Northern Dancer

History's most successful thoroughbred sire, is depicted now and during various stages of his life by the famous West German artist, Klauss Philipp.

Northern Dancer was bred, raced and owned by E. P. Taylor of Toronto, Ontario. When the small colt, sired by Neartic and out of Natalama, did not attract its reserve bid as a yearling, Mr Taylor's Winfield Farm raced him successfully, winning both the Kentucky Derby and the Preakness.

As a sire, Northern Dancer has 544 foals during 19 seasons at stud. He has sired more stake winners (120) and champion race horses (23) than any other horse. At 25, the horse is still standing at stud at Winfield Farm in Maryland.

JOURNAL OF REPRODUCTION AND FERTILITY

SUPPLEMENT No. 35

EQUINE

REPRODUCTION

IV

PROCEEDINGS OF THE FOURTH INTERNATIONAL

SYMPOSIUM ON EQUINE REPRODUCTION

HELD AT THE

UNIVERSITY OF CALGARY

AUGUST 1986

EDITED BY

BARBARA J. WEIR,

I. W. ROWLANDS, W. R. ALLEN and P. D. ROSSDALE

Journal of Reproduction & Fertility
1987

First published 1987

ISSN 0449–3087
ISBN 0 906545 13 7

Published by **The Journals of Reproduction and Fertility Ltd.**

Agent for distribution: **The Biochemical Society Book Depot, P.O. Box 32, Commerce Way, Whitehall Industrial Estate, Colchester, Essex CO2 8HP, U.K.**

Printed in Great Britain by
Henry Ling Ltd., at
The Dorset Press, Dorchester, Dorset

CONTENTS

METHODOLOGY

THE STALLION

THE NON-PREGNANT MARE

Contents

THE PREGNANT MARE

POSTER ABSTRACTS

SUMMATION

SPONSORS

The International and local Organizing Committee express their sincere appreciation to the following organizations for their generous support of the Fourth International Symposium on Equine Reproduction:

Alberta Agriculture—Farming for the Future
Alberta Agriculture—Horse Industry Branch
Alberta Recreation Parks and Wildlife Foundation
Christie Foundation
International Equine Reproduction Trust
Natural Sciences and Engineering Research Council, Canada
Spruce Meadows
Taylor Foundation
Western Canadian Equine Practitioners Association

Dedication to Mr E. P. Taylor

E. P. Taylor is McGill University's first Gold Medal winner for distinguished service to Canadian Society, industrialist, financier and one of the world's most successful breeders of Thoroughbred horses. Of all E. P. Taylor's remarkable accomplishments in the international financial community, none has surpassed his contributions to the sport of horse racing.

Although the acquisition and expansion of commercial enterprises has been his lifelong vocation, horse racing became his hobby and it grew into a consuming passion. To the pursuit of his hobby, Mr Taylor brought his boundless energy and his willingness to gamble for high stakes. As always, he declined to settle for anything less than excellence.

A plan began to take form in Mr Taylor's mind 50 years ago when he acquired a Toronto Brewery, the ancillary assets of which included a modest group of Thoroughbred horses which he raced under the name of the Cosgrave Stable. There were seven tracks in Ontario then, and most of them were in physical decline. To say merely that E. P. Taylor rescued Ontario racing from a state of decay is a palpable understatement. He not only saved racing but he is the man responsible for completely revitalizing the sport across this country.

In 1947, he was elected to the Board of the Ontario Jockey Club. He convinced his fellow directors that the Ontario Jockey Club must absorb six independent operators in the province. He planned to concentrate Ontario racing on only three of the seven tracks. His fellow directors, all men of wealth and importance in Ontario, were startled by the scope and projected costs of Taylor's plan. Nevertheless, they knew that he was the one Canadian with the imagination, know-how and access to financial resources who would assure the success of such a venture. Accordingly, they built the magnificent new Woodbine; they completely refurbished and enlarged the plant at Fort Erie and they completely reconstructed the original Woodbine course, the name of which was changed to Greenwood. Today the racing complexes at Woodbine, Fort Erie and Greenwood stand as glittering testimonials to one man's energy.

E. P. Taylor never has been a man who could be satisfied with half-measures. Realizing that the breeding industry provided the backbone of racing, he made his first major incursion in this field in 1950 when he purchased the late Colonel R. S. McLaughlin's complete stock of Thoroughbreds and his breeding nursery. From that purchase emerged Winfields Farms which as become the most successful Thoroughbred breeding establishment in North America. Northern Dancer, history's most successful Thoroughbred sire, was bred and raced under Winfields' colours (see 'Frontispiece').

Racing has been E. P. Taylor's hobby but no man can indulge himself in a hobby interminably unless his self-indulgence is condoned by his wife. The late Mrs Taylor, petite and indomitable, always was very much in the middle of the family Thoroughbred enterprises. Mrs Taylor was an independent thinker who quietly questioned many of the Great Man's rasher pronunciamentos.

E. P. Taylor's contributions to the Thoroughbred industry have been unrivalled. The point was illustrated when the new Woodbine was opened on 12 June 1956 and the very first race over the new track was one by one of Taylors' horses named Landscape. Columnist Red Smith, musing on the fact that Woodbine had been built at Taylor's insistence and the opening race had been won by Taylor's horse, wrote in the New York Herald–Tribune: "E. P. Taylor giveth—and E. P. Taylor taketh away...."

To which countless Canadian horse-racing enthusiasts gratefully would add the appropriate tagline: "Blessed be the name of E. P. Taylor".

James Coleman
Racing Department,
Calgary Exhibition & Stampede

FOREWORDS

The Fourth International Symposium on Equine Reproduction was, by all standards, deemed a technical and scientific success. Held at the University of Calgary, in Calgary, Alberta, Canada, 24–29 August, 1986, the Symposium was attended by biologists and veterinarians from all parts of the world. We were particularly pleased to be able to dedicate the Symposium to Edward P. Taylor and to have Dr Yoshimasa Nishikawa as honorary chairman.

Many more papers were submitted for consideration than could be accepted for presentation at the Symposium. While it is unfortunate that papers had to be rejected, it speaks well for the increased interest and enthusiasm for equine research. In 1974 we wondered if there would be enough new research to support a Symposium on Equine Reproduction in 1978. Today, many worldwide institutions have become involved in research on this subject and the need for this Symposium is even more evident. These International Symposia meetings have contributed in a significant way to the increased interest and participation in equine reproduction research. This interest has resulted in improved scientific quality of papers presented. It is also obvious that many more young scientists are presenting papers at our Symposia and this augurs well for the future.

Among the important dividends to be obtained from the Symposium are the communication among participants and the discussions, which occur in all aspects of equine reproduction. It was apparent from some of the discussions held during this symposium, as in past symposia, that money for research is a continuing problem. With few exceptions, laboratories suffer from lack of funds to conduct research on a long-term basis.

The local Conference Organizing Committee (L. Burwash, R. B. Church, H. Guenter, L. Kokoski), under the chairmanship of Dr Peter F. Flood, did an outstanding job with this Symposium and all of us extend our grateful thanks to this group. The hospitality extended by the Southerns at their barbeque, held at Spruce Meadows, made it an enjoyable experience for everyone in attendance.

Keith Betteridge's summary of papers presented at the Symposium was both eloquent and humorous.

The International Equine Reproduction Symposia Committee continues to function in the best interests of the organization. I sincerely appreciate their help and guidance over the past two years and throughout the conference. I should especially like to acknowledge the efforts of W. R. Allen, P. F. Flood, D. Mitchell, P. D. Rossdale and my administrative assistant Cherie Frost.

J. P. Hughes
Chairman, International
Organizing Committee

It is my pleasure as Chairman of the Local Arrangements Committee to welcome you on behalf of the Committee to the Fourth International Symposium on Equine Reproduction. The previous three Symposia in Cambridge, Davis and Sydney have set standards that place the merit of the symposia beyond doubt and their published proceedings have become works of standard reference in their field. We hope, and have every reason to believe, that this symposium will be as stimulating and productive as its predecessors.

It is with quite another kind of pleasure that I welcome you to Canada, to the West, the prairies and to the mountains, to the Province of Alberta, to the City of Calgary, host of the 1988 Winter Olympic Games, and to the University of Calgary. They need no introduction.

P. F. Flood
Chairman, Local
Arrangements Committee

INTERNATIONAL ORGANIZING COMMITTEE

PARTICIPANTS

Australia
C. F. Chandler
K. F. Dowsett
C. R. E. Halnan
J. Hyland
C. F. P. Irwin
H. Mortimer
Virginia Osborne
D. Pascoe
R. Pascoe
P. Williamson
M. Wylie

Belgium
D. Ladry
M. Vandeplassche

Brazil
A. A. T. A. Carrascoza
A. J. L. Develey
S. Filizzola
M. Henry
U. R. Reiner
C. A. M. Silva

Canada
Roxy Bell
P. Benns
V. G. Bermudez
K. J. Betteridge
F. Bristol
B. Brummelman
B. Buckrell
L. Burwash
W. Burwash
Joan B. Carrick
R. B. Church
G. D. Davis
P. F. Flood
A. K. Goff
Karen Hage
D. S. Irvine
W. H. Johnson
G. Klavano
H. Loewe
Ria Mackay
R. J. Mapletoft
C. Marsan
B. D. Murphy
C. Piche
A. Purvis
K. Rajkumar
A. Rathwell
J. Rhodes

D. Rieger
April Romagnano
N. C. Savage
J. Sirois
A. Smit
Gayle Sommer
E. Tolksdorff
J. D. A. Twidale
R. Wise
Jeanine Woods

China
W. Y. He

Czechoslovakia
Z. Müller

Finland
Minna Liisa Heiskanen
Terttu Katila
E. Koskinen
P. Mäenpää

France
Francoise Chevalier
G. Duchamp
M. Magistrini
E. Palmer
Helene Rousset
Marianne Vidament

India
S. S. Rathor
P. K. Uppal

Indonesia
D. Mitchell

Iran
M. Emady

Ireland
Stefania Bucca
T. Burns
Ursula Fogarty
D. Forde
J. B. Hughes
L. Keenan
D. Leadon
M. Osborne
J. Wade
J. Weld

Italy
Patrizia de Ferrari
G. Lacalandra
P. Minoia

Japan
Y. Nishikawa

Mexico
F. Gonzalez
A. Saltiel

Netherlands
W. van Leeuwen
P. J. de Vries

New Zealand
Susan Alexander
J. B. Grimmett
C. H. G. Irvine

Poland
A. Okólski
M. Tischner

South Africa
B. Penzhorn
D. Volkmann

Spain
E. Alonso

Sweden
A. Darenius
Kerstin Darenius
J. Hollander
A. Madej

Switzerland
F. Barralet
H. P. Meier
R. Waelchli
Christine Winder

United Kingdom
W. R. Allen
M. S. Boyle
E. W. M. Curnow
D. R. Ellis
Louise Foster
Abigail Fowden
D. Jessett
Jennifer Ousey
Jean P. Renton

S. J. Ricketts
P. D. Rossdale
I. W. Rowlands
M. W. Sanderson
Marian Silver
Soozy Smith
Francesca Stewart
S. Tanner
Elaine Watson
P. F. Watson
Barbara J. Weir
Karen Worthy

United States
G. Adams
Carolyn B. Adams
Karen Affleck
R. Amann
D. F. Antczak
M. Arns
A. C. Asbury
C. B. Baker
J. Baldwin
B. A. Ball
J. M. Bowen
Ann Bowling
T. Bowman
R. H. Bradbury
Barbara Brewer
P. Burns
S. J. Burns
C. M. Clay
H. S. Conboy
M. Couto
P. Daels
W. C. Day
W. L. Donaldson
R. H. Douglas
D. Douglas Hamilton
Meg Douglas Hamilton
Francoise Dudan
M. Eliott
A. Enders
J. W. Evans
E. Fallon
J. E. Fay
Barbara Forney

D. B. Galloway
M. Garcia
S. D. Goodeaux
G. J. Haluska
Kay A. Henderson
R. B. Hillman
Katherine Hines
Katrin Hinrichs
D. Holtan
J. P. Hughes
J. P. Hurtgen
W. Jöchle
W. Jones
K. Kasper
J. P. Kastelic
R. M. Kenney
L. Kloppe
Anne M. Koterba
Michelle LeBlanc
D. Lein
D. E. Little
T. V. Little
I. K. M. Liu
W. E. Loch
T. Lock
R. M. Lofstedt
P. Loomis
C. D. Lothrop
J. E. Lowe
R. Loy
J. Madigan
Elizabeth Mank
J. M. Martin
Sue M. McDonnell
Karen J. McDowell
A. O. McKinnon
D. A. Meirs
Leanne Mellbye
M. Memon
C. D. Morrison
D. Neely
Lynn G. Nequin
T. M. Nett
T. J. Newby
Maire O'Connor
L. M. Olsen
J. G. Oriel

Linda Parry-Weeks
R. L. Pashen
Rachel Pemstein
B. W. Pickett
F. Pipers
Jenifer Plummer
Kathleen F. Pool
J. T. Potter
Janet F. Roser
J. Roszel
Jane Safir
Gayla F. Sargent
H. R. Sawyer
D. H. Schlafer
Patricia Sertich
D. C. Sharp
K. Shiner
M. Dawn Shore
J. Shull
P. Silvia
J. D. Smith
E. L. Squires
P. Strzemienski
Tamara A. Tetzke
D. L. Thompson
P. Timoney
D. Varner
M. Villahoz
Martha M. Vogelsang
S. Vogelsang
J. L. Voss
G. Webb
Joanne Weithenauer
J. Wiest
M. Wilson
G. Woods
W. Zent

West Germany
H. Bader
H. Enbergs
J. Gaus
R. Humke
C. Leiding
W. Leidl
M. Merkt
D. Rath

1, R. Naelchi; 2, A. Enders; 3, A. Smit; 4, E. Fallon; 5, J. B. Hughes; 6, P. de Vries; 7, R. Mackay; 8, W. van Leeuwen; 9, S. Rathur; 10, B. Buckrell; 11, R. Mapletoft; 12, F. Bristol; 13, K. Henderson; 14, D. Irvine; 15, K. Betteridge; 16, K. Hage; 17, D. Volkmann; 18, D. Reiger; 19, B. L. Penzhorn; 20, J. Hyland; 21, A. McKinnon; 22, J. Grimmett; 23, J. Voss; 24, A. Asbury; 25, R. Hillman; 26, J. Hurtgen; 27, G. Davis; 28, P. Williamson; 29, J. Rhodes; 30, D. Jessop; 31, W. Loch; 32, H. Enbergs; 33, W. Humke; 34, C. Winder; 35, K. Worthy; 36, K. Shiner; 37, J. Weld; 38, W. Zent; 39, R. Loy; 40, K. McDowell; 41, D. Sharp; 42, V. Osborne; 43, N. Savage; 44, C. Piche; 45, P. Timoney; 46, D. Pascoe; 47, A. Saltiel; 48, J. Wade; 49, J. Woods; 50, C. Morrison; 51, E. Squires; 52, B. Ball; 53, G. Woods; 54, B. Baker; 55, L. Mellbye; 56, D. Schlafer; 57, D. Safir; 58, D. Thompson; 59, T. Tetzke; 60, J. Wiest; 61, T. Nett; 62, H. Sawyer; 63, P. Loomis; 64, C. Clay; 65, K. Hines; 66, S. Vogelsang; 67, M. Arns; 68, T. Potter; 69, W. Evans; 70, R. Pemstein; 71, D. Forde; 72, R. Bell; 73, J. Carrick; 74, G. Sommer; 75, W. Burwash; 76, R. Douglas; 77, M. Villahoz; 78, H. P. Meirer; 79, S. Bucca; 80, M. O'Connor; 81, J. Ousey; 82, J. Wade; 83, D. Ladry; 84, F. Barrelet; 85, F. Dudan; 86, P. Daels; 87, T. Little; 88, D. Galloway; 89, K. Dowsett; 90, A. Bowling; 91, J. Werthenauer; 92, J. Lowe; 93, D. Meirs; 94, G. Webb; 95, J. Roser; 96, K. Pool; 97, M. Vogelsang; 98, M. Wilson; 99, C. Halman; 100, W. Day; 101, J. Smith; 102, L. Werner; 103, E. Curnow; 104, J. Burns; 105, S. Burns; 106, L. Nequin; 107, A. Darenius; 108, J. Heilander; 109, R. Amann; 110, M. Boyle; 111, R. Pashen; 112, H. W. Yi; 113, F. Gonzalez; 114, K. Kasper; 115, M. Osborne; 116, D. Lein; 117, H. Bader; 118, D. Rath; 119, A. Madej; 120, J. Plummer; 121, P. F. Watson; 122, G. Lacalandra; 123, B. Murphy; 124, R. Lofstedt; 125, A. Fowden; 126, T. Katila; 127, M-L. Heiskanen; 128, J. Filizzola; 129, A. Carrascoza; 130, M. Sanderson; 131, G. Duchamp; 132, J. Fay; 133, B. Forney; 134, A. Devely; 135, U. Reiner; 136, T. Burns; 137, W. Johnson; 138, A. Okölski; 139, T. Newby; 140, J. Gaus; 141, A. Purvis; 142, P. Minola; 143, J. Twidale; 144, E. Watson; 145, L. Foster; 146, S. Smith; 147, P. Mäenpää; 148, E. Kostenen; 149, K. Darenius; 150, L. Parry-Weeks; 151, D. Shore; 152, D. Holtan; 153, P. Strzemienski; 154, M. Viadment; 155, M. Henry; 156, C. Silva; 157, J. Roszel; 158, M. Fogarty; 159, F. Stewart; 160, K. Affleck; 161, S. Tanner; 162, W. Jöchle; 163, E. Tolksdorff; 164, D. Neely; 165, W. Donaldson; 166, R. Pascoe; 167, P. Burns; 168, D. Varner; 169, L. Kloppe; 170, S. Ricketts; 171, E. Mank; 172, D. Little; 173, H. Rousset; 174, F. Chevalier; 175, M. Leblanc; 176, J. Martin; 177, M. Magistrini; 178, C. Marsan; 179, M. Garcia; 180, I. K. M. Liu; 181, A. Rathwell; 182, W. Jones; 183, M. Wylie; 184, D. Ellis; 185, D. Antczak; 186, J. Oriol; 187, V. Bermudez; 188, S. Goodeaux; 189, C. Chandler; 190, P. di Ferrarri; 191, D. P. Leadon; 192, J. M. Bowen; 193, M. Couto; 194, M. Elliott; 195, R. Bradbury; 196, P. Silva; 197, E. Alonso; 198, K. Hinrichs; 199, P. L. Sertich; 200, S. McDonnell; 201, Miss Y. Nishikawa; 202, Mrs Y. Nishikawa; 203, B. J. Weir; 204, S. L. Alexander; 205, I. W. Rowlands; 206, J. W. Shull; 207, M. Memon; 208, A. Goff; 209, J. Baldwin; 210, J. Sirois; 211, H. S. Conboy; 212, L. M. Olsen; 213, P. Irwin; 214, C. Irvine; 215, M. Vandeplassche; 216, D. Mitchell; 217, W. R. Allen; 218, P. D. Rossdale; 219, L. Burwash; 220, J. P. Hughes; 221, Y. Nishikawa; 222, R. B. Church; 223, H. Merkt; 224, R. Kenney; 225, E. Palmer; 226, P. F. Flood; 227, B. W. Pickett; 228, M. Tischner.

HONORARY CHAIRMAN

Professor Yoshimasa Nishikawa

Yoshimasa Nishikawa was born on 13 March 1913. After a bright start at school he enrolled as an undergraduate in the Faculty of Agriculture at the University of Tokyo. He graduated in Veterinary Science in 1936 and immediately joined the research staff at the National Institute of Animal Sciences where he remained for the next 20 years, gaining his Ph.D. degree in 1945. In 1957 he moved to the chair of Animal Reproduction at Kyoto University and then, in 1976, was elected President of Obihiro University of Agriculture and Veterinary Medicine. Upon his retirement from academic life in 1984 he was appointed President of Toyama Women's College where he remains at the present time in charge of over 1000 young female students.

Professor Nishikawa's interest in the then emerging science of animal reproduction was first aroused during his undergraduate years. This spark ignited a veritable fire of enthusiasm for the subject that blazed intensely for the next 50 years. It generated the amazing total of 250 original research papers, 171 review papers and 35 book chapters; and it still burns brightly in retirement.

During his 20 years at the National Institute of Animal Sciences, Professor Nishikawa carried out numerous fundamental and applied experiments on many aspects of reproduction in both the male and female of a wide range of laboratory and domestic animal species including the oldest of man-made mammalian hybrids, the mule. But despite this catholic application across the species boundaries, his abiding interest remained with the horse. He embarked upon all manner of exciting and novel experimental approaches during this period, most of which were aimed at elucidating the physiological mechanisms involved in puberty and gametogenesis in mares and stallions, the endocrinological control of oestrus, follicular growth and ovulation in mares, and practical methods for the collection and preservation of stallion semen.

He characterized in great depth the effects of photoperiodicity on reproductive function in mares and stallions from pioneering experiments on the application of artificial lighting. His studies of the effects of the stage of the ovarian cycle and pregnancy on the characteristics of vaginal and cervical mucus provided, in effect, the first practical use of reproductive cytology in the mare and were something of a forerunner of the progesterone assay in years to come. And lastly, Professor Nishikawa and his team completed an exhaustive study of the effects of exogenous oestrogen on reproductive function in mares from which many findings emerged that still have practical application in modern stud management.

Initially the language barrier prevented much of Professor Nishikawa's work becoming known outside Japan but eventually the Japan Racing Association showed great foresight when it financed the publication in English of a collection of Professor Nishikawa's major experimental studies on the horse. The resulting book, entitled *Studies on Reproduction in Horses*, quickly became a standard and widely read text on equine reproduction. To the present day it remains a fascinating read which provides a clear insight into the innovative and enquiring mind of its author.

In addition to his major input into the research side of animal reproduction, Professor Nishikawa has remained a dedicated and respected teacher in this field. Countless veterinary and animal science undergraduates have benefited from his witty and stimulating lectures over the years and a sizeable body of present-day animal reproduction research workers and teachers in Japan were fortunate to receive inspiration and patient guidance from Professor Nishikawa during their early post-graduate training.

It is not surprising that Professor Nishikawa has received many national and international awards and accolades during his long and active career. He holds honorary emeritus professorships in the Universities of Kyoto and Obihiro, has been awarded honorary doctorates by the Universities on Kon-Kok, Korea, and of Munich, West Germany, and he is an honorary member of many Japanese and international societies concerned with animal production and reproduction. He has also been awarded a great many prizes from Japanese agricultural societies and was decorated with The Order of Merit by The Emperor of Japan in 1985.

Professor Yoshimasa Nishikawa is a distinguished scientist, teacher and international elder stateman in the field of animal reproduction. And to those with a special interest in the science of equine reproduction he is a greatly respected friend and collegue. It is with pride and pleasure that the International Equine Reproduction Symposia Committee bestows upon Professor Nishikawa the title of Honorary Chairman of the 4th International Symposium on Equine Reproduction convened in Calgary, Canada, during August 1986.

W. R. Allen

METHODOLOGY

Convener

C. H. G. Irvine

Chairman

B. D. Murphy

J. Reprod. Fert., Suppl. **35** (1987), 1–8

Application of recombinant DNA techniques to structure–function studies of equine protein hormones

Francesca Stewart, Sarah E. A. Leigh and J. A. Thomson

Thoroughbred Breeders' Association Equine Fertility Unit, Animal Research Station, 307 Huntingdon Road, Cambridge CB3 0JQ, U.K.

Summary. Complementary (c)DNA libraries have been made from horse pituitary gland and endometrial cup tissues with the aim of isolating the genes for the horse gonadotrophins (FSH, LH and CG) and growth hormone (GH). Southern (DNA) and Northern (RNA) blotting techniques were used to demonstrate that several heterologous (human and ovine) cDNA probes would be adequate for isolating the horse genes. A human cDNA probe was then used to isolate the horse gonadotrophin α-subunit cDNA from the pituitary and endometrial cup libraries. The nucleotide sequences from both tissue sources were identical, thereby confirming that there is a single gene for the α-subunit in the horse. As soon as the gonadotrophin β-subunit sequences become available, they will be expressed along with the α-subunit to obtain biologically active hormones.

Horse GH cDNA has also been isolated by using a sheep cDNA probe and its DNA sequence has been determined. It was subsequently used to demonstrate that horse endometrial cups, but not the allantochorion taken at the same stage of gestation, contains mRNA for a GH-like protein. This mRNA could be due to a low level of expression of the GH gene or to expression of the gene(s) for a GH-related protein such as prolactin or placental lactogen.

Introduction

Recombinant DNA technology ('genetic engineering') has provided powerful and novel approaches to the study of protein structure and function. It has largely superseded conventional methods for determining the amino acid sequence of a protein and the ability to point mutate DNA sequences, giving rise to specific amino acid alterations in eucaryotic proteins expressed in bacteria (site-directed mutagenesis), has provided a powerful approach to protein structure–function studies. The technology is based on (1) the integration of foreign DNA fragments (genes) into the DNA of a rapidly replicating vector (plasmid or virus) by *recombination*, (2) the subsequent amplification (*cloning*) of the resulting recombinants in bacteria or yeast, (3) the detection of specific cloned genes, usually by binding (*hybridization*) of characterized, radioactively labelled nucleic acid fragments (probes) to those genes, (4) DNA *sequencing* of the isolated genes and (5) *expression* of the genes and subsequent engineering of the gene products by specific or site-directed mutation of the DNA.

The foreign DNA (genes) to be cloned may be obtained from genomic or complementary (c)DNA. Genomic (chromosomal) DNA can be purified from any tissue (usually spermatozoa or liver) and is digested into fragments of suitable size before cloning. On the other hand, cDNA is synthesized from messenger RNA (mRNA) which has to be purified from the tissues that secrete the protein(s) of interest. In both cases, the DNA fragments are cloned to create 'libraries' which are then screened with probes to find the required genes.

We have used cDNA cloning techniques to begin a study aimed at investigating structure–function relationships of several hormones which are involved in reproduction of horses. The equine gonadotrophins (follicle-stimulating hormone (FSH), luteinizing hormone (LH) and chorionic gonadotrophin (CG)) are of particular interest because of their unique biological and structural properties. In particular, unlike all other gonadotrophins studied so far, horse (and to a lesser extent, donkey) LH and CG exhibit potent FSH-like, as well as LH-like, biological activities in non-equine species (Stewart *et al.*, 1977; Licht *et al.*, 1979; Guillou & Combarnous, 1983; Roser *et al.*, 1984). Cloning the cDNA for these hormones will provide accurate amino acid sequences for use in computer analysis studies aimed at defining hormone–receptor interaction sites (Stewart & Stewart, 1977). In addition, the DNA will be mutated to confirm and extend these predictions and to produce hormones that are more potent or which have altered or more desirable properties. Equine gonadotrophins are in great demand for use in stimulating fertility in farm animals and there would be considerable advantage in being able to produce reliable hormone preparations with optimized biological activities by using recombinant DNA technology.

The DNA sequences would also help to answer some intriguing evolutionary questions. For example, how and when the chorionic gonadotrophins have evolved (and why they appear to be limited only to equids and primates) and what are the exact evolutionary relationships of equids. We have begun this study by creating cDNA libraries from horse pituitary glands (the source of LH and FSH) and endometrial cups (the source of CG), and isolating the horse α- and β-subunit sequences with heterologous (human and ovine) probes. The equivalent donkey genes will also be isolated from cDNA libraries and both the horse and donkey cDNAs used as probes to obtain genomic sequences.

We have further used this technology to study horse reproduction by cloning the gene for horse GH and using it as a probe.

Materials and Methods

Tissues. Pituitary glands (the source of LH, FSH and GH) were obtained at slaughter from an abattoir. The glands were exised from mares and geldings within 10 min of death, frozen immediately in liquid nitrogen and stored for up to 3 months in liquid nitrogen before extraction of messenger RNA (mRNA).

Endometrial cups (the source of CG) were obtained at laparotomy under general anaesthesia from pregnant mares between Days 50 and 60 of gestation. The conceptus was removed and the endometrial cups squeezed off the endometrium with forceps, causing as little damage to the uterine tissues as possible. The tissue was weighed and either extracted immediately or at a later date after storage in liquid nitrogen. Occasionally, some of the allantochorion was also frozen for RNA extraction.

Extraction and purification of polyadenylated messenger RNA. All equipment and solutions were carefully sterilized and gloves were worn throughout to minimize contamination with nucleases present on human skin. Glassware was baked at 180°C for 12 h (to destroy RNAses) and disposable plastics were obtained sterile from the suppliers. Aqueous solutions (apart from those containing Tris buffers) were treated with diethylpyrocarbonate (nuclease inhibitor) before sterilization by autoclaving.

Frozen tissues were ground to a fine powder under liquid nitrogen (using a mortar and pestle) whereas fresh tissues were minced with scissors and/or scalpel blades under lysis buffer (0·2 M-Tris–HCl pH 9·0, 50 mM-NaCl, 10 mM-EDTA, 0·5% (w/v) SDS, 1 mg heparin/ml). The tissues were then homogenized in lysis buffer using a polytron homogenizer (Kinematica, Lucerne, Switzerland) and extracted several times with phenol and phenol/$CHCl_3$ mixture (1:1 v/v) as described by Houghton *et al.* (1980). The RNA and DNA in the aqueous phase was precipitated with ethanol and the resulting pellet was washed in 80% ethanol, dried and redissolved in 10 mM-Hepes, pH 7·4. Total RNA was selectively precipitated by addition of solid NaCl to 3 M and the polyadenylated mRNA was isolated using oligo (dT)-cellulose affinity chromatography (Aviv & Leder, 1972).

In-vitro translations. The quality of the mRNA preparations was assessed by translating them into proteins using the rabbit reticulocyte lysate in-vitro translation system with [^{35}S]methionine as tracer (Jackson & Hunt, 1983). Routinely, 0·3 μg poly A mRNA was used in each translation reaction and the proteins were visualized by SDS-polyacrylamide gel electrophoresis followed by autoradiography.

DNA probes. Because the amino acid sequences of the mammalian gonadotrophins (and also growth hormones) are highly homologous, there was a good chance that hormone genes obtained from other species could be used as probes to isolate the horse genes. Accordingly, we tested the following cDNA sequences for their suitability as probes:

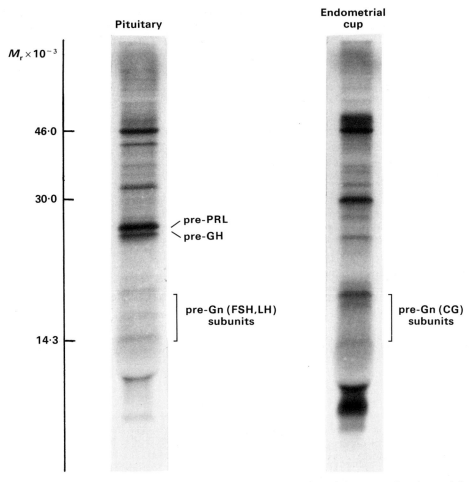

Fig. 1. Autoradiograph of the in-vitro translation products from pituitary and endometrial cup messenger RNA separated by SDS-polyacrylamide gel electrophoresis. Protein bands corresponding to the pre-prolactin (PRL), Growth Hormone (GH) and gonadotrophin (Gn) subunits are indicated.

the human α- and β-subunit CG cDNA probes (kindly provided by J. C. Fiddes (Fiddes & Goodman, 1980)), a bovine α-subunit cDNA (kindly provided by J. H. Nilson (Nilson *et al.*, 1983)) and an ovine GH cDNA probe (kindly provided by J. M. Warwick (Warwick & Wallis, 1984)). The cDNAs were all cloned in plasmid pBR 322 and microgram quantities were routinely prepared (using standard methods, e.g. Maniatis *et al.*, 1982) for use as probes. The preferred method of radio-labelling involved digestion with exonuclease III (New England Biolabs, Bishops Stortford, Herts, U.K.) followed by 'filling in' with [α-^{32}P]dCTP or dGTP (Amersham International, Aylesbury, Bucks, U.K. or NEN, Du Pont (UK) Ltd, Stevenage, Herts, U.K.) and unlabelled nucleotides (Boehringer Corp Ltd, Lewes, Sussex, U.K.) via the action of *E. coli* DNA polymerase I Klenow fragment (Biochemistry Department, Cambridge University or Boehringer Corp Ltd, U.K.).

Southern and Northern hybridizations. To test whether the probes would be adequate for identifying the horse genes, they were hybridized to horse DNA and mRNA using standard Southern and Northern blotting techniques. Horse DNA was prepared from testes or lymphocytes, digested with a variety of restriction enzymes, electrophoresed in 1% agarose, denatured and blotted on to a nylon transfer membrane (usually Gene Screen; NEN, Southampton, U.K.) before hybridization ('Southern blot'; Southern, 1975). The poly A mRNA was first denatured, electrophoresed under denaturing conditions in a 1% agarose/50% formaldehyde gel, and was blotted and hybridized in a similar way to the DNA ('Northern blot'; Thomas, 1980).

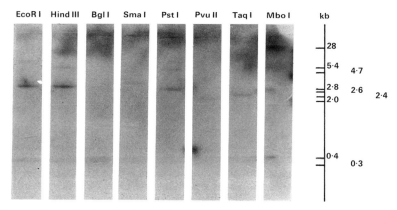

Fig. 2. Autoradiograph of a Southern blot. Genomic horse DNA was digested with several restriction enzymes, electrophoresed in 1% agarose, blotted on to a nylon membrane and hybridized to the ^{32}P-labelled β-hCG cDNA probe. kb = kilo base pairs.

cDNA synthesis and cloning. cDNA was synthesized from the poly A mRNA by the Gübler & Hoffman (1983) method with most of the modifications described by Reinach & Fischman (1985). First-strand synthesis was achieved with oligo dT priming and AMV reverse transcriptase (Anglian Biotechnology Ltd, Colchester, U.K.), followed by second-strand synthesis with RNase H, *E. coli* DNA polymerase I (Anglian Biotechnology) and *E. coli* DNA ligase (Pharmacia Ltd, Milton Keynes, U.K.). The EcoR 1 sites were methylated with EcoR 1 methylase (New England Biolabs) and the ends of the cDNA were blunted or 'polished' with T$_4$ DNA polymerase (Anglian Biotechnology Ltd). EcoR 1 linkers (Pharmacia) were attached and the cDNA ligated to EcoR 1-digested plasmid (pUC8, 18, 19 or 13; Vieira & Messing, 1982), or blunt-end ligation was used to insert the cDNA into the Sma I site of the plasmid. The recombinant plasmids were transformed into *E. coli* host cells; either HB101 (Boyer & Roulland-Dussoix, 1969) or TG1 (Gibson, 1984). The resultant libraries were stored in glycerol broth in microtitre trays at −20°C as described by Gergen *et al.* (1979).

Screening cDNA libraries. The libraries were plated on to nutrient agar (containing 200 μg ampicillin/ml) as described by Gergen *et al.* (1979) and were transferred to either Gene Screen Plus transfer membrane (NEN, Du Pont (UK) Ltd, Stevenage, Herts, U.K.) or Whatman 541 filter paper. Hybridizations with ^{32}P-labelled probes were carried out in 50% formamide at 42°C for Gene Screen and at 37°C for Whatman 541 papers. Clones that hybridized with the probes (positives) were identified by autoradiography and their inserts were isolated and characterized by restriction enzyme digestion and further Southern hybridization experiments.

DNA sequencing. The positive inserts were sub-cloned into the sequencing vector bacteriophase M13 (mp 8, 9, 18 or 19; Messing & Vieira, 1982) and were sequenced by the Sanger dideoxy termination method (Sanger *et al.*, 1977). The resulting sequences were identified by comparison with previously published amino acid and DNA sequences.

Results

In-vitro translations

The translated horse pituitary mRNA gave a characteristic gel electrophoresis pattern with pre-prolactin and pre-GH being readily identifiable (example shown in Fig. 1). The pre-gonadotrophin subunits were tentatively identified according to their molecular weights. As expected, the endometrial cup mRNA also produced protein bands corresponding to the gonadotrophin subunits but not to prolactin and GH (Fig. 1). The quantity (intensity) and quality (clarity) of the protein bands gave a good indication of the quality of the mRNA preparations which, in turn, correlated well with the quality of cDNA synthesized from them.

Southern and Northern hybridizations

Southern hybridizations of the horse DNA with the human and ovine gonadotrophin probes produced distinct bands of various sizes, depending on the restriction enzyme(s) used. An example

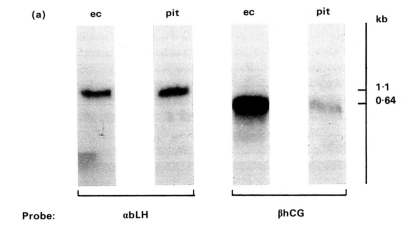

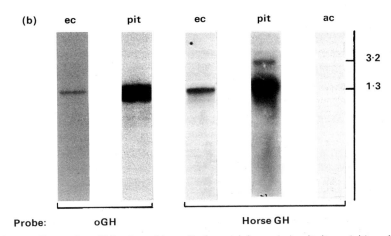

Fig. 3. Autoradiographs of Northern blots. Endometrial cup (ec), pituitary (pit) and allanto-chorion (ac) mRNA was electrophoresed in 1% agarose, blotted on to a nylon membrane and hybridized to (a) the ^{32}P-labelled bovine LH α-subunit (αbLH) and human β-subunit CG (βhCG) probes and (b) the ovine growth hormone (oGH) and horse GH probes. kb = kilo base pairs.

of a horse genomic Southern blot which was hybridized to the β-hCG probe is shown in Fig. 2. Donkey genomic DNA produced very similar results (not shown), suggesting that the donkey and horse CG β-subunit genes are very similar in structure. These results also showed that the heterologous probes would be adequate for isolating the horse genes. The Northern hybridization results (Fig. 3a) confirmed this by giving distinct bands at approximately 1·0 kilo base pairs with the α probe and 0·6 kilo base pairs with the β probe (the expected sizes of the subunit mRNAs). Hybridization of the β-hCG probe with the pituitary mRNA was consistently less than that obtained with endometrial cup mRNA, thereby suggesting a weaker homology between the horse and human LH β-subunits compared to the horse and human CG β-subunits. The ovine GH probe gave a very strong band of hybridization with the horse pituitary mRNA (at around 1·2 kilo base pairs) and a faint band of the same size with endometrial cup mRNA (Fig. 3b). The latter result was unexpected and was investigated further (see below).

cDNA libraries and nucleotide sequencing

Routinely, a yield of 100–800 ng double-stranded cDNA was obtained from 2–3 µg poly A mRNA from which libraries of several thousand recombinants were obtained. Many of the gonadotrophin and GH cDNA clones which were selected by screening and subsequently sequenced were incomplete, probably due to incomplete second-strand synthesis and/or internal restriction enzyme sites. However, most of the gonadotrophin α-subunit has been sequenced (only the sequence for the signal peptide is missing) from the libraries for pituitary and endometrial cup cDNA. The translated amino acid sequence agrees with the published protein sequences of Moore *et al.* (1979) and Bousfield & Ward (1982) but not with that of Rathnam *et al.* (1978). The nucleotide sequences from both sources (pituitary and endometrial cup) were identical, thereby confirming that there is a single gene for the α-subunit in the horse. The translated horse GH amino acid sequence also agrees with that determined by conventional protein-sequencing methods (Zakin *et al.*, 1976). The DNA and deduced amino acid sequences will be published elsewhere.

GH-like protein in endometrial cup tissue

The faint band of hybridization obtained with the ovine GH probe and endometrial cup mRNA (Fig. 3b) suggested that endometrial cup tissue contained mRNA for a GH-like protein. This was investigated further by using the horse GH cDNA to probe the pituitary, endometrial cup and allantochorionic mRNA preparations (Fig. 3b). The allantochorionic mRNA was negative and, as expected, the pituitary mRNA gave the strongest signal (Fig. 3b). The signal with endometrial cup mRNA was similar to that obtained with the ovine GH probe, thereby confirming the presence of a GH-like mRNA.

Discussion

Our results have clearly demonstrated that there is considerable homology between the horse, human and ovine protein hormone genes and that the human and ovine gonadotrophin and GH genes have proved to be adequate probes for the identification and isolation of the horse genes.

Our Southern hybridization experiments have shown that the structural genes for the gonadotrophin subunits are very similar in horses and donkeys, despite differences in the biological activity of the hormones (Stewart *et al.*, 1977; Roser *et al.*, 1984). In addition, the relative simplicity of the β-subunit restriction digest patterns (Fig. 1b), compared to the human (Boorstein *et al.*, 1982), suggests that equids probably do not possess a complex family of CG β-subunit genes. The human CG β-subunit genes have evolved via an initial duplication of the LH β-subunit gene, followed by a single nucleotide deletion giving rise to a C-terminal extension, followed by further gene duplications (Fiddes & Goodman, 1980; Boorstein *et al.*, 1982). These events have clearly not occurred (at least, not exactly) in equids since horse LH (Bousfield *et al.*, 1985) and possibly donkey LH (Roser *et al.*, 1984) already possess a C-terminal extension and are therefore very similar in size and structure to the CGs in both equine species. The exact DNA sequences of the horse and donkey β-subunit genes should provide answers to these questions, as well as providing accurate amino acid sequences for computer analysis aimed at locating those residues responsible for specificity of receptor binding. This will provide the basis for future studies (using site-directed mutagenesis) to alter the biological activities of the hormones. Expression will be initially tried in yeast but, to achieve full biological activity, a mammalian cell such as that used for the expression of human CG (Reddy *et al.*, 1985) will probably be used.

The complete horse α-subunit cDNA sequence has confirmed that the most recent amino acid sequences of Moore *et al.* (1979) and Bousfield & Ward (1982) are correct and that the earlier sequence of Rathnam *et al.* (1978) was incorrect. Furthermore, our finding that the sequences from the pituitary and endometrial cup cDNA libraries are identical has confirmed that there is a single gene for the gonadotrophin α-subunit in horses.

The horse cDNA clones we have obtained will ultimately be used as probes to obtain the genomic sequences for the various hormone genes. This should help to answer some of the evolutionary questions, as well as those relating to gene linkage and control of gene expression.

Our finding that endometrial cup mRNA hybridizes with the sheep and horse GH cDNA probes indicates that endometrial cup cells secrete a GH-like protein. The degree of hybridization was weak compared to that obtained with an equivalent amount of pituitary mRNA, suggesting either a low level of expression of GH or expression of a related protein such as prolactin or placental lactogen. Attempts have been made to detect placental lactogen in the pregnant mare (Forsyth *et al.*, 1975) but all of the samples in the study were taken late in pregnancy, long after the endometrial cups have disappeared. Furthermore, the assays were designed to measure lactogenic rather than GH activity. We hope to isolate and sequence the cDNA for this protein which should provide positive identification. We shall also assay tissue and blood samples for GH and prolactin activities to determine its biological activity.

The function of a GH-like protein in endometrial cup cells is unclear. In ruminants and primates, placental lactogen is secreted in increasing quantities during pregnancy and appears to be important for stimulating mammogenesis and fetal growth in late pregnancy. This seems unlikely in the mare, since endometrial cups have disappeared by mid-pregnancy and the allantochorion (which forms the true placenta) does not appear to secrete the protein. However, the endometrial cup cells are analogous to the cells which secrete placental lactogen in primates and ruminants. In all cases, the secretory cells invade the maternal endometrium and undergo differentiation to form bi- or multinucleate cells. Hoshina *et al.* (1983) have shown that human placental lactogen is expressed only in fully differentiated trophoblast, the syncytium, which is also the tissue that secretes CG. This may also be the case in equids in which the endometrial cups could be considered as the fully differentiated component of the trophoblast layer. The observation suggests a further interesting similarity between primate and equine placentations.

We thank W. R. Allen and M. S. Boyle for collecting the horse tissues and Julia M. Warwick for help in the initial stages of the project. This work was financed by the Thoroughbred Breeders' Association of Great Britain, the Agricultural and Food Research Council (AG63/171) and the Commission of the European Communities (GB1-2-088-UK).

References

Aviv, H. & Leder, P. (1972) Purification of biologically active globin messenger RNA by chromatography on oligothymidylic and cellulose. *Proc. natn. Acad. Sci. U.S.A.* **69**, 1408–1412.

Boorstein, W.R., Vamvakopoulos, N.C. & Fiddes, J.C. (1982) Human chorionic gonadotropin β-subunit is encoded by at least eight genes arranged in tandem and inverted pairs. *Nature, Lond.* **300**, 419–422.

Bousfield, G.R. & Ward, D.N. (1982) A re-examination of the amino acid sequence of equine FSH alpha. *Endocrinology* **110**, Suppl. p. 129, Abstr.

Bousfield, G.R., Sugino, H. & Ward, D.N. (1985) Demonstration of a COOH-terminal extension on equine lutropin by means of a common acid-labile bond in equine lutropin and equine chorionic gonadotropin. *J. biol. Chem.* **260**, 9531–9533.

Boyer, H.W. & Roulland-Dussoix, D. (1969) A complementation analysis of the restriction and modification of DNA in *Escherichia coli*. *J. molec. Biol.* **41**, 459–463.

Fiddes, J.C. & Goodman, H.M. (1980) The cDNA for the β-subunit of human chorionic gonadotropin suggests evolution of a gene by readthrough into the 3'-untranslated region. *Nature, Lond.* **286**, 684–687.

Forsyth, I.A., Rossdale, P.D. & Thomas, C.R. (1975) Studies on milk composition and lactogenic hormones in the mare. *J. Reprod. Fert., Suppl.* **23**, 631–635.

Gergen, P., Stern, R.H. & Wensink, P.C. (1979) Filter replicas and permanent collections of recombinant DNA plasmids. *Nucleic Acids Res.* **7**, 2115–2136.

Gibson, T.J. (1984) *Studies on the Epstein–Barr virus genome.* Ph.D. thesis, University of Cambridge.

Gübler, U. & Hoffman, B.J. (1983) A simple and very efficient method for generating cDNA libraries. *Gene* **25**, 263–269.

Guillou, F. & Combarnous, Y. (1983) Purification of equine gonadotropins and comparative study of their acid-dissociation and receptor-binding specificity. *Biochim. Biophys. Acta* **755**, 229–236.

Hoshina, M., Hussa, R., Pattillo, R. & Boime, I. (1983) Cytological distribution of chorionic gonadotropin subunit and placental lactogen messenger RNA in neoplasms derived from human placenta. *J. biol. Chem.* **97**, 1200–1206.

Houghton, M., Stewart, A.G., Doel, S.M., Emtage, J.S., Eaton, M.A.W., Smith, J.C., Patel, T.P., Lewis, H.M., Porter, A.G., Birch, J.R., Cartwright, T. & Carey, N.H. (1980) The amino terminal sequence of human fibroblast interferon as deduced from reverse transcripts obtained using synthetic oligonucleotide primers. *Nucleic Acids Res.* **3**, 1913–1931.

Jackson, R.J. & Hunt, T. (1983) Preparation and use of nuclease-treated rabbit reticulocyte lysates for the translation of eukaryotic messenger RNA. *Methods Enzymol.* **96**, 50–74.

Licht, P., Bona-Gallo, A., Aggarwal, B.B., Farmer, S.W., Castelino, J.B. & Papkoff, H. (1979) Biological and binding activities of equine pituitary gonadotrophins and Pregnant Mare Serum Gonadotrophin. *J. Endocr.* **83**, 311–322.

Maniatis, T., Fritsch, E.F. & Sambrook, J. (1982) *Molecular Cloning, a Laboratory Manual.* Cold Spring Harbor Laboratory, Cold Spring Harbor.

Messing, J. & Vieira, J. (1982) A new pair of M13 vectors for selecting either DNA strand of double-digest restriction fragments. *Gene* **19**, 269–276.

Moore, W.T., Ward, D.N. & Burleigh, B.D. (1979) Primary structure of Pregnant Mare Serum Gonadotropin alpha subunit. *Fedn Proc. Fedn Am. Socs exp. Biol.* **38**, 462–463.

Nilson, J.H., Thomason, A.R., Cserbak, M.T., Moncman, C.L. & Woychik, R.P. (1983) Nucleotide sequence of a cDNA for the common α-subunit of the bovine pituitary glycoprotein hormones. *J. biol. Chem.* **258**, 4679–4682.

Rathnam, P., Fujiki, Y., Landefeld, T.D. & Saxena, B.B. (1978) Isolation and amino acid sequence of the alpha-subunit of Follicle Stimulating Hormone from equine pituitary glands. *J. biol. Chem.* **253**, 5355–5362.

Reddy, V.B., Beck, A.K., Garramone, A.J., Velucci, V., Lustbader, J. & Bernstine, E.G. (1985) Expression of human choriogonadotropin in monkey cells using a single simian virus 40 vector. *Proc. natn. Acad. Sci. U.S.A.* **82**, 3644–3648.

Reinach, F.C. & Fischman, D.A. (1985) Recombinant DNA approach for defining the primary structure of monoclonal antibody epitopes. *J. molec. Biol.* **181**, 411–422.

Roser, J.F., Papkoff, H., Murthy, H.M.S., Chang, Y.S., Chloupek, R.C. & Potes, J.A.C. (1984) Chemical, biological and immunological properties of pituitary gonadotropins from the donkey (*Equus asinus*): comparison with the horse (*Equus caballus*). *Biol. Reprod.* **30**, 1253–1262.

Sanger, F., Nicklen, S. & Coulson, A.R. (1977) DNA sequencing with chain terminating inhibitors. *Proc. natn. Acad. Sci. U.S.A.* **74**, 5463–5467.

Southern, E.M. (1975) Detection of specific sequences among DNA fragments separated by gel electrophoresis. *J. molec. Biol.* **98**, 503–517.

Stewart, F., Allen, W.R. & Moor, R.M. (1977) Influence of foetal genotype on the Follicle-Stimulating Hormone: Luteinizing Hormone ratio of Pregnant Mare Serum Gonadotrophin. *J. Endocr.* **73**, 419–425.

Stewart, M. & Stewart, F. (1977) Constant and variable regions in glycoprotein hormone beta subunit sequences: implications for receptor binding specificity. *J. molec. Biol.* **116**, 175–179.

Thomas, P.S. (1980) Hybridization of denatured RNA and small DNA fragments transferred to nitrocellulose. *Proc. natn. Acad. Sci. U.S.A.* **77**, 5201–5205.

Vieira, J. & Messing, J. (1982) The pUC plasmids, an M13mp7-derived system for insertion mutagenesis and sequencing with synthetic universal primers. *Gene* **19**, 259–268.

Warwick, J.M. & Wallis, M. (1984) Characterization of mRNA sequences encoding sheep somatotropin (growth hormone) by cloning of pituitary complementary DNA. *Biochem. Soc. Trans.* **12**, 247–248.

Zakin, M.M., Poskus, E., Langton, A.A., Ferrara, P., Santome, J.A., Dellacha, J.M. & Palladini, A.C. (1976) Primary structure of equine Growth Hormone. *Int. J. Pep. Prot. Res.* **8**, 435–444.

J. Reprod. Fert., Suppl. **35** (1987), 9–18

Comparison by three different radioimmunoassay systems of the polymorphism of plasma FSH in mares in various reproductive states

S. L. Alexander, C. H. G. Irvine and Julie E. Turner

Veterinary Science Department, Lincoln College, Canterbury, New Zealand

Summary. FSH was measured in the pituitary, and in pituitary venous and jugular blood collected at frequent intervals from mares in various reproductive states, using 3 validated and highly specific radioimmunoassay systems based on different antibodies, 'o', 'h' and 'e'.

In the pituitary, 4 forms of FSH were found which differed in isoelectric point and relative potency in the 3 assays. In jugular blood, mean FSH concentrations and short-term patterns depended on the assay used and the reproductive state of the mare. In pituitary venous blood, although FSH concentrations were greatly elevated above jugular values, the relationship amongst the 3 assays was similar to that in jugular blood. However, the pulsatility of FSH secretion at oestrus (1–2 pulses per h in 4 of 5 mares) could be observed only in pituitary blood. When the rate of folliculogenesis was slow (acyclicity, oestrus), FSH concentrations measured in assay 'o' were relatively high and showed smaller fluctuations than in assays 'h' and 'e'. By contrast, when folliculogenesis was rapid (transitional phase, dioestrus), FSH tended to be lowest in assay 'o' and pulses had a similar magnitude in all assays.

These results suggest that plasma FSH is polymorphic and that changes in the circulating form may have functional importance. Furthermore, since FSH polymorphism affects immunoactivity, the choice of a radioimmunoassay for clinical or research use is difficult without further FSH structure–function studies to determine which form of the hormone should be measured in a given situation.

Introduction

Horse pituitary FSH, like LH, is a glycoprotein of marked polymorphism (Braselton & McShan, 1970; Reichert, 1971) which is due at least in part to variable sialylation (Aggarwal & Papkoff, 1981). Horse plasma LH appears also to be polymorphic, circulating in different forms depending upon the reproductive state (Alexander & Irvine, 1982, 1984; Adams *et al.*, 1986). The nature of plasma FSH in the horse has not been studied but in other species circulating FSH shows polymorphism (Blum & Gupta, 1985), which is steroid-modulated (Bogdanove *et al.*, 1975) and is thought to have functional importance.

Because desialylation can increase the potency of horse FSH in an homologous radioimmunoassay (RIA) (Aggarwal & Papkoff, 1981), it is possible that the variably sialylated FSH iso-hormones differ in relative affinity for different FSH antibodies. If so, then the nature of plasma FSH could be studied by comparing potency estimates in several RIAs. This would offer an alternative to the more commonly used tools of radioreceptor assay and in-vitro bioassay, which even when sensitive enough may be difficult to apply to the measurement of plasma FSH (van Damme *et al.* 1979; Combarnous & Hengé, 1981) or in which horse LH may cross-react (Papkoff, 1984). On the other hand, if FSH polymorphism affected immunoactivity, then the choice of an RIA system for clinical or research use could influence the results. In that case, FSH structure–function relationships would need to be determined to decide which form of the hormone ought to be measured in a given set of circumstances.

In this paper, we describe the use of 3 different RIA systems to measure horse FSH in the pituitary, secreted by the pituitary, and in blood in mares in various reproductive states.

Materials and Methods

Assays

The 'human system'. This RIA used an antiserum against human FSH (Batch 5, NIADDK), human FSH for iodination (NIADDK) and horse FSH standards (E219 B, gift from Dr H. Papkoff, University of California, San Francisco, CA, U.S.A.), as described previously (Evans & Irvine, 1976). At a final dilution of 1:135 000 the antibody specifically bound 18·4% of added ^{125}I-labelled human FSH (30 000 c.p.m./tube, sp. act. $\sim 100\,\mu$Ci/μg). Non-specific binding was 1·8% of total radioactivity. The minimum detectable concentration (i.e. zero plus 2 s.d.) was 0·59 ng/ml plasma. The mean within-assay coefficient of variation (CV) in the concentration range in which most plasma samples fell was 9·2%. Mean between-assay CV was 7·3%.

The 'ovine system'. This RIA used an antiserum against sheep FSH (H-31, NIADDK), human FSH for iodination (NIADDK) and horse FSH standards (E219 B). A similar assay except that horse FSH is labelled has been validated for use in the horse (Urwin, 1983). At a final dilution of 1:28 000 the antibody specifically bound 16·7% of added ^{125}I-labelled human FSH (15 000 c.p.m./tube). Non-specific binding was 1·0% of total radioactivity. The minimum detectable concentration was 1·5 ng/ml plasma. The antibody showed negligible cross-reaction with horse LH added in amounts up to 10 times those found at the peak of the ovulatory LH surge. Likewise, TSH did not seem to react in the assay, since when 3 dioestrous mares were given thyrotrophin-releasing hormone (150–300 μg i.v., which is sufficient to raise plasma TSH concentration by at least 100%; Thompson *et al.*, 1983) apparent FSH concentrations were not elevated in blood samples collected at 15, 30 and 60 min after injection. Dilutions of plasma from oestrous and dioestrous mares were parallel to the standard curve as assessed by F-test. Mean within- and between-assay coefficients of variation were 4·6% and 10·0%, respectively.

The 'equine system'. This RIA used an antiserum against horse FSH (RB 530, gift from Dr H. Papkoff) and horse FSH (E219 B) for iodination and standards, as described previously (Irvine *et al.*, 1985). At a final dilution of 1 : 175 000 the antibody specifically bound 30·0% of added ^{125}I-labelled horse FSH (15 000 c.p.m./tube, sp. act. $\sim 100\,\mu$Ci/μg). Non-specific binding was 2·5% of total radioactivity. The minimum detectable concentration was 0·50 ng/ml plasma. Mean within- and between-assay coefficients of variation were 4·1% and 8·6%, respectively.

Assays were performed in triplicate using a partly randomized design. To correct for the effect of between-assay variation on results, aliquants of 1 or 2 replicate serum standards were measured in every assay.

Experiments

Isoelectric focussing (IEF) of pituitaries. Fresh horse pituitaries (0·45 kg) were homogenized in 0·025 M-phosphate buffer, pH 7·3, and extracted with 40% ethanol and then acetone as described by Braselton & McShan (1970). The precipitate was recovered by centrifugation, dialysed, lyophilized and reconstituted in 0·025 M-phosphate, pH 7·3. Inert protein was removed by precipitation with phosphoric acid, and the supernatant was dialysed against distilled water then centrifuged and brought to pH 7·4 with 0·05 M-phosphate buffer. The preparation was concentrated to 50 ml by ultrafiltration, 30 ml was loaded onto an IEF column (LKB 8100-1, LKB Produkter AB, Bromma, Sweden), and focussed in a sucrose density gradient between pH 2 and 11 for 21 h. Fifty 10-ml fractions were collected, the pH was measured and the fractions were dialysed against distilled water and then lyophilized. After reconstitution in 1 ml 0·01 M-phosphate buffer, pH 7·0, fractions with a pH < 7·7 were assayed in each of the 3 FSH RIAs using a 3 × 3 assay design.

Collection of jugular blood samples. Two series of experiments were performed in consecutive years to compare jugular FSH concentrations at various reproductive states in (1) the ovine and human systems, and (2) the ovine and equine systems. The series could not be run concurrently because of a temporary shortage of anti-human FSH. Jugular cannulae (16-gauge, 11·6 cm; Angiocath: Deseret Co., Sandy, UT, U.S.A.) were inserted in all mares and blood samples were collected into tubes containing EDTA (1 mg/ml blood).

Five seasonally acyclic mares with inactive ovaries and 3 with ovarian follicles > 15 mm in diameter were bled at 30-min intervals for 12 h. The last 3 mares ovulated for the first time of the season within a month of blood sampling and therefore were considered to have been in the transitional period between non-breeding and breeding seasons at the time of the experiment. To compare endogenously generated FSH pulses with GnRH-induced ones, 4 mares with inactive ovaries were given a physiologically sized dose of GnRH (10–50 μg i.v.; Alexander & Irvine, 1986) and bled at −2, −1, −0·5, 0, 0·25, 0·5, 0·75 and 1 h from injection. All samples were assayed in the ovine and human systems. The GnRH stimulation test was repeated in another 4 mares with inactive ovaries and all samples were assayed in the ovine and equine systems.

Two oestrous mares were bled at 30-min intervals for 12 h and samples assayed in the ovine and human systems. Four oestrous mares were bled at 10-min intervals for 3 h and samples assayed in the ovine and equine systems. Two of these 4 mares were also bled at 10-min intervals for 1 h, 12 days after ovulation.

Collection of pituitary venous blood. To compare the forms of FSH secreted by the pituitary with those in circulation, relatively undiluted pituitary venous blood was collected from unanaesthetized horses using the non-surgical method of Irvine & Huun (1984). Briefly, a 100-cm No. 7 Swan-Ganz cannula was inserted into the facial vein and was manipulated along a venous pathway until its tip entered the intercavernous sinus (ICS) so as to lie near the outlet of the pituitary veins. The ipsilateral jugular vein was also cannulated as described above. Pituitary and jugular blood were collected from mares in seasonal acyclicity (N = 2; pituitary blood every 5 min, jugular every 15 min for 12 h), in oestrus (N = 5; pituitary blood every 5 min, jugular every 5 or 10 min for 3–6 h) and in dioestrus (N = 3; pituitary blood every 5 min, jugular every 15 min for 24 h). The bleeding schedules were chosen on the basis of gonadotrophin pulse frequencies at these 3 reproductive states (Alexander & Irvine, 1986) to observe at least 1 pulse per experiment. All samples were assayed for LH, and then samples encompassing 1–2 h around secretory episodes were assayed in each of the 3 FSH assay systems. Occasionally there was insufficient plasma to do all assays, in which case measurements in the equine system were omitted. The order in which FSH assays were performed on each set of samples was varied.

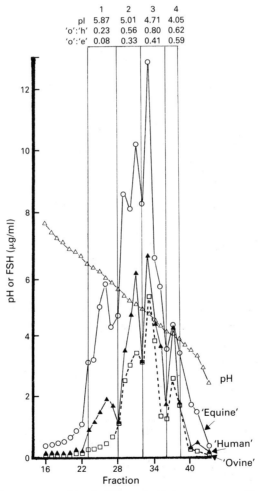

Fig. 1. FSH concentrations measured by 3 different radioimmunoassay (RIA) systems, 'ovine' (□--□), 'equine' (○—○) and 'human' (▲—▲) in fractions eluting at different pH (△—△) after isoelectric focussing of a crude extract of horse pituitaries. Vertical lines represent arbitrary boundaries between FSH peaks. The pI of each peak fraction and the ratios of its potency estimates measured in the ovine and human systems ('o':'h') and in the ovine and equine systems ('o':'e') are shown at the top of the graph.

Statistical analysis

Comparisons between assay systems were made by Student's paired *t* test except as noted in the text. Ratios were converted to logarithms before analysis, and means of ratios are geometric. Data are expressed as means ± s.e.m.

Jugular FSH pulses were identified using criteria based on assay statistics which have been described previously (Alexander & Irvine, 1986).

Results

Isoelectric focussing

After isoelectric focussing of a crude extract of pituitaries, FSH eluted in 3 major peaks with isoelectric points (pI) of 5·01, 4·71 and 4·05, respectively. A minor peak occurred at a pI of 5·87 (Fig. 1). The distribution of FSH in these 4 peaks differed in the 3 assay systems so that marked variation was found when potency estimates for peak fractions were compared between pairs of assays (Fig. 1). In particular, the form of FSH with a pI of 5·87 was poorly recognized by the ovine system in comparison to the other 2 systems.

Jugular and pituitary venous FSH concentrations in acyclic mares

In jugular blood, mean between-pulse FSH concentrations were higher in the ovine than in the

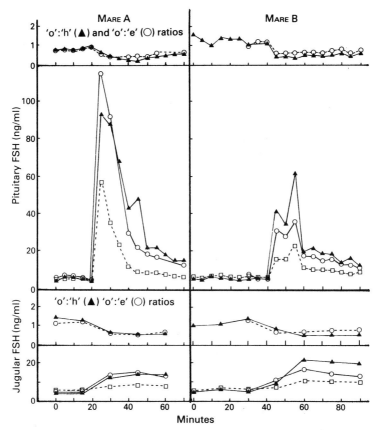

Fig. 2. FSH concentrations measured in the ovine (□--□), equine (○—○) and human (▲—▲) RIA systems in pituitary and jugular blood collected at frequent intervals from 2 seasonally acyclic mares (A and B). The 'o':'h' (▲—▲) and 'o':'e' (○--○) ratios in each sample are shown above each graph of FSH concentrations.

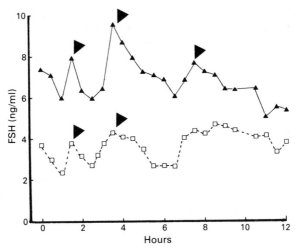

Fig. 3. FSH concentrations measured in the ovine (□--□) and human (▲—▲) RIA systems in jugular blood collected every 30 min from a mare in the transition between non-breeding and breeding seasons. Pulses are marked by arrow heads.

equine system ($P < 0.01$) and slightly but not significantly higher in the ovine than in the human system (Table 1). However, the increase in jugular concentrations during concurrent pulses of secretion was greater when measured in the human than in the ovine system, leading to a drop ($P < 0.05$, 5 mares bled every 30 min for 12 h) in the ratio between concentrations measured in the ovine and human systems ('o':'h' ratio) from 1.18 ± 0.11 to 0.77 ± 0.06 at the peak of the pulse. The relatively small size of fluctuations measured in the ovine system led to fewer increases meeting the criteria for classification as pulses and thus pulse frequency was lower ($P < 0.05$) when determined by the ovine (0.08 ± 0.0 pulses/h) than by the human system (0.2 ± 0.04 pulses/h). The response to a small dose of GnRH was also smaller ($P < 0.05$) in the ovine (before injection: 5.3 ± 1.3 ng/ml; increase after GnRH: 5.4 ± 0.63 ng/ml) than in the human system (before injection: 5.1 ± 1.3 ng/ml; increase: 43.8 ± 9.5 ng/ml). The 'o':'h' ratio therefore fell ($P < 0.05$) from 1.02 ± 0.12 before GnRH to 0.13 ± 0.01 after injection. Likewise, the ovine to equine ('o':'e') ratio decreased ($P < 0.02$) from 1.79 ± 0.11 before GnRH to 0.97 ± 0.08 after injection.

Results in the 2 acyclic mares with pituitary venous cannulae were consistent with these observations (Fig. 2). In both mares, the increase in jugular FSH concentrations during the pulse was least as measured by the ovine system. During the pulse, FSH concentrations in intercavernous sinus blood were greatly elevated over those in the jugular (Fig. 2), confirming the proximity of the cannula to the pituitary. Nevertheless, changes during the pulse in the 'o':'h' and 'o':'e' ratios were similar to those measured in jugular blood (mean 'o':'h', before pulse—pituitary 1.06, jugular 1.26; at pulse's peak—pituitary 0.49, jugular 0.52; mean 'o':'e', before pulse—pituitary 0.92, jugular 1.19; at peak—pituitary 0.49, jugular 0.59). In Mare A, the duration of secretion of FSH as indicated by a positive concentration gradient between pituitary venous and jugular blood was markedly shorter when measured in the ovine than in the other 2 systems (Fig. 2).

In 3 mares in the transition between non-breeding and breeding seasons, mean jugular FSH concentrations were higher ($P < 0.01$) in the human (6.6 ± 0.6 ng/ml) than in the ovine system (3.6 ± 0.41 ng/ml). During pulses, both ovine and human systems measured comparable fractional increases (Fig. 3), so that the 'o':'h' ratio during concurrent pulses was not different from basal values (before pulse, 0.52 ± 0.01; at peak, 0.51 ± 0.01; 5 pulses). Consequently, the number of FSH pulses detected was similar in both systems (pulses per h, ovine: 0.23 ± 0.06; human: 0.19 ± 0.03).

Jugular and pituitary venous FSH concentrations in cyclic mares

At oestrus, mean jugular FSH concentrations were higher ($P < 0.05$) in the ovine than in the human system, and slightly but not significantly higher in the ovine than in the equine system (Table 1). One mare had extraordinarily high FSH when measured in the ovine system (mean 'o':'h', 7·9) and her data were not included in the means shown in Table 1. In all systems, concentrations were relatively stable and pulses were not detected consistently in any of the 3 systems, even when mares were bled at 10-min intervals (pulses per h, ovine: 0·09 ± 0·06 (6 mares); human: 0·21 ± 0·12 (N = 2), equine: 0·14 ± 0·08 (N = 4)). Nevertheless, collection of pituitary venous

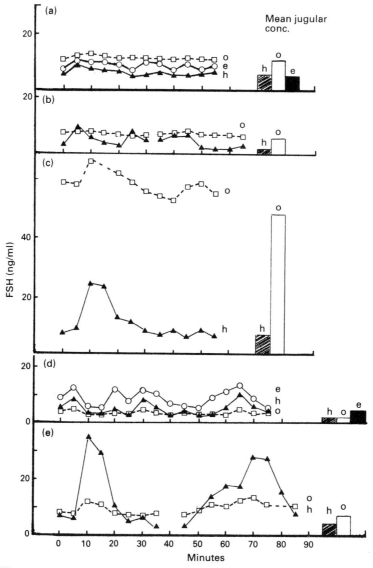

Fig. 4. FSH concentrations measured by the ovine (□--□), human (▲—▲) and, in some cases, equine (○—○) RIA systems in pituitary blood collected every 5 min from oestrous mares before ovulation (a,b,c) or on the day of ovulation (d,e). Graphs (b) and (e) show data from the same mare at these 2 times. The bars represent mean jugular FSH concentrations in the 3 assays.

Table 1. FSH concentrations measured by 3 different radio-immunoassay systems, 'human', 'ovine', and 'equine' in jugular blood from mares in various reproductive states

| Group | FSH conc. (ng/ml) | | |
	'Human' RIA	'Ovine' RIA	'Equine' RIA
Acyclic			
1	5·1 ± 0·72 (7)	6·6 ± 0·31 (7)	—
2	—	5·8 ± 0·28* (7)	3·6 ± 0·36 (7)
3	4·2 ± 0·10 (2)	5·8 ± 0·35 (2)	4·8 ± 0·15 (2)
Oestrous			
4	3·5 ± 0·53 (6)	6·6 ± 1·1* (6)	—
5	—	5·7 ± 0·82 (4)	4·6 ± 0·24 (4)
6	4·0 ± 1·2 (2)	6·6 ± 3·8 (2)	4·8 ± 0·10 (2)
Dioestrous			
7	—	6·8 ± 0·92 (5)	8·8 ± 1·1 (5)
8	11·8 ± 4·5 (3)	7·1 ± 1·3 (3)	9·7 ± 1·8 (3)

Values are mean ± s.e.m. for the no. of mares indicated in parentheses.
*$P < 0.05$ when compared within group by paired t test.

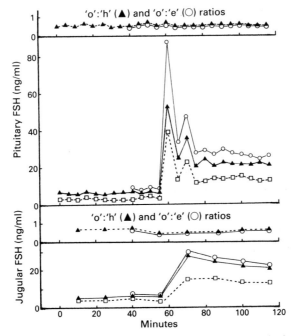

Fig. 5. FSH concentrations measured in 3 different RIA systems (ovine, □--□; human, ▲—▲; equine, ○—○) in pituitary and jugular blood collected at frequent intervals from a typical dioestrous mare. The 'o':'h' (▲--▲) and 'o':'e' (○—○) ratios in each sample are shown above each graph of FSH concentrations.

blood showed that the secretion of the hormone was pulsatile, particularly when measured in the human or the equine systems (Fig. 4) with a frequency of 1–2 pulses per h in 4 of the 5 mares. Therefore, during secretion, the 'o':'h' ratio in pituitary blood fell ($P < 0.01$) to 1·49 ± 0·24 from a mean pre-pulse value of 2·65 ± 0·38. Likewise, in both mares in which equine system measurements could be made, the 'o':'e' ratio dropped during secretion from a mean value of 0·99 to 0·69.

In dioestrous mares, mean between-pulse FSH concentrations were slightly but not significantly higher in the equine and human systems than in the ovine system (Table 1). During pulses, the fractional increase in jugular concentrations was similar in the 3 assay systems, and therefore the 'o':'h' and 'o':'e' ratios remained constant throughout the sampling period (mean 'o':'h'; before pulse 0·62 ± 0·06; at peak 0·55 ± 0·05; during pulse 0·60 ± 0·06; mean 'o':'e'; before pulse 0·67 ± 0·05; at peak 0·61 ± 0·02; during pulse 0·70 ± 0·04; not significantly different, 2-way analysis of variance, N = 3 mares). Attempts to collect pituitary venous blood were unsuccessful in 1 of the 3 dioestrous mares as was shown by the absence of a 'pituitary' to jugular concentration gradient even during a pulse. In the remaining 2 mares, the 'o':'h' and 'o':'e' ratios during a pulse were similar in pituitary and jugular blood (mean 'o':'h'; before pulse—pituitary 0·47, jugular 0·51; at peak—pituitary 0·45, jugular 0·45; mean 'o':'e'; before pulse—pituitary 0·51, jugular 0·58; at peak—pituitary 0·53, jugular 0·57) (Fig. 5).

Discussion

These results show that horse pituitary FSH comprises a range of forms with isoelectric points from 4·0 to 5·9, as has been reported earlier (Braselton & McShan, 1970; Reichert, 1971; Irvine, 1979). Although the 3 RIA systems measured similar FSH patterns after isoelectric focussing, the distribution of immunoactivity within the 4 peaks observed varied with assay system. Thus, the ratio between potency estimates obtained in the ovine and equine systems ('o':'e') varied 7-fold from a minimum at the most alkaline peak (pH 5·87) to a maximum at the most acid (pH 4·01). Likewise, the ovine to human ratio ('o':'h') covered a 4-fold range and was lowest at pH 5·87 but highest at pH 4·57. These differences between systems could not be explained by varied cross-reaction with LH or TSH, which was negligible in the ovine (this paper; Urwin, 1983) and human (Evans & Irvine, 1976) systems, or with free alpha subunits, which would not be expected to bind to heterospecific antibodies (Vaitukaitis, 1978). On the other hand, the equine antibody does show cross-reaction (12% at 50% binding; Irvine *et al.*, 1985) with a highly purified preparation of horse pituitary LH (E98A; Licht *et al.*, 1979) but not with serum LH (Irvine *et al.*, 1985). Preparation E98A has a relatively low sialic acid content of 5·8% (Licht *et al.*, 1979) and therefore could be expected to have an isoelectric point exceeding 5·0 (Irvine, 1979). The very low 'o':'e' ratio in the alkaline range may therefore have been due to the presence of this form of LH. However, the similarity in FSH profiles between the equine and human systems, and the dissimilarity between LH and FSH isoelectric focussing patterns (Irvine, 1979), suggest that LH cross-reaction did not contribute substantially to the equine system results. It therefore appears that the various forms of pituitary FSH differed in relative binding affinity to the 3 antibodies used.

In jugular blood, the mean FSH concentrations and short-term patterns observed depended on the assay system used. Furthermore, the relationship between assay systems varied with reproductive state. In acyclic mares with inactive ovaries and in oestrous mares, FSH concentrations measured in the ovine system were relatively high and, in acyclic mares, showed smaller fluctuations than those measured in the other 2 systems. By contrast, in mares in active folliculogenesis (i.e. transitional phase and dioestrus), FSH concentrations tended to be lowest when measured in the ovine system and fluctuations had a similar magnitude in all 3 assay systems. These observations suggest that plasma FSH, like pituitary FSH, is polymorphic and that the forms in circulation vary with reproductive state. In other species, microheterogeneity of plasma FSH has been reported for the rat (Blum & Gupta, 1985), rhesus monkey (Peckham *et al.*, 1973) and human (Wide, 1982). Since the reproductive state of the animal can modulate the forms of FSH within the pituitary (rat: Robertson *et al.*, 1982; hamster; Ulloa-Aguirre & Chappel, 1982; monkey: Chappel *et al.*, 1984) and in circulation (rat: Bogdanove *et al.*, 1975; monkey: Peckham & Knobil, 1976;

human: Wide, 1982), it has been suggested that changes in the quality as well as the quantity of FSH secreted constitute an important part of the pituitary's signal to the gonads. Our results in the horse support this concept, since changes in the relationship amongst the 3 RIA systems were associated with the rate of folliculogenesis.

In all reproductive states studied, FSH concentrations in intercavernous sinus blood were elevated above jugular values during pulses of secretion demonstrating the proximity of the cannula to the pituitary. In oestrous mares, the pulsatile nature of FSH secretion could be detected in pituitary but not jugular blood. Jugular LH pulses are not easily detected at oestrus because of their small fractional amplitude (Alexander & Irvine, 1986). This is probably because the long circulatory half-life of horse gonadotrophins (~ 5 h; Irvine, 1979) acts to damp peripheral fluctuations particularly when pulse frequency is high as at oestrus.

In oestrous and acyclic mares, the form of FSH secreted was not identical to that in circulation as assessed from the significant decreases in the ovine:human ratio which occurred in pituitary blood during a pulse. Between peaks of secretion, the 'o':'h' ratio was higher, particularly at oestrus, than that observed for any form of FSH isolated from the pituitary, suggesting that the secreted hormone may have been modified peripherally as has been reported for rat LH (Campbell *et al.*, 1978). A peripheral origin for this form of FSH is supported by the small response to endogenous or exogenous GnRH measured in the ovine system at these reproductive states. Alternatively, since the focussed pituitary pool was collected primarily from geldings, it is possible that oestrous and acyclic mares synthesize a form of FSH having a very high 'o':'h' ratio. Presumbly, this molecule would have a long circulatory half-life to explain its preferential accumulation in plasma. A long half-life for the ovine-active molecule is suggested by the slow decay of jugular pulses in acyclic mares and by the apparent preference of the ovine antibody for more acidic FSH forms, which probably contain more sialic acid residues and hence have a longer half-life (Ulloa-Aguirre *et al.*, 1984). Further work is needed to distinguish between these possible explanations.

The situation was simpler in dioestrous and transitional phase mares, in which the form of FSH secreted appeared to be identical to that in circulation and had an 'o':'h' ratio in the range of forms isolated from the pituitary.

The polymorphism of plasma FSH and its effect on immunoactivity make the choice of an FSH RIA system for diagnostic or research use difficult. To decide which form of the hormone should be measured, the physiological role of the various isohormones needs to be determined. That the various FSH isohormones do subserve different functions is suggested by the marked difference in circulating forms between mares with and without developing follicles. Further study is needed before RIA results can be confidently accepted as interpreting the FSH signal of the pituitary to the ovary.

We thank the Grayson Foundation, Inc., Lexington, KY, U.S.A., for funding the intercavernous sinus cannulation studies; and Dr Harold Papkoff and the National Pituitary Agency for gifts of assay reagents.

References

Adams, T.E., Horton, M.B., Watson, J.G. & Adams, B.M. (1986) Biological activity of luteinizing hormone (LH) during the estrous cycle of mares. *Dom. Anim. Endocr.* **3**, 69–78.

Aggarwal, B.B. & Papkoff, H. (1981) Relationship of sialic acid residues to in vitro biological and immunological activities of equine gonadotropins. *Biol. Reprod.* **24**, 1082–1087.

Alexander, S.L. & Irvine, C.H.G. (1982) Radioimmunoassay and in-vitro bioassay of serum LH throughout the equine oestrous cycle. *J. Reprod. Fert., Suppl.* **32**, 253–260.

Alexander, S.L. & Irvine, C.H.G. (1984) Alteration of relative bio-immunopotency of serum LH by GnRH and estradiol treatment in the mare. *Proc. 10th Int. Congr. Anim. Reprod. & A.I., Urbana-Champaign* Vol. II, 3, Abstr.

Alexander, S.L. & Irvine, C.H.G. (1986) Effect of graded doses of gonadotrophin-releasing hormone on serum LH concentrations in mares in various reproductive states: comparison with endogenously generated LH pulses. *J. Endocr.* **110**, 19–26.

Blum, W.F.P. & Gupta, D. (1985) Heterogeneity of rat FSH by chromatofocusing: studies on serum FSH, hormone released in vitro and metabolic clearance rates of its various forms. *J. Endocr.* **105**, 29–37.

Bogdanove, E.M., Nolin, J.M. & Campbell, G.T. (1975) Qualitative and quantitative gonad-pituitary feedback. *Recent Prog. Horm. Res.* **31**, 567–626.

Braselton, W.E. & McShan, W.H. (1970) Purification and properties of follicle-stimulating and luteinizing hormones from horse pituitary glands. *Archs Biochem. Biophys.* **139**, 45–58.

Campbell, G.T., Nansel, D.D., Meinzer, W.M., Aiyer, M.S. & Bogdanove, E.M. (1978) Prolonged infusion of rat luteinizing hormone alters its metabolic clearance patterns: indirect evidence for postsecretory mutation of luteinizing hormone. *Endocrinology* **103**, 683–693.

Chappel, S.C., Bethea, C.L. & Spies, H.G. (1984) Existence of multiple forms of follicle-stimulating hormone within the anterior pituitaries of Cynomolgus monkeys. *Endocrinology* **115**, 452–461.

Combarnous, Y. & Hengé, M.-H. (1981) Equine follicle-stimulating hormone: purification, acid dissociation, and binding to equine testicular tissue. *J. biol. Chem.* **256**, 9567–9572.

Evans, M.J. & Irvine, C.H.G. (1976) Measurement of equine FSH and LH: response of anestrous mares to gonadotropin releasing hormone. *Biol. Reprod.* **15**, 477–484.

Irvine, C.H.G. (1979) Kinetics of gonadotrophins in the mare. *J. Reprod. Fert., Suppl.* **27**, 131–141.

Irvine, C.H.G. & Huun, R. (1984) Long-term collection of pituitary venous blood in the unrestrained horse. *N.Z. med. J.* **97**, 735, Abstr.

Irvine, C.H.G., Alexander, S.L. & Hughes, J.P. (1985) Sexual behavior and serum concentrations of reproductive hormones in normal stallions. *Theriogenology* **23**, 607–617.

Licht, P., Bona-Gallo, A., Aggarwal, B.B., Farmer, S.W., Castelino, J.B. & Papkoff, H. (1979) Biological and binding activities of equine pituitary gonadotrophins and pregnant mare serum gonadotrophin. *J. Endocr.* **83**, 311–322.

Papkoff, H. (1984) Interaction of equine luteinizing hormone with gonadal binding sites of luteinizing hormone and follicle-stimulating hormone. In *Hor-mone Receptors in Growth and Reproduction*, pp. 55–61. Ed. B. B. Saxena. Raven Press, New York.

Peckham, W.D. & Knobil, E. (1976) The effects of ovariectomy, estrogen replacement, and neuraminidase treatment on the properties of the adenohypophysial glycoprotein hormones of the rhesus monkey. *Endocrinology* **98**, 1054–1060.

Peckham, W.D., Yamaji, T., Dierschke, D.J. & Knobil, E. (1973) Gonadal function and the biological and physicochemical properties of follicle-stimulating hormone. *Endocrinology* **92**, 1660–1666.

Reichert, L.E. (1971) Electrophoretic properties of pituitary gonadotropins as studied by electrofocusing. *Endocrinology* **88**, 1029–1044.

Robertson, D.M., Foulds, L.M. & Ellis, S. (1982) Heterogeneity of rat pituitary gonadotropins on electrofocusing; differences between sexes and after castration. *Endocrinology* **111**, 385–391.

Thompson, D.L., Godke, R.A. & Nett, T.M. (1983) Effects of melatonin and thyrotropin releasing hormone on mares during the nonbreeding season. *J. Anim. Sci.* **56**, 668–677.

Ulloa-Aguirre, A. & Chappel, S.C. (1982) Multiple species of follicle-stimulating hormone exist within the anterior pituitary gland of male golden hamsters. *J. Endocr.* **95**, 257–266.

Ulloa-Aguirre, A., Miller, C., Hyland, L. & Chappel, S.C. (1984) Production of all follicle-stimulating hormone isohormones from a purified preparation by neuraminidase digestion. *Biol. Reprod.* **30**, 382–387.

Urwin, V. (1983) The use of heterologous radioimmunoassays for the measurement of follicle-stimulating hormone and luteinizing hormone concentrations in horse and donkey serum. *J. Endocr.* **99**, 199–209.

Vaitukaitis, J.L. (1978) Glycoprotein hormones and their subunits: immunological and biological characterization. In *Structure and Function of the Gonadotropins*, pp. 339–360. Ed. K. W. McKerns. Plenum Press, New York.

van Damme, M.-P., Robertson, D.M., Marana, R., Ritzen, E.M. & Diczfalusy, E. (1979) A sensitive and specific in vitro bioassay method for the measurement of follicle-stimulating hormone activity. *Acta endocr., Copenh.* **91**, 224–237.

Wide, L. (1982) Male and female forms of human follicle-stimulating hormone in serum. *J. clin. Endocr. Metab.* **55**, 682–688.

J. Reprod. Fert., Suppl. **35** (1987), 19–24

Measurement of free cortisol and the capacity and association constant of cortisol-binding proteins in plasma of foals and adult horses

C. H. G. Irvine and S. L. Alexander

Veterinary Science Department, Lincoln College, Canterbury, New Zealand

Summary. A direct method for measuring the capacity of the high-affinity binding protein, CBG, based on charcoal adsorption, was validated for use in the horse.

Several unique aspects of cortisol binding in the horse were observed: (1) CBG content at birth was the lowest of any species studied, (2) CBG concentration increased with age whereas in other species it decreases, (3) the plasma of the new born foal has a binding protein, not reported for other species, which binds as much cortisol as does CBG. Its capacity and affinity are intermediate between albumin and CBG. It may be involved prenatally in increasing dam to fetus transfer of cortisol, and post-natally in buffering the effects of the huge stress-related peripartum release of cortisol.

Introduction

The neonate of the horse is the most mature among the domestic species (Irvine & Evans, 1975) yet if born only 2–3 weeks premature its survival rate is very low. In many species cortisol plays a major role in preparing the fetus for adjustment to extrauterine existence (Liggins, 1976). Also, in some species the fetal adrenocortical axis is pivotal in the induction of parturition (Liggins *et al.*, 1973). A large percentage of cortisol in plasma is firmly bound to the specific binding protein, cortisol-binding globulin (CBG), which prevents it leaving plasma and entering cells (Westphal, 1969). However, the small amount which is free in plasma enters cells rapidly and it is this which influences their metabolism (Ballard *et al.*, 1982). Not surprisingly the amount of free, rather than total, cortisol in plasma has been found to correlate with clinical status (Kaufman *et al.*, 1984). Since no data are available on the concentration of free cortisol in horses of any age, the present study was undertaken to devise techniques for: (1) measuring free cortisol in horses of all ages, and (2) determining the capacity and association constants of horse cortisol binding proteins.

Materials and Methods

Animals

Eleven foals, Thoroughbred or Pony, were bled by jugular venepuncture within 4 h of birth and 8 foals were bled at 3 months of age. Samples were taken also from 11 non-pregnant, untrained horses.

Free cortisol measurement

Total cortisol was measured by radioimmunoassay using magnetic antibody (Rossdale *et al.*, 1982). Free cortisol was measured by a modification of a free thyroxine method (Irvine, 1974), which uses hydrated Sephadex beads as small dialysis sacs allowing small molecules such as cortisol to enter freely but excluding proteins, e.g. CBG.

Because the affinities of cortisol for its binding proteins and for Sephadex differ from those for the thyroxine system, the free thyroxine method required several modifications. The procedure developed was as follows (c.p.m. and cortisol concentration for a typical assay are given in parentheses).

Distribution of cortisol between the Sephadex grains (Sephadex bound) and the excluded volume in the absence of binding protein. Tritiated cortisol (0·3 ng [1,2,6,7-^{3}H]cortisol; sp.act.80 Ci/mol; The Radiochemical Centre, Amersham, U.K.) was diluted in 2·5 ml of a buffer containing 0·04 M-Tris, 0·1 M-NaCl and sufficient HCl to bring the pH to 7·40 (buffer + tracer mixture): 0·2 ml was counted (6722 c.p.m. = 33 608 c.p.m./ml). Duplicate 1·0-ml samples were added to exactly 175 mg Sephadex G-25 coarse (Pharmacia Fine Chemicals, Uppsala, Sweden) in 10 × 75 mm glass tubes in a water bath at 37°C, stirred intermittently for 10 min then allowed to settle for 1 min: 0·2 ml of the supernatant was counted (4964 c.p.m.). Since the excluded volume in this system is 0·674 ml, as determined by addition of a marker not absorbable to Sephadex such as ^{125}I-labelled albumin, and c.p.m. in the excluded volume are unbound, i.e. free, c.p.m. free = 4964 × 0·674/0·2 = 16 728 c.p.m.; Sephadex bound = 33 608 − 16 728 = 16 880 c.p.m. and free/Sephadex bound = 0·990. This fraction is not changed by the addition of protein to the excluded volume (Irvine, 1974).

Determination of the concentration of free cortisol in serum. Buffer + tracer (2·0 ml) was added to 0·5 ml serum which in this example contained 133·9 nmol cortisol/l; 0·2 ml was counted (5682 c.p.m. = 28 411 c.p.m./ml = 27·1 pmol/ml). In duplicate, 1·0 ml was added to exactly 175 mg Sephadex G-25 coarse and equilibrated as above. The supernatant (0·2 ml) was counted (6958) and the c.p.m. in the excluded volume was 6958 × 0·674/0·2 = 23 448, and the Sephadex bound was 28 411 − 23 448 = 4963. Since free/Sephadex bound = 0·990, free c.p.m. in the excluded volume = 4914 (= 4914/28 411 × 27·1 pmol in 0·674 ml = 6·95 nmol/l). The fraction protein bound in the excluded volume = 23 448 − 4914 = 18 534 c.p.m. (18 534/4914 × 6·95 = 26·2 nmol/l).

To determine non-specific binding the assay was repeated in tubes in which 5000 ng cortisol had been dried down before addition of reagents. The quantity of 5000 ng was chosen because the resulting free/bound fraction was high (∼ 3) and identical when amounts from 2000 to 10 000 ng cortisol were added, showing that after addition of 5000 ng cortisol all measurable binding was non-specific, i.e. to high-capacity, low-affinity binding proteins. In the example calculated above, after addition of 5000 ng cortisol, 0·2 ml supernatant gave 4790 c.p.m. and so the excluded volume contained 4790 × 0·674/0·2 = 16 142 c.p.m. and the Sephadex-bound fraction was 28 411 − 16 142 = 12 269 c.p.m. Since free/Sephadex bound = 0·990, free c.p.m. in the excluded volume = 12 146. The protein-bound fraction in the excluded volume = 16 142 − 12 146 = 3996 c.p.m. Free/non-specific counts in this dilute serum was 12 146/3996 = 3·04, irrespective of the amount of cortisol added.

In this example, therefore, in which no cortisol was added, the protein bound fraction was 26·2 nmol/l, the free fraction was 6·95 nmol/l, non-specific binding was 6·95/3·04 = 2·29 nmol/l and the specifically bound fraction was 26·2 − 2·29 = 23·91 nmol/l.

Determination of the association constant (K) and capacity (C) of cortisol binding protein(s). The assay was repeated in tubes in which 2·5, 5, 10, 20 and 40 ng cortisol had been dried down before addition of reagents. Free and specifically bound cortisol concentrations were plotted to obtain *K* and *C* values (Scatchard, 1949).

Measurement of CBG by charcoal adsorption

CBG was measured by the charcoal adsorption method described by Ballard *et al.* (1982) for use in the sheep. Briefly, duplicate 0·1 ml aliquants of plasma were added to tubes in which 0·3 ng [^{3}H]cortisol and 30 ng unlabelled cortisol had been dried down. After vortexing, the tubes were incubated for 30 min at 37°C, vortexed and then incubated at 2°C for 30–120 min. To each tube 0·03 ml acid-washed, activated charcoal solution (85 mg/ml in 0·04 M-Tris pH 7·4) was added. The tubes were vortexed for 3 sec and after 3 min were centrifuged for 5 min at 6000 *g* in the cold. Immediately 0·05 ml was removed for counting. CBG capacity was calculated as the total cortisol in the system multiplied by the fraction bound in the supernatant. When it became apparent that the nature and amounts of proteins to which cortisol was bound in the horse neonate differed greatly from that in sheep, the validity of the method was investigated by: (a) doubling and halving the concentration of charcoal used, (b) varying the amount of unlabelled cortisol added from 20 to 52 ng/0·1 ml plasma, (c) increasing the incubation time with charcoal, (d) heating plasma to 60°C for 15 min to reduce CBG affinity.

Data are expressed as mean ± s.e.m. unless otherwise noted.

Results

Free cortisol assay

When the quantities and volumes used in the original method (Irvine, 1974) were modified as described in 'Materials and Methods' the distribution of [^{3}H]cortisol between the phases, free, specifically bound, non-specifically bound and Sephadex bound was optimal to enable accurate calculations to be made. However, the measurements were made in diluted serum; furthermore, cortisol was added to serum as carrier for the tracer, and removed by Sephadex. These changes in the ratio of the components, i.e. cortisol, proteins and water, in a system containing low-capacity binding proteins such as CBG, would alter the percentage free, and thus the estimate of free cortisol obtained. It was necessary to correct for these factors to determine the concentration of free cortisol in unchanged serum. This correction was made by calculating K_a(CBG) from the values of free cortisol, specifically and non-specifically bound cortisol obtained for a 0·2 in 0·674 ml dilution of each unknown serum sample as described above, using the mass action equation:

$$K_a = [\text{cortisol bound to CBG}]/([\text{free cortisol}][\text{free CBG}])$$

Free CBG was obtained by subtracting the amount of cortisol bound to CBG from the total CBG capacity. Total CBG, determined by charcoal adsorption as described in 'Materials and Methods', was 207·55 nM for the undiluted specimen serum, or 61·59 nM for the 0·2/0·674 serum dilution used to obtain the other data needed to obtain K_a in diluted serum. Using these values (in nM)

$$K_a = \frac{23·91}{6·95\,(61·59 - 23·91)} = 0·091\,\text{nM}^{-1}$$

Since K_a is independent of concentration, free and specifically bound cortisol could be determined for unchanged serum by substitution of the following known values into the mass action equation.

Total cortisol concentration in the serum sample was 133·9 nM consisting of free [F] + specifically bound [SB] + non-specifically bound [NSB] in which the fraction [F]/[NSB] in a 0·2/0·674 dilution of serum was 3·04, or 0·2/0·674 × 3·04 = 0·90 in undiluted serum. Therefore, 133·9 = [F] + [SB] + 1·1[F]. Substituting:

$$0·091 = \frac{[\text{SB}]}{\{(133·9 - [\text{SB}])/2·1\}\{207·55 - [\text{SB}]\}}$$

In undiluted serum, therefore, [SB] = 108·64 nM, and [F] = 12·03 nM.

The concentrations of total, free, specifically bound and non-specifically bound cortisol were 135·8 ± 17·4, 10·5 ± 2·6, 113·6 ± 14·0 and 11·7 ± 2·8 nM respectively in 11 adult horses, 109·3 ± 13·8, 27·3 ± 4·8, 27·3 ± 4·9 and 27·3 ± 4·2 nM in 11 newborn foals and 31·8 ± 6·6, 3·32 ± 0·8, 24·82 ± 6·3, and 3·69 ± 0·7 nM in 8 3-month-old foals. K_a values for adults, newborn foals and 3-month-old foals were 0·091 ± 0·008, 0·054 ± 0·005 and 0·090 ± 0·009 nM^{-1} respectively.

Measurement of CBG capacity by charcoal adsorption

Preliminary experiments with foal and adult plasma showed that CBG measurements were unaffected by: (a) doubling (mean ± s.d., 96·0 ± 3·2% of standard) or halving (98·1 ± 2·7%) the amount of charcoal used, (b) varying the amount of unlabelled cortisol (110% of standard at twice the added cortisol), or (c) increasing the incubation time with charcoal from 3 to 5 min (5 min estimate 97·7% of 3-min estimate, although there was a marked decline by 10 min). In 2 trials, heating plasma to 60°C for 15 min decreased CBG capacity estimates to 31·8% or 45·4% of pre-heating values. The method as originally described gave highly reproducible results with neonatal or adult serum with complete removal of cortisol from non-specifically bound proteins and insignificant stripping from specifically bound proteins. Within-assay coefficient of variation was 2·04% ($n = 27$) and between-assay coefficient of variation was 4·25% ($n = 12$). CBG capacity was 233·7 ± 14·7 nM in 11 adults, 45·9 ± 2·2 nM in 11 newborn foals and 107·7 ± 7·3 nM in 8 3-month-old foals.

Measurement of the association constant (K_a) and capacity (C) of cortisol-binding protein(s) by sequential cortisol displacement

In 11 adult horses Scatchard plots of SB/F, against [SB], gave a good fit to a straight line from which a K_a value of 1·01 ± 0·066 × 10^8 M^{-1} and a capacity of 71·8 ± 7·06 nM in diluted serum or 242.0 ± 23.8 nM in undiluted serum were obtained. The correlation between the charcoal adsorption and cortisol displacement methods for CBG measurement was 0.90.

Since [F]/[NSB] was 0.90 in undiluted serum and mean [F] was 10.5 nM, [NSB] was 11.7 nM. Assuming that this is bound to albumin, which is widely believed to be the only protein other than CBG capable of binding cortisol, and that the albumin concentration was 4.4×10^{-4} M, substitution into the equation:

$$K_{a(alb)} = \frac{[albB]}{[F]([alb\ capacity] - [albB])} = (in\ nM)\frac{11.7}{10.5(4.4 \times 10^5 - 11.7)}$$

and $K_{a(alb)} = 2.5 \times 10^3$ M^{-1} which is almost identical to that reported for other species (Dalle *et al.*, 1980; Ballard *et al.*, 1982).

In 11 newborn foals the ratio of [free]:[NSB] estimated by addition of excess cortisol (5000 ng/tube) to diluted serum was 1.69:1 compared with 3.04:1 in adult horses. Furthermore, [NSB] was approximately twice that calculated from [albumin] and $K_{a(albumin)}$. Displacement with increasing amounts of cortisol generated curvilinear Scatchard plots, after deduction of NSB to albumin, which could be resolved into one component with a K_a of 0.60×10^8 M^{-1} and a capacity of 40.1 ± 3.6 nM, and another component with an affinity of the order of 10^6 M^{-1}. Its capacity could not be determined from Scatchard analysis because it intercepted the baseline at a very obtuse angle. However, when the sum of the concentrations found to be CBG bound, albumin bound and free (27.3, 27.3 and 27.3 nM respectively) were subtracted from the total of 109.3, the remainder was 27.3 nM. The amount of this neonatal component declined steadily but was still detectable at 3 months of age.

Discussion

There is considerable evidence that the physiological activity of cortisol resides with the unbound steroid, rather than the bound forms (Ballard *et al.*, 1982). Fluctuations in the concentration of CBG and free cortisol have been described in neonates of several species; however, there are no data for the horse. There is little uniformity in the methods used to measure free cortisol or in the criteria for a valid assay. The present method for measuring free cortisol was developed to meet the following criteria.

(1) Appropriate correction must be made for: (a) changes in the amount of cortisol caused by removal by Sephadex and addition with tracer, and (b) changes in the volume of aqueous phase caused by removal by Sephadex and addition with diluent.

(2) Since the K_a values for specific binding proteins, and hence the free fraction, are strongly dependent on temperature, pH and ionic composition of the medium, any assay system used must preserve physiological conditions without introducing interference in equilibria. For example, buffers with supraphysiological phosphate concentrations have been shown to give spurious data on interactions between hormones and specific binding proteins (Irvine, 1974) yet are universally used as diluents in free cortisol assays. The Tris buffer used in the present method gave the same results as a totally physiological system using bicarbonate buffer. Since various non-physiological factors may affect cortisol binding to each protein to a different degree, the concentrations obtained with non-physiological conditions bear no constant relationship to those occurring *in vivo* (Irvine, 1974).

(3) During the prolonged incubation required for conventional equilibrium dialysis microbial growth and pH changes must be prevented without incorporating substances which affect binding equilibria. A rapid microdialysis sytem using hydrated microspheres, such as Sephadex, overcomes this problem.

The method described here appears to be the first free cortisol assay which fulfils the above criteria.

There was good correlation between estimates of CBG capacity by charcoal adsorption and by the changes in free cortisol induced by sequential displacement analysis. Since these estimates depend on quite different techniques this provides support for the validity of the free cortisol method employed.

While the mean $K_{a(CBG)}$ was approximately $0.9 \times 10^8 \text{ M}^{-1}$ in the adult horse it was only $0.5 \times 10^8 \text{ M}^{-1}$ in the neonate. Values for $K_{a(CBG)}$ similar to that observed in the adult horse have been reported for adults of other species (in M^{-1}, human: 0.7×10^8 (Westphal, 1969); monkey: 0.93×10^8 (Westphal, 1969); guinea-pig: 1.0×10^8 (Dalle *et al.*, 1980). In the guinea-pig, unlike the horse, the affinity of CBG for cortisol is similar in neonates and adults (Dalle *et al.*, 1980).

In the few species studied there is a marked decrease in CBG capacity from birth through infancy. For example, in the guinea-pig CBG falls steadily from birth to Day 20 (Dalle *et al.*, 1980) and in the sheep from birth until the last measurement which was on Day 21 (Ballard *et al.*, 1982). In both species adult values were similar to the last infancy values. In contrast, in the horse the values increased markedly from 45·9 nM at birth to 107·7 nM at 3 months and 233·7 nM in adults. The CBG capacity in the horse neonate is by far the lowest of any species (in nM, horse: 45·9; sheep: 314·6 (Ballard *et al.*, 1982); monkey: 138–276 (Beamer *et al.*, 1973); baboon: 580 (Oakey, 1975); human: 250 (De Moor *et al.*, 1962); guinea-pig: 2498 (Dalle *et al.*, 1980); rat: 400 (Van Baelen *et al.*, 1977). However, the binding of cortisol in the horse neonate is supplemented by an unidentified protein which has not been recorded in any other species. This protein has a capacity and affinity intermediate between cortisol and albumin, and binds as much cortisol as does CBG. Its effect will be to increase the fetal cortisol pool and also the potential rate of transfer from dam to fetus. It may serve to damp the rise in plasma free cortisol induced by the sudden increase in cortisol secretion which begins immediately before full term delivery (M. Silver, personal communication). By 30–60 min after birth total plasma cortisol concentration reaches a peak of about 400 nmol/l (Silver *et al.*, 1984), which is ∼9 times the CBG capacity reported in this paper. The foal therefore appears to be unique in this respect since total plasma cortisol levels rarely exceed CBG capacity in any species in any circumstances, and, indeed, without this extra binding protein the free cortisol concentration in plasma, and in tissues in equilibrium with it, would reach extremely high levels. This neonatal protein may therefore compensate for the low levels of CBG in the newborn foal.

We thank Robert Cash and Peter Rossdale of Beaufort Cottage Stables, Newmarket, for assistance; Marian Silver of the Physiological Laboratory, University of Cambridge, for help and facilities; and the Wellcome Trust for support.

References

Ballard, P.J., Kitterman, J.A., Bland, R.D., Clyman, R.I., Gluckman, P.D., Platzker, A.C.G., Kaplan, S.L. & Grumbach, M.M. (1982) Ontogeny and regulation of corticosteroid binding globulin capacity in plasma of fetal and newborn lambs. *Endocrinology* **110**, 359–366.

Beamer, N., Hagemenas, F.C. & Kittinger, G.W. (1973) Development of cortisol binding in the Rhesus monkey. *Endocrinology* **93**, 363–368.

Dalle, M., El Hani, A. & Delost, P. (1980) Changes in cortisol binding and metabolism during neonatal development in the guinea-pig. *J. Endocr.* **85**, 219–227.

De Moor, P., Heirwegh, K., Hereman, J. & Declerck-Raskin, M. (1962) Protein binding of corticoids studied by gel filtration. *J. clin. Invest.* **41**, 816–824.

Irvine, C.H.G. (1974) Measurement of free thyroxine in human serum by a Sephadex binding method. *J. clin. Endocr. Metab.* **38**, 468–476.

Irvine, C.H.G. & Evans, M.J. (1975) Post natal changes in total and free thyroxine and triiodothyronine in foal serum. *J. Reprod. Fert. Suppl.* **23**, 709–715.

Kaufman, H., Bruhis, S., Roitman, A., Baumann, B. & Laron, Z. (1984) Evidence of familial transcortin deficiency. *Proc. 7th Int. Congr. Endocr. Quebec.* Abstr. 1117.

Liggins, G.L. (1976) Adrenocortical related maturational events in the fetus. *Am. J. Obstet. Gynec.* **126**, 931–939.

Liggins, G.C., Fairclough, R.J., Grieves, S.A., Kendall, J.Z. & Knox, B.S. (1973) The mechanism of initiation of parturition in the ewe. *Recent Prog. Horm. Res.* **29**, 111–150.

Oakey, R.E. (1975) Serum cortisol binding capacity and

cortisol concentration in the pregnant baboon and its fetus during gestation. *Endocrinology* **97**, 1024–1029.

Rossdale, P.D., Burguez, P.N. & Cash, R.S.G. (1982) Changes in blood neutrophil/lymphocyte ratio related to adrenocortical function in the horse. *Equine vet. J.* **14**, 293–298.

Scatchard, G. (1949) The attraction of proteins to small molecules and ions. *Ann. N.Y. Acad. Sci.* **41**, 660–672.

Silver, M., Ousey, J.C., Dudan, F.E., Fowden, A.L., Knox, J., Cash, R.S.G. & Rossdale, P.D. (1984) Studies on equine prematurity 2: postnatal adrenocortical activity in relation to plasma adrenocorticotrophic hormone and catecholamine levels in term and premature foals. *Eq. vet. J.* **16**, 286–291.

Van Baelen, H., Vandoren, G. & De Moor, P. (1977) Concentration of transcortin in the pregnant rat and its foetuses. *J. Endocr.* **75**, 427–431.

Westphal, U. (1969) Assay and properties of corticosteroid-binding globulin and other steroid-binding serum proteins. *Meth. Enzymol.* **XV**, 761–796.

J. Reprod. Fert., Suppl. **35** (1987), 25–31

Printed in Great Britain
© 1987 Journals of Reproduction & Fertility Ltd

Assessment of fertility and semen evaluations of stallions

H. Rousset, Ph. Chanteloube*, M. Magistrini* and E. Palmer*

*Institut du Cheval, 19230 Arnac-Pompadour, and *I.N.R.A., Reproductive Physiology,
37380 Nouzilly, France*

Summary. (1) Various estimations of motility (subjective appreciation, count on video record, Doppler laser apparatus and optic-microcomputer analyser) or percentage of live spermatozoa (eosin–nigrosin staining) showed a higher repeatability for measurements of a same sample than for straws of a same ejaculate: the values were high (respectively > 0.78 and > 0.69) except for the optical analyser and staining.

(2) Semen samples were collected from 80 stallions 5 times at 24-h intervals. The repeatability varied from 0.37 to 0.69 for gel-free volume, concentration, total sperm number, % of motile spermatozoa in raw and extended semen at collection or after storage. Therefore, to assess a stallion, 2–7 ejaculates, according to the characteristic being measured, are needed to get a repeatability of 0.80 for the mean value.

(3) A study on 79 stallions serving mares over 4–152 oestrous cycles per year showed that 17 seasons are necessary to measure fertility based on the pregnancy rate per oestrous cycle with the same accuracy. If they served more than 25 cycles every year, 6 seasons only were required. In both cases, it was less than for % of live foals per season.

Introduction

Examination of stallion semen can be practised on many occasions, e.g. purchase, suspicion of low fertility or selection for artificial insemination. However, the real semen characteristics or the real fertility of a sire are never known because many sources of variation modify the estimation from one measurement to another. Some fixed sources of variation can be taken into account, e.g. season (Pickett *et al.*, 1976), age (Amann *et al.*, 1979) or frequency of ejaculation (Sullivan & Pickett, 1975), whereas others are due to biological phenomena of random variations or unavoidable error at any measurement (Palmer & Fauquenot, 1985).

Practical estimates of the semen characteristics or fertility of a given stallion might therefore be considered to be a problem. The number of ejaculates examined varies from 2 collected 1 h apart (Pickett *et al.*, 1976) to 7 consecutive daily collections (Gebauer *et al.*, 1974). Fewer recommendations are given on the number of repeated measurements for a sample. These numbers have never been established on a statistical basis. The objective of this study was therefore to determine for each variation level how many repetitions are necessary to assess each characteristic with a given precision.

Materials and Methods

Experiment 1

In a study on the effect of ejaculation frequency on frozen semen (Magistrini *et al.*, 1987), 336 ejaculates were collected from 6 stallions (56 collections each) and frozen. Their vitality, i.e. percentage of live or motile spermatozoa, was estimated according to different techniques after thawing for 30 sec at $+37°C$, dilution 1:4 (v/v) with I.N.R.A. 82 extender (final concentration: 20×10^6 spermatozoa/ml) and incubation for 10 min.

Counts on video record and subjective estimation. The video method was modified from that of Katz & Overstreet (1981) and was as described by Magistrini *et al.* (1987). For subjective measurement, a technician estimated the percentage of motile spermatozoa at the screen of the monitor.

Eosin–nigrosin staining. Diagonal scans of 150 spermatozoa smeared over each slide were used to detect live (not coloured) or dead (coloured) spermatozoa (Ortavant, 1952).

Optical motility analyser. A computerized electro-optical system, Semen Analyser (SEM Israel Ltd, P.O. Box 1545, Ramat-Gan, Israel) provided, after an interval of 30 sec, an index of sperm motility (S.M.I., Bartoov *et al.*, 1981). The sample was placed in a microcapillary tube and inserted in a thermoregulated holder at $+37°C$.

Spermokinesimeter. Based on light diffusion by moving cells (Dopler–Fizeau effect), the SORO spermokinesimeter measured the diffused light spectrum integral for a still or moving sample (50 µl droplet of extended semen). Motility was directly achieved by the ratio of the two figures (Dubois *et al.*, 1975).

To measure repeatability within samples, these different estimations were repeated ($n = 2$–4) in one straw of the 56 ejaculates of a stallion which showed a great variability among ejaculates (Exp. 1a). To estimate repeatability within ejaculates, 2 straws of the 336 ejaculates were examined (except SORO, Exp. 1b).

Experiment 2

Stallions of different breeds (23 Thoroughbred, 8 French Trotters, 9 French Saddle and 40 Breton draught stallions), ranging in age from 2 to 11 years, were examined during the years 1980–1984 in two national studs (Le Pin, Hennebont) during the purchase procedures (October to February) or at the end of the breeding season (July to August).

Five successive ejaculates were collected at daily intervals. Immediately after collection, semen was filtered and sperm concentration was measured with a spectrophotometer. Motility (percentage of motile spermatozoa) was microscopically noted by visual estimation just after collection (raw and extended semen) and after 24 h storage at $+4°C$ (extended semen). The percentage of abnormal spermatozoa was finally determined, for the draught stallions only, in a haemocytometer cell. No distinction was made between the different morphological characteristics, since some percentages were assessed on very small numbers. Gel volume was not considered as, in most cases, the response was only quantal (presence or absence).

Experiment 3

This study was conducted over 5 breeding seasons (1981–1985) at the Hennebont National Stud within a population of 79 Breton draught stallions, aged from 4 to 16 years. For each season, pregnancy rate per oestrous cycle was determined as follows:

$$(P + 0·7 \times NR)/(\text{all cycles except those of unknown results})$$

where: P = pregnant cycles (echography diagnosis) and NR = cycles followed by non-return to oestrus (i.e. more than 21 days after oestrus). Based on large statistical numbers, in this type of management, 70% of non-return values are pregnancies.

These stallions were mainly used for natural matings.

Selection was also made of a subpopulation of 49 stallions which were used for more than 25 oestrous cycles per season.

Statistical analysis

Data collected from all these experiments were submitted to the same type of analysis.

First, variables for which the χ^2 goodness-of-fit test rejected the normality hypothesis were treated as follows: individuals were ranked and then split into classes. Limits were chosen so that the repartition of individuals among classes fitted a normal distribution. Each individual was given the number of its class as the transformed data. The number of classes was 5 for vitality estimations (cumulated frequencies of each class equal to 7, 30, 70, 97 and 100%) and 10 for other variables (cumulated frequencies equal to 3, 7, 15, 30, 50, 70, 85, 93, 97 and 100%). This transformation did not concern pregnancy rate data that were normally distributed.

In Exps 1 and 3, the statistical model was a simple one, with random hierarchical effects for samples and measures within samples (samples are drops or smears according to the technique; Exp. 1a) for ejaculates and straws within ejaculates (Exp. 1b), for stallions and years within stallions (Exp. 3).

In Exp. 2, the model was mixed, with fixed effects (breed, rank of ejaculate) and random hierarchical effects (stallions within breeds and ejaculates within stallions).

We were then able to calculate the between ($\sigma^2 b$) and within ($\sigma^2 w$) class-variance at any random level (stallion, ejaculate, or sample) and the repeatability or intra-class correlation ($r = \sigma^2 b/\sigma^2 b + \sigma^2 w$)), which is also the proportion of total variance of an observation at the level being considered.

Finally, we determined the number of repetitions (measures, straws, ejaculates or years) necessary to get a repeatability of 0·80 for their mean value. The repeatability of the average of N repeated measurements is:

$$\frac{N \times r}{1 + (N-1) \times r} \quad (r = \text{repeatability of individual measurements})$$

Therefore, N is defined as:

$\frac{4(1-r)}{r}$, where r is the previously calculated repeatability.

Results

Tables 1–4 present the mean and standard deviation, between and within class variance, repeatability and N: original and transformed data are presented for each variable.

Table 1. Different vitality assessments for 2–4 measurements from 56 samples (drop or smear) of stallion semen

Motility assessment	Units	No. of observations	Mean ± s.d.	Variance Between samples	Variance Within samples	Repeatability	N‡
Subjective estimation	%	2	36·7 ± 13·3	146·4	31·7	0·82	1
			3·06 ± 1·06	0·66	0·28	0·70	1·7
Count on video record	%	2	34·8 ± 11·6	122·5	12·8	0·91	1
			3·12 ± 0·97	0·93	0·20	0·82	1
Staining	%	4	70·5 ± 7·1	34·3	16·9	0·67	2·0
			3·04 ± 1·01	0·62	0·41	0·60	2·7
Optical analyser	SMI†	3	60·7 ± 45·4	1313·3	768·1	0·63	2·3
			2·77 ± 1·23	0·99	0·54	0·65	2·2
Spermokinesimeter*	%	2	41·2 ± 12·4	119·7	34·1	0·78	1·1
			3·07 ± 0·99	0·61	0·37	0·62	2·5

Results are given, for each estimation, with the original (first line) and the transformed (2nd line) data.
*On only 53 ejaculates.
†SMI = sperm motility index (see text).
‡See text.

Table 2. Different vitality assessments for 2 straws from 336 frozen ejaculates of stallions

Motility assessment	Units	Mean ± s.d.	Variance Between ejaculates	Variance Within ejaculates	Repeatability	N‡
Subjective estimation	%	46·1 ± 13·6	256·7	58·0	0·69	1·8
		3·07 ± 1·03	0·65	0·42	0·60	2·7
Count on video record	%	41·9 ± 11·3	90·7	36·6	0·71	1·6
		3·05 ± 1·02	0·62	0·42	0·60	2·7
Staining	%	68·7 ± 8·9	48·2	31·3	0·61	2·6
		2·97 ± 1·01	0·51	0·52	0·50	4·0
Optical analyser*	SMI†	110·0 ± 60·0	1438·6	2100·4	0·41	5·8
		3·02 ± 1·01	0·43	0·58	0·42	5·5

Results are given, for each estimation, with the original (first line) and the transformed (2nd line) data.
*On only 280 ejaculates (5 stallions).
†SMI = sperm motility index (see text).
‡See text.

Experiment 1

In both parts of this experiment, eosin–nigrosin staining gave a higher percentage of live spermatozoa than did video counting or subjective estimation of motile spermatozoa. It seems that there are some spermatozoa that are non-motile but still alive.

There was a significant variation amongst ejaculates and samples. Repeatability was always higher for counting and subjective estimation and lower for the optical analyser. Moreover, it was higher for observations of the same sample (from 0·63 to 0·91 for original data and from 0·60 to 0·82 for transformed ones; Table 1) than for straws of the same ejaculate (from 0·41 to 0·71 for original data and from 0·42 to 0·69 for transformed ones, Table 2). N therefore varies from 1 to 2 for numbers of measurements per sample and from 3 to 4 for numbers of straws per ejaculate, except for the optical analyser (respectively 2 and 6).

Experiment 2

The results of successive collections from the 80 stallions displayed a significant variance amongst stallions within breeds for all the data, significant effect of breed for all the data except percentage of abnormal spermatozoa and of rank of ejaculate for concentration, volume and total sperm number (Table 3).

Repeatabilities ranged from 0·3 to 0·6 for the transformed data and from 0·4 to 0·7 for original ones (0·37, total sperm number; 0·41, motility in raw semen; 0·45, sperm concentration; 0·46, gel-free volume; 0·62 and 0·69, motility in extended semen; 0·68, abnormal spermatozoa). Values for transformed data did not differ more than 0·10 except for concentration (0·58 *versus* 0·45) and total sperm number (0·58 *versus* 0·37).

Table 3. Characteristics of semen collected from 80 stallions 5 times at 24-h intervals

Semen characteristic	Mean ± s.d.	Significant effects*	Variance Between stallion	Within stallion	Repeatability	N†
Concentration (10^{-6}/ml)	168·8 ± 108·6	B, S, R	3904·1	4767·6	0·45	4·9
	5·58 ± 1·98		1·71	1·24	0·58	2·9
Gel-free volume (ml)	44·1 ± 27·8	B, S, R	298·8	353·4	0·46	4·7
	5·53 ± 1·99		1·43	1·80	0·44	5·1
Total sperm no. ($\times 10^{-9}$)	5·8 ± 4·8	B, S, R	6·5	11·2	0·37	6·9
	5·52 ± 1·99		1·83	1·25	0·58	2·9
Abnormal spermatozoa (%)‡	28·6 ± 12·8	S	113·2	52·7	0·68	1·9
	5·18 ± 2·38		3·84	1·80	0·68	1·9
Motility of raw semen at collection (%)§	53·3 ± 17·3	B, S	105·8	152·0	0·41	5·8
	3·05 ± 0·95		0·29	0·49	0·37	6·8
Motility of extended semen at collection (%)¶	56·0 ± 15·3	B, S	150·6	92·0	0·62	2·4
	3·07 ± 1·04		0·66	0·40	0·62	2·4
Motility of extended semen stored 24 h at +4°C (%)‖	33·3 ± 19·7	B, S	251·4	111·3	0·69	1·8
	3·11 ± 1·06		0·68	0·39	0·64	2·2

Results are given, for each estimation, with the original data (first line) and the transformed data (2nd line).
*B = breed, S = stallion within breed, R = rank of the ejaculate (1–5).
†See text.
‡For only 40 Breton draught stallions.
§For only 79 stallions.
¶For only 72 stallions.
‖For only 4 ejaculates (1–4) and 77 stallions.

At collection, extended semen had a higher motility repeatability than did raw semen for both sets of data (original, 0·62 *versus* 0·41; transformed, 0·62 *versus* 0·37). Depending on the characteristics and the data, N ranged from 2 to 7.

Experiment 3

Stallions of the first and second groups had been serving during an average of 3·9 and 2·8 seasons, with 26·7 ± 14·9 and 36·5 ± 14·8 oestrous cycles per season respectively. The average pregnancy rate per oestrous cycle was respectively 38·5% (0–100%) and 36·9% (10–72%) (Table 4).

There was a significant stallion effect in the two groups. Nevertheless, repeatability was lower for non-selected stallions (0·19) than for those that were used for more than 25 oestrous cycles per season (0·41).

Table 4. Pregnancy rate per oestrous cycle (Breton draught stallions)

| Population studied | Mean ± s.d. | Variance | | Repeatability | N* |
		Between stallions	Within stallions		
79 stallions 3·9 seasons/♂ 26·7 ± 14·9 cycles/season	38·5 ± 14·5	39·1	172·2	0·19	17·1
49 stallions 2·8 seasons/♂ 36·5 ± 14·8 cycles/season	36·9 ± 12·8	67·6	96·0	0·41	5·6

*See text.

Therefore, according to the same criteria as previously, 17 seasons are necessary to estimate the fertility of a stallion by the pregnancy rate per cycle, and only 6 if this figure is determined every year on more than 25 cycles.

Discussion

On the one hand, our statistical model presents the inconvenience of not expressing the real effect of the rank of ejaculate which is proportional to the mean value rather than additive, especially for volume and total sperm number. Since each stallion is examined within a short period (5 days), the season effect is confused with that for the stallion.

On the other hand, data were split into classes of limited numbers which weaken the influence of extreme values. In fact, transformation for 5 or 10 classes gives respectively means of 3 ± 1 and $5·5 \pm 2$ and, therefore, the maximum deviations are respectively 2 (i.e. 2σ) and 4·5 (i.e. $2·25\sigma$).

These remarks may explain why, in most cases, repeatabilities of original data were slightly higher (the difference does not exceed +0·10) than for transformed data, except for concentration and total sperm number (difference, −0·13 and −0·21). Thus, the difference for N does not usually exceed −1·4, except for concentration (+2·0) and total sperm number (+4·0).

In the comparison of the various techniques used to estimate vitality, the optical analyser was the less repeatable since a greater part of the variation was due to the error term. This can result either from a lack of experience with this device used for the first time in our laboratory or from its high sensitivity to sperm concentration variations (Bartoov *et al.*, 1981), which can happen in spite of normalized filling procedures for the straws. The spermokinesimeter SORO does not improve repeatability compared to subjective estimation but, in practice, it could maintain, in different laboratories, a repeatability equal to that of visual estimation by the same well-trained technician. Finally, counting on video record gives the best repeatability. However, the improvement is weak compared to the time spent on the technique. Eosin–negrosin staining, which takes as much time as counting, is less repeatable than subjective estimation. This last technique is therefore at once the simplest and the more repeatable.

Repeatability was higher within measures of the same sample than within straws of an ejaculate. It can be explained by other sources of variability between straws, e.g. absence of homogeneity when the ejaculate is split into straws and variations during the freezing and thawing procedures. So, in practice, any sample should be observed microscopically twice and 3 straws at least are necessary to estimate correctly the motility of spermatozoa in a frozen ejaculate.

In Exp. 2, motility showed higher repeatability for extended semen than for raw semen. Visual estimation is actually influenced by sperm concentration and standardization of this measure, at a concentration of 20×10^6 spermatozoa/ml, enhances repeatability.

Repeatabilities of stallion semen characteristics amongst ejaculates have been estimated in previous studies (Pattie & Dowsett, 1982; Palmer & Fauquenot, 1985). Nevertheless, direct comparisons of these results are impossible since uncontrolled sources of variation are different from one study to another. Age and season were not taken into account in our experiment or in that of Palmer & Fauquenot (1985). However, samples were collected from stallions by the same method in our study and in that of Palmer & Fauquenot (1985): 5 ejaculations at daily intervals, whereas Pattie & Dowsett (1982) recorded data with numbers of collections varying from 1 to 47 and with intervals between collections from 1 h to 1 year. Our results are higher than those of Pattie & Dowsett (1982) (range from 0·26 to 0·53 for gel-free volume, total sperm number, sperm concentration and non-motile spermatozoa), perhaps due to the short period of collections. We may only roughly compare the rank of each characteristic for each study.

In all experiments, characteristics with higher repeatabilities are concentration, extended semen motility at collection and, to a lesser extent, total sperm number, whereas volume has low repeatability. The reason could be that the volume may be dependent upon sexual excitement which is quite difficult to control. We therefore suggest that, in field practice, semen should be collected from a stallion 5 times at daily intervals, corresponding to the weekly scale of work organization.

A previous study on the foaling rate of 357 stallions of different studs and breeds also showed a low repeatability (from 0·15 to 0·18 according to the fertility parameter used) when there was no stallion selection and a slight improvement (values varying from 0·23 to 0·30) for stallions serving more than 25 mares per season (from 0·23 to 0·30) (Langlois, 1977). According to our criteria, it would take at least 9 years to estimate foaling rate.

Foaling rate is a rough estimation since it mixes up real fertility and management of the mare (control of ovulation, number of cycles per mare) and of the stallion (frequency of ejaculation). Dowsett & Pattie (1982) prefer the percentage of pregnancy per service which is also dependent on the numbers of services allowed at each oestrus. Pregnancy rate per oestrous cycle is not much better since it considers several successive cycles of a same mare as independent. However, each estimation attributes to a stallion all that may come from its management.

Since even pregnancy rate per cycle requires at least 6 years for precise assessment of the fertility of a stallion, it is better to look for a relation between fertility estimation and seminal characteristics. This search may be performed according to two different ways. The first is to obtain evidence of a significant correlation, e.g. between pregnancy rate and sperm concentration (Kenney *et al.*, 1971) or threshold levels for one or several characteristics taken individually which may impair fertility, e.g. motility and percentage of abnormal spermatozoa (Aamdal, 1979; Palmer & Fauquenot, 1985), concentration (Dowsett & Pattie, 1982; Palmer & Fauquenot, 1985), or volume, total sperm number and percentage of live spermatozoa (Dowsett & Pattie, 1982). The second is to consider that only a combination of factors can account for the result of a complex phenomenon, as Kenney *et al.* (1971) did with pregnancy rate and concentration, volume, percentage of morphologically normal spermatozoa and motility of extended semen. This last method needs a greater amount of data to find a reliable prediction and, anyway, both of them presuppose a precise assessment of semen characteristics.

In conclusion, we recommend that duplicate measurements are made on semen samples, that at least 3 straws are checked for estimation of frozen semen and 5 ejaculates are examined for the judgement of the stallion and that any conclusion from pregnancy rates based on small numbers of mares is considered with caution.

References

Aamdal, J. (1979) Fertility and infertility in stallions. *Proc. 30th Meeting Euro. Assoc. Anim. Prod.*, Harrogate, England.

Amann, R.P., Thompson, D.L., Jr, Squires, E.L. & Pickett, B.W. (1979) Effects of age and frequency of ejaculation on sperm production and extragonadal sperm reserves in stallions. *J. Reprod. Fert., Suppl.* **27**, 1–6.

Bartoov, B., Kalay, D. & Mayevsky, A. (1981) Sperm motility analysor (SMA), a practical tool of motility and cell concentration determinations in artificial insemination centers. *Theriogenology* **15**, 173–182.

Dowsett, K.F. & Pattie, W.A. (1982) Characteristics and fertility of stallion semen. *J. Reprod. Fert., Suppl.* **32**, 1–8.

Dubois, M., Jouannet, P., Berge, P., Volochine, B., Serres, C. & David, G. (1975) Méthode et appareillage de mesure objective de la mobilité des spermatozoïdes humains. *Annls Phys. Biol. Méd.* **9**, 19–41.

Gebauer, M.R., Pickett, B.W. & Swiestra, E.E. (1974) Reproductive physiology of the stallion. II. Daily production and output of sperm. *J. Anim. Sci.* **39**, 732–736.

Katz, D.F. & Overstreet, J.W. (1981) Sperm motility assessment by videomicrography. *Fert. Steril.* **35**, 188–193.

Kenney, R.M., Kingston, R.S, Rajamannon, A.H. & Ramberg, C.F., Jr (1971) Stallion semen characteristics for predicting fertility. *Proc. 17th Ann. Cong. Am. Ass. equine Pract.*, pp. 53–67.

Langlois, B. (1977) Influence de l'étalon sur la fertilité dans l'élevage des chevaux de sang. *Annls Zootech.* **26**, 329–344.

Magistrini, M., Chanteloube, Ph. & Palmer, E. (1987) Influence of season, frequency of ejaculation on stallion sperm production for freezing. *J. Reprod. Fert., Suppl.* **35**, 127–133.

Ortavant, R. (1952) Contribution à l'étude de la différenciation des spermatozoïdes morts et des spermatozoïdes vivants dans le sperme de taureau. *Annls Zootech.* **1**, 5–12.

Palmer, E. & Fauquenot, A. (1985) Mesure et prédiction de la fertilité des étalons. Etude méthodologique. In *Le Cheval. Reproduction. Sélection. Alimentation*, pp. 113–126. INRA, Paris.

Pattie, W.A. & Dowsett, K.F. (1982) The repeatability of seminal characteristics of stallions. *J. Reprod. Fert., Suppl.* **32**, 9–13.

Pickett, B.W., Faulkner, L.C., Seidel, G.E., Jr, Berndtson, W.E. & Voss, J.L. (1976) Reproductive physiology of the stallion. VI. Seminal and behavioral characteristics. *J. Anim. Sci.* **43**, 617–625.

Sullivan, J.J. & Pickett, B.W. (1975) Influence of ejaculation frequency of stallions on characteristics of semen and output of spermatozoa. *J. Reprod. Fert., Suppl.* **23**, 29–34.

J. Reprod. Fert., Suppl. **35** (1987), 33–38

Evaluation of cellulose acetate/nitrate filters for the study of stallion sperm motility

P. J. Strzemienski, P. L. Sertich, D. D. Varner* and R. M. Kenney

Section of Reproductive Studies, University of Pennsylvania—New Bolton Center, School of Veterinary Medicine, Kennett Square, Pennsylvania 19348, U.S.A.

Summary. Stallion semen was diluted in a Hepes-supplemented buffer (CM) (10^6 spermatozoa/ml) and placed in the upper well of a Sykes-Moore chemotaxis chamber. Chambers were incubated in a humidified atmosphere (5% CO_2 in air) at 37°C for 1 and 2 h and spermatozoa were allowed to swim through filters with a mean pore size of 3, 5 or 8 μm. Spermatozoa entered filters of all three pore sizes. Distance travelled was greater for each increase in pore size ($P < 0.01$) but did not differ ($P > 0.05$) between 1 and 2 h of incubation. Extended semen from stallions of different fertility was analysed for the minimal concentration of spermatozoa needed to enter filters with a 3 μm pore size. Sperm progression into the filter reflected the motility of the ejaculate. Assuming that sperm motility is a good indicator of fertility, this method may be useful for estimating the fertility of a stallion. Progression of spermatozoa into filters with a pore size of 3 μm was hampered by supernatants from overnight cultures of *Streptococcus zooepidemicus* and *Enterobacter aerogenes*. Motility decreased after 2 h of incubation in supernatant from *S. zooepidemicus* diluted 1:5 and *E. aerogenes* supernatant diluted 1:5 and 1:10 in culture medium. In contrast, the bacterial supernatants were chemokinetic to horse neutrophils and did not affect their viability.

Introduction

Characterization of sperm motility is useful in assessing an animal's potential fertility. Motility can be determined by visual estimates with a microscope, by analysis of photomicrographic recordings (Lustig & Lindahl, 1970; Panidis *et al.*, 1982; Van Huffel *et al.*, 1985), by spectrophotometer determinations (Atherton *et al.*, 1978; Levin *et al.*, 1984) and by cinematography (Fray *et al.*, 1972; Katz & Yanagimachi, 1980; Holt *et al.*, 1985). Correlation of these characteristics to sperm motility behaviour *in vivo* and fertilization potential remains unresolved (Katz & Yanagimachi, 1980). Study of sperm progression into cervical mucus has provided additional data on motility (Katz & Overstreet, 1980) but its use may be restricted to particular species (Gaddum-Rosse *et al.*, 1980). In addition, batches of cervical mucus are also difficult to standardize. The use of a readily obtainable matrix may be of help in resolving some of these restrictions.

Recently, polycarbonate filters have been used to study the action of pharmacological agents (Hong *et al.*, 1981) and chemoattractants (Gnessi *et al.*, 1985) on sperm motility. Because of the filter thickness (10 μm) this method relies on the motility of the spermatozoa for passage through a filter of defined pore size. Motility is evaluated by counting the number of spermatozoa swimming through the filter pore.

In this study, cellulose acetate/nitrate filters are evaluated for their use in studying stallion sperm motility.

---

*Present address: Department of Large Animal Medicine and Surgery, Texas A & M University, College Station, TX 77843, U.S.A.

Materials and Methods

Sperm migration test

Semen was collected with a Nishikawa artificial vagina (Scott Medical Products, Haywood, CA). The semen was immediately filtered with a milk filter (Premium Milk Filter, Agway Inc., Syracuse, NY) and the percentage of total and progressively motile spermatozoa were estimated visually by two trained observers. Sperm concentration was determined by a haemocytometer count. Dilutions of spermatozoa were made with culture medium (Whittingham, 1971) containing 0·01 M-Hepes (4-(2-hydroxyethyl)-1-piperazineethanesulphonic acid) and bovine serum albumin (Sigma Chemical Co., St Louis, MO; 12 mg/ml) (pH 7·4). Semen was kept at room temperature (21·5–24°C) during all manipulations.

The bottom well of a Sykes-Moore chemotaxis chamber (Bellco Glass Co., Vineland, NJ) was filled with culture medium. The top well was separated from the bottom well by a cellulose acetate/nitrate filter (Millipore Corp., Danvers, MA). After securing the filter, 1 ml sperm suspension was placed in the top well of the chamber and incubated at 37°C in humidified 5% CO_2 in an air atmosphere. Humidity was controlled by bubbling the gas mixture into a gas washing bottle containing distilled water at 37°C. After incubation, the filters were removed and immediately fixed in 10% neutral buffered formalin overnight. The filters were then washed twice, stained in haematoxylin, washed, and placed with the upper surface in contact with a clean glass slide. After drying, the filters were mounted with a glass coverslip.

Distance of sperm travel was measured by a method used to evaluate neutrophil migration (Zigmond & Hirsch, 1972) utilizing the micrometer scale on the fine focus adjustment knob of the microscope. Measurements were made from the top of the filter to a plane within the filter containing the two most forward cells in focus using a × 40 objective. Ten readings were taken in a straight line bisecting each filter. Two filters were used for a treatment. Formalin-fixed or freeze-thawed spermatozoa did not enter the filter, indicating that spermatozoa needed to be motile to enter the filter.

Experiment 1

Stallions were individually stabled and fed a daily grain and hay ration. Four stallions (two Trakehner, one Standardbred and one Dutch warm-blood) were selected from visual estimations of progressive sperm motility (> 75%) and their success in a commercial breeding programme. Semen was diluted to 10^6 spermatozoa/ml and placed over a filter with a pore size of 3, 5, or 8 µm. Filters were removed after 1 or 2 h of incubation.

Experiment 2

Stallion A (Standardbred) and Stallions B and C (Dutch warmbloods) were classified as fertile or sub-fertile based on the recovery of embryos, 7 days after artificial breeding. Pregnancy rates for Stallions A, B and C were 75% (18/24), 50% (7/14) and 12% (1/8) respectively. Stallion D (Thoroughbred) was referred to the Hofmann Center clinic at New Bolton Center because of subfertility. The breeding history of Stallion E (Standardbred) was not known. Semen was diluted to a concentration of 10^7, 10^6, 10^5, 10^4 and 10^3 spermatozoa/ml and placed over a filter with a 3 µm pore size. Chambers were incubated for 2 h.

Experiment 3

Two ejaculates and peripheral blood neutrophils obtained from Stallion A were utilized to study the effect of bacterial supernatants on sperm and neutrophil movement. *Streptococcus zooepidemicus* isolated from a mare with an active endometritis and *Enterobacter aerogenes* isolated from stallion semen were grown overnight in RMPI culture medium (Sigma). The bacterial suspensions were then centrifuged at 1500 *g* for 30 min, filter sterilized (0·22 µm filter) and frozen at −20°C in aliquants. Fresh horse serum was activated with zymosan to activate complement components (Strzemienski *et al.*, 1984). On the day of assay, an aliquant of the bacterial supernatant was thawed, diluted appropriately in culture medium for use with spermatozoa or neutrophils. Filters were fixed in formalin at 2 h of incubation for spermatozoa and after 45 min of incubation for neutrophils.

Statistical analysis

The results for Exp. 1 were analysed by a two-way analysis of variance with 3 levels of pore size and 2 levels of time. The effects of bacterial supernatants on sperm migration in Exp. 3 were analysed by a two-way analysis of variance with 2 levels of supernatant and 8 levels of dilution. A two-way analysis of variance was also used to compare neutrophil response to zymosan-activated serum and bacterial fluids using 3 levels of fluid and 7 levels of dilution. Comparison between means was done by the Least Significant Difference test (Steel & Torrie, 1960).

Results

Experiment 1

Although there was a difference in the distances travelled by spermatozoa into the filters of different pore size (Table 1) the distance travelled in filters of each size did not change ($P > 0.05$) after 1 h of incubation. Infrequently, spermatozoa were interpreted to have turned within the filter toward the point of entry.

Table 1. Comparison of pore size and length of incubation on sperm travel (μm) in cellulose acetate/nitrate filter

Incubation time	Filter pore size		
	3 μm	5 μm	8 μm
1 h	18 ± 3[a]	34 ± 4[b]	59 ± 12[c]
2 h	19 ± 4[a]	38 ± 6[b]	64 ± 11[c]

Values are mean ± s.d.
Values not sharing the same superscript are different ($P < 0.01$).

Experiment 2

After incubation for 2 h, samples from two ejaculates from Stallion A and from one ejaculate of Stallion B had entered a 3 μm filter when diluted to 10^4 spermatozoa/ml (Table 2). The percentage of progressive motility did not appear to change during this time. One ejaculate from Stallion B was damaged by heating during processing, as indicated by the initial progressive/total motility values, and did not penetrate the filter when at a 10^6 dilution. Dilution of both ejaculates of Stallion C beyond 10^6 spermatozoa/ml prevented spermatozoa from entering the filters. The percentage of progressively motile spermatozoa from both ejaculates from Stallion C was reduced by 2 h of incubation. Ejaculates from Stallions D and E were also limited in the ability of their spermatozoa to enter the filter at lower concentrations. Progressive/total motility values did not appear to change for Stallion D, after 2 h of incubation, while most spermatozoa from Stallion E had considerable head-to-head agglutination.

Table 2. Effect of sperm dilution on penetration (μm) into a filter with a 3 μm pore size

Stallion	Sperm conc.					Motility (progressive/total)	
	10^7	10^6	10^5	10^4	10^3	0 min	120 min
A	20 ± 5	17 ± 4	11 ± 6	0	ND	75/80	80/85
A	19 ± 4	20 ± 3	17 ± 4	17 ± 2	14 ± 3	75/80	85/90
A	21 ± 2	20 ± 3	18 ± 2	18 ± 4	ND	85/90	77/85
B*	19 ± 9	6 ± 7	0	0	0	5/10	3/3
B	26 ± 9	13 ± 2	18 ± 4	17 ± 2	ND	39/42	30/35
C	16 ± 2	6 ± 2	0	0	ND	25/30	10/25
C	13 ± 4	13 ± 4	0	0	0	30/50	10/45
D	16 ± 3	0	0	0	ND	30/50	30/50
E	16 ± 4	6 ± 3	0	0	ND	50/60	70†

*Semen damaged during handling.
†Motile spermatozoa were agglutinated.

Table 3. Distance travelled by spermatozoa or neutrophils (µm) in response to supernatants from cultures of *Streptococcus zooepidemicus* and *Enterobacter aerogenes*

Above filter/ below filter*	Spermatozoa			Neutrophils	
				Supernatant	
	S. zooepidemicus	*E. aerogenes*	Complement	*S. zooepidemicus*	*E. aerogenes*
B/B	14 ± 4	14 ± 4	38 ± 3	38 ± 3	38 ± 3
B/5	13 ± 4	11* ± 2	50* ± 5	49* ± 3	39 ± 3
B/10	12 ± 2	11* ± 3	ND	40 ± 3	42 ± 3
5/5	12 ± 3	11* ± 2	41 ± 3	ND	ND
10/10	12 ± 5	15 ± 5	—‡	—‡	—‡
10/5	11* ± 2	10* ± 3	ND	47 ± 3	52* ± 3
B/100	13 ± 3	16 ± 6	ND	32* ± 3	47 ± 2
B/1000	14 ± 3	12 ± 4	ND	40 ± 3	43 ± 3

Values are means ± s.d. for 2 replicates of Stallion A spermatozoa.
†B/5 = culture medium/reciprocal of dilution used.
‡Not tested.
*Different from controls, B/B ($P < 0.01$).

Experiment 3

The effect of supernatants from *S. zooepidemicus* and *E. aerogenes* on the penetration of spermatozoa from Stallion A into 3 µm filters is presented in Table 3. Progression was less ($P < 0.01$) when spermatozoa were suspended in a 1:10 dilution of *S. zooepidemicus* or *E. aerogenes* supernatant and placed over a well containing a 1:5 dilution of the same supernatant. With the exception of spermatozoa placed in a 1:5 dilution of *E. aerogenes* supernatant, only dilutions of 1:5 or 1:10 inhibited sperm progression into the filter ($P < 0.01$).

The average motility for Stallion A at the beginning of the experiments was 80/85% (progressively motile spermatozoa/total motility) in culture medium alone. After 2 h of incubation motility was not changed. Sperm motility decreased in *S. zooepidemicus* supernatant diluted 1:5 (70/65) but was not diminished when the dilution was increased to 1:10. Sperm motility decreased for 1:5 and 1:10 dilutions of *E. aerogenes* supernatant to 45/50% and 60/70% motility respectively at 2 h of incubation.

As a positive control, peripheral blood neutrophils were also tested (Table 3). Both bacterial supernatants and zymosan-activated serum were stimulatory to neutrophils obtained from Stallion A. Zymosan-activated serum diluted 1:5 increased neutrophil movement above that of controls ($P < 0.01$). Migration increased slightly ($P > 0.01$) when the top and bottom chambers contained a 1:5 dilution of the serum. Neutrophil migration was greater ($P < 0.01$) than in controls only when supernatant from *S. zooepidemicus* was diluted 1:5. Other dilutions with this supernatant in the bottom well (B/10, 10/5) suggest that the stimulatory action is due to a chemokinetic effect. In contrast, migration of neutrophils to *E. aerogenes* supernatant diluted 1:5 or 1:10 was not different ($P > 0.01$) from controls. However, neutrophil migration was greater ($P < 0.01$) when a 1:5 dilution of *E. aerogenes* supernatant was used in the lower chamber and a 1:10 dilution of the supernatant was used in the upper chamber, suggesting chemokinetic activity. Neutrophil viability remained at 98% live, as determined by trypan blue exclusion, after 2 h of incubation in a 1:10 dilution of each bacterial supernatant.

Discussion

These results have shown that stallion spermatozoa can travel through a filter matrix having a pore size of 3 µm. The ability of stallion spermatozoa to travel into this matrix would suggest either a compressing of the sperm head membrane or loss of some membrane components since the stallion sperm head has a width of 3–3·9 μm and a length of 6·41–7·0 μm (Cummins & Woodall, 1985). Direct measurements of spermatozoa entering the filter with light microscopy is made difficult due to orientation and some distortion of the spermatozoa when in the membrane. Sperm heads did not appear to be uniform in size; smaller head sizes were not always associated with forward moving spermatozoa.

Increased sperm penetration in filters having 5 and 8 μm pore sizes indicates more efficient travel. In this assay, sperm penetration into the filter may be influenced by gravity although formalin-fixed and freeze–thawed spermatozoa do not enter the filter, indicating the necessity for motility. The matrix construction of the filter, due to the union of the filter pores, will allow spermatozoa to turn to the point of origin, also indicating the dependence on motility for sperm progression into the filter. It appears that progression into the filter is due to the movement of the spermatozoa and that the total distance travelled into the filter reflects the progressive nature of the sperm sample studied.

Lack of significant sperm progression after 1 h of incubation is difficult to interpret. Lodging of spermatozoa is the most likely explanation due to the tortuosity and diameter variations of the filter. A build-up of waste products or fatigue due to an increased workload on the spermatozoa in negotiating the filter pore matrix may also be tenable. The rigid matrix of the filter could prevent efficient removal of waste products or an efficient supply of metabolizable substrate. Studies examining the water binding capacity of cervical mucus could not eliminate an impairment of diffusable waste products or nutritive substances by the cervical mucus (Katz & Singer, 1978). The Hepes-containing buffer used in this study was selected for its supportive influence on stallion sperm motility (37°C for 2 h) and its semi-defined composition. It is possible that the requirements for free swimming spermatozoa are different from those for spermatozoa challenged by the restrictive environment of the filter membrane structure. The use of the present filter system should allow the use of other sperm-supportive media to investigate their effect on filter penetration.

Although the sample of stallions in this study is small, our results suggest a possible means of screening stallion semen. As would be expected motility appears to play a major role in the concentration of spermatozoa needed to enter a filter. Data from Stallion B in Exp. 2 also suggest that certain motility characteristics may be needed for entrance of spermatozoa into the filter. Up to 8 motility patterns of stallion spermatozoa have been shown photomicrographically (Van Huffel *et al.*, 1985). Correlation of these patterns to sperm negotiation into the filter may be of use in defining the relative contribution of each characteristic.

Polycarbonate filters have been used to study the influence of pharmacological agents (Hong *et al.*, 1981) and have shown the influence of chemotactic agents on human sperm motility (Gnessi *et al.*, 1985). These assays utilize the thin nature of the filter by counting the number of spermatozoa passing through the pores and therefore monitor sperm progressive motility for a short period of time. On the other hand, cellulose acetate/nitrate filters present a constant if not increasing resistance to sperm travel. Weakening of sperm motility mechanisms by toxic bacterial products may not be noticed with the use of polycarbonate filters while the increased workload placed on spermatozoa by the cellulose acetate/nitrate filters may uncover those weakened mechanisms.

Both bacterial supernatants reduced sperm motility over the 2 h incubation period, which is the most probable cause for reduced travel into the filter. However, it was also observed that the supernatant for *E. aerogenes* initially appeared to increase forward swimming velocity as well as accentuate sperm motility. Both bacterial supernatants used in this study are crude extracts and therefore may contain a number of components with diversified functions. To our knowledge, no data are available on the interaction of micro-organism by-products and gamete function. Our results have shown that conditions favourable to neutrophil activity are not optimal for sperm function. Since both male and female gametes are dependent upon their environment for proper function, any alteration of this environment by micro-organisms will modulate their function. Elucidation of those bacterial products which may influence sperm motility mechanisms may increase our understanding of the disease process and its relationship to fertility.

P. J. Strzemienski et al.

We thank Thornbrook Farm and Mr A. W. Berry for support; Mark Cahoon for assistance with the animals; and Pat Brewer for secretarial assistance.

References

Atherton, R.W., Radany, E.W. & Polakoski, K.L. (1978) Spectrophotometric quantitation of mammalian spermatozoan motility 1. *Human. Biol. Reprod.* **18**, 624–628.

Cummins, J.M. & Woodall, P.F. (1985) On mammalian sperm dimensions. *J. Reprod. Fert.* **75**, 153–175.

Fray, C.S., Hoffer, A.P. & Fawcett, D.W. (1972) A re-examination of motility patterns of rat epididymal spermatozoa. *Anat. Rec.* **173**, 301–307.

Gaddum-Rosse, P., Blandau, R.J. & Lee, W.I. (1980) Sperm penetration into cervical mucus *in vitro*. 1. Comparative studies. *Fert. Steril.* **33**, 636–643.

Gnessi, L., Ruff, M.R., Fraioli, F. & Pert, C.B. (1985) Demonstration of receptor-mediated chemotaxis by human spermatozoa. *Expl Cell. Res.* **161**, 219–230.

Holt, H.V., Moore, H.D.M. & Hillier, S.G. (1985) Computer-assisted measurement of sperm swimming speed in human semen; correlation of results with *in vitro* fertilization assays. *Fert. Steril.* **44**, 112–119.

Hong, C.Y., Chaput de Saintonge, D.M. & Turner, P. (1981) A simple method to measure drug effects on human sperm motility. *Br. J. clin. Pharmac.* **11**, 385–387.

Katz, D.F. & Singer, J.R. (1978) Water mobility within bovine cervical mucus. *Biol. Reprod.* **17**, 843–849.

Katz, D.F. & Overstreet, J.W. (1980) Mammalian sperm movement in the secretions of the male and female genital tract. In *Testicular Development, Structure and Function.* pp. 481–489. Eds A. Steinberger & E. Steinberger. Raven Press, New York.

Katz, D.F. & Yanagimachi, R. (1980) Movement characteristics of hamster spermatozoa within the oviduct. *Biol. Reprod.* **22**, 759–764.

Levin, R.M., Hypolite, J.A. & Wein, A.J. (1984) Clinical use of the turbidimetric analysis of sperm motility: An update. *Andrologia* **16**, 434–438.

Lustig, G. & Lindahl, P.E. (1970) Activation of motility in bull and rabbit spermatozoa by ultrasonic treatment recorded by a photographic method. *Int. J. Fert.* **15**, 135–142.

Panidis, D., Tarlatzis, B. & Papaloucos, A. (1982) Objective evaluation of spermatozoal motility in fertile males with the multiple exposure photography (MEP) method. *Acta endocr., Suppl.* **248**, 19–20.

Steel, R. & Torrie, J. (1960) *Principles and Procedures of Statistics.* McGraw-Hill, New York.

Strzemienski, P.J., Do, D. & Kenney, R.M. (1984) Antibacterial activity of mare uterine fluid. *Biol. Reprod.* **31**, 303–311.

Van Huffel, X.M., Varner, D.D., Hinrichs, K., Garcia, M.C., Strzemienski, P.J. & Kenney, R.M. (1985) Photomicrographic evaluation of stallion spermatozoal motility characteristics. *Am. J. vet. Res.* **46**, 1272–1275.

Whittingham, D.G. (1971) Culture of mouse ova. *J. Reprod. Fert., Suppl.* **14**, 7–21.

Zigmond, S.H. & Hirsch, J.G. (1972) Leukocyte locomotion and chemotaxis: new methods for evaluation and demonstration of a cell-derived chemotactic factor. *J. exp. Med.* **137**, 387–410.

THE STALLION

Convener

B. W. Pickett

Chairmen

B. W. Pickett

H. Merkt

J. Reprod. Fert., Suppl. **35** (1987), 39–43

Fertility of pasture bred mares in synchronized oestrus

F. Bristol

*Department of Herd Medicine and Theriogenology, Western College of Veterinary Medicine,
University of Saskatchewan, Saskatoon, Saskatchewan, Canada S7N 0W0*

Summary. Oestrus was synchronized in 220, 300 and 272 mares in 1983, 1984 and 1985 respectively. Mares were given two injections of 250 µg fenprostalene 15 days apart except in 1983 and 1984 when 56 and 53 of the synchronized mares were given 1–10 daily injections of 150 mg progesterone and 10 mg oestradiol-17β to delay and synchronize post-partum oestrus. At 2 days after the second PG injection or 7 days after the last progesterone + oestradiol treatment, mares were divided into groups of 15–21, and each group was placed in a separate pasture with a stallion for 7 weeks. Pregnancy rates were 87·7, 93·7 and 97·1%, and foaling rates were 72·3, 89·7 and 94·1% in 1983, 1984 and 1985 respectively. The number of abortions occurring mainly between 3 and 6 months of gestation varied from 34 (17·8%) in 1983 to 12 (4·3%) in 1984 and 8 (3·0%) in 1985.

Introduction

Fertility of horses is generally considered to be low when compared with other species of domestic livestock (Hutton & Meacham, 1968; Merkt *et al.*, 1979; Rowlands, 1981). While fertility of hand-mated horses has received a great deal of attention there are very few reports on the fertility of pasture-bred or free-running horses. Foaling rates in feral horses have been reported to vary from 23 to 82% (Tyler, 1968; Salter & Hudson, 1982; Keiper & Houpt, 1984; Berger, 1983). Age, quality of forage (Tyler, 1968; Keiper & Houpt, 1984), and male takeovers of harems (Berger, 1983) have been suggested as reasons for this variation.

Little information is available on fertility of domestic horses under free-running pasture conditions. Most workers have reported that pregnancy or foaling rates for pasture-bred horses range from 95 to 100%, but these reports involved small numbers of horses (Ickes, 1965; Hutton & Meacham, 1968; Collery, 1974). Ginther *et al.* (1983) reported mean pregnancy rates of 51 and 54% in small herds of Ponies (20 mares per stallion) while in large herds (60 mares per stallion), the rate was 48% during a 48-day breeding season. In free-ranging Icelandic Toelter horses, conception and foaling rates were estimated to be 82·1% and 81·1% respectively (Hugason *et al.*, 1985).

While synchronization of oestrus with artificial insemination has been successfully used for mares (Palmer & Jousset, 1975; Hyland & Bristol, 1979; Voss *et al.*, 1979), there is only one report on the use of synchronization of oestrus in a small number of pasture-bred mares with one stallion (Bristol, 1982). The aim of this study was to evaluate the effect of synchronization of oestrus on the fertility of a large number of mares bred at pasture in an attempt to concentrate the foaling period.

Animals and Methods

The data for this study were obtained from the breeding records of a pregnant mare urine farm for 3 breeding seasons from 1983 to 1985. Cross-bred draught type mares in good body condition between 3 and 15 years of age were used. The stallions were Belgian or Belgian crossbreds ranging in age from 3 to 13 years. The number of mares and stallions used in each of the years is shown in Table 1. Oestrus was synchronized by 2 s.c. injections of 250 µg fenprostalene, a prosta-glandin (PG) F-2α analogue (Synchrocept B: Syntex Agribusiness, Mississauga, Ontario, Canada), given 15 days apart in lactating and non-lactating mares that had foaled on or before 12 May (Hyland & Bristol, 1979). In 1983 and 1984

Table 1. The distribution of stallions and mares between years and method of synchronization of oestrus

Year	No. of stallions	No. of mares treated with PG	No. of mares treated with progesterone + oestradiol	Total no. of mares
1983	11	164	56	220
1984	15	247	53	300
1985	14	272	0	272

those mares that foaled after 12 May were synchronized by delaying post-partum oestrus with 1–10 daily i.m. injections of 150 mg progesterone and 10 mg oestradiol-17β beginning on the day of parturition (Bristol *et al.*, 1983). Groups of 15–21 mares were placed in separate brome or mixed-grass pastures that varied in size from 56 to 120 hectares. On 1 June, i.e. 2 days after the second PG injection or 7 days after the last progesterone + oestradiol treatment, one stallion was placed in each of the pastures for 7 weeks. After the stallions were removed, the mares remained on pasture until the 2nd week of September. A semen sample had been collected from each stallion and evaluated 2 weeks before breeding began.

The mares were examined for pregnancy 7 weeks after the beginning of the breeding season with a linear array ultrasound scanner and re-examined by rectal palpation 8 weeks later. This was immediately before they were put in the urine collection barns during the 3rd week of September. Mares were considered to have had an embryonic death if they were pregnant at the first examination and non-pregnant at the second examination. Most non-pregnant mares were culled from the herd. While the mares were stabled, they were examined every day for any signs of abortion.

Results

The overall pregnancy and foaling rates for the 3 years of the study are shown in Table 2. the pregnancy rate increased from 87·7 to 97·1% over the 3-year period, and a similar increase was observed for the foaling rate. There was very little variation in fertility between individual stallions, and their pregnancy rates were 70–95% in 1983 to 85–100% in 1984 and 1985. The stallions suffered no injuries, and with the exception of 1 stallion in 1985 they had normal semen. In all 5 ejaculates examined the progressive motility was never greater than 10% in the apparently abnormal stallion but his pregnancy rate was 100%.

Pasture conditions were very good in 1983 and 1985, but in 1984 there was a drought during the summer months and some of the mares had to be given supplementary hay during August and September.

The pregnancy losses for the 3 years of study are shown in Table 3. Embryonic deaths were low in all years while number of abortions was highest in 1983. Most of the abortions occurred shortly after the pregnant mares were put in the barn: over the 3 years of the study, 40 (74·1%) abortions occurred during the first 10 weeks of stabling.

The earliest parturition occurred on 23 April in 1984. Overall, 70·2% of the mares foaled during the first 22 days of the foaling period (Table 4). During the course of the study, 11 mares died in the pasture or in the barn and were excluded from the data.

Table 2. Reproductive performance of pasture-bred mares after synchronization of oestrus with fenprostalene or progesterone and oestradiol-17β

Year	No. pregnant (%)	No. foaling (%)	Total
1983	193 (87·7)	159 (72·3)	220
1984	281 (93·7)	269 (89·7)	300
1985	264 (97·1)	256 (94·1)	272

Table 3. Pregnancy losses in pasture-bred mares after synchronization of oestrus with fenprostalene or progesterone and oestradiol-17β

Year	No. of embryonic deaths (%)	No. of abortions (%)
1983	5 (2·6)	34 (17·8)
1984	3 (1·1)	12 (4·3)
1985	3 (1·2)	8 (3·0)

Table 4. The distribution of foaling in pasture-bred mares after synchronization of oestrus with fenprostalene or progesterone and oestradiol-17β

Year	No. of mares foaling between 23 April and 15 May	No. of mares foaling after 15 May	Total
1983	116 (73·0)	43 (27·0)	159
1984	187 (69·5)	82 (30·5)	269
1985	177 (69·1)	79 (30·9)	256

Discussion

The pregnancy rates in this study are higher than those for hand bred horses, estimated to be 60–70% with a foaling rate of about 60% (see Rowlands, 1981) and are similar to those observed for small groups of non-synchronized free running horses exposed to a stallion during the breeding season (van Rensburg & van Heerden, 1953; Hutton & Meacham, 1968; Ickes, 1965; Collery, 1974). Ginther (1983) reported a lower pregnancy rate of 71% in 3 groups of 20 light horse mares, but they were only exposed to the stallion for 30 days in September, the end of the breeding season in the northern hemisphere. In Icelandic Toelter horses bred from May to July in herds of 5 to 25 mares the pregnancy rate was estimated to be 82·1% (Hugason *et al.*, 1985). Most studies on pasture-bred horses have been carried out during the summer months (van Rensburg & van Heerden, 1953; Bristol, 1982; Hugason *et al.*, 1985); in contrast, hand breeding usually begins in late winter and generally ends by mid-summer. This may contribute to the high fertility of pasture-mated mares.

Most horses are selected for conformation or athletic performance while selection for fertility is rarely practised. Until 1983 the herd had been undergoing rapid expansion and very few mares that failed to conceive were culled. After the 1983 breeding season and in subsequent breeding seasons, almost all of the non-pregnant mares were removed from the herd. This selection for fertility with rigorous culling of most mares that failed to conceive probably contributed to the high fertility of the mares in this study and the increase in pregnancy rate between 1983 and 1985. Other authors have suggested that pasture breeding has a beneficial effect because many mares that fail to conceive with hand mating will readily become pregnant when bred at pasture (van Rensburg & van Heerden, 1953; Ickes, 1965). Even the one stallion with less than 10% progressively motile spermatozoa before the breeding season had a pregnancy rate of 100%. A Quarter-horse stallion with low fertility when used for hand breeding was also reported to have 100% pregnancy rate when used at pasture (Bowen *et al.*, 1983). Drought conditions and poor pastures in 1984 did not appear to have an effect on fertility. Possibly there was sufficient grass growth at the beginning of the season to provide adequate nutrition for successful breeding. Poor body condition during late pregnancy and early lactation reduces fertility but mares in good body condition or gaining weight around the time of breeding have high fertility (Schryver & Hintz, 1980; Henneke *et al.*, 1984). In this study the mares were in good body condition at parturition and only lost weight after foaling.

Stallions herd their band of mares closely and tease them frequently. Continuous social interaction between the stallion and mares may contribute to the high fertility observed in pasture-bred horses (Bristol, 1982). Running mares in a pasture with a stallion having a retroverted penis before the breeding season appeared to hasten the onset of normal oestrous cycles (Belonje, 1956). Bowen *et al.* (1983) demonstrated that more mares conceive early in the breeding season under pasture breeding conditions than with closely supervised hand breeding. Courting behaviour of the boar has beneficial effect on the fertility of sows at natural and artificial matings; allowing sows housed in pairs to be courted by a boar increased conception rate significantly from 62·1 to 86·7% (Hemsworth *et al.*, 1978) and exposing sows to a vasectomized boar 15 min after a fertile mating also improved conception rate (Mah *et al.*, 1985). Likewise, ewes penned with a vasectomized ram after artificial insemination have a higher conception rate than those isolated after insemination (Restall, 1961). Multiple matings may also improve fertility and stallions at pasture mate most mares a number of times during each oestrus (Bristol, 1982; unpublished). Ewes mated more than once during oestrus by the same rams were more likely to conceive than those mated only once (Mattner & Braden, 1967).

The incidence of embryonic death in horses between about 30 and 90 days of gestation was estimated to range from 5 to 14% (Ginther, 1979) but recent studies with ultrasound scanners have demonstrated that the incidence of early embryonic death is higher and occurs much earlier than previously reported (Villahoz *et al.*, 1985; Ginther *et al.*, 1985). In this study, the incidence between Days 14 and 110 of gestation ranged from 1·1 to 2·6% but this may be an underestimate since some early pregnancies may not have been detected and subsequently lost. Other mares may have already lost the embryo by the time pregnancy diagnosis was performed. However, in view of the very high pregnancy and foaling rates, it is assumed that the incidence of embryonic death was low. In managed free running horses, the incidence of pregnancy loss has been estimated to be only 1·1% (Hugason *et al.*, 1985).

The abortion rate between 3 months and term ranged from 17·8 to 3·0%. The stress of a sudden change in diet and environment associated with stabling may precipitate these abortions. Mitchell (1971) reported a 12% incidence of abortions on pregnant mare urine farms with most of them occurring during the first 2 months of stabling. Roberts (1971) observed 5 abortions at 4–5 months of gestation in 2-year-old pregnant mares that had been shipped by rail for 7 days. An abattoir survey of recently transported, stressed mares showed positive evidence of abortion in 50% of the pregnant mares (Osborne, 1975). Stressful conditions such as pain, infectious diseases and weaning cause a marked drop in plasma progestagen concentration in pregnant mares that may lead to pregnancy failure (van Niekerk & Morgenthal, 1982).

In each of the years, the majority of mares foaled within a 22-day period, indicating that synchronization of oestrus was effective in concentrating the breeding season and foaling period. It is therefore probable that most mares conceived at the first synchronized oestrus of the breeding season. This is consistent with previous reports in which conception rates at first oestrus in artificially inseminated and hand-mated mares were 63·8% and 64% respectively (Palmer & Jousset, 1975; Hyland & Bristol, 1979), while in 20 synchronized pasture-bred mares it was 85% (Bristol, 1982).

Rowlands (1981) speculated that low fertility is likely to exist in all equids and may be a possible reflection of evolutionary misdirection. This study demonstrates that fertility can be very high in oestrus-synchronized horses when they are served at pasture during the natural breeding season and there is selection for mares that conceive readily.

I thank Dr P. Doig of Syntex Agribusiness for the Synchrocept B; Mr F. Clement, owner, for the use of the horses; and Mrs J. Deschner for manuscript preparation. The study was supported by the Western College of Veterinary Medicine Equine Health Research Fund.

References

Belonje, C.W.A. (1956) The influence of running a stallion with non-pregnant Thoroughbred mares. *J. S. Afr. vet. med. Ass.* **27**, 57–60.

Berger, J. (1983) Induced abortion and social factors in wild horses. *Nature, Lond.* **303**, 59–61.

Bowen, J.M., Wigington, S. & Bergeron, H. (1983) Advancing the equine breeding season naturally. *Proc. 8th Eq. Nutr. & Physiol. Symp.* pp. 90–95.

Bristol, F. (1982) Breeding behaviour of a stallion at pasture with 20 mares in synchronized oestrus. *J. Reprod. Fert., Suppl.* **32**, 71–77.

Bristol, F., Jacobs, K.A. & Pawlyshyn, V. (1983) Synchronization of estrus on post-partum mares with progesterone and estradiol 17β. *Theriogenology* **19**, 779–785.

Collery, L. (1974) Observations of equine animals under farm and feral conditions. *Equine vet. J.* **6**, 170–173.

Ginther, O.J. (1979) *Reproductive Biology of the Mare: Basic and Applied Aspects*, pp. 248–250. Cross Plaines, Wisconsin.

Ginther, O.J. (1983) Sexual behavior following introduction of a stallion into a group of mares. *Theriogenology* **19**, 877–886.

Ginther, O.J., Scraba, S.T. & Nuti, L.C. (1983) Pregnancy rates and sexual behavior under pasture breeding conditions in mares. *Theriogenology* **20**, 333–345.

Ginther, O.J., Bergfelt, D.R., Leith, G.S. & Scraba, S.T. (1985) Embryonic loss in mares: incidence and ultrasonic morphology. *Theriogenology* **24**, 73–86.

Hemsworth, P.H., Beilharz, R.G. & Brown, W.J. (1978) The importance of the courting behaviour of the boar on the success of natural and artificial matings. *Appl. Anim. Ethol.* **4**, 341–347.

Henneke, D.R., Potter, G.D. & Kreider, J.L. (1984) Body condition during pregnancy and lactation and reproductive efficiency of mares. *Theriogenology* **21**, 897–910.

Hugason, K., Arnason, T. & Jonmundsson, J.V. (1985) A note on the fertility and some demographical parameters of Icelandic Toelter horses. *Livestock Prod. Sci.* **12**, 161–167.

Hutton, C.A. & Meacham, T.N. (1968) Reproductive efficiency on fourteen horse farms. *J. Anim. Sci.* **27**, 434–438.

Hyland, J.H. & Bristol, F. (1979) Synchronization of oestrus and timed insemination of mares. *J. Reprod. Fert., Suppl.* **27**, 251–255.

Ickes, M.W. (1965) Reproduction panel. *Proc. 11th Ann. Conv. Am. Assoc. eq. Pract.* pp. 339–342.

Keiper, R. & Houpt, K. (1984) Reproduction in feral horses: an eight year study. *Am. J. vet. Res.* **45**, 991–995.

Mah, J., Tilton, J.E., Williams, G.L., Johnson, J.N. & Marchello, M.J. (1985) The effect of repeated mating at short intervals on reproductive performance of gilts. *J. Anim. Sci.* **60**, 1052–1054.

Mattner, P.E. & Braden, A.W.H. (1967) Studies in flock mating of sheep. 2. Fertilization and prenatal mortality. *Aust. J. exp. Agric. Anim. Husb.* **7**, 110–116.

Merkt, H., Jacobs, K-O., Klug, E., & Aukes, E. (1979) An analysis of stallion fertility rates (foals born alive) from the breeding documents of the Landgestüt Celle over a 158-year period. *J. Reprod Fert., Suppl.* **27**, 73–77.

Mitchell, D. (1971) Cited by Roberts, p. 142.

Osborne, V.E. (1975) Factors influencing foaling percentages in Australian mares. *J. Reprod. Fert., Suppl.* **23**, 477–451.

Palmer, E. & Jousset, B. (1975) Synchronization of oestrus in mares with a prostaglandin analogue and HCG. *J. Reprod. Fert., Suppl.* **23**, 269–274.

Restall, B.J. (1961) Artificial insemination of sheep. VI. The effect of post inseminal coitus on percentage of ewes lambing to a single insemination. *Aust. vet. J.* **37**, 70–72.

Roberts, S.J. (1971) *Veterinary Obstetrics and Genital Diseases.* Ithaca, New York.

Rowlands, I.W. (1981) Perspectives in perissodactyls. *Equine vet. J.* **13**, 85–87.

Salter, R.E. & Hudson, R.J. (1982) Social organization of feral horses in Western Canada. *Appl. Anim. Ethol.* **8**, 207–223.

Schryver, H.F. & Hintz, H.F. (1980) Nutrition and reproduction in the horse. In *Current Therapy in Theriogenology*, pp. 768–777. Ed. D. A. Morrow. W. B. Saunders, Philadelphia.

Tyler, S. (1968) The behaviour and social organization of the New Forest ponies. *Anim. Behav. Monogr.* **5**, 87–196.

van Niekerk, C.H. & Morgenthal, J.C. (1982) Fetal loss and the effect of stress on plasma progestagen levels in pregnant Thoroughbred mares. *J. Reprod. Fert., Suppl.* **32**, 453–457.

van Rensburg, S.W.J. & van Heerden, J.S. (1953) Infertility in mares caused by ovarian dysfunction. *Ondestepoort J. vet. Res.* **26**, 285–313.

Villahoz, M.D., Squires, E.L., Voss, J.L. & Shideler, R.K. (1985) Some observations on early embryonic death in mares. *Theriogenology* **23**, 915–924.

Voss, J.L., Wallace, R.A., Squires, E.L., Pickett, B.W. & Shideler, R.K. (1979) Effects of synchronization and frequency of insemination of mares. *J. Reprod. Fert., Suppl.* **27**, 257–261.

J. Reprod. Fert., Suppl. **35** (1987), 45–49

Printed in Great Brita
© 1987 Journals of Reproduction & Fertility L

Pharmacological manipulation of sexual behaviour in stallions

S. M. McDonnell, M. C. Garcia and R. M. Kenney

University of Pennsylvania, School of Veterinary Medicine, New Bolton Center, Kennett Square, Pennsylvania 19348, U.S.A.

Summary. Series of experiments and clinical trials were conducted to evaluate the effects of psychoneurotropic agents on sexual behaviour of stallions. The benzodiazepine derivative, diazepam (Valium), effectively reversed experimentally suppressed pre-copulatory arousal and response. Diazepam treatment also blocked the negative effect of novel environment on sexual response. The dibenzazepines imipramine and clomipramine induced erection, masturbation, and ejaculation in the absence of a sexual stimulus.

Introduction

Sexual behaviour dysfunction remains one of the most difficult problems in equine reproduction. About one fourth of the stallions evaluated each year by the University of Pennsylvania Hofmann Center Reproduction Clinic involves a fertility-limiting behaviour problem. Although there is no reliable way to translate these figures to the stallion population at large, the costs represented are enormous.

Advances in the neurophysiology and neuropharmacology of sexual behaviour have brought a multitude of potential approaches to the control of male sexual arousal and response. This paper briefly reviews our recent experimental and clinical studies of the effects of psychopharmacological agents on sexual behaviour of stallions. Sexual response includes at least two principal interdependent neural processes. The first is the more cerebrally mediated process of goal-directed interest and arousal. The second is the more peripherally mediated ability to achieve and maintain erection, copulate and ejaculate. Accordingly, our work towards understanding and manipulating sexual behaviour of stallions has had two areas of focus corresponding to these processes.

Sexual Interest and Arousal

Following work done in other species suggesting effects of various CNS-active agents on general sexual interest and arousal, we are systematically evaluating the effects of several drugs on sexual response in experimental horse and pony stallions and geldings, as well as in clinical case stallions.

GnRH

A number of studies in mammals suggests that the hypothalamic decapeptide, known as gonadotrophin-releasing hormone (GnRH) for its role in regulating the pituitary–testicular axis-mediated endocrine events, may have a direct CNS-mediated role in the regulation of sexual behaviour (Moss, 1978). When GnRH challenge tests (500 µg i.v.; Cystorelin: CEVA Laboratories, Overlook Park, KS) were administered to clinical case stallions with low–normal androgen levels in order to evaluate the integrity of the pituitary–gonadal axis, we observed Flehmen response as well as sudden improvement (subjectively assessed) in sexual interest and arousal. We conducted an experiment to evaluate the effects of synthetic GnRH (25 µg s.c. every 3 h for 3 weeks; Cystorelin: CEVA Laboratories) on sexual behaviour of long-term pony geldings (N = 12) with or without androgen replacement (testosterone propionate, 200 µg/kg every 48 h for 3 weeks; Sigma Chemical Company, St Louis, MO). Preliminary evaluation of performance in a series of sexual behaviour trials (4-min teasing exposure to an ovariectomized oestrogen-primed stimulus mare, 3 times per week, 2 weeks of baseline, 3 weeks of treatment, 3 weeks post-treatment) indicated a significant effect of GnRH treatment on olfactory investigation responses in geldings with androgen replacement. There were no significant effects of GnRH treatment on aggression, erection, or mounting responses.

Yohimbine

Yohimbine, an indolealkylamine alkaloid similar to reserpine, is a centuries-old purported aphrodisiac which is currently under renewed scientific scrutiny for its effects on sexual behaviour. Work in rats has shown that this alpha-2 adrenergic receptor antagonist enhances sexual interest and arousal (Clark *et al.*, 1984). However, similar effects have been difficult to demonstrate in other animal species or in man. In a preliminary experiment, we administered yohimbine (0·15 mg/kg s.c. or i.v. 15–20 min before sexual exposure; Sigma) or vehicle to long-term pony geldings (N = 12) with and without low level androgen replacement (testosterone propionate, 50 µg/kg s.c. 3 times/week at 16 h before a sexual behaviour test; Sigma). After intravenous yohimbine administration the animals exhibited moderately heightened general activity and excitability. Several bucked and frolicked on their way to and while in the sexual behaviour test pen. However, treatment had no apparent effects on precopulatory investigatory behaviour, aggressive responses, erection, mounting, or thrusting measured during standardized sexual behaviour trials (4-min teasing exposure to an ovariectomized oestrogen-primed stimulus mare, 3 times weekly for 2 weeks).

PCPA

Parachlorophenylalanine, a CNS-active selective serotonin synthesis inhibitor has been shown to influence sexual behaviour in rats (Sachs & Dewsbury, 1971), cats (Ferguson *et al.*, 1970), monkeys (Redmond *et al.*, 1971), and men (Sicuteri *et al.*, 1975). In a preliminary experimental trial, we administered PCPA (4 mg/kg i.m., twice/day for 1 week; Pfizer, Groton, CT) to a single pony stallion. This animal became extremely depressed and anorexic, and treatment was terminated. In spite of this adverse effect similar to that reported in cats and primates, sexual response (attention time, erection latency, erection time, mount frequency and olfactory investigation in weekly 4-min exposures to an ovariectomized oestrogen-primed stimulus mare) was greater than before treatment.

Naloxone

Endogenous opiate peptides have been implicated in an inhibitory role in regulation of sexual motivation. Several opiate antagonists have had facilitatory effects on male sexual arousal and response in some species (Sachs *et al.*, 1981). In preliminary evaluation of two pony stallions, we have seen no effect of the opiate antagonist naloxone hydrochloride (4 mg i.v. 2 min before sexual exposure: Narcan, Dupont Pharmaceuticals, Manati, Puerto Rico) on sexual arousal or response in standardized 4-min teasing trials as above.

Benzodiazepines

In the domestic stallion, considerable clinical evidence suggests that dysfunction may result from man-made experiences which inadvertently or by design inhibit sexual interest or arousal. Recent experimental findings confirm that inadequate sexual behaviour can result from negative experiences similar to those encountered by domestic stallions (McDonnell *et al.*, 1985, 1986; McDonnell, 1985). Therefore, a considerable portion of our work towards pharmacological manipulation of sexual behaviour in stallions has focussed on agents known to reverse or block the effects of negative experience on goal-directed behaviour. The benzodiazepines, or minor tranquilizers, have distinct anxiolytic or antidepressant effects at dose levels which do not impair ongoing motor or goal-directed behaviour. They reverse or block negative experience suppression of goal-directed behaviour. Among the benzodiazepine derivatives, diazepam (Valium) remains the most intensively studied for its ability to attenuate effects of negative experience. We have studied the effects of diazepam on several types of experimentally-induced sexual suppression in stallions, as well as on a small number of spontaneously 'inhibited' clinical case stallions.

One type of experience-related sexual suppression that we have successfully modelled in the stallion is 'response-contingent aversive conditioning'. This paradigm involves direct punishment alone or in combination with negative reinforcement of sexual interest, arousal, or response. This results in a rapid, nearly complete suppression of sexual response. Using the appropriate contingencies it is possible to model specific variations of this type of suppression, including mare- or situation-specific suppression similar to that reported in clinical cases with spontaneous dysfunction. Conditioned as well as direct aversive stimuli are effective in suppressing response. In all variations of response-contingent aversive suppression that we have modelled, diazepam (0·05 mg/kg slow i.v., 5–7 min before sexual exposure; Valium, Hoffmann-LaRoche, Nutley, NJ) has effectively reversed or blocked the suppression. For example, stallions that showed no interest in the stimulus after aversive conditioning returned to normal sexual behaviour when given diazepam, while saline-treated controls remained suppressed. In addition, the administration of diazepam during the aversive conditioning procedures appeared to delay the ensuing suppression. In several clinical case stallions with spontaneous dysfunction that may have been related to negative experience, diazepam treatment has been judged by clinicians to be one of the key factors in the successful retraining. These clinical cases have included shy, novice breeding stallions, experienced natural service breeding stallions with apparent aversion to collection by artificial vagina, and stallions that exhibit normal arousal and mounting but suddenly dismount before ejaculation.

Erection and Ejaculation

Several vasoactive or neuromyotropic agents have been used to facilitate erection and ejaculation, principally to enhance smooth muscle activity in the reproductive tract. Several beta-antagonists, prostaglandins and oxytocin have been administered to stallions, with inconsistent results. We have completed preliminary studies of agents with both central and peripheral effects on neurotransmission.

Dibenzazepines

The tricyclic antidepressants have been implicated in effecting changes in erection and ejaculation. Human patients taking tricyclic antidepressants for manic-depressive or obsessive-compulsive personality disorders have reported both disturbance and facilitation of libido, erection and ejaculation (Quirk & Einarson, 1982; McLean *et al.*, 1983; Beeley, 1984). Despite inconsistency of effects and poorly understood mechanism of action, these drugs have been used to treat retrograde ejaculation (Brookes *et al.*, 1980) and premature ejaculation (Mitchell & Popkin, 1983).

We have studied the effects of imipramine on sexual behaviour in horses. Imipramine was administered to 5 male horses (400–500 kg body wt): one was an inexperienced young stallion, two were mature normal breeding stallions, one was a 5-year-old stallion with erection and ejaculatory dysfunction, and one a long-term castrated male horse. Oral treatment (100–600 mg, twice/day; imipramine hydrochloride: Rugby Laboratories, Rockville Centre, Long Island, NY) led to frequent erection and masturbation while at rest in the stall in a non-sexual context. Intravenous treatment, over a range of doses (50–1000 mg, imipramine hydrochloride in sterile water; Sigma), similarly induced erection and masturbation in all animals. Typically, erection occurred within 10 min after injection, and the erection and masturbation continued intermittently for 1–2 h. These erections proceeded as during sexual excitement to a normal firmness and eventual engorgement of the glans penis. Two stallions ejaculated while masturbating. Mild ataxia and drowsiness appeared at the higher doses, but the animals remained responsive to auditory, visual and tactile stimuli. Erection and masturbation were often interrupted by activities about the barn or the approach of the handler, suggesting cortical inhibitory control of the erection. When tested in a sexual context, the two mature breeding stallions mated normally immediately after i.v. treatment (500 mg). The 5-year-old stallion, which had not ejaculated over several months of breeding attempts, spontaneously ejaculated after i.v. imipramine treatment. Subsequently, this stallion has ejaculated during copulation while on low dose oral (100 mg twice/day) imipramine treatment. Plasma total androgens increased during treatment in these stallions. The long-term castrate showed erection and masturbation after i.v. imipramine treatment, suggesting that the effect of imipramine is not testosterone-dependent. Prolonged imipramine treatment (6 months, 100 mg orally twice/day and intermittent i.v. injection) had no adverse effects on thyroid and adrenal function or serum chemistry and haemogram.

Preliminary trials have been conducted using the closely related tricyclic compound, clomipramine. Oral clomipramine treatment (100–600 mg twice/day) resulted in frequent erection and masturbation in the stall, with no adverse effects on thyroid and adrenal function or serum chemistry and haemogram.

We observed consistently positive effects of imipramine and clomipramine on erection and ejaculation in the horses studied. These penis drop and erection responses were clearly not an automatic peripheral neurovascular response, as they were under cortically-mediated inhibitory control. The erection and masturbation responses seemed to occur within the context of a general sexual motivational state induced by the drug treatment.

Discussion

As we learn more about the physiology and pharmacology of sexual behaviour it is becoming increasingly apparent that a variety of agents will emerge as aids to therapy. Some of these drugs may generally enhance sexual arousal, but many may have rather specific effects on one or a cluster of sexual responses. To treat sexual behaviour dysfunction in stallions effectively we must continue to work towards diagnosis of the specific nature of the dysfunction and accordingly select an appropriate agent or combination of agents for therapy. In this process, traditional evaluation and behavioural retraining procedures will remain the most important part of the overall therapeutic strategy.

Although most of our recent work follows similar work in other species, we view the stallion as an excellent model for study of human sexual behaviour and dysfunction, including pharmacological manipulation. Stallions, like primates, are single-mount, multiple-thrust ejaculators, with the musculocavernous penis being similar to that of man. In addition, the effects of experience on stallion sexual response closely resemble purported effects of experience in men. For these reasons, we believe that work in the stallion may lead the way to better understanding of the control of male sexual behaviour in mammals in general.

We thank Mrs A. Riggs, Mr A. Berry, Hamilton Research Laboratory, Mrs Laura Thorn, Hoffmann-LaRoche, CEVA Laboratories and Pfizer for financial support; and Nancy Diehl, Karen Heaps and Alexandra Stockwell for help with portions of this work.

References

Beeley, L. (1984) Drug-induced sexual dysfunction and infertility. *Adv. Drug React. Ac. Pois. Rev.* **3**, 23–42.

Brookes, M.E., Berezin, M. & Braf, Z. (1980) Treatment of retrograde ejaculation with imipramine. *Urology* **15**, 4.

Clark, J.T., Smith, E.R. & Davidson, J.M. (1984) Enhancement of sexual motivation in male rats by yohimbine. *Science, N.Y.* **225**, 847–849.

Ferguson, J., Henriksen, S., Cohen, H., Mitchell, G., Barchas, J. & Dement, N. (1970) Hypersexuality and behavioural changes in cats caused by the administration of p-chlorophenylalanine. *Science, N.Y.* **168**, 499–501.

McDonnell, S.M. (1985) *Stallion sexual behavior: negative experience effects and pharmacological manipulation.* Doctoral Dissertation, University of Delaware.

McDonnell, S.M., Kenney, R.M., Meckley, P.E. & Garcia, M.C. (1985) Conditioned suppression of sexual behavior in stallions and reversal with diazepam. *Physiol. & Behav.* **34**, 951–956.

McDonnell, S.M., Kenney, R.M., Meckley, P.E. & Garcia, M.C. (1986) Novel environment suppression of sexual behaviour in stallions and effects of diazepam. *Physiol. & Behav.* **37**, 503–505.

McLean, J.D., Forsythe, R.G. & Kapkin, I.A. (1983) Unusual side effects of clomipramine associated with yawning. *Can. J. Psychiat.* **28**, 569–570.

Mitchell, J.E. & Popkin, M.K. (1983) Antidepressant drug therapy and sexual dysfunction in men: a review. *J. clin. Psychopharmacol.* **3**, 76–79.

Moss, R.L. (1978) Effects of hypothalamic peptides on sex behaviour in animal's and man. In *Psychopharmacology: a Generation of Progress*, pp. 431–440. Eds M. A. Lipton, A. DiMascio & K. F. Killam. Raven Press, New York.

Quirk, K.C. & Einarson, T.R. (1982). Sexual dysfunction and clomipramine. *Can. J. Psychiatry* **27**, 228–231.

Redmond, D.E., Mass, J.W., Kling, A., Graham, C.W. & Dekirmenjian, H. (1971) Social behavior of monkeys selectively depleted of monoamines. *Science, N.Y.* **174**, 428–431.

Sachs, B.D. & Dewsbury, D.A. (1971) P-chlorophenylalanine facilitates copulatory behaviour in male rats. *Nature, Lond.* **232**, 400–401.

Sachs, B.D., Valcourt, R.J. & Flagg, H.C. (1981) Copulatory behavior and sexual reflexes of male rats treated with naloxone. *Pharmacol. Biochem. & Behav.* **14**, 251–253.

Sicuteri, F., del Bene, E. & Anselmi, B. (1975) Aphrodisiac effect of testosterone in parachlorophenylalanine-treated sexually deficient men. In *Sexual Behavior: Pharmacology and Biochemistry*, pp. 335–339. Eds M. Sandler & G. L. Gessa. Raven Press, New York.

J. Reprod. Fert., Suppl. **35** (1987), 51–58

Circadian, circhoral and seasonal variation in patterns of gonadotrophin secretion in geldings

L. S. Hoffman, T. E. Adams and J. W. Evans*

Department of Animal Science, University of California, Davis, California 95616, U.S.A.

Summary. Blood samples were obtained from 5 mixed-breed, long-term castrated geldings during five 24-h periods between May 1984 and April 1985. Blood samples were collected, beginning at 09:00 h, at 15-min intervals for 8 h and hourly for the remaining 16 h. Plasma concentrations of LH and FSH were determined by RIA. Seasonal changes in hormone concentrations and frequency and amplitude of secretory pulses were evaluated. No diurnal variation in either LH or FSH secretion was observed: however, marked circhoral fluctuations in LH and FSH secretion were noted. Mean LH and FSH concentrations in these long-term castrated geldings were comparable to those measured in intact stallions. The mean LH concentration significantly varied with time of year, with a peak value in late April and a nadir in November. FSH pulse frequency also significantly varied with season, with a peak in August and a nadir in May. The amplitude of secretory episodes of LH and FSH did not fluctuate with season. We conclude that: (1) the long-term castrated gelding does exhibit pulsatile gonadotrophin secretion, and (2) gonadotrophin secretion is subject to seasonal variation in the absence of gonadal steroids.

Introduction

There is growing evidence that the reproductive status or fertility of the stallion is subject to seasonal variation. Seminal volumes (total, gel and gel-free) and sperm concentration decrease during the winter months (Pickett *et al.*, 1976; Harris *et al.*, 1983), seminal osmolality and pH increase (Pickett *et al.*, 1976), while sperm motility remains unchanged (Pickett *et al.*, 1976; Harris *et al.*, 1983).

Morphological and physical features of the testis appear to change with season; testicular weight, parenchymal weight and seminiferous tubular volume per testis increase by over 30% between April and May in the northern hemisphere (Johnson & Neaves, 1981) and the number of Sertoli cells per testis increases between June and July (Johnson & Thompson, 1983). Additionally, libido, which is also important in determining the fertility of the stallion, has been shown to increase during the breeding season (Pickett *et al.*, 1976; Thompson *et al.*, 1977; Turner & Kirkpatrick, 1982).

The gonadal steroids, androgens and oestrogens, promote male fertility by supporting spermatogenesis, maintaining the secondary sex tissues and stimulating sexual behaviour. In turn, the gonadal steroids are under the control of the pituitary gonadotrophins, luteinizing hormone (LH) and follicle-stimulating hormone (FSH). There have been reports that serum concentrations of LH and, possibly, FSH concentrations fluctuate with season in the stallion. Peripheral blood LH concentration peaks during the early summer months and declines to a low in the winter (Thompson *et al.*, 1977; Harris *et al.* 1983). Serum FSH concentration may also peak in the summer (Harris *et al.*, 1983; Johnson & Thompson, 1983).

Considerable research has been conducted to study the feedback control of the gonadal steroids on LH and FSH secretion; the male castrate is an animal model frequently used to study this

---

*Present address: Department of Animal Science, Texas A & M University, College Station, Texas 77480, U.S.A.

control. Castration in the stallion results in an immediate decline in peripheral testosterone and oestradiol concentrations and an immediate rise in LH and FSH concentrations (Thompson *et al.*, 1979). Using steroid replacement therapy on acutely castrated stallions, Thompson *et al.* (1979) found that administration of testosterone plus oestradiol-17β restored LH concentration to intact levels while oestradiol-17β treatment partly restored FSH concentration.

Before the gelding is further used to study the gonadal control of gonadotrophin secretion in the male horse, it would be useful to have a clear understanding of the normal patterns of LH and FSH secretion. The purpose of this study was to examine LH and FSH secretion in the gelding over the period of a year and the following questions were specifically addressed. (1) Does the gelding display a circadian rhythm in gonadotrophin release? (2) Does the gelding display pulsatile LH and/or FSH release? (3) Are the characteristics of gonadotrophin secretion, concentration, pulse frequency and pulse amplitude, subject to seasonal fluctuation?

Materials and Methods

Animals and housing. Five geldings, of mixed breeds, were used. The geldings were between 4 and 10 years old and had been castrated for a minimum of 1 year before start of the study. They were housed in two outdoor pens and fed alfalfa pellets and watered *ad libitum*. They were allowed to acclimatize to their housing for at least 2 months before the onset of the study. They were also vaccinated for common diseases and dewormed.

On the day preceding a blood sampling session, each gelding was shaved for catheterization and placed in an individual stall which was exposed to the outside light. The stall lights were turned off at sunset and back on at sunrise to duplicate the naturally occurring photoperiod. A 25-W red light bulb was turned on in each stall at night to permit blood sampling during the dark hours.

Experimental protocol. Each gelding was fitted with an indwelling angiocatheter, inserted into the jugular vein. During each sampling session, plasma samples were taken over a period of 24 h at a frequency of once every 15 min for the first 8 h and then hourly for the remaining 16 h.

Five 24-h blood sampling sessions were performed over a period of 1 year. The dates of each session were: 8 May 1986, 1 August 1984, 6 November 1984, 5 February 1985 and 23 April 1985.

One of the original 5 geldings died of an injury (unrelated to the study) after the first sampling session. An additional gelding was used for the remaining four sessions. This discrepancy was accounted for in the statistical analysis of the data.

Hormone assays. Plasma LH concentrations were determined via the RIA system validated and described by Geschwind *et al.* (1975) and Evans *et al.* (1979) and using the RIA procedure described by Nett *et al.* (1975) with the modifications described below. An ovine LH preparation, donated by Dr H. Papkoff at the University of California at San Francisco (Papkoff IV 28 BP) was labelled with ^{125}I using the Iodogen method of Moberg *et al.* (1981). The concentration of tracer used was 12 000 c.p.m. per 100 μl. A rabbit anti-ovine LH serum, provided by Dr Gordon Niswender (Niswender, No. 15), was used at a dilution of 1:60 000. The horse LH standard was provided by Dr H. Papkoff (Papkoff, E98A). Goat anti-rabbit serum (obtained from Antibodies Inc., Davis, CA) was used as a second antibody at a dilution of 1:120. The total incubation time for the assay was 5 days (120 h) and incubation was conducted at room temperature (25°C).

An average of 33·2% of the tracer was bound in the absence of unlabelled hormone. The sensitivity of the assay was 1·92 ng/ml. The average intra- and inter-assay coefficients of variation were 4·36% and 11·87%, respectively.

Plasma FSH concentrations were determined via an RIA system validated in this laboratory (described below) and using the RIA procedure detailed by Nett *et al.* (1975). A horse FSH preparation, provided by Dr H. Papkoff (Papkoff, E219B), was labelled with ^{125}I using the Iodogen method of Moberg *et al.* (1981). The concentration of tracer used was 12 000 c.p.m. per 100 μl. A rabbit anti-human FSH serum (NIAMDD; AFP-005) was used at a dilution of 1:20 000. The horse FSH standard used was the same preparation as that used for the tracer (Papkoff, E219B). Goat anti-rabbit serum was used as a second antibody at a dilution of 1:120. The total incubation time for the assay was 5 days (120 h) and the incubation was conducted at 25°C (room temperature).

An average of 40·2% of the tracer was bound in the absence of unlabelled hormone. The sensitivity of the assay ranged from 8·75 to 8·92 ng/ml. The average intra- and inter-assay coefficients of variation were 4·57% and 11·00%, respectively.

The FSH antibody showed no cross-reactivity with purified preparations of horse GH, horse LH, horse CG, human CG or sheep FSH (Fig. 1). The percentage binding values for increasing volumes of a serum sample containing a high concentration of FSH formed a curve which was parallel to the standard curve (Fig. 2).

A control stallion sample, which was obtained from one horse in June 1984, had an LH concentration of 8·91 ng/ml and an FSH concentration of 13·14 ng/ml. Stanfield (1983), working in this laboratory and using the same RIA system, obtained values for LH in a group of control stallions which ranged from 3·67 to 13·13 ng/ml.

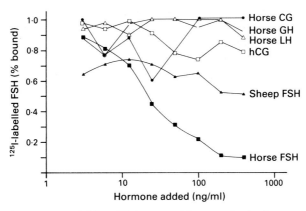

Fig. 1. FSH antibody cross-reactivities with horse and human chorionic gonadotrophins, horse growth hormone and LH, and ovine FSH.

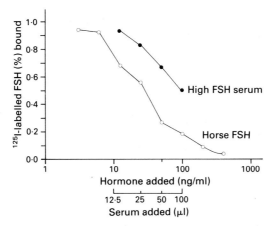

Fig. 2. FSH radioimmunoassay standard curve and curve for serial dilution of horse serum containing a high FSH concentration.

Analysis of data. An LH or FSH pulse was defined as a peak value which occurred within two samples of a nadir and which was greater than the nadir value by a determined factor of variability (Δ). This variability factor was defined as two times the standard deviation (Steel & Torrie, 1980). In other words:

$$x_{\text{peak}} - x_{\text{nadir}} > \Delta$$

where $x =$ the concentration of LH or FSH (ng/ml) in a sample; $\Delta = 2\sigma$; and

$$\sigma = \sqrt{\sum_{j=i}^{\infty} \frac{(\text{mean } x\% \text{ CV})}{k}}$$

where k = the total no. of samples taken in a given sampling session for a particular horse; mean = the average hormone concentration for a particular horse and sampling session; $j =$ the sample number; and %CV = the percentage coefficient of variation for a given sample from a particular sampling session and horse.

Statistical methods. The following values were determined: (1) average LH and FSH concentrations for each horse and blood sampling session; (2) pulse frequency for LH and FSH secretion for each horse for the first 8 h of each sampling session; (3) mean pulse amplitude for LH and FSH secretion for each horse for the first 8 h of each sampling session where: pulse amplitude = value of a pulse (ng/ml) − mean hormone conc. (ng/ml).

The average LH and FSH concentrations, pulse frequencies and mean pulse amplitudes were analysed as a function of horse and time of bleeding session using Analysis of Variance and a Repeated Measures model.

Results and Discussion

LH and FSH concentrations for hourly samples were plotted for each 24-h blood sampling session for each horse (see Fig. 3). From the statistical analysis and these plots, it was observed that there were no diurnal fluctuations in LH or FSH secretion.

LH and FSH concentrations for the 15-min samples were plotted for the first 8 h of each 24-h blood sampling session for each horse (see Fig. 4). It can be observed from these plots that both LH and FSH were secreted in a pulsatile manner. Average LH pulse frequency ranged from 0.63 ± 0.2 (s.e.m.) to 0.97 ± 0.1 pulses/h and FSH pulse frequency ranged from 12.88 ± 1.5 to 16.36 ± 0.6 pulses/h (Table 1).

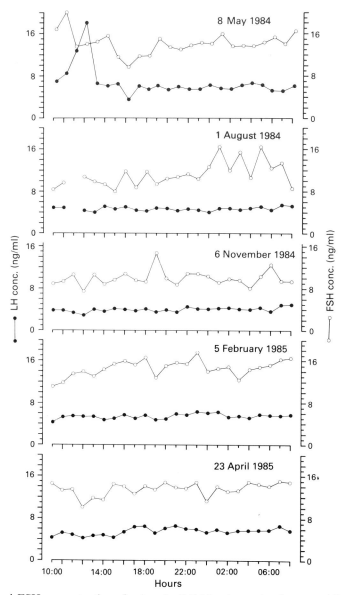

Fig. 3. LH and FSH concentrations for hourly (24) blood samples from a gelding sampled 5 times during a year.

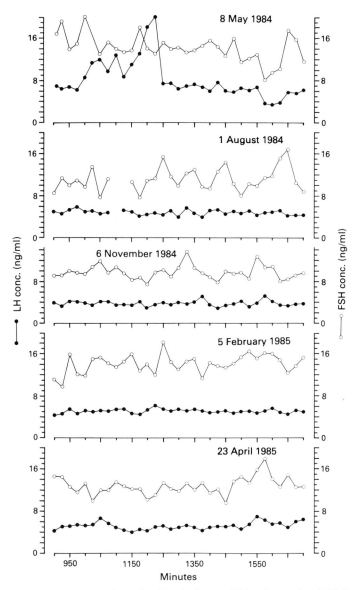

Fig. 4. LH and FSH concentrations for 15-min interval blood samples (2 h) from a gelding sampled 5 times during a year.

The values for each measure of hormone secretion, concentration, pulse frequency and pulse amplitude were averaged over all horses for each sampling session (Table 1; Fig. 5). LH concentration significantly varied with time ($P < 0.009$) with a peak value occurring in May (8.40 ± 1.3 ng/ml) and a nadir in November (6.00 ± 1.1 ng/ml). FSH pulse frequency also significantly varied with time ($P < 0.01$) with a peak value occurring in August (1.23 ± 0.1 pulses/h) and a nadir occurring in May (0.88 ± 0.1 pulses/h). The remaining variables, LH pulse frequency, LH pulse amplitude, FSH concentration and FSH pulse amplitude, did not significantly vary with time. We found no evidence for a diurnal pattern of LH or FSH secretion in this study and none has been reported in the literature for either the gelding or the stallion. There are conflicting reports as to whether testosterone secretion in stallions is under the control of a circadian rhythm. Ganjam & Kenney

Table 1. Seasonal influences on LH and FSH concentrations and pulse release measures for 5 geldings sampled at different times of the year

Sample periods	LH			FSH		
	Plasma conc.* (ng/ml)	Pulse frequency (pulses/h)	Pulse amplitude (ng/ml)	Plasma conc. (ng/ml)	Pulse frequency† (pulses/h)	Pulse amplitude (ng/ml)
8 May 1984	8·4 ± 1·3	0·6 ± 0·2	1·9 ± 0·2	14·8 ± 0·3	0·9 ± 0·1	3·2 ± 0·01
1 August 1984	7·1 ± 1·3	1·0 ± 0·1	1·8 ± 0·5	12·9 ± 1·5	1·2 ± 0·1	3·5 ± 0·1
6 November 1984	6·0 ± 1·1	0·9 ± 0·1	1·3 ± 0·2	16·3 ± 1·7	1·1 ± 0·1	4·3 ± 0·7
5 February 1985	7·7 ± 1·2	0·8 ± 0·1	1·0 ± 0·1	16·4 ± 0·6	0·9 ± 0	3·0 ± 0·3
23 April 1985	8·0 ± 1·4	0·9 ± 0·1	1·2 ± 0·1	15·2 ± 0·5	1·1 ± 0·1	2·7 ± 0·2

Values are mean ± s.e.m.
Values vary significantly over time: *$P < 0.009$; †$P < 0.01$.

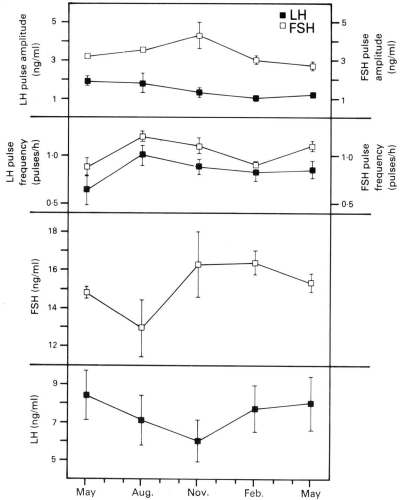

Fig. 5. Seasonal variations in mean (± s.e.m.) concentrations of LH and FSH and mean pulse frequencies and amplitudes for 5 geldings sampled 5 times during a year.

(1975) observed that total androgen concentration peaks at 06:00 h and Sharma (1976) observed a general tendency for testosterone concentration to peak at 08:00 h. However, Ganjam (1979) found that, although total androgen concentration displayed episodic fluctuation, there was no clear diurnal pattern. Stanfield (1983), working in this laboratory, observed that testosterone concentration peaked in the afternoon.

The observed range of values for LH and FSH peripheral plasma concentrations were comparable to those found in intact stallions, using the same assay system. Castration, in most male species, results in a long-term rise in gonadotrophin concentrations. However, Irvine & Alexander (1982) reported that long-term castration in stallions may actually result in a return to precastration concentrations of LH and FSH. The results from this study may add support to this hypothesis.

The criteria we used to define an LH and FSH pulse are similar to those employed by others (Goodman & Karsch, 1980; Fitzgerald *et al.*, 1983). However, the factor of variability which we used to isolate pulses incorporates the average percentage coefficient of variation for each assay. We therefore accounted for the variability which occurred between duplicate measurements of a given sample.

The observed values for LH pulse frequency in this study ranged from 0·25 to 1·00 pulses per hour. Fitzgerald *et al.* (1983), who observed pulsatile LH release in ovariectomized mares, found that pulse frequency ranged from 0·05 to 0·55 pulses per hour. In the intact mare, pulsatile LH activity was not observed in the oestrous phase of the post-partum cycle, but was observed in the luteal phase (Fitzgerald *et al.*, 1985). Pulsatile LH activity has been observed in the ram at a range of 0·05 to 0·52 pulses per hour (Lincoln & Short, 1980). Pulsatile LH activity occurs and ranges from 1 to 2 pulses per hour in the ewe (Karsch *et al.*, 1984). The values we observed for LH pulse frequency in the gelding therefore appear to fall within the spectrum of values observed for the ram, ewe and ovariectomized mare.

The observed values for FSH pulse frequency were surprisingly high, ranging from 0·63 to 1·40 pulses per hour. Pulsatile FSH activity has not been previously reported for the gelding, stallion, mare, ram or ewe.

Variations in LH and FSH concentrations, pulse frequency and pulse amplitude were assessed as a function of time of year. Of these values, LH concentration and FSH pulse amplitude significantly varied with time. LH concentration peaked in late April and declined to a low point in November. FSH pulse frequency peaked in August and declined to a low between February and May. Although the other values did not significantly vary with time in this study, we cannot rule out the possibility that they are also subject to seasonal fluctuations. For example, in 3 of 5 horses, FSH concentration clearly peaked in May and declined in November (exactly opposite to the pattern of LH secretion).

There is general agreement that LH concentration fluctuates as a function of season in the stallion. Thompson *et al.* (1977) found a peak LH concentration of 80–90 ng/ml in May–June and a low concentration of 18 ng/ml in November. Harris *et al.* (1983), who did their work in the southern hemisphere, observed a peak of over 50 ng/ml in November (which corresponds to May in the northern hemisphere) and a low concentration of 13 ng/ml in June–July (which corresponds to December–January).

There is less certainty as to whether FSH concentration fluctuates with season in the stallion. Harris *et al.* (1983) reported that FSH concentration ranges from 10 to 18 ng/ml, with peak concentrations occurring during the spring (September–November in the southern hemisphere; March–May in the northern hemisphere). Johnson & Thompson (1983) also found a 'tendency' for elevated FSH concentrations to occur during the breeding season.

Our observation of a fluctuation in LH concentration with season in the long-term gelding is in contrast with the results of Irvine & Alexander (1982) who did not report any seasonal fluctuation. However, there have been reports that seasonal fluctuation in gonadotrophin secretion can occur in gonadectomized animals. Fitzgerald *et al.* (1983) observed that both LH concentration and pulse frequency rose between February and May in a group of ovariectomized mares. They speculated that a 'gonadal-independent modulation of tonic LH secretion' may occur.

In summary, it was observed that, in the long-term castrated stallion, LH and FSH secretion are not under the control of a circadian rhythm, but are subject to pulsatile activity. Further, at least certain measures of gonadotrophin secretion are subject to seasonal variation in the absence of gonadal steroids.

We thank Dr Harold Papkoff for the gifts of purified LH and FSH, and the National Institute of Arthritis, Diabetes and Digestive and Kidney Diseases for the gift of FSH antibody.

References

Evans, J.W., Hughes, J.P., Neely, D.P., Stabenfeldt, G.H. & Winget, C.M. (1979) Episodic LH secretion patterns in the mare during the oestrous cycle. *J. Reprod. Fert., Suppl.* **27**, 143–150.

Fitzgerald, B.P., I'Anson, H., Loy, R.G. & Legan, S.J. (1983) Evidence that changes in LH pulse frequency may regulate the seasonal modulation of LH secretion in ovariectomized mares. *J. Reprod. Fert.* **69**, 685–692.

Fitzgerald, B.P., I'Anson, H., Legan, S.J. & Loy, R.G. (1985) Changes in patterns of luteinizing hormone secretion before and after the first ovulation in the postpartum mare. *Biol. Reprod.* **33**, 316–323.

Ganjam, V.K. (1979) Episodic nature of the Δ^4-ene and Δ^5-ene steroidogenic pathways and their relationship to the adreno-gonadal axis in stallions. *J. Reprod. Fert., Suppl.* **27**, 67–71.

Ganjam, V.K. & Kenney, R.M. (1975) Androgens and oestrogens in normal and cryptorchid stallions. *J. Reprod. Fert., Suppl.* **23**, 67–73.

Geschwind, I.I., Dewey, R., Hughes, J.P., Evans, J.W. & Stabenfeldt, G.H. (1975) Plasma LH levels in the mare during the oestrous cycle. *J. Reprod. Fert., Suppl.* **23**, 207–212.

Goodman, R.L. & Karsch, F.J. (1980) Pulsatile secretion of luteinizing hormone, differential suppression by ovarian steroids. *Endocrinology* **107**, 1286–1290.

Harris, J.M., Irvine, C.H.G. & Evans, M.J. (1983) Seasonal changes in serum levels of FSH, LH and testosterone and in semen parameters in stallions. *Theriogenology* **19**, 311–317.

Irvine, C.H.G. & Alexander, S. (1982) Importance of testicular hormones in maintaining the annual pattern of LH secretion in the male horse. *J. Reprod. Fert., Suppl.* **32**, 97–102.

Johnson, L. & Neaves, W.B. (1981) Age-related changes in the Leydig cell population, seminiferous tubules, and sperm production in stallions. *Biol. Reprod.* **24**, 703–712.

Johnson, L. & Thompson, D.L., Jr (1983) Age-related and seasonal variation in the Sertoli cell population, daily sperm production and serum concentrations of follicle-stimulating hormone, luteinizing hormone and testosterone in stallions. *Biol. Reprod.* **29**, 777–789.

Karsch, F.J., Bittman, E.L., Foster, D.L., Goodman, R.L., Legan, S.J. & Robinson, J.E. (1984) Neuroendocrine basis of seasonal reproduction. *Recent Prog. Horm. Res.* **40**, 185–225.

Lincoln, G.A. & Short, R.V. (1980) Seasonal breeding, nature's contraceptive. *Recent Prog. Horm. Res.* **36**, 1–43

Moberg, G.P., Watson, J.G. Stoebel, D.P. & Cook, R. (1981) Effect of cortisol and dexamethasone on the oestrogen-induced release of luteinizing hormone in the anoestrous ewe. *J. Endocr.* **90**, 221–225.

Nett, T.M., Holtan, D.W. & Estergreen, V.L. (1975) Levels of LH, prolactin and oestrogens in the serum of post-partum mares. *J. Reprod. Fert., Suppl.* **23**, 201–206.

Pickett, B.W., Faulkner, L.C., Seidel, G.E., Jr, Berndtson, W.E. & Voss, J.L. (1976) Reproductive physiology of the stallion. VI. Seminal and behavioral characteristics. *J. Anim. Sci.* **43**, 618–625.

Sharma, O.P. (1976) Diurnal variations of plasma testosterone in stallions. *Biol. Reprod.* **15**, 158–162.

Stanfield, J.M. (1983) *Seasonal responses of the stallion testis to an hCG challenge.* Master's thesis in Animal Science, University of California at Davis.

Steel, R. & Torrie, J. (1980) *Principles and Procedures of Statistics. A Biometrical Approach.*, 2nd edn, p. 27. McGraw Hill, New York.

Thompson, D.L., Jr, Pickett, B. W., Berndtson, W.E., Voss, J.L. & Nett, T.M. (1977) Reproductive physiology of the stallion. VIII. Artificial photoperiod, collection interval and seminal characteristics, sexual behavior and concentrations of LH and testosterone in serum. *J. Anim. Sci.* **44**, 656–664.

Thompson, D.L., Jr, Pickett, B.W., Squires, E.L. & Nett, T.M. (1979) Effect of testosterone and estradiol-17β alone and in combination on LH and FSH concentrations in blood serum and pituitary of geldings and in serum after administration of GnRH. *Biol. Reprod.* **21**, 1231–1237.

Turner, J.W. & Kirkpatrick, J.F. (1982) Androgens, behaviour and fertility control in feral stallions. *J. Reprod. Fert., Suppl.* **32**, 79–87.

J. Reprod. Fert., Suppl. **35** (1987), 59–65

Relationship of age and season and consumption of *Senecio vulgaris* to LH/hCG receptors in the stallion testis

J. W. Evans*, J. Stanfield, L. S. Hoffman, and C. Slaussen

Department of Animal Science, University of California, Davis, CA 95616, U.S.A.

Summary. Testes were obtained from 70 colts and stallions and were pooled according to age (4 months to 23 years) to determine the relationship of age to LH/hCG receptor kinetics. The receptor concentration (R_t) increased from 0.069×10^{-11} M/mg crude membrane fraction (CMF) for the 4–14-month pools to 0.464×10^{-11} M for the 2–3-year-old pools. A 10-fold increase in testicular size also occurred, and so the total number of receptors per testis was significantly increased. A further increase to 1.237×10^{-11} M/mg CMF was observed for stallions older than 5 years. No differences in binding affinities (K_a) were observed for the various age groups: K_a values varied from 0.19 to 2.19×10^{11} M^{-1}.

A seasonal effect was not ($P < 0.28$) observed for R_t [0.74×10^{-11} M *vs* 1.15×10^{-11} M/mg CMF, winter (N = 5) *vs* summer (N = 4)] or for K_a (0.259×10^{11} M^{-1} *vs* 0.393×10^{11} M^{-1}, winter *vs* summer).

Immediately before death resulting from inadvertent long-term consumption of *Senecio vulgaris*, testes were removed from 5 stallions and pooled. An R_t of 0.046×10^{-11} M/mg CMF was observed. Therefore, a decrease to 2.7% was observed when compared with values for normal stallions. The K_a, 0.036×10^{11} M^{-1}, was decreased to < 10%. Stallions consuming sublethal amounts may have altered reproductive functions.

Introduction

Numerous reports have outlined the secretory patterns of hormones related to reproduction in stallions, but very little information is available regarding the relationship between plasma gonado-trophins (Wesson & Ginther, 1981), testicular gonadotrophin receptors and androgens (Hamilton *et al.*, 1984) during sexual development of colts. The influence of season on attainment of puberty has been reported by Wesson & Ginther (1981). In contrast, numerous reports have described physiological mechanisms and differences in their control during sexual development in other species. Temporal patterns and changes in testicular LH/hCG (rats: Ketelslegers *et al.*, 1978; bulls: Sundby *et al.*, 1981; sheep: Barenton *et al.*, 1983; boars: Sundby *et al.*, 1981, 1983; Peyrat *et al.*, 1981; Vandalem & Hennen, 1981; hamsters: Stallings *et al.*, 1985), FSH (bulls: Dias & Reeves, 1982) and steroid (ram: Monet-Kuntz *et al.*, 1984) receptors have been reported. Evans *et al.* (1982), Stewart & Allen (1979), Matteri *et al.* (1986) and Licht *et al.* (1979) have reported on the kinetics of LH/hCG receptors in the mature stallion.

The effects of age and season on many seminal characteristics in the stallion have been well documented. These differences have been observed to be associated with changes in Leydig and Sertoli cell populations, seminiferous tubules and serum concentrations of FSH, LH and testosterone (Johnson & Thompson, 1983; Johnson & Neaves, 1981).

---

*Present address: Department of Animal Science, Texas A&M University, College Station, TX 77843, U.S.A.

The aims of this investigation were to study the changes in LH/hCG receptor kinetics in the stallion testis during sexual development and ageing and the influence of season of the year and a poisonous plant, *Senecio vulgaris*, on those kinetics.

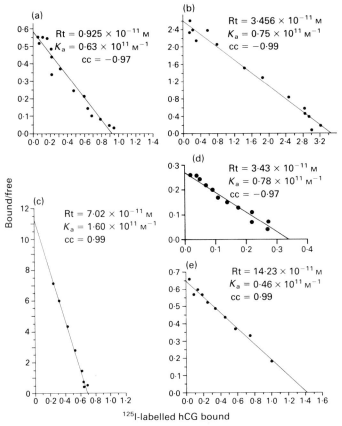

Fig. 1. Scatchard plots illustrating age-related LH/hCG receptor kinetics: (a) 12 months of age, values at 10^{-11} M; (b) 24 months of age, values at 10^{-11} M; (c) 6 years of age, values at 10^{-10} M; (d) 15·5 years of age, values at 10^{-10} M; (e) 23 years of age, values at 10^{-10} M.

Materials and Methods

Testes were obtained from 70 colts and stallions of various breeds (Appaloosa, Arabian, Paint, Quarter Horse and Thoroughbred) and ranging in age from 4 months to 23 years (Table 1) for the age-related study, from 9 stallions for the study on seasonal effects (Table 2) and from 5 stallions for the *Senecio vulgaris* study. Testes were processed according to methods previously described by Evans *et al.* (1982). Crude membrane fractions (CMF) obtained from colts or stallions of similar ages were combined, mixed and placed in aliquants into preweighed plastic vials immediately before Scatchard (1949) analysis (Fig. 1). Two or three pools were made for the 2-, 3-, 5- and 6-year-old age categories since the testes in each pool were collected and assayed at different times (Table 1). All testes for the *Senecio vulgaris* study were processed as a pool. One testis from each stallion, used for the seasonal-effect study was processed separately and an individual Scatchard analysis was performed.

Highly purified hCG (CR-121, Dr R. E. Canfield) was radiolabelled with ^{125}I or ^{131}I using a modification of the lactoperoxidase method (Gospodarowicz, 1973; Evans *et al.*, 1982). Scatchard plots were performed according to the methods of Evans *et al.* (1982). The quantity of CMF was 14 mg for the 4- and 5-months-old pools and 7 mg for all other pools. Protein content of the CMF was determined with the Bio-Rad protein assay (Bio Rad Laboratories, Richmond, CA).

The receptor concentrations were calculated without any corrections, per mg CMF and per µg protein in the CMF. Differences amongst means were compared with Tukey's studentized range test (SAS Institute Inc., Cary, NC). A *P* value of 0·05 or less was considered significant.

Table 1. Age-related changes in LH/hCG receptor concentrations (R_t) and binding affinities (K_a) in horse testes

| Age | No. | Assay | Receptor conc. (R_t) $\times 10^{-11}$ M | | | K_a ($\times 10^{11}$ M^{-1}) | Mean testicular wt (g) |
			R_t	R_t/mg CMF	R_t/µg protein		
4 months	3	A*	0·407	0·027	0·001	0·72	3·1
5 months	5	A	0·314	0·021	0·001	2·19	3·2
11 months	4	A	0·390	0·056	0·001	1·38	9·5
12 months	7	B	0·925	0·132	0·009	0·63	
13 months	2	A	0·761	0·109	0·003	1·34	10·5
14 months	5	A	0·497	0·071	0·002	0·58	9·6
Mean			0·549[a]	0·069[a]	0·0028[a]	1·14	7·2[c]
±s.e.m.			±1·122	±0·160	±0·0072	±0·19	±15·3
2 years	6	A	2·034	0·291	0·002	0·63	109·4
	4	B	3·456	0·494	0·002	0·75	
	7	C	3·870	0·553	0·002	0·76	
2·5 years	2	A	1·064	0·151	0·006	1·24	128·0
3 years	4	A	1·117	0·160	0·0007	1·33	108·8
	3	B	6·974	0·996	0·005	0·32	
	5	D	4·207	0·601	0·013	0·19	
Mean			3·264[b]	0·464[b]	0·0044[b]	0·75	115·4[d]
±s.e.m.			±1·039	±0·148	±0·0067	±0·18	±19·7
5 years	1	A	5·879	0·840	0·011	0·75	102·0
	2	D	11·179	1·597	0·029	0·29	
6 years	1	A	11·956	1·708	0·009	0·65	125·0
	1	B	7·020	1·038	0·047	1·60	
	2	D	7·264	1·038	0·021	0·36	
8 years	1	A	4·854	0·692	0·087	1·10	161·0
10 years	1	B	13·543	1·934	0·055	0·34	
15·5 years	1	A	3·434	0·491	0·006	0·78	136·6
21 years	1	A	9·339	1·334	0·017	1·03	236·0
20+ years	1	A	14·236	2·034	0·020	0·46	97·0
23 years	1	A	6·308	0·901	0·006	1·05	117·0
Mean			8·637[b]	1·237[b]	0·028[b]	0·76	139·2[d]
±s.e.m.			±0·829	±0·118	±0·005	±0·14	±12·9

*Pools with same letter were assayed at the same time.
Within columns a < b, $P < 0.001$ and c < d, $P < 0.0001$. Values are least square means ± s.e.m.

Results

Testicular growth was observed to occur in 3 phases (Table 1). During the 4–14-month period, the testis weights varied from 3·1–10·5 g and increased to 109·4 g at 2 years. No further increases in size were observed. The changes in receptor concentration followed a different pattern (Table 1). Between 14 and 24 months, the peripubertal period, a 6-fold increase ($P < 0.0001$) in concentration from 0.69×10^{-12} M/mg CMF to 4.46×10^{-12} M/mg CMF occurred. During the 2- to 3-year period, the receptor concentration [4·64 (± 1.48) $\times 10^{-12}$ M/mg CMF] remained constant (range 1.51–9.96×10^{-12} M/mg CMF). In stallions 5 years of age and older, the receptor concentration was 12·37 (± 1.18) $\times 10^{-12}$ M/mg CMF and the values ranged from 4.9 to 20.34×10^{-12} M/mg CMF. Binding affinities for the different age groups were not different: 4–14 months, $1.14 (\pm 0.19) \times 10^{11}$ M^{-1}; 2–3 years, $0.75 (\pm 0.18) \times 10^{11}$ M^{-1} and 5–23 years, $0.76 (\pm 0.14) \times 10^{11}$ M^{-1}.

The LH receptor concentration in testes removed from stallions consuming lethal amounts of *Senecio vulgaris* was 0.456×10^{-12} M/mg CMF (3.19×10^{-12} M as assayed and 1.2×10^{-14} M/µg protein) and the binding affinity was 3.6×10^{13} M^{-1}.

Table 2. Seasonal effect on testicular LH/hCG receptor kinetics in horses

| | | | Receptor conc. $(R_t) \times 10^{-11}$ M | | | | Mean |
Season	Horse	Age (years)	R_t	R_t/mg CMF	R_t/μg protein	K_a $(\times 10^{11}$ M$^{-1})$	testicular wt (g)
Winter	1	6	9·15	1·31	0·030	0·247	225
(Feb. 1982)	2	6	5·38	0·77	0·011	0·424	177
	3	3	6·66	0·95	0·025	0·130	77
	4	3	1·42	0·20	0·003	0·439	67
	5	3	3·18	0·45	0·008	0·054	119
Mean			5·16	0·74	0·015	0·259	133
±s.e.m.			±1·63	±0·23	±0·004	±0·063	±33·7
Summer	6	5	10·76	1·54	0·030	0·344	167
(July 1982)	7	5	11·60	1·66	0·028	0·232	256
	8	3	7·77	1·11	0·018	0·160	134
	9	3	2·01	0·29	0·009	0·183	51
Mean			8·04	1·15	0·021	0·230	152
±s.e.m.			±1·82	±0·26	±0·005	±0·070	±37·6

Values are least square means ± s.e.m.

No changes ($P > 0.247$) in receptor concentration were observed when the winter values ($0.74 \pm 0.23 \times 10^{-11}$ M/mg CMF) were compared with the summer values ($1.15 \pm 0.26 \times 10^{-11}$ M/mg CMF) (Table 2).

Discussion

Puberty in the male horse has been defined as the age at which an ejaculate will contain a minimum of 100×10^6 total spermatozoa with at least 10% progressive motility (Cornwell *et al.*, 1973). In Quarter Horse colts, the onset of puberty was 68.8 ± 12.7 (s.e.) weeks. Skinner & Bowen (1968) observed the appearance of spermatozoa in ejaculates of Welsh ponies at 13 months (11·5–14·5 months) which agrees with the findings of Horie & Nishikawa (1955) for the Anglo-Norman breed. Attainment of puberty may be affected by season of birth (Wesson & Ginther, 1981). Ponies born during the spring reached puberty by the end of their second summer (~ 16 months of age) whereas of those born during the summer, only 1 of 4 had reached puberty at the end of their second summer (~ 12–13 months of age).

There are marked differences in the temporal pattern of plasma hormones and LH receptors between species during development. In hamsters, there was a 10-fold increase in LH receptor concentration between 23 and 35 days of age; these receptor changes preceded testicular growth between 35 and 47 days of age; and increases in testosterone concentrations occurred between Days 40 and 58 of age (Stallings *et al.*, 1985). In the lamb, the total number of LH receptors increased 33-fold from 10 to 120 days of age (Barenton *et al.*, 1983). However, the receptor concentration continually decreased from birth to 120 days of age as a result of the volume of seminiferous tubules growing at a faster rate than the volume of interstitial tissue during that period. They observed a large increase in the number of receptors between 80 and 120 days of age and that their concentration stopped decreasing during that period. During this phase of development, plasma LH concentrations did not increase but the release of testosterone was much greater following an LH pulse, suggesting an increase in the sensitivity or responsiveness of the Leydig cells to LH. In rats, Ketelslegers *et al.* (1978) observed a progressive rise in plasma LH concentration between Days 36 and 51 and a marked increase in total numbers of binding sites, beginning at 15–38 days of age and continuing to

Day 60. These were correlated to testicular growth and a continuous increase in receptor concentration from 5 days of age to Day 60. The prepubertal rise in plasma testosterone occurred about 15 days after testicular LH receptors began to increase and was coincident with the continuing rise in LH receptor concentration and the progressive rise in LH during the 35–55-day period. In the pig, specific LH binding per mg protein was highest at 2–3 weeks and lowest at 7 weeks of age when observed for 22 weeks after farrowing (Sundby *et al.*, 1983). These changes occurred in accordance with morphological and functional changes of the Leydig cells. Vandalem & Hennen (1981) observed that the total number of LH receptors per testis in the pig increased progressively from 2 to 6 months of age; however, the concentration of receptor sites decreased from birth to adult age. In prepubertal bulls, LH binding was higher than in older bulls (Sundby *et al.*, 1981).

Very little information is available on the stallion concerning age-related changes in plasma gonadotrophin and androgen concentrations and changes in testicular morphology and functions. Testicular growth in the stallion occurs in 3 phases, similar to those of the hamster (Stallings *et al.*, 1985). A 3-fold increase in testicular size (3–10 g) was observed during the 5- to 14-month period and a further 10-fold increase (10 to 100 g) occurred during the 14- to 24-month period (Table 1). These changes are in agreement with values and patterns of growth published by Arthur & Allen (1972), Johnson & Neaves (1981), Johnson & Thompson (1983), and Thompson *et al.* (1979). Sertoli cell numbers increase until 4–5 years of age due to increases in testicular weight and then the number remains constant. In male horses less than 4 years of age, the number of Sertoli cells per gram of parenchyma is greater than in older stallions (Johnson & Thompson, 1983). In contrast, stallions experience an increase in total number of Leydig cells and in the number per gram of parenchyma with advancing age beyond 2 years of age (Johnson & Neaves, 1981). A 2-fold increase in the total number occurs between post-puberty and adulthood. Associated with the changes in Sertoli and Leydig cells are changes in concentrations of LH, FSH and testosterone and daily sperm production (Wesson & Ginther, 1981; Johnson & Thompson, 1983; Hamilton *et al.*, 1984). The age-related temporal patterns of gonadotrophins, androgens and testicular morphological changes are consistent with the observed changes in LH/hCG receptor concentrations in this report (Table 1). Between 14 and 24 months, the peri-pubertal period, a 6-fold increase ($P < 0.001$) in concentration from 0.69×10^{-12} to 4.46×10^{-12} M/mg CMF occurred. A 15-fold increase ($P < 0.001$) in testicular size also occurred so that the total number of receptors per testis was significantly increased. During the 2- to 3-year period, the receptor concentration (4.46×10^{-12} M/mg CMF) remained constant as did testicular size. A further increase in receptor concentration to $12.37 (\pm 1.18) \times 10^{-12}$ M/mg CMF was observed for stallions 5 years of age and older. Therefore, the age-related changes in endocrine and in morphological events are related. The prepubertal colt has low plasma LH concentrations and low numbers of LH receptors, which are correlated with low plasma testosterone concentrations. During the peripubertal period, plasma LH remains low but the Leydig cell (Ketelslegers *et al.*, 1978; Peyrat *et al.*, 1981) could be responsible for the 2-fold increase in testosterone concentration that occurs during this period (Johnson & Thompson, 1983). After adulthood, the plasma LH concentration increases with age (Johnson & Thompson, 1983), the LH receptor concentration remains constant and testicular size increases so that the total number of receptors per testis increases. This combination of events may therefore explain the age-related increasing testosterone concentration observed by Johnson & Thompson (1983).

No age-related changes were observed for the affinity constants (K_a) which ranged in value from 0.19 to 2.19×10^{11} M^{-1} (Table 1; Fig. 1). Similar observations have been reported for rats (Ketelslegers *et al.*, 1978), lambs (Barenton *et al.*, 1983) and pigs (Vandalem & Hennen, 1981; Sundby *et al.*, 1983). The values observed for the stallion (Table 1; Evans *et al.*, 1982) are similar to those reported for other species by the above authors.

A reduction in LH receptor concentration to 2·7% (4.5×10^{-12} M/mg CMF) and a reduction in binding affinity to 10% (3.6×10^9 M^{-1}) was observed for testes removed from stallions immediately before death resulting from inadvertent long-term consumption of alfalfa hay contaminated with *Senecio vulgaris* when compared with values for older stallions (Table 1). Stallions consuming sublethal amounts may have altered reproductive functions. This may explain some cases of lowered fertility or infertility.

The influence of season on reproduction in the stallion has received considerable attention, especially with regard to changes in seminal characteristics (Pickett & Voss, 1972; Harris *et al.*, 1983), circulating hormones (Johnson & Thompson, 1983; Stanfield *et al.*, 1984; Clay *et al.*, 1985; Thompson *et al.*, 1985), behaviour (Pickett & Voss, 1972) and photoperiodic stimulation (Burns *et al.*, 1982; Clay *et al.*, 1985). No changes ($P > 0.247$) in receptor concentration (Table 2) were observed relative to season in contrast to the season-related changes discussed above. No changes in testicular weights were observed in this small number of observations (Table 2). Since other authors have reported stallions to have increased testicular weights during the breeding season (Johnson & Thompson, 1983), the total number of LH/hCG binding sites would be increased during the breeding season. Similar results have been reported for other seasonally breeding species. During the non-breeding season for rams, reduction in testicular weight and function is associated with a decrease in the number of Leydig cells and an increase in the number of LH receptors per Leydig cell (Barenton & Pelletier, 1983). In golden hamsters, an initial decrease in LH receptor concentration occurs during the nonbreeding season and either remains constant or increases to above the value observed during the breeding season (Bartke *et al.*, 1982). Therefore in the ram and hamster and probably in the stallion, the total number of LH receptor sites is reduced during the non-breeding season. Photoperiod has no significant effect on binding affinity of LH receptors of other species (Bartke *et al.*, 1982; Barenton & Pelletier, 1983; Tähkä & Rajaniemi, 1985) and results for the stallions in this study are in agreement with that concept (Table 2).

We thank Dr R. E. Canfield for the hCG (CR-121); and Gene Mikuckis for his technical expertise.

References

Arthur, G. H. & Allen, W. E. (1972) Clinical observations on reproduction in a pony stud. *Equine vet. J.* **4**, 109–117.

Barenton, B. & Pelletier, J. (1983) Seasonal changes in testicular gonadotropin receptors and steroid content in the ram. *Endocrinology* **112**, 1441–1446.

Barenton, B., Hochereau-de Reviers, M.T. & Saumande, J. (1983) Changes in testicular gonadotropin receptors and steroid content through postnatal development until puberty in the lamb. *Endocrinology* **112**, 1447–1453.

Bartke, A., Klemcke, H.G., Amador, A. & Van Sickle, M. (1982) Photoperiod and regulation of gonadotropin receptors. *Ann. N.Y. Acad. Sci.* **383**, 122–131.

Burns, P.J., Jaward, M.J., Edmundson, A., Cahill, C., Boucher, J.K., Wilson, E.A. & Douglas, R.H. (1982) Effect of increased photoperiod on hormone concentrations in Thoroughbred stallions. *J. Reprod. Fert., Suppl.* **32**, 103–111.

Clay, C.M., Squires, E.L., Amann, R.P. & Pickett, B.W. (1985) The influence of photoperiod on the seasonal reproductive cycle of stallions. *Proc. 9th Equine Nutr. and Physiol. Soc.* pp. 296–301.

Cornwell, J.C., Hauer, E.P., Spillman, T.E. & Vincent, C.K. (1973) Puberty in the Quarter Horse colt. *J. Anim. Sci.* **36**, 215, Abstr.

Dias, J.A. & Reeves, J.J. (1982) Testicular FSH receptors and affinity in bulls of various ages. *J. Reprod. Fert.* **66**, 39–45.

Evans, J.W., Mikuckis, G.M. & Roser, J.F. (1982) Comparison of the interaction of equine LH and human chorionic gonadotrophin to equine testicular receptors. *J. Reprod. Fert., Suppl.* **32**, 113–121.

Gospodarowicz, D. (1973) Properties of the luteinizing hormone receptor of isolated bovine corpus luteum plasma membranes. *J. biol. Chem.* **248**, 5042–5079.

Hamilton, M.-J., Hughes, I.M. & Hegreberg, G.A. (1984) Serum testosterone levels in young normal horses. *Theriogenology* **22**, 417–421.

Harris, J.M., Irvine, C.H.G. & Evans, M.J. (1983) Seasonal changes in serum levels of FSH, LH and testosterone and in semen parameters in stallions. *Theriogenology* **19**, 311–322.

Horie, T. & Nishikawa, Y. (1955) Studies on the development of the testis and epididymis of the horse. II. The development of the epididymis and the properties of the sperm collected from the epididymis. *Bull. natn. Inst. Agric. Sci., Tokyo, Ser. G.* **10**, 351–353.

Johnson, L. & Neaves, W.B. (1981) Age-related changes in the Leydig cell population, seminiferous tubules, and sperm production in stallions. *Biol. Reprod.* **24**, 703–712.

Johnson, L. & Thompson, D.L. (1983) Age-related and seasonal variation in the Sertoli cell population, daily sperm production and serum concentrations of follicle stimulating hormone, luteinizing hormone and testosterone in stallions. *Biol. Reprod.* **29**, 777–789.

Ketelslegers, J.M., Hetzel, W.D., Sherins, R.J. & Catt (1978) Developmental changes in testicular gonadotropin receptors: plasma gonadotropins and plasma testosterone in the rat. *Endocrinology* 103, 212–222.

Licht, P., Bona-Gallo, A., Aggarwal, B.B., Farmer, S.W., Castelino, J.B. & Papkoff, H. (1979) Biological and binding activities of equine pituitary gonadotrophins and pregnant mare serum gonadotrophin *J. Endocr.* 83, 311–322.

Matteri, R.L., Papkoff, H., Ng, D.A., Swedlow, J.R. & Chang, Y.-S. (1986) Isolation and characterization of three forms of luteinizing hormone from the pituitary gland of the horse. *Biol. Reprod.* 34, 571–578.

Monet-Kuntz, C., Hochereau-de Reviers, M.T. & Terqui, M. (1984) Variations in testicular androgen receptors and histology of the lamb testis from birth to puberty. *J. Reprod. Fert.* 70, 203–210.

Peyrat, J.P., Meusy-Dessolle, N. & Garnier, D. (1981) Changes in Leydig cells and luteinizing hormone receptors in porcine testis during postnatal development. *Endocrinology* 108, 625–631.

Pickett, B.W. & Voss, J.L. (1972) Reproductive management of the stallion. *Proc. 18th Ann. Conv. Am. Ass. equine Pract.* pp. 501–531.

Scatchard, G. (1949) The attraction of proteins to small molecules and ions. *Ann. N Y. Acad. Sci.* 51, 660–672.

Skinner, J.D. & Bowen, J. (1968) Puberty in the Welsh stallion. *J. Reprod. Fert.* 16, 133–135.

Stallings, M.H., Matt, K.S., Amador, A., Bartke, A., Siler-Khodr, T.M., Soares, M.J. & Talamantes, F. (1985) Regulation of testicular LH/hCG receptors in golden hamsters (*Mesocricetus auratus*) during development. *J. Reprod. Fert.* 75, 663–670.

Stanfield, J.M., Hoffman, L. & Evans, J.W. (1984) Responses of stallions to hCG during winter vs summer. *J. Anim. Sci.* 59 (Suppl. 1), 492–493.

Stewart, F. & Allen, W.R. (1979) The binding of FSH, LH and PMSG to equine gonadal tissues. *J. Reprod. Fert., Suppl.* 27, 431–440.

Sundby, A., Torjesen, P. & Hansson, V. (1981) Testicular LH and FSH receptors during development in bulls. Characterization of testicular receptor in bull and pig. HCG-induced down-regulation of the testicular LH receptor in pig. *Acta endocr., Copenh.* 97, *Suppl.* 243, Abstr. 386.

Sundby, A., Torjesen, P. & Hansson, V. (1983) Developmental pattern and sensitivity to down-regulation of testicular LH receptors in the pig. *Int. J. Androl.* 6, 194–200.

Tähkä, K.M. & Rajaniemi, H. (1985) Periodic manipulation of testicular LH receptors in the bank vole (*Clethrionomys glareolus*). *J. Reprod. Fert.* 75, 513–519.

Thompson, D.L., Pickett, B.W., Squires, E.L. & Amann, R.P. (1979) Testicular measurements and reproductive characteristics in stallions. *J. Reprod. Fert., Suppl.* 27, 13–17.

Thompson, D.L., St. George, R.L., Jones, L.S. & Garza, F. (1985) Patterns of secretion of luteinizing hormone, follicle stimulating hormone and testosterone in stallions during the summer and winter. *J. Anim. Sci.* 60, 741–747.

Vandalem, J.L. & Hennen, G. (1981) Testicular gonadotrophic receptors in the maturing pig. *Acta endocr., Copenh.* 97, *Suppl.* 432, Abstr. 388.

Wesson, J.A. & Ginther, O.J. (1981) Puberty in the male pony: gonadotropin concentrations and the effect of castration. *Anim. Reprod. Sci.* 4, 165–175.

J. Reprod. Fert., Suppl. **35** (1987), 67–70

Effects of month and age on prolactin concentrations in stallion serum

D. L. Thompson, Jr, L. Johnson* and J. J. Wiest

*Department of Animal Science, Louisiana Agricultural Experiment Station, Louisiana State University Agricultural Center, Baton Rouge, Louisiana 70803 and *Department of Cell Biology, University of Texas Health Science Center at Dallas, Dallas, Texas 75235, U.S.A.*

Summary. Prolactin concentrations in stallion serum were measured by a newly developed radioimmunoasay based on anti-dog prolactin serum and radiolabelled horse prolactin. Samples of serum from a total of 444 stallions were obtained at a commercial abattoir monthly from April to the following March. Ages of stallions were estimated from eruption and wear patterns of incisors. In the analysis of variance, both month ($P < 0.01$) and age ($P < 0.05$) were significant sources of variation whereas there was no interaction between these factors. Monthly means for prolactin concentrations were greatest between May and August with peak concentrations occurring in July (7.1 ± 0.5 ng/ml). Lowest mean concentrations of prolactin (1.2 ng/ml) were observed during the winter months (December and January). Concentrations of prolactin were lowest in stallions $\leqslant 5$ years old and were highest in stallions > 10 years old. It appears that prolactin secretion in the stallion is strongly influenced by season, as it is in the mare and in other seasonally breeding animals. Moreover, prolactin secretion in the stallion tends to increase as the stallion ages.

Introduction

Stallions exhibit seasonal patterns in seminal and behavioural characteristics and in secretion of LH and testosterone (Pickett *et al.*, 1970, 1976; Berndtson *et al.*, 1974; Thompson *et al.*, 1977; Johnson & Thompson, 1983). Moreover, some of these characteristics can be stimulated by additional light applied in the winter (Thompson *et al.*, 1977). With the development of an homologous radioimmunoassay for horse prolactin (Roser *et al.*, 1984), Johnson (1986) was able to demonstrate that prolactin secretion in mares was seasonal and responded to added light in winter (Johnson & Malinowski, 1986). The purpose of the present study was to determine whether serum prolactin concentrations in stallions are (1) affected by season as they are in mares and (2) affected by age in a manner as are concentrations of LH, FSH and testosterone (Johnson & Thompson, 1983).

Materials and Methods

Blood samples were collected monthly from a total of 444 stallions at a commercial abattoir in Fort Worth, Texas, U.S.A. from April 1984 through to March 1985. Blood was collected at exsanguination and was placed on ice until centrifugation. Serum was harvested and was stored at $-15°C$. The age of each stallion was estimated from patterns of eruption and wear of the lower and upper incisors (Ensminger, 1977). Stallions were grouped into five age categories similar to those described previously (Johnson & Thompson, 1983).

Prolactin concentrations were measured with a heterologous radioimmunoassay based on antiserum generated against dog prolactin (reagents for the dog–dog assay supplied by Dr A. F. Parlow, Director, Pituitary Hormones & Antisera Center, Harbor-UCLA Medical Center, 1000 West Carson Street, Torrance, CA 90509, U.S.A.) and radio-labelled horse prolactin. The antiserum was initially diluted 1:10 000 and was used at 200 μl/tube. Horse prolactin used for radioiodination was prepared in our laboratory as described by Li & Chung (1983) except that the starting material was the 0.2 M-HPO_3 precipitate described by Papkoff (1976). After purification, polyacrylamide gel electrophoresis of the preparation resulted in a single band of stainable protein that was coincident with the prolactin immunoreactivity and with radiolabelled dog prolactin. The horse prolactin was radioiodinated by general procedures described by Greenwood *et al.* (1963) and was subsequently purified by gel filtration.

In the heterologous radioimmunoassay, parallel inhibition curves were produced by horse prolactin, horse pituitary extract and sera of horses of various reproductive states. Electrophoresis of an aliquant of equine pituitary extract showed that only one peak of immunoreactivity was present which was coincident with radiolabelled horse prolactin. Moreover, prolactin concentrations were similar ($r = 0.95$) for samples of serum assayed in the heterologous assay and in the homologous assay of Roser *et al.* (1984; reagents supplied by Dr H. Papkoff, Hormone Research Institute, University of California, San Francisco, CA 94143, U.S.A.). In addition, concentrations of prolactin in serum of horses treated with gonadotrophin-releasing hormone were unchanged whereas concentrations of both LH and FSH were elevated. Moreover, prolactin concentrations were elevated in sera from mares treated with thyrotrophin-releasing hormone (TRH; Thompson *et al.*, 1986). Intra- and interassay coefficients of variation averaged 6 and 8%, respectively. Sensitivity of the assay was 0·1 ng/tube.

Data were grouped by month and age and were analysed by least-squares analysis of variance with a 5 × 12 factorial arrangement of treatments (SAS, 1982). Differences amongst means were assessed for significance by Duncan's multiple-range test (Steel & Torrie, 1960).

Results and Discussion

For measurement of prolactin in serum and pituitary extracts of horses, the dog–horse radioimmuno-assay described herein was essentially equivalent to the homologous assay (Roser *et al.*, 1984) and to a horse–dog assay described previously (Thompson *et al.*, 1986). Sensitivity of the homologous assay (Roser *et al.*, 1984) was about 2-fold greater than the dog–horse or horse–dog assays and about 4-fold greater than the dog–dog assay. In each of the 4 types of assay, (1) purified horse prolactin and horse tissues produced parallel inhibition curves, (2) a single peak of immunoreactivity in polyacrylamide gels was detected after electrophoresis of horse pituitary extract, (3) prolactin concentrations in serum samples obtained in summer were consistently higher than those obtained in winter and (4) the response in prolactin concentrations to TRH injection (Thompson *et al.*, 1986) was detected. Therefore, depending on availability of reagents, any of these 4 assays could potentially be used for studying prolactin concentrations in horse tissues.

Concentrations of prolactin in serum of stallions were affected by both month ($P < 0.01$) and age ($P < 0.05$) as assessed by analysis of variance. However, there was no interaction between these two factors. As shown in Fig. 1, mean concentrations of prolactin were highest during July

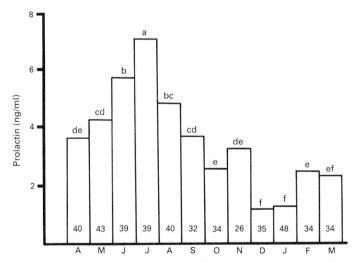

Fig. 1. Mean concentrations of prolactin in stallion serum collected monthly at an abattoir beginning in April. Values within each bar indicate the number of stallions sampled that month. Means with no common superscript (a–f) differ ($P < 0.05$).

Table 1. Mean concentrations of prolactin in serum of stallions of various ages

Age (years)	No. of stallions	Prolactin (ng/ml)
<2	64	$3 \cdot 3 \pm 0 \cdot 4$[a]
2–3	163	$3 \cdot 5 \pm 0 \cdot 2$[a]
4–5	91	$3 \cdot 2 \pm 0 \cdot 3$[a]
6–10	58	$3 \cdot 9 \pm 0 \cdot 4$[ab]
>10	68	$4 \cdot 6 \pm 0 \cdot 4$[b]

Values are means ± s.e.m. Means with no common superscript differ ($P < 0 \cdot 05$).

($7 \cdot 1 \pm 0 \cdot 5$ ng/ml) and were lowest during winter ($\sim 1 \cdot 2$ ng/ml). These seasonal changes in prolactin concentrations are similar to those reported for mares by Johnson (1986), for which prolactin concentrations were also highest in July, but averaged $\sim 3 \cdot 5$ ng/ml compared with $7 \cdot 1$ ng/ml reported herein for stallions; concentrations of prolactin in mares in winter (~ 1 ng/ml) were similar to those for stallions.

Prolactin concentrations were highest in stallions >10 years old and lowest in stallions ≤5 years old (Table 1). This effect of age on prolactin concentrations was similar to that observed previously for concentrations of LH, FSH and testosterone in a different group of stallions (Johnson & Thompson, 1983). The simultaneous increase in concentrations of testosterone and gonadotrophins as stallions age may indicate a reduction in sensitivity of the hypothalamic–pituitary axis to negative feedback by testicular products. Other explanations are that these changes in prolactin and gonadotrophin secretion represent (1) a maturing of the horse pituitary up to about 5 years of age or (2) a decrease in clearance rates of these hormones as the stallions age.

Although prolactin concentrations in mares and stallions are profoundly influenced by season (or photoperiod), the physiological significance of this influence is unknown. In other seasonally breeding species, prolactin has been implicated in the regulation of fat metabolism and deposition (Meier, 1975) and regulation of accessory sex gland and testicular function (Horrobin, 1974) in addition to its involvement in lactation and behaviour (Friesen & Forsbach, 1981). In various vertebrates and invertebrates, prolactin appears to be a major regulator of osmotic balance (Ensor & Ball, 1972; Nicoll, 1981). These are therefore potentially fertile areas for future research on prolactin effects in the horse.

This research was partly funded by Grant HD16773 from the National Institutes of Health, U.S.A. We thank Beltex Corp., Fort Worth, Texas, U.S.A. for horse tissues; and Dr Harold Papkoff and Dr A. F. Parlow for antisera and purified hormones.

References

Berndtson, W.E., Pickett, B.W. & Nett, T.M. (1974) Reproductive physiology of the stallion. IV. Seasonal changes in the testosterone concentration of peripheral plasma. *J. Reprod. Fert.* **39**, 115–118.

Ensminger M.E. (1977) *Horses and Horsemanship*, 5th edn. Interstate, Danville, IL.

Ensor, D.M. & Ball, J.N. (1972) Prolactin and osmoregulation in fishes. *Fedn Proc. Fedn Am. Socs exp. Biol.* **31**, 1615–1623.

Friesen, H. & Forsbach, G. (1981) Prolactin secretion during pregnancy and lactation. In *Prolactin*, pp. 167–180. Ed. R. B. Jaffe. Elsevier-North Holland, New York.

Greenwood, F.C., Hunter, W.M. & Glover, G.S. (1963) The preparation of ^{131}I-labelled human growth hormone of high specific activity. *Biochem. J.* **89**, 114–123.

Horrobin, D.F. (1974) *Prolactin 1974*. Medical and Technical Publishing Co. Ltd., Lancaster.

Johnson, A.L. (1986) Serum concentrations of prolactin,

thyroxine and triiodothyronine relative to season and the estrous cycle in the nonpregnant mare. *J. Anim. Sci.* **62**, 1012–1020.

Johnson, A.L. & Malinowski, K. (1986) Daily rhythm of cortisol, and evidence for a photo-inducible phase for prolactin secretion in nonpregnant mares housed under non-interrupted and skeleton photoperiods. *J. Anim. Sci.* **63**, 169–175.

Johnson, L. & Thompson, D.L., Jr (1983) Age-related and seasonal variation in the Sertoli cell population, daily sperm production and serum concentrations of follicle-stimulating hormone, luteinizing hormone and testosterone in stallions. *Biol. Reprod.* **29**, 777–789.

Li, C.H. & Chung, D. (1983) Studies on prolactin 48: isolation and properties of the hormone from horse pituitary glands. *Archs Biochem. Biophys.* **220**, 208–213.

Meier, A.H. (1975) Chronophysiology of prolactin in the lower vertebrates. *Amer. Zool.* **15**, 905–916.

Nicoll, C.S. (1981) Role of prolactin in water and electrolyte balance in vertebrates. In *Prolactin*, pp. 127–166. Ed. R. B. Jaffe. Elsevier-North Holland, New York.

Papkoff, H. (1976) Canine pituitary prolactin: isolation and partial characterization. *Proc. Soc. exp. Biol. Med.* **153**, 498–500.

Pickett, B.W., Faulkner, L.C. & Sutherland, T.M. (1970) Effect of month and stallion on seminal characteristics and sexual behavior. *J. Anim. Sci.* **31**, 713–728.

Pickett, B.W., Faulkner, L.C., Seidel, G.E., Jr, Berndtson, W.E. & Voss, J.L. (1976) Reproductive physiology of the stallion. VI. Seminal and behavioral characteristics. *J. Anim. Sci.* **43**, 617–625.

Roser, J.F., Chang, Y.-S., Papkoff, H. & Li, C.H. (1984) Development and characterization of a homologous radioimmunoassay for equine prolactin. *Proc. Soc. exp. Biol. Med.* **175**, 510–517.

SAS (1982) *SAS User's Guide: Statistics.* Statistical Analysis System Institute Inc., Cary.

Steel, R.G.D. & Torrie, J.H. (1960) *Principles and Procedures of Statistics.* McGraw-Hill Book Co., New York.

Thompson, D.L., Jr, Pickett, B.W., Berndtson, W.E., Voss, J.L. & Nett, T.M. (1977) Reproductive physiology of the stallion. VIII. Artificial photoperiod, collection interval and seminal characteristics, sexual behavior and concentrations of LH and testosterone in serum. *J. Anim. Sci.* **44**, 656–664.

Thompson, D.L., Jr, Wiest, J.J. & Nett, T.M. (1986) Measurement of equine prolactin with an equine-canine radioimmunoassay: seasonal effects on the prolactin response to thyrotropin releasing hormone. *Dom. Anim. Endocrinol.* **3**, 247–252.

J. Reprod. Fert., Suppl. **35** (1987), 71–78

In-vitro biosynthesis of C_{18} neutral steroids in horse testes

S. J. Smith, J. E. Cox, E. Houghton*, M. C. Dumasia* and M. S. Moss*

*Division of Equine Studies and Farm Animal Surgery, Department of Veterinary Clinical Science, University of Liverpool, Leahurst, South Wirral, L64 7TE, and *Horseracing Forensic Laboratory Limited, P.O. Box 15, Snailwell Road, Newmarket, Suffolk CB8 7DT, U.K.*

Summary. Deuterium, ^{14}C- and ^{3}H-labelled steroid substrates were incubated with minced testicular tissue from stallions of different ages. After extraction and separation of the neutral and phenolic fractions the metabolites were identified by gas chromatography–mass spectrometry. The presence of the expected C_{19} neutral and C_{18} phenolic steroids was confirmed. An isomer of 5(10)-oestrene-3,17-diol was also identified.

Introduction

The C_{18} neutral steroids oestrane-3,17α-diol (Fig. 1a), 5(10)-oestrene-3,17-diol (Fig. 1b), and 19-nortestosterone (Fig. 1c) (Houghton *et al.,* 1984a) and 19-norandrostenedione (Fig. 1d) have been detected in normal stallion urine (M. C. Dumasia & E. Houghton, unpublished observations). Steroid profile studies in stallion testicular tissue (Dumasia *et al.,* 1985) have shown that these steroids are produced in the testis; oestrane-3,17α-diol, 19-nortestosterone and 19-norandrostenedione are present in the unconjugated fraction and four isomers of 5(10)-oestrene-3,17-diol have been detected in the sulpho-conjugated fraction. The presence of these steroids in both urine and testicular tissue suggests that they may be intermediates or by-products in the biosynthesis of oestrogens, a theory originally proposed by Dorfman & Ungar (1965). The in-vitro biosynthesis of these steroids has, therefore, been studied with particular reference to their role in the formation of oestrogens in the stallion testis.

Previous incubation studies on stallion testicular tissue have been carried out by a number of workers. Baggett *et al.* (1959), using [^{14}C]testosterone as substrate, showed some conversion to oestrone and oestradiol. Perfusion studies by Nyman *et al.* (1959) and Savard & Goldzieher (1960) using [^{14}C]acetate demonstrated conversion to both androgens and oestrogens. Bedrak & Samuels (1969) have incubated a number of radiolabelled androgens and progestagens with stallion testicular homogenates and demonstrated that steroidogenesis in the stallion testis occurs by the Δ^4 and Δ^5 pathways: 19-norandrostenedione was formed with 17α-hydroxyprogesterone as substrate and 19-nortestosterone with testosterone as substrate. Oh & Tamaoki (1970) have used progesterone, 17α-hydroxyprogesterone, androstenedione, testosterone and dehydroepiandrosterone (DHA) in incubation studies with cell-free homogenates from stallion testis and again demonstrated conversion to oestrogens.

In the present study a mixture of substrates with radioactive and stable isotope labels has been used to facilitate the detection of metabolites and their identification by combined gas chromatography–mass spectrometry (GC–MS) as their methoxylamine–trimethylsilyl (MO–TMS) or trimethylsilyl (TMS) derivatives.

Materials and Methods

[16,16-^{2}H$_2$]Dehydroepiandrosterone (^{2}H$_2$-DHA) was synthesized according to the method of Gaskell & Finlay (1979). [16,16-^{2}H$_2$]Androst-5-ene-3,17-diol (Δ^5-diol) was synthesized by reduction of [^{2}H$_2$]DHA with sodium borohydride; similarly [4-^{14}C]Δ^5-diol was synthesized from [4-^{14}C]DHA. Radiolabelled [4-^{14}C]DHA and [7-^{3}H]DHA

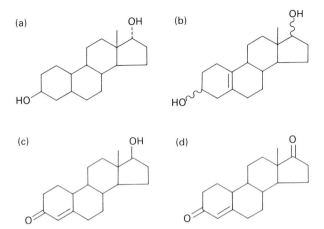

Fig. 1. C_{18} neutral steroids isolated from stallion urine and testicular tissue: (a) oestrane-3,17α-diol, (b) 5(10)-oestrene-3,17-diol, (c) 19-nortestosterone, (d) 19-norandrostenedione.

sulphate were purchased from NEN Research Products (Dupont, U.K. Ltd, Stevenage, Herts) and [4-^{14}C]19-nortestosterone from Amersham International plc, Amersham, Bucks, U.K.).

Triethylaminohydroxypropyl Sephadex LH-20 (TEAP LH-20), which was used for separation of neutral and phenolic steroids, was prepared according to the method of Axelson *et al.* (1981).

Testes were obtained from horses castrated under general anaesthesia and were used fresh. Decapsulated testes were minced and wet tissue (1 g) was incubated with a mixture of radiolabelled and deuterium-labelled substrates (50 μg) in phosphate buffer (0·1 M, pH 7·4, 10 ml). The buffer contained nicotinamide (30 mM), magnesium chloride (4 mM), glutathione (8 mM), glucose 6-phosphate (6 mM), glucose 6-phosphate dehydrogenase (2 units) and nicotinamide adenine dinucleotide phosphate (NADP; 0·1 mM). Incubations were performed at 37°C in a shaking water bath for 15 min in an atmosphere of 95% O_2 and 5% CO_2, and then terminated by addition of methanol (90 ml). Controlled incubations were performed with minced tissue but without added steroid substrates. The procedure for isolation, purification and identification of the steroids is shown in Fig. 2. The efficiency of the analytical procedure was monitored at each stage by removing aliquants for scintillation counting to determine radioactive content. Radiolabelled steroids were located on TLC plates using a radiochromatogram spark chamber (Panax, Betagraph, Bournemouth). Derivatization and GC–MS analyses were carried out as described previously (Houghton *et al.*, 1984b).

Results and Discussion

Classical methods of identification of metabolites in in-vitro steroid biosynthesis studies using radiolabelled substrates have been based upon chromatographic purification, microchemical reactions and crystallization to constant activity. However, identification is then dependent upon the availability of carrier steroids which have the correct stereochemistry. In steroid biosynthesis studies there is a clear requirement to distinguish between endogenous material and substrate-derived species. Radiolabelled substrates permit the continuous monitoring of the efficiency of analytical procedures whilst stable isotope labelling provides a characteristic mass shift or isotopic clusters for unequivocal identification of transformation products by GC–MS (Braselton *et al.*, 1973). The ideal substrate should, therefore, contain a radiolabel for isolation purposes and a stable isotope label for identification of substrate-related species by GC–MS.

The use of ^{13}C (which labels the substrate in the steroid nucleus) will minimize any metabolic isotope effects but synthesis of ^{13}C-labelled materials is expensive and difficult. However, a number of techniques exist for the facile incorporation of deuterium into steroids (Dehennin *et al.*, 1980) but care must be taken to introduce the deuterium into a site remote from metabolism. In the present study [16,16-^{2}H$_2$]DHA was synthesized as a substrate for use in conjunction with

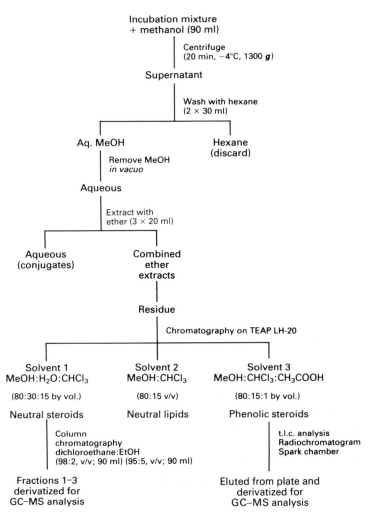

Fig. 2. Flow chart of the analytical procedures for the extraction, purification and identification of steroids after in-vitro incubations with minced stallion testes.

[4-^{14}C]DHA in in-vitro biosynthesis studies and also for use as a precursor for the synthesis of other deuterium-labelled steroids. As the deuterium atoms were not introduced at a primary metabolic site associated with the conversion of androgens to oestrogens it was felt that any isotope effects upon metabolism would be minimized, and the results support this assumption. The advantage of using [16,16-^{2}H$_2$]DHA for identification purposes by GC–MS is demonstrated in Figs 3(a) and 3(b), the mass spectrum of the MO–TMS derivative of the deuterated species showing a characteristic shift of two mass units in certain fragment ions.

Table 1 shows the results of incubations of testes from two stallions. Testicular tissue from a 2-year-old stallion was incubated in May with a mixture of [4-^{14}C]DHA and [16,16-^{2}H$_2$]DHA and from a 3-year-old stallion in July with (i) a mixture of [4-^{14}C]DHA and [16,16-^{2}H$_2$]DHA, (ii) a mixture of [4-^{14}C]Δ^5-diol and [16,16-^{2}H$_2$]Δ^5-diol, (iii) [7-^{3}H]DHA sulphate and (iv) [4-^{14}C]19-nortestosterone.

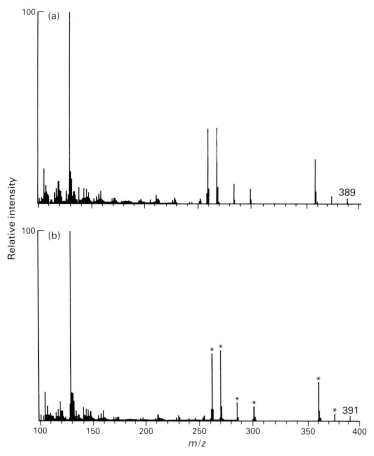

Fig. 3. Mass spectra of the MO–TMS derivatives of (a) DHA and (b) $[^2H_2]$DHA. Ions showing two mass unit shifts are marked (*).

The metabolic profile obtained with DHA (Fig. 4) was similar to that obtained by Bedrak & Samuels (1969) and Oh & Tamaoki (1970). In addition the conversion of DHA to Δ^5-diol and testosterone, a pathway suspected by Bedrak & Samuels (1969), was confirmed. The present study also demonstrated the conversion of DHA to the C_{18} neutral steroid 5(10)-oestrene-3,17-diol; the mass spectra of the *bis*-TMS derivatives of the standard steroid and the deuterium-labelled isolate are shown in Fig. 5.

The incubation with Δ^5-diol gave a similar metabolic profile to that for DHA (Table 1). The interconversion of both Δ^5-diol and DHA, and also the formation of testosterone and androstenedione in both incubations, demonstrates the presence of the Δ^4 and Δ^5 pathways. The C_{18} neutral steroid, 5(10)-oestrene-3,17-diol, was again detected along with 19-hydroxytestosterone and 19-hydroxydehydroepiandrosterone. 19-Hydroxysteroids have previously been identified in in-vitro incubation studies by Bedrak & Samuels (1969) and Oh & Tamaoki (1970).

The ratio of oestrone to oestradiol formed from Δ^5-diol was 1:6 compared with to 6:1 for DHA (2-year-old stallion tissue). Incubations with DHA and Δ^5-diol showed the presence of 17β-hydroxysteroid dehydrogenase activity in stallion testis. However, the ratios of oestrone to oestradiol formed from Δ^5-diol and DHA indicate that the 3β-hydroxysteroid dehydrogenase with Δ^{5-4} isomerase is the more predominant pathway in oestrogen biosynthesis.

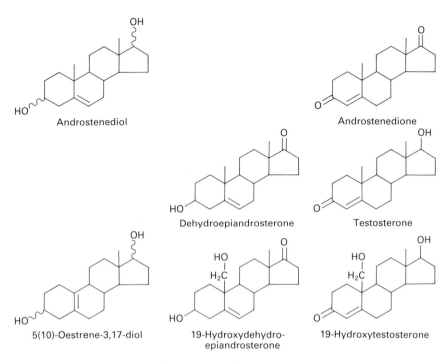

Fig. 4. Structures of steroids identified from in-vitro incubations.

After termination of the DHA sulphate incubation the radioactivity remained in the aqueous fraction. Following isolation of the conjugates by solid-phase extraction, solvolysis and separation on TEAP LH-20, no phenolics were detected. In the neutral fraction DHA and a trace amount of Δ^5-diol were identified by t.l.c. As only conjugates were isolated the result demonstrated that under the experimental conditions used DHA sulphate was not an intermediate in the formation of C_{18} neutral steroids.

After isolation of the metabolites from the 19-nortestosterone incubation, oestrone and oestradiol were identified in the phenolic fraction by t.l.c. Column chromatography of the neutral fraction resulted in the isolation of three radioactive bands. Location of the radioactivity after chromatography on t.l.c. showed that the radiolabelled components in two of the bands had R_f values similar to those of 19-nortestosterone and oestranediol. The radiolabelled component in the third band was more polar, having an R_f value similar to 19-hydroxytestosterone, indicating possible hydroxylation of 19-nortestosterone in the A-ring. In-vitro aromatization of C_{18} neutral steroids has previously been reported for equine placental tissue (Starka *et al.*, 1965) and 1β-hydroxy-19-norandrostenedione has been identified from in-vitro incubations with human placental tissue (Townsley & Brodie, 1966).

In the present study incubation times were deliberately kept short to reduce the conversion to oestrogens and so increase the yield of C_{19} and C_{18} neutral steroids. The isolation of an isomer of the neutral C_{18} steroid 5(10)-oestrenediol from both DHA and Δ^5-diol is of particular importance because four isomers of this steroid have been detected in stallion testicular tissue and are the major components in the neutral sulphate fraction (Dumasia *et al.*, 1985). C_{18} neutral steroids have been reported from horse (Short, 1960; Silberzhan *et al.*, 1985) and pig (Khalil & Walton, 1985) follicular fluid, from stallion (Dumasia *et al.*, 1985) and pig (Ruokonen & Vihko, 1974) testicular tissue and stallion urine (Houghton *et al.*, 1984a; Courtot *et al.*, 1984).

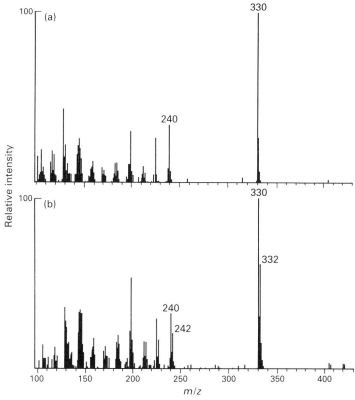

Fig. 5. Mass spectra of the *bis*-TMS derivatives of (a) 5(10)-oestrene-3,17-diol and (b) deuterium-labelled isolate.

Table 1. Steroids searched for and identified (+ to + + + +) in extracts of testicular incubations with deuterium-labelled substrates using the procedure outlined in Fig. 2

| | 3-year-old | | 2-year-old |
	DHA	Δ_5-diol	DHA
Ether*	83%	89%	95%
Neutral steroids*	55%	60%	70%
Fraction 1			
Dehydroepiandrosterone	+ + + +	+ + +	+ +
Androstenedione	+ +	+	+
19-Nortestosterone			
Testosterone	+ +	+ + +	
Androstenediol	+ +	+	
Fraction 2			
Androstenediol		+ + + +	
19-Hydroxyandrostenedione			
5(10)-Oestrene-3,17-diol	+	+	
Fraction 3			
19-Hydroxydehydroepiandrosterone		+	+
19-Hydroxytestosterone		+	
Phenolic steroids*	−2%†	20%	11%
Ratio of oestrone:oestradiol		1:6	6:1

*Recovery of radioactivity.
†Part of this fraction was lost.

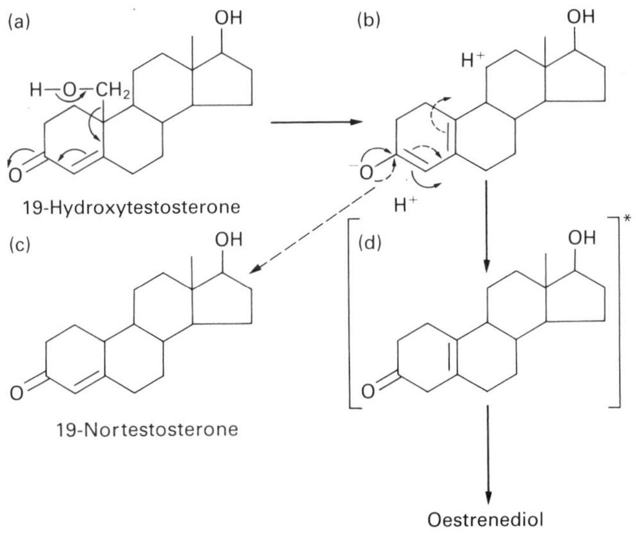

Fig. 6. Proposed mechanism for the formation of C_{18} neutral steroids. *Compound not isolated.

Although Dorfman & Ungar (1965) have postulated that C_{18} neutral steroids could be formed during aromatization of androgens to oestrogens, little experimental evidence is available concerning their formation *in vivo* and *in vitro*. A possible mechanism for their formation is shown in Fig. 6. It is well established that the initial step in aromatization of Δ^4-3-oxo C_{19} steroids is hydroxylation at C-19 (Fig. 6a). Deformylation of the 19-hydroxysteroid via a reverse aldol-type condensation yields the intermediate (Fig. 6b). Protonation of this intermediate at one of two sites can then yield either a Δ^4-3-oxosteroid (Fig. 6c; protonation at C-10) or a $\Delta^{5(10)}$-3-oxosteroid (Fig. 6d; protonation at C-4). The $\Delta^{5(10)}$-3-oxosteroid would be readily isomerized to a Δ^4-3-oxosteroid, whereas the reverse reaction is improbable; $\Delta^{5(10)}$-3-oxosteroid is also a probable precursor of 5(10)-oestrene-3,17-diol. The reduction of the 3-oxo group to form 5(10)-oestrene-3,17-diol stabilizes the A-ring against aromatization and the detection of significant amounts of sulphated 5(10)-oestrenediol in stallion testicular tissue would be consistent with this theory.

We thank the Horserace Betting Levy Board for financial support; the local veterinary practices of Newmarket for the supply of testes; and Mark Dunnett and Phil Teale of Horseracing Forensic Laboratory Limited for technical assistance.

References

Axelson, M., Sahlberg, B.L. & Sjovall, J. (1981) Analysis of profiles of conjugated steroids in urine by ion-exchange separation and gas chromatography–mass spectrometry. *J. Chromatogr.* **224,** 355–370.

Baggett, B., Engel, L.L., Balderas, L. & Lanman, G. (1959) Conversion of C^{14}-testosterone to C^{14} oestrogenic steroids by the endocrine tissues. *Endocrinology* **64,** 600–608.

Bedrak, E. & Samuels, L.T. (1969) Steroid biosynthesis by the equine testis. *Endocrinology* **85,** 1186–1195.

Braselton, W.E., Orr, J.C. & Engel, L.L. (1973) The twin ion technique for detection of metabolites by gas chromatography-mass spectrometry: intermediates in oestrogen biosynthesis. *Analyt. Biochem.* **53,** 64–65.

Courtot, D., Guyot, J. & Benoit, E. (1984) Mise en evidence de l'elimination urinaire de la 19-nortestosterone d'origine endogene chez le cheval male. *C. r. hebd. Séanc. Acad. Sci., Paris* **299,** Serie III, **6,** 139–141.

Dehennin, L., Reiffsteck, A. & Scholler, R. (1980) Simple methods for the synthesis of twenty different, highly enriched deuterium labelled steroids, suitable as internal standards for isotope dilution mass spectrometry. *Biomed. Mass Spectrom.* **7,** 493–499.

Dorfman, R.J. & Ungar, F. (1965) *Metabolism of Steroid Hormones*, pp. 132–133. Academic Press, New York.

Dumasia, M.C., Houghton, E. & Jackiw, M. (1985) Identification of endogenous C_{18} neutral steroids in stallion testis tissue by GC–MS. *J. Endocr.* **107** (Suppl.) Abstract No. 135.

Gaskell, S.J. & Finlay, E.M.H. (1979) A simple rapid procedure for the preparation of deuterium-labelled testosterone. *J. Label. Compound Radiopharm.* **17,** 861–869.

Houghton, E., Copsey, J., Dumasia, M.C., Haywood, P.E., Moss, M.S. & Teale, P. (1984a) The identification of C_{18} neutral steroids in normal stallion urine. *Biomed. Mass Spectrom.* **11,** 96–99.

Houghton, E., Teale, P. & Dumasia, M.C. (1984b) Improved capillary gas chromatographic-mass spectrometric method for the determination of anabolic steroid and corticosteroid metabolites in horse urine using on-column injection with high-boiling solvents. *Analyst* **109,** 273–275.

Khalil, M.W. & Walton, J.S. (1985) Identification and measurement of 4-oestrene-3,17-dione (19-nor-Δ4) in porcine follicular fluid using high performance liquid chromatography and capillary gas chromatography-mass spectrometry. *J. Endocr.* **107,** 375–381.

Nyman, M.A., Geiger, J. & Goldzieher, J.W. (1959) Biosynthesis of oestrogen by the perfused stallion testis. *J. biol. Chem.* **234,** 16–18.

Oh, R. & Tamaoki, B. (1970) Steroidogenesis in equine testis. *Acta endocr., Copenh.* **16,** 1–16.

Ruokonen, A. & Vihko, R. (1974) Steroid metabolism in testis tissue; concentrations of unconjugated and sulphated neutral steroids in boar testis. *J. Steroid Biochem.* **5,** 33–38.

Savard, K. & Goldzieher, J.W. (1960) Biosynthesis of steroids in stallion testis tissue. *Endocrinology* **66,** 617–624.

Short, R.V. (1960) Identification of 19-norandrostenedione in follicular fluid. *Nature, Lond.* **188,** 232.

Silberzahn, P., Dehennin, L., Zwain, I. & Reiffsteck, A. (1985) Gas chromatography-mass spectrometry of androgens in equine ovarian follicles at ultrastructurally defined stages of development. Identification of 19-nortestosterone in follicular fluid. *Endocrinology* **177,** 2176–2181.

Starka, L., Breuer, J. & Breuer, H. (1965) Biogenese von Oestrogenen in der Plazenta des Pferdes. *Naturwissenschaften* **52,** 540–541.

Townsley, J.D. & Brodie, J. (1966) Studies on the mechanism of oestrogen biosynthesis. *Biochem. J.* **101,** 25–26.

J. Reprod. Fert., Suppl. **35** (1987), 79–86

Distribution of spermatozoa in the mare's oviduct

M. S. Boyle, D. G. Cran*, W. R. Allen and R. H. F. Hunter†

Thoroughbred Breeders' Association Equine Fertility Unit and *AFRC Institute of Animal Physiology & Genetics Research, Animal Research Station, 307 Huntingdon Road, Cambridge CB3 0JQ and †School of Agriculture, University of Edinburgh, West Mains Road, Edinburgh EH9 3JG, U.K.

Summary. The morphology of the uterotubal junction (UTJ) and caudal isthmus during the peri-ovulatory period, and the distribution of spermatozoa within the region, were studied in 10 Pony mares. The proximal tip of the uterine horn and caudal 1–2 cm of the isthmus were removed during oestrus or shortly after ovulation from animals mated or artificially inseminated within the previous 24 h. The tissues were incised longitudinally and fixed for scanning electron microscopy.

Analysis of micrographs showed deep longitudinal and oedematous folds in the preovulatory samples. After ovulation, much of the folding and oedema disappeared. There was a regional arrangement of ciliated and nonciliated cells and the cilia showed evidence of directional orientation. Occasional spermatozoa were seen in some specimens deep in furrows between folds of the UTJ.

Introduction

In many species there is a close temporal relationship between oestrus and ovulation which ensures that spermatozoa are deposited in the female reproductive tract at the ideal time to fertilize the ovum. In equids, however, this relationship is less precise. Oestrus may last for 7 or more days with ovulation occurring at any time from several days before, to 48 h after, the end of oestrus (Ginther, 1979). Although most mares ovulate before the end of oestrus, a proportion cease being receptive to the stallion some hours before ovulation. In these cases, therefore, the ability of spermatozoa to survive for long periods within the mare's reproductive tract may be important if fertilization is to occur.

When male equids are given free access to a group of females, multiple mating during oestrus appears to be normal (Klingel, 1969; Bristol, 1982). In modern equine stud management, on the other hand, it is common practice to mate or artificially inseminate mares only once every 48 h during oestrus without any apparent decrease in fertility. Indeed, Day (1942) recorded conception occurring in mares last inseminated as much as 6 days before ovulation. This ability of stallion spermatozoa to survive for long periods within the female tract is in marked contrast to their life of a few hours in ejaculated semen maintained *in vitro* at body temperature.

Also, in equids, semen is deposited directly into the uterus at copulation. The only other large domestic animal in which this occurs is the pig. The result is that, in these species, spermatozoa and male accessory sex gland secretions are rapidly distributed to the uterine horns. Therefore, unlike many other domestic animals in which capacitation and selection of spermatozoa take place during transport from the cervix through the uterus, these functions in the mare and sow must occur at the level of the uterotubal junction (UTJ). Furthermore, it appears likely that the UTJ may act as a filter to prevent the passage of excessive sperm numbers to the site of fertilization in the ampulla and thus reduce the risks of polyspermy.

---

‡Present address: AFRC Institute of Animal Physiology & Genetics Research, Babraham, Cambridge CB2 4AT, U.K.

These features of equine reproduction raise fundamental questions about the structure and function of the mare's UTJ and the caudal portion of the oviduct with regard to the storage and regulation of spermatozoa. Recent studies in the sow (Fléchon & Hunter, 1981; Hunter, 1984) and ewe (Hunter & Nichol, 1983) indicate that the UTJ and caudal isthmus in these species are specialized for these purposes, enabling spermatozoa to be maintained close to the site of fertilization and beyond the population of uterine polymorphonuclear leucocytes until ovulation is imminent. A related question concerns the fate of male accessory sex gland secretions introduced into the uterus at the time of mating. Evidence from a number of other species indicates that seminal plasma contains an ovicidal factor and must, therefore, be kept away from the site of fertilization (Chang, 1949). On the other hand, Mann *et al.* (1956b) produced biochemical evidence to suggest that stallion seminal plasma may enter the mare's oviduct soon after mating although the extent of its distribution remains to be clarified.

The present study was undertaken to examine the distribution of spermatozoa within the mare's reproductive tract following mating within variable times shortly before or after ovulation. Special attention was paid to ultrastructural changes occurring in the UTJ and caudal isthmus of the oviduct and to the storage of spermatozoa in these regions.

Materials and Methods

Ten young adult Welsh Pony, Shetland and crossbred mares were used in the study. They were teased regularly and jugular vein blood samples were collected daily before and during oestrus for measurement of plasma progesterone concentrations in an enzyme-linked amplified immunoassay (Stanley *et al.*, 1986). Follicular development was monitored by manual palpation and ultrasound scanning of the ovaries *per rectum*. The mares were inseminated daily during oestrus. In most cases this was achieved by natural mating to a fertile Pony stallion but, on two occasions, mares were artificially inseminated directly into the uterus with a split ejaculate of freshly collected, undiluted semen. Care was taken to ensure that the final insemination of each mare took place not more than 24 h and not less than 3 h before removal of the oviduct.

At predetermined times in relation to ovulation, the tips of both uterine horns with their associated oviducts and ovaries were removed from the 10 mares. In 5 animals this was done via midline laparotomy under general anaesthesia. In the other 5, the mares were shot and the whole reproductive tract was removed immediately. The samples were wiped clean of blood and the follicular status of the ovaries was recorded. Each oviduct was then dissected free from its enveloping mesosalpinx and, with its UTJ, was separated from the uterine horn. The caudal 1–2 cm of each oviduct was opened longitudinally with fine, clean scissors by cutting in a caudal direction. In 6 cases, the incision was extended to open out the UTJs on each side but in the remaining 4 mares, one UTJ was treated in a similar manner while on the other side the transection was ended just anterior to the UTJ which was left intact and dissected free from surrounding endometrial tissue. A portion of the oviduct in the region of the isthmic–ampullary junction was also transected longitudinally in 3 mares.

Samples from the caudal isthmus, the isthmus adjacent to the ampulla and the UTJ (both transected and in plan view) were prepared for scanning electron microscopy by rinsing in phosphate-buffered Dulbecco's saline before fixation in 4% glutaraldehyde in 0·2 M-cacodylate buffer (pH 7·2). The fixed tissues were subsequently dehydrated through 50%, 70%, 90% and 100% ethanol and immersed in amyl acetate. After critical-point drying, the tissues were coated with gold and examined at an accelerating voltage of 30 kV using a Philips 501 scanning electron microscope.

At the time of recovery of the tissues, 4 of the mares were still in oestrus with a large, preovulatory follicle in their ovaries. The other 6 mares had ovulated between 2 and 36 h earlier.

Results

Morphology of the UTJ and distal oviduct

In mares before ovulation, it was difficult to distinguish the UTJ from the oedematous surface of the endometrium. Within a few hours after ovulation, however, the entrance to the oviduct could be readily discerned as a discrete, pale area, 1–2 mm in diameter, at the tip of the uterine horn. By 24 h after ovulation, 8–9 deep folds could be seen radiating outwards from the centre of the UTJ (Fig. 1). When the distal oviduct and UTJ were opened longitudinally, these radial folds were seen to be continuations of longitudinal furrows originating on the luminal surface of the oviduct.

Before ovulation, the epithelial surface of the UTJ looked turgid and oedematous and was contorted by many additional folds (Fig. 2). After ovulation, however, the oedema appeared to subside and the transverse and secondary furrows in the luminal surface disappeared leaving only the longitudinal folds (Fig. 3). This pattern was repeated in the caudal oviduct adjoining the UTJ; during oestrus, the luminal surface of the oviduct was oedematous and exhibited both longitudinal and transverse furrows (Fig. 4) but after ovulation, the oedema disappeared to leave a rather featureless surface in which vestiges of the longitudinal furrows could just be distinguished (Fig. 5).

Uterine histology

Two basic types of epithelial cells were recognized, ciliated and non-ciliated, but within this broad classification, wide variations were evident. In some areas, epithelial cells were devoid of surface protrusions. More often, however, the cells were polygonal in outline and showed a distinctly convex surface. Individual cells were frequently separated by deep clefts (Fig. 6) and many possessed dense accumulations of microvilli which were frequently associated with cilia of markedly differing types and density. Generally, the cilia appeared to originate from the clefts between individual cells (Fig. 6) and they ranged from short, sparse tufts to dense mats obscuring the underlying surface of the cells (Fig. 10).

Regional distribution of cells

The luminal surface of endometrial cells in the vicinity of the UTJ was generally rather featureless, with few microvilli and a complete absence of cilia (Fig. 7). At the UTJ, the cells in the areas between the troughs and furrows were also featureless (Fig. 8). However, the major indentations of this region were lined by cells exhibiting microvilli and, typically, cilia which appeared growing from intercellular clefts as described above (Figs 8 & 9). Cell types in the caudal oviduct were a continuation of those in the UTJ but the impression was gained that the cilia in the distal oviduct were denser than those in the UTJ (Fig. 10). In views of the anterior oviduct close to the ampulla, microvillous and ciliated cells were seen but these were more sparsely located than in the caudal portions of the duct. No marked difference in cell type or distribution could be distinguished between samples fixed before or after ovulation. However, when viewing the ciliated cells, the cilia frequently appeared to have a directional orientation, possibly indicating a regional synchronization of their action (Fig. 11).

In all but one of the specimens, very few spermatozoa were seen either at the UTJ or within the isthmus whereas they were much in evidence on the endometrial surface close to the UTJ. Occasional spermatozoa were seen deep within the clefts between the longitudinal folds of the surface of the UTJ (Fig. 12) but this could not be taken as conclusive evidence of their deliberate sequestration in this region. The only samples in which spermatozoa were observed in the isthmus were in mares that had ovulated before removal of the oviduct. But even in these cases, the number of cells seen was very low. Spermatozoa were not seen to be agglutinated but, in some instances, both the visible spermatozoa and the surface of the UTJ and oviduct were coated with a mucus-like substance which could have been either an oviducal secretion or a component of seminal plasma. Also, the entire surface of individual spermatozoa, including the acrosomal cap and the equatorial region, appeared homogeneous as though a coating was still present on the cells. No signs of phagocytosis were seen, nor were leucocytes observed in any of the preparations.

Discussion

By ligating the caudal oviducts of mated, oestrous gilts at various times before ovulation, Hunter (1984) showed that viable spermatozoa capable of fertilizing ova seldom pass beyond the caudal

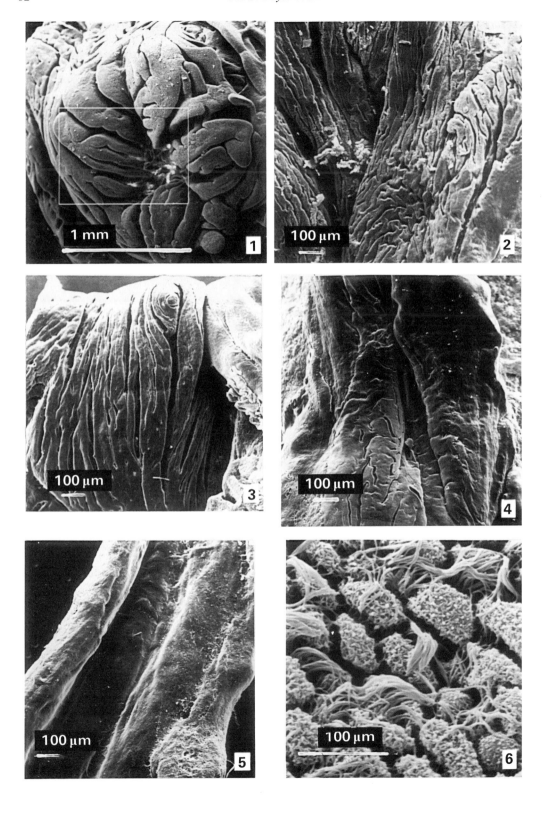

1–2 cm of the isthmus until close to ovulation. An earlier ultrastructural study of the UTJ and caudal isthmus of the sow by Fléchon & Hunter (1981) described how these regions of the reproductive tract may be capable of acting as a storage site for spermatozoa and as a valve controlling access to the upper reaches of the duct. In view of the similarity between the mare and sow with regard to the direct deposition of semen into the uterus at mating, the present study paid particular attention to the structure and possible functions of the UTJ and caudal isthmus in the mare.

The prominent polypoid processes which are a feature of the sow's UTJ during oestrus were less evident in preparations taken from mares. However, the extensive distension and folding of the epithelial surfaces of both the UTJ and caudal oviduct of the sow before ovulation were also apparent in oestrous mares suggesting that, in this species too, these structures may be capable of acting as a mechanical valve to occlude the lumen of the oviduct. Were the lumen of the oviduct to be blocked before ovulation by this process of oedema and surface folding, the resulting longitudinal furrows would constitute the only pathway for cells or other particulate matter. Furthermore, these furrows would seem to constitute ideal sites for sperm storage in the mare, analogous to the sperm reservoirs located in the pig by Rigby (1966). Specimens recovered from mares between 12 and 24 h after ovulation showed a marked reduction in oedema and general smoothing of the surface contours similar to the situation described in the sow. However, the timing of sample recovery in this study was not sufficiently precise to determine when, in relation to ovulation, the changeover occurs in the mare.

Spermatozoa were notable mainly for their scarcity, even in mares that had been mated within 3 h of sampling. In a few preparations of the UTJ, spermatozoa were seen on the epithelial surface, often aggregated in clumps and associated with mucoid secretions of unknown origin. These groups of surface spermatozoa resembled the spermatozoa found in a preparation of the endometrium adjacent to the UTJ and it seems likely that they do not form part of the population of viable spermatozoa which would eventually be available for fertilization. The location of this population of biologically significant spermatozoa is suggested by the few specimens that showed sequestration of some spermatozoa deep in the furrows between the folds of the UTJ. The fact that

---

Fig. 1. General plan view of UTJ as seen from the uterine horn 24 h after ovulation. Terminal folds radiate from the entrance to the oviduct.

Fig. 2. Portion of UTJ opened out. Mare in mid-oestrus. Longitudinal folds and ridges are swollen and oedematous. Epithelial surface is contorted by many secondary folds.

Fig. 3. Portion of oviduct opened out. Taken from mare 12 to 24 h after ovulation. The swelling has subsided, there is no secondary folding and the longitudinal folds are no longer oedematous.

Fig. 4. Opened out length of caudal oviduct adjacent to the UTJ taken from mare in late oestrus. Surface is swollen and oedematous.

Fig. 5. Opened out length of caudal oviduct adjacent to the UTJ taken from a mare that had ovulated 12–24 h earlier. The surface has lost its tumescent appearance and only vestiges of longitudinal folds can be seen.

Fig. 6. Epithelial cells from area bordering longitudinal folds of the UTJ of a mare that had ovulated 12–24 h earlier. Cells are polygonal with convex surfaces. Individual cells are densely covered by microvilli and separated by deep clefts. Long tufts of cilia present and apparently originating from the inter-cellular clefts.

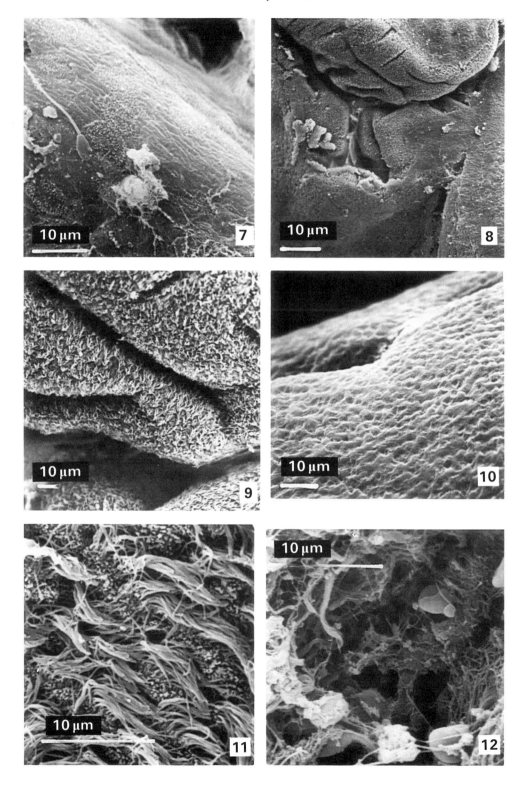

no spermatozoa were noted in the caudal isthmus from any mare still in oestrus, but were present in samples from mares that had ovulated some hours earlier, gives support to the suggestion that movement of spermatozoa from the UTJ to the oviduct is closely controlled in relation to ovulation. However, the number of specimens involved in the present study is too small to be conclusive and the evidence obtained is contradicted to some extent by the previous experiments of Parker *et al.* (1975). These authors flushed the oviducts of groups of mares inseminated 12 h before slaughter at various stages of the oestrous cycle and recovered relatively large numbers of spermatozoa from the oviducts at all stages of the cycle. It may, nevertheless, be significant that the largest numbers of spermatozoa recovered by Parker *et al.* (1975) were from the groups that contained the highest proportion of recently ovulated mares. Bader (1982) also attempted to recover spermatozoa from the tracts of mares inseminated shortly after ovulation. Spermatozoa had reached the ampulla in some mares inseminated 2 h before the oviducts were removed but, in most cases, transport to the upper portions of the oviduct did not appear to be complete until 4 to 6 h after insemination.

Other, earlier experiments have examined the question of the selectivity of the UTJ in allowing the passage of cells versus fluids from the uterus into the oviduct. Polge (1978) inseminated barium sulphate into the uterus of a sow in oestrus 1 h before slaughter. Radiographs of the excised uterus showed that the radio-opaque material had become uniformly distributed throughout the uterus including the tips of both uterine horns, yet no radio-opacity could be detected in either oviduct. Similarly, Mann *et al.* (1956b) had earlier undertaken a biochemical analysis of the flushings from various parts of the reproductive tracts of recently mated gilts and found no evidence that constituents of seminal plasma had entered the oviducts. Conversely, however, in a concurrent experiment on a mated Pony mare, the same authors found both citric acid and ergothionine in oviducal washings recovered 50 min after mating (Mann *et al.*, 1956a). Likewise, Einarsson *et al.* (1980) found radioactivity in all parts of the oviducts of gilts inseminated 1 h before slaughter with seminal plasma containing radiolabelled molecules of various sizes.

These two experiments indicate that components of seminal plasma may gain entry to the oviducts in both the sow and the mare. Such an hypothesis would also seem to be supported by the experiments of Hunter & Nichol (1986) who found a greater temperature difference between the ampullary and uterine ends of the oviducts of mated than of unmated gilts. It is therefore reasonable to speculate that the presence of seminal plasma at the UTJ in both species may be a necessary prerequisite to stimulate physiological changes within the oviduct directed towards enhancing and/or governing the storage and transport of spermatozoa and ova within this tube.

Fig. 7. Epithelial surface of endometrium adjacent to UTJ. Featureless, slightly domed epithelial cells some of which are sparsely covered by short microvilli. Mare about 6 h after ovulation.

Fig. 8. View of UTJ showing distribution of epithelial cells. Bordering the major folds, the cells are densely ciliated but the intervening areas are covered with non-ciliated cells. Mare in oestrus.

Fig. 9. Higher magnification of layer of ciliated epithelial cells lining the longitudinal folds of the UTJ. Mare in oestrus.

Fig. 10. Dense mat of ciliated cells lining the surface of the oviduct of a mare in oestrus which had been mated 3 h earlier. No spermatozoa were apparent.

Fig. 11. Ciliated cells from UTJ of mare in oestrus. Cilia showing directional orientation.

Fig. 12. View into fold in UTJ showing possible sites for sperm sequestration.

In the present study, the presence of superficial spermatozoa and the nature of the underlying epithelial surfaces, both at the UTJ and in the caudal oviduct, were frequently obscured by a surface coating of mucoid material. While this did not appear to be a local exocrine secretory substance, it was none the less impossible to be certain whether it contained elements of seminal plasma. The sites of deposition and accumulation of seminal plasma during the peri-ovulatory period in the mare, and its possible roles in the storage of spermatozoa and achievement of fertilization, therefore remain important topics for further investigation.

I thank Mr R. Turner, AFRC Food Research Institute, Norwich, for assistance.

References

Bader, H. (1982) An investigation of sperm migration into the oviduct of the mare. *J. Reprod. Fert., Suppl* **32**, 59–64.

Bristol, F. (1982) Breeding behaviour of a stallion at pasture with 20 mares in synchronized oestrus. *J. Reprod. Fert., Suppl.* **32**, 71–77.

Chang, M.C. (1949) The effect of seminal plasma on fertilized rabbit ova. *Proc. natn. Acad. Sci. U.S.A.* **36**, 291–300.

Day, F.T. (1942) Survival of spermatozoa in the genital tract of the mare. *J. agric. Sci., Camb.* **32**, 108–111.

Einarsson, S., Jones, B., Larsson, K. & Viring, S. (1980) Distribution of small and medium-sized molecules within the genital tract of artificially inseminated gilts. *J. Reprod. Fert.* **59**, 453–457.

Fléchon, J-E. & Hunter, R.H.F. (1981) Distribution of spermatozoa in the utero-tubal junction and isthmus of pigs, and their relationship with the luminal epithelium after mating: a scanning electron microscope study. *Tissue & Cell* **13**, 127–139.

Ginther, O.J. (1979) *Reproductive Biology of the Mare: Basic and Applied Aspects*, Ch. 6. Cross Plaines, Wisconsin.

Hunter, R.H.F. (1984) Pre-ovulatory arrest and peri-ovulatory redistribution of competent spermatozoa in the isthmus of the pig oviduct. *J. Reprod. Fert.* **72**, 203–211.

Hunter, R.H.F. & Nichol, R. (1983) Transport of spermatozoa in the sheep oviduct: preovulatory sequestering of cells in the caudal isthmus. *J. exp. Zool.* **228**, 121–128.

Hunter, R.H.F. & Nichol, R. (1986) A preovulatory temperature gradient between the isthmus and ampulla of pig oviducts during the phase of sperm storage. *J. Reprod. Fert.* **77**, 599–606.

Klingel, H. (1969) Reproduction in the plains zebra, *Equus burchelli bochmi:* behaviour and ecological factors. *J. Reprod. Fert., Suppl.* **6**, 339–345.

Mann, T., Leone, E. & Polge, C. (1956a) The composition of the stallion's semen. *J. Endocr.* **13**, 279–290.

Mann, T., Polge, C. & Rowson, L.E.A. (1956b) Participation of seminal plasma during the passage of spermatozoa in the female reproductive tract of the pig and horse. *J. Endocr.* **13**, 133–140.

Parker, W.G., Sullivan, J.J. & First, N.L. (1975) Sperm transport and distribution in the mare. *J. Reprod. Fert., Suppl.* **23**, 63–66.

Polge, C. (1978) Fertilization in the pig and horse. *J. Reprod. Fert.* **54**, 461–470.

Rigby, J.P. (1966) The persistence of spermatozoa at the utero-tubal junction of the sow. *J. Reprod. Fert.* **11**, 153–155.

Stanley, C.J., Paris, F., Webb, A.E., Heap, R.B., Ellis, S.T., Hamon, M., Worsfold, A. & Booth, J.M. (1986) Use of a new and rapid milk progesterone assay to monitor reproductive activity in the cow. *Vet. Rec.* **118**, 664–667.

J. Reprod. Fert., Suppl. **35** (1987), 87–94

Printed in Great Britain
© 1987 Journals of Reproduction & Fertility Ltd

Ultrasonography of accessory sex glands in the stallion

T. V. Little and G. L. Woods*

Department of Clinical Sciences, New York State College of Veterinary Medicine, Ithaca, New York 14853, U.S.A.

Summary. The accessory sex glands of 10 stallions were examined by transrectal ultrasonography. Seminal vesicles were $26·4 \pm 5·2$ (s.d.) mm in width and $9·2 \pm 3·1$ mm in height. Ampullae were $16·3 \pm 3·6$ mm in width and $12·9 \pm 3·9$ mm in height. Bulbourethral glands were $19·7 \pm 4·6$ mm in width and $32·4 \pm 6·7$ mm in length. Prostate lobes exceeded 34 mm in width and were $23·5 \pm 5·7$ mm in height. The prostatic isthmus was $6·0 \pm 1·4$ mm in height. The seminal colliculus, masculine uterus, and deferent ducts were also identified and characterized.

Five of these stallions were killed to compare transrectal results with water bath ultrasonograms and gross dissections of the isolated accessory sex glands. Transrectal ultrasonograms were anatomically and acoustically similar to water bath ultrasonograms. The anatomical relationships, physical dimensions and acoustic characteristics of the stallion's accessory sex glands were accurately represented by transrectal ultrasonography.

Introduction

The accessory sex glands of the stallion are the seminal vesicles, ampullae, prostate gland and bulbourethral glands (Setchell, 1977; Cooper, 1979; Thompson *et al.*, 1980). Their secretions constitute 95% of the stallions's ejaculate volume (Pickett *et al.*, 1983) and are thought to be beneficial, though not essential to fertility (Barker & Gandier, 1957; Amann, 1981). Proposed mechanisms for enhancing fertility in various species include optimizing urethral pH, providing energy substrates, maintaining osmotic balance, and enhancing sperm transport in the female. Their secretions may also serve antioxidant, bacteriocidal and immunoprotective functions (Mann, 1975; Rodger, 1975; Spring-Mills & Hafez, 1980).

Direct evaluation of the stallion's accessory sex glands by rectal palpation has been a routine part of clinical fertility examinations of the stallion (Kenney, 1975; Neely, 1980; Kenney *et al.*, 1983). Thorough palpation of these glands is difficult because of their small size, soft consistency and thick connective tissue coverings (Nickel *et al.*, 1973; Neely, 1980). Transrectal ultrasonography has proved a valuable adjunct to rectal palpation of the reproductive tract of the mare (Ginther, 1986). Similar application to the stallion's accessory sex glands seemed possible because of a soft tissue nature and an anatomical location similar to that of the mares' reproductive tract. Ultrasonographic analysis of the male accessory sex glands is routine in man (Rifkin *et al.*, 1983) but there are no reports of this application in stallions. The objectives of this study were (1) to describe the technique of transrectal ultrasonography of the accessory sex glands of stallions and (2) to compare the anatomical relationships, physical dimensions and acoustic characteristics of these transrectal ultrasonograms to water bath ultrasonograms and gross dissections of isolated stallion accessory sex glands.

---

*Present address: Department of Veterinary Clinical Medicine and Surgery, University of Idaho, Moscow, Idaho 83843, U.S.A.

Materials and Methods

Seven horses and three Pony stallions aged 3–25 years were studied in October and November of 1985. Transrectal ultrasonography of the seminal vesicles, prostate gland, bulbourethral glands and ampullae of each stallion was performed. The stallion's masculine uterus, prostatic ducts, pelvic urethra dilatation and seminal vesicle necks were examined when visualized. The ultrasound machine was a B mode, real time unit (Model 210 DX, Technicare: Johnson & Johnson Ultrasound Inc., Ramsey, NJ, U.S.A.) with a 7·5 MHz, linear-array transducer.

Each stallion had been sexually rested for at least 1 week and sexual stimulation before examinations was avoided. Stallions were restrained in stocks without sedation and the rectum was cleared of faecal material. The transducer was well lubricated with coupling gel and inserted into the stallion's rectum. The accessory glands were scanned over their entire length, first longitudinally and then transversely to their long axes (Fig. 1). Ultrasonograms were recorded on videotape and simultaneously narrated by the examiner. Photographs of still images were made with a 35 mm camera monitor on Plus X Panchromatic film. Acoustic characteristics were described as echogenic (white), echolucent (black), or intermediate to those extremes. Anatomical relationships of observed structures were noted and physical dimensions were retrieved from videotape recordings. Bilateral height and width values were recorded for all glands except the bulbourethral glands for which length and width were the recorded dimensions.

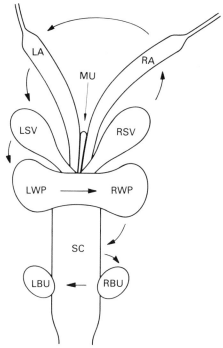

Fig. 1. Examination sequence for transrectal and water bath ultrasonography. RSV, right seminal vesicle; RA, right ampulla; LA, left ampulla; LSV, left seminal vesicle; LWP, left prostate; RWP, right prostate; SC, seminal colliculus; RBU, right bulbourethral gland; LBU, left bulbourethral gland; the masculine uterus (MU) was also detectable.

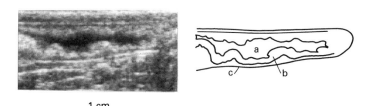

1 cm

Fig. 2. Seminal vesicle fundus in longitudinal section by transrectal ultrasonography: (a) lumen, (b) folded mucosa, (c) capsular border.

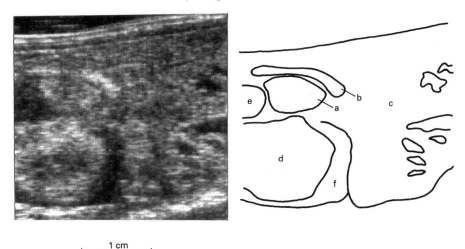

1 cm

Fig. 3. Prostatic isthmus in transverse section by transrectal ultrasonography: (a) ampulla, (b) seminal vesicle neck, (c) prostatic parenchyma, (d) bladder neck, muscosal layer, (e) masculine uterus, (f) bladder neck, muscular layer.

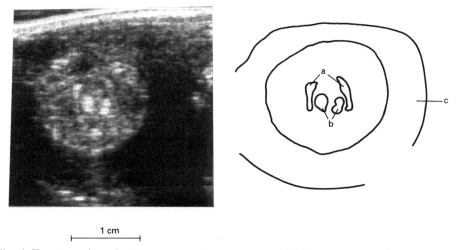

1 cm

Fig. 4. Excretory ducts in transverse section by transrectal ultrasonography, just proximal to the seminal colliculus. (a) Excretory ducts of the seminal vesicles, (b) excretory ducts of the ampullae, and (c) the surrounding urethralis muscle.

Immediately after the transrectal ultrasonogram, 2 horse and 3 Pony stallions were killed and their accessory sex glands were removed for direct measurement. After direct measurement, the isolated glands were placed in a tepid water bath and the scanning sequence was repeated. Physical dimensions were again determined from narrated videotape recordings of the water bath ultrasonograms.

Results

Transrectal ultrasonography

The fundic portions of the seminal vesicles were located dorsolateral to the ampullae. Their thick walls were slightly more echogenic than surrounding viscera and their lumina, when distended with secretions, were echolucent (Fig. 2). These fundic portions were compressed dorsoventrally and were $26{\cdot}4 \pm 5{\cdot}2$ (s.d.) mm in width and $9{\cdot}2 \pm 3{\cdot}1$ mm in height. The neck of each seminal vesicle passed ventral to the prostatic isthmus accompanied ventromedially by the ipsi-lateral deferent duct (Fig. 3). The paired seminal vesicles and ampullae tapered to excretory ducts, and together with the pelvic urethra passed within the tubular urethralis muscle. The excretory duct of each seminal vesicle was observed as a laterally compressed echogenic crescent lateral to the excretory deferent duct (Fig. 4).

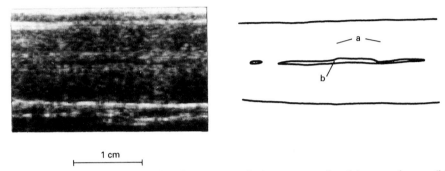

1 cm

Fig. 5. Ampulla in longitudinal section by transrectal ultrasonography: (a) parenchyma, (b) lumen.

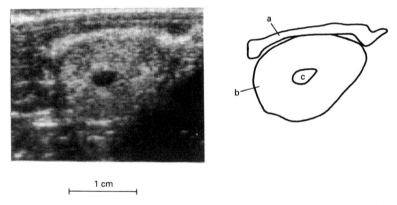

1 cm

Fig. 6. Ampulla in transverse section by transrectal ultrasonography. (a) Seminal vesicle neck, (b) parenchyma, (c) lumen containing fluid. The echolucent area in the lower right is the urinary bladder.

The ampullae were located dorsolateral to the urinary bladder. They were 16.3 ± 3.6 mm in width and 12.9 ± 3.9 mm in height and were readily observed in transverse or longitudinal section (Figs 5 & 6). In 1 of 10 stallions both ampullae were to the right of the urinary bladder. The ampullae converged, passed ventral to the prostatic isthmus and tapered to excretory ducts. The excretory duct of each ampulla accompanied the excretory duct of each seminal vesicle to the seminal colliculus (Fig. 4). Excretory duct length from the prostatic isthmus to the seminal colliculus was approximately 4 cm.

The lumen of the ampulla was echolucent when it contained fluid. When the lumen did not contain fluid, it appeared as a broken echogenic line. The entire length of one stallion's ampulla was fluid filled. The ampullar parenchyma, which varied greatly in echogenicity, was homogeneous in 8 stallions and non-homogeneous or mottled in 2 stallions (Fig. 7).

The prostate lobes and isthmus were located just below the rectum and dorsal to the pelvic musculature. They bordered the excretory seminal vesicles, ampullae and bladder neck on three sides (Fig. 3). Wings of the prostate extended approximately 4 cm laterally from the midline and measured $23\cdot5 \pm 5\cdot7$ mm in height (Fig. 8). The isthmus of the prostate was a maximum of 8 mm in height (Fig. 9). The prostate lobes had echolucent spaces throughout their parenchyma which at gross dissection were confirmed as accumulated secretions.

The bulbourethral glands were uniformly echogenic structures which were located dorsolateral to the pelvic urethra at the ischial arch. These ovoid glands were surrounded by a thin echolucent border which corresponded to the bulboglandularis muscle (Fig. 9). They measured $19\cdot7 \pm 4\cdot6$ mm in width and $32\cdot4 \pm 6\cdot7$ mm in length. Echolucent areas were occasionally observed throughout their parenchyma. Excretory ducts, which were observed inconsistently, passed from the caudal portion of each gland to the dilatation of the pelvic urethra.

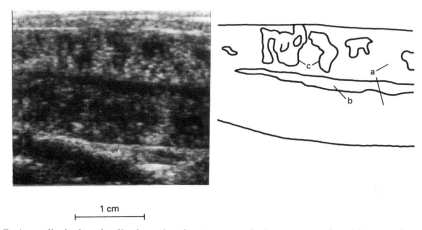

1 cm

Fig. 7. Ampulla in longitudinal section by transrectal ultrasonography: (a) parenchyma, (b) lumen containing fluid, (c) lucencies in parenchyma.

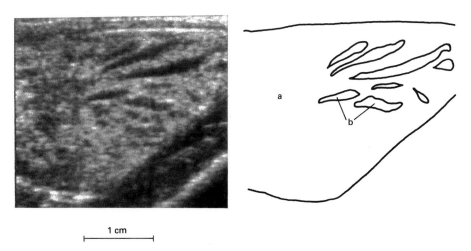

1 cm

Fig. 8. Lobe of prostate gland in transverse section by transrectal ultrasonography: (a) prostate parenchyma, (b) echolucencies in parenchyma.

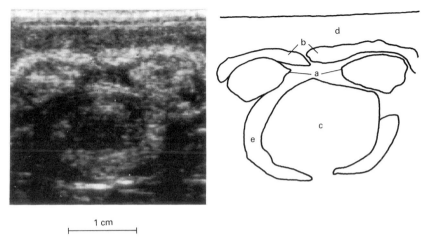

1 cm

Fig. 9. Transverse section at prostatic isthmus by transrectal ultrasonography: (a) ampullae, (b) seminal vesicle necks, (c) bladder neck, lumen, (d) prostatic isthmus, (e) bladder neck, muscular layer.

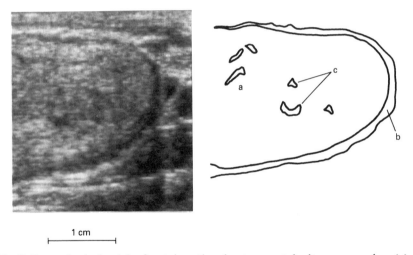

1 cm

Fig. 10. Bulbourethral gland in frontal section by transrectal ultrasonography: (a) gland parenchyma, (b) bulboglandularis muscle, (c) echolucencies in parenchyma.

The masculine uterus and the seminal colliculus were also visualized during transrectal ultrasound exams. The masculine uterus appeared in 8 stallions as a round structure between the ampullae (Fig. 3). Their maximal transverse diameters ranged from 2 to 9 mm. The masculine uterus was identified in all stallions by dissection. The seminal colliculus was identified as a less echogenic area at the termination of the ampullar and seminal vesicular excretory ducts.

Post-mortem studies

Water bath ultrasonograms were acoustically similar to transrectal ultrasonograms except that the parenchyma of the seminal vesicles and the capsular surfaces of all accessory sex glands were more echogenic by water bath ultrasonography (Fig. 11). Physical dimensions of left and right accessory glands, which were not significantly different, were combined (Table 1). The only difference in physical dimensions between transrectal ultrasonography, water bath ultrasonography and direct measurement was that water bath measurements of the seminal vesicles were greater in height than were transrectal measurements of the same glands (Table 1). The mean dimensions by all methods were in general agreement with published dimensions for the stallions' accessory sex glands (Ellery, 1971; Nickel *et al.*, 1973; Gebauer *et al.*, 1974).

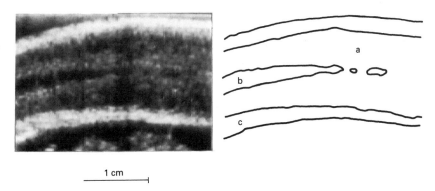

1 cm

Fig. 11. Ampulla in longitudinal section by water bath ultrasonography: (a) ampullar parenchyma, (b) lumen containing fluid, (c) hyperechogenic surface.

Table 1. Dimensions of accessory sex glands determined by transrectal ultrasound, water bath ultrasound and direct measurement

	Transrectal ultrasound	Direct measurement	Water bath ultrasound
Seminal vesicles			
Width (mm)	23·7 ± 5·5	23·0 ± 4·1	18·7 ± 4·2
Height (mm)	9·7 ± 3·6[a]	12·7 ± 4·2	16·2 ± 3·4[b]
Ampulla			
Width (mm)	15·1 ± 2·9	16·5 ± 4·7	14·7 ± 3·9
Height (mm)	11·1 ± 3·0	13·5 ± 4·2	13·2 ± 4·4
Prostate lobe			
Width (mm)	>34	39·4 ± 4·7	>34
Height (mm)	22·0 ± 4·9	19·8 ± 4·4	23·4 ± 2·6
Prostate isthmus			
Height (mm)	5·0 ± 1·1	4·2 ± 0·4	6·2 ± 0·8
Bulbourethral gland			
Width (mm)	19·0 ± 6·3	20·1 ± 4·8	18·8 ± 4·5
Length (mm)	28·5 ± 7·4	33·1 ± 7·5	28·1 ± 5·2

Values are mean ± s.d. for 5 stallions, left and right sides combined.
Values with different superscripts are significantly different ($P < 0.01$).

Discussion

Anatomical relationships of the accessory sex glands of stallions were directly and accurately demonstrated with transrectal ultrasound. This included anatomical structures such as the excretory ducts and the seminal colliculus which were inaccessible by rectal palpation. The difficulty in discerning seminal vesicles by palpation was explained by their compressible nature and the ease by which their secretions were redistributed throughout the gland.

The main difference in physical dimensions among methods was in seminal vesicle measurements by transrectal ultrasonography compared to water bath ultrasonography. This difference was caused by contraction of the seminal vesicles in the water bath thereby resulting in a round cross-sectional shape to the glands. A similar trend toward a symmetrical cross-section was seen in ampullar height and width by the water bath technique. A contributing factor to dorsoventral compression of both ampullae and seminal vesicles during transrectal ultrasonograms may have been the downward pressure necessary to achieve transducer contact.

Acoustic characteristics of the accessory sex glands were also accurately represented by transrectal ultrasonography. Increased echogenicity of the seminal vesicles in the water bath was in part a result of gland contraction. Increased surface echogenicity was the result of a greater difference in acoustic impedance across the water-soft tissue interface compared to the soft tissue-soft tissue interface of transrectal ultrasonograms.

This study demonstrates that transrectal ultrasonography accurately characterizes the accessory sex glands of the stallion. Further ultrasound studies are indicated to define normal measurements as well as to identify acoustic and dimensional changes associated with age, season, sexual stimulation and ejaculation. With normal characteristics established, abnormalities of the accessory sex glands and excurrent duct system may be assessed more directly by transrectal ultrasonography.

We thank the Harry M. Zweig Memorial Trust and Thornbrook Farms for financial support; Pitman-Moore Inc., New Jersey, for providing ultrasound machines; and C. B. Baker, B. Ball, R. Hillman and D. Vanderwall for technical assistance.

References

Amann, R.P. (1981) A review of anatomy and physiology of the stallion. *Eq. Vet. Sci.* (May/June) 83–105.

Barker, C.A.V. & Gandier, J.C.C. (1957) Pregnancy in a mare resulting from frozen epididymal spermatozoa. *Can. J. comp. Med.* **21,** 47–51.

Cooper, W.L. (1979) Methods of determining the site of bacterial infections in the stallion reproductive tract. *Proc. Soc. Theriogenology Mobile*, pp. 1–3.

Ellery, J.C. (1971) *Spermatogenesis, accessory sex gland histology and the effects of seasonal change in the stallion.* Ph.D. thesis, University of Minnesota, St. Paul.

Gebauer, M.R., Pickett, B.W. & Swierstra, E.E. (1974) Reproductive physiology of the stallion. III. Extragonadal transit time and sperm reserves. *J. Anim. Sci.* **39,** 737–742.

Ginther, O.J. (1986) Ovaries and uterus. In *Ultrasonic Imaging and Reproductive Events in the Mare*, pp. 115–194. Equiservices, Cross Plains, Wisconsin.

Kenney, R.M. (1975) Clinical fertility evaluation of the stallion. *Proc. 21st A. Conv. Am. Ass. equine Pract.* pp. 336–355.

Kenney, R.M., Hurtgen, J., Pierson, R., Witherspoon, D. & Simms, J. (eds) (1983) Physical Examination of the Genital Organs. In *Manual for Clinical Fertility Evaluation of the Stallion*, pp. 26–29. Society for Theriogenology, Hastings, Nebraska.

Mann, T. (1975) Biochemistry of stallion semen. *J. Reprod. Fert., Suppl.* **23,** 47–52.

Neely, D.P. (1980) Physical examination and genital

diseases of the stallion. In *Current Therapy in Theriogenology*, pp. 694–706. Ed. D. A. Morrow. W. B. Saunders, Philadelphia.

Nickel, R., Schummer, A. & Seiferle, E. (1973) Male genital organs of the horse. In *Viscera of the Domestic Mammals*, pp. 341–344. Eds A. Schummer, R. Nickel & W. O. Sack. Springer-Verlag, New York.

Pickett, B.W., Voss, J.L. & Squires, E.L. (1983) Fertility evaluation of the stallion. *Compend. Cont. Educ. Pract. Vet.* **5,** 194–202.

Rifkin, M.D., Kurtz, A.B., Choi, H.Y. & Goldberg, B.B. (1983) Endoscopic ultrasonic evaluation of the prostate using a transrectal probe: prospective evaluation and acoustic characterization. *Radiology* **149,** 265–271.

Rodger, J.C. (1975) Seminal plasma, an unnecessary evil? *Theriogenology* **3,** 237–247.

Setchell, B.P. (1977) Male reproductive organs and semen. In *Reproduction in Domestic Aminals*, pp. 229–256. Eds H. H. Cole & P. T. Cupps. Academic Press, New York.

Spring-Mills, E. & Hafez, E.S.E. (1980) Functional anatomy. In *Male Accessory Sex Glands*, pp. 68–69. Eds E. Spring-Mills & E. S. E. Hafez. Elsevier/North Holland Biomedical Press, Amsterdam.

Thompson, D.L., Jr, Pickett, B.W., Squires, E.L. & Nett, T.M. (1980) Sexual behaviour, seminal pH and accessory sex gland weights in geldings administered testosterone and(or) estradiol-17 beta. *J. Anim. Sci.* **51,** 1358–1366.

J. Reprod. Fert., Suppl. 35 (1987), 95–102

The carrier state in equine arteritis virus infection in the stallion with specific emphasis on the venereal mode of virus transmission

P. J. Timoney, W. H. McCollum, T. W. Murphy, A. W. Roberts, J. G. Willard* and G. D. Carswell*

Department of Veterinary Science, University of Kentucky, Lexington, KY 40546, and *Department of Agriculture and Natural Resources, Morehead State University, Morehead, KY 40351, U.S.A.

Summary. The carrier state has been confirmed virologically in Thoroughbred and non-Thoroughbred stallions naturally infected with equine arteritis virus (EAV). Short-term or convalescent and long-term carriers occur. The frequency rate of the long-term carrier state in Thoroughbreds was high, averaging 33·9% among the three groups of stallions under study. While the convalescent carrier state only lasted a few weeks after clinical recovery, the long-term carrier state could persist for years. There was evidence, however, that not all such carriers might remain persistently infected for life. Carrier stallions appeared to shed EAV constantly in the semen but not in the respiratory secretions or urine. It also could not be demonstrated in the buffy coat of the blood. All of the carrier stallions have continued to maintain moderate to high neutralizing antibody titres to EAV in serum. Virus was associated with the sperm-rich and not the pre-sperm fraction of semen. There was relatively little variation in virus concentration between sequential ejaculates from the same stallion. Transmission of EAV infection by long-term carrier stallions would appear to occur solely by the venereal route. Such carriers are thought to play an important epidemiological role in the dissemination and perpetuation of the virus.

Introduction

The outbreak of equine viral arteritis in central Kentucky during the second half of the 1984 breeding season (Timoney, 1985) marked the first occasion this disease had been recorded in the Thoroughbred population of North America since the causative virus was isolated and identified as a new viral pathogen of the horse in 1953 (Doll *et al.*, 1957). A significant outcome of the investigations into that outbreak has been the finding that the carrier stallion probably plays a major epidemiological role in the dissemination and perpetuation of equine arteritis virus (EAV) from year to year (Timoney & McCollum, 1985). Up to that point, little information was available about the carrier state in this disease.

There are reports in the older veterinary literature of the transmission of a clinically similar disorder 'epizootic cellulitis-pinkeye' or 'equine influenza' by apparently healthy stallions to mares at time of breeding (Pottie, 1888; Clark, 1892; Bergman, 1913; Schofield, 1937). Based on clinical observations and the pattern of transmission of the infection to mares, Pottie (1888) and Clark (1892) postulated that the causative organism was present in the semen of certain stallions for an undefined period after they had clinically recovered from the disease.

Very little is currently known, however, about the characteristics of the carrier state in stallions after natural exposure to EAV, except that it occurred at a relatively high frequency rate in stallions on certain farms involved in the 1984 outbreak of equine viral arteritis (Timoney, 1985). In the absence of such information, it is difficult to formulate effective control measures for the management of carrier stallions without running the risk of being excessively restrictive in respect of isolation requirements.

The present study was undertaken to elucidate in part the nature of the carrier state in the stallion persistently infected with EAV. Specific aims were as follows: (a) to study the duration of the carrier state and to ascertain if long-term shedding stallions can, in time, cease to be carriers of the virus; (b) to define the route(s) whereby the virus can be transmitted by a carrier stallion; (c) to characterize the nature of the shedding pattern and to determine whether EAV is shed constantly or intermittently in the semen; (d) to establish whether there are major fluctuations over time in the amount of virus being shed in the semen; and (e) to investigate the possible circulation of EAV in the buffy coat of the blood of carrier stallions. Confirmation of the carrier state in naturally acquired EAV infection in the stallion has been the subject of a brief report (Timoney *et al.*, 1986).

Materials and Methods

Stallions

Investigation of the characteristics of the carrier state in the stallion was undertaken in a group of 18 stallions, all of them identified as long-term shedders of EAV on the basis of the results of a test breeding programme (Timoney *et al.*, 1986). Confirmation of carrier status was contingent upon the successful transmission of infection to susceptible mares, i.e. the demonstration of seroconversion or the development of neutralizing antibody titres to EAV in the test mares within 28 days of their last breeding date.

The stallions screened for evidence of the carrier state were, with few exceptions, all Thoroughbreds. Initially, they comprised the 25 stallions known to have been involved in the 1984 outbreak of equine viral arteritis (Group 1), all of which have continued to maintain moderate to high antibody titres to the virus. Stallions investigated since then were those in Group 2 containing 18 seropositive Thoroughbred stallions that were detected during the mandatory sero-logical survey undertaken in Kentucky in January 1985 to establish the state-wide prevalence of EAV infection in the Thoroughbred stallion population and to identify any further carriers of the virus, and those in Group 3, containing 4 Thoroughbred stallions, 5 Standardbred stallions and a Tennessee walking horse stallion that were serologically tested for the first time during the second half of 1985 or early in 1986 and were found to be seropositive for antibodies to EAV. While none of the stallions known to have been vaccinated previously against equine viral arteritis, 6 of the 25 stallions in Group 1 had received vaccine 4 days before the onset of clinical signs of the disease. None of these 6 stallions, however, subsequently remained long-term shedders of the virus. With one exception, a Thoroughbred stallion in Group 3, all of the confirmed carriers were located in Kentucky. The 18 long-term shedding stallions that were investigated ranged in age from 7 to 17 (mean 11·8) years.

Experimental procedures

Sampling. Much of the data included in this study were derived from an in-depth investigation of the carrier state in 2 of the carrier stallions in Group 3, one a 7 year-old Thoroughbred (Stallion 15) and the other a 12-year-old Tennessee walking horse stallion (Stallion 18). Each of these stallions was located at a separate experimental facility and was freely available for sampling at regular intervals over an extended period of time. Stallion 15 was sampled 3 times a week for 3 months from mid-October through mid-January and Stallion 18 was sampled concurrently twice a week for about 13 weeks.

For a variety of reasons, more importantly the time of semen collection in relation to the breeding season and the ability of a particular stallion to breed artificially, it was not possible to monitor the majority of the remaining carrier stallions at regular much less the same frequency intervals. Consequently, the number of samples examined from individual stallions varied greatly from 1 to as many as 37. Similarly, the sampling interval was also variable, ranging from a few days to several months. For 3 stallions, however, the interval between sample collections was extended to 12 to 14 months.

The following specimens were taken on each sampling date for attempted isolation of EAV.

(a) Naturally voided urine was collected using a large, wide-mouth container. With the intention of avoiding possible contamination with virus-containing semen, efforts were made to obtain a mid-stream specimen of urine at least 30 min before attempting a semen collection.

(b) A nasopharyngeal swab was taken and transferred immediately into a vial of viral transport medium.

(c) A 20-ml sample of blood obtained by jugular venepuncture was collected into acid–citrate–dextrose anti-coagulant. In addition to repeated in-vitro attempts to isolate EAV from the buffy coat of carrier stallions, efforts were also made to recover virus by the inoculation of susceptible horses. At some point in the sampling period, 80–100 ml citrated blood were collected for this purpose from each of a randomly selected group of 8 Group-1 stallions. Although not for virus isolation, an additional blood sample for serum harvest and antibody determination to EAV was taken on each sampling date.

(d) Semen, preferably an entire ejaculate, was the last of the specimens to be collected. It was obtained using a closed ended artificial vagina (modified Colorado type) or a condom and a teaser or dummy mare. When it was not possible to obtain a semen collection by this means, a dismount sample was obtained at time of breeding. No antiseptics or disinfectants were used in the washing of the external genitalia of the stallion before collection. On a number of occasions, pre-sperm or pre-ejaculatory fluid was collected initially from Stallion 15 and tested separately for the presence of virus. The gel fraction was also harvested several times from this stallion and similarly processed for virus isolation. Care was taken at all times to avoid possible thermal inactivation of any virus present in the semen through inadvertent contact of the sample with the heated liner of the artificial vagina.

Immediately after collection, all of the specimens taken for virus isolation were refrigerated in crushed ice or on freezer packs. Samples were transported to the laboratory with minimal delay and in the vast majority of cases, they were processed and inoculated into cell culture on the day of collection.

Virus isolation. The procedures used to isolate EAV from nasopharyngeal swabs and buffy coats derived from citrated blood samples were essentially similar to those of McCollum *et al.* (1971) with the modification that isolation was attempted in monolayer cultures (25 cm^2 flasks) of RK-13 cells as described by Timoney *et al.* (1986). Recovery of virus from urine specimens was attempted following preliminary clarification of the samples by centrifugation at 800 *g* for 15 min. Semen samples were pre-treated before inoculation by short-term sonication (3 × 15 sec cycles) followed by centrifugation at 1000 *g* for 10 min to sediment the spermatozoa. Monolayer cultures of RK-13 cells were inoculated with serial 10-fold dilutions of the sonicated seminal plasma (Timoney *et al.*, 1986).

In attempting to transmit EAV infection to susceptible horses, buffy coats were harvested from the larger volumes of citrated blood as described by McCollum *et al.* (1971), washed and resuspended in 10 ml phosphate-buffered saline pH 7·2. Horses seronegative for antibodies to EAV were inoculated intravenously with buffy coat suspensions from the 8 stallions to undergo this screening procedure, using 1 recipient per carrier stallion. Daily clinical observations were taken for 6 days as well as nasopharyngeal swabs and citrated blood samples for virus isolation. Each horse was killed on the 6th day after inoculation, necropsied and a wide range of tissues (selected thoracic and abdominal lymphatic glands, lung, liver, spleen and kidney) and body fluids (peritoneal, pleural and pericardial) were harvested. Isolation of EAV from body fluids and clarified 10% tissue suspensions was attempted in cell culture as previously described.

The identity of isolants of EAV was confirmed in a one-way plaque-reduction assay using rabbit antiserum prepared against the Bucyrus strain of the virus.

Serology. Determination of neutralizing antibody titres to EAV was carried out using a microneutralization test in RK-13 cells in the presence of guinea-pig complement (Miles Scientific, Naperville, IL, U.S.A.).

Results

The carrier state was confirmed in 18 (17 Thoroughbreds and a Tennessee walking horse) out of 53 stallions seropositive for antibodies to EAV on the basis of successful transmission of infection to test mares and/or isolation of the virus from semen (Table 1). All of the carrier stallions were

Table 1. Frequency of the carrier state amongst 53 stallions seropositive for antibodies to equine arteritis virus

Group no.	No. of stallions	Breed*	No. of confirmed carriers†		Frequency rate (%) of long-term carriers
			Short-term	Long-term	
1	25	TB	5	9	36
2	18	TB	—	5	27·7
3	4	TB	—	3	
	5	STB	—	0	40
	1	TWH	—	1	
Total	53		5	18	33·9

*TB, Thoroughbred; STB, Standardbred; TWH, Tennessee Walking Horse.
†Confirmation of carrier status based on successful transmission of equine arteritis virus infection to test mares and/or isolation of the virus in monolayer cultures of RK-13 cells.

asymptomatically infected with the virus. In some instances, infection of the mares was reportedly associated with the development of clinical signs of variable severity characteristic of equine viral arteritis. This was observed more often in mares mated with certain stallions than with others. Both short-term or convalescent and long-term or chronic carriers were shown to occur. With the exception of 8 of the confirmed shedders resultant from the 1984 outbreak of equine viral arteritis in Kentucky which were located on Farm A, all of the remaining carriers were located on different premises.

Frequency rate

The frequency rate of the long-term carrier state among the stallions included in this study averaged 33·9%, ranging from 27·7% in Group 2 to 40% in Group 3. Whereas 25% of the 25 stallions in Group 1 became short-term carriers, 36% remained carriers of the virus for at least 1 year. Occurrence of the long-term carrier state varied greatly amongst affected stallions on different farms, ranging from 47% of 17 stallions on Farm A to 0% of 7 stallions on Farm B.

Duration of carrier state

The convalescent carrier state persisted for a period of weeks after clinical recovery from the disease. Of 7 stallions with equine viral arteritis on Farm B and of 17 stallions on Farm A, 3 and 2 respectively shed detectable virus in the semen for 1–2 and 4–5 weeks respectively. Duration of the long-term carrier state exceeded 1 year or more in all cases when sequential sampling of the stallions was possible (Table 2). On circumstantial evidence, 3 of the chronic shedders have almost

Table 2. Results of attempted isolation of equine arteritis virus from the nasopharynx, buffy coat, urine and semen of 18 long-term carrier stallions

| Group no. | Stallion no. | Age (years) | Breed | No. of isolations† (no. of specimens examined) | | | | Sampling period (months) | Duration of carrier state (years) |
				N/P swab	Buffy coat	Urine	Semen		
1	1	13	TB	0 (2)	0 (2)	—†	11 (11)	7	2·1
	2	13	TB	0 (2)	0 (2)	—‡	16 (17)	10	2·1
	3	11	TB	0 (2)	0 (2)	—‡	28 (34)	6	2·1
	4	9	TB	0 (2)	0 (2)	—‡	20 (24)	10	2·1
	5	8	TB	0 (2)	0 (2)	—‡	11 (11)	6	2·1
	6	12	TB	—‡	—‡	0 (1)	0 (1)	—	✶1·8§
	7	14	TB	0 (2)	0 (2)	—‡	15 (15)	10	2·1
	8	12	TB	0 (2)	0 (2)	—‡	1 (30)	8	✶1·6
	9	8	TB	0 (2)	0 (2)¶	0 (2)	9 (9)	9	≥1·8§
2	10	14	TB	0 (5)	0 (6)¶	0 (6)	3 (3)	14	≥5
	11	12	TB	0 (2)	0 (2)¶	—‡	2 (2)	14	≥5
	12	17	TB	0 (3)	0 (3)¶	0 (2)	1 (2)**	2	≥5
	13	17	TB	0 (1)	0 (1)¶	—‡	6 (6)	10	≥1
	14	7	TB	0 (3)	0 (4)¶	0 (4)	3 (3)	16	≥3
3	15	8	TB	0 (37)	0 (38)¶	0 (12)	37 (37)	9	≥1
	16	16	TB	0 (1)	0 (1)	0 (1)	1 (1)	—	≥1
	17	9	TB	—‡	—‡	—‡	1 (1)	—	≥0·3
	18	13	TWH	0 (24)	0 (25)¶	0 (24)	34 (34)	6	≥3

*TB, Thoroughbred; TWH, Tennessee Walking Horse.
†Isolation of equine arteritis virus attempted in monolayer cultures of RK-13 cells.
‡Not sampled.
§Dead.
¶Virus isolation attempted once by i.v. inoculation of buffy coat into a susceptible horse.
**Virus-negative semen collection was pre-ejaculatory fluid.

certainly been carriers of the virus for at least 5 years. This is based on breeding data for 1982 and the demonstration of a 95–100% prevalence of antibodies to EAV in the sera of mares 3–5 weeks after being served by these stallions in that year. The carrier status of 8 of the affected stallions in 1984 has continued to be monitored over the intervening 26 months, principally by means of a test breeding programme. Transmission of EAV infection has been demonstrated in 48 out of 51 mares mated during this time. The 3 mares that did not become infected had been covered by 1 stallion. The first 5 mares served by him in the 18-month interval since his clinical recovery from the disease became infected with EAV, but the last 3 mares did not become infected.

Route(s) of virus shedding

As previously indicated, much of the data on the route(s) of shedding of EAV by long-term carrier stallions was obtained from an in-depth investigation of the carrier state in two stallions (Nos 15 and 18) in Group 3 (Table 2). These results together with additional information on the remaining carriers revealed that, whereas EAV was isolated from the majority of semen samples, it could not be detected in nasopharyngeal secretions, urine or buffy coat specimens from any of the stallions. Furthermore, none of the horses inoculated with buffy coat samples from 8 of the carrier stallions became infected with virus. Isolations of EAV from semen were made on first passage in cell culture with visible cytopathic effect appearing after 3–5 days. All of the carrier stallions continued to maintain moderate to high serum neutralizing antibody titres to EAV during the sampling period.

Pattern of viral shedding

Investigation of the pattern of viral shedding by stallions that are long-term carriers of EAV showed that virus was shed virtually constantly in the semen (Table 2; Fig. 1). Although most of the semen samples examined from 8 of the 9 stallions in Group 1 were dismount specimens, the

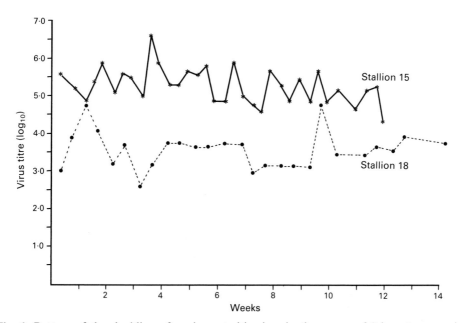

Fig. 1. Pattern of the shedding of equine arteritis virus in the semen of 2 long-term carrier stallions. Stallion 15 was sampled three times weekly and Stallion 18 twice weekly over a 3-month period. Virus titres are expressed as $\log_{10}$ plaque-forming units per $1 \cdot 0$ ml seminal plasma.

remainder of the stallions were screened using entire ejaculates. Based on the results of a test breeding programme together with virus isolations from semen, there was no evidence that any of the long-term carrier stallions included in this study were intermittent shedders of EAV.

During the 3 months of intensive sampling of Stallions 15 and 18, the amount of virus detected in the semen of each stallion ranged from 2.5×10^4 to 4.4×10^6 and from 3.7×10^2 to 5.8×10^4 plaque-forming units (pfu) respectively per ml of sonicated seminal plasma (Fig. 1). In each stallion, there was less than one log of variation in the infectivity titre of 88% of the ejaculates tested. No major fluctuations in virus concentration were detected in either stallion. The infectivity titre of semen from Stallion 15 tended to fall in the range 9×10^4 to 9×10^5 pfu/ml, but the corresponding titre for Stallion 18 was nearly 2 logs lower, ranging from 1×10^3 to 1×10^4 pfu/ml. There was evidence of considerably greater variation in virus concentration between sequential dismount semen samples obtained from 6 of the 9 stallions in Group 1.

Attempts to isolate EAV from 6 samples of pre-ejaculatory fluid collected from Stallion 15 were unsuccessful, as was an attempt from a specimen of the pre-sperm fraction of an ejaculate from Stallion 12 in Group 2 (Table 2). Several gel fractions of the ejaculate collected from Stallion 15 yielded EAV on culture, with virus concentrations falling within the previously stated range for entire ejaculates from this stallion.

Discussion

Stallions may become short-term or convalescent or long-term carriers after natural exposure to EAV. Existence of the two types of the carrier state has been demonstrated in this study, the findings of which corroborate the earlier observations of Pottie (1888), Clark (1892) and Bergman (1913). Existence of convalescent as well as long-term carriers of the virus has not been reported previously.

The frequency rate of the long-term carrier state in Thoroughbreds was surprisingly high, averaging 33·9% of the three groups of stallions under study. There was evidence of considerable disparity in the carrier rate amongst stallions on the same farm and between groups of stallions on different farms. The reasons why as many as 8 (47%) of a group of 17 stallions on Farm A that were primarily infected with EAV in May 1984 became long-term shedders of the virus remain unclear. All of the affected group experienced disease of approximately equivalent clinical severity and duration. The mean period of sexual rest after clinical recovery was shorter for the non-carrier group (7·7 days) than for those that later became long-term shedders (9·3 days). In sharp contrast to this was the complete absence of the long-term carrier state among 7 stallions on Farm B that was linked epidemiologically to the stallions on Farm A (Timoney, 1985). It is possible that the period of 2 weeks of enforced sexual rest that immediately followed initial diagnosis of the disease on the farm may have been largely instrumental in preventing establishment of the long-term carrier state in this group of stallions. Other possible factors that could have had a contributory influence include the fact that these stallions were less severely affected than those on Farm A and 6 of them had been vaccinated with an experimental modified live equine viral arteritis vaccine 4 days before the onset of clinical signs of the disease. Since the strain of EAV was similar to that present on Farm A, the difference in host response among the stallions on Farm B may also have been dose related.

Duration of the carrier state in the stallion after natural infection with EAV can vary from a period of weeks to years. It would appear that some long-term carriers may even remain infected for life. Pottie (1888), Bergman (1913) and Schofield (1937) reported that certain stallions that had experienced 'epizootic cellulitis-pinkeye' or 'influenza' could continue to transmit the disease to mares they served in succeeding breeding seasons, in one instance for as long as 6·5 years (Bergman, 1913). The results of the present study would indicate that not all long-term shedding stallions will remain permanent carriers of EAV. There is considerable evidence that Stallion 8 in Group 1, that had been a confirmed carrier for 18 months, had not shed virus in the semen for the ensuing

7–8-month period. Continued monitoring of this stallion will be required to confirm that he remains a non-shedder and therefore no longer a carrier of EAV. An additional stallion, No. 6 in Group 1, may also have ceased to shed virus in the semen at the time of his death. Examination of a dismount sample taken 1 week previously, together with exhaustive testing of a wide range of tissues obtained at necropsy, failed to demonstrate the presence of EAV. Neither stallion had undergone any change in positive serological status for EAV during this time. Although further studies are required to confirm these findings, they provide grounds for optimism that perhaps more of the current long-term shedders of the virus may in time cease to be carriers.

In contrast to the importance of transmission of EAV by infective aerosolised nasal secretions during the acute phase of the infection (McCollum *et al.*, 1971; McCollum & Swerczek, 1978), stallions appear to shed this virus solely by the venereal route during the chronic carrier phase of the disease. Based on the earlier findings of McCollum *et al.* (1971) who recovered virus from the kidney and urine of horses up to 19 days after experimental challenge, it was considered initially that EAV might remain localized in the kidneys or elsewhere in the urinary tract of the shedding stallion. This has not been supported by the results of the present study, however, which indicate that long-term carrier stallions harbour the virus at some immunologically sheltered site(s) in the genital tract, confirming the observation of Pottie (1888) in respect of stallions affected with 'epizootic cellulitis-pinkeye' that "it is the seminal fluid which propagates the disease". While repeated attempts to isolate EAV from the buffy coat were unsuccessful, this does not totally preclude the possibility of the circulation of low levels of virus in the blood of long-term carrier stallions from time to time, even in the face of persisting moderate to high homologous neutralizing antibody titres in all of the carrier stallions.

The results of the in-depth investigation of 2 of the carrier stallions, confirming constancy of virus shedding in the semen, are supported by epidemiological field data which indicate that there was virtually 100% transmission of EAV to mares mated by long-term shedding stallions in previous breeding seasons (Timoney & McCollum, 1985). It is readily apparent that clinical evidence of equine viral arteritis in these mares may be mild or very often absent. While establishment of clinical infection in mares mated to certain stallions may be dose related, it is not always the case. There is evidence to indicate that many of the long-term carrier stallions are shedding large quantities of EAV in the semen and that the strains of virus involved are of such modified virulence that the vast majority of mares mated to these stallions under normal breeding conditions become asymptomatic cases of infection (P. J. Timoney & W. H. McCollum, unpublished observations). It would appear from the results of this study that the concentration of virus being shed in the semen is not subject to major fluctuations over time and, also, that it is characteristic for a particular stallion.

Assessment of the shedding status of a putative carrier stallion by means of attempted isolation of EAV from semen is best undertaken with entire ejaculates rather than dismount samples. It is evident that, while there was virtually complete agreement between the results of a test breeding programme and isolations of virus from complete collections, the isolation rate from some dismount samples was not of equivalent accuracy. Preliminary screening of stallions using dismount semen samples is nevertheless a feasible, practical procedure. Current indications are that EAV is present in the sperm-rich fraction of the ejaculate and that a false negative isolation result could arise from testing a specimen of the pre-sperm fraction.

The frequency of the long-term carrier state in Thoroughbred stallions exposed to EAV was surprisingly high, emphasizing the significant epidemiological role the carrier stallion probably plays in the widespread dissemination of the virus at time of breeding. Vaccination with a modified live vaccine against equine viral arteritis (Arvac; Fort Dodge Labs, Fort Dodge, IA, U.S.A.) has been used extensively in the Thoroughbred stallion population in Kentucky since early 1985 as an effective means of protection against the disease and possible establishment of the carrier state in additional stallions. Determination of the site(s) of virus localization and the mode of persistence of EAV in the long-term shedding stallion are some of the aspects of the carrier state that currently need to be resolved.

We thank the veterinarians, farm owners and staff who participated in this study; and the Kentucky state veterinarians, Dr R. I. Hail and Dr M. J. McDonald. This work was supported in part by a grant from the Grayson Foundation, Inc., and is published as paper No. 86-4-183 by permission of the Dean and Director, College of Agriculture and Kentucky Agricultural Experiment Station.

References

Bergman, A.M. (1913) Beitrage zur kenntnis der virus-trager bei rotlaufseuche, influenza erysipelatosa, des pferdes. *Z. Infektkrankrh. Parasit. Krankh. Hyg. Haust.* **13**, 161–174.

Clark, J. (1892) Transmission of pink-eye from apparently healthy stallions to mares. *J. comp. Path.* **5**, 261–264.

Doll, E.R., Bryans, J.T., McCollum, W.H. & Crowe, M.E.W. (1957) Isolation of a filterable agent causing arteritis of horses and abortion of mares. Its differentiation from the equine abortion (influenza) virus. *Cornell Vet.* **47**, 3–41.

McCollum, W.H. & Swerczek, T.W. (1978) Studies of an epizootic of equine viral arteritis in racehorses. *J. eq. Med. Surg.* **2**, 293–299.

McCollum, W.H., Prickett, M.E. & Bryans, J.T. (1971) Temporal distribution of equine arteritis virus in respiratory mucosa, tissues and body fluids of horses infected by inhalation. *Res. vet. Sci.* **2**, 459–464.

Pottie, A. (1888) The propagation of influenza from stallions to mares. *J. comp. Path.* **1**, 37–38.

Schofield, F.W. (1937) A report of two outbreaks of equine influenza due to virus carriers (stallions). *Ann. Rpt Dept. Agric., Ontario,* p. 15.

Timoney, P.J. (1985) Clinical, virological and epidemiological features of the 1984 outbreak of equine viral arteritis in the thoroughbred population in Kentucky, USA. In *Proc. Grayson Foundation International Conference of Thoroughbred Breeders Organizations on Equine Viral Arteritis,* pp. 24–33. Grayson Foundation Inc., Lexington, Kentucky.

Timoney, P.J. & McCollum, W.H. (1985) The epidemiology of equine viral arteritis. *Proc. 31st Ann. Conv. Am. Assoc. Equine Practnr* pp. 545–551.

Timoney, P.J., McCollum, W.H., Roberts, A.W. & Murphy, T.W. (1986) Demonstration of the carrier state in naturally acquired equine arteritis virus infection in the stallion. *Res. vet. Sci.* **41**, 279–280.

J. Reprod. Fert., Suppl. **35** (1987), 103–107

Motility and ATP content of extended equine spermatozoa in different storage conditions*

Minna-Liisa Heiskanen, Arja Pirhonen†, E. Koskinen and P. H. Mäenpää†

The State Horse Breeding Institute, SF-32100 Ypäjä, and †Department of Biochemistry, University of Kuopio, P.O. Box 6, SF-70211 Kuopio, Finland

Summary. The role of various environmental conditions on sperm motility and ATP content was investigated by incubating raw and washed spermatozoa collected with an open-ended artificial vagina from 10 stallions in various biological and artificial media under different atmospheric conditions. Spermatozoa did not survive for more than 12 h when kept unextended in the original seminal fluid in any circumstances. The most favourable media tested for long-term sperm survival were Kenney's medium or Kenney's medium supplemented with 10 mM-theophylline and 10 mM-Hepes, pH 7·2. Centrifugation and slow cooling to 5–7°C improved the survival as did incubation in atmosphere containing 5% CO_2 or in a closed plastic bag with no air-space. In the most favourable circumstances, spermatozoa could stay alive, in some instances, for up to 4–5 days. The pregnancy rates 16 days after oestrus in mares inseminated with extended and cooled spermatozoa stored for 24 h were 82% ($n = 11$) and 70% ($n = 10$) per first oestrous cycle for Kenney's medium and the supplemented Kenney's medium, respectively.

Introduction

Ejaculated spermatozoa kept *in vitro* survive to a lesser extent than those retained within the female genital tract. The preservation of equine spermatozoa without freezing for prolonged periods of time would provide an efficient means for utilizing stallions and for breeding mares at convenient locations. The fertility of stallion spermatozoa extended in various extenders and stored at different temperatures tends to be in general relatively low although developments in cooling technology have improved pregnancy rates markedly (Douglas-Hamilton *et al.*, 1984).

The report by Makler *et al.* (1984) on the effects of incubation of human spermatozoa in various biological and artificial media in atmospheres containing 5% CO_2 lead us to study stallion spermatozoa under similar conditions. Our aim was also to compare sperm survival under those conditions with that obtained by extending the spermatozoa with the dry-milk extender described by Kenney *et al.* (1975). In addition, the cooling container for stallion spermatozoa (Douglas-Hamilton *et al.*, 1984) was tested to determine the most favourable conditions for sperm storage for at least 24 h.

Materials and Methods

The study was conducted at the State Horse Breeding Institute in south-western Finland over the years 1984–86. Semen was collected from 9 Finnish-bred and 1 German riding horse stallions using an open-ended artificial vagina. Usually the three first fractions of the ejaculate were collected and pooled for further studies. Immediately after collection, semen was evaluated for volume, sperm concentration and motility. Aliquants were frozen in liquid nitrogen

---

*Reprint requests to Dr P. Mäenpää.

for DNA and ATP determinations and the rest of the pooled sample was stored as such or extended with various biological and cell culture media or with Kenney's dry-milk extender (Kenney *et al.*, 1975) (49 g glucose, 24 g instant non-fat dry milk, 1.5×10^6 i.u. penicillin and 1.5 g streptomycin in 1 litre) or with the same medium supplemented with 10 mM-theophylline and 10 mM-Hepes, pH 7.2, to achieve 1:5 dilution (semen to extender). Usually one half of the extended sperm sample was washed by centrifugation for 3 min at 700 *g* and the soft pellet resuspended with the respective extender to the original volume.

In the first series of experiments, the raw semen or extended sperm samples were maintained in plastic Petri dishes in an incubator at 25–27°C for up to 24 h. The atmosphere was air or 95% O_2 + 5% CO_2. In the second series of experiments, the sperm samples were transferred to polyethylene bags which were stored open or closed, with no air-space, at room temperature or in cooling containers (Equitainer or Retainer, Hamilton Equine Systems, Wenham, MA, U.S.A.). Aliquants of the samples were evaluated at designated times for motility and ATP content. ATP was determined by a bioluminescence assay (ATP Monitoring Reagent, LKB-Wallac Oy, Turku, Finland) using an LKB-Wallac Luminometer 1250. DNA was determined according to Schneider (1957) and the motility evaluation was performed microscopically with a Makler chamber (Sefi Medical Instruments, Haifa, Israel). Sperm density was estimated with a Bürker haemocytometer.

The biological and cell culture media were purchased from Flow Laboratories, McLean, VA, U.S.A. The following media were tested: Ham's F-10 medium, Eagles Minimal Essential Medium, Ovum Culture Medium, phosphate-buffered saline and tryptose phosphate broth. Hepes-buffer concentrate, fetal bovine serum, newborn bovine serum and penicillin-streptomycin concentrate were also from Flow Laboratories. Bovine serum albumin and theophylline were from Sigma Chemical Co., St Louis, MO, U.S.A.

In the third series of experiments, 21 mares (4 maiden, 13 foaling and 4 barren mares) aged 3–16 years were inseminated during one cycle with spermatozoa ($\sim 1500 \times 10^6$ spermatozoa) extended with Kenney's medium or Kenney's medium supplemented with 10 mM-theophylline and 10 mM-Hepes, pH 7.2, slowly cooled in a container (Equitainer or Retainer) and stored for 24 h. The pregnancy rates were determined by rectal palpation and ultrasonic examination at 16 days using an Aloka SSD-210 DX scanner (Aloka Co. Ltd, Tokyo, Japan).

Results

Experiment 1

When extended stallion spermatozoa were incubated at room temperature in Petri dishes in various biological and artificial media, only incubation in an atmosphere containing 5% CO_2 or addition of 10 mM-theophylline seemed to prolong sperm survival (data not shown). Even in the most favourable circumstances, however, the progressive motility was relatively low at 24 h and bacterial growth occurred occasionally in spite of the presence of penicillin-streptomycin in the extenders. Since the survival of spermatozoa extended in Kenney's medium was always better, only that medium was further studied in the next series of experiments.

Experiment 2

The progressive motility of spermatozoa was well maintained for at least 24 h and in some instances for up to 4–5 days when spermatozoa extended with Kenney's medium or Kenney's medium supplemented with theophylline and Hepes were slowly cooled to 5–7°C (Fig. 1). Washing prolonged the long-term survival especially during the 2nd and 3rd day of storage. Determination of the ATP content of spermatozoa indicated that, when expressed per 10^6 cells (Fig. 2a) or per DNA (Fig. 2b), centrifugation and supplementing the extender with theophylline and Hepes were beneficial for sperm survival. After storage for 48 h, the difference between the washed and unwashed spermatozoa in motility was about 2-fold (Fig. 3). The difference was significant for Kenney's extender ($P < 0.01$) and the supplemented Kenney's extender ($P < 0.001$). When the different atmospheric conditions were tested during slow cooling of the sperm sample extended in Kenney's extender, the survival diminished in the following order: closed bag with no air-space → 95% O_2 + 5% CO_2 → air.

The DNA content of the spermatozoa in sperm samples was closely correlated with cell density ($r = 0.929$, $P < 0.001$). However, the ATP content of spermatozoa was significantly correlated with motility only in semen samples with unwashed spermatozoa extended with Kenney's medium

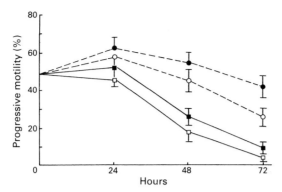

Fig. 1. Mean values (±s.e.m.) of percentage motile stallion spermatozoa in different extenders: □—□, Kenney's extender; ■—■, Kenney's extender supplemented with theophylline and Hepes; ○—○, Kenney's extender, washed spermatozoa; ●—●, Kenney's extender supplemented with theophylline and Hepes, washed spermatozoa.

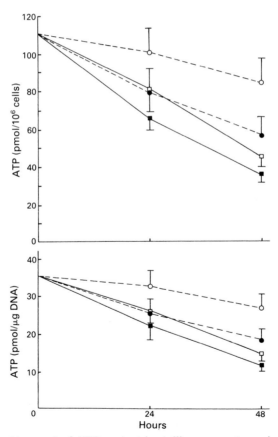

Fig. 2. Mean values (±s.e.m.) of ATP content in stallion spermatozoa in different extenders: □—□, Kenney's extender; ■—■, Kenney's extender supplemented with theophylline and Hepes; ○-○, Kenney's extender, washed spermatozoa; ●—●, Kenney's extender supplemented with theophylline and Hepes, washed spermatozoa.

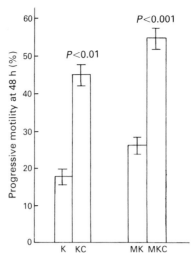

Fig. 3 Mean values (±s.e.m.) of percentage motile spermatozoa in different extenders after slow cooling and storage at 5–7°C for 48 h: K, Kenney's extender; KC, Kenney's extender, washed spermatozoa; MK, modified Kenney's extender (supplemented with theophylline and Hepes); MKC, modified Kenney's extender, washed spermatozoa.

Table 1. Effect of extenders on first cycle pregnancy rates at 16 days in mares after storage of spermatozoa for 24 h

Stallion	No. of mares pregnant/ no. inseminated (%)			
	Kenney's extender		Modified Kenney's extender	
A	3/3	(100)	3/3	(100)
B	5/5	(100)	1/4	(25)
C	1/3	(33)	3/3	(100)
Total	9/11	(82)	7/10	(70)

($r = 0.743, P < 0.001$) or the supplemented Kenney's medium ($r = 0.530, P < 0.05$). The correlation was not significant with washed spermatozoa.

Experiment 3

When 21 mares were inseminated (within 30 h before to 6 h after ovulation) with spermatozoa extended with Kenney's medium or with Kenney's medium supplemented with theophylline and Hepes (washed spermatozoa), slowly cooled to 5–7°C and stored for 24 h, the pregnancy rates with both methods of extension and storage were satisfactory at 16 days after insemination (Table 1).

Discussion

The present results for stallion semen confirm the findings reported for human semen by Makler *et al.* (1984) that storage of spermatozoa in atmospheres containing 5% CO_2 prolongs the long-term survival of spermatozoa. However, storage of unextended or extended stallion semen in a closed polyethylene bag with no air-space (Douglas-Hamilton *et al.*, 1984) was found to be most favourable for survival. The metabolic activity of spermatozoa may produce enough CO_2 to result in better survival of the stored spermatozoa. The mechanism by which high partial pressure of CO_2 influences sperm survival remains to be determined.

From the various media tested, only Kenney's medium (Kenney *et al.*, 1975) or Kenney's medium supplemented with theophylline and Hepes were judged as satisfactory for long-term survival of stallion spermatozoa. Hepes ensures the maintenance of the pH in sperm samples at or near the physiological pH. The mechanism by which methylxanthines, such as theophylline and caffeine, stimulate sperm motility is thought to involve the inhibition of phosphodiesterase activity and the subsequent accumulation of cyclic nucleotides, especially cyclic AMP, within sperm cells (Garbers *et al.*, 1971). The elevated intracellular cAMP concentration is probably involved in controlling the calcium ion flux across the sperm plasma membrane (Peterson *et al.*, 1979). Maintenance of low intracellular Ca^{2+} levels seems to be important in many species for optimum sperm motility (Tash & Means, 1983). Caffeine, however, has been reported to have teratogenic potential (Nishimura & Nakal, 1960) and other methylxanthines could have teratogenic properties, limiting their use in improving sperm motility and fertilizing ability.

Separation of seminal plasma from spermatozoa by centrifugation prolonged the long-term survival of extended spermatozoa. This is in accord with previous results (Pickett *et al.*, 1975). The results on the ATP content of raw stallion spermatozoa generally agree with earlier reports (see Pfetsch, 1979). However, correlation with motility was significant only with unwashed, extended spermatozoa. Nevertheless, our results indicate that determination of sperm ATP content and motility can supplement each other in evaluation of stallion spermatozoa during storage.

In the insemination trial, spermatozoa extended with Kenney's extender or with Kenney's extender supplemented with theophylline and Hepes, slowly cooled to 5–7°C and stored for 24 h before insemination, resulted in satisfactory pregnancy rates when examined at 16 days after the first cycle insemination. In comparison, the overall foaling percentage in Finland of Finnish-bred mares was 51·9% (total number of mares 2671) in 1985.

We thank Marja Reittonen, Eeva-Kaisa Titov, Arja Mansala, Arja Kalliojärvi, Iikka Huttunen and Niilo Lyytikäinen for excellent technical assistance and students of veterinary medicine Helena Kuutsi, Heli Lindeberg, Anu Saikku and Dr Katri Helminen, DVM, for help with inseminations. This study was supported in part by grants from the Academy of Finland and the Erkki Rajakoski Foundation, Finland.

References

Douglas-Hamilton, D.H., Osol, R., Osol, G., Driscoll, D. & Noble, H. (1984) A field study of the fertility of transported equine semen. *Theriogenology* **22**, 291–304.

Garbers, D.L., Lust, W.D., First, N.L. & Lardy, H.A. (1971) Effect of phosphodiesterase inhibitors and cyclic nucleotides on sperm respiration and motility. *Biochemistry, N.Y.* **10**, 1825–1831.

Kenney, R.M., Bergman, R.V., Cooper, W.L. & Morse, G.W. (1975) Minimum contamination techniques for breeding mares: technique and preliminary findings. *Proc. 21st Meeting Am. Ass. eq. Pract.*, pp. 327–336.

Makler, A., Fisher, M., Murillo, O., Laufer, N., DeChernay, A. & Naftolin, F. (1984) Factors affecting sperm motility. IX. Survival of spermatozoa in various biological media and under different gaseous compositions. *Fert. Steril.* **41**, 428–432.

Nishimura, M.B. & Nakal, K. (1960) Congenital malformations in offspring of mice treated with caffeine. *Proc. Soc. exp. Biol. Med.* **104**, 140–142.

Peterson, R.N., Seyler, D., Bundman, D. & Freund, M. (1979) The effect of theophylline and dibuturyl cAMP on the uptake of radioactive calcium and phosphate ions by boar and human spermatozoa. *J. Reprod. Fert.* **55**, 385–390.

Pfetsch, J. (1979) *Zur ATP-Bestimmung im Sperma einiger Haustierarten mit dem Biolumineszenzverfahren.* Dissertation, University of München.

Pickett, B.W., Sullivan, J.J., Byers, W.W., Pace, M.M. & Remmenga, E.E. (1975) Effect of centrifugation and seminal plasma on motility and fertility of stallion and bull spermatozoa. *Fert. Steril.* **26**, 167–174.

Schneider, W.C. (1957) Determination of nucleic acids in tissues by pentose analysis. *Methods Enzymol.* **3**, 680–684.

Tash, J.S. & Means, A.R. (1983) Cyclic adenosine 3′,5′-monophosphate, calcium and protein phosphorylation in flagellar motility. *Biol. Reprod.* **28**, 75–104.

J. Reprod. Fert., Suppl. **35** (1987), 109–112

Printed in Great Britain
© 1987 Journals of Reproduction & Fertility Ltd

Influence of chlorhexidine on seminal patterns in stallions

D. Rath, C. Leiding, E. Klug and H.-C. Krebs

Klinik für Andrologie und Besamung der Haustiere and Institut für Chemie, Tierärztliche Hochschule Hannover, Bischofsholer Damm 15, 3000 Hannover, Federal Republic of Germany

Summary. Ejaculates were collected at 3-day intervals before, during and after a washing procedure with chlorhexidine (2%). Semen motility and pathology were determined before and after deep-freezing. Blood samples were taken before and within 1 h after washing procedures and then extracted in ether. This was followed by HPL chromatography. Chlorhexidine concentrations in blood and seminal plasma were distinctly higher in the treated stallions than in control groups. Concentrations in the control groups were below the detection limit of the column. Significant correlations between decreasing semen quality (% immotile cells, % morphological aberrations) and chlorhexidine treatment were found. A change of the present requirements for imports into North America should be considered to improve the semen quality of deep-frozen ejaculates.

Introduction

Taylorella (*Haemophilus*) *equigenitalis* causes contagious equine metritis (CEM). After the appearance of this disease in Europe, the U.S. Department of Agriculture imposed a general embargo on horses older than 731 days coming from this area, in order to protect the national population. After intervention by European breeding associations and veterinary officials, an exception was agreed upon for stallions in January 1980 and for mares in May 1981. A permit was given only after strict investigations and pretreatment procedures. These requirements for stallions are: washing the penis and prepuce on 5 consecutive days with a 2% chlorhexidine solution, followed by application of 0·2% nitrofurazone ointment. The washing procedure has to be effected after erection and emission of the penis, which easily happens in the presence of a mare in heat. At 7 days after the last chlorhexidine treatment, microbiological swab samples have to be collected from the urethra, the fossa glandis and the free part of the penis (Klug *et al.*, 1980). Sampling has to be repeated on the 14th and 21st days after the last chlorhexidine washing. For transportation of the samples, Amies or Stuart Medium is suitable. All samples must be free from the organism and export must take place within 30 days after last sampling. Otherwise the whole procedure has to be repeated (Lüning *et al.*, 1983).

In the Hannoverian Stallion Centre in Celle/Lower Saxony, horse semen has been deep frozen on a regular commercial basis for several years. In 1983/84 a higher percentage of ejaculates which did not fulfil the minimal requirements for the deep-freezing process arose in stallions treated with chlorhexidine according to the U.S. demands. In such samples the sperm characteristics were lower than for normal fresh semen or the results after thawing did not satisfy the minimal requirements for insemination (Klug, 1982). In 1984, 55·8% (N = 111) ejaculates from groups of stallions without chlorhexidine preparation were considered freezable but only 50·1% (N = 104) from stallions with the treatment could be frozen, i.e. about 6% more ejaculates did not reach the accepted criteria for freezing. Huston *et al.* (1982) reported irreversible changes in cell membranes and enzyme metabolism by protein precipitation which have been attributed to the cationic nature of chlorhexidine. Figure 1 presents the chemical structural formula.

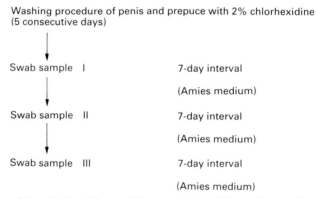

Fig. 1. The structural formula of chlorhexidine (from Huston et al., 1982).

The following experiments were undertaken to determine the effect of chlorhexidine treatment on the freezability of stallion semen.

Materials and Methods

First of all, the results of semen evaluations in 363 ejaculates of Hannoverian stallions collected in 1983/84 and 1984/85 were compared between one group (N = 160) without and another (N = 203) with chlorhexidine treatment. The percentages of motility and morphological aberrations have been verified for this breed.

Additionally, a trial with two Hannoverian stallions was executed in the Clinic for Andrology. Both stallions were washed in a manner similar to the export conditions (Fig. 2) and semen was collected by means of an artificial vagina (Hannover Type) and a mare in heat. In the first phase, 2 ejaculates were taken weekly for 6 weeks from both stallions before the washing period started. In the second phase semen collection started 60 min after the washing procedure with chlorhexidine (Synopharm, Hamburg, F.R.G.) and was repeated for both animals twice a week, for 8 consecutive weeks. For the third step, after export preparations had finished, semen was collected again for both stallions twice a week, for 10 consecutive weeks. Immediately after semen collection, intensive investigation of sperm motility and of the percentage of morphological aberrations was carried out. Additionally, some doses of each ejaculate were deep-frozen in liquid nitrogen, following the Hannover method using big straws (Makrotüb: Minitüb, Landshut, F.R.G.) (Martin & Klug, 1979). After thawing, the motility of the semen was evaluated as before.

Washing procedure of penis and prepuce with 2% chlorhexidine
(5 consecutive days)

Swab sample   I   7-day interval

                                   (Amies medium)

Swab sample   II   7-day interval

                                   (Amies medium)

Swab sample   III   7-day interval

                                   (Amies medium)

Fig. 2. Diagram of the chlorhexidine washing procedure. The swab samples were taken from the urethra, fossa glandis and free part of the penis.

To prove that chlorhexidine may directly or indirectly cause damage to spermatozoa, a difficult but sensitive method similar to that described by Huston et al. (1982) was established. The content of chlorhexidine was determined in blood and seminal plasma using high-performance liquid chromatography (HPLC). Heparinized blood samples were collected from the jugular vein. After centrifugation at 1500 g for 10 min, plasma was extracted with NaOH and diethyl ether. Mechanical shaking and sedimentation was followed by decantation of the organic layer. This was then evaporated to dryness and redissolved to a mobile phase to give a suitable sample for injection. A high-performance liquid chromatography (HPLC) column with a detection limit of 0·2 µg/ml was filled with an aliquant of the mobile phase. Chlorhexidine concentration in blood plasma was determined by comparison with a standard curve, established by 5 different concentrations. The same procedure was done with seminal plasma from ejaculates collected 60 min after washing the penis and prepuce.

Results

Table 1 demonstrates the results from semen evaluation of 363 ejaculates, collected from stallions in 1984 and 1985. The average of forward movement in the untreated group was significantly higher than that in ejaculates from washed stallions. The opposite is obvious in the percentage of non-motile spermatozoa. There were no differences in morphological aberrations in the two groups.

Table 1. Semen characteristics of ejaculates from Hannovarian stallions with or without the chlorhexidine washing procedure in 1984 and 1985

	Without chlorhexidine	With chlorhexidine
Forward movement (%)	41·7 ± 13·9	33·7 ± 16·7*
Non-motile spermatozoa (%)	34·3 ± 14·1	43·4 ± 19·4*
Morphological aberrations (%)	39·6 ± 11·5	41·0 ± 17·6

Values are mean ± s.d. for 363 ejaculates.
*$P < 0.01$ compared to value without chlorhexidine treatment.

Table 2. Semen characteristics of 2 Hannoverian stallions before, during and after the chlorhexidine washing procedure

	Before	During	After
Fresh semen			
Forward movement (%)			
Stallion I	[a]33·8 ± 7·7 (13)	[b]22·9 ± 9·1 (14)	[c]13·0 ± 8·0 (20)
Stallion II	[b]32·7 ± 16·7 (16)	[c]11·1 ± 9·8 (14)	[c]10·5 ± 13·1 (15)
Nonmotile cells (%)			
Stallion I	[a]31·5 ± 9·9 (13)	[b]41·4 ± 11·1 (14)	[c]56·2 ± 16·2 (20)
Stallion II	[b]30·7 ± 7·0 (15)	[c]39·3 ± 9·2 (14)	[c]41·6 ± 7·6 (19)
Morphological aberrations (%)			
Stallion I	[b]30·2 ± 4·9 (12)	[ca]45·9 ± 10·6 (14)	[cb]56·6 ± 13·1 (20)
Stallion II	[b]29·9 ± 4·6 (14)	[c]42·5 ± 12·7 (14)	[c]48·8 ± 10·9 (19)
Frozen semen			
Forward movement (%)			
Stallion I	[b]28·2 ± 7·8 (11)	[c]11·9 ± 7·2 (13)	[c]7·5 ± 2·7 (6)
Stallion II	[b]30·4 ± 8·1 (12)	[b]25·4 ± 11·4 (13)	[bc]19·3 ± 5·3 (7)

Values are mean ± s.d. for the no. of observations in parentheses.
Within rows (i.e. values for each stallion), values with different superscripts are significantly different; a,b $P < 0.05$; a,c and b,c $P < 0.01$.

The results for the two stallions tested in trials in the clinic are shown in Table 2. Progressive motility decreased in both stallions during and after the washing procedure, while the percentages of non-motile cells and morphological aberrations increased during the tests. Morphological aberrations occurred mainly as acrosomal defects and plasma droplets in the middle-piece area. The results for forward motility in frozen–thawed samples are given in Table 2. Both stallions responded to the washing procedure with a significant deterioration in sperm motility. The chlorhexidine content (µg/ml) in blood plasma, detected by means of HPLC, at 30 min, 60 min and 24 h after washing was 4·6 ($n = 5$), 2·6 ($n = 4$) and 1·3 ($n = 5$) in Stallion I and 6·8 ($n = 5$), 2·5 ($n = 5$) and 2·1 ($n = 5$) in Stallion II. The chlorhexidine content in seminal plasma, collected 60 min after the washing procedure, was 12·8 ± 10·5 (s.d.) µg/ml ($n = 4$ for 2 stallions).

Discussion

The present results confirmed that the percentage of immotile spermatozoa was significantly greater in the horses exposed to the chlorhexidine washing procedure. Experiments with two stallions showed that, after chlorhexidine washing, both forward motility and percentage of non-motile spermatozoa deteriorated significantly. Additionally, a significant increase of morphological aberrations was obvious in both stallions, comparing the abnormalities in fresh ejaculates from the beginning and the end of the investigation.

The same effects were observed in frozen semen. Forward motility decreased by more than half in samples from stallions washed with chlorhexidine. High-performance liquid chromatography showed that chlorhexidine can be detected in blood and seminal plasma, strengthening the suspicion that it penetrates mucous membranes. This substance may therefore be responsible for the observed aberrations and the low semen quality. In view of the negative effects of chlorhexidine on membranes and enzyme metabolism, mentioned in the literature, long-term alterations in spermatogenesis may be induced. Even in the present experiments semen quality did not improve during the 3 months after the last washing. The extremely low incidence of CEM (Eckstein *et al.*, 1983), e.g. in the Federal Republic of Germany, does not seem to justify a prophylactic treatment with such an influence on semen quality.

References

Eckstein, K., Merkt, H., Kirpal, G. & Klug, E. (1983) Erhebungen über Verbreitung und Bedeutung der übertragbaren Gebärmutterentzündung der Pferde (CEM '77) in der Bundesrepublik Deutschland. *Prakt. Tierarzt* **12**, 1096–1104.

Huston, C.E., Wainwright, P., Cooke, M. & Simpson, R. (1982) High-performance liquid chromatography method for the determination of chlorhexidine. *J. Chromatog.* **237**, 457–464.

Klug, E. (1982) Untersuchungen zur klinischen Andrologie des Pferdes. *Habil.-Schr., Hannover*, 175–190.

Klug, E., Merkt, H., Kirpal, G. & Flüge, A. (1980) Maßnahmen zur Eindämmung eines akuten Auftretens der kontagiösen equinen Metritis (CEM 77) im Bereich einer staatlichen Deckstelle. *Dtsch. tierärztl. Wschr.* **87**, 158–163.

Lüning, I., Klug, E., Gaus, J. & Mattos, R. (1983) Exportvorbereitung von Pferden nach Nordamerika gemäß CEM-Bestimmungen der Importländer. *Prakt. Tierarzt* **6**, 485–492.

Martin, J.C. & Klug, E. (1979) Zur Samenübertragung beim Pferd—Spermakonservierung in Kunststoffröhrchen. *Prakt. Tierarzt* **3**, 196–200.

J. Reprod. Fert., Suppl. **35** (1987), 113–120

Proteins in stallion seminal plasma

R. P. Amann, M. J. Cristanelli and E. L. Squires

Animal Reproduction Laboratory, Colorado State University, Fort Collins, Colorado 80523, U.S.A.

Summary. Motility and fertility of frozen–thawed semen differs greatly amongst stallions. Differences in seminal plasma might be one cause of this variation. For 8 ejaculates from each of 17 stallions, seminal plasma was saved at $-20°C$ and spermatozoa were cryopreserved. Based on post-thaw sperm motility, seminal plasma samples from 7 stallions (2 good, 3 variable, 2 poor sperm motility) were selected for measurement of electrolytes, protein content and analysis by sodium dodecylsulphate gel electrophoresis (10% gel, Coomassie blue stain). Variation in seminal plasma was significant ($P < 0.05$) amongst stallions for concentrations of Na, K, Ca and Cl, but not for P or protein. A total of 27 proteins, of 13 000 to 122 000 molecular weight, were detected in seminal plasma. There was a difference ($P < 0.05$) amongst stallions in the proportion of ejaculates containing 13 of the 27 proteins. However, only for Proteins 23 and 26 was there a significant correlation between the relative amount of protein and sperm motility (0, 30 or 60 min after thawing; $r = 0.42–0.50$). These correlation coefficients, or those with concentration of K, were too low to be of predictive value. Therefore, although an effect of a minor protein can not be excluded, variation in the relative amounts of major proteins in seminal plasma probably is not a cause of differences in post-thaw motility of stallion spermatozoa.

Introduction

The best procedures for freezing spermatozoa from most stallions yield 1-cycle pregnancy rates <75% as good as those obtained with fresh semen from the same stallion. Despite recent research (Loomis *et al.,* 1983; Cochran *et al.,* 1984; Cristanelli *et al.,* 1984), there continues to be variability in post-thaw motility of ejaculates from a given stallion. Such variation could reflect inconsistencies in the procedure used for cryopreservation of stallion spermatozoa, innate differences in the spermatozoa before emission, or differences in seminal plasma which result in differences in the spermatozoa after admixture with the seminal plasma. Although the nuances of any procedure for cryopreservation should not be overlooked, it was considered that ejaculate-to-ejaculate variation in seminal plasma was likely to be greater than differences in the spermatozoa *per se.* Therefore, experiments were designed to test the hypothesis that differences in the post-thaw motility of stallion spermatozoa might be due to relative differences in the concentration of one or more electrolytes or proteins in seminal plasma.

Spermatozoa from the cauda epididymidis are functionally different from ejaculated spermatozoa. Consequently, the admixture of spermatozoa with seminal plasma at the time of emission and ejaculation could influence subsequent sperm function, susceptibility to damage during centrifugation, cooling of freeze–thawing, or fertility of spermatozoa in undiluted (neat) semen, extended semen, or cryopreserved semen. Seminal plasma apparently has a role in the processes of reproduction beyond serving as a fluid medium for the transport of spermatozoa (Chang, 1947; Rozin, 1960; Mann, 1964; Marden & Werthessen, 1956; Lindholmer, 1974). Vesicular gland fluid from bulls (Matousek, 1968) and boars (Moore *et al.,* 1976) contains proteins which contribute to seminal plasma and sensitize spermatozoa to make them less resistant to cold shock. Although stallion spermatozoa are more resistant to cold shock than are boar or bull spermatozoa (Watson &

Plummer, 1985), stallion seminal plasma may contain proteins that adversely affect sperm survival. Stallion seminal plasma reduces the motility of frozen–thawed spermatozoa (Smith & Polge, 1950; Rajamannan et al., 1968; Nishikawa et al., 1968; Nishikawa, 1972). Stallion seminal plasma also has an adverse effect on sperm motility and fertility (Marden & Werthessen, 1956; Pickett et al., 1975). However, excessive dilution of semen or removal of seminal plasma by centrifugation also can be deleterious to the spermatozoa (Pickett et al., 1975).

There is no comprehensive study of the electrolytes or proteins in seminal plasma from the stallion or jackass, although limited data are available (Mann, 1964; Gebauer et al., 1976). Comprehensive and comparative studies of the electrophoretic characteristics of seminal plasma proteins of other domestic animals, laboratory animals and humans have revealed a multiplicity of proteins. Some originate from a particular gland and others are common to several organs. Many proteins in seminal plasma become bound to spermatozoa. However, correlations between concentrations of proteins in seminal plasma and the motility, or fertility, of fresh or frozen–thawed semen have not been reported.

The objectives of this experiment were to determine (a) the components of variance for concentrations of electrolytes and protein in seminal plasma and post-thaw sperm motility, and (b) the relationships among concentrations of electrolytes, protein, or the relative amounts of specific proteins in seminal plasma, and post-thaw sperm motility.

Materials and Methods

Eighteen stallions of light horse breeds were assigned to two replicates, of 9 stallions each, although 1 stallion subsequently was deleted from the experiment because of low sperm concentration. Before starting the experiment, semen was collected by artificial vagina from each stallion 4 times within 3 days and then the stallions were given 3 days of sexual rest. Starting on either 9 March or 12 April, for the two replicates, one ejaculate was collected every 3rd day until a total of 8 ejaculates had been processed and frozen for each stallion. The gel-free semen was evaluated for volume, sperm concentration, and percentage of progressively motile spermatozoa (Pickett & Back, 1973). A 10-ml aliquant of semen from each ejaculate was centrifuged at 1100 g for 5 min followed by 12 000 g for 10 min to provide sperm-free seminal plasma which was stored at $-20°C$. The remaining semen was considered suitable for processing if it contained $\geq 20\%$ progressively motile spermatozoa, $\geq 50 \times 10^6$ spermatozoa/ml, and $\geq 4 \times 10^9$ spermatozoa.

Each ejaculate processed was diluted to 50×10^6 spermatozoa/ml with a centrifugation medium and centrifuged (Cochran et al., 1984) to concentrate the spermatozoa. The supernatant was aspirated and the spermatozoa were resuspended at 200×10^6 progressively motile spermatozoa per 1·0 ml of extender containing 4% glycerol (Cochran et al., 1984) and packaged in 0·5-ml polyvinyl chloride straws. The straws were cooled at 10°C/min from ambient to $-15°C$ and at 25°C/min from -15 to $-120°C$, using a programmable liquid nitrogen freezer, and then plunged into liquid nitrogen (Cochran et al., 1984). Semen was stored at $-196°C$ for ≥ 7 days before thawing and evaluation.

Semen was thawed by immersing the straw in 75°C water for exactly 7 sec and immediately transferring the straw into water at 37°C for ≥ 5 sec (Cochran et al., 1984). To evaluate the quality of the frozen–thawed spermatozoa, the contents of two straws representing each ejaculate were pooled and a 150-µl aliquant of the semen was diluted with 1·5 ml of a clarified (5 000 g for 20 min) extender without glycerol and incubated at 37°C. The percentage of progressively motile spermatozoa was evaluated subjectively ($\times 200$ phase-contrast microscopy) 0, 30, 60 and 90 min later by two technicians unaware of the identity of the samples. Values for each time point were averaged.

It was deemed impractical to perform electrophoretic analyses on all 136 ejaculates. Consequently, ejaculates were ranked, within stallion, on the basis of survival of progressively motile spermatozoa between the initial evaluation and the post-thaw evaluations. Selections were made to provide seminal plasma samples from stallions with good, average, and poor sperm motility after thawing (N = 2, 3 and 2, respectively).

The protein concentration in seminal plasma from each ejaculate ($n = 56$) was determined using the Bio-Rad assay (Bio-Rad Laboratory, Richmond, CA) and bovine serum albumin as the standard. An aliquant of each seminal plasma sample was thawed at room temperature and centrifuged at 15 600 g for 5 min. Then, 10 µl seminal plasma were diluted with 10 ml deionized water and 800 µl of the diluted sample were mixed with 200 µl Bio-Rad reagent. Absorbance was measured at 595 nm. Another aliquant of seminal plasma was analysed for Na and K using a Nova Model 1 Analyzer (Nova Biomedical, Newton, MA) and for Ca, P and Cl using a Roto-chem Model II-A Analyzer (American Instruments Co., Silver Spring, MD).

A third aliquant of seminal plasma was thawed at room temperature and diluted to 7 mg/ml with deionized water. Diluted seminal plasma then was mixed 1:1 with treatment buffer containing 4% sodium dodecylsulphate (SDS) and 10% 2-mercaptoethanol; the mixture was boiled for 5 min. The separating gel contained (10% (w/v) polyacrylamide and 0·63% (w/v) bisacrylamide and the stacking gel contained 4% polyacrylamide and 0·035% bisacrylamide. Each gel (1·5 mm thick) contained ten 1-cm wells and 50 µl denatured seminal plasma protein, or molecular weight standard, was loaded into each well. Electrophoresis was at 20°C using 30 amp/gel with a Tris–glycine buffer (0·25 M-Tris, 0·192 M-glycine, 0·01% (w/v) SDS; pH 8·3) and required about 3 h. Molecular weight standards were β-lactoglobulin, 18 400; trypsin inhibitor, 20 100; ovalbumin, 45 000; bovine serum albumin, 66 000; phosphorylase β-subunit, 97 400; and β-galactosidase subunit, 116 000. Proteins in the gels were fixed and concurrently stained with 0·125% (w/v) Coomassie blue in methanol:acetic acid:deionized water (50:10:40 by vol.) for 10–14 h and destained for 4–8 h in a similar solution lacking Coomassie blue.

A photograph (Polaroid-type 55 film) was taken of each gel. Each lane of Coomassie blue-stained proteins as imaged on the negative was scanned using a laser densitometer (Biomed Instruments, Inc., Chicago, IL) coupled to a strip-chart recorder. The resulting graphs were used to calculate the relative mobility of each molecular weight standard and also the relative amounts of each protein detected. The molecular weight of each protein was calculated by comparison with a standard curve derived from the relative mobilities of the 6 standard proteins. The amount of protein in each peak was calculated by multiplying the width at the half-height by the total height of the peak. Values for the relative amount of the protein were normalized by dividing the area of each peak in a sample by the total area for all peaks within each sample to give a relative amount (percentage of total protein).

A one-way analysis of variance (with stallion as a random variable) was used to determine whether there were significant differences amongst stallions for (a) progressive motility of spermatozoa at 0, 30, 60, or 90 min after thawing, (b) concentrations of electrolytes, and (c) concentration of protein in seminal plasma. In addition, the variance was partitioned into two components representing the variation amongst stallions and the variation amongst ejaculates within stallions (Snedecor & Cochran, 1978). Potential differences among stallions in the proportion of ejaculates (within a stallion) having a particular seminal plasma protein were evaluated by χ^2 tests. Stepwise multiple regression (Steel & Torrie, 1980) was used to estimate the predictive value of independent variables (namely, the absolute or relative concentration of each electrolyte and the 27 seminal plasma proteins, and volume, sperm concentration of total sperm number in the semen at initial evaluation) for predicting the population mean percentage of progressively motile spermatozoa at 0, 30, 60 or 90 min after thawing. Simple correlations also were calculated.

Results

Post-thaw sperm motility varied ($P < 0·05$) amongst stallions at 0, 30, 60 and 90 min after thawing, although the variation associated with stallion was less than that associated with ejaculates within stallions (Table 1). Immediately after thawing (0 min), only 20% of the variance was associated with stallions, although at 30, 60 or 90 min this component accounted for 31–43% of the variance. Therefore, when screening stallions on the basis of post-thaw motility of their spermatozoa, semen

Table 1. Mean value and estimates of variance components for the progressive motility of stallion spermatozoa at 0, 30, 60, or 90 min after thawing and for the concentrations of electrolytes or protein in seminal plasma*

		Variance components	
Characteristic	Mean	Amongst stallions	Ejaculates within stallions
Motility at:			
0 min	43	33·7 (20%)	134·1 (80%)
30 min	42	53·8 (31%)	118·5 (69%)
60 min	39	106·0 (43%)	138·1 (58%)
90 min	36	52·8 (31%)	120·0 (69%)
Concentration of:			
Protein	12·1	1·5 (4%)	32·3 (96%)
Na	2·59	3407 (39%)	5260 (61%)
K	0·77	11·5 (68%)	5·5 (32%)
Ca	0·13	26·4 (34%)	54·0 (66%)
P	0·04	0·0 (0%)	6·5 (100%)
Cl	4·04	2969 (41%)	4274 (59%)

*Based on data for 8 ejaculates from each of 17 stallions for motility or 7 stallions for protein concentration.

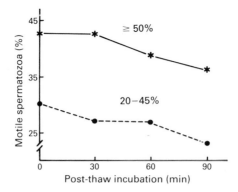

Fig. 1. Mean percentage of progressively motile spermatozoa at 0, 30, 60, and 90 min after thawing. The upper line represents the mean for 100 ejaculates for which $\geq 50\%$ of the spermatozoa were progressively motile immediately after collection, and the lower line represents data for 36 ejaculates for which 20–45% of the spermatozoa were progressively motile immediately after collection.

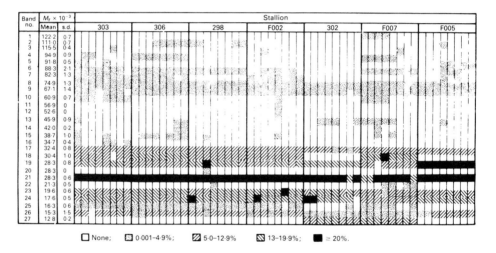

☐ None; ▨ 0·001–4·9%; ▧ 5·0–12·9%; ▨ 13–19·9%; ■ $\geq 20\%$.

Fig. 2. A comparison of the proteins detectable by SDS-gel electrophoresis of seminal plasma from 8 ejaculates from each of 7 stallions selected because post-thaw motility of the spermatozoa was good (Stallions 303 and 306), average (Stallions 298, F002 and 302) or poor (Stallions F007 and F005). The relative amount of each seminal plasma protein is depicted.

should be incubated 30 or 60 min before evaluation. The percentage of progressively motile spermatozoa declined between 0 and 90 min of incubation at 37°C. Three stallions consistently ejaculated semen in which < 50% of the spermatozoa were progressively motile at initial evaluations immediately after collection. A similar decline occurred for spermatozoa from ejaculates containing $\geq 50\%$ progressively motile spermatozoa immediately after collection and samples containing 20–45% progressively motile spermatozoa at the initial evaluation (Fig. 1). When the 17 stallions were ranked on the basis of sperm survival between the initial evaluation and immediately after thawing, averaged across the 8 ejaculates, the 3 stallions with low initial motility ranked 8th, 15th

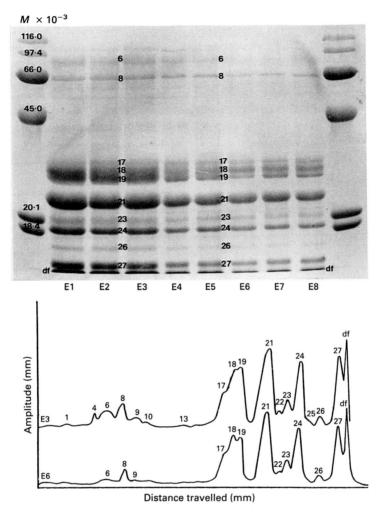

Fig. 3. Electrophoretogram (upper) and laser densitometer scans (lower) of proteins in seminal plasma separated by SDS-gel electrophoresis. The samples are from Stallion F-007 and the laser scans are for Ejaculates 3 and 6. The numbers designate the seminal plasma proteins depicted in Fig. 2.

and 17th. This might be evidence that semen containing a relatively low percentage of progressively motile spermatozoa upon initial evaluation ($< 50\%$) does not survive the procedures involved with cryopreservation and thawing as well as spermatozoa in samples containing $\geq 50\%$ progressively motile spermatozoa at initial evaluation.

The mean protein concentration in seminal plasma did not differ ($P > 0.05$) amongst stallions and 96% of the variation was associated with ejaculates within stallions (Table 1). Similarly, for all electrolytes, except K, more variation was associated with ejaculates within stallion than amongst stallions (Table 1). Correlations between concentrations of Na and K, Ca, P and Cl were -0.82, -0.62, 0.01 and 0.70 and those of K with Ca, P and Cl were 0.53, 0.04 and -0.56. The concentration of any electrolyte (except K at collection or 60 min), or total protein, in seminal plasma was not significantly correlated with the percentage of motile spermatozoa immediately after collection or in samples incubated for 0–60 min after thawing (r = -0.29 to 0.37).

Table 2. Summary of stepwise multiple regression analyses of the relationship of sperm motility after thawing and a pool of independent variables*

Step	Multiple R^2	Independent variable	Beta†
0 min after thawing			
1	0·22	% Protein 27	−0·47
2	0·39	P conc.	−0·42
3	0·47	% Protein 26	0·28
30 min after thawing			
1	0·25	% Protein 23	0·50
2	0·44	P conc.	−0·44
3	0·57	% Protein 26	0·39
4	0·63	Spermatozoa/ejaculate	0·26
60 min after thawing			
1	0·21	% Protein 23	0·46
2	0·36	P conc.	−0·39
3	0·48	% Protein 26	0·39
4	0·61	Spermatozoa/ejaculate	0·38
90 min after thawing			
1	0·24	% Protein 23	0·49
2	0·37	P conc.	−0·37
3	0·50	Spermatozoa/ejaculate	0·36
4	0·61	% Protein 26	0·39

*All multiple correlation coefficients (R) were significant ($P < 0.001$).
†Partial regression coefficient.

Altogether 27 proteins were detected in the 56 samples of seminal plasma subjected to electrophoresis (Fig. 2). Data for a typical stallion are presented in Fig. 3. Based on χ^2 analysis, there was a difference ($P < 0.05$) in the proportion of ejaculates within a stallion having seminal plasma proteins 1, 2, 4, 5, 6, 7, 9, 10, 13, 15, 16, 18, 22, 25 and 27 (Fig. 2). If the relative amount of a protein was <5% of the total protein present, quantification of the absence or relative amount of that protein was difficult; many of the proteins (1, 2, 4, 5, 6, 7, 9, 13, 15, 16) present at <5% were those for which the relative amount differed significantly amongst ejaculates within stallions. Consequently, data for these proteins should be interpreted with caution. Furthermore, it may have been concluded erroneously (based on χ^2 analysis) that the relative amounts of seminal plasma proteins 3, 11, 12, 14 and 20 did not differ among ejaculates within a stallion because these proteins were undetectable in many ejaculates. There were numerous significant correlations amongst the relative amounts of the 27 seminal plasma proteins, but the biological significance of these relationships remains unknown.

Stepwise multiple regression revealed that 47–63% of the variation in motility at 0, 30, 60 or 90 min after thawing could be accounted for by the stepwise addition of 3 or 4 independent variables (Table 2). Although the multiple regression coefficients were significant ($P < 0.01$), using this pool of independent variables probably is not useful in predicting the post-thaw motility of stallion spermatozoa because of the time required to prepare and analyse SDS-gels of seminal plasma.

For each of the 27 seminal proteins, its relative concentration in each of the 56 samples analysed was correlated with the percentage of progressively motile spermatozoa at 0, 30, 60 and 90 min after thawing, as well as immediately after collection. Only for Proteins 23 and 26 was there a small ($r = -0.39$ to 0.50; $P < 0.01$) but significant linear relationship (Fig. 4).

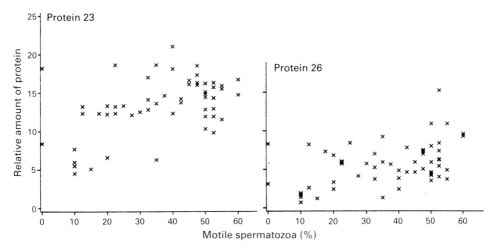

Fig. 4. Relative amounts (% of total protein) of seminal plasma proteins (Nos 23 and 26) and the percentage of progressively motile spermatozoa after incubation at 37°C for 60 min after thawing ($r = 0.46$ and 0.42, respectively; $n = 56$).

Discussion

We postulated that ejaculate-to-ejaculate differences in post-thaw motility of stallion spermatozoa might be attributable to relative differences in the concentration of one or more electrolytes or proteins in seminal plasma. This proved to be correct in that the concentration of K or relative amount of Proteins 23 and 26 was correlated ($r = -0.39$ to 0.50) with sperm motility 0–90 min after thawing. Using additional independent variables, multiple R-values of 0.69–0.79 were obtained. Because of inherent limitations of quantifying proteins in a stained gel, including incomplete resolution and non-linearity of staining, we expressed the data on a relative, rather than absolute, basis; a similar mass of protein was loaded into each gel well. Possibly these limitations of the procedure altered the magnitude of the observed correlations. Because quantification of seminal plasma proteins is time consuming, even though the multiple correlation matrix accounted for 47–63% of the variation in the post-thaw motility of stallion spermatozoa, this approach is not practicable as a predictive test of sperm quality.

The variability amongst stallions in post-thaw motility of spermatozoa confirms classic fertility studies (Nishikawa, 1972). This variation cannot be attributed to consistent differences in the concentration of protein in seminal plasma, since this attribute did not differ significantly amongst stallions. However, ejaculate-to-ejaculate variation within stallions was considerable. The observation (Table 1) that there is considerable variation among ejaculates in post-thaw motility of spermatozoa also is consistent with many studies with bulls. The relatively small variance associated with stallions (20–43%), as compared to that for ejaculates within stallions, may be a consequence of rejection of stallions producing abnormal semen from this project. Nevertheless, 3 of the 17 stallions studied consistently produced semen which resulted in poor post-thaw survival of spermatozoa.

This study did not reveal a simple criterion, based on evaluation seminal plasma, for selecting ejaculates for cryopreservation. However, selection of stallions consistently producing semen in which a higher percentage of spermatozoa are progressively motile immediately after seminal collection should increase the probability of obtaining satisfactory post-thawing motility and fertility of cryopreserved spermatozoa.

References

Chang, M.C. (1947) Effects of testis hyaluronidase and seminal fluids on the fertilizing capacity of rabbit spermatozoa. *Proc. Soc. exp. Biol. Med.* **66**, 51–54.

Cochran, J.D., Amann, R.P., Froman, D.P. & Pickett, B.W. (1984) Effects of centrifugation, glycerol level, cooling to 5°C, freezing rate and thawing rate on the post-thaw motility of equine sperm. *Theriogenology* **22**, 25–38.

Cristanelli, M.J., Squires, E.L., Amann, R.P. & Pickett, B.W. (1984) Fertility of stallion semen processed, frozen and thawed by a new procedure. *Theriogenology* **22**, 39–45.

Gebauer, M.R., Pickett, B.W., Faulkner, L.C., Remmenga, E.E. & Berndtson, W.E. (1976) Reproductive physiology of the stallion. VII. Chemical characteristics of seminal plasma and spermatozoa. *J. Anim. Sci.* **43**, 626–632.

Lindholmer, C.H. (1974) The importance of seminal plasma for human sperm motility. *Biol. Reprod.* **10**, 533–542.

Loomis, P.R., Amann, R.P., Squires, E.L. & Pickett, B.W. (1983) Fertility of unfrozen and frozen stallion spermatozoa extended in EDTA-lactose-egg yolk and packaged in straws. *J. Anim. Sci.* **56**, 687–693.

Mann, T. (1964) *The Biochemistry of Semen and of the Male Reproductive Tract*, pp. 343–350. Methuen, London.

Marden, W. & Werthessen, N.T. (1956) Influence of seminal fluid on sperm motility. *Fert. Steril.* **7**, 508–515.

Matousek, J. (1968) Seminal vesicle fluid substance of bulls sensitizing spermatozoa to cold shock. *Proc. 6th Int. Congr. Anim. Reprod. & A.I., Paris* **1**, 553–555.

Moore, H.D.M., Hall, G.A. & Hibbitt, K.G. (1976) Seminal plasma proteins and the reaction of spermatozoa from intact boars and from boars without seminal vesicles to cooling. *J. Reprod. Fert.* **47**, 39–45.

Nishikawa, Y. (1972) Motility and fertilizing ability of frozen horse spermatozoa. *Proc. 7th Int. Symp. Zootech.*, 155–167.

Nishikawa, Y., Wade, Y. & Shinomiya, S. (1968) Studies on deep freezing of horse spermatozoa. *Proc. 6th Int. Congr. Anim. Reprod. & A.I., Paris* **2**, 1589–1591.

Pickett, B.W. & Back, D.G. (1973) Procedures for preparation, collection, evaluation and insemination of stallion semen. *Colorado State Univ. Exp. St. Gen. Ser.* **935**.

Pickett, B.W., Sullivan, J.J., Byers, W.W., Pace, M.M. & Remmenga, E.E. (1975) Effect of centrifugation and seminal plasma on motility and fertility of stallion and bull spermatozoa. *Fert. Steril.* **26**, 167–174.

Rajamannan, A.H.J., Zemjanis, R. & Ellery, J. (1968) Freezing and fertility studies with stallion semen. *Proc. 6th Int. Congr. Anim. Reprod. & A.I., Paris* **2**, 1601–1604.

Rozin, S. (1960) Studies on seminal plasma: I. The role of seminal plasma in motility of spermatozoa. *Fert. Steril.* **11**, 278–285.

Smith, A.U. & Polge, C. (1950) Survival of spermatozoa at low temperatures. *Nature, Lond.* **166**, 668–669.

Snedecor, G.W. & Cochran, W.G. (1978) *Statistical Methods,* 6th edn. Iowa State University Press, Ames.

Steel, R.G. & Torrie, J.H. (1980) *Principles and Procedures of Statistics. A Biometrical Approach.* McGraw-Hill, New York.

Watson, P.F. & Plummer, J.M. (1985) The responses of boar sperm membranes to cold shock and cooling. In *Deep Freezing of Boar Semen*, pp. 113–127. Eds L.A. Johnson & K. Larsson. Swedish University of Agricultural Science, Uppsala.

J. Reprod. Fert., Suppl. **35** (1987), 121–125

Practicalities of insemination of mares with deep-frozen semen

Z. Müller

Research Station for Horse Breeding Slatinany, Centre of Horse Insemination, Pardubice-Mnetice, 533 01, Czechoslovakia

Summary. From 341 stallions examined for sperm quality, 61% of warm-blooded stallions and 47% of cold-blooded stallions fulfilled the pre-existing criteria for their occasional use in insemination. From these stallions 51–71% of acceptable ejaculates were obtained. Altogether 959 mares were inseminated in an average of 1·36 oestrous cycles. For the insemination of one mare in one oestrous cycle on the average 2·2 insemination doses were used. These inseminations were carried out by 41 cattle insemination technicians trained in mare insemination. A pregnancy rate of 56% and a foaling rate of 48% were achieved. Differences were found between the fertility of individual stallions (29–62%), and the pregnancy rate of mares after insemination carried out by individual technicians (0–79%).

Introduction

While insemination with frozen semen is, in some species of agricultural animals, the chief method of reproduction and breeding, it is rarely practised in horse breeding. Apart from restrictions for registration legalities, physiological peculiarities of reproduction in stallions and mares have been a hindrance. However, interest in the insemination of horses with frozen semen has increased. New methods of working and freezing stallion spermatozoa with fairly good results have been developed (Cochran *et al.*, 1983; Cristanelli *et al.*, 1984). So far the practical methods of insemination, i.e. assessment of the usefulness of individual stallions for freezing spermatozoa and the proper accomplishment of insemination of mares, are not fully documented.

In Czechoslovakia, the solution of this problem has started mainly from experiences of Soviet and Polish authors (Naumenkov & Romanikova, 1970; Tischner *et al.*, 1974). Their methods have been adapted to our conditions (Müller, 1982). We have been trying to work out a selection system for stallions for insemination, the freezing procedures for the semen of individual stallions, and methods for insemination of mares under practical conditions existing in agricultural farms.

Materials and Methods

Stallion selection. The stallions enrolled in the insemination programme were those for which the spermatozoa showed minimum motility of 60%, a maximum of morphologically abnormal spermatozoa of 30%, and from this number at the most 10% of morphological changes of a primary character. Another condition for the use of a given stallion has been a good freezability of his spermatozoa.

The quality of the ejaculate was verified between 1981 and 1985 in 341 stallions, the freezability sperm examination was carried out in 105 stallions always a short time before the planned use for insemination. The stallions (3–19 years of age) were mostly Czech warm-blooded animals, with some being Czech cold-blooded and very few Thoroughbreds.

Semen collection. Semen was collected at the Stallion Central Insemination Station at Mnetice throughout a whole year. During the stay of stallions at this station (4–18 months), they were used only for the collection of ejaculates for freezing. Then they were again used for natural mating.

Only the sperm fraction of the ejaculate (the first 2–4 ejaculate fractions) was collected by means of artificial vagina, as described by Tischner *et al.* (1974). Ejaculates < 10 ml in volume containing < 70% of morphologically normal spermatozoa, and having motility < 60% and a density < 200 × 10⁶ per ml, were discarded.

Deep freezing and thawing of semen. The semen was treated after the method of Müller (1982). The diluent described by Naumenkov & Romanikova (1970) was used (redistilled water, 100 ml; lactose, 11 g; EDTA, 100 mg; sodium citrate, 89 mg; sodium bicarbonate, 8 mg; egg yolk, 1·6 g; glycerol, 3·5 ml). Initially, the semen was diluted at a ratio 1:3–4, but since 1983 it has been diluted according to volume, density of spermatozoa and expected motility after thawing, so that each insemination dose should contain 300 × 10⁶ progressively motile spermatozoa after thawing. Spermatozoa were frozen in hermetically closed aluminium tubes in special equipment. At 7–10 days after freezing the control sample from each frozen ejaculate was thawed and if the following conditions were not met, i.e. a motility minimum of 30%, survivability at 2–4°C for more than 120 h, and if one dose contained < 200 × 10⁶ progressively motile spermatozoa, then all insemination doses from the ejaculate were discarded. Ejaculates in which pathogenic microorganisms were found were also discarded.

Thawing was carried out in a 45–50°C water-bath for about 20 sec, but at the maximum temperature of the insemination apparatus, i.e. at the environmental temperature.

Formation of stores of frozen semen. Insemination doses corresponding to the above-mentioned criteria were handed over to the Central Bank for frozen stallion semen, where they were stored. From each stallion a secure store of insemination doses was formed; they were or will be used only after the death of the stallion. A working store was also established for the insemination of mares over a period of several years.

The stores of frozen semen were formed first of all from the following groups of stallions: stallions selected according to the quality of their offspring or their own efficiency; imported stallions; young prospective stallions.

Insemination of mares. In the given period, 959 mares of various ages (3–21 years) were inseminated. The mares were mainly of the Czech warm-blooded breed, with a few of the Czech cold-blooded type. The mares were inseminated in the surroundings of 34 insemination centres located over the whole of Bohemia. Most mares were inseminated in the period from February until June, a smaller number of mares in the period from February until October.

The insemination of mares was carried out by insemination technicians of cattle, who were theoretically trained in the insemination of mares. In the first year of their practical activity they were supervised by a more experienced worker. The number of insemination technicians increased from 7 in 1981 to 41 in 1985.

Insemination was carried out *per vaginam*. The semen was placed into the last third of the cervix or into the uterine body. The determination of the insemination time was based, first of all, on the rectal control of follicular development. Judgement of the viscosity of the cervical mucus was also used and sometimes the duration of oestrus could be taken into consideration. The examination of mares for pregnancy was carried out about 2 months after the insemination by means of uterine palpation.

Results and Discussion

Choice of stallions for insemination

Altogether, 341 stallions, > 75% of all stallions mating in the Bohemian horse-breeding area, have been examined. The fixed criteria for the enrolment of each stallion for the insemination programme were fulfilled by 35% of the stallions; a further 26% of stallions were on the borderline of these criteria (Table 1). The worst quality of semen was found with stallions of the cold-blooded type (Group 2), for which only a maximum of 47% of the stallions could be selected for possible use in the insemination programme. With the stallions for which quality of the ejaculate was not adequate for the required demands (Group C), a low motility of spermatozoa was found in 9% of cases, a high rate of abnormal spermatozoa in 24% of cases and a low motility and a high rate of abnormal spermatozoa as well in 67% of cases.

Because of low freezing capability, 14% of the stallions were excluded, although the main cause of low freezability of the ejaculate of stallions in Groups B and C was the poor quality of the ejaculate. The freezibility of the ejaculate of stallions whose ejaculate was of a good quality (Group A) was very good. From this group only 5% of stallions have been excluded (Table 2).

Freezing of spermatozoa

A great difference was found in the motility of spermatozoa between individual groups of stallions (Table 2). The average value of 31% has, however, been greatly influenced by the number

Table 1. Examination of stallion sperm quality in relation to the given criteria for use in the insemination programme

Stallion group * †		No. of stallions (%)	Ejaculate volume (ml)	No. of spermatozoa ($\times 10^{-6}$/ml)	Sperm motility (%)	Morphologically abnormal spermatozoa (%)	Protoplasmic droplets (%)
A	1	7 (25)	26·6	436·2	66·6	26·8	5·3
	2	14 (19)	30·3	475·3	62·5	23·5	4·5
	3	99 (41)	31·6	507·1	66·3	22·8	6·2
	Mean	120 (35)	31·2	501·0	66·1	23·2	6·1
B	1	9 (32)	37·4	508·0	56·0	41·0	5·6
	2	20 (28)	26·8	384·6	57·3	33·2	6·1
	3	57 (24)	31·8	485·8	57·1	30·2	7·4
	Mean	86 (25)	31·2	465·5	57·0	32·1	6·9
C	1	12 (43)	30·3	372·0	42·3	58·2	9·8
	2	38 (53)	32·0	418·8	38·0	42·0	8·0
	3	85 (35)	32·3	421·6	38·4	52·1	10·1
	Mean	135 (40)	32·0	416·3	38·7	49·9	9·5
Mean 1		28 (100)	31·6	431·6	52·8	45·4	7·4
Mean 2		72 (100)	30·8	416·2	46·8	38·6	7·3
Mean 3		241 (100)	31·9	472·2	54·3	35·4	7·9
Mean A, B, C		341	31·6	457·7	52·7	36·8	7·7

*A, stallions meeting the given criteria, good ejaculate quality; B, stallions on the verge of the given criteria, average ejaculate quality; C, stallions not meeting the given criteria, poor ejaculate quality.
†1, Thoroughbred stallions; 2, cold-blooded stallions; 3 warm-blooded stallions.

Table 2. Characteristics of stallion semen used for freezing

Stallion group	No. of stallions examined	Stallions with semen used for freezing (%)	Ejaculates used for freezing (%)	Sperm motility (%)	Survival of spermatozoa (h)	Ejaculates used for storage* (%)
A	63	95	81	33·4	210	88
B	34	82	69	27·0	168	74
C	8	25	40	18·3	121	68
Mean	105	86	76	31·1	197	83

*Ejaculates used for storage/ejaculates used for freezing.

of stallions examined in each group and is not necessarily a characteristic of the whole population of stallions.

After semen collection, 14% of ejaculates were excluded, and after freezing and thawing of the control sample a further 17% of frozen ejaculates were excluded. There were 71% acceptable ejaculates for the stallions of Group A, 51% for Group B and 27% for Group C. The discarding of individual ejaculates before freezing was due to low motility of spermatozoa (Group A, 9%), or to low motility and higher number of abnormal spermatozoa as well (Group B, 31%; Group C, 60%) and to a low concentration of spermatozoa in the ejaculate (Group A, 10%; Group B, 7%). The discarding of ejaculates after thawing was mainly caused by the low motility, as can be seen in Table 2.

Comparison with the results of other authors is difficult. Loomis *et al.* (1983) found a sperm motility of >30% in 31% of frozen ejaculates from 3 stallions used for the insemination of mares,

Cochran *et al.* (1983) found 62% of acceptable ejaculates and Cristanelli *et al.* (1984) reported 51% of acceptable ejaculates from 3 stallions with an average sperm motility of 38%. Tischner (1979) examined the frozen ejaculates of 36 stallions and reported a motility of >40% with 20% of stallions, a motility of 20–40% with 60% of stallions and <20% motility with 20% of stallions. Considering the number of stallions examined, their selection, various evaluaton criteria and other influences, it can be said that the results of the present work are not contradictory.

Fertility of frozen semen

In the given period 959 mares were inseminated. In the past 2 years the number of inseminated mares has reached more than 5% of the total number of mares used for breeding in Bohemia. A pregnancy rate of 56% and a foaling rate of 48% have been reached (Table 3). The pregnancy rates in the first, second and third oestrous cycles are similar to those of 56·3, 56 and 29% given by Nishikawa & Shinomyia (1976), Cristanelli *et al.* (1984) and Loomis *et al.* (1983). The comparison of foaling rate is even more difficult because of a lack of references in the literature.

Differences in the pregnancy rate were found amongst the stallions, with the 46 in Group A giving 43% (37–61%), 18 in Group B giving 36% (29–50%) and the 2 in Group C giving a 13% rate. These results more or less correspond to the data of the above-mentioned authors. The greatest differences in the pregnancy rate of mares were found amongst individual insemination technicians (Table 4).

On average each mare required 1·36 oestrous cycles for insemination. From the number of mares inseminated in the first cycle that remained non-pregnant only 47% were inseminated in the next cycle, and from the number of non-pregnant mares inseminated in the second cycle only 49% were further inseminated. For the insemination of one mare in one oestrous cycle the average insemination dose was 2·2.

The results obtained in this study have been gained in practical conditions, which has both its advantages and disadvantages. Some data, mainly those concerning the results dealing with the fertility of mares after insemination have been influenced by various factors, and so they have not been obtained in standard conditions. However, there are two indicators of how the pregnancy rate of mares can be improved after insemination with frozen semen. These are the differences among pregnancy rates of mares achieved by individual inseminators and after insemination with the semen of individual stallions.

Table 3. Pregnancy rate and foaling rate (% in parentheses) after artificial insemination of mares in 1981–1985

	1981	1982	1983	1984	1985	Total
Mares pregnant						
1st oestrus	29/56 (52)	34/104 (33)	78/159 (49)	131/309 (42)	142/331 (43)	414/959 (43)
2nd oestrus	6/13 (46)	12/40 (30)	19/37 (51)	33/92 (36)	31/82 (38)	101/264 (38)
3rd and subsequent oestrous periods	0/1 (0)	5/12 (42)	4/14 (29)	5/22 (23)	8/31 (24)	22/80 (27)
Total	35/56 (63)	51/104 (49)	101/159 (64)	169/309 (55)	181/331 (55)	537/959 (56)
Resorptions	0/56 (0)	4/104 (4)	9/159 (6)	18/309 (6)	—	31/628 (5)
Abortions	5/56 (9)	5/104 (5)	4/159 (3)	12/309 (4)	—	26/628 (4)
Foals born	30/56 (54)	42/104 (40)	88/159 (55)	139/309 (45)	—	299/628 (48)

Table 4. The pregnancy rate of mares assessed according to skill of insemination technicians

Technician group	No. of technicians	% Pregnancy rate (range)
Good*	19	68 (50–79)
Average†	11	57 (40–64)
Poor‡	11	32 (0–44)
Mean	41	56 (0–79)

*Technicians very well skilled with additional practice.
†Technicians well skilled but with little practice.
‡Technicians poorly skilled without practice.

For the insemination of mares it is very important to determine the most suitable time for the insemination so that it precedes ovulation slightly. The prediction of ovulation is not a simple affair and requires, besides theoretical assumptions, a lot of practical experience of the worker treating the mare. A simple and exact ovulation detection test would be of great help for this purpose. Another important item in the insemination is the accomplishment of the insemination without coarse treatment applied to the mare and without other stress factors. It is also important to feign the friction movement of the stallion in order to make the uterus contract during the insemination.

For freezing of the ejaculate, the selection of stallions and the selection of ejaculates will be necessary, and simple and practical examination methods to determine characteristics correlating with the fertility of the spermatozoa are needed. It will then be necessary to work out methods of increasing the concentration of spermatozoa and, if possible, of separating motile and immotile spermatozoa, and, further, to solve the question of suitable protective media and processes for freezing and thawing stallion spermatozoa.

The advantages of the method of cryopreservation of stallion semen described in this study are: a simple method, especially for the insemination of the mare; prevention of secondary bacterial contamination of the insemination doses; and clear identifiability of insemination doses. The disadvantage is a relatively large volume of diluted semen (15 ml) which is frozen and stored. Although this method may not be a final solution of the problem, it is at present suitable for practical purposes.

I thank my co-workers and all insemination technicians for participating in this study.

References

Cochran, J.D., Amann, R.P., Squires, E.L. & Pickett, B.W. (1983) Fertility of frozen thawed stallion semen extended in lactose-EDTA-egg yolk extender and packaged in 1·0-ml straws. *Theriogenology* **20**, 735–741.

Cristanelli, M.J., Squires, E.L., Amann, R.P. & Pickett, B.W. (1984) Fertility of stallion semen processed, frozen and thawed by a new procedure. *Theriogenology* **22**, 39–45.

Loomis, P.R., Amann, R.P., Squires, E.L. & Pickett, B.W. (1983) Fertility of stallion spermatozoa frozen in an EDTA-Lactose-egg yolk extender and packaged in 0·5-ml straws. *J. Anim. Sci.* **56**, 687–693.

Müller, Z. (1982) Fertility of frozen equine semen. *J. Reprod. Fert., Suppl.* **32**, 47–51.

Naumenkov, A.L. & Romanikova, N.K. (1970) Razbavitel semini. *Konievodstvo i Konnyj sport* **5**, 33–34.

Nishikawa, Y. & Shinomyia, S. (1976) Results of conception tests of frozen horse semen during the past ten years. *Proc. 8th Int. Congr. Anim. Reprod. & A.I., Krakow* **1**, 184.

Tischner, M. (1979) Evaluation of deep-frozen semen in stallions. *J. Reprod. Fert., Suppl.* **27**, 53–59.

Tischner, M., Kosiniak, K. & Bielanski, W. (1974) Analysis of the pattern of ejaculation in stallions. *J. Reprod. Fert.* **41**, 329–335.

J. Reprod. Fert., Suppl. **35** (1987), 127–133

Influence of season and frequency of ejaculation on production of stallion semen for freezing

M. Magistrini, Ph. Chanteloube and E. Palmer

I.N.R.A., 37380 Monnaie, France

Summary. In an attempt to define optimal season and ejaculation frequency for frozen semen, semen was collected from 6 stallions (3 horses and 3 ponies) 3 times per week or every day, alternating every week, for 1 year. The semen was evaluated and frozen. All the samples were thawed at the end of the experiment. At collection, fresh semen evaluations showed that winter (as opposed to spring and summer) was associated with low sexual behaviour, small volumes of spermatozoa and gel, high sperm concentration and lower motility. The high ejaculation frequency yielded a decreased volume, concentration of spermatozoa in the ejaculate and slightly improved motility. The quality of thawed semen was analysed by video and microscope estimations for motility and by two staining methods for vitality. No variation was observed according to the ejaculation frequency; the best freezability was obtained in winter but the difference was small compared to between-stallion variability and optimization of frequency and season did not change a 'bad freezer' into a good one.

Introduction

Since studies conducted by Nishikawa (1959) on stallion semen, many authors (Pickett *et al.*, 1970, 1975a, b, 1976; Cornwell *et al.*, 1972; Gebauer *et al.*, 1974; Thompson *et al.*, 1975, 1977; Sullivan & Pickett, 1975) have observed variations of seminal characteristics throughout the year. The analysis of these variations was important to determine the reproductive function of the stallions with more efficiency. The above-mentioned authors have largely reported the influence of season, stallion and frequency of ejaculation on sperm characteristics but very few data (Nishikawa *et al.*, 1968; Nishikawa & Shinomiya, 1972; Oshida *et al.*, 1972) are available about the best period of the year for freezing semen. The aim of the present experiments was therefore to determine whether there were favourable periods for freezing semen during the breeding and non-breeding seasons. We have also tried to determine whether frequency of ejaculation influenced the quality and the freezability of stallion spermatozoa.

Materials and Methods

From February 1984 through February 1985, semen was collected from 6 stallions, 3 horses and 3 ponies, ranging in age from 10 to 17 years. The semen collection was done alternately according to a low frequency, 3 times a week (Monday, Wednesday, Friday), and a high frequency, 5 times a week (Monday–Friday inclusive). The year was partitioned into 14 periods and observations and freezing were assessed for each 4-week period on 4 ejaculates per stallion (the 2 last ones from the low frequency and the 2 last ones from the high frequency collections).

Observation at collection

The sexual behaviour of the 6 stallions was studied by measuring ejaculation time and number of mounts per ejaculate for each collection analysed. Ejaculation time was the total time for the visual contact with the mare to ejaculation in the artificial vagina.

The seminal characteristics of each ejaculate analysed, filtered semen (semen is filtered after collection, ml), presence of gel or not, concentration (optical density: 10^6 spermatozoa/ml) were measured. Total number of spermatozoa per ejaculate was calculated from filtered volume and concentration.

Finally, the semen was extended (1:3 (v/v) in INRA 82 + 2% (v/v) egg yolk) and mobility was estimated subjectively on different microscope fields; diluted semen was then frozen in 0·5 ml straws (Palmer, 1984) and stored at − 196°C in liquid nitrogen until the end of the experiment.

Observations on frozen semen

All the observations on the 14 periods × 6 stallions × 2 frequencies × 2 ejaculates × 2 straws thawed and analysed were grouped at the end of the experiment.

Motility of frozen–thawed spermatozoa was estimated by 2 methods. (1) Video estimations: the method was slightly modified from that of Katz & Overstreet (1981). Ten microscope fields were recorded on 2 drops (10 µl) of diluted thawed spermatozoa pipetted onto a slide and covered by coverslips; the videotape was then analysed and 10 spermatozoa per field, distributed uniformly on the grid screen of the monitor were noted as motile or not by moving the recorder backwards and forwards. (2) Subjective estimations: a technician estimated the percentage of motile spermatozoa on the screen of the monitor.

Vitality of frozen semen was evaluated by eosin–nigrosin staining (Ortavant *et al.*, 1952) for one slide per straw and 150 spermatozoa counted per slide. The analysis of morphology was assessed on the same slides and sperm head, midpiece and tail anomalies were estimated.

The presence of the acrosome on sperm heads was analysed by the triple stain described by Talbot & Chacon (1981) for one slide per straw and 300 spermatozoa counted per slide. This staining was completed only for a part of the ejaculates (14 periods × 6 stallions × 2 frequencies × 1 ejaculate × 1 straw).

The results were subjected to factorial analysis of variance. Effect of frequency or period were considered significant with the frequency × stallion or period × stallion interactions as error terms so that the conclusion could be extrapolated to a large population of horses. When a significant period effect was found, a seasonable pattern was tested by adjustment to a sinusoidal function with a period of 1 year by the least squares method and accepted only when the correlation between data and the sinusoidal function was significant. The results are given with the amplitude of the function, the period at which it is maximum and the correlation.

The patterns present the mean data of the 6 stallions for each period: 24 observations per period for the criteria analysed at collection and 48 observations per period for the frozen semen except for the triple stain.

Results and Discussion

All the variables analysed from fresh and frozen semen are presented in Table 1.

Patterns of sexual behaviour (Fig. 1a) measured by ejaculation time and number of mounts per ejaculate showed significant ($P < 0.01$) effect of season (maximum in period 1 for ejaculation time and period 13 for a number of mounts). The 2 criteria were lower in the breeding season and the collections for all the stallions were much more rapid and with reduced number of mounts. However, frequency did not modify the behaviour parameters.

Filtered volume (Fig. 1d) and presence or not of gel in the ejaculates (Fig. 1e) were influenced by the season ($P < 0.01$; maximum on period 5 for the 2 criteria). However, only filtered volume was modified by frequency ($P < 0.01$); the mean filtered was 31·6 ml in the slow frequency compared with 27·2 ml in the rapid one, and these observations agree with previous results (Pickett *et al.*, 1970, 1975b, 1976; Sullivan & Pickett, 1975). While Thompson *et al.* (1977) reported a significant difference of gel volume for ejaculates collected at frequencies of 1 or 24 h, in our experiment a 24- or 48-h interval did not affect the presence of gel in the ejaculates. We therefore deduce that the seminal vesicles which secrete the gelatinous material can replenish the secretions more quickly than a 24-h ejaculation frequency can empty them.

The pattern of sperm concentration (Fig. 1h) was influenced by the season ($P < 0.01$, maximum in period 12) and was inverted compared to filtered volume; a high ejaculation frequency decreased the semen concentration from 209×10^6 to 133×10^6 spermatozoa per ml ($P < 0.01$). The concentration was significantly higher during period 12 as reported by Oshida *et al.* (1972) but showed a different pattern compared to Pickett *et al.* (1976); our results for the frequency of ejaculation agree with the previous investigations of Pickett *et al.* (1975a) and Sullivan & Pickett (1975).

Total number of spermatozoa (Fig. 1i) was influenced by period ($P < 0.05$) but no clear

Table 1. Variations of variables analysed for semen according to stallion, collection frequency, period and season effects

Variable	Overall mean	Stallion effect		Frequency effect		Period effect	Season effect (sinusoidal adjustment)¶			
		P	s.d.*	P	H-L†	P	P	Amplitude (max.–min.)	Period‡	r§
Fresh semen										
Ejaculation time (min)	6·2	0·01	3·5	NS		0·01	0·01	2·7	1	0·16
Mounts/ejaculate	1·4	0·01	0·5	NS		0·01	0·01	0·4	13	0·19
Filtered volume (ml)	29·4	0·01	9·4	0·001	−4·4	0·01	0·01	13·1	5	0·27
% ejaculate with gel	17·2	0·01	9·7	NS		0·01	0·01	30·0	5	0·27
Sperm conc. (10^{-6}/ml)	171·6	0·01	42·9	0·01	−76	0·01	0·01	76·8	12	0·29
Total number (10^{-9})	4·7	0·01	2·1	0·01	−2·6	0·05	NS			
Motility (%)	64·9	0·01	4·8	0·05	2·1	0·01	0·01	5·1	8	0·20
Frozen–thawed semen										
Motility (video counts) (%)	41·9	0·001	5·9	NS		0·01	0·01	5·7	1	0·20
Motility (subjective) (%)	46·0	0·01	9·0	NS		0·01	0·05	5·0	1	0·14
Live spermatozoa (eosin–nigrosin) (%)	68·7	0·001	5·5	0·01	2	0·01	NS			
Normal live spermatozoa (eosin–nigrosin) (%)	64·5	0·001	7·5	0·01	2·1	0·05	NS			
Live spermatozoa with intact acrosomes (triple stain) (%)	37·0	0·01	4·4	0·05	3·2	0·01	0·01	4·1	8	0·18

*Standard deviation of stallion means.

†Mean of high frequency − mean of low frequency.

‡Period with highest mean.

§Correlation between adjusted function and measurements.

¶$X = \bar{x} + \dfrac{A}{2}\cos 2\pi\left(\dfrac{P-PM}{13}\right)$, where x = dependent variable; P = Period: independent variable; A = Amplitude of seasonal variation and PM = Period of maximum of adjusted function (adjusted parameters). The periodicity of 1 year was the fixed parameter.

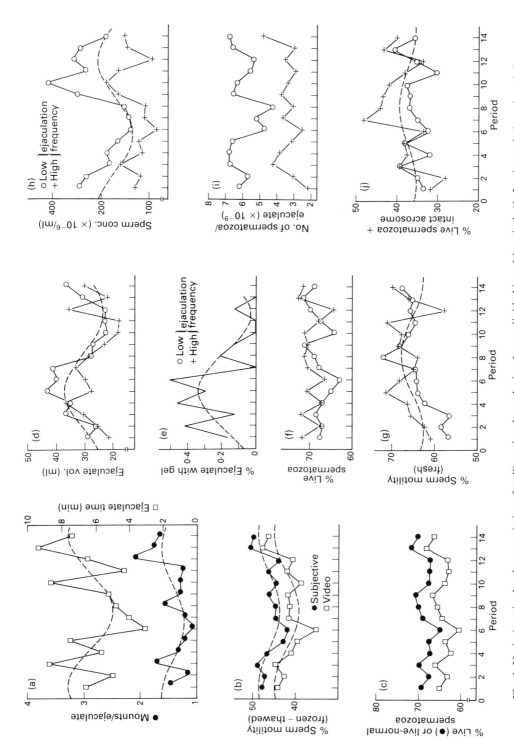

Fig. 1. Variations in the characteristics of stallion ejaculates during 1 year divided into 14 periods (1–3, winter; 3–6, spring; 6–9, summer; 9–12, autumn; 12–14, winter). The broken lines in (a), (b), (d), (e), (g), (h), and (j) represent the calculated sinusoidal curve for the data.

seasonal pattern could be identified. A high ejaculation frequency induced a 44% decrease of total spermatozoa per ejaculate ($P < 0.01$; 6.09×10^9 *vs* 3.4×10^9 spermatozoa). We have not observed a higher number during the breeding season as reported by Pickett *et al.* (1970, 1976) and Cornwell *et al.* (1972). However, Thompson *et al.* (1977) found an increase in number of spermatozoa in March followed by a decrease in April and May. In our experiment, total number of spermatozoa was higher in period 14 compared to period 1. This fact can perhaps be explained by the following: (1) after regular semen collection for 1 year, the stallions were more experienced at the end of the experiment than at the beginning; (2) sperm production varied amongst the years.

If there is modification of the number of spermatozoa per ejaculate, there must be variation in spermatogenesis. Nishikawa (1959) did not report any modification but Berndtson *et al.* (1983) reported an increase of spermatogenesis at the onset of the breeding season. The small seasonal variation of total number of spermatozoa in our experiment did not allow us to detect fluctuations of spermatogenesis during the year.

The pattern of sperm motility in diluted semen (Fig. 1g) varied significantly ($P < 0.01$) with the season (maximum in period 8) and was slightly ($P < 0.05$) improved by a high frequency of ejaculation (65·9% *vs* 63·8%). Previous studies on stallion seminal characteristics have never presented seasonal variation of sperm motility at collection, except for Cornwell *et al.* (1972). In our experiment, motility at collection was significantly correlated ($P < 0.01$ when $r > 0.148$) with other variables such as: filtered volume ($r = -0.344$), frozen semen motility estimations ($r = 0.197$, $r = 0.202$, subjective and video counts respectively) and frozen live spermatozoa ($r = 0.408$).

The estimations of motility of spermatozoa by subjective and video counts (Fig. 1b) showed seasonal variation ($P < 0.05$ and $P < 0.01$, respectively, with maximum in period 1 for both) but there was no effect of frequency; there was a high correlation between the 2 methods ($r = 0.861$). Motility of frozen–thawed spermatozoa is significantly higher (winter) when motility of fresh spermatozoa is lower, concentration of high and volume low. We can suppose an influence of the quality of the seminal plasma on the freezability of the semen. Volume is mostly influenced by the secretions of accessory glands, and so, during winter, these secretions may be better for best freezing of spermatozoa.

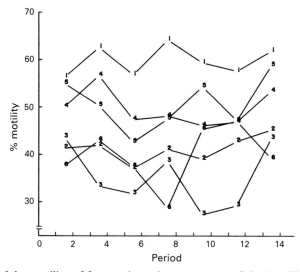

Fig. 2. Variation of the motility of frozen–thawed spermatozoa of the 6 stallions (Ponies: 1, 2 and 3; horses: 4, 5 and 6) during the 14 periods of the experiment. The pattern for each stallion is the mean value for the two ejaculation frequencies.

The vitality evaluated by eosin–nigrosin staining demonstrated a significant ($P < 0.01$) variation of live spermatozoa (Fig. 1f) with the period and the frequency ($P < 0.01$) but no clear seasonal pattern was identified. A high collection frequency improved the percentage of live spermatozoa but the difference was low (69·6% *vs* 67·6% for a low frequency). When compared, the 2 patterns of live and live normal spermatozoa showed a parallel evolution (Fig. 1c); live normal spermatozoa are also influenced by the period ($P < 0.05$) and presented a slight improvement ($P < 0.01$) with a high ejaculation frequency (65·5% *vs* 63·4% with the low frequency). This is probably because of the low variation of the total percentage of anomalies during the seasons and with the two frequencies of ejaculation. Percentages of the 2 categories of stained spermatozoa were highly correlated ($r = 0.967$) and percentage of live spermatozoa was also correlated with frozen motility estimations ($r = 0.321$ and $r = 0.402$ for subjective and video counts respectively).

As with eosin–nigrosin, the triple stain can evaluate vitality of semen. The percentage of live spermatozoa was significantly influenced by season ($P < 0.05$, maximum in period 9) and frequency did not interfere. Patterns of live spermatozoa with intact acrosomes (Fig. 1j) showed a seasonal variation ($P < 0.01$, maximum in period 8) and a high frequency improved this class ($P < 0.05$, 38·5% *vs* 35·4% with the low frequency).

All the results analysed in this experiment demonstrated variations of the seminal characteristics amongst periods, season and frequency of ejaculation, but the greater factor of variability remains the stallion. Indeed, stallion effect variation is significant for all variables analysed (Table 1). Figure 2 illustrates the variability of motility of frozen spermatozoa from the 6 stallions during the 14 periods of the experiment (seasonal effect, $P < 0.05$). All the stallions were significantly different ($P < 0.01$) from each other and collection frequency slightly influenced the motility of frozen spermatozoa. Of the 6 stallions, 3 (nos 2, 3 & 6) remained the worst whatever the period of the year or frequency of ejaculation. Optimization of season and frequency of ejaculation does not change a bad 'freezer' into a good one.

Such results could alter the programmes for managing reproductive activity of stallions. Natural service or artificial insemination with fresh semen could be used during the breeding season and collection and freezing out of season.

We thank A. Fauquenot and F. Lescot for technical assistance; and N. Chauvin and D. Tanzi for typing the manuscript.

References

Berndtson, W.E., Squires, E.L. & Thompson, D.L., Jr. (1983) Spermatogenesis, testicular composition and the concentration of testosterone in the equine testis as influenced by season. *Theriogenology* **20**, 449–456.

Cornwell, J.C., Gutrie, L.D., Spillman, T.E., McCraine, S.E., Haver, E.P. & Vincent, C.K. (1972) Seasonal variation in stallion semen. *J. Anim. Sci.* **34**, 353, Abstr.

Gebauer, M.R., Pickett, B.W. & Swierstra, E.E. (1974) Reproductive physiology of the stallion. II. Daily production and output of sperm. *J. Anim. Sci.* **39**, 732–736.

Katz, D.F. & Overstreet, M.D. (1981) Sperm motility assessment by videomicrography. *Fert. Steril.* **35**, 188–193.

Nishikawa, Y. (1959) *Studies on Reproduction in Horses.* Japan Racing Assn, Shiba Tamuracho Minatoku, Tokyo.

Nishikawa, Y. & Shinomiya, S. (1972) Freezability of horse semen collected during the non breeding season. *Proc. 7th Int. Congr. Anim. Reprod. & A.I., Munich,* **2**, 1539–1543.

Nishikawa, Y., Waide, Y. & Shinomiya, S. (1968) Studies on deep freezing of horse spermatozoa. *Proc. 6th Int. Congr. Anim. Reprod. & A.I., Paris,* **2**, 1589–1591.

Ortavant, R., Dupont, S., Pauthe, H. & Roussel, G. (1952) Contribution à l'étude de la différenciation des spermatozoïdes morts et des spermatozoïdes vivants dans le sperme de taureau. *Annls Zootech.* **1**, 5–12.

Oshida, H., Tomizuka, T., Masaki, J., Hanada, A. & Nagase, H. (1972) Some observations on freezing stallion semen. *Proc. 7th Int. Congr. Anim. Reprod. & A.I., Munich,* **2**, 1934–1937

Palmer, E. (1984) Factors affecting stallion semen survival and fertility. *Proc. 10th Int. Congr. Anim. Reprod. & A.I. Urbana-Champaign* **3**, 377, Abstr.

Pickett, B.W., Faulkner, L.C. & Sutherland, T.M. (1970) Effect of month and stallion on seminal characteristics and sexual behavior. *J. Anim. Sci.* **31**, 713–728.

Pickett, B.W., Faulkner, L.C. & Voss, J.L. (1975a) Effect of season on some characteristics of stallion semen. *J. Reprod. Fert., Suppl.* **23**, 25–28.

Pickett, B.W., Sullivan, J.J. & Seidel, G.E., Jr (1975b) Reproductive physiology of the stallion. V. Effect of frequency of ejaculation on seminal characteristics and spermatozoa output. *J. Anim. Sci.* **40**, 917–923.

Pickett, B.W., Faulkner, L.C., Seidel, G.E., Jr, Berndtson, W.E. & Voss, J.L. (1976) Reproductive physiology of the stallion. VI. Seminal and behavioral characteristics. *J. Anim. Sci.* **43**, 617–625.

Sullivan, J.J. & Pickett, B.W. (1975) Influence of ejaculation frequency of stallions on characteristics of semen and output of spermatozoa. *J. Reprod. Fert., Suppl.* **23**, 29–34.

Talbot, P. & Chacon, R.S. (1981) A triple stain technique for evaluating normal acrosome reactions of human sperm. *J. exp. Zool.* **215**, 201–208.

Thompson, D.L., Pickett, B.W. & Voss, J.L. (1975) Equine seminal patterns and collection interval. *J. Anim. Sci.* **41**, 382, Abstr.

Thompson, D.L., Jr, Pickett, B.W., Berndtson, W.E., Voss, J.L. & Nett, T.M. (1977) Reproductive physiology of the stallion. VIII. Artificial photoperiod, collection interval and seminal characteristics, sexual behaviour and concentration of LH and testosterone in serum. *J. Anim. Sci.* **44**, 656–664.

J. Reprod. Fert., Suppl. **35** (1987), 135–141

Printed in Great Britain
© 1987 Journals of Reproduction & Fertility Ltd

Use of different nonglycolysable sugars to maintain stallion sperm viability when frozen or stored at 37°C and 5°C in a bovine serum albumin medium

M. J. Arns, G. W. Webb, J. L. Kreider*, G. D. Potter and J. W. Evans

Texas Agricultural Experimental Station, Department of Animal Science, Texas A & M University, College Station, Texas 77843, U.S.A.

Summary. Bovine serum albumin (BSA) diluents containing lactose, raffinose or sucrose were not different ($P > 0.05$) in their ability to maintain stallion sperm viability, as determined by percentage motile spermatozoa (PMS) and their rate of forward movement (RFM), when stored at 37 or 5°C for 24 h. These diluents did promote a higher ($P > 0.05$) PMS and RFM, when compared with BSA diluents containing arabinose or galactose. The BSA–arabinose and BSA–galactose diluents did not differ ($P < 0.05$) in their ability to support sperm viability and were detrimental to spermatozoa. The fertility of freshly collected and diluted spermatozoa was not different ($P > 0.05$) when extended in BSA diluents containing lactose, raffinose or sucrose. There was no difference ($P > 0.05$) in PMS and RFM of frozen–thawed stallion spermatozoa when the spermatozoa were frozen, thawed and incubated at 37°C for 180 min in BSA diluents containing lactose, raffinose or sucrose. Spermatozoa from 6 of 8 stallions did not survive the freezing process. A one-cycle conception rate of 32% was obtained from frozen–thawed spermatozoa extended in BSA diluents containing lactose, raffinose or sucrose. This rate was 78% of the conception rate obtained when the same mares were inseminated with fresh semen in a subsequent study (41%).

Introduction

The addition of serum albumin as a constituent of semen diluents has been shown to prolong motility in stallion (Kreider *et al.*, 1985), bull (Brown & Senger, 1980), boar (Dixon *et al.*, 1980), ram (Harrison *et al.*, 1982), mouse (Fraser, 1985) and human (Suter *et al.*, 1979) spermatozoa.

A BSA–sucrose diluent enhanced stallion sperm motility through 8 h of incubation at 37°C (Klem *et al.*, 1983). The role of non-glycolysable sugars in promoting stallion sperm motility is unknown. Nishikawa (1959) reported that a sucrose solution maintained a higher percentage of motile spermatozoa when stallion spermatozoa were stored at 4°C for 24 h than did a lactose solution. Lapwood & Martin (1966) found that ram spermatozoa had higher motility scores when stored at 37°C in diluents containing glucose, mannose, fructose or sucrose. In comparison, ram spermatozoa had higher motility scores when stored at 5°C in diluents containing arabinose, galactose, ribose, xylose or lactose. Differences in the cryoprotective abilities of different sugars on spermatozoa have also been identified for the bull (Unal *et al.*, 1978) and ram (Jones & Martin, 1965).

This study investigated the abilities of BSA diluents containing arabinose, galactose, lactose, raffinose or sucrose to maintain or enhance stallion sperm viability when frozen or stored at 37 or 5°C.

---

*Present address: Dean Lee Research Station, Alexandria, LA 71301, U.S.A.

Materials and Methods

Experiment I

Eighteen ejaculates, 3 from each of 6 stallions, were utilized to investigate the abilities of BSA diluents containing nonglycolysable sugars to maintain sperm viability after storage for 24 h at 37°C or 5°C. Semen from 2 stallions was collected every 3 days with an artificial vagina. The gel-free ejaculates were evaluated initially for concentration, volume, percentage motile spermatozoa and rate of forward movement. After initial evaluation, each ejaculate was split into 5 equal parts and each diluted with 1 of 5 diluents (Table 1) to give a final concentration of 100×10^6 motile spermatozoa/ml. One 4–5 ml aliquot was taken from each of the 5 aliquots and stored at 37°C. A second 4–5 ml aliquot was placed in a 200 ml flask containing water at 37°C. The flask, containing 5 culture tubes, was placed in a cold room (5°C) for cooling and storage. Samples were taken from each tube at 0, 1, 2, 3, 4, 8, 16 and 24 h after dilution for motility measurements. The percentage motile and rate of forward movement were evaluated as described by Kreider (1983) in a double-blind procedure. Each of the cold samples was warmed for 2 min before evaluation by placing a drop of the sample on a microscope slide and placing it in a 37°C incubator.

The BSA diluents containing arabinose or galactose did not support stallion sperm viability and therefore these media were omitted from the following experiments.

Experiment II

Fertility of stallion spermatozoa diluted in extenders containing BSA and nonglycolysable sugars was investigated in this experiment. Mares (N = 65) were randomly assigned to 1 of 3 treatment groups. Each group was inseminated with spermatozoa diluted in BSA–lactose, BSA–raffinose or BSA–sucrose. Diluted semen (1:1 v/v) containing 500×10^6 motile spermatozoa was used for insemination once a preovulatory follicle (> 30 mm in diameter and/or a change in turgidity) was detected on the mare's ovary. Follicular changes were determined by rectal palpation and/or ultrasonography. Pregnancies were determined at 18–21 days and at 30 days after ovulation by ultrasonography.

Experiment III

Ejaculates, 3 from each of 8 stallions, were collected from the stallions as in Exp. I. After initial evaluation, semen was split into three aliquots and extended in a one-to-one ratio. The diluted semen was split into 20–30-ml volumes for centrifugation. Following careful mixing of the diluted semen in 50-ml centrifugation tubes, 0·25 ml of a 30% (w/v) BSA solution was placed beneath the diluted semen. The aliquots were then centrifuged at 400 *g* for 15 min at 22°C. After centrifugation, the supernatant was aspirated, leaving ~0·25 ml with the sperm pellet. The pellet was then resuspended to a concentration of 1×10^9 cells/ml in a glycerolated BSA–nonglycolysable sugar diluent (Table 1). Glycerol was added at 5% (w/v) such that the final concentration of glycerol in the extended semen was 3·5%. Once the pellet was resuspended, the semen was packaged and frozen in liquid nitrogen vapour as described by Cochran *et al.* (1984).

For evaluation, straws were taken from the liquid nitrogen storage tank and immediately submerged into a 37°C water bath for 30 sec. The semen was then diluted about 10-fold (to 5 ml) with the corresponding BSA diluent. The post-thaw viability of the stallion spermatozoa was evaluated for motility and rate of forward movement, as in Exp. I, at 0, 30, 60, 90, 120 and 180 min after thawing.

Table 1. Composition of BSA diluents containing arabinose (A), galactose (G), lactose (L), raffinose (R) or sucrose (S)

Ingredient	BSA–A*	BSA–G*	BSA–L*	BSA–R*	BSA–S*
Arabinose (g)	4·385	—	—	—	—
Galactose (g)	—	5·266	—	—	—
Lactose (g)	—	—	10·000	—	—
Raffinose (g)	—	—	—	17·372	—
Sucrose (g)	—	—	—	—	10·001
BSA† (g)	3·002	3·002	3·002	3·000	3·001
Gentocin‡ (ml)	1	1	1	1	1
Osmolality (mosmol)	290	290	310	337	320
pH§	7·1	7·2	7·1	7·2	7·2

*Diluted to 100 ml with deionized water.
†Sigma (A-4503) Fraction V powder.
‡Schering Corp. (gentamicin sulphate, 50 mg/ml).
§pH adjusted with 1 N-KOH.

Experiment IV

To investigate the fertility of frozen–thawed stallion spermatozoa which had been frozen in BSA diluents containing lactose, raffinose or sucrose, semen from 3 stallions was processed as in Exp. III and 22 mares were inseminated every day once a preovulatory follicle was present. Straws were thawed at 37°C for 30 sec and deposited into the mare via a bovine artificial insemination gun.

Statistical analysis

The percentage motility and rate of forward movement variables were analysed as a split plot in time by analysis of variance (ANOVA) procedures of the Statistical Analysis System (SAS, 1985). When significant time–treatment interactions were detected by the ANOVA procedure, treatments were compared within times, and times within treatments (Gill, 1978). Because of the high correlation associated with repeated sampling, conservative degrees of freedom were used. In addition the error term for the variation due to stallion was the ejaculate-within-stallion error mean square. The error term for treatments and the time–treatment interaction was the ejaculate-within-stallion-treatment interaction error mean square (Geisser & Greenhouse, 1958). The Student–Neuman–Keuls procedure (SAS, 1985) was used to evaluate differences between treatments when analysis of variance procedures detected significant treatment effects of variation. Pregnancy rates were examined by χ^2 analysis using the BSA–sucrose treatment values as the expected values (Steel & Torrie, 1960).

Results and Discussion

Experiment I

When stallion spermatozoa were incubated at 37°C for 24 h in BSA diluents containing arabinose, galactose, lactose, raffinose or sucrose, there was no difference ($P > 0.05$) in the ability of the diluents to promote stallion sperm viability (Table 2). These diluents promoted a higher ($P < 0.01$) sperm viability than did the BSA–arabinose or BSA–galactose diluents which were not different ($P > 0.05$) from one another.

The percentage motility and rate of forward movement values for stallion spermatozoa incubated in BSA–lactose, BSA–raffinose and BSA–sucrose diluents are lower than those reported by Klem *et al.* (1983) and Kreider *et al.* (1985) for stallion spermatozoa incubated at 37°C in different BSA diluents.

A decline ($P < 0.05$) in the percentage of motile stallion spermatozoa occurred at each observation through 16 h after dilution for all diluents. A time–diluent interaction existed ($P < 0.01$) for the rate of forward movement variable. The decline over time in percentage motility for stallion spermatozoa diluted in the BSA–sucrose diluent in this study appeared greater than that for stallion spermatozoa incubated in BSA diluents as reported by Klem *et al.* (1983) and Kreider *et al.* (1985).

Table 2. Mean percentage of motile stallion spermatozoa and mean rate of forward movement of spermatozoa incubated at 37°C in BSA diluents containing arabinose (A), galactose (G), lactose (L), raffinose (R) or sucrose (S)

Time (h)	Percentage motile spermatozoa					Rate of forward movement*				
	BSA–A	BSA–G	BSA–L	BSA–R	BSA–S	BSA–A	BSA–G	BSA–L	BSA–R	BSA–S
0	31[ax]	23[bx]	70[ct]	72[cv]	73[cv]	2.3[av]	2.0[bv]	3.3[ct]	3.5[cw]	3.5[cu]
1	5[ay]	5[ay]	48[bu]	51[bw]	51[bw]	1.1[aw]	1.1[aw]	2.8[bu]	2.8[bx]	3.0[bw]
2	2[ay]	3[ay]	38[bv]	37[bx]	41[bx]	0.7[ax]	0.7[ax]	2.6[bvw]	2.7[bx]	2.7[bw]
3	1[ay]	1[ay]	35[bvw]	32[bxy]	34[by]	0.5[axy]	0.5[axy]	2.7[bvw]	2.6[bx]	2.7[bw]
4	1[ay]	1[ay]	31[bvw]	28[by]	29[by]	0.3[ayz]	0.3[ayz]	2.4[bw]	2.6[bx]	2.6[bw]
8	0[ay]	0[ay]	10[bx]	8[bz]	8[bz]	0[az]	0[az]	1.7[bx]	1.6[by]	1.8[bx]
16	0[ay]	0[ay]	4[by]	4[bz]	4[bz]	0[az]	0[az]	1.0[by]	0.9[bz]	1.0[by]
24	0[ay]	0[ay]	2[by]	2[bz]	2[abz]	0[az]	0[az]	0.7[bz]	0.7[bz]	0.6[bz]

*Values are subjective and range from 0 to 4 with 4 being the most progressive.
[a,b,c]Means within rows which have different superscripts differ ($P < 0.05$).
[t,u,v,w,x,y,z]Means within columns which have different superscripts differ ($P < 0.05$).

When stored at 5°C, the BSA diluents containing lactose, raffinose or sucrose promoted higher ($P < 0.05$) sperm viability at each observation throughout the 24-h storage period as compared with BSA media containing arabinose or galactose. There was no difference ($P > 0.05$) in the ability of the BSA diluents with lactose, raffinose or sucrose, or with galactose or arabinose, to maintain viability when stored at 5°C for 24 h (Table 3). The decline in percentage motility and rate of forward movement when stored at 5°C was more gradual than for stallion spermatozoa incubated at 37°C. A decline ($P < 0.05$) in percentage motility occurred after 1 h of storage at 5°C for spermatozoa diluted in the BSA–sucrose and BSA–lactose and after 2 h of storage in the BSA–raffinose diluent. Motility for spermatozoa diluted in BSA media containing lactose, raffinose or sucrose declined ($P < 0.05$) during the 2nd and 3rd hour of storage and then was maintained ($P < 0.05$) through 8 h of storage. Motility continued to decline ($P < 0.05$) between 8 and 16 h of storage and then was maintained through 24 h (Table 3).

Rate of forward movement of stallion spermatozoa diluted in BSA–lactose, BSA–raffinose and BSA–sucrose was maintained ($P < 0.05$) through 4 h of storage and then declined ($P < 0.05$) through 16 h (Table 3). Stallion and time–diluent interaction differences ($P < 0.05$) were detected for both measures.

These results indicate an adverse reaction of stallion spermatozoa to BSA–arabinose and BSA–galactose diluents, as they failed to support the viability of stallion spermatozoa during storage at 37 or 5°C, while the BSA–sucrose, BSA–lactose and BSA–raffinose supported viability during this time. These results contrast earlier work with stallion spermatozoa in which a 10% sucrose solution promoted a higher percentage motility than did a 13% lactose solution (47% *vs* 23%) for stallion spermatozoa stored at 4°C for 24 h (Nishikawa, 1959).

A difference between sugars has been detected in their ability to maintain ram sperm viability over time (Lapwood & Martin, 1966), and the temperature at which the spermatozoa were stored (37 or 5°C) had an effect on which sugar solution provided the optimal medium. In addition, no sugar solution appeared to be detrimental to the ram spermatozoa. A toxic effect of some sugars on sperm motility, as was seen for the stallion spermatozoa diluted with the BSA media containing arabinose or galactose, has also been reported for bee spermatozoa (Poole & Edwards, 1972) exposed to diluents containing different sugars.

Experiment II

No difference ($P > 0.05$) was detected between mean pregnancy rates during first, second and over-all oestrous cycles for lactating or non-lactating mares inseminated with stallion spermatozoa diluted in BSA–lactose, BSA–raffinose or BSA–sucrose (Table 4).

Experiment III

There was no difference ($P > 0.05$) in the post-thaw viability of stallion spermatozoa diluted in BSA–lactose, BSA–raffinose or BSA–sucrose diluents. This finding is similar to those of Nagase *et al.* (1968) and Unal *et al.* (1978) who observed no difference between solutions of raffinose or lactose in their ability to enhance recovery of bull spermatozoa after freezing. This does not support the earlier work of Nagase *et al.* (1964) on bull spermatozoa or that of Sanatarius & Giersch (1983) on thylakoid membranes (Sanatarius & Giersch, 1983) in which freezing was enhanced when sugars of high molecular weight were included in the freezing medium. Both studies found that the cryoprotective ability decreased as the molecular weight of the sugar decreased. Sanatarius & Giersch (1983) found that, even though the main action of the sugars is in their colligative action, not all of the protective action of the sugars could be accounted for by this mechanism. They concluded that a membrane–sugar interaction could exist. This may account for the fact that there was no difference between the disaccharides (lactose and sucrose) and the trisaccharide (raffinose) in their ability to enhance recovery of frozen–thawed stallion spermatozoa

Table 3. Mean percentage of motile stallion spermatozoa and rate of forward movement of spermatozoa incubated at 5°C in BSA diluents containing arabinose (A), galactose (G), lactose (L), raffinose (R) or sucrose (S)

Time (h)	Percentage motile spermatozoa					Rate of forward movement*				
	BSA–A	BSA–G	BSA–L	BSA–R	BSA–S	BSA–A	BSA–G	BSA–L	BSA–R	BSA–S
0	14[ax]	9[ax]	51[bv]	51[bu]	58[bw]	1.7[ax]	1.3[aw]	2.6[bx]	2.7[bx]	3.0[bx]
1	7[ay]	6[ay]	44[bw]	48[buv]	50[bx]	1.1[ay]	1.1[aw]	2.8[bx]	2.9[bx]	3.0[bx]
2	7[ay]	5[ayz]	40[bw]	44[bvw]	46[bx]	1.1[ay]	1.1[bx]	2.7[cx]	2.8[cx]	3.0[cx]
3	6[ay]	5[ayz]	33[bxy]	38[bw]	38[by]	1.1[ay]	0.7[bxy]	2.6[cx]	2.8[cx]	2.8[cx]
4	4[ayz]	4[ayz]	38[bxw]	40[bw]	40[by]	0.9[ay]	0.6[by]	2.6[cx]	2.7[cx]	2.8[cx]
8	2[az]	2[ayz]	30[by]	33[bx]	34[by]	0.5[az]	0.4[az]	2.2[by]	2.3[by]	2.4[by]
16	2[az]	2[az]	23[bz]	27[by]	26[bz]	0.4[az]	0.4[az]	1.9[bz]	2.0[bz]	2.1[bz]
24	2[az]	2[az]	20[bz]	21[bz]	23[bz]	0.4[az]	0.3[az]	1.8[bz]	1.9[bz]	2.0[bz]

*Values are subjective and range from 0 to 4 with 4 being the most progressive.
[a,b,c]Means within rows which have different superscripts differ ($P < 0.05$).
[u,v,w,x,y,z]Means within columns which have different subscripts differ ($P < 0.05$).

Table 4. Mean conception rates (%) of mares inseminated with stallion spermatozoa diluted with BSA–lactose, BSA–raffinose or BSA–sucrose diluents

	Treatment		
	BSA–lactose	BSA–raffinose	BSA–sucrose
Non-lactating mares			
First oestrus	50 (3/6)	71 (5/7)	86 (12/14)
Second oestrus	67 (2/3)	50 (1/2)	50 (1/2)
Total	83 (5/6)	86 (6/7)	93 (13/14)
Lactating mares			
First oestrus	60 (9/15)	78 (11/14)	67 (6/9)
Second oestrus	83 (5/6)	67 (2/3)	67 (2/3)
Total	93 (14/15)	93 (13/14)	89 (8/9)
All mares combined			
First oestrus	57 (12/21)	76 (16/21)	78 (18/23)
Second oestrus	78 (7/9)	60 (3/5)	60 (3/5)
Total	90 (19/21)	90 (19/21)	91 (21/23)

No difference ($P > 0.05$) was detected between treatments for mean pregnancy rates.

in this study. It is also possible that, if a higher recovery of motile spermatozoa had been obtained, a difference would have been detected.

A decline in percentage motility and rate of forward movement occurred after the initial recovery and was significantly ($P < 0.05$) decreased at each observation from 60 min after thawing until 180 min after thawing (Table 5). The post-thaw motilities reported in this study are lower than those reported by Cochran *et al.* (1984) and Cristanelli *et al.* (1984). Moreover, the motility was maintained through 90 min of post-thaw incubation in those studies which is different from the decline seen in the present study (Table 5).

Spermatozoa from 6 of 8 stallions did not survive freezing and ejaculates from these stallions were not utilized. A difference ($P < 0.05$) was detected in the percentage of motile spermatozoa from the 2 stallions which provided semen which did survive the freezing process.

The recovery of spermatozoa during the centrifugation process was 90% (80–100%). This is similar to the recovery rate reported by Cochran *et al.* (1984).

Table 5. Mean motility (%) and rate of forward movement of frozen–thawed stallion spermatozoa incubated at 37°C in BSA diluents containing lactose (L), raffinose (R) or sucrose (S)

Time (min)	Percentage motility			Rate of forward movement*		
	BSA–L	BSA–R	BSA–S	BSA–L	BSA–R	BSA–S
0	18^{vw}	18^{vw}	18^v	1·9^{xy}	2·2^x	1·8^{xy}
30	21^w	20^w	17^v	2·0^x	2·0^{xy}	2·0^x
60	16^v	18^v	15^w	1·8^{xy}	2·0^{xy}	1·8^{xy}
90	15^x	15^x	13^x	1·7^y	1·9^y	1·5^y
120	10^y	10^y	9^y	1·2^z	1·3^z	1·1^z
180	5^z	6^z	5^z	0·9^z	1·1^z	1·0^z

*Values are subjective and range from 0 to 4 with 4 being the most progressive.

No difference ($P < 0.05$) was detected between diluents in their ability to maintain post-thaw motility or rate of forward movement.

^{v,w,x,y,z}Means within columns which have different superscripts differ ($P < 0.05$).

The use of a 'cushion' during the centrifugation process (Cochran *et al.*, 1984) appeared to reduce the time required for resuspension of the sperm pellet during this study.

Experiment IV

Conceptions rates obtained with thawed spermatozoa which had been frozen in BSA diluents containing lactose, raffinose or sucrose were 28% (2/7), 28% (2/7) and 38% (3/8), respectively. Statistical analysis was not performed between treatments due to the low conception rates within each group. The combined conception rate resulting from frozen stallion semen was 32%; this was 78% of the conception rate obtained when these same mares were inseminated with fresh semen during a subsequent study. There was an average of 958×10^6 ($706–1203 \times 10^6$) spermatozoa/ml and the average number of motile spermatozoa per straw was 82×10^6, based on an average post-thaw motility of 18% (10–30%). Two straws were used for each insemination (164×10^6 motile spermatozoa per insemination).

Several reasons may exist for the poor conception rates in this study. First, the mares were of unknown fertility, and secondly the recovery of motile spermatozoa is lower in this study than in that of Cristanelli *et al.* (1984), in which, moreover, the values were not different from those obtained with fresh semen. However, Cristanelli *et al.* (1984) culled ejaculates with post-thaw motilities of 35% or less. In the present study, all ejaculates were used regardless of the post-thaw quality. Even though conception rates were low in this study, the fact that pregnancies did result is encouraging and suggests that poor quality ejaculates may be used for insemination provided that enough straws are used per insemination to provide an adequate number of viable spermatozoa.

We thank John Labore and Dr Charles Gates for assistance in the statistical analysis of these results. Technical article No. 22003 of Texas Agricultural Experiment Station.

References

Brown, D.V. & Senger, P.L. (1980) Influence of homologous blood serum on motility and head-to-head agglutination in nonmotile ejaculated bovine spermatozoa. *Biol. Reprod.* **23**, 271–275.

Cochran, J.D., Amann, R.P., Froman, D.P. & Pickett, B.W. (1984) Effects of centrifugation, glycerol level, cooling to 5°C, freezing rate and thawing rate on post-thaw motility of equine sperm. *Theriogenology* **22**, 25–38.

Cristanelli, M.J., Squires, E.L., Amann, R.P. & Pickett, B.W. (1984) Fertility of stallion semen processed, frozen and thawed by a new procedure. *Theriogenology* **22**, 39–45.

Dixon, K.E., Songy, E.A., Thrasher, D.M. & Kreider, J.L. (1980) Effect of bovine serum albumin on the evaluation of boar spermatozoa and their fertility. *Theriogenology* **13**, 437–444.

Fraser, L.R. (1985) Albumin is required to support the acrosome reaction but not capacitation in mouse spermatozoa *in vitro. J. Reprod. Fert.* **74**, 185–196.

Geisser, S. & Greenhouse, S.W. (1958) An extension of Box's results on the use of the F distribution in multivariate analysis. *Annl Math. Stat.* **29**, 885–891.

Gill J.L. (1978) *Design and Analysis of Experiments in the Animal and Medical Sciences*, vol. 2, pp. 209–210. Iowa State University Press, Ames.

Harrison, R.A.P., Dott, H.M. & Foster G.C. (1982) Bovine serum albumin, sperm motility, and the 'dilution effect'. *J. exp. Zool.* **222**, 81–88.

Jones, R.C. & Martin, I.A.C. (1965) Deep freezing ram spermatozoa: the effects of milk, yolk citrate and synthetic diluents containing sugar. *J. Reprod. Fert.* **10**, 413–423.

Klem, M.E., Gorman-McAdams, J., Potter, G.D. & Kreider, J.L. (1983) Motility of equine spermatozoa incubated in a sucrose BSA medium. *Proc. 8th Equine Nutr. and Physiol. Symp.*, Lexington, pp. 72–74.

Kreider, J.L. (1983) Stallion management and artificial insemination, Anim. Sci. Dept. Horse Breeders School Manual, pp. 46–57. Texas A & M University.

Kreider, J.L., Tindall, W.C. & Potter, G.D. (1985) Inclusion of bovine serum albumin in semen extenders to enhance maintenance of stallion sperm viability. *Theriogenology* **23**, 399–408.

Lapwood, K.R. & Martin, I.A.C. (1966) The use of monosaccharides, disaccharides, and trisaccharides in synthetic diluents for storage of ram spermatozoa at 37°C and 5°C. *Aust. J. biol. Sci.* **19**, 655–671.

Nagase, H., Niwa, T., Yamishita, S. & Irie, S. (1964) Deep freezing bull semen in concentrated pellet form. II. Protective action of sugars. *Proc. 5th Int. Congr. Anim. Reprod. & A.I., Trento* **4**, 498–502.

Nagase, H., Yamishita, S. & Irie, S. (1968) Protective action of sugars against freezing injury of bull spermatozoa. *Proc. 6th Int. Congr. Anim. Reprod. & A.I., Paris* **2**, 1111–1113.

Nishikawa, Y. (1959) *Studies on Reproduction in Horses.* Japan Racing Ass., Tokyo.

Poole, H.K. & Edwards, J.F. (1972) Effect of toxic and non-toxic sugars on motility of honey bees (*Apis mellifera* L.). *Experientia* **28**, 235.

Sanatarius, K.A. & Giersch, C. (1983) Cryopreservation of spinach chloroplast membranes by low-molecular-weight carbohydrates. II. Discrimination between colligative and noncolligative protection. *Cryobiology* **20**, 90–99.

Statistical Analysis System (SAS) (1985) Users Guide. SAS Institute Inc., Cary.

Steel, R.G.D. & Torrie, J.H. (1960) *Principles and Procedures of Statistics*, pp. 352–365. McGraw Book Co., New York.

Suter, D., Chow, P.Y.W. & Martin, I.A.C. (1979) Maintenance of motility in human spermatozoa by energy derived through oxidation phosphorylation and addition of albumin. *Biol. Reprod.* **20**, 505–510.

Unal, M.B., Berndtson, W.E. & Pickett, B.W. (1978) Influence of sugars with glycerol on post-thaw motility of bovine spermatozoa frozen in straws. *J. Dairy Sci.* **61**, 83–89.

J. Reprod. Fert., Suppl. **35** (1987), 143–148

Fertility of stallion semen frozen in 0·5-ml straws

D. H. Volkmann and Deborah van Zyl

Department of Genesiology, Faculty of Veterinary Science, University of Pretoria, Private Bag X04, Onderstepoort, 0110 Republic of South Africa

Summary. Semen of 2 pony stallions was frozen by 2 methods in 0·5 ml PVC straws. The fertility of the frozen-thawed semen was evaluated by inseminating 60 mares during 69 oestrous cycles. An overall single cycle pregnancy rate of 55% was achieved. Freezing method, stallion, insemination during steroid-synchronized oestrus or insemination only every 2nd day during oestrus did not significantly influence pregnancy rates. Pregnancy rates were significantly improved from a mean 44% to a mean 73% when the mean number of progressively motile spermatozoa per insemination was increased from 175×10^6 to 249×10^6. It is concluded that the simpler freezing technique will yield satisfactory pregnancy rates when semen with a post-thaw progressive motility of 30% is used for AI. Starting when ovulation is anticipated to occur within the next 48 h mares should be inseminated every 2nd day with at least 220×10^6 progressively motile spermatozoa per insemination until ovulation is confirmed.

Introduction

The potential advantages of using frozen stallion semen for the artificial insemination (AI) of mares have been discussed by Amann & Pickett (1984). Colorado workers were the first to successfully freeze stallion semen in 0·5 ml straws (Cochran *et al.*, 1984) and use it for AI in mares (Cristanelli *et al.*, 1984). The freezing method used by Cochran *et al.* (1984) involved the layering of 0·25 ml of a hyperosmotic medium below extended semen before centrifugation. Studies carried out in this laboratory (unpublished) have shown that this technically somewhat complicated procedure could be omitted without harmful effects on progressive motility or acrosomal integrity of frozen-thawed stallion semen. It was then decided to determine the fertility of frozen–thawed semen that was processed either with or without the hyperosmotic cushion during centrifugation. Cristanelli *et al.* (1984) only used frozen–thawed semen with a progressive motility of at least 40% for AI. Discarding all semen that does not show a progressive motility of 40% results in high semen losses. We therefore investigated the possibility of using frozen–thawed semen with a progressive motility of at least 30% for AI. Simultaneously we also evaluated the effects of insemination times, frequency of inseminations, sperm numbers, acrosomal integrity and synchronization of oestrus on pregnancy rates of mares inseminated with semen frozen by each method.

Materials and Methods

Semen of two Nooitgedacht pony stallions (1 and 2) of known fertility was collected and frozen by two methods (A and B) during the breeding season. Method A was based on the method described by Cochran *et al.* (1984). It involved the extension of fresh semen in an isosmotic centrifugation medium; layering 0·25 ml of a hyperosmotic EDTA/glucose solution below the extended semen in 50-ml centrifugation tubes; centrifugation at 400 *g* for 15 min; aspiration of the supernatant; resuspension of the sperm pellet in lactose–EDTA–egg yolk freezing medium to a concentration of 750×10^6 spermatozoa/ml; packaging of the semen in 0·5 ml PVC straws and freezing them at $-60°C$/min in liquid nitrogen vapour. From the first extension until freezing the semen was maintained at 20°C. Method B was the same as Method A, but the hyperosmotic EDTA/glucose cushion was omitted during centrifugation.

Semen was thawed at 75°C for 6 sec and then rapidly transferred to a waterbath at 37°C (Cochran *et al.*, 1984). For evaluation semen was thawed 2–7 days after freezing. The progressive motility of the frozen–thawed semen was determined by phase-contrast microscopy after extending 0·25 ml thawed semen in 2·5 ml 5% (w/v) glucose and 5% skim milk powder (w/v) diluent (Pickett *et al.*, 1981) at 37°C and incubating it for 30–40 min. The sperm count per straw was also recorded for each batch of frozen semen. Acrosomal integrity was assessed using the Spermac stain (Stain Enterprises, P.O. Box 12421, Onderstepoort, 0110 Republic of South Africa) (Oettlé, 1986).

For insemination, mares were prepared using accepted procedures to ensure maximum hygiene precautions. Semen was stored in liquid nitrogen for 2–58 weeks before it was used for AI. Straws were thawed at 75°C for 6 sec and then rapidly transferred to a waterbath at 37°C. They were dried and loaded into bovine AI pistolettes which had been extended by 12 cm. Apart from the normal sterile pistolette sheath a second flexible sterile plastic sleeve (as used for bovine embryo transfer) was pulled over the assembled pistolette. First an arm length glove was put on, and then a sterile surgical glove was pulled on over the long glove. The tip of the first pistolette was buried in the gloved fist. The fist was lubricated sparingly with sterile lubricant and then advanced to the external cervical os. The index finger was then extended to locate the cervical os. By pulling on the caudal end of the outer plastic sleeve and simultaneously advancing the pistolette the outer plastic sleeve was penetrated and the tip of the pistolette advanced through the cervix to about 2 cm into the uterine body where the semen was expelled. The pistolette was removed, the fist closed and pulled back to the level of the vulva. Here it was opened slightly to spread the vulvar lips and receive the second pistolette. The second insemination was then carried out in the same manner as the first.

Experiment 1. Nooitgedacht pony mares (300–500 kg) were used in the 1984–85 breeding season. Breeding records and a thorough clinical examination were used to exclude any mares of doubtful reproductive status from the trial. The mares were randomly assigned to one of two groups: those in Group A (N = 11) were inseminated with semen frozen by Method A and those in Group B (N = 11) were inseminated with semen frozen by Method B. Semen from Stallion 1 only was used.

Mares were teased daily. As soon as a mare showed oestrus (excluding foal heat), she was rectally examined every morning to monitor follicular maturation. Based on the size and degree of fluctuation of the follicle, the intensity of oestrous symptoms and the lengths of previously recorded oestrous periods, the anticipated day of ovulation was predicted. Every mare was inseminated once or twice only. If she ovulated within 24 h of the first insemination she was not inseminated again. If she had not ovulated within 24 h a second insemination was performed 24 h after the first. No further inseminations were performed even if the mare had not ovulated within 24 h after the second insemination. Pregnancy was diagnosed at 20 and 30 days after ovulation by rectal palpation. Mares that were not pregnant after the first cycle were inseminated again during a second cycle. Thus, 38 cycles were utilized. These included 5 cycles during which a single insemination was carried out during ovulation or within 10 h after ovulation and 2 cycles during which insemination was carried out more than 48 h before ovulation (Fig. 1).

Experiment 2. The mares inseminated in Exp. 1 required daily attendance by a veterinarian for 4 months. This intensive veterinary attention would largely prohibit the use of frozen semen for AI under farm conditions. This prompted the synchronization of oestrus in three groups of mares (A, B and C) during the 1985/86 breeding season. The selection of mares, insemination technique and pregnancy diagnosis were performed as in Exp. 1. For synchronization each mare was injected with a daily dose of 150 mg progesterone and 10 mg oestradiol-17β in oil intramuscularly for 10 days (Loy *et al.*, 1981; Bristol *et al.*, 1983). As recommended by Loy *et al.* (1981) a single dose of 5 mg prostaglandin F-2α was given on Day 10.

Group A mares (N = 15) were synchronized early, Group B mares (N = 18) in the middle of and Group C mares (N = 15) late in the breeding season. As the mares showed oestrus they were examined as in Exp. 1. When ovulation was expected to occur within the next 48 h, each mare was inseminated every 2nd day until ovulation was confirmed. In this manner 13 mares in Groups A and B and 12 mares in Group C were inseminated: 25 mares received semen from Stallion 2 frozen by Method A (13 mares) or Method B (12 mares) and the other 13 mares were inseminated with semen from Stallion 1 frozen by Method A (7 mares) or Method B (6 mares).

Z-tests on normal approximations of binomial data were used to test the significance of the effects of freezing method (A or B), stallion (1 or 2), acrosomal damage (≤ 50% or > 50%), number of progressively motile spermatozoa per insemination ($\leq 210 \times 10^6$ of 220×10^6) and timing of insemination (1–23 h or 24–47 h before ovulation) on single cycle pregnancy rates in the mares. The same test was used to establish whether the pregnancy rates in the three groups of synchronized mares were significantly different and to compare the pregnancy rate after AI during synchronized oestrus to the pregnancy rate after AI during natural oestrus.

Results

The progressive motility and acrosomal damage of semen used in these trials are given in Table 1. Figure 1 illustrates the time of insemination relative to the time of ovulation and gives the number of pregnancies per cycle of mares in Exp. 1. Table 2 shows the first and second cycle pregnancy rates of mares in Exp. 1 inseminated with semen frozen by Methods A and B respectively. There were no

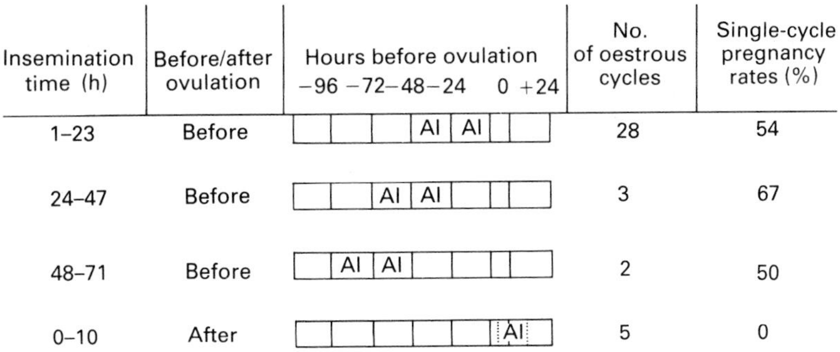

Insemination time (h)	Before/after ovulation	Hours before ovulation −96 −72 −48 −24 0 +24	No. of oestrous cycles	Single-cycle pregnancy rates (%)
1–23	Before	AI AI	28	54
24–47	Before	AI AI	3	67
48–71	Before	AI AI	2	50
0–10	After	AI	5	0

Fig. 1. Insemination times relative to ovulation and single-cycle pregnancy rates in 22 mares that were inseminated with frozen semen during 38 cycles in Exp. 1.

Table 1. Mean ± s.d. progressive motility and damaged acrosomes in post-thaw semen of Stallions 1 and 2 frozen by Methods A and B

		Post-thaw progressive motility (%)	Post-thaw damaged acrosomes (%)
Stallion 1	Method A	32·7 ± 3·6	51·7 ± 5·4
	Method B	36·1 ± 5·5	46·0 ± 6·8
Stallion 2	Method A	35·5 ± 6·4	46·8 ± 4·2
	Method B	34·0 ± 6·5	46·0 ± 11·5

Table 2. First and second cycle pregnancy rates of mares inseminated 1–47 h before ovulation with semen frozen by Methods A and B

Freezing method	Cycle of mare	No. of mares inseminated	No. of mares pregnant	Pregnancy rate per cycle (%)
A	First	11	7	64
	Second	4	2	50
	Total	15	9	60
B	First	11	6	55
	Second	5	3	60
	Total	16	9	56
A + B	First	22	13	59
	Second	9	5	56
	Total	31	18	58

differences between methods or between cycles. There was also no significant difference in pregnancy rates of mares inseminated with semen that contained >50% or ≤50% damaged acrosomes.

In Exp. 2, 9 of 13 mares in Group A, 6 of 13 mares in Group B and 6 of 12 mares in Group C became pregnant. These pregnancy rates were not statistically different. Pooling the results of all 3 groups showed a mean pregnancy rate of 55% in Exp. 2. Of the 13 mares inseminated with semen from Stallion 1 and of 25 mares inseminated with semen from Stallion 2, 6 (46%) and 13 (60%) respectively became pregnant (not significant). Of 18 mares inseminated with semen frozen by

Table 3. The effects of timing of insemination and number of progressively motile spermatozoa per insemination on single cycle pregnancy rates in mares inseminated with frozen–thawed semen

Progressively motile spermatozoa per AI ($\times 10^{-6}$)		Time of insemination before ovulation (h)		
		1–23	24–47	Total
$\leqslant 210$	No. of cycles	33	10	43
Range 137–210	% pregnancy	42[a]	50	44[a]
Mean 175				
s.d. 23·1				
> 220	No. of cycles	16	10	26
Range 222–334	% pregnancy	81[b]	60	73[b]
Mean 249				
s.d. 29·3				
Total	No. of cycles	49	20	69
Range 137–334	% pregnancy	55	55	55
Mean 203				
s.d. 44·3				

Values within a column with different superscript differ ($P < 0.01$).

Method B and of 20 (50%) mares inseminated with semen frozen by Method A, 11 (61%) and 10 (50%) became pregnant (not significant).

The mean 55% pregnancy rate achieved in Exp. 2 was not significantly different from the mean of 59% for first cycle pregnancy rate or the 56% pregnancy rate in mares inseminated during a second cycle in Exp. 1. The 5 post-ovulatory inseminations and the 2 inseminations carried out more than 48 h before ovulation in Exp. 1 were not included in this calculation because no mares in Exp. 2 were inseminated at such times during oestrus.

All the results from Exps 1 and 2 (excluding the 7 inseminations in Exp. 1 mentioned above) were pooled to evaluate the effects of timing of inseminations and numbers of progressively motile spermatozoa per insemination on pregnancy rates (Table 3). Insemination 1–23 h before ovulation did not yield significantly different pregnancy rates from insemination at 24–47 h before ovulation. Pregnancy rates were significantly different between mares that were inseminated with more than 220×10^6 and mares that received fewer than 210×10^6 progressively motile spermatozoa per insemination. This effect was independent of the timing of AI.

Discussion

Various methods of freezing stallion semen have been used but single-cycle pregnancy rates have seldom reached more than 50%. For ease of comparison the single-cycle pregnancy rates given in 9 different reports published since 1971 have been extracted and tabulated in Table 4. Using the freezing method of Cochran *et al.* (1984), Cristanelli *et al.* (1984) were the first to report pregnancy rates of over 50% with semen frozen in 0·5 ml straws. The overall single-cycle pregnancy rates reported here reflect much the same result. The results also show that there was no effect of the stallion providing the semen on pregnancy rates.

The fact that the layering of a hyperosmotic cushion below extended semen can be omitted without ill effects on pregnancy rates is in agreement with previous observations (unpublished) that post-thaw progressive motility and acrosomal integrity were not affected by the omission of this technically somewhat complicated procedure. Method B is technically less complicated and therefore more practical.

Table 4. Selected data on the fertility of frozen stallion semen extracted and adapted from 9 previous publications

Authors	Semen packaging	Progressively motile spermatozoa per AI ($\times 10^{-6}$)	Time of AI	No. of cycles used for insemination	Single-cycle pregnancy rates (%)
Ellery *et al.* (1971)	Plastic sachets	60–400	Every 48 h before ovulation	17	41
Dockhorn *et al.* (1971)	Pellets	100–150	Daily before and after ovulation	(a) 30 (b) 34	(a) 33 (b) 18
Blobel & Klug (1975)	Pellets	?	Before and after ovulation	52	46
Klug *et al.* (1975)	Pellets	100	?	219	23
Nishikawa (1975)	Straws, 1 ml	350–900*	Before ovulation	(a) 273 (b) 299†	(a) 47 (b) 65‡
Martin *et al.* (1979)	Straws, 4 ml	200	6 h after ovulation	19	63
Müller (1982)	Aluminium tubes, 15 ml	?	Daily before ovulation	246	37
Loomis *et al.* (1983)	Straws, 0·5 ml	100–130	Daily before ovulation	120	29
Cristanelli *et al.* (1984)	Straws, 0·5 ml	161–418	Daily before ovulation	108	56

*Total number, not only progressively motile spermatozoa.
†Number of mares, not cycles.
‡More than one oestrus used per mare.

Also of great significance is the finding that frozen–thawed semen with a progressive motilitiy of more than 30% could be used to achieve very satisfactory pregnancy rates. This result in a significant decrease in the number of ejaculates that have to be discarded after freezing. Cristanelli *et al.* (1984) obtained a progressive motility of at least 40% in only 49% of ejaculates frozen. In contrast, 82% of all ejaculates frozen for this trial showed a progressive motility of at least 30%.

The number of insemination doses and the number of opportunities to carry an infection into the uterus during an oestrous period are significantly reduced when a mare is only inseminated every 2nd day. The number of inseminations per oestrous period can be reduced even further when the first insemination is delayed until a very ripe follicle is present in the ovary. There was a non-significant reduction in pregnancy rates from 81 to 60% in the mares inseminated with more than 220×10^6 progressively motile spermatozoa 1–23 h or 24–47 h before ovulation respectively. This tendency was not evident in the mares that were inseminated with fewer than 210×10^6 progressively motile spermatozoa. Trials with larger groups of mares are required to investigate this aspect more extensively.

The finding that frozen–thawed spermatozoa obviously remain viable and able to fertilize an ovum that has ovulated up to 47 h after insemination indicates that inseminations need not be performed after ovulation as has been previously advocated (Dockhorn *et al.*, 1971; Blobel & Klug, 1975; Martin *et al.*, 1979) (Table 4). Indeed, all 5 mares that were inseminated during or within 10 h after ovulation in this trial failed to conceive. A possible explanation for this finding would be that the freezing techniques used by Cochran *et al.* (1984) and in this trial lessened the degree of acrosome damage occurring before and during freezing. The semen used in this trial was safely stored in liquid nitrogen within 2 h of collection. During processing it was kept at the sub-physiological, yet relatively harmless, temperature of 20°C (Province *et al.*, 1985) to inhibit most biological activity (acrosome changes, energy consumption) as much as possible.

Mean pregnancy rates per cycle were improved from 44 to 73% by increasing the mean number of progressively motile spermatozoa per insemination from 175×10^6 to 249×10^6 (Table 3). It may be assumed that a proportion of progressively motile spermatozoa in post-thaw semen may have suffered some degree of acrosomal damage during freezing and may therefore not be able to fertilize. The percentage of damaged acrosomes in post-thaw semen used in this trial was between 46 and 53% as compared to a mean 13% in fresh semen. This observation may account, in part at least, for the high number of progressively motile spermatozoa required to achieve satisfactory conception rates in mares inseminated with frozen–thawed semen. Some of the poorer pregnancy rates reported by others (Table 4) may have been due to the low numbers of progressively motile spermatozoa per insemination used in those trials. Due to the narrow range of acrosomal damage in the semen used in this trial the effect thereof on fertility could not be evaluated reliably.

The synchronization óf oestrus with ovarian steroids and prostaglandin F-2α did not affect pregnancy rates in these trials. This means that veterinary attention can be concentrated on a group of mares for a relatively short period during the breeding season when frozen semen is used for AI.

We thank Professor R. I. Coubrough for help in preparing the manuscript; and the staff of the Genesiology Department, numerous students of the BVSc VI class and Mrs C. R. Volkmann for enthusiastic help during the insemination trials. These projects were liberally supported by the S A Medical Services.

References

Amann, R.P. & Pickett, B.W. (1984) An overview of frozen equine semen: procedures for thawing and insemination of frozen equine spermatozoa. Animal Reproduction Laboratory, Colorado State Univ. Special Series 33, Fort Collins, Colorado.

Blobel, K. & Klug, E. (1975) Tiefgefriersamen-Übertragung beim Pferd—Ein Versuch unter praktischen Gestütsbedingungen. *Berl. Münch. Tierärztl. Wschr.* **88,** 465–785.

Bristol, F., Jacobs, K.A. & Pawlyshyn, V. (1983) Synchronization of estrus in post-partum mares with progesterone and estradiol-17β. *Theriogenology* **19,** 779–785.

Cochran, J.D., Amann, R.P., Froman, D.P. & Pickett, B.W. (1984) Effects of centrifugation, glycerol level, cooling to 5°C, freezing rate and thawing rate on the post-thaw motility of equine semen. *Theriogenology* **22,** 25–38.

Cristanelli, M.J., Squires, E.L., Amann, R.P. & Pickett, B.W. (1984) Fertility of stallion semen processed, frozen and thawed by a new procedure. *Theriogenology* **22,** 39–45.

Dockhorn, W., Schmidt, D.R. & Schützler, H. (1971) Konservierung und Einsatz von Hengstperma in Pelletform, 2. Mitteilung: Einsatz von pelletiertem Hengst-sperma zur künstlichen Besamung von PMS-Stuten. *Fortpfl. Haust.* **7,** 100–108.

Ellery, J.C., Graham, E.F. & Zemjanis, R. (1971) Artificial insemination of pony mares with semen frozen and stored in liquid nitrogen. *Am. J. vet. Res.* **32,** 1693–1698.

Klug, E., Treu, H., Hillmann, H. & Heinze, H. (1975) Results of insemination of mares with fresh and frozen stallion semen. *J. Reprod. Fert., Suppl.* **23,** 107–110.

Loomis, P.R., Amann, R.P., Squires, E.L. & Pickett, B.W. (1983) Fertility of unfrozen and frozen stallion spermatozoa extended in EDTA-lactose-egg yolk and packaged in straws. *J. Anim. Sci.* **56,** 687–693.

Loy, R.G., Pemstein, R., O'Canna, D. & Douglas, R.H. (1981) Control of ovulation in cycling mares with ovarian steroids and prostaglandin. *Theriogenology* **15,** 191–200.

Martin, J.C., Klug, E. & Günzel, A.-R. (1979) Centrifugation of stallion semen and its storage in large volume straws. *J. Reprod. Fert., Suppl.* **27,** 47–51.

Müller, Z. (1982) Fertility of frozen equine semen. *J. Reprod. Fert., Suppl.* **32,** 47–51.

Nishikawa, Y. (1975) Studies on the preservation of raw and frozen horse semen. *J. Reprod. Fert., Suppl.* **23,** 99–104.

Oettlé, E.E. (1986) Using a new acrosome stain to evaluate sperm morphology. *Vet. Med.* 263–266.

Pickett, B.W., Voss, J.L., Squires, E.L. & Amann, R.P. (1981) *Management of the stallion for maximum reproductive efficiency.* Animal Reproduction Laboratory, Colorado State Univ., Fort Collins, Colorado.

Province, C.A., Squires, E.L., Pickett, B.W. & Amann, R.P. (1985) Cooling rates, storage temperatures and fertility of extended equine spermatozoa. *Theriogenology* **23,** 925–934.

THE NON-PREGNANT MARE

Conveners

J. P. Hughes

M. Vandeplassche

Chairmen

R. M. Kenney

M. Vandeplassche

D. Mitchell

F. Bristol

J. Reprod. Fert., Suppl. **35** (1987), 149–155

An update of chromosomal abnormalities in mares

Ann Trommershausen Bowling, L. Millon and J. P. Hughes

Serology Laboratory and Department of Reproduction, School of Veterinary Medicine, University of California, Davis, CA 95616, U.S.A.

Summary. Chromosomal abnormality was detectable in 98 of 180 mares aged 3 years or over with gonadal dysgenesis. The most common abnormality was X monosomy (63,X). The second most common abnormality was a karyotype indistinguishable by G- or C-banding from that of a male horse (64,XY).

Two mares demonstrated structural abnormality of one X chromosome [64,X,del(Xp)] which has not previously been reported in horses. One of these foaled a filly with the same karyotype as her dam. Blood typing confirmed parentage of the foal. This is the only example in our experience of fertility in a mare with gonadal dysgenesis and chromosomal abnormality.

Chromosomal abnormalities were also found in 4 yearling fillies investigated solely because of small size, poor conformation and lack of vigour. One was 63,X; one was 63,X/64,XX; one was 64X,del(Xp) and the 4th had an autosomal trisomy, tentatively 64,XX,i(?26), which demonstrated a second new abnormal karyotype of the horse.

Introduction

Since the first reports of mare infertility associated with sex chromosome aneuploidy (e.g. Payne *et al.*, 1968; Chandley *et al.*, 1975; Hughes *et al.*, 1975a, b) the occurrence of karyotype abnormalities in infertile mares has continued to be documented (e.g., Blue, 1976; Kieffer *et al.,* 1976; Hughes & Trommershausen-Smith, 1977; Blue *et al.*, 1978; Trommershausen-Smith *et al.*, 1979; Walker & Bruere, 1979; Metenier *et al.*, 1979; Miyake *et al.*, 1979; Buoen *et al.*, 1983; Halnan *et al.*, 1985; Halnan & Irwin, 1985; Schroeder, 1986). As for the analogous human condition of female gonadal dysgenesis (Turner syndrome), the most commonly reported chromosomal abnormality is X monosomy, an aneuploid karyotype in which one of the sex chromosomes is lacking. In addition to 63,X, the aneuploid condition 65,XXX as well as various chimaeric/mosaic combinations such as 63,X/64,XX, 63,X/64,XY and 63,X/64,XY/65,XXY have been reported. Mare infertility has also been associated with the 64,XY euploid karyotype of the male horse.

The purpose of this report is to update the accounting of karyotype results from this laboratory for gonadal dysgenesis in mares and thereby confirm the association of that phenotype with chromosomal abnormality. In addition, studies of 4 small, unthrifty yearling fillies suggest that chromosome analysis might occasionally be useful for identifying abnormal females before breeding age. As well, 2 new euploid abnormal karyotypes are described.

Materials and Methods

Between 1974 and 1986 we karyotyped 180 mares with gonadal dysgenesis aged 3 years or over. The presenting problems included chronic infertility, failure to cycle regularly, very small ovaries (0·5 × 0·5 × 1·0 cm) lacking follicular activity and a flaccid uterus and cervix. Usually, but not always, the mares were reported to be of small size. In addition 4 yearling fillies were referred because of small size, poor conformation and lack of vigour. Most of the cases were not locally available for examination so peripheral blood samples were collected in acid citrate-dextrose anticoagulant and sent by referring veterinarians to the laboratory via rapid courier. The methods used for culture of lymphocytes from peripheral blood, for G- and C-banding of chromosomes, for polymorphonuclear (PMN) leucocyte drumstick analysis and blood typing were as described previously (Trommershausen-Smith *et al.*, 1979).

Results

A summary of karyotyping results for mares aged 3 years or over with primary infertility and gonadal dysgenesis is presented in Table 1. Chromosomal abnormality was detected in 98 of 180 cases (54·4%).

Table 1. Primary infertility in 180 mares aged 3 years and older

Karyotype	Breed*									Total
	AP	AR	MH	PP	QH	ST	TB	TWH	XX	
63,X	0	20	1	2	12	3	16	1	1	56
63,X/64,XX	1	3	0	0	0	1	1	0	0	6
63,X/64,XY	0	2	0	0	0	0	2	0	0	4
65,XXX	0	1	0	0	0	1	3	0	0	5
64,XX/65,XXY	0	0	0	0	1	0	1	0	1	3
64,X,del(Xp)	0	1	0	0	1	0	0	0	0	2
64,XY	1	5	2	0	6	0	7	0	1	22
64,XX	2	41	2	1	13	4	14	1	4	82

AP, Appaloosa; AR, Arabian; MH, Morgan; PP, Peruvian Paso; ST, Standardbred; TB, Thoroughbred; TWH, Tennessee Walking Horse; XX, Crossbred or other breed.

Gonadal dysgenesis with X, X/XX or X/XY

The most commonly encountered abnormality was 63,X which accounted for 31% of cases. One yearling Arabian also demonstrated the 63,X karyotype.

Pedigree information was available for 18 Arabians. Most of these horses were produced by matings between unrelated animals, although a few showed moderate inbreeding. Two affected mares were sired by the same stallion. The average maternal age at conception was 8·5 years (range 3–18, median 8). The average paternal age at conception was 10·8 years (range 2–26, median 10).

In another 10 (5·6%) infertile mares more than one cell line was detected, including a euploid line, either 64,XX or 64,XY, as well as a 63,X line. One of the yearlings, a Thoroughbred, also demonstrated a 63,X/64,XX karyotype. No evidence for red cell chimaerism was found in 5 mares blood-typed for 7 systems of blood group markers (A, C, D, K, P, Q, U) and 2 loci of red cell proteins (*Hb, PGD*). Pedigree information was available for the 5 Arabians. The average maternal age at conception was 6·6 years (range 3–10, median 7). The average paternal age at conception was 13 years (range 3–27, median 14).

After chromosome analysis one of the 63,X mares was listed as the dam of two registered foals, but blood typing studies invalidated this possibility (Bowling, 1985). As far as we are aware, none of the other mares with a 63,X line has been reported to have produced a foal.

Gonadal dysgenesis with X,del(Xp)

Two mares of this study, 1 Arabian and 1 Quarter Horse, demonstrated a euploid karyotype with a structural abnormality of one of the X chromosomes, namely a deletion of its short arms [64,X,del(Xp)]. The same abnormality was additionally observed in an Arabian that had been reported by us to have an autosomal deletion, diagnosed as a yearling (Trommershausen-Smith *et*

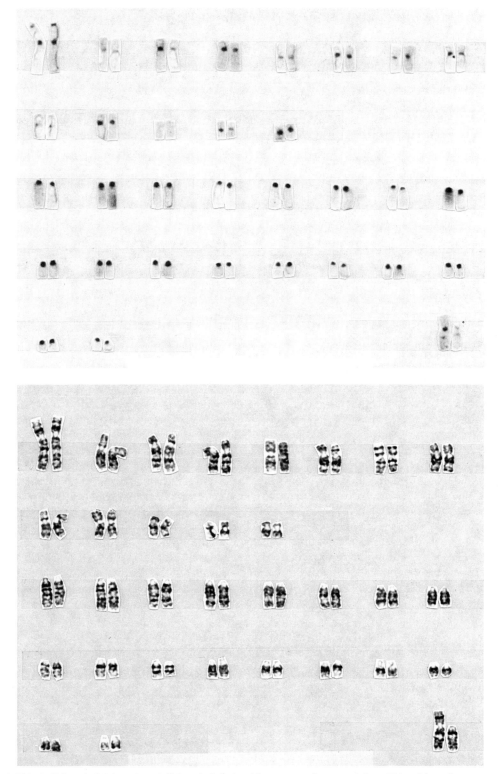

Fig. 1. C-banded (above) and G-banded (below) karyotype from peripheral blood lymphocytes of an Arabian mare showing, in the lower right corner, a normal X chromosome and a second X with the short arms deleted [64,X,del(Xp)].

al., 1979). Improved resolution of C- and G-bands clearly showed the abnormality to involve an X chromosome (Fig. 1). Two chromosomes with interstitial C-bands characteristic of the X were present, one a sub-metacentric and one an acrocentric. G-banding confirmed the acrocentric to be an X with the short (p) arms deleted. Both Arabians showed about 3% PMN drumstick counts, characteristic of an XX female.

The 2 Arabians were available for further study. Both mares were small in size and considered to be of poor conformation, particularly of the hind legs ('sickle-hocked' and 'cow-hocked'), characteristics commonly described for non-mosaic 63,X gonadal dysgenesis. Upon rectal palpation both had infantile, inactive reproductive structures although one mare exhibited occasional oestrous cycles, twice developed follicles at 4 years of age, was mated and ovulated, but failed to conceive. The second mare was not closely followed after an intensive initial examination failed to detect evidence of oestrus or follicle development, but at 5 years of age, after pasture service to a pony stallion, foaled a filly with the same 64,X,del(Xp) karyotype as her dam. Blood typing confirmed parentage of the foal. This is the only example in our experience of fertility in a mare with gonadal dysgenesis and chromosomal abnormality.

The 2 Arabian mares were unrelated. Each was the first foal of her dam, conceived when the dam was a 3-year-old. For one the sire was 7 years old at the time of conception, for the other, the sire was 11 years of age.

Gonadal dysgenesis with XY

Twenty-two mares (12·2%) with primary infertility and gonadal dysgenesis had only 64,XY line, indistinguishable by G- or C-banding from that of a male horse. These cases occurred in several breeds. The Shetland Pony and one of the Arabians, a mare born co-twin with a colt, were previously reported (Trommershausen-Smith *et al.*, 1979). Often rectal palpation could not distinguish between the 63,X condition and 64,XY. The pony mare did not exhibit masculine behaviour and the gonads were typical for 63,X gonadal dysgenesis. In contrast, one of the Thoroughbreds was a large mare that exhibited masculine behaviour and upon rectal palpation was found to lack a uterus and cervix. The gonads were histologically determined to be testes. This mare matched the characteristics of a Quarter Horse mare with testicular feminization previously described by Kieffer *et al.* (1976).

In a karyotyping study separate from the 184 animals considered for the rest of this paper, a series of female offspring of a stallion who had sired more than one infertile 64,XY female was karyotyped. Many but not all in the series were weanlings or yearlings whose reproductive status had not yet been determined. According to the stud book record the stallion had sired 215 foals (65 males, 150 females). Of the 67 daughters karyotyped, 22 (32·8%) were 64,XY, confirmed by C- and G-banding.

Gonadal hypoplasia with 65,XXX or 64,XX/65,XXY

Five mares were non-mosaic aneuploids with an extra X chromosome and 3 were mosaic aneuploids. All were karyotyped from mailed samples and none was available for detailed study.

Gonadal hypoplasia with 64,XX

For 46% of the breeding-age mares with primary infertility and gonadal hypoplasia no chromosomal abnormality could be detected. It is possible that some had undetected chimaerism or small deletions and that karyotyping of solid tissue might have allowed detection of an abnormality, but at least 9 of these mares were known to have produced foals after their referral for karyotyping.

Autosomal structural abnormality 64,XX,i(?26)

An autosomal structural abnormality was found in a yearling Thoroughbred filly referred for karyotyping because of poor conformation and lack of vigour. An extra metacentric in a 64,XX karyotype was identified by G-banding to be an isochromosome of one of the smaller acrocentrics, tentatively pair 25 or 26 (Fig. 2). Upon rectal palpation gonads and tubular structures appeared normal. She was conceived when her sire was 5 and her dam 3 years of age.

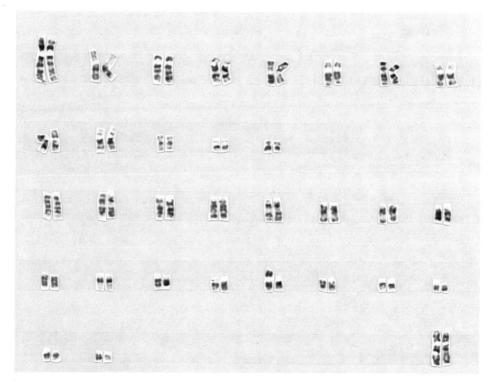

Fig. 2. G-banded karyotype from peripheral blood lymphocyte of a Thoroughbred mare showing an isochromosome (4th row, 5th pair) for one of the small metacentrics [64,XX,i(?25)].

Discussion

Chromosomal abnormality was detectable in more than half of mares referred for karyotyping because of gonadal dysgenesis. Not all mares with small gonads had detectable chromosome abnormality and some of these produced foals after the karyotyping study.

Of the 71 mares with detectable X monosomy, 57 (80·3%) presented the classical 63,X karyotype; 11 (15·5%) were mosaic for a second cell line with structurally normal sex chromosomes and 3 (4·2%) presented non-mosaic Xp deletion. These percentages can be compared to data from 275 human Turner syndrome patients in which 55% had the classical 45,X karyotype and 12·7% were mosaic for the monosomic line and a second line with structurally normal sex chromosomes (Fryns *et al.*, 1983).

The prognosis for fertility is poor for any condition involving X monosomy. Blood typing studies cast doubt on the fertility reported for one 63,X mare alleged to have produced two foals after the karyotyping studies (Bowling, 1985). Although fertility has been demonstrated in a limited number of Turner syndrome patients, the human syndrome is associated with high frequency of reproductive failure and high fetal loss. Collected data reported by Fryns *et al.* (1983) showed 15 early miscarriages and 4 stillbirths in 56 pregnancies. Of the 37 live-borns, 3 presented autosomal trisomy, and 5 a 45,X Turner syndrome.

Only a limited number of human patients with 'pure' Xp deletion without detectable 45,X mosaicism have been reported. In the series of 275 Turner syndrome patients above, the Xp deletion was found in only 2 patients, both in a mosaic condition with an X monosome line. Fryns *et al.* (1983) found 12 females to be partly monosomic for the X chromosome [del(Xp)] amongst a series of 32 930 patients. Three of the 12 Xp deletion patients had primary amenorrhoea and streak gonads but in the others secondary sexual development was normal. Of the 12 patients, 2 produced live-born children. One Xp deletion female with a normal phenotype had 2 daughters, one with a normal karyotype and the other with the same Xp deletion as her mother. A second patient with Xp deletion had 2 miscarriages before giving birth to a normal boy with 46,XY karyotype.

Although the Xp deletion genotype would seem to be relatively less rare in horses than humans, the occurrence of live-born offspring has now been documented for both.

As yet no compelling evidence has been found to indicate an inherited tendency for the production of X monosomy in humans or horses. Although two 63,X Arabian mares of this study had the same sire, that stallion is the most prolific sire of the breed to date. Rather than suggesting a genetic component, the two mares may reflect the statistical odds of such a chromosomal error arising at random, or may correlate with breeding methods for managing large numbers of mares. However, in various human studies (Warburton, 1983) consistent evidence has been shown for a small inverse maternal age effect of 45,X live births. Both the Xp deletion Arabians as well as two of the 18 63,X Arabians were first foals for their dams, conceived when the dams were 3-year-olds. A comparable control group has not yet been identified for comparison with the affected mares to ascertain whether the inverse maternal age effect is seen in horses as well as humans.

Warburton (1983) summarizes human data showing increased risk for XXX with maternal age, but we did not have pedigree data for our mares and could not ascertain the possibility of a similar situation in horses.

Unlike the sex chromosome aneuploid conditions, a genetic aetiology must be considered for XY females. The case of Kieffer *et al.* (1976) suggests occurrence of X-linked testicular feminization, at least in Quarter Horses. The XY Shetland Pony described by us may be analogous to XY gonadal dysgenesis which may be inherited in humans (Simpson, 1976).

A familial hypothesis must also be considered for the stallion siring XY females. This family does not fit a pattern of X-linked inheritance. The high incidence of the abnormality in the family as well as the predominant lack of mating to related females, suggest a dominant mode of inheritance for this defect. One hypothesis to explain the mechanism of the effect is a translocation of Y chromosome material to an autosome. Phenotypic males sired by this stallion would have received both his Y chromosome and the Y part translocated to an autosome. If the chromosomes segregate at random in meiosis I then it would be anticipated that half of his offspring which receive a Y chromosome would fail to receive the translocated counterpart and the genetic rearrangement/ deficit would fail to produce phenotypic males. The stallion would have a higher ratio of daughters to sons and about one third of his daughters would be XY and infertile. We did not observe any structural abnormalities of the chromosomes in this sire family, but did not perform high-resolution banding or use probes for Y chromosome DNA which might be far more sensitive than our methods for detecting such a translocation.

These studies confirm the role of chromosome abnormalities in primary infertility of the mare and demonstrate the advisability of obtaining a karyotype analysis to identify the nature of the problem in the event it may be of a hereditary nature. It also may be worthwhile to karyotype young females which are very small and unthrifty to identify at an early age those which may have a karyotype abnormality that limits reproductive usefulness.

We thank veterinarians and owners throughout the U.S. for the blood samples and information about infertile mares; and Dr Clive Halnan for valuable assistance with banding techniques.

References

Blue, M. (1976) *A clinical cytogenetical study of the horse.* Ph.D. Thesis, Massey University, Palmerston North, New Zealand.

Blue, M.G., Bruere, A.N. & Dewes, H.F. (1978) The significance of the XO syndrome in infertility of the mare. *N.Z. vet. J.* **26**, 137–141.

Bowling, A.T. (1985) Blood typing invalidates alleged evidence of fertility of XO mare. *Theriogenology* **24**, 203–210.

Buoen, L.C., Eilts, B.E., Rushmer, A. & Weber, A. (1983) Sterility associated with an XO karyotype in a Belgian mare. *J. Am. vet. med. Ass.* **182**, 1120–1121.

Chandley, A.C., Fletcher, J., Rossdale, P.D., Peace, C.K., Ricketts, S.W., McEnery, R.J., Thorne, J.P., Short, R.V. & Allen, W.R. (1975) Chromosome abnormalities as a cause of infertility in mares. *J. Reprod. Fert., Suppl.* **23**, 377–383.

Fryns, J.P., Kleczkowska, A. & Van den Berghe, H. (1983) The X chromosome and sexual development: clinical aspects. In *Cytogenetics of the Mammalian X Chromosome, Part B: X Chromosome Anomalies and Their Clinical Manifestations*, pp. 115–126. Ed. A. A. Sandberg. Liss, New York.

Halnan, C.R.E. & Irwin, C.F.P. (1985) Dysgenesis of the tubal system in an infertile mare karyotype 63X:64,XY:65,XXY. *J. Equine vet. Sci.* **5**, 328–329.

Halnan, C.R.E., Hutchins, D.R. & Brownlow, M. (1985) Failure to conceive in a Standardbred mare: karyotype 64,XX:64,XY. *J. Equine vet. Sci.* **5**, 161–162.

Hughes, J.P. & Trommershausen-Smith, A. (1977) Infertility in the horse associated with chromosomal abnormalities. *Aust. vet. J.* **53**, 253–257.

Hughes, J.P., Benirschke, K. & Kennedy, P.C. (1975a) XO-gonadal dysgenesis in the mare (Report of two cases). *Equine vet. J.* **7**, 109–112.

Hughes, J.P., Benirschke, K., Kennedy, P.C. & Trommershausen-Smith, A. (1975b) Gonadal dysgenesis in the mare. *J. Reprod. Fert., Suppl.* **23**, 385–390.

Kieffer, N.M., Burns, S.J. & Judge, N.G. (1976) Male pseudohermaphroditism of the testicular feminizing type in a horse. *Equine vet. J.* **8**, 8–41.

Metenier, L., Driancourt, M.A. & Cribiu, E.P. (1979) An XO chromosome constitution in a sterile mare (*Equus caballus*). *Annls Genet. Sel. Anim.* **11**, 161–163.

Miyake, Y.-I., Ishikawa, T. & Kawata, K. (1979) Three cases of mare sterility with sex-chromosomal abnormality (63,X). *Zuchthygiene* **14**, 145–150.

Payne, H.W., Ellsworth, K. & DeGroot, A. (1968) Aneuploidy in an infertile mare. *J. Am. vet. med. Ass.* **158**, 1293–1299.

Schroeder, W.G. (1986) X-Y sex reversal syndrome in the mare. *J. Equine vet. Sci.* **6**, 26–27.

Simpson, J.L. (1976) *Disorders of Sexual Differentiation.* Academic Press, San Francisco.

Trommershausen-Smith, A., Hughes, J.P. & Neely, D.P. (1979) Cytogenetic and clinical findings in mares with gonadal dysgenesis. *J. Reprod. Fert., Suppl.* **27**, 271–276.

Walker, K.S. & Bruere, A.N. (1979) XO condition in mares. *N.Z. vet. J.* **27**, 18–19.

Warburton, D. (1983) Parental age and X-chromosome aneuploidy. In *Cytogenetics of the Mammalian X Chromosome, Part B: X Chromosome Anomalies and Their Clinical Manifestations*, pp. 23–33. Ed. A. A. Sandberg. Liss, New York.

J. Reprod. Fert., Suppl. **35** (1987), 157–167

Fine structure of the follicular oocyte of the horse

M. M. Vogelsang*, D. C. Kraemer*†, G. D. Potter* and G. G. Stott‡

*Departments of *Animal Science, †Veterinary Physiology and Pharmacology, and ‡Veterinary Anatomy, Texas Agricultural Experiment Station, Texas A & M University, College Station, Texas 77843, U.S.A.*

Summary. Oocytes recovered by follicular aspiration were evaluated by light and transmission electron microscopy. Of the 22 oocytes, 4 exhibited characteristics of degeneration, and the remaining 18 were in various stages of meiotic development. Of the non-degenerate oocytes, 14 were in the germinal vesicle stage, 2 had undergone nuclear membrane disintegration, 1 displayed chromosomes in late metaphase I–early anaphase I, and 1 oocyte was in the process of extrusion of the first polar body. Although some oocytes retained complete cumulus cell investments, oocytes were predominantly enclosed only by the corona radiata. Ultrastructural evaluation revealed that follicular oocytes of the horse are similar to those of other mammals. An abundance of vesicular agranular endoplasmic reticulum in horse oocytes was the most obvious difference. Variation in intracellular organization seemed to be primarily dependent upon stage of meiotic development.

Introduction

Follicular oocytes of several mammalian species have been extensively studied. Oocytes of the rat (Sotelo & Porter, 1959; Odor, 1960), guinea-pig (Anderson & Beams, 1960; Adams & Hertig, 1964), rabbit (Blanchette, 1961; Zamboni & Mastroianni, 1966), mouse (Merchant & Chang, 1971; Wassarman & Josefowicz, 1978) and man (Hertig & Adams, 1967; Baca & Zamboni, 1967) have been examined in great detail. The ultrastructural morphology of oocytes from laboratory animals and women has served as the foundation for comparative studies on oocytes of domestic livestock (cow: Fleming & Saacke, 1972; sheep: Cran *et al.,* 1980; pig: Motlik & Fulka, 1976; Cran, 1985). The increased interest in in-vitro manipulation and fertilization of the oocytes of domestic animals has prompted continued investigation into changes that occur within oocytes during maturation before fertilization. However, there have been no studies reported that examined specifically the ultrastructure of the horse follicular oocyte. Characterization of the fine structure of such oocytes would facilitate advanced studies on fertilization and research in the areas of genetic engineering, endocrinology and equine infertility. The purpose of this research was to evaluate oocytes from normal cyclic mares and to describe and characterize the fine structure of the oocyte.

Experimental procedures

Twenty-two oocytes were recovered from mares by the procedures described by Vogelsang (1986) or from ovaries obtained at a local abattoir. Twelve of the oocytes were obtained from mares treated with hCG, GnRH, or horse anterior pituitary extract for stimulation of follicular and oocyte maturation; however, these were not found to differ from oocytes recovered from naturally cyclic, unstimulated mares. Oocytes were placed in buffered 2·5% glutaraldehyde immediately after recovery, and the diameter, overall appearance and state of cumulus cell compactness were recorded. Oocytes were washed in cacodylate buffer and post-fixed in 1·0% osmium tetroxide for 1 h. Oocytes were further subjected to staining *en bloc* in 1·0% uranyl acetate for 4 h and then washed in cacodylate buffer and glass-distilled water before dehydration. Dehydration was accomplished by transferring oocytes through a graded alcohol series and into 100% propylene oxide. Infiltration was achieved by placing oocytes in a 1:1 ratio of propylene oxide and Maraglas resin for 2 h. Embedding was completed by placing oocytes in 100% resin and leaving them in a drying oven 52°C for 24–36 h.

Thick sections of oocytes were mounted on glass slides and stained with toluidine blue for light microscopy. Ultrathin sections were mounted on copper mesh grids for transmission electron microscopy. Sections were stained for 2 min in lead citrate to promote contrast in the section.

Results and Discussion

Oocytes examined in this study were in all stages of cumulus cell investment. Some (10%) were encompassed within a dense compact cluster of cumulus cells (Fig. 17), while others (49%) were within a loosely arranged cumulus mass (Fig. 13). Of the 6 oocytes (35%) collected which were surrounded only by the corona radiata, 3 were degenerate. Two oocytes (6%) were completely denuded of cumulus cells; 1 of these was degenerate and 1 appeared to be in early metaphase I as indicated by a fragmented and wavy nuclear membrane. Diameter of the whole-mounted oocytes was 125 μm, exclusive of zona pellucida and corona radiata. Oocytes were characteristically round, but some of those deemed to be degenerate were more oval in shape.

Light and transmission electron microscopy indicated that cells of the corona radiata were closely packed (Fig. 17). Cell-to-cell communication was suggested by junctional complexes. Junctions between follicular cell processes and the oocyte surface were also evident when the corona was intact (i.e. not dispersed). Cellular debris, possibly remnants of corona cell processes, was found within the perivitelline space of oocytes in which cumulus expansion had begun. No attachment of these remnants to the oocyte surface was observed, which is similar to the findings of Fleming & Saacke (1972) and Cran *et al.* (1980).

The zona pellucida varied in appearance. In most specimens, the zona pellucida appeared as homogeneous material of low density, which was located between the oocyte and corona radiata (Fig. 10). The only structures found within the matrix of the zona pellucida were microvilli from the oocyte surface and corona cell processes from the follicular cells. A two-layered zona pellucida was observed in two of the oocytes. The layer closest to the oocyte surface was the more densely staining of the two layers. This layer went through a gradual transition to a very lightly-stained external band which was the wider of the two layers. Width of the zona pellucida was 7–10 μm as measured from cell membrane of the oocyte to the base of the corona radiata.

Fig. 1. Transmission electron micrograph of intracellular organization of a horse oocyte. Mitochondria (M), some with cristae visible, are in close association with agranular endoplasmic reticulum (AER). Golgi complexes (G) are found in lower part of micrograph. × 12 000.

Fig. 2. Transmission electron micrograph of mitochondria (M) with blebbing from outer membrane (arrows) from a horse oocyte. × 30 000.

Fig. 3. Transmission electron micrograph showing cristae configuration within mitochondria (M) of a horse oocyte. These are in close apposition to lipid droplets (L). × 19 000.

Fig. 4. Transmission electron micrograph taken near periphery of a horse oocyte. Mitochondria (M) and agranular endoplasmic reticulum (AER) are evident as well as a group of cortical granules (CG) which are apparently migrating to the oocyte surface. × 7000.

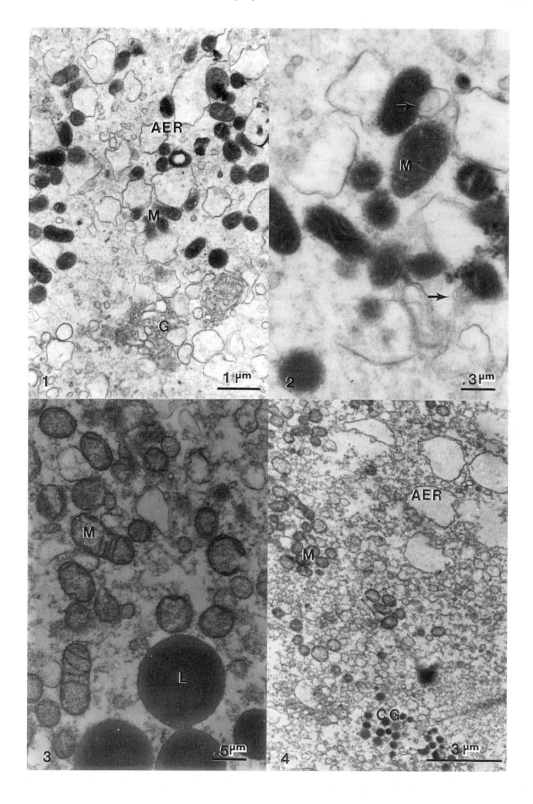

Microvilli were a prominent feature of the oocyte surface (Fig. 8). They were quite uniform in length and width and did not contain other cytoplasmic structures. Although most of the oocytes in this study exhibited a very regular and continuous spacing of microvilli, some oocytes contained a sparse distribution of microvilli over their entire surface. There was some evidence of pinocytotic activity at the oocyte surface. In most instances, this occurred near the base of the microvilli. One oocyte exhibited increased pinocytotic activity as shown by actual discontinuity of the membrane.

Mitochondria were observed throughout the cytoplasm and were seen in peripheral and juxta-nuclear positions. Most commonly, mitochondria were in a maturative state. Dumb-bell shaped mitochondria (Fig. 3) were found in all oocytes. Some mitochondria exhibited blebbing of mitochondrial membranes, while others were vacuolated (Fig. 1). Both of these characteristics were believed to signify maturation of mitochondria. The vesicular blebs arising from the outer membrane of the mitochondria (Fig. 2) were attached in such a way that cytoplasm did not appear to be incorporated into the inner mitochondrial matrix. However, the mitochondrial blebs did contain a very fine filamentous matrix. Cristae were evident in many of the mitochondria. They were seen to cross the organelle longitudinally or transversely. Some cristae appeared apposed to the limiting membrane of the mitochondrion, and others protruded inwardly from the surface of the mitochondrion and ended abruptly in the centre of the organelle (Fig. 5). Mitochondria were occasionally found near large groups of lipid droplets (Fig 3).

Mitochondria of the horse follicular oocyte had no preferential distribution within the cell as was also observed by Suzuki *et al.* (1981) and Sundstrom *et al.* (1985) for human oocytes. Mitochondria did seem to proliferate around and were closely associated with vesicular agranular endoplasmic reticulum located throughout the cytoplasm (Fig. 5). In some cases, membranes of the endoplasmic reticulum appeared to be continuous with the mitochondria (Fig. 1).

Endoplasmic reticulum in the horse oocyte was almost exclusively of the agranular vesicular type (Fig. 1). Smaller oval-shaped vesicles and larger irregular vesicles were observed. As commonly reported in the literature, endoplasmic reticulum of these types predominated and did not conform to the tubular and lamellar systems (Thorpe, 1984) usually observed in other cell types. Within the endoplasmic reticulum there was a very fine, slightly granular matrix which is common to rabbit (Zamboni & Mastroianni, 1966) and human (Baca & Zamboni, 1967) oocytes. Appearance of the inner endoplasmic reticulum varied within the oocyte as some vesicles exhibited a more granular matrix than others. There was predominance of endoplasmic reticulum at the periphery of the oocyte. Occasionally, tubular endoplasmic reticulum was observed in close association with membranes of compound aggregates (Fig. 6). Endoplasmic reticulum was found close to, but not in direct communication with, the nuclear membrane (Fig. 9).

The relationship of agranular endoplasmic reticulum to the Golgi complex observed in the horse oocyte (Fig. 5) has been described for human oocytes by Sathananthan *et al.* (1985). Golgi complexes did not assume the typical configuration of a layer of smooth-membraned tubules found in other cell types. There was some indication of layered tubules but Golgi were primarily observed as one or two tubular cisternae to which sac-like vesicles were attached. Numerous relatively small vesicles (probably tubules in cross-section) were in the immediate vicinity of these sac-like

Fig. 5. Transmission electron micrograph of typical intracellular distribution of organelles of a horse oocyte. Golgi complexes (G) are evident and are in close proximity to agranular endoplasmic reticulum (AER). Mitochondria (M) are also closely associated with AER. × 21 000.

Fig. 6. Transmission electron micrograph of a compound aggregate (CA) in a horse oocyte. Vesicles (V) of various sizes are found within the CA and the organelle is partly encircled by tubular agranular endoplasmic reticulum (arrow). × 33 000.

Fig. 7. Transmission electron micrograph of nucleoplasm of a horse oocyte with a compact two-zoned nucleolus (Nu). Clumps of heterochromatin (H) are visible. × 9500.

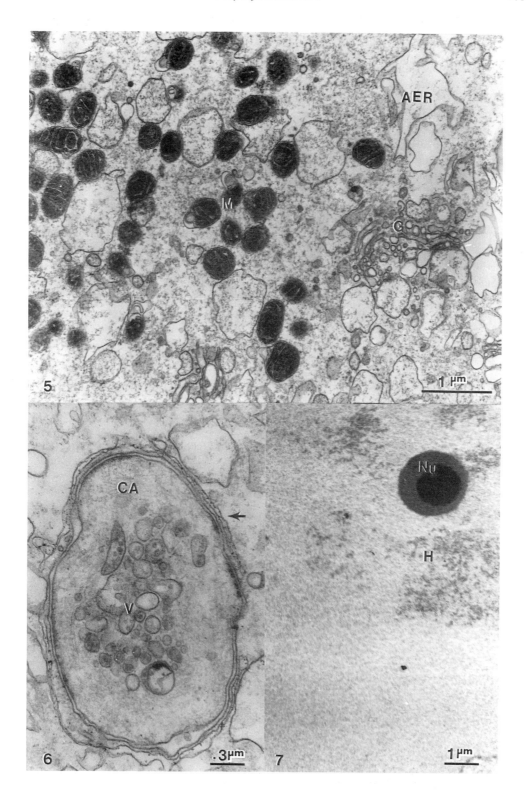

vesicular–tubular structures. The vesicles appeared most frequently on the concave secretory face of the Golgi apparatus. This association describes the secretory, production-orientated role of the Golgi (Thorpe, 1984). Blebbing, or budding, from the Golgi complexes contributed to the large number of membrane-bound vesicles in the cytoplasm of the oocyte.

Golgi complexes were not confined to the juxtanuclear region of the oocyte but were evident throughout the cytoplasm as well as within peripheral areas. This pattern of Golgi migration has been described in other mammalian oocytes to coincide with oocyte maturation (Adams & Hertig, 1964; Baca & Zamboni, 1967; Anderson, 1974; Tesoriero, 1981).

Cortical granules were occasionally observed in close proximity to Golgi complexes, but they were more frequently found close to or at the oocyte surface. An assemblage of these structures that appeared to be in the process of migration to the cell surface was observed in some oocytes (Fig. 4). The relatively small, very dense structures were membrane-bound and quite uniform in size. In some oocytes the cortical granules appeared to be aligning themselves along the oocyte border. These were probably migrating from the peripherally-located Golgi complexes. Apparently, at the stages of development of oocytes studied in this project, the number of cortical granules is not sufficient to demonstrate the very close apposition of these structures, as observed by Cran *et al.* (1980) and Sathananthan *et al.* (1985).

The compound aggregate (Fig. 6) was not a frequently observed structure in the horse pre-ovulatory oocyte, which corresponds with the findings of Sundstrom *et al.* (1985) for human preovulatory oocytes. These organelles were occasionally surrounded by agranular endoplasmic reticulum. The inner matrix was quite similar to the general cytoplasmic matrix. Within the compound aggregate there was a widely varying number of small vesicular structures of various shapes and densities. Most internal structures were vesicles; dense unbound structures were not observed in the compound aggregates of horse oocytes.

Lipid droplets were present in all oocytes recovered from preovulatory follicles of mares (Fig. 3). The large dense spherical droplets tended to be found in groups randomly located throughout the cell. The non-membrane bound lipid droplets were observed in close apposition to mitochondria or intimately associated with agranular endoplasmic reticulum. Although not quantified, there was a large volume of lipid in horse oocytes. Lipid droplets were also found clustered together, as described by Norrevang (1968), and it was difficult to distinguish their boundaries.

In the oocytes which had an intact nucleus, the structure was located eccentrically within the cell (Fig. 11). The nuclear membrane had a very characteristic appearance (Fig. 8) with denser staining of the inner membrane relative to the outer membrane. Nuclear pores were quite prominent and appeared circular or as constricted areas of the double membrane (Fig. 9), which occurred at regularly spaced intervals. Proliferation of nuclear pores has been suggested as an indication that meiotic activity is resuming (Sotelo & Porter, 1959; Odor, 1965; Baca & Zamboni, 1967; Anderson, 1974). Oocytes examined in the current study suggest that this proliferation is also characteristic of preovulatory horse oocytes. The nuclear membrane presented a wavy or undulating outline in most cases, but had a very tortuous and folded outline in 2 oocytes (Fig. 10). Masui & Clarke (1979) reported that this phenomenon occurred shortly before germinal vesicle breakdown.

Heterochromatin was visible in the intact nucleus of some oocytes (Fig. 7) indicating these cells were in diakinesis (Baca & Zamboni, 1967). Two of the oocytes exhibited this characteristic in conjunction with the compacting of the nucleolus into a two-zoned structure as described by Fleming & Saacke (1972). This two-zoned nucleolus (Fig. 7) had a very dense inner zone and an outer zone of lesser density. Also seen in this micrograph is the heterochromatin, some of which is associated with the nucleolus. Generally, much of the nucleoplasm could be described as having a fine granular texture and was homogeneous throughout the greater portion of the nucleus, except for the previously described structures.

A few of the oocytes displayed nuclear characteristics indicative of later stages of maturity. A cluster of very small vesicles (Fig. 12), which were centrally located, was found in 2 oocytes. These correspond to the vesiculation of the nuclear membrane described by Masui & Clarke (1979).

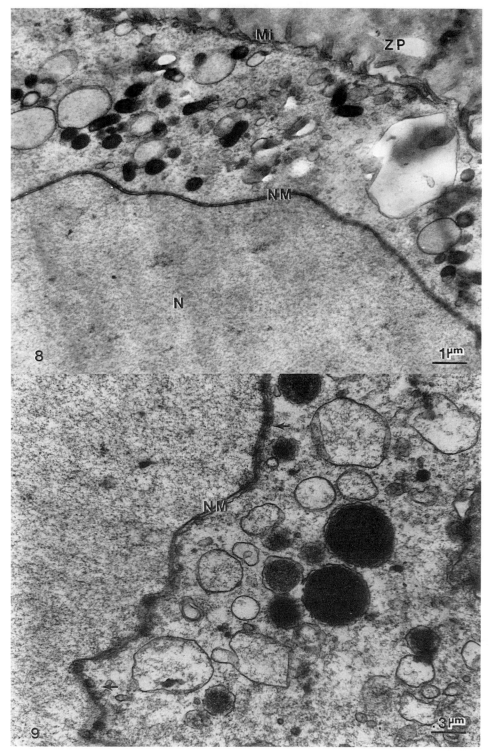

Fig. 8. Transmission electron micrograph of a horse oocyte illustrating zona pellucida (ZP), nucleus (N) with intact nuclear membrane (NM) and cytoplasmic organelles close to the oocyte periphery. Uniformity of surface microvilli (Mi) is also evident. × 15 000.

Fig. 9. Transmission electron micrograph of nuclear membrane (NM) of a horse oocyte and juxtanuclear organelles. Nuclear pores (arrows) are in evidence at intervals along the membrane. × 32 000.

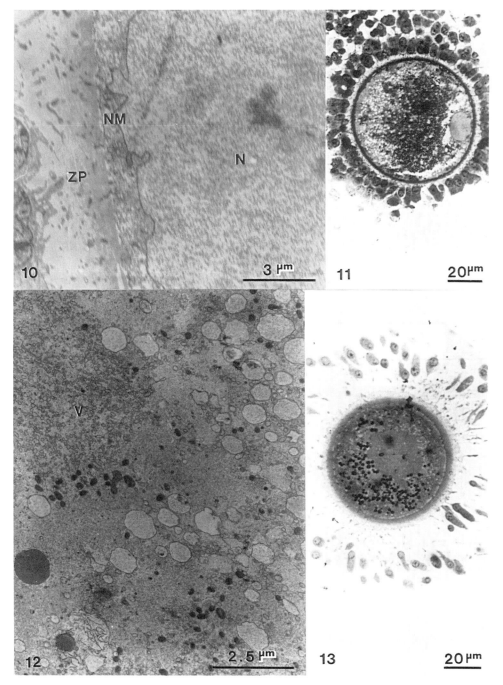

Fig. 10. Transmission electron micrograph showing undulating nuclear membrane (NM) of eccentrically located nucleus (N) of a horse oocyte. The zona pellucida (ZP) contains corona cell processes which traverse the zona to the oocyte surface. × 3000.

Fig. 11. Light micrograph of a horse oocyte (Fig. 10) with eccentrically positioned nucleus and intact corona radiata. × 500.

Fig. 12. Transmission electron micrograph of a horse oocyte in which nuclear membrane had fragmented and a region of small smooth-surfaced vesicles (V) had developed. Other cellular organelles are not located within this region but are found around it. × 9000.

Fig. 13. Light micrograph of a horse oocyte with dispersing corona radiata and indistinct, undulating nuclear membrane. × 500.

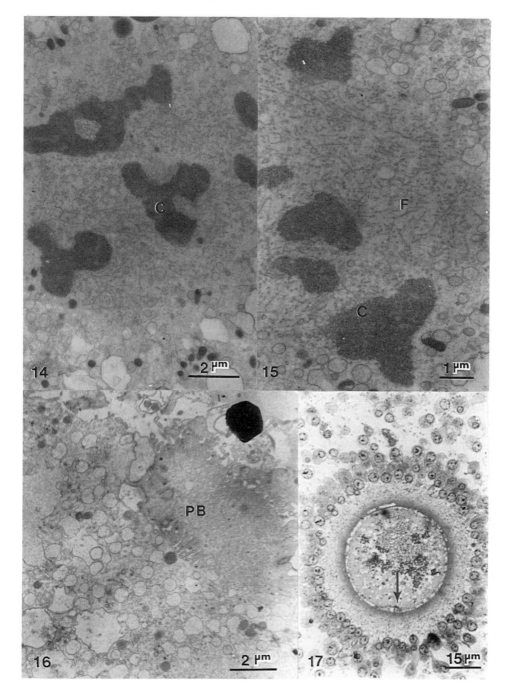

Fig. 14. Transmission electron micrograph of condensed chromosomes (C) of a horse oocyte in late metaphase I–early anaphase I of meiosis. × 7000.

Fig. 15. Transmission electron micrograph of condensed chromosomes (C) of horse oocytes. Fibrils (F) of the meiotic spindle are also evident. × 10 000.

Fig. 16. Transmission electron micrograph of a horse oocyte in the process of polar body (PB) extrusion. × 6600.

Fig. 17. Light micrograph of a horse oocyte showing corona radiata, zona pellucida and an indented area (arrow) depicting polar body extrusion. × 650.

Hadek (1965) interpreted this as the initial stages of metaphase I. Other cytoplasmic organelles were not found within this localized vesicular region, but were close to it. Nucleolar or chromosomal material was not evident within this area. Baca & Zamboni (1967) found a similar arrangement in the human follicular oocyte. The membranous structures within this oocyte appeared normal and intact such that this observation was not believed to have been an artefact of fixation. Another oocyte, in which the nucleolus was not evident, exhibited a more profuse proliferation of vesicular structures of similar size. In this particular oocyte, there was a localized area of these vesicles towards the periphery of the cell, which also contained small spherical immature mitochondria. These small vesicles appeared to have migrated throughout the cell and many were observed close to the oocyte surface.

In one oocyte the chromosomes of late metaphase I or early anaphase I of meiosis were observed (Fig. 14). It appeared that the chromosome pairs had begun to separate and move along the spindle toward opposite poles. The entire meiotic spindle is not shown, but close observation (Fig. 15) revealed the presence of the microtubular structure of the spindle fibrils as described by Zamboni & Mastroianni (1966) and Baca & Zamboni (1967). The chromosomes were seen as dense, granular, irregularly shaped aggregates which were located in an area devoid of other organelles.

Observed at the periphery of this same oocyte was an indented or 'pinched-in' area similar to that described by Odor & Renninger (1960) and Zamboni & Mastroianni (1967), indicating initial stages of polar body formation. As reported by Odor & Renninger (1960), this indention was seen in one area of the oocyte membrane; however, plane of section may have been the reason for this observation. One area of another oocyte (nucleus not intact) was found in which a section of cytoplasm appeared to be in the process of extrusion (Fig. 16). Although not positively identified as such, it is possible that this represents a portion of the first polar body before complete separation from the secondary oocyte. Sectioned in a perpendicular plane, the entire dimension of this polar body would have been visible.

As shown by these observations, the non-degenerate preovulatory horse oocyte is not unlike other mammalian oocytes at similar stages of maturation. Major similarities include (1) appearance of the zona pellucida, (2) numerous surface microvilli, (3) large numbers of mitochondria in all stages of development, (4) cortical granule migration and (5) nucleus characteristics of the germinal vesicle oocyte. However, horse oocytes differed from those of other species in (1) the absence of granular endoplasmic reticulum, (2) the lack of lamellar organization of the Golgi complex, (3) the absence of mitochondrial clustering around a dense matrix, and (4) the lack of a reticulate nucleolus.

We thank Dr M. J. Bowen for assistance in collecting some of the oocytes used in this study, and Dr T. Caceci and K. Neck for technical assistance in the preparation of tissue for electron microscopy.

References

Adams, E. & Hertig, A. (1964) Studies on guinea pig oocytes. I. Electron microscopy observations on the development of cytoplasmic organelles in oocytes of primordial and primary follicles. *J. Cell Biol.* **21**, 397–427.

Anderson, E. (1974) Comparative aspects of the ultrastructure of the female gamete. *Int. Rev. Cytol., Suppl.* **4**, 1–70.

Anderson, E. & Beams, H. (1960) Cytological observations on the fine structure of the guinea pig ovary with special reference to the oogonium, primary oocyte and associated follicular cells. *J. Ultrastruct. Res.* **3**, 432–446.

Baca, M. & Zamboni, L. (1967). The fine structure of human follicular oocytes. *J. Ultrastruct. Res.* **19**, 354–381.

Blanchette, E. (1961) A study of the fine structure of the rabbit primary oocyte. *J. Ultrastruct. Res.* **5**, 349–363.

Cran, D. (1985) Qualitative and quantitative structural changes during pig oocyte maturation. *J. Reprod. Fert.* **74**, 237–245.

Cran, D., Moor, R. & Hay, M. (1980) Fine structure of the sheep oocyte during antral follicle development. *J. Reprod. Fert.* **59**, 133–143.

Fleming, W. & Saacke, R. (1972) Fine structure of the bovine oocyte from the mature Graafian follicle. *J. Reprod. Fert.* **29**, 203–213.

Hadek, R. (1965) The structure of the mammalian egg. *Int. Rev. Cytol.* **18**, 29–71

Hertig, A. & Adams, E. (1967) Studies on the human oocyte and its follicle. I. Ultrastructural and histochemical observations on the primordial follicle stage. *J. Cell Biol.* **34**, 647–675.

Masui, Y. & Clarke, H. (1979) Oocyte maturation. *Int. Rev. Cytol.* **57**, 185–282.

Merchant, H. & Chang, M. (1971) An electron microscopic study of mouse eggs matured in vivo and in vitro. *Anat. Rec.* **171**, 21–38.

Motlik, J. & Fulka, J. (1976) Breakdown of the germinal vesicle in pig oocytes in vivo and in vitro. *J. exp. Zool.* **198**, 155–162.

Norrevang, A. (1968) Electron microscopic morphology of oogenesis. *Int. Rev. Cytol.* **23**, 113–186.

Odor, D. (1960) Electron microscopic studies on ovarian oocytes and unfertilized tubal ova in the rat. *J. Biophys. Biochem. Cytol.* **7**, 567–574.

Odor, D. (1965) The ultrastructure of unilaminar follicles of the hamster ovary. *Am. J. Anat.* **116**, 493–522.

Odor, D. & Renninger, D. (1960) Polar body formation in the rat oocyte as observed with the electron microscope. *Anat. Rec.* **137**, 13–23.

Sathananthan, A., Ng, S., Chia, C., Law, H., Edirisinghe, W. & Ratnam, S. (1985) The origin and distribution of cortical granules in human oocytes with reference to golgi, nucleolar and microfilament activity. *Am. N.Y. Acad. Sci.* **442**, 251–264.

Sotelo, J. & Porter, K. (1959) An electron microscopic study of the rat ovum. *J. Biophys. Biochem. Cytol.* **5**, 327–341.

Sundstrom, P., Nilsson, B., Liedholm, P. & Larsson, E. (1985) Ultrastructure of maturing human oocytes. *Ann. N.Y. Acad. Sci.* **442**, 324–331.

Suzuki, S., Kitai, H., Tojo, R., Seki, K., Oho, M., Fujiwara, T. & Iizuka, R. (1981) Ultrastructure and some biologic properties of human oocytes and granulosa cells cultured in vitro. *Fert. Steril.* **35**, 142–148.

Tesoriero, J. (1981) Early ultrastructural changes of developing oocytes in the dog. *J. Morph.* **168**, 171–179.

Thorpe, N. (1984) *Cell Biology.* John Wiley & Sons, New York.

Vogelsang, M. (1986) *Collection and evaluation of the preovulatory equine oocyte.* Ph.D. thesis, Texas A & M University.

Wassarman, P. & Josefowicz, W. (1978) Oocyte development in the mouse: an ultrastructural comparison of oocytes isolated at various stages of growth and meiotic competence. *J. Morph.* **156**, 209–236.

Zamboni, L. & Mastroianni, L. (1966) Electron microscopic studies on rabbit ova. I. The follicular oocyte. *J. Ultrastruct. Res.* **14**, 95–117.

J. Reprod. Fert., Suppl. **35** (1987), 169–181

Changes in thecal and granulosa cell LH and FSH receptor content associated with follicular fluid and peripheral plasma gonadotrophin and steroid hormone concentrations in preovulatory follicles of mares*

J. E. Fay† and R. H. Douglas‡

Department of Veterinary Science, University of Kentucky, Lexington, Kentucky 40546–0076, U.S.A.

Summary. Individual antral follicles from 11 horse mares were studied at three stages of the oestrous cycle to determine the characteristics of the presumptive ovulatory follicle. Mares were ovariectomized (ovex) during the late luteal phase on Day 14 after ovulation (Group 1) and on the 1st (Group 2) or 4th (Group 3) day of oestrus. Every follicle >5 mm in diameter was dissected from each ovary; follicles ⩾15 mm in diameter were analysed separately while others were pooled by size for subsequent analyses.

The presumptive ovulatory follicle possessed the following characteristics: in Groups 2 and 3 they were the largest in size; in all groups they contained the largest amount of protein in the granulosa cell component; they were the most vascular; they contained the highest follicular fluid oestradiol concentration; and they possessed the highest granulosa cell LH/hCG receptor content. In 9 mares, there was one such follicle present. In 2 mares, there were 2 such follicles present and one of these mares had a history of repeated double ovulations. The presumptive ovulatory follicle of Group 1 mares contained more thecal LH receptor than did non-ovulatory follicles. There were no differences in granulosa FSH receptor content amongst follicles, although smaller follicles tended to have increased FSH receptor content. Follicular fluid FSH and LH reflected peripheral hormone concentrations with follicular fluid LH being highest in presumptive ovulatory follicles of Group 3 mares. Follicular fluid progesterone concentrations tended to be higher in presumptive ovulatory follicles when compared with small follicles. Within a class of follicle, follicular fluid androstenedione and testosterone concentrations were lower in follicles of Group 1 mares than in follicles from Group 2 and 3 mares, reflecting endogenous stimulation of androgen biosynthesis by LH. Elevated follicular-fluid oestradiol occurred only in the presumptive ovulatory follicle with concentrations at least 30–50-fold higher than in non-ovulatory follicles. These results demonstrate that the presumptive ovulatory follicle in horse mares has been selected by at least Day 14 after ovulation and can be identified at this time by its various biochemical characteristics. Only after the first day of oestrus were physical characteristics such as size accurately able to predict ovulatory follicle status.

These results clearly demonstrate the relationship between LH stimulation of androgen biosynthesis *in vivo* regardless of follicle size or status (ovulatory or non-ovulatory). The increase in thecal LH receptor in presumptive ovulatory follicles of Group 1 mares may be important in the development of ovulatory follicle status since elevated LH receptor in the theca at this time would allow for increased responsiveness to basal LH concentrations and would permit an increase in androgen biosynthesis

---

*Reprint requests to R. H. Douglas.
†Present address: Colorado State University, School of Veterinary Medicine, Fort Collins, CO 80521, U.S.A.
‡Present address: BET Farms Inc. 6174 Jacks Creek Road, Lexington, KY 40515, U.S.A.

which could then be aromatized to oestrogen, a critical step in the final maturation of ovulatory follicles.

Introduction

Accurate identification of preovulatory follicles in the mare would permit further in-depth investigation of not only the time course of the selection process but the mechanism of that process. With this information, a more complete understanding of the mare reproductive cycle might be possible.

Little is known about follicular dynamics and function in the mare. In the ewe and in the cow, preovulatory follicles have been identified during the peri-ovulatory period by the following characteristics: (1) their large size; (2) high follicular fluid oestradiol concentrations; and (3) an increased ability of the granulosa and thecal cells to bind LH/hCG (Dufour *et al.*, 1972; England *et al.*, 1981; Webb & England, 1982a, b; Staigmiller *et al.*, 1982). In addition, the number of follicles demonstrating these characteristics is equal to the ovulation rate of the breed studied (Webb & England, 1982a) in each of two breeds of sheep with different ovulation rates. Collectively, these data suggest that these physiological markers are useful in the identification of the ovulatory follicle with a high degree of accuracy.

Kenney *et al.* (1979) removed ovaries from mares in oestrus and reported that high oestradiol concentrations in the follicular fluid occurred in histologically viable follicles and that granulosa cells from these follicles bound iodinated hCG. This preliminary information would suggest that in the mare, as in the ewe and the cow, the presence of increased granulosa and thecal LH/hCG receptors as well as an increased follicular fluid concentration of oestradiol would be useful physiological markers for the identification of the presumptive ovulatory follicle.

With these data in mind, we designed the present experiment to determine the LH/hCG receptor content of the granulosa and thecal cell layers, the FSH receptor content of the granulosa, and the follicular fluid concentrations of various steroid and gonadotrophic hormones in individual follicles removed from mares at three well-defined stages of the oestrous cycle. It was our intent not only to identify the characteristics of the presumptive ovulatory follicle, but to compare these characteristics with those of non-ovulatory follicles throughout the follicular phase. In addition, we hoped to identify the onset of the development of characteristics which identified the presumptive ovulatory follicle from non-ovulatory follicles in order to determine the potential mechanism of this process.

Materials and Methods

Animals

Twelve horse mares of mixed breeding, 6–15 years of age, were teased daily with a vigorous stallion to determine oestrous behaviour at the onset of the breeding season. Ovaries were palpated *per rectum* approximately every 3rd day during dioestrus and daily during oestrus to monitor ovarian follicular development and ovulation. When the mares had exhibited one normal interovulatory interval (18–24 days) they were bled daily until the next ovulation. If this interval was also of normal length, the mares were randomly assigned as they ovulated to one of three treatment groups.

One mare ovulated on the morning of scheduled surgery and was eliminated, and so Group 3 has only 3 mares in it. Group 1 (N = 4) mares were ovariectomized during dioestrus on Day 14 after ovulation; Group 2 mares (N = 4) were ovariectomized on the 1st day of oestrus; and Group 3 mares (N = 3) were ovariectomized on the 4th day of oestrus. Mares were again bled daily from assignment to groups until the day of ovariectomy.

Surgical procedures

Mares were fasted for 24 h before surgery. On the day of surgery, mares were restrained in a palpation chute, tranquilized with promazine, and given xylazine intravenously as an analgesic agent. Epidural anaesthesia was induced with 10 ml 2% lidocaine given between coccygeal vertebrae 1 and 2. The vagina was then rinsed with dilute (3%) povidone-iodine (Betadine) and the perineal area was prepared for surgery. The actual ovariectomy was performed through the anterior vaginal wall according to the technique of Scott & Kunze (1977). An incision was made through the anterior vaginal wall dorsal to the external os of the cervix through which a gauze pad soaked in 2% lidocaine was passed and placed around the ovarian pedicle to accomplish local anaesthesia. This was removed and a chain ecraseur was passed through the incision and placed over the ovary. The chain loop was slowly closed to separate the ovary from the pedicle. The ovary and the ecraseur were removed and the process repeated for the second ovary.

Specimen collection

Daily blood samples were collected into heparinized vacutainers from the jugular vein of each animal. Blood was immediately placed on ice and brought to the laboratory where it was centrifuged at 2000 *g*. Plasma was harvested and stored for future hormone analyses.

Ovaries obtained at ovariectomy were immediately trimmed free of fat and connective tissue and gross morphological characteristics were identified and recorded. The ovaries were placed in sterile bags in an ice bath to cool to 4°C and then stored at -70°C.

Radioimmunoassay of plasma and follicular fluid hormones

LH. Plasma LH concentrations were determined utilizing rabbit anti-ovine LH serum (GDN No. 15) in an assay previously validated in our laboratory for use with horse plasma (Burns *et al.*, 1982). Highly purified horse LH (McShan g100 LH f2) was used for iodination and standards. The sensitivity of the assay was 0·075 ng LH per assay tube. Intra-assay and interassay variation were 7% and 9%, respectively. The assay was considered valid for the measurement of follicular fluid LH when parallelism was exhibited between dilutions of 50–300 µl and the standard curve. In addition, the recovery of known amounts of exogenous LH when added to the follicular fluid was between 95 and 105% of the amount added after endogenous concentrations were subtracted.

FSH. Plasma FSH concentrations were determined utilizing rabbit antihuman FSH serum (NIAMDD) in an assay described by Freedman *et al.* (1979) and validated in our laboratory by Burns & Douglas (1981) for use with horse plasma. Highly purified horse FSH (Braselton & McShan, 1970) was used for iodination and standards. Follicular fluid assay procedures and validation steps were as described for LH. Intra-assay and interassay coefficients of variation were calculated as 8 and 10%, respectively. The assay sensitivity was 0·125 ng FSH/tube.

Progesterone. Plasma progesterone concentrations were determined utilizing a commercially available solid-phase ^{125}I-labelled progesterone RIA kit (Diagnostic Products, Los Angeles, CA). The kit was designed for use as a direct (non-extraction method) assay of progesterone in human serum. The kit was modified and validated in our laboratory such that progesterone in horse plasma could be measured directly (Fay, 1983). Intra- and inter-assay variations were <10%. The sensitivity of the assay was 100 pg/ml or 10 pg/tube. Plasma samples from 12 dioestrous mares were assayed in both the direct assay and in a previously validated progesterone RIA which used ether extraction of sample, [^{3}H]progesterone-labelled antigen, and charcoal–dextran separation of bound from free hormone. Means for the 12 samples were 12·17 ng/ml for the standard progesterone assay and 12·28 ng/ml for the direct assay. Validations of follicular fluid progesterone concentrations were conducted as described above. Results from these experiments indicated that follicular fluid could not be assayed directly in the system outlined and therefore extraction with 10 volumes of ethyl ether followed by reconstitution in plasma from an ovariectomized mare was necessary before follicular fluid progesterone could be assayed in the direct-assay system utilized for plasma samples.

Androstenedione, testosterone and oestradiol-17β. Plasma androstenedione, testosterone and oestradiol-17β concentrations were determined by RIA utilizing a method described by Jawad & Wilson (1982) for the measurement of serum oestriol concentrations. The characteristics of the androstenedione antibody utilized have been described previously by Abraham *et al.* (1975). [^{3}H]Androstenedione (androst-4-ene-3,17-dione; NET 469) from New England Nuclear (Boston, MA) was utilized for the labelled antigen. There were no differences in plasma or follicular fluid samples which were extracted with 10 volumes of ethyl ether and reconstituted or were extracted and chromatographed on Sephadex LH:20 coulumns with hexane:benzene:methanol (85:5:5 by vol.) as the solvent system (Carr *et al.*, 1971). The chromatography step was therefore eliminated. The sensitivity of the assay was 2 pg/tube and the intra-and inter-assay variations were 8%.

The antibody used for the measurement of testosterone concentrations in plasma and follicular fluid was provided by Dr Gordon Niswender at Colorado State University. Characteristics of this antibody have been described previously (Niswender *et al.*, 1975). [^{3}H]Testosterone (NET 370) from New England Nuclear was used as the labelled antigen. As for androstenedione, there were no differences in samples that had been extracted or extracted and then chromatographed, and so the chromatography step was eliminated. Sensitivity of the assay was 2 pg/tube and the intra- and inter-assay variations were <8%.

The antibody used for the determination of oestradiol concentrations in plasma and follicular fluids was provided by Dr R. L. Butcher at West Virginia University. Characteristics of this antibody have been described previously (Butcher *et al.*, 1974). The labelled antigen was [^{3}H]oestradiol (NET 517) and was purchased from New England Nuclear. Plasma samples were extracted with 10 volumes of ethyl ether and then chromatographed (LH-20 columns) with methylene chloride:methanol (95:5 v/v) as the solvent system as described by Butcher *et al.* (1974). There were no differences in follicular fluid samples that were extracted and then chromatographed and samples that had been diluted 1:1000 in assay diluent and then assayed directly. Therefore all follicular fluid samples were assayed directly after dilution in assay buffer. Sensitivity of the assay was 1 pg/tube and the intra- and inter-assay variations were < 7%.

Calculation of RIA data

Data from all RIAs were analysed with a microcomputer program in which a logit-log transformation yielded linear standard curves ($r = 0.99$). Curve fit of the standard curve, maximum binding, non-specific binding and quality control samples were assessed before unknowns were computed with linear regression analysis. Appropriate corrections were made for steroid assays in which losses due to extraction and chromatography were noted.

Iodination of hormones for RIA and for receptor quantification

A lactoperoxidase enzymic iodination technique (Morrison & Bayse, 1970; Miyachi et al., 1972) was used for the iodination of hCG and LH. Separation of ^{125}I-labelled hormone from free ^{125}I was accomplished with chromatography on biogel P-60 utilizing 0.01 M-phosphate, 1% BSA as the eluting buffer. ^{125}I-labelled FSH for both RIA and receptor quantification was prepared with the minimum time, temperature and chloramine T technique as described by Leidenburger & Reichert (1972). The specific activity of these preparations was determined with a self-displacement radioreceptor assay as described by Diekman et al. (1978). Specific activity of the labelled preparations varied from 20 to 50 $\mu Ci/\mu g$ hormone depending on the iodination. Only one batch of labelled FSH or hCG was used to assess FSH and LH receptor concentrations. Maximum bindability was 25% for the FSH preparation and 50% for the hCG preparation.

Preparation of tissue for receptor quantification

Pairs of ovaries which were frozen at ovariectomy were thawed overnight at 4°C. Upon thawing, ovaries were photographed, weighed, measured and assigned an identification number. Every follicle greater than 5 mm was then dissected from each ovary and was also assigned an identification number. Follicles < 15 mm in diameter from one ovary were pooled by size for subsequent analyses. Every follicle ⩾ 15 mm was analysed separately. Follicle measurements included size, weight, volume of follicular fluid removed and scores for vascularity (a subjective evaluation with 1 being least vascular and 4 being most vascular). Follicular fluid was aspirated from individual or pools of follicles with a sterile syringe and 18-gauge needle. This follicular fluid was then centrifuged at 2000 g to collect granulosa cells suspended in the follicular fluid. Follicular fluid was then placed into individually labelled polypropylene vials and stored at $-20°C$ for future hormone analyses. Individual follicles were then incised and the thecal shell was completely scraped with a scalpel blade and rinsed with Tris buffer (0.05 M-Tris–HCl, 1 mM-CaCl$_2$, 1 mM-NaN$_3$ and 0.1% BSA) to collect granulosa cells. The thecal shell was rinsed again and the granulosa cells from each step were pooled and centrifuged at 2000 g. The resulting pellet was resuspended in Tris buffer and homogenized in a ground-glass tissue homogenizer. The thecal shell, after the granulosa cells were removed, was homogenized in a Sorvall motor driven tissue grinder. This mixture was then further homogenized in a ground glass tissue homogenizer and then filtered through four layers of cheesecloth to remove debris. The homogenates (granulosa and thecal cell) were centrifuged at 30 000 g for 30 min to recover a crude membrane preparation. The resulting pellet was diluted in the original volume of Tris buffer and resuspended in a ground-glass tissue homogenizer.

Receptor assay methodology

All receptor assays were conducted in Tris incubation buffer. When granulosa and thecal cell preparations were prepared, 200 μl of the membrane suspension was pipetted in duplicate into clean, disposable glass 12 × 75 mm culture tubes, and 100 μl additional buffer were also added to these tubes. Non-specific binding for every duplicate pair was determined by adding 1000-fold excess of unlabelled hormone to an additional set of tubes containing the sample. The numbers of FSH and LH binding sites were assessed by either Scatchard analysis, in which various quantities of labelled hormone (1.3–50.3 fmol) were added to a constant amount of sample, or by single-point saturation analysis in which a single dose of labelled hormone (a dose which was mid-point on the Scatchard curve) was added. Labelled hormones were prepared such that all doses were delivered in a 100 μl volume. Total volume of the reaction was then 400 μl. Tubes were mixed and allowed to incubate for 24 h at 27°C (a time and temperature previously demonstrated to achieve equilibrium conditions). At the end of incubation, 3 ml cold phosphate buffered saline (PBS) were added and the tubes were centrifuged at 2000 g for 20 min. Supernatant was decanted and an additional 3 ml PBS were added and centrifuged as before. All tubes were then counted in a Beckman auto-gamma counter with a counting efficiency of > 75%. Nonspecific binding was 0.5–1.5% for ^{125}I-labelled hCG and 1.0–2.0% for ^{125}I-labelled FSH assays. The remaining granulosa and thecal cell homogenates from each sample were then assayed for protein content by the method of Bradford (1976). Protein per tube ranged from 50 to 150 mg and all results are expressed per unit of protein.

Statistical analysis of data

All data were examined by analysis of variance (ANOVA). Amongst-mean comparisons of plasma hormone concentrations were made using Dunnett's t test (Gill, 1978), Duncan's multiple range test (Steele & Torrie, 1960) or Bonferronni's t test (Gill, 1978) when appropriate. Amongst-mean comparisons (among groups within follicle class and among classes of follicle within group of mares) of the data obtained from ovarian follicles were made with Bonferronni's t statistic. Due to heterogeneity of variance, some of the data (follicular fluid testosterone and LH concentrations) were transformed (log base 10) before statistical analysis. All data are presented in figures and tables are represented as mean ± s.e.

Results

Classification of ovulatory and non-ovulatory follicles

Follicles were classified as presumptive ovulatory or non-ovulatory based on the concentration of LH/hCG receptor in the granulosa cell component. For all data analyses and for presentation in the figures the follicles were divided into four classes.

Class 1: presumptive ovulatory follicles. These follicles contained significantly increased ($P < 0.001$) content of LH/hCG receptor in the granulosa cell layer and increased ($P < 0.001$) concentration of oestradiol in the follicular fluid.

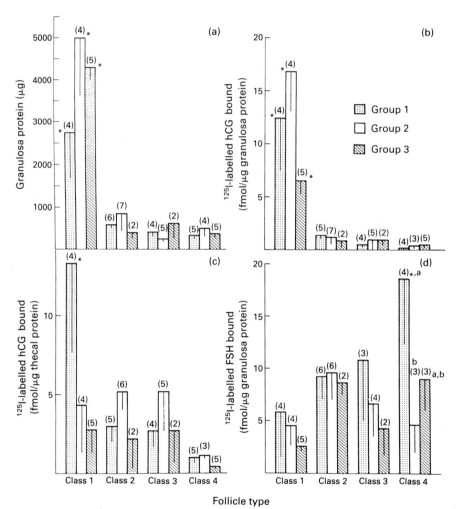

Fig. 1. Individual histograms representing (a) protein content of the granulosa cell layer, (b) binding of ^{125}I-labelled hCG per unit of granulosa cell protein, (c) binding of ^{125}I-labelled hCG per unit of thecal protein, and (d) binding of ^{125}I-labelled FSH per unit of granulosa cell protein in presumptive ovulatory (Class 1) and non-ovulatory follicles (Classes 2, 3 & 4; see text) removed from mares in Group 1, Group 2 or Group 3. Values are mean ± s.e.m. for the number of follicles indicated in parentheses. a, b. $P < 0.05$ for group means within a follicle class; *$P < 0.05$ for follicle class within a group.

Class 2: non-ovulatory follicles > 25 mm in diameter. These follicles were large in size but did not demonstrate the characteristics of the presumptive ovulatory follicle.

Class 3: non-ovulatory follicles 15–25 mm in diameter. These follicles were of intermediate size and did not possess any of the characteristics of the presumptive ovulatory follicle.

Class 4: non-ovulatory follicles < 15 mm in diameter. These were small follicles which were pooled for analysis and did not demonstrate any of the characteristics of the presumptive ovulatory follicles.

Binding of ^{125}I-labelled hCG to granulosa and thecal cell homogenates

In all mares there was one follicle which demonstrated an increased ($P < 0.001$) content of LH/hCG receptor in the granulosa cell component of the follicle (Fig. 1b). In 2 mares there were 2 such follicles present and one of these mares had a history of double ovulations. There was no significant difference in the content of LH receptor per unit protein of the presumptive follicles amongst the different groups of mares. However, receptor binding tended to increase from Day 14 after ovulation to Day 1 of oestrus and then fall sharply by the 4th day of oestrus. Linear Scatchard plots of the binding data revealed one homogeneous class of receptor whose affinity constant did not change significantly over the course of the experiment.

The presumptive ovulatory follicle also contained more ($P < 0.05$) protein in the granulosa cell layer when compared with other non-ovulatory follicles (Fig. 1a). The more mature presumptive ovulatory follicles of mares in Groups 2 and 3 tended to have more protein than those of Group 1. When total binding of ^{125}I-labelled hCG (by taking into account the increase in LH/hCG receptor/ unit protein as well as the increased protein content) in presumptive ovulatory follicles is compared with non-ovulatory follicles, the relative difference becomes even larger.

There was an increased ($P < 0.05$) amount of LH/hCG receptor in the thecal cell homogenate of presumptive ovulatory follicles of mares ovariectomized on Day 14 after ovulation when compared with the amount of thecal LH/hCG receptor present in other follicles from this group of mares (Fig. 1c). There were no differences amongst follicles in thecal LH/hCG receptor content in follicles from mares in Groups 2 or 3.

Binding characteristics of receptor for ^{125}I-labelled FSH

The binding characteristics to the granulosa cell homogenates are depicted in Fig. 1(d). Presumptive ovulatory follicles tended to have the lowest amount of FSH receptor with small follicles tending to have the highest amount of FSH receptor. Small non-ovulatory follicles (Class 4) from Group 1 mares had more ($P < 0.05$) FSH receptor when compared with other follicles from this group of mares or when compared with Class 4 follicles from Group 2 or 3 mares. There was no demonstrable FSH binding to the thecal component of any follicle.

Peripheral hormone concentrations

Plasma LH concentrations became elevated when plasma progesterone concentrations fell below 1 ng/ml. LH concentrations continued to increase until maximum concentrations were achieved at, or on the day after ovulation. On the day of ovariectomy, mean plasma LH concentration in Group 1 mares was 1.5 ± 0.2 ng/ml. LH concentrations were higher ($P < 0.01$) on the day of ovariectomy in Group 2 mares (6.9 ± 1.2 ng/ml) and were still increasing by the 4th day of oestrus in Group 3 mares (8.9 ± 2.1 ng/ml; Fig. 2a, inset).

Plasma FSH concentrations became elevated during the late oestrus and early dioestrus and maximum concentrations (10 ng/ml) between Days 5 and 10 after ovulation. Mean plasma FSH concentration was 6.6 ± 0.8 ng/ml in Group 1 mares on the day of ovariectomy and declined

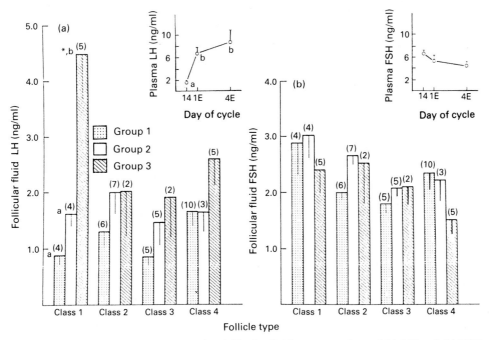

Fig. 2. Individual histograms representing follicular fluid concentrations of (a) LH and (b) FSH in presumptive ovulatory (Class 1) and non-ovulatory follicles (Classes 2, 3 & 4; see text) removed from mares in Groups 1, 2 and 3. Plasma concentrations of these hormones are represented in the insets. Values are mean ± s.e.m. for the number of follicles indicated in parentheses. a, b $P < 0.05$ for group means within a follicle class; *$P < 0.05$ for follicle class within a group.

gradually thereafter to levels approaching baseline concentrations as seen during pretreatment cycles (Fig. 2b, (inset).

Plasma progesterone concentrations became elevated above 1 ng/ml by 1 day after ovulation and rose sharply to reach maximum concentrations of about 10–12 ng/ml between Days 8 and 10 after ovulation. Progesterone concentrations declined gradually thereafter and fell below 1 ng/ml between Days 15 and 16 after ovulation. In Group 1 mares, plasma progesterone concentrations were still elevated (7.3 ± 2.4 ng/ml) on the day of ovariectomy whereas in Group 2 and 3 mares, progesterone concentrations were below 1 ng/ml on the day of ovariectomy (Fig. 3a, inset).

Plasma androstenedione concentrations were measured in selected daily samples. Androstenedione concentrations were measured on Days 13 and 14 after ovulation, Days −1 and 1 of oestrus, and the 3rd and 4th days of oestrus during the pretreatment and ovariectomy cycle. Androstenedione concentrations remained constant throughout late dioestrus and early oestrus with concentrations approaching 200 pg/ml (Fig. 3b, inset). Androstenedione concentrations increased ($P < 0.05$) by Day 4 of oestrus to levels approaching 450 pg/ml.

Plasma testosterone concentrations were also measured in selected daily samples. Testosterone concentrations were elevated (40–50 pg/ml) during mid–late dioestrus, fell gradually to reach baseline concentrations (10–20 pg/ml) at the beginning of oestrus with concentrations again becoming elevated during late oestrus (Fig. 3c, inset).

Oestradiol concentrations were low (4 pg/ml) during late dioestrus, becoming elevated at the onset of oestrus with significant elevations ($P < 0.05$) occurring by the 4th day of oestrus (Fig. 3d, inset).

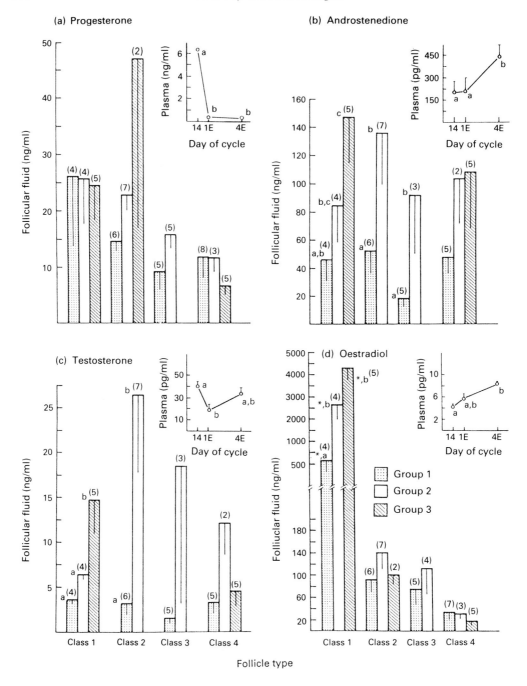

Fig. 3. Individual histograms representing follicular fluid concentrations of (a) progesterone, (b) androstenedione, (c) testosterone and (d) oestradiol in presumptive ovulatory (Class 1) and non-ovulatory follicles (Classes 2, 3 & 4; see text) removed from mares in Groups 1, 2 and 3. There were no data available in some classes of small follicles from Group 3 mares (progesterone, Class 2; androstenedione, Classes 2 & 3; testosterone, Classes 2 & 3; oestradiol, Class 3). Plasma concentrations of these hormones are presented in the insets. Values are mean $\pm$ s.e.m. for the number of follicles indicated in a parentheses. a, b, c $P < 0.05$ for group means within a follicle class; *$P < 0.05$ for follicle class within a group.

Follicular fluid hormone concentrations

The presumptive ovulatory follicle in all three groups of mares contained greater ($P < 0.001$) concentrations of oestradiol when compared with non-ovulatory follicles (Fig. 3d). Presumptive ovulatory follicles of Group 1 mares contained lower ($P < 0.01$) oestradiol concentrations than did the presumptive ovulatory follicles of mares in Groups 2 and 3.

There were no differences in follicular fluid androstenedione or testosterone concentrations between presumptive ovulatory and non-ovulatory follicles when compared within a group of mares (Fig. 3b, c). When androgen concentrations were examined within a class of follicle amongst groups of mares (i.e. with respect to time of ovariectomy and stage of follicle maturity), concentrations of both androstenedione and testosterone were significantly ($P < 0.05$) lower in the presumptive ovulatory and non-ovulatory follicles of Group 1 mares when compared with the same classes of follicles of Group 3 mares (and in some cases Group 2 mares). Androstenedione concentrations in Group 1 presumptive ovulatory follicles did not differ from non-ovulatory follicles from this group of mares, but when compared with presumptive ovulatory follicles of Group 3 mares, they were significantly ($P < 0.05$) lower. This relationship also held true for Class 2 and 3 follicles. Similar relationships also existed for testosterone concentrations. In small Class 4 follicles there were no differences in follicular fluid androstenedione concentration. Likewise, there were no differences in testosterone concentrations in Class 3 and 4 follicles.

Follicular fluid progesterone concentrations did not differ amongst groups within a class of follicle or amongst class of follicle among groups of mares (Fig. 3a). There was a trend, which approached significance, for large follicles to have more progesterone than smaller follicles.

FSH was detectable in the follicular fluid of all follicles (Fig. 2b). There were no differences in the concentration of FSH amongst groups of mares within a class of follicle or amongst classes of follicles in mares of a group.

LH was detectable in all follicles in concentrations below those seen in the peripheral plasma (Fig. 2a). Follicular fluid LH concentrations seemed to reflect peripheral plasma concentrations with follicular fluid concentrations rising from Day 14 after ovulation (Group 1 mares) to the 4th day of oestrus (Group 3 mares) within each class of follicle. Follicular fluid LH concentrations were significantly ($P < 0.01$) higher in the presumptive ovulatory follicles of Group 3 mares when compared with the presumptive ovulatory follicles of Group 1 and 2 mares. In addition, follicular fluid LH concentrations of presumptive ovulatory follicles of Group 3 mares were significantly higher ($P < 0.05$) than those in non-ovulatory follicles within this group of mares.

Physical properties of presumptive ovulatory follicles

The presumptive ovulatory follicle of Group 1 mares was similar in size to other large follicles present in 2 of the mares. In the other 2 mares it was the largest follicle present. In Group 2 and 3 mares, the presumptive ovulatory follicle was the largest follicle present ($P < 0.01$). The presumptive ovulatory follicle of Group 2 and 3 mares was larger ($P < 0.05$) than that of Group 1 mares. The size of the presumptive ovulatory follicle did not differ between mares in Groups 2 and 3 (Table 1).

Table 1. Physical characteristics of the ovulatory follicle

	Group 1	Group 2	Group 3
Size (mm)	28.7 ± 1.5^a	40.7 ± 2.8^b	36.7 ± 1.0^b
Follicular fluid volume (ml)	17.3 ± 4.0^a	35.5 ± 13.3^b	24.8 ± 1.7^{ab}
Vascularity	2.25^a	3.25^a	3.0^a

Means with different superscripts are significantly different ($P < 0.05$).

The presumptive ovulatory follicle from all groups contained the largest volume of follicular fluid. Follicles >15 mm in diameter demonstrated a significant linear relationship between the volume of follicular fluid recovered and follicle size.

The presumptive ovulatory follicle in all groups was more vascular ($P < 0.05$) when compared with non-ovulatory follicles. There were no differences in vascularity scores of presumptive ovulatory follicles when compared amongst groups of mares.

The influence of presumptive ovulatory follicle growth on ovarian size and weight measurements can be seen in Table 2. Within Groups 2 and 3, when the ovary containing the presumptive ovulatory follicle was compared with the opposite ovary, the ovary containing the presumptive ovulatory follicle was significantly ($P < 0.05$) larger and heavier. This relationship was not significant for Group 1 mares, in which the ovary containing the presumptive follicle was similar in size to the opposite ovary.

Table 2. Physical characteristics of ovaries

	Mean size (cm)		Mean weight (g)	
	Ovary with ovulatory follicle	Ovary without ovulatory follicle	Ovary with ovulatory follicle	Ovary without ovulatory follicle
Group 1	5.4 ± 0.28^a	4.6 ± 0.23^a	109.4 ± 9.6^a	76.9 ± 10.0^a
Group 2	6.5 ± 0.56^a	5.0 ± 0.41^b	164.5 ± 41.3^a	78.8 ± 19.4^b
Group 3	6.3 ± 0.45^a	5.1 ± 0.55^b	148.0 ± 28.9^a	87.8 ± 28.5^b

Means with different superscripts were significantly different from one another ($P < 0.05$).

Discussion

The results of this experiment clearly demonstrate that in the mare, as in the ewe and the cow, the presumptive ovulatory follicle can be identified during the peri-ovulatory period by several physiological markers which include: (1) the presence of an increased number of LH/hCG receptors in the granulosa cell layer, and (2) the presence of high follicular fluid oestradiol concentrations. In addition, these follicles were the most vascular and by the first day of oestrus were the largest follicles present in the ovaries. The presence of an increased granulosa cell layer protein content in presumptive ovulatory follicles presumably reflects an increased number of granulosa cells, also characteristic of healthy, metabolically active follicles (Peters & McNatty, 1980). These data also suggest that the presumptive ovulatory follicle had been selected by Day 14 after ovulation since a single large follicle demonstrating all the characteristics defined above was present in all mares at this time.

In 9 of the 11 mares studied, there was only one follicle present which demonstrated the characteristics of the ovulatory follicle. In 2 mares, there were two such follicles present and one of these mares had a history of repeated double ovulations. Therefore the number of 'activated' or presumptive ovulatory follicles was equal to the expected ovulation rate for this population of mares. This phenomenon has also been reported for sheep (Webb & England, 1982a) and cattle (Staigmiller *et al.*, 1982; Ireland & Roche, 1983).

LH/hCG receptor content/unit protein in the presumptive ovulatory follicle was not different among groups, but a tendency towards decreased numbers of LH/hCG receptors in Group 3 mares approached significance. This phenomenon would support the concept of down-regulation, since, by the 4th day of oestrus, mares had been exposed to 3–4 days of increased concentrations of LH (4–5-fold increased above baseline) although maximum LH concentrations in the plasma had not been attained. In the ewe and the cow (Webb & England, 1982b; Staigmiller *et al.*, 1982) the binding of labelled hCG to granulosa and thecal cells of ovulatory follicles decreased significantly when data from animals ovariectomized after the LH surge were compared with results from animals which were ovariectomized before or during the ascending limb of the LH surge. In the rat, prolonged elevations in plasma LH have been shown to decrease the total numbers of LH receptors per unit tissue by mechanisms other than occupation (Richards *et al.*, 1976; Rao *et al.*, 1977).

The increased content of LH/hCG receptor in the thecal component of the presumptive ovulatory follicle of Group 1 mares when compared with non-ovulatory follicles from this same group of mares may be indicative of the mechanism whereby one follicle becomes dominant. An increased thecal LH/hCG receptor content in this follicle would allow for an increased ability to produce aromatizable substrate for oestradiol production when compared with non-ovulatory follicles. Bogovich *et al.* (1981) suggested that presumptive ovulatory follicles have a requirement for LH which acts to increase thecal cell function and androgen synthesis, thereby selectively permitting follicles to produce oestradiol, a critical step in the final maturation of follicles destined to ovulate. By the first day of oestrus, there were no differences in thecal LH/hCG receptor content when ovulatory and non-ovulatory follicles were compared.

LH and FSH concentrations in horse follicular fluid have not been previously reported. In the present study, LH and FSH were detectable in the follicular fluid of all follicles in concentrations at or below those present in peripheral plasma. In particular, follicular fluid concentrations of LH seemed to reflect peripheral plasma values with concentrations rising from Day 14 after ovulation through to Day 4 of oestrus in all groups of follicles. Highest follicular fluid LH concentrations occurred in follicles from Group 3 mares which had been exposed to elevated peripheral LH concentration for 3–4 days. The marked elevation in LH concentrations in presumptive ovulatory follicles of Group 3 mares presumably is a result of the increased vascularity of these follicles when compared with non-ovulatory follicles. These results agree with published data from other species in which follicular fluid concentrations of gonadotrophins were examined (McNatty *et al.*, 1981; Fortune & Hansel, 1985).

Follicular fluid steroid hormone patterns demonstrated consistent cyclic changes which reflected the health and physiological maturity of the follicles studied. As the presumptive ovulatory follicle matured from Day 14 after ovulation to the 4th day of oestrus, follicular fluid oestradiol concentrations continued to rise. Follicular fluid concentrations of oestradiol in presumptive ovulatory follicles from Group 3 mares were 9 times higher than those found in presumptive ovulatory follicles of Group 1 mares. Maximum oestradiol concentrations in presumptive ovulatory follicles were 30–50-fold higher than maximum concentrations attained in non-ovulatory follicles. The values for both presumptive ovulatory and non-ovulatory follicle follicular fluid oestradiol concentrations are consistent with previously reported values for the mare (YoungLai, 1971; Kenney *et al.*, 1979; Silberzahn *et al.*, 1985) as well as for other species (Moor *et al.*, 1978; McNatty *et al.*, 1981; Webb & England, 1982a, b; Ireland & Roche, 1983; Fortune & Hansel, 1985). Collectively, these data indicate that follicular fluid oestradiol concentrations are a reliable indicator of follicular growth and maturity in a number of species including the mare.

Follicular fluid androgen concentrations (androstenedione and testosterone) did not differ between presumptive ovulatory and non-ovulatory follicles within each group studied. When follicular fluid androgen concentrations were compared over the course of the experiment, the relationship between LH stimulation of androgen biosynthesis became clearly evident. Follicular fluid androgen concentrations in follicles removed from mares in Groups 2 and 3 were significantly elevated when compared with concentrations present in the follicular fluid of Group 1 mares. Peripheral plasma concentrations of LH also became elevated during this period. The phenomenon of LH-dependent androgen biosynthesis has been well studied *in vivo* and *in vitro* (Fortune & Armstrong, 1977) and results from the present in-vivo experiment agree well with the previously published data. Kenney *et al.* (1979) reported elevated androgen concentrations in viable follicles and concluded that the increased androgen was not causing atresia. Collectively, these data would indicate that increased androgen concentrations in the follicular fluid are a consequence of increased LH stimulation and, at least in the mare, do not influence atresia. As was evident in the present experiment, non-ovulatory follicles are also capable of responding to LH stimulation with resultant increased follicular fluid androgen concentrations. In these follicles, there seems to be a lack of conversion from androgen to oestrogen, presumably from a deficiency of aromatase activity.

Follicular fluid progesterone concentrations in the present experiment did not differ among groups. There was a trend for large follicles (primarily the presumptive ovulatory follicle) to have higher progesterone concentrations which approached significance. It seems reasonable to suggest that progesterone concentrations may have continued to increase had we carried the experiment closer to the time of ovulation.

In summary, the results of the present experiment clearly demonstrate that the presumptive ovulatory follicle can be identified by various physiological and biochemical markers as early as Day 14 after ovulation. In addition, a consistent pattern of follicular fluid hormones which reflect the maturity and physiological state of the follicle has been defined and allows comparison of presumptive ovulatory and non-ovulatory follicles.

We thank Dr O. J. Ginther, Dr G. D. Niswender, Dr R. Butcher and Dr G. Abraham as well as the NIAMDD for the generous supply of antisera and purified hormones; the farm crew of the Department of Veterinary science for assistance with animal care throughout the course of the experiment; Dr M. J. Evans, Dr R. G. Loy, Dr S. J. Legan and Dr P. J. Burns for thoughtful comments during the design of this experiment; and Ms Lelia Garrison and Ms Susan Hipkins for assistance in the laboratory. Supported in part by a grant from the Grayson Foundation.

References

Abraham, G.E., Manlimos, F.S., Solis, M. & Wickman, A.C. (1975) Combined radioimmunoassay of four steroids in one ml of plasma. 11. Androgens. *Clin. Biochem.* **8,** 374–378.

Bogovich, K., Richards, J.S. & Reichert, L.E. (1981) Obligatory role of luteinizing hormone (LH) in the initiation of preovulatory follicle growth in the pregnant rat: specific effects of human chorionic gonadotrophin and follicle stimulating hormone on LH receptors and steroidogenesis in theca, granulosa and luteal cells. *Endocrinology* **109,** 860–867.

Bradford, M.M. (1976) A rapid and sensitive method for the quantification of microgram quantities of protein utilizing the principle of protein-dye binding. *Analyt. Biochem.* **72,** 248–254.

Braselton, W.E. & McShan, W.H. (1970) Purification and properties of follicle stimulating and luteinizing hormones from horse pituitary glands. *Archs Biochem. Biophys.* **139,** 45–58.

Burns, P.J. & Douglas, R.H. (1981) Effects of daily administration of estradiol on follicular growth, ovulation and plasma hormones in mares. *Biol. Reprod.* **24,** 1026–1031.

Burns, P.J., Jawad, M.J., Edmundson, A., Cahill, C., Boutcher, J.K., Wilson, E.A. & Douglas, R.H. (1982) Effect of increased photoperiod on hormone concentrations in Thoroughbred stallions. *J. Reprod. Fert., Supp.* **32,** 103–111.

Butcher, R.L., Collins, W.E. & Fugo, N.W. (1974) Plasma concentrations of LH, FSH, Prolactin, progesterone and estradiol throughout the 4-day estrous cycle of the rat. *Endocrinology* **94,** 1704–1708.

Carr, B.R., Mikhail, G. & Flickinger, G.L. (1971)

Column chromatography of steroids on Sephadex LH-20. *J. clin. Endocr. Metab.* **33,** 358–359.

Diekman, M. A., O'Callaghan, P., Nett, T.M. & Niswender, G.D. (1978) Quantification of occupied and unoccupied luteal receptors for LH throughout the estrous cycle and early pregnancy in ewes. *Biol. Reprod.* **19,** 999–1009.

Dufour, J., Whitmore, H.L., Ginther, O.J. & Casida, L.E. (1972) Identification of the ovulating follicle by its size on different days of the estrous cycle in heifers. *J. Anim. Sci.* **34,** 85–91.

England, B.G., Webb, R. & Dahmer, M.K. (1981) Follicular steroidogenesis and gonadotropin binding to ovine follicles during the estrous cycle. *Endocrinology* **109,** 881–887.

Fay, J.E. (1983) *The presumptive ovulatory follicle in horse mares: changes in thecal and granulosa cell luteinizing hormone and follicle stimulating hormone receptor content associated with follicular fluid and peripheral plasma hormone concentrations.* Ph.D. dissertation, University of Kentucky, Lexington.

Fortune, J.E. & Armstrong, D.T. (1977) Androgen production by theca and granulosa isolated from proestrus rat follicles. *Endocrinology* **100,** 1341–1347.

Fortune, J.E. & Hansel, W. (1985) Concentrations of steroids and gonadotropins in follicular fluid from normal heifers and heifers primed for superovulation. *Biol. Reprod.* **32,** 1069–1079.

Freedman, L.J., Garcia, M.C. & Ginther, O.J. (1979) Influence of photoperiod and ovaries on seasonal reproductive activity in mares. *Biol. Reprod.* **20,** 567–574.

Gill, J.L. (1978) *Design and Analysis of Experiments in* the *Animal and Medical Sciences*, vol. 1. Iowa State University Press, Ames.

Ireland, J.J. & Roche, J.F. (1983) Development of non-ovulatory antral follicles in heifers: changes in steroids in follicular fluid and receptors for gonadotropins. *Endocrinology* **112**, 150–156.

Jawad, M.J. & Wilson, E.A. (1982) Effect of dextran-coated charcoal on radioimmunoassays of unconjugated estriol in serum. *J. Steroid Biochem.* **16**, 797–800.

Kenney, R.M., Condon, W., Ganjam, V.K. & Channing, C. (1979) Morphological and biochemical correlates of equine ovarian follicles as a function of their state of viability or atresia. *J. Reprod. Fert., Suppl.* **27**, 163–171.

Leidenburger, F. & Reichert, L.E., Jr (1972) Evaluation of a rat testis homogenate radioligand receptor assay for human pituitary LH. *Endocrinology* **91**, 901–909.

McNatty, K.P., Gibb, M., Dobson, C., Thurley, D.C. & Findlay, J.K. (1981) Changes in the concentration of gonadotrophic and steroidal hormones in the antral fluid of ovarian follicles throughout the oestrous cycle of the sheep. *Aust. J. biol. Sci.* **34**, 67–80.

Miyachi, Y., Vaitukaitis, J.L., Nieschlag, E. & Lipsett, M. (1972) Enzymic radioiodination of gonadotropins. *J. clin. Endocr. Metab.* **34**, 23–28.

Moor, R.M., Hay, M.F., Dott, H.M. & Cran, D.G. (1978) Macroscopic identification and steroidogenic function of atretic follicles in sheep. *J. Endocr.* **77**, 309–318.

Morrison, M. & Bayse, G.D. (1970) Catalysis of iodination by lactoperoxidase. *Biochemistry, N.Y.* **9**, 2995–3000.

Niswender, G.D., Nett, T.M., Meyer, D.L., & Hagerman, D.D. (1975) Factors influencing the specificity of antibodies to steroid hormones. In *Steroid Immunoassay*, pp. 61–66. Eds E. H. D. Cameron, S. G. Hillier & K. Griffiths. Apha Omega Alpha Publishing, Cardiff.

Peters, H. & McNatty, K.P. (1980) *The Ovary: a Correlation of Structure and Function in Mammals*, 86 pp. University of California Press, Berkeley.

Rao, M.C., Richards, J.S., Midgley, A.R., Jr & Reichert L.E., Jr (1977) Regulation of gonadotropin receptors by LH in granulosa cells. *Endocrinology* **101**, 512–523.

Richards, J.S., Ireland, J., Rao, M.C., Bernath, G.A., Midgley, A.R., Jr & Reichert, L.E., Jr (1976) Ovarian follicular development in the rat: hormone receptor regulation by estradiol, follicle stimulating hormone and luteinizing hormone. *Endocrinology* **99**, 1562–1576.

Scott, E.A. & Kunze, D.J. (1977) Ovariectomy in the mare: presurgical, surgical and post-surgical considerations. *J. equine Med. Surgery* **1**, 5–12.

Silberzahn, P., Almahbobi, G., Dehennin, L. & Merouane, A. (1985) Estrogen metabolites in equine ovarian follicles: gas chromatographic-mass spectrometric determinations in relation to follicular ultrastructure and progestin content. *J. Steroid Biochem.* **22**, 501–505.

Staigmiller, R., England, B.G., Webb, R., Short, R.E. & Bellows, R.A. (1982) Estrogen secretion and gonadotropin binding by individual bovine follicles during estrus. *J. Anim. Sci.* **55**, 1473–1482.

Steele, R.E. & Torrie, J.H. (1960) *Principles and Procedures of Statistics*. McGraw-Hill Book Co., New York.

Webb, R. & England, B.G. (1982a) Identification of the ovulatory follicle in the ewe: associated changes in follicular size, thecal and granulosa LH receptors, antral fluid steroids and circulating hormones during the periovulatory period. *Endocrinology* **110**, 873–881.

Webb, R. & England, B.G. (1982b) Relationship between LH receptor concentrations in thecal and granulosa cells and in-vivo and in-vitro steroid secretion by ovine follicles during the preovulatory period. *J. Reprod. Fert.* **66**, 169–180.

YoungLai, E.V. (1971) Steroid content of the equine ovary during the reproductive cycle. *J. Endocr.* **50**, 589–597.

J. Reprod. Fert., Suppl. **35** (1987), 183–189

Influence of follicular status on twinning rate in mares

R. R. Pascoe, D. R. Pascoe and M. C. Wilson

Oakey Veterinary Hospital, Oakey, Queensland 4401, Australia

Summary. Between 1982 and 1985, 1015 mares were evaluated using the following parameters: age, mare status (maiden, barren, lactating), Caslick index, Caslick operation, ovarian cycle, ovarian and follicular size, treatments (hCG and intrauterine infusions), number of ovulations after mating (184 mares), number of conceptuses present, dimensions of embryonic vesicles, and pregnancy status 45 days after mating. Mares were examined ultrasonographically between Days 12 and 25 to detect the presence of embryonic vesicles and measure the dimensions of each vesicle. The data were analysed by a stepwise logistic regression method.

Mare and follicular status were significant ($P < 0.005$) predictors of pregnancy outcome. Results from 2949 mare cycles (268 maiden; 1047 lactating; 618 barren) for twin conceptus rates were 15·3%, 8·8% and 14·0% respectively. Based on follicle status, twin-conceptus rates with a single palpable follicle >25 mm in diameter were 11% in maiden, 11% in barren and 5·3% in lactating mares. In 23·7% of cycles (22·4%, 28·8% and 20·7%) 2 palpable follicles >25 mm were recorded, resulting in twin conception rates of 30%, 23·3% and 22% in maiden, barren and lactating mares, respectively.

Twin conceptus rate for all mares with a single follicle >25 mm was 7·6% and with 2 follicles >25 mm 23·6%. Mares with 2 large follicles (>40 mm) at mating had a 38·2% (29/76) twin conception rate. Mares with known synchronous double ovulations had a 40% (22/55) twin conception rate.

An analysis of co-variance was used to calculate the daily embryonic growth rate between Days 10 and 16 of gestation for 11 Standardbred mares. A quadratic growth model was predicted ($P < 0.01$) with an estimated mean daily growth rate of $4·14 \pm 0·15$ mm for a Day-14 embryonic vesicle. The calculated age difference between twin embryonic vesicles based on the size difference between vesicles suggests that 96·5% (55/57) of the embryonic vesicles resulted from synchronous ovulations.

Introduction

'Twin pregnancy' remains a serious problem for all horse breeders, particularly those concerned with Thoroughbred and Standardbred mares. Various management practices have been used to attempt to reduce the incidence of twin conceptions. Commonly, mating of the mare was avoided when 2 or more large follicles were detected on rectal palpation. Alternatively, when synchronous ovulations (<48 h) apart) were encountered, mating was delayed for 18–24 h after the first follicle ovulated (Ginther *et al.,* 1982a). Both management systems result in up to 17% loss of available oestrus cycles in a breeding season (Ginther, 1982a,b) and twin conceptions still occurred in about 4·5% of the mares that were mated (Ginther *et al.,* 1982a).

Recent studies have indicated that twin conceptuses occurred in 20% of 127 mares with 2 asynchronous ovulations (>48 h apart) and none occurred in 99 mares with synchronous ovulations (<48 h apart) (Ginther *et al.,* 1982b; Ginther, 1983a,b). In a similar study with Thoroughbred and Standardbred mares, twin conceptuses occurred in 0% (0/57) of mares with synchronous ovulations and 7% (2/31) of mares with asynchronous ovulations (Woods *et al.,* 1983). Ginther *et al.* (1982b) have postulated that the reason for the absence of detectable twin conceptuses after synchronous ovulations, as well as the low percentage of twin births or twin abortions, is probably due to a biological embryo-reduction mechanism.

Ginther *et al.* (1982a) and Woods *et al.* (1983) have shown that the single pregnancy rate was significantly higher in mares with double synchronous ovulations (83% and 74%, respectively) than in mares with single ovulations (54% and 54%, respectively). Furthermore, these authors suggested that, because of the higher single conception rate with synchronous double ovulations, a more efficient method of breeding would be to mate mares regardless of the number of large follicles present.

In breeding practice, however, mares with twin conceptuses are of financial significance because of the eventual high loss associated with abortion or stillbirth. Jeffcott & Whitwell (1973) reported a 64·5% abortion or stillbirth rate for twin conceptuses with only 18 of the 124 (14·5%) possible foals reaching the 2nd week of neonatal life. Pascoe (1983), in another study of 130 mares carrying twins, reported that only 11% produced viable foals and, furthermore, the same group of mares produced only 38% viable foals the following year. These results suggest that twinning significantly reduced foaling not only in the twinning year, but also in the subsequent mating season.

Diagnostic ultrasonography has allowed recognition of twin conceptuses before Day 30 of gestation (Palmer & Driancourt, 1980). In addition, the identification of a 'mobile phase' (Day 10–Day 16) (Ginther, 1983a) and a 'fixation period' (> Day 16), for the early embryonic vesicles as well as the changes within the embryonic vesicle as the fetus develops (> Day 21) (Ginther, 1983b) has afforded veterinarians the opportunity to identify and dispose of one or both twin conceptuses before Day 36 of gestation.

The purpose of the present study was to record all palpable follicle sizes (⩾ 25 mm in diameter) before mating. Irrespective of the number of follicles present at the time, mares were mated and, in conjunction with ultrasonography, pregnancy data were obtained to derive: (1) the probability of single or twin conceptuses occurring in the mare, knowing the follicular status of the mare at the time of mating; (2) the probability of twin conceptuses occurring in the mare with known synchronous ovulations; and, based on the size difference between each vesicle in a pair of twin conceptions, (3) the likelihood that synchronous ovulations occurred.

Materials and Methods

Two studies were conducted during the 1982–85 breeding seasons on 1026 Thoroughbred and Standardbred mares aged 2–24 years. The mares were maintained on 12 broodmare farms which were visited on alternate days for routine fertility management. All mares were grazed under open paddock conditions on pasture crops (barley and oats in early spring, and alfalfa and millet in late spring and summer). The first study, on 1015 mares, investigated the probability of single or twin conceptuses occurring regardless of the mare's follicular status at the time of mating, the probability that twin conceptuses occurred from synchronous double ovulations (184 pregnant mares), and the likelihood that 2 synchronous ovulations occurred based on the size difference between each vesicle in a pair of twin conceptuses (57 mares). The second study on 11 mares (Standardbred) investigated the growth rate of single conceptuses between Days 10 and 16 of gestational age.

Study 1. Maiden and barren mares were palpated at the onset of the first detected oestrous period and lactating mares at the first oestrous period after foaling, and then all mares were palpated on alternate days to detect the number and size of all ovarian follicles. Most mares were treated with hCG (5000 i.u., i.v. in September; 2500 i.u., i.v., October–December) (Burns Veterinary Supply, Oakland, CA) 12 h before mating and were usually not examined for ovulation. Failure to show oestrus 48 h after the last examination was used as an indicator of ovulation. However, 184 pregnant mares were examined for ovulation, and double ovulations were recorded as synchronous if they occurred within 48 h of the last rectal examination.

A total of 2949 breeding cycles was evaluated using the following information: age, mare status, Caslick index, Caslick operation, ovarian cycle, ovarian and follicular sizes, examination and mating dates, treatments (hCG, intra-uterine antibiotics), number of ovulations after mating (184 pregnant mares), number of conceptuses detected, presence and location of endometrial cysts, dimensions of embryonic vesicles (ultrasound), treatment with progesterone after mating, and pregnancy status 45 days after mating.

Study 2. Eleven Standardbred mares maintained on one farm, were examined daily by ultrasonography and rectal palpation of the genital tract from the first day of oestrus. The mares were inseminated on alternate days during the oestrous period and the day of ovulation was recorded as Day 0.

Ultrasonography. Diagnostic ultrasonography (in 1982–83; Fisher Vetscan, 3·5 MHz; Solus Xray, Brisbane; since 1984, Technicare 210, 5 MHz; Pitman Moore, Englewood, CO, U.S.A) was used to determine the number of conceptuses in the uterus. Embryonic motility (Ginther, 1983a) was used to differentiate embryos from endometrial cysts between Days 12 and 16 of gestation. Uniformity in embryonic size, location of conceptuses within the uterus, and lack of known endometrial cysts from previous ultrasonographic examinations were used to determine the presence of conceptuses between Days 18 and 21 of gestation. After Day 21 of gestation, the presence of developing fetal structures within the embryonic vesicle confirmed pregnancy (Ginther, 1983b).

Mares were examined by emptying the rectum and advancing a lubricated (ultrasound gel) probe to the level of the uterine bifurcation. The probe was moved along the left uterine horn to the left ovary and, in a slow continual sweep, passed over the uterine horns from the left ovary to the right ovary, and then back down the right uterine horn to the uterine bifurcation. The uterine body was examined using a slow sigmoid sweep with the probe from the bifurcation to the cervix. All non-echogenic circular areas were evaluated and measured.

In Study 1, ultrasonography was performed on all mares between Day 17 and Day 25 (1982–84) and between Day 12 and Day 25 (1985) of gestation to detect pregnancy. All pregnant mares were re-examined after Day 42 of gestation.

In Study 2, ultrasonography was used to confirm ovulation. Each mare was examined and scanned daily from Day 10 to Day 16 to detect the presence of an embryonic vesicle. The size of each embryonic vesicle (width, height) was recorded.

Statistical analysis. Data in Study 1 were analysed using a stepwise logistic regression program (Engleman, 1985). Pregnancy status (single or twin) was used as the dependent variable, and step selection to enter or remove a variable was based on an asymptotic covariance estimate (ACE) with a P value to remove a variable set at 0·15 and P value to enter a variable set at 0·1.

Recordings for each cycle were transformed to fit into 6 categories, based on the number of follicles (>25 mm) present at the time of examination and size of the individual follicles: 1 follicle <40 mm in size (Category 1); 1 follicle >40 mm in size (Category 2); 2 follicles, each between 25 and 40 mm in size and ≤10 mm difference in size (Category 3); 2 follicles both >40 mm in size and ≤10 mm difference in size (Category 4); 2 follicles (>25 mm) with a >10 mm and <20 mm size difference between the 2 follicles (Category 5); 2 follicles (>25 mm) with >20 mm difference in size between the two follicles (Category 6).

Data in Study 2 were analysed using an analysis of covariance (Jeanrich *et al.*, 1985). Day of gestation was used as the covariate and embryonic vesicle size (mm) was elected as the dependent variable.

Results

Study 1

The stepwise logistic regression analysis produced a statistical model which predicted the probable pregnancy outcome (single or twin conceptuses) based on the mare's status and the follicular status of the mare's ovaries at the last examination before mating. Mare and follicular status, were significant ($P <0.005$ level) predictors of pregnancy outcome. The observed and predicted twin conceptus rate for mares with a single follicle palpated at the last examination before mating are shown in Table 1. Similarly, the observed and predicted twin conceptus rate for mares with 2 follicles palpated at the last examination before mating are shown in Table 2.

Table 1. Probability of twin conceptuses occurring from known mare status with single follicles of > 25 mm in diameter

Mare status	Follicle diameter	Twin conceptus rate	
		Observed* %	Expected† %
Maiden	Single <40 mm	16·7 (12/72)	12·3 (9/72)
	Single >40 mm	8·1 (11/136)	10·4 (14/136)
Barren	Single <40 mm	13·0 (18/137)	10·6 (15/137)
	Single >40 mm	10·0 (29/289)	8·9 (26/289)
Lactating	Single <40 mm	4·2 (10/241)	6·9 (17/241)
	Single >40 mm	5·8 (34/589)	5·8 (34/589)

*Actual number of twin conceptuses detected.
†Closest approximate number of twin conceptuses expected based on the total number of mares observed.

Table 2. Probability of twin conceptuses occurring from known mare status with two follicles > 25 mm in diameter

Mare status	Follicle diameter	Twin conceptus rate	
		Observed* %	Expected† %
Maiden	Two, > 25 mm, < 40 mm and ≤ 10 mm difference	22·7 (5/22)	31·0 (7/22)
	Two, > 40 mm and ≤ 10 mm difference	44·4 (4/9)	47·6 (4/9)
	Two, > 25 mm, and > 10 mm and < 20 mm difference	16·7 (1/6)	21·3 (1/6)
	Two, > 25 mm and > 20 mm difference	34·8 (8/23)	24·0 (6/23)
Barren	Two, > 25 mm, < 40 mm and < 10 mm difference	23·6 (17/72)	27·5 (20/72)
	Two, > 40 mm and ≤ 10 mm difference	42·9 (12/28)	43·4 (12/28)
	Two, > 25 mm, and > 10 mm and < 20 mm difference	12·5 (3/24)	18·6 (4/24)
	Two, > 25 mm and > 20 mm difference	16·7 (8/48)	21·0 (10/48)
Lactating	Two, > 25 mm, < 40 mm and ≤ 10 mm difference	24·4 (21/86)	19·1 (16/86)
	two, > 40 mm and ≤ 10 mm difference	33·3 (13/39)	32·2 (13/39)
	Two, > 25 mm, and > 10 mm and < 20 mm difference	19·2 (5/26)	12·5 (4/26)
	Two, > 25 mm and > 20 mm difference	13·6 (9/66)	14·3 (9/66)

*Actual number of twin conceptuses detected.
†Closest approximate number of twin conceptuses expected based on the total number of mares observed.

Synchronous double ovulations were recorded in 55 mares with 40% (22/55) of the mares carrying twin conceptuses. Single ovulations were recorded in 129 mares and 10·9% (14/129) of these mares were carrying twin conceptuses.

The overall pregnancy rate per cycle was 68·2% (268/393) for maiden mares, 66·8% (598/895) for barren mares, and 63% (1047/1661) for lactating mares. In all mares the combined occurrence of twin conceptuses for a single follicle was 7·8% (114/1464) and for 2 follicles was 23·6% (106/449). During this 4-year study, 207 mares conceived twins and only 14 (6·8%) of these mares conceived twins at subsequent matings, 4 in the same year, 6 the following year and 2 each at 2 and 3 years after the 1982 breeding season.

Study 2

The analysis of covariance considered polynomial growth functions and indicated that, between 10 and 16 days of gestation, a quadratic growth model provided the best fit. This model ($P < 0.01$)

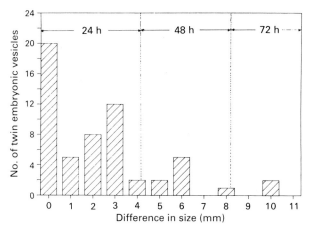

Fig. 1. Size difference between individual pairs of embryonic vesicles of a gestational age between Days 13 and 16. Each 24-h period represents the calculated age difference between each individual pair of embryos based on a daily growth rate of 4·14 ± 0·15 mm (see text).

provided a significantly better fit than a linear growth model, but no additional improvement was obtained with a cubic model. The estimated mean (±s.e.m.) daily growth rate for a Day-14 embryonic vesicle was 4·14 ± 0·15 mm.

Twin conceptuses between Day 13 and Day 16 were detected in 57 of the mares. The size difference (mm) between each embryonic vesicle of a pair is represented in Fig. 1. For each mare, the age difference in hours (hi) calculated as hi = (d/slope) × 24 h where hi is the age (h) difference between each embryonic vesicle, d is the difference in size (mm) between the 2 embryonic vesicles, and slope is the estimated mean growth rate of 4·14 ± 0·15 mm for a Day-14 embryonic vesicle, derived from the quadratic model. The age difference between twin embryonic vesicles was <48 h in 96·5% (55/57) of mares and in the remaining 2 mares the embryonic vesicles were <72 h different in age (Fig. 1).

Discussion

From these results it can be concluded that, for this population of pregnant mares, the majority of twin conceptuses occur in 23·6% of mares which had 2 follicles >25 mm at the time of the last examination before mating. The presence of 2 large follicles (>40 mm) at the time of mating greatly increases the chances of twin conceptuses (38·2%, 29/76) occurring regardless of the synchrony of the ovulations. Twin conceptuses occurred in 40% (22/55) of mares which had synchronous double ovulations and, using the estimated age difference between each embryonic vesicle of a pair between Day 13 and Day 16 of gestation, 96·5% (55/57) of the twin conceptuses which resulted originated from synchronous double ovulations.

The logistic regression provided a good model to predict the probable pregnancy outcome in this population of mares when the status and follicular activity of the mares are known. The model predicts the probable twin conceptus rate regardless of the time or number of ovulations. The incidence of twin conceptuses occurring was approximately 50% lower in lactating mares (5·3%) than for maiden (11%) or barren (11%) mares when a single follicle was palpated on the ovaries (Table 1). However, 2 follicles were recorded in 20·7% of the cycles for lactating mares, in 22·4% of the cycles for maiden mares, and in 28·8% of the cycles for barren mares and there were associated twin conception rates of 22%, 23·3% and 30%, respectively (Table 2). These results suggest that the presence of 2 follicles and the occurrence of multiple ovulations as judged by the twin conception rate was not markedly altered by the mare's status, whereas the decreased incidence of twin conceptuses in lactating mares with single follicles may be related to the suppressive action of nursing on reproductive function which occurs in mares (Ginther et al., 1972).

The occurrence of repeated twin conceptuses in individual mares for this population appears to be sporadic with only 4 mares producing twin conceptuses in the same season. Although the familial incidence of twin conceptuses was not examined, the low incidence of individual mares (6·8%) which conceived twins more than once suggests that twinning is not a strong genetic trait within individual mares in this population.

The mating of mares irrespective of the number of follicles present at the time of breeding and the use of hCG to induce ovulation after mating appear to be an efficient management system. It has been shown that the use of hCG before mating decreased the interval between the first and last ovulations (Woods et al., 1982). Therefore, provided mares were judiciously teased 48 h after mating, a rectal examination to check for ovulation was not then necessary. Although the frequency of twin conceptuses was increased using this management system, careful ultrasonography 14–16 days after mating should detect all twin conceptuses. One report on large numbers of mares shows that the use of embryonic vesicle rupture at this time is a highly effective method of effecting embryonic reduction between Days 12 and 16 (49/50) (Pascoe, 1986).

In previous studies of broodmare farm records, twin conceptuses apparently result from asynchronous ovulations (>48 h) and not from synchronous ovulations (Ginther et al., 1982a; Woods et al., 1983). In the current study, ovulation data were obtained for 189 mares. In this group, 55 mares had synchronous double ovulations which resulted in 22 of the 55 (40%) mares carrying twin conceptuses, as demonstrated by ultrasonography performed between Days 12 and 25 of gestation. This finding differed from previous reports of 0% for synchronous double ovulations (Ginther et al., 1982a; Ginther, 1983a; Woods et al., 1983). However, the twin conceptus rate for known single ovulations, 10·9%, was similar to the range (4–7%) of twin conceptus rates previously reported for Thoroughbreds and Standardbreds (Hughes et al., 1982; Woods et al., 1983).

Calculating the effective age difference between each embryonic vesicle of the 57 sets of twin conceptuses detected by ultrasonography indicated that the maximum age difference between any embryonic vesicle of a given pair was <72 h. There was <48 h age difference between 96·7% of the embryonic vesicles of a given pair when an average daily growth rate of 4·14 ± 0·15 mm (Day 14) was used for the calculation. The calculated growth rate was slightly higher than the growth rate of 3·4 mm previously reported when averaging the width of single embryonic vesicles in 40 mares (Ginther, 1983b). The small age difference (<48 h) between each 2 embryonic vesicles would suggest that they originated from 2 ovulations which occurred <48 h apart (synchronous).

A biological or natural embryo-reduction mechanism has been postulated to occur between Day 6 and Day 20 based upon the single and twin conceptus rate of superovulated mares and the pregnancy rate obtained from analysing broodmare records from various farms (Ginther et al., 1982a). For the population of mares in the current study, the data indicate that the biological embryo-reduction mechanism proposed by Ginther et al. (1982b) was not effective in causing single embryonic reduction in mares which had synchronous double ovulations, because 96·7% of the twin conceptuses were estimated to be <48 h different in age, and, when synchronous double ovulations were recorded, 40% of the mares conceived twins.

This raises the question as to the fertility of ova produced by superovulating mares with horse pituitary extract and the viability of these embryos during the first 20 days of gestation. One study on the viability of Day-7 embryos transferred from superovulated donors to recipient mares reported a 33% (7/21) (8 donor mares) pregnancy rate per embryo transferred compared to a 50% (4/8) pregnancy rate for single ovulating donors (controls) (Woods & Ginther, 1984). The above study suggests that decrease viability of the superovulated embryos was due to a biological embryo-reduction mechanism within the uterus.

The high twin conception rate in the current study indicates that the embryo-reduction mechanism was not effective in reducing twin conceptuses in the "prefixation" period (<Day 16). Further

investigation is needed to prove the viability of superovulated embryos by attempting to transfer the embryos from the oviduct (Days 2–4 after ovulation) or by using in-vitro fertilization.

In conclusion, the results of this study provide a useful model to predict the probability of twin conceptuses occurring when the mare's status and follicular status are known. The high frequency of twin conceptuses from synchronous and estimated synchronous ovulations found in this study indicates the need to examine further the embryo-reduction mechanism previously postulated (Ginther *et al.*, 1982b).

We thank the Queensland Equine Research Foundation for funding of equipment used for the ultrasound examinations reported in this project.

References

Engleman, L. (1985) Stepwise logistic regression. In *BMDP Statistical Software*, pp. 330–344. Ed. W. J. Dixon. University of California Press, Berkeley.

Ginther, O.J. (1982a) Twinning in mares: a review of recent studies. *J. equine vet. Sci.* **2**, 127–135.

Ginther, O.J. (1982b) Effect of reproductive status on twinning and on side of ovulation and embryo attachment in mares. *Theriogenology* **20**, 383–388.

Ginther, O.J. (1983a) Mobility of the early equine conceptus. *Theriogenology* **19**, 603–611.

Ginther, O.J. (1983b) Fixation and orientation of the early equine conceptus. *Theriogenology* **19**, 613–623.

Ginther, O.J., Whitmore, H.L. & Squires, E.L. (1972) Characteristics of estrus, diestrus, and ovulation in mares and effects of season and nursing. *Sm. J. vet. Res.* **33**, 1935–1939.

Ginther, O.J., Douglas, R.H. & Lawrence, J.R. (1982a) Twinning in mares. A survey of veterinarians and analysis of Theriogenology records. *Theriogenology* **18**, 333–341.

Ginther, O.J., Douglas, R.H. & Woods, G.L. (1982b) A biological embryo-reduction mechanism for the elimination of excess embryos in mares. *Theriogenology* **18**, 475–485.

Hughes, J.P., Stabenfeldt, G.H. & Evans, J.W. (1982) Clinical and endocrine aspects of the estrous cycle of the mare. *Proc. 28th Ann. Conv. Am. Assoc. equine Pract.*, pp. 119–150.

Jeanrich, R., Sampson, P. & Frane, J. (1985) Analysis of variance and covariance including repeated measures. In *BMDP Statistical Software*, pp. 359–387. Ed. W. J. Dixon. University of California Press, Berkeley.

Jeffcott, L.B. & Whitwell, K.T. (1973) Twinning as a cause of foetal and neonatal loss in the Thoroughbred mare. *J. comp. Path.* **83**, 91–106.

Palmer, E. & Driancourt, M.A. (1980) Use of ultrasonic echography in equine gynecology. *Theriogenology* **13**, 203–216.

Pascoe, D.R. (1986) *Single embryonic reduction in the mare with twin conceptuses: studies of hormone profiles and drug therapies using a physiological model, and manual and surgical reduction techniques* in vivo. PhD. dissertation, University of California, Davis.

Pascoe, R.R. (1983) Methods for the treatment of twin pregnancy in the mare. *Equine vet. J.* **15**, 40–42.

Woods, G.L. & Ginther, O.J. (1984) Collection and transfer of multiple embryos in the mare. *Theriogenology* **21**, 461–469.

Woods, G.L., Scraba, S.T. & Ginther, O.J. (1982) Prospects for induction of multiple ovulations and collection of multiple embryos in the mare. *Theriogenology* **17**, 61–72.

Woods, G.L., Sprinkle, T.A. & Ginther, O.J. (1983) Prevention of twin pregnancy in the mare. *Proc. Ann. Conf. Am. Soc. Theriogenol.*, pp. 194–202.

J. Reprod. Fert., Suppl. **35** (1987), 191–196

Evaluation of mare oocyte collection methods and stallion sperm penetration of zona-free hamster ova

A. Okólski, P. Babusik*, M. Tischner and W. Lietz

Department of Animal Reproduction, Academy of Agriculture, 30-059 Kraków, Al. Mickiewicza 24/28, Poland, and *Institute of Animal Production, Nitra, Czechoslovakia

Summary. Comparisons were made between 2 methods of oocyte recovery from the ovarian follicles of slaughtered mares: 500 oocytes (3 per mare) were obtained by aspiration of follicular fluid from ovaries of 162 mares, and 120 oocytes (8 per mare) by isolation and rupture of follicles from ovaries of 14 mares. In the oocytes recovered after rupture of follicles, 89·2% were morphologically unchanged, in comparison to 29·3% obtained by aspiration of follicular fluid.

Stallion spermatozoa capacitated *in vitro* were tested on zona-free hamster oocytes. The stallion spermatozoa were washed in TCM-199 and preincubated for 4–5 h at a concentration of $0·5–1·0 \times 10^9$/ml. Of the 305 hamster oocytes fertilized by spermatozoa obtained from 8 stallions, 77·8% penetration was confirmed on the basis of the presence of enlarged heads of spermatozoa and male pronuclei with their corresponding tails. Spermatozoa from one of the stallions were tested 10 times, 8 months later. The percentage of penetrated oocytes averaged 52·8% (39·1–76·9%). In the control group of oocytes preincubated in the absence of spermatozoa, spontaneous activation and female pronuclei were found in 12 out of a total of 57.

Introduction

In studies of in-vitro fertilization, the basic problems involve those of oocyte recovery, as well as preparation of semen for fertilization. Despite the fact that to this time attempts at in-vitro fertilization of mare oocytes have been unsuccessful, the preliminary stages of study in this field have been carried out. Primarily they concern the successful trials of culturing mare oocytes *in vitro* (Fulka & Okólski, 1981; Desjardins *et al.*, 1985), as well as studies evaluating fertilizing ability of stallion spermatozoa.

In a review of the current literature, there are no reports of successful oocyte recovery from ovarian follicles in mares. Furthermore, studies evaluating the fertilizing ability of stallion spermatozoa are limited to one account (Brackett *et al.* 1982b) in which workers using the zona-free hamster ova penetration test showed that stallion spermatozoa capacitated in in-vitro conditions are capable of penetrating hamster oocytes. Advances in research are significantly greater in the utilization of this test to evaluate sperm fertilizing ability in human medicine (Yanagimachi *et al.*, 1976; Barros *et al.*, 1979; Quinn *et al.*, 1982), as well as in other species of domestic animals such as cattle (Bousquet & Brackett, 1982; Brackett *et al.*, 1982a), and pigs (Imai *et al.*, 1977; Pavlok, 1981; Babusik & Majerciak, 1983).

The aims of the present study were (1) to compare 2 methods of oocyte recovery from mare ovarian follicles and (2) to evaluate the penetration of fresh spermatozoa into zona-free hamster ova.

Materials and Methods

Mare oocyte collection

The experiment was carried out on ovaries removed from different breeds of slaughtered mares, ranging in age from 4 to 20 years. Two different methods of oocyte recovery were used. Immediately after slaughter, follicular fluid containing oocytes was aspirated with a syringe fitted with a 21-gauge × 3·3 cm needle, from the ovaries of 162 mares. The samples were then transported to the laboratory in containers maintained at 20–25°C. There, under a stereomicroscope, the oocytes were separated from the follicular fluid.

In the second method, the entire ovaries of 14 mares were transported to the laboratory, where the follicles were isolated, and separately placed on watch glasses in PBS medium. The follicles were then ruptured with needles or incised under a stereomicroscope. Oocytes were collected from antral follicles of different sizes (> 5 mm in diameter), and were examined morphologically. Criteria for evaluating oocyte quality included the presence and the state of cumulus cell layers, as well as morphology of the oocyte cytoplasm. A three-stage classification system was used: (1) oocytes with intact cytoplasm, surrounded by compact and dense cumulus layers, (2) oocytes with intact cytoplasm, surrounded by dispersed layers or remnants of cumulus cells, (3) oocytes with degenerative changes in the cytoplasm.

Hamster egg penetration assay by stallion spermatozoa

Semen preparation. Fresh ejaculates, obtained from 4 Polish ponies of the Tarpan type, and 4 half-breed stallions, ranging from 1·5 to 15 years of age, were evaluated for fertilizing ability. The reproductive history of 2 of the stallions was well known (Nos 1 and 2, Table 2). They were used regularly with donor mare embryos, and several mares mated with these stallions bore offspring. The remainder of the stallions were not used for mating, although, from the start of sexual maturity, their semen was regularly examined; sperm motility and routine laboratory values were found to be within normal limits.

Stallion semen was collected with an open artificial vagina (Tischner *et al.*, 1974). Only the medial spermatic fraction of the ejaculate was collected, with 4–5 ml samples of the fresh ejaculate being diluted with TCM-199 medium at a ratio of 1:1 (v/v). Osmolality of the medium was approximately 280 mosmol. The composition of the culture medium was similar to that described by Pavlok & McLaren (1972) and modified by Motlik & Fulka (1974), except that bovine or fetal calf serum (20% final concentration) was substituted for freeze-dried calf serum proteins. After thoroughly stirring the semen with medium, the spermatic suspension was centrifuged for 2 min at 200 g to eliminate any contamination. The spermatic supernatant (4–5 cm) was centrifuged for 10 min at 2500 g, poured off, and then 1 ml fresh medium was added to the dense mass of spermatozoa. The contents were again thoroughly mixed, centrifuged for 10 min at 2500 g, the supernatant discarded, and spermatozoa in the remainder of the fluid were disrupted. The prepared spermatozoa were preincubated for 4–5 h in tissue culture medium (0·1 ml), covered with paraffin oil on a watch glass and placed in a Petri dish. Sperm concentration during preincubation was 0·5–1·0 × 10^9/ml. Sperm preincubation was carried out at 37·5°C, under artificial atmospheric conditions (5% CO$_2$ in air). At a concentration of 0·5–1·0 × 10^6/ml, preincubated stallion spermatozoa were added to the cultivation drop with zona-free hamster ova.

Sperm motility was evaluated by standard methods, directly after collection, before preincubation and before addition to oocytes.

Preparation of zona-free hamster ova. Golden hamsters were superovulated with 50 i.u. PMSG (Antex-Leo), i.p., and 72 h later, 50 i.u. hCG (Biomed) were administered i.p. Cumulus cells were recovered 15–17 h after the last injection and treated with hyaluronidase (190 TRU; 1 ml) and α-chymotrypsin (2 mg/ml) to disperse the cumulus cells and to remove the zona pellucida, respectively. After flushing, the ova were transferred to sterile watch glasses containing 0·2 ml medium under paraffin oil. Each droplet contained 10–30 zona-free hamster ova. These experiments were carried out in 2 stages. In the first stage, the semen collected from 8 stallions (Table 2) was used, with preincubation time of oocytes with spermatozoa ranging from 12 to 16 h. In the second stage, semen was collected for each trial from one stallion, whose reproductive history was well documented (Stallion 1, Table 2). The prepared zona-free hamster ova inseminated with preincubated spermatozoa comprised the experimental group, while oocytes without the presence of spermatozoa were similarly incubated as controls. The incubation period for both groups was 3·5–4·5 h.

After incubation, the ova were transferred to slides, fixed in ethanol/acetic acid (3:1 v/v), and stained in 1% orcein in 45% acetic acid. The critiera used to evaluate sperm penetration of the hamster ova were the presence of an enlarged sperm head or sperm pronuclei and the corresponding sperm tail in the ovum.

Results

Recovery of oocytes from mares

The results are shown in Table 1. The percentage of oocyte recovery, in relation to follicles and number of morphologically unchanged oocytes, was higher ($P < 0·01$, comparison by frequency U-test) when the method of rupturing the ovarian follicles was used.

Stallion sperm penetration rate

The results of the interaction between zona-free hamster ova and preincubated stallion spermatozoa are presented in Tables 2 and 3. In the first stage of this experiment, with spermatozoa

Table 1. Recovery and morphological evaluation of mare oocytes

Method of oocyte recovery	No. of mares	No. of follicles		No. of oocytes		Morphological evaluation			
								With degenerative changes	
		Total	Average per mare	Total (%)	Average per mare	Total no.	Intact (%)	Cytoplasm (%)	Cumulus (%)
Aspiration of follicular fluid	162	1471	9·08	500 (33·9)*	3·08	365	107 (29·3)	36 (9·8)	222 (60·8)*
Rupture of isolated follicles	14	120	8·57	112 (93·3)*	8·00	112	100 (89·2)	9 (8·0)	73 (2·8)*

*P < 0·01 comparison by frequency U-test.

Table 2. Penetration of zona-free hamster ova by spermatozoa from 8 stallions

Stallion	Age (years)	Breed*	No. of ova		Average no. of male pronuclei per ovum
			Inseminated	Penetrated (%)	
1	15	Pp	52	38 (73·0)	1·94
2	15	Pp	77	59 (76·6)	1·33
3	4	hb	36	31 (86·1)	1·51
4	5	hb	61	46 (75·4)	1·60
5	2·5	Pp	27	22 (81·5)	1·54
6	1·5	Pp	55	46 (83·6)	1·82
7	1·7	hb	64	49 (76·7)	2·04
8	1·6	hb	20	14 (70·0)	1·64
Total			392	305 (77·8)	1·68

*Pp, Polish pony; hb, Anglo-Arab half-breed.

from 8 stallions, the penetration rate was 78%. The breed and age of stallion had no significant effect on the penetration rate (Table 2).

In the second stage of this experiment, semen was collected from only one of the stallions (Stallion 1, Table 3). The mean penetration rate of 53% was lower than that for the study with 8 stallions but the highest value (76·9%) was similar.

In this study, sperm motility rate after collection was 60–80%, changed in value in subsequent trials after washings and centrifuging to 40–80%, and, following 4–5 h incubation, the rate was maintained at a level of 10–60% (Table 3).

In the control group, spontaneous activation, as indicated by presence of female pronuclei, was seen in 12 (21·05%) of the 57 ova preincubated without spermatozoa.

Discussion

Aspiration of follicular fluid containing oocytes is the prime method of extracting immature oocytes in different species of animals. Utilizing this method in cattle, 30–60% follicular oocytes can be recovered (Kanagawa, 1979; Leibfried & First, 1979). In our studies, only 33·9% of mare follicular oocytes were recovered, of which a large number exhibited degenerative changes. The reason for the low effectiveness of this method may be due to difficulty in separating the oocytes from the remainder of the follicle, particularly in the case of small follicles. In general, the cumulus cells are regularly found to be damaged during extraction of oocytes from follicles. Rupturing isolated follicles proved to be 3 times more effective in recovering oocytes from mares, and 89% of oocytes were morphologically unchanged. Oocytes were absent in only 8 of the 120 isolated follicles. Kątska (1984), applying this method to secure ovarian oocytes in cattle, obtained 155 oocytes from a total of 154 follicles. The mare ovarian follicles we examined did not contain more than one oocyte. Despite the relatively high recovery rate of oocytes by rupture of ovarian follicles, this method cannot be recommended for application for large numbers of animals, considering the high labour and time consumption required to recover an oocyte. For development of a method of in-vitro fertilization of mare oocytes, this method should find particular application in oocyte recovery from valuable mares, since attempts to induce superovulation in this species produce only very limited results. Such studies have been carried out for other species (Brackett et al., 1982b), and in-vitro fertilization of oocytes in cattle, by bull spermatozoa capacitated in vitro, followed by transfer of the embryos at the 4-cell stage, resulted in the recipient giving birth to a normal bull calf. The present studies with stallion spermatozoa capacitated in vitro have demonstrated their ability to penetrate zona-free hamster ova.

In the first study carried out in June, on the semen of 8 stallions, a very high penetration rate (78%) was obtained. The percentage of penetration was more than 20% lower in the second experiment, when the same test was repeated several months later, between March and May, using semen from just one stallion (Stallion 1, Table 2). In the second experiment, incubation time of the oocytes together with spermatozoa was reduced from 12–16 h to 3·5–4·5 h. It is possible that this factor could have influenced the different results, although such a time frame is routinely used in the zona-free hamster ova penetration test in other animal species. Other factors which may have affected the outcome include use of a different series of hamsters, medium and agents. In the current literature, only Brackett et al. (1982b) conducted trials of penetration of zona-free hamster ova by stallion spermatozoa capacitated in vitro. The percentage of interaction was evaluated on the basis of oocytes with spermatozoa attached, with penetrated spermatozoa and with male pronuclear formation. The rate was 82·6% when spermatozoa were preincubated in BWW medium for 18–26 h, but 100% with preincubation of spermatozoa for 10 min in DM medium with higher osmolality, and subsequently for 1 h in isotonic DM medium. The presence of enlarged sperm heads and male pronuclei with their corresponding tails was evidence of penetration in our studies. The average number of male pronuclei per oocyte was 1·68 and 1·59 in the first and second

Table 3. Penetration of zona-free hamster ova by spermatozoa from Stallion 1 only

| Attempt | No. of ova | | Average no. of male pronuclei per ovum | No. of ova | | % motile spermatozoa | | |
	Inseminated	Penetrated (%)		Incubated without spermatozoa	With spontaneous activation	Fresh semen	Preincubated Before	Preincubated After
1	23	9 (39·1)	1·22	10	5	70	50	10
2	42	25 (59·5)	1·28	8	3	60	40	20
3	—	—	—	6	0	—	—	—
4	13	10 (76·9)	1·60	2	0	70	40	40
5	25	12 (48·0)	1·75	7	1	80	—	30
6	16	7 (43·7)	2·10	—	—	60	60	20
7	4	3 (75·0)	1·67	5	0	80	60	30
8	19	10 (52·6)	1·60	3	0	70	60	20
9	18	9 (50·0)	1·55	11	3	70	50	30
10	16	8 (50·0)	2·25	5	0	80	80	60
Total	176	93 (52·8)	1·59	57	12			

experiments, respectively. There was no conclusive relationship between sperm motility and the percentage of penetrated oocytes. In the study by Brackett *et al.* (1982), the results showed 100% interaction, 41% penetration by spermatozoa obtained from stallions with questionable fertility, and 100% penetration of 8 oocytes from 2 fertile stallions.

It remains to be seen whether the zona-free ova penetration test will find application for evaluation of fertility in other domestic male animals, or in stallions. Bousquet & Brackett (1982), comparing the ability of frozen semen of 2 bulls to interact with zona-free hamster ova, concluded that the interaction rate was correlated with the 60-day non-return data for cows inseminated with semen from these same bulls. On the same premise, this test should similarly find application in evaluation of fertility, particularly in young stallions before prospective mating with exceptionally valuable mares.

Studies conducted on methods of recovery and culture of mare ovarian oocytes, as well as preparation of stallion spermatozoa for penetration, should lead to successful trials of fertilization of mare oocytes in in-vitro conditions.

This work was supported by grant No. CPBP 0506314. We thank Mrs M. Kolanos for help with the preparation of the MS.

References

Babusik, P. & Majerciak, P. (1983) Fertilizing capacity of fresh and frozen boar ejaculates according to the penetration test. *Polnohospodarstvo* **12**, 1061–1070.

Barros, C., Gonzalez, J., Herrera, E. & Bustos-Obregon, E. (1979) Human sperm penetration into zona-free hamster oocytes as a test to evaluate the sperm fertilizing ability. *Andrologia* **11**, 197–210.

Bousquet, D. & Brackett, G. (1982) Penetration of zona-free hamster ova as a test to assess fertilizing ability of bull sperm after frozen storage. *Theriogenology* **2**, 199–213.

Brackett, B.G., Bousquet, D., Boice, M.L., Donawick, W.J., Evans, J.F. & Dressel, M.A. (1982a) Normal development following in vitro fertilization in the cow. *Biol. Reprod.* **27**, 147–158.

Brackett, B.G., Cofone, M.A., Boice, M.L. & Bousquet, D. (1982b) Use of zona-free hamster ova to assess sperm fertility ability of bull and stallion. *Gamete Res.* **5**, 217–227.

Desjardins, M., King, W.A. & Bousquet, D. (1985) In vitro maturation of horse oocytes. *Theriogenology* **23**, 187, Abstr.

Fulka, J. & Okólski, A. (1981) Culture of horse oocytes in vitro. *J. Reprod. Fert.* **61**, 213–215.

Imai, H., Niwa, K. & Iritani, A. (1977) Penetration *in vitro* of zona-free hamster eggs by ejaculated boar spermatozoa. *J. Reprod. Fert.* **51**, 495–497.

Kanagawa, H. (1979) Recovery of unfertilized ova from slaughtered cattle. *Jpn. J. vet. Res.* **27**, 72–76.

Katska, L. (1984) Comparison of two methods for recovery of ovarian oocytes from slaughter cattle. *Anim. Reprod. Sci.* **7**, 461–463.

Leibfried, L. & First, N.L. (1979) Characterization of bovine follicular oocytes and their ability to mature in vitro. *J. Anim. Sci.* **48**, 76–86.

Motlik, J. & Fulka, J. (1974) Fertilization of pig follicular oocytes cultivated *in vitro*. *J. Reprod. Fert.* **36**, 235–237.

Pavlok, A. (1981) Penetration of hamster and pig zona-free eggs by boar ejaculated spermatozoa preincubated *in vitro*. *Int. J. Fert.* **26**, 101–106.

Pavlok, A. & McLaren, A. (1972) The role of cumulus cells and the zona pellucida in fertilization of mouse eggs *in vitro*. *J. Reprod. Fert.* **29**, 91–97.

Quinn, P., Barros, C. & Whittingham, D.G. (1982) Preservation of hamster oocytes to assay the fertilizing capacity of human spermatozoa. *J. Reprod. Fert.* **66**, 161–168.

Tischner, M., Kosiniak, K. & Bielański, W. (1974) Analysis of the pattern of ejaculation in stallions. *J. Reprod. Fert.* **41**, 329–335.

Yanagimachi, R., Yanagimachi, H. & Rogers, B.J. (1976) The use of zona-free animal ova as a test-system for the assessment of the fertilizing capacity of human spermatozoa. *Biol. Reprod.* **15**, 471–476.

J. Reprod. Fert., Suppl. 35 (1987), 197–209

Extraspecific donkey-in-horse pregnancy as a model of early fetal death

W. R. Allen, Julia H. Kydd, M. S. Boyle and D. F. Antczak*

*Thoroughbred Breeders' Association Equine Fertility Unit, Animal Research Station,
307 Huntingdon Road, Cambridge CB3 0JQ, U.K. and *James Baker Institute for Animal Health,
New York State College of Veterinary Medicine, Cornell University, Ithaca, New York 14853,
U.S.A.*

Summary. Transfer of donkey embryos to horse mares provides a useful model of early fetal death. Endometrial cups do not develop in this one type of extraspecific pregnancy and 80% of donkey fetuses are aborted between Days 80 and 100 of gestation in conjunction with abnormal implantation and an intense accumulation of leucocytes in the endometrium of the surrogate mare. Treatment of mares carrying donkey conceptuses with progestagen (allyl trenbolone) or purified horse chorionic gonadotrophin does not prevent abortion. However, passive immunization with serum from mares carrying intraspecific horse fetuses, or active immunization with donkey lymphocytes, causes a marked increase in fetal survival rate and the birth of live foals. Furthermore, both cytotoxic (rejection type) and immunoprotective maternal immune responses to the xenogeneic donkey fetus can be recalled in mares carrying repeated donkey-in-horse pregnancies.

We suggest that the endometrial cup reaction in normal equine pregnancy provides a vital and temporally important antigenic stimulus which results in the mare mounting an immunoprotective response towards her allogeneic fetus *in utero*.

Introduction

Many past surveys have given widely varying figures for the rate of fetal mortality in mares during the first 100 days of gestation. These range from as low as 2·9% in Arabian mares in North America (Hutton & Meacham, 1968), through 6·5% and 9% respectively in barren/maiden and lactating Thoroughbred mares in the United Kingdom (M. W. Sanderson & W. R. Allen, unpublished data) to an overall figure of 12·5% in lactating Thoroughbred mares in West Germany with a sharp rise to 17% in mares mated in the first post-partum oestrus (Merkt, 1966). Figures of this type are influenced by such factors as breed and mare age, the methods and frequency of data collection and, perhaps most important, the accuracy of pregnancy diagnosis, the stage of gestation when it is first carried out and the frequency of repeat examinations.

There are many possible causes of early pregnancy loss in the mare, including lethal chromosomal abnormalities in the fetus, chronic endometritis, luteal insufficiency and nutritional stress. However, no accurate figures have been recorded as to the relative occurrence and importance of any particular cause of fetal death within the overall loss rate for the population or breed as a whole. Nor are there meaningful data on the morphological, biochemical and endocrinological changes associated with fetal death. A major reason for this paucity of information stems from the fact that the diagnosis of fetal death is, by the nature of the condition, usually made after the event, not before or during its occurrence.

A model of pregnancy in which a high rate of early fetal death occurs spontaneously at a predictable time in gestation has obvious advantages as a tool for studying the inter-relationship of maternal and fetal factors required to maintain the pregnancy state. The extraspecific transfer of donkey embryos to horse mares provides such a model of pregnancy failure in which >80% of conceptuses die and are aborted between Days 80 and 100 (Allen, 1982). This paper reviews the gross, histological and endocrinological changes associated with this example of equine fetal death and summarizes the results of experiments undertaken to define the relative roles played by genetic, developmental, endocrinological and immunological factors in its cause.

Equine placentation and the kinetics of the endometrial cup reaction

By Day 36 after ovulation the equine trophoblast is clearly divided into its invasive and non-invasive components. The former comprises a discrete, pale, annulate band that surrounds the spherical conceptus in the region where the enlarging allantoic and regressing yolk sac membranes abut (Ewart, 1897). This chorionic girdle consists of finger-like projections of rapidly multiplying trophoblast cells which separate from the fetal membranes and invade the underlying maternal endometrium between Days 36 and 38 (Allen *et al.*, 1973). Once in the endometrial stroma the girdle cells undergo a dramatic transformation; they cease to divide, become sessile, enlarge greatly, acquire a second nucleus and begin secreting CG. The cells remain clumped together in the endometrium to form the endometrial cups which are seen as a series of ulcer-like endometrial protuberances in the gravid horn arranged in a circle around the conceptus (Cole & Goss, 1943; Clegg *et al.*, 1954). The cups have a lifespan of 60–100 days throughout which increasing numbers of lymphocytes, macrophages, plasma cells and eosinophils accumulate in the surrounding endometrial stroma. These maternal leucocytes form a solid band of cells that walls off the fetal cup from the adjacent maternal tissues (Fig. 1a). After Day 80, the leucocytes begin to invade the cup tissue and appear to destroy the individual cup cells. Eventually, between Days 100 and 140, the whole necrotic cup is sloughed from the surface of the endometrium in a manner histologically similar to the rejection of a skin allograft (Allen, W. R., 1975).

The non-invasive trophoblast covers the remainder, and hence the great majority, of the conceptus at Day 36. From Day 42–45 it begins to establish a stable, microvillous contact with the adjacent endometrial epithelium. The whole conceptus begins to expand and elongate so that, by Day 90–100, the allantochorion is firmly attached to endometrium over the entire internal surface of the uterus. Increasingly complex and branching allantochorionic villi protrude into corresponding endometrial crypts, thus greatly enlarging the total area for placental exchange (Samuel *et al.*, 1975). However, despite this vast area of intimate materno–fetal epitheliochorial contact, no accumulation of leucocytes occurs at the trophoblast–endometrium interface, as it does around the endometrial cups (Fig 1b; Allen, 1979).

Maternal immune responses in pregnancy

There is clear evidence of strong humoral and cell-mediated maternal immune responses to foreign antigens expressed by the horse conceptus. On the humoral side, >90% of primigravid mares exhibit high titres of anti-paternal lymphocytotoxic antibody in their serum in early gestation. This is first detected between Days 44 and 72, soon after invasion of the chorionic girdle cells to form the endometrial cups at Day 36 (Allen, 1979; Antczak *et al.*, 1984). Antibody persists throughout gestation, shows strong allospecificity for the mating stallion and is absent in mares carrying fetuses that are histocompatible with regard to Class I Major Histocompatibility Complex (MHC) antigens (Antczak *et al.*, 1984). An anamnestic rise in titre occurs in mares carrying second pregnancies by the same stallion and in primigravid mares that had previously received skin grafts from an MHC-incompatible mating sire (Crump *et al.*, 1987).

The leucocyte response to the endometrial cups has the histological appearance of a cytotoxic reaction to foreign antigens expressed by the fetal cup cells. The large numbers of lymphocytes that

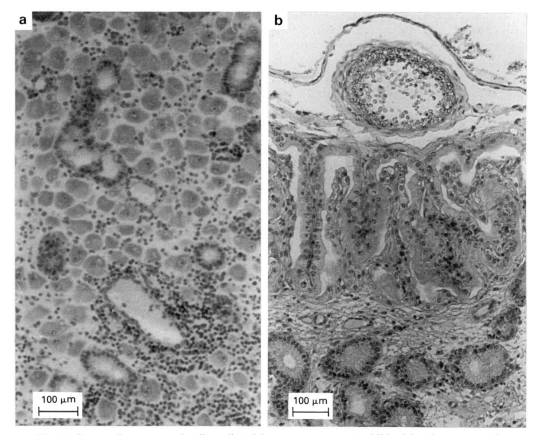

Fig. 1. Contrasting maternal cell-mediated immune responses exhibited by the mare to the invasive and non-invasive components of intra-species horse trophoblast: (a) leucocytes accumulated around the invasive trophoblast at the base of an endometrial cup at Day 65; (b) the allantochorion–endometrium interface at Day 83 showing the absence of leucocytes associated with the non-invasive trophoblast.

accumulate at the border of the cups begin actively to invade the cup tissue beyond Day 70–80 and they penetrate and destroy individual cup cells. This concept, that the leucocyte reaction hastens the death and desquamation of the cups, is supported by two other experiments. First, chorionic girdle cells recovered before invasion of the endometrium at Day 36 differentiate *in vitro* to form stable monolayers that survive and continue to secrete CG into the culture medium for > 200 days, more than twice the average lifespan of endometrial cups *in vivo* (Allen & Moor, 1972). Second, in mares mated to their co-twin brothers to whom, as a result of cellular exchange from placental fusion during co-twin life *in utero*, they are sufficiently tolerant to accept skin grafts, serum CG concentrations remain elevated to beyond Day 200 of gestation, thereby reflecting prolonged lifespan of the endometrial cups (Bouters *et al.*, 1978).

These and other findings suggest that cytotoxic antibody production and the endometrial cup leucocyte reaction represent, respectively, the humoral and cell-mediated arms of the maternal immune response to paternally-derived fetal alloantigens expressed on endometrial cup cells (Kydd *et al.*, 1982). However, the results of other experiments cast some doubts on this hypothesis. First, it has not yet been possible, using well characterized anti-MHC monoclonal antibodies in an immunoperoxidase labelling system, to demonstrate Class I MHC antigens on mature endometrial cup cells or non-invasive trophoblast, although the progenitor tissue of the cups, the chorionic girdle, stains heavily (Crump *et al.*, 1987). Second, in mares carrying Class I MHC-compatible fetuses, maternal serum CG concentrations are not higher than those in mares carrying histo-incompatible fetuses and CG activity does not persist for any longer in pregnancy. Furthermore, a leucocyte reaction that is of equivalent intensity to the reaction observed in normal, histo-incompatible pregnancy (Allen *et al.*, 1984) occurs around the endometrial cups. An even more dramatic effect of fetal genotype on the cell-mediated response is seen in interspecies mating between horses and donkeys. In mares and jenny donkeys carrying, respectively, hybrid mule and hinny conceptuses, the lymphocyte accumulation around the endometrial cups is greatly increased compared to that seen during normal intraspecies pregnancy in each species (Allen, W. R., 1975).

Despite the inability to demonstrate Class I MHC antigens on mature endometrial cup cells by histochemical means, the strong affinity of anti-Class I MHC sera for chorionic girdle tissue (Crump *et al.*, 1987) and the close temporal association between cytotoxic antibody production and formation of the endometrial cups, still support the invasive component of the trophoblast as the source of alloantigen in equine pregnancy. On the other hand, the development of a typical leuco-cyte response to the cups in Class I MHC histocompatible pregnancy indicates that other types of fetal antigens are also expressed on cup cells. These may be minor histocompatibility antigens or, more likely, antigens that are specific for trophoblast tissue. Furthermore, the great increase in the leucocyte response to the endometrial cups observed in hybrid pregnancies indicates that species-related antigens are also expressed on at least the invasive portion of the trophoblast.

Successful extraspecific pregnancy in equids

The use of artificial insemination and embryo transfer has demonstrated an amazing capacity of female equids to carry to term a variety of inter- and extra-specific offspring that differ markedly in genotypic and phenotypic characteristics from the surrogate mother (Table 1; Allen, 1982; Davies *et al.*, 1985; Kydd *et al.*, 1985). A common feature of all successful intra-, inter- and extra-species

Table 1. Endometrial cup development and fetal survival in intra-, inter- and extra-specific equine pregnancies

Fetal genotype	Mare genotype	Fetal:maternal karyotype (2n =)	Endometrial cup development and lifespan*		Outcome of pregnancy
Intraspecific mating					
Horse	Horse	64:64	+ + +	L	Term
Donkey	Donkey	62:62	+	L	Term
Grant's zebra	Grant's zebra	46:46	+	L	Term
Interspecific mating					
Mule	Horse	63:64	+	S	Term
Hinny	Donkey	63:62	+ + + +	L	Term
Extraspecific embryo transfer					
Przewalski's horse	Horse	66:64	+ +	L	Term
Horse	Donkey	64:62	+ + + +	L	Term
Horse	Mule	64:63	+ + +	L	Term
Donkey	Mule	62:63	+	L	Term
Grant's zebra	Donkey	46:62	+	S	Term
Grant's zebra	Horse	46:64	±	VS	Term
Donkey	Horse	62:64	—		80% abort at Day 80–95

*Endometrial cups persist for: L = >60 days; S = 15–30 days; VS = <10 days.

pregnancies is invasion of fetal chorionic girdle cells into the endometrium at Day 36–38. The result-ing endometrial cups vary greatly between the types of pregnancy with regard to: (a) the amount of cup tissue that develops, as demonstrated by serum CG concentrations and reflecting the breadth and overall development of the chorionic girdle and its success in invading the endometrium; and (b) the lifespan of the cups, demonstrated by the duration of CG activity in serum and, possibly, reflecting the intensity and success of the maternal leucocyte reaction in hastening the death of the cups. As shown in Table 1, endometrial cup development and lifespan ranges from large and rela-tively long-lived cups in mares carrying horse conceptuses and in donkeys carrying hinny and horse conceptuses, to very small and transient cups in mares carrying zebra conceptuses. It appears that the paternal genome has a marked influence on the growth of the progenitor chorionic girdle and, hence, on the amount of endometrial cup tissue that develops after invasion; regardless of changes in uterine environment resulting from embryo transfer, a much broader and more active chorionic girdle develops on the conceptus when a horse is the sire than when a donkey or Grant's zebra is the sire. But despite these big variations, any conceptus that gives rise to any degree of endometrial cup development in the maternal uterus seems able to implant normally and proceed safely to term.

The donkey-in-horse model of pregnancy failure

One type of extraspecific equine pregnancy that differs markedly from the others is that resulting from the transfer of donkey embryos to horse mares. In this instance the donkey chorionic girdle fails completely to invade the endometrium of the surrogate mare at Day 36 and so no endometrial cups are formed. Despite this early developmental deficiency and the resulting absence of CG in maternal blood, the donkey conceptus continues to grow normally during the next 25–30 days. The expanding donkey allantochorion is closely applied to the endometrium but there is minimal or no microvillous attachment between fetal and maternal epithelial surfaces and the endometrium does not undergo the architectural modifications that would normally provide crypts to accommodate the allantochorionic villi during implantation. Between Days 65 and 80 the fetus becomes increas-ingly congested and wasted in appearance and the amniotic fluid assumes a claret colour due to haemolysis of fetal red blood cells (Fig. 2a). During the same period, increasing numbers of lymphocytes and other leucocytes congregate in the supepithelial endometrial stroma throughout the entire area of endometrium in contact with donkey allantochorion. Appreciable quantities of exocrine material, secreted by the endometrial glands and the trophoblast cells, accumulates between the endometrial epithelium and the allantochorion, thereby further preventing the normal process of attachment and implantation (Fig. 2b). Eventually, between Days 80 and 95 in most cases, the accumulated lymphocytes begin to pass through the endometrial epithelium and actively attack the xenogeneic donkey trophoblast. This finally leads to abortion of the dead fetus and pale, autolysing placenta when progesterone concentrations fall to < 1 ng/ml, so allowing the cervix to relax and the mare to return to oestrus (Allen, 1982).

Hormonal treatment of mares carrying donkey conceptuses

Secondary or accessory corpora lutea only occasionally develop after Day 40 in a small proportion of mares carrying donkey conceptuses. In the majority, the primary corpus luteum is the only source of progesterone with the result that maternal plasma progesterone levels remain low but relatively constant at 3–8 ng/ml between Days 40 and 70. They tend to decline steadily after this time in those mares in which fetal degeneration and death is taking place, finally reaching basal values of < 1 ng/ml coincidentally with abortion. In mares in which the donkey conceptus is implanting normally and is destined to continue to term, on the other hand, plasma progesterone concentrations begin to rise slowly after about Day 100, presumably as a result of increasing quantities of progesterone secreted by the placenta.

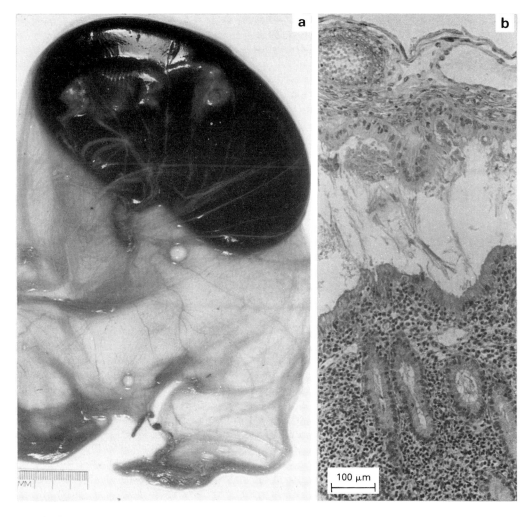

Fig. 2. Extraspecific donkey-in-horse conceptuses. (a) Aborted at Day 87 of gestation: the fetus shows generalized muscle wasting and the amniotic fluid is claret-coloured due to haemolysis of fetal red blood cells. (b) Allantochorion–endometrium interface in a mare at Day 79: exocrine secretion is accumulated between maternal and fetal epithelial layers and the endometrial stroma is filled with lymphocytes and other leucocytes.

Two experiments were undertaken to determine whether endocrinological deficiencies might play a significant part in the process of fetal death and abortion that occurs in the majority of donkey-in-horse pregnancies. In the first, 4 mares carrying donkey fetuses were each given a daily oral dose of 44 mg allyl trenbolone (Regumate: Hoechst Animal Health, Milton Keynes, U.K.) between Days 35 and 84; lower doses of this potent, synthetic progestagen had previously been shown to maintain pregnancy in mares bilaterally ovariectomized at Day 35 (Shideler *et al.*, 1982). Plasma progesterone levels in the allyl trenbolone-treated mares fell to baseline within 8 to 21 days after starting treatment and, hence, much earlier in gestation than in the untreated mares. Despite the lack of endogenous progesterone, the conceptuses continued to develop normally to around Day 75 in all 4 mares. Beyond this stage, however, they underwent the typical pattern of fetal regression leading to death and resorption between Days 85 and 105.

Table 2. Endocrinological treatments given to mares carrying extraspecific donkey conceptuses

Treatment	No. of pregnancies	Outcome		Fetal survival rate (%)
		Aborted before Day 110	Remained pregnant to > 150 days	
None	15	12	3	20
Oral allyl trenbolone (44 mg daily)	4	4	0	0
Partly purified horse CG (20 000–300 000 i.u.)	7	6	1*	14
Total	26	22	4	15

*Fetus viable and allantochorion implanted normally at Day 83.

In the second experiment, 7 mares carrying donkey conceptuses were given sporadic or regular intravenous injections of various doses of partly purified horse CG (20 000–300 000 i.u.), between Days 37 and 84. These doses of CG were high enough to give serum CG concentrations of 1–12 i.u./ml in the treated mares for 5–8 days after each injection. However, despite the large quantities of exogenous gonadotrophin, sharp rises in plasma progesterone concentrations, indicating the occurrence of secondary luteal development, occurred in only 1 of the 7 treated mares. Furthermore, in 6 mares the donkey conceptus was either aborted at the expected time, or was showing the typical degenerative changes leading to abortion when removed surgically between Days 65 and 87 (Table 2). In the 7th mare, however, the fetus was well developed and viable when removed at Day 83 and the allantochorion was firmly implanted in the endometrium. It seems likely that this pregnancy would have continued to term had it been left *in situ* (Allen, 1982).

As summarized in Table 2, neither the high doses of allyl trenbolone nor the injections of exogenous horse CG showed any evidence of increasing the survival rate of xenogeneic donkey-in-horse fetuses.

Immunization of mares carrying donkey fetuses

The absence of endometrial cups in the donkey-in-horse pregnancy model coupled with: (a) the success of other extraspecific equine pregnancies in which some endometrial cup development occurs; (b) the failure of hormonal therapy to increase the survival rate of donkey-in-horse fetuses; and (c) the dense leucocyte response in the mare's endometrium associated with the death of the donkey fetus, all suggested some form of immunological deficiency as the underlying cause of this example of early pregnancy loss. Three experiments were undertaken to test the hypothesis.

Passive immunization. Six mares carrying donkey conceptuses were given i.v. infusions, every 4–5 days between Days 40 and 85 of gestation, of 800–1200 ml serum recovered from mares carrying normal intraspecies horse pregnancies at equivalent stages of gestation. Two other control mares carrying donkeys were given similar infusions of serum from non-pregnant mares. Both control mares and 3 of the mares infused with pregnant mares' serum aborted their donkey conceptuses during the characteristic period (Days 84–110). However, the other 3 pregnant mares' serum-treated mares gave birth of live donkey foals, thereby raising fetal survival rate to 50% (Table 3).

Active immunization. Nine mares carrying donkey conceptuses were given a mixture of i.v. and intramuscular injections, 4 times between Days 25 and 54, of 200×10^6 mixed, washed donkey peripheral blood lymphocytes. Six mares received lymphocytes recovered from the genetic sire and dam of the donkey embryo transferred to them and the other 3 received lymphocytes from unrelated male and female donkeys. As shown in Table 3, the conceptus implanted normally and

Table 3. Immunological treatments given to mares carrying extraspecific donkey conceptuses and expression of the immunoprotective and cytotoxic (rejective) facets of maternal immune memory in successive pregnancies

| Treatment | No. of pregnancies | Outcome | | Fetal survival rate (%) |
		Aborted before Day 110	Remained pregnant to > 150 days	
Immunological treatments				
Passive immunization				
Infused serum of pregnant mares	6	3	3	50
Infused serum of non-pregnant mares	2	2	0	0
Active immunization				
Injected parental donkey lymphocytes	6	2	4	67
Injected unrelated donkey lymphocytes	3	1	2	67
Immune memory				
Mare previously carried first donkey fetus to term	13	0	13	100
Mare previously aborted first donkey fetus	12	12	0	0

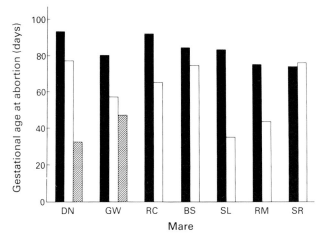

Fig. 3. Gestational age when fetal death occurred in 7 mares carrying first (closed bars), second (open bars) and third (hatched bars) successive extraspecific donkey pregnancies, illustrating anamnestic recall of a primary immunorejective type of maternal immune response towards the xenogeneic donkey fetus.

continued to term in 4 of the 6 mares immunized with parental, and in 2 of the 3 mares immunized with unrelated, donkey lymphocytes. This gave an overall fetal survival rate of 67% in the lymphocyte-immunized mares, compared to the 'control' survival rate of only 20% in untreated mares.

Expression of immune memory. Anamnestic recall of a previous immune response upon subsequent exposure to the same antigens is a fundamental concept of immunology. The success of

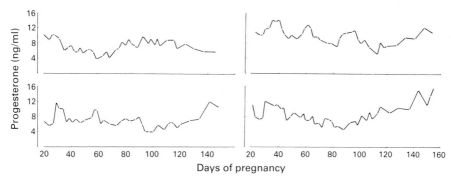

Fig. 4. Plasma progesterone profiles during the first 100 days of gestation in 4 mares each carrying a second successive donkey conceptus. These mares were immunized against donkey lymphocytes in their first successful donkey-in-horse pregnancies, thereby illustrating recall of the maternal immunoprotective response towards the donkey fetus that was generated at that time. Plasma progesterone concentrations in these mares were low and there are no signs of secondary luteal development.

the serum infusion and lymphocyte immunization experiments created two groups of horse mares that enabled an examination of maternal immune memory of fetal antigens. In the first group of 9 mares that had all aborted a first donkey pregnancy during the characteristic period of gestation, 9 second and 3 third consecutive donkey-in-horse pregnancies were established in subsequent years. All of these 12 repeat pregnancies were aborted at either the same (1 case) or earlier (11 cases) stages of gestation compared to the time of loss of the first pregnancy (Table 3; Fig. 3). However, 9 second and 4 third consecutive donkey-in-horse pregnancies established in 9 mares that had carried their first donkey pregnancies to term all implanted successfully and were carried to Day 150 without any further immunizations or other treatments (Fig. 4). Abortion was induced deliberately at this stage in most of these mares so that they could be re-used more rapidly in other experiments.

The findings in these animals therefore demonstrated clearly that both the rejection (cytotoxic) and immunoprotective types of maternal responses to the fetus can be effectively recalled in subsequent pregnancies.

Discussion

A deficiency of the steroid and other hormones necessary to maintain the uterus and related organs in 'the pregnancy state', or a disturbance of the feto–maternal interactions that normally induce maternal immune tolerance towards the antigenically foreign fetus throughout its intrauterine life, are just two of the many possible causes of early fetal death in viviparous mammals. The curious action of the equine fetus in injecting a small portion of its trophoblast tissue into the maternal endometrium before implantation, and the very obvious endocrinological and immunological facets of the resulting endometrial cup reaction, make equids suitable for study of these two aspects of pregnancy wastage. The complete failure of endometrial cup development in mares carrying extraspecific donkey conceptuses coupled with the high, closely-timed and partly reversible rate of fetal death, makes the donkey-in-horse pregnancy model particularly valuable for dissecting and differentiating the relative roles and importance of hormones and maternal immune responses in the maintenance of pregnancy.

On the hormonal side, some basic conclusions can be drawn from the donkey-in-horse experiments. First, it is clear that CG *per se* is not essential for the maintenance of equine pregnancy; 14 donkey foals have been born live and a further 10 fetuses carried to >150 days of gestation in the complete absence of this hormone. Second, the secondary luteal structures which normally develop after Day 40 in intraspecies equine pregnancy and the resulting sharp rises in maternal plasma progesterone concentrations are also not essential; 17 donkey-in-horse pregnancies in which thrice-weekly plasma progesterone assays have shown no evidence of accessory luteal development have been maintained to >150 days.

A third endocrinological finding of interest in the donkey-in-horse experiments was the inability of exogenous CG to stimulate secondary ovulations. In the 7 pony mares given a total of 29 separate injections of large doses of horse CG, a definite rise in plasma progesterone concentrations, indicating secondary luteal development, occurred only twice in the same mare. On all other occasions, progesterone values remained steady or continued to decline slowly after each injection of gonadotrophin. Conversely, sharp rises in plasma progesterone concentrations between Days 40 and 80 have been observed on 5 isolated occasions in the total of 14 untreated donkey-in-horse pregnancies monitored, thereby demonstrating that secondary ovulations can sometimes occur in the absence of chorionic gonadotrophin. These findings support the results of other experiments (Squires & Ginther, 1975; Evans & Irvine, 1975; Allen, W.E., 1975; Stewart & Allen, 1979; Garcia et al., 1979) which, when considered together, cast doubt upon the earlier suggestions of Cole (1936) and Amoroso et al. (1948) that CG is solely responsible for stimulating secondary luteal development in equine pregnancy. It now seems likely that pituitary FSH, not CG, induces secondary follicular growth with only the LH-like component of CG acting to ovulate and/ or luteinize these follicles when they become sufficiently mature to respond to the stimulus.

One further feature of the donkey-in-horse experiments may be a valuable pointer in relation to the common veterinary practice of administering exogenous progesterone or synthetic progestagens to pregnant mares that have histories of previous early pregnancy loss. The close correlation between falling plasma progesterone concentrations and fetal degeneration in mares aborting a donkey conceptus suggests that, when low maternal plasma progesterone concentrations are found in conjunction with fetal death in mares carrying normal intraspecies horse pregnancies, it is most likely to be a case of a defective fetus causing luteal regression, rather than a defective corpus luteum resulting in the loss of an otherwise healthy conceptus. Furthermore, the more rapid than normal decline in endogenous plasma progesterone concentrations noted in all 4 of the mares carrying donkey conceptuses treated with allyl trenbolone suggests some form of negative feedback effect that caused suppression of either a fetal luteotrophin or, more likely, a fetal anti-luteolysin, that was acting to prolong the lifespan and secretory function of the primary corpus luteum. Whether a similar feedback effect would occur in progestagen-treated pregnant mares that have adequate levels of CG in their blood has yet to be determined. But in view of the risk it is clear that, if progestagen therapy is to be instituted during early pregnancy, sufficient exogenous hormone should be given to permit maintenance of the pregnancy in the absence of any contribution from the ovaries. Treatment should also continue until at least Day 100 of gestation so that the placenta can mature sufficiently to secrete enough progesterone to maintain pregnancy when therapy with exogenous hormone is ceased.

In addition to its endocrinological features, the donkey-in-horse pregnancy model has given valuable information concerning the necessity of feto–maternal immunological interactions in early gestation if the antigenically foreign fetus is to be protected from potentially lethal maternal immune responses. The absence of endometrial cups in this one unsuccessful model of extraspecies pregnancy and the marked improvement in fetal survival rate achieved by immunization of the surrogate mothers suggest that, in normal equine pregnancy, the invasion of chorionic girdle cells into the endometrium to form the endometrial cups provides a vital and temporally important antigenic stimulus to the mother that results in her becoming tolerant of any foreign fetal antigens that may subsequently be expressed at the feto–maternal interface.

The precise nature of any foreign antigens expressed on the invasive and non-invasive components of equine trophoblast have yet to be determined. But already it is clear that paternal Class I MHC antigens are present in high concentrations on the progenitor tissue of the endometrial cups and that these alloantigens are recognized by the mother during development of the cups (Antczak *et al.*, 1984; Crump *et al.*, 1987). Likewise, histochemical studies with a panel of monoclonal antibodies raised against endometrial cup cells have isolated two antibodies that show specific, selective and reciprocal staining patterns on the invasive and non-invasive components of horse trophoblast (Antczak *et al.*, 1987). It therefore becomes reasonable to postulate that both paternally-derived alloantigens and trophoblast-specific molecules are presented for maternal immune recognition by the development of the endometrial cups and that both types of antigens are concerned in eliciting the humoral and cell-mediated responses observed.

The success of lymphocyte immunizations in raising fetal survival rate in the donkey-in-horse pregnancy model parallels similar studies in human and murine pregnancy. Mowbray *et al.* (1985), Beer *et al.* (1986) and others have reported a large increase in live birth rates in women with a history of recurrent spontaneous first trimester abortion, by immunizing with peripheral blood lymphocytes from either the husband or an unrelated third party donor. Likewise, Kiger *et al.* (1985) have demonstrated a significant decrease in the number of resorptions in the CBA × DBA/2 model of mouse pregnancy after immunization in the CBA mothers with lymphocytes from BALB c mice that share the same $H-2^d$ histocompatibility locus with the DBA/2 fathers. Although the mechanisms involved in generating the fetal immunoprotection induced by lymphocyte immuniz-ation are not yet understood, available evidence points towards involvement of the so-called trophoblast–lymphocyte cross-reacting antigen (TLX) that has been identified on human (McIntyre & Faulk, 1982) and rat (Gill & Repetti, 1979) placentae. In view of the similarly beneficial effects of this form of therapy in the human, murine and equine models of pregnancy failure, the possibility must be considered that a TLX-type of antigen is expressed on the invasive trophoblast cells of the endometrial cups in the mare.

The immunoprotection afforded to the equine fetus could take the form of enhancing or blocking antibodies, as described by Voisin & Chaouat (1974) and others in the mouse, or it may involve the generation of specific suppressor cells in the uterus. Clark *et al.* (1984) have described the presence of small, granulated non-T lineage cells with potent suppressor activity in the deciduoma surrounding viable mouse fetuses, and a considerable reduction in the numbers of these cells around conceptuses undergoing resorption. The results of preliminary experiments designed to determine the functional activities of the lymphocytes that accumulate around the endometrial cups are of considerable significance. When recovered between Days 55 and 70 of gestation, these endometrial cup lymphocytes exhibit a potent ability to suppress the normal mitogenic response of peripheral blood lymphocytes from the dam and unrelated horses (J. Kydd & W. R. Allen, unpublished data). It therefore appears that at least a proportion of the leucocytes

Fig. 5. Three surrogate pony mares with their transferred extraspecific Przewalski's horse, donkey and Grant's zebra foals.

attracted to the endometrial cups have suppressor, rather than cytotoxic, capabilities. This, in turn, suggests the likelihood that stimulation of specific uterine suppressor cell activity is a major function of the endometrial cup reaction in normal equine pregnancy. These cells could then act to inhibit potentially harmful maternal responses to any foreign antigens that may be expressed on the non-invasive trophoblast of the allantochorion during implantation.

In the overall search to understand the mechanisms that provide immunoprotection for the mammalian fetus *in utero*, the mare and her relatives provide a valuable and often contrasting adjunct to the more intensively studied murine and primate species. In particular, her ability to carry to term extraspecific fetuses of such widely differing genotype and phenotype as the donkey, Przewalski's horse and Grant's zebra (Fig. 5) demonstrate that, if the fetus supplies the correct type of antigenic signal to the mother at the appropriate stage of gestation, it receives a strikingly effective immune shield in return.

This study was financed by The Thoroughbred Breeders' Association of Great Britain, The Horserace Betting Levy Board, The Harry M. Zweig Memorial Fund and the National Institutes of Health (HD 15799).

References

Allen, W.E. (1975) Ovarian changes during early pregnancy in Pony mares in relation to PMSG production. *J. Reprod. Fert., Suppl.* **23,** 425–428.

Allen, W.R. (1975) Immunological aspects of the equine endometrial cup reaction. In *Immunology of Trophoblast*, pp. 217–253. Eds R. G. Edwards, C. W. S. Howe & M. H. Johnson. Cambridge University Press.

Allen, W.R. (1979) Maternal recognition of pregnancy and immunological implications of trophoblast-endometrium interactions in equids. In *Maternal Recognition of Pregnancy* (Ciba Fdn Series No. 64), pp. 323–352. Excerpta Medica, Amsterdam.

Allen, W.R. (1982) Immunological aspects of the endometrial cup reaction and the effects of xenogeneic pregnancy in horses and donkeys. *J. Reprod. Fert., Suppl.* **31,** 57–94.

Allen, W.R. & Moor, R.M. (1972) The origin of equine endometrial cups. I. Production of PMSG by foetal trophoblast cells. *J. Reprod. Fert.* **29,** 313–316.

Allen, W.R., Hamilton, D.W. & Moor, R.M. (1973) The origin of equine endometrial cups. II. Invasion of the endometrium by trophoblast. *Anat. Rec.* **117,** 475–501.

Allen, W.R., Kydd, J.H., Miller, J. & Antczak, D.F. (1984) Immunological studies on feto-maternal relationships in equine pregnancy. In *Immunological Aspects of Reproduction in Mammals*, pp. 183–193. Ed. D. B. Crighton. Butterworths, London.

Amoroso, E.C., Hancock, J.L. & Rowlands, I.W. (1948) Ovarian activity in the pregnant mare. *Nature, Lond.* **161,** 355–356.

Antczak, D.F., Miller, J. & Remick, L.H. (1984) Lymphocyte alloantigens of the horse. II. Antibodies to ELA antigens produced during equine pregnancy. *J. Reprod. Immunol.* **6,** 283–287.

Antczak, D.F., Oriol, J.G., Donaldson, W.L., Poleman, C., Stenzler, L., Volsen, S.G. & Allen, W.R. (1987) Differentiation molecules of the equine trophoblast. *J. Reprod. Fert., Suppl.* **35,** 371–378.

Beer, A.E., Quebbeman, J.F. & Zhu, X. (1986) Non-paternal leukocyte immunization in women previously immunized with paternal leukocytes. Immune responses and subsequent pregnancy outcome. In *Reproductive Immunology 1986*, pp. 261–268. Eds D. A. Clark & B. A. Croy. Elsevier, Amsterdam.

Bouters, R., Spincemaille, J., Vandeplassche, M., Bonte, P. & Coryn, M. (1978) Implantatie, foetale outwikkeling en PMSG-productie bij tweeking-merries na paring met kun chimaere tweelingbroeder. *Verh. K. vlaam. Acad. Geneesk. Belg.* **40,** 253–256.

Clark, D.A., Slapsys, R.M., Croy, B.A., Kreek, J. & Rossant, J. (1984) Local active suppression by suppressor cells in the decidua: a review. *Am. J. Reprod. Immunol.* **5,** 78–83.

Clegg, M.T., Boda, J.M. & Cole, H.H. (1954) The endometrial cups and allantochorionic pouches in the mare with emphasis on the source of equine gonadotrophin. *Endocrinology* **54,** 448–463.

Cole, H.H. (1936) On the biological properties of mare gonadotrophic hormone. *Am. J. Anat.* **59,** 299–332.

Cole, H.H. & Goss, H. (1943) The source of equine gonadotropin. In *Essays in Honor of Herbert M. Evans*, pp. 107–119. University of California Press, Berkeley.

Crump, A., Donaldson, W.L., Miller, J., Kydd, J., Allen, W.R. & Antczak, D.F. (1987) Expression of MHC antigens on equine trophoblast. *J. Reprod. Fert., Suppl.* **35,** 379–388.

Davies, C.J., Antczak, D.F. & Allen, W.R. (1985) Reproduction in mules: embryo transfer using sterile recipients. *Equine vet. J., Suppl.* **3,** 63–67.

Evans, M.J. & Irvine, C.H.G. (1975) Serum concentrations of FSH, LH and progesterone during the oestrous cycle and early pregnancy in the mare. *J. Reprod. Fert., Suppl.* **23,** 193–200.

Ewart, J.C. (1897) *A Critical Period in the Development of the Horse.* Adam & Charles Black, London.

Hutton, C.A. & Meacham, T.N. (1968) Reproductive efficiency on fourteen horse farms. *J. Anim. Sci.* **27,** 434–438.

Garcia, M.C., Freedman, L.J. & Ginther, O.J. (1979) Interaction of seasonal and ovarian factors in the regulation of LH and FSH secretion in the mare. *J. Reprod. Fert., Suppl.* **27,** 103–111.

Gill, T.J. & Repetti, C.F. (1979) Immunologic and genetic factors influencing reproduction. A Review. *Am. J. Path.* **95,** 465–570.

Kiger, N., Chaouat, G., Kolb, J-P., Wegmann, T.G. & Guenet, J-L. (1985) Immunogenetic studies of spontaneous abortion in mice. Pre-immunization of females with allogeneic cells. *J. Immunol.* **134,** 2966–2970.

Kydd, J., Miller, J., Antczak, D.F. & Allen, W.R. (1982) Maternal anti-fetal cytotoxic antibody responses of equids during pregnancy. *J. Reprod. Fert., Suppl.* **32,** 361–369.

Kydd, J.H., Boyle, M.S., Allen, W.R., Shephard, A. & Summers, P.M. (1985) Transfer of exotic equine embryos to domestic horses and donkeys. *Equine vet. J., Suppl.* **3,** 80–83.

McIntyre, J.A. & Faulk, W.P. (1982) Allotypic trophoblast-lymphocyte cross-reactive (TLX) cell surface antigens. *Human Immunol.* **4,** 27–35.

Merkt, H. (1966) Fohlenrosse und Fruchtresorption. *Zuchthygiene* **1,** 102–108.

Mowbray, J.F., Gibbings, C., Liddell, H., Reginald, P.W., Underwood, J.L. & Beard, J.W. (1985) Controlled trial of treatment of recurrent spontaneous abortion by immunization with paternal cells. *Lancet* **1,** 941–943.

Samuel, C.A., Allen, W.R., Steven, D.H. (1975) Ultrastructural development of the equine placenta. *J. Reprod. Fert., Suppl.* **23,** 575–578.

Shideler, R.K., Squires, E.L., Voss, J.L., Eikenberry, D.J. & Pickett, B.W. (1982) Exogenous progestagen therapy of ovariectomized pregnant mares. *J. Reprod. Fert., Suppl.* **32,** 459–464.

Squires, E.L. & Ginther, O.J. (1975) Follicular and luteal development in pregnant mares. *J. Reprod. Fert., Suppl.* **23,** 429–433.

Stewart, F. & Allen, W.R. (1979) The binding of FSH, LH and PMSG to equine gonadal tissues. *J. Reprod. Fert., Suppl.* **27,** 431–440.

Voisin, G.A. & Chaouat, G. (1974) Demonstration, nature and properties of maternal antibodies fixed on placenta and directed against paternal antigens. *J. Reprod. Fert., Suppl.* **21,** 81–103.

J. Reprod. Fert., Suppl. **35** (1987), 211–220
Printed in Great Britain
© 1987 Journals of Reproduction & Fertility Ltd

Infusion of gonadotrophin-releasing hormone (GnRH) induces ovulation and fertile oestrus in mares during seasonal anoestrus

J. H. Hyland, P. J. Wright, I. J. Clarke*, R. S. Carson*, D. A. Langsford and L. B. Jeffcott

*Department of Veterinary Clinical Sciences, University of Melbourne, Werribee, Victoria 3030 and *Medical Research Centre, Prince Henry's Hospital, St Kilda Road, Melbourne, Victoria 3004, Australia*

Summary. In Exp. 1, 30 Standardbred mares in deep seasonal anoestrus were divided into 3 equal groups and treated with 0, 50 (G50) or 100 (G100) ng GnRH $kg^{-1} h^{-1}$ for 28 days via osmotic minipumps. Ovulation occurred in 0/10, 3/10 and 7/10 mares respectively ($P < 0.05$). Plasma GnRH profiles (Days −6, 0, 2, 6, 12, 20, 28 and 34 relative to pump insertion) were dose-dependent ($P < 0.01$) and peaked on Day 12 of infusion. Mean daily plasma LH concentrations were biphasic in treated mares that ovulated, with LH peaks occurring around Day 6 and Days 16–20. By contrast, in treated mares that did not ovulate the initial LH rise was followed by a steady decline to the end of the experiment. LH pulse frequency in treated mares increased between Day 0 and Day 21 of the experiment. LH pulse frequency in G100 mares was higher ($P < 0.05$) than in G50 and control mares on Day 3, and higher than the controls on Days 7 and 21 of the experiment. There were no significant differences in LH pulse amplitude between the groups on the days studied.

In Exp. 2, 27 Standardbred mares in shallow seasonal anoestrus received no treatment (N = 13) or a subcutaneous infusion of GnRH (100 ng $kg^{-1} h^{-1}$) via osmotic minipump for 28 days (N = 14). Mares were served by a stallion during oestrus. Day of ovulation was earlier in treated than in control mares (18·6 ± 2 *vs* 41·9 ± 6 days; $P < 0.001$). Likewise, time of conception was earlier in treated than in control mares (25·2 ± 6 *vs* 49·1 ± 9 days; $P < 0.05$). One mare in the control group failed to conceive while one treated mare conceived to an undetected ovulation.

The results show that constant GnRH infusion induces ovulation and fertile oestrus in mares during deep and shallow seasonal anoestrus.

Introduction

Seasonal anoestrus in mares presents severe management and economic difficulties for the horse breeding industry. The problem is compounded by the imposition of an arbitrary birthday for performance horses, which is 1 August and 1 January in the southern and northern hemispheres, respectively. Any method which reliably induced ovulation in mares during deep or shallow anoestrus could have significant commercial implications for the horse breeding industry by encouraging more efficient use of stallions and stud farm personnel.

Reproductive activity in the mare is stimulated by increasing photoperiod, and so artificial lighting, given to simulate long days, can be used to advance the breeding season in mares (Burkhardt, 1946; Kooistra & Ginther, 1975; Palmer *et al.*, 1982). The effects of photoperiod on reproductive activity in seasonally breeding species appear to be mediated by the pineal gland in ferrets (Thorpe & Herbert, 1975), hamsters (Reiter, 1981), ewes (Bittman & Karsch, 1984) and perhaps mares (Grubaugh *et al.*, 1982). It has been suggested (Ginther, 1979) that melatonin, produced by the pineal gland, modifies secretion of GnRH by the hypothalamus. This can be inferred from one study (Strauss *et al.*, 1979) which showed that the hypothalamic GnRH content of ovariectomized mares was lower during winter months than during summer months.

Indirect evidence of a reduction in GnRH secretion during seasonal anoestrus comes from experiments in which plasma LH concentrations were estimated over 1 year in 11 pony mares (Ginther, 1979) and for 4 months in 5 ovariectomized horse mares (Fitzgerald *et al.*, 1983). Plasma LH concentrations were low during winter, but were high during summer. Furthermore, the number of GnRH pulses measured in blood leaving the pituitary gland of mares is lower during seasonal anoestrus than during the ovulatory season (Irvine & Alexander, 1984). In addition, GnRH pulses, measured in the pituitary portal blood of conscious sheep, appear to be temporally associated with LH pulses occurring in peripheral blood (Clarke & Cummins, 1982). Further evidence supporting the role of GnRH secretion in seasonal reproduction comes from a study in sheep in which a continuous infusion of GnRH induced oestrus and ovulation during seasonal acyclicity (Wright *et al.*, 1983).

Ovulation has been induced in anoestrous mares by use by multiple injections of equine pituitary extract (Woods & Ginther, 1982), GnRH (Evans & Irvine, 1979) or hCG (Bour & Palmer, 1984). These methods have the disadvantage that they are inconvenient to use on large numbers of horses.

The hypothesis to be tested in these experiments was that a constant infusion of GnRH would induce ovulation and fertile oestrus in mares during deep and shallow seasonal anoestrus.

Materials and Methods

The study was divided into 2 experiments. In the first study (Exp. 1) we assessed the ability of constant infusions of GnRH for inducing ovulation during deep (June/July) seasonal anoestrus. In the second study (Exp. 2), this work was extended to examine whether GnRH infusions could be used to induce fertile oestrus during shallow (September/October) seasonal anoestrus.

Experiment 1

Experimental animals. Mares were defined as being in deep seasonal anoestrus when plasma progesterone concentrations were < 1 ng/ml in all blood samples collected every 3 or 4 days for 4 weeks, and when palpation of the ovaries *per rectum* revealed them to be small, hard and with little or no palpable follicular activity.

The 30 maiden Standardbred mares, 3–7 years old and weighing 344–530 kg were divided into 3 equal groups. They were fed about 4 kg of pasture hay in the morning and 6 kg of a grain and chaff mixture in the evening and kept in yards for the duration of the GnRH infusion and for 6 days before the start of infusion; otherwise, they remained in paddocks. The mares were vaccinated against strangles and tetanus and received regular anthelmintic treatments.

Experimental procedures. Treated groups received an infusion of synthetic GnRH (Novachem P/L., Victoria, Australia) at a dose rate of 50 (Group G50) or 100 (Group G100) ng kg^{-1} h^{-1}, while the remaining mares acted as untreated controls. The GnRH was dissolved in sterile 0·15 M-saline and delivered via osmotic minipumps (model 2ML4; Alza Corp., Palo Alto, CA, U.S.A.) which, according to the manufacturer's instructions, were functional for 28 days. The minipumps were loaded with GnRH solution and placed in a beaker of sterile 0·15 M-saline overnight at room temperature to initiate pump activity. Next day (Day 0) the pumps were implanted subcutaneously in the side of the neck of the treated mares, using aseptic procedures.

Blood collection. On Day −7 selected mares were fitted with indwelling jugular catheters. Blood (10 ml) was collected into heparinized tubes every hour, from 07:00 to 18:00 h between Day −6 and Day 28. At weekends blood was collected every 3 h between 09:00 and 18:00 h. Additional samples were taken every 15 min (intensive bleeds) between 09:00 and 17:00 h on Days −4, 0, 3, 7, 14, 21 and 28. Blood samples were held in ice for up to 1 h before plasma was removed and maintained at −20°C until assayed for LH and progesterone.

Additional blood samples were collected for plasma GnRH measurement on Days −6, 0, 2, 6, 12, 20, 28 and 34. Blood was collected into 2·5 ml of 0·15 M-saline containing bacitracin on ice.

Assessment of ovarian activity. Ovarian activity was monitored 3 times weekly by palpation of the ovaries *per rectum*. If a follicle >3 cm in diameter was detected, ovarian palpation was carried out daily until ovulation or follicular regression occurred. On the same days that ovarian palpations were carried out, the mares were tested with a stallion for behavioural oestrus.

Experiment 2

Experimental animals. Standardbred mares, in good condition and weighing 376–527 kg were maintained in paddocks at the Veterinary Clinical Centre and received about 5 kg pasture hay twice per day. The 27 mares were diagnosed as being in shallow anoestrus by (i) plasma progesterone concentrations < 1 ng/ml in blood collected every 3 or 4 days for the 4 weeks before start of the experiment, (ii) one or more ovarian follicles > 2 cm in diameter and (iii) irregular displays of behavioural oestrus when exposed to a stallion every 2 or 3 days.

Experimental procedures. The mares were assigned to control (N = 13) or treated (N = 14) groups so that ovarian activity was similar between the two groups when treament began. Treated mares received an infusion of GnRH in saline at a dose rate of 100 ng kg^{-1} h^{-1} for a period of 28 days, delivered by subcutaneously implanted minipump (as above) and the control mares remained untreated. The pumps were implanted (Day 0) between 10 September and 21 September in 1984 and on 2 October in 1985.

Blood collection. Blood samples were collected into heparin by jugular venepuncture every 3 or 4 days before implantation of the minipumps. Samples were then taken every 2 or 3 days until about Day 30 of pregnancy. Blood was held on ice for up to 2 h before plasma was removed by centrifugation, and stored at −20°C until assayed for LH and progesterone.

Assessment of ovarian activity. Ovarian palpations *per rectum* were carried out every 3 or 4 days for 1 month before implantation of the pumps, and then every 2 or 3 days or daily when a follicle > 3 cm in diameter was detected. The day of ovulation was recorded as the first day on which ovulation was detected. Using these methods, we failed to detect ovulation in one mare. Mares were mated to a fertile stallion every other day during oestrus until ovulation occurred.

Assay procedures

Plasma LH concentrations were determined by radioimmunoassay established in this laboratory (J. H. Hyland & L. B. Jeffcott, unpublished). Briefly, the assay utilized iodinated purified sheep LH (Papkoff (G3222B) as tracer and purified horse LH (Landefield & McShan, 1974) as standard. Rabbit anti-ovine LH (GDN15) was the antibody, features of which have been published elsewhere (Nett *et al.,* 1975). The sensitivity of the assay was 0·2 ng/ml and intra- and inter-assay coefficients of variation did not exceed 20%. Purified horse LH was recovered quantitatively after addition to plasma from an anoestrous mare at concentrations of 1, 2·5, 5 and 10 ng/ml ($y = 0.95x + 0.29$; $r = 0.96$, $n = 54$). Parallelism was observed between the standard curve and increasing dilutions of plasma from a dioestrous mare.

Plasma progesterone concentrations were determined by using commercial radioimmunoassay kits. In 1984, we used a kit containing a solid phase antibody (DPC Coat-a-Count, Dist. Bio-Mediq Australia, Doncaster, Victoria, Australia). According to the manufacturer's literature, the major cross-reactants with the antibody are deoxycortico-sterone (1·7%), 20α-dihydroprogesterone (2·0%), deoxycortisol (2·4%) and 5α-pregnandione (1·3%). The assay was validated for horse plasma in our laboratory by preparing standards in anoestrous mare plasma (100 μl). Parallelism between the standard curve and increasing dilutions of dioestrous mare plasma was observed. Sensitivity of the assay (minimum concentration significantly different from zero; Burger *et al.,* 1972) was about 0·1 ng/ml. Recovery of progesterone added to anoestrous mare plasma was quantitative with a regression of $y = 1.05x + 0.04$ and $r = 0.999$ ($n = 27$). The within-assay coefficient of variation for this experiment was 6·1% ($n = 3$). Samples collected during 1985 were analysed using a double-antibody method in kit form (Farmos Diagnostica, Turku, Finland). According to the manufacturer's literature the major cross-reactants to the antiserum are 11-hydroxyprogesterone (80%), 5α-dihydroprogesterone (8·8%) and 5β-dihydroprogesterone (7·1%). The assay was validated in our laboratory for the measurement of progesterone in horse plasma. Recovery of progesterone added to anoestrous mare plasma was quantitative, with a regression of $y = 0.93x + 0.72$ ($r = 0.98$, $n = 21$). The intra- and inter-assay coefficients of variation were 8·8% and 6·6%, respectively ($n = 7$). Assay sensitivity, measured as the lowest concentration significantly different from zero (Burger *et al.,* 1972), was 0·1 ng/ml. Parallelism was obtained between the standard curve, prepared in anoestrous horse plasma, and increasing dilutions of plasma from a dioestrous mare.

The progesterone concentration in plasma from a dioestrous mare was measured in both assay systems. The mean progesterone concentration (± s.e.m.) was 6·3 ± 0·3 ng/ml ($n = 25$) and 4·9 ± 0·1 ng/ml ($n = 21$) by the DPC Coat-a-Count and Farmos methods, respectively ($P < 0.001$).

Plasma GnRH concentrations were measured by the method of Jonas *et al.* (1975), after extraction of GnRH from horse plasma with acidified (0·01 M-glacial acetic acid) methanol. Recovery was 64·9 ± 1·7% ($n = 4$). Features of the antibody have been described previously (Nett *et al.,* 1973). In our system, assay sensitivity (Burger *et al.,* 1972) was about 0·6 pg/ml and intra- and inter-assay coefficients of variation were less than 20%. Recovery of GnRH added to serum from an hypophysectomized sheep was quantitative, with a regression equation of $y = 1.2x + 1.3$ ($r = 0.97$, $n = 12$).

Statistical methods

A plasma LH pulse was defined by the criteria of Fitzgerald *et al.* (1983). These were: (i) a peak concentration must occur within 3 samples of the previous nadir, (ii) the amplitude of the pulse (peak minus nadir) must be at least twice the sensitivity of the assay, and (iii) the LH concentration at the peak must exceed the 95% confidence limits of the LH concentration at both the preceding and subsequent nadirs. Confidence limits for the nadirs were calculated by the method of Burger *et al.* (1972).

Mean daily plasma LH concentrations were calculated by averaging hourly plasma LH concentrations between 07:00 and 18:00 h between Day −6 and Day 28. Blood samples were collected 3 times each week for 4 weeks after the end of the GnRH infusion and mean weekly plasma LH concentrations were calculated.

The proportions of mares ovulating in each group in Exp. 1 were compared by χ^2 tests. Analysis of variance for continuous data over time (Gill & Hafs, 1971) was used to analyse plasma GnRH profiles and mean daily LH profiles. LH pulse frequency and LH pulse amplitude data were analysed by analysis of variance and Student's *t* test using a pooled variance. Student's *t* test was used to compare mean day of ovulation and conception between control and treated mares in Exp. 2. Values are given as the mean and s.e.m.

Results

Experiment 1

Ovulation data. Ovulations occurred in 3 of 10 Group G50 mares, and 7 of 10 Group G100 mares, whereas none of the control mares ovulated. Ovulation occurred 20·3 ± 2 days and 19·9 ± 1·7 days after insertion of the pump in mares in Groups G50 and G100 respectively. Growth of the ovulating follicle was calculated from the time the follicle was first detected to the day of ovulation. Mean growth time was 7·7 days (range 6–10 days) for Group G50 mares and 10·2 days (range 7–14 days) in Group G100 mares.

Hormone profiles. There was a dose-related rise in mean plasma GnRH concentrations (Fig. 1), which was evident within 2 days of implantation of the pump. This rise reached a maximum of 32·7 pg/ml and 61·7 pg/ml on Day 12 in mares in Groups G50 and G100, respectively. Concentrations then declined to baseline by Day 34 when the pump had ceased to function. There were significant treatment ($P < 0·01$) and day ($P < 0·001$) effects of GnRH infusions on mean plasma GnRH concentrations.

There was a significant ($P < 0·001$) treatment effect on mean daily LH concentrations (Fig. 2). Plasma LH concentrations were higher in Group G100 mares than in Group G50 mares or control mares between Day 0 and Day 28. Mean daily LH profiles were biphasic in the treated mares, with peaks occurring on Days 6 and 18 in Group G50 mares and on Days 7 and 21 in Group G100 mares. No trends in plasma LH concentrations were seen in control mares, which remained around baseline for the duration of the experiment. In treated mares, plasma LH concentrations returned to baseline concentrations within 1 week of the end of GnRH infusion and remained low for 3 weeks. By the 4th week after GnRH infusion, mean weekly plasma LH concentrations were higher in all groups, indicating that the reproductive system of the mares may have been responding to increasing daylength at this time.

Examination of mean daily plasma LH profiles in treated mares which ovulated, and treated mares which did not ovulate (Fig. 3) shows a biphasic release of LH in ovulating mares, the second peak of which is higher than the first. Ovulations occurred in association with this second peak. By contrast, mean daily LH concentrations in the treated mares which did not ovulate showed only a single peak. In these mares, mean daily LH concentrations rose in parallel with ovulating mares, to peak around 6 days after pump insertion, but then declined slowly to reach levels slightly higher than baseline 24–28 days after pump insertion.

There was a significant ($P < 0·05$) effect of treatment on LH pulse frequency (Fig. 4a) on Days 3, 7 and 21. Mean LH pulse frequency (pulses/8 h) increased in the treated groups between Day −4 and Day 7 (G50, range 0–0·9; G100, range 0·2–1·5), while values for control mares were variable (range 0–0·5). Maximum LH pulse frequencies in Group G50 mares (1·7 pulses/8 h) occurred on Days 21 and 28, while Group G100 mares had a maximum of 2·1 pulses/8 h on Day 21. Pulse frequency was lower in all groups on Day 14 compared to Day 7, but increased on Day 21 and Day 28, except in mares in Group G100, in which there was a decline in LH pulse frequency between Day 21 and Day 28. In 1984, a number of blood samples collected during the intensive bleeding period on Day 14 were lost. Consequently, data on LH pulse frequency and LH pulse amplitude on Day 14 of the 1984 experiment are not included. Therefore, N = 5, 5 and 6 for controls, and mares in Groups G50 and G100 respectively on Day 14.

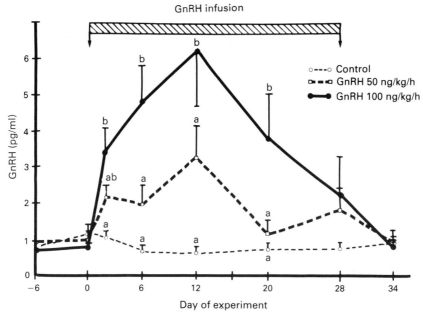

Fig. 1. Plasma GnRH concentrations in mares given GnRH infusions via minipumps during deep seasonal anoestrus. Values are mean $\pm$ s.e.m. for 10 mares. Values with different superscript letters are significantly different ($P < 0.05$).

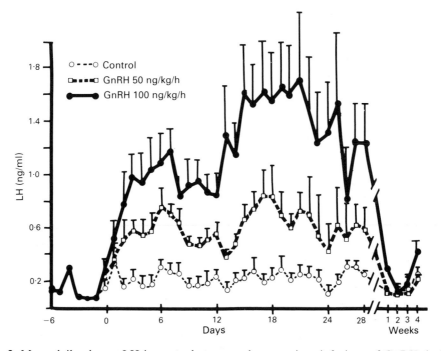

Fig. 2. Mean daily plasma LH in control mares and mares given infusions of GnRH during deep seasonal anoestrus. There is a significant treatment effect ($P < 0.001$). Values are mean $\pm$ s.e.m. for 5–10 mares.

In mares in Group G100 LH pulse frequency was significantly ($P < 0.05$) higher than in the other 2 groups on Days 3 and 7. On Day 21, LH pulse frequency was higher ($P < 0.05$) in Group G100 mares than in the controls. By Day 28, LH pulse frequency in the low-dose GnRH group was significantly higher than in the other 2 groups.

There were no significant trends in LH pulse amplitude (Fig. 4b) within or between treatment groups.

Progesterone concentrations increased within 1 or 2 days of ovulation. The length of the luteal phase was calculated as the day of the first sustained rise in progesterone concentration associated with ovulation, to the first day that progesterone concentrations fell below 1 ng/ml. Using these criteria, luteal life-span ranged from 6 days to > 24 days. Luteal life-span was negatively correlated with day of ovulation relative to day of pump insertion ($r = 0.66$; $P < 0.05$; $n = 10$), as was peak height of plasma progesterone profiles ($r = -0.93$; $P < 0.001$; $n = 10$).

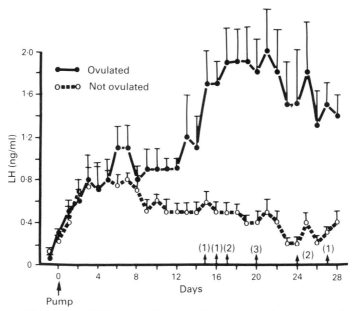

Fig. 3. Mean daily plasma LH in treated mares that ovulated (arrow, no. of mares) and in treated mares that did not ovulate in response to GnRH infusion during deep seasonal anoestrus. Values are mean $\pm$ s.e.m. for 5–10 mares.

Experiment 2

Control (N = 13) and treated (N = 14) mares ovulated 41.9 ± 5.5 days and 18.6 ± 3.0 days, respectively, after implantation of the pump or assignment to the experiment ($P < 0.001$). The period of growth of the ovulating follicle was calculated as the time it was first detected by palpation to the day of ovulation. Mean growth time was 6.9 ± 1.1 and 8.3 ± 0.6 days in control and treated mares, respectively. Day of conception (calculated from the day of pump implantation) was later in control mares (49.1 ± 8.7 days) than in the treated mares (25.2 ± 6.0 days; $P < 0.05$): 9 of 13 control mares and 11 of 14 treated mares conceived to the first ovulation after pump implantation. One of the control mares failed to conceive and one treated mare conceived to an undetected ovulation, about 6 days after service. Data from these mares have been excluded from the conception results. Two mares resorbed embryos after conceiving to a GnRH-associated ovulation in 1985.

Plasma LH concentrations in all but 1 treated mare rose within 2 days of pump implantation. This initial elevation of LH was followed by a distinct LH surge in 11 of the 14 treated mares during

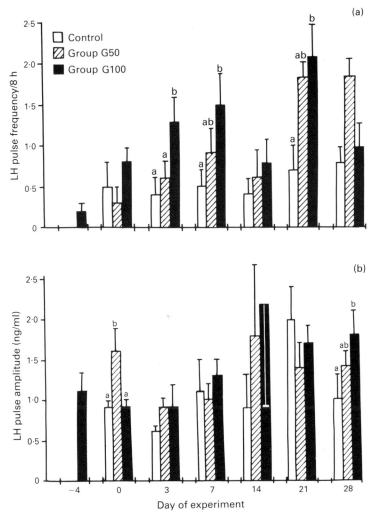

Fig. 4. LH pulse frequency (a) and LH pulse amplitude (b) in control mares and mares treated with GnRH infusions (50 or 100 ng/kg/h) during deep seasonal anoestrus. GnRH infusions began on Day 0. Values are mean ± s.e.m. for 10 mares/group except on Day 14, when N = 5, 5 and 6 for controls, and Groups G50 and G100 respectively. Values with different letters are significantly different (*P* < 0·05).

the period the pumps were functional. These LH surges resulted in ovulation in all but 1 mare, in which a second LH surge was associated with ovulation on Day 33 after pump insertion. In the 3 mares in which surges were not observed during GnRH infusion, small LH rises occurred and were associated with ovulation in 2 mares, 10 and 25 days after insertion of the pump.

Mares were examined by ultrasound echography to monitor pregnancy until Day 45 of gestation. Growth of the vesicle and embryo were within normal limits up to this time except in the 2 mares which resorbed embryos.

Discussion

The results of Exp. 1 clearly demonstrate that a continuous infusion of GnRH, given to mares during deep seasonal anoestrus, induced ovulation in 50% of the treated mares. Ovulation has been induced in seasonally anoestrous ewes with multiple (every 2 h), low-dose (250, 500 or 1000 ng) GnRH injections, given for 8 days (McLeod *et al.*, 1982). This translates into a dose of about $2 \cdot 5$–10 ng kg^{-1} h^{-1}. In addition, Wright *et al.* (1983) induced fertile oestrus in ewes pre-treated with progesterone implants and given continuous GnRH infusions for 7 days at approximate dose rates of $1 \cdot 25$–$2 \cdot 5$ ng kg^{-1} h^{-1}. The dose rate (50 or 100 ng kg^{-1} h^{-1}) required to induce ovulation in this experiment and the duration of infusion in our mares were considerably higher than those for similar studies in ewes. The reasons for the differences in response to GnRH treatment between the 2 species are not readily apparent, but may relate to the rate of growth of the preovulatory follicle. In the ewe, this period is about 2 days (Driancourt & Cahill, 1984), while in pony mares growth of the preovulatory follicle takes 2–11 days, with an average of $5 \cdot 7$ days (Ginther, 1979). Despite the fact that plasma LH concentrations increased within 1 or 2 days of pump insertion in mares which eventually ovulated, ovulation occurred 15–27 days later, suggesting that maturational processes within the pituitary gland may need to occur in deeply anoestrous mares before surge release of LH can occur in response to exogenous GnRH. Alternatively, follicular oestradiol may be required to effect positive feedback on pituitary LH release (Richards, 1980). Evans & Irvine (1979) suggested that the failure of deeply anoestrous mares to ovulate in response to GnRH and progesterone may have been due to failure of an LH surge because of inadequate oestradiol priming. In our study, oestradiol concentrations were $< 0 \cdot 5$ nmol/l (data not shown) at all times sampled, despite ovulation and the apparent formation of normal corpora lutea in some mares. All mares which ovulated showed oestrus, so presumably oestrus can occur in mares despite low plasma oestradiol concentrations.

Infusions of GnRH via the minipump did not produce linear increases in plasma GnRH concentrations. The reason for this is not known, but may be related to the stability of GnRH in solution, or to pooling of GnRH solution at the site of the implant.

Plasma LH concentrations increased within hours of pump insertion, indicating that there was a readily releasable pool of pituitary LH (Stelmasiak & Galloway, 1977) responsive to GnRH stimulation. The biphasic nature of mean daily LH concentrations suggests that the pituitary gland may have initially become refractory to GnRH due to down-regulation of GnRH receptors, as demonstrated in ewes (Nett *et al.*, 1981) and monkeys (Knobil, 1980). It has also been suggested that continuous exposure of the pituitary to low levels of GnRH may serve a primary function for subsequent LH release (Lincoln *et al.*, 1985). Our results confirm this observation, as the initial decline in mean daily plasma LH concentrations was followed by a secondary rise which was associated with ovulation in 10 of the treated mares. The biphasic nature of LH release was especially obvious in mean daily LH profiles of treated mares which ovulated (Fig. 3). By comparison, in treated mares which did not ovulate, the initial increase in LH concentrations was followed by a steady decline for the rest of the experimental period. Whether the decline in mean daily LH concentrations occurred as a result of pituitary refractoriness, pump failure or breakdown of GnRH is not known.

The pulsatile nature of LH secretion associated with the preovulatory LH surge, which has been observed in a number of species (women: Midgley & Jaffe, 1971; cow: Rahe *et al.*, 1980; rat: Gallo, 1981; ewe: Baird, 1978) is thought to be generated by the pulsatile release of GnRH from the hypothalamus. In this study, continuous GnRH administration resulted in an increase in LH pulse frequency between Day 0 and Day 7, suggesting that constant GnRH infusion increased the sensitivity of the pituitary gland to endogenously generated GnRH pulses. Support for this hypothesis comes from the work of Hoff *et al.* (1979) who showed in women that low-dose GnRH infusions resulted in greater LH release in response to a single injection of GnRH at the end of the infusion, compared to controls. However, GnRH pulse frequency in hypophysial portal blood of mares is reduced during seasonal anoestrus (Irvine & Alexander, 1984), suggesting that constant GnRH infusions may change the frequency of operation of the LH pulse generator.

The LH pulse amplitude in treated and control mares showed no consistent trends over time, although pulse amplitude tended to be higher at the end of the experiment. This suggests that the rise in mean daily LH concentration was due to an increased pulse frequency. A combination of increased LH pulse frequency and LH pulse amplitude is thought to be responsible for the pre-ovulatory LH surge in the cow (Rahe *et al.*, 1980), but in the ewe, LH pulse frequency is considered more important than LH pulse amplitude in production of the preovulatory LH surge (Baird, 1978).

Ovulation was followed by an increase in plasma progesterone concentrations in all mares. There was a tendency for the duration of the post-ovulatory progesterone rise to be shorter in mares which ovulated later. This may have been due to a lack of luteal support, because progester-one secretion is dependent on basal LH concentrations in sheep (Denamur *et al.*, 1973; Baird & McNeilly, 1981). Furthermore, LH antiserum has been shown to reduce luteal life-span in ewes and cows (Hansel *et al.*, 1973) and in mares (Pineda *et al.*, 1973). The above studies suggest that a minimum concentration of LH is necessary after ovulation for maintenance of the corpus luteum. Mean daily plasma LH concentrations and plasma progesterone concentrations fell to baseline within several days of the pump ceasing to function, indicating that circulating LH concentrations were insufficient to maintain viability of the newly-developed corpora lutea in mares which ovulated later.

As a result of GnRH infusion, the time to ovulation and conception in Exp. 2 was significantly reduced during shallow anoestrus. Previous attempts to induce ovulation in seasonally anoestrous mares using GnRH have met with limited success (Evans & Irvine, 1977, 1979). However, Bour & Palmer (1984) successfully induced fertile ovulation in mares during seasonal anoestrus by multiple i.m. injections of hCG. In the present experiment, fertile ovulation was induced in 11 of 14 mares, an average of 23 days earlier than in the controls when data for the 2 years were analysed. This difference would have been greater if the 1985 experiment had been started earlier than 2 October. At this time of year in Victoria, mares are in late transitional oestrus and the beneficial effects of GnRH treatment are likely to be reduced. In 1984, the experiment began on 10 September and the mean day of ovulation in the treated mares was 36 days earlier than the controls.

That GnRH infusion was not detrimental to conception was shown by the fact that the mean day of conception in treated mares occurred 15 days earlier than in the controls, using combined data for the 2 years. In 9 of 13 treated mares, conception occurred to the ovulation induced by GnRH infusion. One treated mare became pregnant on Day 33 after pump insertion and the other 2 treated mares became pregnant 40 and 45 days after the ovulation induced by the GnRH infusion.

The results of these experiments show that mares in shallow seasonal anoestrus are more likely to respond to GnRH infusion than are mares in deep seasonal anoestrus. These findings have significant implications for the horse breeding industry in that selected mares can be treated early in the breeding season with GnRH to induce ovulation. This will allow stud masters to arrange breed-ing schedules to conserve stallions and improve the efficiency of stud-farm personnel.

We thank Dr G. D. Niswender for the supply of LH antiserum; Dr L. Reichert for LH standards; Kathy Burman for carrying out the GnRH assay; and Garry Anderson for statistical analyses and advice. This study was supported in part by the Australian Equine Research Foundation.

References

Baird, D.T. (1978) Pulsatile secretion of LH and ovarian estradiol during the follicular phase of the sheep estrous cycle. *Biol. Reprod.* **18**, 359–364.

Baird, D.T. & McNeilly, A.S. (1981) Gonadotrophic control of follicular development and secretion in the sheep oestrous cycle. *J. Reprod. Fert., Suppl.* **30**, 119–133.

Bittman, E.L. & Karsch, F.J. (1984) Nightly duration of pineal melatonin secretion determines the repro-ductive response to inhibitory daylength in the ewe. *Biol. Reprod.* **30**, 585–593.

Bour, B. & Palmer, E. (1984) Stimulation of seasonally anovulatory pony mares with repeated low doses of hCG. *Proc. 10th Int. Cong. Anim. Reprod. & A.I.* Urbana-Champaign, Abstr. 310.

Burger, H.G., Lee, V.W.K. & Rennie, G.C. (1972) A generalized computer programme for the treatment of data from competitive protein-binding assays including radioimmunoassays. *J. Lab. Clin. Med.* **80**, 302–312.

Burkhardt, J. (1946) Transition from anoestrus in the mare and the effects of artificial lighting. *J. agric. Sci., Camb.* **37**, 64–68.

Clarke, I.J. & Cummins, J.T. (1982) The temporal relationship between gonadotrophin releasing hormone (GnRH) and luteinizing hormone (LH) secretion in ovariectomized ewes. *Endocrinology* **111**, 1737–1743.

Denamur, R., Martinet, J. & Short, R.V. (1973) Pituitary control of the ovine corpus luteum. *J. Reprod. Fert.* **32**, 207–220.

Driancourt, M.A. & Cahill, L.P. (1984) Preovulatory follicular events in sheep. *J. Reprod. Fert.* **71**, 205–211.

Evans, M.J. & Irvine, C.H.G. (1977) Induction of follicular development, maturation and ovulation by gonadotrophin releasing hormone administration to acyclic mares. *Biol. Reprod.* **16**, 452–462.

Evans, M.J. & Irvine, C.H.G. (1979) Induction of follicular development and ovulation in seasonally acyclic mares using gonadotrophin-releasing hormone and progesterone. *J. Reprod. Fert., Suppl.* **27**, 113–121.

Fitzgerald, B.P., l'Anson, H., Loy, R.G. & Legan, S.J. (1983) Evidence that changes in LH pulse frequency may regulate the seasonal modulation of LH secretion in ovariectomized mares. *J. Reprod. Fert.* **69**, 685–692.

Gallo, R.V. (1981) Pulsatile LH release during the ovulatory surge on proestrus in the rat. *Biol. Reprod.* **24**, 100–104.

Gill, J.L. & Hafs, H.D. (1971) Analysis of repeated measurements of animals. *J. Anim. Sci.* **33**, 331–336.

Ginther, O.J. (1979) *Reproductive Biology of the Mare: Basic and Applied Aspects.* Equiservices, Cross Plains, Wisconsin.

Grubaugh, W., Sharp, D.C., Berglund, L.A., McDowell, K.J., Kilmer, D.M., Peck, L.S. & Seamans, K.W. (1982) Effects of pinealectomy in pony mares. *J. Reprod. Fert., Suppl.* **32**, 293–295.

Hansel, W., Concannon, P.W. & Lukaszewska, J.H. (1973) Corpora lutea of the large domestic animals. *Biol. Reprod.* **8**, 222–245.

Hoff, J.D., Lasley, B.L. & Yen, S.S.C. (1979) The functional relationship between priming and releasing actions of luteinizing hormone-releasing hormone. *J. clin. Endocr. Metab.* **49**, 8–11.

Irvine, C.H.G. & Alexander, S.L. (1984) LH and gonadotrophin releasing hormone (GnRH) concentrations in pituitary venous blood collected by a novel technique for long term sampling in the unrestrained, ambulatory horse. *Proc. Endocrine Soc. Aust.* **27**, Suppl. 7, Abstr. 7.

Jonas, H.A., Burger, H.G., Cumming, I.A., Findlay, J.K. & de Kretser, D. (1975) Radioimmunoassay for luteinizing hormone releasing hormone (LHRH): its application to the measurement of LHRH in ovine and human plasma. *Endocrinology* **96**, 406–415.

Knobil, E. (1980) The neuro-endocrine control of the menstrual cycle. *Recent Prog. Horm. Res.* **36**, 53–82.

Kooistra, L.H. & Ginther, O.J. (1975) Effect of photoperiod on reproductive activity and hair in mares. *Am. J. vet. Res.* **36**, 1413–1419.

Landefield, T.D. & McShan, W.H. (1974) Equine luteinizing hormone and its subunits. Isolation and physicochemical properties. *Biochemistry, N.Y.* **13**, 1389–1393.

Lincoln, D.W., Fraser, H.M., Lincoln, G.A., Martin, G.B. & McNeilly, A.S. (1985) Hypothalamic pulse generators. *Recent Prog. Horm. Res.* **41**, 369–419.

McLeod, B.J., Haresign, W. & Lamming, G.E. (1982) The induction of ovulation and luteal function in seasonally anoestrous ewes treated with small-dose multiple injections on GnRH. *J. Reprod. Fert.* **65**, 215–221.

Midgley, A.R., Jr & Jaffe, R.B. (1971) Regulation of human gonadotrophin. X. Episodic fluctuation of LH during the menstrual cycle. *J. clin. Endocr. Metab.* **33**, 962–969.

Nett, T.M., Akbar, A.M., Niswender, G.D., Hedlund, M.T. & White, W.F. (1973) A radioimmunoassay for gonadotrophin-releasing hormone (GnRH) in serum. *J. clin. Endocr. Metab.* **36**, 880–885.

Nett, T.M., Holtan, D.W. & Estergreen, V.L. (1975) Oestrogens, LH, PMSG and prolactin in serum of pregnant mares. *J. Reprod. Fert., Suppl.* **23**, 457–462.

Nett, T.M., Crowder, M.E., Moss, G.E. & Duello, T.M. (1981) GnRH-receptor interaction. V. Downregulation of pituitary receptors for GnRH in ovariectomized ewes by infusion of homologous hormone. *Biol. Reprod.* **24**, 1145–1155.

Palmer, E., Driancourt, M.A. & Ortavant, R. (1982) Photoperiodic stimulation of the mare during winter anoestrus. *J. Reprod. Fert., Suppl.* **32**, 275–282.

Pineda, M.H., Garcia, M.C. & Ginther, O.J. (1973) Effect of antiserum against an equine pituitary fraction on corpus luteum and follicles in mares during diestrus. *Am. J. vet. Res.* **34**, 181–183.

Rahe, C.H., Owens, R.E., Fleeger, J.L., Newton, J.T. & Harms, P.G. (1980) Pattern of plasma luteinizing hormone in the cyclic cow: dependence upon the period of the cycle. *Endocrinology* **107**, 498–503.

Reiter, R.J. (1981) The mammalian pineal gland: structure and function. *Am. J. Anat.* **162**, 287–313.

Richards, J.S. (1980) Maturation of ovarian follicles: actions and interactions of pituitary and ovarian hormones on follicular cell differentiation. *Physiol. Rev.* **60**, 51–89.

Stelmasiak, T. & Galloway, D.B. (1977) A priming effect of LH-RH on the pituitary in rams. *J. Reprod. Fert.* **51**, 491–493.

Strauss, S.S., Chen, C.L., Kabra, S.P. & Sharp, D.C. (1979) Localization of gonadotrophin-releasing hormone (GnRH) in the hypothalamus of ovariectomized pony mares by season. *J. Reprod. Fert., Suppl.* **27**, 123–129.

Thorpe, P.A. & Herbert, J. (1975) Studies on the duration of the breeding season and photorefractoriness in female ferrets pinealectomized or treated with melatonin. *J. Endocr.* **70**, 255–262.

Woods, G.L. & Ginther, O.J. (1982) Ovarian response, pregnancy rate, and incidence of multiple fetuses in mares treated with an equine pituitary extract. *J. Reprod. Fert., Suppl.* **32**, 415–421.

Wright, P.J., Clarke, I.J. & Findlay, J.K. (1983) The induction of fertile oestrus in seasonally anoestrous ewes using a continuous low dose administration of gonadotrophin releasing hormone. *Aust. vet. J.* **60**, 254–255.

J. Reprod. Fert., Suppl. **35** (1987), 221–228

Printed in Great Britain
© 1987 Journals of Reproduction & Fertility Ltd

Alternative solutions to hCG induction of ovulation in the mare

G. Duchamp, B. Bour*, Y. Combarnous and E. Palmer

INRA, 37380 Nouzilly, France

Summary. Injection of hCG (2000–2500 i.u., i.v.) to mares when a follicle reaches 35 mm induces ovulation between 24 and 48 h. However, repeated injections induce antibodies against hCG. We report attempts to induce ovulation without this inconvenience. We called 'response' an ovulation between 24 and 48 h after treatment. The typical response to hCG was obtained in 73% (N = 145) of treated mares. After immunization against hCG, the response (0%, N = 10) was less than in non-immunized controls (100%, N = 9). Simultaneous injection of dexamethasone and hCG resulted in induction of ovulation (71%, N = 14). However, simultaneous hCG + dexamethasone resulted in antibody formation similar to that induced by hCG alone, when injected repeatedly every 21 days. Neither GnRH (2 mg i.m.) nor partly purified pig LH (26 mg i.v. or s.c.) induced ovulation consistently (40%, N = 30 and 31%, N = 16). Crude horse gonadotrophin (60 mg i.v. or s.c.) induced ovulation (86%, N = 14). Fertility was not different from control (61%, N = 13 *vs* 40%, N = 10). Crude horse gonadotrophin also induced ovulation in mares previously immunized against hCG (78%, N = 9): 50 or 25 mg gave satisfactory response (86%, N = 29 and 57%, N = 40). We conclude that crude horse gonadotrophin is a good alternative to hCG for the induction of ovulation in mares.

Introduction

Modern techniques of reproduction such as embryo transfer, AI of frozen semen or in-vitro fertilization need a precise prediction of the time of ovulation. Rectal palpation or ultrasonography are not accurate enough for this purpose. In many species the interval between the LH peak and ovulation is very constant (Cumming *et al.*, 1971). The control of the LH signal can give a precise timing of ovulation in the mare. Human chorionic gonadotrophin (hCG) is widely used for induction of ovulation in the mare but it has the inconvenience of inducing formation of antibodies which reduce its efficiency after several successive injections. The present experiments were undertaken to study the efficiency of hCG in relation to its immunogenic activity. We also tried other methods by altering the immune reaction to hCG or by using other compounds such as gonadotrophin-releasing hormone (GnRH), pig LH or crude horse gonadotrophins.

Materials and Methods

Animals. Different types of animals were used in the successive experiments: Trotter and saddle mares (400–600 kg) of unknown conditions at stud for Exp. 1; Welsh pony mares (150–300 kg) from the INRA laboratory with regular checks of the level of antibodies against hCG in Exps 2, 4, 6 & 7; Saddle mares from the INRA laboratory (400–600 kg) in Exps 3 & 7; and Shetland ponies (males, geldings or females, 150–250 kg) of a riding pony club for Exp. 5.

---

*Present address: Haras National, 60200 Compiègne, France.

Assessment of ovarian function. Cyclicity was usually verified before use of 'cyclic' females by weekly analysis of blood progesterone. Follicular growth and ovulation were checked by rectal palpation (Exp. 1) or ultrasound (Exps 2, 3, 4, 6 & 7) at least twice a week, every other day as soon as a 25-mm follicle was detected, and every day when the follicle was > 30 mm in diameter (Palmer & Driancourt, 1980). Ovulation was usually confirmed by the increase of blood progesterone occurring 24–48 h later in daily blood samples.

Measurements of antibodies against hCG. Measurements were performed according to the technique of Roser *et al.,* (1979) with minor modification (dioxane final concentration: 50·8%). Results are expressed as percentage precipitated radioactive hCG. When data from several assays were pooled, they were expressed as % of the hyperimmune control.

Compounds used. The commercial preparation of hCG (Chorulon) was kindly donated by Intervet Company (4 9000 Angers, France). The commercial preparation of GnRH (Fertagyl) was donated by Intervet Company. A long-acting preparation of dexamethasone salts (1 mg sodium phosphate/ml + 2 mg phenylpropionate/ml) (Dexafort) was donated by Invervet Company. Pig LH was the preparation CY 1300 from our laboratory containing approximately 20% pure LH (0·3 × NIH-LH-S1) and <0·2% FSH.

Horse pituitaries were supplied commercially and the first step in the preparation of crude horse gonadotrophin was as described by Guillou & Combarnous (1983). Briefly, pituitaries were ground and extracted twice in 40% ethanol, and the extract precipitated by 80% ethanol at −20°C. The precipitate was then dialysed against ammonium bicarbonate and lyophilized. The resulting preparation (CEG) contained approximately 6% pure horse LH and 2–4% pure horse FSH as measured by a pig radioreceptor assay.

Experiment 1. This was a field trial of the use of hCG in large mares designed to define the 'typical response' to hCG. Large mares were injected i.v. with 2500 i.u. when a follicle > 30 mm in diameter was detected. The ovaries were palpated daily to find the day of ovulation.

Experiment 2. The animals of experiments published elsewhere (Bour & Palmer, 1984) were used to test the response of mares immunized against hCG.

In Exp. 2A, Welsh pony mares (N = 4) in deep anoestrus received 20 daily intramuscular injections of 200 i.u. hCG; ovulation was not induced but the antibody titre was raised (Bour & Palmer, 1984). At 32 ± 12 days later, when the first preovulatory follicle (30 mm soft or 35 mm) of the season was detectable, they received an intravenous injection of 2000 i.u. hCG. The 3 control mares (without the previous immunizing protocol) received the i.v. injection of 2000 i.u. hCG when they had their first preovulatory follicle of the season.

In Exp. 2B, 6 Welsh pony mares were treated daily during the breeding season with 200 i.u. hCG i.m. during one cycle (from ovulation 1 to ovulation 2). This treatment raised antibodies against hCG. At the subsequent cycle (ovulation 3) they received one i.v. injection of 2000 i.u. hCG when a follicle reached preovulatory size (30 mm soft or 35 mm). The 6 control Welsh pony mares did not receive the daily injection but received the preovulatory i.v. injection of 2000 i.u. hCG at the same period of the year.

The results of Exps 2A and 2B were pooled to compare response of immunized *vs* non-immunized mares to an ovulatory injection of hCG.

Experiment 3. The possible use of GnRH instead of hCG to induce ovulation of a preovulatory follicle in cyclic large mares was tested. Large mares of the INRA experimental herd and Le Pin National stud were used: 31 mares received 2500 i.u. hCG i.v. when the follicle was > 34 mm in diameter and the mare was in oestrus; 30 mares received 2 mg GnRH i.m. using the same criteria; and 32 control mares received no injection but the interval from 'follicle larger than 34 mm' was used for the comparison.

Experiment 4. Welsh pony mares (N = 16) were used to test three other possible ways of inducing ovulation: (1) hCG + corticoid, assuming the corticoid could reduce the immune response without reducing the ovulatory response; (2) pig LH, a preparation which may become available as a by-product of the preparation of FSH for superovulation in cattle; (3) crude horse gonadotrophin (CEG) which contains more LH than FSH. Its injection in attempts to induce superovulation in mares raises the measurable LH in the blood to levels similar to the LH peak (Palmer, 1985). The protocol of the experiment was to use 4 cycles of each animal and a different treatment in each cycle (control, hCG + corticoid, pig LH or CEG). The order of the treatments was different in each animal so as to suppress seasonal effect or delayed effects in the response to the different treatments.

At each cycle, mares were usually inseminated with fresh semen every other day. On Day 7, embryo collection was attempted to measure fertility. On Day 9, an injection of a prostaglandin analogue was given to induce rapid return to oestrus.

Injection was decided when a mare reached oestrus and had a soft follicle > 30 mm in diameter or a follicle > 34 mm whatever its consistency.

Compounds and doses injected in the different groups were:
(a) control, 2 ml saline (9 g NaCl/l) i.v.;
(b) 2000 i.u. hCG (Chorulon) i.v. + 20 ml corticoid (20 mg sodium phosphate of dexamethasone + 40 mg phenylpropionate of dexamethasone) i.m. (Dexafort);
(c) 26 mg pig LH ('CY1300') injected i.v. to 8 animals or s.c. to 8 animals in 2 ml saline; and
(d) 60 mg of the CEG preparation injected in 2 ml saline, i.v. to 8 mares and i.m. to 8 mares.

Experiment 5. The effect of corticoid on the immune response to successive i.v. injections of hCG every 21 days was tested in two groups of ponies: the 'control' group received only hCG (200 i.u., i.v.) 5 times at 21-day intervals. The 'corticoid' group received i.v. hCG + 20 ml corticoid i.m. on 5 occasions.

Binding of radioactive hCG to the serum immunoglobulins of these animals was measured in 10 blood samples collected every 21 days, starting on the day of first injection.

Experiment 6. The effect of CEG (60 mg i.v.) on females selected for a high level of antibodies against hCG was tested in 9 pony mares. No control group was included in this experiment.

Experiment 7. A dose effect study of induction of ovulation by CEG was undertaken with 43 pony mares (200–300 kg) and 26 large mares (400–600 kg). Two doses (25 mg and 50 mg) of CEG were injected i.m. when the animals were in oestrus and had a follicle > 32 mm (ponies) or 34 mm (large mares) in diameter.

Statistical analysis and presentation of data. According to the literature and the results of Exp. 1, we consider that ovulation is induced when it occurs between 24 and 48 h after an LH stimulus. Therefore, all statistical comparisons between groups consist of a χ^2 test on the proportion of mares ovulating 24–48 h after the stimulation. However, to give more complete information, the figures will present histograms of 'day of ovulation' *vs* % of mares. Ovulations occurring more than 6 days after the stimulus are included in the Day 6 column. When a mean calculation was made, mean ± s.e.m. values are presented.

Results

Experiment 1

As shown in Fig. 1, a 'typical' response to hCG ovulation between 24 and 48 h after injection, was obtained in 145 (73%) of the mares.

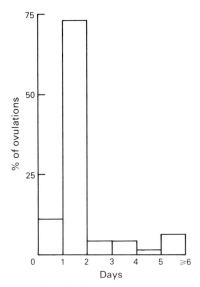

Fig. 1. Histogram of percentage of ovulation occurring in successive 24-h intervals after i.v. injection of 2500 i.u. hCG in a field trial on 145 large mares. hCG was injected on the day when a follicle > 30 mm in diameter was detected.

Experiment 2

There was no difference in the results of the spring and summer experiments which were pooled in Fig. 2. The ovulatory response to hCG was obtained in all 9 of the non-immunized females and in 0 of 10 of the immunized mares ($P < 0.01$).

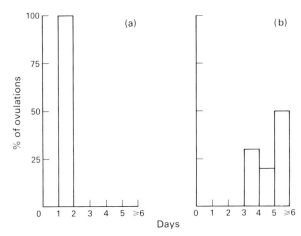

Fig. 2. Histogram of percentage of ovulation occurring in successive 24-h intervals after i.v. injection of 2000 i.u. hCG in (a) non-immunized (N = 9) or (b) immunized (N = 10) Welsh pony mares. Immunization by 20–30 daily i.m. injections of 200 i.u. hCG was terminated 20–50 days before the induction of ovulation with hCG.

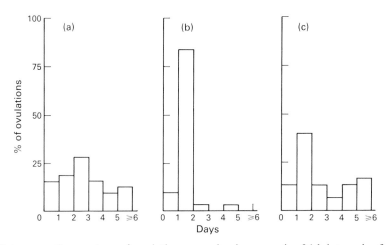

Fig. 3. Histogram of percentage of ovulation occurring in successive 24-h intervals after (a) no treatment (N = 32), (b) 2500 i.u. hCG i.v. (N = 31) or (c) 2 mg GnRH i.m. (N = 30) when the follicle was > 34 mm in diameter.

Experiment 3

As shown in Fig. 3, hCG induced ovulation in 26/31 (84%) of the mares, GnRH in 12/30 (40%), and in only 6/32 (18%) of the untreated control mares by Day 2. The hCG group was different ($P < 0.01$) from both GnRH and control groups which were not different from one another ($P > 0.05$).

Experiment 4

A few cycles are missing because one mare returned to anoestrus during the experiment and one ovulated twice before injection. Results of i.m. or i.v. injection of CEG and of s.c. or i.v. injection of pig LH were similar. Consequently, the routes of injection are not distinguished in Fig. 4.

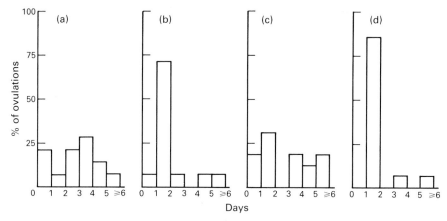

Fig. 4. Histogram of percentage of ovulation occurring in successive 24-h intervals in Welsh pony mares after (a) saline (N = 14), (b) 2000 i.u. hCG + 20 ml dexamethasone salts (N = 14), (c) 26 mg pig LH (N = 16) or (d) 60 mg crude horse gonadotrophins (CEG; N = 14), injected when the follicle was 30 mm and soft or 35 mm in diameter.

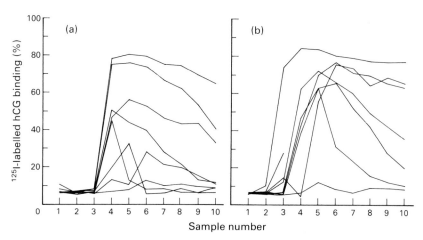

Fig. 5. Individual profiles of radioactive hCG binding in the serum of ponies injected 5 times at 21-day intervals with (a) hCG (N = 8) or (b) hCG + corticoids (N = 8). Injections were performed immediately after Samples 1–5.

Both CEG and hCG + corticoid induced ovulation ($P < 0.05$) but pig LH was less effective in comparison to the saline injection.

The fertility showed an overall statistical difference ($P < 0.05$) but none of the treated groups differed significantly from the control: embryos collected per inseminated mare were 8/13, 8/11, 4/10 and 2/14 for CEG, hCG + corticoid, control and pig LH treatments respectively.

Experiment 5

The successive injections of hCG or hCG + corticoids at 21-day intervals induced formation of antibodies in 14/16 animals. Figure 5 presents the hCG-binding profile of animals of both groups. A few samples are missing because one horse in the hCG + corticoid group died and sometimes the ponies were not available for the sampling.

No statistical difference was found between the groups, in spite of a possible earlier rise in the hCG + corticoid group.

The injections of corticoid did not suppress the immune response to hCG and this approach was abandoned.

Experiment 6

Of 9 mares with antibodies against hCG, 6 (78%) ovulated on Day 2 after injection of CEG (1 on Day 1 and 1 on Day 6) and, in spite of the absence of controls, we consider that CEG is efficient in inducing ovulation in this type of mare. These mares had received 4.2 ± 1.0 injections of hCG in their life and bound $65 \pm 8\%$ of radioactive hCG compared to the hyperimmune controls.

Experiment 7

The dose–response effects of CEG injected i.m. showed no difference due to breed (pony *vs* large mares) but a significant ($P < 0.05$) difference of response between 50 mg (80%, N = 29) and 25 mg (57%, N = 40) (Fig. 6).

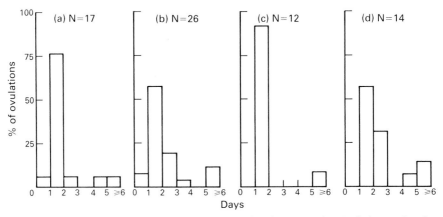

Fig. 6. Histograms of percentage ovulations occurring in successive 24-h intervals after an injection of (a,c) 50 or (b,d) 25 mg CEG to (a,b) Welsh pony mares or (c,d) large mares.

Discussion

After Day (1939) the use of hCG has been well documented (Davidson, 1947; Loy & Hughes, 1966; Sullivan *et al.*, 1973; Voss *et al.*, 1974). A 'typical' response to hCG is ovulation between 24 and 48 h after injection. Consequently, adequate statistical test of its effect is not analysis of variance of interval to ovulation to show hastening of ovulation, but reduction of the variability of time to ovulation or percentage of mares ovulating at the expected time of 24–48 h. Our results confirm those of previous studies.

Concerning antibody formation and reduction of efficiency of hCG, the literature is contra-dictory. Loy & Hughes (1966) found no reduction of effect during successive injections whereas Sullivan *et al.* (1973) did. Using animals with high antibody titres ($80.4 \pm 6.7\%$ of hyperimmune controls) we have shown clearly that hCG induction of ovulation is impaired. However, our immunizing hCG pretreatment (200 i.u./day for 20–30 days) is not the usual way of administration of hCG. Roser *et al.* (1979) showed that antibodies may appear after 2–5 injections but failed to show refractoriness. Our study confirms the heterogeneity of immune response to 5 successive injections, 14/16 animals showing a rise of measurable antibodies after 3 injections and 2/16 showing no rise at all.

Fauci (1979) reports that dexamethasone depresses antibody formation and increases their catabolism. James & Merlin (1982) showed that glucocorticoids depress lymphocyte concentrations and consequently antibody formation. We therefore tried to give dexamethasone at the same time as hCG but antibody formation was not reduced. However, only one time and one dose of corticoid were used and our failure does not mean that glucocorticoid cannot suppress the immune response. Nevertheless, we preferred to try other approaches because suppression of the immune system may have other major consequences on health of the animals.

Asa & Ginther (1982) showed that dexamethasone administered chronically may suppress gonadotrophins, follicular growth, oestrus and ovulation. In our experiment, one injection had no effect on immediate ovulation or on the subsequent cycle and did not depress fertility.

The use of GnRH as an alternative to hCG has been proposed by Irvine *et al.* (1975). However, several authors (Ginther & Wentworth, 1974; Noden & Oxender, 1976) have not found an effect on time of ovulation. Obviously, one injection of a high dose (2 mg in our study) is not a 'physiological' treatment and is not as efficient as hCG in inducing ovulation.

The use of other gonadotrophins is an obvious possibility. However, development is limited to availability of the compounds. Our studies have shown the usefulness of CEG in the induction of multiple follicular growth in mares. Its preparation is simple and we hope that it will be available in the future. Radioreceptor assays performed in our laboratory (Guillou & Combarnous, 1983; Y. Combarnous, unpublished) using horse testis and ^{125}I-labelled horse LH-A showed relative potencies of: (1) hCG = 2·4 × horse LH-A (Guillou & Combarnous, 1983); (2) horse LH-A = 2·7 × horse LH E98A (from H. Papkoff); and (3) horse LH E98A = pure pLH, and respective purities are: pure hCG = 15 000 i.u./mg, CEG = 5% horse LH-A, pig LH CY1300 = 20% pure pig LH. From these data, we can consider that our 60 mg CEG is about equivalent to 40 mg pig LH or to 18 750 i.u. hCG. Such a calculation does not take into account the very different half-lives of the hormones (Irvine, 1979). Assays in different systems show sometimes very different potencies. Theoretical calculation can give only an indication of the range of doses to be tested. We cannot say whether the failure of 26 mg pig LH CY1300 to induce ovulation is due only to an insufficient dose. Nevertheless, the two pituitary hormone preparations have been tried at quantities about 10 times the dose of hCG necessary for induction of ovulation.

In our experiment, some mares had 'anaphylactic-like' shocks after i.v. injection of pig LH; we therefore decided to give the pig LH subcutaneously to half of the animals and none showed this type of shock. However, at least at the dose tested, pig LH seemed a poor ovulation inducer. In addition, this preparation, being a foreign protein, may raise the same problems of antibody formation as does hCG.

Induction of ovulation with CEG seems to be the best alternative to hCG: being an homologous preparation, it should not raise antibodies (but this should be verified). At 60 mg (Exp. 4) and 50 mg (Exp. 7), it induces ovulation consistently. The fertility of induced ovulation was not impaired (ultrasound diagnosis of pregnancy). It was efficient in mares which already had high antibody titres against hCG. For practical purposes, we thought that the i.m. route should be tested and found that the effective dose was 25–50 mg. CEG is now used routinely at the INRA laboratory instead of hCG for induction of ovulation in the mare.

We thank Intervet for the generous gifts of compounds used.

References

Asa, C.S. & Ginther, O.J. (1982) Glucocorticoid suppression of oestrus, follicles, LH and ovulation in the mare. *J. Reprod. Fert., Suppl.* **32**, 247–251.

Bour, B. & Palmer, E. (1984) Stimulation of seasonally anovulatory pony mares with repeated low doses of hCG. *Proc. 10th Int. Congr. Anim. Reprod. & A.I., Urbana-Champaign,* p. 310, Abstr.

Cumming, I.A., Brown, J.M., Blockey, M.A. de, Winfield,

G.G., Baxter, R. & Goding, J.R. (1971) Constancy of interval between LH release and ovulation in the ewe. *J. Reprod. Fert.* **24,** 134–136.

Day, F.T. (1939) Ovulation and the descent of the ovum in the fallopian tube of the mare after treatment with gonadotrophic hormones. *J. agric. Sci., Camb.* **29,** 459–469.

Davidson, W.F. (1947) The control of ovulation in the mare with reference to insemination with stored sperm. *J. agric. Sci., Camb.* **37,** 287–290.

Fauci, A.S. (1979) Immunosuppressive and anti-inflammatory effect of glucocorticoides. In *Glucocorticoid Hormone Action,* pp 449–465. Eds J.D. Baxter & G.G. Rousseau. Springer-Verlag, Berlin.

Ginther, O.J. & Wentworth, B.C. (1974) Effect of a synthetic GnRH on plasma concentration of luteinizing hormone in ponies. *Am. J. vet. Res.* **35,** 79–81.

Guillou, F. & Combarnous, Y. (1983) Purification of equine gonadotrophins and comparative study of their acid dissociation and receptor-binding specificity. *Biochem. Biophys. Acta* **755,** 229.

Irvine, C.H.G. (1979) Kinetics of gonadotrophins in the mare. *J. Reprod. Fert.,* Suppl. **27,** 131–141.

Irvine, D.S., Downey, B.R., Parker, W.G. & Sullivan, J.J. (1975) Duration of oestrus and time of ovulation in mares treated with synthetic GnRH (AY 24031). *J. Reprod. Fert., Suppl.* **23,** 279–283.

James, A.R. & Merlin, L. (1982) Effect of glucocorticoids on the bovine immune system. *J. Am. vet. med. Ass.* **180,** 896–897.

Loy, R.G. & Hughes, J.P. (1966) The effects of human chorionic gonadotrophin on ovulation, length of oestrus and fertility in the mare. *Cornell Vet.* **56,** 41–50.

Noden, P.A. & Oxender, W.D. (1976) LH and ovulation after GnRH in mares. *J. Anim. Sci.* **42,** 1360.

Palmer, E. (1985) Recent attempts to improve synchronization of ovulation and to induce superovulation in the mare. *Equine vet. J., Suppl.* **3,** 11–18.

Palmer, E. & Driancourt, M.A. (1980) Use of ultrasonic echography in equine gynecology. *Theriogenology* **13,** 203–216.

Roser, J.F., Kiefer, B.L., Evans, J.W., Neely, D.P. & Pacheco, C.A. (1979) The development of antibodies to human chorionic gonadotrophin following its repeated injection in the cyclic mare. *J. Reprod. Fert., Suppl.* **27,** 173–179.

Sullivan, J.J., Parker, W.G. & Larson, L.L. (1973) Duration of estrus and ovulation time in non lactating mares given human chorionic gonadotrophin during three successive estrous periods. *J. Am. vet. med. Ass.* **162,** 895–898.

Voss, J.L., Pickett, B.W., Burwash, L.D. & Daniels, W.H. (1974) Effect of human chorionic gonadotrophin on duration of estrous cycle and fertility of normally cycling, not lactating mares. *J. Am. vet. med. Ass.* **165,** 704–706.

J. Reprod. Fert., Suppl. **35** (1987), 229–237

Inhibition of ovulation in the mare by active immunization against LHRH*

J. M. Safir, R. G. Loy and B. P. Fitzgerald

Department of Veterinary Science, University of Kentucky, Lexington, KY 40546–0076, U.S.A.

Summary To investigate the hypothesis that the onset of the breeding season in the mare may be due to a daylength-induced seasonal increase in LHRH pulse frequency, 5 mares were immunized against LHRH. Beginning 1 December, 5 immunized and 5 untreated control mares were exposed to an abrupt, artificial increase in daylength (16L:8D) to advance the onset of the breeding season. In control mares ovulation occurred 49·6 ± 3·5 (s.e.m.) days later (18 January), whereas in 3/5 immunized mares ovulation had not occurred by 1 April. In the remaining 2 mares, although ovulation occurred once (Mare 79) or twice (Mare 72) during February, a booster immunization restored acyclicity for the duration of the study (No. 72) or to 30 March (No. 79). The absence or occurrence of ovulation in LHRH-immunized mares appeared to be related to antibody titre, such that the highest antibody titres were observed in those mares that remained anovulatory throughout the experimental period, while low titres were seen in the 2 mares that ovulated.

In 3 acyclic immunized mares, LH pulses were not observed in blood samples collected frequently at 2-week intervals from November to January. In contrast, LH pulse frequency in control mares, and in the 2 immunized mares that ovulated, increased from 0/12 h (November) to 2–5/12 h (January). These results confirm our previous observation that, in the mare, the onset of the breeding season is associated with an increase in LH pulse frequency. Furthermore, the results suggest that the increase in LH pulse frequency reflects an increase in pulsatile LHRH release from the hypothalamus.

Introduction

In the horse mare, the onset of the breeding season occurs in response to an increase in daylength, and can be advanced by exposure of anoestrous mares to an artificial increase in photoperiod (Sharp & Ginther 1975). In the mare, transition into the breeding season is associated with an increase in the frequency of pulsatile luteinizing hormone (LH) secretion (Murdoch *et al.*, 1985). Furthermore, the timing of the increase in LH pulse frequency can be advanced when mares are exposed to an artificial extension of daylength to advance the onset of the breeding season (Fitzgerald *et al.*, 1987). In consideration of these studies, therefore, it may be proposed that, in the mare, a daylength-mediated seasonal increase in LH pulse frequency may play a role in the initiation of the breeding season. Further, in consideration of the role of luteinizing hormone-releasing hormone (LHRH) as the proposed stimulus for pulsatile LH secretion (Levine *et al.*, 1982) the aforementioned hypothesis may be extended to propose that the onset of the breeding season reflects an increase in LHRH pulse frequency.

In several species (e.g. rats, sheep, pigs, monkeys) the role of LHRH in the control of LH secretion has been extensively investigated using the technique of immunoneutralization (Fraser &

*Reprint requests to Dr B. P. Fitzgerald.

Baker, 1978; Fraser *et al.*, 1984, 1986; McNeilly *et al.*, 1984; Esbenshade & Britt, 1985). It is evident from these studies that both active and passive immunization of males and females against LHRH results in a suppression of pulsatile LH secretion and ultimately an inhibition of reproductive function. In view of these studies, therefore, the purpose of the present study was to examine by active immunization the role of LHRH in the mare during the transition into the breeding season. In this regard, we investigated the hypothesis that immunoneutralization of LHRH would suppress the occurrence of pulsatile LH secretion, an index of LHRH release, and ultimately inhibit the onset of the breeding season.

Materials and Methods

Animals and treatments

Ten horse mares of mixed breeding were maintained on pasture in Lexington (38°2′N). Animals were given a supplement of a small amount of hay and oats when housed in individual stalls during experimental procedures. Water was available *ad libitum*.

Beginning 1 December 1985, 5 mares which had previously been immunized against LHRH (see below) and 5 anoestrous mares which served as non-immunized controls were exposed to an abrupt artificial increase in daylength to advance the onset of the breeding season. Mares were kept in individual stalls between 15:00 and 07:00 h daily and exposed to artificial lighting between 16:00 and 23:00 h every evening. Daylength was thereby extended to at least 16 h light per day.

Follicular development, oestrus and ovulation

Follicular development was assessed by palpation of the ovaries *per rectum* every other day unless significant follicular growth was noted (⩾ 25 mm), after which palpation was undertaken daily. The occurrence of ovulation was indicated by the collapse of a follicle. Oestrous behaviour was monitored by daily teasing with a stallion.

Immunization against LHRH

The immunogen was prepared by conjugating LHRH (Wyeth Pharmaceuticals, Philadelphia, PA, U.S.A.) to human serum albumin (HSA, Fraction V; Sigma Chemical Co., St Louis, MO, U.S.A.) using the carbodiimide reaction (Fraser *et al.*, 1974). On 31 July 1985, 4 mares received a primary immunization consisting of 2 mg LHRH coupled to 2 mg HSA emulsified in Freund's complete adjuvant (2:3 ratio) (Clarke *et al.*, 1978). The immunogen was injected i.m. at 4 sites. A fifth mare (No. 81) was immunized 1 year previously and had received booster immunizations on 4 occasions between September 1984 and June 1985, the most recent occurring on 5 July 1985. After the primary immunization, all mares (including Mare 81) received booster immunizations 4, 14, 20 and 29 weeks later. Booster immunizations consisted of 2 mg LHRH conjugated to 2 mg HSA emulsified in Freund's incomplete adjuvant, and were injected at multiple (4–10) s.c. sites on the back.

Antibody titre to LHRH was determined in serial dilutions of sera (1:100 to 1:100 000) by incubating 200 µl diluted serum with ^{125}I-labelled LHRH (∼ 10 000 c.p.m.; New England Nuclear, Boston, MA, U.S.A.) in a final volume of 300 µl. After incubation for 24 h at 4°C, separation of antibody-bound and free ^{125}I-labelled LHRH was achieved by adding 5 volumes of ice-cold absolute ethanol, centrifuging for 30 min at 4°C, and decanting the supernatant. Antibody-bound ^{125}I-labelled LHRH was estimated by counting the pellet. Antibody titre was expressed as the initial dilution of serum which bound 33% of a constant amount of ^{125}I-labelled LHRH.

Collection of blood samples

Changes in the pattern of LH secretion were determined in blood samples (5 ml) collected via an indwelling jugular catheter. Samples were collected every 15 min for 12-h periods before (27 November) and 2 weeks after exposure to an abrupt increase in daylength. Thereafter, periods of frequent sample collection were performed every 2 weeks until each control mare ovulated or until 30 January 1986. All blood samples were kept overnight at 4°C. Sera were collected and stored frozen until assayed.

Radioimmunoassay of LH

Luteinizing hormone concentrations were measured in duplicate 10–200 µl samples of serum using a radioimmunoassay procedure described previously (Loy *et al.*, 1982). An equine pituitary LH preparation (E98A), kindly supplied by Dr H. Papkoff, was used as standard. This preparation has 2·97 times the activity of NIH-LH-S1 according to the ovarian ascorbic acid depletion assay (Licht *et al.*, 1979). The sensitivity of the radioimmunoassay was 0·1 ng/ml for 200 μl serum. Intra-assay and inter-assay coefficients of variation were 13 and 16% respectively for a serum pool that produced a 50% inhibition of binding of radiolabelled LH to antibody (19 assays).

Analysis of data

Changes in mean serum LH concentrations before ovulation were analysed using a repeated measures analysis of variance. The occurrence of an LH pulse was identified as described previously (Fitzgerald *et al.*, 1985). Data are presented as mean ± s.e.m.

Results

In 5 untreated control mares, ovulation occurred 49·6 ± 3·5 days after 1 December, corresponding to 18 January. In each mare, ovulation was preceded by prolonged oestrous behaviour which averaged 18 days in duration (range 11–37 days). In 3 mares, one or two large preovulatory-size follicles (> 35 mm diameter) were identified during the period of prolonged oestrus but these follicles regressed before development of the follicle which ovulated. In the remaining 2 mares, the first follicle identified as > 35 mm diameter ovulated.

The 5 mares immunized against LHRH included one animal (Mare 81) which had been immunized for about 1 year and anovulatory for about 1·5 years, a period including seasonal anoestrus. However, in this mare ovulation unexpectedly occurred on 25 November and again on 17 December, an interovulatory interval of 23 days. After ovulation on 25 November, introduction of the teaser stallion evoked a passive behavioural response and oestrus occurred within 10 days. The second ovulation occurred 3 days before a booster immunization which suppressed the occurrence of further ovulations such that Mare 81 remained anovulatory for the remainder of the study (18 December–31 March).

The occurrence of ovulation was also arrested in 2 of the 4 mares which received primary immunization in July 1985. In both mares (Mares 77 and 85), introduction of a teaser stallion evoked a passive behavioural response. Furthermore, rectal examination of the reproductive tract of both mares, and of Mare 81 during the anovulatory period of January–March, revealed small, hard ovaries without discernible structures, an atonal uterus and a relaxed cervix.

In the remaining 2 mares (Mares 72 and 79), which received primary immunization in July 1985, ovulation occurred during the experimental period. In Mare 72, ovulation occurred 64 days (2 February) after exposure to an abrupt increase in daylength. The occurrence of ovulation was unaccompanied by an alteration in sexual behaviour and oestrus continued to be exhibited until 22 February when a second ovulation occurred, an interovulatory interval of 20 days. A booster immunization on 21 February blocked the occurrence of the next anticipated ovulation and Mare 72 remained anovulatory for the remainder of the study (36 days). The final mare of the group (Mare 79) ovulated on 11 February which was followed by typical aggressive behaviour in response to introduction of a teaser stallion. On 21 February, a booster immunization blocked the occurrence of the next anticipated ovulation but the delay was temporary and ovulation again occurred 47 days later (30 March).

In control mares, serum LH concentrations in November were low (0·20 ± 0·02 ng/ml, N = 5) and in all mares the pattern of secretion during a 12 h period was non-pulsatile (Table 1).

Table 1. Changes in mean (± s.e.m.) serum LH concentrations, LH pulse frequency and amplitude in 5 non-immunized control mares exposed to an abrupt increase in daylength beginning 1 December

	27 November	19 December	2 January	16 January
Serum LH (ng/ml)	0·20 ± 0·02 (2)	0·70 ± 0·10 (5)	1·26 ± 0·30 (5)	1·93 ± 0·31 (2)
LH pulse frequency/12 h	0·00	1·00 ± 0·31	1·60 ± 0·24	4·00 ± 1·00
LH pulse amplitude (ng/ml)	—	1·58 ± 0·43	1·45 ± 0·18	2·25 ± 0·16

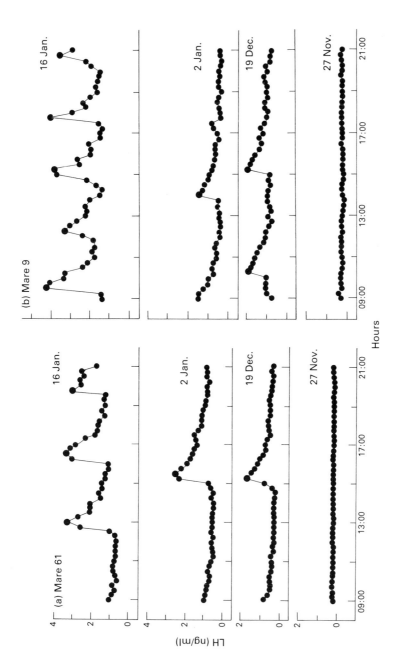

Fig. 1. Profile of the pattern of LH secretion in control mares (a, Mare 61; b, Mare 9) 2 weeks before (27 November) and after exposure to an abrupt increase in photoperiod beginning 1 December. Blood samples were collected frequently (every 15 min) on the dates indicated. The peak of an LH pulse is indicated by the larger symbol.

Subsequently, 19 days after exposure to an abrupt increase in daylength, a pulsatile pattern of LH secretion was observed in 4/5 mares (mean pulse frequency $1 \cdot 0 \pm 0 \cdot 31/12$ h, range 0–2). All 5 mares exhibited LH pulses during a 12-h period on 2 January (Table 1) and mean serum LH values were higher ($P < 0 \cdot 05$) than in November. Profiles of the pattern of LH secretion in 2 mares which had not ovulated before 16 January are illustrated in Fig. 1. In both mares, frequent LH pulses were observed during a 12 h period which corresponded to Day -7 (Mare 61, 3 pulses/12 h; Fig. 1a) and Day -14 (Mare 9, 5 pulses/12 h; Fig. 1b), before ovulation, respectively. In both mares, and for the group overall, mean LH pulse amplitude did not change ($P > 0 \cdot 05$) during the experimental period (Table 1).

Table 2. Changes in mean serum LH concentrations, LH pulse frequency and amplitude in 5 mares immunized against LHRH (antibody titre is indicated in parentheses)

		Time of year					
	Mare	27 Nov.	19 Dec.	2 Jan.	16 Jan.	30 Jan.	31 Mar.
Serum LH (ng/ml)	72	0·14 (<100)	0·12 (<100)	0·22 (<100)	1·27 (<100)	4·48 (<100)	(110)
	77	0·14 (33 000)	0·18 (16 000)	0·16 (14 000)	0·29 (5 200)	0·11 (6 600)	(5 400)
	79	0·43 (3 700)	0·20 (2 900)	0·24 (14 000)	0·71 (5 600)	1·17 (4 300)	(3 200)
	81	1·64* (2 100)	4·64* (900)	0·40 (5 200)	0·21 (3 700)	0·72 (3 000)	(4 000)
	85	0·10 (5 200)	0·10 (6 800)	† (37 000)	† (26 000)	† (30 000)	(15 000)
LH pulses 12 h	72	0	0	0	2	0	
	77	0	0	0	0	0	
	79	2	0	0	2	2	
	81	0	0	0	0	0	
	85	0	0	0	0	0	
LH pulse amplitudes (ng/ml)	72	—	—	—	1·92,0·95	—	
	77	—	—	—	—	—	
	79	0·31,1·51	—	—	0·43,0·45	1·35,1·02	
	81	—	—	—	—	—	
	85	—	—	—	—	—	

*Blood samples collected 2 days after ovulation.
†Below sensitivity of assay.

In mares immunized against LHRH, the pattern of LH secretion and mean serum LH concentrations appeared to be related to antibody titre (Table 2). The highest antibody titre was detected in Mares 85 and 77, both of which remained anovulatory throughout the experimental period. As shown in Fig. 2(b), serum LH concentrations in Mare 85 remained close to or below the limits of hormone detection and LH pulses were not observed. In the 2 mares which were anovulatory throughout the experimental period (Mare 77) or between January and March (Mare 81), serum LH remained low and the pattern of secretion was non-pulsatile (Table 2). Finally, in the remaining 2 mares (Mares 72 and 79), both of which ovulated and exhibited low LHRH antibody titres, 2 LH pulses/12 h were detected in both mares on 16 January and in Mare 79 on 30 January, Day -12 before ovulation. As shown in Fig. 2(a), serum LH concentrations in Mare 72 were elevated ($4 \cdot 0$ ng/ml) on 30 January, Day -2 before ovulation, but LH pulses were not identified.

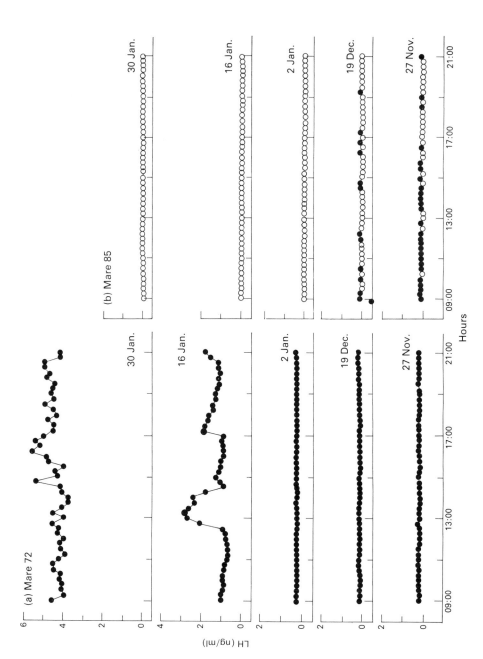

Fig. 2. Profile of the pattern of LH secretion in LHRH-immunized mares (a, Mare 72; b, Mare 85) 2 weeks before (17 November) and after exposure to an abrupt increase in photoperiod beginning 1 December. Blood samples were collected frequently (every 15 min) for 12 h on the dates indicated. Serum LH concentrations below assay sensitivity are indicated (○), and the peak of an LH pulse is indicated by the larger symbol.

Discussion

The objective of the present study was to examine the hypothesis that, in the mare, the onset of the breeding season is associated with, and may be caused by, an increase in frequency of pulsatile secretion of LHRH from the hypothalamus. In this study, the occurrence of pulsatile LH secretion, as measured in blood samples collected from the jugular vein, was used as an index of pulsatile LHRH secretion (Levine *et al.*, 1982). That the occurrence of ovulation was blocked in 3/5 mares immunized against LHRH, and in those mares LH pulses were not observed during the anovulatory period, supports the aforementioned hypothesis. However, absence of occurrence of follicular development and ovulation appeared to be related to LHRH antibody titre. Two mares (Nos 77 and 85), which remained anovulatory throughout the duration of the experiment, had the highest titres of the group and the lowest mean serum LH concentrations (Table 2). This observation suggests that circulating antibodies intercepted LHRH released from the hypothalamus, thus blocking stimulation of the pituitary gland by LHRH (Takabashi *et al.*, 1978). The presence of a high LHRH antibody titre most probably resulted in a suppression of serum LH concentrations, therefore providing inadequate stimulation of the ovaries which is necessary for follicular maturation and ovulation.

Although one mare immunized against LHRH (No. 81) ovulated in November and again in December, after an anovulatory period of more than 1 year, a booster immunization in late December resulted in a more than 5-fold increase in LHRH antibody titre in this animal. As a result, LH concentrations remained low and ovulation was arrested in Mare 81 for the duration of the study. The mare with the lowest antibody titre (Mare 72) showed pulses of LH and increased serum LH concentrations before ovulation in February. Booster immunizations administered to Mare 72 failed to raise LHRH antibody titre to any significant extent. However, a booster administered on 21 February blocked the occurrence of the next anticipated ovulation, such that Mare 72 remained anovulatory through the end of the experiment (36 days). The effect was temporary, however, as ovulation again occurred on 12 April, 60 days later (B. P. Fitzgerald, unpublished observations). A similar phenomenon was observed in Mare 79, although antibody titre appeared to be relatively high when compared with the antibody titres of the 3 anovulatory mares (Mares 77, 81 and 85). Ovulation occurred in Mare 79 on 11 February, preceded by the occurrence of LH pulses and an increase in mean serum LH concentrations. As in Mare 72, a booster immunization administered to Mare 79 on 21 February blocked the occurrence of the next anticipated ovulation, but 47 days later (31 March) ovulation again occurred.

There may be several explanations for the occurrence of ovulation in LHRH-immunized Mares 72 and 79. Low antibody titre most probably accounts for failure to block ovulation in Mare 72. The occurrence of ovulation in Mare 79, however, may be due to incomplete neutralization of LHRH, resulting in a measurable antibody titre but one not sufficient to block completely circulating LHRH from reaching the pituitary.

The present results confirm previous observations from this laboratory that, in the mare, LH pulse frequency increases during the transition from seasonal anoestrus to the breeding season (Murdoch *et al.*, 1985; Fitzgerald *et al.*, 1987). As detected by an increase in LH pulse frequency, the response to an increase in daylength appeared to be rapid. In 5 untreated control mares, LH pulses were not observed during a 12-h period of sample collection in November, before exposure to an abrupt increase in daylength beginning 1 December. However, within 3 weeks of exposure to a stimulatory daylength, LH pulses were observed in 4/5 mares. This apparent rapid response to an alteration in photoperiod, in terms of a change in the pattern of LH secretion, is in contrast to the apparently slower response observed in another seasonal breeder, namely the ewe. In this species, exposure to a stimulatory daylength leads to an increase in LH pulse frequency 40–50 days later (Walton *et al.*, 1980). However, the seemingly rapid alteration in the pattern of LH secretion observed in the mare is slower than the response seen in the quail, in which an increase in serum LH, which may reflect an increase in LH pulse frequency, is evident within 24 h of exposure to a stimulatory daylength (Follett, 1978). Thus, in terms of an alteration in the pattern of LH secretion, considerable diversity appears to exist between individual species of seasonal breeders.

Based on the aforementioned considerations we propose that one aspect of the mechanisms governing the onset of the breeding season of the mare is a daylength-mediated, seasonal increase in LHRH pulse frequency. The present results suggest that immunoneutralization of LHRH suppresses the occurrence of pulsatile LH secretion; however, changes in the pattern of secretion of follicle-stimulating hormone (FSH) were not studied. The regulation of FSH secretion during the period of transition is important for several reasons. First, we have observed the occurrence of coincident LH and FSH pulses during the transition period (Hines *et al.*, 1986). Second, in that study we reported that serum FSH concentrations progressively decrease during the transition period and may reflect a decrease in FSH pulse amplitude. Third, towards the completion of the present study we examined follicular development in anovulatory LHRH-immunized mares using the technique of ultrasonic echography. Our unpublished observations indicate that although serum LH concentrations were undetectable, follicles (< 5 mm diameter) could be identified. This remarkable observation suggests that in the mare antral follicular development may be independent of serum LH concentrations. This concurs with studies undertaken in LHRH-immunized female rats, in which the ovaries of those animals with high antibody titres and low serum concentrations of LH and FSH contained some small antral follicles (Fraser & Baker, 1978). Therefore, these observations emphasize the necessity to examine changes in the pattern of secretion of FSH in mares actively immunized against LHRH.

We thank the staff of the University Farm for care of the animals; Dr T. R. Nett, Dr L. E. Reichert, Jr, Dr H. Papkoff and Dr L. Edgerton for reagents used in the radioimmunoassays; Dr F. Bex of Wyeth Pharmaceuticals for LHRH; Dr H. Fraser for technical advice; and Ms Lelia Garrison and Ms Kathy Williams for assistance with the radioimmunoassays.

This study was supported by the Grayson Foundation, Inc. The investigation reported in this paper (No. 86-4-172) was in connection with a project of the Kentucky Agricultural Experiment Station and is published with approval of the director.

References

Clarke, I.J., Fraser, H.M. & McNeilly, A.S. (1978) Active immunization of ewes against luteinizing hormone-releasing hormone, and its effects on ovulation and gonadotrophin, prolactin and ovarian steroid secretion. *J. Endocr.* **78**, 39–47.

Esbenshade, K.K. & Britt, J.H. (1985) Active immunization of gilts against gonadotropin-releasing hormone: effects on secretion of gonadotropins, reproductive function, and responses to agonists of gonotropin-releasing hormone. *Biol. Reprod.* **33**, 569–577.

Fitzgerald, B.P., I'Anson, H., Legan, S.J. & Loy, R.G. (1985) Changes in patterns of luteinizing hormone secretion before and after the first ovulation in the postpartum mare. *Biol. Reprod.* **33**, 316–323.

Fitzgerald, B.P., Affleck, K.J., Barrows, S.P., Murdoch, W.L., Barker, K.B. & Loy, R.G. (1987) Changes in LH pulse frequency and amplitude in intact mares during the transition into the breeding season. *J. Reprod. Fert.* **79**, 485–493.

Follett, B.K. (1978) Photoperiodism and seasonal breeding in birds and mammals. In *Control of Ovulation*, pp. 267–293. Eds G. E. Lamming & D. B. Crighton. Butterworth Press, London.

Fraser, H.M. & Baker, T.G. (1978) Changes in the ovaries of rats after immunization against luteinizing hormone releasing hormone. *J. Endocr.* **77**, 85–93.

Fraser, H.M., Gunn, A., Jeffcoate, S.L. & Holland, D.T. (1974) Effect of active immunization to luteinizing hormone-releasing hormone on serum and pituitary gonadotrophins, testes and accessory sex organs in the male rat. *J. Endocr.* **63**, 399–406.

Fraser, H.M., McNeilly, A.S. & Popkin, R.M. (1984) Passive immunization against LHRH: elucidation of the role of LH-RH in controlling LH and FSH secretion and LH-RH receptors. In *Immunological Aspects of Reproduction in Mammals*, pp. 399–418. Ed. D. B. Crighton. Butterworth, London.

Fraser, H.M., McNeilly, A.S., Abbott, M. & Steiner, R.A. (1986) Effect of LHRH immunoneutralization on follicular development, the LH surge and luteal function in the stumptailed macaque monkey (*Macaca arctoides*). *J. Reprod. Fert.* **76** 299–309.

Hines, K.K., Affleck, K.J., Barrows, S.P., Murdoch, W.L., Barker, K.B., Loy, R.G. & Fitzgerald, B.P. (1986) Evidence that a decrease in follicle stimulating hormone (FSH) pulse amplitude accompanies the onset of the breeding season in the mare. *Biol. Reprod.* **34**, Supp. 1, 270, Abstr.

Levine, J.E., Pau, K.-Y.F., Ramirez, V.D. & Jackson, G.L. (1982) Simultaneous measurement of luteinizing hormone-releasing hormone and luteinizing hormone release in unanesthetized, ovariectomized sheep. *Endocrinology* **111**, 1440–1455.

Licht, P., Bona Gallo, A., Aggarwal, B.B., Farmer, S.W., Castelino, J.B. & Papkoff, H. (1979) Biological and binding activities of equine pituitary gonadotrophins and pregnant mare serum gonadotrophin. *J. Endocr.* **83**, 311–332.

Loy, R. G., Evans, M.J., Pemstein, R. & Taylor, T.B. (1982) Effects of injected ovarian steroids on reproductive patterns and performance in postpartum mares. *J. Reprod. Fert., Suppl.* **32**, 199–204.

McNeilly, A.S., Fraser, H.M. & Baird, D.T. (1984) Effect of immunoneutralization of LH releasing hormone on LH, FSH and ovarian steroid secretion in the preovulatory phase of the oestrous cycle in the ewe. *J. Endocr.* **101**, 213–219.

Murdoch, W.L., Affleck, K.J., Barrows, S.P., Loy, R.G. & Fitzgerald, B.P. (1985) Evidence of an increase in luteinizing hormone (LH) pulse frequency in intact mares during the transition from anestrus to the breeding season. *Biol. Reprod.* **32**, *Suppl.* 1, 266, Abstr.

Sharp, D.C. & Ginther, O.J. (1975) Stimulation of follicular activity and estrous behaviour in anestrous mares with light and temperature. *J. Anim. Sci.* **41**, 1368–1372.

Takabashi, M., Ford, J.J., Yoshinaga, K. & Greep, R.O. (1978) Active immunization of female rats with luteinizing hormone releasing hormone (LHRH). *Biol. Reprod.* **17**, 754–761.

Walton, J.S., Evins, J.D., Fitzgerald, B.P. & Cunningham, F.J. (1980) Abrupt decrease in daylength and short-term changes in the plasma concentrations of FSH, LH and prolactin in anoestrous ewes. *J. Reprod. Fert.* **59**, 163–171.

J. Reprod. Fert., Suppl. **35** (1987), 239–243

Printed in Great Britain
© 1987 Journals of Reproduction & Fertility Ltd

Induction of ovulation in cyclic mares by administration of a synthetic prostaglandin, fenprostalene, during oestrus

N. C. Savage and R. M. Liptrap*

*Departments of Clinical Studies and *Biomedical Sciences, Ontario Veterinary College,
University of Guelph, Guelph, Ontario, Canada N1G 2W1*

Summary. Fenprostalene (250 µg) or saline was given at 60 h after the onset of oestrus in alternate oestrous periods to 8 mares for 4 cycles. Onset of oestrus and stage of cycle were determined by daily teasing, palpation and ultrasonography until time of treatment when follicular development was monitored at 12-h intervals to confirm ovulation. Serum progesterone concentrations were monitored daily.

The interval from treatment to ovulation was significantly decreased (41·25 *vs* 73·50 h; $P = 0.001$) as was the duration of oestrus (5·63 *vs* 6·88 days; $P = 0.005$). There was no significant difference in the duration of dioestrus (15·88 *vs* 15·75 days; $P = 0.94$). There were no significant differences between serum progesterone concentrations at onset of oestrus, at treatment or at examination before ovulation. Follicular size was not significantly different at onset of oestrus or before ovulation. There was no statistically significant difference in follicle size at the time of treatment when a possible horse effect was controlled for, although a significant difference ($P < 0.05$) was seen when the horse effect was not taken into account. Clinically, 81% of treated mares had ovulated by 48 h after treatment compared to 31% of the control mares (χ^2; $P < 0.025$). The results demonstrate that fenprostalene can effectively hasten time of ovulation in cyclic mares.

Introduction

The process of ovulation involves final maturation of the follicle and the ovum within, rupture of the follicle with ovum release and the initial development of the corpus haemorrhagicum. These biochemical and morphological changes are modulated and integrated by physiological control mechanisms and the involvement of prostaglandins (PGs) in the process has been demonstrated in numerous species. In rats in which ovulation has been blocked by indomethacin or by pentobarbitone treatment, final maturation of the ovum and ovulation can be induced by administration of PGE-2 (Armstrong & Grinwich, 1972). Brown & Poyser (1984) further demonstrated a preferential increase in PGE production in the rat ovary in late pro-oestrus that was associated with an increase in ovarian cyclo-oxygenase activity. In rabbits that are mated or treated with an ovulatory dose of hCG, PGF and PGE show a marked increase only in follicles that actually progress to ovulation (Yang *et al.* 1973, 1974) and the intrafollicular injection of indomethacin or antiserum against PGF-2α effectively prevented ovulation (Armstrong *et al.*, 1974). These observations suggest an important role for PGF-2α in the ovulation process in the rabbit. A role for PGF-2α in follicular development, maturation and ovulation has also been demonstrated in the guinea-pig (Sharma, *et al.*, 1976; Tam & Roy, 1982). In follicular fluid of the prepubertal gilt treated with PMSG and hCG, levels of PGF and PGE increased markedly towards ovulation (Tsang *et al.*, 1979). Treatment with indomethacin prevented this ovarian prostaglandin synthesis and ovulation (Ainsworth *et al.*, 1979) while administration of PGF-2α overcame the ovulatory inhibition (Downey & Ainsworth, 1980). Increases in concentrations of PGE-2, PGF-2α and 6-keto-PGF-1α in theca and granulosa cells of ovaries of prepubertal gilts treated with PSM/hCG provided additional evidence for their role in ovulation (Evans *et al.,* 1983; Ainsworth *et al.,* 1984). In the oestrous ewe, PGE-2α and PGF-2α concentrations are elevated in ovarian follicles (Murdoch & Dailey, 1979) and Murdoch & Myers (1983) further demonstrated that treatment of oestrous ewes with indomethacin caused a reduction in ovarian blood supply and failure of ovulation. In the human female, a significant increase in PGF-2α concentrations occurs in ovarian follicular fluid just before ovulation, suggesting a role for this prostaglandin in the ovulation process (Darling *et al.,* 1982; Lumsden, *et al.,* 1986).

There is little information on the role of prostaglandins in ovulation in the horse. The objectives of this study were to evaluate the use of the PGF-2α analogue, fenprostalene, for induction of ovulation in the oestrous mare and to determine its effect on the subsequent luteal phase of the cycle. Co-incidently, a study of the dynamics of follicular development during the oestrous cycle was made using real-time ultrasonography.

Materials and Methods

Eight mature Standardbred and Quarterhorse mares in good body condition were used between 1 May and 31 August 1985. They were stabled but turned out for exercise as weather permitted. All mares were teased, palpated and examined by rectal ultrasonography daily until Day 3 of oestrus, from which time ovarian examination was performed every 12 h until ovulation was detected. The sizes and positions of follicular and luteal structures were recorded. Ultrasonic examinations were made with a portable linear array realtime ultrasound system with a 5 mHz transducer (Equisonic 310; Equisonic Inc., Bensonville, IL, U.S.A.).

Fenprostalene or saline (9 g NaCl/l) control injections were used alternately during four successive cycles, allocating mares to treatment or control group randomly in the initial cycle. Injections (1·0 ml) were given subcutaneously during the afternoon 60 h after the onset of oestrus. Fenprostalene was given at a dose of 250 μg diluted 1:1 (v/v) with normal saline (9 g NaCl/l) to reduce the possibility of reaction to the polyethylene glycol base (Bosu *et al.,* 1983). Ovulation was considered to have occurred when the absence of a previously recorded large follicle was associated with an ultrasonographic image of a corpus haemorrhagicum (Ginther & Pierson, 1984). Mares were considered to be in dioestrus at the first sign of non-receptivity. No fertility studies were undertaken.

Blood samples, collected by jugular venepuncture daily before teasing, were centrifuged within 1 h of collection and the separated sera were frozen until assayed. Progesterone values were determined by radioimmunoassay (Abraham *et al.,* 1971) using anti-progesterone-11-(succinyl)–BSA (Steranti Research Limited, St Albans, Herts, U.K.) and [1,2,6,7,16,17-³H]progesterone (Amersham International Ltd., Bucks, U.K.). Mean recovery of progesterone added to plasma of a gelding was 85·7% ($n = 6$). The mean coefficient of variation within assays was 5·15% ($n = 7$) and that between assays was 7·08% ($n = 6$). The sensitivity of the assay was 35 pg.

The effects of fenprostalene treatment on the interval from treatment to ovulation, duration of oestrus, duration of dioestrus, follicular development and progesterone concentrations were assessed. Data were analysed using analysis of variance, with horse effects being the random variable and treatment effects the fixed variable. The χ^2 test was used to determine the proportion of mares ovulating by 48 h after treatment.

Results

The interval from treatment to ovulation and the durations of oestrus and dioestrus are shown in Table 1. In mares treated with fenprostalene the interval to ovulation was significantly reduced ($P < 0.001$) as was the duration of oestrus ($P < 0.005$). The duration of dioestrus was not significantly different between the two treatment groups although there was a significant horse effect. Two mares consistently had longer dioestrous periods than was expected (mean ± s.d. 19·0 ± 1·15 days for Mare A; 17·0 ± 0·82 days for Mare B confirmed by progesterone concentrations).

Table 2 shows the serum progesterone concentrations and follicular diameters measured at the onset of oestrus, at the time of treatment and at the last examination before ovulation.

There was a significant horse effect ($P < 0.10$) for serum progesterone concentration at onset of oestrus but no significant effect was found due to treatment. Serum progesterone concentrations were not significantly different in either group at the time of treatment or at the last examination time.

Table 1. Effect of fenprostalene treatment on the treatment–ovulation interval, duration of oestrus and duration of dioestrus in mares

	Fenprostalene	Saline
No. of mare cycles	16	16
Interval to ovulation (h)	$41 \cdot 5 \pm 21 \cdot 00^a$	$73 \cdot 50 \pm 30 \cdot 31^b$
Duration of oestrus (days)	$5 \cdot 63 \pm 1 \cdot 02^a$	$6 \cdot 88 \pm 1 \cdot 31^b$
Duration of dioestrus (days)	$15 \cdot 88 \pm 1 \cdot 99$ ^{a}	$15 \cdot 75 \pm 2 \cdot 18†$

Values are mean ± s.d.
Means with different superscripts within rows are significantly different $P < 0 \cdot 01$.
†Significant horse effect.

Table 2. Serum progesterone concentrations and follicular size at onset of oestrus, at treatment and at the last examination before ovulation in mares

	Fenprostalene	Saline
No. of mare cycles	16	16
Serum progesterone conc. (ng/ml)		
At onset of oestrus	$1 \cdot 72 \pm 1 \cdot 70$	$1 \cdot 64 \pm 1 \cdot 71†$
At treatment	$0 \cdot 73 \pm 0 \cdot 71$	$0 \cdot 49 \pm 0 \cdot 44$
Before ovulation	$0 \cdot 81 \pm 0 \cdot 81$	$0 \cdot 90 \pm 0 \cdot 74$
Follicular diameter (mm)		
At onset of oestrus	$34 \cdot 3 \pm 5 \cdot 4$	$30 \cdot 6 \pm 7 \cdot 4$
At treatment	$42 \cdot 8 \pm 4 \cdot 6^a$	$38 \cdot 2 \pm 7 \cdot 5^b†$
Before ovulation	$47 \cdot 1 \pm 4 \cdot 8$	$47 \cdot 3 \pm 5 \cdot 8†$

Values are mean ± s.d.
Means with different superscripts within rows are significantly different, $P < 0 \cdot 05$.
†Significant horse effect.

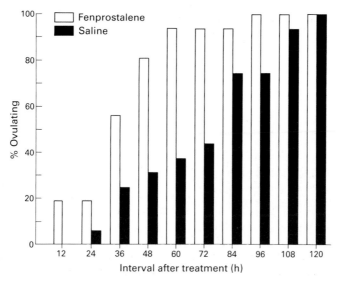

Fig. 1. Cumulative percentage of mares ovulating after treatment with fenprostalene or saline.

Follicular size at the onset of oestrus was not significantly different in either group. A significant difference ($P < 0.05$) was found in follicular size at treatment. However, when horse effect was controlled for there was no significant difference due to treatment. A significant horse effect was found for follicle size at examination before ovulation but no treatment effect existed when horse effect was controlled.

The cumulative percentages of mares that had ovulated at successive 12-h intervals after treatment are recorded in Fig. 1. By 48 h after treatment (108 h after onset of oestrus) 81% of fenprostalene-treated mares had ovulated compared to 31% of those treated with saline.

Discussion

This study provides evidence for the ability of an analogue of PGF-2α (fenprostalene) to hasten follicular maturation and time of ovulation in the mare when injected during oestrus. The durations of oestrus and dioestrus during control cycles in this study were in agreement with those reported by others (Ginther, 1979).

Fenprostalene, the synthetic prostaglandin analogue used, was chosen because of its relatively long duration of action; the half-life for fenprostalene is approximately 24 h in cattle (Herschler & Reid, 1982). However, whether this influenced its ability to hasten ovulation in the mare cannot be decided from this study.

The 'mapping' of follicles using ultrasonography allowed changes in development to be monitored although growth rates of individual follicles were not determined. After the onset of oestrus, the number of follicles >20 mm in diameter decreased as the largest follicle increased in size. Numerous small follicles (<10 mm diam.) were found in the ovaries throughout the oestrous cycle. The follicle that eventually ovulated was identifiable in the first 2–3 days of oestrus. Similar changes were reported by Ginther (1979).

Serum progesterone concentrations at the onset of oestrus were slightly higher than might be expected in mares showing oestrous behaviour (Ginther, 1979), as they were at the time of treatment (mid-oestrus) and preceding ovulation. Mid-dioestrous concentrations of progesterone were also high (about 15–20 ng/ml). However, these discrepancies may be due to individual variation or the assay procedure used.

Neely *et al.* (1975) reported ovulation of dioestrous follicles after intrauterine saline infusion, suggesting a role for PGF-2α in ovulation. In dioestrous mares with large follicles and treated with PGF-2α, a reduction in the interval to ovulation was reported if follicles were larger than 35 mm diameter (Neely *et al.,* 1983). Ovulation of large (>40 mm diam.) follicles within 5 days of treatment with PGF-2α in 62% of mares with persistent CL was reported by Loy *et al.* (1979). These authors also found a similar response in post-partum mares. In this study, follicular diameter of the largest follicle at the time of treatment was 42·8 mm and 38·2 mm in fenprostalene- and saline-treated mares respectively. Diameters of follicles at the examination preceding ovulation were 47·1 and 47·3 mm for fenprostalene- and saline-treated mares respectively. This is in agreement with the data of others (Ginther, 1979). The exact mechanism by which fenprostalene was able to hasten ovulation is uncertain. Further investigation is required to determine whether it acts on the hypothalamo–pituitary axis (Noden *et al.,* 1978; Nett, *et al.,* 1979) or at the ovarian level. The relatively larger size and volume of the preovulatory horse follicle makes it ideal for such research.

Clinically, the ability of fenprostalene to induce ovulation in 81% of mares by 48 h after treatment provides an additional agent for use in the management of the time of ovulation.

Financial support for this research was provided by the Ontario Racing Commission (N.C.S.) and the Natural Sciences and Engineering Research Council (R.M.L.). We thank Dr P. Doig of Syntex for the Synchrocept B and for encouragement (N.C.S.); Ms P. Huether for excellent technical assistance; Mrs M. Montgomery for statistical analysis; and Miss Kim Austin and Miss Rozanne MacLean for manuscript preparation.

References

Abraham, G.E., Swerdloff, R., Tulchinsky, D. & Odell, W.D. (1971) Radioimmunoassay of plasma progesterone. *J. clin. Endocr. Metab.* **32,** 619–624.

Ainsworth, L., Tsang, B.K., Downey, B.R., Baker, R.D., Marcus, G.J. & Armstrong, D.T. (1979) Effects of indomethacin on ovulation and luteal function in gilts. *Biol. Reprod.* **21,** 401–411.

Ainsworth, L., Tsang, B.K., Marcus, G.J. & Downey, B.R. (1984) Prostalglandin production by dispersed granulosa and theca interna cells from porcine preovulatory follicles. *Biol. Reprod.* **31,** 115–121.

Armstrong, D.T. & Grinwich, D.I. (1972) Blockade of spontaneous and LH-induced ovulation in rats by indomethacin, an inhibitor of prostaglandin biosynthesis. *Prostaglandins* **1,** 21–28.

Armstrong, D.T., Grinwich, D.I., Moon, Y.S. & Zamecnik, J. (1974) Inhibition of ovulation in rabbits by intrafollicular injection of indomethacin and prostaglandin F antiserum. *Life Sci.* **14,** 129–140.

Bosu, W.T.K., McKinnon, A.O., Lissemore, K. & Kelton, D. (1983) Clinical and luteolytic effects of fenprostalene (a prostaglandin $F_{2\alpha}$ analogue) in mares. *Can. vet. J.* **24,** 347–351.

Brown, C.G. & Poyser, N.L. (1984) Studies on ovarian prostaglandin production in relation to ovulation in the rat. *J. Reprod. Fert.* **72,** 407–414.

Darling, M.R.N., Jogee, M. & Elder, M.G. (1982) Prostaglandin $F_{2\alpha}$ levels in the human ovarian follicle. *Prostaglandins* **23,** 551–556.

Downey, B.R. & Ainsworth, L. (1980) Reversal of indomethacin blockade of ovulation in gilts by prostaglandins. *Prostaglandins* **19,** 17–22.

Evans, G., Dobias, M., King, G.J. & Armstrong, D.T. (1983) Production of prostaglandins by porcine preovulatory follicular tissues and their roles in intrafollicular function. *Biol. Reprod.* **28,** 322–328.

Ginther, O.J. (1979) *Reproductive Biology of the Mare: Basic: and Applied Aspects.* McNaughton and Gunn, Inc., Ann Arbor.

Ginther, O.J. & Pierson, R.A. (1984) Ultrasonic anatomy of equine ovaries. *Theriogenology* **21,** 471–483.

Herschler, R.C. & Reid, J.F.S. (1982) New prostaglandins: present studies and future. In *Veterinary Pharmacology and Toxicology,* pp. 401–402. Le Congress European, Toulouse.

Loy, R.G., Buell, J.R., Stevenson, W. & Hamm, D. (1979) Sources of variation in response intervals after prostaglandin treatment in mares with functional corpora lutea. *J. Reprod. Fert., Suppl.* **27,** 229–239.

Lumsden, M.A., Kelly, R.W., Templeton, A.A., Van Look, P.F.A., Swanston, I.A., & Baird, D.T. (1986) Changes in the concentration of prostaglandins in preovulatory human follicles after administration of hCG. *J. Reprod. Fert.* **77,** 119–124.

Murdoch, W.J. & Dailey, R.A. (1979) Prostaglandins E_2 and $F_{2\alpha}$ in ovine follicles before and during oestrus. *Biol. Reprod.* **20** (Suppl. 1), 51A, Abstr. 80.

Murdoch, W.J. & Myers, D.A. (1983) Effect of treatment of oestrous ewes with indomethacin on the distribution of ovarian blood to the periovulatory follicle. *Biol. Reprod.* **29,** 1229–1232.

Neely, D.P., Hughes, J.P., Stabenfeldt, G.H. & Evans, J.W. (1975) The influence of intrauterine saline infusion on luteal function and cyclic ovarian activity in the mare. *Equine vet. J.* **6,** 150–157.

Neely, D.P., Lui, I.K.M. & Hillman, R.B. (1983) In *Equine Reproduction,* pp. 26–27. Hoffman-LaRoche, Nutley.

Nett, T.M., Pickett, B.W. & Squires, E.L. (1979) Effects of Equimate (ICI-81008) on levels of luteinizing hormone, follicle stimulating hormone and progesterone during the oestrous cycle of the mare. *J. Anim. Sci.* **48,** 69–75.

Noden, P.A., Oxender, W.D. & Hafs, H.D. (1978) Early changes in serum progesterone, estradiol and LH during $PGF_{2\alpha}$-induced luteolysis in mares. *J. Anim. Sci.* **47,** 666–671.

Sharma, S.C., Wilson, C.M.W. & Pugh, D.M. (1976) *In vitro* production of prostaglandins E and F by the guinea pig ovarian tissue. *Prostaglandins* **11,** 555–568.

Tam, W.H. & Roy, R.J.J. ((1982) A possible role of prostaglandin F-2α in the development of ovarian follicles in guinea-pigs. *J. Reprod. Fert.* **66,** 277–282.

Tsang, B.K., Ainsworth, L., Downey, B.R. & Armstrong, D.T. (1979) Preovulatory changes in cyclic AMP and prostaglandin concentrations in follicular fluid of gilts. *Prostaglandins* **17,** 141–148.

Yang, N.S.T., Marsh, J.M. & LeMaire, W.J. (1973) Prostaglandin changes induced by ovulatory stimuli in rabbit Graafian follicles. The effect of indomethacin. *Prostaglandins* **4,** 395–404.

Yang, N.S.T., Marsh, J.M. & LeMaire, W.J. (1974) Post ovulatory changes in the concentration of prostaglandins in rabbit Graafian follicles. *Prostaglandins* **6,** 37–44.

J. Reprod. Fert., Suppl. 35 (1987), 245–252

Patterns of oxytocin secretion during the oestrous cycle of the mare

T. A. Tetzke*, S. Ismail, G. Mikuckis and J. W. Evans*

Department of Animal Science, University of California, Davis, CA 95616, U.S.A.

Summary. From a group of 11 cyclic mares, blood samples were collected at 3-min intervals for 2 h and at 15-min intervals for an additional 6 h during four stages of the oestrous cycle. Mean plasma oxytocin concentrations (pg/ml, LSM $\pm$ se) were greater on Day 15 after ovulation ($169 \cdot 9 \pm 17 \cdot 6$) than on Day 0 ($82 \cdot 6 \pm 17 \cdot 6$; $P < 0 \cdot 01$), Day 3 ($97 \cdot 2 \pm 20 \cdot 4$, $P < 0 \cdot 01$) and Day 7 after ovulation ($104 \cdot 0 \pm 25 \cdot 0$, $P < 0 \cdot 05$).

Oxytocin was secreted in a pulsatile manner throughout the oestrous cycle, with short (0–29 min), medium (30–89 min) and long (>90 min) duration rhythms. No differences in pulse frequencies were observed throughout the oestrous cycle. Pulse amplitudes for rhythms of short and medium duration were greater on Day 7 after ovulation than on Day 0 ($P < 0 \cdot 01$ for short and $P < 0 \cdot 05$ for medium rhythms), Day 3 ($P < 0 \cdot 01$) and Day 15 after ovulation ($P < 0 \cdot 05$ for short rhythms). Fluctuations in baseline values for oxytocin over the 8-h sampling period were indicative of a possible circadian rhythm.

These results are consistent with previous observations for the mare and other species that oxytocin is secreted in a pulsatile manner and may be involved in the regulation of the oestrous cycle by promoting luteolysis via the synthesis and release of uterine PGF-2α.

Introduction

Oxytocin is synthesized and secreted along with its neurophysin from the posterior pituitary. Oxytocin has been identified in the corpora lutea of some mammals (Wathes & Swann, 1982; Wathes *et al.*, 1983; Khan-Dawood *et al.*, 1984). In the cow (Wathes *et al.*, 1983) and ewe (Wathes & Swann, 1982), high concentrations of oxytocin are synthesized and secreted from the ovary during the oestrous cycle. In circulation, oxytocin binds to specific binding sites, causing smooth muscle contractions in the uterus, mammary gland (Soloff, 1979) and oviduct (Soloff, 1975). In addition, oxytocin binding sites have been identified in the endometrial portion of the uterus in some species (Roberts *et al.*, 1976; Fuchs *et al.*, 1985; Stull & Evans, 1986).

Circulating oxytocin concentrations vary throughout the oestrous cycle. In ruminants, oxytocin concentrations are greater during the luteal phase (Flint & Sheldrick, 1983; Schams, 1983), but in primates they are greater in mid-cycle (Falconer *et al.*, 1980; Amico *et al.*, 1981). A previous investigation in the mare reported higher oxytocin concentrations during oestrus and early dioestrus than at any other stage of the cycle (Burns *et al.*, 1981). It has been suggested that oxytocin may play a role in luteolysis and/or ovulation.

Accumulating evidence supports the idea that oxytocin is involved in luteolysis. Administration of oxytocin to cattle caused premature luteolysis (Hansel & Wagner, 1960), and immunization against oxytocin in sheep and goats interfered with luteolysis (Sheldrick *et al.*, 1980; Cooke & Homeida, 1985). Oxytocin may act via a direct interaction with its endometrial receptor in the late

---

*Present address: Department of Animal Science, Texas A&M University, College Station, TX 77843, U.S.A.

luteal phase, stimulating the synthesis and release of uterine PGF-2α. Addition of oxytocin to sheep (Roberts *et al.*, 1976) and horse (King & Evans, 1984) endometrial tissue collected during the luteal phase and incubated *in vitro* caused the synthesis and release of PGF-2α. Furthermore, administration of oxytocin caused a rise in uterine venous PGF-2α concentration in heifers (Newcomb *et al.*, 1977) and a rise in PGFM in ewes (Fairclough *et al.*, 1980).

Steroids mediate oxytocin activity. Negoro *et al.* (1973) found that paraventricular (PVN) neurone activity varies throughout the oestrous cycle, oestrogen exciting and progesterone lowering PVN unit activity. Steroids also regulate the number of uterine oxytocin binding sites. Oestrogens increase and progesterone inhibits the formation of uterine oxytocin receptors as observed in the ewe (McCracken *et al.*, 1981) and the rabbit (Nissenson *et al.*, 1978). In the mare, Stull & Evans (1986) found a greater number of uterine oxytocin binding sites at the time of luteolysis which is a period of increasing oestrogen and decreasing progesterone concentrations.

The pulsatile secretion of oxytocin has been studied extensively during suckling and parturition (Wakerley *et al.*, 1973; Mitchell *et al.*, 1982). It appears that pituitary oxytocin is secreted in 'spurts' due to an acceleration in the firing rate of PVN neurones (Summerlee *et al.*, 1985). Pulsatile secretion of oxytocin has been reported during the oestrous cycle in the cow and ewe (Mitchell *et al.*, 1982; Walters *et al.*, 1984). Walters *et al.* (1984) found that oxytocin and progesterone are secreted concomitantly from the ovary in a pulsatile fashion during the mid-luteal phase of the cycle. No investigations have been made into the secretion pattern of oxytocin during the oestrous cycle in the mare. The objective of the present study was to determine plasma oxytocin concentrations and patterns of secretion during different stages of the horse oestrous cycle in relation to other hormonal and physiological events.

Materials and Methods

Sample collections. Jugular vein blood samples were collected from July through to September 1983 from 12 normally cyclic mares. Ovarian status was determined by rectal palpation and verified with plasma progesterone concentrations. Ovulation was designated as Day 0. Samples were collected at 3-min intervals for 2 h and at 15-min intervals for an additional 6 h during each of four stages of the oestrous cycle: Day 0 (N = 8), Day 3 (N = 6), Day 7 (N = 4) and Day 15 (N = 8) after ovulation. One mare was classified as having a 38-day spontaneously prolonged corpus luteum as evidenced by the progesterone concentrations, failure to show oestrus and failure to ovulate: no further details of this mare are reported. After collection, blood samples were immediately poured into test-tubes containing a phenanthroline solution (10 μl/10 ml blood) and refrigerated until plasma was collected after centrifugation. Plasma samples were stored at −20°C until assayed.

Oxytocin assay. Oxytocin was assayed using a double-antibody RIA developed in this laboratory. Briefly, the assay procedure was as follows. Oxytocin (Ferring Pharmaceuticals, Malmo, Sweden) was iodinated with ^{125}I (Amersham Corporation, Arlington Heights, IL, U.S.A.) using the chloramine-T method (Greenwood *et al.*, 1963). The labelled hormone was separated from the free iodine by Rexyn 202 (Fisher Scientific, Springfield, NJ, U.S.A.) and gel filtration (Sephadex G-25 fine).

Oxytocin was extracted from plasma samples by adding 0·25 ml 1 N-HCl and 0·5 ml Fullers earth (mesh size 100–200, Sigma Corporation, St Louis, MO, U.S.A.; 100 mg/ml water) to 13 × 100 borosilicate culture tubes (Fisher Scientific), vortexed for 3 min and then centrifuged (1700 g) for 10 min. Then 1 ml 80% acetone–1 N-HCl was added to the pellet, vortexed for 2 min and centrifuged (1700 g) for 10 min. The supernatant was decanted into a 13 × 100 borosilicate culture tube, washed with 0·5 ml petroleum ether (nanograde; Mallinckrodt, St Louis, MO, U.S.A.) by vortexing for 1 min and the petroleum ether aspirated off. The extracts were dried and the tubes were sealed with parafilm. The extracts were stored at 20°C up to a maximum of 2 weeks before assay. The extraction efficiency was 70·7 ± 0·9% (mean ± s.e.m.).

In preparation for the assay, extracts were reconstituted with 0·25 ml 0·01 M-phosphate buffer (pH 7·5; 372 mg EDTA, 48 mg L-cystine, 4 mg phenanthroline, 0·1 g thimerosal and 5 g bovine serum albumin per litre), vortexed for 2 min, allowed to stand for 1 h, vortexed for 1 min and then centrifuged (1700 g) for 10 min. Aliquants (100 μl) were assayed in duplicate.

Antisera to oxytocin were produced in rabbits by injection of oxytocin (Calbiochem–Behring, San Diego, CA, U.S.A.) conjugated to bovine serum albumin (Sigma) at 2-week intervals for 20 weeks. The antiserum produced cross-reacted <1% with arginine vasopressin, vasopressin and angiotensin I and II and was used at a final dilution of 1:250 000. Goat anti-rabbit antibody (Antibodies Inc., Davis, CA, U.S.A.) was used as the second antibody.

After a 24-h incubation of the sample and antiserum, labelled oxytocin was added (7000 c.p.m.). After 72 h, 400 μl goat anti-rabbit antibody (1:45 dilution) and 100 μl normal rabbit serum (1:20 dilution) were added and the tubes were incubated for an additional 24 h. After centrifugation (1700 g) for 20–30 min, the supernatant was aspirated and the pellets counted in a Micromedic Assay Compucenter. The lower limit of sensitivity was 20 pg/ml. Intra-assay coefficients of variation were $3 \pm 0.7\%$, $3.1 \pm 0.5\%$ and $2.7 \pm 0.5\%$ for 3 pools of mare plasma. Interassay coefficients of variation were 9·7, 14·8 and 20% for quality control points ED_{20}, ED_{50} and ED_{80}, respectively.

Statistical analysis. Oxytocin concentrations for each mare were plotted. From these graphs, pulse amplitudes, durations and periods were determined. Rhythms were divided into three categories according to duration: short (1–29 min), medium (30–89 min) and long (>90 min). Pulse amplitude and duration observations were coupled for analysis, while pulse period data were dealt with as single entity. A one-way analysis of variance and least square mean test (SAS Institute Inc., Cary, NC 27511, copyright 1985) were used to test for significant differences in oxytocin concentrations, pulse amplitudes, pulse periods and pulse durations during different stages of the oestrous cycle. The results are expressed as least square means $\pm$ s.e. of the least square mean unless otherwise stated.

Results

Oxytocin concentrations fluctuated throughout the oestrous cycle (Table 1). Mean plasma oxytocin concentrations were greater on Day 15 than on Day 0 ($P < 0.01$), Day 3 ($P < 0.01$) and Day 7 ($P < 0.05$) after ovulation. No significant differences ($P > 0.05$) were found between remaining groups. The variability amongst groups was attributed to differences ($P < 0.0001$) amongst mares within a group (Table 1).

Table 1. Mean plasma oxytocin concentrations for individual mares at different stages of the oestrous cycle (days after ovulation)

	Day 0		Day 3		Day 7		Day 15
Mare no.	Oxytocin (pg/ml)	Mare no.	Oxytocin (pg/ml)	Mare no.	Oxytocin (pg/ml)	Mare no.	Oxytocin (pg/ml)
1	57.2 ± 1.8^a	1	$87.1 \pm 1.9^{c*}$	1	80.8 ± 5.5^b	1	$169.0 \pm 2.6^{d*}$
2	55.8 ± 1.7^a	2	$93.6 \pm 1.9^{d*}$	6	176.1 ± 5.5^a	3	144.0 ± 2.7^b
3	76.7 ± 1.8^b	4	57.8 ± 1.8^c	8	80.5 ± 5.5^b	4	$160.5 \pm 2.6^{c*}$
4	$69.3 \pm 1.8^{c*}$	7	85.3 ± 1.9^c	10	78.5 ± 5.6^b	6	104.0 ± 2.6^a
5	112.0 ± 1.8^d	8	153.5 ± 1.9^b			7	138.0 ± 2.5^b
6	48.7 ± 1.9^e	9	106.3 ± 1.9^e			9	330.4 ± 3.0^e
7	$75.2 \pm 1.8^{b*}$					10	156.8 ± 2.5^c
2†	165.7 ± 1.8^f					11	156.6 ± 2.6^c
Overall mean	82.6 ± 13.7		97.2 ± 13.0		$104 \quad \pm 24.0$		170.0 ± 24.0

Values are least square means $\pm$ s.e. Superscript letters denote differences at the $P < 0.01$ level amongst mares on a day after ovulation.
*Values that are different at $P < 0.05$ on a particular day.
†Second oestrous cycle for Mare 2.

Rapid fluctuations in oxytocin concentrations were observed during all stages of the oestrous cycle and are indicative of a pulsatile secretion of oxytocin. The pulsatile secretion occurred in short, medium and long rhythmic patterns. Mean period and duration values did not differ within rhythms at various stages of the oestrous cycle and so all values were grouped (Table 2). Mare 4 on

Table 2. Mean $\pm$ s.e.m. amplitudes, periods and durations for short, medium and long rhythms throughout the oestrous cycle

Rhythm	Mean amplitude (pg/ml)	Mean period (min)	Mean duration (min)
Short	26.3 ± 2.2	12.9 ± 0.7	12.8 ± 0.6
Medium	38.3 ± 5.1	49.8 ± 2.4	53.4 ± 2.0
Long	43.9 ± 3.9	124.0 ± 7.9	128.8 ± 8.5

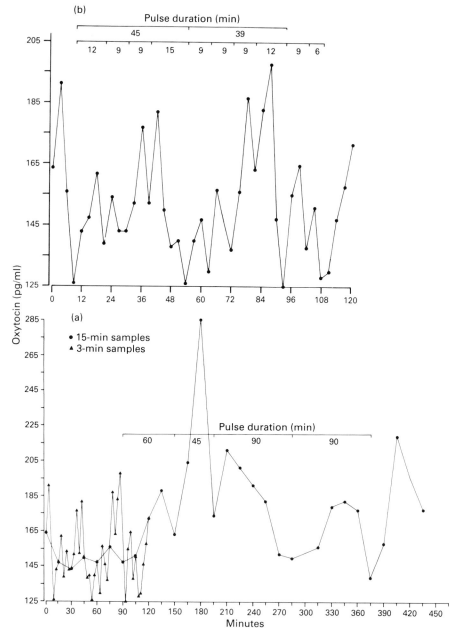

Fig. 1. Oxytocin concentrations in a mare (Day 15 after ovulation) sampled at 15-min (a) and 3- and 15-min (a, b) intervals. Pulse durations are denoted by a bar and time in minutes; (a) demonstrates long-duration pulses and (b) short- and medium-duration pulses.

Day 15 after ovulation demonstrates the presence of the three rhythmic patterns (Figs 1a & b). The short and medium rhythms were most pronounced during the 3-min sampling period. The 15-min sampling interval was not sufficient to describe this pulse activity, but it did demonstrate the longer, more sustained secretions (Fig. 1a). Long-duration rhythms were sometimes difficult to detect and quantitate, although pulsatile secretions were evident.

Mean amplitudes for short and medium rhythms (Table 3) were greater on Day 7 than on Day 0 ($P < 0.01$ for short and $P < 0.05$ for medium rhythms), Day 3 ($P < 0.01$) and Day 15 after ovulation ($P < 0.05$ for short rhythms). No significant differences were found amongst the remaining pulse data.

Large-amplitude, short-duration pulses of oxytocin referred to as 'spurt release', were observed in some mares for each stage of the oestrous cycle (Fig. 2). The largest amplitudes, ranging from 58

Table 3. Mean amplitudes (pg/ml; least squares $\pm$ s.e. for no. of cycles in parentheses) of short, medium and long rhythms at different days after ovulation

Day after ovulation	Short rhythm	Medium rhythm	Long rhythm
Day 0	19.6 ± 3.2^b (8)	30.5 ± 8.1^c (7)	36.2 ± 6.8 (8)
Day 3	22.8 ± 3.7^b (6)	21.8 ± 8.7^b (6)	38.5 ± 9.6 (4)
Day 7	43.8 ± 5.2^a (3)	63.5 ± 10.7^a (4)	46.8 ± 11.1 (3)
Day 15	27.9 ± 3.2^c (8)	42.3 ± 8.7 (6)	52.6 ± 7.3 (7)

[a,b] $P < 0.01$ within rhythm group.
[a,c] $P < 0.05$ within rhythm group.

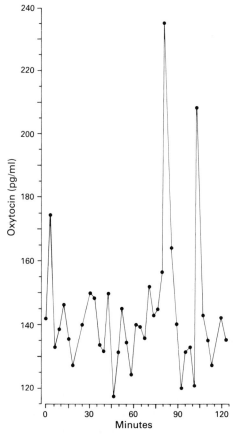

Fig. 2. Oxytocin concentrations for 3-min interval samples collected from a mare on Day 15 after ovulation. This figure demonstrates the 'spurt' release of oxytocin.

to 586 pg/ml, were observed in Mares 1 and 8 on Day 7 after ovulation. These pulses were 7–16 times greater in amplitude than the overall mean amplitude.

The baseline about which the oxytocin pulses were regulated varied over the 8-h sampling period during all stages of the oestrous cycle. Of the 27 observations, 18 showed an increase in the baseline activity during the sampling period (Fig. 3a), 2 mares showed a decrease in baseline activity (Fig. 3b) and 7 no baseline movement. The fluctuation in the baseline is indicative of a possible circadian rhythm.

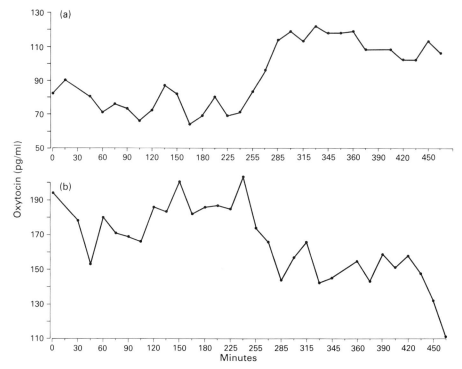

Fig. 3. Oxytocin concentrations for 15-min interval samples, demonstrating (a) an increase (Day 3 after ovulation) and (b) a decline (Day 15 after ovulation) in the oxytocin baseline over time.

Discussion

The results demonstrated cyclic variations in plasma oxytocin concentrations throughout the oestrous cycle of the mare. Although oxytocin concentrations were higher in the mare, the trend is similar to that observed in ruminants (Sheldrick & Flint, 1981; Schams, 1983) in that highest concentrations were found during the mid- and late luteal phases and were low during ovulation and early dioestrus. These findings are not in agreement with those previously reported by Burns *et al.* (1981) for the mare, who observed higher concentrations during oestrus and early dioestrus than during mid- and late luteal phases.

The rise in oxytocin concentrations observed during the luteal phase of the cycle is consistent with the possibility that oxytocin may have a role in the synthesis and release of uterine PGF-2α in a dose-dependent manner (Roberts *et al.*, 1976; McCracken *et al.*, 1981). Previous findings in the mare are in support of this idea: (1) uterine PGF-2α was luteolytic and peaked between Days 14 and 17 (Douglas & Ginther, 1972); (2) the number of endometrial oxytocin receptors was greater on Days 14–17 after ovulation (Stull & Evans, 1986); and (3) addition of oxytocin to horse endometrial tissue incubated *in vitro* caused an increase in PGF-2α synthesis and secretion in the late luteal phase (King & Evans, 1984). It therefore appears that in the mare an increase in circulating oxytocin and an increase in its endometrial receptor in the late luteal phase may act in concert to stimulate the synthesis and secretion of endometrial PGF-2α.

Pulsatile oxytocin secretion has been observed in the ewe (Mitchell *et al.*, 1982) and cow (Walters *et al.*, 1984). In this study, oxytocin was secreted in a pulsatile manner which occurred in short, medium and long rhythms. The 3-min sampling interval was sufficient to detect 'spurt release' as suggested by Gibbens *et al.* (1973). These 'spurts' in oxytocin may result from an increased high frequency activity between PVN neurones as observed during suckling in rats (Lincoln *et al.*, 1973). Mean pulse amplitudes were greater during Day 7 after ovulation which corresponds to the initial rise in mean oxytocin concentrations observed at that time. The pronounced increase in baseline oxytocin on Day 15 was not due to a change in pulse frequency or amplitude but possibly due to an increase in the set point about which homeostatic regulation occurs.

Circulating oxytocin concentrations are mediated by ovarian steroids—oestrogen stimulates and progesterone inhibits pituitary oxytocin secretion (Roberts & Share, 1969; Negoro *et al.*, 1973). In the present study, the rise in mean oxytocin concentration coincides with the time of progesterone withdrawal and a preovulatory rise in oestrogen concentrations (Hughes *et al.*, 1972; Pattison *et al.*, 1972). Significantly lower oxytocin concentrations were observed at a time of increased progesterone and decreased oestrogen concentrations. The low concentrations observed at oestrus and early dioestrus therefore resulted from a decrease in secretion. The increase in pulse amplitude on Day 7 may be due to a changing hormonal environment. Although progesterone concentrations are at a peak on Day 7 after ovulation it may be losing its inhibitory effect. Since oestrogen has an excitatory effect on PVN neurones, PVN neurones may be more sensitive to a slight increase in oestrogen occurring at mid-dioestrus in the mare. These findings support the idea that an increase in oxytocin secretion is due to a stimulation by oestrogen and a loss of inhibition to progesterone at the time of luteolysis in the mare.

The fluctuations observed in the baseline during the sampling period are indicative of a possible circadian rhythm. Further investigations need to be conducted to clarify the existence of any long-term rhythms of oxytocin in the mare.

References

Amico, J.A., Seif, S.M. & Robinson, A.G. (1981) Elevation of oxytocin and oxytocin-associated neurophysin in the plasma of normal women during mid-cycle. *J. clin. Endocr. Metab.* **53**, 1229–1232.

Burns, P.J., Kumaresan, P. & Douglas, R.J. (1981) Plasma oxytocin concentrations in cyclic mares and sexually aroused stallions. *Theriogenology* **16**, 531–539.

Cooke, R.G. & Homeida, A.M. (1985) Suppression of prostaglandin F2α release and delay of luteolysis after active immunization against oxytocin in the goat. *J. Reprod. Fert.* **75**, 63–68.

Douglas, R.H. & Ginther, O.J. (1972) Effect of prostaglandin F2α on length of diestrus in mares. *Prostaglandins* **2**, 265–268.

Fairclough, R.J., Moore, L.G., McGowen, L.R., Peterson, A.J., Smith, J.F., Tervit, H.R. & Watkins, W.B. (1980) Temporal relationship between plasma concentrations of 13, 14-dihydro-15-keto prostaglandin F2α and neurophysin I/II around luteolysis in sheep. *Prostaglandins* **20**, 199–208.

Falconer, J., Mitchell, M.D., Mountford, L.A. & Robinson, J.S. (1980) Plasma oxytocin concentrations during the menstrual cycle in the rhesus monkey (*Macaca mulatta*). *J. Reprod. Fert.* **59**, 69–72.

Flint, A.P.F. & Sheldrick, E.L. (1983) Evidence for a systemic role for ovarian oxytocin in luteal regression in sheep. *J. Reprod. Fert.* **67**, 215–225.

Fuchs, A., Fuchs, F. & Soloff, M.S. (1985) Oxytocin receptors in nonpregnant human uterus. *J. clin. Endocr. Metab.* **60**, 37–41.

Gibbens, D., Boyd, N.R.H. & Chard, T. (1973) Spurt release of oxytocin during human labour. *J. Endocr.* **53**, liv–lvi, Abstr.

Greenwood, F.C., Hunter, W.M. & Glover, J.S. (1963)

The preparation of ^{131}I-labelled human growth hormone of high specific radioactivity. *Biochem. J.* **89**, 114–123.

Hansel, W. & Wagner, W.C. (1960) Luteal inhibition in the bovine as a result of oxytocin injections, uterine dilatation, and intrauterine infusions of seminal and preputial fluids. *J. Dairy Sci.* **43**, 796–805.

Hughes, J.P., Stabenfeldt, G.H. & Evans, J.W. (1972) Clinical and endocrine aspects of the estrous cycle of the mare. *Proc. 18th A. Meeting Am. Assoc. Equine Pract.* 119–151.

Khan-Dawood, F.S., Marut, E.L. & Dawood, M.Y. (1984) Oxytocin in the corpus luteum of the cynomolgus monkey (*Macaca fascicularis*). *Endocrinology* **115**, 570–574.

King, S.S. & Evans, J.W. (1984) Equine endometrial PFG2α production in response to oxytocin and arachadonic acid during the normal estrous cycle and spontaneously prolonged corpus luteum syndrome. *Proc. 10th Int. Congr. Anim. Reprod. & A. I., Urbana-Champaign* **1**, 483, Abstr.

Lincoln, D.W., Hill, A. & Wakerley, J.B. (1973) The milk-ejection reflex of the rat: an intermittent function not abolished by surgical levels of anaesthesia. *J. Endocr.* **57**, 459–476.

McCracken, J.A., Schramm, W. Barcikowski, B. & Wilson, Jr., L. (1981) The identification of prostaglandin F2α as a uterine luteolytic hormone and the hormonal control of its synthesis. *Acta vet. scand., Suppl.* **77**, 71–88.

Mitchell, M.D., Kraemer, D.L., Brennecke, S.P. & Webb, R. (1982) Pulsatile release of oxytocin during the estrous cycle, pregnancy and parturition in sheep. *Biol. Reprod.* **27**, 1169–1173.

Negoro, H., Visessuwan, S. & Holland, R.C. (1973) Unit activity in the paraventricular nucleus of female rats at different stages of the reproductive cycle and after ovariectomy, with or without oestrogen or progesterone treatment. *J. Endocr.* **59**, 545–558.

Newcomb, R., Booth, W.D. & Rowson, L.E.A. (1977) The effect of oxytocin treatment on the levels of prostaglandin F in the blood of heifers. *J. Reprod. Fert.* **49**, 17–24.

Nissenson, R., Flouret, G. & Hechter, O. (1978) Opposing effects of estradiol and progesterone on oxytocin receptors in rabbit uterus. *Proc. natn. Acad. Sci. U.S.A.* **75**, 2044–2048.

Pattison, M.L., Chen, C.L. & King, S.L. (1972) Determination of LH and estradiol-17β surge with reference to the time of ovulation in mares. *Biol. Reprod.* **7**, 136, Abstr.

Roberts, J.S. & Share, L. (1969) Effects of progesterone and estrogen on blood levels of oxytocin during vaginal distention. *Endocrinology* **84**, 1076–1081.

Roberts, J.S., McCracken, J.A., Gavagan, J.E. & Soloff, M.S. (1976) Oxytocin-stimulated release of prostaglandin F2α from ovine endometrium *in vitro*: Correlation with estrous cycle and oxytocin-receptor binding. *Endocrinology* **99**, 1107–1114.

Schams, D. (1983) Oxytocin determination by radioimmunoassay. III. Improvement to subpicogram sensitivity and application to blood levels in cyclic cattle. *Acta endocr., Copenh.* **103**, 180–183.

Sheldrick, E.L. & Flint, A.P.F. (1981) Circulating concentrations of oxytocin during the estrous cycle and early pregnancy in sheep. *Prostaglandins* **22**, 631–635.

Sheldrick, E.L., Mitchell, M.D. & Flint, A.P.F. (1980) Delayed luteal regression in ewes immunized against oxytocin. *J. Reprod. Fert.* **59**, 37–42.

Soloff, M.S. (1975) Oxytocin receptors in rat oviduct. *Biochem. Biophys. Res. Commun.* **66**, 671–677.

Soloff, M.S. (1979) Minireview: Regulation of oxytocin action at the receptor level. *Life Sci.* **25**, 1453–1460.

Stull, C.L. & Evans, J.W. (1986) Oxytocin binding in the uterus of the cycling mare. *J. equine Vet. Sci.* **6**, 114–119.

Summerlee, A.J.S., Paisley, A.C., O'Byrne, K.T. & Fletcher, J. (1985) Oxytocin neuron activity and plasma oxytocin during reflex milk ejection in rabbits. *Neurosci. Lett. Suppl.* **22**, 5597.

Wakerley, J.B., Dyball, R.E.J. & Lincoln, D.W. (1973) Milk ejection in the rat: the result of a selective release of oxytocin. *J. Endocr.* **57**, 557–558.

Walters, D., Schams, D. & Schallenberger, E. (1984) Pulsatile secretion of gonadotrophins, ovarian steroids and ovarian oxytocin during the luteal phase of the oestrous cycle in the cow. *J. Reprod. Fert.* **71**, 479–491.

Wathes, D.C. & Swann, R.W. (1982) Is oxytocin an ovarian hormone? *Nature, Lond.* **297**, 225–227.

Wathes, D.C., Swann, R.Q., Birkett, S.D., Porter, D.G. & Pickering, B.T. (1983) Characterization of oxytocin, vasopressin, and neurophysin from the bovine corpus luteum. *Endocrinology* **113**, 693–698.

J. Reprod. Fert., Suppl. **35** (1987), 253–260

Printed in Great Britain
© 1987 Journals of Reproduction & Fertility Ltd

Oxytocin stimulation of plasma 15-keto-13,14-dihydro prostaglandin F-2α during the oestrous cycle and early pregnancy in the mare

A. K. Goff, D. Pontbriand and J. Sirois

Centre de recherche en reproduction animale, Faculté de médicine vétérinaire, Université de Montréal, C.P. 5000, St-Hyacinthe, Québec, Canada J2S 7C6

Summary. In Exp. 1, 4 mares were given oxytocin intravenously (10 i.u./500 kg body wt) daily between Days 9 and 14 (Day 0 = day of ovulation) when pregnant and on Days 9–14, 16, 18, 20 when non-pregnant (not inseminated). In the non-pregnant mares the increase in plasma PGFM response to oxytocin was greater at Day 13 (235 ± 54 pg/ml) than at Day 11 (113 ± 38 pg/ml; $P < 0.05$) and was maximum at Day 16. However, these animals did not return to oestrus and plasma progesterone did not fall below 4 ng/ml. There was no significant increase in response to oxytocin between Days 9 and 14 in the pregnant animals. In Exp. 2, when these same mares were challenged with oxytocin on alternate day (Days 9, 11 and 13 for pregnant mares, Days 9, 11, 13, 15, 17 and 19 for non-pregnant mares) there was a significant difference in the response between non-pregnant and pregnant mares by Day 13 (383 ± 19 pg/ml *vs* 88 ± 9 pg/ml; $P < 0.005$). Plasma progesterone concentrations declined normally and the mares returned to oestrus. During oestrus the response to oxytocin decreased dramatically in mares receiving oxytocin on alternate days, and no response was seen by Day 19. The response also declined after Day 16 in the non-pregnant mares that had daily injections of oxytocin even though plasma progesterone remained elevated. The decreased response coincided with the increase in plasma oestrogen concentrations, suggesting that oestrogens play a role in the control of the response. These results show that the release of PGF-2α in response to oxytocin changes throughout the oestrous cycle, being maximum at the time of luteolysis. This increase in response is suppressed in the pregnant mare and indicates that maternal recognition of pregnancy occurs between Days 11 and 13.

Introduction

As in the other large domestic animals, prostaglandin (PG) F-2α is thought to be responsible for luteolysis in the mare. The mare differs from the cow and sheep in the fact that unilateral hysterectomy ipsilateral to the corpus luteum does not prolong the luteal phase (Ginther & First, 1971), suggesting that PGF-2α acts systemically and not locally in this species.

A considerable amount of work has been done in recent years in sheep (Wathes & Swann, 1982; Flint & Sheldrick, 1983; Sheldrick *et al.*, 1980) and cattle (Fields *et al.*, 1983; Schams, 1983; Schams *et al.*, 1983; Wathes *et al.*, 1983), to clarify the role of oxytocin in luteolysis. It was first shown by Armstrong & Hansel (1959) that exogenous oxytocin is able to shorten the luteal phase of the oestrous cycle in the cow. This luteolytic action of oxytocin is mediated via the release of uterine PGF-2α (Armstrong & Hansel, 1959; Ginther *et al.*, 1967). There is, however, very little information on the possible role of oxytocin in luteolysis in the mare. It was originally thought that oxytocin was not involved since administration of exogenous oxytocin between Days 3 and 8 of the cycle did not result in premature luteolysis (Neely et al., 1979). However, it has now been shown that oxytocin stimulates the release of PGF-2α at around the time of luteolysis (Betteridge et al., 1985). This, together with the fact that the mare is very sensitive to the luteolytic action of PGF-2α, suggests that oxytocin may play a role in luteolysis. The profile of plasma oxytocin during the oestrous cycle appears to be different from that of the cow and sheep and more similar to that of the human, i.e. high during the follicular phase and low during the luteal phase (Burns et al., 1981). Plasma concentrations of oxytocin have not, however, been measured during luteolysis.

The peaks in plasma PGFM concentrations, seen during luteolysis, are not seen in the pregnant mare (Douglas & Ginther, 1976; Kindahl et al., 1982), suggesting that pregnancy is maintained by an inhibition of the release of PGF-2α. The release of PGF-2α in response to oxytocin is inhibited in the cow (Lafrance & Goff, 1985) and the sheep (Fairclough et al., 1984). The objectives of the present study were to determine whether (1) there are changes in the response to oxytocin throughout the oestrous cycle in the mare; and (2) the response to oxytocin is modified in the pregnant animal.

Materials and Methods

Animals

The experimental animals were 6 light horses, weighing 300–450 kg, that had exhibited normal oestrous cycles and were of proven fertility before the experiments. They were part of a larger experimental herd of which the general and reproductive management have been previously described (Betteridge et al., 1985). Mares in the non-pregnant groups were not mated. Pregnancy diagnosis and monitoring of conceptus growth were performed daily between Days 9 and 14 with a real-time linear array ultrasound scanner equipped with a 5 MHz intra-rectal probe (model LS-300, Equisonics Inc., Elk Grove, IL, U.S.A.). On Day 14, the blastocyst was recovered by non-surgical uterine flush (Betteridge et al., 1985). Mares that were inseminated but found to be non-pregnant were not used in the study.

Blood sampling and oxytocin injection procedures

Daily blood samples for progesterone and oestrogen determinations were taken from the jugular vein into heparinized vacutainers (Becton-Dickinson, Rutherford, NJ, U.S.A.). The frequent serial samples for PGFM determination were taken by syringe through an indwelling catheter (Angiocath, Deseret Medical Inc., Sandy, Utah, U.S.A.) in the jugular vein and then transfered into heparinized tubes. The catheter was kept patent by flushing with 2 ml heparinized saline (40 i.u./ml; 10 000 units Hepalean/ml: Organon Canada Ltd, Toronto) after each sampling. For the preliminary experiment, blood samples were collected at intervals of 5 min for 30 min, the oxytocin was injected via the catheter, and then samples were taken every 5 min for 30 min and finally every 15 min for the next 60 min. For the other experiments, blood samples were taken every 5 min for 15 min before and then every 5 min for 15 min after oxytocin injection. All blood samples (7 ml) were immediately kept at 4°C and centrifuged within 45 min, and the plasma was stored at −20°C until used for hormone analysis. The oxytocin injections were of purified pituitary oxytocin in aqueous solution (Austin Laboratories Ltd, Joliette, Quebec, Canada) containing 20 i.u./ml (USP).

Experimental protocol

A preliminary experiment was conducted using 2 mares to examine whether the response to oxytocin changed during the cycle and if this was modified in the same animals when pregnant. Oxytocin (20 i.u./500 kg) was injected on Days 5, 7, 9, 11 and 13 after ovulation in the cyclic and pregnant animals. Oxytocin was also given during oestrus in the cyclic animals.

Experiment 1 was performed in 4 mares to examine in further detail the increase in response to oxytocin with respect to the decline in progesterone concentrations. Oxytocin (10 i.u./500 kg) was injected daily from Day 9 to Day 14 in non-pregnant mares and also in these same mares when pregnant. In the pregnant mares, the embryo was collected on Day 14 and the experiment terminated. In the non-pregnant mares oxytocin was also injected on Days 16, 18 and 20.

Experiment 2 was designed to compare the effect of daily injection of oxytocin to that of injection of oxytocin on alternate days. The same 4 mares as in Exp. 1 were used. Oxytocin (10 i.u./500 kg) was injected on Days 9, 11 and 13 in non-pregnant and pregnant mares, and continued on Days 15, 17 and 19 in the non-pregnant mares.

Radioimmunoassays

Aliquants (300 μl) of plasma were assayed for concentrations of 15-keto-13,14-dihydro PGF-2α (PGFM) by the specific radioimmunoassay and method described by Kindahl et al. (1976, 1982). The sensitivity of the assay was 40 pg/ml plasma. The intra- and inter-assay coefficients of variation were 14·4% and 14·6% respectively.

Aliquants of 50 μl plasma were extracted with petroleum ether and the progesterone content was measured by radioimmunoassay as described previously (Betteridge *et al.*, 1985). Recovery rates were between 80% and 94%. The antibody used was raised in rabbits against 11-hemisuccinate/bovine serum albumin conjugate (Sigma Chemical Co., St Louis, MO, U.S.A.). The cross-reactivity of this antiserum at 50% displacement is: 45% for 5α-pregnan-3,20-dione; 2·2% for 5α-pregnan-3α-ol-20-one; 0·9% for pregnenolone; 0·4% for 17α-hydroxyprogesterone, oestradiol-17β and oestrone; 0·2% for 20α-dihydroprogesterone; and 0·05% for testosterone, androstenedione and hydrocortisone. The sensitivity of the assay was 240 pg/ml plasma. The intra- and inter-assay coefficients of variation were 12·2% and 12·3% respectively.

Total oestrogen concentration was measured by radioimmunoassay in 50 μl unextracted plasma using an antibody (Lot No. RI PA) obtained from Radioimmunoassay Systems Laboratories Inc. (Carson, California, U.S.A.). The antibody was raised against oestrone-3-carboxymethylether–BSA and was used at a final concentration of 1:140 000. It showed, at 50% displacement, a 100% cross-reaction for oestrone sulphate, oestrone and oestradiol-17β, 55% for equilin, 11% for oestradiol-17α 0·44% for oestriol, 0·03% for androstenedione and <0·01% for cortisol, pregnenolone, progesterone and testosterone. The intra- and inter-assay coefficients of variation were 10% and 13% respectively. The sensitivity of the assay was 120 pg/ml.

Statistical analysis

Results from Exps 1 and 2 were analysed by polynomial linear regression analysis (Snedecor & Cochran, 1967) of the data for Days 9 to 14. Comparisons between individual means were performed using Duncan's New Multiple Range Test (Duncan, 1955).

Results

Preliminary experiment

The increase of PGFM in response to oxytocin at different days after ovulation in the pregnant and non-pregnant mares is shown in Fig. 1. The response to oxytocin was low (1·7 ± 0·33-fold

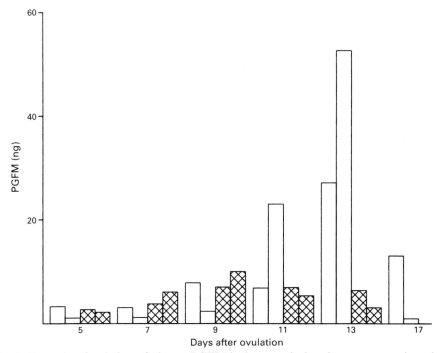

Fig. 1. Oxytocin stimulation of plasma PGFM in 2 mares during the oestrous cycle and early pregnancy. The total PGFM was calculated as the area under the curve for the PGFM concentration in the blood samples taken during the 90 min after oxytocin injection. The open bars represent non-pregnant mares and the cross-hatched bars represent these same two mares when pregnant.

increase in PGFM) on Days 5, 7 and 9 and there was no difference between the pregnant and non-pregnant mares. On Day 11 there was a marked increase in PGFM after oxytocin in one non-pregnant mare (9·2-fold) and by Day 13 both non-pregnant mares exhibited large increases (21- and 11-fold). During oestrus there was a dramatic decline in the response to oxytocin. However, when these mares were pregnant the increase in PGFM at Days 11 and 13 was low, similar to that seen at Days 5, 7 and 9.

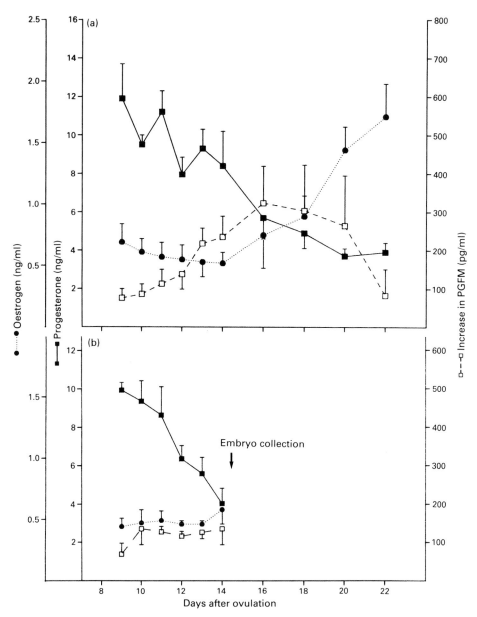

Fig. 2. Effect of oxytocin on plasma hormones in (a) 4 non-pregnant mares and (b) the same 4 mares when pregnant. Oxytocin (10 i.u./500 kg) was injected daily on Days 9–14 (a, b) and then on Days 16, 18, 20, 22 (a) (Exp. 1). The values are mean ± s.e.m. (for PGFM in samples taken 5, 10 and 15 min after injection of oxytocin, basal concentrations < 50 pg/ml).

Experiment 1

The plasma concentrations of progesterone and oestrogen are correlated with the increase in PGFM in response to oxytocin in the non-pregnant mare in Fig. 2. Since the plasma PGFM concentrations peaked 5–10 min after oxytocin injection in the preliminary experiment, the increase in PGFM in Exps 1 and 2 is represented as the mean PGFM concentration at 5, 10 and 15 min after oxytocin. Basal concentrations were <50 pg/ml and did not vary with day or pregnancy.

As in the preliminary experiment the response to oxytocin was low at Day 9, but increased with time and by Day 13 was significantly greater ($P < 0.05$) than that seen at Day 11. The maximum response was reached by Day 16 but it then declined and by Day 22 the response was similar to that

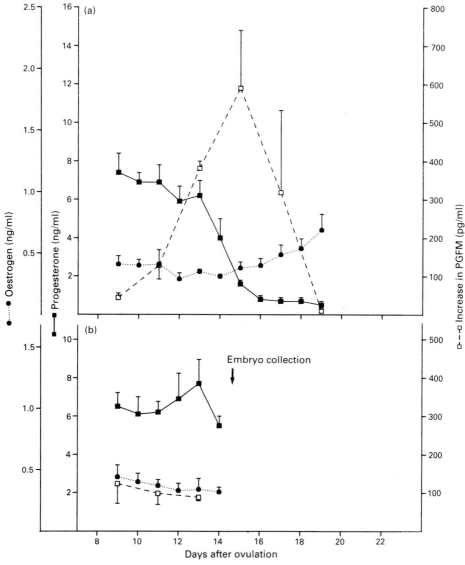

Fig. 3. Effect of administration of oxytocin (10 i.u./500 kg) on plasma hormones in (a) 4 non-pregnant mares and (b) the same 4 mares when pregnant. Injections were given on alternate days (Days 9, 11, 13, 15, 17 and 19 for (a) and Days 9, 11 and 13 for (b)) (Exp. 2). Values are mean ± s.e.m. (for PGFM in samples taken 5, 10 and 15 min after injection of oxytocin, basal concentrations <50 pg/ml).

at Day 9. Plasma progesterone decreased slowly throughout the treatment period, but the concentration did not fall below 4 ng/ml and none of the mares returned to oestrus. Plasma oestrogen concentrations remained stable between Days 9 and 14 and then gradually increased; by Day 18 the level was significantly greater than that at Day 14 ($P < 0.05$). This increase in oestrogen concentration corresponded to the growth of the large antral follicle as detected by rectal palpation and ultrasound scanning. Although the increase in plasma oestrogen occurred during the decline in the response to oxytocin, there was no significant correlation between them.

The results obtained when these same mares were pregnant are shown in Fig. 2. There was a slight ($P > 0.1$) increase in the response to oxytocin between Days 9 and 10, but then no further change. Polynomial linear regression analysis showed that there was a significant ($P < 0.025$) effect of pregnancy on the stimulation of PGFM concentrations by oxytocin. Plasma progesterone concentration declined during the treatment period, with the concentration at Day 13 significantly less ($P < 0.05$) than at Day 9. Although the decrease in plasma progesterone appeared to be greater in the pregnant animals, statistical analysis showed no effect of pregnancy on the decline in progesterone concentration.

Experiment 2

As in Exp. 1, in the non-pregnant mares the response to oxytocin increased with day of the cycle (Fig. 3), by Day 13 the response to oxytocin was greater than that at Day 11 ($P < 0.05$). The maximum response was seen at Day 15 and by Day 19 (3 days after plasma progesterone had dropped below 1 ng/ml and 2 days after the start of oestrus) there was no increase in PGFM in response to oxytocin. In contrast to Exp. 1, all mares returned to oestrus and ovulated at the normal time. There was a significant interaction between experiment and the response to oxytocin, showing that daily administration of oxytocin results in a decreased stimulation of PGFM. The plasma progesterone and oestrogen profiles were similar to those found in normal cyclic mares, i.e. progesterone started to decline by Day 14 and was below 1 ng/ml by Day 16 and plasma oestrogens started to increase after Day 16. When these same animals were pregnant, there was no significant change in the response to oxytocin between Day 9 and 13 (Fig. 3); there was therefore a significant effect of pregnancy on the response to oxytocin ($P < 0.005$).

Discussion

Oxytocin is now thought to play an important role in luteolysis in the cow and the sheep, whereas in the mare little information is available concerning oxytocin during this period. The present results show a similar pattern of response to that reported previously for the cow (Lafrance & Goff, 1985) and the sheep (Fairclough *et al.*, 1984). The response to oxytocin is greatest around the time of luteolysis, suggesting that oxytocin may also be responsible for the pulsatile release of PGF-2α in the mare. However, since plasma profiles of oxytocin during luteolysis have not yet been reported for the mare, no definite conclusions can be made. The increase in response to oxytocin is probably due to the action of progesterone since progesterone priming of an ovariectomized mare for 14 days resulted in induction of responsiveness to oxytocin (Betteridge *et al.*, 1985).

The decline in the response to oxytocin, seen after Day 16, does not appear to be due to the decreased progesterone concentrations. During the initial decline in progesterone there was no decrease in response to oxytocin, and furthermore, in Exp. 1 progesterone concentrations remained at around 4 ng/ml whereas the response to oxytocin continued to decline. Increasing total oestrogen concentration that occurs before ovulation may be involved in turning off the response to oxytocin but further work is necessary to establish this. When progesterone concentrations fell below 1 ng/ml and the animals returned to oestrus (Exp. 2) the decline in the response was more dramatic, indicating a need for decreasing progesterone and increasing oestradiol to inhibit quickly the response to oxytocin as seen in normal circumstances. Thus, a possible cause of pseudopregnancy, which is often seen in the mare, could be a premature rise in plasma oestrogen concentrations which would inhibit the ability of oxytocin to stimulate the release of prostaglandin necessary for complete luteolysis. This would agree with the results of Berg & Ginther (1978) that oestrogen administration prevents luteolysis and causes pseudopregnancy.

The reason why daily injection of oxytocin prevented luteolysis is not clear, although a similar phenomenon has been observed in the sheep (Flint & Sheldrick, 1985) and could be due to refractoriness of the uterus to oxytocin either by down regulation of the oxytocin receptor or by a decrease in the precursors or enzyme activity necessary for prostaglandin synthesis.

The previous observations that the peaks of PGFM normally seen during luteolysis are abolished during early pregnancy suggest that the corpus luteum is maintained by an inhibition of the release of PGF-2α (Douglas & Ginther, 1976; Kindahl *et al.*, 1982). Pregnancy decreases the response to oxytocin in the cow (Lafrance & Goff, 1985) and the sheep (Fairclough *et al.*, 1984). Our present results show a similar pattern in the pregnant mare, in which the release of PGF-2α in response to oxytocin is also significantly reduced. Hershman & Douglas (1979) have determined the time of maternal recognition of pregnancy (i.e. the time at which the embryo maintains progesterone secretion by the corpus luteum) in the mare to be between Days 14 and 16. They arrived at this conclusion by demonstrating that the earliest time that embryos could be removed from the uterus, and result in a prolongation of the luteal phase, was between Days 14 and 16. Although this method has been used successfully in the cow and sheep, in the mare the results are complicated by the fact that uterine flushing itself can prolong the luteal phase (Betteridge *et al.*, 1985). This is in contrast to the situation in sheep (Moor, 1968) and cows (Northey & French, 1980) in which flushing the uterus has no effect on cycle length in non-pregnant animals. The increase in the response to oxytocin between Days 11 and 13 in the non-pregnant mares, and the fact that this increase was abolished in the pregnant mares, suggest that maternal recognition of pregnancy occurs between Days 11 and 13.

The exact mechanism by which the embryo maintains progesterone production, and the nature of the embryonic signal involved, are not well understood. In the pig, the embryonic signal is thought to be oestradiol (Flint *et al.*, 1979). A large amount of oestradiol is released from horse conceptus membranes (Zavy *et al.*, 1979) and the intact conceptus (Goff *et al.*, 1983), and so oestradiol has also been postulated to be the embryonic signal in the mare. However, exogenous oestradiol has been shown to have different effects on cycle length in the mare (Berg & Ginther, 1978; Woodley *et al.*, 1979), possibly because the timing and the dose given may be critical. Our observations that increasing plasma oestrogen concentrations correspond to decreased responsiveness to oxytocin suggest that oestrogens produced by the embryo may maintain the corpus luteum in a similar fashion. The stimulation of PGF-2α release by oxytocin may provide a more quantitative approach by which embryonic secretions can be examined for their ability to maintain the corpus luteum.

We thank Dr H. Kindahl, Swedish University of Agricultural Sciences, for the generous gift of the antibody against 15-keto-13,14-dihydro PGF-2α. This work was supported by grants from NSERC and MRC Canada.

References

Armstrong, D.T. & Hansel, W. (1959) Alteration of the bovine estrous cycle with oxytocin. *J. Dairy Sci.* **42**, 533–542.

Berg, S.L. & Ginther, O.J. (1978) Effect of estrogens on uterine tone and life span of the corpus luteum in mares. *J. Anim. Sci.* **47**, 203–208.

Betteridge, K.J., Renard, A. & Goff, A.K. (1985) Uterine

prostaglandin release relative to embryo collection, transfer procedures and maintenance of the corpus luteum. *Equine vet. J., Suppl.* **3**, 25–33.

Burns, P.J., Kumaresan, P. & Douglas, R.H. (1981) Plasma oxytocin in cyclic mares and sexually aroused stallions. *Theriogenology* **16**, 531–539.

Douglas, R.H. & Ginther, O.J. (1976) Concentration of

prostaglandins F in uterine venous plasma of anesthetized mares during the estrous cycle and early pregnancy. *Prostaglandins* **11**, 251–260.

Duncan, D.B. (1955) Multiple range and multiple F tests. *Biometrics* **11**, 1–9.

Fairclough, R.T., Moore, L.G., Peterson, A.J. & Watkins, W.B. (1984) Effect of oxytocin on plasma concentrations of 13,14-dihydroprostaglandin F and the oxytocin-associated neurophysin during the estrous cycle and early pregnancy in the ewe. *Biol. Reprod.* **31**, 36–43.

Fields, P.A., Elridge, R.K., Fuchs, A.-R., Roberts, R.F. & Fields, M.J. (1983) Human placental and bovine corpora luteal oxytocin. *Endocrinology* **113**, 1544–1546.

Flint, A.P.F., Burton, R.D., Gadsby, J.E., Saunders, P.T.K. & Heap, R.B. (1979) Blastocyst oestrogen synthesis and the maternal recognition of pregnancy. In *Maternal Recognition of Pregnancy* (Ciba Fdn Symp. N.S. 64), pp. 209–228. Ed. J. Whelan. Excerpta Medica, Amsterdam.

Flint, A.P.F. & Sheldrick, E.L. (1983) Evidence for a systemic role for ovarian oxytocin in luteal regression in sheep. *J. Reprod. Fert.* **67**, 215–225.

Flint, A.P.F. & Sheldrick, E.L. (1985) Continuous infusion of oxytocin prevents induction of uterine oxytocin receptor and blocks luteal regression in cyclic ewes. *J. Reprod. Fert.* **75**, 623–631.

Ginther, O.J., Woody, C.O., Mahajan, S., Janakiraman, K. & Casida, L.E. (1967) Effect of oxytocin administration on the oestrous cycle of unilaterally hysterectomized heifers. *J. Reprod. Fert.* **14**, 225–229.

Ginther, O.J. & First, N.L. (1971) Maintenance of the corpus luteum in hysterectomized mares. *Am. J. vet. Res.* **32**, 1687–1691.

Goff, A.K., Renard, A. & Betteridge, K.J. (1983) Regulation of steroid synthesis in pre-attachment equine embryos. *Biol. Reprod.* **28**, Suppl. 1, 137, Abstr.

Hershman, L. & Douglas, R.H. (1979) The critical period for the maternal recognition of pregnancy in pony mares. *J. Reprod. Fert., Suppl.* **27**, 395–401.

Kindahl, H., Edqvist, L.-E., Granstrom, E. & Bane, A. (1976) The release of prostaglandin $F_{2\alpha}$ as reflected by 15-keto-13,14-dihydro-prostaglandin $F_{2\alpha}$ in the peripheral circulation during normal luteolysis in heifers. *Prostaglandins* **11**, 871–878.

Kindahl, H., Knudsen, O., Madej, A. & Edqvist, L.-E. (1982) Progesterone, prostaglandin $F_{2\alpha}$, PMSG and oestrone sulphate during early pregnancy in the mare. *J. Reprod. Fert., Suppl.* **32**, 353–359.

Lafrance, M. & Goff, A.K. (1985) Effect of pregnancy on oxytocin-induced release of prostaglandin $F_{2\alpha}$ in heifers. *Biol. Reprod.* **33**, 1113–1119.

Moor, R.M. (1968) The effect of the embryo on corpus luteum function. *J. Anim. Sci.* **27**, 97–118.

Neely, D.P., Stabenfeldt, G.H. & Sauter, C.L. (1979) The effect of exogenous oxytocin on luteal function in mares. *J. Reprod. Fert.* **55**, 303–308.

Northey, D.L. & French, L.R. (1980) Effect of embryo removal and intrauterine infusion of embryonic homogenates on the lifespan of the bovine corpus luteum. *J. Anim. Sci.* **50**, 298–302.

Schams, D. (1983) Oxytocin determination by radioimmunoassay III. Improvement to subpicogram sensitivity and application to blood levels in cyclic cattle. *Acta endocr., Copenh.* **103**, 180–183.

Schams, D., Walters, D.L., Schallenberger, E., Bullermann, B. & Karg, H. (1983) Ovarian oxytocin in the cow. *Acta endocr., Copenh.* **102**, Suppl. 253, 147–148.

Sheldrick, E.L., Mitchell, M.D. & Flint, A.P.F. (1980) Delayed luteal regression in ewes immunized against oxytocin. *J. Reprod. Fert.* **59**, 37–42.

Snedecor, G.W. & Cochran, W.G. (1967) Factorial experiments. In *Statistical Methods*, pp. 349–354. Iowa State University Press, Ames.

Stabenfeldt, G.H., Hughes, J.P., Evans, J.W. & Neely, D.P. (1974) Spontaneous prolongation of luteal activity in the mare. *Equine vet. J.* **6**, 158–163.

Wathes, D.C. & Swann, R.W. (1982) Is oxytocin an ovarian hormone? *Nature, Lond.* **297**, 225–227.

Wathes, D.C., Swann, R.W., Birkett, S.D., Porter, D.G. & Pickering, B.T. (1983) Characterization of oxytocin, vasopressin and neurophysin from the bovine corpus luteum. *Endocrinology* **113**, 693–698.

Woodley, S.L., Burns, P.J., Douglas, R.H. & Oxender, W.D. (1979) Prolonged interovulatory interval after oestradiol treatment in mares. *J. Reprod. Fert., Suppl.* **27**, 205–209.

Zavy, M.T., Mayer, R., Vernon, M.N., Bazer, F.W. & Sharp, D.C. (1979) An investigation of the uterine luminal environment of non-pregnant and pregnant pony mares. *J. Reprod. Fert., Suppl.* **27**, 403–411.

J. Reprod. Fert., Suppl. **35** (1987), 261–267

Release of LH, FSH and GnRH into pituitary venous blood in mares treated with a PGF analogue, luprostiol, during the transition period

W. Jöchle*, C. H. G. Irvine†, S. L. Alexander† and T. J. Newby‡

10 Old Bontoon Road, Denville, NJ 07834, U.S.A.; †Lincoln College, Canterbury, New Zealand; and ‡Norden Laboratories, Lincoln NE 68501, U.S.A.

Summary. Nine mares received cannulae to collect blood from the pituitary venous outflow in the intercavernous sinus (ICS) and the jugular vein; in 4 mares, only jugular cannulae were used. Those 4 mares and 3 of the mares with cannulae in both positions received 7·5 mg luprostiol i.m. and 1 mare with both cannulae was treated with 3·75 mg uprostiol i.v. Blood samples were kept before and after treatment at 2-, 5- or 10-min intervals and concentrations of LH, FSH and GnRH were determined by RIA. Treatments resulted in an immediate sharp rise of LH and FSH in ICS and jugular blood samples within 2–10 min, with ICS concentrations rising earlier, and with peak levels of LH 8 to 100 times higher, respectively.

In ICS samples, GnRH was elevated consistently only after LH and FSH had reached peak levels. At both locations, LH and FSH concentrations remained elevated 60–120 min after treatment, but had returned to baseline by 240 min. In 5 untreated mares with cannulae at both locations, sampling at 5-min intervals for 12 or 24 h revealed no pulses of LH or FSH in 3 mares, and only one pulse a day, preceded by several small rises of GnRH during the hour before the pulses, in 2 mares.

Introduction

The luteolytic effects of prostaglandin (PG) F-2α and its analogues in horses, cattle, sheep and pigs have been extensively documented and are widely used for therapy and biotechnology. In the horse, PGF-2α and its analogues are firmly established luteolysins for the shortening of individual oestrous cycles, the treatment of persistently dioestrous mares, the routine induction of a 'second' foal heat, and the interruption of early gestations.

Clinical observations, combined with plasma progesterone determinations, indicated that even in mares with baseline plasma progesterone concentrations treatment with a PG analogue may induce oestrus and/or ovulations without the preceding luteolysis (Lamond *et al.*, 1975). These observations were verified in clinical studies in Brazil (Reiner & Jöchle, 1981) and Australia (Howey *et al.*, 1983) in 55 mares with baseline plasma progesterone values which had failed to begin cycling during the first half of the season and were treated once or twice with alfaprostol and 47·3 and 60% respectively, responded with oestrus and/or fertile ovulations. In Germany, luprostiol was given to 18 transitional-phase mares with multiple small follicles and 13 (72·2%) responded with oestrus and ovulations (Güngerich, 1980).

In the absence of elevated plasma progesterone concentrations, PGF-2α and its analogues cause a sustained release of LH (Hafs *et al.*, 1975) in cows and heifers and/or testosterone in bulls (Haynes *et al.*, 1977, 1978; Müller, 1980) in peripheral blood, similar to the response achieved with GnRH injections. Similar responses of jugular plasma LH and FSH to a luteolytic dose of alfaprostol occur in transition-phase mares (unpublished data). With the development of a technique to study directly in the venous outflow of the pituitary the release of pituitary hormones in response to hypothalamic releasing factors and any other agents (Irvine & Huun, 1984), this PGF-related phenomenon can be studied in detail. The purpose of this study was to determine effects of a PGF analogue, luprostiol, on the release of GnRH, LH and FSH in response to i.m. or i.v. administration in mares during the transition period, and on the onset and duration of hormone release, and on the sequence and magnitude of response.

Materials and Methods

Mares and treatment. The 8 mares selected were aged 5–14 years and weighed 500–600 kg. All were in late anoestrus or the transitional phase as assessed by an irregular pattern of response to 3 times weekly teasing, a positive behavioural response to 2·5 mg oestradiol benzoate i.m. and the presence of some follicular activity without development to the preovulatory stage. In 4 mares, a 100-cm Swan-Ganz 7 French cannula was inserted into one external maxillary vein and manipulated along venous pathways and into the intercavernous sinus until its tip was in apposition to the outlet of major pituitary veins, as described by Irvine & Huun (1984). The cannula was kept patent with heparinized saline while the mares were allowed to roam freely on pasture in a large enclosure. Jugular cannulae (Angiocath 16 gauge, 13 cm: Deseret Co., Sandy, UT, U.S.A.) were placed in all 8 mares. They were given no other treatment. After 24 h, 3 mares with intercavernous sinus cannulae and 4 mares with only jugular cannulae were given 7·5 mg luprostiol (Norden Laboratories, Lincoln, NE, U.S.A.) i.m. and one mare with an intercavernous sinus cannula was given 3·75 mg luprostiol by slow i.v. injection. Blood samples were collected from both cannulae before and after luprostiol administration at the intervals shown in 'Results'. Samples were placed in chilled heparinized tubes, immediately centrifuged in a refrigerated centrifuge and stored at −20°C until assayed.

In addition, blood samples were taken at 5-min intervals for 12 or 24 h from 5 untreated mares in the transition phase, which were fitted with similar cannulae.

Assays. Luteinizing hormone was measured by radioimmunoassay (RIA) using an antiserum to ovine LH (GDN15), ovine LH for iodination (LER 1374A) and horse LH standards (E98A; gift from Dr H. Papkoff, University of California, San Francisco, CA, U.S.A.) as described previously (Irvine *et al.*, 1985). The minimum detectable dose (i.e. zero plus 2 × s.d.) was 13 pg/tube. The mean within-assay coefficients of variation at low, moderate and high LH concentrations were 7·5%, 3·3% and 5·6%, respectively. The mean between-assay coefficient of variation, determined from repeated measurements ($n = 25$) of a plasma sample containing moderate LH concentrations, was 11·2%.

GnRH was measured by RIA using an antibody raised and characterized by Dr A. Caraty, INRA, Nouzilly, France (Caraty *et al.*, 1980). The synthetic decapeptide (Hoechst Op L 056; gift from Hoechst AG, Frankfurt, FRG) was used as standard and for iodination by the lactoperoxidase method followed by purification on QAE-Sephadex to obtain mono-iodo-GnRH (Nett & Adams, 1977). At a final dilution of 1:272 000 the antibody specifically bound approximately 40% of the 4000 c.p.m. of ^{125}I-labelled GnRH added. The minimum detectable dose in this assay was 0·28 pg/tube. Before assay, plasma samples were extracted in 5 volumes of nanograde methanol. Mean ± s.d. recovery after extraction and assay of GnRH added to horse jugular plasma was 72·6 ± 7% ($n = 8$), which was similar to the recovery after extraction only of a tracer amount of ^{125}I-labelled GnRH (77·4 ± 1·5, $n = 11$). The slope of the regression line fitted to logit–log transformed displacement curves was similar whether standards had been diluted in buffer, or diluted and extracted from horse plasma (mean of 11 slopes ± s.d., serum: −2·25 ± 0·21; buffer: −2·27 ± 0·22, not significantly different, paired *t* test). The mean (±s.d.) GnRH concentration measured in 100 μl methanol-extracted horse jugular blood was 0·30 ± 0·12 pg ($n = 68$ samples) which was very close to the detection limit of the assay. The mean within-assay coefficient of variation in the low concentration range in which all plasma samples fell was 10·0%. Mean between-assay coefficient of variation determined from repeated assay of jugular blood spiked to a moderate GnRH concentration was 11·0%.

FSH was measured by radioimmunoassay using an antibody against human FSH (Batch 5, NIADDK), human FSH for iodination (NIADDK) and horse FSH standards (gift from Dr L. Nuti, University of Wisconsin, Madison, WI, U.S.A.) as described previously (Evans & Irvine, 1976). At a final dilution of 1:135 000 the antibody specifically bound 18·4% of added FSH (30 000 c.p.m./tube, sp. act. ~100 μCi/μg). Non-specific binding was 1·8% of total radioactivity. The minimum detectable concentration (i.e. zero plus 2 s.d.) was 59 pg/tube. The mean within-assay coefficient of variation was 8·7%. Mean between-assay coefficient of variation was 7·8%.

Results

In the mares with intercavernous sinus cannulation, LH peaks in intercavernous sinus plasma were 8, 95 and 104 times the concurrent levels in jugular plasma, and were always followed by an increase in LH in jugular plasma. Similar patterns were observed with FSH, but with intercavernous sinus peak levels only 3–5 times more than jugular levels, respectively (Figs 1 & 2). This indicates that the cannula was in close apposition to the pituitary venous outflow. This was

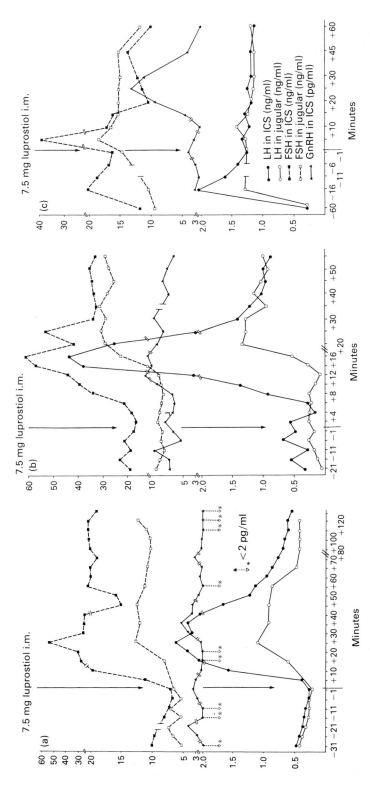

Fig. 1. Effects of treatment with 7·5 mg luprostiol, i.m., on LH, FSH and GnRH concentrations in pituitary venous blood (ICS) and in the jugular vein of 3 mares. In (c) the mare had an endogenous surge of gonadotrophins between −16 and −60 min; the values at −60 min are normal for this time of the year.

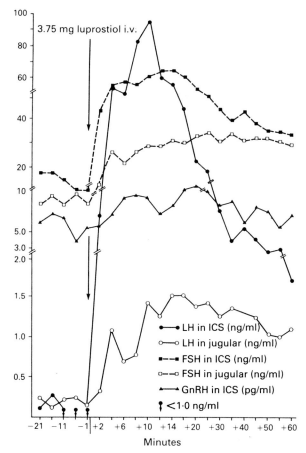

Fig. 2. Effects of treatment with 3·75 mg luprostiol, i.v., on LH, FSH and GnRH concentrations in pituitary venous blood (ICS) and in the jugular vein of a mare.

supported by the finding of measurable GnRH concentrations in intercavernous sinus blood, but not in jugular blood.

In 7 out of the 8 mares given luprostiol there was an increase in gonadotrophin concentrations in jugular blood by 5 or 10 min after injection, or by 15 min when neither 5- nor 10-min samples were collected (Figs 1–3). In these mares, rises of LH and FSH occurred even earlier in intercavernous sinus blood, after 2–10 min, with substantially higher peak levels (Figs 1 & 2). In one mare (Fig. 1c) there was an episode of LH and FSH release into pituitary blood peaking between 16 and 60 min before luprostiol injection. This mare showed only a brief (5–10 min) increase in gonadotrophins after luprostiol, and similar concentrations in intercavernous sinus and jugular blood.

In all other mares, gonadotrophins in intercavernous sinus blood peaked between 8 and 20 min, and in jugular blood always slightly later, after 10 to 25 min (Figs 1–3). Gonadotrophins showed little decline by the end of the sampling period at 60 or 120 min. At 50 min after treatment pituitary venous blood values for gonadotrophins remained higher than those in jugular blood, in 2 of 4 mares, indicating that pituitary secretion was continued for that period.

There was no difference in the gonadotrophin profile in jugular blood between mares with and without intercavernous sinus cannulae (Figs 1–3), indicating that cannulation did not affect the drainage of pituitary blood.

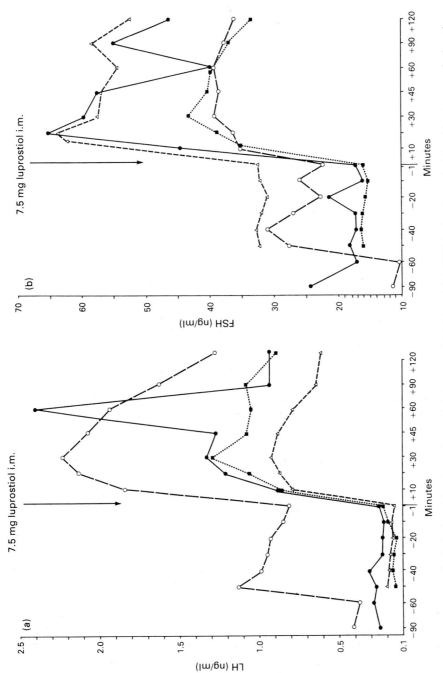

Fig. 3. Effects of treatment with 7·5 mg luprostiol, i.m., on (a) LH and (b) FSH in jugular vein blood in 4 mares during the transition phase.

After i.m. or i.v. injection of luprostiol a peak of GnRH occurred in pituitary blood, with increases of 2–4 times over pretreatment levels (Figs 1 & 2). The onset of the increase and the peak followed those in gonadotrophins. Those peaks were clearly seen also in the mare (Fig. 1c) which had obviously experienced a release within the hour before treatment.

In the 5 mares with intercavernous sinus and jugular cannulae which remained as untreated controls, samplings at 5-min intervals did not reveal any releases of LH, FSH or GnRH in 2 mares sampled for 24 h and in 1 mare sampled for 12 h. In the other 2 mares, one pulse of LH and FSH in 24 h was recorded which was preceded by several small peaks of GnRH at 10–20-min intervals, which occurred within 1-h before the gonadotrophins began to rise.

Although the results are not shown, all elevations of gonadotrophins in response to luprostiol had returned to baseline levels by 4 h after treatment.

Discussion

Luprostiol treatment resulted in GnRH release in all mares with intercavernous sinus cannulation. This was preceded in all mares which were in the transition phase by marked increases of LH and of FSH in pituitary venous blood as well as in jugular blood. An exception was one mare in which a pulse of both gonadotrophins had been recorded within an hour before treatment.

The release of LH and FSH seen with luprostiol surpassed in duration, but not in amplitude, the pulsatile releases seen at about 10–12-h intervals in pituitary and peripheral blood of cyclic mares. Contrary to the sequence of events in these pulsatile releases when LH and FSH are preceded by GnRH by about 5 min in pituitary blood (C. H. J. Irvine & S. L. Alexander, unpublished), luprostiol caused an immediate release within 2–3 min after i.v., and within 10 min after i.m., administration. This was followed by GnRH release, which occurred 20–25 min after treatment; the duration in minutes and the amplitude of its pulse-like appearance were different from those occurring during spontaneous pulses: the duration was longer and the amplitude only about 40% of that seen during oestrus. The pattern of LH concentrations in plasma after luprostiol treatment in mares resembles LH and/or testosterone increases seen in peripheral blood of bulls treated with PGF-2α (Haynes *et al.*, 1977, 1978) or with alfaprostol or cloprostenol (Müller, 1980). What may have caused this immediate LH and FSH release remains to be clarified. The delayed, but consistent, release of GnRH might be responsible for the sustained production and/or release of LH and FSH observed in our studies.

Attempts to gain insight into hormonal release patterns during the transition period in 5 untreated mares fitted with intercavernous sinus cannulae yielded inconclusive results. Pulses of GnRH and of LH and FSH release were very infrequent and did not occur at all in 2 mares during 24 h and 1 mare during 12 h of frequent sampling. In 2 mares, one surge of LH and FSH in 24 h was seen in each of them, preceded by several small GnRH pulses. Bursts of GnRH, followed by sharp increases of gonadotrophins as seen in cyclic mares (C. H. G. Irvine & S. L. Alexander, unpublished) were missing. The more striking was the rather uniform response to luprostiol and the reversal in the pattern of occurrence of hormone releases, with GnRH following and not preceding LH and FSH.

The response to luprostiol treatment may explain the clinical observations (see 'Introduction') when mares with very low progesterone concentrations responded to treatments with PGF analogues with oestrus and/or ovulations. It also helps to understand results from a 4-year field study (U. R. Reiner & W. Jöchle, unpublished) of 192 mares which did not start to cycle during the first 10 weeks of the season and were treated once or twice with PGF analogues alfaprostol (N = 177), cloprostenol (N = 9) or dinoprost (N = 6): while only 35·3% responded with oestrus, 67·7% ovulated within 10 days. In 68 mares not showing oestrus but follicular growth and ovulations, as determined by once or twice daily rectal palpation, timely forced mating resulted in a pregnancy rate of 55·9%. Similar attempts by Peterson (1985) in Minnesota and Martin *et al.* (1981) in

Northern Germany, using alfaprostol for the treatment of transition-phase mares or mares in anoestrus during the early part of the season, were unsuccessful.

It has been reported that treatment of mares on Day 2 or 3 of oestrus with a PGF analogue shortens the time to ovulation (Arbeiter & Arbeiter, 1983, 1985). This effect has not been confirmed in a population of problem mares (Hackmann, 1982), but enquiries with practitioners in North America indicate that this practice is widespread. All these observations indicate that the LH (and FSH) releasing property of luprostiol, and of other PGF analogues, deserves greater attention and requires clinical and endocrinological evaluation of limiting and permissive factors.

References

Arbeiter, K. & Arbeiter, E. (1983) Anwendung von Prostaglandin $F_{2\alpha}$-Analoga bei der Stute: I. Alfaprostol zur Ovulationsausloesung waehrend der Rosse und II. Tiaprost zur Brunstinduktion nach uebergangener Fohlenrosse. *Dtsch tieraerztl. Wschr.* **90**, 386–388.

Arbeiter, K. & Arbeiter, E. (1985) Alfaprostol zur Ovulationsausloesung waehrend der Rosse (2. Mitteilung). *Dtsch tieraerztl. Wschr.* **92**, 87–88.

Caraty, A., de Reviers, M.-M., Pelletiers, J. & Dubois, P.M. (1980) Reassessment of LRF radioimmunoassay in the plasma and hypothalamic extracts of rats and rams. *Reprod., Nutr. Develop.* **20**, 1489–1531.

Evans, M.J. & Irvine, C.H.G. (1976) Measurement of equine FSH and LH: response of anestrous mares to gonadotropin releasing hormone. *Biol. Reprod.* **15**, 477–484.

Güngerich, B. (1980) *Untersuchungen zur Anwendung eines Prostaglandin F_2 alpha-analogs beim Pferd unter Praxisbedingungen bei gleichzeitiger Erweiterung des klinischen Indikationsbereichs.* Vet. Med. Dissertation, Tieraerztl. Hochschule Hannover, FRG.

Hackmann, F. (1982) *Zur Oestrus- und Ovulationsterminierung beim Pferd. Einfluss eines Prostaglandin F_2 alpha Analogs (K 11941) sowie der Palpation of klinisch bedeutsame Reproduktionsparameter bei der Stute.* Vet. Med. Dissertation. Tieraerztl. Hochschule Hannover, FRG.

Hafs, H.D., Louis, T.M., Stellflug, J.N., Convey, E.M. & Britt, J.H. (1975) Blood LH after $PGF_{2\alpha}$ in diestrous and ovariectomized cattle. *Prostaglandins* **10**, 1001–1009.

Haynes, N.B., Kiser, T.E., Hafs, H.D. & Marks, J.D. (1977) Prostaglandin $F_{2\alpha}$ overcomes blockade of episodic LH secretion with testosterone, melengestrol acetate or aspirin in bulls. *Biol. Reprod.* **17**, 723–728.

Haynes, N.B., Collier, R.J., Kiser, T.E. & Hafs, H.D. (1978) Effects of prostaglandin E_2 and $F_{2\alpha}$ on serum luteinizing hormone, testosterone and prolactin in bulls. *J. Anim. Sci.* **47**, 923–926.

Howey, W.P., Jöchle, W. & Barnes, W.J. (1983) Evaluation of clinical and luteolytic effects of a novel prostaglandin analogue in normal and problem mares. *Aust. vet. J.* **60**, 180–183.

Irvine, C.H.G. & Huun, R. (1984) Long-term collection of pituitary venous blood in the unrestricted horse. *N.Z. Med. J.* **95**, 735.

Irvine, C.H.G., Alexander, S.L. & Hughes, J.P. (1985) Sexual behavior and serum concentrations of reproductive hormones in normal stallions. *Theriogenology* **23**, 607–617.

Lamond, D.R., Buell, J.R. & Stevenson, W.S. (1975) Efficacy of a prostaglandin analogue in reproduction in the anoestrous mare. *Theriogenology* **3**, 77–85.

Martin, J.C., Klug, E., Merkt, H., Himmler, V. & Jöchle, W. (1981) Luteolysis and cycle synchronization with a new prostaglandin analog for artificial insemination in the mare. *Theriogenology* **16**, 433–446.

Müller, U.M. (1980) *Die Testosteronsekretion des zuchtreifen Stieres unter verschiedenen Bedingungen.* Vet. Med. Dissertation, Universitaet Zürich, Switzerland.

Nett, T.M. & Adams, T.E. (1977) Further studies on the radioimmunoassay of gonadotropin releasing hormone: effect of radioiodination, antiserum and unextracted serum on levels of immunoreactivity in serum. *Endocrinology* **101**, 1135–1144.

Peterson, D.E. (1985) Gonadotrophin releasing effects of alfaprostol in seasonally anoestrous/oestrous mares. *Equine vet. Sci.* **5**, 331–334.

Reiner, U.R. & Jöchle, W. (1981) Effectiveness of a novel prostaglandin analog in postpartum mares and mares with early season anoestrum. *Zuchthygiene* **16**, 110–115.

J. Reprod. Fert., Suppl. 35 (1987), 269–276

Printed in Great Britain
© 1987 Journals of Reproduction & Fertility Ltd

Plasma prolactin concentrations in non-pregnant mares at different times of the year and in relation to events in the cycle

K. Worthy, K. Colquhoun*, R. Escreet*, M. Dunlop*, J. P. Renton*
and T. A. Douglas

*Departments of Veterinary Clinical Biochemistry and *Reproduction and Surgery, University of Glasgow Veterinary School, Garscube Estate, Bearsden, Glasgow G61 1QH, U.K.*

Summary. Plasma prolactin concentrations were measured in mares using an homologous radioimmunoassay. An annual rhythm in plasma prolactin was found, with concentrations higher during the summer than during the winter. In addition to this seasonal pattern, occasional high concentrations of prolactin were seen when concentrations were otherwise basal. Blood samples taken from mares during an oestrous cycle in October–November showed that prolactin values were basal for most of the cycle, with a marked rise in prolactin shortly before the onset of oestrus. This prolactin peak was associated with an increase in the size of the largest follicle, and with a peak of PGFM in some mares, but did not appear to be related to the LH surge. In oestrous cycles in March and May–June, there was a wide variation in the baseline of prolactin secretion, in accordance with the seasonal pattern already mentioned. However, the peak of prolactin seen around oestrus in October–November was less obvious in March and May–June. Post-partum mares showed a high but irregular profile of prolactin concentrations with no clear-cut pattern in relation to the oestrous cycle.

Introduction

Reports of plasma prolactin concentrations in the non-pregnant mare are very scarce, and the importance of this hormone to the oestrous cycle and in relation to seasonal breeding in mares is unknown. Johnson (1986) reported a marked rise in plasma prolactin concentrations in Standardbred mares during summer, and low prolactin concentrations in winter. Similar patterns have been found in other seasonally breeding mammals (Karg & Schams, 1974; Webster & Haresign, 1983), and appear to be largely dictated by changes in the photoperiod (Walton *et al.*, 1980) or ambient temperature (Peters & Tucker, 1978; Hill & Alliston, 1981). The functional significance of this annual rhythm in prolactin concentrations has not yet been established for any species.

In contrast to this seasonal pattern, Johnson (1986) reported that there was no change in plasma prolactin concentrations in mares during an oestrous cycle. This is in marked contrast to the large surge in prolactin concentrations found at or around the time of oestrus in the ewe, cow, sow and rat (Kann & Denamur, 1974; Karg & Schams, 1974; Dusza & Krzymowska, 1979).

The data presented here are the results of a preliminary investigation into the patterns of plasma prolactin in the oestrous cycle of Pony mares. These patterns were related to important events in the oestrous cycle, such as luteolysis, follicular development, and oestrous behaviour, as well as to changes in plasma concentrations of progesterone, PGFM and LH. Oestrous cycles during different months of the year and after foaling were also studied. In addition, seasonal changes in plasma prolactin concentrations were measured.

Materials and Methods

Animals

Six Pony mares of various ages and proven fertility were used. Some of the mares were studied at several different times of the year, allowing comparison of blood hormone concentrations and clinical changes both within and between animals. All blood samples were taken in the morning by jugular venepuncture.

Two mares were blood-sampled twice weekly throughout the year, to reveal any seasonal changes in plasma prolactin concentrations. During this period they experienced urban lighting.

Several mares were studied during oestrous cycles in March (spring), May–June (summer) and October–November (winter) in 1985–1986. Blood samples were taken daily from Day 1 to Day 14 of the cycle, and more frequently from Day 14 until the beginning of oestrus. During the cycle studied in October–November, a uterine biopsy was performed on Mares 3 and 4 to compare an induced oestrus with a natural oestrus in Mare 5. Daily samples were also available from 4 of the mares for the first few weeks *post partum* (1984/85), during which time all of these mares were suckling a foal.

Regular assessments of reproductive status by means of rectal palpation of the tract were made in all mares. Daily assessments of oestrous behaviour were also made throughout the year by teasing with a stallion, except during the summer when for practical reasons this was not possible. Plasma progesterone concentrations were determined in all cases to confirm cyclic activity.

Mares were housed from October to May and fed on a ration of concentrates with ad-libitum access to hay and water. They were kept at grass for the summer months.

Hormone determinations

Prolactin. Plasma prolactin concentrations were measured using an homologous double-antibody radioimmunoassay with a highly purified horse prolactin as tracer and standard, and a specific antiserum raised against this standard. The assay validation and methodology have been detailed previously (Worthy *et al.*, 1986). Within this study the inter-assay coefficient of variation, calculated from plasma pools included in all assays, was 19%; the intra-assay CV for randomly selected duplicate pairs was 6%. The limit of sensitivity of the assay, defined as the value at twice the standard deviation of blank values, was 1·5 ng/ml.

Luteinizing hormone. Plasma LH concentrations were estimated by means of an homologous double-antibody radioimmunoassay using a purified preparation of horse LH (E 98A) which has previously been characterized (Licht *et al.*, 1979). A specific antiserum against horse LH was used (UCB Bioproducts, Brussels, Belgium) which had previously shown cross-reactions of 23% for PMSG and <1% for horse FSH. The assay methodology was as described by Farmer & Papkoff (1979) with the following modifications. The hormone was iodinated by means of a solid-phase lactoperoxidase system (Habener *et al.*, 1979), giving a stable tracer with 25–30% binding at an initial dilution of the primary antiserum of 1:20 000. The plasma samples (100 µl) were initially made up to 300 µl with assay buffer; 100 µl antiserum were added and the assay incubated for 24 h at 4°C before addition of 100 µl tracer (10 000 c.p.m.). The bound fraction of the ^{125}I-labelled LH was precipitated by means of a polyethylene glycol (PEG)-assisted double-antibody system. For this, PEG (final concentration 2·5%) was added and the assay then incubated at room temperature for 2 h before centrifugation at 2000 *g* and separation of the supernatant.

An inhibition curve for a pooled mare plasma at three dilutions was parallel to the standard curve. Known amounts of the horse LH standard were added to a mare plasma pool and assayed. After adjustment for the original concentration of LH in the pool, the recovery figure was 106 ± 9%. The limit of sensitivity of the assay was 1·3 ng/ml, and the inter- and intra-assay coefficients of variation were 16% and 11% respectively.

Progesterone. Plasma concentrations of progesterone were estimated by the radioimmunoassay method of Munro *et al.* (1979). The inter- and intra-assay CVs were 20% and 11%, and the limit of sensitivity was 0·3 ng/ml.

PGFM. Plasma concentrations of PGFM (13,14-dihydro 15-keto prostaglandin F-2α) were determined by means of an assay based on the method of Dobson (1983), with the following modifications. All plasma samples were heat-treated at 40°C for 30 min, rather than extracted. This technique has been validated by Stewart *et al.* (1984) for measurement of prostaglandin metabolites, and did not affect the standards in this study. The antiserum used had previously been shown to have cross-reactions of 8% with 15-keto PGF-2α, and <1% with 13,14-dihydro-PGF-2α, PGF-1, PGE-1, and PGE-2 (Mitchell *et al.*, 1976). The bound and free fractions were separated by charcoal (0·5 g in 100 ml Tris–HCl buffer). The limit of sensitivity of the assay was 20 pg/ml, and the inter- and intra-assay CVs were 17% and 20% respectively.

Results

Seasonal changes in prolactin concentrations

Weekly mean prolactin concentrations in Mares 2 and 6 throughout the year are depicted in

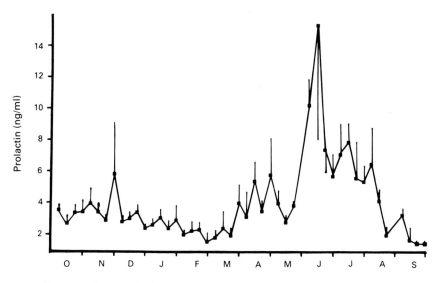

Fig. 1. Weekly mean (±s.e.m.) plasma prolactin concentrations in Mares 2 and 6 throughout the year.

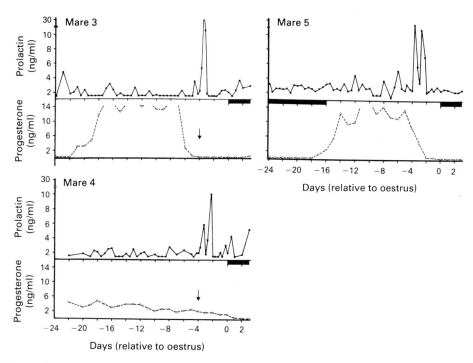

Fig. 2. Plasma concentrations of prolactin and progesterone in Mares 3, 4 and 5 throughout an oestrous cycle during October–November. The bars indicate the days when oestrus was shown; the arrows indicate the time of uterine biopsy in Mares 3 and 4.

Fig. 1. Prolactin concentrations were generally low (1–3 ng/ml) during the winter, and rose in late spring to a peak (8–15 ng/ml) in summer. However, some peaks of prolactin were seen during the winter, when levels were otherwise basal.

Prolactin concentrations during the oestrous cycle

As in previous years at this stud (Colquhoun, 1984), most of the mares continued to cycle throughout the winter months.

October–November. Blood-sampling was carried out 1–3 times daily during a complete cycle in 3 mares. As shown in Fig. 2, plasma prolactin concentrations were basal for most of the cycle. Shortly before oestrus a marked rise in prolactin concentrations, lasting for 1–2 days, was seen in all the mares. Prolactin concentrations then returned to basal. This peak of prolactin occurred when progesterone concentrations were falling. It appears that in Mare 3 progesterone concentrations had already begun to fall before the uterine biopsy was performed, but in Mare 4 the biopsy induced a return to oestrus from a state of prolonged dioestrus.

Figure 3 illustrates the concurrent changes in other hormones for Mares 3 and 5. The peak of prolactin was accompanied by a large pulse of plasma PGFM in Mare 5, in which a natural oestrus

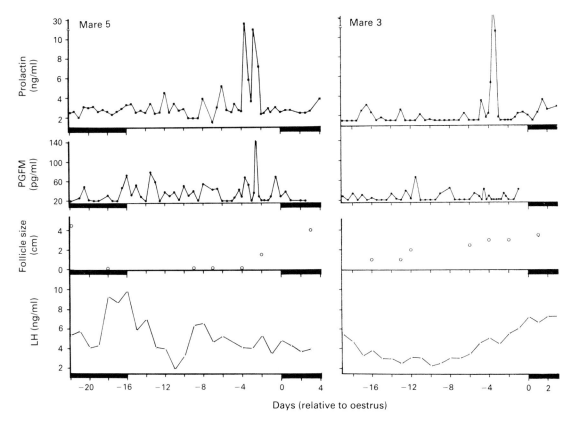

Fig. 3. Plasma concentrations of prolactin, PGFM and LH in Mares 3 and 5 during October–November, together with the diameter of the largest follicle palpable per rectum. The bars indicate the days when oestrus was shown.

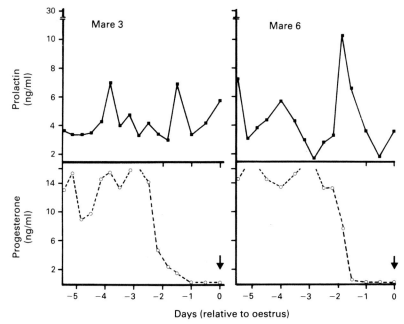

Fig. 4. Plasma concentrations of prolactin and progesterone in Mares 3 and 6 during luteolysis in March. The arrows indicate the first positive signs of oestrus.

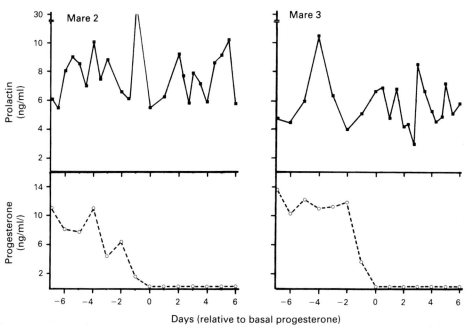

Fig. 5. Plasma concentrations of prolactin and progesterone in Mares 2 and 3 during luteolysis in May–June.

occurred, but a significant increase in PGFM was not detected in Mares 3 and 4 in which a uterine biopsy was performed. Increases in the size of the largest follicle palpable *per rectum* began shortly after prolactin levels returned to basal. LH concentrations rose in all the mares during oestrus, and

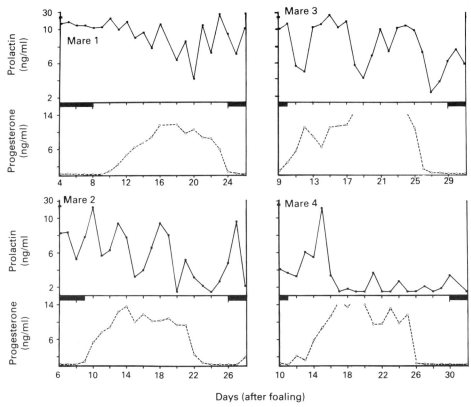

Fig. 6. Plasma concentrations of prolactin and progesterone in Mares 1, 2, 3 and 4 during the oestrous cycle after the first post-partum oestrus. The bars indicate the days that oestrus was shown.

reached a low point in mid-dioestrus; there did not appear to be any relationship with changes in plasma prolactin concentrations.

March. Plasma prolactin concentrations were again elevated in Mares 3 and 6 at the time of declining progesterone concentrations, before oestrus. However, the peaks of prolactin appeared to be generally smaller (Fig. 4).

May–June. Plasma prolactin concentrations in Mares 2 and 3 showed an elevated baseline at this time, compared to the basal values seen during the winter (Fig. 5). No clear-cut patterns of prolactin secretion were seen, although at the time of luteolysis both mares showed some rise in prolactin concentrations.

Post-partum cycle. During the first post-partum oestrus, prolactin values were high in Mares 1, 2 and 3, but were quite low in Mare 4 (Fig. 6). In Mares 1, 2 and 3 there also appeared to be a rise in prolactin concentrations around the time of the second post-partum oestrus, but in Mare 4 no large increase was seen. The baseline level of prolactin was generally high during the post-partum period.

Discussion

Plasma prolactin concentrations in Pony mares varied with the season, being high during the summer and low during the winter. This agrees with findings for Standardbred mares (Johnson,

1986) and for stallions (K. Hardman & K. Worthy, unpublished observations). Prolactin concentrations were therefore highest during the period of maximal reproductive activity. However, it seems that for most of the oestrous cycle only a minimum quantity of prolactin need be present, since the mares continued to show oestrous cycles in the face of a wide variation in the baseline prolactin concentrations at different times of year.

The exact pattern of prolactin concentrations in the oestrous cycle of the mare is unclear. In an oestrous cycle during October–November, prolactin values showed a striking increase before the onset of oestrus in mares with both natural and induced oestrus (Fig. 2). Oestrous cycles were also studied in March and May–June and in post-partum mares. In these cycles, a rise in plasma prolactin before oestrus was demonstrated in a number of mares, but there was a great deal of between-animal variation (Figs 4, 5 & 6). Also, Johnson (1986) found little change in prolactin concentrations in daily samples taken during an oestrous cycle in May–June, although the current results suggest that daily sampling may not detect all changes in prolactin secretion.

In addition, the role which prolactin plays in the oestrous cycle of the mare is obscure. In other species, it is thought that the surge of prolactin found around oestrus may be related to the increased oestrogen production by preovulatory follicles, since oestrogen is known to stimulate prolactin secretion (McNeilly, 1980). Preliminary work in the mare suggests that the peak of prolactin is indeed associated with increasing oestrogens (K. Worthy, unpublished observations). Also, increases in the size of the largest follicle began around the time of the prolactin surge. However, the timing of the LH surge seemed to be unrelated to changes in prolactin concentrations.

In the ewe, rat and pig, prolactin is required for the formation and/or maintenance of the corpus luteum (McNeilly *et al.*, 1982). Conversely, there is evidence of an involvement of prolactin in luteolysis in the rat and mouse (Ensor, 1972). In the current study, the peak of prolactin before oestrus was in some cases associated with a peak of PGFM (Fig. 3 and unpublished data). In the cow and rat, PGF-2α can stimulate a marked increase in prolactin concentrations (Louis *et al.* 1974; Deis & Vermouth, 1975). However, it seems unlikely that a rise in prolactin is critical for luteolysis in the mare, since the peak of prolactin did not always coincide with the decline in progesterone concentrations at luteolysis. Studies using agents such as bromocriptine to eliminate any rise in prolactin are needed to elucidate the role of prolactin in the oestrous cycle of the mare.

We thank the Horserace Betting Levy Board for financial support, and Professor Li and Professor H. Papkoff for generous donation of reagents.

References

Colquhoun, K.M. (1984) *A study of certain factors affecting reproduction in the mare*. Ph.D. thesis, University of Glasgow.

Deis, R.P. & Vermouth, N.T. (1975) Effect of prostaglandin $F_{2α}$ on serum prolactin and LH concentrations in male and female rats. *J. Reprod. Fert.* **45**, 383–387.

Dobson, H. (1983) Liquid-phase radioimmunoassay for PGFM. In *A Radioimmunoassay Laboratory Handbook*, pp. 60–63. Liverpool University Press, Liverpool.

Dusza, L. & Krzymowska, H. (1979) Plasma prolactin concentrations during the oestrous cycle of sows. *J. Reprod. Fert.* **57**, 511–514.

Ensor, D.M. (1972) *Comparative Endocrinology of Prolactin*. Chapman & Hall, London.

Farmer, S.W. & Papkoff, H. (1979) Immunochemical studies with pregnant mare serum gonadotrophin. *Biol. Reprod.* **21**, 425–431.

Habener, J.F., Rosenblatt, M., Dee, P.C. & Potts, J.T., Jr (1979) Cellular processing of preproparathyroid hormone involves rapid hydrolysis of leader sequence. *J. biol. Chem.* **254**, 10596–10606.

Hill, T.G. & Alliston, C.W. (1981) Effects of thermal stress on plasma concentrations of luteinizing hormone, progesterone, prolactin and testosterone in the cycling ewe. *Theriogenology* **15**, 201–209.

Johnson, A.L. (1986) Serum concentrations of prolactin, thyroxine and triiodothyronine relative to season and the estrous cycle in the mare. *J. Anim. Sci.* **62**, 1012–1020.

Kann, G. & Denamur, R. (1974) Possible role of prolactin during the oestrous cycle and gestation in the ewe. *J. Reprod. Fert.* **39**, 473–483.

Karg, H. & Schams, D. (1974) Prolactin release in cattle. *J. Reprod. Fert.* **39**, 463–472.

Licht, U.L., Bona-Gallo, A., Aggarval, B.B., Farmer,

S.W., Castelino, J.B. & Papkoff, H. (1979) Biological and binding activities of equine pituitary gonadotrophins and PMSG. *J. Endocr.* **83**, 311–322.

Louis, T.M., Stellflug, J.N., Tucker, H.A. & Hafs, H.D. (1974) Plasma prolactin, growth hormone, luteinizing hormone and glucocorticoids after prostaglandin $F_{2\alpha}$ in heifers. *Proc. Soc. exp. Biol. Med.* **147**, 128–133.

McNeilly, A.S. (1980) Prolactin and the control of gonadotrophin secretion in the female. *J. Reprod. Fert.* **58**, 537–549.

McNeilly, A.S., Glasier, A., Jonassen, J. & Howie, P.W. (1982) Evidence for direct inhibition of ovarian function by prolactin. *J. Reprod. Fert.* **65**, 559–569.

Mitchell, M.D., Flint, A.P.F. & Turnbull, A.C. (1976) Plasma concentrations of 13,14-dihydro 15-keto prostaglandin $F_{2\alpha}$ during pregnancy in sheep. *Prostaglandins* **11**, 319–329.

Munro, C.D., Renton, J.P. & Butcher, R. (1979) The control of oestrous behaviour in the mare. *J. Reprod. Fert., Suppl.* **27**, 217–227.

Peters, R.R. & Tucker, H.A. (1978) Prolactin and growth hormone response to photoperiod in heifers. *Endocrinology* **103**, 229–234.

Stewart, D.R., Kindahl, H., Stabenfeldt, G.H. & Hughes, J.P. (1984) Concentrations of 15-keto-13,14-dihydro prostaglandin $F_{2\alpha}$ in the mare during spontaneous and oxytocin induced foaling. *Equine vet. J.* **16**, 270–274.

Walton, J.S., Evins, J.D., Fitzgerald, B.P. & Cunningham, F.J. (1980) Abrupt decrease in daylength and short-term changes in the plasma concentrations of FSH, LH and prolactin in anoestrous ewes. *J. Reprod. Fert.* **59**, 163–171.

Webster, G.M. & Haresign, W. (1983) Seasonal changes in LH and prolactin concentrations in ewes of two breeds. *J. Reprod. Fert.* **67**, 465–471.

Worthy, K., Escreet, R., Renton, J.P., Eckersall, P.D., Douglas, T.A. & Flint, D.J. (1986) Plasma prolactin concentrations and cyclic activity in pony mares during parturition and early lactation. *J. Reprod. Fert.* **77**, 569–574.

J. Reprod. Fert., Suppl. **35** (1987), 277–280

Prolactin response to thyrotrophin-releasing hormone stimulation in normal and agalactic mares

C. D. Lothrop, Jr, J. E. Henton*, B. B. Cole and H. L. Nolan

*Departments of Environmental and * Rural Practice, College of Veterinary Medicine,
University of Tennessee, Knoxville, Tennessee 37901, U.S.A.*

Summary. Serum prolactin concentration was determined before and after TRH administration to normal mares at 10 months of gestation, 2 and 4 months *post partum* and during a -7- to $+14$-day peri-parturient period. The serum prolactin concentration increased significantly ($P < 0.05$) at 15, 30 and 60 min after TRH administration in the normal mares regardless of the season of the year, pregnancy or lactation status. However, during the periparturient period, the basal prolactin concentration was increased 4-fold and there was only a marginal increase after TRH administration. Of 9 agalactic mares, 6 had a decreased basal serum prolactin concentration; the other 3 agalactic mares had a normal to increased basal concentration suggestive of a peripheral resistance to prolactin.

Introduction

Prolactin, a large polypeptide hormone synthesized in the pituitary lactotrophs, is the major lactogenic hormone in most mammalian species and necessary for all phases of lactation (Carroll, 1980; Clemens & Sharr, 1980; Henton *et al.*, 1983). The synthesis and release of prolactin is regulated in a positive manner by serotonin and negatively by dopamine. Thyrotrophin releasing hormone (TRH) stimulates the release of thyrotrophin (TSH) and prolactin in many species (Jacobs *et al.*, 1971; Convey *et al.*, 1973; Davis & Borger, 1972; Kelly *et al.*, 1973; Lothrop *et al.*, 1984; Lothrop & Nolan, 1986). However, two studies have suggested that TRH may not stimulate prolactin release in the mare (Thompson *et al.*, 1983; Thompson & Nett, 1984). Thyrotrophin was increased significantly after TRH administration in the same mares which showed no increase in prolactin (Thompson *et al.*, 1983; Thompson & Nett, 1984).

Horse prolactin has been isolated and purified and an homologous radioimmunoassay has been developed (Li & Chung, 1983; Roser *et al.*, 1984). Both dog and pig prolactin had significant inhibition with cross-reactivities of $\sim 20\%$ with horse prolactin in an homologous RIA (Roser *et al.*, 1984). Only slight inhibition and non-parallelism was observed with sheep prolactin.

Previous investigations of prolactin physiology in the mare have used the heterologous sheep prolactin RIA for serum prolactin measurements (Thompson *et al.*, 1983; Thompson & Nett, 1984; Nett *et al.*, 1975a, b). Since sheep prolactin cross-reacts minimally in a horse prolactin RIA it seemed possible that small increases in the serum prolactin concentration during pregnancy and lactation or after TRH administration might not be detectable with the sheep prolactin assay. The purpose of this investigation was to determine whether small but possibly significant increases in serum prolactin, as determined with a heterologous RIA for dog prolactin, occurred after TRH administration to normal mares. Additionally, serum prolactin concentrations were determined in 9 agalactic mares before and after TRH administration.

Materials and Methods

Adult horses and Pony mares of apparent normal health and lactation were maintained on pasture and were given supplementary grass hay and sweet feed. Only animals with known mating dates were used as the control group in this study. Agalactic mares were either brought to the clinic for examination and treatment or seen by one of us (J.E.H.) on an ambulatory basis.

The TRH response test was performed as previously described (Lothrop & Nolan, 1986). Mares were given 1 mg TRH i.v. and blood samples, for serum collection, were taken immediately before and 15, 30 and 60 min after TRH administration. The blood was allowed to clot at room temperature, centrifuged and the serum stored at $-20°C$ until hormone analysis.

Prolactin was quantified with a heterologous antiserum to dog prolactin obtained from Dr A. F. Parlow (Papkoff, 1976) and/or an homologous antiserum to horse prolactin obtained from Dr H. Papkoff (Roser *et al.*, 1984). Roser *et al.* (1984) have shown that dog prolactin cross-reacts significantly with horse prolactin in an homologous assay for horse prolactin. Preliminary investigations in our laboratory based on the work of Roser *et al.* (1984) confirmed that a heterologous assay to dog prolactin could be used to estimate serum prolactin concentrations in horses.

The antisera to dog and horse prolactin were both used at final tube dilutions of 1/50 000. Dog prolactin was iodinated by the chloramine T and metabisulphite method and used as the radioligand in both assays. Assay conditions were as suggested by Dr A. F. Parlow for the dog prolactin radioimmunoassay.

Validation of the heterologous assay to dog prolactin and the homologous assay to horse prolactin demonstrated linearity with sample volumes ranging from 25 to 100 μl. The intra-assay CV ranged from 5·5 to 13·0% for the dog prolactin assay and 1·9 to 5·3% for the horse prolactin assay. The inter-assay CV was approximately twice the intra-assay CV for both antisera. A purified horse prolactin preparation was not available for these experiments. The calculated sensitivity with each antiserum was 0·1 ng with dog prolactin. The antiserum to dog prolactin under-estimates the absolute serum prolactin concentration because of incomplete cross-reactivity with horse prolactin, but appears to be a more sensitive assay than the previously described heterologous sheep prolactin assay. Significance was determined with Student's *t* test.

Results

The measured serum prolactin concentration was dependent on the specificity of the antiserum used in the radioimmunoassay. When the homologous antiserum to horse prolactin was used, the prolactin value was 2–2·5 times greater than that measured with the heterologous antiserum to dog prolactin (Samples 1 and 2, 6 replicates each, mean $\pm$ s.e.m.: 42·2 $\pm$ 2·2 and 155·8 $\pm$ 2·9 ng/ml for the homologous assay, and 17·6 $\pm$ 0·97 and 50·5 $\pm$ 6·6 ng/ml for the heterologous assay). The difference in the serum prolactin concentration as determined with the two antisera is consistent with the previously reported partial cross-reactivity of dog prolactin in the horse prolactin assay. Although the heterologous dog prolactin assay is not as sensitive as the homologous horse prolactin assay, we conclude that the dog prolactin assay does provide a reasonable estimate of the serum prolactin concentration in horses.

Serum prolactin concentrations were measured before and after TRH administration in a group of normal mares to determine whether TRH stimulated prolactin release in horses (Table 1). The

Table 1. Effect of seasonality on prolactin response (ng/ml) to TRH administration in normal horses

Season*	Group	No. of mares	Before TRH	After TRH		
				15 min	30 min	60 min
Winter	1	4	3·2 ± 0·4	10·1 ± 3·4	13·8 ± 6·3	9·0 ± 1·8
	2	6	4·6 ± 0·9	10·4 ± 8·3	8·1 ± 2·1	7·1 ± 1·3
Spring	3	5	2·9 ± 0·8	8·2 ± 1·3	8·3 ± 1·4	7·4 ± 1·1
Summer	4	8	3·0 ± 0·8	10·3 ± 0·4	11·5 ± 3·9	10·7 ± 3·4
	5	7	6·0 ± 1·9	17·7 ± 4·5	17·2 ± 5·5	15·6 ± 4·5

Values are mean ± s.e.m.
*Horses were tested in late December (Group 1), early January (Group 2), mid-March (Group 3), mid-June (Group 4) and early August (Group 5).

serum prolactin concentration increased significantly ($P < 0.05$) at 15, 30 and 60 min after TRH administration in all mares. There was no apparent influence of seasonality on the basal serum prolactin concentration or prolactin response to TRH administration.

The serum prolactin response to TRH administration was determined in pregnant (10 months gestation), peri-parturient and lactating (2 and 4 months *post partum*) mares (Table 2). Basal and TRH-stimulated prolactin concentrations were similar in pregnant and post-partum mares. However, at the time of parturition, the basal serum prolactin concentration was significantly ($P < 0.05$) elevated compared to that at 10 months gestation and *post partum* in the same mares. At parturition there was only a marginal increase in prolactin after TRH administration.

Table 2. Prolactin response (ng/ml) to TRH administration in pregnant and lactating mares

				After TRH		
Group	Reproductive status	No. of mares	Before TRH	15 min	30 min	60 min
1	Pregnant (10 months)	8	4·3 ± 1·3	11·5 ± 1·9	12·8 ± 2·6	11·4 ± 1·9
2	Peri-parturient (−7 days to +14 days)	12−	15·3 ± 3·0	20·6 ± 3·9	21·4 ± 4·3	23·5 ± 5·7
3	*Post partum* (2 months)	8	2·2 ± 0·4	14·2 ± 3·1	14·0 ± 1·9	12·1 ± 2·6
4	*Post partum* (4 months)	5	7·4 ± 2·3	20·2 ± 6·0	19·6 ± 7·4	18·0 ± 4·8

Values are mean ± s.e.m.

Basal serum prolactin concentrations and those after TRH administration were determined in 9 agalactic mares at the time of parturition (Table 3). Of the 9 mares, 6 had a low basal serum prolactin concentation with a significant increase in the serum prolactin after TRH administration (Group 1). The other 3 mares had an elevated basal serum prolactin concentration with a partial rise in the serum prolactin concentration after TRH administration (Group 2).

Table 3. Prolactin response (ng/ml) to TRH administration in normal and agalactic mares

				After TRH		
Mares		No. of mares	Before TRH	15 min	30 min	60 min
Normal*		12−	15·3 ± 3·0	20·6 ± 3·9	21·4 ± −4·3	23·5 ± −5·7
Agalactic†	Group 1	6	−2·6 ± 0·8	12·3 ± 2·2	10·2 ± −2·9	10·2 ± −1·7
	Group 2	3	23·5 ± 3·5	37·6 ± 9·5	41·1 ± 10·8	42·1 ± 13·9

Values are mean ± s.e.m.
*Group 2 in Table 3.
†Divided into 2 groups based on the basal prolactin concentration.

Discussion

The results of this study suggest that a heterologous RIA for dog prolactin can be used to estimate serum prolactin concentrations in normal mares. Since dog prolactin has only a 20% cross-reactivity to horse prolactin, the concentrations reported in this study are probably minimal estimates of the actual serum prolactin concentration.

Previous investigations (Nett *et al.*, 1975a, b; Thompson *et al.*, 1983; Thompson & Nett, 1984) of horse prolactin physiology have been hampered by the absence of a sensitive assay for horse prolactin, as shown by Roser *et al.* (1984).

Serum prolactin concentrations rose significantly after TRH administration in a group of normal mares in this study, but the increase was not affected by seasonality, pregnancy status or lactation except during the periparturient period. At the periparturient interval, the basal serum prolactin concentration was significantly increased and there was only a marginal rise after TRH administration. Previous studies (Thompson *et al.*, 1983; Thompson & Nett, 1984) suggesting TRH may not be an effective stimulus of prolactin in the mare may have been hampered by the use of the less sensitive sheep prolactin RIA.

The increased basal prolactin concentrations during the peri-parturient period may indicate an increased demand for prolactin associated with production of colostrum and early milk. The absence of a marked increase in prolactin during this period may indicate that pituitary prolactin production is maximal at this time, although a decreased rate of prolactin degradation could not be ruled out by these studies.

The 6 agalactic mares with a decreased basal prolactin concentration may have been agalactic from the lack of endogenous serotonergic and/or TRH stimulation. Alternatively, excessive inhibitory override by dopamine or dopamine-like molecules could also have caused a decrease in the serum prolactin concentration. The 3 mares with an increased basal serum prolactin concentration may have been agalactic from a peripheral resistance to prolactin.

We thank Dr H. Papkoff and the antiserum to horse prolactin; and Dr A. F. Parlow and the National Hormone and Pituitary Program for the purified dog prolactin and antiserum to dog prolactin.

References

Carroll, E.J. (1980) *Veterinary Endocrinology and Reproduction.* Ed. L. E. McDonald. Lea & Febiger, Philadelphia.

Clemens, J.A. & Sharr, C.J. (1980) Control of prolactin secretions in mammals. *Fedn Proc. Fedn Am. Socs exp. Biol.* **39,** 2588–2591.

Convey, E.M., Tucker, H.A., Smith, V.G. & Zolman, J. (1973) Bovine prolactin, growth hormone, thyroxine and corticoid response to thyrotropin-releasing hormone. *Endocrinology* **92,** 471–476.

Davis, S.L. & Borger, M.L. (1972) Prolactin secretion stimulated by TRH. *J. Anim. Sci.* **35,** 239, Abstr.

Henton, J.E., Lothrop, C.D., Dean, D. & Waldrop, V. (1983) Agalactia in the mare: a review and some new insights. *Theriogenology* **20,** 203–221.

Jacobs, L.S., Snyder, P.J., Wilber, J.B., Utiger, R.D. & Daughaday, W.H. (1971) Increased serum prolactin after administration of synthetic thyrotropin-releasing hormone (TRH) in man. *J. clin. Endocr. Metab.* **33,** 996–998.

Kelly, P.A., Bedirian, K.N., Baker, R.D. & Friesen, H.G. (1973) Effect of synthetic TRH on serum prolactin, TSH and milk production in the cow. *Endocrinology* **92,** 1289–1293.

Li, C.H. & Chung, D. (1983) Studies on prolactin: isolation and properties of the hormone from horse pituitary glands. *Arch. Biochem. Biophys.* **220,** 208–213.

Lothrop, C.D. & Nolan, H.L. (1986) Equine thyroid function assessment with the thyrotropin-releasing hormone response test. *Am. J. Vet. Res.* **47,** 942–944.

Lothrop, C.D., Tamas, P.M. & Fadok, V.A. (1984) Canine and feline thyroid function assessment with the thyrotropin-releasing hormone response test. *Am. J. vet. Res.* **45,** 2310–2313.

Nett, T.M., Holtan, D.W. & Estergreen, V.L. (1975a) Oestrogens, LH, PMSG and prolactin in serum of pregnant mares. *J. Reprod. Fert., Suppl.* **23,** 457–462.

Nett, T.M., Holtan, D.W. & Estergreen, V.L. (1975b) Levels of LH, prolactin and oestrogens in the serum of post-partum mares. *J. Reprod. Fert., Suppl.* **23,** 201–206.

Papkoff, H. (1976) Canine pituitary prolactin: Isolation and partial characterization. *Proc. Soc. exp. Biol. Med.* **153,** 498–500.

Roser, J.F., Chang, Y.S., Papkoff, H. & Li, C.H. (1984) Development and characterization of a homologous radioimmunoassay for equine prolactin (41829). *Proc. Soc. exp. Biol. Med.* **175,** 510–517.

Thompson, D.L. & Nett, T.M. (1984) Thyroid stimulating hormone and prolactin secretion after thyrotropin releasing hormone administration to mares: Dose response during anestrus in winter and during estrus in summer. *Domestic Anim. Endocr.* **1,** 263–268.

Thompson, D.L., Godke, R.A. & Nett, T.M. (1983) Effects of melatonin and thyrotropin releasing hormone on mares during the nonbreeding season. *J. Anim. Sci.* **56,** 668–677.

J. Reprod. Fert., Suppl. **35** (1987), 281–288

Reproductive patterns in cyclic and pregnant thyroidectomized mares

J. E. Lowe, R. H. Foote*, B. H. Baldwin*, R. B. Hillman and F. A. Kallfelz

*Department of Clinical Studies, New York State College of Veterinary Medicine and *Department of Animal Science, New York State College of Agriculture and Life Sciences, Cornell University, Ithaca, New York 14853, U.S.A.*

Summary. Three Quarter-horse mares were thyroidectomized at about 1·5 years of age. Three similar intact mares served as controls. The study continued through two breeding seasons. The thyroidectomized mares were lethargic, rear limbs were oedematous and hair coats were coarse. They displayed a tranquil oestrous behaviour when exposed to a stallion and were only mildly antagonistic when not in oestrus. Length of oestrous cycles varied but most often they were 19–24 days long. Duration of oestrus (mean $\pm$ s.e.m.) for the control and thyroidectomized mares was $12\cdot9 \pm 2\cdot9$ and $11\cdot7 \pm 2\cdot2$ days respectively ($P > 0\cdot05$). The peak of LH during oestrus was as high as 60 ng/ml blood serum with no difference between the two groups. Peak progestagen on Day 7 after ovulation for controls was $9\cdot0 \pm 1\cdot6$ ng/ml and was not different ($P > 0\cdot05$) from the peak of $6\cdot3 \pm 1\cdot7$ ng/ml for thyroidectomized mares on Day 8. Pregnancy was achieved in both groups of mares, including the use of semen from a thyroidectomized stallion. Thyroxine was detectable in one pregnant thyroidectomized mare during the last two-thirds of pregnancy only.

Introduction

Few studies have been conducted with thyroidectomized horses. Previous research from our laboratories (Lowe *et al.*, 1974) revealed multiple changes in growth, pelage and metabolic characteristics in thyroidectomized horses, but other than an effect on libido, semen characteristics and testicular histology in stallions was normal (Lowe *et al.*, 1975). Simultaneously with these studies on reproductive performance in stallions, reproductive patterns in a group of thyroidectomized mares was compared with controls. The objectives of the studies with mares were to examine the effects of thyroidectomy on (1) the reproductive organs, (2) the oestrous cycle, (3) on luteinizing hormone (LH) and progestagens and (4) on fertility.

Materials and Methods

Animals and thyroidectomy

At about 1·5 years of age, 3 Quarter-horse mares were surgically thyroidectomized. Three similar intact mares served as controls (Table 1). One of the controls had to be removed from the study because of colic and abdominal adhesions.

Surgery was performed with the mares under halothane anaesthesia using a ventral approach to the thyroid lobes. If present, isthmus thyroid tissue was also removed. Vessels were ligated as necessary as the thyroid lobe was dissected free. No attempt was made to save the parathyroid glands because a lower pair is present at the thoracic inlet (Krook & Lowe, 1964). All 3 mares recovered uneventfully.

Total serum thyroxine in controls was 10–30 ng/ml throughout the 67-week study. These values were normal (Thomas, 1978). After thyroidectomy the total serum thyroxine declined to zero, as reported previously (Lowe *et al.*, 1974) and this is considered evidence for complete thyroidectomy.

Table 1. Age of Quarter-horse mares during experimental periods

Animal no.	Treatment	Age at thyroidectomy (months)	Age during exp. (months)	
			1st year	2nd year
137*	Control	—	20	—
141	Control	—	20–27	32–36
167	Control	—	20–27	32–36
138	Thx	17	19–26	31–35
173	Thx	18	20–27	32–36
174	Thx	19	21–28	33–37

*Excluded from the experiment due to illness.

Measures of reproductive performance

Reproductive organs. Mares were palpated regularly *per rectum* from January until May or until the mares were seen in oestrus. Then they were palpated daily. Palpation continued until August during the first year and was repeated from January until May during the 2nd year. The ovaries were examined for the presence and location of significant structures. Uterine tone was classified as poor, fair, good or excellent. The cervix was graded as having no definable structure, or soft and flabby or as having a definite firm structure.

Oestrous cycle. Mares were teased with one or more stallions starting about 1 month before the palpations *per rectum* were instituted. This test for oestrous behaviour was conducted several times per week until it appeared that the mare was approaching oestrus. Then teasing was conducted daily until pregnancy or anoestrus was confirmed.

Progestagen and LH. Each week after the morning feeding, 20 ml blood were taken from the jugular vein. Daily samples were taken as soon as oestrous cycles began. Blood was allowed to clot and aliquants of serum were placed in several tubes and frozen until assayed.

Progestagens were assayed by the competitive protein binding assay of Murphy (1967). All samples were assayed in triplicate, with two sets of standards (0–10 ng pure progesterone) run with each assay. Appropriate blanks were run with each assay. Extraction efficiency was repeatably 88–90%. Sensitivity was <0·1 ng/ml. Intra-assay coefficient of variation was <10% while the inter-assay coefficient of variation ranged from 12 to 14%. While most of the steroid measured was probably progesterone, small amounts of 17α-hydroxyprogesterone (Ginther, 1979) and particularly 20α-dihydroprogesterone during the second part of pregnancy may have been present (Seren *et al.*, 1981). The term progestagen is therefore used to refer to the steroids measured by this assay.

Luteinizing hormone (LH) was measured by Dr G. D. Niswender, using the assay of Niswender *et al.* (1969). Samples were assayed in duplicate and any duplicates differing by more than 10% were rerun. All appropriate blanks, standards and other steps to assure a valid assay were routinely employed in that laboratory, but specific characteristics of each assay are no longer available.

Fertility. All mares were mated during oestrus when palpation of the ovaries indicated that ovulation was imminent. Fresh semen collected from stallions (Lowe *et al.*, 1975) was used for insemination, or, if it was not a regular semen collection day, other stallions were used for natural mating. Mares were inseminated only once during oestrus. Pregnancy was determined by palpation *per rectum* and also by the immunopregnancy test of Chak & Bruss (1968). After the 67-week experimental period ended the mares were kept through foaling and non-experimental mating was carried out the following year.

Results and Discussion

General characteristics

The thyroidectomized mares were lethargic and slow moving. An example of a control and a thyroidectomized mare is shown in Fig. 1. Oedema of the rear limbs and a 'meaty' thickened appearance of the face occurred. The hair coats were coarse, rough and dull. Hair coats were 4–6 weeks later in shedding compared to controls. Sensitivity to cold was observed often in thyroidectomized animals. In cold weather, during daily paddock exercise periods, thyroidectomized animals would shiver continually while controls did not shiver. Thyroidectomized mares ate about one-half the amount of diet available *ad libitum* compared to controls.

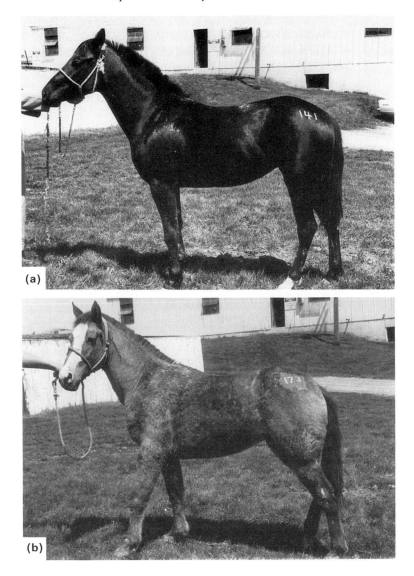

Fig. 1. Photograph of a control mare (a) and a thyroidectomized mare 21 weeks after thyroidectomy (b).

Reproductive characteristics

Reproductive behaviour. Behavioural signs of oestrus were reduced in thyroidectomized mares. The thyroidectomized mares were placid about everything. When in standing oestrus they did not become alert and interested in the approaching teaser as did the controls by whinnying and moving about. They would raise their tails, 'wink' and stand for the stallion but in a more tranquil way. By the same token, when in dioestrus, the thyroidectomized mares would ignore the teaser and only kick if the stallion bit them or tried to mount. Even then, the kicking was in slow motion. The controls on the other hand would lay their ears back and kick rather violently at the same teaser, letting the stallion recognize explicitly that they were not in oestrus.

Reproductive organs. Although individual mares differed, no consistent differences were found between the two groups of mares. They followed normal patterns as previously described (Ginther *et al.*, 1972; Ginther, 1979) for mares during the period of transition from the anovulatory season until oestrous cycles were established. Accompanying oestrus, the uterine walls became more flaccid and the cervix more relaxed. The ovaries became larger as large preovulatory follicles developed. The 2 controls were first observed in oestrus at the end of December and in March the first year. The 3 thyroidectomized mares were first observed in oestrus in January (2) and in May.

Oestrous cycles. These were variable in both groups the first breeding season. The intervals between ovulations for 21 oestrous cycles were generally 19–24 days, but one interval was 8 days, two were 59 days (1 control and 1 thyroidectomized mare) and 1 thyroidectomized mare had one 113-day interval. The average lengths of oestrus ± the standard errors for control mares was 12·9 ± 2·9 days *versus* 11·7 ± 2·2 days for the thyroidectomized mares (*P* > 0·05). Oestrus terminated within 2 days after ovulation except once in one thyroidectomized mare. Characteristic examples of these relationships and the associated hormonal changes can be seen in Figs 2 and 3.

Figure 2 illustrates the LH and progestagen patterns associated with oestrus and ovulation in a control mare. No oestrus was observed in Mare 167 during July, nor was an LH peak detected with the daily blood sampling, but the progestagen profile indicates that an ovulation followed by formation of a corpus luteum occurred.

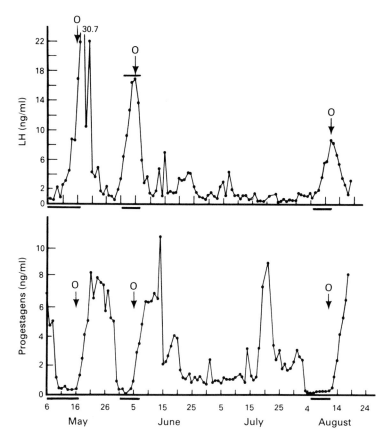

Fig. 2. Luteinizing hormone (LH) and progestagen concentration in the blood serum of a control mare (No. 167) with recurring oestrous cycles. Solid bars beneath each horizontal axis indicate length of oestrus; O = ovulation.

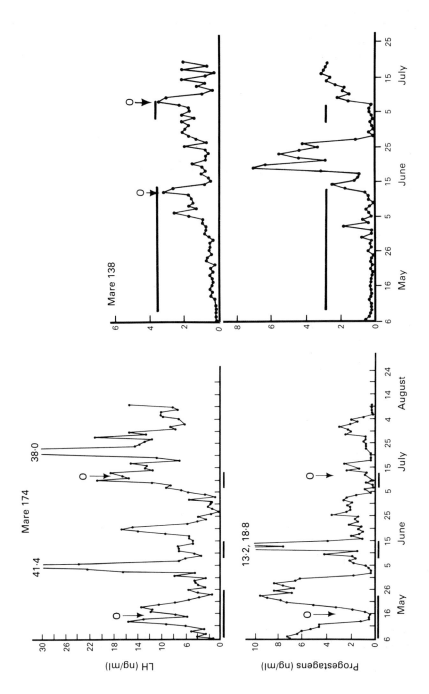

Fig. 3. Luteinizing hormone (LH) and progestagen concentration in the blood serum of two thyroidectomized mares showing contrasting oestrous cycle patterns. Horizontal bars indicate oestrus; O = ovulation.

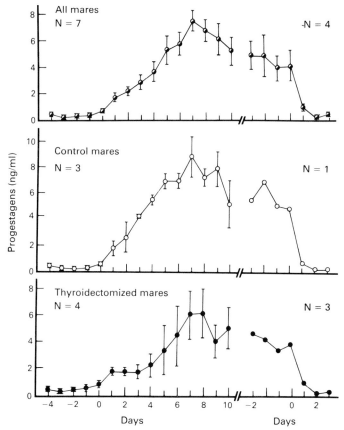

Fig. 4. Progestagen concentrations for control and thyroidectomized mares during the oestrous cycle. The first 0 represents ovulation and the 0 at the right is the last day of appreciable luteal activity. The graphs are broken because of variable lengths of dioestrus.

The thyroidectomized mare No. 174 (Fig. 3a) followed an irregular cyclic pattern similar to the control mare, No. 167, except that Mare 174 had a long oestrus in May with two LH peaks and a mid-cycle ovulation. The pattern for thyroidectomized mare No. 138 is shown in Fig. 3(b). This mare had a long behavioural oestrus and by far the lowest LH concentration in the blood serum of any mare. Davis & Borger (1973) reported low LH values for thyroidectomized lambs. However, all values are within the extreme range of values reported in the literature for LH in mares (Whitmore *et al.*, 1973; Nett *et al.*, 1976; Urwin & Allen, 1983). In a summary of the literature, Ginther (1979) attributed this wide variation to different assay procedures and purity of LH standards.

The progestagen concentrations (Fig. 4) and patterns also were consistent with the other reproductive observations and previous reports (Plotka *et al.*, 1972; Sharp & Black, 1973; Whitmore *et al.*, 1973; Holtan *et al.*, 1975, Nett *et al.*, 1976; Hunt *et al.*, 1978; Urwin & Allen, 1983). Most of the progestagen during the oestrous cycle is progesterone (Ginther, 1979). The tendency for thyroidectomized animals to have lower progestagen values than controls was not significant ($P > 0.05$) and is attributable to variation amongst individual mares. The actual peak value for the controls was 9.0 ± 1.6 ng/ml on Day 7 and for thyroidectomized mares was 6.3 ± 1.7 ng/ml on Day 8.

The right side of Fig. 4 shows the decline in progestagens as luteal function decreases. Day 0 represents the last day of luteal activity, as measured by significant concentrations of serum

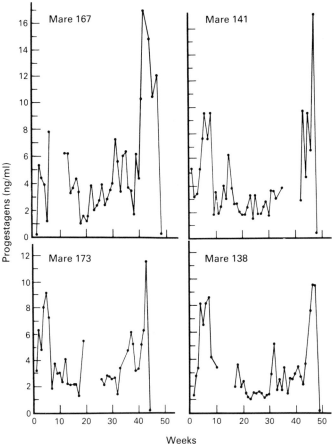

Fig. 5. Progestagen patterns during pregnancy in control (Nos 167 & 141) and thyroidectomized (Nos 173 & 138) mares.

progestagen. The decline follows a pattern similar to that described by Stabenfeldt *et al.* (1972) and summarized by Ginther (1979).

Mating and pregnancy. During the two breeding seasons one control mare was mated in both years and one was only mated in the first year. All first inseminations resulted in a pregnancy, including one to semen from a thyroidectomized stallion (Lowe *et al.*, 1975). In two seasons, 3 pregnancies in the thyroidectomized group of mares resulted from one or two matings, including one from semen from another thyroidectomized stallion (Lowe *et al.*, 1975). Two thyroidectomized stallions therefore impregnated a control and a thyroidectomized mare. One mare (No. 174; Fig. 4) did not conceive during either year after mating during two oestrous periods.

The progestagen pattern of 4 pregnant mares is illustrated in Fig. 5. The typical patterns (Ginther, 1979) of an initial rise in progestagens followed by a decline to about 2–4 ng/ml and then a marked rise the last few weeks of pregnancy are evident from Fig. 5. The progestagen concentration declined to zero in both groups at parturition. Significant amounts of 17α-hydroxyprogesterone are present the first 2–3 months of pregnancy with high concentrations of 20α-dihydroprogesterone during the last month of pregnancy (Seren *et al.*, 1981).

Values for LH were as high as 60 ng/ml during the first 15–17 weeks of pregnancy, then declined to non-detectable concentrations by 6 months and remained low during the remainder of pregnancy. During this time the placental unit is the source of progestagens. There was no difference in

LH values in the two groups although one thyroidectomized mare (No. 173) presented a more erratic pattern than did the other mares. This mare had a normal progestagen profile (Fig. 5).

One distinct event occurred during the pregnancy of thyroidectomized mare No. 138. This mare had a low detectable thyroxine concentration in blood serum after about 4 months of gestation which persisted only until parturition. This result is interpreted to indicate some placental transfer of thyroxine. During this time the mare shed its rough coat which was replaced with a normal hair coat. The scaly skin and swelling around the eyes and oedema of the legs disappeared. This mare, during the following year, did not develop the superficial characteristics of a thyroidectomized mare despite the lack of detectable circulating thyroxine. This 'carryover' effect was also seen to some extent in stallions after treatment with iodinated casein (Lowe *et al.*, 1974). Thyroidectomized mare No. 173 never had detectable thyroxine in its blood serum and retained the appearance of a thyroidectomized animal.

These studies confirm those with thyroidectomized stallions in that behavioural characteristics of reproductive performance are diminished but not eliminated in the absence of detectable thyroxine. The thyroidectomized mares were somewhat erratic in oestrous cycle patterns but not more than was true for the limited number of controls, or is found in young mares generally. The LH and progestagen concentrations and patterns during the oestrous cycles were not different in thyroidectomized and control mares, and the former were capable of normal pregnancy.

We thank Dr G. D. Niswender for the LH assays.

References

Chak, R. & Bruss, M. (1968) The MIP test for the diagnosis of pregnancy in mares. *Proc. 14th A. Conv. Am. Assoc. Equine Prac.* pp. 53–55.

Davis, S.L. & Borger, M.L. (1973) The effect of thyroidectomy on the secretion of prolactin and on plasma levels of thyrotropin, luteinizing hormone and growth hormone in lambs. *Endocrinology* **92**, 1736–1739.

Ginther, O.J. (1979) *Reproductive Biology of the Mare.* McNaughton and Gunn, Inc., Ann Arbor.

Ginther, O.J., Whitmore, H.L. & Squires, E.L. (1972) Characteristics of estrus, diestrus and ovulation in mares and effects of season and nursing. *Am. J. Vet. Res.* **33**, 1935–1939.

Holtan, D.W., Nett, T.M. & Estergreen, V.L. (1975) Plasma progestins in pregnant, postpartum and cycling mares. *J. Anim. Sci.* **40**, 251–259.

Hunt, B., Lein, D.H. & Foote, R.H. (1978) Monitoring of plasma and milk progesterone for evaluation of postpartum estrous cycles and early pregnancy in mares. *J. Am. vet. med. Assoc.* **172**, 1298–1302.

Krook, L. & Lowe, J.E. (1964) Nutritional secondary hyperparathyroidism in the horse. *Path. Vet.* **1**, Suppl. 1, 1–98.

Lowe, J.E., Baldwin, B.H., Foote, R.H., Hillman, R.G. & Kallfelz, F.A. (1974) Equine hypothyroidism: The long term effects of thyroidectomy on metabolism and growth in mares and stallions. *Cornell Vet.* **64**, 276–295.

Lowe, J.E., Baldwin, B.H., Foote, R.H., Hillman, R.B. & Kallfelz, F.A. (1975) Semen characteristics in thyroidectomized stallions. *J. Reprod. Fert., Suppl.* **23**, 81–86.

Murphy, B.E.P. (1967) Some studies of the protein-binding of steroids and their application to the routine micro and ultramicro measurement of various steroids in body fluids by competitive protein-binding radioassay. *J. clin. Endocr. Metab.* **27**, 973–990.

Nett, T.M., Pickett, B.W., Seidel, G.E., Jr. & Voss, J.L. (1976) Levels of luteinizing hormone and progesterone during the estrous cycle and early pregnancy in mares. *Biol. Reprod.* **14**, 412–415.

Niswender, G.D., Reichert, L.E., Jr., Midgley, A.R., Jr., & Nalbandov, A.V. (1969) Radioimmunoassay for bovine and ovine leutinizing hormone. *Endocrinology* **84**, 1166–1173.

Plotka, E.D., Witherspoon, D.M. & Foley, C.W. (1972) Luteal function in the mare as reflected by progesterone concentrations in peripheral blood plasma. *Am. J. vet. Res.* **33**, 917–920.

Seren, E., Tamanini, C, Gaiani, R. & Bono, G. (1981) Concentrations of progesterone, 17α-hydroxyprogesterone and 20α-dihydroprogesterone in the plasma of mares during pregnancy and at parturition. *J. Reprod. Fert.* **63**, 443–448.

Sharp, D.C. & Black, D.L. (1973) Changes in peripheral plasma progesterone throughout the oestrous cycle of the pony mare. *J. Reprod. Fert.* **33**, 533–538.

Stabenfeldt, G.H., Hughes, J.P. & Evans, J.W. (1972) Ovarian activity during the estrous cycle of the mare. *Endocrinology* **90**, 1380–1384.

Thomas, C.L. (1978) Radioimmunoassay of equine serum for thyroxine: Reference values. *Am. J. vet. Res.* **39**, 1239.

Whitmore, H.L., Wentworth, B.C. & Ginther, O.J. (1973) Circulating concentration of luteinizing hormone during estrous cycle of mares as determined by radioimmunoassay. *Am. J. vet. Res.* **34**, 631–635.

Urwin, V. & Allen, W.R. (1983) Follicle stimulating hormone, luteinizing hormone and progesterone concentrations in the blood of Thoroughbred mares exhibiting single and twin ovulations. *Equine vet. J.* **15**, 325–329.

J. Reprod. Fert., Suppl. **35** (1987), 289–296

Use of push–pull perfusion techniques in studies of gonadotrophin-releasing hormone secretion in mares

D. C. Sharp and W. R. Grubaugh

Department of Animal Science, University of Florida, Gainesville, FL 32611, U.S.A.

Summary. Push–pull perfusion was used to study GnRH secretory ability of the hypothalamus in anoestrous, transitional, dioestrous and oestrous Pony mares. The technique involved placement of a concentric (tube within a tube) cannula into the area of the medial basal hypothalamus and perfusing a carrier medium (artificial cerebrospinal fluid) through the inner tube whilst aspirating from the outer tube so that the flow rate within the hypothalamic tissue was essentially constant. The perfusion rate was 0·5 ml/10 min and samples were collected at 10-min intervals for 10–15 h. The carrier medium, which contained GnRH, was acidified and frozen until measurement by radioimmunoassay. Blood samples were taken from the jugular vein simultaneously for measurement of LH. Nineteen mares were perfused, representing anoestrus (N = 3), transition (N = 6), dioestrus (N = 7), and oestrus (N = 3). GnRH secretion was minimal during anoestrus, with the majority of samples being below the level of detection of the assay (1·9 pg/tube). During the transition phase, the overall mean secretory rate was increased, and there appeared to be more secretory episodes. During the breeding season, GnRH secretion was markedly increased over anoestrus and transition, with oestrous mares secreting significantly greater amounts of GnRH than dioestrous mares. There were no significant differences among groups in number of secretory peaks, or the interval between peaks. There was, however, a significant ($P < 0.0001$) difference among groups in overall mean secretory rate. These results suggest that GnRH secretion is increased in early transition, thus accounting for the increased FSH secretion that takes place then, but LH is not increased. During the breeding season, GnRH is markedly increased, and this is reflected in elevated LH during oestrus.

Introduction

Studies of the mechanisms regulating the secretion of gonadotrophins have been hampered in horses, as in other species, by our inability to monitor the afferent signal for gonadotrophin secretion, gonadotrophin-releasing hormone (GnRH). This decapeptide is known to be synthesized in neural perikarya within the anterior and medial basal hypothalamus of most species, and transported to the anterior pituitary via the hypothalamo–hypophysial portal system (Yen, 1986; Moore, 1986). Rapid degradation of the decapeptide prevents measurement in plasma distal to the pituitary and, until recently, the most frequently utilized method of studying GnRH has been to monitor tissue concentrations within the hypothalamus (Strauss *et al.,* 1979; Hart *et al.,* 1984). While this technique has produced useful and reproducible results it does not offer vital information about secretory dynamics. These studies were therefore undertaken to devise techiques by which GnRH secretory dynamics could be studied, and to examine the changes in GnRH secretion during different phases of the annual reproductive cycle.

Materials and Methods

Perfusion device. The technique of push–pull perfusion, as described by Meyers (1977) and by Levine & Ramirez (1982), was adapted for use in the mare. The technique utilized a stainless-steel plate which was affixed to the dorsal calvarium with bone screws and a stainless-steel, 13-gauge guide cannula which was inserted into the brain to about the level of the thalamus. At the time of perfusion, a concentric (tube within a tube) perfusion cannula was inserted through the guide cannula to the region of interest and artificial cerebrospinal fluid (aCSF; Levine & Ramirez, 1982) was introduced through the inner cannula (21-gauge). Simultaneously, aCSF was withdrawn from the outer cannula (19 gauge) via the side port (see Fig. 1). The effect of the simultaneous infusion and withdrawal was to create a small (3–4 mm) droplet of aCSF within the neuropil at the tip of the cannula into which GnRH and other neurosecretions were deposited. The perfusate was collected at 10-min intervals, the volume measured, and immediately added to cold 2 N-acetic acid (for GnRH and/or endogenous opioid peptides) or to 0·8 N-perchloric acid (1:1, v/v; for HPLC analysis of catecholamines).

Perfusion of aCSF was carried out using a Sage Syringe pump, model 355 (Orion), modified with the 'pull' syringe fixed to the back end of the push bar so that infusion and withdrawal rates were identical. The rate of infusion was 0·5 ml/10 min. Infusions were carried out at 10-min intervals for 10–15 h.

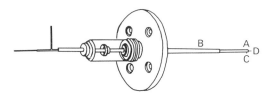

Fig. 1. Diagram of the push–pull perfusion device. The perfusion cannula (A) is shown in place inserted through the guide cannula (B). The perfusion cannula is constructed of stainless-steel hypodermic tubing, 21-gauge inside 19-gauge tubing. The tip of the inner perfusion cannula (C) protrudes from the outer cannula (D) by 1 mm. The reinforcing collar is added at the time of perfusion to help guide and reinforce the perfusion cannula. During times of non-perfusion, the reinforcing collar is removed and a rounded cap (not shown) is fastened, using the same screw threads, to seal to device.

Surgical placement. Our previous results (Strauss *et al.*, 1979) indicated that the GnRH-rich portion of the hypothalamus of the horse extends from the pre-optic, anterior hypothalamus through the medial basal hypothalamus and thus represents a corridor of GnRH-rich neurones that is nearly 1·5 cm in rostral–caudal dimension. Similarly, the medio-lateral dimension of GnRH-rich neurones extends nearly 1·0 cm from midline.

Pony mares weighing an average of 180 kg were anaesthetized with thiopentone sodium (Boehringer Ingelheim, St Louis, MO) and maintained at surgical plane with halothane inhalant anaesthesia (Fluothane: Fort Dodge Lab, Inc., Fort Dodge, IA). The mares were positioned in sternal recumbency with head fixed in an upright position, and the dorsal surface was prepared for surgery. A curving incision was made exposing the parietal bone and the insertion of the interscutularis muscle. The periostium was removed, and a portion of the interscutularis insertion was reflected laterally to permit fixation of the device. A burr hole was placed through the parietal bone about 3–5 mm lateral to midline and approximately equidistant from the rostral border of the zygomatic arch and the atlas. Bone wax was used to effect a seal between the parietal bone and the stainless-steel skull plate which was affixed by 4 stainless-steel bone screws. The guide cannula was inserted and a lateral radiograph taken to verify placement. The rostral portion of the sella turcica, which could be visualized in this view, was considered to be the desirable target site. Adjustment in the rostral–caudal plane could be made, up to about 5 mm, by tightening or loosening the appropriate screws holding the skull plate, rotating the angle of entry slightly. The radiograph also served the important function of estimating depth that the perfusion cannula needed to be inserted at the time of push–pull perfusion. Measurements were made from the ventral-most portion of the guide cannula to an area about 2–4 mm dorsal to the sella and allowances were made for radiographic magnification. The device was sealed by a neoprene-gasketed cap which could be removed at the time of push–pull perfusion.

Experimental animals. Twenty-four Pony mares were used for these studies representing the following phases of the annual reproductive cycle; anoestrus (Group A, N = 5); transition from anoestrus to the breeding season (Group B, N = 7); oestrus (Group C, N = 5); and dioestrus (Group, D, N = 7). Anoestrous mares were perfused during November and December, transition mares were perfused during February through April and cyclic mares (Groups C and D) were perfused during May through August. The reproductive status of mares was monitored by palpation of the ovaries *per rectum* and by teasing with a vigorous stallion as previously reported (Sharp *et al.*, 1975, 1979a). After perfusion, mares were killed and the position of the perfusion cannula was verified. Only those mares in which the perfusion cannula had been within the above described location were used for data analysis.

Radioimmunoassay. GnRH was measured in perfusate within 24 h of collection by a radioimmunoassay developed by Dr M. S. A. Kumar (present address: Tufts University, Boston, MA) and validated for use in equine brain tissue and perfusate in this laboratory. Sensitivity of the assay was 5·04 ± 2·01 pg/tube and the mean slope of all standard curves represented in these data was $-1.00 + 0.04$. Addition of increasing volumes of perfusate resulted in GnRH estimates that were linear and that were parallel to the standard curve and addition of authentic GnRH to aCSF, or perfusate from non-GnRH rich areas of the brain resulted in accurate estimates of the hormone.

Luteinizing hormone was measured in peripheral plasma by radioimmunoassay using antiserum generously donated by Dr G. D. Niswender (GDN No. 15, anti-ovine LH antiserum) and horse LH generously donated by Dr H. Papkoff for iodination and standard curves. Sensitivity of the LH assay was 0.69 ± 0.06 ng per tube and the mean slope of all standard curves represented in these data was $-1.31 + 0.13$. Quantification of authentic horse LH added to plasma from pituitary stalk-sectioned Pony mares (Sharp *et al.*, 1982) resulted in linear and accurate estimates of the hormone.

Statistical analysis. Mean concentration of GnRH and LH in perfusate and peripheral plasma were tested for statistical differences amongst mares, amongst season groups, and over time by least squares analysis of variance with attention to all interactions. Single degree of freedom orthogonal comparisons were used to describe differences among season groups. Regression analyses were performed to test for significant time trends in the two hormones within and among season groups. A modification of the method of Christian *et al.* (1978) was used to identify 'peak' events. Basically, the method involved repetitive tests for skewness with ranked data and sequential deletion of highest values until the data were normalized (no longer skewed). Peaks were defined as concentrations of hormone that exceeded the mean of the normalized data by two standard deviations.

Results

Placement of the cannula was inappropriate in only 5 mares; of the remaining 19 mares, 3 were in Group A, 6 in Group B, 7 in Group C, and 3 in Group D. The efficiency of placement of the device, coupled with the extended data collection time possible, suggests that the technique of push–pull perfusion can be a very useful tool for studying hypothalamic GnRH secretion rate in conscious horses.

Table 1. Peak analysis for GnRH and LH endpoints* in mares in Groups A, B, C and D

	GnRH		LH	
Group	No. of peaks	IPI (min)	No. of peaks	IPI (min)
A (anoestrous)	2.3 ± 1.5	143.0 ± 67.0	1.3 ± 0.9	25.0 ± 25.0
B (transitional)	3.2 ± 0.5	158.6 ± 57.0	1.8 ± 0.6	88.7 ± 40.5
C (dioestrous)	2.0 ± 0.7	165.0 ± 57.7	1.2 ± 0.6	80.0 ± 52.9
D (oestrous)	3.0 ± 0.0	65.0 ± 15.0	2.5 ± 0.5	141.2 ± 66.2

*Number of peaks, and inter-peak interval (IPI) expressed per 10 h perfusion time to standardize amongst mares.

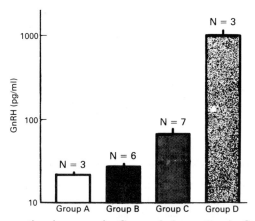

Fig. 2. Mean GnRH secretion in mares in Group A (anoestrous), Group B (transitional), Group C (dioestrous), and Group D (oestrous). The overall mean ($\pm$s.e.m.) is shown for all mares within each group. There was a significant ($P < 0.01$) difference amongst groups in mean GnRH secretion. Note that the scale on the ordinate is a log scale.

There was a significant ($P < 0.001$) difference among season groups in the concentration of GnRH that was characterized by increasing mean GnRH secretory rate as mares progressed from anoestrus to oestrus of the breeding season. Figure 2 demonstrates the mean secretory rate by season group. Of the 3 mares in Group A, 1 did not have any detectable GnRH, and the other 2 had only occasional events of detectable GnRH contrasted against a background of non-detectable values (see Fig. 3a). The mean GnRH secretory rate in Group B was elevated slightly and was characterized by more frequent and higher amplitude secretory events than observed in Group A (see Figs 3b & 3c) although the number of episodes or peaks that were detected did not differ significantly among season groups (see Table 1). GnRH was considerably higher in mares in Groups C and D during the breeding season, with GnRH elevated to an even greater extent in the latter (Figs 3d, e, f). Overall, there were about 65% as many LH peaks identified as there were GnRH peaks. Furthermore, when LH and GnRH were viewed concurrently, there was not always

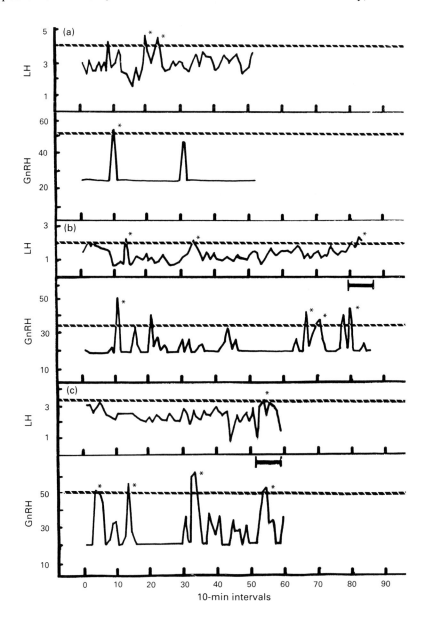

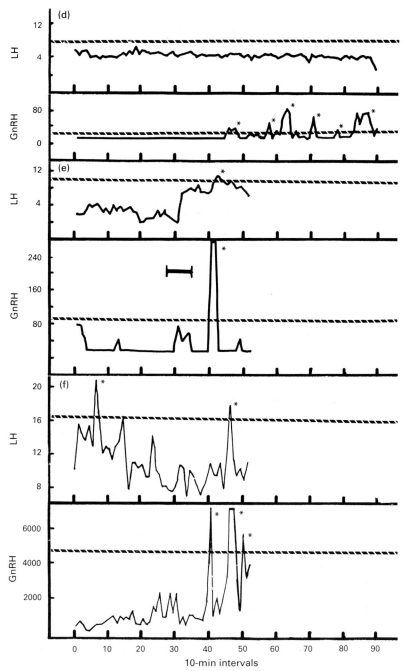

Fig. 3. Concentrations of LH (ng/ml) in jugular vein plasma and GnRH (pg/ml) in the push–pull perfusate in representative mares from (a) Group A (anoestrous), (b, c) Group B (transitional), (d, e) Group C (dioestrous) and (f) Group D (oestrous). The broken horizontal line in the panels indicates the cut-off value for detection of peak events. The asterisks indicate concentrations of hormone that were considered to be peaks. The horizontal bar indicates the inclusion of 15 mM-KCl in the perfusate in (b) and (c) and naloxone in (e).

temporal agreement between the two hormones. That is, occasionally a peak of GnRH was seen unaccompanied by an LH peak and similarly, occasionally an LH peak was seen unaccompanied by a GnRH peak (Fig. 3).

In representative mares, 15 mM-KCl was added to the aCSF perfusate to test for GnRH secretion by stimulus–secretion coupling (Figs 3b, C) and in 2 dioestrous mares the opioid receptor inhibitor, naloxone (Sigma Chemical Company, St Louis, MO), was included in the aCSF at 1 mg/ml (Fig. 3e). When such a stimulation was imposed, the result was a significant coordinated increase in both GnRH and LH.

Discussion

These results indicate that it is feasible to monitor the secretion of GnRH directly in the hypothalamus of the horse and thus to assess GnRH secretory ability. The salient points of this study are as follows. (1) The mean secretory rate of GnRH increased from the anoestrous to the transition period, and further increased once the breeding season had been reached. Furthermore, the secretion of GnRH during oestrus of the breeding season was considerably higher than during any other reproductive state tested. (2) The secretion of GnRH, as monitored at 10-min intervals, was best described as 'irregularly episodic' but not regularly pulsatile. That is, the occasional secretory events that were classified as peaks did not appear to occur with any obvious regularity, and certainly were not of the frequency described as circhoral by Dierschke *et al.* (1970) and Clarke & Cummins (1982). (3) The secretion of LH and GnRH was not always in temporal agreement. There were 1·5 times as many GnRH peaks as there were LH peaks, suggesting that not every GnRH secretory event initiates a comparable LH response. It is more difficult to explain the occurrence of LH secretory episodes that were unaccompanied by GnRH secretion. A possible explanation is that, if the position of the push–pull perfusion device were 'upstream' from a significant number of GnRH neurones, the additive firing of the neurones 'downstream' to the device, and therefore undetected at the position of the device, could account for the LH response.

These results agree well with our previous report of increased hypothalamic GnRH concentrations in ovariectomized Pony mares during the equivalent breeding season when compared with ovariecto-mized Pony mares during the equivalent anoestrus (Strauss *et al.*, 1979). These results have been corroborated by Hart *et al.* (1984). The measurement of GnRH concentrations in hypothalamic tissue is a useful technique that has produced valuable information (Wheaton *et al.*, 1975; Kumar *et al.*, 1977; Estes *et al.*, 1977; Strauss *et al.*, 1979; Kalra & Kalra, 1983; Hart *et al.*, 1984), but suffers the criticism of offering a static measurement, and does not provide information as to the secretory dynamics of GnRH. The present results suggest that our former conclusions of increased GnRH secretory activity during transition and the breeding season were correct, and that there is a seasonally-mediated increase in GnRH secretion shortly after the shortest day of the year.

The seasonal pattern of gonadotrophin secretion in mares has been well characterized (Ginther, 1979; Garcia *et al.*, 1979; Sharp, 1980) and it is accepted that circulating LH declines in the fall and winter. However, in the spring, LH concentrations remain low throughout most of the transition period into the breeding season proper. We have reported that, although transitional mares undergo marked folliculogenesis, and display constant or sporadic oestrous behaviour, there is little or no increase in circulating LH until 3–5 days before the first ovulation of the year (Sharp *et al.*, 1979b). On the other hand, there is a significant elevation in circulating FSH that accompanies the renewed follicular development, and FSH concentrations remain elevated, albeit irregular in concentration, throughout the early portion of the transition phase (Ginther, 1979; S. D. Davis, W. R. Grubaugh & D. C. Sharp, unpublished data). The present results confirm that there is a slight, but likely critical, increase in the secretion of GnRH during this time and we propose that the increased GnRH secretion during very early transition is responsible for the increased FSH secretion which, in turn, initiates follicular redevelopment. The long time between the first stages of follicular redevelopment and the first ovulation of the year (60–90 days in the Pony mare; Sharp, 1980) is a time that was characterized in these studies by little or no obvious change in GnRH secretion, except for increased mean secretory rate when compared with that of anoestrous mares. The lack of LH secretion could possibly be explained by reduced pituitary LH stores, as reported by Hart *et al.* (1984), and we have observed increased pituitary LH concentrations in ovariectomized, oestradiol-treated Pony mares during the equivalent anoestrus (D. C. Sharp & W. R. Grubaugh, unpublished observations). The steroid-treated mares had significantly elevated circulating LH concentrations and, more importantly, responded to exogenous GnRH administration with a significantly greater amount of LH secretion. The significance of these observations is that, as previously reported, during transition into the breeding season, mares exhibit marked follicular development, but do not ovulate until several dominant follicles have appeared and regressed (Sharp, 1980; Seamans & Sharp, 1982). Furthermore, the early transitional follicles do not appear to have normal steroidogenic patterns and concentrations of peripherally circulating oestrogens remain low until shortly before the first ovulation of the year.

We therefore propose that, shortly after the shortest day, increasing daylength, in some manner as yet undetermined, causes increased GnRH secretion from the hypothalamus. The available pituitary FSH stores and reduced pituitary LH stores (Hart *et al.*, 1984) permit increased secretion of FSH which, in turn, causes increased follicular development, but there is no increase in LH because of the reduced LH stores. The follicles which develop do not secrete sufficient concentrations of oestrogens to affect pituitary LH concentrations. In time, driven by mechanisms that are also not yet understood, the population of developing follicles becomes competent, and secretion of oestradiol increases pituitary LH stores so that an ovulatory surge can take place. After the consequent ovulation the hypothalamic–pituitary–ovarian axis is free to undergo the normal cyclic feedbacks that take place during the breeding season.

Supported by the National Institutes of Health Grant No. 10862. University of Florida Institute of Food and Agricultural Sciences, Florida Agricultural Experiment Station Journal Series No. 7784.

References

Christian, L.E., Everson, D.O. & Davis, S.L. (1978) A statistical method for detection of hormone secretory spikes. *J. Anim. Sci.* **46**, 699–706.

Clarke, I.J. & Cummins, J.T. (1982) The temporal relationship between gonadotropin releasing hormone (GnRH) and luteinizing hormone (LH) secretion in ovariectomized ewes. *Endocrinology* **111**, 1737–1739.

Dierschke, D.J., Bhattacharya, A.N., Atkinson, L.E. & Knobil, E. (1970) Circhoral oscillations of plasma LH levels in the ovariectomized rhesus monkey. *Endocrinology* **87**, 850–853.

Estes, K.S., Padmanabhan, V. & Convey, E. (1977) Localization of gonadotropin-releasing hormone within the bovine hypothalamus. *Biol. Reprod.* **17**, 706–711.

Garcia, M.C., Freedman, L.J. & Ginther, O.J. (1979) Interaction of seasonal and ovarian factors in the regulation of LH and FSH secretion in the mare. *J. Reprod. Fert., Suppl.* **27**, 103–111.

Ginther, O.J. (1979) *Reproductive Biology of the Mare.* McNaughton and Gunn, Ann Arbor.

Hart, P.J., Squires, E.L., Imel, K.J. & Nett, T.M. (1984) Seasonal variation in hypothalamic content of gonadotrophin-releasing hormone (GnRH), pituitary receptors for GnRH, and pituitary content of luteinizing hormone and follicle stimulating hormone in the mare. *Biol. Reprod.* **30**, 1055–1062.

Kalra, S.P. & Kalra, P.S. (1983) Neural regulation of luteinizing hormone secretion in the rat. *Endocr. Rev.* **4**, 311–351.

Kumar, M.S.A., Chen, C.L. & Kalra, S.P. (1977) Distribution of LH-RH in canine hypothalamus. *Endocrinology* **100**, Suppl. 59, Abstr. 293.

Levine, J.E. & Ramirez, V.D. (1982) Luteinizing hormone-releasing hormone release during the rat estrous cycle and after ovariectomy, as estimated with push-pull cannulae. *Endocrinology* **111**, 1439–1455.

Meyers, R.D. (1977) Chronic methods: intraventricular infusion, cerebrospinal fluid sampling, and push-pull perfusion. In *Methods in Psychobiology*, vol. 3, pp. 381–313. Ed. R. D. Meyers. Academic Press, New York.

Moore, R.Y. (1986) Neuroendocrine mechanisms: cells and systems. In *Reproductive Endocrinology*, pp. 3–32. Eds S. C. C. Yen & R. B. Jaffe. W. B. Saunders Co., Philadelphia.

Seamans, K.W. & Sharp, D.C. (1982) Changes in equine follicle aromatase activity during transition from winter anoestrus. *J. Reprod. Fert., Suppl.* **32**, 225–233.

Sharp, D.C. (1980) Environmental influences on reproduction in horses. *Vet. Clinics of North America: Large Animal Practice* **2**, 207–225.

Sharp, D.C., Kooistra, L. & Ginther, O.J. (1975) Effects of artificial light on the oestrous cycle of the mare. *J. Reprod. Fert., Suppl.* **24**, 241–246.

Sharp, D.C., Garcia, M.C. & Ginther, O.J. (1979a) Luteinizing hormone during sexual maturation in pony mares. *Am. J. vet. Res.* **40**, 584–586.

Sharp, D.C., Vernon, M.W. & Zavy, M.T. (1979b) Alteration of seasonal reproductive patterns following superior cervical ganglionectomy. *J. Reprod. Fert., Suppl.* **27**, 87–93.

Sharp, D.C., Grubaugh, W.R., Bergland, L.A., McDowell, K.J., Kilmer, D.M., Peck, L.S., Seamans, K.W. & Chen, C.L. (1982) Effects of pituitary stalk transection on endocrine function in Pony mares. *J. Reprod. Fert., Suppl.* **32**, 297–302.

Strauss, S.S., Chen, C.L., Kalra, S.P. & Sharp, D.C. (1979) Localization of gonadotrophin releasing hormone (GnRH) in the hypothalamus of ovariectomized Pony mares by season. *J. Reprod. Fert., Suppl.* **27**, 123–129.

Wheaton, J.W., Krulich, L. & McCann, S.M. (1975) Localization of luteinizing hormone-releasing hormone in the pre-optic area and hypothalamus of the rat using radioimmunoassay. *Endocrinology* **97**, 30–38.

Yen, S.S.C. (1986) Neuroendocrine control of hypophyseal function. In *Reproductive Endocrinology*, pp. 33–74. Eds S. C. C. Yen & R. B. Jaffe, W. B. Saunders Co., Philadelphia.

J. Reprod. Fert., Suppl. **35** (1987), 297–303

Clinical and endocrine aspects of the oestrous cycle in donkeys (*Equus asinus*)

M. Henry, A. E. F. Figueiredo, M. S. Palhares and M. Coryn*

*Department of Veterinary Clinics and Surgery, School of Veterinary Medicine, University of Minas Gerais, P.B. 567 Pampulha, 31.270 Belo Horizonte, Brazil and *Faculty of Veterinary Medicine, State University, 9000 Ghent, Belgium*

Summary. Oestrous behaviour and occurrence of ovulation was studied in 13 jenny asses (3–18 years) during a 15-month period. Teasing was performed daily and follicular growth, ovulation and changes of the genitalia were checked every other day by rectal palpation during dioestrus or the anovulatory season or daily during oestrus and when a follicle ⩾3 cm in diameter was present. Plasma oestrogens and progesterone levels were measured by radioimmunoassay (RIA).

Of the 13 jennies, 6 showed a typical seasonal ovarian activity, 1 stopped cycling only at the end of the experiment, 2 ovulated regularly and 4 did not cycle. Mean (±s.d.) length of the anovulatory season in the 6 jennies with seasonal cyclic ovarian activity was $166·3 \pm 63·2$ days (74–263 days). Irregular oestrous periods were detected during the anovulatory season of these jennies. Irregular signs of oestrus and follicular growth were detected in the 4 jennies which remained acyclic during the whole experiment. Mean (±s.d.) length of the ovulatory season was $197·8 \pm 63·4$ days (102–291 days). Overall mean ±s.d. lengths for the oestrous cycle and dioestrous and oestrous periods were $25·9 \pm 2·7$ ($n = 74$), $18·2 \pm 2·3$ ($n = 70$) and $7·9 \pm 2·5$ ($n = 79$) days. Ovulatory oestrous periods tended to be longer at the start and end of the season than during the summer months. Dioestrous length was not affected by season. Ovulations occurred near the end of oestrus and were more frequent from the left (61%, $n = 77$) than on the right (39%, $n = 49$) ovary ($P < 0·02$). The incidences of single, double, triple and quadruple ovulations were, respectively, 62·8% ($n = 54$), 25·5% ($n = 22$), 10·5% ($n = 9$) and 1·1% ($n = 1$).

One period of spontaneous prolongation of the luteal activity was detected and lasted for 80 days. Progesterone concentrations were <1 ng/ml from $1·75 \pm 0·88$ ($n = 8$) days before the onset of oestrous behaviour up to the day after ovulation. Corpus luteum lifespan was $15·7 \pm 2·1$ days (range 13–20 days, $n = 9$). Oestrogen values peaked 1–3 days before ovulation and reached dioestrous levels up to 3 days after ovulation.

Introduction

The oestrous cycle in donkeys has received little attention from researchers dealing with equine reproduction. The characteristics of the oestrous cycle have been reported (Nishikawa & Yamazaki, 1949a, b) and data on follicular growth and gonadotrophin changes during the oestrous cycle of jennies have been published (Vandeplassche *et al.*, 1981). However, information on donkey reproduction is still sparse.

The purpose of this study was to determine the effect of season on the behavioural pattern and ovarian activity in 13 jennies and to gather some information on the endocrinology of the oestrous cycle.

Materials and Methods

Reproductive behaviour and occurrence of ovulation were studied in 13 jenny asses from the beginning of April 1984 to the end of June 1985, at 19°51'06" latitude south and 043°57'03" longitude west. The jennies were of different breeds (5 Pêgas and 8 crossbreds), purchased from donkey breeders in the neighbourhood of the Veterinary School. The age, estimated by eruption and wear of the incisor teeth, ranged from 3 to 18 years, weight from 140 to 307 kg, and height from 1·10 to 1·33 m. Previous reproductive histories were not known.

The jennies were kept outdoors in a paddock and fed forage twice a day. The animals were confined in a group rather than individually. There was no control of the amount of forage ingested per jenny per day. Quality of food consumed was influenced by climatic changes and grain concentrate supplementation was not given in any period of the experiment. Salt and trace minerals were given *ad libitum*. Weight was monitored throughout the period, nothing being done to maintain the jennies in a constant body condition.

Daily individual teasing of the jennies was done in rotation by 3 jack asses. Courting was initiated through frontal contact followed by flank and eventually by posterior presentation of the jenny. The jack was allowed to mount the jenny at any time since the penis was laterally deviated to prevent copulation. Courting lasted for about 2 min and whenever teasing was doubtful, it was checked by using another teaser jackass. Jennies were considered to be in oestrus when 2 or more of the following signs were observed: posturing, mouth clapping, ear touching the neck, urinating and tail raising. Mouth clapping and ear touching the neck, generally associated, were the most evident signs of oestrus. The periods of non-receptivity were characterized by signs such as tail switching, kicking and moving.

Follicular growth, ovulation and changes of the genitalia were detected by rectal palpation once a day during oestrus or when a follicle of $\geqslant 3$ cm diameter was found, and every other day during the anovulatory season and dioestrus. Size of the largest follicles was estimated.

Blood samples were collected by venepuncture of the jugular, daily and once a week during the ovulatory and anovulatory seasons respectively. After centrifugation, plasma samples were stored at $-20°$C until assayed. Progesterone concentrations were measured by RIA after extraction with petroleumether (b.p. 30–45°C) and without further purification. The antiserum used, raised in sheep against progesterone-11-hemisuccinate–BSA cross-reacted with 17α-hydroxyprogesterone (1·15%), testosterone and cortisol (<0.03%). No cross-reaction was found for oestrone, oestradiol-17β, oestriol and pregnanediol. A satisfactory standard curve was obtained over the range 0·005–1 ng progesterone using 0·2 ml antiserum at a dilution of 1:50 000, binding 35% of the added [1,2,6,7-³H] progesterone at 0 ng progesterone. Sensitivity of the assay was 0·005 ng. The regression equation was $y = 0.005 + 0.890x$. Precision was 0.625 ± 0.025 ng (mean $\pm$ s.d., $n = 9$), 0.470 ± 0.015 ng, 0.320 ± 0.035 ng, 0.110 ± 0.020 ng, 0.035 ± 0.005 ng. The intra- and inter-assay coefficients of variation were 7·05 and 8·75% respectively.

Oestradiol-17β concentrations were measured by radioimmunoassay after extraction with diethylether and without further purification.The antiserum used, raised in sheep against 17β-oestradiol-3-hemisuccinate–BSA, cross-reacted with oestrone (9·2%), oestriol (2·3%) and oestradiol-17α (0·9%). A suitable standard curve (range 5–320 pg) was obtained with 0·5 ml antiserum (1:70 000), binding 40% of the added [2, 4, 6, 7-³H] oestradiol-17β at 0 pg oestradiol-17β. When known amounts of oestradiol-17β, added to 1 ml plasma, were assayed a regression line with the equation $y = 19 \pm 0.85x$ (y = pg found, x = pg added) was obtained. The precision of the assay was determined by the repeated assay of different plasma samples (1 ml): 124 ± 6 pg (mean $\pm$ s.d.), 104 ± 4 pg, 74 ± 2 pg, 57 ± 3 pg, 23 ± 2 pg. The intra- and inter-assay coefficients of variation were 5·75 and 8·30% respectively. Sensitivity was 5 pg.

Results and Discussion

The anovulatory season

Of the 13 jenny asses, 10 were acyclic at the start of the study; 6 went through a period of cyclic ovarian activity and 4 (Nos 1–4) remained acyclic during the whole experiment. Jenny 11 showed quiescence of the ovaries only at the end of the 1985 season. During the anovulatory season, one or more of the signs characteristic of oestrus were detected in 5 of the 6 jennies with seasonal cyclic ovarian activity (Table 1). The anovulatory oestrous periods were very irregular in length (range 1–21 days) and the interanovulatory oestrous intervals were quite variable (see Fig. 1). An effect of year on the occurrence of anovulatory oestrus at the end of the ovulatory season could be observed. More jennies showed anovulatory oestrous periods at the end of the 1984 than the 1985 season. The occurrence of oestrus during the anovulatory season has been previously reported for jennies (Nishikawa *et al.*, 1949a), as well for horses (van Niekerk, 1967a; Ginther, 1974).

During this period, follicular growth was found in all but one jenny. Most of the time, polyfollicular ovaries were present, with the number of palpable follicles ranging from 1 to 8 per ovary. Ovulation of follicles which apparently did not luteinize, as indicated by lack of a soft recent corpus

Table 1. Characteristics of the anovulatory and ovulatory seasons in donkeys

	Mean ± s.d. (*n*)	Range
Anovulatory season		
Length (days)	166·3 ± 63·2 (6)	74–263
Anovulatory oestrus		
No. of periods/jenny	4·2 ± 1·7	0– 10
No. of days/period	1·8 ± 1·9	1– 8
Interval between anovulatory oestrous		
periods (days)	6·5 ± 6·7	1– 22
Ovulatory season		
Length (days)	197·8 ± 63·4 (6)	102–291
No. of oestrous cycles/jenny	8·1 ± 1·7 (6)	5– 10
Length of oestrus (days)	7·9 ± 2·6 (9)	4– 22
Length of dioestrus (days)	18·2 ± 2·4 (7)	14– 30
Length of oestrous cycle (days)	25·9 ± 2·7 (7)	20– 40
No. of 'silent' oestrous periods (%)	2 (2)	
No. of split oestrous periods (%)	7 (7·7)	
No. of anovulatory oestrous periods (%)	3 (3)	
No. of suboestrous periods (%)	5 (5·4)	

luteum at rectal palpation and/or persistence of the external oestrous behaviour, was found 4 times in 3 jennies (Nos 1, 9 & 10).

Of 4 jennies which remained acyclic for the whole year, 3 showed irregular signs of oestrus. The number of anovulatory oestrous periods per jenny varied from 2 periods in Jenny 4 to 41 in Jenny 1 and the lengths of these oestrous periods ranged from 1 to 24 days. Follicles up to 2 cm diameter were found in all jennies during the whole period, but follicles of ovulatory size, which remained in the ovaries for long periods before regression, were found in Jennies 2 and 3 up to January and in Jennies 1 and 4 for almost the whole period. It was not clear why follicles of ovulatory size did not ovulate; presumably the final stimulus for ovulation was lacking. 'Silent' ovulation was detected only once in Jennies 2, 3 and 4 and twice in Jenny 1. The lack of resumption of cyclic ovarian activity could not be related to age or breed.

The ovulatory season

The characteristics of the ovulatory season are given in Table 1. The first ovulations of the season occurred between August and November and the last between February and April. Mean oestrous cycle length is similar to that reported for donkeys by Berliner *et al.* (1938) and Vandeplassche *et al.* (1981) although slightly longer than the findings of Nishikawa *et al.* (1949a) (22·8 ± 0·1 days). In comparison to a mean oestrous cycle length calculated for horses from an average of many reports (21·7 ± 3·5 days) (Ginther, 1979), the oestrous length in donkeys is clearly longer.

The mean lengths of components of the oestrous cycle for separate months of the year are shown in Table 2. Ovulatory oestrous periods tended to be longer at the beginning and end of the season as compared to the summer months. Comparison between means was significant only between some months. Dioestrous length was not affected by season. In horses, ovulatory oestrous periods are longer for the spring and autumn than for the summer months (van Niekerk, 1967b; Hughes *et al.*, 1972; Ginther, 1974). Similar results for the dioestrous length among months have been found in the mare (Hughes *et al.*, 1972). No trend to longer dioestrous periods during summer

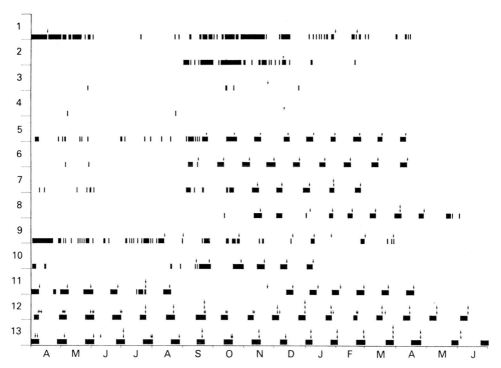

Fig. 1. Occurrence of oestrus (solid blocks) and ovulation (arrows) in jenny donkeys (Nos 1–13) over a 15-month period.

Table 2. Donkey oestrous cycle parameters among months

	Oestrous cycle (days)	Ovulatory oestrous period (days)	Dioestrous period (days)
July	28·0 ± 1·6 (4)	9·3 ± 2·8^{bf} (3)	18·0 ± 1·0 (4)
August	26·0 ± 1·7 (3)	8·7 ± 1·1^{bdf} (3)	18·3 ± 2·5 (3)
September	26·5 ± 2·1 (2)	13·0 ± 5·9^{a} (5)	19·0 ± 1·9 (5)
October	27·3 ± 2·4 (6)	8·8 ± 1·5^{bde} (6)	17·0 ± 1·7 (5)
November	25·0 ± 1·4 (6)	7·5 ± 1·0^{cd} (8)	17·6 ± 2·6 (7)
December	24·9 ± 2·1 (8)	6·6 ± 0·5^{ce} (9)	18·1 ± 1·3 (8)
January	25·0 ± 1·9 (10)	6·3 ± 1·1^{c} (11)	18·3 ± 2·5 (10)
February	24·3 ± 2·3 (8)	6·7 ± 1·8^{cef} (7)	17·4 ± 1·7 (7)
March	24·5 ± 2·2 (7)	6·6 ± 1·0^{ce} (7)	18·5 ± 1·5 (8)
April	26·0 ± 2·4 (6)	7·0 ± 1·4^{cef} (6)	22·0 ± 7·0 (3)
May	33·5 ± 9·2 (2)	7·0 ± 0·0^{cef} (2)	24·5 ± 7·8 (2)
June	29·0 ± 2·8 (2)	7·5 ± 0·7^{cef} (2)	23·0 ± 1·4 (2)

Values are mean ± s.d. for the no. of observations in parentheses.

Oestrous cycle, oestrous and dioestrous lengths are associated with the month in which they began. Within a column, means without a common superscript letter are significantly different ($P < 0.05$, Student's *t*-test).

than spring and autumn, as was found in a group of large ponies or crosses between ponies and horses (Ginther, 1974), was observed.

Silent oestrus involving quiet ovulation was found twice (Jennies 9 and 11) during the breeding season (2%). Suboestrus, characterized by the presence of some signs of oestrus of low intensity, lasting only 1–3 days and followed by an ovulation approximately at the expected time, was found

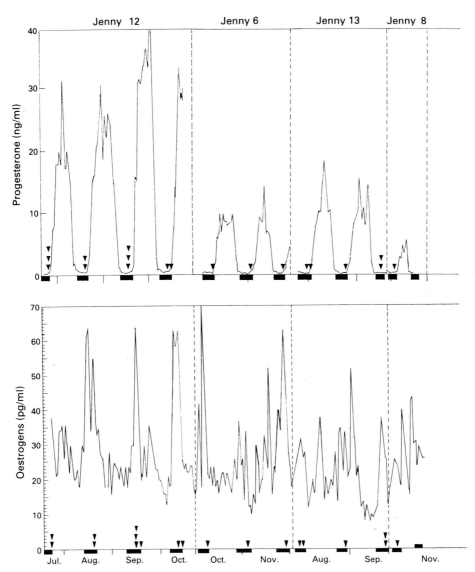

Fig. 2. Progestagen and oestrogen concentrations in the blood of non-pregnant jennies. Periods of oestrus are indicated by solid blocks and ovulations by arrows.

5 times (5·4%) in Jennies 8 and 9. Split oestrous periods, herein defined as non-oestrous behaviour during a period of ovulatory oestrus, was found in 5 jennies during 7 (7·7%) oestrous periods (3 in Jenny 9, 1 in each of Jennies 5, 6, 10 and 11). Rejection of the teaser occurred at any time within the oestrous period and lasted for 1–3 consecutive days. Eight of the 14 abnormal oestrous behaviours were related to Jenny 9 and the remaining 6 cases occurred in 5 jennies, suggesting that irregular oestrous behaviour during the breeding season may be more frequent in some individuals. Irregularities of oestrous behaviour have been found previously for jennies (Nishikawa *et al.*, 1949a) and in mares (Hughes *et al.*, 1972; Ginther, 1979) and the incidence of silent and split oestrus in mares has been shown to be 1 and 2%, respectively.

Anovulatory oestrous periods during the breeding season occurred 3 times (3%) in 2 jennies (2 in No. 9, 1 in No. 8). On one occasion in Jenny 9 and once in Jenny 8, one or more follicles of

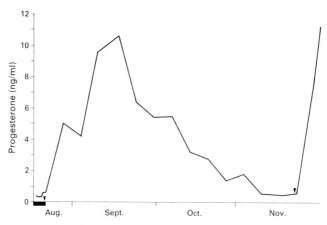

Fig. 3. Progestagen concentrations in the blood of a non-pregnant jenny during a prolonged luteal phase. Blood samples were taken once a week. The period of oestrus is indicated by the solid block and ovulations by arrows.

$\leqslant 2.5$ cm in diameter were palpated in one of the ovaries. The follicles remained palpable for the next 7 days after the end of oestrus. On the second occasion in Jenny 9, a follicle reached ovulatory size at the end of oestrus, regressing gradually thereafter. On all 3 occasions, an ovulatory oestrus was detected at the expected time, before and after these periods. It has been suggested that in horses luteinization of follicles occurred when an anovulatory oestrus was detected during the breeding season (Ginther, 1974). This may also have occurred in jennies, because rectal palpation was done only at 24-h intervals and so it cannot be excluded that an ovulation was not detected.

Forty (51·4%) of the ovulations were detected on the last day of oestrus, 19 (24·4%) at 1, 4 (5·1%) at 2, 3 (3·9%) at 3 and 1 (1·3%) at 4 days before the end of oestrus. The incidence of ovulations 1, 2 and 3 days after the end of oestrus was, respectively, 10·2 (N = 8), 2·6 (N = 2) and 1·3% (N = 1) of the ovulatory oestrous periods. The occurrence of ovulation near the end of oestrus is in agreement with previous reports for jennies (Nishikawa *et al.*, 1949b; Vandeplassche *et al.*, 1981) and horses (Hughes *et al.*, 1972; Ginther, 1974).

Ovulations were more frequent on the left (61%, n = 77) than on the right (39%, n = 49) side ($P < 0.02$). In a previous study, it was not concluded, due to a limited number of observations, whether ovulation was more frequent in one of the ovaries (Vandeplassche *et al.*, 1981). Nishikawa & Yamazaki (1949b) reported that, as in horses (Ginther, 1979), there was a slightly greater frequency of ovulation from the left ovary in jennies. In the present study the incidence of single, double, triple and quadruple ovulations was, respectively, 62·8% (n = 54), 25·5% (n = 22), 10·5% (n = 9) and 1·1% (n = 1): 12 (37·5%) of the multiple ovulations involved only the left, 5 (15·6%) only the right and 15 (46·9%) both ovaries. Incidences of multiple ovulations of 5·3% (Vandeplassche *et al.*, 1981) and 31·8% (Nishikawa & Yamazaki, 1949b) have been reported. In our study, 90% of the multiple ovulations occurred in 2 jennies (Jenny 12 ovulated more than one follicle in 16 and Jenny 13 in 13 of the 16 ovulatory oestrous periods studied). This suggests that occurrence of multiple ovulations may be more frequent in some individuals, as has been reported for horses (Hughes *et al.*, 1972). If so, the lack of such animals in a studied group could greatly change the observed incidence of multiple ovulations. Breed seems to be one of the determining factors in the occurrence of multiple ovulations in horses (Henry *et al.*, 1982), but this still needs to be studied in donkeys. The intervals between ovulations were: 24 (57·1%) on the same day, 9 (21·4%) at 1 day, 2 (14·3%) at 2 days, and 1 (2·4%) at 3,4 and 7 days later. As the time from the first ovulation increased, the incidence of a second ovulation decreased.

Progesterone concentrations were low ($\leqslant 1$ ng/ml from 1.75 ± 0.88 ($n = 8$) days before the onset of oestrous behaviour up to the day after the ovulation (Fig. 2). Dioestrous values reached a plateau between Days 4 and 14 after ovulation and decreased from the plateau to oestrous levels within 2 days. Progesterone concentrations were >2 ng/ml for a period of 15.7 ± 2.1 days (range 13–20 days, $n = 9$). Maximal progesterone values during dioestrus were influenced by the number of ovulations: concentrations with 1, 2 and 4 ovulations were, respectively, 15.6, 37.0 and 51.4 ng/ ml (see Fig. 2). Oestrogen concentrations began to rise just after the decrease of progesterone, peaked 1–3 days before ovulation and reached dioestrous levels on the day of to 3 days after ovulation (see Fig. 2).

Spontaneous prolonged luteal function was observed only once in Jenny 11: progesterone concentrations remained above 2 ng/ml for about 80 days (see Fig. 3). At least one follicle of 2–3 cm diameter was present in all examinations during the whole period. The largest follicle palpated was 4 cm diameter between 17 and 24 days before complete regression of the corpus luteum and thereafter regressed. No ovulation occurred during this period. A dry, close and pale cervix and a tense uterine tonus was found during this phase. The jenny was not mated during the normal preceding ovulatory oestrus, but contracted piroplasmosis during the expected dioestrus. The ovulatory oestrus succeeding the prolonged luteal phase was silent. The same phenomenon, described as pseudopregnancy, has been found in unmated mares (Ginther, 1979). The persistence of the corpus luteum for periods of about 2 months has been reported for mares (Hughes *et al.*, 1972).

References

Berliner, V.R., Sheets, W.E., Means, R.H. & Cowart, F.C. (1938) Oestrous cycle of jenneys and sperm production of jacks. *Proc. Am. Soc. Anim. Prod.* pp. 295–298.

Ginther, O.J. (1974) Occurrence of anestrus, estrus, diestrus and ovulation over a 12-month period in mares. *Am. J. vet. Res.* **35**, 1173–1179.

Ginther, O.J. (1979) *Reproductive Biology of the Mare: Basic and Applied Aspects.* McNaughton and Gunn, Ann. Abor.

Henry, M., Coryn, M. & Vandeplassche, M. (1982) Multiple ovulation in the mare. *Zentbl. VetMed. A* **29**, 170–184.

Hughes, J.P., Stabenfeldt, G.H. & Evans, J.W. (1972) Clinical and endocrine aspects of the oestrous cycle of the mare. *Proc. Ann. Conv. Am. Ass. equine Pract.,* pp. 119–148.

Nishikawa, Y. & Yamazaki, Y. (1949a) Studies on reproduction in asses. I. Breeding season, oestrous cycle and length of heat. *Jap. J. Zootech. Sci.* **19**, 119–124.

Nishikawa, Y. & Yamazaki, Y. (1949b) Studies on reproduction in asses. II. Growth of follicles in ovaries and ovulation. *Jap. J. Zootech. Sci.* **20**, 29–32.

Vandeplassche, M., Wesson, J.A. & Ginther, O.J. (1981) Behavioral, follicular and gonadotropin changes during the estrous cycle in donkeys. *Theriogenology* **16**, 239–249.

van Niekerk, C.H. (1967a) Pattern of the oestrous cycle of mares. I. The breeding season. *J. S. Afr. vet. med. Assoc.* **38**, 295–298.

van Niekerk, C.H. (1967b) Pattern of the oestrous cycle. II. The duration of the oestrous cycle and the oestrous period. *J. S. Afr. vet. med. Assoc.* **38**, 299–307.

Vermeulen, A. & Verdonck, L. (1976) Radio-immuno-assay of 17-hydroxy-5-androstan-3-one, 4-androstene-3,17-dione, dehydroepiandrosterone, 17-hydroxy progesterone and progesterone and its application to human male plasma. *J. Steroid Biochem.* **7**, 1–5.

J. Reprod. Fert., Suppl. **35** (1987), 305–309

Printed in Great Britain
© 1987 Journals of Reproduction & Fertility Ltd

Cervico-endometrial cytology and physiological aspects of the post-partum mare

A. Saltiel*, A. Gutierrez, N. de Buen-Llado† and C. Sosa‡

Departamentos de Reproducción, Histología† y Genética‡, Facultad de Medicina Veterinaria y Zootecnia, Universidad Nacional Autónoma de México, Ciudad Universitaria, 04510, Mexico

Summary. After parturition, Thoroughbred mares were mated at the first post-partum oestrus (N = 24) or at a subsequent oestrus (N = 12). All mares were examined daily for: oestrous detection, palpation *per rectum* of the genital tract, vaginoscopic examination and cervico-endometrial cytology. Pregnancy diagnosis was carried out at Days 18, 35 and 45 after mating. An identical first service conception rate of 50% was found in both groups. The number of neutrophils followed a descending profile to only scattered cells at the first post-partum oestrus and in Group II mares remained at this very low level during the period of study. The percentage of histiocytes and eosinophils increased on Days 10 and 17, and 5 and 6 *post partum*, respectively. The percentage of lymphocytes remained low and constant during the period of study. Bacterial flora decreased from Days 2 to 9 and increased from Days 13 to 17 *post partum*. Cellular necrosis and erythrocytes decreased and ciliocytopholia increased as mares approached the first post-partum oestrus. A positive correlation was found between amount, colour, viscosity and turbidity of secretions and all cellular types, ciliocytopholia, cellular necrosis and bacterial flora. The number of neutrophils was positively correlated with the percentage of eosinophils, bacterial flora and cellular necrosis but had a negative association with the presence of ciliocytopholia. Two mares that did not re-establish cyclic ovarian activity after parturition had delayed uterine involution. Mares not conceiving at the first post-partum oestrus exhibited a more prolonged presence of cellular necrosis and erythrocytes and an increased presence of bacterial flora and lymphocytes as compared to mares conceiving at this period.

Introduction

Despite a number of studies on mating during the first post-partum oestrus (Ginther, 1979; Loy, 1980), this method is still considered controversial, probably due to the lack of objective parameters on which to base a decision. Although the bacteriological environment and uterine involutionary characteristics have been investigated (Andrews & McKenzie, 1941; Gygax *et al.*, 1979), cytological studies of the uterus at first post-partum oestrus have received little attention. The purpose of the present study was to characterize and correlate age and parity of the mare and uterine cytology, oestrous behaviour, palpation *per rectum* of the reproductive tract and vaginoscopic findings of the genital tract with the fertility of the first post-partum oestrus.

Materials and Methods

A total of 36 pregnant Thoroughbred mares was used in this study. As the first 24 mares foaled, they were alternately assigned to two groups. The remainder were incorporated into Group I due to management requirements of the

---

*Present address: Criadero de Caballos Deportivos La Silla, Hidalgo 1911 Poniente, Colonia Obispado, Monterrey, Nuevo León 64060, Mexico.

breeding farm where the study was performed. Group I mares were mated with the proven stallions during the first post-partum oestrus and subsequent oestrous periods when necessary. Group II mares were also mated with the same stallions beginning at the second post-partum oestrus, which was induced by prostaglandin treatment (Tolskdorff *et al.*, 1976) on Day 6 of the first post-partum dioestrous period.

From Day 1 *post partum* until they were mated for the first time, mares in both groups were examined daily for their behavioural response to a stallion; palpation *per rectum* of the genital tract; and a vaginoscopic examination through a glass speculum and sampling for cervico-endometrial cytology (introduction of a sterile swab previously moistened with distilled water through the cervix into the uterus). A smear of the fluid was obtained on a glass slide, air-dried and subsequently stained with Giemsa stain (Luna, 1968). Altogether, 542 slides were evaluated cytologically. The percentage of inflammatory cell types was determined by counting 200 cells. Due to the large numbers of neutrophils observed, this cell type was also appraised on a scale of 1–5, i.e. from very slight to very abundant, respectively. Additionally, the presence or absence of erythrocytes, cellular necrosis, ciliocytopholia and intra- and/or extracellular bacterial flora was evaluated. Secretions as seen during vaginoscopic examination were categorized according to amount, colour, viscosity, and turbidity. Pregnancy diagnosis was performed by palpation *per rectum* of the uterus at 18, 35 and 45 days after mating. Data were arranged by mare and group, according to age and parity (Neter & Nasserman, 1974). Normal values were postulated for every variable (Daniel, 1979) and correlations between variables were estimated (Weisberg, 1980).

Results and Discussion

Reproductive parameters

An identical first service conception rate (50%) was achieved by mares in both groups. Average intervals from parturition to oestrus, ovulation and conception were 7·2 and 7·9 days, 10·9 and 11·5 days and 26·5 and 44·3 days for Groups I and II, respectively. Despite the shortened dioestrous phase due to prostaglandin treatment in Group II mares, the 17·8 days difference found between groups in the average interval from parturition to conception supports the observation of Loy (1980) that mating at the first post-partum oestrus is a reasonable choice in attempting to shorten the foaling interval in a time-limited breeding season.

Follicular activity

An increase in follicle diameter to > 35 mm was observed beginning at Day 4 *post partum* and reached a peak between Days 7 and 10 *post partum*. This coincides with oestrous behaviour (Fig. 1). No significant differences were found between the activity of left and right ovaries. This finding agrees with results previously reported for cyclic mares (Saltiel *et al.*, 1982). The chronology of ovarian events documented in this study is comparable to that observed by others (Ganjam *et al.*, 1975; Ginther, 1979).

Physical characteristics of the uterus

The physical characteristics of the uterus in mares mated at the second post-partum oestrus (Group II) are shown in Fig. 1. Uterine tone and tubularity increased during the first 5 days *post partum*. As mares entered first post-partum oestrus, these characteristics decreased with a subsequent increase after ovulation, probably due to the establishment of luteal activity (Ganjam *et al.*, 1975). Uterine involution based on size and evaluated by palpation *per rectum* was not correlated with the age of the mare, parity, year of last parturition or conception rate. These results suggest that physical characteristics of the genitalia have a doubtful predictability in terms of uterine ability to support a pregnancy when first post-partum oestrus is used for mating, a finding which is in agreement with those of other authors (Gygax *et al.*, 1979; Loy, 1980). The uterus tended to involute more quickly in mares that exhibited an early post-partum oestrus than in 2 mares in which mares foaled but failed to establish follicular activity, a situation similar to a winter-like anoestrus.

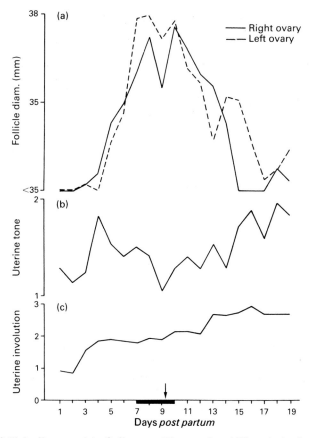

Fig. 1. Average follicle diameter (a) of all mares (Groups I and II) and physical characteristics of the uterus of mares not mated at the first post-partum oestrus (Group II); (b) average uterine tone and tubularity (1 = moderate, 2 = good); and (c) average uterine involution (1 = slight, 2 = moderate, 3 = good). Bar indicates days of oestrus and arrow the mean day of ovulation.

Cervico-endometrial cytology

The cervico-endometrial cytology of mares mated at the second post-partum oestrus (Group II) is shown in Fig. 2. The number of neutrophils decreased *post partum* to the onset of the first oestrus and remained at this value during the period of study. Positive correlations ($P < 0.01$) were established between neutrophil numbers and percentage of eosinophils, presence of bacterial flora and cellular necrosis, and negative correlations ($P < 0.01$) with the presence of ciliocytopholia. The percentage of neutrophils showed a biphasic profile that decreased on Days 7 and 16 *post partum*. These fluctuations were not significant, indicating that the estimated number of neutrophils more efficiently reflected the dynamics of this cellular type. The percentage of histiocytes and eosinophils showed a sharp rise on Days 10 and 17, and 5 and 6 *post partum*, respectively. No explanation is offered for these responses, although a relationship between eosinophils and the entrance of air into the genital tract of the mare has been suggested (Slusher *et al.*, 1984). However, Beeson & Bass (1977) have found receptors for oestrogens in eosinophils of the genital tract of the rat and suggested an oestrogen-transport function of this cell type. Percentage of lymphocytes remained low and constant during the period of study. The large amount of bacterial flora found in the present study is comparable with results published by Gygax *et al.* (1979) based on bacteriological observations. These results confirm that it is possible for post-partum mares to become pregnant at

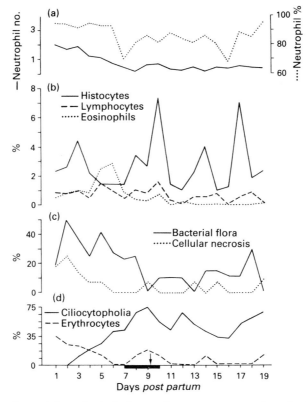

Fig. 2. Cervico-endometrial cytological changes in mares not mated at the first post-partum oestrus (Group II): (a) neutrophils (1 = very slight, 2 = slight, 3 = moderate); (b) histocytes, lymphocytes and eosinophils; (c) bacterial flora and cellular necrosis; and (d) ciliocytopholia and erythrocytes. Bar indicates days of oestrus and the arrow the mean day of ovulation.

an early post-partum mating in spite of an acute surface bacterial contamination. In the present study, cellular necrosis occurred both in epithelial and inflammatory cells and was positively correlated ($P < 0.01$) with neutrophil numbers and percentage of lymphocytes. Cellular necrosis decreased drastically before the appearance of the first post-partum oestrus, which suggests that cellular regeneration takes place rapidly after parturition. This cellular regeneration is undoubtedly the basis for the mare's ability to conceive at the first post-partum oestrus. Ciliocytopholia, which refers to the appearance of apical segments of ciliated endocervical and endometrial cells, is considered a degenerative change of the genital epithelia in other species (Hollander & Gupta, 1974). The ciliocytopholia profile showed a marked increase as mares entered the first post-partum oestrus and was positively correlated ($P < 0.01$) with uterine involution. In women, ciliogenesis is considered an indicator of oestrogenic stimulation (Ferenczy, 1981). Erythrocytes were found in most post-partum mares in this study; the presence of erythrocytes decreased as mares entered the first post-partum oestrus. A positive correlation ($P < 0.01$) was found between the presence of this cell type and the number of neutrophils and percentage of histiocytes. Uterine involution was retarded in the presence of erythrocytes ($P < 0.01$).

Secretions

The amount, colour, viscosity and turbidity of secretions decreased as mares entered the first post-partum oestrus. Positive correlations ($P < 0.01$) were found between these variables and the number of neutrophils, percentage of lymphocytes and eosinophils, the presence of erythrocytes, cellular necrosis and bacterial flora, and a negative correlation ($P < 0.01$) was found with the presence of ciliocytopholia.

Fertility of mares mated at first post-partum oestrus

Mares not conceiving at the first post-partum oestrus had a more prolonged presence of cellular necrosis and erythrocytes than did mares conceiving during the first post-partum oestrus. Bacterial flora were present in relation to mating (Days 9–11 *post partum*) in more mares not conceiving during the first post-partum oestrus than in mares conceiving at the same period. A greater percentage of lymphocytes also characterized these mares. These results suggest that the lack of conception at the first post-partum oestrus could be due to a slower and/or less effective uterine involutionary process and to reduced uterine defence mechanisms against bacteria, as suggested by Asbury *et al.* (1984), for cyclic mares.

This study was supported in part by a grant from Evelyn Aluja, in whose memory we dedicate this paper.

References

Andrews, F.N. & McKenzie, F.F. (1941) Estrus, ovulation and related phenomena in the mare. *Res. Bull. Mo. agric. Exp. Stn*, No. **329**, 1–117.

Asbury, A.C., Gorman, N.T. & Foster, G.W. (1984) Uterine defense mechanisms in the mare: serum opsonins affecting phagocytosis of *Streptococcus zooepidemicus* by equine neutrophils. *Theriogenology* **21**, 375–385.

Beeson, P.B. & Bass, D.A. (1977) The eosinophil. In *Major Problems in Internal Medicine*, pp. 154–158, Ed. L. H. Smith, Jr., W. B. Saunders Co., Philadelphia.

Daniel, W.W. (1979) *Bioestadística. Bases para el Análisis de las Ciencias de la Salud*. Limusa, México.

Ferenczy, A. (1981) The ultrastructural dynamics of endometrial hyperplasia and neoplasia. In *Advances in Clinical Cytology*, pp. 1–43. Eds L. G. Koss & D. V. Coleman. Butterworths, London.

Ganjam, V.K., Kenney, R.M. & Flickinger, G. (1975) Plasma progestagens in cycling, pregnant and *post-partum* mares. *J. Reprod. Fert., Suppl.* **23**, 441–447.

Ginther, O.J. (1979) *Reproductive Biology of the Mare: Basic and Applied Aspects*. McNaughton & Gunn, Inc., Ann Arbor.

Gygax, A.P., Ganjam, V.K. & Kenney, R.M. (1979) Clinical, microbiological and histological changes associated with uterine involution in the mare. *J. Reprod. Fert., Suppl.* **27**, 571–578.

Hollander, D.H. & Gupta, P.K. (1974) Detached ciliary tufts in cervico-vaginal smears. *Acta cytol.* **18**, 367–369.

Loy, R.G. (1980) Characteristics of *post-partum* reproduction in mares. *Vet. Clin. N.A.* **2**, 345–359.

Luna, G.L. (1968) *Manual of Histologic Staining Methods of the Armed Forces Institute of Pathology*, 3rd edn, McGraw-Hill Book Co., New York.

Neter, J. & Nasserman, W. (1974) *Applied Linear Statistical Models*. R. D. Irwin, Inc., Illinois.

Saltiel, A., Calderon, A., Garcia, N. & Hurley, D.P. (1982) Ovarian activity in the mare between latitude 15° and 22°N. *J. Reprod. Fert., Suppl.* **32**, 261–267.

Slusher, S.H., Freeman, K.P. & Roszel, J.F. (1984) Eosinophils in equine uterine cytology and histology specimens. *J. Am. vet. med. Ass.* **184**, 665–670.

Tolksdorff, E., Jochle, W., Lamond, D.R., Klug, E. & Merkt, H. (1976) Induction of ovulation during the *post-partum* period in the Thoroughbred mare with a prostaglandin analogue. *Theriogenology* **6**, 403–409.

Weisberg, M.S. (1980) *Applied Linear Regression*. John Wiley and Sons, New York.

J. Reprod. Fert., Suppl. **35** (1987), 311–316

Effects of susceptibility of mares to endometritis and stage of cycle on phagocytic activity of uterine-derived neutrophils

A. C. Asbury and P. J. Hansen

Department of Reproduction, College of Veterinary Medicine, University of Florida, Gainesville, Florida 32610, U.S.A.

Summary. Fourteen mares, 7 susceptible and 7 resistant to bacterial endometritis, were used to provide circulating and uterine-derived neutrophils. Uterine neutrophils were recruited by inoculating cell-free filtrates of *Streptococcus zooepidemicus*, or control vehicle. Mares were assigned to schedules for collection of neutrophils at oestrus or dioestrus. Phagocytic activity of circulating and uterine cells was evaluated by an assay for chemiluminescence after addition of opsonized streptococci. Chemiluminescence generated by circulating neutrophils was greater ($P < 0.05$) for susceptible mares (28 ± 4.9 V) than for resistant mares (13.4 ± 2.8 V), but was unaffected by stage of cycle or by the interaction. Chemiluminescence by uterine-derived neutrophils from susceptible mares was greater ($P < 0.10$) than for resistant mares. There was an interaction ($P < 0.05$) with stage of oestrous cycle. Uterine cells from resistant mares in oestrus produced more chemiluminescence than did those from resistant mares in dioestrus (11.5 ± 4.1 *vs* 7.1 ± 2.1 V). The activity of uterine-derived cells of susceptible mares was unaffected by stage of cycle. Susceptibility to endometritis was not associated with a defect in the phagocytic function of uterine neutrophils. Also the function of uterine cells from resistant mares was greater during oestrus than dioestrus.

Introduction

Phagocytosis by neutrophils in the uterine lumen is an important component of the defence mechanisms against bacterial endometritis, a major cause of equine infertility (Hughes & Loy, 1969; Peterson *et al.*, 1969). Various aspects of the phagocytic process have been studied in an attempt to associate some degree of failure with susceptibility to infection and subsequent infertility. Two recent publications have suggested that the neutrophil itself loses some of its function during the migration to the uterus of mares susceptible to endometritis (Liu *et al.*, 1985; Cheung *et al.*, 1985). The first report suggested that chemotactic responsiveness and cell deformability both suffered during the migration to the uterine lumen in such mares. In the second study, the ability of neutrophils harvested from the uterus of susceptible mares to ingest and kill organisms was greatly reduced. Phagocytosis and killing were measured using a fluorochrome assay to count ingested *Candida albicans* blastospores and to evaluate viability to the ingested organisms.

The study reported here was designed to evaluate phagocytic function of uterine neutrophils by a different assay and under somewhat different conditions. Neutrophils were recruited to the uterus using a cell-free filtrate of bacterial growth rather than a bacterial inoculum. Phagocytosis was evaluated by an assay for chemiluminescence using opsonized streptococci as the substrate.

Materials and Methods

Animals. Fourteen cyclic mares of light breeds were kept in paddocks, fed hay and grain and teased daily with a stallion to determine sexual receptivity. Ovulation was determined by palpation of the ovaries *per rectum*. Seven mares were classified as resistant to endometritis based on nulliparous status and freedom from any signs of uterine inflammation as determined by physical examination and endometrial culture, cytology and biopsy. Seven mares were classified as susceptible based on a documented history of persistent clinical endometritis, histological evidence of uterine inflammation and some evidence of active endometritis at the onset of the experiment.

Uterine inoculum. A cell-free filtrate of cultures of *Streptococcus zooepidemicus* was used as an inoculum to initiate the inflammation necessary for the recruitment of neutrophils to the uterine lumen. The filtrate was prepared as described elsewhere (Couto & Hughes, 1985). Briefly, *S. zooepidemicus* of horse uterine origin was grown in Brain-Heart Infusion Broth, and filtered through an 0·2 µm filter. All of the filtrate required for the experiment was prepared in one batch and frozen in 50-ml aliquants. Non-inoculated broth was similarly filtered and frozen to be used as a control inoculum.

Experimental design. Within classification (susceptible or resistant), mares were randomly assigned to inoculation with filtrate or with the control vehicle and to one of two experimental schedules (Fig. 1). In Schedule A, neutrophil function (see below) was evaluated 7 or 8 days after evolution during the first oestrous cycle and again on Day 2 or 3 of the third oestrous period. In Schedule B, evaluations were on Days 2 or 3 of the second oestrous period and 7 or 8 days after ovulation had occurred during the third oestrous period. Since the procedures involved uterine infusions and potential endogenous prostaglandin release, mares in Schedule B were treated with 10 mg prostaglandin F-2α injected intramuscularly on Day 7 of the second dioestrus to mimic the effect of the neutrophil collection procedure done at dioestrus in Schedule A.

To ensure retention of uterine catheters during the neutrophil collection procedures, a purse string suture was placed in the cervix of each mare by a technique previously described (Asbury *et al.*, 1984). Sutures were placed on Day 2 (Schedule A) or Day 10 (Schedule B) of dioestrus. The cervical canal was closed to a diameter that would allow free placement or removal of a Foley catheter. The sutures remained in place for the duration of the experiment.

Neutrophil collection procedures. A 24 French Foley catheter with a 75-ml balloon (C. R. Bard, Murray Hill, NJ) was inserted through the cervix and the balloon inflated. The uterus was washed with 80 ml sterile saline (9 g NaCl/l) by repeated injection and aspiration via a syringe. If the flushings appeared grossly cellular, 2 or 3 litres of saline were infused and recovered to remove accumulated neutrophils. Then 50 ml inoculum (filtrate or control vehicle) were placed into the uterus and the exterior end of the catheter was sealed to exclude air.

At 2 and 4 h after inoculation the uterus was again flushed with 80 ml saline to remove material that had accumulated in response to the inoculum. The flushing at 4 h was used for the preparation of uterine neutrophils. The recovered material was centrifuged at 2000 *g* and the pellet of cells was then treated to prepare neutrophils for phagocytic assay as described below. At the same time as the uterus was flushed, circulating blood was collected into heparinized tubes by jugular venepuncture to provide circulating neutrophils.

Neutrophil preparation. Circulating neutrophils were prepared for chemiluminescence assay by a technique described previously (Asbury *et al.*, 1982). Briefly, erythrocytes were sedimented in 6% dextran. Leucocytes were removed from the remaining supernatant by centrifugation (400 *g*, 12 min). The resultant pellet was washed and suspended in saline, and treated with ammonium chloride and EDTA to lyse any remaining erythrocytes. Isotonicity was restored by the addition of citrate–saline, and centrifugation was repeated (400 *g*, 7 min). The pellet was washed again with saline.

Neutrophils were counted in a haemocytometer under phase-contrast microscopy, allowing easy differentiation of neutrophils from lymphocytes by shape and conformation. The cell concentrate was then diluted with Hanks Buffered Saline Solution (HBSS; GIBCO, NY) to a neutrophil concentration of 1×10^6/ml. Uterine neutrophils were prepared by washing the original pellet from the uterine flush four times with HBSS, counting neutrophils as described above, and adjusting the concentration to 1×10^6/ml. Previous trials had shown no difference in phagocytic rate when uterine neutrophils were prepared in this manner in comparison to preparing them as the circulating cells were handled.

Reagents for chemiluminescence. Luminol (5-amino-2,3-dihydro-1,4-phthalazinedione) solutions were prepared by previously reported methods (Asbury *et al.*, 1982). Concentrated solutions were prepared 2 or 3 days before any assay, and dilutions (1/240) for use in a particular run were made on the day before the assay. All solutions of luminol were stored in the dark until used.

Streptococci, the substrate for phagocytosis in the assay, were prepared as previously described (Asbury *et al.*, 1982). Briefly, cultures of *S. zooepidemicus* were grown in Trypticase Soy Broth, concentrated by centrifugation, and washed twice with phosphate-buffered saline (PBS). The bacteria were then killed by autoclaving, washed again in PBS and the concentration adjusted with PBS so that when 20 µl were diluted in 5 ml PBS the absorbance at 525 nm was 0·3. Bacteria were opsonized in the serum of the mare whose neutrophils were being tested. A 0·1 ml aliquant of the stock suspension of streptococci was incubated in 1 ml serum at 37°C for 20 min. The combination was then centrifuged at 2000 *g* for 10 min, and the resulting pellet was resuspended in 2·5 ml saline.

Chemiluminescence assay. Chemiluminescence was measured in a luminometer (LKB, Model 1251, Gaithersburg, MD), controlled by a custom program in a personal computer (Apple Computer, Inc, Cupertino, CA). The luminometer was run in the integration mode (time = 5 sec) with continuous mixing at 37°C. Five tubes were used in each assay, one control (circulating neutrophils, no streptococci) and duplicates of circulating (Tubes 2 and 3) and uterine (Tubes 4 and 5) neutrophils plus opsonized streptococci. To each of the first 3 tubes, 0·5 ml circulating neutrophil preparation was added and an equal volume of uterine neutrophils was added to Tubes 4 and 5. Then 0·25 ml of the final dilution

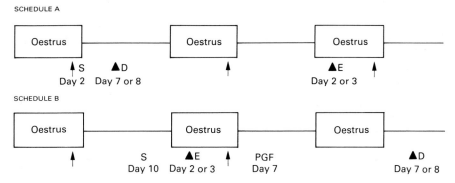

Fig. 1. Diagram of the experimental design indicating sequence of procedures relative to oestrous cycles. ↑ = ovulation; S = purse string suture applied to cervix; ▲E = collection of neutrophils at oestrus; ▲D = collection of neutrophils at dioestrus; PGF = prostaglandin injection.

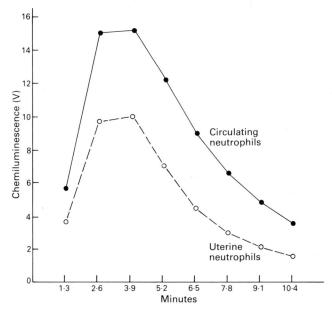

Fig. 2. Chemiluminescence curves for circulating and uterine neutrophils from Mare J, Schedule A, collection at oestrus. The shape of the curves is typical for the results obtained throughout the experiment.

of luminol was added to all tubes, and the cells and neutrophils were allowed to equilibrate. When the readings from all 5 tubes were stable and below 25 mV per counting interval (5 sec), the assay was started by pipetting 0·5 ml HBSS to Tube 1 (control) and 0·5 ml opsonized streptococci to Tubes 2, 3, 4 and 5. These reagents were added at 15-sec intervals, the time required for the luminometer to advance and count one tube. With 5 tubes in the chamber the total time elapsed in counting each set once was 80 sec (5 sec required to restart each set). Points on the chemiluminescence curve (Fig. 2) are therefore 80 sec apart. Each set of 5 tubes was counted 8 times without interruption.

Analysis of data. Data on peak chemiluminescence produced by uterine neutrophils were analysed separately from data on chemiluminescence produced by circulating neutrophils. Both sets of data were analysed by analysis of variance with main effects of status (susceptible *versus* resistant), treatment (filtrate *versus* vehicle control) and schedule (A *versus* B). Stage of cycle effect (oestrus *versus* dioestrus) was nested within animal.

Results

A typical set of chemiluminescence response data is presented in Fig. 2. The curve for circulating and uterine neutrophils is representative of the shapes of all chemiluminescence responses obtained in the experiment. For each tube, peak chemiluminescence was invariably noted at either the second or third reading, thus ensuring that the peak was recorded in each case. Values for Tube 1 (control) were essentially zero in all readings and were therefore not subtracted from the counts obtained for Tubes 2–5.

Mean ($\pm$ s.e.m.) peaks of chemiluminescence generated by circulating and uterine neutrophils in this study are presented in Table 1. For circulating and uterine neutrophils there were no significant differences between filtrate and control vehicle, nor any significant differences between Schedule A or B, so data are presented as pooled across these classifications. Circulating cells from susceptible

Table 1. Values (mean $\pm$ s.e.m.) for chemiluminescence (V) of neutrophils

	Susceptible mares	Resistant mares
Circulating neutrophils	$28\cdot0 \pm 4\cdot9$*	$13\cdot0 \pm 2\cdot8$*
Uterine neutrophils	$11\cdot1 \pm 2\cdot5$†	$8\cdot8 \pm 2\cdot3$†

*$P < 0\cdot05$; †$P < 0\cdot10$.

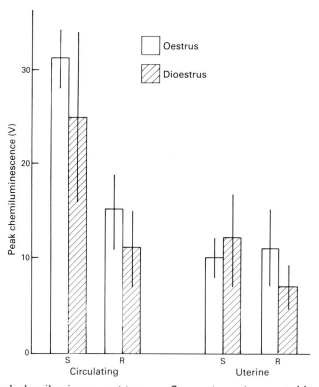

Fig. 3. Mean peak chemiluminescence ($\pm$ s.e.m., 7 mares/group) generated by circulating and uterine neutrophils, analysed for effect of oestrus and dioestrus, and for effect of resistance (R) and susceptibility (S) to endometritis.

mares produced a greater ($P < 0.05$) response than did those from resistant mares. There was no effect of the stage of cycle or the interaction on chemiluminescence production by these cells.

Uterine cells from susceptible mares produced a greater ($P < 0.10$) response than those from resistant mares. An interaction with stage of cycle was noted with the uterine-derived cells ($P < 0.05$) and is illustrated in Fig. 3. Uterine neutrophils from resistant mares in oestrus produced more CL than did those cells from mares in dioestrus. No difference between oestrus and dioestrus was noted in the function of cells derived from the uteri of susceptible mares.

Discussion

The increased activity of uterine-derived neutrophils from resistant mares during oestrus compared with dioestrus is logical in the scheme of uterine defence mechanisms. Naturally occurring contamination of the reproductive tract is an oestrus-related event. A similar effect of cycle on uterine resistance has been noted by Washburn *et al.* (1982) who reported that mares treated with progesterone were more likely to develop endometritis when the uterus was inoculated with bacteria than were mares treated with oestrogen. The absence of a stage of cycle effect on uterine cells from susceptible mares is difficult to evaluate. In fact the positive effect of susceptibility to endometritis on phagocytic function of both uterine and circulating neutrophils warrants further study.

A major inference of this study is that neutrophils harvested from the uterine lumen of susceptible mares are as capable of luminol-dependent chemiluminescence as are those cells derived from resistant mares. Our interpretation is in disagreement with the conclusions of other investigators (Cheung *et al.*, 1985). Several differences in our approaches to this question should be noted, the first of which is the use of a chemiluminescence assay to evaluate neutrophil function. Chemiluminescence production has been closely correlated with oxidation of glucose via the hexose monophosphate shunt, and with the subsequent production of superoxide anion and singlet oxygen (Allen *et al.*, 1972; Allen *et al.*, 1974). These workers associated the sharp burst in chemiluminescence from neutrophils with the initiation of phagocytosis, and the oxidative metabolism with microbicidal activity. The generation of chemiluminescence is therefore a measure of cell metabolism during the phagocytic process. Given the limitations of any in-vitro evaluation of a complex biological process, we consider the assay for chemiluminescence to be related to in-vivo phagocytosis.

Neutrophils in this study were recruited from the uterus with a cell-free initiator of inflammation. This method delays any phagocytic activity by the neutrophils until they contact the bacterial substrate in the chemiluminescence assay. It is possible that early phagocytic activity by the neutrophils during recruitment with a bacterial inoculum could influence the results of a subsequent phagocytic assay. Another potential difference between this experiment and the one reported by Cheung *et al.* (1985) is the microorganisms used as a target for phagocytosis.

Selection of susceptible mares for this trial was based on known history of endometritis, and on clinical and laboratory evidence of uterine inflammation. There was no requirement for these mares to be classified as Category III by the system of Kenney (1978), and in fact only 2 of 7 were so classified. Although there may be subjective differences in the assignment of classification by different individuals, the major feature of Category III is perceived to be endometrial fibrosis.

In summary, the results presented here suggest that a defect in the neutrophil itself is not a factor in decreased resistance to bacterial endometritis. We previously reported no differences in the numbers of neutrophils migrating to the uterine lumen between resistant and susceptible mares, and no differences in amounts of protein transported (Asbury *et al.*, 1982). We suggest that the ability of the neutrophil to exhibit phagocytosis in the uterus is independent of the state of the mare's resistance. Other factors affecting phagocytotic function, such as chemotaxis, opsonization and adherence, along with mechanical evacuation of uterine contents, may be more important in determining the success of uterine defence.

We thank E. Chalmers, M. Cowart, K. Fincher, M. Fuller, A. Sage and S. Sams for invaluable assistance with this study. Supported in part by a Research Development Award from the Division of Sponsored Research, University of Florida. Published as Florida Agricultural Experimental Station Journal Series No. 7510.

References

Allen, R.C., Stjernholm, R.L. & Steele, R.H. (1972) Evidence for the generation of an electronic excitation state(s) in human polymorphonuclear leukocytes and its participation in bactericidal activity. *Biochem. Biophys Res. Commun.* **47**, 679–684.

Allen, R.C., Yevich, S.J., Orth, R.W. & Steele, R.H. (1974) The superoxide anion and singlet molecular oxygen: their role in the microbicidal activity of the polymorphonuclear leukocyte. *Biochem. Biophys. Res. Commun.* **60**, 909–917.

Asbury, A.C., Schultz, K.T., Klesius, P.H., Foster, G.W. & Washburn, S.M. (1982) Factors affecting phagocytosis by neutrophils in the mare's uterus. *J. Reprod. Fert., Suppl.* **32**, 151–159.

Asbury, A.C., Bloomer, R.J. & Foster, G.W. (1984) A technique for long-term collection of uterine contents from mares. *Theriogenology* **21**, 367–374.

Cheung, A.T.W., Liu, I.K.M., Walsh, E.M. & Miller, M.E. (1985) Phagocytic and killing capacities of uterine-derived polymorphonuclear leukocytes from mares resistant and susceptible to chronic endometritis. *Am. J. vet. Res.* **46**, 1938–1940.

Couto, M.A. & Hughes, J.P. (1985) Intra-uterine inoculation of a bacteria-free filtrate of *Streptococcus zooepidemicus* in clinically normal and infected mares. *J. Eq. vet. Sci.* **5**, 81–86.

Hughes, J.P. & Loy, R.G. (1969) Investigations on the effect of intrauterine inoculation of *Streptococcus zooepidemicus* in the mare. *Proc. 15th Ann. Conv. Am. Assoc. equine Pract., Houston,* pp. 289–292.

Kenney, R.M. (1978) Cyclic and pathological changes of the mare endometrium as detected by biopsy, with a note on early embryonic death. *J. Am. vet. Med. Ass.* **172**, 241–262.

Liu, I.K.M., Cheung, A.T.W., Walsh, E.M., Miller, M.E. & Lindenberg, P.M. (1985) Comparison of peripheral blood and uterine-derived polymorphonuclear leukocytes from mares resistant and susceptible to chronic endometritis: chemotactic and cell elastimetry analysis. *Am. J. vet. Res.* **46**, 917–920.

Peterson, F.B., McFeely, R.A. & David, J.S.E. (1969) Studies on the pathogenesis of endometritis in the mare. *Proc. 15th Ann. Conv. Am. Assoc. equine Pract., Houston,* pp. 279–287.

Washburn, S.M., Klesius, P.H., Ganjam, V.K. & Brown, B.G. (1982) Effect of estrogen and progesterone on the phagocytic response of ovariectomized mares infected in utero with β-hemolytic streptococci. *Am. J. vet. Res.* **43**, 1367–1370.

J. Reprod. Fert., Suppl. **35** (1987), 317–325

Printed in Great Britain
© 1987 Journals of Reproduction & Fertility Ltd

Dynamics of the acute uterine response to infection, endotoxin infusion and physical manipulation of the reproductive tract in the mare

P. Williamson, S. Munyua, R. Martin and W. J. Penhale

School of Veterinary Studies, Murdoch University, Murdoch, Western Australia 6150, Australia

Summary. The uterine responses after the infusion of saline (PBS), a bacterial suspension, or lipopolysaccharide derived from *Escherichia coli*, and after stimulation of the reproductive tract were compared. All infusions provoked a response involving both serum proteins and leucocytes. Protein levels peaked within a few hours of infusion, whereas leucocyte concentration peaked later at around 6 h. Bacterial recovery from the uterus followed a similar pattern, with recovery falling dramatically by 12 h. In mares known to be susceptible to infection large numbers of bacteria were again recovered after 24 h. No differences were apparent between resistant and susceptible mares in protein or leucocyte concentrations. Stimulation of the cervix and uterus resulted in a protein and neutrophil response. In contrast, vaginal stimulation failed to provoke the uterine defences.

Introduction

The uterus of the mare is normally resistant to infection, and invading bacteria are rapidly eliminated (Peterson *et al.*, 1969; Hughes & Loy, 1969). At present the factors which prevent the establishment of infection are not well understood, but research to date has shown that the reaction of the uterus to bacterial invasion has both soluble and cellular components (Peterson *et al.*, 1969; Hughes & Loy, 1969, 1975; Asbury *et al.*, 1980; Liu *et al.*, 1986). These investigators considered that an infiltration of neutrophils into the endometrium and uterine lumen was responsible for the early removal of invading bacteria in normal mares. In addition to the neutrophil infiltration, an influx of serum proteins occurs after infection (Strzemienski & Kenney, 1984) and the endometrium also has the ability to mount a typical mucosal immune response, with the selective production of IgA and IgG (Mitchell *et al.*, 1982; Widders *et al.*, 1985).

While most mares are able to withstand infection, certain mares are prone to recurrent uterine infection, despite proper treatment and veterinary supervision. This suggests that these mares have impaired uterine defences. Asbury *et al.* (1980) found no difference between mares known to be resistant or susceptible to uterine infection in the concentrations of neutrophils recovered from the uterus after the infusion of large numbers of pathogenic bacteria.

The present study was undertaken to characterize the neutrophil and protein influx into the uterus during the acute response to bacterial invasion, and to determine the pattern of bacterial recovery from the uterus after experimental infection. The uterine reaction to infection was compared in mares known to be resistant or susceptible to uterine infection. In addition, non-specific stimuli were applied to the uterus to determine whether the acute uterine response was similar for all noxious stimuli, or was specific for infectious agents.

Materials and Methods

Animals

A total of 64 Standardbred or Thoroughbred mares, aged between 2 and 16 years, were used in the experiments. All mares were judged healthy and were run on irrigated pasture and drenched periodically for parasite control.

Mares were initially classified as being resistant or susceptible to uterine infection on the basis of their having a history of recurrent endometritis, or having active infection or inflammation on swabbing and biopsy at the preliminary examination (Williamson *et al.*, 1983). This classification system was later modified for mares involved in the bacterial challenge experiments. These mares were judged as being susceptible to infection if bacteria were still recovered from the uterus 4 days after experimental infection with *Streptococcus zooepidemicus*. The rationale for adopting this criterion was based on consideration of the critical time allowable for mares to clear bacteria which contaminated the uterus at service, with regard to the arrival and subsequent survival of embryos in the uterus. In addition, the results of swabbing 20 of these mares at 10 days after infection resulted in 17 of the 20 mares having the same bacteriological status, indicating that swabs taken at 4 days give a reasonably accurate prediction of the persistence of infection.

Experiment 1: uterine challenge with Streptococcus zooepidemicus. An active inoculum of $> 10^9$ *Streptococcus zooepidemicus* (originally isolated from the uterus of a clinically infected mare) in 2 ml phosphate-buffered saline (PBS) was infused into the uterus by catheter and washed through 10 ml isotonic mannitol. Isotonic mannitol was selected to facilitate a continuing study of the ionic balance of uterine fluids during infection. Mannitol is an inert sugar and the bacteria were shown to survive in it for a period similar to that in PBS (W. J. Penhale, unpublished observations). In this experiment, 43 mares in groups of 5–6, with sampling extending through all seasons over several years, were used.

The mares were divided into two broad treatment groups. In Trial 1 (30 mares) serial flushings of the uterus were collected after bacterial challenge (27 mares) or infusion of 10 ml isotonic mannitol (3 mares). The timing of the serial sampling was staggered between groups to enable samples to be available for analysis at short intervals after infection, while minimizing trauma which could distort the findings. A guarded uterine swab for bacteriology was obtained before each uterine flush. The times of serial collections for different groups were: 0, 2, 4, 6, 24, 48 and 96 h in Group 1 (12 mares); 0, 3, 6, 9, 12, 15, 24 and 96 h in Group 2 (8 mares); and 0, 2, 6, 24, 48 and 96 h in Group 3 (10 mares).

The uterus was again swabbed and flushed at 10 days after bacterial challenge for 22 of these mares (Groups 1 and 3).

Uterine flushings were analysed for leucocyte concentration, protein concentration and composition, the concentration of immunoglobulins A (IgA) and G (IgG), and were cultured for bacterial growth.

In Trial 2 (13 mares), serial swabbing of the uterus after bacterial challenge was undertaken to determine the pattern of recovery of bacteria without the possible distortions caused by serial flushings. The times of serial swabbing of the two groups were: 0, 1, 2, 4, 6, 12, 24 and 96 h in Group 4 (6 mares) and 0, 1, 3, 6, 9, 12, 18, 24, 48 and 96 h in Group 5 (7 mares).

Experiment 2: uterine infusion of an endotoxin, lipopolysaccharide Escherichia coli. A group of 21 mares was used in this experiment, the same mares being used a number of times for infusion and sample collection, with an interval of at least 2 weeks between successive use. A solution containing 500 µg lipopolysaccharide *Escherichia coli* 0127:B8 (Sigma Chemicals, St Louis, MO, U.S.A.) per 20 ml PBS, pH 7·4, was infused into the uterus via a modified Foley catheter. Uterine samples were collected before infusion and at 2, 3, 6, 9 or 24 h after infusion of lipopolysaccharide. An equal number of samples was collected from the uteri of mares infused with PBS alone, to provide control mare data. A minimum of 3 mares was sampled at each of the designated times after treatment. Uterine flushings were subjected to the same analysis as those collected in Exp. 1.

Experiment 3: response of the endometrium to manipulation of different segments of the reproductive tract. The uterine response to physical manipulation of the uterus, cervix or vagina was assessed by collecting uterine flushings 6 h after physical manipulation, and 9 mares were utilized for stimulation of each of the different segments of the tract.

Uterine stimulation was effected by infusing 60 ml PBS into the uterus by catheter and then vigorously massaging the uterus *per rectum* for 3 min. It is probable that during this treatment the whole tract was disturbed. For cervical stimulation a Foley catheter was passed into the cervix, with care being taken to ensure that uterine contact did not occur. The cuff of the catheter was inflated and the cervix was then massaged *per rectum* for 3 min. In the course of this procedure it was assumed that the vagina was also disturbed. In the third group of mares the vagina was massaged for 3 min *per rectum* after the positioning in it of an inflated glove.

Control mares (2) had 60 ml PBS infused gently into the uterus, without massaging, or were subjected to rectal massage of the uterus without PBS infusion.

Methods

Routine bacteriology and cytology. Guarded endometrial swabs for bacteriology were obtained at each of the collection times detailed in Exp. 1 and from mares in Exps 2 and 3 before the trials. Swabs were inoculated onto agar gel containing 10% sheep blood as described by Williamson *et al.* (1983). A second swab from the uterus at each collection was rolled on to gelatin-coated glass slides for cytological examination (Wingfield-Digby, 1978). Smears for cytology were stained with Wright's stain, coverslipped and examined under × 400 magnification for neutrophils. In addition, when uterine washings were obtained from mares in the bacterial challenge experiments, a loopful of the washings containing around 20 µl fluid was streaked onto blood agar plates for bacteriology.

Inoculated blood agar plates were incubated for 24–48 h at 37°C and the number of colony forming units (CFU) of the infused organism (*Streptococcus zooepidemicus*) was estimated. When a heavy growth of bacteria occurred it was impossible to count bacterial colonies accurately, and the estimated number of colonies was recorded as being greater than 1000. Random samples of the bacteria recovered from the uterus of mares in the bacterial challenge experiments were serotyped to ensure that the organism infused was causing the observed response.

Uterine flushing procedure. Uterine flushings were obtained using a modified Foley catheter (Zavy *et al.*, 1977). In sampling each mare, 50 ml of isotonic mannitol (bacterial challenge experiments) or phosphate-buffered saline (PBS) (endotoxin and manipulation experiments) was instilled into the uterus and the uterus was gently massaged *per rectum* for 4 min to ensure adequate mixing. Fluid was then allowed to flow back along the catheter from the uterus into sterile containers, until sufficient fluid for testing (20 ml) had been collected. The procedure was than halted, to minimize the time and trauma of collection. Samples (2 ml) of each flushing were decanted for estimation of leucocyte concentration. The remaining fluid was immediately centrifuged at 2400 *g* for 10 min, and the supernatant was decanted, placed on ice, then frozen and stored at −20°C until analysed for proteins.

Enumeration of leucocytes in uterine washings. Leucocyte concentrations were estimated on aliquants of fresh washings using a Coulter electronic particle counter. The proportion of neutrophils, the dominant leucocyte type present, was then estimated from a differential count performed on the uterine smears made just before each flushing. In the few samples in which significant numbers of erythrocytes were present, a correction factor based on the approximate number of erythrocytes, was used to estimate true neutrophil concentrations (Coles, 1980).

Protein assays on uterine flushings. Protein concentration in the uterine flushings were determined by spectro-photometry, using the technique described by Bradford (1976) and Spector (1978).

Electrophoretic analysis was performed on undiluted uterine fluids collected 12 h after infection, and on uterine flushings collected throughout the sampling period, using mare serum for comparison and a Gelman semimicro-electrophoresis chamber and Gelman sepraphore (Clifford Instruments Inc. 1969). The cellulose acetate strip was soaked in Gelman high-resolution buffer for 10 min before 10 µl of the sample were loaded. After 25 min of electro-phoresis the acetate strips were stained with Ponceux S and then dried before scanning. Immunoglobulin levels (IgA, IgG and IgM) assayed in the uterine washings were measured using the radiol diffusion technique described by Mancini *et al.* (1965). Commercial antiserum (Flow Laboratories, Sydney) was used against IgG. Antiserum prepared in our laboratories by vaccinating rabbits was used against IgA.

Statistics. The results of the neutrophil, protein and immunoglobulin concentration studies in uterine flushings are presented as group means ± standard errors. With the bacteriological counts, estimates of the number of colonies of *Streptococcus zooepidemicus* (CFU) growing on each bacterial plate were made by counting a number of sectors on each plate. When the approximate number of CFUs exceeded 1000, it was impossible to count accurately their numbers, and a total of > 1000 was recorded. This made it impossible to calculate group means or apply conventional statistical analysis to the data. The results are therefore represented as schematic curves, fitted as closely as possible to the estimated median value of each group for each sampling time. The curves indicate the relative changes in bacterial recovery rates over the sampling time. Student's *t* test was used to determine the significance of differences between group means for protein and neutrophil concentrations.

Results

Experiment 1

Uterine challenge with Streptococcus zooepidemicus. Mares used in this trial all returned negative bacteriological growth from pre-treatment uterine swabs. All of the mares challenged with *Streptococcus zooepidemicus* infusion showed a clinical response characterized by a mucopurulent discharge from the vulva within 12 h of infection. Differential counts on the uterine smears prepared from swabs taken 1 h after bacterial challenge showed that a rapid inflammatory reponse occurred, which was predominantly neutrophilic. Neutrophils remained the predominant type of leucocyte present throughout the remainder of the sampling period in mares in which inflammation persisted. The mean neutrophil concentrations in uterine washings recovered at different intervals after bacterial challenge are shown in Fig. 1. Neutrophil numbers rose dramatically within 2 h of infusion of the bacteria, with levels fluctuating but remaining elevated at each of the subsequent sampling times, up to 4 days after challenge. By 10 days after challenge neutrophil concentrations had fallen but still remained above the pretreatment levels. No difference in either the pattern of recovery or the concentrations of neutrophils was seen between the resistant mares and the suscept-ible mares still infected at 4 days after infection. The unchallenged control mare showed a milder, predominantly neutrophilic response which followed a pattern similar to that of the infected mares.

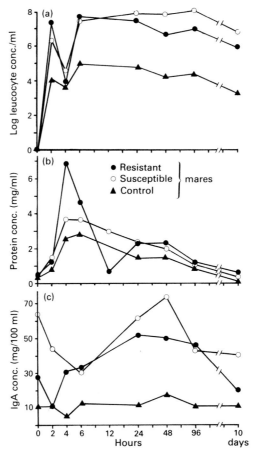

Fig. 1. The mean leucocyte (a), protein (b) and IgA (c) concentrations in uterine flushings collected from mares at the specified times after uterine challenge with an active inoculum of *Streptococcus zooepidemicus.*

Protein concentration in uterine washings. The protein concentrations in pre-challenge uterine washings obtained from mares susceptible to uterine infection were similar to those in the washings from resistant and control mares (0.45 ± 0.18 mg/ml compared with 0.50 ± 0.30 and 0.35 ± 0.10 mg/ml respectively). The control mares and challenge mares showed a marked increase in protein concentrations in uterine flushings collected within the first 6 h of infusion of the bacterial suspension or isotonic mannitol (Fig. 1). There were no significant differences in the initial responses of mares in the 3 groups. By 12 h after challenge, protein levels in the washings had begun a steady decline in the susceptible and control mares, a decline which continued at all subsequent sampling times. In the resistant mares, the fall in protein levels appeared greater than with the susceptible and control mares at 12 h, but by 24 h concentrations had risen again to levels similar to those measured in the other groups. Protein levels then gradually fell in parallel in the resistant, susceptible and control mares.

The results of densitometer analyses of a typical 12-h sample are presented in Table 1. All major serum proteins were present in the uterine fluids. The relative concentration of albumin was generally lower in uterine fluid than in serum, but relative concentrations of α-2, β-1 and β-2 globulins were higher in uterine fluid and washings.

Table 1. Concentrations of protein components in serum and uterine fluid from a typical mare obtained 6 h after bacterial infusion into the uterus, calculated from a densitometer tracing after cellulose acetate strip electrophoresis

	Uterine fluid (undiluted)		Serum	
	Concentration (mg/ml)	Relative %	Concentration (mg/ml)	Relative %
Total protein	68		56	
Albumin	11·9	26·4	28·6	51·1
α1	1·5	2·2	3·9	6·9
α2	8·7	12·8	4·4	7·9
β1	16·0	23·5	5·54	9·9
β2	9·6	14·2	4·81	8·9
γ	14·2	21·0	8·18	15·7

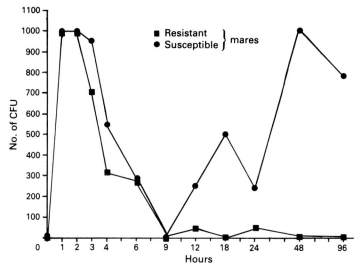

Fig. 2. Mean CFU of *Streptococcus zooepidemicus* recovered from uterine swabs taken at different intervals after bacterial challenge by infusion of $> 10^9$ bacteria into the uterus.

Immunoglobulins in uterine washings. Immunoglobulins IgA and IgG were detected in most of the uterine washings tested. The minimum levels of immunoglobulins detected in the assays were 22 mg IgA/100 ml, 15 mg IgG/100 ml and 10 mg IgM/100 ml. IgA and IgG were present in the uterus of all groups of mares at the time of first sampling, before bacterial challenge. Levels rose to a peak in samples collected at 24–48 h after bacterial challenge (Fig. 1). At 96 h after challenge levels of IgG had declined in washings from susceptible and resistant mares. In susceptible mares IgA values remained elevated in washings collected up to 10 days after infection. Immunoglobulin levels in washings from control mares rose over the sampling period in parallel to the profiles from the resistant mares, but with lower levels present. IgM was detected in only 4 uterine washing samples, from 4 different mares.

Bacterial recovery from the uterus after challenge. The results of bacterial culture at 96 h for 32 mares for which an accurate history was available matched up perfectly with the anticipated resistance or susceptibility of these mares based on that history. A great deal of variation occurred between mares in the pattern of recovery of bacteria, especially in those mares that were subjected to concurrent uterine swabbing and flushing. Recovery patterns differed between the mares still infected at 4 days after challenge (susceptible) and those that had cleared the infection (resistant). These patterns are illustrated in Fig. 2 based on the results obtained from Trial 2 mares which were not flushed over the period when swabs were collected. Bacterial recovery was high at the initial swabbing, with most mares having in excess of 1000 CFU growing on inoculated culture plates. The numbers of bacteria recovered fell dramatically in almost all mares (from both the resistant and susceptible groups) in samples collected between 4 and 24 h after challenge. Recovery remained low in resistant mares at subsequent swabbings, but in the susceptible mares large numbers of bacteria were again recovered at 48 or 96 h of sampling. The pattern was essentially the same for Trial 1 mares, with the initial recovery of high numbers of bacteria, the rapid fall in recovery rates within 12–24 h of challenge and their reappearance at 48 and 96 h in susceptible mares. All resistant mares in Trial 1 had cleared the infection by 48 h after challenge.

Experiment 2: uterine infusion of endotoxin

The concentrations of leucocytes in serial uterine flushings collected before and after lipopolysaccharide or PBS infusion are shown in Fig. 3. Concentrations rose within 2 h of infusion, peaking at 6 h in both groups. Concentrations of leucocytes were higher in flushings from the lipopolysaccharide group at every collection interval. At 9 h after infusion, leucocyte concentrations

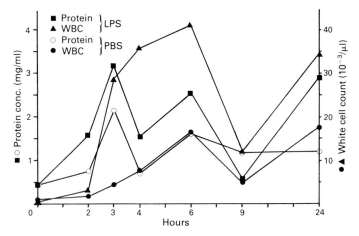

Fig. 3. Concentrations of white blood cells and protein in uterine flushings collected at the specified intervals after infusion of lipopolysaccharide (LPS) or PBS into the uterus.

Table 2. Protein and leucocyte concentrations in uterine flushings collected before (0 h) and 6 h after mechanical stimulation of different parts of the reproductive tract

	0 h	6 h		
		Vaginal stimulation	Cervical stimulation	Uterine stimulation
Protein conc. (mg/ml)	0·39 ± 0·16	0·49 ± 0·37	1·66 ± 1·32	1·75 ± 1·3
White cell count (× 10⁻³/µl)	0·49 ± 0·17	0·81 ± 0·34	19·2 ± 23·2	17·2 ± 12·7

Values are mean ± s.e.m.

decreased in both groups, but had risen again to nearly peak levels in the 24-h samples from all mares except for one of the mares infused with lipopolysaccharide.

Results of protein measurements on uterine flushings collected serially after the infusions are also presented in Fig. 3. Before infusion the mean concentration of protein in uterine fluid was 0.44 ± 0.28 and 0.43 ± 0.26 mg/ml in the PBS and lipopolysaccharide groups respectively. Mean protein concentration rose significantly over the 3 h after infusion, then, after falling at 4 h, rose again in both groups at 6 h. Levels fluctuated at subsequent samplings, but still remained elevated above baseline values at 24 h.

Experiment 3: physical stimulation of the reproductive tract

The results of analysis of uterine flushings collected 6 h after physical stimulation of different segments of the reproductive tract are presented in Table 2. There was a significant increase in both leucocyte and protein concentrations ($P < 0.05$) in the 6 h samples from mares subjected to uterine and cervical stimulation. This increase did not occur in the mares subjected only to vaginal stimulation. The control procedures caused only mild, transient increases in protein and neutrophil levels in washings.

Discussion

The adverse stimuli applied to the uterus in the present study provoked a response typical of that reported after experimental infection, with an increase in leucocyte and protein concentrations in uterine washings (Peterson *et al.*, 1969; Hughes & Loy, 1969; Blue *et al.*, 1982). In the present study serial sampling during the acute phase of the response gave a clearer picture of the dynamics of the uterine reaction. Leucocytes (predominantly neutrophils) poured into the uterus within a few hours of bacterial or lipopolysaccharide invasion. A biphasic pattern of influx emerged, with the concentrations peaking at around 6 h after bacterial or lipopolysaccharide infusion, dipping at 12 h, and then increasing again by 24 h. Samples collected after bacterial infusion showed that these increased levels of leucocytes persisted until at least 4 days after challenge, then fell by 10 days in both the resistant and susceptible groups of mares, despite the persistence of infection in susceptible mares. A similar pattern occurred with protein concentrations after infusion with β-haemolytic streptococcus and lipopolysaccharide, with peak levels occurring 3–4 h after infusion. The pattern of change in leucocyte and protein concentrations in flushings from mares sampled serially over the observation period (Exp. 1) was similar to that from mares collected singly, at set time intervals after infusion (Exp. 2). This indicates that the serial flushing of the uterus at 1 or 2 h intervals did not markedly distort the response of the uterus to the initial infusion. The similarity of bacterial recovery patterns from the uteri of flushed and non-flushed mares also provides evidence that the flushing procedure did not grossly distort the response pattern.

Increases in protein and neutrophil concentrations were also seen 6 h after the physical stimulation of the cervix and uterus. The combined evidence of the present experiments suggests that an acute, non-specific reaction was triggered when the uterus and cervix was disturbed by infusion or manipulation. The reaction to saline infusion into the uterus during the luteal phase has been shown to shorten the cycle by releasing prostaglandin (Neely *et al.*, 1974). It therefore appears that an inflammatory reaction or physical disturbance of the cervix or uterus can precipitate the non-specific defence mechanism of the uterus.

The secondary influx of neutrophils and the increased levels of IgA seen several days after infection, especially evident in those mares that remained infected beyond 48 h, was not paralleled by continued high levels of protein in the uterine washes. This indicates that the initial protein influx, rich with serum proteins, was not maintained with persistent infection, and that the continuing IgA response was due to local production of antibody, directed more specifically at the invading pathogen, in accord with the conclusions of Kenney & Khaleel (1975) and Mitchell *et al.* (1982).

No differences were apparent in the acute uterine reaction between susceptible and resistant mares. However, the pattern of recovery of bacteria from the uterus of groups of susceptible and resistant mares, initially similar, diverged after the 24-h swabbing. Bacterial recovery from both groups fell dramatically 3–6 h after infection and remained low at all further swabbings in the resistant mares. In the susceptible mares, however, large numbers of bacteria were again recovered from 48 or 96 h, and in subsequent swabs. These results indicate that the acute defence mechanisms of the uterus are initially effective in combating the invading organisms in both groups of mares, but that this affect cannot be sustained in the susceptible mares. The copious mucopurulent discharge observed at the vulva of all mares after infection may form part of the initial defences, mechanically flushing some of the invading organisms from the uterus. Those bacteria that remain in the suscept- ible mares appear to become established in the uterus. Liu *et al.* (1986) showed that neutrophils from the uterus of mares with severe endometrial degeneration had poor chemotaxis and deform- ability when harvested 12 h after intrauterine infusion of β-haemolytic streptococcus. This degeneration of neutrophil function occurred earlier in these mares than in mares with normal endometria. It is generally considered that mares with degenerate endometria are more susceptible to endometritis than normal mares, and thus equate to the susceptible mares of the present study. The combined findings of these 2 studies suggest that early loss of neutrophil function in susceptible mares, despite the high concentration of neutrophils present, may be associated with the break- down of the uterine defence mechanisms in susceptible mares. Bacteria may become established at the time of neutrophil malfunction and multiply, leading to their re-emergence in large numbers at 48–96 h, as noted in the present study. Liu *et al.* (1986) also found that neutrophils recovered from the uterus of susceptible mares at 15 and 20 h after infection again showed increased chemotactic activity, and suggested that this renewed activity indicated continuing recruitment of neutrophils due to the persistence of bacteria in the uteri of these mares. More research is necessary to clarify this relationship between infection and uterine neutrophil function. It seems likely that the impaired function of uterine neutrophils is due to the local uterine environment, as Asbury *et al.* (1980) have shown that circulating neutrophils remain functional throughout experimental uterine infection in both resistant and susceptible mares. Asbury *et al.* (1984) suggested that bacterial opsonization was impaired in the uterus of mares susceptible to endometritis, interfering with phagocytosis. In the present study it was shown that all major serum proteins, including the β-globulins which contain the principal opsonins, were present in the initial influx into the uterus. It is possible that their opsonizing activity was not sustained, or that the opsonizing agents could not function properly in the uterine environment of susceptible mares.

This study has clarified the nature of the uterine reaction to infection and to other adverse stimuli. The initial non-specific response seems effective in mares both susceptible to, and resistant to uterine infection. This defence cannot be maintained in susceptible mares and infection becomes established.

We thank Mr J. Murray, Ms M. Sharpe, J. O'Connor and A. Dunning for expert technical assistance and advice, and Mr D. Brockway and his staff for management of the mares. This work was funded by the Australian Equine Research Foundation and by the University's Special Research Grant Fund.

References

Asbury, A.C., Halliwell, R.E.W., Foster, G.W. & Longino, S.J. (1980) Immunoglobulins in uterine secretions of mares with differing resistance to endometritis. *Theriogenology* **14**, 299–308.

Asbury, A.C., Gorman, N.T. & Foster, G. (1984) Uterine defense mechanisms in the mare: serum opsonins affecting phagocytosis of *Streptococcus zooepidemicus* by equine neutrophils. *Theriogenology* **21**, 375–385.

Blue, M.G., Brady, A.A., Davidson, J.W. & Kenney, R.M. (1982) Studies on the composition and anti- bacterial activity of uterine fluid in mares. *J. Reprod. Fert., Suppl.* **32**, 143–146.

Bradford, M.M. (1976) A rapid and sensitive method for quantitation of microgram quantities of protein using the principle of protein dye binding. *Analyt. Biochem.* **72**, 248–254.

Coles, E.H. (1980) *Veterinary Clinical Pathology*, 2nd edn, pp. 392–393. W.B. Saunders Co., Burgess Hill.

Hughes, J.P. & Loy, R.G. (1969) Investigations on the effect of intra-uterine inoculation of *Streptococcus zooepidemicus* in the mare. *Proc. 15th Am. Assoc. Equine Pract.* 289–291.

Hughes, J.P. & Loy, R.G. (1975) The relation of infection to infertility in the mare and stallion. *Equine vet. J.* **7**, 155–159.

Kenney, R.M. & Khaleel, A. (1975) Bacteriostatic activity of the mare uterus: a progress report on immunoglobulins. *J. Reprod. Fert., Suppl.* **23**, 357–358.

Liu, I.K.M., Cheung, A.T.W., Walsh, W.M. & Ayin, S. (1986) The functional competence of uterine-derived polymorphonuclear neutrophils (PMN) from mares resistant and susceptible to chronic uterine infection: a sequential migration analysis. *Biol. Reprod.* **35**, 1168–1174.

Mancini, G., Carbonara, A.D. & Heremans, J.F. (1965) Immunochemical quantitation of antigens by single radial immunodiffusion. *Immunochemistry* **2**, 235–254.

Mitchell, G., Liu, I.K.M., Perryman, L.E., Stabenfeldt, G.H. & Hughes, J.P. (1982) Preferential production and secretion of immunoglobulins by the equine endometrium—a mucosal immune system. *J. Reprod. Fert., Suppl.* **32**, 161–166.

Neely, D.P., Hughes, J.P., Stabenfeldt, G.H. & Evans, J.W. (1974) The influence of intrauterine saline infusion on luteal function and cyclic ovarian activity in the mare. *Equine vet. J.* **6**, 150–157.

Peterson, F.B., McFeely, R.A. & David, J.S.E. (1969) Studies on the pathogenesis of endometritis in the mare. *Proc. 15th Am. Assoc. Equine Pract.* 279–287.

Spector, T. (1978) Refinement of the Coomassie Blue method of protein quantitation. *Analyt. Biochem.* **86**, 142–147.

Strzemienski, P.J. & Kenney, R.M. (1984) Effect of stage of cycle, sampling frequency and recovery of micro-organisms on total protein content of mare uterine flushings. *J. Reprod. Fert.* **70**, 327–332.

Widders, P.R., Stokes, C.R., David, J.S.E. & Bourne, F.J. (1985) Immunohistological studies on the local immune system in the reproductive tract of the mare. *Res. Vet. Sci.* **38**, 88–95.

Williamson, P., Dunning, A., O'Connor, J. & Penhale, W.J. (1983) Immunoglobulin levels, protein concentrations and alkaline phosphatase activity in uterine flushings from mares with endometritis. *Theriogenology* **19**, 441–448.

Wingfield-Digby, N.J. (1978) The techniques and clinical application of endometrial cytology in mares. *Equine vet. J.* **10**, 167–170.

Zavy, M.T., Bazer, F.W., Sharp, D.C. & Wilcox, C.J. (1977) Uterine luminal proteins in the cycling mare. *Biol. Reprod.* **20**, 689–698.

J. Reprod. Fert., Suppl. **35** (1987), 327–334

Printed in Great Britain
© 1987 Journals of Reproduction & Fertility Ltd

Factors affecting uterine clearance of inoculated materials in mares

M. J. Evans*, J. M. Hamer, L. M. Gason and C. H. G. Irvine

Veterinary Science Department, Lincoln College, Canterbury, New Zealand

Summary. Twelve acyclic mares of various ages (2–29 years) and parity (maiden–multiparous) were given oestradiol-17β i.m. (winter 1982) or progesterone i.m. (winter 1983) to induce changes in the endometrium consistent with oestrus and dioestrus, respectively. After hormone treatment, mares were inoculated intrauterine with 50 ml saline containing 5×10^5 *Streptococcus zooepidemicus* bacteria, ^{51}Cr-labelled 15-μm microspheres, and 500 mg charcoal (Groups E + B and P + B) or microspheres and charcoal only (Groups EC and PC). At 5 h after inoculation uteri were flushed with 50 ml saline containing tracer amounts of ^{125}I-labelled HSA. In Group E + B inoculated materials were cleared more rapidly in younger than in older mares, and there was a significant positive correlation between age and bacterial concentration, total numbers of bacteria, and amounts of microspheres and charcoal in the uterus and also between age and the WBC concentration and total numbers of WBC. In Group EC there was a significant positive correlation between age and the amounts of microspheres and charcoal remaining in the uterus, and between age and WBC concentration and total numbers of WBC. We suggest that in the oestrogen-dominated uterus physical drainage may be a factor in determining whether a mare is 'resistant' or 'susceptible' to bacterial challenge of the uterus. Physical clearance is increased in younger (resistant) mares in the presence and absence of an antigenic stimulus. In mares in Group P + B there was no correlation between age and the values measured. In Group PC all mares had bacteria present in the flush although bacteria were not included in the inoculum. There was no correlation between age and the other values measured. It appears that progesterone treatment results in previously resistant mares becoming susceptible to bacterial challenge; impaired physical clearance may contribute to this increased susceptibility. Physical clearance may therefore have an important role in the early elimination of bacteria and may contribute to the resistance of mares to the establishment of uterine infections.

Introduction

Although bacterial contamination of the uterus of the mare is known to occur frequently (Bryans, 1962; Gygax *et al.,* 1979), it appears that while most mares are 'resistant' to the establishment of infection, there are mares which are 'susceptible' to this contamination, and infection can become established (Hughes & Loy, 1969, 1975; Peterson *et al.,* 1969). There are many factors which may contribute to this susceptibility (Hughes & Loy, 1975), and these include increased age and parity (Hughes *et al.* 1966) and progesterone domination of the uterus (Washburn *et al.,* 1982; Evans *et al.,* 1986).

---

*Present address: Department of Endocrinology, The Princess Margaret Hospital, Christchurch 2, New Zealand.

In this study, we have assessed the effects of progesterone, oestradiol, age and parity, in determining the differences which exist between the response of the 'susceptible' and 'resistant' uterus to bacterial challenge.

Materials and Methods

The 4 parts of this experiment were carried out at intervals during 2 non-breeding seasons. The same 12 mares were used in each treatment section, except for 2 substitutions which were age and parity matched with the mares they replaced. The mares used were acyclic, and were representative of the cross-section of animals which might be presented for breeding at stud. While the youngest mares (2–3 years) were known to be maiden at the beginning of the experiment, the older mares had histories which included variable parity, repeat breedings and intrauterine antibiotic treatment, in previous seasons. Although only age was used as a definitive parameter in the analysis, it should be noted that parity generally increased with increasing age and, associated with this, an increase in invasive procedures to the reproductive tract. It was anticipated that this group of mares would include some animals which would be resistant and some susceptible to infection of the reproductive tract (Hughes *et al.*, 1966).

The 4 studies were carried out in the following sequence.

Group E + B

Mares were pre-treated, i.m., for 2 days with 1·5 mg oestradiol-17β (Intervet: Sydney, NSW) in oil, followed by 2 days of 5 mg oestradiol-17β. On the day of treatment a uterine culture swab was obtained using a guarded swab, after which mares were given an intrauterine inoculation of 50 ml isotonic saline (9 g NaCl/l) containing 5×10^5 *Streptococcus zooepidemicus* bacteria, ^{51}Cr-labelled 15 μm microspheres and 500 mg charcoal, as described previously (Evans *et al.*, 1986).

Group EC

Mares were treated as for Group E + B except that bacteria were omitted from the inoculum.

Group P + B

Mares were pre-treated, i.m., for 3 days with 150 mg progesterone (Sigma: St. Louis, MO, U.S.A.) in oil. On the day of treatment a uterine swab was obtained, then mares were given an intrauterine inoculation as for Group E + B.

Group PC

Mares were treated as for Group P + B except that bacteria were omitted from the inoculum.

We have previously shown that the oestradiol or progesterone treatments induce changes in the endometrium consistent with oestrus and dioestrus respectively (Hamer *et al.*, 1985). The inoculum contained bacteria commonly associated with infection of the horse uterus at a concentration well within the range frequently occurring in stallion semen (Burns *et al.*, 1975). The microspheres are cleared only by physical removal through the cervix and the charcoal is cleared by cellular defence mechanisms and physical removal.

At 5 h after inoculation, the uterus of each mare was infused ('flushed') with 50 ml sterile saline (9 g NaCl/l) containing tracer amounts of ^{125}I-labelled human serum albumin (HSA), using a two-way catheter with an inflatable cuff (Evans *et al.*, 1986). After manual manipulation of the uterus *per rectum*, the saline was collected by gravity flow.

The concentrations of bacteria, white blood cells, microspheres and charcoal were calculated for each flush as described previously (Evans *et al.*, 1986). From the dilution of the ^{125}I-labelled HSA in the flush solution, the volume of fluid present in the uterus before the flush was determined (Evans *et al.*, 1986), and so the total amounts of bacteria, WBC, microspheres and charcoal present in the uterus at the time of the flush were calculated. The amounts of microspheres and charcoal were expressed as a percentage of that inoculated.

Results were analysed by 2-way least squares analysis of variance to determine the effect of treatment group and age on the measured end-points. For this analysis, mares were categorized as 2–5, 11–16 or 18–29 years of age.

The effect of treatment on each end-point was also assessed using paired *t* tests, with each mare as her own pair. The correlation of age with each end-point within each treatment was also calculated, and the level of significance determined.

Results

The results are given in Tables 1 and 2. Analysis of variance showed that treatment group and age category had significant effect ($P < 0·05$) for all end-points, except the effect of age on the amount of charcoal in the flush.

Table 1. Bacteria, white blood cells, microspheres and charcoal recovered in a flush 5 h after intrauterine inoculation of microspheres and charcoal together with bacteria (Group E + B) or without bacteria (Group EC) from mares pre-treated with oestradiol

Mare	Age[a]	Parity[b]	Fluid in uterus (ml)[c]		Bacteria/ml[d]		Total bacteria[e]		WBC/ml[f]		Total WBC[g]		Microspheres[h]		Charcoal[i]	
			E + B	EC	E + B	EC	E + B	EC	E + B	EC	E + B	EC	E + B	EC	E + B	EC
1	2	0	0	9	0	0	0	0	6·08	5·30	7·77	7·07	0	10·9	14·9	2·9
2	2	0	6	6	0	2·48	0	4·22	5·54	5·56	7·29	7·31	0	2·0	0	0
3A/3B[j]	3/4	0/0	10	9	0	0	0	0	4·43	4·85	6·21	6·62	4·8	7·5	0·3	30·7
4	4	0	0	10	0	0	0	0	6·51	5·45	8·21	7·23	0	15·6	0	0·9
5	11	1	5	43	3·18	2·78	4·92	4·75	6·34	5·11	8·08	7·08	17·8	41·4	23·0	28·1
6	11	0	9	10	0	0	0	0	4·80	5·43	6·57	7·21	0	24·8	0	13·8
7A/7B[j]	13/18	5/4	14	16	0	0	0	0	4·43	6·14	6·24	7·96	0	6·4	0	2·0
8	15	2	37	45	2·70	3·78	4·64	0	5·15	5·68	7·08	7·66	0	35·6	12·3	39·6
9	17	3	298	140	6·00	0	8·54	6·06	8·00	5·85	12·54	8·12	36·0	93·1	47·9	62·1
10	22	4	117	75	6·26	0	8·48	0	8·00	6·86	10·22	8·95	3·2	73·5	4·8	100·0
11	23	3	24	62	6·00	0	7·87	0	10·00	5·75	11·87	7·80	12·2	53·1	17·8	78·6
12	28	9	15	26	4·34	5·00	6·15	6·88	10·00	6·10	11·81	7·98	58·7	27·2	100·0	95·8

[a] Age in years, first year of experiment.
[b] Number of foals born at time of experiment.
[c] Volume of fluid present in uterus before flush; calculated from dilution of ^{125}I-labelled HSA in flush (see text).
[d] Number of bacteria/ml recovered flush solution (see text) expressed as $\log_{10}$.
[e] Total number of bacteria in uterus at time of flush (see text) expressed as $\log_{10}$.
[f] Number of white blood cells/ml of recovered flush solution (see text) expressed as $\log_{10}$.
[g] Total number of white blood cells in uterus at time of flush (see text) expressed as $\log_{10}$.
[h] Counts due to ^{51}Cr-labelled microspheres in uterus expressed as % of number inoculated.
[i] Amount of charcoal in uterus expressed as % of amount inoculated.
[j] Mares 3A and 7A used only in Group E + B.

Table 2. Bacteria, white blood cells, microspheres and charcoal recovered in a flush 5 h after intrauterine inoculation of microspheres and charcoal together with bacteria (Group P + B) or without bacteria (Group PC) from mares pre-treated with progesterone

Mare	Age[a]	Parity[b]	Fluid in uterus (ml)[c]		Bacteria/ml[d]		Total bacteria[e]		WBC/ml[f]		Total WBC[g]		Microspheres[h]		Charcoal[i]	
			P + B	PC	P + B	PC	P + B	PC	P + B	PC	P + B	PC	P + B	PC	P + B	PC
1	3	0	64	25	4·14	3·86	6·20	5·74	4·70	4·78	6·76	6·65	63·6	†	3·5	8·4
2	3	0	24	4	2·90	3·51	4·77	5·24	5·40	4·95	7·26	6·69	23·8	†	0	1·4
3B	4	0	33	47	3·53	4·40	5·45	6·38	5·23	4·00	7·15	5·98	46·0	†	2·9	39·2
4	5	0	46	71	3·78	3·04	5·76	5·12	5·73	4·70	7·72	6·78	66·3	†	0	0
5	12	1	211	111	2·28	3·54	4·69	5·75	4·84	4·48	7·26	6·68	96·0	†	52·9	17·0
6	12	0	43	85	3·21	1·95	5·18	4·08	4·48	4·60	6·45	6·81	52·9	†	29·1	11·1
7B	18	4	43	13	1·62	2·89	3·59	4·69	5·16	0	7·13	0	65·6	†	1·9	3·7
8	16	2	35	49	3·00	2·79	4·93	4·79	4·48	5·20	6·41	7·20	16·8	†	100·0	11·2
9	18	3	28	27	3·60	6·05	5·49	7·93	6·06	6·51	7·95	8·39	30·3	†	50·5	100·0
10	23	4	300	70	8·18	3·00	10·63	5·08	7·11	4·90	9·57	6·98	100·0	†	56·8	100·0
11	24	3	84	13	4·58	2·81	6·71	4·61	5·81	4·84	7·94	6·65	67·8	†	28·9	5·9
12	29	9	6	41	1·60	2·49	3·35	4·45	5·13	4·48	6·87	6·43	23·0	†	64·5	22·9

[a]Age in years, second year of experiment.
[b]Number of foals born at time of experiment.
[c]Volume of fluid present in uterus before flush; calculated from dilution of ^{125}I-labelled HSA in flush (see text).
[d]Number of bacteria/ml recovered flush solution (see text) expressed as $\log_{10}$.
[e]Total number of bacteria in uterus at time of flush (see text) expressed as $\log_{10}$.
[f]Number of white blood cells/ml of recovered flush solution (see text) expressed as $\log_{10}$.
[g]Total number of white blood cells in uterus at time of flush (see text) expressed as $\log_{10}$.
[h]Counts due to ^{51}Cr-labelled microspheres in uterus expressed as % of number inoculated.
[i]Amount of charcoal in uterus expressed as % of amount inoculated.
†No data.

Groups E + B and EC

Culture of uterine swabs obtained immediately prior to uterine inoculation showed absence of bacterial contamination in all mares in Group E + B except Mares 9 and 12 (light growth of *S. zooepidemicus*) and 10 (moderate growth of *S. zooepidemicus*), each with a history of intermittent uterine infection. When bacteria, microspheres and charcoal were inoculated together, all 3 markers were cleared rapidly (Table 1). The bacteria were cleared more rapidly in younger than in older mares, and there was a significant positive correlation between age and bacterial concentration ($P < 0.01$) and total bacteria ($P < 0.01$). Similarly, there was a significant positive correlation between age and the amount of microspheres ($P < 0.05$) and charcoal ($P < 0.05$) remaining in the uterus at the time of the flush, and also between age and the WBC concentration ($P < 0.01$) and total WBC number ($P < 0.01$).

Culture of uterine swabs obtained before inoculation with microspheres and charcoal showed absence of bacterial contamination in all mares in Group EC. Although 4 mares had some bacteria present at the time of the flush (Table 1) there was no correlation with age. There was a significant positive correlation with age and the amount of microspheres ($P < 0.05$) and charcoal ($P < 0.01$) remaining in the uterus, and also between age and WBC concentration and total WBC number ($P < 0.01$).

The rapid clearance of inoculated bacteria in Group E + B, particularly in the younger mares, resulted in no significant differences between the groups for bacterial concentration and total number of bacteria. However, the amounts of microspheres and charcoal were greater in mares in Group EC than Group E + B ($P < 0.01$ and $P < 0.05$ respectively).

Groups P + B and PC

Pre-inoculation uterine swabs showed that mares in Group P + B were free of bacterial contamination of the uterus except for Mare 4 (light growth of *S. zooepidemicus*) and Mares 9 and 10 (light mixed growth including *Enterobacter aerogenes*). Considerable amounts of at least one marker were retained by each mare at the time of the flush; the clearances of bacteria and microsphere were not correlated with the age of the mare, nor were the amounts of WBC. The amount of charcoal, however, was positively correlated with the age of the mare ($P < 0.05$).

All mares in Group PC were free of bacterial contamination of the uterus immediately before the flush except for Mare 9 (heavy growth of *E. aerogenes*). Although only microspheres and charcoal were inoculated, all mares had bacteria present in the flush (Table 2). Neither bacterial concentration nor total numbers of bacteria were correlated with age. There were no results for microspheres in this group due to insufficient radioactive counts, but all but one mare retained some charcoal. The clearance of charcoal was not correlated with age, nor were the concentrations of WBC or total numbers of WBC.

Although bacteria were only included in the inoculum for mares in Group P + B, the apparent contamination by bacteria during the inoculation process, and their subsequent multiplication, resulted in all mares in Group PC having bacteria present in the flush. Consequently, there were no differences between the groups for bacterial counts. There were also no differences between the groups for the amount of charcoal, the WBC concentration and total numbers of WBC.

Discussion

Our findings are consistent with previous work in which inoculation of the same number of *S. zooepidemicus* organisms into the oestrogen-dominated uterus of maiden mares elicits a rapid response, including drainage of material through the cervix within 2 h (Evans *et al.*, 1986), as also occurs within 5 h in oestrous or dioestrous maiden fillies inoculated with much higher numbers of the same organism (Hughes & Loy, 1969).

The correspondingly rapid clearance from younger mares of inoculated microspheres, which, because of their size, cannot be removed other than by physical drainage, suggests that this mechanism may be a factor in determining whether a mare is resistant or susceptible to bacterial challenge to the uterus. The finding that the clearance of microspheres was positively correlated with age when there were no bacteria in the inoculum further suggests that one aspect of the clearance from resistant mares is independent of any antigenic stimulus or polymorphonuclear neutrophil (PMN) activity (although transient bacterial contamination during inoculation cannot be discounted) and may therefore be a passive event dependent on the conformation or function of the reproductive tract. However, since over the whole group, microspheres were lower when bacteria were in the inoculum, an antigenic stimulus may enhance this physical clearance.

The finding that it is the young nulliparous mares which appear both the most resistant to bacterial challenge, and efficient in the clearance of inoculated materials, agrees with the observations of Hughes & Loy (1969) that the healthy uterus of young mares has a marked capacity to handle infectious organisms rapidly, in contrast to the situation in older mares and in some mares after a number of pregnancies (Hughes *et al.*, 1966).

The correlation between increasing age and the number of WBC suggests, that, irrespective of whether the inoculum contained an antigenic stimulus, increased susceptibility to uterine infection is not the result of inadequate numbers of WBC. This is consistent with the finding of Liu *et al.* (1986) who reported that the numbers of uterine PMNs were not higher in mares deemed resistant than in mares considered susceptible to induced uterine infection, measured at 5 and 12 h after uterine inoculation with *S. zooepidemicus* (1×10^9 organisms). Similarly, Asbury *et al.* (1982) found no differences in the numbers of neutrophils in resistant and susceptible mares 8 h after inoculation with 7×10^9 *S. zooepidemicus*. Much higher numbers of organisms were inoculated in these two studies than in the present experiment.

We have previously shown that, when maiden mares were treated with progesterone, and the same numbers of *S. zooepidemicus* bacteria were inoculated with microspheres and charcoal, bacteria were still present in 3 of 4 mares after 3 days (Evans *et al.*, 1986). Moreover, microspheres and charcoal were also present in 2 of 4 mares at this time.

The absence of correlation between age and the amount of bacteria and microspheres in the flush suggests that progesterone treatment results in previously resistant mares becoming susceptible to bacterial challenge. Since the microspheres can only be removed from the uterus by drainage through the cervix, one factor contributing to the loss of resistance may involve a decrease in the physical elimination of inoculated material.

There is evidence that it is the functional role of neutrophils which is of major importance in differentiation between the response of resistant and susceptible mares (Cheung *et al.*, 1985; Liu *et al.*, 1985, 1986). Although there are marked differences in the functional competence of PMNs 12 h after intrauterine inoculation of *S. zooepidemicus* bacteria into luteal-phase mares, there are no differences observed at 5 h after inoculation (Liu *et al.*, 1986). It is therefore difficult to predict whether differing functional competence of the PMNs contributed to the more rapid elimination of bacteria from young mares in the present experiment, since such clearance of bacteria was complete at 5 h in the oestradiol pre-treated young resistant mares.

When bacteria were inoculated with microspheres and charcoal, some differences in measured endpoints due to steroid pretreatment were observed, although some of these effects were compounded by the differing age (and parity) of the mares. Young mares were clearly more efficient at clearing inoculated bacteria when pre-treated with oestradiol than with progesterone (Tables 1 and 2); however, the older mares did not show this effect. Therefore, there were no significant overall differences in bacterial concentration between these groups. The total, however, was significantly higher ($P < 0.05$) in mares in Group P + B than in Group E + B. Similarly, the amount of microspheres present in Group P + B mares was higher than in Group E + B ($P < 0.01$), but there were no differences in the amount of charcoal. In comparison, WBC concentration, but not the total number of WBC, was higher in Group E + B.

The present results are consistent with those obtained in maiden mares 7 days after similar inoculation (Evans *et al.,* 1986) in which progesterone-treated mares had higher numbers of bacteria, and a greater volume of purulent fluid in the uterus than did oestradiol-treated mares. In addition, retention of the non-antigenic markers tended to be higher in the progesterone-treated mares 3 days after inoculation, particularly in the older maiden mares (Evans *et al.,* 1986). Similarly, Washburn *et al.* (1982) and Ganjam *et al.,* (1982) found higher numbers of *S. zooepidemicus* bacteria in progesterone-treated than in oestradiol-treated mares 7 days after inoculation of this organism into ovariectomized mares (age not specified). However, the results from the present study suggest that the most pronounced effects due to steroid treatment are observed in the uterine response of young mares classed as resistant to bacterial challenge.

When bacteria were excluded from the inoculum in Groups EC and BC, bacterial values were markedly higher in Group PC than in Group EC ($P < 0.01$), which resulted from apparent contamination during the inoculation of the PC mares. This contamination occurred using the identical inoculation procedure as used in Group EC. Only one Group PC mare showed bacteria in the uterus before the inoculation. This finding illustrates the undesirability of invasion of the reproductive tract of any mare during a progesterone-dominated period. The WBC concentration was higher in Group EC (as occurred when bacteria were present in the inoculum); there were no differences in total numbers of WBC or charcoal between mares in Groups EC or PC.

In the oestrogen-treated mare, clearance of non-antigenic markers inoculated alone appears more efficient in young (resistant) mares. Our attempt to determine whether physical clearance of inoculated materials is similarly more efficient in young mares treated with progesterone, or whether it is compromised, was negated by the contamination during inoculation. We conclude from this study that, when oestrogen dominated, the uteri of young (resistant) mares eliminate bacteria and non-antigenic markers rapidly (within 5 h); this elimination does not occur in aged (susceptible) mares. Physical clearance, which occurs with or without antigenic stimulus, is more efficient in young mares and may contribute to this rapid elimination. When progesterone dominated, the uteri of young mares appear to become less resistant to bacterial challenge and respond in similar manner to aged mares; impaired physical clearance may contribute to this apparent increase in susceptibility.

Physical clearance may therefore have an important role in the early elimination of bacteria, and contribute to the resistance of mares to intrauterine bacterial challenge.

This project was funded by the New Zealand Equine Research Foundation, with support also from Lincoln College Special Research Grant and the Norman Cunningham Fellowship Trust.

We thank C. Graham, Ministry of Agriculture and Fisheries, Lincoln, and J. M. Aitken, Department of Microbiology, The Princess Margaret Hospital, Christchurch, for performing bacteriological and cell counting procedures; Anne Lawson for technical assistance; and Trish Riddell for typing the manuscript.

References

Asbury, A.C., Schultz, K.T., Klesius, P.H., Foster, G.W. & Washburn, S.M. (1982) Factors affecting phagocytosis of bacteria by neutrophils in the mare's uterus. *J. Reprod. Fert., Suppl.* **32**, 151–159.

Bryans, J.T. (1962) Research on bacterial diseases of horses. *Lectures, Stud Managers Course,* Lexington, Kentucky, 153 pp.

Burns, S.J., Simpson, R.B. & Snell J.R. (1975) Control of microflora in stallion semen with a semen extender. *J. Reprod. Fert., Suppl.* **23**, 139–142.

Cheung, A.T.W., Liu, I.K.M., Walsh, E.M. & Miller, M.E (1985) Phagocytic and killing capacities of uterine-derived polymorphonuclear leukocytes from mares resistant and susceptible to chronic endometritis. *Am. J. vet. Res.* **46**, 1938–1943.

Evans, M.J., Hamer, J.M., Gason, L.M., Graham, C.S.,

Asbury, A.C. & Irvine, C.H.G. (1986) Clearance of bacteria and non antigenic markers following intra uterine inoculation into maiden mares: effect of steroid hormone environment. *Theriogenology* **26**, 37–50.

Ganjam, V.K., McLeod, C., Klesius, P.H., Washburn, S.M., Kwapien, R., Brown, B. & Fazeli, M.H. (1982) Effect of ovarian hormones on the phagocytic response of ovariectomized mares. *J. Reprod. Fert., Suppl.* **32**, 169–174.

Gygax, A.P., Ganjam, V.K. & Kenney, R.M. (1979) Clinical, microbiological and histological changes associated with uterine involution in the mare. *J. Reprod. Fert., Suppl.* **27**, 571–578.

Hamer, J.M., Taylor, T.B., Evans, M.J., Gason, L.M. & Irvine, C.H.G. (1985) Effect of administration of estradiol and progesterone, and bacterial contamination, on endometrial morphology of acyclic mares. *Anim. Reprod. Sci.,* **9**, 317–322.

Hughes, J.P. & Loy, R.G. (1969) Investigations on the effect of intra uterine inoculations of Streptococcus zooepidemicus in the mare. *Proc. 15th A. Conv. Am. Ass. equine Pract,* pp. 289–292.

Hughes, J.P. & Loy, R.G. (1975) The relation of infection to infertility in the mare and stallion. *Equine vet. J.* **7**, 155–159.

Hughes, J.P., Loy R.G., Asbury, A.C. & Burd, H.E. (1966) The occurrence of Pseudomonas in the reproductive tract of mares and its effect on fertility. *Cornell Vet.* **56**, 595–610.

Liu, I.K.M., Cheung, A.T.W., Walsh, E.M., Miller, M.E. & Lindenberg, P.M. (1985) Comparison of peripheral blood and uterine-derived polymorphonuclear leukocytes from mares resistant and susceptible to chronic metritis: chemotactic and cell elastimetry analysis. *Am. J. vet. Res,* **46**, 917–920.

Liu, I.K.M., Cheung, A.T.W., Walsh, E.M. & Ayin, S. (1986) The functional competence of uterine-derived polymorphonuclear neutrophils (PMN) from mares resistant and susceptible to chronic uterine infection: a sequential migration analysis. *Biol. Reprod.* **35**, 1168–1174.

Peterson, F.B., McFeely, R.A. & David, J.S.E. (1969) Studies on the pathogenesis of endometritis in the mare. *Proc. 15th A. Conv. Am. Ass. equine Pract.* pp. 279–287.

Washburn, S.M., Klesius, P.H., Ganjam, V.K. & Brown, B.G. (1982) Effect of estrogen and progesterone on the phagocytic response of ovariectomized mares infected in utero with β-hemolytic streptococci. *Am. J. vet. Res.* **43**, 1367–1379.

J. Reprod. Fert., Suppl. **35** (1987), 335–342

Comparison of progesterone and progesterone + oestrogen on total and specific uterine proteins in pony mares*

K. J. McDowell†, D. C. Sharp and W. Grubaugh

Department of Animal Science, Bldg 459, Shealy Drive, University of Florida, Gainesville, Florida 32611, U.S.A.

Summary. Eight ovariectomized pony mares were used to test the effect of various doses of progesterone (0, 50, 150, 450 mg/day, in oil, i.m., for 10 days) on progesterone and LH in the peripheral circulation, and on total protein and uteroferrin in uterine secretions. Progesterone increased uteroferrin, but there were no differences amongst doses of progesterone. Progesterone treatment decreased LH, and tended to increase total protein.

Eighteen ovariectomized mares were given vehicle, oestradiol (10 mg/day, in oil, i.m.), progesterone or progesterone + oestradiol for 28 days. Both the last two steroid treatments significantly increased total protein and uteroferrin in the uterine secretions, compared to vehicle or oestradiol alone. Progesterone + oestradiol increased utero-ferrin, but not total protein compared to progesterone. Nine ovariectomized progesterone-primed mares were used to compare systemic and intraluminal adminis-tration of oestradiol. There were no differences between routes of administration of oestradiol.

In conclusion, administration of progesterone increased total protein and utero-ferrin in uterine secretions, and progesterone + oestradiol increased them further.

Introduction

It is well established that early embryonic mortality is a major source of reproductive inefficiency in domestic animals (Short, 1969). This is true for horses, in which embryonic loss may be as high as 25% by Day 25. In two recent studies using frequent ultrasound examinations of mated mares, 25·7% of the conceptuses observed on Day 11 were lost by Day 25 (Ginther, 1985), and 77·1% of all early embryonic deaths observed between 15 and 50 days occurred before Day 35 (Villahoz *et al.*, 1985). In addition, Chevalier & Palmer (1982) reported that the most frequently observed abnormality among early conceptuses was the small size of the vesicle in relation to the stage of gestation; 78% of those conceptuses were subsequently lost. While the reasons for these early conceptus losses are not known, it is probable that deficiencies in uterine secretions, either qualita-tive or quantitative, may lead to inadequate nourishment for the conceptus and result in poor conceptus growth or death.

The ability of the endometrium to provide adequate amounts of appropriate secretions depends to a large extent on the ovarian steroid progesterone. The necessity for ovaries or progesterone for successful pregnancy has been well documented (Holtan *et al.*, 1979; Shideler *et al.*, 1982; Hinrichs *et al.*, 1985), indicating that progesterone is the primary limiting ovarian factor necessary for pregnancy maintenance. In addition, the horse conceptus synthesizes oestrogens in large amounts (Heap *et al.*, 1982; Zavy *et al.*, 1979b, 1984).

---

*Reprint requests to D. C. Sharp.
†Present address: Colorado State University, Department of Physiology, Ft Collins, CO 80523, U.S.A.

Total protein content of uterine secretions increases with advancing luteal function and decreases abruptly as progesterone declines in non-pregnant mares. However, with luteal maintenance and continued progesterone production in pregnant mares, the protein content of the uterine secretions remains elevated (Zavy *et al.,* 1979a, b). One such progesterone-induced protein is uteroferrin (M_r 35 000, pI 9·7), a cationic acid phosphatase with many physical, chemical and immunological properties similar to pig uteroferrin (Zavy *et al.,* 1979b; McDowell *et al.,* 1982; Ketcham *et al.,* 1985). Uteroferrin may serve to carry iron to the developing horse conceptus in a manner similar to that described for pigs (Renegar *et al.,* 1982). The acid phosphatase enzymic activity of uteroferrin is easily monitored and can be used as an indicator of proteins synthesized and secreted by the progestational uterus.

Collectively, these data suggest that survival of the conceptus may depend, in part, upon the ability of the endometrium to respond to steroids with appropriate uterine secretions. The objectives of the present studies therefore were: (1) to test whether there is a threshold level of progesterone and/or a dose response to increasing levels of progesterone relative to the total protein and uteroferrin contents of the uterine secretions; (2) to compare the influence of progesterone versus oestrogen or the combination of the two on protein and uteroferrin contents of the uterine secretions; and (3) to compare routes of administration of oestrogen (systemic or intrauterine [intraluminal]) on protein and uteroferrin contents of the uterine secretions.

Materials and Methods

Animals. All of the pony mares used in the following experiments were ovariectomized at least 10 months before the onset of the experiments. None of the mares was assigned to more than one experiment.

Experiment 1. Mares were used to test the effect of various levels of progesterone on protein content and uteroferrin (monitored via its acid phosphatase activity) in uterine secretions, and on progesterone and LH concentrations in the peripheral circulation. The experiment was a 4 × 4 Latin square, with 2 mares in each cell. Each treatment occurred once during each period, and each mare received each treatment once during the course of the experiment. There was a 14-day interval (no treatment) between each period. Treatments were 0, 50, 150 or 450 mg progesterone in 3 ml sesame oil administered i.m. daily for 10 days. On Day 10 of each period, a blood sample and uterine secretions were collected from each mare. On the last day of the final period, blood samples were obtained from each mare every 4 h for 24 h.

Experiment 2. Mares were used to test the effects of progesterone, oestradiol, or the combination of these steroids on protein content and uteroferrin in uterine secretions. The following treatments were in 3 ml sesame oil and delivered i.m. daily for 28 days: vehicle (N = 4, Group 1), 10 mg oestradiol benzoate (N = 5, Group 2), 150 mg progesterone (N = 4, Group 3) or progesterone + oestradiol (N = 5, Group 4). The experiment was repeated four times during the year (November, February, May and July). At the end of each 28-day treatment period, uterine secretions were collected, and protein content and acid phosphatase activity were determined.

Experiment 3. Mares were used to compare the route of administration of oestrogen on protein content and uteroferrin in the uterine secretions. The experiment was a 3 × 3 Latin square, with 3 mares in each cell. Each treatment occurred once during each period, and each mare received each treatment once during the course of the experiment. There was a 10-day interval (no treatment) between each period. The design was to simulate the oestrous cycle or early pregnancy by administering oestrogen (i.m.), followed by progesterone (i.m.), with either vehicle or oestrogen (simulating conceptus oestrogen synthesis during the first 20 days of pregnancy; Zavy *et al.,* 1979b, 1984) administered intraluminally or systematically.

Each treatment consisted of systemic injection of oestradiol benzoate (10 mg/day, Days 1–4) followed by progesterone (150 mg/day, Days 5–19). In addition, on Days 13–19, mares received vehicle intraluminally (treatment 1), oestrogens intraluminally (3 µg oestrone and 6 µg oestradiol per day) (treatment 2) or vehicle intraluminally and oestradiol (10 mg per day) systemically. Oestrone and oestradiol were dissolved in 10% (v/v) ethanol in 0·9% (w/v) saline. Systemic hormone treatments were delivered in 3 ml sesame oil administered i.m. daily. Intraluminal treatments were delivered by use of sterile Alzet Osmotic Pumps (model 2001, Alza Corp, Palo Alto, CA) sutured to the serosal surface of one uterine horn. Each pump was fitted with a 5-cm catheter which was introduced into the uterine lumen near the utero-tubal junction. At the end of each period, pumps were removed through a small (2–3 cm) flank incision with the aid of a laparoscope. Surgeries were performed under general anaesthesia.

At the end of each period, uterine secretions were obtained, and protein content and acid phosphatase activity were determined.

Collection of blood samples and uterine secretions. Blood samples were collected into heparinized tubes by jugular venepuncture and immediately placed on ice. Uterine secretions were collected via transcervical flush. Sterile 0·9%

(w/v) saline (60 ml) was introduced into the uterine lumen through a Foley catheter, mixed with the luminal contents and withdrawn into a sterile syringe. The flushings were placed on ice for transport to the laboratory where they were centrifuged and filtered.

Protein content in the uterine flushings was measured by the method of Lowry *et al.* (1951) and acid phosphatase activity was determined as described by Ketcham *et al.* (1985).

Radioimmunoassays. Progesterone was measured in plasma by a solid-phase radioimmunoassay (COAT-A-COUNT, Diagnostic Products Corp., Los Angeles, CA) validated in our laboratory. Sensitivity of the assay was 0·01 ng. All samples (100 μl each) were run in one assay utilizing three standard curves. Intra-assay coefficient of variation (CV) was 6·8%.

Luteinizing hormone (LH) was measured in heparinized samples utilizing a double-antibody RIA described elsewhere (Niswender *et al.*, 1969; Nett *et al.*, 1975) and validated in our laboratory. Rabbit anti-ovine LH (No. 15, donated by Dr G. D. Niswender, Colorado State University, Ft Collins, CO) was the first antibody, and sheep anti-rabbit gamma globulin (Sigma, St Louis, MO) was used as the second antibody. The horse LH standard was donated by Dr H. Papkoff (University of California, Davis). Sensitivity of the assay was 0·04 ng. All samples (100 μl each) were run in one assay utilizing three standard curves. Intra-assay CV was 9·23%.

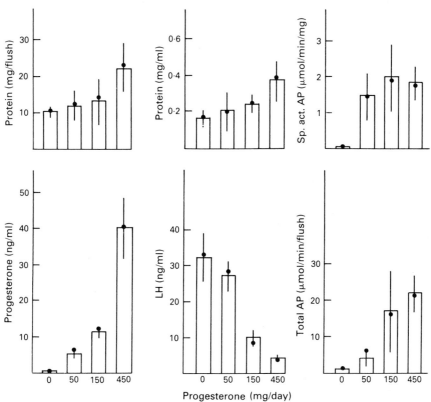

Fig. 1. Means ± s.e.m. (histograms) and LSM (●) for plasma progesterone and LH concentrations, total protein (mg/flush), protein concentration (mg/ml), acid phosphatase (AP) total activity (μmol *p*-nitrophenol released/min/flush), and acid phosphatase specific activity (μmol *p*-nitrophenol released/min/mg protein) in uterine secretions of mares in Exp. 1 (8 mares/dose of progesterone).

Significance levels for orthogonal comparisons:

Contrasts	Sp. act. AP	Total AP	Progesterone	LH
0 *vs* 50, 150, 450	$P < 0.05$	$P < 0.05$	$P < 0.005$	$P < 0.001$
50 *vs* 150, 450	$P > 0.5$	$P > 0.1$	$P < 0.001$	$P < 0.005$
150 *vs* 450	$P > 0.5$	$P > 0.5$	$P < 0.001$	$P > 0.1$

Statistical analyses. Least squares analysis of variance was performed utilizing the General Linear Models Procedures of the Statistical Analysis System (SAS, 1979) to test for treatment effects. Comparisons among means of treatment groups were made by orthogonal contrasts. When mares were sampled every 4 h (Exp. 2), time was considered as a continuous independent variable, which was analysed by polynomial regression. Tests for homogeneity of regression were used to detect differences in time trends for progesterone and LH responses among groups.

Results

Experiment 1

Data from one mare were not obtained in the 4th period (450 mg treatment) because of cervical adhesions. There was no effect of period on any parameter analysed; therefore, data from all periods were combined.

Arithmetic means ($\pm$ s.e.m.) and least squares means (LSM) are shown in Fig. 1. Increasing doses of progesterone increased ($P < 0.001$) concentrations of progesterone and decreased ($P < 0.001$) LH in the peripheral circulation, although the highest dose of progesterone (450 mg/day) decreased LH to only 5.49 ± 0.85 (s.e.m.) ng/ml. Orthogonal comparisons (Fig. 1, legend) showed that peripheral plasma progesterone increased significantly and peripheral plasma LH decreased significantly with each increasing dose of progesterone administered, except between 150 and 450 mg/day for LH.

Orthogonal comparisons indicated that progesterone administration significantly increased acid phosphatase total activity and acid phosphatase specific activity but there were no differences among doses of progesterone greater than 0 (Fig. 1). Although progesterone administration tended to increase total protein and protein concentration in the uterine flushings, these apparent increases were not significant.

When blood samples were obtained at 4-h intervals for 24 h, there were significant differences among treatment means ($P < 0.001$) for peripheral plasma LH. However, when time was considered as a continuous independent variable, the interactions of treatment and time were not significant ($P > 0.5$; data not shown). There were differences amongst treatment means ($P < 0.001$) for peripheral plasma progesterone. When time was considered as a continuous independent variable, there were significant interactions of treatment and time (Fig. 2; $P < 0.001$) that were best described by fourth-order polynomial regression equations (not shown). Tests for homogeneity of regression indicated that the progesterone curves were not parallel ($P < 0.005$). Peripheral plasma progesterone concentration was elevated by 4 h after injection, remained elevated for 4 h, then returned to baseline by 24 h (Fig. 2).

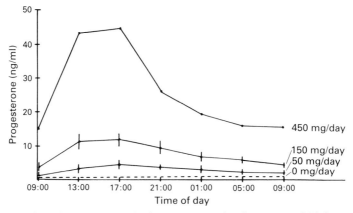

Fig. 2. Concentrations (mean $\pm$ s.e.m.) of progesterone in plasma over 24 h in ovariectomized mares in Exp. 1 treated with different doses of progesterone (2 mares/dose except for 1 mare at 450 mg/day). Values were at or below assay sensitivity for mares treated with 0 mg/day.

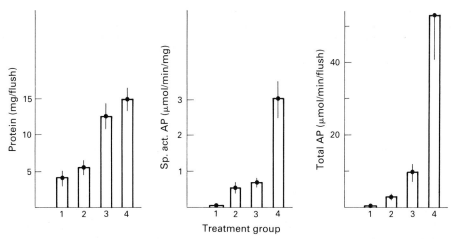

Fig. 3. Means ± s.e.m. (histograms) and LSM (●) for total protein (mg/flush), acid phosphatase (AP) specific activity (μmol *p*-nitrophenol released/min/mg protein), and acid phosphatase total activity (μmol *p*-nitrophenol released/min/flush) in uterine secretions of mares in Exp. 2 (Group 1, vehicle, N = 4; Group 2, 10 mg oestradiol benzoate, N = 5; Group 3, 150 mg progesterone, N = 4; Group 4, progesterone + oestradiol, N = 5).

Significance levels for orthogonal comparisons:

Contrasts	Total protein	Sp. act. AP	Total AP
1 *vs* 2, 3, 4	$P < 0.05$	$P < 0.001$	$P < 0.001$
2 *vs* 3, 4	$P < 0.001$	$P < 0.001$	$P < 0.001$
3 *vs* 4	$P > 0.1$	$P < 0.001$	$P < 0.001$

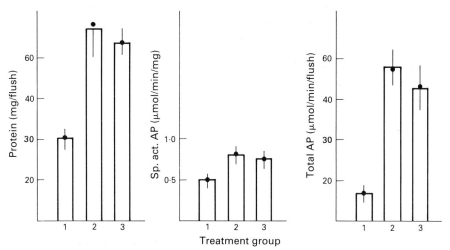

Fig. 4. Means ± s.e.m. (histograms) and LSM (●) for total protein and acid phosphatase (AP) specific and total activity in uterine secretions of mares in Exp. 3 (see Fig. legend 3 for units; see text for treatments; 3 mares/group).

Significance levels for orthogonal comparisons:

Contrasts	Total protein	Sp. act. AP	Total AP
1 *vs* 2, 3	$P < 0.001$	$P = 0.08$	$P < 0.001$
2 *vs* 3	$P = 0.50$	$P = 0.9$	$P = 0.40$

Experiment 2

There was no effect of month or the interaction of treatment and month on any parameter studied in uterine secretions. Therefore, data for all months were combined.

Arithmetic means ($\pm$ s.e.m.) and LSM are shown in Fig. 3. There were significant effects of treatment ($P < 0.001$) for total protein and specific and total acid phosphatase activity. Progesterone + oestradiol significantly increased specific and total acid phosphatase but not total protein, when compared with progesterone alone (Fig. 3).

Experiment 3

There were no effects of period in any analysis; therefore, data from all periods were combined. Arithmetic means ($\pm$ s.e.m.) and LSM are shown in Fig. 4. There were effects of treatments on total protein ($P < 0.001$) and total acid phosphatase ($P < 0.001$), but not on specific acid phosphatase ($P = 0.20$). Results obtained from orthogonal comparisons are shown in Fig. 4, legend. Treatments 2 and 3 (oestrogen intraluminally or systemically) clearly increased total protein ($P < 0.001$) and total acid phosphatase ($P < 0.001$) above Treatment 1, and increased specific acid phosphatase ($P = 0.08$) to a lesser degree. However, there was no difference between oestrogen administered intraluminally or systemically (Treatments 2 and 3).

Discussion

In Exp. 1, total recoverable uterine protein and total acid phosphatase activity generally increased with increasing doses of progesterone. In Exp. 2, the combination of progesterone + oestradiol increased total recoverable protein and greatly increased total acid phosphatase activity, indicating that these hormones are synergistic in stimulating secretion of uterine specific proteins. Geisert *et al.* (1982) proposed that oestrogen production by pig blastocysts may stimulate secretion of uterine proteins.

Progesterone insufficiency is frequently implicated as a cause of abortion in women (Hernandez Horta *et al.*, 1977; Hensleigh & Fainstat, 1979; Gautray *et al.*, 1981), cattle (Sreenan & Diskin, 1983), sheep (Wilmut *et al.*, 1985) and mares (Squires, *et al.*, 1981). Preliminary data suggest that short interovulatory intervals occur repeatedly in some mares and may be associated with early embryonic loss (Ginther, 1985); progesterone therapy is often administered to abortion-prone mares (Shideler *et al.*, 1982). Cause of short interovulatory intervals other than luteal dysfunction may include premature prostaglandin F secretion due to endometritis (Neely *et al.*, 1979) or neuro-endocrine disorders (Gautray *et al.*, 1978). For a recent review concerning mechanisms associated with subnormal luteal function, see Garverick & Smith (1986).

It is well established that severe periglandular fibrosis is associated with pregnancy wastage (Kenney, 1978). It is not unlikely, therefore, that some embryonic loss may be due to the inability of the endometrium to provide appropriate quantity or quality of histotrophe. In a preliminary study (K. J. McDowell, D. C. Sharp & A. C. Asbury, unpublished), exogenous progesterone (200 mg/day for 2 weeks) was given to 6 mares, one with a Grade I biopsy score (characteristic of a healthy uterus; Kenney, 1978) and 5 with Grade III biopsy scores (characteristic of severe periglandular fibrosis). Endometrial explants from each mare were incubated *in vitro* in the presence of [^{3}H]leucine. The percentage incorporation of the radiolabel into non-dialysable macromolecules was less for the Grade III mares than for the Grade I mare (9.92 ± 0.62 *vs* 16.61%). In addition, acid phosphatase activity in the uterine secretions was less for the Grade III mares than for the Grade I mare (1.76 ± 0.68 *vs* 11.20 μmol/min/mg protein). These preliminary results suggest that severe periglandular fibrosis may render the uterus incapable of responding to progesterone with increased secretion of specific uterine proteins.

Progesterone has a negative feedback on LH in mares (Garcia & Ginther, 1978). In addition, Pineda *et al.* (1972) demonstrated luteal regression in mares treated with an antiserum against horse pituitary fraction having LH activity, demonstrating the necessity of LH for normal luteal function. The possibility is raised, therefore, that exogenous progesterone may depress pituitary LH through negative feedback, which may result in decreased endogenous progesterone secretion and induced luteal insufficiency if the exogenous progesterone is terminated. In the studies presented here, LH decreased with each increasing dose of progesterone administered, but only to about 4·5 ng/ml at the highest dose of progesterone given (450 mg/day). Weithenauer *et al.* (1986) have reported significant differences among progesterone concentrations in mares given progesterone (150 mg/day), progesterone + oestradiol (150 and 10 mg/day) or vehicle on Days 1–25 of gestation. There were also significant differences in diameter and area of the embryonic vesicles as measured by ultrasonic scanning. Embryonic vesicles of the progesterone-treated mares had a faster rate of development. Although treatment was discontinued on Day 25, the pregnancies were not disrupted. It may be that there is a critical level of LH that is not under progesterone negative feedback control.

In conclusion, exogenous progesterone increased total protein content in uterine secretions, and also increased secretion of uterine specific proteins, as monitored by uteroferrin. In addition, oestrogen acted synergistically with progesterone to increase both total protein and uteroferrin in uterine secretions of mares.

Supported by The Morris Animal Foundation and the Grayson Foundation. University of Florida Institute of Food and Agricultural Sciences, Journal Series No. 7782.

References

Chevalier, F. & Palmer, E. (1982) Ultrasonic echography in the mare. *J. Reprod. Fert., Suppl.* **32**, 423–430.

Garcia, M.C. & Ginther, O.J. (1978) Regulation of plasma LH by estradiol and progesterone in ovariectomized mares. *Biol. Reprod.* **19**, 447–453.

Garverick, H.A. & Smith, M.F. (1986) Mechanisms associated with subnormal luteal function. *J. Anim. Sci.* **62**, Suppl. 2, 92–105.

Gautray, J.P., Jalivet, A., Goldenberg, F., Tajchner, G. & Eberhard, A. (1978) Clinical investigations of the menstrual cycle. II. Neuroendocrine investigation and therapy of the inadequate luteal phase. *Fert. Steril.* **29**, 275–281.

Gautray, J.P., de Brux, J., Tajchner, G., Robel, P. & Mouren, M. (1981) Clinical investigation of the menstrual cycle. III. Clinical, endometrial and endocrine aspects of luteal defect. *Fert. Steril.* **35**, 296–303.

Geisert, R.D., Renegar, R.H., Thatcher, W.W., Roberts, R.M. & Bazer, F.W. (1982) Establishment of pregnancy in the pig. I. Interrelationships between pre-implantation development of the pig blastocyst and uterine endometrial secretions. *Biol. Reprod.* **27**, 925–939.

Ginther, O.J. (1985) Embryonic loss in mares: incidence, time of occurrence, and hormonal involvement. *Theriogenology* **23**, 77–89.

Heap, R.B., Hamon, M. & Allen, W.R. (1982) Studies on oestrogen synthesis by the preimplantation equine conceptus. *J. Reprod. Fert., Suppl.* **32**, 343–352.

Hensleigh, P.A. & Fainstat, T. (1979) Corpus luteum dys-

function: serum progesterone levels in diagnosis and assessment of therapy for recurrent and threatened abortion. *Fert. Steril.* **32**, 396–400.

Hernandez Horta, J.L., Gardillo, F.J., Soto de Leon, B. & Cartes-Gallegas, V. (1977) Direct evidence of luteal insufficiency in women with habitual abortion. *Obstet. Gynecol.* **49**, 705–708.

Hinrichs, K., Sentich, P.L., Cummings, M.R. & Kenney, R.M. (1985) Pregnancy in ovariectomized mares achieved by embryo transfer: a preliminary study. *Equine vet. J., Suppl.* **3**, 74–75.

Holtan, D.W., Squires, E.L., Lapin, D.R. & Ginther, O.J. (1979) Effect of ovariectomy on pregnancy in mares. *J. Reprod. Fert., Suppl.* **27**, 457–463.

Kenney, R.M. (1978) Cyclic and pathologic changes in the mare endometrium as detected by biopsy, with a note on early embryonic death. *J. Am. vet. med. Ass.* **172**, 241–262.

Ketcham, C.M., Baumbach, G.A., Bazer, F.W. & Roberts, R.M. (1985) The type 5, acid phosphatase from spleen of humans with hairy cell leukemia. *J. biol. Chem.* **260**, 5768–5776.

Lowry, O.H., Rosebrough, N.C., Farr, A.L. & Randall, R.J. (1951) Protein measurement with the Folin phenol reagent. *J. biol. Chem.* **193**, 265–275.

McDowell, K.J., Sharp, D.C., Fazleabas, A.T., Roberts, R.M. & Bazer, F.W. (1982) Partial characterization of the equine uteroferrin-like protein. *J. Reprod. Fert., Suppl.* **32**, 329–334.

Neely, D.P., Kindahl, H., Stabenfeldt, G.H., Edqvist, L.E.

& Hughes, J.P. (1979) Prostaglandin release patterns in the mare: physiological, pathophysiological and therapeutic responses. *J. Reprod. Fert., Suppl.* **27**, 181–189.

Nett, T.M., Holtan, D.W. & Estergreen, V.L. (1975) Levels of LH, prolactin and oestrogens in the serum of post-partum mares. *J. Reprod. Fert., Suppl.* **23**, 201–206.

Niswender, G.D., Reichert, L.E., Midgely, A.R. & Nalbandov, A.V. (1969) Radioimmunoassay for bovine and ovine luteinizing hormone. *Endocrinology* **84**, 1166–1173.

Pineda, H.M., Ginther, O.J. & McShan, W.H. (1972) Regression of corpus luteum in mares treated with an antiserum against equine pituitary fraction. *Am. J. vet. Res.* **33**, 1767–1773.

Renegar, R.H., Bazer, F.W. & Roberts, R.M. (1982) Placental transport and distribution of uteroferrin in the fetal pig. *Biol. Reprod.* **27**, 1247–1260.

SAS (1979) *SAS User's Guide.* Statistical Analysis System, Inc., Cary, NC.

Shideler, R.K., Squires, E.L., Voss, J.L., Eikenberry, D.J. & Pickett, B.K. (1982) Progestagen therapy of ovariectomized pregnant mares. *J. Reprod. Fert., Suppl.* **32**, 459–464.

Short, R.V. (1969) When a conceptus fails to become a pregnancy. In *Foetal Autonomy* (Ciba Fdn Symp.), pp. 2–26. Eds G. E. W. Wolstenholme & M. O'Connor. J. & A. Churchill, London.

Sreenan, J.M. & Diskin, M.G. (1983) Early embryonic mortality in the cow: its relationship with progesterone concentration. *Vet. Rec.* **112**, 517–521.

Squires, E.L., Webel, S.K., Shideler, R.K. & Voss, J.L. (1981) A review on the use of altrenogest for the broodmare. *Proc. 27th Ann. Conv. Am. Assoc. equine Pract.* 221–231.

Villahoz, M.D., Squires, E.L., Voss, J.L. & Shideler, R.K. (1985) Some observations on early embryonic death in mares. *Theriogenology* **23**, 915–924.

Weithenauer, J., McDowell, K.J., Davis, S.D. & Rothman, T.K. (1986) Effect of exogenous progesterone and estrogen on early embryonic growth in pony mares. *Biol. Reprod.* **34**, Suppl. 1, Abstr.

Wilmut, I., Sales, D.I. & Ashworth, C.J. (1985) The influence of variation in embryo stage and maternal hormone profiles on embryo survival in farm animals. *Theriogenology* **23**, 107–119.

Zavy, M.T., Bazer, F.W., Sharp, D.C. & Wilcox, C.J. (1979a) Uterine luminal proteins in the cycling mare. *Biol. Reprod.* **20**, 689–698.

Zavy, M.T., Mayer, R., Vernon, M.W., Bazer, F.W. & Sharp, D.C. (1979b) An investigation of the uterine luminal environment of non-pregnant and pregnant pony mares. *J. Reprod. Fert., Suppl.* **27**, 403–411.

Zavy, M.T., Vernon, M.W., Sharp, D.C. & Bazer, F.W. (1984) Endocrine aspects of early pregnancy in pony mares: a comparison of uterine luminal and peripheral plasma levels of steroids during the estrous cycle and early pregnancy. *Endocrinology* **115**, 214–219.

J. Reprod. Fert., Suppl. **35** (1987), 343–351

Role of anaerobic bacteria in equine endometritis

S. W. Ricketts and M. E. Mackintosh

*Beaufort Cottage Stables, High Street, Newmarket, Suffolk CB8 8JS and Animal Health Trust,
Bacteriology Unit, Lanwades Park, Newmarket, Suffolk CB8 7DW, U.K.*

Summary. This study, performed over 3 breeding seasons, surveyed anaerobic and aerobic bacterial isolates from 362 clitoral and endometrial swabs and uterine washes from 263 Thoroughbred maiden, foaling, foal heat and barren mares, and from 113 urethral, urethral fossa, preputial and pre-ejaculatory fluid swabs from 29 Thoroughbred stallions. The significance of isolates was determined by their association with acute endometritis, as determined by concurrent endometrial smear results and by consideration of age and reproductive status before and after the survey.

The results suggest that the horse uterus may harbour obligate anaerobes as surface commensals. These organisms normally inhabit the external genital surfaces of mares and stallions and are periodically introduced into the uterus at coitus or in association with genital pathology, e.g. pneumovagina or vagino/cervical injury. They may act as opportunist pathogens when there is epithelial damage, e.g. during the post-partum involutionary period. Synergism with aerobic bacteria may result in mixed infection and active endometritis.

In the mare, the predominant uterine anaerobic species is *Bacteroides fragilis*. This species is predominantly penicillin- and aminoglycoside-resistant, a feature which is pertinent when intrauterine antimicrobial therapy for endometritis is considered. Detailed anaerobic diagnoses are unavoidably time consuming and as mixed infections are common, nitrofurantoin or metronidazole should be included in antibiotic mixtures. Failure to do this may account for some cases of aerobe-negative persistent endometritis.

Introduction

Transient endometritis is an inevitable sequal to coitus in mares. Ejaculation occurs through the open oestrous cervix, contaminating the uterine lumen with semen, environmental and external genital microorganisms and debris (Kenney *et al.*, 1975; Simpson *et al.*, 1975). The normal, genitally healthy mare produces an efficient transient post-coital acute endometritis, which resolves within 48–72 h (Hughes & Loy, 1969; Peterson *et al.*, 1969), leaving the endometrium in a satisfactory state to receive the fertilized ovum from the oviduct at about 5 days after ovulation. Mares with genital abnormality, e.g. pneumovagina, vulval, recto-vaginal or cervical injury, or mares that have impaired local endometrial defence mechanisms (Hughes & Loy, 1975), produce an inefficient, persistent acute endometritis, which invariably results in conception failure or early embryonic loss. Such mares are treated with intrauterine antimicrobial medication (Ricketts, 1986) and, when appropriate, are treated before and/or after coitus with a variety of prophylactic techniques (Asbury, 1986).

Transient endometritis is an inevitable sequal to parturition in mares. The uterine lumen is exposed to environmental microorganisms during second- and third-stage labour and during the involutionary period, when movement of the mare causes the enlarged, pendulous uterus to aspirate air through the relaxed vulva, vagina and cervix. This occurs at a time when the endometrium is showing signs of histological 'repair', with luminal epithelial damage, microcotyledonary crypt resolution, stromal haemorrhage, oedema and polymorphonuclear cell infiltration (Ricketts, 1978; Gygax *et al.*, 1979). Severe acute post-parturient endometritis may follow parturient trauma and/or placental retention and may be associated with delayed uterine involution. Even when parturition and uterine involution proceed normally, bacteriological and histological examinations suggest a high incidence of surface bacterial contamination (Gygax *et al.*, 1979).

It has been suggested that the normal horse uterus is bacteriologically sterile or has a temporary, non-resident microflora rather than a resident microflora (Woolcock, 1980), but these comments are based on studies which have traditionally ignored anaerobic cultural techniques. There have been many surveys of aerobic bacteria associated with endometritis in mares (Shin *et al.*, 1979) and mycotic endometritis (Blue, 1986) is well recognized. Some strains of *Klebsiella pneumoniae* and *Pseudomonas aeruginosa* can cause epidemic venereal endometritis (Rossdale & Ricketts, 1980), but in most bacterial (most commonly *Streptococcus zooepidemicus*, *Escherichia coli* and *Staphylococcus aureus*) and fungal infections, individual predisposing factors (see above) are an essential pre-requisite. In 1977 an epidemic venereal endometritis was first recognized (David *et al.*, 1977) which was associated with bacteria that could be seen on Gram-stained smears but could not be cultured aerobically. Several strains of *Bacteroides fragilis* were isolated but these were considered to be opportunist rather than primary pathogens (Powell, 1980). Microaerophilic culture on chocolated blood agar identified *Taylorella equigenitalis* as the cause (Taylor *et al.*, 1978). Endometrial cytological (Wingfield Digby, 1978; Wingfield Digby & Ricketts, 1982) and histological examinations (Ricketts, 1978) have demonstrated that endometritis can occur without a growth of microorganisms following standard aerobic and microaerophilic cultural techniques. Scanning electron microscopic studies (Ricketts *et al.*, 1978; Samuel *et al.*, 1979) have demonstrated epithelial surface organisms in such cases. Special cultural techniques are required for anaerobic bacteriology (Baker & Kenney, 1980) and no comprehensive studies have been reported to date. We have therefore investigated the role of anaerobic bacteria in endometritis in horses.

Materials and Methods

Swab samples (475) were collected from 29 Thoroughbred stallions (Table 1) and 263 Thoroughbred mares (Table 2). Swabs from stallions were collected from four penile sites (David *et al.*, 1977) (Table 1) after pre-moistening with transport medium. The erect penis was held by a gloved hand during teasing with an oestrous mare. These samples were collected during January and early February, before the start of the breeding season, during routine screening for potential venereal disease bacteria.

Samples from the clitoral fossa and sinus (Medical Wire and Equipment Co., Ltd, Corsham, Wiltshire, U.K.) were collected (Powell *et al.*, 1978), after pre-moistening in the urethral opening. These samples were collected during routine pre-coital screening for potential venereal disease bacteria.

Endometrial swabs were collected by two methods. Non-guarded swabs (Medical Wire and Equipment Co., Ltd) were collected via a sterile disposable vaginal speculum (Barker and Barker Ltd, Yardley, Birmingham, U.K.) through the open, early oestrous cervix (Ricketts, 1981), using standard hygienic techniques. These swabs were taken throughout the breeding season during routine pre-coital screening. Guarded swabs (Kallajan Industries Inc., Long Beach, CA, U.S.A.) were collected via a sterile disposable speculum, as above.

Uterine washes were collected via 24 French, 30-ml cuff, Foley catheters (Warne Surgical Products Ltd, Lurgan, Co. Armagh, N. Ireland) by methods described by Watson *et al.* (1986), from 12 barren mares at the end of the breeding season.

Table 1. Swab samples collected from stallions

Stallions	No.	Average age (years)	Urethra	Urethral fossa	Smegma	Pre-ejaculatory fluid	Total
Maiden	1	5	—	1	—	—	1
Mature	20	9·8	21	26	20	13	80
Teaser	8	11·4	7	9	8	8	32
Total	29	10·1	28	36	28	21	113

Table 2. Swab samples collected from mares

Mares	No.	Average age (years)	Clitoris	Non-guarded swab	Guarded swab	Uterus wash	Total
				Uterus			
Maiden	33	4	2	29	12	0	43
Foaling	86	18	7	96	26	0	129
Foal heat	82	10	0	70	16	0	86
Barren	62	12	6	56	30	12	104
Total	263	10	15	251	84	12	362

All swabs were placed into Amies' charcoal transport medium immediately after collection and were plated out for aerobic, microaerophilic and anaerobic bacteriological examinations within 24 h.

Aerobic cultures were performed at 37°C on blood and McConkey's agar, and microaerophilic cultures at 37°C on chocolated blood agar in 10% carbon dioxide, using standard techniques (Mackintosh, 1981).

Anaerobic cultures were performed, by direct inoculation, on two plates of Wilkins–Chalgren agar (Oxoid) containing 5% horse blood. One plate contained 136 µg neomycin sulphate/ml (Sigma, Poole, Dorset, U.K.) (equivalent to 100 µg neomycin/ml). On each plate a disc containing 5 µg metronidazole/ml was placed on the area of the heavy inoculum. The plates were incubated for 48 h at 37°C in an environment of 10% hydrogen, 10% carbon dioxide and 80% nitrogen in an anaerobic incubator (Gallenkamp). Swabs were also incubated anaerobically in an enrichment medium of Robertson's meat granules (Lab M) rehydrated in GLC broth (Lab M) for 24 h before inoculation on Wilkins–Chalgren agar with 5% horse blood and a metronidazole disc followed by incubation as above. Obligate anaerobes were differentiated from the facultative anaerobes by their growth inhibition around the metronidazole disc. Isolates were purified and identified by methods described by Willis (1977). Identification of volatile fatty acids produced by growth of pure cultures in GLC broth was made by gas–liquid chromatography (Holdeman & Moore, 1972).

Endometrial smears, for cytological examinations, were taken concurrently with endometrial swab samples, processed and interpreted using techniques described by Wingfield Digby (1978) and Wingfield Digby & Ricketts (1982). The uterine washings were centrifuged at 125 g for 15 min and the leucocytes were counted using a haemocytometer chamber. Stained smears were examined for the presence of endometrial epithelial cells and leucocytes. The proportion of polymorphonuclear leucocytes (PMNLs) seen in each smear was classified as − (no PMNLs seen), + − (<0·5%), 1+ (0·5–5%), 2+ (5–30%) and 3+ (>30%). Acute endometritis was defined by ⩾1+ PMNLs.

Ages and details of reproductive status before and after the survey were obtained from stud-farm records and from the General Stud Book (Wetherbys, Ltd, Wellingborough, Northamptonshire, U.K.).

Results

Neither *T. equigenitalis* nor *Ps. aeruginosa* was isolated from any of the swabs in this survey. *K. pneumoniae* was not isolated from any of the stallion swabs or mare clitoral swabs, but one isolate (capsule type 7) was obtained from one non-guarded endometrial swab collected from an 8-year-old foaling mare.

Stallion swabs

As shown in Table 3, 149 anaerobic isolations were made from 96% of the 113 stallion swabs. *Peptococcus* spp. was the most frequent isolate overall and from the urethra, urethral fossa and pre-ejaculatory fluid. *Clostridium perfringens* was the second most frequent isolate overall and was the most frequent isolate from the penile smegma. *B. fragilis* was the third most frequent isolate overall and was the second most frequent isolate from the urethral fossa. There was no obvious difference in the anaerobic isolates with age and between the mature stallions and the maiden and teaser stallions. None of the survey stallions exhibited abnormal fertility and none was associated with signs of epidemic venereal disease.

Table 3. Anaerobes isolated from stallion swabs

	Urethra	Urethral fossa	Smegma	Pre-ejaculatory fluid	Total
	No. (%)	No. (%)	No. (%)	No. (%)	No. (%)
Total no. of swabs	28	36	28	21	113
Anaerobes isolated	27 (96)	35 (97)	27 (96)	19 (90)	108 (96)
Peptococcus spp.	23 (82)	25 (69)	10 (36)	13 (62)	71 (63)
Clostridium perfringens	4 (14)	4 (11)	14 (50)	7 (33)	29 (26)
Bacteroides fragilis	3 (11)	14 (39)	4 (14)	2 (10)	23 (20)
Fusobacterium mortiferum	1 (4)	6 (17)	3 (11)	2 (10)	12 (11)
Clostridium sporogenes	1 (4)	2 (6)	5 (18)	1 (5)	9 (8)
Fusobacterium spp.	2 (7)	1 (3)	0	0	3 (3)
Clostridium cadaveris	0	0	1 (4)	1 (5)	2 (2)
Total isolates	34	52	37	26	149

Clitoral swabs

Of the 15 clitoral swabs, 26 anaerobic isolations were made from all of them (100%).

B. fragilis was the most frequent isolate (100% swabs), followed by *F. mortiferum* (47%), *Cl. perfringens* (13%), *Cl. sporogenes* (7%) and *B. melaninogenicus* (7%). There was no obvious difference in the anaerobic isolates with age and between the barren, foaling and maiden mares. None of the survey mares was associated with epidemic venereal disease. At the end of the season after collection 10 mares (67%) were pregnant.

Endometrial swabs

As shown in Table 4, 198 anaerobic isolations were made from 42% of the 335 endometrial swabs. *B. fragilis* was the most frequent isolate from all the swabs, followed by *F. mortiferum*, *Peptostreptococcus anaerobius*, *Peptococcus* spp. and *Cl. perfringens*. Altogether, 62 swabs (18%) were associated with cytological evidence of acute endometritis. Of these 28/94 (45%) grew *B. fragilis*, 7/18 (11%) grew *Cl. perfringens* and 6/30 (10%) grew *F. mortiferum*. The isolations from foaling mares, mares at the foal heat and maiden and barren mares are given in Table 5. None of the isolations in maiden mares was associated with acute endometritis, but cytological evidence was found in the other categories of mares.

Uterine washes

Anaerobic isolates were obtained from 7 (58%) of the 12 washes (Table 6).

Overall cytological and bacteriological correlations

Of the 347 uterine samples examined overall, 26% showed cytological evidence of acute endometritis (Table 7). Of the 276 (74%) samples with no cytological evidence of acute endometritis, 46% produced no aerobic or anaerobic growth. From the 18 samples with cytological evidence of acute endometritis and anaerobic but no aerobic growth, 10 pure cultures (7 *B. fragilis*, 2 *Peptococcus* spp. and 1 *Cl. perfringens*) and 15 mixed cultures (with *F. mortiferum*, *Cl. sporogenes*, *Cl. cadaveris* and *Cl. histolyticum*) were isolated.

Table 4. Anaerobes isolated from endometrial swabs with cytological findings

	Non-guarded swabs	Guarded swabs	Total
	No. (AE)*	No. (AE)*	No. (AE)*
Total no. of swabs	251 (52)	84 (10)	335 (62)
Anaerobes isolated	119 (35)	23 (4)	142 (39)
Bacteroides fragilis	80 (25)	14 (3)	94 (28)
Fusobacterium mortiferum	25 (6)	5 (0)	30 (6)
Peptostreptococcus anaerobius	21 (5)	0	21 (5)
Peptococcus spp.	15 (5)	2 (0)	17 (5)
Clostridium perfringens	13 (6)	5 (1)	18 (7)
Fusobacterium spp.	5 (1)	0	5 (1)
Clostridium sporogenes	5 (2)	0	5 (2)
Bacteroides melaninogenicus	3 (1)	0	3 (1)
Clostridium cadaveris	2 (2)	0	2 (2)
Clostridium histolyticum	1 (1)	0	1 (1)
Clostridium spp.	1 (1)	0	1 (1)
Veillonella sp.	1 (0)	0	1 (0)
Total isolates	172	26	198

*No. of swabs with cytological evidence of acute endometritis, i.e. $\geqslant 1 +$ PMNL score.

Table 5. Anaerobes isolated from endometrial swabs with cytological findings correlated with reproductive status

	Maiden mare	Foaling mare	Foal heat mare	Barren mare
	No. (AE)*	No. (AE)*	No. (AE)*	No. (AE)*
Total no. of swabs	41 (2)	122 (23)	86 (27)	89 (15)
Anaerobes isolated	21 (0)	39 (15)	38 (18)	44 (11)
Bacteroides fragilis	17 (0)	26 (10)	24 (12)	27 (6)
Fusobacterium mortiferum	4 (0)	3 (1)	11 (4)	11 (1)
Peptostreptococcus anaerobius	1 (0)	4 (3)	7 (1)	9 (1)
Peptococcus spp.	3 (0)	2 (0)	6 (4)	6 (1)
Clostridium perfringens	0	9 (2)	4 (3)	5 (2)
Fusobacterium spp.	2 (0)	2 (0)	1 (1)	0
Clostridium sporogenes	0	2 (0)	2 (2)	1 (0)
Bacteroides melaninogenicus	0	3 (0)	0	1 (1)
Clostridium cadaveris	0	0	1 (1)	1 (1)
Clostridium histolyticum	0	1 (1)	0	0
Clostridium spp.	0	0	1 (1)	0
Veillonella sp.	0	0	0	1 (0)
No anaerobes isolated	20 (2)	83 (8)	48 (9)	45 (4)

*No. of swabs with cytological evidence of acute endometritis, i.e. $\geqslant 1 +$ PMNL score.

Table 6. Uterine wash results

Mare no.	Age (years)	Cytology (PMNL score)*	Aerobic culture	Anaerobic culture
1	14	3+	*E. coli*	No growth
2	7	3+	No growth	*B. fragilis*
3	18	1+	No growth	*B. fragilis*
4	9	1+	No growth	*Peptococcus* spp.
5	11	3+	No growth	*B. fragilis*
6	5	1+	No growth	*B. fragilis*
7	8	3+	No growth	No growth
8	12	−	No growth	No growth
9	16	+−	No growth	*B. melaninogenicus* + *F. mortiferum*
10	14	−	No growth	*B. fragilis* + *F. mortiferum*
11	9	+−	No growth	No growth
12	17	+−	No growth	No growth

*−, none; +−, <0·5%; 1+, 0·5–5%; 2+, 5–30%; 3+, >30%.

Table 7. Cytological and bacteriological correlations for endometrial samples

Growth permutations	Acute endometritis		No acute endometritis	
	No.	(%)	No.	(%)
O₂ no growth + NO₂ no growth	7	(10)	128	(46)
O₂ growth + NO₂ no growth	18	(25)	47	(17)
O₂ no growth + NO₂ growth	18	(25)	53	(19)
O₂ growth + NO₂ growth	28	(40)	48	(18)
Total samples	71		276	

O_2 = aerobic culture; NO_2 = anaerobic culture.

Correlations with reproductive status after the survey

At the end of the stud seasons during which the 335 endometrial swab samples were collected, 220 (66%) of the associated mares were pregnant. Pregnancy occurred in 89/142 (63%) of mares with anaerobic isolations and in 131/193 (68%) of mares without such isolations.

Discussion

Potentially pathogenic obligate anaerobic bacterial species are found abundantly in the environment and on human and animal skin and mucosal surfaces (Kohn, 1978; Moss, 1983), where they occupy a commensal role as a constituent of the normal microbiological flora. Up to 70% of post-partum women have *Clostridium* spp. in their uteri, with no evidence of clinical disease. *Bacteroides* spp. were isolated from the cervix and vagina in 65% normal healthy women, attending a family planning clinic, the most common species being of the *B. melaninogenicus/oralis* group. *Bacteroides* spp. were isolated from the vagina and uterus of mature bitches during different cyclic stages (Baba *et al.*, 1983).

The present study suggests that the external genitalia of the normal mare and stallion, irrespective of age and reproductive status, possess a mixed obligate anaerobic microflora. From stallion swabs, *Peptococcus* spp. was the most frequent isolate from the urethra, urethral fossa and pre-ejaculatory fluid. *Cl. perfringens* was the most frequent isolate from the penile smegma. *B. fragilis* was the second most frequent isolate from the urethral fossa. From clitoral swabs, *B. fragilis* was the most frequent isolate, followed by *F. mortiferum*, *Cl. perfringens*, *Cl. sporogenes* and *B. melaninogenicus*. In horses, as in most animals, the external genital microflora reflects, to an important extent, environmental and faecal organisms.

The endometrial swab results from maiden mares with no cytological evidence of acute endometritis suggest that obligate anaerobic bacteria can inhabit the normal horse uterus as surface commensals. *B. fragilis* was the most frequent isolate followed by *F. mortiferum*, *Peptococcus* sp. and *Pst. anaerobius*.

Bacteria of the *Bacteroides–Fusobacterium* group are significant anaerobic pathogens in many human clinical conditions (Duerden *et al.*, 1976; Wilks *et al.*, 1984), including pelvic infections in women. *B. fragilis*, *Peptostreptococcus* and *Clostridium* spp. have been implicated in endometritis in women (Kohn, 1978). *Bacteroides* spp. may play an important role, together with *Gardnerella vaginalis*, in the pathogenesis of human non-specific vaginitis and balanoposthitis (Jones *et al.*, 1982; Moss, 1983), and have been isolated from 19% of affected women.

Obligate anaerobes have been isolated in surveys of samples from non-specific veterinary infections (Prescott, 1978; Hirsh *et al.*, 1979), *Bacteroides* spp. being the major group. *Bacteroides* spp., *Clostridium* spp. and *Fusobacterium* spp. are common isolates from cow uterine samples in cases of post-partum metritis (Olson *et al.*, 1984). *Fusobacterium* spp. can cause perineal and peri-rectal abscesses in horses (Kohn, 1978).

The present study suggests that *B. fragilis*, *F. mortiferum*, *Pst. anaerobius* and *Cl. perfringens* may be associated with acute endometritis in foaling, foal heat or barren mares. *B. fragilis* was the most frequent isolate overall and, in foal heat mares, this species was associated with acute endometritis in 50% of the isolates.

These studies confirm that bacteriological examinations of endometrial swab samples, taken with whatever equipment and by whatever techniques, are not accurately interpretable without concurrent cytological data (Wingfield Digby & Ricketts, 1982).

Of the cases of acute endometritis sampled, 40% grew anaerobes and aerobes, suggesting that, as in other species, mixed infections are common. Pathogenic synergy has been demonstrated (Kelly, 1978; Brook *et al.*, 1984; Brook, 1985) in that the growth rate of facultative and aerobic bacteria is enhanced in the presence of anaerobes, especially *Bacteroides* spp. *Corynebacterium pyogenes* acts synergistically with *F. necrophorum* and *B. melaninogenicus* to cause pyometra and metritis in cows (Olson *et al.*, 1984). Anaerobe inhibition of aerobe leucocyte phagocytosis (Ingham *et al.*, 1977) appears to be one of the mechanisms responsible, possibly via competition for serum opsonins (Rodloff *et al.*, 1986). The efficiency of *Streptococcus zooepidemicus* uterine leucocyte phagocytosis is depressed in mares with persistent endometritis (Asbury *et al.*, 1982) and in mares with chronic endometrial disease (Cheung *et al.*, 1985). This depression can be reduced or abolished by treatment with opsonins in the form of homologous plasma, *in vitro* and *in vivo*. Clearly, more research must be conducted into the role that anaerobic bacteria may play in the pathogenesis of persistent endometritis in mares.

In human medicine, it is recognized that, as anaerobic bacterial technology is time consuming, appropriate treatment must begin before a specific diagnosis is available (Leigh & Simmons, 1977). Within the constraints of a horse industry based upon an arbitrarily and unphysiologically shortened breeding season, similar procedures are adopted by veterinary clinicians. Cytological diagnosis of acute endometritis may be made within minutes of sample collection and aerobic bacterial results may be available overnight, but anaerobic cultures and identification can take several days. Anaerobic species are frequently identified by their antibiotic sensitivity patterns (Duerden *et al.*, 1976) and most strains of *B. fragilis* are resistant to penicillin and the aminoglycoside antibiotics. Nitrofurantoin and metronidazole have been shown to be bactericidal for *B. fragilis* and *Cl. perfringens* (Ralph, 1978). 'First choice' intra-uterine antimicrobial treatments should therefore be deliberately broad spectrum, to take account of mixed infections, and a water-soluble powder containing 3 g sodium benzylpenicillin (Crystapen: GlaxoVet, Ltd, Uxbridge, U.K.) and 1 g neomycin, 40 000 i.u. polymixin B and 600 mg furaltadone (Intra-Uterine Wash: Univet Ltd, Bicester, U.K.) would appear to cover for the Gram-positive and -negative aerobic bacteria and anaerobic bacteria commonly isolated from cases of endometritis in horses.

We thank Mrs K. Gifkins and Mrs A. Cheese of Beaufort Cottage Laboratories and Miss M. Blackett of the Animal Health Trust for technical assistance; and the Home of Rest for Horses for financial support (M.E.M.).

References

Asbury, A.C. (1986) Failure of uterine defence mechanisms in mares. In *Current Therapy in Equine Medicine,* 2nd edn, pp. 508–511. Ed. N.E. Robinson. W. B. Saunders Co., Philadelphia.

Asbury, A.C., Schultz, K.T., Klesius, P.H., Foster, G.W. & Washburn, S.M. (1982) Factors affecting phagocytosis of bacteria by neutrophils in the mare's uterus. *J. Reprod. Fert., Suppl.* 32, 151–159.

Baba, E., Hata, H., Fukata, T. & Arakawa, A. (1983) Vaginal and uterine microflora of adult dogs. *Am. J. vet. Res.* 44, 606–609.

Baker, C.B. & Kenney, R.M. (1980) Systematic approach to the diagnosis of the infertile or subfertile mare. In *Current Therapy in Theriogenology,* pp. 721–735. Ed. D. A. Morrow. W. B. Saunders Co., Philadelphia.

Blue, M.G. (1986) Fungal endometritis. In *Current Therapy in Equine Medicine,* 2nd edn, pp. 511–513. Ed. N. E. Robinson. W. B. Saunders Co., Philadelphia.

Brook, I. (1985) Enhancement of growth of aerobic and facultative bacteria in mixed infections with bacteroides species. *Infection and Immunity* 50, 929–931.

Brook, I., Hunter, V. & Walker, R.I. (1984) Synergistic effect of *Bacteroides, Clostridium, Fusobacterium,* anaerobic cocci and aerobic bacteria on mortality and induction of subcutaneous abscesses in mice. *J. Infect. Diseases* 149, 924–928.

Cheung, A.T.W., Liu, I.K.M., Walsh, E.M. & Miller, M.E. (1985) Phagocytic and killing capacities of uterine-derived polymorphonuclear leukocytes from mares resistant and susceptible to chronic endometritis. *Am. J. vet. Res.* 46, 1938–1940.

David, J.S.E., Frank, C.J. & Powell, D.G. (1977) Contagious Metritis 1977. *Vet. Rec.* 101, 189–190.

Duerden, B.I., Holbrook, W.P., Collee, J.G. & Watt, B. (1976) The characterization of clinically important Gram negative anaerobic bacilli by conventional bacteriological tests. *J. appl. Bact.* 40, 163–188.

Gygax, A.P., Ganjam, V.K. & Kenney, R.M. (1979) Clinical, microbiological and histological changes associated with uterine involution in the mare. *J. Reprod. Fert., Suppl.* 27, 571–578.

Hirsh, D.C., Biberstein, E.L. & Jang, S.S. (1979) Obligate anaerobes in clinical veterinary practice. *J. clin. Microbiol.* 10, 188–191.

Holdeman, L.V. & Moore, W.E.C. (1972) *Anaerobe Laboratory Manual.* Virginia Polytechnic Institute and State University Anaerobe Laboratory, Blacksburg.

Hughes, J.P. & Loy, R.G. (1969) Investigations on the effect of intrauterine inoculations of *Streptococcus zooepidemicus* in the mare. *Proc. 15th Ann. Conv. Am. Ass. equine Pract.* pp. 289–292.

Hughes, J.P. & Loy, R.G. (1975) The relation of infection to infertility in the mare and stallion. *Equine vet. J.* 7, 155–159.

Ingham, H.R., Sisson, P.R., Tharagonnet, D., Skelton, J.B. & Codd, A.A. (1977) Inhibition of phagocytosis *in vitro* by obligate anaerobes. *Lancet* ii, 1252–1254.

Jones, B.M., Kinghorn, G.R. & Duerden, B.I. (1982) An overview of the diagnosis and treatment of *Gardnerella vaginalis* and *Bacteroides* associated vaginitis. *Eur. J. clin. Microbiol.* 10, 320–325.

Kelly, M.J. (1978) The quantitative and histological demonstration of pathogenic synergy between *Escherichia coli* and *Bacteroides fragilis* in guinea-pig wounds. *J. med. Microbiol.* 11, 513–523.

Kenney, R.M., Bergman, R.V., Cooper, W.L. & Morse, G.W. (1975) Minimal contamination techniques for breeding mares: techniques and preliminary findings. *Proc. 21st Ann. Conv. Am. Ass. equine Pract.* pp. 327–336.

Kohn, C. (1978) Management of anaerobic infections. *Proc. 2nd Equine Pharmacol. Symposium,* pp. 109–113.

Leigh, D.A. & Simmons, K. (1977) Identification of non-sporing anaerobic bacteria. *J. clin. Path.* 30, 991–992.

Mackintosh, M.E. (1981) Bacteriological techniques in the diagnosis of equine genital infection. *Vet. Rec.* 108, 52–55.

Moss, S. (1983) Isolation and identification of anaerobic organisms from the male and female urogenital tracts. *Br. J. vener. Dis.* 59, 182–185.

Olson, J.D., Ball, L., Mortimer, R.G., Farin, P.W., Adney, W.S. & Huffman, E.M. (1984) Aspects of bacteriology and endocrinology of cows with pyometra and retained fetal membranes. *Am. J. vet. Res.* 45, 2251–2255.

Peterson, F.B., McFeely, R.A. & David, J.S.E. (1969) Studies on the pathogenesis of endometritis in the mare. *Proc. 15th Ann. Conv. Am. Ass. equine Pract.* pp. 279–288.

Powell, D.G. (1980) Contagious Equine Metritis. In *Current Therapy in Theriogenology*, pp. 779–783. Ed. D. A. Morrow. W. B. Saunders Co., Philadelphia.

Powell, D.G., David, J.S.E. & Frank, C.J. (1978) Contagious Equine Metritis. The present situation reviewed and a revised code of practice for control. *Vet. Rec.* **103**, 399–402.

Prescott, J.F. (1978) Identification of some anaerobic bacteria in nonspecific anaerobic infections in animals. *Can. J. comp. Med.* **43**, 194–199.

Ralph, E.D. (1978) The bactericidal activity of nitrofurantoin and metronidazole against anaerobic bacteria. *J. antimicr. Chemother.* **4**, 177–184.

Ricketts, S.W. (1978) *Histological and histopathological studies of the endometrium of the mare.* Fellowship thesis R.C.V.S., London.

Ricketts, S.W. (1981) Bacteriological examinations of the mare's cervix: Techniques and interpretation of results. *Vet. Rec.* **108**, 46–51.

Ricketts, S.W. (1986) Uterine abnormalities. In *Current Therapy in Equine Medicine,* 2nd edn, pp. 503–508. Ed. N. E. Robinson. W. B. Saunders Co., Philadelphia.

Ricketts, S.W., Rossdale, P.D. & Samuel, C.A. (1978) Endometrial biopsy studies of mares with Contagious Equine Metritis 1977. *Equine vet. J.* **10**, 160–166.

Rodloff, A.C., Becker, J., Blanchard, D.K., Klein, T.W., Hahn, H. & Friedman, H. (1986) Inhibition of macrophage phagocytosis by *Bacteroides fragilis, in vivo* and *in vitro. Infection and Immunity* **52**, 488–492.

Rossdale, P.D. & Ricketts, S.W. (1980) *Equine Stud Farm Medicine*, 2nd edn. Bailliere Tindall, London.

Samuel, C.A., Ricketts, S.W., Rossdale, P.D., Steven, D.H. & Thurley, K.W. (1979) Scanning electron microscope studies of the endometrium of the cyclic mare. *J. Reprod. Fert., Suppl.* **27**, 287–292.

Shin, S.J., Lein, D.H., Aronson, A.L. & Nusbaum, S.R. (1979) The bacteriological culture of equine uterine contents, in-vitro sensitivity of organisms isolated and interpretation. *J. Reprod. Fert., Suppl.* **27**, 305–315.

Simpson, R.B., Burns, S.J. & Snell, J.R. (1975) Microflora in stallion semen and their control with a semen extender. *Proc. 21st Ann. Conf. Am. Ass. equine Pract.* **00**, 255–261.

Taylor, C.E.D., Rosenthal, R.G., Brown, D.F.J., Lapage, S.P., Hill, L.H. & Legross, R.M. (1978) The causative organism of Contagious Equine Metritis, 1977—proposal for a new species to be known as Haemophilus equigenitalis. *Equine vet. J.* **10**, 136–144.

Watson, E.D., Stokes, C.R., David, J.S.E., Bourne, F. J. & Ricketts, S.W. (1986) Concentrations of uterine luminal prostaglandins in mares with acute and persistent endometritis. *Equine vet. J.* **19**, 31–37.

Wilks, M., Thin, R.N. & Tabaqchali, S. (1984) Quantitative bacteriology of the vaginal flora in genital disease. *J. med. Microbiol.* **18**, 217–231.

Willis, A.T. (1977) *Anaerobic Bacteriology: Clinical and Laboratory Practice* 3rd edn. Butterworths, London.

Wingfield Digby, N.J. (1978) The technique and clinical application of endometrial cytology in mares. *Equine vet. J.* **10**, 167–170.

Wingfield Digby, N.J. & Ricketts, S.W. (1982) Results of concurrent bacteriological and cytological examinations of the endometrium of mares in routine studfarm practice 1978–1981. *J. Reprod. Fert., Suppl.* **32**, 181–185.

Woolcock, J.B. (1980) Equine bacterial endometritis: diagnosis, interpretation and treatment. *Vet. Clinics N. Am.; Large Animal Practice* **2**, 241–251.

THE PREGNANT MARE

Convener

E. Palmer

Chairmen

D. C. Sharp

W. R. Allen

K. J. Betteridge

J. Reprod. Fert., Suppl. **35** (1987), 353–361

Analysis of X-chromosome inactivation in horse embryos*

A. Romagnano†‡, C. L. Richer†‡, W. A. King‡ and K. J. Betteridge‡§

†*Departement d'Anatomie, Faculté de Médecine, and* ‡*Centre de Recherche en Reproduction Animale, Faculté de Médecine Vétérinaire, Université de Montréal, C.P. 6128, Succursale A, Montréal, Québec, Canada H3C 3J7*

Summary. To define the time of X-chromosome inactivation in the horse, 122 conceptuses were collected transcervically between Days 6 and 28 (ovulation = Day 0) and subjected to cytogenetic analysis: 59 of the embryos were divided and in 41 of these separate cytogenetic analysis of the embryonic disc and remaining tissues was possible. Conceptuses were measured and photographed before capsule removal, culture in the presence of 5-bromodeoxyuridine and subsequent fixation for cytogenetic analysis. On average, 15 slides were prepared per conceptus. C-banding was used to determine the sex of each conceptus and, in the females, the effects of 5-bromodeoxyuridine incorporation were revealed by u.v. irradiation which produced R-bands and allowed identification of the inactive X-chromosome. Inactivation of the X chromosome began gradually in the trophoblastic ($\pm$ attached endoderm and mesoderm) cells around Day 7·5 and in the embryonic disc around Day 11·5. Cells with an inactive X-chromosome predominated in the trophoblast by Day 10·5 and in the disc by Day 12·5.

Introduction

The original X-inactivation hypothesis (Lyon, 1961) proposed that one of the two X chromosomes in female mammalian embryos becomes inactivated early in embryogenesis, and that the inactive X is stably inherited in daughter cells. The presence of an inactive X-chromosome can be recognized in several ways: by the appearance of the sex chromatin, by the reduction of X-linked gene products, and by visualization of late replication (the inactive X begins its replication late in the second half of the cell cycle after its homologue and the autosomes have completed theirs). In the embryos of most domestic animals, including the horse, the dense granular nature of the nucleus masks the sex chromatin or Barr body (Rowson, 1974). Furthermore, late replication is a more reliable indicator of X-inactivation than is the appearance of sex chromatin (Lyon, 1972). The inactive X is readily revealed by differential staining after incorporation of an analogue of thymidine, 5-bromodeoxyuridine (BrdU), during the last half of the S-phase of the cell cycle. It is then incorporated not only into the C-bands and all the G-bands, but also into the late replicating R-bands of the inactive X-chromosome.

Since 1961, X-inactivation has been studied extensively by cytological, cytogenetic and biochemical means in early embryos of the mouse (Martin, 1982; Sugawara *et al.*, 1983, 1985), and, to a lesser extent, in other eutherian mammals including man and macaque (Park, 1957), rat (Zybina, 1960), cat (Austin, 1966), dog (Austin, 1966), rabbit (Melander, 1962; Kinsey, 1967; Issa *et al.*, 1969; Plotnick *et al.*, 1971), golden hamster (Hill & Yunis, 1967), pig (Axelson, 1968) and roe deer (Aitken, 1974). In a recent study in man, the phenomenon has been investigated in the chromosomes of cells taken from first-trimester spontaneous abortions and newborn infant placentae (Migeon *et al.*, 1985). The somatic cells of metatherian (marsupial) and some monotreme mammals

---

*Reprint requests to Dr C.L. Richer.
§Present address: Department of Biomedical Sciences, Ontario Veterinary College, University of Guelph, Guelph, Ontario, Canada N1G 2W1.

have also been studied, but, to date, their embryos have been too scarce for developmental X-inactivation studies (VanderBerg, 1983; VanderBerg *et al.*, 1983).

The size and accessiblity of embryos have also limited cytogenetic X-inactivation studies, even in the mouse. For example, the embryos of both mouse and man have short preimplantation periods and the mouse embryo is embedded in the uterine wall by the time the X-chromosome of the embryonic ectoderm (which gives rise to the ectoderm, mesoderm and endoderm of the fetus) is inactivated. Mechanical removal of the embryo from the uterine wall can lead to inadvertent mixing of cells from these various lineages as well as an overall loss of cells. The loss of cells adds to the difficulties of accomplishing cytogenetic studies of the inactive X-chromosome in very small amounts of tissue with a low mitotic index.

Horse conceptuses are more accessible than those of the mouse; attachment is superficial in the horse and migration of chorionic cells into the endometrium does not begin until about 37 days gestation (Allen *et al.*, 1973). For at least the first 3 weeks of this preattachment period, the horse conceptus is enclosed in an acellular capsule which replaces the zona pellucida soon after the blastocyst arrives in the uterus, about 5·5 days after ovulation (Marrable & Flood, 1975; Betteridge *et al.*, 1982; Flood *et al.*, 1982; Denker *et al.*, 1987). Embryogenesis therefore proceeds to relatively advanced stages while the entire conceptus remains separate from the maternal tissue and is readily recoverable through the cervix.

In mice, there is a 2-fold difference in the levels of X-linked enzyme activities between female and male embryos before one of the X-chromosomes in the females is inactivated (Epstein *et al.*, 1978; Monk, 1981). This fact has been exploited to diagnose the sex of mouse embryos biochemically during transfer (Williams, 1986), raising the possibility of using similar techniques in the domestic species (Rieger, 1984). With these facts in mind, the objective of this study was to determine the time of X-inactivation in the horse by detection of the late replicating X-chromosome in embryonic tissues.

Materials and Methods

Animals and their management. The experimental herd of 10–20 light horse or pony-type mares, ranging from 200 to 500 kg in weight, were fed hay and concentrates both during winter when they were housed, and between the months of May and October when they were run outdoors. They were teased daily and their reproductive tracts were palpated daily during oestrus for the diagnosis of ovulation which was designated as Day 0. Since ovulation could have occurred at any time during the 24 h preceding the detection of a ruptured follicle, the day on which the latter was diagnosed was taken as Day 0.5 ± 0.5. Mares were mated or artificially inseminated using one of 3 stallions as previously described (Betteridge *et al.*, 1985).

Conceptuses. Although the term 'embryo' is retained in the title of this paper for conformity with similar studies in other mammals, the term 'conceptus' more aptly describes the range of developmental stages that we have studied in the horse.

Over a period of 3 years, 122 conceptuses, including 5 sets of twins and 1 set of triplets, were collected during 113 transcervical collections between Days 6 and 28. The collection techniques, which have been described elsewhere (Betteridge *et al.*, 1985), permitted the recovery of conceptuses which usually remained intact if under 23 mm in diameter but had to be crushed if larger. The flushing medium consisted of a phosphate-buffered saline (PBS) adjusted to pH 7·2, warmed to 38°C, and containing freshly added penicillin (100 i.u./ml) and streptomycin (0·1 µg/ml).

After collection, the conceptuses were photographed and, if intact, the outer diameters (including the capsule or zona pellucida) were measured with a calibrated eyepiece and microscope or with a ruler for those too large for the microscope. Dimensions of 5 conceptuses that were ruptured during recovery were available from ultrasonic examination made *in situ* just before collection. Before culture, capsules were removed under a dissection microscope using two 25-gauge needles on tuberculin syringes, or for one Day-6·5 blastocyst still invested with traces of zona pellucida as well, with the aid of micromanipulators (Picard *et al.*, 1985).

When possible, the embryonic disc was dissected from the rest of the conceptus immediately after capsule removal using the same two tuberculin syringes fitted with 25-gauge needles. These tissues were then cultured separately to assess X-inactivation in each. Although it is recognized that the trophectoderm of the conceptus from which the disc had been removed could have endodermal and sometimes mesodermal cells attached to it in various amounts, depending on the stage of development (see Ginther, 1979), the tissue devoid of disc will be referred to as 'trophoblast' for convenience. In the two largest conceptuses (Nos 25 & 26, Table 1) the recognizable embryo proper was cultured after being cut into small pieces (~ 1 mm^3) with iridectomy scissors.

Table 1. The proportions (percentages) of R-banded metaphases that revealed an inactive X chromosome in each of the female conceptuses studied cytogenetically by the direct technique

| | Conceptus | | Proportion (%) of metaphases with an inactive X in: | | |
| | Day of collection (±0·5 days) | Diameter (mm) | Whole conceptus | Disc | Trophoblast |
No.					
1	6·5	0·20	0/30 (0)		
2	7·5	n.m.	0/63 (0)		
3	7·5	0·69	0/20 (0)		
4	7·5	0·71	0/10 (0)		
5	7·5	0·65	1/50 (2)		
6	8·5	0·82	0/20 (0)		
7	8·5	1·33	0/30 (0)		
8	8·5	1·17	2/50 (4)		
9	9·5	0·6	0/20 (0)		
10	9·5	n.m.	2/10 (20)		
11	9·5	3·00		0/10 (0)	4/10 (40)
12	10·5	5·00		0/10 (0)	5/10 (50)
13	10·5	n.m.		0/10 (0)	3/10 (30)
14	10·5	5·00			7/10 (70)
15	11·5	6·00	10/20 (50)		
16	11·5	7·71	5/30 (17)		
17	11·5	5·24		3/10 (30)	4/10 (40)
18	12·5	9·00		6/10 (60)	7/10 (70)
19	12·5	9·25		4/10 (40)	
20	12·5	11		6/10 (60)	
21	13·5	16		14/20 (70)	
22	14·5	19			7/10 (70)
23	14·5	18		5/10 (50)	
24	14·5	22		6/10 (60)	2/3 (67)
25	19·5	n.m.		6/10 (60)*	
26	22·5	16		5/10 (50)*	

n.m. = not measured.
*Embryo proper in these advanced conceptuses.

Culture and fixation. The 26 conceptuses were processed by the direct technique described by Romagnano *et al.* (1985); 2 were studied following long-term culture requiring a minimum of 2 weeks before chromosomes could be harvested (see below).

In preparation for the direct technique, after capsule removal whole conceptuses were cultured for up to 10 h in MEM containing 0·25 mM-Hepes buffer and Hanks salts to which 1% L-glutamine, 10% charcoal-treated fetal calf serum (FCS), antibiotics, and 50 µg BrdU/ml had been added (Romagnano *et al.*, 1985). However, this study was carried out concurrently with other investigations and therefore 12 conceptuses had a preincubation of less than 4 h in medium containing serum of a different source and, in 4 cases, radioactive steroid substrates (Marsan *et al.*, 1987). After this preincubation, embryos were decapsulated, cultured and fixed in the same manner as those embryos not preincubated. When dissection was possible, the embryonic disc was cultured as a whole, while the trophoblastic tissue was first cut into small pieces using the same needles on tuberculin syringes. In the course of the study, the direct technique was modified slightly: fetal bovine serum was omitted from the culture medium and a vortex was employed for a few seconds after the addition of 50% acetic acid. On average, 10 slides (range 7–20) were prepared from each whole conceptus, and 20 slides (range 12–30) from those that had been dividied (10 each from the embryonic disc and trophoblast).

In the 2 conceptuses studied after a prolonged culture, embryonic disc fragments only were cultured in gelatinized T-25 culture flasks (Falcon Plastics, Becton-Dickinson, Mississauga, Ontario) containing 5 ml MEM (with 0·25 mM-Hepes buffer and Hanks salts) supplemented with 1% L-glutamine, 20% FCS and antibiotics as previously described (Murer-Orlando *et al.*, 1982). This technique was slightly modified however; at 7 h before harvest, BrdU was added to the flasks at a final concentration of 50 µg/ml. Each flask yielded enough cells for the preparation of 15–20 slides.

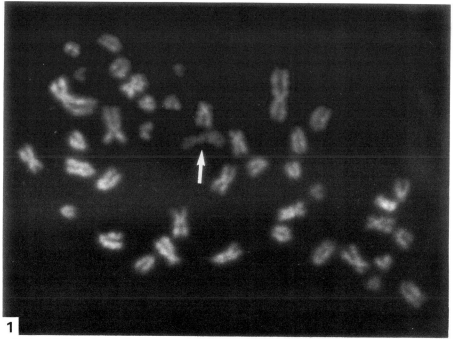

Fig. 1. Partial RBA-banded metaphase from an 8·5 day embryo prepared by the direct technique. The dark chromosome, indicated by the arrow, is the inactive X. According to the criteria of RBA staining the inactive X is reliably revealed as a red chromosome both under the microscope and on colour prints.

Staining. A portion of the slides prepared from each conceptus larger than 0·82 mm were CBG banded (Sumner, 1972) and the remainder were RBA banded (Dutrillaux *et al.*, 1973) or RB–FPG banded (Romagnano & Richer, 1984). Slides from the smaller embryos, which produced fewer preparations, were first RBA banded and then CBG banded. From each conceptus, 10–63 well-stained RBA banded metaphases, and a similar number of CBG and RB–FPG banded metaphases, were examined.

The following criteria were used to identify the sex chromosomes. The Y chromosome, a small acrocentric, is darkly stained after CBG-banding, dull red after RBA-banding and pale after RB–FPG banding. The active X-chromosome, a large submetacentric, has a centromeric and an interstitial band after CBG banding, fluorescent yellow R-bands after RBA and dark R-bands after RB–FPG banding. The inactive X-chromosome, a large submetacentric, has a centromeric and an interstitial band after CBG banding, and is predominantly dull red after RBA and pale after RB–FPG banding.

A few additional slides were simply Giemsa stained and the mitotic index was calculated. The mitotic index (cells in metaphase/total cells counted, expressed as a percentage) of all cultured conceptus tissues (irrespective of sex) was determined when the nuclei of at least 200 cells (range 200–1000) could be clearly identified.

Results

Of the 120 conceptuses examined by the direct technique, 92 (77%) were sexed after C-banding: 53 were males, including all three triplets, and 39 females. This is not a significant variation from the expected 50:50 sex ratio ($\chi^2 = 2\cdot14$). All 122 conceptuses examined were judged karyotypically normal, as no numerical or sex chromosome anomalies were found.

Separation of the embryonic disc from the trophoblast proved possible only in conceptuses measuring over 3 mm in diameter on Day 9·5 or later. Four conceptuses were not measured intact: Nos 2 and 13 in error; Nos 10 and 25 because of damage during collection (Table 1).

Of the 39 known females, 13 possessed too few (ranging from 0 to 9) well spread RBA-banded metaphases to be further analysed. Twenty-six conceptuses (12 entire; 14 divided into trophoblast and embryonic disc) revealed 10–63 well-spread R-banded metaphases in which the presence of the inactive X-chromosomes could be reliably ascertained (Fig. 1). The numbers of inactive X-chromosomes detected in these are recorded in relation to their developmental stage in Table 1. As can be seen, in the whole conceptuses no inactive Xs were observed in the Day-6·5 blastocyst (0·20 mm), or in 5 of the 7 studied on Days 7·5 and 8·5 and ranging from 0·65 to 1·17 mm in diameter. Only 1 inactive X was observed amongst the 143 R-banded metaphases examined in Day-7·5 conceptuses (0·7%) and only 2 in the 100 R-banded metaphases from the 3 whole conceptuses studied on Day 8·5 (2%). While 1 of the 2 whole Day-9·5 conceptuses still displayed no inactive X, the other, as well as the 2 examined on Day 11·5 displayed an increase in the frequency of X-inactivation. When disc and trophoblastic tissues could be analysed separately, X-inactivation was found exclusively in the trophoblast in the 3 youngest conceptuses examined (Nos 11, 12 & 13; Days 9·5–10·5; Table 1) and

Table 2. The distribution of conceptuses according to their mitotic indices after culture with or without dissection of the embryonic disc from the trophoblastic tissue

No. of conceptuses	Type of culture	Mitotic index		
		<1%	1–10%	>10%
54	Whole	36 (67%)	15 (28%)	3 (5%)
41	Dissected	28 (68%)	13 (32%)	0 (0%)

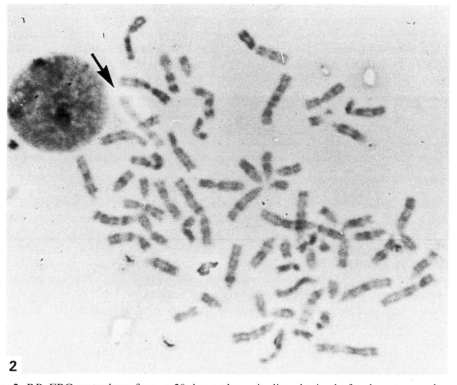

2

Fig. 2. RB–FPG metaphase from a 28-day embryonic disc obtained after long-term culture. The pale chromosome, indicated by the arrow, is the inactive X.

was first detected in the disc on Day 11·5, after which time it affected all conceptuses examined (Table 1). The mitotic index was unaffected by dissection; it was less than 1% in about two-thirds of the conceptuses, and very seldom exceeded 10% (Table 2). In the 41 dissected conceptuses, the mitotic index was higher in the embryonic disc than in the trophoblast in 22 (54%), was the same in both tissues in 6 (15%), and was higher in the trophoblast than in the disc in 13 (31%).

The two embryonic discs examined after prolonged culture were dissected from embryos aged 12·5 and 28 days. Both were female and revealed inactive Xs in all R-banded metaphases studied. Figure 2 depicts an RB–FPG banded metaphase from the 28-day embryonic disc culture. The inactive X is very pale and easily recognizable (see arrow). Both cultures remained viable for up to 6 months and retained their state of X-inactivation during up to 4 subcultures and harvests.

Discussion

The advantage of cytogenetic techniques over biochemical methods is that the former gives direct evidence of whether or not the X-chromosome is active within individual cells (Rastan, 1981). The results of this study therefore indicate that X-inactivation occurs in some cells of the early blasto-cyst in the horse and is gradually detected in more cells as the embryo develops. In conceptuses studied whole, the earliest inactive X was detected on Day 7·5 (diameter of 0·65 mm), but this X-chromosome was the only one found in the inactive state in all the 143 metaphases examined on Day 7·5. The youngest conceptuses divided (Day 9·5, diameter 3·00 mm) had an inactive X in 40% of its trophoblastic cells and 0% of its embryonic disc cells. This pattern was repeated in both Day-10·5 conceptuses studied, suggesting that the inactive Xs found in whole conceptuses on or before Day 10·5 are of trophoblastic origin.

X-inactivation in the embryonic disc was first observed in a Day-11·5 conceptus of 5·24 mm in diameter. By Day 12·5, about 50% of the embryonic disc cells bore an inactive X. This was demon-strated directly by the majority of Conceptuses 18–26 (Table 1) and indirectly in the long-term cultures from a Day-12·5 embryonic disc and a Day-28 embryo in which X-inactivation was also shown to be maintained after up to 4 subcultures.

In mouse embryos, the process of X-inactivation is also a gradual one. It begins in the trophectoderm on Day 3·5, in the primitive endoderm on Day 4·5, in the embryonic ectoderm on Day 6·0 and finally in the oocytes of the embryonic ovary on Day 11·5 (Martin, 1982; Takagi, 1983). There is general agreement that X-inactivation has not occurred in the morula, but occurs concurrently with its first differentiation which gives rise to the trophectoderm (Martin, 1982). According to most authors, X-inactivation is concomitant with tissue differentiation in the mouse (Monk & Kathuria, 1977; Kratzer & Gartler, 1978; Monk & Harper, 1979), although Sugawara *et al.* (1985) challenge this belief and propose that overt cell differentiation precedes the appearance of the asynchronously replicating X-chromosome. According to our study, X-inactivation appears to begin in the cells of the early horse blastocyst about 2 days after the trophectoderm has differen-tiated. This would suggest that, in the horse, X-inactivation is not totally concomitant with cell differentiation, in agreement with the contention of Sugawara *et al.* (1985).

As yet, X-inactivation has been studied cytogenetically in the early embryos of very few other species. In the cow, 27%, 30% and 46% of cells examined were found to contain inactive X chromosomes at Days 12, 13 and 14 respectively (W. A. King, A. Romagnano, C. L. Richer & K. J. Betteridge, unpublished data).

In our horse studies the percentage of metaphases with an inactive X in trophoblastic or embryonic disc cells varied between 0 and 70%. These numbers probably reflect a true hetero-geneity of inactivation in the tissue. However, the fact that an inactive X was never found in more than 70% of the cells could have been caused by the natural asynchrony found in embryonic cells; BrdU may not have been properly incorporated in all metaphases and hence some inactive Xs may not have been detected. This raises the question of what other staining techniques could be used to quantify X-inactivation in the horse.

A technique that is apparently specific to mice (Kanda, 1973) was adapted to murine embryonic cells after preliminary studies on adult bone-marrow cells (Rastan, 1981). These studies showed a darkly staining chromosome in 73·1% of the bone-marrow metaphases examined in female mice, in agreement with the figure (72%) originally described by Kanda (1973). However, when applied to the embryonic cells the percentages were higher, and a darkly stained inactive X was revealed in over 80% (range 74·5–91·6%) of metaphase cells from Day-13·5 embryos (Rastan, 1981). Rastan (1982) also examined the timing of X-inactivation in whole mouse embryos on Days 5·5, 6·5 and 7·5, and in epiblasts isolated from Day-6 embryos, in which inactive Xs were seen in 86·9, 85·8, 83·5 and 90·9% of metaphase cells respectively. Unfortunately, this technique has not proved applicable to horse chromosomes prepared from leucocytes or embryos (A. Romagnano & C. L. Richer, unpublished results).

Other staining methods used to study X-inactivation in early mouse embryos were Q-banding by quinicrine mustard, G-banding by acetic saline Giemsa staining (Takagi & Oshimura, 1973) and finally R-banding following BrdU incorporation and staining with acridine orange (BrdU–AO; Takagi & Sasaki, 1975). The last technique was considered the most useful as it has been proposed that the inactive X-chromosome is not only late replicating during embryogenesis, but early replicating as well (Takagi, 1974). Hence not only are dull red X-chromosomes scored as being inactive, but brightly fluorescent (green–yellow) Xs are as well (Takagi, 1974; Sugawara *et al.*, 1983, 1985). The usefulness of the BrdU–AO method was judged by being able to identify the allocyclic X (the X-chromosome which replicates out of cycle, be it early or late) in more than 80% of metaphases from Day-7·5 embryos as the dullest red or the brightest fluorescent complement (Takagi, 1974). According to Sugawara *et al.* (1983) the early replicating X-chromosome shifts to become a late replicating one in the embryo proper on Day 6·4. As interesting as this hypothesis is, one cannot totally eliminate the possibility of an uneven BrdU incorporation in asynchronous cells. Indeed, a late replicating X appears brightly fluorescent in a cell which has incorporated BrdU in the first half of its S-phase and is thus G-banded. When unsynchronized cells are placed in the presence of BrdU, we have observed, as have Dutrillaux *et al.* (1973), that not all cells are R-banded, but that a few are G-banded and a few others present intermediary banding. In the present study only well R-banded metaphases containing one dull red X-chromosome were scored as positive for an inactive X, while those with 2 well banded X chromosomes were scored as negative and were assumed to contain two active Xs. Although the R-banding pattern was consistent between the homologues in the negatively scored metaphases, it was not always as consistent between the two Xs. However, in clearly R-banded metaphases, we have never observed a distinct bright fluorescent allocyclic X-chromosome as described by others (Takagi, 1974; Sugawara *et al.*, 1983, 1985). We feel that, if fluorescent early replicating X chromosomes existed in horse embryos, we would have noticed them.

In agreement with previous studies on mouse embryos (Takagi, 1978; Ilgren, 1981) the majority of horse embryonic discs displayed a higher mitotic index than did trophoblastic tissue after in-vitro culture. The lowered mitotic index of the trophoblastic cells further decreased the number of successfully R-banded metaphases, so that, despite the great difference in cell numbers in the embryonic disc versus the trophoblast in early conceptuses, the disc was more easily studied.

Another interesting finding was the complete lack of karyotypically abnormal conceptuses; of the 92 analysed after C-banding, not one was found to be aneuploid. This is rather surprising considering their early ages. Extensive studies on human spontaneous and induced abortions as well as limited ones on human and domestic animal embryos, strongly suggest that the younger the embryo the higher the expected percentage of anomalies (Hassold & Jacobs, 1984; Rudak *et al.*, 1984; King, 1985), so our material would be a high risk group.

In conclusion our results show that X-inactivation in early horse conceptuses is a gradual process which begins in the trophoblast on Day 7·5 and then in the disc around Day 11·5, but not concomitantly with their differentiation. By Day 10·5 in the trophoblast and Day 12·5 in the disc,

X-inactivation appears to have spread to a good proportion of the cells. In horses, therefore, any exploitation of X-linked factors for distinguishing male from female embryos would need to be directed at blastocysts collected before Day 7, very soon after their arrival in the uterus.

We thank Dr Arnaud Renard and Dr Jean Sirois and the Animal care staff of the CRRA for help with the embryo collections; Dr Louis Picard for micromanipulation; Dr Alan Goff and Charles Marsan for help with short-term culture preparation; Louise Laquere for excellent technical assistance and maintenance of long term cultures; Jean Leveille for skilful photography; and the FCAC and NSERC for financial support.

References

Aitken, R.J. (1974) Sex chromatin formation in the blastocyst of the roe deer (*Capreolus capreolus*) during delayed implantation. *J. Reprod. Fert.* **40**, 235–239.

Allen, W.R., Hamilton, D.W. & Moor, R.M. (1973) The origin of equine endometrial cups. II. Invasion of the endometrium by trophoblast. *Anat. Rec.* **117**, 485–502.

Austin, C.R. (1966) Sex chromatin in embryonic and fetal tissue. In *The Sex Chromatin*, pp. 241–254. Ed K. L. Moore. Saunders, Philadelphia.

Axelson, M. (1968) Sex chromatin in early pig embryos. *Hereditas* **60**, 347–354.

Betteridge, K.J., Eaglesome, M.D., Mitchell, D., Flood, P.F. & Bériault, R. (1982) Development of horse embryos up to twenty-two days after ovulation: observations on fresh specimens. *J. Anat.* **135**, 191–209.

Betteridge, K.J., Renard, A. & Goff, A.K. (1985) Uterine prostaglandin release relative to embryo collection, transfer procedures and maintenance of the corpus luteum. *Eq. vet. J., Suppl.* **3**, 25–33.

Denker, H.-W, Betteridge, K.J. & Sirois, J. (1987) Shedding of the 'capsule' and proteinase activity in the horse. *J. Reprod. Fert., Suppl.* **35**, 703.

Dutrillaux, B., Laurent, C., Couturier, J. & Lejeune, J. (1973) Coloration par L'acridine orange des chromosomes préalablement traités par le 5-bromodeoxyuridine (BrdU). *C. r. hebd. Scéanc. Acad. Sci., Paris* **276**, 3179–3181.

Epstein, C.J., Smith, S., Travis, B. & Tucker, G. (1978) Both X chromosomes function before visible X-chromosome inactivation in female mouse embryos. *Nature, Lond.* **274**, 500–502.

Flood, P.F., Betteridge, K.J. & Diocee, M.S. (1982) Transmission electron microscopy of horse embryos 3–16 days after ovulation. *J. Reprod. Fert., Suppl.* **32**, 319–327.

Ginther, O.J. (1979) *Reproductive Biology of the Mare. Basic and Applied Aspects*, pp. 255–263. O. J. Ginther, Cross Plaines, Wisconsin.

Hassold, T.J. & Jacobs, P.A. (1984) Trisomy in man. *Ann. Rev. Genet.* **18**, 69–97.

Hill, R.N. & Yunis, J.J. (1967) Mammalian X-chromosomes: change in patterns of DNA replication during embryogenesis. *Science, N.Y.* **155**, 1120–1121.

Ilgren, E.B. (1981) On the control of the trophoblastic

giant-cell transformation in the mouse: homotypic cellular interactions and polyploidy. *J. Embryol. exp. Morph.* **62**, 183–202.

Issa, M., Blank, C.E. & Atherton, G.W. (1969) The temporal appearance of sex chromatin and of the late-replicating X chromosome in blastocysts of the domestic rabbit. *Cytogenetics* **8**, 219–237.

Kanda, N. (1973) A new differential technique for staining heteropycnotic X-chromosome in female mice. *Expl Cell Res.* **80**, 463–467.

King, W.A. (1985) Intrinsic embryonic factors which may affect survival after transfer. *Theriogenology* **23**, 161–174.

Kinsey, J.D. (1967) X-chromosome replication in early rabbit embryos. *Genetics, Princeton* **55**, 337–343.

Kratzer, P.G. & Gartler, S.M. (1978) HGPRT activity changes in preimplantation mouse embryos. *Nature, Lond.* **274**, 503–504.

Lyon, M.F. (1961) Gene action in the X-chromosome of the mouse (*Mus musculus* L.). *Nature, Lond.* **190**, 372–373.

Lyon, M.F. (1972) X-chromosome inactivation and developmental patterns in mammals. *Biol. Rev.* **47**, 1–35.

Marrable, A.W. & Flood, P.F. (1975) Embryological studies on the Dartmoor pony during the first third of gestation. *J. Reprod. Fert., Suppl.* **23**, 499–502.

Marsan, C., Goff, A.K., Sirois, J. & Betteridge, K.J. (1987) Steroid secretion by two different cell types of the horse conceptus. *J. Reprod. Fert., Suppl.* **35**, 363–369.

Martin, G.R. (1982) X-chromosome inactivation in mammals. *Cell* **29**, 721–724.

Melander, Y. (1962) Chromosomal behaviour during the origin of sex chromatin in the rabbit. *Hereditas* **48**, 645–661.

Migeon, B.R., Wolf, S.F., Axelman, J., Kaslow, D.C. & Schmidt, M. (1985) Incomplete dosage compensation in chorionic villi of human placenta. *Proc. natn. Acad. Sci. U.S.A.* **82**, 3390–3394.

Monk, M. (1981) A stem-line model for cellular and chromosomal differentiation in early mouse development. *Differentiation* **19**, 71–76.

Monk, M. & Harper, M.I. (1979) Sequential X chromosome inactivation coupled with cellular differentiation in early mouse embryos. *Nature, Lond.* **281**, 311–313.

Monk, M. & Kathuria, H. (1977) Dosage compensation for an X-linked gene in preimplantation mouse embryos. *Nature, Lond.* **270**, 599–601.

Murer-Orlando, M., Betteridge, K.J. & Richer, C.L. (1982) Cytogenetic sex determination in cultured cells of pre-attachment horse embryos. *Ric. Sci. Educaz. Permanent, Suppl.* **24**, 372–378.

Park, W.W. (1957) The occurrence of sex chromatin in early human and macaque embryos. *J. Anat.* **91**, 369–373.

Picard, L., King, W.A. & Betteridge, K.J. (1985) Production of sexed calves from frozen-thawed embryos. *Vet. Rec.* **117**, 603–608.

Plotnick, F., Klinger, H.P. & Kosseff, A.L. (1971) Sex-chromatin formation in preimplantation rabbit embryos. *Cytogenetics* **10**, 224–253.

Rastan, S. (1981) *Aspects of X-chromosome inactivation in mouse embryogenesis* Ph.D. thesis, University of Oxford.

Rastan, S. (1982) Timing of X-chromosome inactivation in postimplantation mouse embryos. *J. Embryol. exp. Morph.* **71**, 11–24.

Rieger, D. (1984) The measurement of metabolic activity as an approach to evaluating viability and diagnosing sex in early embryos. *Theriogenology* **21**, 138–149.

Romagnano, A. & Richer, C.L. (1984) R-banding of horse chromosomes: The fluorescence-photolysis-Giemsa technique after bromodeoxyuridine incorporation. *J. Hered.* **75**, 269–272.

Romagnano, A., King, W.A., Richer, C.L. & Perrone, M.-A. (1985) A direct technique for the preparation of chromosomes from early equine embryos. *Can. J. Genet. Cytol.* **27**, 365–369.

Rowson, L.E.A. (1974) The role of research in animal production. *Vet. Rec.* **95**, 276–280.

Rudak, E., Dor, J., Mashiach, S., Nebel, L. & Goldman, B. (1984) Chromosome analysis of multipronuclear human oocytes fertilized *in vitro. Fert. Steril.* **41**, 538–545.

Sugawara, O., Takagi, N. & Sasaki, M. (1983) Allocyclic early replicating X chromosome in mice: genetic inactivity and shift into a late replicator in early embryogenesis. *Chromosoma* **88**, 133–138.

Sugawara, O., Takagi, N. & Sasaki, M. (1985) Correlation between X-chromosome inactivation and cell differentiation in female preimplantation mouse embryos. *Cytogenet. Cell Genet.* **39**, 210–219.

Sumner, A.T. (1972) A simple technique for demonstrating centromeric heterochromatin. *Expl Cell Res.* **75**, 304–306.

Takagi, N. (1974) Differentiation of X chromosomes in early female mouse embryos. *Expl Cell Res.* **86**, 127–135.

Takagi, N. (1978) Preferential inactivation of the paternally derived X chromosome in mice. In *Genetics and Chimeras of Mammals*, pp. 341–359. Ed. A. McLaren, Plenum Press, New York.

Takagi, N. (1983) Cytogenetic aspects of X-chromosome inactivation in mouse embryogenesis. In *Cytogenetics of the Mammalian X Chromosome Part A: Basic Mechanisms of X Chromosome Behavior*, pp. 21–50. Ed. A. A. Sandberg, A. R. Liss, Inc., New York.

Takagi, N. & Oshimura, M. (1973) Fluorescence and Giemsa banding studies of the allocyclic X chromosome in embryonic and adult mouse cells. *Expl Cell Res.* **78**, 127–135.

Takagi, N. & Sasaki, M. (1975) Preferential inactivation of the paternally derived X chromosome in the extraembryonic membranes of the mouse. *Nature, Lond.* **256**, 640–642.

VanderBerg, J.L. (1983) Developmental aspects of X chromosome inactivation in eutherian and metatherian mammals. *J. exp. Zool.* **228**, 271–286.

VanderBerg, J.L., Johnston, P.G., Cooper, D.W. & Robinson, E.S. (1983) X-chromosome inactivation and evolution in marsupials and other mammals. *Isozymes: Curr. Top. Biol. Med. Res.* **9**, 201–218.

Williams, T.J. (1986) A technique for sexing mouse embryos by a visual colorimetric assay of the X-link enzyme, glucose 6-phosphate dehydrogenase. *Theriogenology* **25**, 733–739.

Zybina, E.V. (1960) Sex-chromatin in the trophoblast of young rat embryos. *Dokl. Akad. Nauk., SSSR* **130**, 633–635.

J. Reprod. Fert., Suppl. 35 (1987), 363–369

Printed in Great Britain
© 1987 Journals of Reproduction & Fertility Ltd

Steroid secretion by different cell types of the horse conceptus

C. Marsan, A. K. Goff, J. Sirois and K. J. Betteridge*

Centre de recherche en reproduction animale, Faculté de médecine vétérinaire, Université de Montréal, C.P. 5000, St-Hyacinthe, Québec, Canada J2S 7C6

Summary. Horse conceptuses were recovered non-surgically at Day 12–Day 15 and were dissociated with collagenase. Separation of the cells on a 31·8% Percoll gradient gave two bands of cells and indirect evidence suggests that the low density cells (LDC) are endoderm and the higher density cells (HDC) are trophectoderm. Each band was incubated for 24 h in Minimum Essential Medium and concentrations of oestradiol and progesterone in the medium were measured by radioimmunoassay (RIA). The LDC secreted predominately progesterone (log oestradiol/progesterone = − 0·994 ± 0·141; N = 15) whereas the HDC secreted mainly oestradiol (log oestradiol/progesterone = 0·522 ± 0·135). When incubated for a further 24 h in fresh medium the oestradiol production decreased by 85% and 84% whereas the progesterone production decreased by 57% and 50% for LDC and HDC respectively. When cells from each band were incubated for 24 h with [³H]pregnenolone, separation of the radiolabelled metabolites by reversed-phase high-pressure liquid chromatography revealed a different steroid profile for each cell type. Therefore, in the intact embryo, an interaction between cells may be important for oestrogen synthesis and thus the maintenance of pregnancy.

Introduction

The embryonic signal involved in the prolongation of the life-span of the corpus luteum has been studied in several species (see Heap *et al.*, 1981). Evidence for a role of steroid hormones in early pregnancy was initially provided by Huff & Eik-Nes (1966) who demonstrated conversion of acetate to pregnenolone and metabolism of other steroids in 6-day-old rabbit blastocysts. In the pig, histochemical (Flood, 1974; King & Ackerley, 1985) and biochemical (Perry *et al.*, 1973) evidence has shown that steroid metabolism in the early blastocyst begins before the time when luteolysis would occur in the non-pregnant female (see Perry *et al.*, 1976). Because the production of oestrogen is much higher in the pig than in many other species (Gadsby *et al.*, 1980) and exogenous oestrogen is luteotrophic in this species (Gardner *et al.*, 1963), oestrogen is thought to be involved in maternal recognition of pregnancy in the pig.

The horse conceptus has also been shown to produce considerable amounts of oestrogens as well as some androgens (Zavy *et al.*, 1979; Flood *et al.*, 1979) which have therefore been postulated to be involved in embryonic signalling in this species. It is uncertain, however, whether oestrogens from the conceptus act in similar ways in these two species since there is a controversy about whether exogenous oestrogens are luteotrophic (Berg & Ginther, 1978; Allen, 1979) in mares. Oestrogen production rises during the period between Days 12 and 20 after ovulation, but may begin as early as Day 6 (Paulo & Tischner, 1985). This encompasses the time of maternal recognition of pregnancy which, in the mare, was believed to occur between Days 14 and 16 (Hershman & Douglas, 1979), but may be as early as Day 11–13 (Betteridge *et al.*, 1985; Goff *et al.*, 1987).

---

*Present address: Department of Biomedical Sciences, Ontario Veterinary College, University of Guelph, Guelph, Ontario, Canada N1G 2W1.

Goff *et al.* (1983) have reported a decrease in oestradiol secretion after rupture of the intact horse blastocyst which suggests some kind of regulation of steroidogenesis by the early conceptus. At Day 12–15, the horse blastocyst consists of the inner cell mass as well as trophoblastic ectoderm, endoderm and some mesoderm (Ginther, 1979). In other steroidogenic tissues an interaction between different cell types has been shown to be essential for normal function (Armstrong & Dorrington, 1977; Fitz *et al.*, 1982). It is therefore possible that an interaction between different cell types within the intact conceptus is necessary for the normal secretion of oestradiol. The objective of this study was to investigate the steroidogenic potential of different cell types from horse conceptuses collected between Days 12 and 15.

Materials and Methods

Embryo collection. Conceptuses were collected between mid-May and early October during 3 consecutive breeding seasons from an experimental herd of light horse and pony types weighing between 200 and 500 kg. The mares were teased daily and the ovaries were palpated each morning during oestrus for the diagnosis of ovulation. Ovulation was defined as occurring midway between two examinations which revealed rupture of a follicle, so that the day on which this was detected was Day 0.5 ± 0.5. Mares were naturally mated or artificially inseminated on alternate days beginning on the 2nd day of oestrus.

The conceptuses were collected by transcervical uterine flushing between Days 12·5 and 15·5 as previously described by Betteridge *et al.* (1982). The flushing medium consisted of a phosphate-buffered saline (PBS) adjusted to pH 7·2, warmed to 38°C, and containing freshly added penicillin (100 i.u./ml) and streptomycin (0·1 µg/ml).

After collection, each conceptus was photographed and, if intact, its diameter was measured with a ruler (accurate to within 1 mm). The dimensions of 5 conceptuses that were ruptured during recovery were obtained by the use of realtime ultrasound scanning examination *in situ* just before collection.

Cell dissociation and incubation. After removal of the capsule, each conceptus was cut into small fragments with two tuberculin syringes fitted with 25-gauge needles. The embryonic disc plus a small amount of trophoblast tissue was removed and the remaining tissue was dissociated with 0·25% collagenase type I (200 units per mg; Sigma St Louis, MO, U.S.A.) in Hanks balanced salt solution (HBSS: GIBCO, Burlington, Canada) without Ca^{2+} and Mg^{2+}. During the 20-min period of dissociation at 37°C the fragments were frequently disrupted with a siliconized Pasteur pipette.

After dissociation the cells were transferred with 300 µl HBSS to the top of a 31·8% continuous Percoll gradient (Pharmacia Fine Chemicals, Dorval, Canada). This gradient was then centrifuged at 400 *g* for 15 min. The separated populations of cells that resulted were washed and resuspended in Eagle's Minimal Essential Medium (MEM: GIBCO) and the numbers of live cells were counted with a haemocytometer after staining with trypan blue.

Cells were incubated for 24 h in MEM containing 25 mM-Hepes buffer, Hanks salts, 1% L-glutamine and 1% penicillin/streptomycin after which the medium was removed and frozen. The cells from 7 conceptuses were incubated for a further 24 h in fresh medium.

Experiments. In a preliminary experiment with a Day-13·5 conceptus, the density gradient was divided into twelve 1-ml fractions. In 15 subsequent experiments using separate conceptuses, only the two visible bands of cells in the density gradient were collected with a siliconized Pasteur pipette.

For indirect identification of these two cells populations, a part of the endodermic layer from 8 of the 15 conceptuses was removed using a non-enzymic procedure of cell dissociation (Dziadek, 1981) which involves incubating the ruptured blastocysts in MEM containing 1 M-glycine for 15 min at 37°C. This technique causes the separation of tissues with poor cell-to-cell contact. The endodermic cells that detached from the inner surface of the blastocyst were further dissociated with collagenase before incubation. On two occasions these endodermic cells were subjected to fractionation on a density gradient.

The aromatase potential of both types of cell were compared using 6 conceptuses. Half of the cells of each type were incubated for 24 h in MEM alone (controls) and the other half in MEM containing 16 ng androstenedione/ml. In all of these experiments, progesterone and oestradiol-17β were measured in the medium by radioimmunoassay (RIA).

For the last experiment both types of cell from 3 conceptuses were incubated for 8 h in 1 ml MEM containing 10^6 d.p.m. of tritiated androstenedione or pregnenolone. These precursors had been checked for purity by high-pressure liquid chromatography (HPLC). The specific activities of [1,2,6,7-³H(N)]androstenedione and [7-³H(N)]-pregnenolone (NEN Research Products, Lachine, Quebec, Canada) were 85 Ci/mmol and 19·3 Ci/mmol respectively. After incubation, half of the medium was separated into conjugated, phenolic and neutral fractions as described by Gadsby *et al.* (1980). The other half of the medium was passed through a Sep-Pak cartridge from which steroids were eluted with acetonitrile. The steroids in each fraction were then separated by HPLC.

Radioimmunoassay. Oestradiol-17β and progesterone were measured in 100 µl samples of unextracted medium by radioimmunoassay. The antibody used for the progesterone assay was raised in rabbits against 11-hydroxyprogesterone 11-hemisuccinate–bovine serum albumin conjugate (Sigma). The cross-reactivity of this antiserum at 50% displacement is 45% for 5-pregnane-3-ol-20-one; 0·9% for pregnenolone; 0·4% for 17α-hydroxyprogesterone, oestradiol-17β and oestrone; 0·2% for 20α-dihydroprogesterone and <0·05% for testosterone, androstenedione and hydrocortisone. The intra- and inter-assay coefficients of variation were 8% and 16% respectively. The sensitivity was 120 pg/ml.

Oestradiol antibody was supplied by Dr D. T. Armstrong and the assay was performed as described by Dorrington & Armstrong (1975). The sensitivity of the assay was 60 pg/ml, and the intra- and interassay coefficients of variation were 6% and 9% respectively.

High-pressure liquid chromatography. A Water's Associates HPLC system with a Nova Pak reverse-phase radial compression cartridge, mounted in a radial compressor (Water's Model RCM 100), was used. After filtration through 0·22 μm microfilter a volume of 500 μl was injected and steroids were eluted with a gradient of 12% diethyl ether in acetonitrile:water from 30:70 (v/v) to 80:20 (v/v) over 40 min with a flow rate of 1 ml/min. Five internal unlabelled standards (hydrocortisone, testosterone, androstenedione, 20α-dihydroprogesterone and progesterone) were included with each run. The retention times of standards using this system were 2·03 min for oestrone sulphate, 8·97 min for oestriol, 9·38 min for hydrocortisone, 19·30 min for oestradiol-17β, 20·30 min for testosterone, 20·62 min for oestradiol-17α, 22·46 min for 17α-hydroxyprogesterone, 22·68 min for oestrone, 23·00 min for androstenedione, 30·56 min for 20α-dihydroprogesterone, and 36·86 min for progesterone. The elution of standards was monitored by u.v. absorption at 254 or 280 nm.

Fractions of 0·5 or 0·25 ml were collected directly into 7-ml scintillation vials. After evaporation of the diethyl ether 5·5 ml HB/b (Beckman Ltd) scintillation fluid were added and the samples were counted in a Beckman 3801 scintillation counter. Radioactivity was expressed as d.p.m.

Statistical analysis was performed on log-transformed data by the paired *t* test.

Results

The 19 conceptuses used varied from 1·4 to 2·3 cm in diameter. The preliminary experiment showed that it was possible to separate different populations of cells from a Day-13·5 horse blastocyst on a density gradient. The low-density cells (LDC) which migrated in the first 4 fractions produced mainly progesterone and very little oestradiol, whereas the high-density cells (HDC) found in fractions 6 to 11 produced predominantly oestradiol (Table 1). This was confirmed when only the two visible bands of cells were used (Table 2). The LDC (density = 1·03 g/ml) produced more progesterone than oestradiol (1167 ± 530 and 203 ± 158 pg/10^4 cells respectively; $P < 0·01$) whereas the HDC (density = 1·055 g/ml) produced more oestradiol than progesterone (886 ± 334 and 365 ± 201 pg/10^4 cells respectively; $P < 0·02$). When expressed as the logarithm of the ratio

Table 1. Oestradiol (E$_2$) and progesterone (P$_4$) production during 24-h culture of the different cellular fractions separated by Percoll density gradient (the embryonic disc was not removed before cell dissociation)

Steroids measured	Fraction number*											
	1	2	3	4	5	6	7	8	9	10	11	12
E$_2$ (pg/10^4 cells)	<24	<25	<25	<25	260	376	441	197	255	238	622	379
P$_4$ (pg/10^4 cells)	386	845	753	456	539	107	148	69	82	201	403	1910
log E$_2$/P$_4$	−1·2	−1·5	−1·5	−1·3	−0·3	0·5	0·5	0·5	0·5	0·07	0·2	−0·7

*Fraction 1 is the top of the gradient (lowest density) and Fraction 12 is the bottom of the gradient (highest density).

Table 2. Oestradiol and progesterone production by low density cells (LDC) and high density cells (HDC) incubated for 24 and 48 h

Type of cells	No. of cells (× 10^{-4})	Oestradiol-17β (pg/10^4 cells)		Progesterone (pg/10^4 cells)	
		1st day (*n* = 15)	2nd day (*n* = 7)	1st day (*n* = 15)	2nd day (*n* = 7)
LDC	3·44 ± 0·71	203 ± 158	31 ± 18	1167 ± 530	505 ± 161
HDC	8·54 ± 1·68	886 ± 334	139 ± 171	365 ± 201	181 ± 77

Each value represents the mean ± s.e.m.

Table 3. Oestradiol and progesterone production by different cell types incubated alone or with 16 ng androstenedione/ml for 24 h

Type of cells	No. of cells ($\times 10^{-4}$)	Oestradiol-17β (pg/10^4 cells)		Progesterone (pg/10^4 cells)	
		Control	Androstenedione	Control	Androstenedione
LDC	2·82 ± 0·96	460 ± 389	1364 ± 536	2713 ± 1119	2403 ± 877
HDC	11·72 ± 2·59	686 ± 151	23420 ± 3262	104 ± 54	133 ± 67

Each value represents the mean ± s.e.m., $n = 6$.

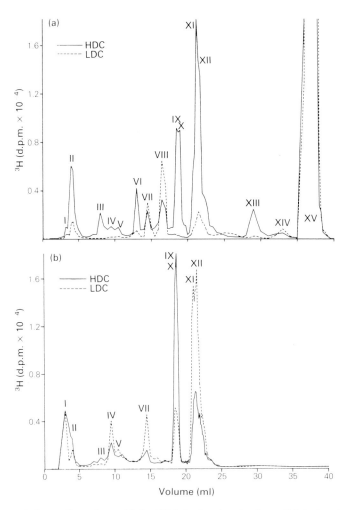

Fig. 1. Characteristic profiles of steroids by HPLC after incubation of high density cells (HDC) and low density cells (LDC) with (a) [^{3}H]pregnenolone or (b) [^{3}H]androstenedione for 8 h at 37°C. The cells were separated on a 31·8% Percoll gradient from one Day-14 conceptus after removal of the embryonic disc.

of oestradiol/progesterone there was a significant difference between the LDC and the HDC (-0.994 ± 0.141 and 0.522 ± 0.135; $P < 0.01$). During the second 24-h culture period the oestradiol production decreased by 85% for LDC and by 84% for HDC whereas the progesterone production decreased by 57% and 50% for LDC and HDC respectively (Table 2). After this reduced steroid output, the LDC still produced more progesterone and less oestradiol than the HDC.

The incubation of both cell types with androstenedione (Table 3) demonstrated that the oestradiol production by the LDC increased but there was a much greater increase in oestradiol production by the HDC. The addition of androstenedione had no effect on progesterone production (Table 3).

The endodermic cells dissociated by the glycine and collagenase treatment migrated to the same portion of the 31·8% Percoll gradient as did the LDC showing that they were of the same density. Furthermore, when the endodermal cells were incubated in MEM they produced 67 ± 37 pg oestradiol/10^4 cells and 1247 ± 357 pg progesterone/10^4 cells ($n = 8$). The log oestradiol/progesterone ratio of -1.276 ± 0.177 was not significantly different from the LDC ratio of -0.944 ± 0.141.

Separation of radiolabelled metabolites

The control incubation of medium without cells showed that there was no significant degradation of radiolabelled precursors during culture, freezing and Sep-Pak treatment.

When incubated with [^{3}H] pregnenolone for 8 h, the HDC showed a steroid profile of 14 radiolabelled metabolites (Fig. 1a). It was not possible to get a good separation of pregnenolone (Peak XV) and progesterone with the technique used, and so the results obtained by RIA of progesterone production cannot be confirmed by HPLC. There were, however, differences in the profiles for the HDC and LDC: Peaks VI, IX, X and XI were produced mainly by HDC whereas Peaks VII and VIII were the main metabolites of the LDC. Peaks IX and XI had the same retention times as oestradiol and androstenedione respectively.

When another aliquant of the same cells was incubated with [^{3}H]androstenedione, Peaks VI and VIII disappeared whereas Peaks IX and X remained elevated (Fig. 1b). The separation of phenolic from neutral steroids showed that Peaks VII and VIII were neutral whereas Peaks III, IV, V, IX and X seemed to be phenolic.

The conjugated steroid fraction eluted in the first 5 ml and constituted the first and second peaks.

Discussion

These results demonstrate that it is possible to separate the cells from the horse conceptus into two populations using a Percoll density gradient. These two types of cells seem to have different steroidogenic potential with the lower density cells producing mainly progesterone and the higher density cells producing oestradiol. The variability seen in the results is probably due to a certain amount of cell contamination between types as well as to differences between embryos. The decrease in steroid production seen during the second 24-h incubation can be explained in part by the fact that some cells were lost during the withdrawal of the first culture medium. It may also be due to the depletion of steroid precursors or enzyme cofactors. The decrease in oestradiol production by both cell types was greater than that observed for progesterone and may suggest an inactivation or destruction of the aromatase complex during the culture as observed with granulosa cells in culture (Kettelberger *et al.*, 1986).

That these two cells types are able to synthesize different steroids, and the results obtained with radioimmunoassay of steroids are not simply due to the release of steroids stored within the cells, has been confirmed by demonstrating that these two cell types have different steroidogenic profiles after incubation with radiolabelled precursors. It has not, however, been possible to confirm the production of progesterone by the cells since it cannot be adequately separated from pregnenolone with the procedure used. Like tissues from conceptuses at later stages (Heap *et al.*, 1982) both types of cells produced conjugated steroids, Peaks I and II, which are possibly sulphated and/or glucuronidated steroids.

Several metabolites were produced by both types of cells but there were also distinct differences between the profiles of the two cell types, confirming that there is a difference in their steroidogenic ability. No formal identification of the steroids has yet been performed, but Peaks IX and XI had the same retention times as oestradiol and androstenedione respectively. Peaks VI and VIII are probably C_{21} steroids since they disappeared when the cells were incubated with labelled androstenedione. Peak VII seems to be a neutral steroid and probably an androgen because it is a metabolite of androstenedione.

The exact location of these two cell types within the intact conceptus is not known at the present time. However, since the embryo at this stage consists of the embryonic disc, the ectoderm and the endoderm (Ginther, 1979) it is possible that one is of ectodermal and the other of endodermal origin. The non-enzymic isolation of endodermic cells gives indirect evidence that the LDC consist mainly of endoderm cells because these cells migrated to the same position as the LDC in the Percoll gradient and they produced oestradiol and progesterone in the same ratio as the LDC.

These results suggest that there is some mechanism, as in ovarian follicles (Fortune & Armstrong, 1978), by which intercellular interactions are important in the synthesis and secretion of steroids by the young conceptus. One possibility is that the endoderm cells produce mainly progesterone which is then transferred to the trophectodermic cells for oestradiol production. The large increase in oestradiol production by HDC compared with LDC when incubated with exogenous androstenedione and the incorporation of [³H]pregnenolone into phenolic compounds (Peaks IX and X) support our hypothesis that the aromatase complex is mainly active in HDC. The previous observation that rupture of the conceptus leads to a decrease in oestradiol production (Goff *et al.*, 1983) may be explained by trauma to the endoderm–trophectoderm interaction and subsequent decrease in transfer of steroids between the cells. Since production of steroids, especially oestrogen, is probably important for maintenance of an adequate uterine environment and the corpus luteum, it seems likely that any such disturbance of steroidogenesis could be related to early embryonic mortality.

We thank Dr D. T. Armstrong for the antiserum against oestradiol-17β. This work was supported by the MRC of Canada.

References

Allen, W.R. (1979) Maternal recognition of pregnancy and immunological implications of trophoblast–endometrium interactions in equids. In *Maternal Recognition of Pregnancy*, pp. 323–346. Ed. J. Whelan. Excerpta Medica, Amsterdam.

Armstrong, D.T. & Dorrington, J.H. (1977) Estrogen biosynthesis in the ovaries and testes. In *Advances in Sex Hormone Research*, Vol. III, pp. 217–258. Eds J. A. Thomas & R. L. Singhal. University Park Press, Baltimore.

Berg, S.L. & Ginther, O.J. (1978) Effect of estrogens on uterine tone and life span of the corpus luteum in mares. *J. Anim. Sci.* **47**, 203–208.

Betteridge, K.J., Eaglesome, M.D., Mitchell, D., Flood, P.F. & Bériault, R. (1982) Development of horse embryos up to twenty-two days after ovulation: observations on fresh specimens. *J. Anat.* **135**, 191–209.

Betteridge, K.J., Renard, A. & Goff, A.K. (1985) Uterine prostaglandin release relative to embryo collection, transfer procedures and maintenance of the corpus luteum. *Equine vet. J., Suppl.* **3**, 25–33.

Dorrington, J.H. & Armstrong, D.T. (1975) Follicle-stimulating hormone stimulates oestradiol-17β synthesis in culture Sertoli cells. *Proc. natn. Acad. Sci. U.S.A.* **72**, 2677.

Dziadek, M.A. (1981) Use of glycine as a non-enzymic procedure for separation of mouse embryonic tissues and dissociation of cells. *Expl Cell Res.* **133**, 383–393.

Fitz, T.A., Mayan, M.H., Sawyer, H.R. & Niswender, G.D. (1982) Characterization of two steroidogenic cell types in the ovine corpus luteum. *Biol. Reprod.* **27**, 703–711.

Flood, P.F. (1974) Steroid metabolizing enzymes in the early pig conceptus and in the related endometrium. *J. Endocr.* **63**, 413–414.

Flood, P.F., Betteridge, K.J. & Irvine, D.S. (1979) Oestrogens and androgens in blastocoelic fluid and cultures of cells from equine conceptuses of 10–22 days gestation. *J. Reprod. Fert., Suppl.* **27**, 414–420.

Fortune, J. & Armstrong, D.T. (1978) Hormonal control of 17β estradiol biosynthesis in proestrous rat follicles: estradiol production by isolated theca versus granulosa. *Endocrinology* **102**, 227–235.

Gadsby, J.E., Heap, R.B. & Burton, R.D. (1980) Oestrogen production by blastocyst and early embryonic tissue of various species. *J. Reprod. Fert.* **60**, 409–417.

Gardner, M.L., First, N.L. & Casida, L.E. (1963) Effect of exogenous estrogens on corpus luteum maintenance in gilts. *J. Anim. Sci.* **22**, 132–134.

Ginther O.J. (1979) In *Reproductive Biology of the Mare*, ch. 9, pp. 255–267. McNaughton and Gunn, Michigan.

Goff, A.K., Renard, A. & Betteridge, K.J. (1983) Regulation of steroid synthesis in pre-attachment equine embryos. *Biol. Reprod.* **28**, Suppl. **1**, 137, Abstr.

Goff, A.K., Pontbriand, D. & Sirois, J. (1987) Oxytocin stimulation of plasma 15-keto-13,14-dihydro prostaglandin F-2α during the oestrous cycle and early pregnancy in the mare. *J. Reprod. Fert., Suppl.* **35**, 253–260.

Heap, R.B., Flint, A.P.F. & Gadsby, J.E. (1981) Embryonic signals and maternal recognition. In *Cellular and Molecular Aspects of Implantation*, pp. 311–326. Eds S. R. Glasser & D. W. Bullock. Plenum Press, New York.

Heap, R.B., Hamon, M. & Allen, W.R. (1982) Studies on oestrogen synthesis by the preimplantation equine conceptus. *J. Reprod. Fert., Suppl.* **32**, 343–352.

Hershman, L. & Douglas, R.H. (1979) The critical period for the maternal recognition of pregnancy in Pony mares. *J. Reprod. Fert., Suppl.* **27**, 395–401.

Huff, R.L. & Eik-Nes, K.B. (1966) Metabolism *in vitro* of acetate and certain steroids by six-day-old rabbit blastocysts. *J. Reprod. Fert.* **11**, 57–63.

Kettelberger, D.M., Hoover, D. & Anderson, L. (1986) Loss of aromatase activity in porcine granulosa cells during spontaneous luteinisation *in vitro*. *Biol. Reprod.* **34**, Suppl. 1, 160, Abstr.

King, G.J. & Ackerley, C.A. (1985) Demonstration of oestrogens in developing pig trophectoderm and yolk sac endoderm between Days 10 and 16. *J. Reprod. Fert.* **73**, 361–367.

Paulo, E. & Tischner, M. (1985) Activity of 5α-hydroxysteroid dehydrogenase and steroid hormones content in early preimplantation horse embryos. *Folia Histoch. Cytobiol.* **23**, 81–84.

Perry, J.S., Heap, R.B. & Amoroso, E.C. (1973) Steroid hormone production by pig blastocysts. *Nature, Lond.* **245**, 45–47.

Perry, J.S., Heap, R.B., Burton, R.D. & Gadsby, J.E. (1976) Endocrinology of the blastocyst and its role in the establishment of pregnancy. *J. Reprod. Fert., Suppl.* **25**, 85–104.

Zavy, M.T., Mayer, R., Vernon, M.W., Bazer, F.W. & Sharp, D.C. (1979) An investigation of the uterine luminal environment of non-pregnant and pregnant pony mares. *J. Reprod. Fert., Suppl.* **27**, 403–411.

J. Reprod. Fert., Suppl. **35** (1987), 371–378

Differentiation molecules of the equine trophoblast

D. F. Antczak, J. G. Oriol, W. L. Donaldson, C. Poleman, L. Stenzler,
S. G. Volsen* and W. R. Allen*

*James A. Baker Institute for Animal Health, New York State College of Veterinary Medicine,
Ithaca, NY 14853, U.S.A. and *Thoroughbred Breeders' Association Equine Fertility Unit,
307 Huntingdon Road, Cambridge CB3 0JQ, U.K.*

Summary. Monoclonal antibodies raised against horse placenta were tested using an indirect immunoperoxidase-labelling technique for reactivity with a panel of tissues from adult horses and conceptuses of various gestational ages. The pattern of reactivity of 4 of the antibodies (F67.1, F71.3, F71.7, F71.14) on trophoblastic tissues described unique antigenic phenotypes for the non-invasive trophoblast of the allantochorion, the invasive trophoblast of the chorionic girdle, and the mature endometrial cup cells, which are derived from the chorionic girdle. Two of the monoclonal antibodies (F67.1 and F71.3) reacted only with chorionic girdle and the endometrial cups. Antibody F71.7 labelled strongly the non-invasive allantochorion from Day 29 of gestation to term. However, F71.7 failed to label mature endometrial cups and stained chorionic girdle only weakly, suggesting that the ability of the girdle cells to synthesize the molecule identified by F71.7 was gradually lost after development of the girdle. Antibody F71.14 reacted with trophoblastic tissues from all stages of gestation tested, with the exception of chorionic girdle. The other 3 anti-trophoblast monoclonal antibodies (F71.1, F71.2 and F71.8) labelled trophoblast-derived tissues from all stages tested. When the monoclonal antibodies were tested on cultured fetal and placental cells from Day 33 conceptuses recovered non-surgically from pregnant mares, the reactivities of the monoclonal antibodies on cultured cells were mostly identical to their reactivities *in situ* on tissue samples of similar gestational age; F67.1 and F71.3 were strong, specific markers for chorionic girdle cells, and F71.7 labelled allantochorion weakly *in vitro*, but failed to label chorionic girdle cells. The monoclonal antibodies described here are useful for identifying the origin and lineage of equine placental cells in culture and may be markers for molecules which play important roles in the differentiation of the equine trophoblast *in vivo*.

Introduction

The placenta is a highly specialized organ with diverse functions that include fetal nutrition, respiration and waste removal, as well as ensuring that the fetus does not succumb to a lethal maternal immune response directed against foreign fetal or placental antigens. The diversity of placental types (Amoroso, 1952) indicates that these functions can be performed successfully in a variety of ways.

The trophoblast of the horse is of particular interest because of the manner in which it differentiates during the first third of pregnancy (Ewart, 1897). The major portion of the trophoblast, which forms the outermost layer of the allantochorionic membrane, forms non-invasive villi that interdigitate with corresponding crypts in the epithelium of the maternal endometrium. Villus formation begins after Day 40 and the villi become increasingly complex as pregnancy proceeds. However, little maternal cellular response is generated towards this non-invasive trophoblast, which is characteristic of the epitheliochorial type of placentation.

The trophoblast of equids also consists of an invasive portion called the endometrial cups. The cups are formed by the invasion of the endometrium by the specialized trophoblast cells of the chorionic girdle between Days 36 and 38. The endometrial cups secrete chorionic gonadotrophin (CG), which can be detected in maternal blood between Days 40 and 150 of gestation. The endometrial cups are the target of an intense, progressive accumulation of maternal mononuclear cells that results in the destruction of the cups and their sloughing from the endometrium (Allen, 1975).

Recently, 19 monoclonal antibodies have been produced which label various cells and tissues of the horse placenta and endometrium (Antczak *et al.*, 1986). The aim of that work was to develop antibody reagents capable of distinguishing between the invasive and non-invasive forms of trophoblast. In the present study a subgroup of those reagents was tested on somatic and placental tissues of various ages.

Materials and Methods

Recovery of fetal, placental and maternal tissues. Tissues were obtained from pregnant and non-pregnant mares at various stages of gestation. Conceptuses aged 29–37 days were recovered as described below. Endometrial cups, allantochorion, endometrium and fetuses from conceptuses aged 47–100 days were obtained during surgical hysterotomy and term placentae were obtained at spontaneous foaling. Various adult tissues and organs were obtained from horses at necropsy in the Pathology Department of New York State College of Veterinary Medicine. Small pieces of these tissues (~ 10 mm^3) were snap frozen in OCT embedding medium and stored at -70°C until use.

Non-surgical recovery of Day 29–37 conceptuses. Horse conceptuses were recovered non-surgically by uterine lavage at various times between 29 and 37 days after ovulation. A Foley-type catheter (Veterinary Concepts Inc., Spring Valley, WI, U.S.A.) was passed through the cervix after digital dilatation and the cuff was inflated in the uterine lumen. Sterile Dulbecco's phosphate-buffered saline (Flow Laboratories Ltd, Rickmansworth, Herts, U.K.) with 2×10^{-5} g kanamycin sulphate/ml (Sigma, St Louis, MO, U.S.A.) was infused by gravity flow until the mare showed signs of mild discomfort. The catheter was removed with the cuff inflated to dilate the cervix further and the fluid was recovered in a sterile dish by manual massage of the uterus *per rectum*. The conceptus was usually recovered after a single infusion and although the allantochorion was ruptured by the process, the yolk sac and amnion remained intact in most instances. Small pieces of chorionic girdle, allantochorion, and fetus were snap-frozen for immunohistochemistry.

In-vitro culture of fetal and placental tissues. Methods and culture conditions similar to those described by Allen & Moor (1972) were used. Conceptuses recovered non-surgically between Days 29 and 37 after ovulation as described above were dissected in a laminar-flow hood with the aid of a low-power microscope. Three tissues were obtained for in-vitro culture: chorionic girdle (invasive trophoblast), allantochorion (non-invasive trophoblast plus allantoic mesoderm and endoderm), and fetal skin as a source of fibroblasts.

The chorionic girdle was dissected free of adjacent allantochorion by gentle dissociation using watchmaker's forceps and a scalpel blade. The small fragments of girdle tissue were passed through a 22-gauge needle in 10 ml culture medium (DMEM plus 10% heat-inactivated fetal calf serum, 0·005% sodium pyruvate, 1% non-essential amino acids, 2 mM-L-glutamine, 100 i.u. penicillin/ml, 100 mg streptomycin/ml, $2·5 \times 10^{-5}$ M-2-mercaptoethanol, 5×10^{-4} mg vitamin C/ml and 4×10^{-4} mg insulin/ml). This produced small clumps of girdle cells of uniform size which settled to the bottom of a standing test tube in less than 10 min at 4°C. The bottom 2 ml of the contents that contained these clumps was resuspended in 10 ml culture medium and 100 μl samples of the suspension were plated into tissue-culture slide chambers (Miles Scientific, Naperville, IL, U.S.A.). The cells were allowed to adhere to the slides overnight and a further 400 μl culture medium were added to each chamber the following day. The medium was changed as required during the culture period. A piece of allantochorion, approximately 10×15 mm, was cut into small pieces with scissors and stirred at room temperature in 50 ml 0·25% trypsin for 30 min. The cells in suspension were centrifuged at 800 *g* for 10 min at 4°C, resuspended in 2·0 ml culture medium, counted using a phase-contrast microscope, and cultured in slide chambers at densities ranging from 4×10^4 to 8×10^5 cells/ml in a volume of 400 μl. Small pieces of fetal skin were dissociated and the fibroblasts cultured in the same manner as for the allantochorion.

All cultures were incubated at $37·5^{\circ}$C in a humidified atmosphere containing 5% CO_2. After culture periods ranging from 1 day to 8 weeks, the slides were washed free of medium, fixed in acetone, and stored at -20°C until used for immunohistochemistry.

Immunohistochemistry. A panel of 7 monoclonal antibodies raised in rats immunized with equine placental tissue was used in this study (Table 1). The reactivities of these antibodies and the technique of indirect immunoperoxidase labelling employed has already been described (Antczak *et al.*, 1986). Briefly, 5–7 μm acetone-fixed sections were incubated with monoclonal antibodies as the first-stage reagent and a horseradish peroxidase-conjugated goat anti-rat Ig antiserum (heavy and light chain specific; Cappel Laboratories, Cochranville, PA, U.S.A.) as the second-stage reagent. The slides were developed with amino-ethyl-carbazole (AEC) and counterstained with haematoxylin.

Table 1. Monoclonal antibodies to horse trophoblast

		Specificity*			
Monoclonal antibody	Ig isotope	Endometrial cup (invasive trophoblast)	Allantochorion (non-invasive trophoblast)	Maternal endometrial tissues	Maternal non-endometrial tissues
F67.1	IgG2a	+	−	−	−
F71.1	IgG2a	+	+	+	−
F71.2	IgG2a	+	+	+	+
F71.3	IgG2a	+	−	+	+
F71.7	IgG2c	−	+	+	+
F71.8	IgG2a	+	+	−	+
F71.14	IgM	+	+	−	±

*The detailed specificity of these monoclonal antibodies for maternal and placental tissues has been described (Antczak *et al.*, 1986). This table summarizes the reactivity of the antibodies on sections of endometrial cup plus associated allantochorion from Day-60 horse pregnancies: + = labelling of tissue with monoclonal antibody; − = failure to label; ± = weak labelling.

Results

The monoclonal antibodies used in the study fell into three groups: (1) F67.1, a monoclonal antibody reactive with endometrial cup tissue only; (2) 5 monoclonal antibodies reactive with antigens shared by the endometrial cups and various placental and adult tissues (F71.1, F71.2, F71.3, F71.8 and F71.14); and (3) F71.7, a monoclonal antibody reactive with the non-invasive trophoblast of the allantochorion and some maternal tissues, but not with the endometrial cups. These monoclonal antibodies were tested in a variety of ways designed to elucidate some of the properties of the molecules which they recognized.

F67.1, a monoclonal antibody to horse CG

F67.1 was the only monoclonal antibody produced in three cell hybridization experiments that reacted exclusively with the endometrial cups and chorionic girdle, and no other fetal, placental, or maternal tissues. Gonadotrophic activity in serum from pregnant mares was removed by incubation of the serum with F67.1. Furthermore, the reaction pattern of F67.1 on equine tissue sections was identical to that of a rabbit antiserum (MSIIB) raised against highly purified horse CG (MSII, 16 000 i.u./mg, M. J. Stewart, Cambridge; Urwin & Allen, 1983).

Temporal expression of trophoblast antigens

The panel of monoclonal antibodies reactive with trophoblast was tested on invasive and non-invasive trophoblast from placentae of various gestational ages (Table 2). F71.1, F71.2 and F71.8 all labelled invasive and non-invasive trophoblast at all stages that were examined. However, these antibodies gave different labelling patterns on maternal tissues (Table 1), and therefore probably do not recognize the same molecule. F71.14 reacted with endometrial cup cells and with the non-invasive trophoblast of the allantochorion of specimens collected after Day 47 (Fig. 1d). F71.14 also labelled the non-invasive trophoblast of the chorion between Days 29 and 37 (Fig. 1e), but it failed to label the rapidly dividing trophoblast of the chorionic girdle (Fig. 1f). F71.7, on the other hand, reacted strongly with all samples of non-invasive trophoblast, but only weakly with Day 33 chorionic girdle and not at all with endometrial cups of any age. Monoclonal antibodies F67.1 and F71.3 (Fig. 1a) were specific markers for the invasive trophoblast. Since F67.1 recognizes horse CG, it was expected to react only with the invasive trophoblast of the chorionic girdle, the progenitor

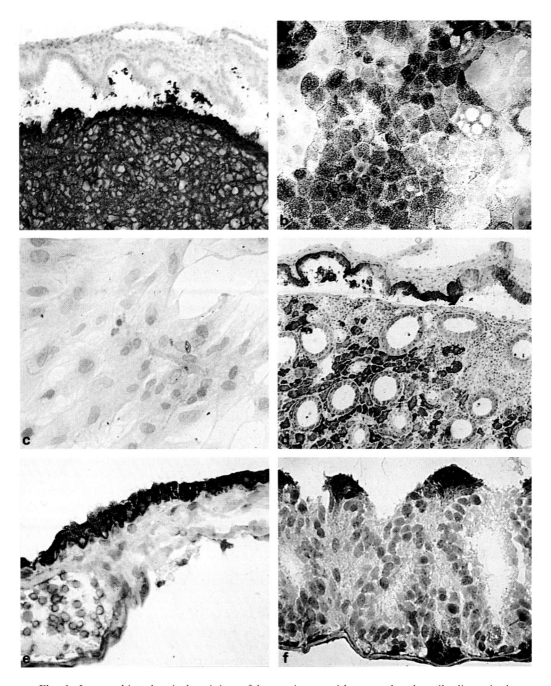

Fig. 1. Immunohistochemical staining of horse tissues with monoclonal antibodies raised against horse trophoblast. (a) F71.3 on Day-60 endometrial cup and allantochorion ($\times$ 130); (b) F71.3 on chorionic girdle cells *in vitro* ($\times$ 230); (c) F71.3 on allantochorion cells *in vitro* ($\times$ 400); (d) F71.14 on Day-60 endometrial cup allantochorion ($\times$ 136); (e) F71.14 on Day-33 allantochorion ($\times$ 350); (f) F71.14 on Day-33 chorionic girdle ($\times$ 350).

Table 2. Reactivity of monoclonal antibodies to horse trophoblast at several stages of gestation*

Type of trophoblast	Stage of pregnancy	F71.1 F71.2 F71.8	F71.14	F71.7	F67.1 F71.3
Invasive	Chorionic girdle (Day 33)	+	−	±	+
	Endometrial cup (Days 47, 60 and 100)	+	+	−	+
Non-invasive (allantochorion)	Days 29, 33 and 37†	+	+	+	−
	Days 47, 60 and 100‡	+	+	+	−
	Term placenta	+	+	+	−

*All monoclonal antibodies were tested with an indirect immunoperoxidase assay using acetone-fixed frozen tissue sections: + = labelling with monoclonal antibody; − = failure to label; ± = weak labelling.
†Before villus formation and implantation.
‡After villus formation and implantation.

Table 3. Reactivity of anti-trophoblast monoclonal antibodies with fetal and placental cells *in vitro*

Monoclonal antibody	Reactivity with cell or tissue type*		
	Chorionic girdle	Allantochorion	Fetal fibroblasts
F67.1	+	−	−
F71.1	+	±	−
F71.2	+ +	+ +	+ +
F71.3	+ +	−	−
F71.7	−	±	−
F71.8	+ +	±	+
F71.14	−	−	−

*Chorionic girdle, allantochorion and fibroblasts were derived from the same horse conceptus, recovered non-surgically at Day 33 and cultured *in vitro* for 7, 14 or 21 days, respectively, before fixation for immunohistochemistry: + + = strong labelling of cells with monoclonal antibody; + = labelling; − = failure to label; ± = weak labelling or labelling of some cells only within a culture.

tissue of the endometrial cups (Allen *et al.*, 1973). Reactivity of F67.1 with Day 33 chorionic girdle demonstrates that CG synthesis is occurring before invasion of the girdle and maturation of the endometrial cups. Because F71.3 did not label the allantochorion from Day 29, it appears that the molecule identified by F71.3 is gained by the invasive trophoblast at the time of girdle formation, and is not lost from the non-invasive chorion.

Antigen expression in vitro

Chorionic girdle, allantochorion and fetal fibroblasts grew at different rates and displayed distinctly different morphologies *in vitro*. The chorionic girdle cells grew rapidly in a radial pattern from small clumps of girdle tissue for about 7 days and then remained sessile for the duration of the culture period. The mature girdle cells were large, binucleated and contained numerous intra-cytoplasmic vacuoles. During the first 2 weeks of culture most cells had the typical morphology previously described by Allen & Moor (1972). After longer culture periods, however, slower growing contaminating cells of cuboidal morphology began to appear in some culture vessels and gradually these produced confluent monolayers reminiscent of cultures of 'paving stone' epithelium. An identical pattern was produced by parallel cultures of the allantochorion from the same conceptus, thereby suggesting that the contaminating cells in the chorionic girdle cultures were derived from the allantochorion. The fetal fibroblast preparations rapidly produced confluent monolayers of cells with the spindle morphology characteristic of fibroblasts.

The anti-trophoblast monoclonal antibodies labelled cultured chorionic girdle, allantochorion, and fetal fibroblasts with staining patterns that closely paralleled those observed *in situ* in tissues of similar gestational age (Table 3). For example, F71.3 strongly labelled chorionic girdle (Fig. 1b), but failed to label allantochorion (Fig. 1c) or fetal fibroblasts. Likewise, F67.1 (anti-horse CG) only labelled cells in cultures of chorionic girdle. F71.7 labelled allantochorion cultures weakly and it also labelled the aforementioned contaminating cells in chorionic girdle cultures that were not stained by F67.1 or F71.3.

F71.14, which failed to label the chorionic girdle cells *in situ*, did not label 7-day-old cultures of chorionic girdle cells derived from Day 33 conceptuses. F71.14 also failed to label in-vitro cultures of allantochorion. This cannot be explained, since the non-invasive trophoblast of the allantochorion was labelled by this antibody on in-situ sections from Days 29 to 37 (Fig. 1e).

Discussion

Monoclonal antibody technology is a powerful tool for the investigation of biological systems. In a consideration of the equine placenta between Days 29 and 60 of gestation, it was our aim to produce monoclonal antibodies that might identify molecules unique to the invasive endometrial cups or the non-invasive trophoblast of the allantochorion; such molecules could be responsible for the distinctly different functions of these two types of trophoblast. A second aim was to identify cell surface molecules of the endometrial cups that could serve as the target for the maternal cell-mediated attack against the cup cells (Allen, 1975).

Only a small number of molecules specific for, or highly restricted to, the placenta have to date been identified using monoclonal antibody technology. These include human chorionic gonadotrophin (Kofler *et al.*, 1982), placental-type alkaline phosphatase (Johnson *et al.*, 1981; Sunderland *et al.*, 1981), and a few other molecules whose gene products have not been well defined (Johnson *et al.*, 1981; Sunderland *et al.*, 1981; Loke *et al.*, 1985; Gogolin-Ewens *et al.*, 1986; Mueller *et al.*, 1986). This apparent paucity of placental-specific molecules (or their lack of immunogenicity) was emphasized at a recent World Health Organization Workshop on Trophoblast Antigens in Toronto, Canada (report in preparation), at which only 5 of 40 monoclonal antibodies submitted for analysis demonstrated reactivity restricted primarily to the trophoblast. However, many of these monoclonal antibodies do distinguish between various cell types in the placenta (Bulmer & Johnson, 1985), thereby making them useful for separating trophoblast subpopulations.

In three fusions in which several hundred antibody-producing hybrids were screened, we identified to monoclonal antibodies specific for horse trophoblast, with the exception of F67.1, which recognizes horse CG. Two other monoclonal antibodies distinguished the non-invasive trophoblast of the allantochorion from the invasive trophoblast of the endometrial cups; F71.7 reacted only with the former and F71.3 only with the latter type of trophoblast. However, even these monoclonal antibodies showed cross-reactivity on several different maternal tissues (Table 1; Antczak *et al.*, 1986). The other trophoblast-reactive monoclonal antibodies (F71.1, F71.2, F71.8 and F71.14) reacted with both the invasive endometrial cups and the non-invasive trophoblast of the allantochorion from Day 47 onwards. Additional molecules that differentiate invasive from non-invasive horse trophoblast may exist. However, these have not been easy to identify using monoclonal antibody technology, perhaps because of their low concentration in the cell or their weak immunogenicity.

It would be biologically economical for a tissue or cell type to 'borrow' molecules used for different purposes by other cells types. This could be accomplished if the function required two or more molecules to effect it. From the immunological point of view, borrowing molecules from adult tissues by the trophoblast would result in a cell surface that would not be recognized as 'non-self' by the maternal immune system. It is possible that horse trophoblast achieves its unique functions in such a fashion. For example, endometrial cup cells express the molecules identified by monoclonal antibodies F71.1, F71.2, F71.3, F71.8 and F71.14 whereas no adult tissue yet identified expresses precisely that constellation of molecules. The non-invasive trophoblast was labelled by these same monoclonal antibodies but with the exception of F71.3. In addition, it expresses the molecule recognized by F71.7. The non-invasive chorion therefore also has a unique molecular phenotype.

Testing placental tissues of various gestational ages for expression of the molecules recognized by the panel of monoclonal antibodies revealed several interesting points. First, F71.7 reacted with trophoblast of the allantochorion before the chorionic girdle developed (Day 29). Sections of chorionic girdle obtained between Days 33 and 35 stained weakly with this antibody, suggesting that expression of the antigen by the invasive trophoblast was gradually lost; fully developed endometrial cups were never positive for the antigen. Second, reactivity with F71.3 was first detected when the chorionic girdle had become visible macroscopically (after Day 30), thereby indicating coincidental appearance of the antigen with the development of invasive trophoblast. It is possible that this molecule is important in the process by which the girdle cells invade the endometrium. Reactivity of F67.1 (anti-horse CG) correlated with the reactivity of F71.3. Third, F71.14 reacted with trophoblast of all gestational ages except the invasive chorionic girdle. The reason for lack of expression of the F71.14 antigen on the chorionic girdle is not known, but it may be that the metabolic requirements of the rapidly dividing invasive trophoblast do not permit expression of that molecule. The remaining trophoblast-reactive antibodies (F71.1, F71.2 and F71.8) labelled both invasive and non-invasive trophoblast and from all stages that were tested.

The invasive and non-invasive portions of the equine trophoblast are easily separated by microdissection between Days 33 and 36 of gestation. However, it is much more difficult to identify and purify chorionic girdle cells from conceptuses of younger gestational ages. The monoclonal antibodies were therefore used for identification of the origin of cultured cells and antibody F71.7 was especially valuable to monitor contamination of cultures of chorionic girdle with allantochorion.

One discrepancy between in-situ and in-vitro labelling patterns was observed with F71.14, which failed to label either chorionic girdle or allantochorion cells *in vitro*. It is possible that cultures of chorionic girdle kept *in vitro* for longer than 7 days would begin to express the F71.14 antigen. The failure to label allantochorion could be due to scarcity of trophoblast cells in those cultures. The allantochorion membrane which was trypsinized before culture consists of trophoblast, mesoderm and allantoic endoderm, the last two layers not being labelled by the antibody *in situ* (Fig. 1e). Although the 'paving stone' morphology of these cultures suggested an epithelial origin of the cells, this was not confirmed using histochemical markers for epithelia. A second possibility is that monoclonal antibody F71.14 fails to label cells *in vitro* because it is of the IgM isotype, the only example of this isotype in the panel of antibodies tested. It may have a weak affinity for its antigen and could therefore have produced binding below the level of detection in our system when tested on in-vitro cultures.

The monoclonal antibodies described here may identify molecules critical to placental function. In particular, the modulation of some of these molecules during development of the chorionic girdle suggests a functional role related to the cell–cell interactions required for girdle invasion. In addition, the unique antigenic phenotypes of the invasive trophoblast of the endometrial cups and the non-invasive trophoblast of the allantochorion may account for the very different biological behaviour of these closely related tissues.

This research was supported by USPHS grant HD-15799, the Dorothy Russell Havemeyer Foundation, Inc., the Zweig Memorial Fund for Equine Research in New York State, USDA grant 85-CRCR-1-1861, the Horserace Betting Levy Board, and the Thoroughbred Breeders' Asociation of Great Britain. We thank Miss L. Dyess and Mrs A. Hesser for secretarial assistance.

References

Allen, W.R.(1975) Immunological aspects of the equine endometrial cup reaction. In *Immunobiology of the Trophoblast*, pp. 217–253. Eds R. G. Edwards, C. Howe & M. H. Johnson. Cambridge University Press, Cambridge.

Allen, W.R. & Moor, R.M. (1972) The origin of the equine endometrial cups. I. Production of PMSG by fetal trophoblast cells. *J. Reprod. Fert.* **29**, 313–316.

Allen, W.R., Hamilton, D.W. & Moor, R.M. (1973) The origin of equine endometrial cups. II. Invasion of the endometrium by trophoblast. *Anat. Rec.* **177**, 485–501.

Amoroso, E.C. (1952) Placentation. In *Marshall's Physiology of Reproduction*, 3rd edn, Vol. 2, pp. 127–297. Ed. A. S. Parkes. Longmans Green, London.

Antczak, D.F., Poleman, J.C., Stenzler, L.M., Volsen, S.G. & Allen, W.R. (1986) Monoclonal antibodies to equine trophoblast. *Trophoblast Research* (in press).

Bulmer, J.M. & Johnson, P.M. (1985) Antigen expression by trophoblast populations in the human placenta and their possible immunobiological relevance. *Placenta* **6**, 127–140.

Ewart, J.C. (1897) *A Critical Period in the Development of the Horse.* Adam and Charles Black, London.

Gogolin-Ewens, K.J., Lee, C.S., Mercer, W.R., Moseby, D.M. & Brandon, M.R. (1986) Characterization of a sheep trophoblast-derived antigen first appearing at implantation. *Placenta* **7**, 243–255.

Johnson, P.M., Cheng, H.M., Molloy, C.M., Stern, C.M.M. & Slade, M.B. (1981) Human trophoblast-specific surface antigens identified using monoclonal antibodies. *Am. J. Reprod. Immun.* **1**, 246–254.

Kofler, R., Berger, P. & Wick, G. (1982) Monoclonal antibodies against human chorionic gonadotropin (hCG): I. Production, specificity and intramolecular binding sites. *Am. J. Reprod. Immun.* **2**, 212–216.

Loke, Y.W., Butterworth, B.H., Margetts, J.J. & Burland, K. (1985) Identification of cytotrophoblast colonies in cultures of human placental cells using monoclonal antibodies. *Placenta* **7**, 221–231.

Mueller, U.W., Hawes, C.S. & Jones, W.R. (1986) Cell surface antigens of human trophoblast: definition of an apparently unique system with a monoclonal antibody. *Immunology* **59**, 135–138.

Sunderland, C.A., Redman, C.W.G. & Stirrat, G.M. (1981) Monoclonal antibodies to human syncytio-trophoblast. *Immunology* **43**, 541–546.

Urwin, V.F. & Allen, W.R. (1983) Follicle stimulating hormone, luteinising hormone and progesterone concentrations in the blood of Thoroughbred mares exhibiting single and twin ovulations. *Equine vet. J.* **15**, 325–329.

J. Reprod. Fert., Suppl. **35** (1987), 379–388

Expression of major histocompatibility complex (MHC) antigens on horse trophoblast

A. Crump, W. L. Donaldson, J. Miller, J. H. Kydd*, W. R. Allen* and D. F. Antczak

James A. Baker Institute for Animal Health, New York State College of Veterinary Medicine, Ithaca, NY 14853, U.S.A. and *Thoroughbred Breeders' Association Equine Fertility Unit, 307 Huntingdon Road, Cambridge CB3 0JQ, U.K.

Summary. Antibodies to fetal major histocompatibility complex (MHC) antigens are routinely detected in the serum of pregnant mares some 2–4 weeks after formation of the endometrial cups at Day 36–38 after ovulation. Several experimental approaches were taken to determine whether paternal MHC antigens are expressed on horse placental tissues. First, absorption of anti-paternal MHC antisera with a large volume of endometrial cup cells removed antibody activity in only 2 of 4 experiments. Second, repeated immunization of horses with endometrial cup tissue recovered from a mare on Day 47 of pregnancy failed to induce the formation of anti-MHC antibodies. Third, a potent anti-MHC antiserum, raised in a pregnant mare which had previously received skin grafts from the MHC homozygous mating stallion, labelled chorionic girdle, but not normal allantochorion, when tested in an indirect immunoperoxidase labelling assay on tissues bearing the MHC antigens of the stallion. These results indicate that the rapidly dividing cells of the chorionic girdle, the progenitor tissue of the equine endometrial cups, express high levels of paternal MHC antigen, and may serve as the alloantigenic stimulus for cytotoxic antibody production by pregnant mares. Conversely, the mature, CG-secreting endometrial cup cells have a much reduced expression of paternal MHC antigen.

Introduction

Pregnant mares regularly make cytotoxic antibody responses against the foreign paternally-inherited major histocompatibility complex (MHC) antigens carried by their fetuses. These responses are unusual for several reasons. First, antibody is detectable in the sera of over 90% of mares carrying MHC-incompatible fetuses (Bright *et al.*, 1978; DeWeck *et al.*, 1978). Second, antibody is detected regularly by Day 60 of gestation in maiden mares carrying their first pregnancies (Allen, 1979; Antczak *et al.*, 1984). Third, mares with persistent levels of serum antibody, raised as a result of one pregnancy, exhibit a rise in titre, indicative of an anamnestic response, when exposed to the same MHC antigens in subsequent pregnancies (Antczak *et al.*, 1984). These observations are striking when compared to other species in which (a) the frequency of sensitization seldom exceeds 30% in multiparous females; (b) antibody is usually detected only in the final stages of gestation; and (c) antibody titres are low (Antczak & Allen, 1984).

Anti-MHC antibodies have not been detected in the sera of unimmunized stallions, geldings or maiden mares and there exists a strong temporal correlation between the development of the endometrial cups at Day 36–38 of pregnancy and the subsequent appearance of cytotoxic antibody in the mare's blood 2–4 weeks later. These observations, coupled with the strong maternal leucocyte response mounted against the invasive trophoblast cells of the endometrial cup, but not against the non-invasive trophoblast of the allantochorion (Allen, 1975), suggested that the endometrial cup cells might express paternal MHC antigens and serve as the source of antigen for maternal sensitization. This paper describes several experiments undertaken to test this hypothesis.

Materials and Methods

Animals and tissues. Horses and ponies of mixed breeding, age and reproductive status from the experimental herds of the Cornell Equine Genetics Center and the Thoroughbred Breeders' Association Equine Fertility Unit, were used in this study as donors of serum and/or lymphocytes, and for planned matings. Peripheral blood lymphocytes (PBL) were isolated from heparinized jugular vein blood by density gradient centrifugation using Ficoll-Isopaque as described by Bright *et al.* (1978).

Samples of fetal, placental and adult tissues from various organs were obtained at *post mortem* or during surgical hysterotomy. In addition, Day-33 fetal sacs were recovered non-surgically as described by Antczak *et al.* (1987) as a source of chorionic girdle and allantochorion. The tissues were either fixed in Bouin's fluid for light microscopy or embedded in OCT medium and snap frozen in liquid N_2.

ELA (MHC) typing. Horses were typed for equine lymphocyte antigens (ELA) of the major histocompatibility complex using a complement-mediated lymphocyte microcytotoxicity assay (Antczak *et al.*, 1982) and a panel of alloantisera characterized in international workshops (Antczak *et al.*, 1986a; Bernoco *et al.*, 1987).

Production of alloantisera. Four previously undescribed alloantisera were used in these experiments, two of which (Nos 2471 and 2488) were produced by skin grafting. Four 10-mm diameter punch biopsies of full thickness skin from a Thoroughbred stallion homozygous for the ELA-A2 antigen (No. 0834) were sutured into prepared biopsy sites on the side of the neck of two young crossbred pony mare recipients which did not carry the ELA-A2 antigen. The mares were rested for 6 months and then mated to the stallion that donated the skin grafts. Pregnancy was diagnosed by ultrasound examination of the uterus at Day 18 after ovulation, after which daily blood samples were taken from the mares and assayed for cytotoxic antibody titre. Titres rose to > 1:500 in both mares by Day 45 and sera recovered between Days 45 and 60 were used as alloantisera in the experiment.

The other two antisera (Nos 1863 and 2184) were produced by immunizing two male horses with endometrial cup tissue recovered from a mare on Day 47 of gestation. The cups were dissected free of overlying allantochorion and surrounding endometrium, minced into small pieces using a scalpel and scissors and forced through a fine nylon mesh. The resulting tissue slurry was washed twice by centrifugation in Dulbecco's phosphate-buffered saline (PBS) and resuspended at a concentration of 0·05 g wet weight/ml PBS for immunization. Each horse received a primary immunization of 5 ml of the endometrial cup cell immunogen intramuscularly on Day 0, followed by further 5-ml booster immunizations on Days 23, 31 and 55. The immunogen was stored at $-70°C$ between immunizations.

Absorption of antisera. Alloantisera were absorbed with washed and packed PBL or washed and packed endometrial cup cells prepared as described above. For lymphocyte absorptions, 100 µl serum were absorbed twice at 4°C with 100×10^6 lymphocytes. The first absorption lasted for 1 h and the second overnight. For absorption with endometrial cup cells, a total volume of tissue equivalent to 20 times the volume of lymphocytes was used, since preliminary experiments showed that smaller amounts than this had no effect.

Table 1. Absorption of alloantibodies to ELA (MHC) antigens by endometrial cup cells

Antiserum	Absorbing tissue*	% Cytotoxicity for lymphocytes of mating stallion*			
		Dilution of antiserum			
		Neat	1/2	1/4	1/8
0122	None	100	100	50	0
(anti-ELA-A1)	'OT' endometrial cup cells	100	50	0	0
	'LS' endometrial cup cells	0	0	0	0
0030	None	95	0	0	0
(anti-ELA-A3)	'OT' endometrial cup cells	0	0	0	0
	'LS' endometrial cup cells	60	0	0	0

*Endometrial cup cells were obtained from Mares OT and LS on Days 45 and 60 of gestation, respectively, after pregnancy was established by the mating stallion. The stallion was heterozygous for the ELA-A1 and ELA-A3 antigens. The mares were maidens and did not carry either ELA antigen of the stallion.

Table 2. Specificity of alloantisera produced by mares after surgical removal of endometrial cups for absorption of anti-paternal ELA (MHC) antisera*

Mare	Day after ovulation	Horse no. (ELA type)‡	
		0007 (*A1*,A2)	0039 (A2,*A3*)
OT	30	0	1
	45†	0	90
	47	0	100
	66	40	90
LS	30	0	40
	60†	0	0
	85	100	0

*Mares OT and LS were donors of the endometrial cups used for the absorption experiment described in Table 1.

†Day of embryo removal.

‡Figures show the percentage of lymphocytes killed from horses carrying the A1 (Horse 0007) or A3 (Horse 0039) ELA antigens of the mating stallion when tested against sera from Mares OT and LS in the microcytotoxicity assay.

Elution of antibody from endometrial cups. Small pieces of endometrium, endometrial cup, allantochorion and fetal liver were obtained at Day 60 of gestation from 2 mares that had been mated to an MHC homozygous stallion. The tissues were minced, forced through nylon mesh and washed in PBS as described previously. Endometrial cup material that did not pass through the nylon mesh was also saved and this fraction, as well as the material that passed through the mesh, was used in the elution experiments.

Elution of the various tissues was performed by incubating 0·5 ml packed tissue slurry with 1 ml glycine–HCl buffer (pH 3·0) for 30 min at 4°C. The mixtures were then centrifuged at 13 000 *g* for 5 min. The supernatants were decanted, adjusted to pH 7·0 by the addition of 1 ml Tris buffer and dialysed in 1 litre volumes of PBS with 3 changes over 2 days. The resulting eluates were stored at −20°C until used.

Indirect immunoperoxidase staining of tissues. Cryostat sections (5–7 μm) of various equine fetal and placental tissues were labelled in an indirect immunoperoxidase assay using a modification of the technique described previously by Antczak *et al.* (1986b). Sections of each tissue were mounted on albumin-coated microscope slides, fixed in acetone for 10 min, covered with a thin film (∼ 50 μl) of the test alloantiserum diluted 1:150 or greater in Tris-buffered saline (pH 7·6–7·7) containing 10% (v:v) normal goat serum (TBS-NGS) and incubated at 22°C in a humidified chamber for 30 min. The slides were then given three 5-min washes in TBS and incubated in TBS-NGS for 15 min before the sections were covered with a thin film of the second-stage reagent, a peroxidase-conjugated goat anti-horse Ig (heavy and light chain specific) antiserum (Cappel Laboratories, Cochranville, PA, U.S.A.). After incubation at 22°C for 30 min, the slides were again given three 5-min washes in TBS. The slides were then flooded with the substrate, amino-ethyl carbazole (AEC), and incubated at 22°C for 40 min before final washing in TBS and counterstaining with haematoxylin.

Results

Absorption of alloantisera with endometrial cup tissue

Suspensions of endometrial cup tissue from 4 pregnancies were used to absorb antisera to paternal ELA antigens. The stallions and mares used in these experiments shared no ELA antigens, giving rise to ELA-incompatible pregnancies. The anti-paternal ELA antisera which were absorbed did not cross-react with the ELA antigens of the mares. The first 2 pregnancies were established by mating 2 mares to an ELA heterozygous stallion that carried the alleles A1 and A3. Cup tissue from one pregnancy (Mare LS, Table 1) absorbed cytotoxic activity from an anti-A1 antiserum, but not an anti-A3 antiserum. Consistent with this observation was the production of anti-A1 antibody by the mare (Table 2), indicating that the fetus had inherited the ELA-A1 antigen of the stallion. Cup tissue from the second mare mated to the same stallion absorbed cytotoxic activity from an anti-A3 serum but not an anti-A1 serum (Mare OT, Table 1). Likewise, the mare made antibody to the A3 allele (Table 2), demonstrating that, in this case, the fetus inherited the ELA-A3 allele from the stallion.

In the two subsequent experiments an ELA-A2 homozygous stallion was mated to 2 maiden mares that did not carry ELA-A2. Various antisera specific for (a) the stallion's ELA antigens, (b) the ELA antigens of the mares, or (c) the antigens of the unrelated 'third party' horses were absorbed with suspensions of endometrial cups, endometrium, allantochorion, fetal liver, or with lymphocytes from the stallion, the mare, or the 'third party' horses. In these experiments there was no evidence for specific absorption of antibodies to paternal ELA antigens by incubation of the antisera with the suspension of endometrial cups or allantochorion. Incubation with fetal liver reduced the titre of antisera to the sire's or the dam's ELA antigens. Incubation with endometrial cup cells removed antibody to maternal ELA antigens. This was to be expected, however, due to contamination of the endometrial cup suspension with maternal endometrial tissues and leucocytes that express high levels of ELA antigens. In only 2 of the 4 experiments, therefore, was the expression of paternal histocompatibility antigens in endometrial cup tissue demonstrated by the ability of that tissue to specifically absorb cytotoxic antibody activity from anti-paternal MHC antisera.

Table 3. Elution of antibody from endometrium and endometrial cups of a pregnant mare*

Eluate or antiserum from Mare 1536	Dilution of antiserum or eluate				
	Neat	1/2	1/4	1/8	1/16
Eluate					
Endometrial cup cell	100	5	0	0	0
Endometrial cup debris	100	100	100	90	0
Endometrium	85	0	0	0	0
Antiserum	100	100	85	90	0

*Mare 1536 (ELA-A9) was mated to Stallion 0834 (ELA-A2). Figures show the percentage of stallion lymphocytes killed by the various eluates or the antiserum.

Table 4. Allo-specificity of anti-paternal antibody eluted from endometrial cups and allantochorion

	Lymphocyte donor no. (ELA type)‡					
	0834* (A2)		1538 (A2, A10)		2446 (A10)	
Eluate†	Neat	1:2	Neat	1:2	Neat	1:2
Endometrial cup cells	90	90	85	25	5	5
Allantochorion	70	10	60	5	10	5
Endometrium	20	5	40	0	5	0
Fetal liver	10	0	0	0	0	0

*No. 0834 was the mating stallion.

†All eluates were prepared from tissue recovered from Mare 1537 (ELA-A6) at Day 60 of gestation.

‡Figures show percentage of lymphocytes killed when tested against the various eluates.

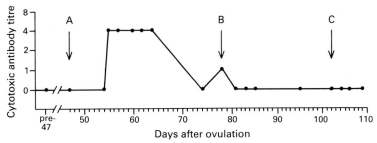

Fig. 1. Cytotoxic antibody response to paternal ELA antigens in a pregnant mare immunized with endometrial cup tissues. Plot of cytotoxic antibody titre versus days after ovulation. Arrows indicate, A: day of surgery for recovery of endometrial cups (Day 47); B and C: days of booster immunizations with endometrial cup cells (Days 78 and 102). The antiserum from the mare was cytotoxic for lymphocytes from horses carrying the A10, but not the A2, ELA antigen of the sire.

Table 5. Donor–recipient combinations used for allo-immunization with endometrial cup tissue

| | | | Equine lymphocyte antigens* | | |
| | | | ELY-1 | | ELY-2 |
Horse†	Identification no.	ELA	.1	.2	.1
Donor sire	1538	A2, A10	+	+	−
Donor dam	1535	A3	+	−	−
Recipient	1863	A3, A5	+	+	−
Recipient	2184	A3, W15	+	+	−

*The ELY-1 (Byrns *et al.*, 1987) and ELY-2 (Antczak, 1984) genetic systems control the expression of lymphocyte antigens not encoded by the ELA region.

†Dam 1535 was homozygous for the ELA-A3 allele. Both recipients were males that had never received blood transfusions.

Elution of antibody from endometrial cups

Antibody eluates were prepared from tissues recovered from 2 mares carrying MHC-incompatible fetuses after mating to the same stallion that was homozygous for ELA-A2. The eluates and serum recovered from each mare at the time of surgical removal of the conceptus were tested against lymphocytes from the mating stallion and other ELA typed horses in the microcytotoxicity assay. For the first mare (No. 1536, ELA-A9), low levels of cytotoxic antibody activity were detected in the eluates from both the endometrial cups and endometrium. Higher levels, equivalent to those in the mare's serum, were detected in the eluate derived from the endometrial cup cell debris that had not passed through the nylon mesh during preparation (Table 3). For the second mare (No. 1537, ELA-A6), a larger range of tissues was used for elution, including fetal liver, endometrium from the non-gravid horn, normal allantochorion and endometrial cups. As found previously, low levels of antibody were eluted from endometrial cups. Very weak antibody activity was detected in the eluates from the allantochorion and endometrium, but there was no activity in the fetal liver eluate. The antibody activity was specific for the ELA-A2 antigen of the mating stallion (Table 4), thereby indicating the expression of paternally-derived MHC antigens on the specialized trophoblast cells of the endometrial cups.

Allo-immunization against endometrial cup tissue

Endometrial cup cell suspensions prepared from tissues recovered surgically at Day 47 from Mare 1535 were used to immunize the mare herself and 2 other horses. The mating and donor-recipient combinations were chosen to ensure that the animals which received cup suspension would be tolerant to the maternal, but not the paternal ELA (MHC) antigens (Table 5). The experiment consisted of several parts.

First, serum samples were obtained from Mare 1535 3 times weekly between Days 20 and 110 after ovulation and tested for cytotoxic antibody to lymphocyte antigens of the mating stallion. Before the experiment this mare had not carried a pregnancy beyond Day 8, although she had previously donated the embryo that developed into one of the horses immunized against her endometrial cups (Recipient 2184, Table 5). In response to her current pregnancy she made antibody specific for the ELA-A10 haplotype of the stallion, demonstrating that the fetus and endometrial cups carried the gene for the A10 antigen. The antibody was first detected on Day 55, 8 days after surgical removal of endometrial cups. When immunized subsequently on two occasions with the suspension of her own endometrial cups there was no anamnestic rise in titre of serum antibody cytotoxic for lymphocytes of the mating stallion (Fig. 1).

Second, two male horses (Nos 1863 and 2184, Table 5) were immunized 4 times with endometrial cup tissue over a 60-day period. Thrice weekly serum samples recovered from these recipients failed to show any cytotoxic antibody for lymphocytes of the mating stallion.

Third, 75 days after the donor mare and the 2 immunized recipients had last been boosted with endometrial cup suspension, the donor mare (No. 1535) and one recipient horse (No. 2184) were immunized with lymphocytes from the mating stallion. Serum samples from the mare showed that, at the time of lymphocyte immunization, the titre of antibody to the ELA-A10 antigen induced by pregnancy had dropped to zero. By 8 days later, however, serum from the mare became cytotoxic for lymphocytes carrying either the ELA-A10 or -A2 antigens, with the titre being much higher against lymphocytes carrying ELA-A10. Weak cytotoxic antibody was detected in the serum of recipient horse No. 2184 at 13 days after immunization, but this was directed against ELA-A2 and not -A10 (Table 6).

Immunoperoxidase labelling of placental tissues

Two technical problems were encountered during indirect immunoperoxidase labelling of horse tissues with horse alloantisera. First, many tissues contained large amounts of immunoglobulin

Table 6. Cytotoxic antibody levels produced by lymphocyte immunization in a mare previously immunized by pregnancy and a horse previously immunized with endometrial cup tissue

| | | Cytotoxic antibody titre of serum from:* | | | |
| | | Mare 1535 | | Recipient horse 2184 | |
Horse no.	ELA type	Day 0	Day 13	Day 0	Day 13
1538	A2, A10	0	≥ 1:32	0	1:2
0197	A6, A10	0	≥ 1:32	0	0
0982	A10, ITH-14	0	≥ 1:32	0	0
0834	A2	0	1:2	0	1:2
2392	A6, A7	0	1:4	0	Neat

*Cytotoxic titres measured on Day 0 or Day 13 after immunization with 1 × 10^8 peripheral blood lymphocytes from Horse 1538, the stallion used to establish pregnancy in Mare 1535. Horse 2184 had been previously immunized 4 times with endometrial cup tissue from that pregnancy.

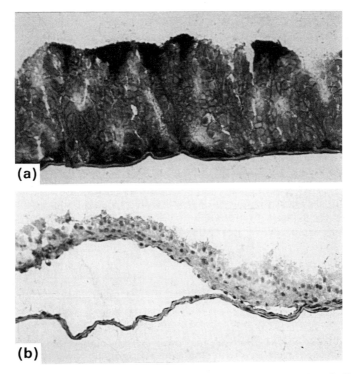

Fig. 2. Indirect immunoperoxidase staining of Day-33 chorionic girdle (a) and allantochorion (b) labelled with horse antiserum 2488 (anti-ELA-A2) at a dilution of 1:1000 as the first-stage reagent. The chorionic girdle shows positive staining while the allantochorion is negative. × 200.

from serum and transudates that were labelled by the peroxidase-conjugated second-stage anti-equine Ig reagent. This was true for the maternal leucocytes accumulated around the endometrial cups and for the cup cells themselves. Placental membranes recovered non-surgically at Day 33, however, did not carry endogenous immunoglobulin. The second problem involved non-specific binding of horse Ig in whole serum to the cryostat sections. This occurred when using horse allo-antisera at dilutions of < 1:150, so that only alloantisera with very high titres could be used at the high dilution required to provide specificity. Most horse anti-ELA (MHC) antisera used regularly for ELA typing have titres of only 1:4–1:8, and could not, therefore, be used in this assay.

Two potent alloantisera (Nos 2471 and 2488) raised against the ELA-A2 antigens by skin grafting and subsequent pregnancy were tested for their ability to bind acetone-fixed, frozen sections of placental tissues obtained from pregnancies in which the sire of the fetus was Horse 0834, an ELA-A2 homozygote. Antiserum 2488, at a dilution of 1:1000, labelled intensely the chorionic girdle of the Day-33 horse conceptus but did not label adjacent normal allantochorion (Fig. 2). Antiserum 2471 showed similar but weaker staining patterns on the tissues. Neither of the antisera raised by repeated immunization of horses with endometrial cup suspension (Nos 1863 and 2184, Table 5) labelled Day 33 chorionic girdle or allantochorion.

The specificity of the labelling exhibited by antisera 2488 and 2471 was demonstrated in several ways. First, serum prepared from Mare 2488 before skin grafting failed to label any tissues at dilutions of > 1:150. Second, absorption of antiserum 2488 with lymphocytes from the stallion against which the antiserum was raised removed all cytotoxic and binding activities. Third, negative control absorptions of the antisera with lymphocytes from unrelated horses that expressed ELA (MHC) antigens different from those against which the antisera were raised removed neither cytotoxic nor binding activity.

Discussion

The intense accumulation of maternal lymphocytes around the endometrial cups, but not the non-invasive allantochorion, coupled with the temporal association between cup development and the subsequent appearance of alloantibody to paternal MHC antigens in the serum of pregnant mares, suggested that the endometrial cups, but not the non-invasive trophoblast, might express paternally inherited MHC antigens (Allen, 1979; Antczak *et al.*, 1984). This is an attractive hypothesis, but evidence to support it has been difficult to obtain.

MHC-matched matings were undertaken for which the fetus inherited no detectable paternal MHC antigens that were not carried by the mother. These mares could not, therefore, make antibody to fetal MHC antigens because of the immunological principle of self-tolerance. In such MHC-compatible pregnancies, however, the intensity of the maternal leucocyte reaction around the endometrial cups at Day 60 of gestation was not reduced compared to the response observed in MHC-incompatible pregnancies, even though no alloantibody was detectable in the serum of the mares carrying MHC-compatible fetuses (Allen *et al.*, 1984).

Absorption of anti-paternal MHC alloantisera with endometrial cup tissue resulted in specific removal of antibody activity in only 2 of 4 experiments (Tables 1 and 2). A volume of packed endometrial cup suspension equivalent to about 20 times that of packed PBL was required to effect antibody absorption. Elution experiments (Tables 3 and 4) indicated that small quantities of cytotoxic antibody may be bound to endometrial cups *in vivo*. Taken together, these results indicate that the level of expression of MHC antigens on endometrial cup cells may be very low compared to expression on adult lymphocytes.

Immunization of naive recipients with endometrial cup tissue failed to induce cytotoxic antibody to paternal MHC antigens, or to prime the recipients for a subsequent response induced by immunization with lymphocytes (Table 6). Similarly, booster immunizations of a mare with endometrial cup tissue from her own pregnancy failed to increase the titre of cytotoxic anti-paternal antibody raised by 'maternal' immunization during pregnancy (Fig. 1). In contrast, immunization of the same pregnancy-sensitized mare with paternal lymphocytes resulted in a strong anamnestic response to the ELA antigen to which the mare had been sensitized during pregnancy (Table 6). These results also suggest that, if endometrial cup cells express paternal MHC antigen, it is not expressed in a form that is highly immunogenic. Furthermore, the production of cytotoxic antibody after lymphocyte immunization demonstrated that immunization with endometrial cup cells had not induced a state of tolerance or non-responsiveness in the immunized recipient horses or the pregnancy-sensitized mare.

If the endometrial cups used for immunization were coated with maternal antibody to MHC antigens, as suggested by the elution experiments, this could provide an explanation for their failure to stimulate an immune response when injected into the recipient horses. However, in the immunization experiments the endometrial cups were recovered on Day 47 of gestation, 8 days before antibody to paternal MHC antigen was detected in the mare's serum (Fig. 1).

In the mouse, monoclonal antibodies to paternal Class 1 MHC antigens were absorbed from the maternal circulation and shown to bind specifically to labyrinthine trophoblast, but not spongiotrophoblast or trophoblast giant cells (Wegmann *et al.*, 1980; Chatterjee-Hasrouni & Lala, 1982). In women, Butterworth *et al.* (1985) have demonstrated binding of monoclonal antibody W6/32 (anti-HLA-A,B,C) to villous and extravillous cytotrophoblast although Sunderland *et al.* (1981) detected MHC antigen only on villous stroma and not on trophoblast. In rodents and humans, therefore, there is evidence for a selective expression of MHC antigens on some types of trophoblast. However, controversy persists concerning precisely which types of trophoblast express MHC antigens, which MHC loci are expressed, and the role of these antigens in inducing maternal immune responses.

In the present experiments, the use of indirect immunoperoxidase labelling of horse placental tissues provided conclusive evidence for expression of paternal MHC antigens by the specialized trophoblast cells of the chorionic girdle. Specific, high-titred, anti-MHC alloantisera stained Day-33 chorionic girdle intensely whereas staining of the normal trophoblast of the allantochorion from the same conceptus could not be detected (Fig. 2).

The absorption and elution experiments described here suggested that endometrial cups might express small quantities of paternal MHC antigen. However, it was not clear how exposure of such a low level of MHC antigen to the maternal immune system in early pregnancy could provide a sufficiently strong stimulus to result in the high levels of alloantibody produced by most mares by Day 60 of pregnancy. The immunoperoxidase labelling experiments offer an explanation: chorionic girdle cells labelled very strongly with anti-MHC antisera, while allantochorion did not. Hence, it seems possible that only the rapidly dividing chorionic girdle cells express high levels of MHC antigen, and once they have completed their invasion of the endometrium and differentiated to become endometrial cup cells, the expression of MHC antigen declines sharply. Murine trophoblast cells cultured *in vitro* with interferon have been shown to increase their level of expression Class 1 MHC antigens (Zuckerman & Head, 1986). An equivalent type of stimulus may modulate temporarily the expression of MHC antigens on chorionic girdle cells in equids.

The functional importance of this proposed regulation of MHC antigen expression by equine trophoblast is unknown. However, it could act as an immunoregulatory signal to the mare to induce the provision of an appropriate environment for implantation during this critical stage of pregnancy. The slow demise of the endometrial cups over a 60-day period is more reminiscent of rejection of skin grafts across minor histocompatibilty antigen barriers than MHC barriers, with which rejection is usually effected within 14 days. Decreased expression of paternal MHC antigens on the endometrial cups might render the cup cells less susceptible to killing by an anti-MHC response.

The proposed modulation of MHC antigens on chorionic girdle and endometrial cup cells could also regulate the composition of maternal leucocytes that surround the endometrial cups. It is possible that these leucocytes contain populations of uterine suppressor cells, as described in the mouse by Clark *et al.* (1984). Such an explanation is consistent with our findings in mares carrying transferred extraspecific donkey conceptuses which do not form endometrial cups. The majority of these conceptuses are aborted at Day 80–95 of gestation by a mechanism that includes a strong, maternal cell-mediated reaction directed against the xenogeneic donkey allantochorion (Allen *et al.*, 1987).

This research was supported by USPHS grant HD-15799, The Zweig Memorial Fund for Equine Research in New York State, The Dorothy Russell Havemeyer Foundation, Inc., The Horserace Betting Levy Board and The Thoroughbred Breeders' Association of Great Britain. We thank Miss Lori Dyess and Mrs Anita Hesser for secretarial assistance.

References

Allen, W.R. (1975) Immunological aspects of the equine endometrial cup reaction. In *Immunobiology of the Trophoblast*, pp. 217–253. Eds R. G. Edwards, C. Howe & M. H. Johnson. Cambridge University Press, Cambridge.

Allen, W.R. (1979) Maternal recognition of pregnancy and immunological implications of trophoblast-endometrium interactions in equids. In *Maternal Recognition of Pregnancy* (Ciba Fdn Series No. 64 (new series)), pp. 323–352. Excerpta Medica, Amsterdam.

Allen, W.R., Kydd, J., Miller, J.M. & Antczak, D.F. (1984) Immunological studies on fetomaternal relationships in equine pregnancy. In *Immunological Aspects of Reproduction in Mammals*, pp. 183–193. Ed. D. B. Crighton. Butterworths, London.

Allen, W.R., Kydd, J. & Antczak, D.F. (1987) Maternal immunological response to the trophoblast in xenogeneic equine pregnancy. In *Proc. 2nd Banff Conference on Reproductive Immunology*, Eds T. J. Gill & T. Wegmann. (in press).

Antczak, D.F. (1984) Lymphocyte alloantigens of the horse III. ELY-2.1 A lymphocyte alloantigen not coded by the MHC. *Anim. Blood Groups and Biochemical Genetics* **15**, 103–115.

Antczak, D.F. & Allen, W.R. (1984) Invasive trophoblast in the genus *Equus. Annls Institut Pasteur, Paris* **135**, 325–331, 341–342.

Antczak, D.F., Bright, S.M., Remick, L.H. & Bauman, B.E. (1982) Lymphocyte alloantigens of the horse. 1. Serologic and genetic studies. *Tissue Antigens* **20**, 172–187.

Antczak, D.F., Miller, J. & Remick, L.H. (1984) Lymphocyte alloantigens of the horse II. Antibodies to ELA antigens produced during equine pregnancy. *J. Reprod. Immunol.* **6**, 283–297.

Antczak, D.F., Bailey, E., Barger, B., Bernoco, D., Bull, R.W., Byrns, G., Guerin, G., Lazary, S., McClure, J., Mottironi, V.D., Symons, R., Templeton, J. & Varewyck, H. (1986a) Joint Report of the Third International Workshop on Lymphocyte Alloantigens of the Horse, Kennett Square. *Animal Genetics* **17**, 363–373.

Antczak, D.F., Poleman, J.C., Stenzler, L.M., Volsen, S.G. & Allen, W.R. (1986b) Monoclonal antibodies to equine trophoblast. *Trophoblast Research* (in press).

Antczak, D.F., Oriol, J.G., Donaldson, W.L., Poleman, C., Stenzler, L., Volsen, S.G. & Allen, W.R. (1987) Differentiation molecules of the equine trophoblast. *J. Reprod. Fert., Suppl.* **35**, 371–378.

Bernoco, D., Antczak, D.F., Bailey, E., Bell, K., Bull, R.W., Byrns, G., Guerin, G., Lazary, S., McClure, J., Templeton, J. & Varewyck, H. (1987) Joint Report of the Fourth International Workshop on Lymphocyte Alloantigens of the Horse, Lexington. *Animal Genetics* (in press).

Bright, S., Antczak, D.F. & Ricketts, S. (1978) Studies on equine leukocyte antigens. *J. Equine Med. Surg.,* Suppl. 1., 229–236.

Butterworth, B.H., Khong, T.Y., Loke, Y.W. & Robertson, W.B. (1985) Human cytotrophoblast populations studied by monoclonal antibodies using single and double biotin-avidin-peroxidase immunocytochemistry. *J. Histochem. Cytochem.* **10**, 977–983.

Byrns, G., Crump, A.L., Lalonde, G., Bernoco, D. & Antczak, D.F. (1987) The ELY-1 locus controls a di-allelic alloantigenic system of equine lymphocytes. *J. Immunogenetics* (in press).

Chatterjee-Hasrouni, S. & Lala, P.K. (1982) Localization of paternal H-2K antigens on murine trophoblast cells *in vivo. J. exp. Med.* **155**, 1679–1689.

Clark, D.A., Slapsys, R., Croy, B.A., Krcek, J. & Rossant, J. (1984) Local active suppression by suppressor cells in the decidua: a review. *Am. J. Reprod. Immunol.* **4**, 78–83.

DeWeck, A.L., Lazary, S., Bullen S., Gerber H. & Meister, U. (1978) Determination of leukocyte histocompatibitiy antigens in horses by serological techniques. In *Equine Infectious Diseases IV. Proceedings of the Fourth International Conference on Equine Infectious Diseases,* pp. 221–227. Eds J. T. Bryans & H. Gerber. Verterinary Publications, Inc., Princton, NJ.

Sunderland, C.A., Naiem, M., Mason, D.Y., Redman, C.W.G. & Stirrat, G.M. (1981) The expression of major histocompatibility antigens by human chorionic villi. *J. Reprod. Immunol.* **3**, 323–331.

Wegmann, T.G., Leigh, J.L., Carlson, G.A., Mossmann, T.R., Raghupathy, R. & Singh, B. (1980) Quantitation of the capacity of the mouse placenta to absorb monoclonal anti-fetal H-2K antibody. *J. Reprod. Immunol.* **2**, 53–59.

Zuckermann, F.A. & Head, J.R. (1986) Expression of MHC antigens on murine trophoblast and their modulation by interferon. *J. Immunol.* **3**, 846–853.

J. Reprod. Fert., Suppl. **35** (1987), 389–397

Genetic and temporal variation in serum concentrations and biological activity of horse chorionic gonadotrophin

A. W. Manning, K. Rajkumar, F. Bristol, P. F. Flood and B. D. Murphy

Departments of Veterinary Physiological Sciences, Biology, Anatomy and Herd Medicine and Theriogenology, University of Saskatchewan, Saskatoon, Canada S7N 0W0

Summary. The variation in the quantity of circulating chorionic gonadotrophin (CG) and its follicle-stimulating hormone (FSH) and luteinizing hormone (LH) activity in rodent bioassay systems was investigated. A portion of the variability in total CG could be attributed to the stallion that sired the pregnancy and it was possible to select sires and mares to increase CG production. It was further demonstrated that FSH activity per unit of CG was greater at Days 71 and 104 of gestation than at Day 39. LH activity per unit of CG varied with the sire, but no effect of day of gestation could be shown. It was demonstrated that removal of sialic acid increased LH bioactivity and it is proposed that variation in biological activity between animals and during gestation may be a function of differences in carbohydrate content of CG.

Introduction

The potent gonadotrophin present in mare serum during the first third of pregnancy was first recognized by Cole & Hart (1930). Cole & Goss (1943) demonstrated that the gonadotrophic activity emanated from the endometrial cups, a finding substantiated by the report of Clegg *et al.* (1954). The landmark studies of Allen & Moor (1972) demonstrated that fetal tissues, specifically the cells of the chorionic girdle, secrete the gonadotrophin *in vitro*. This finding was confirmed by the ultrastructural studies of Allen *et al.* (1973) and Hamilton *et al.* (1973) who showed that the chorionic girdle cells invade the endometrium to form the endometrial cups. The original name given to the substance was pregnant mares' serum gonadotrophin. More recently, the term horse chorionic gonadotrophin (CG) has been adopted, which reflects the embryonic origin of the cells producing the molecule.

Horse CG is known to be a glycoprotein hormone consisting of two non-identical subunits (Papkoff, 1974; Moore & Ward, 1980) and its carbohydrate component is large relative to that of other glycoprotein hormones, as it approaches 45% by weight (Papkoff *et al.*, 1978; Moore *et al.*, 1980). The molecule occurs in different forms depending on whether it is isolated from serum or from endometrial cups (Aggarwal *et al.*, 1980b): the circulating form of the molecule appears to be larger, contains more carbohydrate and has different amino terminal residues from that isolated from the endometrial cups.

It has been shown by a number of authors that the biological activity of horse CG in the mare is similar to that of luteinizing hormone (LH) (Squires & Ginther, 1975; Moore & Ward, 1980). In other mammals, horse CG has both LH and follicle-stimulating hormone (FSH) activity, a property shared with horse pituitary hormones (Licht *et al.*, 1979). The relative occurrence of FSH and LH activity, usually expressed as the FSH:LH ratio, has been shown to vary, depending on whether the CG is isolated from cup tissue or serum (Matteri *et al.*, 1986). An early report, in which biological activity was determined by radioreceptor assay, suggested that no variation existed in FSH:LH ratio between mares or between Days 40 and 80 of gestation (Stewart *et al.*, 1976). In contrast, Gonzalez Mencio *et al.* (1978) reported variation between mares and between Days 50 and 90 of gestation while Aggarwal *et al.* (1980b) reported a difference in FSH:LH ratio between preparations of CG from pooled sera containing high and low concentrations of the hormone. Moore & Ward (1980) suggested that differences in biological activity may be attributable to micro-heterogeneity of the CG molecule, particularly to differences in the structure of the subunits. Papkoff (1981) proposed that the variation may occur because of differences in the carbohydrate component of the molecule.

Previous reports have concerned relatively small groups of mares or pooled serum. One purpose of this investigation was to measure the FSH and LH activity of horse CG in a large number of pregnancies to see whether patterns could be detected in the biological activity of horse CG.

Comparison of the horse with other equids indicates that genotype influences CG production. In the donkey CG production is considerably lower than in the horse (reviewed by Allen, 1982) and it is clear from observations on interspecific crosses that the genotype of the fetus is critical. Mares with mule fetuses had pregnancies characterized by CG production one tenth of that found in mares carrying a conspecific fetus (Clegg *et al.*, 1962). The endometrial cups were correspond-ingly small in mule pregnancies (Allen, 1982). Donkeys carrying hinny fetuses had circulating CG at a level associated with horse pregnancy and large, well developed endometrial cups (Allen, 1982).

Evidence for an effect of fetal genotype on CG production in horses was obtained from observations on 227 pregnancies in Belgian-cross mares, each bred to one of 12 sires (Murphy *et al.*, 1985): two groups of sires were identifiable, those producing pregnancies with high and low titres of CG. Bell (1985), reporting on 100 pregnancies in standard-bred mares, was able to separate the 4 stallions in her study into two statistically dissimilar groups, based on the titre of CG of the mares covered.

A second purpose of this study was to investigate the possibility that CG titres could be selected for in a population of horses by breeding mares with high titre CG pregnancies to sires that produced high titre pregnancies, and conversely, low titre mares to low titre sires.

A third objective was to assess the effect of desialylation on LH and FSH like activity of CG and to determine the sialic acid content of CG samples with different LH:FSH ratios.

Materials and Methods

Animals, breeding and sampling procedures. The mares studied were Belgian-cross animals, part of a herd used for collection of urine from pregnant mares on the Bar-C Ranch, Rossburn, Manitoba. In 1983, four groups of 10 randomly chosen mares were treated with 250 µg synthetic prostaglandin F-2α (Fenprostelene: Syntex Agribusiness, Mississauga, Ontario) to synchronize oestrous cycles. Each group was pastured separately with one of 4 stallions; 2 stallions had previously produced pregnancies with high CG titres and the other 2 pregnancies with low titres. After 7 weeks pregnancy was detected using a 3·5 MHz linear array real-time ultrasound scanner (Equiscan 4100, Bion Corporation, Westminster, CO, U.S.A.). Blood samples (10 ml) were collected at weekly intervals between Days 51 and 100 of gestation. Serum CG concentrations were determined by radioimmunoassay as described below.

In 1985, using results obtained in 1983, 15 mares exhibiting the highest mean levels of CG were mated with a sire previously shown to produce pregnancies with high levels of CG, and the 15 mares producing the lowest levels were mated with a stallion associated with pregnancies with low levels of CG. Two groups of 15 mares that had not previously been studied were mated with 2 stallions that had sired previous pregnancies of high and low CG titres. Oestrus was synchronized with PGF-2α, and mares were pastured with the stallions as before. Pregnancy was con-firmed in 55 of the 60 mares as described above. Blood samples were collected from pregnant mares at 2-week intervals between Days 39 and 104 of gestation, and serum samples were assayed for total CG, and FSH and LH activities as described below.

Radioimmunoassay for CG and bioassays for FSH and LH. All the samples collected during the 1983 and 1985 experiments were subject to double-antibody radioimmunoassay to determine the concentrations of CG. The anti-serum (R-290), purified horse CG for iodination and standard horse CG (PM-68-B) were donated by Dr H. Papkoff. Sheep-anti-rabbit-gamma globulin (W-168) acquired from Dr N. C. Rawlings, was used as the second antibody at a dilution of 1:15. The antiserum has previously been characterized and the assay validated (Farmer & Papkoff, 1979). In the present study, serum was diluted by 1:100 in assay buffer and 0·1 ml assayed. The assay protocol has previously been reported (Murphy *et al.*, 1985). Gamma counter output in counts per minute was converted to ng/ml serum by means of a microcomputer program (RIAPC) developed and donated by Dr D. Rieger. The Second International Standard of CG was diluted and run in each assay to determine the inter-assay coefficient of variation which ranged from 6 to 14%. The intra-assay coefficient of variation, calculated between duplicates, ranged from 0·01 to 11%. The sensitivity of the assay was 7·5 ng CG/ml serum.

Serum samples collected on Days 39, 71 and 104 of gestation from all mares in 1985 were subjected to FSH and LH bioassays. Use of untreated serum in in-vitro bioassays proved difficult because of a variable occurrence of cytotoxicity. The CG was therefore extracted from serum by addition of sufficient ethanol to make a final concentration of 50%. The samples were centrifuged, precipitates discarded and the supernatant adjusted to a final concentration of 90% ethanol. The precipitate containing the CG was recovered by centrifugation and resuspended in the original serum volume of distilled water. Recovery of ^{125}I-labelled horse CG added to serum ranged from 84 to 92%. All extracted samples were assayed for CG by the double-antibody radioimmunoassay described above. Comparison of CG concentrations in serum samples and the same samples after extraction revealed no appreciable difference. Together these observations suggest that there was no significant loss of CG that could be attributed to the extraction procedure.

The LH bioassay used was a modification of that of Van Damme *et al.* (1974) as previously described (Gonzalez Mencio *et al.*, 1978). Briefly, testes from mature Swiss mice were decapsulated, mechanically dispersed and filtered through a silk screen to remove tubular elements. The resultant suspension, consisting primarily of Leydig cells, was concentrated by centrifugation and resuspended in tissue culture Medium 199 containing 2% fetal bovine serum (FBS; GIBCO, Grand Island, NY) and subjected to a 1 h preincubation at 34°C under an atmosphere of 95% O_2:5% CO_2. The collected cells were centrifuged, washed in culture medium and dispensed at $1·5 \times 10^5$ interstitial cells per plastic culture tube (Falcon, Becton-Dickinson, Oxnard, CA). The standard curve, carried out in triplicate, comprised NIADDK-oLH-25 at concentrations from 10 pg to 10 ng per tube. Aliquants of extracted CG were added in duplicate. Standards and unknowns were incubated for 3 h at 34°C under 95% O_2:5% CO_2 in a shaking water bath. Incubations were terminated by centrifugation, the supernatants were collected and their testosterone content was determined by a radioimmunoassay previously validated by Cook *et al.* (1982) using antiserum (NCR 98-18-4-78) donated by Dr N. C. Rawlings. The dose–response relationship between LH and testosterone was rectified by plotting the square root of testosterone concentration against the logarithm of the amount of LH standard added (Van Damme *et al.*, 1974). The coefficient of determination for the linearity of a typical standard curve was 0·97. The Second International Standard of CG was included in each bioassay, and the inter- and intra-assay coefficients of variation, based on this preparation, were 8·38 and 16% respectively.

FSH activity was determined using a method modified from Combarnous *et al.* (1984) in which changes in the activity of the enzyme plasminogen activator in Sertoli cell cultures were assessed. In brief, culture wells (Falcon Multiwell, Becton-Dickinson) were coated under sterile conditions with $1·0 \times 10^5$ c.p.m. ^{125}I-labelled fibrinogen ($\sim 7·5$ µg/well) and dried overnight. The wells were incubated for 2 h at 37°C with 0·5 ml Minimum Essential Medium (MEM, GIBCO) containing 10% FBS to convert plated fibrinogen to fibrin. Sertoli cells were prepared from testes removed and decapsulated from 18–22-day-old Sprague–Dawley rats. Testes were placed in Hanks' balanced salt solution (without Ca^{2+} and Mg^{2+}) and digested with 0·25% trypsin (GIBCO) and 0·001% DNase (Sigma, St. Louis, MO, U.S.A.) for 15 min. The tubules were allowed to settle and the supernatant was removed. Proteolysis was terminated by the addition of 0·125% soybean trypsin inhibitor (Sigma). The tubules were then washed three times and resuspended in filter-sterilized MEM, pH 7·2, buffered with Hepes (Sigma) and containing 100 i.u. penicillin/ml, 100 µg streptomycin/ml and 0·25 µg fungizone/ml (GIBCO, Chagrin Falls, OH). The seminiferous tubules were further macerated by repeated aspiration, and filtered through a silk screen. The Sertoli cell-rich suspension was plated onto ^{125}I-labelled fibrin-precoated wells at $5·0 \times 10^5$ cells/well and incubated in a tissue culture incubator for 48 h at 34°C, 95% humidity and 5% CO_2. Medium was then removed from each well and replaced with 0·5 ml medium containing standard FSH (NIADDK-oFSH-16) (in triplicate), or unknown, extracted CG (in duplicate), which was added at concentrations of 1:300, 1:1000 and 1:3000. After incubation for a further 4 h, media were removed from wells and 0·5 ml MEM containing 10% acid-treated FBS was added as a source of plasminogen. Incubation was continued for a further 16 h, after which 0·25 ml aliquants of medium and associated solubilized ^{125}I-labelled fibrin were removed and counted. The dose–response to FSH was linearized by plotting c.p.m. against the logarithm of FSH concentration, resulting in coefficients of determination in the range of 0·91 to 0·97. The three concentrations of unknowns were generally linear, and calculations were made using the value nearest the 50% response point of the standard curve.

All FSH and LH values were converted to NIH-FSH-S1 and NIH-LH-S1 units, respectively. To preclude the possibility that variations in FSH and LH activity were solely, or in part, a function of serum CG concentrations, all values were expressed as ng LH or FSH per ng horse CG.

Determination of sialic acid content. To test the hypothesis that variation in biological activity of horse CG is attributable to variability in the carbohydrate content of the molecule, the effects of removal of various amounts of sialic acid on the LH and FSH activity of CG were determined. Commercial horse CG (Equinex: Ayerst, Montreal, Canada) was incubated with neuraminidase (Sigma) at one enzyme unit per $1·0 \times 10^4$ i.u. horse CG. The sialic acid concentration of the treated horse CG was determined by an enzymic assay kit from Boehringer-Mannheim (Dorval, Quebec, Canada). The neuraminidase reaction was time dependent and various amounts of sialic acid were removed by incubation at room temperature for periods from 15 min to 18 h. Three preparations were chosen for radio-immunoassay of CG and FSH and LH bioassay, i.e. CG with 20% sialic acid removed (15 min incubation), 53% removed (45 min incubation), 80% removed (6 h incubation).

To determine whether variability existed in sialic acid content of horse CG, 30 samples of serum extract that had a wide range of LH activity by bioassay were selected. The CG component of 2 ml of each was separated by chromatography on Sephadex G-25 columns (PD-10, Pharmacia, Montreal, Canada). Tracer amounts of ^{125}I-labelled horse CG were added to each sample and the columns were eluted with phosphate-buffered saline. Effluent (0·5 ml fractions) was monitored for radioactivity and protein concentration, the latter by measuring u.v. absorbance at 260 nm. The fractions containing CG were collected and pooled and assayed for sialic acid as previously described. When the sialic acid contents of the CG peak and of the pooled fractions were compared, more than 75% of the total sialic acid in the extracted samples appeared in the CG peak.

Statistical analysis. Two-way analysis of variance was performed on the 1983 and 1985 results to test the effect of sire and stage of gestation. Comparisons between individual means were made by Student–Neuman–Keuls tests (Sokol & Rohlf, 1981). The correlation between of LH bioactivity and sialic acid content in CG was determined.

Results

Concentrations and biological activity of horse CG

Mean ($\pm$s.e.m.) CG concentrations from groups of 10 mares mated with each of the 4 stallions in the 1983 trial were 15·8 $\pm$ 2·9, 8·0 $\pm$ 1·9, 13·7 $\pm$ 2·5 and 7·4 $\pm$ 1·5 mg/ml, respectively. Statistical analysis revealed that the CG concentrations in the serum of pregnant mares were influenced by the stallion ($P < 0.05$) and that the individual stallion effects were those expected from analysis of the results of the previous year.

Two-way analysis of variance of CG concentrations in the serum of mares selected for high or low titres or randomly chosen for the 1985 trials further demonstrated that sires influenced CG levels in the mare ($P < 0.05$). Individual comparison of means revealed the presence of differences between the high-titre mares mated with a high-titre sire (Sire 1) and low CG titre mares mated with a low-titre sire (Sire 2) from Day 55 to 104 of gestation ($P < 0.01$) (Fig. 1). The remaining 2 sires that were mated to unselected mares did not produce pregnancies in which CG titres were significantly different from those for Sire 1.

The variation in FSH bioactivity per unit of CG between mares and at different stages of gestation was considerable, with activity as low as 0·38 and as high as 18·03 ng FSH (S1 units)/ng CG. Individual profiles selected to demonstrate this variability are presented in Fig. 2. No effect of sire or day of gestation on variation in FSH bioactivity could be demonstrated (Figs 2 & 3).

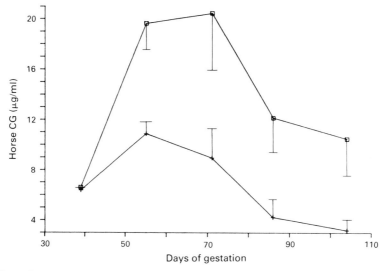

Fig. 1. Mean ($\pm$s.e.m.) CG concentrations from Days 39 to 104 of gestation in 10 mares selected for high titre CG pregnancies and mated with a high producing sire ($\square$) and 10 mares with low titres of CG mated with a 'low' sire ($+$).

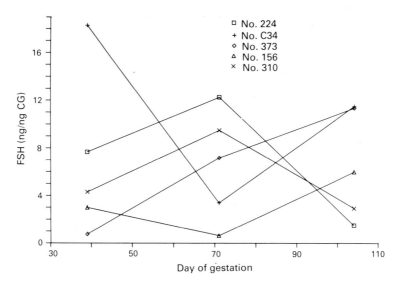

Fig. 2. Profiles of FSH activity per unit of CG in serum samples collected (in 1985) on Days 39, 71 and 104 of gestation in 5 different mares selected to demonstrate the variability in FSH activity.

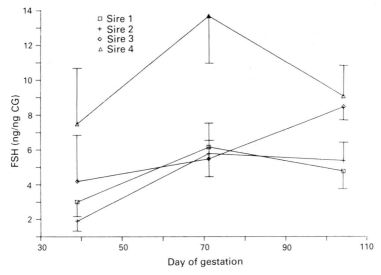

Fig. 3. Mean ($\pm$s.e.m.) FSH activity per unit of CG in serum samples collected (in 1985) on Days 39, 71 and 104 of gestation from mares mated with one of 4 sires.

LH activity of CG also exhibited individual variation, and an effect of day of gestation could not be clearly identified although there appeared to be an upward trend in the Day 71 samples (Fig. 4). There was an effect of sire on LH bioactivity ($P < 0.05$), with LH activity of CG from pregnancies produced by Sire 2 being lower ($P < 0.05$) than that for any of the remaining 3 stallions (Fig. 5).

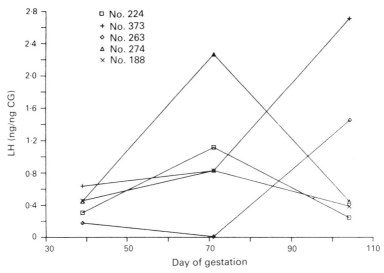

Fig. 4. Profiles of LH activity per unit of CG in serum samples collected (in 1985) on Days 39, 71 and 104 of gestation in mares selected to demonstrate the variability in LH activity.

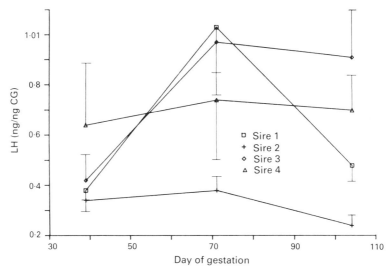

Fig. 5. Mean ($\pm$ s.e.m.) LH activity per unit of CG in serum samples collected (in 1985) on Days 39, 71 and 104 of gestation from mares mated with one of 4 sires.

The ratio of FSH:LH was calculated from bioassay estimates. Mean ratios for all animals at Days 71 and 104 were greater than at Day 39 ($P < 0.05$).

Effects of removal of sialic acid on the biological activity of CG and variation in sialic acid concentration of CG in serum

Neuraminidase treatment of commercial horse CG which removed 20 and 53% of the sialic acid had no effect on the concentrations of CG as measured by radioimmunoassay, or on FSH activity as measured by bioassay. However, removal of 80% sialic acid resulted in a diminution of about 9% in the amount of CG recognized by the Papkoff antibody, and a 22% decrease in FSH

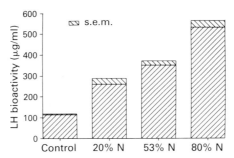

Fig. 6. Augmentation of LH activity of horse CG after incubation with neuraminidase (N) to remove sialic acid. Each bar represents the mean ($\pm$ s.e.m.) of 5 replicates.

bioactivity. Ablation of sialic acid from commercial horse CG resulted in a dose-dependent increase in LH activity (Fig. 6).

There was a demonstrable negative correlation ($r = -0.49$, $P < 0.05$) between the amount of sialic acid present in CG samples from the experimental mares and their potency in the in-vitro LH bioassay. The relationship appeared most pronounced when sialic acid content of the samples was lowest.

Discussion

The results of 1983 trials in which groups of 10 mares were mated with each of four sires confirm the findings of Bell (1985), who demonstrated an effect of sire on the amount of CG produced by mares between Days 50 and 100 of gestation. From the 1985 trial it can be concluded that selection of both sire and mare amplifies the disparity in CG production.

As the sampling regimen differed between years, it is not possible to compare directly the absolute amounts of CG. Comparison of CG titres at Day 70 and 71 in the 1983 and 1985 trials suggested an effect of the mare on the amount of CG produced. This may be a result of the maternal contribution to the embryonic genotype expressed in the amount of CG produced by chorionic cells. Alternatively, some mares could provide a uterine environment more amenable to the production of CG. The development of large, active, endometrial cups and elevated CG concentrations in donkeys that were the recipients of transferred mule or horse embryos (Allen, 1982) argues for an effect of the uterine environment.

The results of the selection trial suggest that it is possible to select mares and sires for CG production, a finding which may be of commercial importance.

The finding that there is variation in both LH and FSH activity between mares and between bleeding dates in the same mare confirms our previous report (Gonzalez Mencio *et al.*, 1978). A consequence of these differences is the fairly wide variation in FSH:LH ratio. The conclusion that there is heterogeneity in biological activity of horse CG preparations has also been reached by Aggarwal *et al.* (1980a) and Papkoff (1981) but contrasts with that of Stewart *et al.* (1976) who, using a radioreceptor assay, concluded that the FSH:LH ratio of horse CG remained constant between Days 40 and 80 of gestation. The failure of these workers to demonstrate the heterogeneity may be due to the methods they used since it has been demonstrated that removal of the carbohydrate components of human chorionic gonadotrophin does not alter its binding to testicular membrane receptors but abolishes its biological activity (Thotakura & Bahl, 1982).

It appears from the current data that some of the variability in biological activity, particularly with respect to LH, may be explained by a sire effect. In particular, pregnancies sired by the stallion associated with low CG titres were also low in LH activity per unit of CG.

Matteri *et al.* (1986) attribute structural heterogeneity of horse CG to variation in glycosylation between preparations isolated from serum and endometrial cups. In the present study it has been shown that differential removal of the sialic acid component of CG causes concomitant elevation in its LH bioactivity. No effect was observed on FSH activity when half of the sialic acid was removed but removal of 80% reduced FSH activity. The former result concurs with the findings of Moore & Ward (1980), who demonstrated enhanced testosterone production by rat Leydig cells *in vitro* with 70% desialated horse CG. The finding that FSH activity was reduced by desialation is in agreement with the response of Sertoli cells *in vitro* to desialated horse CG reported by Farmer & Papkoff (1978). Our results differ somewhat from those of Moore & Ward (1980) who report large increases in the FSH activity of desialated CG, relative to the intact molecule, when tested in a rat granulosa cell plasminogen activator assay. A reconciliation of these disparate findings is not obvious although our trials with rat granulosa cell bioassays for FSH activity revealed the presence of a 5% cross-reaction between LH and FSH (K. Rajkumar & B. D. Murphy, unpublished results). The considerable LH activity of horse CG and the increase in LH activity associated with desialation could have resulted in the conclusion that both LH and FSH activity are increased by removal of sialic acid.

The relationship between LH activity and the sialic acid content of extracted horse CG samples is of some interest because it provides support for the concept that the variation in biological activity of different horse CG samples may reflect differences in the degree of glycosylation of the CG molecule. Removal of sialic acid results in more rapid clearance of the hormone in rats (Aggarwal & Papkoff, 1985). It is therefore possible that some of the variability, not only in the biological activity of CG, but also in its concentration, may be the result of changes in clearance rates rather than production rates in mares.

The molecular basis for variation in sialic acid content of CG between mares and between stages of gestation remains to be discovered. Glycosylation of glycoproteins is a post-transcriptional event, believed to be carried out in Golgi apparatus of the synthesizing cell. As polypeptide determinants can influence protein glycosylations (Henderson, 1984), the variation in the primary structure of horse CG may be responsible for the consequent variation in the quantity of carbohydrate attached. Alternatively, it has been shown that other gene products, including the glycosylating enzymes such as glycosyltransferases and glycohydrolases, may differ between glycoprotein synthesizing cells (Henderson, 1984). Genetic variation in regulation of the synthesis of glycosylating enzymes may be the key to the genetically associated variability in horse CG illustrated by the present study.

We thank the owners and staff of the Bar C Ranch, Rossburn, Manitoba for allowing and aiding this study; J. Malenik, H. Ly, S. Johannesen and S. Emery for excellent technical assistance; P. Schott for data analysis; and S. Johannesen for help in preparation of the manuscript. This study was supported by a National Science and Engineering Research Council Strategic Grant.

References

Aggarwal, B.B. & Papkoff, H. (1985) Plasma clearance and tissue uptake of native and desialylated equine gonadotropins. *Dom. Anim. Endocr.* **2**, 173–181.

Aggarwal, B.B., Farmer, S.W., Papkoff, H. & Seidel, G.E. (1980a) Biochemical properties of equine chorionic gonadotropin from two different pools of pregnant mare sera. *Biol. Reprod.* **23**, 570–576.

Aggarwal, B.B., Farmer, S.W., Papkoff, H., Stewart, F. & Allen, W.R. (1980b) Purification and characterization of the gonadotropin secreted by cultured horse trophoblast cells. *Endocrinology* **106**, 1755–1759.

Allen, W.R. (1982) Immunological aspects of the endometrial cup reaction and the effect of xenogeneic pregnancy in horses and donkeys. *J. Reprod. Fert., Suppl.* **31**, 57–94.

Allen, W.R. & Moor, R.M. (1972) The origin of the equine endometrial cups. I. Production of PMSG by fetal trophoblast cells. *J. Reprod. Fert.* **29**, 313–316.

Allen, W.R., Hamilton, D.W. & Moor, R.M. (1973) The origin of the equine endometrial cups. II. Invasion of the endometrium by trophoblast. *Anat. Rec.* **177**, 485–501.

Bell, R.J. (1985) *Fertility and pregnancy loss following delay of foal estrus with progesterone and estradiol 17β.* M.Sc. thesis, University of Saskatchewan. Saskatoon.

Clegg, M.T., Boda, J.M. & Cole, H.H. (1954) The endometrial cups and allantochorionic pouches in the mare with an emphasis on the source of equine gonadotrophin. *Endocrinology* **54**, 448–463.

Clegg, M.T., Cole, H.H., Howard, C.B. & Pigon, H. (1962) The influence of foetal genotype on equine gonadotrophin secretion. *J. Endocr.* **25**, 245–248.

Cole, H.H. & Goss, H.T. (1943) The source of equine gonadotrophin. In *Essays in Biology in Honor of Herbert M. Evans*, pp. 107–119. University of California Press, Berkeley.

Cole, H.H. & Hart, G.H. (1930) The potency of blood serum of mares in progressive stages of pregnancy in effecting the sexual maturity of the immature rat. *Am. J. Physiol.* **93**, 57–68.

Combarnous, Y., Guillou, F. & Martinat, N. (1984) Comparison of the *in vitro* follicle-stimulating hormone (FSH) activity of equine gonadotrophins (luteinizing hormone, FSH and chorionic gonadotrophin) in male and female rats. *Endocrinology* **115**, 1821–1827.

Cook, S.J., Rawlings, N.C. & Kennedy, R.I. (1982) Quantitations of six androgens by combined high performance liquid chromatography and radioimmunoassay *Steroids* **40**, 369–380.

Farmer, S.W. & Papkoff, H. (1978) Pregnant mare serum gonadotrophin and follicle-stimulating hormone stimulation of cyclic AMP production in rat seminiferous tubule cells. *J. Endocr.* **76**, 391–397.

Farmer, S.W. & Papkoff, H. (1979) Immunochemical studies with pregnant mare serum gonadotrophin. *Biol. Reprod.* **21**, 425–431.

Gonzalez Mencio, F., Manns, J. & Murphy, B.D. (1978) FSH and LH activity of PMSG from mares at different stages of gestation. *Anim. Reprod. Sci.* **1**, 137–144.

Hamilton, D.W., Allen, W.R. & Moor, R.M. (1973) The origin of the equine endometrial cups. III. Light and electron microscopic study of fully developed equine endometrial cups. *Anat. Rec.* **177**, 503–518.

Henderson, E.J. (1984) The role of glycoproteins in the life cycle of the cellular slime mold *Dictyostelium discoideum*. In *The Biology of Glycoproteins*, pp. 371–443 Ed. R. J. Ivatt. Plenum Press, New York.

Licht, P., Bona-Gallo, A., Aggarwal, B.B., Farmer, S.W., Castelino, J.B. & Papkoff, H. (1979) Biological and binding activities of equine pituitary gonadotrophins and pregnant mare serum gonadotrophin. *J. Endocr.* **83**, 311–322.

Matteri, R.L., Papkoff, H., Murthy, H.M.S., Roser, J.F. & Chang, Y-S. (1986) Comparison of the properties of highly purified equine chorionic gonadotropin isolated from commercial concentrates of pregnant mare serum and endometrial cups. *Dom. Anim. Endocr.* **3**, 39–48.

Moore, W.T., Burleigh, B.D. & Ward, D.N. (1980) Chorionic gonadotrophins: comparative studies and comments on relationships to other glycoprotein hormones. In *Chorionic Gonadotropins*, pp. 89–126. Ed. S. J. Segal. Plenum Press, New York.

Moore, W.T. & Ward, D.N. (1980) Pregnant mare serum gonadotropin. An *in vitro* characterization of the lutropin-follitropin dual activity. *J. biol. Chem.* **255**, 6930–6936.

Murphy, B.D., Rajkumar, K., Bristol, F., Flood, P.F. & Chedrese, P.J. (1985) Efecto del genotipo del semental y la edad de la yegua sobre lose niveles de gonadotrofina corionica equina. *Rev. Med. Vet. (Bs. As.)* **66**, 72–76.

Papkoff, H. (1974) Chemical and biological properties of the subunits of pregnant mare serum gonadotrophin. *Biochem. Biophys. Res. Commun.* **58**, 397–404.

Papkoff, H. (1981) Variations in the properties of equine chorionic gonadotropin. *Theriogenology* **15**, 1–11.

Papkoff, H., Bewley, T.A. & Ramachandran, J. (1978) Physicochemical and biological characterization of pregnant mare serum gonadotropin and its subunits. *Biochim. Biophys. Acta* **532**, 185–194.

Sokol, R.R. & Rohlf F.J. (1981) *Biometry*, 2nd edn. W.H. Freeman, New York.

Squires, E.L. & Ginther, O.J. (1975) Follicular and luteal development in pregnant mares. *J. Reprod. Fert., Suppl* **23**, 429–433.

Stewart, F., Allen, W.R. & Moor, R.M. (1976) Pregnant mare serum gonadotrophin: ratio of follicle-stimulating hormone and luteinizing hormone activities measured by radioreceptor assay. *J. Endocr.* **71**, 371–382.

Thotakura, N.R. & Bahl, O.M. (1982) Role of carbohydrate in human chorionic gonadotropin: deglycosylation uncouples hormone-receptor complex and adenylate cyclase system. *Biochem. Biophys. Res. Commun.* **108**, 399–405.

Van Damme, M.P., Robertson, D.M. & Diczfalusy, E. (1974) An improved *in vitro* bioassay for measuring luteinizing hormone (LH) activity using mouse Leydig cells. *Acta endocr., Copenh.* **77**, 65–71.

J. Reprod. Fert., Suppl. **35** (1987), 399–403

Reproductive characteristics of spontaneous single and double ovulating mares and superovulated mares

E. L. Squires, A. O. McKinnon, E. M. Carnevale, R. Morris and T. M. Nett

Animal Reproduction Laboratory, Colorado State University, Fort Collins, Colorado 80523, U.S.A.

Summary. For embryos collected from mares 7 days after ovulation, embryo recovery for single-ovulating mares was 53% compared to 106% for double-ovulating mares. Pregnancy rates 50 days after surgical transfer were 68 and 129%, respectively. Concentrations of LH were similar during the periovulatory period for cycles which included single or double ovulations. Horse pituitary extract given for 5·5 days resulted in ≥ 2 ovulations (mean 3·8) in 26 of 28 mares and 2·0 embryos were recovered per donor compared to 0·65 for controls. Non-surgical pregnancy rates for embryos collected from superovulated mares 6 and 8 days after ovulation (18 of 46; 39%) were identical to those obtained for untreated controls (12 of 31; 39%). The overall pregnancy rate for superovulated donors, which provided 2–4 embryos, was 120%. Since the viability of multiple embryos collected from spontaneous and induced multiple-ovulating mares was equal to that of embryos from single-ovulating mares, collection and transfer of embryos from these mares will increase the efficiency of horse embryo transfer.

Introduction

The production of several foals from genetically valuable mares is one of the major advantages of horse embryo transfer. However, collection and transfer of embryos from single-ovulating mares limits the number of foals that can be produced from a donor in a given year. Based on a 60% embryo recovery rate and a 65% pregnancy rate per embryo transferred, the success of obtaining a pregnancy per oestrous cycle is approximately 40%. The collection and transfer of several embryos during a given oestrous cycle has the potential of increasing the efficiency of an embryo transfer program and decreasing the cost per successful transfer. Woods & Ginther (1982) reported that, of 39 superovulated mares that became pregnant, only one vesicle was detected on Day 28 after ovulation. This natural embryo reduction mechanism in the mare could interfere with programmes to collect and transfer multiple embryos. Data on embryo recovery from spontaneous multiple ovulators and induced superovulated mares, and subsequent pregnancy rate after transfer of several embryos, are limited. Previous studies have suggested that the viability of multiple embryos was reduced when collections were performed on Day 7 (Woods & Ginther, 1984) or Day 8, but not Day 6 (Squires *et al.*, 1985a). Unfortunately, it was not possible to determine whether the reduced viability of the multiple embryos was due to the superovulation treatment or the mare's natural embryo reduction mechanism. Data are presented herein on the characteristics (hormonal, embryo recovery, embryo viability) of spontaneous single- and double-ovulating mares (Exp. 1) and super-ovulated mares (Exp. 2). The objectives were to: (1) compare the viability of embryos from single- and double-ovulating mares, (2) compare the viability of embryos obtained from single and induced multiple-ovulating mares on Days 6 and 8 after ovulation, and (3) determine whether an embryo reduction mechanism occurs in multiple-ovulating mares on Day 6 or 8 after ovulation.

Materials and Methods

Spontaneous single and double ovulators (Exp. 1). Nine European warmbloods and 10 Thoroughbred mares were teased daily and examined during oestrus with ultrasound for detection of follicular activity and ovulation. These mares were selected because of their tendency to ovulate 2 eggs spontaneously. Mares were artificially inseminated every other day during oestrus. Embryo recovery attempts (Imel *et al.*, 1981) were performed 7 days after ovulation, based on the last ovulation. Mares were treated with prostaglandin F-2α after each recovery attempt. All embryos were transferred via flank incision (Squires *et al.*, 1985b) into recipients +1 to −3 days in synchrony with the donor (+1 = recipient ovulated 1 day before donor). Blood samples were collected daily from Day 1 of oestrus until Day 7 after ovulation during 28 oestrous cycles and assayed for LH (Nett *et al.*, 1975), FSH (Nett *et al.*, 1979) and progesterone (Niswender, 1973). Data on concentrations of hormones, embryo recovery and pregnancy rates were grouped as to the number of ovulations per donor (single versus double).

Superovulated mares (Exp. 2). About 70 normally cycling, non-lactating mares, primarily of Quarter Horse and Appaloosa type, were utilized as embryo donors. Ovaries of each mare were examined by ultrasonography daily beginning on Day 12 after ovulation and continuing until all follicles had ovulated or regressed to ≤30 mm in diameter. Due to the unavailability of several recipients at a given time, a greater number of mares served as control (N = 40) than were treated (N = 28) with pituitary extract. For those mares assigned to the treated group, 7·5 mg prostaglandin F-2α were administered on Day 12. Pituitary extract was dissolved in deionized water such that 7 ml contained approximately 750 Fevold Rat units (prepared and supplied by Dr O. J. Ginther). Injections (7 ml) were administered subcutaneously daily until ≥2 follicles of 35 mm diameter were obtained (Woods & Ginther, 1982). At that time, extract treatment was discontinued and 3300 i.u. hCG were given intravenously. Control mares were not treated with placebo but were given hCG during oestrus once a ≥35 mm follicle was obtained. All donor mares were inseminated beginning on Day 2 or 3 of oestrus or once a mare had acquired ≥1 follicle of 35 mm diameter. Inseminations were performed daily with 250×10^6 motile spermatozoa from one stallion. Within groups (treated or control) mares were randomly assigned to embryo recovery on Day 6 or 8 after ovulation. Mares having asynchronous ovulations (>2 days apart) were recycled. Embryo recovery procedures were as described previously (Imel *et al.*, 1981). When only one embryo was recovered from an extract-treated mare, it was not transferred. In addition, a maximum of 4 embryos from any given multiple ovulating mare were transferred. Recipients were from a pool of about 100 non-lactating mares. All transfers were made into recipients non-surgically by one experienced technician using a Cassou gun (Squires *et al.*, 1985b) and only recipients that had ovulated +1 to −3 days in synchrony with the donor were used. An attempt was made to transfer 15–20 Day-6 and -8 embryos from single- and multiple-ovulating mares. This was done to test the hypothesis that Day-8 embryos from multiple-ovulating mares are less viable than Day-6 embryos due to the natural embryo reduction mechanism. Pregnancy detection using ultrasonography was performed on Days 12, 14, 20 and 30 after ovulation.

Daily blood samples were obtained from 10 controls and 10 treated donor mares beginning on Day 12 after ovulation and continuing until the day of embryo recovery. Samples were assayed for progesterone (Niswender, 1973) using a validated radioimmunoassay method.

Statistical analyses. Proportional data were analysed by χ^2 analysis. Quantitative data were analysed by *t* test or analysis of variance (Steel & Torrie, 1980).

Results

Experiment 1

Of the 59 oestrous cycles experienced by this group of mares during the 1985 breeding season, 33 (55·9%) included ≥2 spontaneous ovulations. The majority of the double ovulations occurred on the same day (68·6%) or only one day apart (27%). Embryo recovery 7 days after a single ovulation was 52·8% compared to 106% for spontaneous double-ovulating mares ($P < 0.001$). Embryo recovery was therefore 2-fold with double ovulating mares.

At 50 days after surgical transfer, the pregnancy rate for single-ovulating donors was 68% compared to 129% after transfer of 2 embryos recovered from double-ovulating mares. There was therefore no difference in viability of embryos from single- *versus* double-ovulating mares, at least when collections were performed 7 days after ovulation. In addition, there was no difference in pregnancy rate of embryos from mares that ovulated synchronously (same day) or asynchronously (1 or 2 days apart).

The amount of LH secreted ± 3 days of ovulation (area under the curve) was similar ($P > 0.05$) for mares with single (1032·7 ± 132·5 ng) and double (847·6 ± 95·6 ng) ovulations. In contrast, concentrations of progesterone during the initial 7 days after ovulation (area under the curve) were greater ($P < 0.05$) in double-ovulating (55·2 ± 10·2 ng) than single-ovulating mares

(44·6 ± 11·4 ng). Concentrations of FSH during oestrus and for 7 days after ovulation were similar ($P > 0.05$) during cycles in which mares had single or double ovulations.

Experiment 2

A mean of 5·5 daily injections of pituitary extract was given before 2 follicles of ≥ 35 mm in diameter were obtained (Table 1). The number of ovulations for treated mares ranged from 0 to 10. Of 28 treated mares, 26 had ≥ 2 ovulations. The majority (84%) of ovulations were on the same day. The correlation between the number of treatments and number of ovulation was not significant ($r = 0.306$; $P > 0.05$). The mean number of ovulations for treated mares was greater ($P < 0.05$) than that for controls. The superovulatory response obtained in this study (93% of mares responded, mean 3·8 ovulations per mare) was slightly better than that reported by Woods & Ginther (1983). An average of 2·0 embryos was obtained from each treated mare compared to only 0·65 from controls ($P < 0.05$). The number of embryos recovered from treated mares ($P > 0.05$) ranged from 0 to 6. Embryo recovery from superovulated mares was therefore 3-fold higher than that from untreated controls. Embryo size was similar between treated and control donors when collections were performed at Day 6 (0·177 and 0·213 mm, respectively) but not at Day 8 (0·567 and 0·903 mm, respectively; $P < 0.05$).

There was no difference in the percentage of ovulations that resulted in embryos recovered for treated (54·3%) compared to control mares (54·8%). The correlation between the number of ovulations and number of embryos recovered was not significant ($r = 0.287$). Since embryo recovery per ovulation was approximately 50%, the number of recipients needed per superovulated donor would equal half the number of ovulations.

Non-surgical pregnancy rates were similar ($P > 0.05$) among the four groups (Table 2). Embryos from multiple-ovulating mares were as viable as those from single ovulating mares. In addition, there appeared to be no reduction in pregnancy rates with Day 8 embryos compared to

Table 1. Effect of horse pituitary extract on induction of multiple ovulation and embryo recovery

| Treatment | No. of mares | No. of injections | No. of ovulations | | Mean embryo recovery (no.) | % of ovulations providing embryos |
			All mares	Multiple ovulators		
Extract	28	5·5	3·8	4·0 (N = 26)	2·0	54·3
Control	40	—	1·2	2·0 (N = 5)	0·65	54·8

Table 2. Success of non-surgical transfer of embryos from superovulated and control mares

| | Source of embryo | | | | |
| | Single ovulation | | Multiple ovulations | | |
Day	No. transferred	No. pregnant (%)	No. transferred	No. pregnant (%)	Total (%)
6	16	6 (38)	25	9 (36)	15 (37%)
8	15	6 (40)	21	9 (43)	15 (42%)
Total	31	12 (38)	46	18 (39)	

Day 6 embryos. Further evidence that the multiple embryos were equal in viability to controls was gained by examining the pregnancy rates when 2, 3 or 4 embryos were transferred from a given donor. If the pregnancy rate of 39% is used for embryos from control mares, the expected pregnancy rate for transfer of 2, 3, and 4 embryos from a given donor would be 78, 117 and 156%, respectively. The actual pregnancy rates at Day 30 obtained from transfer of multiple embryos from treated mares were 75, 100 and 180% for 2, 3 and 4 embryos respectively ($P > 0.05$). The overall pregnancy rate per donor, for those in which 2–4 embryos were transferred, was 120%.

As expected, concentrations of progesterone in superovulated mares were higher ($P < 0.05$) than those in single-ovulating mares. Therefore, whether multiple ovulations are spontaneous or induced, concentrations of progesterone are higher when more than one corpus luteum is present.

Discussion

The incidence of spontaneous multiple ovulation in this study was higher than the 12% reported by Ginther *et al.* (1982). Thoroughbreds were reported to have a higher incidence of multiple ovulations than other breeds (Ginther *et al.,* 1982), but data on warmbloods were not available. The increased multiple ovulation rate noted did not appear to be related to concentrations of gonadotrophin. This is similar to a previous study of Urwin & Allen (1983). These investigators collected blood every 3 days from 12 Thoroughbred mares throughout one or two cycles. They were unable to demonstrate a relationship between gonadotrophin concentrations and subsequent ovulation rate. The reason(s) for the higher incidence of twinning for mares in the present study was undetermined. As suggested by Urwin & Allen (1983), it may be that the ovaries of these mares are more sensitive to gonadotrophins. This increased sensitivity could be genetic in origin. Nevertheless, mares that consistently double ovulate are useful in an embryo transfer programme since embryo recovery rates are higher and viability of these multiple embryos equals that of embryos from single-ovulating mares. This is in agreement with previous studies that have shown a higher pregnancy rate in double- *versus* single-ovulating mares (Ginther, 1983). However, one of the embryos is eliminated after it enters the uterus by the mare's natural reduction process.

The results of Exp. 2 differ somewhat from those of previous reports from our laboratory (Squires *et al.,* 1985a) and those of investigators in Wisconsin. Woods & Ginther (1982) reported a higher pregnancy rate and lower embryonic loss in untreated controls than in seasonally anoestrous mares induced to ovulate with pituitary extract. Pregnancy rate was significantly reduced in mares with > 2 ovulations. Of the superovulated mares that did become pregnant, only 1 conceptus was detected on Day 28. In a subsequent study, Woods & Ginther (1983) tested the hypothesis that embryo reduction occurs between Days 7 and 11 after ovulation. More normal embryos were obtained from multiple-ovulating mares on Day 7 (2·9) compared to Day 11 (0·7). These data indicated that embryo reduction occurred between Days 7 and 11. Viability of Day-7 embryos from superovulated and control mares was also evaluated (Woods & Ginther, 1984). The diameter of embryos was less ($P < 0.05$) for the induced multiple-ovulating donors than for single-ovulating donors. Transfer success rate at 49 days tended ($P < 0.10$) to be greater for embryos from single-embryo collections (4 of 8, 50%) than from multiple embryo collections (7 of 21, 33%). Squires *et al.* (1985a) conducted a preliminary trial to determine the pregnancy rates of embryos collected 6 or 8 days after ovulation from untreated, single-ovulating or induced multiple-ovulating mares. The number of transfers in each group was small (9–11). Although there were no significant differences, the lowest pregnancy rate (3 of 11, 27%) was obtained from non-surgical transfer of multiple embryos collected on Day 8. They concluded that additional studies with greater numbers of transfers were needed before firm conclusions could be drawn as to whether multiple embryos are less viable than single embryos and, if so, when and how this embryo reduction mechanism occurs.

The study reported herein was based on a larger number of transfers than in previous reports. The reason for the overall lower non-surgical pregnancy rates in this study was undetermined.

However, the same technician and procedure were used as those that resulted in an overall success rate of 54% in the previous breeding season (Squires *et al.*, 1985a). In any event, based on comparisons to controls, there was no evidence of a reduced viability of multiple embryos collected at Day 6 or Day 8 from superovulated mares. However, embryos from superovulated donors collected 8 days after ovulation were smaller than those from controls but still within the range presented previously for Day-8 embryos (Squires *et al.*, 1985b). Whether this decrease in embryo size was due to the mare's embryo reduction mechanism is uncertain.

Collection of multiple embryos from a given donor mare would greatly improve the efficiency of an embryo transfer programme if the viability of these multiple embryos was equal to that of single embryos. In the present study, embryo reduction had not occurred by Day 7 in naturally occurring multiple-ovulating donors or by Days 6 or 8 in induced superovulated mares. Therefore, collection of multiple embryos from either natural double ovulators or induced multiple ovulators is a method of increasing the efficiency of an embryo transfer programme.

Supported in part by the Van Camp Foundation and Colorado State University Experiment Station.

References

Ginther, O.J. (1983) Effect of reproductive status on twinning and on side of ovulation and embryo attachment in mares. *Theriogenology* **20**, 383–395.

Ginther, O.J., Douglas, R.H. & Lawrence, R. (1982) Twinning in mares: a survey of veterinarians and analyses of theriogenology records. *Theriogenology* **18**, 333–347.

Imel, K.J., Squires, E.L., Elsden, R.P. & Shideler, R.K. (1981) Collection and transfer of equine embryos. *J. Am. vet. med. Assoc.* **179**, 987–991.

Nett, T.M., Holtan, D.W. & Estergreen, V.L. (1975) Levels of LH, prolactin and oestrogens in the serum of post-partum mares. *J. Reprod. Fert., Suppl.* **23**, 201–206.

Nett, T.M., Pickett, B.W. & Squires, E.L. (1979) Effect of Equimate (ICI-81008) on levels of luteinizing hormone, follicle-stimulating hormone and progesterone during the estrous cycle of the mares. *J. Anim. Sci.* **58**, 69–75.

Niswender, G.D. (1973) Influence of the site of conjugation on the specificity of antibodies to progesterone. *Steroids* **22**, 413–424.

Squires, E.L., Garcia, R.H. & Ginther, O.J. (1985a) Factors affecting success of equine embryo transfer. *Equine vet. J., Suppl.* **3**, 92–95.

Squires, E.L., Cook, V.M. & Voss, J.L. (1985b) Collection and transfer of equine embryos. Animal Reproduction Laboratory Bulletin 01, Colorado State University, Ft. Collins, CO 80523.

Steel, R.G. & Torrie, J.H. (1980) *Principles and Procedures of Statistics, A Biometrical Approach.* McGraw-Hill Book Company, New York.

Urwin, V.E. & Allen, W.R. (1983) Follicle stimulating hormone, luteinizing hormone and progesterone concentrations in the blood of Thoroughbred mares exhibiting single and twin ovulations. *Equine vet. J.* **15**, 325–329.

Woods, G.L. & Ginther, O.J. (1982) Ovarian response, pregnancy rate and incidence of multiple fetuses in mares treated with an equine pituitary extract. *J. Reprod. Fert., Suppl.* **32**, 415–421.

Woods, G.L. & Ginther, O.J. (1983) Recent studies relating to the collection of multiple embryos in mares. *Theriogenology* **19**, 101–108.

Woods, G.L. & Ginther, OJ. (1984) Collection and transfer of multiple embryos in the mare. *Theriogenology* **21**, 461–469.

J. Reprod. Fert., Suppl. **35** (1987), 405–417

Ultrastructure of cryopreserved horse embryos

J. M. Wilson*, T. Caceci†, G. D. Potter* and D. C. Kraemer*‡

*Departments of *Animal Science, †Anatomy and ‡Veterinary Physiology and Pharmacology,
Texas A & M University, College Station, Texas 77843, U.S.A.*

Summary. Embryos were recovered non-surgically at about Day 6 after ovulation from 15 Quarter horse-type mares and were evaluated for morphological changes which may occur because of exposure to the cryoprotectant and/or cryopreservation. Electron microscopy was used to elucidate the fine structure of intracellular organelles which, if damaged, could cause cellular death. The horse embryo does not totally re-expand in the 10% glycerol freezing medium, nor will it completely re-expand in the isotonic holding medium following glycerol removal whether or not the embryo has been frozen. Embryos in this study were frozen by the same protocol which had resulted in a 30% pregnancy rate for similarly frozen embryos.

Junctional complexes between trophoblast cells, as well as the plasma and nuclear membranes of trophoblast and inner cell mass (ICM) cells, were intact after treatment in all embryos. Changes in lipid droplets and some mitochondrial degeneration were observed in the ICM cells of the glycerol-treated embryos. The change in the lipid was not observed in the frozen–thawed embryos, but mitochondrial changes were evident in the trophoblast and ICM cells, with the most extensive mitochondrial damage in the ICM cells.

Introduction

Many cryopreserved embryos appear to be viable after cryopreservation, with the majority of the cells intact, but the pregnancy rates following surgical or non-surgical transfer have been considerably lower than those obtained when embryos were transferred soon after collection. These results have been observed using several different cryoprotectants and freezing regimens, regardless of species of embryo frozen.

The reasons why some frozen–thawed embryos, that appear viable at assessment by light microscopy, do not produce pregnancies have not been elucidated.

Mohr & Trounson (1981) described the ultrastructure of cryopreserved cow embryos. In that study cellular damage was observed, but the extent of the damage varied between and within embryos. The Day 5 or younger embryos appeared to sustain the most damage to the plasma membrane, although the junctional complexes appeared to be intact. Day 13 blastocyst trophectoderm cells sustained the most damage, while the undifferentiated embryonic cells appeared to survive the freezing and thawing process. Flood *et al.* (1982) described the ultrastructure of the horse embryo from Day 3 to Day 16, but most attention in their study was devoted to a description of the development of the embryonic capsule of the fresh specimens.

The purpose of the present study was to compare fresh and frozen–thawed horse embryos using electron microscopy to try to elucidate the changes which occur during cryopreservation and subsequent thawing.

Materials and Methods

Embryo recovery. Embryos were recovered non-surgically from 2-year-old maiden fillies using a recovery technique described by Wilson *et al.* (1985). Non-surgical embryo recoveries were performed on Day 6 (∼144 h after ovulation), soon after the embryo enters the uterus (Oguri & Tsutsumi, 1972). The early horse embryo appears to survive cryopreservation better than the expanded blastocyst (Takeda *et al.*, 1984; Slade *et al.*, 1985; Wilson *et al.*, 1986b).

Embryo treatment groups. Embryos were assigned to one of three treatment groups. Group I embryos were placed into a 10% glycerol holding medium, allowed to equilibrate for approximately 30 min, then placed into a 1·08 M-sucrose holding medium for 20 min for glycerol removal.

Once the embryos had contracted and begun to re-expand they were placed into fresh holding medium, allowed to re-expand as much as they would and were then cultured for 1 h at room temperature. Group II embryos were equilibrated in the 10% glycerol medium and frozen–thawed by the method described by Wilson *et al.* (1986b). After thawing, embryos were placed into 10% glycerol holding medium, rehydrated and cultured as described for Group I embryos. Group III embryos were fixed fresh as soon as they had completed the 1 h of culture.

Fixation, sectioning and staining. After culture all embryos were fixed in 2% glutaraldehyde–PBS, post-fixed in 1% osmium tetroxide and washed in PBS. The embryos were dehydrated through ethanol and embedded in Maraglas (Caceci, 1984). Ultrathin sections were stained with uranyl acetate and lead citrate (Hayat, 1981). Thin sections on glass slides were stained with toluidine blue. Ultrathin sections were observed at 60 kV with a Zeiss EM/10C electron microscope and stained thin sections were photographed with a Zeiss inverted light microscope.

Results

Ultrastructural description

The horse embryo appears to be unique among the domestic species. The embryo enters the uterus at Day 6 (~144 h) after ovulation, and is about 150 μm in diameter with an intact zona pellucida (Fig. 1; Oguri & Tsutsumi, 1972). This is the approximate size of the horse oocyte, indicating that all development thus far has been within the confines of the original zona pellucida (Betteridge *et al.*, 1982).

Soon after entering the uterus the embryo begins to develop an acellular capsule between the trophoblast and the zona pellucida (Fig. 2). This sub-zonal layer remains until about Day 20 even though the zona pellucida is shed by late Day 6 (Flood *et al.*, 1982; Wilson *et al.*, 1986a).

The horse embryo enters the uterus as an early blastocyst with a centrally located inner cell mass (ICM) and an expanding blastocoele. The ICM cells had protoplasmic processes which extended from one cell to another. The trophoblast cells had formed a single layer adjacent to the capsule (Figs 3 & 4). Note in these figures that the zona pellucida has thinned and the capsule is evident between the zona pellucida and the trophoblast cells. As the embryo increased in size the ICM cells had begun to coalesce and migrate toward one pole of the blastocoele (Fig. 5). The expanded blastocyst, however, appeared similar to embryos of other species at this developmental stage. The ICM cells were adjacent to the trophoblast cells at one pole. The major difference is that the horse embryo has shed its zona pellucida and is now covered only by the transparent acellular capsule (Fig. 6).

The early blastocyst is covered by a thinning zona pellucida and the capsule. The trophoblast cells appeared more columnar containing many mitochondria and numerous vesicles (Fig. 7). The expanded blastocyst (Day 7 embryo) had no zona pellucida and was covered only by the acellular capsule. Also, the trophoblast cells had become more squamous (Fig. 8). These data agree with those reported by Flood *et al.* (1982) for the developing trophoblast cells. The apical surfaces of trophoblast cells of all developmental stages observed were covered with numerous microvilli. Microvilli were also present between trophoblast cells below junctional complexes which had formed between the trophoblast cells at the apical end (Figs 9 & 10). These junctional complexes were similar to those described for the preimplantation rabbit embryo by Hastings & Enders (1975), and for the horse embryo by Flood *et al.* (1982).

The cellular processes observed in the early blastocysts appeared to be continuous between cells (Fig. 11). Also, the cellular processes between some of the undifferentiated ICM cells appeared to be connected by intercellular bridges (Figs 12 & 13), similar to those described by Weiss (1983) in undifferentiated spermatogonia. Protoplasmic continuity between ICM cells due to the presence of intercellular bridges and cellular processes was also observed (Figs 14 & 15).

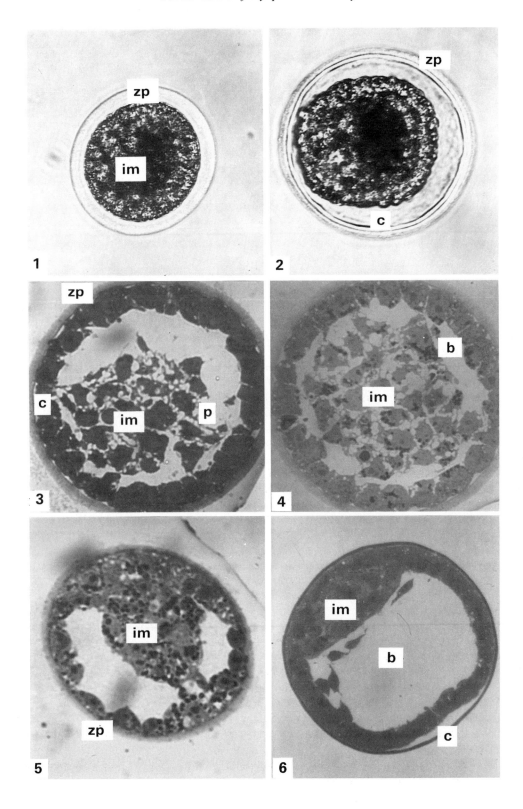

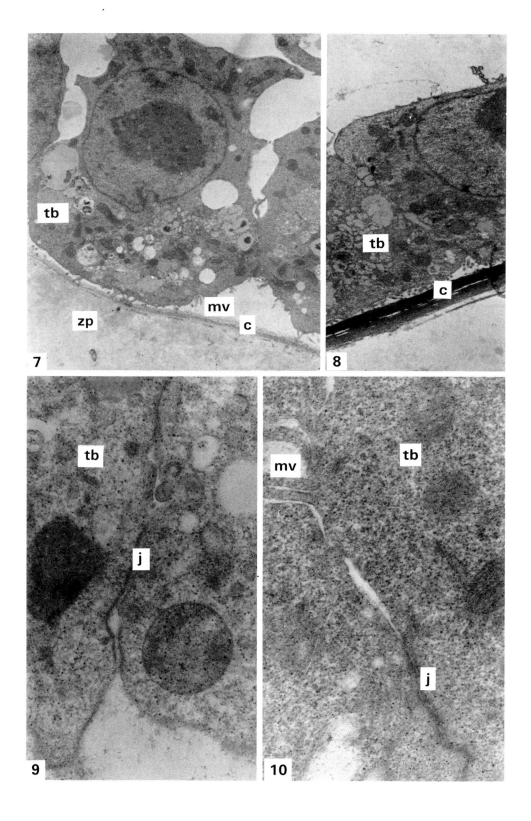

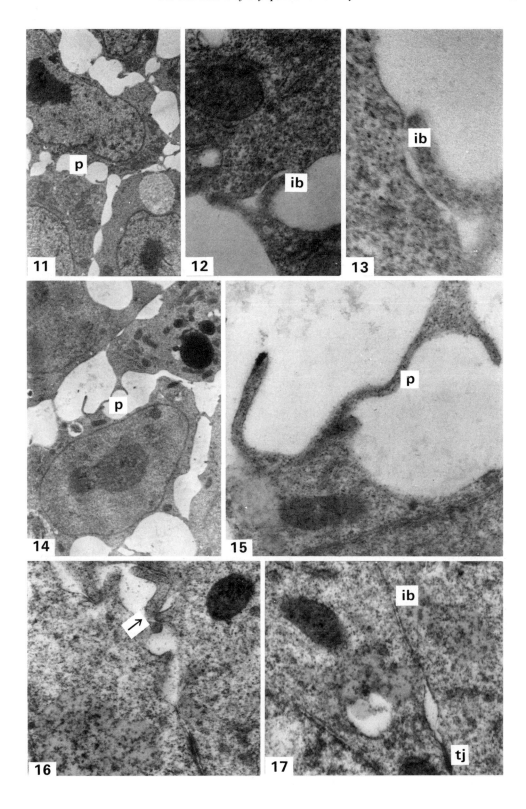

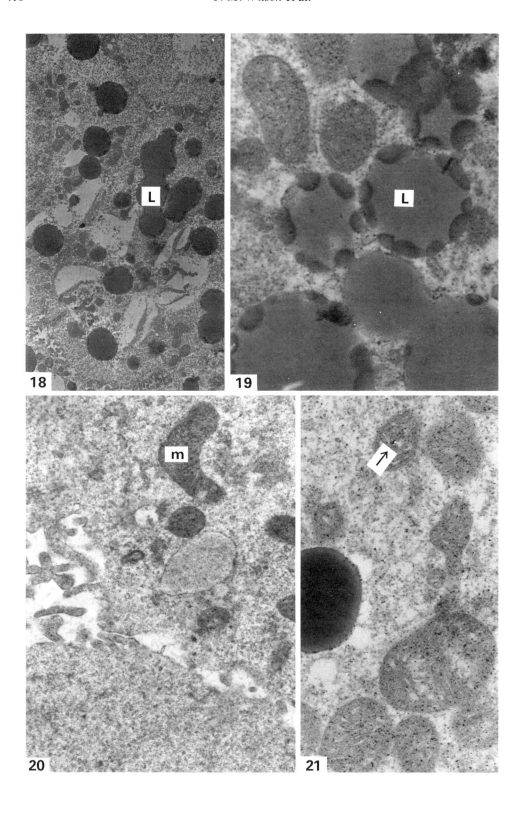

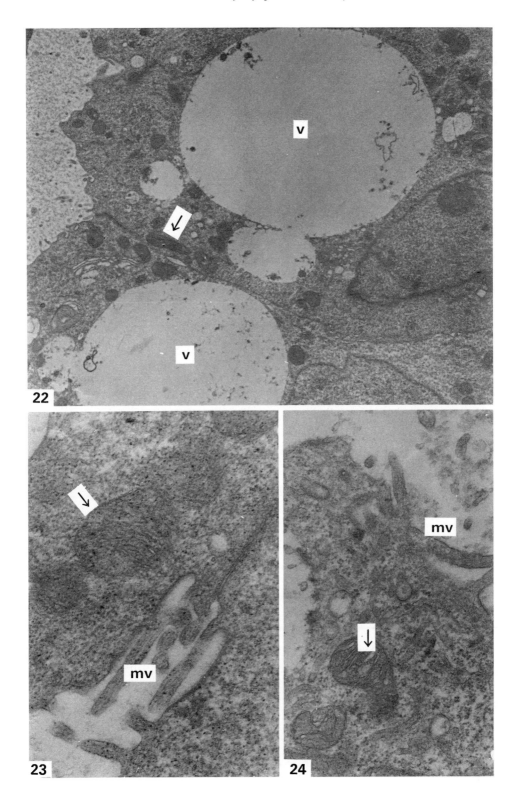

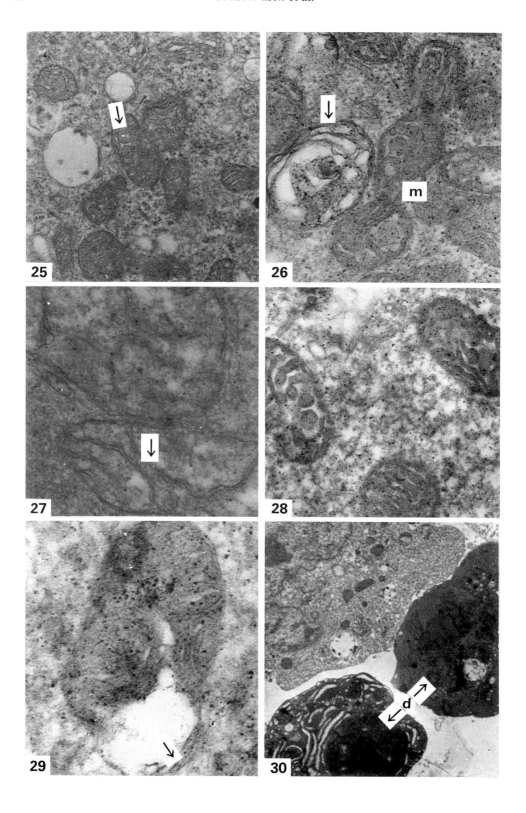

Fig. 1. A 150 μm early blastocyst with intact zona pellucida (zp) and centrally located inner cell mass (im). × 300.

Fig. 2. A 250 μm expanding blastocyst with an intact zona pellucida (zp) and a sub-zonal capsule (c). × 300.

Fig. 3. Light micrograph of a 175 μm early blastocyst with intact zona pellucida (zp) and sub-zonal capsule (c). Note centrally located inner cell mass (im) cells with cellular processes (p). × 750.

Fig. 4. Light micrograph of a 175 μm early blastocyst with centrally located inner cell (im) mass. Note blastocoele (b) surrounding inner cell mass. × 750.

Fig. 5. A 175 μm expanding blastocyst with inner cell mass (im) cells coalescing. Note thinning zona pellucida (zp). × 750.

Fig. 6. A 375 μm expanded blastocyst with intact capsule (c) but the zona pellucida has been shed. Note inner cell mass (im) closely attached at upper most pole of the embryo. Also note cells free in blastocoele (b). × 750.

Fig. 7. Trophoblast (tb) cells of a 175 μm early blastocyst with intact zona pellucida (zp) and developing capsule (c). Note microvilli (mv) on apical surface of trophoblast cells. × 3000.

Fig. 8. Trophoblast (tb) cells of 750 μm expanded blastocyst without zona pellucida. Note sub-zonal capsule (c). × 4000.

Fig. 9. Trophoblast (tb) cells from 175 μm expanding blastocyst with junctional complex (j) at apical end of cell. × 2500.

Fig. 10. Trophoblast (tb) cells from a 250 μm expanding blastocyst with junctional complex (j) and microvilli (mv) between cells below the complex. × 32 000.

Fig. 11. Inner cell mass cells of 250 μm expanding blastocyst. Cells are centrally located in the blastocoele. Note cellular processes (p) between cells. × 3200.

Fig. 12. Intercellular bridge (ib) between two inner cell mass cells. × 25 600.

Fig. 13. Enlargement of intercellular bridge (ib) in Fig. 12. Note continuous cell membranes and protoplasmic continuity. × 56 000.

Fig. 14. Inner cell mass cells connected by intercellular processes (p). × 3000.

Fig. 15. Enlargement of cellular processes (p) in Fig. 14. Note protoplasmic continuity between cells. × 25 600.

Fig. 16. Inner cell mass cells from a 750 μm expanded blastocyst. Note close apposition of cells with microvilli or remnants of cellular processes (arrow). × 25 600.

Fig. 17. Inner cell mass cells from an expanded blastocyst. Note possible intercellular bridge (ib) and tight junction (tj). × 25 600.

Fig. 18. Numerous lipid droplets (l) in inner cell mass cells from glycerol-treated 175 μm early blastocyst. Note staining differences around edges of droplets. × 16 000.

Fig. 19. Enlargement of lipid droplets (l) from glycerol-treated 350 μm expanded blastocyst. Note staining differences. × 32 000.

Fig. 20. Inner cell mass cells of 750 μm expanded blastocyst from control (fresh fixed) embryo. Note mitochondria (m) and microvilli or remnants of cellular processes (arrow). × 16 000.

Fig. 21. Mitochondria in inner cell mass cells of glycerol treated 300 μm expanded blastocyst. Note thickened cristae (arrow). × 32 000.

Fig. 22. Trophoblast cells of frozen–thawed 200 μm expanded blastocyst. Note large vesicles (v) and numerous mitochondria (arrow) which appear to have normal cristae. × 8000.

Fig. 23. Enlargement of trophoblast cells of embryo in Fig. 22. Note intact mitochondria (arrow) with no morphological changes evident. Also note numerous microvilli (mv) below apical junctional complexes (j). × 40 400.

Fig. 24. Trophoblast cells of frozen–thawed 150 μm early blastocyst. Note thickening of mitochondrial cristae (arrow) and numerous microvilli (mv) on apical surfaces of cells. × 32 000.

Fig. 25. Inner cell mass cells of embryo in Fig. 24. Note thickening of mitochondrial cristae (arrow). Most of the mitochondria appear to have suffered morphological damage. × 32 000.

Fig. 26. Inner cell mass cells of frozen–thawed 200 μm expanding blastocyst. Note mitochondria (m) with thickened cristae and early myelin figure (arrow). × 40 400.

Fig. 27. Inner cell mass cells of frozen–thawed 350 μm expanded blastocyst. Note coalescing of inner mitochondrial membranes (arrow). × 64 000.

Fig. 28. Inner cell mass cell of 350 μm frozen–thawed embryo which did not contract in the sucrose medium nor re-expand in the isotonic holding medium. Note mitochondrial damage. × 32 000.

Fig. 29. Blebbing of outer membranes of mitochondria (arrow) in an inner cell mass cell of frozen–thawed 275 μm blastocyst. × 80 000.

Fig. 30. Two degenerative cells (d, arrows) among inner cell mass cells of frozen–thawed 275 μm expanding blastocyst. Embryo was considered viable after thawing by light microscopy. × 3000.

As the embryo increased in diameter the ICM cells became more closely attached and the cellular processes were no longer evident (Fig. 16). However, remnants of the processes or microvilli were observed between ICM cells of expanded blastocysts. Intercellular bridges were also evident in the ICM cells of the expanding blastocysts (Fig. 17). Tight junctions had formed between ICM cells of the expanded blastocyst and intercellular bridges were still evident (Figs 16 & 17). The significance of the cellular processes and intercellular bridges is unknown at this time.

Glycerol-treated embryos

Glycerol-treated embryos appeared normal (partly re-expanded in isotonic holding medium) after the addition and removal of the glycerol in a one-step sucrose medium as described by Wilson *et al.* (1986b). Electron microscopy revealed that all cells had intact plasma and nuclear membranes. The lipid droplets which are in greatest abundance in the ICM cells appeared different from those found in the fresh (control) and the frozen–thawed embryos. Glycerol-treated embryos displayed a different staining contrast of the lipid droplets in the ICM cells compared to controls and frozen–thawed embryos (Figs 18 & 19). Most of the droplets observed had the lobulated pattern shown in Figs 18 & 19. ICM cells of some of the glycerol treated embryos also exhibited morphological changes in the mitochondria. The cristae had become thickened as compared to the cristae of normal mitochondria (compare Figs 20 and 21). These changes in the mitochondria were not as consistent as the lipid changes observed and not all embryos or cells within an embryo displayed the mitochondrial changes.

Frozen–thawed embryos

All but one of the frozen–thawed embryos in this study appeared to survive and partly re-expanded in the isotonic holding medium. However, a 350 μm expanded blastocyst neither condensed in the sucrose medium nor re-expanded in the isotonic holding medium. With the light microscope all cells observed appeared to have intact plasma and nuclear membranes. This was confirmed with the electron microscope. This blastocyst, which would be considered non-viable after thawing, would not have been transferred. Electron microscopy revealed that some of the intracellular vesicles in the trophoblast cells had coalesced (Fig. 22). However, the junctional complexes were intact (Fig. 23). Although most of the mitochondria of the trophoblast cells were normal, a few mitochondria were observed to have a thickening of the cristae (Fig. 24).

Most of the ICM cells of the frozen–thawed embryos did exhibit damage to some of the mitochondria. Several morphological changes were observed in different ICM cells. A thickening of the cristae similar to that observed in the ICM cells of the glycerol-treated embryos was evident in the ICM and trophoblast cells of the frozen–thawed embryos, but the damage was greatest in the ICM cells (Figs 25, 26 & 28). Coalescing of the inner membranes (Fig. 27) and blebbing of the outer mitochondrial membrane (Fig. 29) were also observed.

Although mitochondrial damage was the major defect found in the cells of the frozen–thawed embryos, some degenerate cells were also observed (Fig. 30). This embryo appeared normal before freezing and, with the exception of ICM mitochondrial damage, similar to that observed in all frozen–thawed embryos, the remaining cells appeared normal after thawing.

Discussion

It appears that the early horse embryo may develop differently from that of the other domestic species. The early horse embryo develops a sub-zonal capsule which replaces the zona pellucida (Flood *et al.*, 1982). The zona pellucida is shed at about Day $6\frac{1}{2}$ after ovulation whereas the capsule remains intact until about Day 21 (Betteridge *et al.*, 1982). Several authors (Yamamoto *et al.*, 1982;

Takeda *et al.*, 1984; Slade *et al.*, 1985; Wilson *et al.*, 1986b) have shown that expanded blastocysts (no zona pellucida) do not survive the freeze–thaw procedure as well as do early or expanding blastocysts (zona pellucida intact).

The early embryo also has a centrally located ICM. The undifferentiated ICM cells have intercellular bridges and cellular processes which connect the ICM cells to each other and to the trophoblast cells. These may allow for a synchronous development of these cells. The bridges and processes were evident at least through the developmental stage observed. Bloom & Fawcett (1986) have described the development of intercellular bridges of spermatozoa. In all but the earliest spermatogonial division cytokinesis is incomplete, resulting in groups of conjoined spermatogonia and larger clusters of primary spermatocytes. The protoplasmic continuity which occurs because of the intercellular bridges and cellular processes between ICM cells of the horse embryo may be responsible for a similar syncytium of ICM cells. Loss of this possible synchronous development may contribute to the low survival of embryos frozen–thawed at this developmental stage. As the blastocoele enlarges the ICM appears to migrate to one pole of the embryonic vesicle. The Day 7 horse embryo appears to be similar to the expanded blastocysts of other domestic species with the exception that it has expanded to $\sim 500\ \mu m$.

The early development of the horse embryo appears to differ from that of the cow. These differences may help explain why the ultrastructural changes observed in this study differ from those reported by Mohr & Trounson (1981) for frozen–thawed cow embryos. In that study the plasma membranes of the morulae and early blastocysts sustained the most damage. In the hatched embryos most damage was observed in the trophectoderm cells with the undifferentiated embryonic cells appearing to survive. In the present study it appeared that, with the freeze–thaw protocol used, the trophoblast and ICM cells survive with intact plasma and nuclear membranes.

The differences in freeze–thaw damage between horse and cow embryos may be related to the fact that the cow embryo will re-expand rapidly, after initial shrinkage, when placed in the glycerol medium. However, the horse embryo does not completely re-expand in the glycerol medium, or in the isotonic holding medium after thawing. This would indicate that glycerol does not totally permeate the embryo, leaving the blastocoele partly collapsed during the freeze–thaw process. Some changes must also occur in the trophoblast cells of the horse embryos exposed only to glycerol (not frozen–thawed) because these embryos also remained partly collapsed when placed in the isotonic holding medium. The continued contraction and non-re-expansion is most evident in embryos which do not have a zona pellucida (expanded blastocysts).

The junctional complexes formed between the trophoblast cells of the early embryo could allow for a different milieu to be present between the external environment and the blastocoele. This may be important in the permeation of cryoprotectants, especially in the early blastocysts in which the ICM cells are centrally located and connected only by cellular processes to each other and the trophoblast cells. For a substance to come in contact with the ICM cells it must first pass through the trophoblast cells, and these may somehow limit the permeation of the cryoprotectant and therefore prevent the total re-expansion of the blastocoele.

Embryos which had been exposed only to the cryoprotectant (glycerol) displayed a staining contrast of the lipids (with OsO_4) different from that of fresh (control) and frozen–thawed embryos. McManus & Mowry (1960) have reported that osmium tetroxide will stain unsaturated lipids differently, and the changes observed in the lipids in this study suggest that at least some of the glycerol may have permeated the trophoblast cells and entered the ICM cells.

Some degenerate cells were observed in the ICM of one of the frozen–thawed embryos. However, this was not observed in any of the other frozen–thawed embryos and the degenerate cells were assumed to have been present at the time of collection as it is unlikely that the degeneration could have occurred during the 1 h of culture after thawing.

Most of the damage caused by the freeze–thaw process seemed to occur in the intracellular organelles, especially the mitochondria. The damage observed in the mitochondria of the frozen–thawed ICM cells could be responsible for the death of the cell or render the cells unable to remain in synchrony with the remainder of the syncytia, although the trophoblast cells may survive the freeze–thaw process. Mohr & Trounson (1981) reported that frozen–thawed Day 7 collapsed cow blastocysts with trophoblast cell damage would form another layer of trophoblast cells and a new blastocoele while in culture.

Some of the damage observed in the embryos in this study may be cryoprotectant-induced but most appeared to be from the freeze–thaw process and may be related to the inability of sufficient amounts of the cryoprotectant to permeate the trophoblast cells and enter the ICM cells prior to the beginning of freezing. However, increasing the length of time the embryos are exposed to the cryoprotectant before freezing does not improve permeation and is probably damaging to the embryonic cells. This study suggests that horse embryos may differ in the permeability of the trophoblast cells to glycerol and that, while glycerol appears to protect the plasma and nuclear membranes, it does not protect the mitochondrial membranes from freeze–thaw damage. Further research is indicated to determine which cryoprotectant or combinations of cryoprotectants may be used to protect the horse embryo during cryopreservation.

References

Betteridge, K.J., Eaglesome, M.D., Mitchell, D., Flood, P.F. & Beriault, R. (1982) Development of horse embryos up to twenty-two days after ovulation: observations of fresh specimens. *J. Anat.* **135**, 191–209.

Bloom, W. & Fawcett, D.W. (1986) Male reproduction. In *A Textbook of Histology*, pp. 814–817. Saunders, Philadelphia.

Caceci, T. (1984) A gravimetric formula for Erlandson's Maraglas, D.E.R. 732 embedding medium. *T.S.E.M. Journal* **15**, 26–27.

Flood, P.F., Betteridge, K.J. & Diocee, M.S. (1982) Transmission electron microscopy of horse embryos 3–16 days after ovulation. *J. Reprod. Fert., Suppl.* **32**, 319–327.

Hastings R.A., II & Enders, A.C. (1975) Junctional complexes in the preimplantation rabbit embryo. *Anat. Rec.* **181**, 17–34.

Hayat, M.A. (1981) Cytological staining. In *Principles and Techniques of Electron Microscopy*, pp. 301–436. University Park Press, Baltimore.

McManus, J.F.A. & Mowry, R.W. (1960) Osmium tetroxide staining. In *Staining Methods*, pp. 111–113. Hoeber, New York.

Mohr, L.R. & Trounson, A.O. (1981) Structural changes associated with freezing of bovine embryos. *Biol. Reprod.* **25**, 1009–1025.

Oguri, N. & Tsutsumi, Y. (1972) Non-surgical recovery of equine eggs, and an attempt at non-surgical egg transfer in horses. *J. Reprod. Fert.* **31**, 187–195.

Slade, N.P., Takeda, T., Squires, E.L., Elsden, R.P. & Seidel, G.E., Jr (1985) A new procedure for the cryopreservation of equine embryos. *Theriogenology* **24**, 45–57.

Takeda, T., Elsden, R.P. & Squires, E.L. (1984) In vitro and in vivo development of frozen thawed equine embryos. *Proc. 10th Int. Congr. Anim. Reprod. & A.I., Urbana-Champaign* Vol. II, pp. 246–247.

Weiss, L. (1983) The male reproductive system. In *Histology: Cell and Tissue Biology*, pp. 1000–1051. Elsevier, New York.

Wilson, J.M., Kreider, J.K., Potter G.D. & Pipkin, J.L. (1985) Repeated embryo recovery from two-year-old fillies. *Proc. 9th Symp. Equine Nutr. Physiol.* pp. 398–401.

Wilson, J.M., Caceci, T., Kraemer, D.C., Potter, G.D. & Neck, K.F. (1986a) Hatching of the equine embryo: An electron microscopy study. *Biol Reprod.* **34** (Suppl. 1), 101, Abstr.

Wilson, J.M., Kraemer, D.C., Potter, G.D. & Welsh T.H., Jr (1986b) Non-surgical transfer of cryopreserved equine embryos to pony mares treated with exogenous progestin. *Theriogenology* **25**, 217, Abstr.

Yamamoto, Y., Oguri, N., Tsutsumi, Y. & Hachinohe, Y. (1982) Experiments in the freezing and storage of equine embryos. *J. Reprod. Fert., Suppl.* **32**, 399–403.

J. Reprod. Fert., Suppl. **35** (1987), 419–427

PGF-2α release, progesterone secretion and conceptus growth associated with successful and unsuccessful transcervical embryo transfer and reinsertion in the mare

J. Sirois*, K. J. Betteridge† and A. K. Goff

Centre de recherche en reproduction animale Faculté de médecine vétérinaire, Université de Montréal, C.P. 5000, St-Hyacinthe, Québec, Canada J2S 7C6

Summary. The outcome of 23 collections and reinsertions of conceptuses on Days 10·5–13·5, 4 transfers of Day-10·5, and 13 transfers of Day-6.5 embryos (ovulation = Day 0) was monitored in 30 mares. Blood samples were taken before and after each procedure to measure plasma 15-keto-13,14-dihydroprostaglandin F-2α (PGFM), and then daily for progesterone determinations. Mares were also subjected to daily teasing for detection of oestrus, and to uterine ultrasonography for tracing the development of the conceptus. After the reinsertions, 12/23 conceptuses were detectable immediately after the procedure. Of these, 1 developed into a normal foal but 11 disappeared during the next 4 days. All 4 Day-10·5 conceptuses transferred disappeared within 7 days. Of the 13 Day-6·5 conceptuses transferred, 8 were detectable by Day 11·5 and 5 of these developed into healthy foals. The 3 losses occurred between Days 14 and 25. Small increases in PGFM concentrations occurred in 8/23 collection–reinsertion procedures on Days 10·5–13·5 and after 2/25 Day-6·5 recoveries, but during none of the 13 Day-6·5 transfers. A marked reduction (≥ 40%) in progesterone concentration was noted within 24 h of the procedures in only one Day-10·5 embryo transfer and, in subsequent days, the disappearance of the conceptuses was associated with complete luteolysis in only 1 Day-6·5 transfer, 1 Day-10·5 transfer and 1 Day-12·5 reinsertion. Conceptus growth up to the time of disappearance after reinsertion was normal in 6 mares and retarded in 3; after Day-10·5 transfers it was normal; after the 3 Day-6·5 transfers that were followed by embryonic loss, growth was normal in 2 and retarded in 1 mare. It appears unlikely that PGF-2α release or major reductions in progesterone secretion are implicated in the failures of reinsertions, suggesting that disturbances in the uterine environment or direct adverse effects on the conceptus, which do not usually affect its growth, are more likely to blame.

Introduction

The reasons for wishing to recover conceptuses ('embryos') and then reinsert them into the uterus of the same mare, as well as the conspicuous lack of success in efforts to do so, have been recently discussed (Betteridge *et al.*, 1985; Bowen *et al.*, 1985). An understanding of the reasons for these failures could not only help to counteract them to make reinsertion possible, but could also contribute to our understanding of the pregnancy losses that occur spontaneously and after embryo transfer.

---

*Present address: 829 Veterinary Research Tower, Department of Physiology, Cornell University, Ithaca, NY 14853, U.S.A.
†Present address: Department of Biomedical Sciences, Ontario Veterinary College, University of Guelph, Guelph, Ontario, Canada N1G 2W1.

The course of a pregnancy temporarily interrupted by removal and replacement of the conceptus, or initiated by embryo transfer, can be monitored by the use of β-scan echography (Palmer & Driancourt, 1980; Ginther, 1985). Maternal endocrinological changes can be followed by measuring 15-keto-13,14-dihydroprostaglandin F-2α (PGFM) in paired blood samples to detect release of prostaglandin (PG) F-2α from the uterus during the collection and reinsertion procedures, and by assaying plasma concentrations of progesterone as a measure of luteal activity (Betteridge et al., 1985).

In the present study, all these methods have been used in mares subjected to three procedures: (1) conceptus removal and reinsertion on Days 10·5–13·5 of pregnancy (ovulation = Day 0); (2) transfer of Day-10·5 embryos; (3) transfer of Day-6·5 conceptuses. The first procedure was a further evaluation of the feasibility of conceptus reinsertion, using developmental stages more advanced than hitherto because of the lack of success earlier in pregnancy and because previous work suggested that PGF-2α release was less likely to be induced by uterine flushing of pregnant mares after Day 10·5 than before (Betteridge et al., 1985). The second procedure was to monitor the fate of similarly advanced conceptuses transferred into a uterus that had not been subjected to flushing. The third procedure served two purposes: (1) it determined whether the methods of embryo collection, handling and transfer that were used were adequate to produce pregnancies in more normal circumstances and (2) it allowed comparison of patterns of embryonic loss after normal (Day 6·5) and late (Day 10·5) transfers.

Materials and Methods

Animals and their management

Experiments 1 and 2 (see below) were performed in a small experimental herd of 10–26 mares during three breeding seasons (1984–1986). The general and reproductive management of this herd, composed mainly of light horse and Pony-type mares has been described previously (Betteridge et al., 1985).

Experiment 3 was conducted during 1985 with 5 donor mares, 11 recipients and 1 stallion transported to this laboratory from an equestrian centre especially for embryo transfer. They were Thoroughbred and crossbred horses weighing 475–650 kg. Oestrous cycle synchronization between donor and recipient mares was achieved by subcutaneous injections of 5 mg PGF-2α (Lutalyse; Upjohn, Kalamazoo, MI).

In all mares, oestrus was detected by daily teasing with a vigorous stallion. During oestrus, ovaries were palpated once a day to detect ovulation and donor mares were inseminated naturally or artificially on alternate days, starting when a 2·5-mm follicle was apparent and ending when ovulation had occurred. Mares from which conceptuses were to be recovered after Day 9 were examined by β-scan echography to confirm pregnancy and to locate the position of the conceptus.

Conceptus collections

Mares to be flushed were brought into a chute, sedated with xylazine (Rompun; Haver Lockhart; 0·55 mg/kg i.v.) and their perineal regions were thoroughly washed. All embryos were recovered by transcervical uterine flushes with rigid acrylic catheters as previously described (Betteridge et al., 1985). The flushing medium was phosphate-buffered saline (PBS) pH 7·2, containing 100 i.u. penicillin/ml and 0·1 μg streptomycin/ml and warmed to 38°C before use. This medium was supplemented with 2% heat-inactivated fetal calf serum (FCS, Gibco, Chagrin Falls, OH, U.S.A.) for the recovery of embryos destined for transfer but not when reinsertion was planned. With mares scheduled for a conceptus reinsertion, special care was taken to use as little medium as possible and, if feasible, to confine the flush to the uterine horn containing the conceptus. During collections on Days 10·5–13·5, flushing was stopped as soon as the readily visible conceptus was seen in the recovered medium. The conceptuses that were to be reinserted were kept in the recovery medium whereas those to be transferred were washed in three consecutive changes of fresh medium supplemented with 20% FCS. For Day-6·5 embryo collection, 4–5 samples of 240–480 ml flushing medium were introduced and collected separately in covered beakers. About 90% of the total volume recovered was siphoned off using an inverted filter (pore size; 75 μm) to prevent loss of the embryo. The remaining 10% was decanted into Petri dishes and searched under a stereomicroscope. Once identified, the embryo was washed three times with fresh PBS containing 20% FCS.

Experiments

Experiment 1. Twenty-three conceptuses (12 at Day 10·5, 2 at Day 11·5, 8 at Day 12·5 and 1 at Day 13·5) were transcervically reinserted into their mother shortly (within 20 min) after being recovered. They were reintroduced in

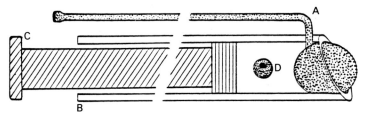

Fig. 1. Catheter used for the reinsertion or transfer of large embryos. (A) modified Foley catheter; (B) rigid acrylic tube; (C) piston fitted with a Silastic plunger; (D) embryo. For details, see text.

the uterus either via the recovery catheter which was left in place (9 mares) or with another acrylic catheter designed for the purpose (Fig. 1). The latter comprised a piston fitted with a Silastic (Dow Corning, Midland, MI) plunger inside an acrylic tube (length: 60 cm, o.d. 13 or 20 mm). At the anterior end of the tube, a 3-mm hole drilled through the wall allowed passage of a 12 FG modified Foley catheter which, when inflated, occluded the tube to retain the conceptus. The loaded catheter was also protected from contamination by a sterile polyethylene sleeve until it was introduced into the cervix under manual intravaginal guidance. With the aid of transrectal manipulation, the catheter was passed from there up to the base of one uterine horn where the cuff was deflated, the Foley catheter withdrawn and the conceptus gently expelled in the uterus. Mares in which the blastocyst was obviously damaged during the procedure were discarded from further consideration and given 5 mg PGF-2α to induce their return to oestrus.

Experiment 2. Four Day-10·5 conceptuses were transcervically transferred into recipient mares on Days 9·5, 10·5 or 11·5 (1, 2 and 1 mare respectively) using the same special reinsertion catheter.

Experiment 3. Thirteen Day-6·5 embryos were transcervically transferred into recipients that had ovulated within 2 days of the donors by using a 0·25-ml straw in a stainless-steel embryo-transfer gun designed for cattle (Hannover cannula; Walter Worrlein, Germany). During its introduction through the vagina and into the cervix, the gun was protected with a rigid plastic sheath to prevent contamination. The embryo was deposited either at the anterior part of the uterine body or at the base of one uterine horn. Recipient mares that did not appear to be pregnant during three echographic examinations, generally on Days 10·5, 11·5 and 12·5, were treated with PGF-2α to induce oestrus.

Monitoring procedures

Uterine prostaglandin release. To monitor possible uterine PGF-2α release induced during embryo recoveries, transfers or reinsertions, the plasma concentrations of its metabolite (PGFM) were measured in paired blood samples taken before and within 2 min after the procedure (Betteridge *et al.*, 1985). For embryo reinsertions, these paired samples spanned both the collection and the reinsertion. PGFM concentrations were considered to have increased significantly when the concentration of the metabolite in the 'after' sample exceeded the mean concentration in all 'before' samples by two standard deviations.

Luteal phase. After a transfer or a reinsertion, recipient mares were teased daily to detect oestrus. Mares were also bled daily for plasma progesterone determinations until either their return to oestrus or at least Day 25. After Day 25, non-pregnant mares were treated with 5 mg PGF-2α to induce oestrus and then assigned to another experiment.

Ultrasonography. Pregnancy diagnosis and the outcomes of reinsertions and transfers were monitored with a real-time linear array ultrasound scanner equipped with a 5 MHz intrarectal probe (model LS 300; Equisonics Inc., Elk Grove, IL, U.S.A.). The diameter, position and development of the conceptus were evaluated daily from Day 9 or 10 to Day 25 and recorded photographically on 35 mm black and white panchromatic film, ASA 125. Growth between Days 10·5 and 15·5 was considered to be retarded if the diameter of a conceptus was more than 2 standard deviations less than the mean diameter of other conceptuses measured during normal pregnancies (Fig. 2).

Blood sampling and hormonal assays. All blood samples (7 ml) were collected from a jugular vein into evacuated tubes containing EDTA as anticoagulant (Vacutainers; Becton-Dickinson, Rutherford, NJ, U.S.A.). They were immersed in an iced water bath until centrifugation which generally occurred within 1 h after collection. The separated plasma was stored at −20°C until radioimmunoassayed.

Aliquants (300 μl) of plasma were assayed for PGFM by the specific radioimmunoassay described by Kindahl *et al.* (1976, 1982). The sensitivity of the assay was 20 pg/ml. The intra- and inter-assay coefficients of variation were 12·9% and 12·4% respectively.

For progesterone determinations aliquants of 30 μl plasma were extracted with petroleum ether (recovery rate 76–92%) and assayed as described previously (Betteridge *et al.*, 1985). In brief, the antibody used was raised in rabbits against 11-hydroxyprogesterone 11-hemisuccinate/bovine serum albumin conjugate (Sigma Chemical Co., St Louis, MO, U.S.A.). The sensitivity of the assay was 240 pg/ml and the intra- and inter-assay coefficients of variation were 13% and 19·3% respectively.

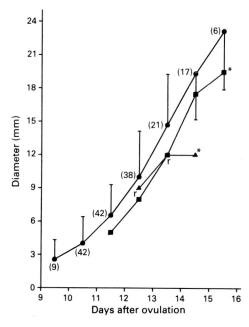

Fig. 2. Two examples of conceptus growth after reinsertion (r). ■, Normal growth after reinsertion of a Day-13·5 conceptus (No. 7 in Table 1); ▲, retarded growth after reinsertion of a Day-12·5 conceptus (No. 1 in Table 1); ●, control growth curve (mean ± 2 s.d.; numbers in parentheses indicate the number of conceptuses for each day). The asterisks denote the last days on which the reinserted conceptuses were detected.

Results

Experiment 1

Of the 23 conceptuses reinserted, 12 (4 Day-10·5, 1 Day-11·5, 6 Day-12·5 and 1 Day-13·5) remained intact during the procedure as confirmed by ultrasonography immediately afterwards. To effect these recoveries, 1, 2 or 5 flushes were required for 6, 5 and 1 mare, respectively, and in only 1 mare did the volume exceed 600 ml (Table 1). The 11 conceptuses that collapsed during the procedure were discarded from the study. Of the 12 intact reinserted conceptuses, only one developed normally and gave rise to a healthy filly. The remaining 11 blastocysts disappeared during the 1st, 2nd, 3rd or 4th subsequent days (2, 3, 5 and 1 mares respectively). Conceptus growth before disappearance was normal in 6 mares and retarded in 3 (Table 1) as exemplified in Figs 2 and 3. Nine of the recipients of these 11 conceptuses were teased and bled daily to determine the time of their return to oestrus and their progesterone profiles; 8 had a prolonged luteal phase (no return to oestrus before Day 24·5) and one had a normal oestrous cycle (interovulatory interval of 20 days). There were three types of progesterone profile associated with prolonged luteal phases: a slow decline of progesterone concentrations (2 mares), a constant high level (4 mares) and a decline followed by an increase without return to oestrus (2 mares) (Fig. 4). None of these mares showed a decrease in progesterone concentrations of 40% or more on the day after the reinsertion.

The mean (±s.d.) concentration of PGFM in all 'before' samples was 24·1 ± 11 pg/ml. PGFM concentrations increased significantly (>46·1 pg/ml) during 8 of the 23 collection-reinsertions (Table 2). In these 8 mares, mean (±s.d.) PGFM concentration in 'after' samples was 62·5 ± 12·6 pg/ml (range 48·5–84·4 pg/ml).

Table 1. Characteristics of the recovery and fate of the 12 conceptuses remaining intact during reinsertion in Exp. 1

Conceptus no.	Mare no.	Day	No. of lavages (vol. ml) for recovery	Diameter (mm)	Growth before disappearance*	Day of disappearance (day after reinsertion)	Oestrous cycle length*
1	9	12·5	2 (600)	9	R	15·5 (3rd)	P
2	11	12·5	1 (300)	11	N	15·5 (3rd)	N
3	26	12·5	5 (1190)	11	N	14·5 (2nd)	P
4	26	11·5	2 (480)	10	—	12·5 (1st)	P
5	75	12·5	2 (400)	11	R	15·5 (3rd)	P
6	9	12·5	2 (480)	12	N	15·5 (3rd)	P
7	9	13·5	1 (240)	12	N	16·5 (3rd)	P
8	78	12·5	1 (300)	4	R	16·5 (4th)	P
9	72	10·5	2 (600)	4	N	12·5 (2nd)	P
10	9	10·5	1 (300)	4	Normal pregnancy established		
11	15	10·5	1 (240)	4	—	11·5 (1st)	PGF-2α
12	74	10·5	1 (240)	4	N	12·5 (2nd)	PGF-2α

*R = retarded; N = normal; P = prolonged.

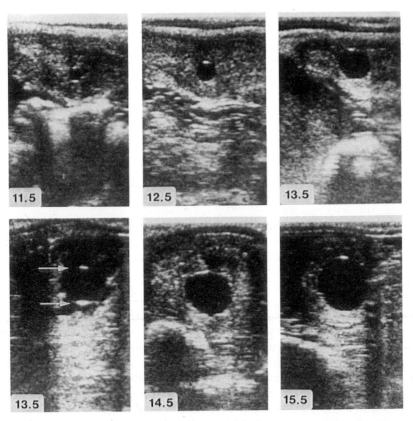

Fig. 3. The growth of conceptus No. 7 (Table 1) before and after reinsertion on Day 13·5. Immediately after reinsertion, the conceptus (delimited by arrows) can be seen to be floating in a small volume of medium left in the uterus after the recovery. Despite normal growth up to Day 15·5, the conceptus had disappeared by Day 16·5 (see also Fig. 2).

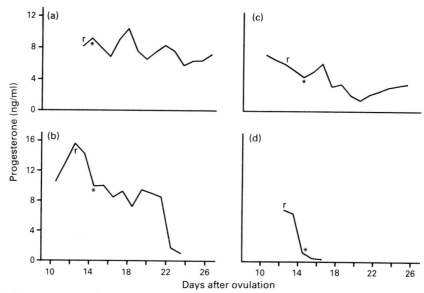

Fig. 4. Four representative plasma progesterone profiles after embryo reinsertion (r). Profiles (a), (b) and (c) were from mares exhibiting prolonged oestrous cycles associated with progesterone concentrations that were maintained at high levels (a, conceptus No. 7 in Table 1), declined slowly (b, conceptus No. 5), or declined and then increased again without a return to oestrus (c, conceptus No. 6). Profile (d) (conceptus No. 2) depicts normal luteolysis followed by a return to oestrus on Day 15·5. The asterisks denote the last days on which the conceptuses were seen.

Table 2. Relationships between PGF-2α release (reflected as increased plasma PGFM concentrations) during embryo transfer and reinsertion and plasma progesterone concentrations 24 h later

| | Day-6·5 embryo | | Day-10·5 embryo | | Day-10·5–13·5 embryo |
	Recovery	Transfer	Recovery	Transfer	Recovery and reinsertion
Proportion of mares with PGF-2α release	2/25	0/13	1/4	0/4	8/23
Proportion of mares with a decrease in progesterone of ⩾40%	—	0/11	—	1/4	0/9

Experiment 2

No normal pregnancy was established following the transfer of four Day-10·5 conceptuses. All 4 were intact at ultrasonography immediately after the transfer but they disappeared during the 1st, 3rd, 4th or 7th day after transfer (Table 3). Growth before disappearance was normal in the 3 mares in which it could be evaluated. Three recipient mares showed a prolonged luteal phase; two with no return to oestrus before Day 25 and one with a return to oestrus at Day 21 and an interovulatory interval of 28 days. In these mares, plasma progesterone concentrations remained relatively constant for at least 6 days after the transfers. In the fourth recipient, a 60% decrease in progesterone concentration was noted 24 h after transfer and this mare had a normal oestrous cycle (interovulatory interval of 21 days). In this 4th mare, the disappearance of the conceptus (on Day 14·5) was associated with luteolysis. A significant increase in PGFM concentration occurred during one recovery and none of the transfers (Table 2).

Table 3. Relationship between recipient–donor synchrony and the detection of conceptuses by ultrasonography on various days after transfer

| Synchrony (days)* | Day-6·5 transfers | | | | | Day-10·5 transfers | | | | |
| | Transfers | Recipients with conceptus on Day: | | | | Transfers | Recipients with conceptus on Day: | | | |
		11·5	15·5	20·5	25·5		11·5	15·5	20·5	25·5
−2	2	1	1	1	1	—	—	—	—	—
−1	4	3	3	3	3	1	1	1	0	0
0	4	1	1	1	1	2	1	0	0	0
+1	1	1	1	1	0	1	1	0	0	0
+2	2	2	1	1	0	—	—	—	—	—
Total	13	8	7	7	5	4	3	1	0	0

*Recipients ovulating after (minus sign) or before the donor (plus sign).

Experiment 3

Of 13 Day-6·5 embryos transferred, 8 (61%) developed and were detected by ultrasonography on Day 11·5. However, of these, 3 were lost on Days 14·5, 22·5 and 24·5 in recipient mares which had ovulated 2, 1 and 2 days respectively before the donors (Table 3). The others continued to term and gave rise to 5 healthy foals. Conceptus mobility (Days 10·5–15·5), fixation (Day 15–16) and orientation for the 5 normal pregnancies were observed to follow the course described by Ginther (1985). In the 3 cases of embryonic loss, growth of the 2 conceptuses that disappeared on Days 14·5 and 22·5 was normal. However, in the one that vanished on Day 24·5, growth was retarded and its diameter was reduced by 7 mm between Days 19·5 and 22·5. Moreover, an embryo was not detected inside this embryonic vesicle on Days 20·5 and 22·5. On Day 24·5, neither vesicle nor liquid was observed inside the uterus. In another case (the Day-22·5 loss), an apparently collapsed embryonic vesicle and free liquid were observed in the uterine lumen on the morning of the loss but had completely disappeared by the afternoon.

Progesterone concentrations remained constant after transfer and up to Day 25 in the 5 mares in which a normal pregnancy was established. In the 2 in which embryonic loss occurred and for which blood samples were available, progesterone remained constant in the one in which pregnancy terminated on Day 22·5; in the mare that lost her conceptus on Day 24·5, progesterone concentrations fell by 30% on the 4th day after the transfer but were nevertheless maintained at about 12 ng/ml until Day 24·5. In the third case of embryonic loss, the mare was in oestrus on the day the conceptus disappeared (Day 14·5), suggesting that luteolysis had occurred although blood samples for progesterone analysis were unfortunately not taken. For the 5 mares in which no pregnancy could be diagnosed by ultrasonography, 4 of them were monitored and showed normal and constant progesterone levels for the 6 days after the transfer, at which time they were treated with PGF-2α to re-induce oestrus. A significant increase in PGFM concentrations was observed after only 2/25 recoveries and during none of the transfers (Table 1).

Discussion

The birth of the first foal after an embryo reinsertion indicates that this technique can be successful. However, the very high failure rate (22/23) will require correction if reinsertion is to become practical. These failures seem not to be due to the general techniques and materials used (such as catheters, media and manipulative abilities) since 5 healthy foals were born after transfers of Day-6·5 embryos via a similar transcervical approach. From PGFM and progesterone assays, it seems unlikely that either uterine PGF-2α release during embryo collection–reinsertion or luteal deficiency is implicated in most cases. A significant but small increase of PGFM occurred in only a minority of mares (8 out of 23) and the levels attained (48·5–84·4 pg/ml) were below the peaks usually associated with luteolysis (200–500 pg/ml) (Kindahl *et al.*, 1982). Furthermore, no marked progesterone decrease (⩾40%) after reinsertion was noted within 24 h in any of the mares monitored and, in subsequent days, the disappearance of the conceptus was associated with complete luteolysis in only one mare. However, it remains possible that pregnancy is vulnerable to more subtle changes in progesterone secretion than would be detected in daily samples. Plasma progesterone concentrations have been reported to differ significantly between pregnant and non-pregnant mares by Days 9–13 (Forde *et al.*, 1987) and embryonic loss associated with uterine inflammation is usually accompanied by an earlier decline in progesterone concentrations than are pregnancy losses without inflammation (Adams *et al.*, 1987). The prolonged luteal phases observed after most (8/9) embryo collections and reinsertions cannot be presumed to be due to the continued presence of a healthy conceptus because similar prolongations and progesterone profiles can follow collection alone, although perhaps less frequently (Betteridge *et al.*, 1985).

Other causes of pregnancy failure could be the direct effects of the disturbance of the intra-uterine environment and/or trauma inflicted on the conceptus during the procedure. Considering that growth of reinserted conceptuses appeared to be normal before disappearance in most cases (6/9), these putative deleterious influences may exert their effects only some days after reinsertion. For example, they might affect steroid production by the conceptus which is thought to be important for the maintenance of pregnancy (Goff *et al.*, 1983; Marsan *et al.*, 1987). As previously suggested (Bowen *et al.*, 1985), removal of histotrophe (Zavy *et al.*, 1982) by the flushing procedure could also critically affect the development of the embryo. When Day-10·5 conceptuses were transferred into 'undisturbed' uteri of recipient mares, the patterns of pregnancy failure resembled those seen after reinsertion. Poor success rates with transfers after Day 8 have been reported previously (Squires *et al.*, 1982) but the delay between transfer and loss has not been described before. The failure of those late transfers is difficult to interpret. Although histotrophe removal can be excluded as a factor, an absence of histotrophe could still be involved if its secretion depends on the presence of an embryo before Day 10·5. It is therefore not possible to resolve whether failure of Day-10·5 reinsertions is due to a direct effect on the uterine environment, the conceptuses or both.

Uterine PGF-2α release was also rare during Day-6·5 embryo recoveries (2/25) and transfers (0/13), which extends a previous report of no release during 8 sham transfers between Days 7 and 9 (Betteridge *et al.*, 1985). Neither did progesterone decrease by ⩾40% in the 24 h after any of the 11 transfers monitored. In 4 transfers in which no pregnancy was diagnosed by ultrasonography, constant progesterone concentrations indicated that luteolysis could not have been responsible for the lack of embryonic development.

The patterns of losses observed by ultrasonography after transfer and reinsertion, such as the disappearance of an apparently normal growing conceptus, reduction of its diameter before the loss, and the lack of development of the embryo proper inside the embryonic vesicle, have been previously reported to occur in normally mated mares (Ginther *et al.*, 1985) and show that ultrasonographic images must be interpreted with care. The increase in the diameter of the early conceptus of the horse might not necessarily reflect its viability. For example, an intact healthy trophoblast might allow blastocyst expansion and maintenance of its spherical shape for some time within the capsule even though the embryonic disc is irreversibly damaged. Trophoblastic vesicles obtained from disc-free cattle and sheep embryos and transferred into recipients can develop and prolong the luteal phase (Heyman *et al.*, 1984). The possibility that the horse conceptus is damaged by the reinsertion procedure could be investigated by recovering it a second time for evaluation at various intervals after reinsertion.

The way in which the conceptuses of normal appearance disappeared during the 24-h interval between two ultrasonic examinations was not clear. It would be surprising if it happened via expulsion through a closed cervix. Rupture of the conceptus and its capsule seems a more likely possibility; the blastocoelic fluid would be rapidly resorbed by the endometrium, leaving tissue that could not be detected by ultrasonography. Uterine flushes after embryo disappearance could be used to test this hypothesis.

Since Day-10·5 reinsertions were largely unsuccessful and since Day-6·5 embryos can be recovered and transferred so successfully, embryo reinsertion at the earlier stage should be re-investigated despite previous failures. The suggested deleterious effect of the uterine flushes might be overcome by extending the interval between recovery and reinsertion to allow the re-establishment of a more optimal uterine environment. During that time, it should be possible to culture the embryo *in vitro* (Clark *et al.*, 1985) or keep at 4°C (Pashen, 1987) for up to 24 h without compromising its viability.

We thank Dr H. Kindahl, Swedish University of Agricultural Sciences, for the generous gift of the antibody against 15-keto-13,14-dihydro PGF-2α; the animal care staff of the CRRA for help with the horses; and Sylvie Lagacé and Hélène Boucher for typing the manuscript. This work was supported by NSERC, Canada.

References

Adams, G.P., Kastelic, J.P., Bergfelt, D.R. & Ginther, O.J. (1987) Effect of uterine inflammation and ultrasonically-detected uterine pathology on fertility in the mare. *J. Reprod. Fert., Suppl.* **35**, 445–454.

Betteridge, K.J., Renard, A. & Goff, A.K. (1985) Uterine prostaglandin release relative to embryo collection, transfer procedures and maintenance of the corpus luteum. *Equine vet. J. Suppl.* **3**, 25–33.

Bowen, J.B., Salsbury, J.M., Bowen, J.M. & Kramer, D.C. (1985) Non-surgical auto-transfer in the mare. *Equine vet. J. Suppl.* **3**, 100–102.

Clark, K.E., Squires, E.L., Takeda, Y. & Seidel, G.E. (1985) Effect of culture media on viability of equine embryos *in vitro. Equine vet. J. Suppl.* **3**, 35.

Forde, D., Keenan, L., Wade, J., O'Connor, M. & Roche, J.F. (1987) Reproductive wastage in the mare and its relationship to progesterone in early pregnancy. *J. Reprod. Fert., Suppl.* **35**, 493–495.

Ginther, O.J. (1985) Dynamic physical interactions between the equine embryo and uterus. *Equine vet. J. Suppl.* **3**, 41–47.

Ginther, O.J., Bergfelt, D.R., Leith, G.S. & Scraba, S.T. (1985) Embryonic loss in mares: incidence and ultrasonic morphology. *Theriogenology* **24**, 73–86.

Goff, A.K., Renard, A. & Betteridge, K.J. (1983) Regulation of steroid synthesis in pre-attachment equine embryos. *Biol. Reprod.* **28**, Suppl. **1**, 137, Abstr.

Heyman, Y., Camous, S., Fèvre, J., Méziou, W. & Martal, J. (1984) Maintenance of the corpus luteum after uterine transfer of trophoblastic vesicles to cyclic cows and ewes. *J. Reprod. Fert.* **70**, 533–540.

Kindahl, H., Edqvist, L.-E., Granström, E. & Bane, A. (1976) The release of prostaglandin $F_{2\alpha}$ as reflected by 15-keto-13,14-dihydro-prostaglandin $F_{2\alpha}$ in the peripheral circulation during normal luteolysis in heifers. *Prostaglandins* **11**, 871–878.

Kindahl, H., Knudsen, O., Madej, A. & Edqvist, L.-E. (1982) Progesterone, prostaglandin F-2α, PMSG and oestrone sulphate during early pregnancy in the mare. *J. Reprod. Fert., Suppl.* **32**, 353–359.

Marsan, C., Goff, A.K., Sirois, J. & Betteridge, K.J. (1987) Steroid secretion by different cell types of the horse conceptus. *J. Reprod. Fert., Suppl.* **35**, 363–369.

Palmer, E. & Driancourt, M.A. (1980) Use of ultrasonic echography in equine gynecology. *Theriogenology* **13**, 203–216.

Pashen, R.L. (1987) Short-term storage and survival of horse embryos after refrigeration at 4°C. *J. Reprod. Fert., Suppl.* **35**, 697–698.

Squires, E.L., Imel, K.J., Iuliano, M.F. & Shideler, R.K. (1982) Factors affecting reproductive efficiency in an equine embryo transfer programme. *J. Reprod. Fert., Suppl.* **32**, 409–414.

Zavy, M.T., Sharp, D.C., Bazer, F.W., Fazleabar, A., Sessions, F. & Roberts, R.M. (1982) Identification of stage-specific and hormonally induced polypeptides in the uterine protein secretions of the mare during the oestrous cycle and pregnancy. *J. Reprod. Fert.* **64**, 199–207.

J. Reprod. Fert., Suppl. **35** (1987), 429–432

Printed in Great Britain
© 1987 Journals of Reproduction & Fertility Ltd

Exogenous hormone regimens to utilize successfully mares in dioestrus (Days 2–14 after ovulation) as embryo transfer recipients

K. F. Pool, J. M. Wilson, G. W. Webb, D. C. Kraemer*, G. D. Potter and J. W. Evans

Texas Agricultural Experiment Station, Departments of Animal Science & *Veterinary Physiology and Pharmacology, Texas A & M University, College Station, TX 77843, U.S.A.

Summary. Two hormone regimens were utilized for recipient mares which were 2–14 days after ovulation at the time of non-surgical embryo transfer. In Exp. I, 20 embryos were transferred non-surgically into recipient mares which had been given 22 mg altrenogest daily starting the day of recipient ovulation. Higher ($P < 0.05$) pregnancy rates (50% *vs* 0%) were obtained in mares which were 2–6 days after ovulation at the time of transfer compared with mares which were 7–12 days after ovulation. In Exp. II, on the day the donor mare ovulated (Day 0), 10 mg PGF-2α were given to the recipient mare at 0 h, 20 mg oestradiol cypionate at 12, 24 and 36 h, 500 mg progesterone in oil at 48 h and then 22 mg altrenogest at 60, 72 and 96 h. Altrenogest (22 mg/day) was continued until Day 28 (fetal heart beat). The pregnancy rate (58% *vs* 10%) was higher ($P < 0.05$) for the 12 recipient mares which were 10–14 days after ovulation than for the 10 mares that were 5–8 days after ovulation when treatment began. Early embryonic loss was detected in 4/8 pregnant recipient mares between Days 21 and 28 of gestation. We suggest that mares which are 2–14 days after ovulation and not in ovulation synchrony with the donor may be used as embryo transfer recipients; mares which are in early dioestrus can be given altrenogest while those in mid- to late dioestrus can be placed on the more complex hormone regimen.

Introduction

A major problem in the development of embryo transfer procedures in horses has been the synchronization of ovulation between donor and recipient mares. Pregnancy rates have been highest in recipient mares that have ovulated on the same day or within 2 days after the donor mare (Douglas, 1982; Iuliano *et al.*, 1985). Because of the difficulties encountered in synchronizing mares, embryo transfer programmes usually attempt to synchronize at least 2–4 recipients each time a donor mare is bred.

The objective of this study was to develop exogenous hormone regimens to utilize as embryo recipients mares whose ovulations were not synchronized with those of the donor mare.

Materials and Methods

Embryos were recovered on Day 6 or 7 after ovulation using a non-surgical technique as described by Wilson *et al.* (1985). Recovered fluid was filtered through an embryo concentrating filter (Immuno Systems, Inc., Kennebunk, ME) and poured into a gridded Petri dish which was scanned with a dissecting microscope at ×10 magnification. Embryos were washed 5 times in sterile (millipore filtered) holding medium (Dulbecco's phosphate-buffered saline supplemented 10% (v/v) with heat-treated gelding serum). Embryos were placed into a sterile 0·5 ml bovine artificial insemination straw as described by Squires *et al.* (1982). Embryos were transferred with a bovine artificial insemination pipette by the guarded method as described by Iuliano *et al.* (1985). Non-surgical transfers were performed immediately after embryo recovery.

Recipient mares were randomly selected from a group of maiden, parous and non-lactating and post-partum Quarterhorse-type mares. All mares were teased individually in stalls, or in a large teasing pen with the stallion confined to a small teasing cage. Rectal palpation and ultrasonography were used to monitor follicular development on a daily basis and to detect ovulations.

Experiment I. Twenty embryos were transferred non-surgically into recipient mares which were given 22 mg altrenogest (Regu-mate: Hoechst, Sommerville, NJ) daily beginning the day the recipient mare ovulated. At 2–12 days after the start of treatment, mares received Day-6 or -7 embryos. Altrenogest (given orally) was continued until pregnancy detection (Days 14–16) by rectal palpation and ultrasonography. If the mare was pregnant, altrenogest treatment was continued until detection of a fetal heartbeat by ultrasonography (Day 28).

Experiment II. Embryos (22) were transferred non-surgically into dioestrous mares (Days 5–14 after ovulation) which had been placed on an exogenous hormone regimen beginning the day the donor mare ovulated. The treatment schedule was as follows: (treatment Day 0) 0 h, 10 mg PGF-2α (Upjohn, Kalamazoo, MI); then, at 12, 24 and 36 h, oestradiol-17β (20 mg oestradiol cypionate: Tech America, Elwood, KS) was given. At 48 h, 500 mg progesterone in oil (Butler, Columbus, OH) were administered intramuscularly, followed by 22 mg altrenogest orally at 60, 72 and 96 h after PGF-2α. Altrenogest (22 mg/day) was continued until initial pregnancy detection (Days 14–16) and, if pregnant, until detection of a fetal heartbeat by ultrasonography (Day 28). Throughout the treatment regimen, mares were observed for oestrous behaviour and monitored for follicular development.

The effect of treatment on pregnancy rates was analysed by χ^2 analysis (Daniel, 1978).

Results and Discussion

The results of Exp. I are summarized in Table 1. Embryos at 6 or 7 days were transferred into recipient mares which were 2–12 days after ovulation and had been maintained on altrenogest since the day of ovulation. A higher ($P < 0.05$) pregnancy rate (50% *vs* 0%) was obtained in mares which were 2–5 days after ovulation than in mares which were 7–12 days after ovulation. These pregnancy rates are in agreement with previous studies reported for ovulation synchronization periods between donor and recipient mares for successful embryo transfer. Previous attempts to utilize mares which ovulated 2 or more days before the donor mare have been unsuccessful (Oguri & Tsutsumi, 1980). Reported pregnancy rates were higher in recipient mares which had ovulated from 1 day before or up to 2 days after the donor mare (Squires *et al.*, 1982; Iuliano *et al.*, 1985). Vogelsang *et al.* (1985) transferred embryos non-surgically into recipient mares which ovulated 3, 4 and 5 days after the donor mare. Pregnancy rates were 60% (6/10), 0% (0/2), and 0% (0/3), respectively. In the present study, with the administration of altrenogest from the day of recipient ovulation, 2 pregnancies were obtained from 5 embryo transfers into mares which ovulated 4 and 5 days after the donor mare (Table 1).

Table 1. Pregnancy rate by day for recipient mares receiving Day-6 or Day-7 embryos after treatment with altrenogest for various numbers of days after ovulation in the recipient

No. of days treated	No. of transfers (% pregnancy)
2	1 (100)
3	4 (25)
4	4 (75)
5	3 (33)
7	1 (0)
8	1 (0)
9	2 (0)
10	1 (0)
11	2 (0)
12	1 (0)

The results of Exp. II are summarized in Table 2. Recipient mares which were 5–14 days after ovulation were started on the hormone regimen the day the donor mare ovulated. During administration of exogenous oestrogen, all treatment mares exhibited overt oestrous behaviour. Multiple follicles of < 20 mm in diameter were found in mares which were 5–9 days after ovulation at the start of treatment. Mares which were 10–14 days after ovulation at the start of treatment developed 1 or 2 follicles ranging from 20 to 40 mm in diameter during the treatment period, but ovulation was not detected.

Pregnancy rates for mares which were 5–14 days after ovulation at the start of treatment are given in Table 2. Pregnancy rates (58% *vs* 10%) were higher ($P < 0.05$) for recipient mares which were more than 9 days after ovulation at the start of treatment than for mares which were less than 9 days after ovulation when treatment began. Early embryonic loss was detected in 4 of the 8 pregnant recipient mares between Days 121 and 28 of gestation.

Table 2. Pregnancy rate by day for recipient mares receiving Day-6 or Day-7 embryos and treated with an exogenous hormone regimen

Days from ovulation to start of treatment	No. of transfers (% pregnancy)
5	1 (0)
6	2 (0)
7	3 (33)
8	4 (0)
10	5 (40)
11	3 (67)
12	1 (100)
13	2 (50)
14	1 (100)

The synchronization of ovulation between the donor mare and recipient mare has been a major problem in the development of equine embryo transfer procedures. The use of prostaglandin F-2α or various prostaglandin analogues to induce rapid luteolysis during dioestrus in the mare has been a common practice for synchronization of oestrus (Douglas & Ginther, 1972; Allen & Rowson, 1973). Several studies have been conducted to examine the use of exogenous progestagens and oestrogens to synchronize oestrus and ovulation in the mare (Loy & Swan, 1966; Palmer, 1979; Holtan *et al.*, 1977; Loy *et al.*, 1981; Bristol *et al.*, 1983). Daily injections of exogenous hormones may be undesirable for practical use. Therefore an oral synthetic progestagen, altrenogest, has been used to control the oestrous cycle (Webel, 1975; Squires *et al.*, 1979; Allen *et al.*, 1980; Webel & Squires, 1982). These treatments were successful in altering the oestrous cycle of the mare, but exact synchrony of ovulation was not always accomplished.

In an attempt to use as embryo recipients mares whose ovulations were not synchronized with those of the donor mares, Hinrichs *et al.* (1985) established pregnancies in ovariectomized recipient mares which had been treated with exogenous progesterone.

In the present study, therefore, a combination of previously described synchronization treatments was used to prepare a mare in any stage of dioestrus for receipt of an embryo. A treatment similar to the one used for the mare in mid- and late dioestrus has been used in ovariectomized ewes which were slaughtered 6–13 days after embryo transfer (Miller *et al.*, 1977). Of 27 treated ewes, 21 carried normal, developing embryos, at the time of slaughter. In those ewes not treated with oestrogen, embryos ceased to develop normally within 1–2 days of transfer. Decreased amounts of oestrogen and progesterone receptors were found in the endometrium at the time of embryo transfer.

In conclusion, it is possible to utilize mares as recipients which are in very early dioestrus or mid- to late dioestrus with one of the two described regimens. This should decrease the number of potential recipient mares which must be maintained and manipulated each time a donor mare is bred.

Paper No. 22001 from the Texas Agricultural Experiment Station, College Station.

References

Allen, W.R. & Rowson, L.E.A. (1973) Control of the mare's oestrous cycle by prostaglandins. *J. Reprod. Fert.* **33**, 539–543.

Allen, W.R., Urwin, V., Simson, D.J., Greenwood, R.E.S., Crowhurst, R.C., Ellis, D.R., Ricketts, S.W., Hunt, M.D.N. & Digby, N.J.W. (1980) Preliminary studies on the use of an oral progestagen to induce oestrus and ovulation in seasonally anoestrous Thoroughbred mares. *Equine vet. J.* **12**, 141–145.

Bristol, F., Jacobs, K.A. & Pawlyshyn, V. (1983) Synchronization of estrus in post-partum mares with progesterone and estradiol-17β. *Theriogenology* **19**, 779–785.

Daniel, W.W. (1978) Chi-square test of independence and homogeneity. In *Applied Non-Parametric Statistics,* Vol. 1, pp. 166–169. Haughton Mifflin Co., Boston.

Douglas, R.H. (1982) Some aspects of equine embryo transfer. *J. Reprod. Fert., Suppl.* **32**, 405–408.

Douglas, R.H. & Ginther, O.J. (1972) Effect of prostaglandin F$_{2\alpha}$ on length of diestrus in mares. *Prostaglandins* **2**, 265–268.

Hinrichs, K., Sertich, P.L., Cummings, M.R. & Kenney, R.M. (1985) Pregnancy in ovariectomized mares achieved by embryo transfer: a preliminary study. *Equine vet. J.* (Suppl.) **3**, 74–75.

Holtan, D.W., Douglas, R.H. & Ginther, O.J. (1977) Estrus, ovulation, and conception following synchronization with progesterone, prostaglandin F$_{2\alpha}$ and human chorionic gonadotrophin in pony mares. *J. Anim. Sci.* **44**, 431–437.

Iuliano, M.F., Squires, E.L. & Cook, V.M. (1985) Effect of age of equine embryos and method of transfer on pregnancy rate. *J. Anim. Sci.* **60**, 258–263.

Loy, R.G. & Swan, S.M. (1966) Effects of exogenous progestogens on reproductive phenomena in mares. *J. Anim. Sci.* **25**, 821–826.

Loy, R.G., Pemstein, R., O'Canna, D. & Douglas, R.H. (1981) Control of ovulation in cycling mares with ovarian steroids and prostaglandin. *Theriogenology* **15**, 191–199.

Miller, B.G., Moore, N.W., Murphy, L. & Stone, G.M. (1977) Early pregnancy in the ewe: effects of oestradiol and progesterone on uterine metabolism and on early embryo survival. *Aust. J. biol. Sci.* **30**, 279–288.

Oguri, N. & Tsutsumi, Y. (1980) Non-surgical transfer of equine embryos. *Arch. Embryol.* **5**, 108–110.

Palmer, E. (1979) Reproductive management of mares without detection of oestrus. *J. Reprod. Fert., Suppl.* **27**, 263–270.

Squires, E.L., Stevens, W.B., McGlothin, D.E. & Pickett, B.W. (1979) Effect of an oral progestin on the estrous cycle and fertility of mares. *J. Anim. Sci.* **49**, 729–735.

Squires, E.L., Imel, K.J., Iuliano, M.F. & Shideler, R.K. (1982) Factors affecting reproductive efficiency in an equine embryo transfer programme. *J. Reprod. Fert., Suppl.* **32**, 409–414.

Vogelsang, S.G., Bondioli, K.R. & Massey, J.M. (1985) Commercial application of equine embryo transfer. *Equine vet. J.* (Suppl.) **3**, 89–91.

Webel, S.K. (1975) Estrous control in horses with a progestin. *J. Anim. Sci.* **41**, 385, Abstr.

Webel, S.K. & Squires, E.L. (1982) Control of the oestrous cycle in mares with altrenogest. *J. Reprod. Fert., Suppl.* **32**, 193–198.

Wilson, J.M., Kreider, J.L., Potter, G.D. & Pipkin, J.L. (1985) Repeated embryo recovery from two-year-old mares. *Proc. 9th Equine Nut. & Phys. Symp.* pp. 398–400.

J. Reprod. Fert., Suppl. 35 (1987), 433–438

Effect of altrenogest on pregnancy maintenance in unsynchronized equine embryo recipients*

L. C. Parry-Weeks and D. W. Holtan

Department of Animal Science, Oregon State University, Corvallis, Oregon 97331, U.S.A.

Summary. Non-surgical embryo recovery attempts were done on Day 7 after ovulation. Embryo recovery rate from mares of varied reproductive histories was 57% (38/67). Non-surgical transfer of these embryos into altrenogest-treated recipient mares that ovulated between 3 days before and 3 days after the donor resulted in a 30-day pregnancy rate of 77% (10/13). Transfer of embryos into altrenogest-treated recipients that ovulated between 4 days before and 6 days after the donor resulted in an overall pregnancy rate of 64% (16/25) at Day 30 of gestation. No recipients that were in oestrus at the start of treatment, nor recipients that ovulated 5 or more days before the donor, maintained pregnancy. Mean plasma progesterone concentrations of pregnant, altrenogest-treated, embryo-recipient mares; pregnant, altrenogest-treated, untransferred mares; and pregnant, untreated, untransferred mares were comparable ($P > 0.05$). Treatment of embryo-recipient mares with altrenogest appears to be beneficial in extending the degree of donor–recipient synchrony required for successful embryo transfer. Altrenogest treatment also seems to be conducive to pregnancy maintenance in recipients experiencing luteal dysfunction.

Introduction

It has been suggested that a major cost associated with embryo transfer in horses is the maintenance of recipient mares (Douglas *et al.*, 1985). Due to the wide variability between mares in duration of oestrus, hormone therapies for ovulation synchronization do not always achieve close synchrony (Ginther, 1979). Previously, most successful non-surgical embryo transfers have been performed when the recipient ovulated between $+1$ to -2 days of the donor (Iuliano *et al.*, 1985), and so it has often been necessary to maintain several recipient mares in hopes that one recipient will ovulate in close synchrony with the donor. This study was conducted to determine whether altrenogest (synthetic progestagen) treatment of recipient mares, at various degrees of asynchrony compared to the donor's day of ovulation, would permit successful non-surgical embryo transfer.

Materials and Methods

Mares and treatments. Ten light horse and 25 pony mares with unknown reproductive histories were used, often repeatedly over a 2-year period, in the following three treatments: Group 1, altrenogest-treated, embryo-recipient mares (N = 38); Group 2, altrenogest-treated, untransferred, pregnant mares (N = 8); and Group 3, untreated, untransferred, pregnant mares (N = 6). All mares were teased and palpated to detect oestrus, follicular development and ovulations. Embryo donor mares for Group 1 were inseminated every other day of oestrus until ovulation was detected. Treated mares were fed the synthetic progestagen, altrenogest (0·044 mg/kg body weight; allyl-trenbolone; Regu-Mate, Hoechst-Roussel Agri-Vet Co., Summerville, NJ, U.S.A.), mixed in their grain, daily. In Group 1, recipients which ovulated or those with suspected follicular luteinization were started on altrenogest treatment on the day the donor ovulated, or the day the recipient ovulated or luteinized, whichever occurred later. Recipients in oestrus were started on altrenogest treatment the day their donor ovulated. In Group 2, mares were inseminated and fed altrenogest daily starting the day that they ovulated. In Group 3, mares were inseminated, but did not receive altrenogest treatment. Jugular blood samples were collected at 2-day intervals in Groups 1 and 2, starting on Day 1 of

---

*Reprint requests to Dr D. W. Holtan.

treatment, and at 4-day intervals in Group 3, starting on the day of ovulation. Plasma was stored at $-20°C$ until assayed for progesterone concentrations.

Embryo recovery. All embryo recovery attempts were performed non-surgically on Day 7 after the donor's day of ovulation. The recovery medium and uterine flushing procedures were similar to those described by Imel *et al.* (1981), except that uteri were flushed a total of 4, rather than 3, times with Dulbecco's phosphate-buffered saline containing 1% antibiotic-antimycotic mixture and 1% heat-inactivated horse serum (Gibco, NY, U.S.A.). Embryos were rinsed three times in medium containing 10% horse serum.

Embryo transfer. The technique used for non-surgical transfer of embryos was an unguarded method, similar to artificial insemination in horses, which is also described by Imel *et al.* (1981). Embryos were deposited in the uterine body. All transfers were performed by the same technician within 2 h of recovery.

Pregnancy diagnosis. In Groups 1 and 2, each mare's uterus was scanned for pregnancy via ultrasonic echography once between Days 14 and 18 after ovulation, by age of the embryo. Those mares with positive ultrasound diagnoses, and mares in Group 3, were examined for pregnancy maintenance by rectal palpation at frequent intervals until Day 30 of gestation.

Hormone analysis. Progesterone concentrations in plasma were measured by the radioimmunoassay method validated and described by Koligian & Stormshak (1977). Mean recoveries of various quantities of progesterone (25, 50, 100 and 200 pg) added to 100 µl pooled serum obtained from geldings were: $27·9 \pm 1·8$, $63·0 \pm 9·4$, $107·5 \pm 4·3$ and $209·8 \pm 13·4$ pg/100·µl (mean + s.e.), respectively. The working range of the standard curve was between 10 and 2000 pg/tube (0 vs 10 pg; $P < 0·05$, $n = 10$). Recoveries of 90% were obtained after ether extraction of 100 µl plasma, and the effective sensitivity of the assay was 0·11 ng/ml. The intra- and interassay coefficients of variation were 6·8 and 15·8%, respectively. The antiserum used did not appear to cross-react with altrenogest.

Results and Discussion

Embryo recovery

A total of 67 embryo recovery attempts yielded 38 transferable embryos (57%). These data include two sets of twin embryos which were recovered from one light horse and one pony mare, both of which had double ovulations. This embryo recovery rate (57%) is comparable to the first cycle pregnancy rate of 58% (24/41) in a similar group of mares.

Table 1. Influence of altrenogest and donor–recipient ovulation asynchrony on non-surgical embryo transfer pregnancy rates (Group 1)

Donor–recipient asynchrony (days)*	No. of embryos transferred	No. of mares pregnant	% Pregnant at 30 days		
In oestrus	7	0	0		
+8*	1	0	0		
+7	1	0	0	0%	
+6	—	—	—		
+5	4	0	0		
+4	2	1†‡	50		
+3	3	2†‡‡	67		
+2	2	2	100		
+1	4	3‡	75		
0	1	1	100	77%	64%
−1	—	—	—		
−2	1	0	0		
−3	2	2	100		
−4	3	1†	33		
−5	5	3	60		
−6	2	1	50		

*Days recipient ovulated before (+) or after (−) donor.
†A recipient's follicle luteinized.
‡A recipient had luteal dysfunction.

Embryo transfer pregnancy rates

Results obtained from the unguarded, non-surgical transfer of embryos are presented in Table 1. None of the recipients that were in oestrus at the start of altrenogest treatment, nor recipients that ovulated 5 or more days before their donor, maintained pregnancy until ultrasound examination at Days 14–18. In the interval of donor–recipient asynchrony that ranged from +4 to −6 days, the overall 30-day pregnancy maintenance rate was 64% (16/25). In the 13 recipients that were +3 to −3 days asynchronous with their donors, the 30-day pregnancy maintenance rate was 77% (10/13). All recipients which were diagnosed pregnant at the time of ultrasound remained pregnant until Day 30 of gestation. These pregnancy maintenance rates compare favourably with those reported by Iuliano *et al.* (1985) who obtained values for non-surgical embryo transfer of 45% (18/40) and for surgical transfer of 72% (31/43) in the +1 to −3 day synchrony range. In 1985, Vogelsang *et al.* transferred embryos into recipients which were treated with altrenogest starting the day of transfer and achieved a 51% (17/33) overall pregnancy rate when recipients ovulated in the −1 to −3 day range, and 0% (0/7) pregnancy when recipients ovulated at 0, −4, −5 and −6 days.

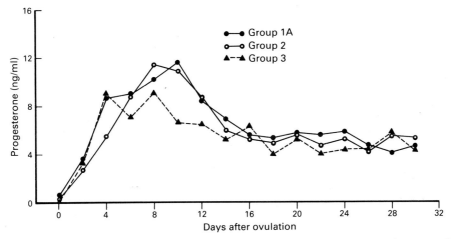

Fig. 1. Mean plasma progesterone profiles of: pregnant, altrenogest-treated, embryo-transfer recipient mares (Group 1A, N = 16); pregnant, altrenogest-treated, untransferred mares (Group 2, N = 8); and pregnant, untreated, untransferred mares (Group 3, N = 6). Pooled s.e.m. = 0·45.

Plasma progesterone concentrations

Mean plasma progesterone concentrations for the first 30 days of gestation were similar ($P > 0.05$: one-way ANOVA) for pregnant mares in Groups 1, 2, and 3 (Fig. 1). Thus, altrenogest alone (Group 2) or altrenogest plus embryo transfer (Group 1) did not affect luteal function when compared to non-transferred, non-treated pregnant controls (Group 3). Mean plasma progesterone profiles of pregnant embryo recipients (Group 1A, N = 16) versus non-pregnant embryo recipients which had ovulated or luteinized (Group 1B, N = 15) differed ($P < 0.05$: one-way ANOVA) at and after Day 14 after ovulation when non-pregnant recipients underwent luteal regression and pregnant recipients generally maintained luteal function (Fig. 2). Progesterone profiles of non-pregnant recipients that were in oestrus at the beginning of altrenogest treatment (Group 1C, N = 7) were very variable between mares (Fig. 3), although in all cases some form of luteinization appeared to occur on or before Day 15 of treatment as indicated by an increase in

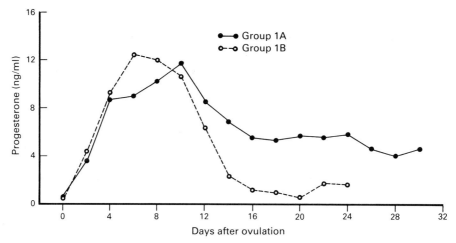

Fig. 2. Mean plasma progesterone profiles of pregnant (Group 1A, N = 16) and non-pregnant (Group 1B, N = 15) altrenogest-treated, embryo-transfer recipient mares. Means differ ($P < 0.05$) at and after Day 14 after ovulation. Pooled s.e.m. = 0·33.

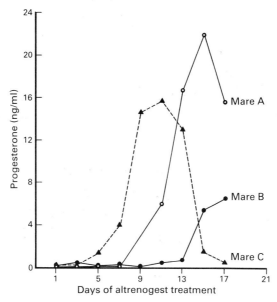

Fig. 3. Three individual plasma progesterone profiles of non-pregnant, altrenogest-treated embryo-transfer recipient mares which were in oestrus at the start of treatment, with presumed luteinization.

plasma progesterone concentration. According to plasma progesterone profiles, these luteinized structures formed and regressed again within 8–10 days in these non-pregnant mares. Within the group of pregnant, altrenogest-treated, embryo recipients (Group 1A), there were 5 atypical individual plasma progesterone profiles. In Fig. 4, Mares D and E are among a total of 4 recipients in this study which appeared to exhibit luteal dysfunction (a plasma progesterone concentration of < 2·5 ng/ml between Days 12 and 20 was used as the critical value to identify mares considered to

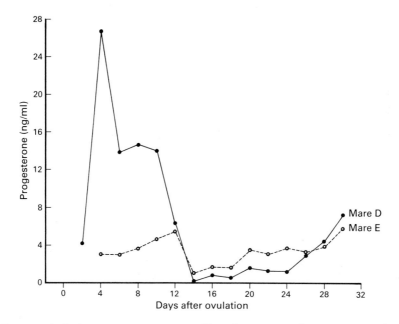

Fig. 4. Two atypical plasma progesterone profiles of pregnant, altrenogest-treated embryo-transfer recipient mares. Mare D had a normal ovulation and luteal dysfunction. Mare E had follicular luteinization and luteal dysfunction; both mares maintained pregnancy.

have luteal dysfunction (Douglas *et al.,* 1985)). Because progesterone is essential for pregnancy maintenance (Shideler *et al.,* 1981), and because altrenogest appears capable of maintaining pregnancy in the absence of a functional corpus luteum during early gestation (Parry & Holtan, 1985), it is our assumption that these 4 recipients with luteal dysfunction maintained their transferred embryos with the aid of altrenogest. The reason for the occurrence of the luteal dysfunction is unknown, but is postulated to be connected with the timing of embryonic signals to the uterus for the maternal recognition of pregnancy (Sharp & McDowell, 1985). In Fig. 4, Mare E represents 1 of 3 pregnant recipients which, in the oestrous period before the start of altrenogest treatment, luteinized rather than ovulated their follicles. Palpation revealed that these follicles would develop normally, but the mares would go out of oestrus without ovulating, then the follicle would become large and firm on palpation. Neely *et al.* (1983) described such structures as "persistent" or "autumn" follicles that were tense, fluid-filled structures, up to 10–15 cm in diameter, formed from ovarian follicles. On histological examination, Neely *et al.* (1983) found a pink–red gelatinous structure in the centre with a thin rim of luteal tissue around its periphery. Plasma progesterone profiles of the 3 pregnant recipients with presumed follicular luteinization differed from ovulating mares in that the progesterone concentrations did not follow the normal Day 6–10 rise and peak pattern, but showed a more gradual and less pronounced increase that fluctuated greatly. Two of these mares with presumed follicular luteinization were amongst those that displayed luteal dysfunction.

Altrenogest treatment therefore appears to: allow for extension of the degree of donor–recipient ovulation synchrony required for successful non-surgical embryo transfer in mares; have no effect on luteal function during pregnancy in mares; and aid in pregnancy maintenance in recipient mares with luteal dysfunction.

The work was supported in part by the Oregon Horsemen's Assoc. Altrenogest was provided by Hoechst–Roussel Agri-Vet Co., Sommerville, NJ, U.S.A.

Technical paper no. 8160, Oregon Agricultural Experiment Station Project No. 347.

References

Douglas, R.H., Burns, P.J. & Hershman, L. (1985) Physiological and commercial parameters for producing progeny from subfertile mares. *Equine vet. J., Suppl.* **3**, 111–114.

Ginther, O.J. (1979) *Reproductive Biology of the Mare: Basic and Applied Aspects.* Published by the author, Cross Plaines.

Imel, K.J., Squires, E.L., Elsden, R.P. & Shideler, R.K. (1981) Collection and transfer of equine embryos. *J. Am. vet. med. Ass.* **179**, 987–991.

Iuliano, M.F., Squires, E.L. & Cook, V.M. (1985) Effect of age of equine embryos and method of transfer on pregnancy rates. *J. Anim. Sci.* **60**, 258–263.

Koligian, K.B. & Stormshak, F. (1977) Nuclear and cytoplasmic estrogen receptors in ovine endometrium during the estrous cycle. *Endocrinology* **101**, 524–533.

Neely, D.P., Liu, I.K.M. & Hillman, R.B. (1983) *Equine Reproduction.* Veterinary Learning Systems Co., Inc., U.S.A.

Parry, L.C. & Holtan, D.W. (1985) Effect of Regu-Mate, a synthetic progestin, on early pregnancy maintenance in the absence of functional corpora lutea in pony mares. *Proc. 9th Equine Nutr. Physiol. Symp., East Lansing,* 382–386.

Sharp, D.C. & McDowell, K.J. (1985) Critical events surrounding the maternal recognition of pregnancy in mares. *Equine vet. J., Suppl.* **3**, 19–22.

Shideler, R.K., Squires, E.L., Voss, J.L. & Eikenberry, D.J. (1981) Exogenous progesterone therapy for maintenance of pregnancy in ovariectomized mares. *Proc. 22nd Ann. Conv. Am. Ass. equine Pract.,* 211–219.

Vogelsang, K.R., Bondioli, K.R. & Massey, J.M. (1985) Commercial application of equine embryo transfer. *Equine vet. J., Suppl.* **3**, 89–91.

J. Reprod. Fert., Suppl. **35** (1987), 439–443

Effect of timing of progesterone administration on pregnancy rate after embryo transfer in ovariectomized mares

K. Hinrichs and R. M. Kenney

Section of Reproductive Studies, University of Pennsylvania School of Veterinary Medicine, New Bolton Center, Kennett Square, Pennsylvania 19348, U.S.A.

Summary. Ovariectomized recipient mares were divided into two groups. Group A mares received 300 mg progesterone in oil i.m. daily starting 5 days before transfer of a 7-day embryo. Group B mares received the same dose of progesterone, but starting at least 4 days before donor ovulation. Presence of an embryonic vesicle was determined by ultrasonography; mares were considered to be pregnant if they had normal vesicle development to Day 18. Pregnancy rates were: Group A, 6/8; Group B, 1/12 ($P < 0.01$). An additional 4 mares in Group B had a vesicle visible at 14 days which degenerated or did not grow. These results indicate that, for establishment of pregnancy after embryo transfer, there is a need for synchrony of donor ovulation with onset of progesterone administration to ovariectomized recipient mares.

Introduction

The requirements for establishment and maintenance of pregnancy in the mare have not been well defined. In light of the success of embryo transfer as a commercial procedure for horses, the events surrounding early pregnancy in the mare take on clinical as well as scientific interest. The mare is similar to most other species in that progesterone alone will maintain pregnancy after ovariectomy in early gestation (review, Adams 1967). Mares ovariectomized at 35 days gestation abort (Holtan *et al.*, 1979), but mares remain pregnant after ovariectomy when treated with progesterone (Terqui & Palmer 1979; Shideler *et al.*, 1982) or the synthetic progestagen, altrenogest (Shideler *et al.*, 1982). Parry & Holtan (1985) gave altrenogest to mares that were treated with a luteolytic dose of prostaglandin at Day 10 of gestation: 7 out of 9 mares were still pregnant at Day 35 on this treatment.

These studies all used mares that had established pregnancy before loss of ovarian hormone stimulation. In some species, the hormone changes associated with oestrus are essential to establishment of pregnancy. In the sheep, as in the mare, progesterone alone will maintain gestation after ovariectomy in early gestation (Moore & Rowson, 1959). However, embryo transfer into long-term ovariectomized ewes is successful only if the ewe has been previously treated to mimic the fall in progesterone, rise in oestrogen, and subsequent rise in progesterone that would normally accompany oestrus (Miller & Moore, 1976). To determine the requirements for establishment and maintenance of pregnancy in long-term ovariectomized mares (ovariectomized before the oestrus associated with pregnancy) a series of experiments was performed.

The first study was conducted to determine whether progesterone alone could prepare ovariectomized mares to establish and maintain pregnancy after embryo transfer (Hinrichs *et al.*, 1985). Recipient mares received progesterone, 300 mg i.m. daily, starting 5 days before transfer of a 7- or 8-day embryo. Of 4 mares, 3 became pregnant after transfer and delivered normal foals within 10 days of the expected foaling date. This study clearly demonstrated that the hormones associated with oestrus are not necessary for the establishment of pregnancy in the mare. A valuable clinical use for these mares as embryo transfer recipients is apparent: ovariectomized recipients do not need to be synchronized before ovulation with the donor, there is no need for teasing or palpation of recipients, and the recipient is dependably in the optimal stage for receiving the embryo.

In a second experiment, the orally active synthetic progestagen, altrenogest, was used to prepare ovariectomized mares as embryo transfer recipients (Hinrichs *et al.*, 1986). Pregnancy rate was 1/6 for mares on 22 mg altrenogest and 2/6 for mares on 66 mg altrenogest. The low pregnancy rate was associated with poor uterine and cervical tone at the time of trans-cervical transfer.

The present experiment was conducted to determine whether the timing of progesterone admin- istration in ovariectomized recipient mares has an effect on pregnancy rate after transfer. When intact recipients are used, best pregnancy rates are reported to be obtained when the recipient ovulates 1 day before to 2 days after the donor (Douglas, 1982; Squires *et al.*, 1982). A similar requirement for synchronization of progesterone treatment with donor ovulation may exist when ovariectomized recipients are used.

There are two possible reasons for the demand for synchrony in intact recipients. First, the embryo must have the ability to prevent luteolysis by Day 15 of gestation, the time of maternal recognition pregnancy in the mare (Hershman & Douglas, 1979). If the recipient ovulates long before the donor, the relatively young embryo may not be mature enough to prevent luteolysis when the recipient is at Day 15. A second possible reason is that the stage of development of the recipient uterus and of the embryo must be matched to have embryo survival.

The existence of a synchrony between the 'age' of the uterus after ovulation (or onset of pro- gesterone stimulation) and the age of the embryo has been established for other species (see review by Wilmut *et al.*, 1985). An important factor in the uterine environment is the protein produced by the uterus, and some uterine proteins appear to be stage specific. In the pig and ewe, uterine proteins change with time through the luteal phase (Geisert *et al.*, 1982; Miller & Moore, 1983). It has been suggested that the proteins secreted 'match' the requirements of the corresponding age embryo and not those of different age embryos.

When progesterone-treated ovariectomized mares are used as recipients, there is no need for embryo signalling for maternal recognition of pregnancy (luteal maintenance). Differences in preg- nancy rate in synchronized and unsynchronized ovariectomized recipients should reflect the effect of timing of progesterone administration on the uterine environment—the effect of uterine 'age' on embryo survival. In this experiment, pregnancy rates after transfer were compared in ovari- ectomized recipients started on progesterone treatment either in synchrony with the donor (starting 2 days after donor ovulation) or not in synchrony with the donor (starting 4 or more days before donor ovulation).

Materials and Methods

Sixteen light horse mares aged 2–10 years with normal reproductive tracts on palpation *per rectum*, no growth on culture of a uterine swab, and Category I or I–II endometria on biopsy (Kenney, 1978) were used as recipients. Mares were ovariectomized at least 21 days before beginning treatment for transfer. Recipient mares were randomly assigned to Group A, to start progesterone treatment 2 days after donor ovulation, or Group B, to start progesterone treat- ment at least 4 days before donor ovulation. Progesterone treatment consisted of 300 mg progesterone in oil i.m. daily (Progesterone Injection; Carter-Glogau Laboratories, Phoenix, AZ).

Twelve light-horse mares aged 4–14 years were used as donors. Donor mares were teased with a stallion daily to detect oestrus. On the first day of oestrus, each donor was matched with a recipient. If the corresponding recipient was in Group B, progesterone treatment was started at this time. Donors in oestrus were palpated *per rectum* and examined by ultrasonography daily to monitor follicular growth. On the basis of follicular size and consistency, mares were inseminated naturally or artificially by a fertile stallion every other day until ovulation. The day of ovulation, as detected by ultrasonography, was designated Day 0. Embryo flush and transfer were performed on Day 7, as described by Douglas (1982) with the modification that the transfer was performed entirely *per vaginam*, without grasping the uterus *per rectum*. If no embryo was recovered for transfer, progesterone treatment of the corresponding recipient was discontinued, and the recipient remained untreated for a minimum of 7 days before starting treatment for a subsequent transfer. If 2 embryos were recovered in the flush from one donor (twins), both embryos were transferred into the corresponding recipient.

After transfer of an embryo, the recipient remained on progesterone treatment until Day 18. Examination by ultrasonography was performed daily from Day 14 to Day 18 to detect the presence of an embryonic vesicle and monitor vesicle growth.

Because of evidence of infection after transfer in 2 mares in Group B (see 'Results'), after the first 5 transfers the transfer technique was modified to include a sterile mitt over the hand and transfer pipette when introducing the pipette into the vagina. This was done to minimize contamination of the pipette or operator's hand by flora in the vulva or vestibule during transfer. Once inside the vagina, the hand and pipette were extruded from the end of the mitt and the cervix was cannulated.

After one embryo transfer, 4 mares from each treatment group were crossed over to the opposite treatment group for a second transfer. Mares infected after transfer were not used again as recipients. Mares were left untreated for a minimum of 7 days after complete disappearance of any vesicle before starting treatment for another transfer.

Recipient mares were considered pregnant if a vesicle was present and had normal development to Day 18. Pregnancy rates for the two groups were compared by Fisher's exact test.

Results

Altogether, 32 flushes of donor mares were performed and 24 embryos were recovered, including 4 sets of twins: 8 transfers were performed in Group A and 12 in Group B. Pregnancy rates (mares with normal vesicle development to Day 18) were 6/8 and 1/12 for the two groups. The difference in pregnancy rate was significant ($P < 0.01$). All mares had good to excellent uterine and cervical tone at the time of transfer, and no difference in uterine or cervical tone was noted between groups.

Because the time of ovulation in the mare cannot be exactly predicted, start of progesterone treatment in mares in Group B varied from 4 to 10 (average 6.7) days before donor ovulation.

The mare in Group B that had a normal pregnancy (Mare SH) was started on progesterone treatment 5 days before donor ovulation. Four additional mares in Group B had a vesicle visible on ultrasound at Day 14 but the vesicles did not grow (Mare HY and one twin in Mare TS), grew to 15 days then degenerated, losing roundness, size and clarity (Mare BL and the other twin in Mare TS), or grew to 15 days and were then lost (Mare VF). These 4 mares were started on progesterone treatment 4–10 days before donor ovulation. All vesicles seen in mares in Group A developed normally to Day 18.

Of 4 sets of twins, 2 sets were transferred into recipients in Group A and developed 2 vesicles in each case. Of 2 sets transferred in Group B, one set did not develop vesicles and the other developed abnormal vesicles as outlined above (Mare TS).

After the first 5 transfers (2 in Group A and 3 in Group B), 2 mares in Group B developed fluid in the uterus, seen on ultrasound. Culture of this fluid yielded *Escherichia coli* in one mare and *Staphylococcus aureus* in the other. Culture of transfer media yielded no growth of bacteria. To minimize contamination, the transfer technique was modified as described in the 'Methods' section. Of the subsequent 9 transfers in Group B, one mare developed fluid in the uterus, but aerobic culture of a uterine swab yielded no growth. No mares in Group A were seen to have uterine fluid.

Discussion

The significant difference in pregnancy rates between Group A and Group B demonstrates that, to obtain optimal pregnancy rates, ovariectomized mares must be started on progesterone treatment in synchrony with donor ovulation. However, the reason for this demand for synchrony is obscured by the tendency of mares on the longer progesterone treatment (Group B) to develop infection after transfer.

Steroid hormones may influence the ability of the endometrium to resist bacterial infection. The longer progesterone treatment of mares in Group B (11–18 days before transfer) may have lowered uterine resistance to bacteria introduced at the time of transfer. Although only 3 mares in Group B developed fluid in the uterus, and so were cultured, it is possible that less dramatic infections may have been present in other mares, and may have interfered with pregnancy. Use of a surgical transfer technique may help to eliminate contamination during transfer and so minimize the effect of infection on pregnancy rate.

Although only 1 mare in Group B had normal vesicle development from Day 14 to Day 18, there were 4 additional mares in Group B that had vesicles of normal appearance at Day 14 which later degenerated. The uterus apparently supported development of these vesicles from Day 7 (the day of transfer) to Day 14, but not afterwards. If the 3 mares which developed uterine fluid are disregarded, this initial pregnancy rate (at 14 days) is 5/9, and is not significantly different from the pregnancy rate in Group A. It would be interesting to flush embryos from these recipients at different times after transfer to assess when the effect of the asynchronous uterus on the embryo begins.

It is possible that oestrogen priming, or oestrogen and progesterone treatment, may influence maintenance of pregnancy in ovariectomized recipient mares. Oestrogen may have an effect by changing uterine protein profiles (McDowell *et al.*, 1987) or resistance to bacteria. The horse embryo has been shown to produce oestrogens *in vitro* as early as 8 days gestation (Flood *et al.*, 1979; Zavy *et al.*, 1979; Heap *et al.*, 1982). Embryonic oestrogen may influence the uterus and cause it to secrete proteins corresponding to the stage of development of the embryo. This has been shown to be true in the pig (Geisert *et al.*, 1982). If this mechanism is acting in the horse, it appears that the uterus must be in the proper stage of progesterone stimulation to respond to the embryo.

This study demonstrates that ovariectomized mares treated with progesterone from 2 days after ovulation in the donor can serve as optimal embryo transfer recipients, as a 75% pregnancy rate was achieved with these mares by a transcervical transfer technique. It is also clear that ovariectomized recipients have poor pregnancy rates (8%) if placed on long-term progesterone treatment before transfer, but the reasons for the difference in embryo survival in these two groups of recipients need to be investigated.

This work was supported by a grant from the Grayson Foundation. We thank Carter-Glogau Laboratories, Inc. for donation of materials; and Ms L. A. Caldwell for excellent technical assistance.

References

Adams, C.E. (1967) Ovarian control of early embryonic development within the uterus. In *Reproduction in Female Mammals*, pp. 532–546. Eds G. E. Lamming & E. C. Amoroso. Butterworths, London.

Douglas, R.H. (1982) Some aspects of equine embryo transfer. *J. Reprod. Fert., Suppl.* **32**, 405–408.

Flood, P.E., Betteridge, K.J. & Irvine, D.S. (1979) Oestrogens and androgens in blastocoelic fluid and cultures of cells from equine conceptuses of 10–22 days gestation. *J. Reprod. Fert., Suppl.* **27**, 413–420.

Geisert, R.D., Renegar, R.H., Thatcher, W.W., Roberts, R.M. & Bazer, F.W. (1982) Establishment of pregnancy in the pig: I. Interrelationships between pre-implantation development of the pig blastocyst and uterine endometrial secretions. *Biol. Reprod.* **27**, 925–939.

Heap, R.B., Hamon, M. & Allen, W.R. (1982) Study on oestrogen synthesis by the preimplantation equine conceptus. *J. Reprod. Fert., Suppl.* **32**, 343–352.

Hershman, L. & Douglas, R.H. (1979) The critical period for the maternal recognition of pregnancy in Pony mares. *J. Reprod. Fert., Suppl.* **27**, 395–401.

Hinrichs, K., Sertich, P.L., Cummings, M.R. & Kenney, R.M. (1985) Pregnancy in ovariectomised mares achieved by embryo transfer. *Equine vet. J. Suppl.* **3**, 74–75.

Hinrichs, K., Sertich, P.L. & Kenney, R.M. (1986) Use of altrenogest to prepare ovariectomized mares as embryo transfer recipients. *Theriogenology* **26**, 455–460.

Holton, D.W., Squires, E.L., Lapin, D.R. & Ginther, O.J. (1979) Effect of ovariectomy on pregnancy in mares. *J. Reprod. Fert., Suppl.* **27**, 457–463.

Kenney, R.M. (1978) Cyclic and pathologic changes of the mare endometrium as detected by biopsy, with a note on early embryonic death. *J. Am. vet. med. Assoc.* **172**, 241–262.

McDowell, K.J., Sharp, D.C. & Grubaugh, W. (1987) Comparison of progesterone and progesterone + oestrogen on total and specific uterine proteins in pony mares. *J. Reprod. Fert., Suppl.* **35**, 335–342.

Miller, B.G. & Moore, N.W. (1976) Effects of progesterone and oestradiol on RNA and protein metabolism in the genital tract and on survival of embryos in the ovariectomized ewe. *Aust. J. biol. Sci.* **29**, 565–573.

Miller, B.G. & Moore, N.W. (1983) Endometrial protein secretion during early pregnancy in entire and ovariectomized ewes. *J. Reprod. Fert.* **68**, 137–144.

Moore, N.W. & Rowson, L.E.A. (1959) Maintenance of pregnancy in ovariectomized ewes by means of progesterone. *Nature, Lond.* **184,** 1410.

Parry, L.C. & Holtan, D.W. (1985) Effect of regumate, a synthetic progestogen, on early pregnancy maintenance in the absence of functional corpora lutea in pony mares. *Proc. 9th equine Nutrition and Physiology Symposium, East Lansing,* pp. 382–385.

Shideler, R.K., Squires, E.L., Voss, J.L., Eikenberry, D.J. & Pickett, B.W. (1982) Progestagen therapy of ovariectomized pregnant mares. *J. Reprod. Fert., Suppl.* **32,** 459–464.

Squires, E.L., Imel, K.J., Iuliano, M.F. & Shideler, R.K. (1982) Factors affecting reproductive efficiency in an equine embryo transfer programme. *J. Reprod. Fert., Suppl.* **32,** 409–414.

Terqui, M. & Palmer, E. (1979) Oestrogen patterns during early pregnancy in the mare. *J. Reprod. Fert., Suppl.* **27,** 441–446.

Wilmut, I., Sales, D.I. & Ashworth, C.J. (1985). The influence of variation in embryo stage and maternal hormone profiles on embryo survival in farm animals. *Theriogenology* **23,** 107–119.

Zavy, M.T., Mayer, R., Vernon, M.W. & Sharp, D.C. (1979) An investigation of the uterine luminal environment in non-pregnant and pregnant Pony mares. *J. Reprod. Fert., Suppl.* **27,** 403–411.

J. Reprod. Fert., Suppl. 35 (1987), 445–454

Effect of uterine inflammation and ultrasonically-detected uterine pathology on fertility in the mare

G. P. Adams, J. P. Kastelic, D. R. Bergfelt and O. J. Ginther

Department of Veterinary Science, University of Wisconsin, 1655 Linden Drive, Madison, Wisconsin 53706, U.S.A.

Summary. The incidence of intrauterine fluid collections during dioestrus (12/43, 28%) and uterine cysts throughout the oestrous cycle (11/73, 15%) found in this study indicates that these ultrasonically detectable abnormalities are prevalent in mares. The hypothesis that uterine cysts do not affect pregnancy was not supported. Intrauterine fluid collections at dioestrus represented the presence of an inflammatory process as indicated by a high biopsy score, reduced progesterone concentrations, and a shorter interovulatory interval. Mares with fluid collections at dioestrus had a lower pregnancy rate at Day 11 and a higher embryonic loss rate by Day 20 than did mares without such collections. The progesterone profile and length of interovulatory interval for mares with uterine inflammation supported the hypotheses that embryonic loss in this herd was due to uterine-induced luteolysis rather than primary luteal inadequacy.

Introduction

The ultrasonic appearance of uterine cysts and free collections of intraluminal uterine fluid has been described (Ginther & Pierson, 1984b), but the relationship of these conditions to infertility is not clear. However, a high incidence of embryonic loss (18%) between Days 11 and 15 (ovulation = Day 0) and small (e.g. 5–20 mm), free collections of uterine fluid was detected ultrasonically during mid-dioestrus in a herd of mares (Ginther *et al.*, 1985a). All 3 mares that had both free intrauterine fluid collections and an embryonic vesicle lost the vesicle by Day 15. Mares with a history of embryonic loss during Days 11–15 and mares with a history of intrauterine fluid collections at some time during the breeding season had the following similarities: (1) reduced pregnancy rate, (2) reduced progesterone concentrations on Days 7 and 11, (3) reduced length of the interovulatory interval both for intervals associated with loss and for periods in which an embryo was not detected, and (4) repeatability of the condition within individuals, indicating a chronic problem. These similarities suggested that the factors responsible for the embryonic loss on Days 11–15 were the same as those responsible for the intrauterine fluid collections. It was suggested that both were attributable to uterine inflammation, and the short interovulatory intervals and reduced progesterone concentrations were due to early activation of the uterine luteolytic mechanism by the inflammatory processes (Ginther *et al.*, 1985a; Ginther, 1986). However, since blood samples for progesterone analyses were not taken before an inflammatory process would be expected to cause luteolysis (before Day 5; Ginther, 1985), early activation of luteolysis could not be differentiated from primary luteal inadequacy.

In a study by Woods *et al.* (1985), subfertile and maiden mares were compared by uterine biopsy and by embryo recovery attempts on Days 7 and 8. Only the subfertile mares had indications of uterine pathology, and more of them had embryonic vesicles that were classified as abnormal. In another study (Ball *et al.*, 1986) the fertilization rate (embryo recovery rate at Day 2) was the same for barren (11/14) and normal (10/14) mares but the estimated loss rate by Day 7 was higher for the barren group (8/11) than the normal group (0/10). The studies indicated that embryonic loss and not fertilization failure was the more important cause of infertility in these groups of mares, and that the embryonic loss was associated with uterine inflammation.

These results from independent and different approaches by two groups of workers implicated uterine inflammatory processes as the cause of infertility in mares.

The purposes of the present studies were to determine the incidence of ultrasonically-detected intrauterine fluid collections and uterine cysts in a group of horse mares (Exp. 1) and to test the following hypotheses: (1) ultrasonically detectable uterine cysts do not affect pregnancy rate (Exp. 2), (2) ultrasonically-detected intrauterine fluid collections are related to uterine inflammation and are associated with poor fertility (Exp. 3), and (3) a high incidence of early embryonic loss in a group of subfertile pony mares was associated with uterine inflammation rather than primary luteal inadequacy (Exp. 3).

Materials and Methods

Transrectal ultrasonic examinations of the reproductive tract were conducted with a real-time, B-mode scanner equipped with a 5 MHz linear-array transducer. The day of ovulation was detected by once daily examinations (Exps 2 and 3) and the presence of a corpus luteum was determined as described by Ginther & Pierson (1983). Examinations for detection of an embryonic vesicle were begun on Day 10 (Exps 2 and 3), and the diameter of the vesicle was measured as described by Ginther & Pierson (1983). The uterus was recorded as being ultrasonically characteristic of dioestrus (homogeneous with poorly-defined endometrial folds) or oestrus (heterogeneous with well-defined endometrial folds) (Ginther & Pierson, 1984b). Uterine cysts and free intrauterine fluid collections were measured and assigned to a location by dividing each uterine horn and the uterine body into three approximately equal segments (total of 9 uterine segments). Cysts were defined as immobile, usually compartmentalized, non-echogenic (fluid-filled) structures with well-defined borders. A cyst score was assigned based on the severity of the cystic condition (no cysts detected, score 0; diameter of the largest cyst 10 mm and fewer than 5 in number, score 1; diameter ≥ 10 mm or more than 5 in number, score 2). Intrauterine fluid collections were defined as mobile non-echogenic areas in the lumen with poorly defined borders. The height of the largest intrauterine fluid collection was determined for each mare. An intrauterine fluid collection score was assigned based on the maximum height of the largest fluid collection during a given daily examination or over a given period of oestrus or dioestrus (no intrauterine fluid detected, score 0; maximum height $\leqslant 10$ mm, score 1; height 11 to 39 mm, score 2; height ≥ 40 mm, score 3). For statistical purposes (Exps 1 and 3) non-pregnant, cyclic mares were defined as being in oestrus or dioestrus on the basis of the ultrasonic appearance of the endometrium and corpus luteum (Ginther & Pierson, 1984a, b; Pierson & Ginther, 1985).

Mares were inseminated artificially (Exps 2 and 3) with at least 100×10^6 progressively motile, morphologically normal spermatozoa. Fresh, non-extended semen was used.

Experiment 1. Horse mares of mixed breeds were examined to determine the types and incidence of ultrasonically-detectable uterine pathology. The 93 mares had been recently acquired from several slaughter-market sales in Iowa, Nebraska and Oklahoma and ranged in age from 1·5 years to > 20 years as estimated by the state of the dentition. The reproductive histories and the reason for sale of the mares were not known. The size and location of intrauterine fluid collections and uterine cysts were recorded and the stage of the oestrous cycle (dioestrus or oestrus) was estimated by ultrasound examination. The mares were examined on only one day and for some mares (N = 32), the age was recorded. The incidence of cysts was determined regardless of the stage of the oestrous cycle. The incidence of intra-uterine fluid collections was determined only for dioestrus. Oestrous mares and those with a newly forming corpus luteum were excluded to avoid the possibility of confusing intrauterine fluid collections with semen or normal fluids at oestrus. Differences in age were evaluated by the Student's *t* test between mares with and without cysts and between dioestrous mares with and without intrauterine fluid collections.

Experiment 2. To determine the relationship of cysts to infertility, data were taken from two separate studies on a total of 22 mares in which ultrasonically detectable uterine cysts were observed. Nine mares were part of the herd studied in Exp. 3 and their data included endometrial biopsy scores. The remaining 13 were from a preliminary study on ultrasonic uterine pathology wherein endometrial biopsies were not examined. Only mares with no other ultrasonically-detected uterine pathology were used. The pregnancy rate on Day 11–12 and on Day 40 was compared by χ^2 analysis amongst mares with a cyst score of 1 (N = 8) or 2 (N = 14) and contemporary herdmates which had no ultrasonically detectable uterine pathology (cyst score, 0; N = 24). To determine the relationship between uterine cysts and endometritis, the biopsy scores were compared between mares with and without cysts by Student's *t* test.

Experiment 3. Fifty-six pony mares were used. The herd was assembled at the end of the previous breeding season on the basis of failing to become pregnant after repeated breedings, embryonic loss before Day 20, or the presence of intrauterine fluid collections.

The uterus was examined by ultrasound for the presence of intrauterine fluid collections during mid-dioestrus (Day 8) preceding the oestrus during which the mares were mated. An endometrial biopsy was taken as described (Kenney, 1978) immediately after the ultrasound examination. The biopsies were examined histologically for indications of inflammation. The extent of inflammation was scored from 1 to 4 according to the classification system (categories, 1, 2a, 2b, 3) described by Kenney (1978) and Kenney & Doig (1986). The biopsy score was used as a reference to determine the relationship of intrauterine fluid collections to endometritis.

The mares were mated during the first oestrus after the biopsy procedure. Ultrasound examinations of the uterus were done on Day 2 and daily from Day 8 until the next ovulation in mares in which an embryonic vesicle was not detected at any time (non-pregnant mares) and in mares that lost the embryonic vesicle. Pregnant mares were scanned until Day 20 and thereafter at 5-day intervals until termination of the experiment on Day 40. The uterus was examined for the presence of intrauterine fluid collections, cysts and embryonic vesicles. The fluid collections were measured as described above. The extent of fluid involvement was scored according to the largest collection detected during Days 8–15. Mares with uterine cysts were assigned to Exp. 2, but mares with both cysts and intrauterine fluid collections were assigned to the appropriate fluid collection group.

First detection of a presumptive embryonic vesicle was defined as the detection of a spherical, non-echogenic structure which was differentiated from a uterine cyst by reference to previous recordings of the number, size, and location of cystic structures. A confirmed embryonic vesicle was defined as a presumptive embryonic vesicle which increased in diameter or changed locations by the following day. Confirmed embryonic loss was defined as failure to detect a previously recorded confirmed embryonic vesicle. The day of embryonic loss was the day of complete disappearance of the vesicle or its remnants. A presumptive embryonic loss was defined as the failure to detect a previously diagnosed presumptive embryonic vesicle. On the day of embryonic loss, the cervix was examined digitally, and a cervix that readily admitted one or more fingers was defined as patent.

A blood sample was collected by jugular venepuncture on Days 2, 5, 8, 11, 15 and 20. Concentrations of progesterone in the plasma were determined using a modification of a solid-phase ^{125}I-radioimmunoassay kit for progesterone (Coat-a-Count Progesterone; Diagnostic Products Corporation, Los Angeles, CA). The kit was provided with progesterone standards prepared in human serum. To determine concentrations of progesterone in mare plasma, the assay has been modified and validated using progesterone standards prepared in plasma from an ovariectomized mare (Bergfelt & Ginther, 1986). The assay coefficients of variation as determined over the 6 assays used herein were 2·7% and 10·3% for within and 3·8% and 10·3% for among assays from pooled plasma samples of dioestrous and oestrous ponies, respectively.

The changes in intrauterine fluid collection scores during the interovulatory interval were characterized using all mares with an interovulatory interval and a fluid collection on any day during the interval, excluding 5 mares with frank pyometra (intrauterine fluid collections > 100 mm). The length of the intervals was normalized by designating the length of the interval as 100% for each mare. Ten points were used by determining the mean of observations within each tenth of the interval. The profile of the resulting mean was characterized by regression analyses. A correlation analysis was done between the fluid collection score obtained on Day 8 before the oestrus during which the mares were mated and the next Day 8 within individuals to obtain a measure of repeatability. The association between fluid collection scores and biopsy scores was examined by a correlation analysis both for fluid collection results obtained on the day of biopsy (mid-dioestrus preceding the oestrus during which mares were mated) and the overall result for Days 8–15.

For further statistical analyses, the mares were grouped according to biopsy score (1–4), intrauterine fluid collection score (0–3), and pregnancy status (non-pregnant, embryo maintained, and embryo lost). Non-pregnant mares were defined as those in which neither a presumptive nor confirmed embryonic vesicle was detected at any time. Only mares with a confirmed embryonic vesicle were used in the analyses of embryonic loss. Mares with a presumptive embryonic loss were evaluated by inspection of data only. The relationships of pregnancy status to biopsy score and fluid collection score over Days 8–15 were examined by χ^2 analyses using the proportion of mares not pregnant, with maintained pregnancy and with embryonic loss for each score. To determine the relationship between infertility and fluid collections at oestrus and dioestrus, pregnancy rate and embryonic loss rate were compared amongst mares which had fluid collections (1) only at dioestrus, (2) only at oestrus, (3) at oestrus and dioestrus, and (4) no fluid collections. The comparisons were made by χ^2 analyses.

Growth rate of the embryonic vesicle was compared between mares that maintained the embryo and mares that lost the embryo by Day 20. The growth rate was characterized by regression analysis and the mean vesicle size on a given day was compared by analysis of variance. The average length of the interovulatory intervals was compared among biopsy groups and fluid collection groups by analysis of variance and multiple comparisons were made by determining the least significant difference (Snedecor & Cochran, 1980). The progesterone concentrations were analysed amongst groups by the General Linear Model of the Statistical Analysis System (SAS, 1985). An overall analysis of variance was performed to test day effect, group effect and day/group interaction. The groups were compared within each sampling day by Duncan's multiple range test (Snedecor & Cochran, 1980). Progesterone concentrations were compared amongst fluid collection groups and amongst biopsy groups. Because 5 of 8 mares with a biopsy score of 1 were pregnant, the pregnant mares were included to preserve a more complete data set for statistical analysis. Statistical inference was not extended past Day 8 since progesterone concentrations for subsequent samples may have been influenced by pregnancy status (Ginther, 1979).

Results

Experiment 1

The results are given in Table 1: 19 mares were pregnant and 74 were non-pregnant. One mare appeared to be anoestrous and was excluded. Of the 12 dioestrous mares with intrauterine fluid collections, 8 had a fluid collection score of 1 (66·7%) and 4 had a score of 2 (33·3%). Of the 11 mares with uterine cysts, 5 had cyst scores of 1 (45·5%) and 6 of 2 (54·5%).

Table 1. Incidence of ultrasonically detected intra-uterine fluid collections (IFCs) and uterine cysts and relationships to age in mares

	No. of mares (%)	Age (years)
Uterine cysts		
Yes	11/73 (15·1)	13·4*
No	62/73 (84·9)	6·9*
IFCs at dioestrus		
Yes	12/43 (27·9)	12·2**
No	31/43 (72·1)	7·5**

*$P < 0.005$; **$P < 0.10$.

Table 2. The pregnancy rates at Day 11–12 and Day 40 and embryonic loss rate to Day 40 in mares with uterine cysts or no uterine pathology

Cyst score*	Pregnancy rate (%)		Embryonic loss rate (%)
	Day 11 or 12	Day 40	
0	14/24 (58·3)	14/24[a] (58·3)	0/14 (0·0)
1	5/8 (62·5)	5/8[ab] (62·5)	0/5 (0·0)
2	5/14 (35·7)	4/14[b] (28·6)	1/5 (20·0)

*0 (no uterine cysts detected); 1 (largest cyst < 10 mm diameter and fewer than 5 in number); 2 (largest cyst ≥ 10 mm or more than 5 in number).
$P < 0.10$ for values with different superscripts.

Experiment 2

There were no differences in pregnancy rates on Days 11–12 among the three cyst groups, but mares in Group 2 tended to have a lower ($P < 0.10$) pregnancy rate at Day 40 than did mares in Groups 0 and 1 (Table 2). No statistical relationship was found between cyst score and biopsy score, since there were no differences in mean biopsy scores amongst the three groups.

There was no difference in the rate of embryonic loss amongst groups; in all three the loss rate was very low (Table 2).

Experiment 3

Presumptive embryonic loss occurred in 2 mares on Days 10 and 11, respectively. The biopsy scores for these two mares were 2 and 3, respectively, and the fluid collection scores were 0 and 1. In another mare, embryonic loss occurred on Day 35. The mare became pseudopregnant (Ginther, 1979). The biopsy and fluid collection scores were 1 and 0, respectively. By design, these 3 mares were excluded and the statistical analyses involved 53 mares.

Intrauterine fluid collections were detected in 30/53 (56·6%) mares during Days 8–15: on all 8 days in 10 mares (33·3%), on 7 days in 1 mare (3·3%), on 6 days in 3 mares (10%), on 5 days in 6 mares (30%), on 4 days in 1 mare (3·3%), on 3 days in 1 mare (3·3%), on 2 days in 4 mares (13·3%), and on 1 day in 4 mares (13·3%). The relationship between height of the largest fluid collection and day of the interovulatory interval is shown in Fig. 1. The changes in size of intrauterine fluid collections best fit a cubic regression ($R^2 = 72·6\%$); all three coefficients contributed significantly, indicating a good fit. The values were minimal at 25% and reached maximal at about 75–80% of the distance between ovulations (Fig. 1).

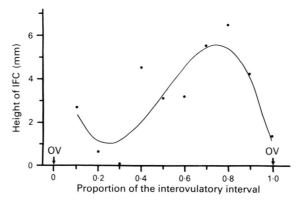

Fig. 1. The changes in size of the largest intrauterine fluid collection over one interovulatory interval. The solid line represents the calculated cubic regression line for the data. The points represent the mean of observations. OV = ovulation.

Table 3. Relationship (no. of mares) between intrauterine fluid collection (IFC) scores (see text) and biopsy scores

IFC score	Biopsy score			
	1	2	3	4
0	8	12	1	1
1	1	11	3	0
2	0	2	5	1
3	0	4	1	2

The presence or absence of an intrauterine fluid collection in individual mares was highly correlated ($r = 0·8$; $P < 0·0001$) between Day 8 before the oestrus during which mares were mated and the next Day 8. The fluid collection scores and biopsy scores on Day 8 before mating (day of biopsy) were also positively correlated ($r = 0·3$; $P < 0·05$). The correlation between biopsy scores and the overall IFC scores for Days 8 to 15 was $r = 0·5$ ($P < 0·0001$) (Table 3).

The pregnancy rate was ($P < 0·005$) for mares which had lower ultrasonically-detected fluid collections during Days 8–15 than for mares without such collections (Table 4). There were no differences in the pregnancy rates among scores of 1, 2 and 3. The pregnancy rate was higher ($P < 0·025$) for mares with biopsy score of 1 than for the other 3 biopsy scores combined (Table 4). There were no differences in pregnancy rates among biopsy scores 2, 3, and 4.

Table 4. Pregnancy rates (Day 11–12) and embryonic loss rates (to Day 40) according to biopsy and intrauterine fluid collection scores (see text)

Technique	Score	Non-pregnant (%)	Pregnancy rate (%)	Embryonic loss rate (%)
Biopsy	1	3/8 (37·5)	5/8 (62·5)	0/5 (0·0)
	2	26/31 (83·9)	5/31 (16·1)	3/5 (60·0)
	3	8/10 (80·0)	2/10 (20·0)	2/2 (100)
	4	3/4 (75·0)	1/4 (25·0)	1/1 (100)
IFC	0	13/23 (56·5)	10/23 (43·5)	3/10 (30·0)
	1	14/15 (93·3)	1/15 (6·7)	1/1 (100)
	2	6/8 (75·0)	2/8 (25·0)	2/2 (100)
	3	7/7 (100)	0/7 (0·0)	0/0

Within each technique and for each endpoint, $P < 0.05$ for biopsy score 1 or IFC score 0 compared with the combined values for the other 3 scores.

Table 5. Relationships between intrauterine fluid collections detected at oestrus and/or dioestrus and infertility in pony mares

	At dioestrus	At oestrus	At oestrus and dioestrus	None
Pregnancy rate (Day 11–12)	1/12[a] (8·3%)	2/3[b] (66·7%)	1/18[a] (5·6%)	9/20[b] (45·0%)
Embryonic loss rate	1/1 (100%)	2/2 (100%)	1/1 (100%)	2/9* (22·2%)

Pregnancy rates with different superscripts are different ($P < 0.05$).
*$P < 0.05$ compared with combined rate for the other 3 groups.

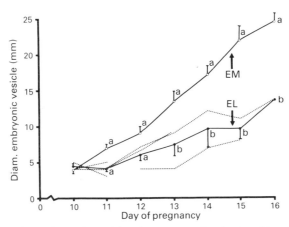

Fig. 2. Growth curve for embryonic vesicles that were maintained to Day 40 (EM) and those that were lost (EL). Individual embryos in the EL group are represented by the broken lines. Within each day diameters with different superscripts are different ($P < 0.05$).

Early embryonic loss occurred in six mares on Days 12, 12, 12, 14, 16 and 17. The loss rate was lower ($P < 0.05$) for mares with no detected fluid collections than for those with such collections during Days 8–15, but there were no differences in the loss rates among scores (Table 4). Similarly, the loss rate was lower ($P < 0.01$) for mares with biopsy score 1 than for mares with biopsy scores 2, 3 or 4, but there were no differences among biopsy scores 2, 3, and 4 (Table 4). Pregnancy rate was not different ($P > 0.50$) between mares with no fluid collections and mares with fluid collections during oestrus only; however, embryonic loss rate was higher ($P < 0.05$) for the latter. Mares with intrauterine fluid collections during dioestrus only and those with collections during dioestrus and oestrus did not differ in pregnancy rate ($P > 0.9$) or embryonic loss rate ($P = 1.0$). Mares in dioestrus with fluid collections (whether or not collections were detected during oestrus) had a lower ($P < 0.025$) pregnancy rate than mares with fluid collections only at oestrus or no collections. Regardless of the stage of the cycle during which an intrauterine fluid collection was observed, all 4 mares that had such collections and became pregnant failed to maintain the embryo beyond 20 days (Table 5).

The slope of the growth curve from Day 10 to Day 16 was different ($P < 0.05$) for embryonic vesicles that were lost than for those maintained; by Day 13 the vesicles that were lost were smaller ($P < 0.05$) than those that were maintained (Fig. 2).

The interovulatory interval for mares that did not become pregnant was shorter ($P < 0.05$) in mares with an intrauterine fluid collection score of 3 than in mares with a score of 0 or 1 (Fig. 3). The interval tended to be shorter ($P < 0.1$) in mares with a score of 2 than in mares with a score of 0 or 1. Similarly, the mean interval was shorter ($P < 0.05$) in mares with a biopsy score of 4, than in mares with a score of 1 or 2 (Fig. 3). The interval tended to be shorter ($P < 0.1$) in mares with a score of 4 than in mares with a score of 3.

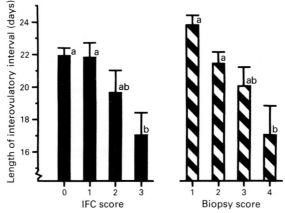

Fig. 3. Effect of intrauterine fluid collection (IFC) score and biopsy score on length of interovulatory interval. Lengths of intervals with different superscripts are different ($P < 0.05$).

The progesterone profiles for mares of different categories are illustrated in Fig. 4. The first significant difference in progesterone concentrations occurred on Day 8; the concentration was lowest ($P < 0.05$) for mares with a fluid collection score of 3 (Fig. 4). By Day 11, mares with a score of 2 or 3 had lower ($P < 0.05$) concentrations than those with a score of 1 or 0. There were no differences among groups on Days 15 and 20. By Day 11, the mean concentration was lower ($P < 0.05$) in the group that lost the embryo than in the group that maintained the embryo. The progesterone concentrations were lower ($P < 0.05$) on Days 8 and 11 in mares with a biopsy score of 4 than in mares with a score of 1.

Discussion

Intrauterine fluid collections have been described in other studies (Ginther & Pierson, 1984b; Ginther

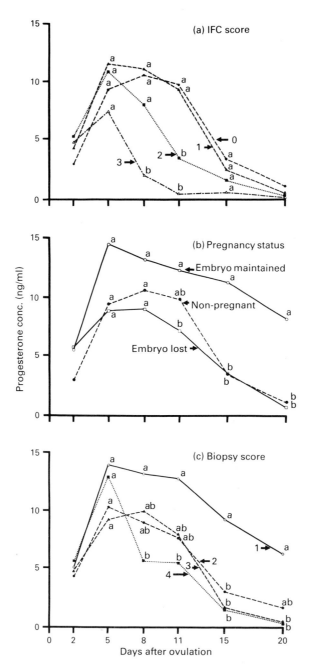

Fig. 4. Progesterone concentrations in mares in relation to (a) IFC score (0 = none; 1 = $\leqslant 10$ mm; 2 = 11–39 mm; 3 = $\geqslant 40$ mm); (b) pregnancy status; and (c) biopsy score (increasing score represents increasing inflammatory change). Within each day, values with different superscripts are different ($P < 0.05$).

et al., 1985a). The incidence found in Exp. 1 (27·9% of dioestrous mares) indicates that such fluid collections may involve a considerable portion of the population. However, the mares were being assembled for slaughter and the group may have included some mares being culled for infertility. The incidence of cysts in Exp. 1 (15·1%) was also high. No previous reports on the incidence of uterine cysts were found; however, they have been described ultrasonically (Ginther & Pierson, 1984b) and by gross and histopathology (Arthur, 1958; Kenney & Ganjam, 1975). According to Kenney (1978) glandular cysts do not become larger than 1 mm in diameter; therefore, the cysts visible by ultra-sound (≥ 3 mm diameter) were not glandular cysts. Cysts vary in size from microscopic to 15 cm (Kenney, 1978) and, because they can be very small, they may be very difficult to appreciate by palpation or gross pathology. Ultrasonic imaging is an ideal tool for studying the extent and the dynamics of the cystic condition.

There appeared to be a quantitative effect of uterine cysts on infertility since a decrease in pregnancy rate was not discernible until the affected mares were separated according to severity. Only the more severely affected group had a tendency to have a reduced pregnancy rate determined on Day 40, although the effect was much less profound than in mares with intrauterine fluid collec-tions. Further studies with greater mare numbers will be needed to clarify the relationship of cysts to infertility. Cysts were more common in older mares, and therefore age must be considered in attempts to attribute infertility problems to the presence of cysts. It has been reported (Ginther & Pierson, 1984b) that intrauterine fluid collections initially detected during dioestrus changed dramatically from one examination to the next. This observation was confirmed in Exp. 3 and the sinusoidal pattern of fluid collection size further defined the dynamic nature of such collections over the oestrous cycle. With reference to a normal 21-day cycle, the intrauterine fluid collections reached a minimum at about Day 5 and a maximum at about Day 16–17. Mares so affected tended to have fluid collections consistently from one cycle to the next and from one day to the next during dioestrus. About 75% of the mares in Exp. 3 had observable fluid collections for 5 or more days between Days 8 and 15. The best time to diagnose this condition therefore would be during mid- to late dioestrus which conveniently coincides with the time of early pregnancy diagnosis by ultrasound.

Intrauterine fluid collections during dioestrus were positively correlated to biopsy score and therefore were indicative of inflammation. Furthermore, higher biopsy scores and scores for fluid collections at dioestrus were associated with shorter interovulatory intervals and an early reduction of progesterone, implicating inflammation as the cause of premature luteal regression, presumably by early activation of uterine-induced luteolysis (Ginther, 1986). An observation of intrauterine fluid at dioestrus therefore has diagnostic importance in the reproductive evaluation of the mare, but the importance of free fluid at oestrus remains unclear. Although 2 of 3 mares with fluid collections only at oestrus became pregnant, neither maintained the embryo past Day 16. During pro-oestrus, the endometrial folds hypertrophy and become more oedematous, and it appears that a higher percent-age of mares have ultrasonically detectable intraluminal fluid at this time. A small amount of luminal fluid may be normal during oestrus but the incidence and extent of involvement relative to infertility needs further investigation. The fluid flux across the endometrium during the oestrous cycle and pregnancy is poorly understood. Perhaps an imbalance between fluid inflow and outflow exists in mares with fluid collections at dioestrus such that fluid builds up during dioestrus and proestrus when the cervix is closed but drains out during oestrus when the cervix is open. Although not critically examined in this study, the fluid from many mares contained inflammatory cells, some had frank pyometra, while still others had fluid collections in which the fluid was pale whitish-opalescent, mucoid and had few cells.

In the mares from Exp. 3, infertility was apparently due to early luteolysis of normally functioning corpora lutea rather than primary luteal insufficiency. The progesterone profiles were consistent for intrauterine fluid collection and biopsy groups and indicated that concentrations rose to normal by Day 5 and by Day 8 began to decline in more severely affected animals. This drop in progesterone was temporally related to the time at which the corpus luteum becomes sensitive to uterine luteolysin (Ginther, 1979). Blockage of uterine-induced luteolysis, on the other hand, starts at Day 11 and is not complete until Day 15 (Betteridge *et al.*, 1985; Ginther, 1986). None of the mares that lost the conceptus before Day 20 became pseudopregnant and embryonic loss preceded or was coincident with a return to oestrus in all cases. Of the 6 losses, 4 occurred during oestrus and the other 2 were lost 3 and 4 days, respectively, before the onset of oestrus. Of the 4 mares in which the cervix was examined on the day of loss, all 4 cervices were patent. No embryonic debris was detected by ultrasonography or by manual exploration of the cervix and vagina, but 2 mares developed intrauterine fluid collections before loss and 2 after loss occurred.

Similar to the previous report (Ginther *et al.*, 1985b), embryonic vesicles which were subsequently lost were smaller than maintained embryonic vesicles. The low pregnancy rate observed in this study was probably due to embryonic loss after entrance of the conceptus into the uterus but before the embryonic vesicles were detectable by ultrasound, rather than to a spermicidal effect (Woods *et al.*, 1985; Ball *et al.*, 1986).

We thank Dr R. M. Kenney and Dr K. Hinrichs for invaluable assistance in the interpretation of the endometrial biopsies; Marcia Campbell, Peggy Hoge and Lisa Kulick for technical assistance; Jacques Carriere for statistical help; and Upjohn Co. for Lutalyse. The work was supported by the College of Agricultural and Life Sciences, University of Wisconsin, Madison. Mares, supplies, facilities, labour and equipment were provided by Equi-Culture, Inc., Cross Plains, Wisconsin.

References

Arthur, G.H. (1958) An analysis of the reproductive function of mares based on post mortem examination. *Vet. Rec.* **70**, 682–686.

Ball, B.A., Little, T.V., Hillman, R.B. & Woods, G.L. (1986) Embryonic loss in normal and barren mares. *Proc. 31st Ann. Conv. Am. Assoc. equine Pract., Toronto*, p. 535–543.

Bergfelt, D.R. & Ginther, O.J. (1986) Follicular populations following inhibition of FSH with equine follicular fluid during early pregnancy in the mare. *Theriogenology* **26**, 733–747.

Betteridge, K.J., Renard, A. & Goff, A.K. (1985) Uterine prostaglandin release in relation to embryo transfer procedures and maintenance of the corpus luteum. *Eq. vet. J., Suppl.* **3**, 15–33.

Ginther, O.J. (1979) *Reproductive Biology of the Mare: Basic and Applied Aspects*, pp. 174–176, 200–202, 335–351. Equiservices, Cross Plains, Wisconsin.

Ginther, O.J. (1985) Embryonic loss in mares; incidence, time of occurrence, and hormonal involvement. *Theriogenology* **23**, 77–89.

Ginther, O.J. (1986) *Ultrasonic Imaging and Reproductive Events in the Mare*, pp. 259–272. Equiservices, Cross Plains, Wisconsin.

Ginther, O.J. & Pierson, R.A. (1983) Ultrasonic evaluation of reproductive tract of the mare: principles, equipment and techniques. *J. equine vet. Sci.* **3**, 195–201.

Ginther, O.J. & Pierson, R.A. (1984a) Ultrasonic anatomy of equine ovaries. *Theriogenology* **21**, 471–483.

Ginther, O.J. & Pierson, R.A. (1984b) Ultrasonic anatomy and pathology of the equine uterus. *Theriogenology* **27**, 505–516.

Ginther, O.J., Bergfelt, D.R., Leith, G.S. & Scraba, S.T. (1985a) Embryonic loss in mares: incidence and ultrasonic morphology. *Theriogenology* **24**, 73–86.

Ginther, O.J., Garcia, M.C., Bergfelt, D.R., Leith, G.S. & Scraba, S.T. (1985b) Embryonic loss in mares: pregnancy rate, length of interovulatory intervals, and progesterone concentrations associated with loss during days 11 to 15. *Theriogenology* **24**, 409–417.

Kenney, R.M. (1978) Cyclic and pathologic changes of the mare endometrium as detected by biopsy, with a note on early embryonic death. *J. Am. vet. med. Ass.* **172**, 241–262.

Kenney, R.M. & Doig, P.A. (1986) Equine endometrial biopsy. In *Current Therapy in Theriogenology*, 2nd edn, pp. 723–729. Ed. D. A. Morrow. W. B. Saunders, Philadelphia.

Kenney, R.M. & Ganjam, V.K. (1975) Selected pathological changes in the mare uterus and ovary. *J. Reprod. Fert., Suppl.* **23**, 335–339.

Pierson, R.A. & Ginther, O.J. (1985) Ultrasonic evaluation of the corpus luteum of the mare. *Theriogenology* **23**, 795–806.

SAS User's Guide (1985) Statistics, Version 5 edn, p. 956. SAS Institute Inc., Cary.

Snedecor, G.W. & Cochran, W.G. (1980) *Statistical Methods*, pp. 233–235. Iowa State University Press, Ames.

Woods, G.L., Baker, C.B., Hillman, R.B. & Schlafer, D.H. (1985) Recent studies relating to embryonic death in the mare. *Equine vet. J., Suppl.* **3**, 104–107.

J. Reprod. Fert., Suppl. **35** (1987), 455–459

Printed in Great Britain
© 1987 Journals of Reproduction & Fertility Ltd

Early pregnancy loss in brood mares

G. L. Woods*, C. B. Baker†, J. L. Baldwin‡, B. A. Ball, J. Bilinski§, W. L. Cooper‡, W. B. Ley¶, E. C. Mank♯ and H. N. Erb

Department of Clinical Sciences, New York State College of Veterinary Medicine, Cornell University, Ithaca, New York 14853; †R. D. 2, Owego, New York 13928; ‡Lana Lobell Farms, Montgomery, New York 12549; §Hudson Valley Equine Center, North Chatham, New York 12132; ¶Virginia-Maryland Regional College of Veterinary Medicine, Blacksburg, Virginia 24061; and ♯Almahurst Farm, Bloomsburg, Pennsylvania 17815, U.S.A.

Summary. During 1985, linear-array ultrasonography was used to study early pregnancy loss in commercial brood mares: 600/1115 (54%) of the cycles resulted in detected pregnancy at Week 2 after ovulation and 80 (13%) of these pregnancies resulted in early pregnancy loss. The pregnancy loss rate was significantly higher ($P < 0.05$) for twin pregnancies (10/41) than for singleton pregnancies (70/559). The pregnancy loss rate was significantly higher ($P < 0.05$) at 2–4 weeks (29/60) than at 6–8 weeks (12/60). The pregnancy rate was significantly less ($P < 0.05$) for post-partum mares inseminated at the foal heat (157/302) than for those inseminated at a subsequent oestrus (203/334). Mares with a history of endometritis had a significantly higher ($P < 0.05$) per cycle pregnancy loss rate (7/26) than did mares with a history of not having endometritis (64/498). When mares that had lost pregnancies were re-mated, 37/75 (49%) detectable pregnancies resulted and 7 (19%) of these pregnancies were again lost. The per cycle pregnancy rate was 56% (153/273), 55% (177/319), 60% (130/216), 51% (72/142), 45% (34/76) and 33% (12/36) for mares aged 2–5, 6–9, 10–13, 14–17, 18–21 and >21 years, respectively. The corresponding per cycle pregnancy loss rates were 12%, 14%, 9%, 14%, 24% and 33%.

Introduction

Transrectal ultrasonography has made possible earlier detection of equine pregnancy, thereby allowing for earlier detection and study of pregnancy loss in the mare (Chevalier & Palmer, 1982). Experienced ultrasonographers, using 5 MHz transducers, can detect pregnancy with 98% accuracy by Week 2 after ovulation (Ginther, 1986). Ultrasonography has been useful in the study of early pregnancy loss in commercial brood mares since many veterinarians routinely use ultrasound machines with 5 MHz transducers and the danger of ultrasonography to pregnancy has been estimated to be minimal (Ginther, 1986).

After initial pregnancy detection at Week 2 in a group of commercial brood mares, it was observed that the pregnancy loss rate for Weeks 2–4 after ovulation was significantly higher than the preganancy loss rate for the two subsequent 2-week periods (Woods *et al.*, 1985). This obser-vation led to the hypothesis that pregnancy losses might be even higher during earlier stages of pregnancy. This hypothesis was supported when Ball *et al.* (1986) determined the pregnancy loss rate before Week 2 for a group of barren mares to be 73%. The objective of the current study was to characterize further early pregnancy loss in commercial brood mares.

*Present address: Department of Veterinary Clinical Medicine and Surgery, WOI Regional Program of Veterinary Medicine, University of Idaho, Moscow, Idaho 83843, U.S.A.

Materials and Methods

Data were collected by 6 veterinarians in New York, Pennsylvania and Virginia during February through August, 1985. Four veterinarians examined Standardbred mares, one veterinarian examined Saddlebred mares and the other examined Thoroughbred mares. Standardbred mares were artificially inseminated with semen from one of 40 Standardbred stallions. Saddlebred mares were artificially inseminated with semen from one of two Saddlebred stallions. Thoroughbred mares were inseminated naturally by one of 18 Thoroughbred stallions. After insemination, ovulations were detected by rectal palpation. Four sequential pregnancy examinations were completed on each mare by ultrasonography at Day 14–18 and Day 24–28 after ovulation and by ultrasonography or rectal palpation at Day 34–38 and Day 44–48 after ovulation. Linear-array ultrasound machines with 5 MHz transducers were used. Some of the pregnancy losses which occurred beyond Day 48 after ovulation were also detected by rectal palpation.

Pregnancy rates were calculated as the percentage of all mare cycles in which pregnancy was detected. Pregnancy loss rates were calculated as the percentage of detected pregnancies in which pregnancy loss was identified. Mares were classified as maiden, foaling or barren. It was noted whether mares had single or twin detected pregnancies and when pregnancy loss occurred. To identify factors which would influence pregnancy and pregnancy loss rates, the mares history of pregnancy loss and history of endometritis were recorded. It also was noted whether mares were mated at the foal heat.

The cycles of those mares that were inseminated after pregnancy loss were also examined. After pregnancy losses were detected, endometrial swabs (Knudsen, 1964) for cytological evaluation and endometrial biopsies (Kenney, 1978) for inflammatory changes were taken from some of the mares. These samples were compared with endometrial swab and biopsy samples from 5 experimental mares in which pregnancy losses were induced with prostaglandin F-2α at Days 32–41 after ovulation. Cytologies were classified as positive for inflammation if more than one neutrophil per high power field ($\times 450$) was observed, and biopsies were classified as positive for inflammation if a moderate to severe inflammatory infiltrate was noted in the endometrial biopsy. Two individuals independently interpreted the endometrial cytologies and biopsies. The pregnancy rates and pregnancy loss rates of mares 2–5, 6–9, 10–13, 14–17, 18–21 and >21 years of age were calculated.

Proportional data were analysed by χ^2 analysis and quantitative data were analysed by Student's t test.

Results

The overall per cycle pregnancy rate at Week 2 was 54% (Table 1). The pregnancy rate was significantly lower ($P < 0.05$) for barren mares than for maiden or foaling mares (Table 1). The pregnancy loss rate was significantly higher ($P < 0.05$) for twin pregnancies than for singleton pregnancies (Table 1). Of the 60 singleton pregnancies which were lost between 2 and 8 weeks after ovulation, more pregnancies tended ($P < 0.1$) to be lost at 2–4 weeks than at 4–6 weeks, and significantly more ($P < 0.05$) pregnancies were lost at 2–4 weeks than at 6–8 weeks.

Mares with a history of pregnancy loss versus mares with a history of not having pregnancy loss (Table 1) had no significant differences ($P > 0.1$) in pregnancy rates or pregnancy loss rates. Mares with a history of endometritis had no significant difference ($P > 0.1$) in pregnancy rates but had a significantly higher ($P < 0.05$) rate of pregnancy loss than did mares with a history of not having endometritis (Table 1). The pregnancy rate (Table 1) was significantly lower ($P < 0.05$) for mares mated at the foal heat than for mares mated after the foal heat, but there was no significant difference ($P > 0.1$) in pregnancy loss rates for these two categories of mare.

Mares were inseminated in 75 cycles after the 80 detected pregnancy losses (Table 1): 37 pregnancies were detected and 7 of these pregnancies were again lost for a 19% repeat pregnancy loss rate. When compared with all mare cycles, the pregnancy rate after pregnancy loss was not significantly less ($P > 0.1$) and the pregnancy loss rate was not significantly higher ($P > 0.1$). Significantly more ($P < 0.05$) pregnancies were lost from reimpregnations after twin pregnancy loss than after loss of a singleton pregnancy.

The results of endometrial swabs for cytological evaluation and endometrial biopsies for evaluation of inflammatory changes after spontaneous or induced pregnancy losses are shown in Table 2. Mares with spontaneous pregnancy losses tended ($P < 0.1$) to have a higher ratio of endometrial swabs which were positive for inflammation and a lower ratio of those that were negative for inflammation by the endometrial swab and the endometrial biopsy.

The pregnancy rates and pregnancy loss rates for the mares of different ages are shown in Table 3. The pregnancy loss rates appeared to be inversely related to the pregnancy rate.

Table 1. Pregnancy and pregnancy loss rates for mares

	Pregnancy rate	Pregnancy loss rate
Classification of mare		
Maiden	76/132 (58%)[a]	9/76 (12%)
Barren	139/293 (47%)[b,a]	19/139 (14%)
Foaling	385/690 (56%)[b]	52/385 (14%)
Total	600/1115 (54%)	80/600 (13%)
Singleton pregnancies		70/559 (13%)
Twin pregnancies		10/41 (24%)
Week of singleton loss		
2–4		29/60 (48%)[a]
4–6		19/60 (32%)[c]
6–8		12/60 (20%)[b]
Mares with a history of pregnancy loss		
Yes	58/125 (46%)	13/58 (22%)
No	158/358 (44%)	36/158 (23%)
Mares with a history of endometritis		
Yes	26/51 (51%)	7/26 (27%)[a]
No	498/932 (53%)	64/498 (13%)
Mares mated at foal heat		
Yes	157/302 (52%)[a]	18/157 (11%)
No	203/334 (61%)[b]	24/203 (12%)
Outcome of cycles after pregnancy loss		
Singleton pregnancy losses	32/64 (50%)	5/32 (16%)[a]
Twin pregnancy losses	5/11 (45%)	2/5 (40%)[b]
Total	37/75 (49%)	7/37 (19%)

[a,b]$P < 0.05$; [a,c]$P < 0.1$.

Table 2. Comparison of uterine inflammation from endometrial swabs and endometrial biopsies from mares after spontaneous pregnancy loss or that induced by prostaglandin F-2α

	Spontaneous pregnancy loss	Induced pregnancy loss
No. of mares with indications of inflammation from:		
Cytology only (%)	8/35[a] (23%)	0/5[c] (0%)
Biopsy only (%)	7/35 (20%)	1/5 (20%)
Both cytology and biopsy (%)	6/35 (17%)	0/5 (0%)
Neither cytology nor biopsy (%)	14/35[a] (40%)	4/5[c] (80%)

[a,c]$P < 0.1$.

Discussion

The overall per cycle pregnancy rate of 54% appeared higher than a 47% per cycle pregnancy rate which was reported by Sullivan *et al.* (1975). The apparently lower rate in the previous study may be in part due to the limitations of pregnancy detection by rectal palpation, since earlier pregnancies which were detected at Week 2 and lost before Week 4 in this study may have been missed by rectal

Table 3. Pregnancy rates and pregnancy loss rates at week 2 after ovulation for 1062 mares of different ages

	Age (years)					
	2–5	6–9	10–13	14–17	18–21	> 21
No. of mares	273	319	216	142	76	36
Pregnancy rate						
No. of mares pregnant	153	177	130	72	34	12
%	56	55	60	51	45	33
Lost pregnancies (%)	18 (12)	24 (14)	12 (9)	10 (14)	8 (24)	4 (33)

palpation in the earlier study. The overall pregnancy loss rate of 13% was similar to an 11% pregnancy loss rate for commercial brood mares studied with a comparable experimental protocol (Woods *et al.*, 1985).

The lower pregnancy rate for barren mares compared to maiden or foaling mares has been reported previously (Sullivan *et al.*, 1975). Although the detected pregnancy loss rate was not statistically higher for barren than for maiden or foaling mares, it is possible that the lower pregnancy rate in barren mares may have been due to a higher undetected pregnancy loss rate (pregnancies lost before the first pregnancy examination at Days 14–18). Ball *et al.* (1986) showed that the fertilization rate of barren mares was similar to that of young reproductively normal mares and that the lower detected pregnancy rate of barren mares was due to a higher rate of pregnancy losses occurring before Week 2.

The higher rate of pregnancy loss in twin versus singleton pregnancies may have been related to the veterinarians' management of twin pregnancies: 2 twin pregnancy losses were induced with prostaglandin F-2α injections and 4 twin pregnancy losses occurred after manual embryo crushing. No mares with singleton pregnancies received prostaglandin F-2α injections or manual embryo crushing.

The higher singleton pregnancy loss rate at 2–4 weeks versus 4–6 or 6–8 weeks after ovulation is consistent with a previous study (Woods *et al.*, 1985). The 60 singleton pregnancy losses which were detected between 2 and 8 weeks after ovulation were possibly the tailing off of a higher rate of undetected pregnancy losses which occurred before Week 2. Ball *et al.* (1986) demonstrated that pregnancy losses were greatest before Week 2 in barren mares. On the assumption that the pregnancy loss rate was highest before the detection of pregnancy with ultrasonography (Week 2), it is possible that detected pregnancy losses represented only a part of the total number of pregnancy losses which occurred.

The significantly higher incidence of pregnancy loss in mares with a history of endometritis suggests that endometritis may have been a cause of some pregnancy losses. The tendency for more endometrial samples from mares in this study to show evidence of inflammation after loss than did endometrial samples from prostaglandin F-2α-treated control mares suggests that some pregnancy losses in this study were related to uterine infections. The lower pregnancy rate for mares inseminated at the foal heat than for mares inseminated after the foal heat has been reported (Caslick, 1937; Jennings, 1941; Trum, 1950). The relatively high pregnancy rate and moderate incidence of observed pregnancy losses in mares re-mated after an initial pregnancy loss suggest that many causes of pregnancy loss were not of a permanent or enduring nature.

Increased age was associated with decreased pregnancy rates. The apparent inverse relationship between pregnancy rates and pregnancy loss rates with age tends to support the hypothesis that some of the undetected pregnancies were actually undetected pregnancy losses which occurred before Week 2. The pregnancy rate was relatively stable from 2 to 13 years, then appeared to decline after 14 years. One explanation for this change in relationship of pregnancy rate to increased mare age would be that multiple factors might cause the pregnancy losses and thereby affect the pregnancy rates. One group of such factors could be intrinsic such as chromosomal abnormalities which appear to increase proportionately with increasing age in other species (Bouie & Bouie, 1976). The older the mare, the greater the chance of these intrinsic factors causing pregnancy loss. A second group might be extrinsic factors such as nutrition and social stress which would not increase proportionately with increasing age. For example, young mares seem to be more susceptible to nutritional deficiencies or social stress than middle-aged mares (Mitchell & Allen, 1975). These factors might interact differently within each of the first three age groups to result in a relatively constant pregnancy rate until 13 years of age. Further work is needed to clarify these diverse factors.

We thank Thornbrook Farms and the Harry M. Zweig Memorial Trust for generous financial support; and A. Guion, K. Roneker, D. Vanderwall and J. Weber for excellent technical assistance.

References

Ball, B.A., Little, T.V., Hillman, R.B. & Woods, G.L. (1986) Pregnancy rates at Days 2 and 4 and estimated embryonic loss rates prior to Day 4 in normal and infertile mares. *Theriogenology* **26,** 611–619.

Bouie, J.G. & Bouie, A. (1976) Chromosome anomalies in early spontaneous abortion—their consequences on early embryogenesis and *in vitro* growth of embryonic cells. *Curr. Top. Pathol.* **62,** 193–208.

Caslick, E.A. (1937) The sexual cycle and its relation to ovulation with breeding records of the Thoroughbred mare. *Cornell Vet.* **27,** 187–206.

Chevalier, F. & Palmer, E. (1982) Ultrasonic echography in the mare. *J. Reprod. Fert., Suppl.* **32,** 423–430.

Ginther, O.J. (1986) *Ultrasonic Imaging and Reproductive Events in the Mare,* 196 pp. Equiservices, Cross Plains, Wisconsin.

Jennings, W.E. (1941) Some common problems in horse breeding. *Cornell Vet.* **31,** 117.

Kenney, R.M. (1978) Cyclic and pathologic changes of the mare endometrium as detected by biopsy, with a note on early embryonic death. *J. Am. vet. med. Ass.* **172,** 241–262.

Knudsen, O. (1964) Endometrial cytology as a diagnostic aid in mares. *Cornell Vet.* **54,** 415–422.

Mitchell, D. & Allen, W.R. (1975) Observations on reproductive performance in the yearling mare. *J. Reprod. Fert., Suppl.* **23,** 531–536.

Sullivan, J.J., Turner, P.C., Self, L.C., Gutteridge, H.B. & Bartlett, D.E. (1975) Survey of reproductive efficiency in the Quarter Horse and Thoroughbred. *J. Reprod. Fert., Suppl.* **23,** 315–318.

Trum, B.F. (1950) The estrous cycle of the mare. *Cornell Vet.* **40,** 17–23.

Woods, G.L., Baker, C.B., Hillman, R.B. & Schlafer, D.H. (1985) Recent studies relating to early embryonic death in the mare. *Equine vet. J., Suppl.* **3,** 104–107.

J. Reprod. Fert., Suppl. **35** (1987), 461–467

Printed in Great Britain
© 1987 Journals of Reproduction & Fertility Ltd

Changes in maternal hormone concentrations associated with induction of fetal death at Day 45 of gestation in mares

L. B. Jeffcott, J. H. Hyland, A. A. MacLean, T. Dyke and
G. Robertson-Smith*

Department of Veterinary Clinical Sciences, University of Melbourne, Werribee, Victoria 3030, Australia

Summary. Pregnant Standardbred mares were allocated to 2 groups. On Day 45 of gestation, 20–45 ml saline (240 g NaCl/l) were injected into the fetal sacs of 10 mares, and the other 10 mares were given sham treatment. Post-operative plasma oestrone sulphate concentrations were lower ($P < 0.01$) on Days 48–55 in saline-treated mares than in sham-treated mares. Mean plasma progesterone profiles were similar in the two groups of mares, although post-operative luteolysis occurred in 4 saline-treated mares. There was no difference in plasma CG profiles between the 2 groups, except that CG concentrations in saline-treated mares were generally lower than those of sham-treated mares. There was generally a post-operative loss of uterine and cervical tone in saline-treated mares.

These results show that the maintenance of maternal plasma oestrone sulphate concentrations requires the presence of a viable embryo and confirm that luteolysis can occur despite high plasma CG concentrations after fetal death.

Introduction

Embryonic death is an important source of economic loss to the horse breeding industry. In Australia, the incidence of embryonic death at less than 5 months gestation has been estimated at 13·5% with 75% occurring before Day 75 (Bain, 1969). The diagnosis of embryonic death is usually made by using the techniques of palpation of the uterus *per rectum* or ultrasound echography (Ginther *et al.*, 1985). A useful addition to these techniques would be a non-invasive method which allowed fetal well-being to be monitored. Measurement of maternal progesterone concentrations is an unreliable diagnosis of embryonic death because pseudopregnancy frequently occurs after loss of the embryo, with maintenance of high progesterone concentrations (Allen, 1978). Likewise, maternal plasma concentrations of chorionic gonadotrophin (CG) indicate only the presence of functional endometrial cups (Allen & Moor, 1972). However, some reports suggest that CG concentrations are lower in commercial stud mares which experience fetal death (Darenius *et al.*, 1982; Fay & Douglas, 1982).

Plasma and urinary oestrone sulphate concentrations in pregnant mares increase from around Day 37 of gestation (Kindahl *et al.*, 1982; Hyland *et al.*, 1984; Evans *et al.*, 1984), and this hormone has been identified in the horse embryo from Day 20 of pregnancy (Heap *et al.*, 1982). The above

---

*Present address: Department of Large Animal Clinics, School of Veterinary Medicine, Purdue University, West Lafayette, IN 47907, U.S.A.

evidence suggests that measurement of maternal oestrone sulphate concentrations could be used to monitor embryonic viability in mares after Day 40 of gestation.

The aim of this experiment was to monitor short-term changes in maternal oestrone sulphate, progesterone and CG in mares, after induction of fetal death on Day 45 of pregnancy.

Materials and Methods

The experiment was carried out during the breeding seasons of 1984 and 1985.

Experimental animals. Maiden Standardbred mares, in good condition and about 3–7 years old were used for the experiment. They were maintained in good body condition by feeding pasture hay twice daily. The mares were kept in outside paddocks until Day 44 of gestation, at which time they were brought into sheltered yards where they were maintained until the end of the experiment. While confined to yards, the mares received lucerne hay in the morning and a grain mixture in the evening.

The mares were teased every 2 or 3 days and mated with a fertile stallion every 2nd day during oestrus, until ovulation was detected by palpation of the ovaries *per rectum*. Ultrasound echography was used to confirm pregnancy 12–15 days after ovulation. The mares were examined every 2 or 3 days with an ultrasound scanner until Day 45 of gestation, from which time examinations were made daily until fetal death was confirmed in treated mares or until Day 55 of gestation in controls.

Blood samples (10 ml) were collected into heparin (10 i.u./ml) every 2 days between Days 10 and 44 of gestation (Day 0 = day of ovulation) and daily from Day 46 to 55 of gestation. On Day 45 of gestation, blood samples were collected at the time of induction of embryonic death or sham operation and at 15, 30, 60, 120, 360 and 540 to 720 min after treatment. Plasma was removed from the blood and stored at $-20°C$ until assayed for oestrone sulphate, progesterone and CG.

Induction of fetal death. Mares were fasted for 24 h before surgery on Day 45 of gestation. Anaesthesia was induced with glyceryl guiacolate and thiopentone sodium. The mares were intubated and anaesthesia was maintained with halothane in oxygen. A ventral midline laparotomy was performed with the mare in dorsal recumbency. The uterus was located and gently lifted through the incision to expose the pregnant horn. In saline-treated mares (N = 10), a 26-gauge × 13 mm hypodermic needle was inserted into the fetal sac. A 50 ml sterile syringe was attached to the needle and fetal fluid (20–45 ml) was withdrawn. An equivalent volume of sterile saline (240 g NaCl/l) was then slowly injected into the fetal sac through the same needle. In control mares (N = 10), the uterus was exposed through the incision for ∼2 min, without further treatment. The uterus was replaced after saline or sham treatment and the incision was closed.

The uterus of each mare was examined by rectal palpation and ultrasound echography at various times before and after saline or sham treatment. The presence or absence of a fetal bulge was noted, as was uterine tone, which was scored from 1 to 5, 5 being maximum turgidity. Observation of the external os of the cervix was done at various times before and after treatment. Cervical tone was scored from $+2$ (cervical folds very tight, consistent with pregnancy) to -2 (cervical folds very loose and oedematous, consistent with oestrus).

Hormone assays. Oestrone sulphate was measured using a direct radioimmunoassay on 20 µl plasma (Hyland *et al.*, 1984). The antibody was raised in sheep against oestrone-3-methyl-phosphothionate coupled to bovine serum albumin (Cox *et al.*, 1979). Major cross-reactants to the antiserum were oestrone (200%), equilin (34%) and equilenin (12%). Oestrone sulphate (Sigma Co., St Louis, MO, U.S.A.) standards were prepared in gelding plasma. Standards and assay samples were run in triplicate. Assay sensitivity (lowest concentration significantly different from zero, Burger *et al.*, 1972) was 0·5 ng/ml. The intra- and inter-assay coefficients of variation, using plasma from a pregnant mare, were 4·9% and 14·7%, respectively (n = 6).

Plasma progesterone concentrations were measured using a commercial radioimmunoassay kit (Farmos Diagnostica, Turku, Finland). The antiserum cross-reacts <1% with all major steroids tested, except for 11-hydroxyprogesterone (80%), 5α-dihydroprogesterone (8·8%) and 5β-dihydroprogesterone (7·1%). Sensitivity of the assay was 0·1 ng/ml. Recovery of progesterone added to plasma from an anoestrous mare at concentrations of 1, 2·5, 5 and 10 ng/ml was quantitative ($y = 0·93x + 0·72$; $r = 0·98$; n = 21). The intra- and inter-assay coefficients of variation, using dioestrous mare plasma, were 8·8% and 6·6% respectively. Parallelism was observed between the standard curve and serial dilutions of plasma from a dioestrous mare.

Chorionic gonadotrophin was measured by heterologous radioimmunoassay. Purified sheep LH (Papkoff G3222B), radioiodinated with iodine-125 (Greenwood *et al.*, 1963), was used as tracer. Rabbit anti-ovine LH serum (GDN15) was diluted 1:50 000 in phosphate-buffered saline (PBS) containing 1% bovine serum albumin at pH 7·4. In the absence of added CG, the antibody bound 28–44% of the ^{125}I-labelled LH (100 µl). A commercial preparation of horse CG (Folligon: Intervet Australia P/L, NSW, Australia) was used as standard (100 µl; 0·005–0·5 i.u./tube) and prepared in anoestrous mare plasma. Standards were added to 100 µl antibody and 100 µl ^{125}I-labelled LH in triplicate tubes. Unknown plasma samples (50 µl of 1:100 dilution in PBS) were added to antibody and tracer and incubated for 48 h at 4°C. Donkey anti-rabbit IgG (Wellcome Research Labs, Beckenham, U.K.) was diluted 1:30 in PBS and added to all tubes, vortexed and incubated at 4°C for a further 24 h. At the end of this period, the tubes were centrifuged at 1500 *g* for 10 min and the supernatant poured off. Bound radioactive hormone was counted in a Packard Auto-Gamma 5780.

Features of the antibody have been published elsewhere (Nett *et al.*, 1975). Sensitivity of the assay (lowest concentration of CG significantly different from zero; Burger *et al.*, 1972) was 0·2 i.u./ml. Intra- and inter-assay coefficients of variation, using plasma from a mare at Day 50 of gestation, were 14·1% and 24·3%, respectively (*n* = 7). Commercial horse CG, added to anoestrous mare plasma at concentrations of 2·5, 5·0, 10 and 25 i.u./ml, was recovered quantitatively, with a regression of $y = 0.95x + 0.8$, $r = 0.997$ (*n* = 20). Parallelism was observed between the standard curve and increasing dilutions of plasma from a pregnant mare.

Statistical analyses. Student's *t* test was used to compare cervical and uterine tone. A method for analysing continuous data over time (Gill & Hafs, 1971) was used to compare hormone profiles in saline- and sham-treated mares. Differences in hormone concentrations between saline- and sham-treated mares at particular sampling times were compared using Student's *t* test.

Results

Mare and fetus examinations

Ultrasound echography showed that the fetal heart beat was absent in 8 mares within 3 h and in 2 mares within 24 h of treatment with hypertonic saline. The fetal bulge remained palpable from 6 h to 8 days after treatment in the saline-treated mares. The first noticeable change was a thickening and softening of the uterine wall in the area of the fetal bulge. This was followed by a reduction in size. Within 1–3 days of detectable changes in the fetal bulge, the fetal sac was not palpable *per rectum*.

Between Days 45 and 46, cervical tone decreased from + 1·2 to 0 in the saline-treated mares and from + 1·3 to + 0·7 in the sham-treated controls. Cervical tone score remaind lower in the saline-treated mares until Day 55, but increased to pretreatment scores in the controls. The crevices of the treated mares were significantly (*P* < 0·05) softer than those of the controls on Days 49–53 of gestation. In one saline-treated mare, the fetal membranes were observed protruding from the cervical os 5 days after operation, but were absent by the 6th post-operative day. Uterine tone was maintained in both groups of mares until 3 days after treatment at which time there was a decrease in uterine tone in treated mares. By 8 days after induction of fetal death, uterine tone was significantly (*P* < 0·05) lower than in mares which underwent sham treatment, and remained this way until the end of the experiment.

Hormone measurements

Oestrone sulphate. There was a significant (*P* < 0·01) treatment effect on mean plasma concentrations. Values were < 1 ng/ml in both groups until Day 38 of gestation (Fig. 1a). Between Days 38 and 45 of gestation there was a rapid rise in plasma oestrone sulphate concentration to 3·8 ng/ml (sham treatment) and 5·1 ng/ml (saline treatment). After saline or sham treatment there was a further marked rise in plasma concentrations, which peaked at 30 min in both groups. Thereafter, there was a gradual decline and, by 24 h after saline or sham treatment, concentrations had reached pretreatment levels. Mean plasma oestrone sulphate concentrations in sham-treated mares rose from 3·8 ng/ml at Day 46 to 4·7 ng/ml by Day 55, when sampling was discontinued. By contrast, mean plasma concentrations in saline-treated mares fell from 4·3 ng/ml at Day 46 to 1·3 ng/ml by Day 55.

Progesterone. Mean plasma progesterone concentrations (Fig. 1b) remained between 4 and 5 ng/ml in both groups from Day 10 to Day 38 of gestation. On Day 38, plasma progesterone concentrations increased and by Day 45 were approximately 6 ng/ml in both groups. There was a marked rise in mean plasma progesterone concentrations which peaked 15 min after saline or sham treatment on Day 45. A post-operative decline in mean plasma progesterone concentrations occurred to below pre-treatment concentrations by 9–12 h. However, progesterone concentrations

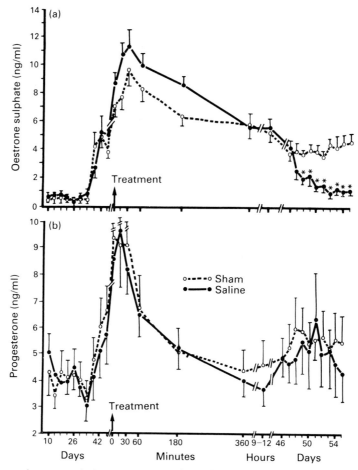

Fig. 1. Changes in maternal plasma concentrations of (a) oestrone sulphate and (b) progesterone in mares in which fetal death was induced by injection of 24% (w/v) saline into the fetal sac, or in mares which were sham-treated (time 0). Values are mean ± s.e.m. for 10 mares per group. *$P < 0.05$ compared with value for sham-treated mares.

increased over the next 3 days in both groups before falling to 5·7 ng/ml (sham treated) and 4·3 ng/ml (saline treated) by Day 55 of gestation. Mean plasma progesterone profiles for sham- and saline-treated mares were similar, with no statistically significant differences between the 2 groups at any time sampled. However, luteolysis apparently occurred in 4 saline-treated mares in which plasma progesterone concentrations declined after treatment to 10–30% of pre-treatment concentrations by the end of the experiment.

Chorionic gonadotrophin. Mean maternal plasma CG concentrations (Fig. 2) were baseline until Day 38 of gestation. Between Days 38 and 45 of gestation there was a sharp rise in plasma CG concentrations, which were higher in the sham-treated than in the saline-treated mares. There was a decline in plasma CG concentrations in both groups after treatment. By 24 h after operation, plasma CG had recovered to pretreatment concentrations. From Day 46 to Day 55, plasma CG concentrations increased rapidly in parallel, and, except for Day 49, were always higher in the sham-treated mares. Mean CG concentrations were significantly ($P < 0.05$) higher in sham-treated mares on Day 42, Day 45 at 3 h and 6 h after treatment and on Day 46 and Day 47. There was no overall treatment effect on plasma CG concentrations.

Discussion

Injection of 24% saline into the fetal sac resulted in fetal death in all treated mares whereas all fetuses survived in the sham-treated group. This method has been used to induce abortion in women (Csapo *et al.*, 1969), and was chosen because of the potential trauma to the uterus by methods such as aspiration of fetal fluids *per vaginam* (Pascoe, 1979) or crushing of the fetal sac *per rectum* (Roberts, 1982). These methods may also alter the secretion of oestrone sulphate and CG from the pregnant uterus and its contents. Prostaglandin F-2α was not used to terminate pregnancy because of its luteolytic effects and because prostaglandins are less effective in causing abortion in mares after formation of the endometrial cups (Ginther, 1979).

Although not critically examined, uterine tone gradually decreased in 5 of 10 saline-treated mares within 4 days of operation. Uterine tone was maintained in the sham-treated mares, all of which retained viable embryos until the end of the experiment. Hoppe (1968) noted that uterine tone was maintained in mares which lost embryos during the second month of gestation, and so it appears that there is some variation between mares as regards maintenance of uterine tone after fetal loss.

Cervical tone was reduced in saline-treated mares within 24 h of surgery. Cervical relaxation, similar to the oestrous condition, occurred in 2 treated mares. Fetal membranes were observed hanging through the external cervical os on Day 50, which coincided with maximum cervical relaxation in one of these mares.

Maternal plasma oestrone sulphate profiles up to Day 45 of gestation were similar to published data (Kindahl *et al.*, 1982; Hyland *et al.*, 1984). Coincident with the time of treatment or sham-treatment there was a sharp rise of mean plasma oestrone sulphate. This rise occurred in both groups of mares and may have been due to anaesthesia or handling the uterus during exposure of

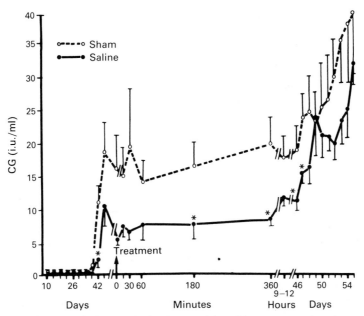

Fig. 2. Changes in maternal plasma chorionic gonadotrophin concentrations in mares in which fetal death was induced by injection of 24% (w/v) saline into the fetal sac, or in mares which were sham-treated (time 0). Values are mean ± s.e.m. for 10 mares per group. *$P < 0.05$ compared with value for sham-treated mares.

the pregnant horn through the abdominal incision. By Day 46 plasma oestrone sulphate concentrations had returned to preoperative levels. In treated mares the concentrations continued to decline, and between Days 48 and 55 were significantly lower than at the same days in sham-treated mares. These observations suggest that maintenance of oestrone sulphate concentrations requires the presence of a viable fetus. Oestrone sulphate production during pregnancy is dependent on a viable fetus in cows (Heap & Hamon, 1979) and sows (Gadsby *et al.*, 1980), although placental oestrogen production in the cow is first detected from Day 150 of gestation. By comparison, oestrogens are produced by the early (Day 15) pig embryo and converted to the sulphate by endometrial enzymes (Bazer & Thatcher, 1977) before release into the maternal circulation. It appears that the horse conceptus is similar to the embryo of the pig, because a sulphotransferase system has been identified in the endometrium of pregnant mares and the yolk sac of horse embryos contains high concentrations of oestrone sulphate between Days 30 and 40 of gestation (Heap *et al.*, 1982). The presence of a living embryo in these species is therefore essential for significant oestrone sulphate production. The post-operative decline in plasma oestrone sulphate concentrations in saline-treated mares in this study confirms these observations.

Terqui & Palmer (1979) found that the sustained rise in 'total oestrogens' which began between Days 35 and 40 in pregnant mares could be abrogated or reduced by ovariectomy at Day 34 or Day 42 of gestation, whereas ovariectomy at Day 56 of gestation made little difference to the increase in plasma oestrone sulphate concentrations before or after this time. These authors concluded that the initial oestrone sulphate rise was ovarian in origin. It is difficult to reconcile their results and the results of this study, unless the ovaries are supplying oestrogen precursors to the developing fetus.

Maternal plasma progesterone profiles to Day 45 of gestation were consistent with published data (Burns & Fleeger, 1975; Holtan *et al.*, 1975). There was a marked rise in mean plasma progesterone concentrations in both groups of mares, which peaked at 0 and 15 min relative to saline or sham treatment, respectively. Progesterone concentrations declined over the next 24 h to reach pretreatment concentrations. Progesterone concentrations in 4 saline-treated mares declined between Days 46 and 52 to <1 ng/ml in 2 mares and around 2 ng/ml in the other 2 mares. By comparison, Allen (1978) observed maintenance of luteal function in 5 ponies undergoing fetal death after Day 36 of gestation. This suggests that CG is not entirely responsible for maintaining progesterone secretion by the ovaries after formation of the endometrial cups. In this experiment, observations were continued until Day 55 of gestation. It is therefore possible that luteolysis could have occurred after the experimental period in some or all of the 6 mares which maintained high progesterone concentrations to Day 55 of gestation.

Maternal plasma CG concentrations were similar to published data (Nett *et al.*, 1975; Fay & Douglas, 1982). Between Days 42 and 55 of gestation, concentrations in saline-treated mares were lower than those in controls, except at Day 49. Despite this difference, CG profiles were similar between the two groups. These data are similar to the results of Mitchell (1971) who reported that fetal loss after formation of the endometrial cups did not appear to affect subsequent secretion of CG. However, Fay & Douglas (1982) observed that plasma CG concentrations at Days 42–45 of gestation were lower in mares that lost fetuses (3·3 i.u./ml) than in mares which foaled normally (19·6 i.u./ml). This suggests that reduced CG production may be detrimental to fetal viability and is an observation worthy of further study.

This experiment has shown that measurement of plasma oestrone sulphate concentrations is a useful method for monitoring fetal viability in mares after Day 40 of gestation. It has also shown the variable short-term change in plasma progesterone concentrations in mares after induction of fetal death at Day 45 of gestation. The rise in plasma CG concentrations despite fetal death confirms previous observations that endometrial cup activity is independent of a viable fetus.

We thank Dr G. D. Niswender for supplying LH antiserum for use in the CG assay; Dr R. Cox for supply of the antiserum to oestrone sulphate; Debra Langsford and Debbie Yeomans for hormone assays; and Garry Anderson for statistical analyses.

This project was supported by The Australian Equine Research Foundation.

References

Allen, W.E. (1978) Some observations on pseudo-pregnancy in mares. *Br. vet. J.* **134**, 263–269.

Allen, W.R. & Moor, R.M. (1972) The origin of the equine endometrial cups. I. Production of PMSG by fetal trophoblast cells. *J. Reprod. Fert.* **29**, 313–316.

Bain, A.M. (1969) Foetal losses during pregnancy in the Thoroughbred mare: A record of 2562 pregnancies. *N.Z. vet. J.* **17**, 155–158.

Bazer, F.W. & Thatcher, W.W. (1977) Theory of maternal recognition of pregnancy in swine based on estrogen controlled endocrine versus exocrine secretion of $PGF_{2\alpha}$ by the uterine endometrium. *Prostaglandins* **14**, 397–406.

Burger, H.G., Lee, V.W.K. & Rennie, G.C. (1972) A generalised computer programme for the treatment of data from competitive protein binding assays including radioimmunoassays. *J. Lab. Clin. Med.* **80**, 302–312.

Burns, S.J. & Fleeger, J.L. (1975) Plasma progestagens in the pregnant mare in the first and last 90 days of gestation. *J. Reprod. Fert., Suppl.* **23**, 435–439.

Cox, R.I., Hoskinson, R.M. & Wong, M.S.F. (1979) Antisera reactive directly to estrone sulphate. *Steroids* **33**, 549–561.

Csapo, A.I., Knobil, E., Pulkkinen, M., Van der Molen, H.J., Sommerville, I.F. & Wiest, W.G. (1969) Progesterone withdrawal during hypertonic saline-induced abortions. *Am. J. Obstet. Gynec.* **105**, 1132–1134.

Darenius, K., Kindahl, H., Knudsen, O., Madej, A. & Edqvist, L.E. (1982) PMSG, progesterone and oestrone sulphate during normal pregnancy and early fetal death. *J. Reprod. Fert., Suppl.* **32**, 625–626.

Evans, K.L., Kasman, L.H., Jughes, J.P., Couto, M. & Lasley, B.L. (1984) Pregnancy diagnosis in the domestic horse through direct urinary estrone conjugate analysis. *Theriogenology* **22**, 615–620.

Fay, J.E. & Douglas, R.H. (1982) The use of radioreceptor assay for the detection of pregnancy in the mare. *Theriogenology* **18**, 430–444.

Gadsby, J.E., Heap, R.B. & Burton, R.D. (1980) Oestrogen production by blastocyst and early embryonic tissue of various species. *J. Reprod. Fert.* **60**, 409–417.

Gill, J.L. & Hafs, H.D. (1971) Analysis of repeated measurements of animals. *J. Anim. Sci.* **33**, 331–336.

Ginther, O.J. (1979) *Reproductive Biology of the Mare: Basic and Applied Aspects.* Equiservices Inc., Wisconsin.

Ginther, O.J., Garcia, M.C., Bergfelt, D.R., Leith, G.S. & Scraba, S.T. (1985) Embryonic loss in mares: pregnancy rate, length of interovulatory intervals, and progesterone concentrations associated with loss during days 11 to 15. *Theriogenology* **24**, 409–417.

Greenwood, F.C., Hunter, W.M. & Glover, J.S. (1963) The preparation of ^{131}I-labelled human growth hormone of high specific radioactivity. *Biochem. J.* **89**, 114–123.

Heap, R.B. & Hamon, M. (1979) Oestrone sulphate in milk as an indicator of a viable conceptus in cows. *Br. vet. J.* **135**, 355–363.

Heap, R.B., Hamon, M. & Allen, W.R. (1982) Studies on oestrogen synthesis by the pre-implantation equine conceptus. *J. Reprod. Fert., Suppl.* **32**, 343–352.

Holtan, D.W., Nett, T.M. & Estergreen, V.L. (1975) Plasma progestins in pregnant, post partum and cycling mares. *J. Anim. Sci.* **40**, 251–260.

Hoppe, R. (1968) The embryonic mortality in the mare. *Proc. 6th Int. Congr. Anim. Reprod. & A.I., Paris,* vol. II, 1573–1576.

Hyland, J.H., Wright, P.J. & Manning, S.J. (1984) An investigation of the use of plasma oestrone sulphate concentrations for the diagnosis of pregnancy in mares. *Aust. vet. J.* **61**, 123.

Kindahl, H., Knudsen, O., Madej, A. & Edqvist, L.E. (1982) Progesterone, prostaglandin F-2α, PMSG and oestrone sulphate during early pregnancy in the mare. *J. Reprod. Fert., Suppl.* **32**, 353–359.

Mitchell, D. (1971) Early fetal death and a serum gonadotrophin test for pregnancy in the mare. *Can. vet. J.* **12**, 41–44.

Nett, T.M., Holtan, D.W. & Estergreen, V.L. (1975) Oestrogens, LH, PMSG and prolactin in serum of pregnant mares. *J. Reprod. Fert., Suppl.* **23**, 457–462.

Pascoe, R.R. (1979) A possible new treatment for twin pregnancy in the mare. *Eq. vet. J.* **11**, 64–65.

Roberts, C.J. (1982) Termination of twin gestation by blastocyst crush in the broodmare. *J. Reprod. Fert., Suppl.* **32**, 447–449.

Terqui, M. & Palmer, E. (1979) Oestrogen pattern during early pregnancy in the mare. *J. Reprod. Fert., Suppl.* **27**, 441–446.

J. Reprod. Fert., Suppl. **35** (1987), 469–478

Induction of ovulation in anoestrous mares with a slow-release implant of a GnRH analogue (ICI 118 630)

W. R. Allen, M. W. Sanderson, R. E. S. Greenwood*, D. R. Ellis*, J. S. Crowhurst*, D. J. Simpson* and P. D. Rossdale†

Thoroughbred Breeders' Association Equine Fertility Unit, Animal Research Station, 307 Huntingdon Road, Cambridge CB3 0JQ; *Reynolds House, 166 High Street, Newmarket, Suffolk CB8 8AH; and †Beaufort Cottage Stables, High Street, Newmarket, Suffolk CB8 8JS, U.K.

Summary. A total of 18 experimental pony and 136 commercial maiden, barren and foaling Thoroughbred mares in seasonal or lactation-related anoestrus were injected subcutaneously with 1 or 2 slow-release D,L-lactide–glycolide co-polymer implants impregnated with 0·9 or 1·8 mg of the potent GnRH analogue, ICI 118 630, to give a daily release of, respectively, 30 or 60 µg analogue for 28 days; 32 of the Thoroughbred mares were also given a daily oral dose of 27·5 mg allyl trenbolone for 5 days after injection of the implant. Thirteen pony (76%) and 120 Thoroughbred (88%) mares ovulated 3–18 days after treatment with ICI 118 630 and the additional treatment with allyl trenbolone did not significantly reduce the considerable variation in the interval between treatment with GnRH analogue and ovulation. Of 100 Thoroughbred mares mated during the GnRH analogue-induced oestrus, (70%) conceived.

The results of this trial demonstrated that low-dose, slow-release formulations of GnRH agonists have considerable potential as a practical method of hastening renewed ovarian cyclicity in anoestrous mares.

Introduction

The disparity between the arbitrary mating period for Thoroughbreds and the physiological breeding season of the mare (Osborne, 1966) necessitates the development of efficient, practical methods for shortening seasonal anoestrus in barren and maiden mares and lactation-related anoestrus in early foaling mares. Many previous studies, notably those of Sharp *et al.* (1975) and Palmer *et al.* (1982), have demonstrated the beneficial effects of artificial lighting in seasonally anoestrous mares. However, increased daylength must be applied for 6–8 weeks from the time of the winter solstice to be effective and there remains considerable variation between individual mares in groups of animals maintained under the same lighting regimen as to the time of the first ovulation.

More recent approaches to anoestrus in mares have included the administration of progesterone or synthetic progestagens with or without oestradiol-17β for 10–15 days, in the form of daily intramuscular (i.m.) injections (Taylor *et al.*, 1982), daily oral dosing (Allen *et al.*, 1980) or a progesterone-releasing intravaginal device (Palmer, 1979; Rutten *et al.*, 1986). This type of progesterone feedback withdrawal therapy relies upon suppression of luteinizing hormone (LH) release from the pituitary gland during treatment followed by a surge-type, rebound release when administration of the drug is ceased. The technique is relatively efficacious but works well only in seasonally anoestrous mares that are already well advanced in the transitional phase between deep anoestrus and renewed cyclic activity; it is much less successful in mares still in deep anoestrus and in those exhibiting lactation-related anoestrus (Allen *et al.*, 1980).

A single injection of small quantities of the hypothalamic decapeptide, gonadotrophin-releasing hormone (GnRH), or its synthetic analogues, stimulates a maximal release of both follicle-stimulating hormone (FSH) and LH from the pituitary gland of the mare (Evans & Irvine, 1976, 1977, 1979). The heights of the GnRH-induced gonadotrophin peaks in peripheral blood, as reflecting the amounts of each hormone released from the pituitary, are dependent upon the stage of the ovarian cycle at the time of treatment. However, regardless of cycle, the duration of gonado-trophin release is relatively short; serum LH concentrations have returned to preinjection values by 4 h, and serum FSH values within 8–12 h, after each administration of GnRH (Evans & Irvine, 1976). A slow but steady rise in serum LH concentrations always precedes, and is an essential stimulus for, the first spontaneous ovulation of the new breeding season in mares (Garcia et al., 1979). This increase in basal LH secretion by the pituitary gland appears to be stimulated by an increase in the frequency of pulsatile releases of GnRH from the hypothalamus (Jöchle et al., 1987).

In their extensive study of the actions of exogenous GnRH in sheep, McLeod et al. (1982a, b) showed that follicular development, ovulation and normal luteal function can be induced in anoestrous ewes by pulsatile administration of small doses of GnRH at 2-h intervals over 3–5 days, delivered by means of a subcutaneous mini-pump. Furthermore, the same results can be achieved by continuous infusion of very low doses of GnRH, thereby eliminating the need for the pulsatile delivery system (McLeod et al., 1983). This discovery opened the way for development of practical methods of chronic low-dose GnRH administration by means of slow-release subcutaneous implant formulations. The present paper describes the successful use of one such preparation for inducing fertile ovulations in anoestrous barren, maiden and lactating pony and Thoroughbred mares.

Materials and Methods

GnRH preparation

A synthetic analogue of GnRH (ICI 118 630), with the peptide arrangement D-Ser(Bu^t)^{6}Azgly NH$_2$ ^{10}-LHRH, and shown to exhibit 100 times the potency of natural GnRH in rats (Dutta et al., 1978), was used in the experiments. The active principle was incorporated into a D,L-lactide–glycolide co-polymer, moulded to give very small tubular implants, measuring 0·25 or 0·5 cm in length and 1·2 mm in cross-sectional diameter and containing 0·9 or 1·8 mg analogue which was released continuously at an average rate of 30 or 60 µg/day for 28 days (ICI Pharmaceuticals Division, Cheshire, U.K.). To give a daily release of 30, 60, or 90 µg ICI 118 630, 1, 2 or 3 implants were injected beneath the skin on the side of the neck of treated mares via a 16-gauge hypodermic needle attached to a disposable syringe. The very small size of the implants meant that they remained impalpable through the skin after insertion.

Animals

A total of 18 experimental Welsh and crossbred anoestrous pony mares aged 2–8 years, and 155 maiden, barren and lactating anoestrous Thoroughbred mares aged 3–21 years, were treated with ICI 118 630 between February and May in each of 1984, 1985 and 1986.

The pony mares, weighing 200–450 kg, and 2 experimental Thoroughbred mares, were maintained in half-covered concrete yards throughout the winter and fed on a small ration of concentrates and good quality hay. They were teased regularly for signs of oestrous behaviour and their ovaries were occasionally palpated or examined by ultra-sound echography per rectum to monitor follicular growth. Jugular vein blood samples were recovered daily, or thrice weekly, from all treated and control mares for measurement of plasma progesterone and serum FSH and LH concentrations. All these experimental mares were judged to be in winter anoestrus at the time of treatment with ICI 118 630 on the basis of long hair coats, an absence of large follicles in the ovaries and plasma progesterone concentrations of < 1 ng/ml in weekly blood samples recovered during the preceding month.

The commercial Thoroughbred mares were maintained at 'public' or 'private' stud-farms in the Newmarket area. Most of the mares had arrived at the stud during the early part of February, having come from a wide variety of home management regimens during the preceding winter months. They were teased daily or every other day and visual inspection of the cervix by vaginoscope and rectal palpation of the ovaries to monitor follicular growth and ovulation was carried out by the attending veterinary clinician as considered necessary and according to the management policy of the study concerned. Jugular blood samples were recovered regularly before and after treatment with ICI 118 630 implants or other hormone preparations, for measurement of plasma progesterone, but not serum FSH and LH concentrations. All treated mares were considered to be in anoestrus on the basis of their failure to show oestrous behaviour, the appearance of the cervix, an absence of large follicles in the ovaries and at least 2 jugular vein plasma samples recovered between 2 and 9 days previously that showed progesterone concentrations of < 1 ng/ml.

Hormone assays

Progesterone. Plasma progesterone concentrations were measured during the first year of the study by radio-immunoassay as described by Urwin & Allen (1982). A goat anti-progesterone-11α-succinyl–bovine serum albumin conjugate (No. 465/7, Specific Antisera Ltd, Cheshire, U.K.) was used as the antibody and 1,2,6,7,(n)-^{3}Hprogesterone (TRK413; sp.act. 81 Ci/mmol; Radiochemical Centre, Amersham, U.K.) as label. The limit of sensitivity of the assay in horse plasma was 0·5 ng and the intra- and interassay coefficients of variation were 7 and 8% respectively. An Amplified Enzyme-linked Immunoassay (IQ (Bio) Ltd, Cambridge, U.K.) available commercially in kit form (Enzynogest; Hoechst Animal Health Division, Milton Keynes, U.K.) was used during the remainder of the trial. This rapid and accurate method utilizes 96-well microtitre plates coated with a mouse monoclonal antibody to pro-gesterone, alkaline phosphatase coupled to progesterone as conjugate, NADP as substrate, alcohol dehydrogenase and lipoamide dehydrogenase as enzyme amplifier and iodo-nitro tetrazolium violet as colour marker (Stanley *et al.,* 1986). The degree of colour change in test wells in the plate is compared with 8 duplicate standards of progesterone dissolved in oestrous mare plasma by measuring wavelength transmission in a Titerek Uniskan plate reader (Flow Laboratories, Rickmansworth, U.K.). Results were calibrated using specially written software (IQ (Bio) Ltd) for a BBC2 micro-computer linked directly to the plate reader. The limit of sensitivity of the assay for measurement of progesterone in horse plasma was 0·2 ng and the intra- and interassay coefficients of variation were both < 10%.

FSH and LH. Serum concentrations of FSH and LH were measured by double-antibody radioimmunoassay as described by Urwin (1983). The heterologous assay for FSH utilized a rabbit anti-ovine FSH serum (H-31, NIAMDD) as first antibody and highly purified horse FSH (E99B; Licht *et al.,* 1979) as radioligand and reference standard. The heterologous assay for LH utilized an anti-horse CG serum (MSIIB) prepared against highly purified horse CG (MSII, 16 000 i.u./mg; M. J. Stewart, Cambridge) and highly purified horse LH (E99A, Licht *et al.,* 1979) as radioligand and standard. A sheep anti-rabbit gamma globulin serum was used as second antibody in both assays. The limits of sensitivity of the assays for measurement of FSH and LH in mare serum were 0·5 and 0·25 ng/ml respectively and the intra- and interassay coefficients of variation were 10 and 12% respectively for FSH and 9 and 9% respectively for LH.

Treatment regimens

Experimental mares. During the first year of the study (1984), 6 pony and 2 Thoroughbred experimental mares were each given a single implant of 1·8 mg ICI 118 630 on the same day (5 March). Eight other pony mares remained as untreated controls. Regular teasing and daily blood sampling was begun for all the mares 3 days before injection of the implant (Day 0) and continued until 10 days after ovulation, or until Day 30, whichever occurred first. In the second year (1985), 5 pony mares were given a single implant of 0·9 mg ICI 118 630 and 7 others were given a single 1·8 mg implant, on the same day (13 March). Eight other anoestrous pony mares acted as untreated controls. As before, teasing and daily blood sampling were continued for 30 days after treatment or until 10 days after ovulation.

Commercial Thoroughbred mares. In each of the 3 years of the trial, individual maiden, barren and lactating mares were injected with one or more implants on the same occasion to give a dose of 0·9, 1·8 or 2·7 mg ICI 118 630; the great majority of mares received a single 1·8 mg implant delivering ∼ 60 µg GnRH analogue/day for 28 days.

Selection of suitable mares for treatment was arbitrary and was influenced by factors such as permission of the owner or failure to respond to prior treatment with oral progestagen (Regumate; Hoechst Animal Health Division) and/or prostaglandin analogue. Nevertheless, all the treated mares were considered to be in anoestrus at the time of treatment (Day 0). Regular teasing continued after treatment and ovarian and cervical changes were monitored twice weekly, or more frequently when mares began showing oestrus. Jugular vein blood sampling was continued 2 or 3 times weekly until a rise in plasma progesterone concentrations confirmed the occurrence of ovulation. Mares that had not ovulated by Day 18 were considered to have failed to respond to the treatment.

In the third year of the trial (1986), 32 mares were given a daily oral dose of 27·5 mg allyl trenbolone (Regumate) for 5 days after injection of the single implant of 1·8 mg ICI 118 630 in an attempt to synchronize better the interval between treatment and ovulation. Six other mares were given a single i.m. injection of 10 mg oestradiol benzoate (Intervet Laboratories, Cambridge, U.K.) between Days 5 and 11 after injection of the GnRH implant, in an attempt to induce the expression of oestrous signs and improve 'ripening' and relaxation of the cervix.

Results

Experimental mares

Of the 13 anoestrous pony mares and the 2 experimental Thoroughbred mares given a single implant of 1·8 mg ICI 118 630 in 1984 and 1985, and of the 5 pony mares given a single implant of

Table 1. Response of anoestrous experimental pony and Thoroughbred mares to a subcutaneous injection of a single implant containing 0·9 or 1·8 mg ICI 118 630

Dose of ICI 118 630 in implant (mg)	No. of mares treated	No. of mares that ovulated in < 18 days	Mean (range) interval in days from treatment to:	
			Onset of oestrus	Ovulation
0·9	5 pony	2	6 (5, 7)	13·5 (12, 15)
1·8	13 pony	12	4·2 (2–7)	11·3 (4–17)
	2 Thoroughbred	1	3	10

0·9 mg ICI 118 630 in 1985, 12, 1 and 2 respectively began to show oestrous behaviour between Days 2 and 7 after treatment (Day 0) and they ovulated between Days 4 and 17. Ovulation did not occur until Days 23–67 in the other 4 treated and the 16 untreated control mares (Table 1).

Serum LH concentrations in the GnRH-treated mares which responded rose steadily from baseline values of 1–3 ng/ml before injection of the implant to reach variable peaks of 8–23 ng/ml associated with ovulation. Serum FSH concentrations, on the other hand, showed an initial slight increase for 1–2 days after treatment, but then declined steeply over the next few days to reach minimum values (18–25 ng/ml) during early oestrus. Plasma progesterone concentrations remained low (< 1 ng/ml) until ovulation but then rose sharply over the next 5–8 days, indicating development of a functional corpus luteum (Fig. 1a–c). Three of the 4 treated pony mares that were mated, and the one Thoroughbred mare that was inseminated artificially with fresh, diluted semen during this first GnRH-induced oestrus, conceived. The other 7 pony mares that ovulated within 18 days after GnRH treatment showed a normal dioestrous interval (15–18 days) and they all continued to cycle normally thereafter.

The Thoroughbred and 4 pony mares that failed to respond to treatment all showed an initial rise in serum LH concentrations and corresponding decline in serum FSH values during the first 6 days after treatment but beyond this stage the serum concentrations of both gonadotrophins tended to drift back towards their starting values (Fig. 1d). The Thoroughbred mare was given a second implant of 1·8 mg ICI 118 630 on Day 21, after which she showed a similar pattern of transitory changes in serum gonadotrophin concentrations (Fig. 1d). This mare did not ovulate until 5 May, 61 days after injection of the first implant and considerably later in the breeding season than expected in view of her good bodily condition and early loss of hair coat.

Table 1 also compares the response rate in the pony mares given a single implant of 0·9 or 1·8 mg ICI 118 630; care was taken to assign equivalent numbers of smaller and larger animals to each dosage group. Whereas 12 of the 13 mares (92%) given a 1·8 mg implant ovulated within 18 days after treatment, only 2 of the 5 mares (40%) injected with a 0·9 mg implant did so. From this result it was concluded that, in view of the limited availability of only two sizes of implant, the implant containing 1·8 mg ICI 118 630 (releasing 60 µg analogue/day) constituted the most practical for general use in all mares.

Commercial Thoroughbred mares

The Thoroughbred mares showed no significant differences in response to treatment in each of the 3 years of the trial and overall results are therefore summarized in Table 2. The 4 mares injected with 2 implants simultaneously to give a daily release of 90 µg ICI 118 630, the 6 mares injected with oestradiol benzoate some days after injection of the GnRH implant, and the 32 mares treated with

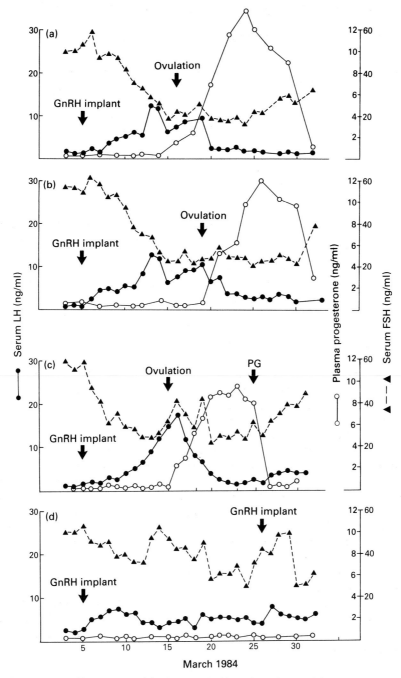

March 1984

Fig. 1. Hormone profiles measured in 2 pony (a, b) and 2 Thoroughbred (c, d) experimental anoestrous mares each injected subcutaneously on 5 March 1984 with a single implant containing 1·8 mg ICI 118 630. The mare shown in (d) was given a second 1·8 mg implant on 26 March and did not ovulate until 12 May. PG = injection of prostaglandin analogue.

allyl trenbolone for 5 days after injection of a single 1·8 mg implant are all included in the overall figures since it was clear that none of these minor changes to the basic treatment regimen caused any significant alteration to the occurrence, or otherwise, of ovulation. However, 3 maiden mares that failed to respond to a single implant of 0·9 mg ICI 118 630 are excluded from the figures in view of the results of the dose–response trial in ponies. Similarly, 16 other mares were excluded from the trial when they were found, on the basis of high plasma progesterone concentrations exhibited on the day when the implant was injected, to have been wrongly diagnosed clinically as being in anoestrus. All these mares had apparently ovulated without showing oestrous behaviour ('silent' ovulation) during the period between the previous 'diagnostic' blood sample showing baseline progesterone values and the day of treatment. They were all treated with prostaglandin analogue when they were judged to be >Day 7 of dioestrus and all underwent luteolysis and subsequently ovulated again spontaneously. However, no fewer than 9 of these 16 mares again failed to show oestrous behaviour and did not exhibit sufficient relaxation of the cervix to enable them to be mated during this second, prostaglandin-induced follicular phase.

Table 2. Overall response of anoestrous commercial Thoroughbred mares treated with implants containing 1·8–2·7 mg ICI 118 630 during 1984–1986

Mare status	No. of mares (%)			
	Given implants	Ovulating in <18 days	Ovulating and mated in <18 days	Conception rate (% mares mated)
Barren	79	69 (87%)	61 (77%)	67
Maiden	25	23 (92%)	16 (64%)	63
Foaling	32	28 (88%)	23 (72%)	86
Total	136	120 (88%)	100 (73%)	70

Of the total of 136 GnRH-treated anoestrous Thoroughbred mares (79 barren, 25 maiden, 32 lactating), 120 (88%) ovulated within 18 days of treatment (Table 2). The intervals from treatment (Day 0) to (a) onset of oestrus, and (b) ovulation, were both rather variable, with only a moderate peak in the treatment–ovulation interval occurring between Days 9 and 11 (Fig. 2a). Some mares ovulated very quickly after treatment (Days 3–4), presumably due to misdiagnosis of the presence of a maturing follicle in their ovaries when the implant was injected. These animals added considerably to the total of 20 mares that ovulated in response to treatment but were not able to be mated (see Table 2). Oral administration of allyl trenbolone for 5 days after injection of the GnRH implant was unsuccessful. This additional therapy tended merely to delay ovulation without narrowing significantly the treatment–ovulation range (Fig. 2b).

Of the 120 mares that ovulated in response to treatment with GnRH, 100 (83%) were mated. However, the remaining 20 mares (15% of all treated mares) failed to show sufficiently strong oestrous behaviour and/or cervical relaxation to enable mating to occur, despite the diagnosis of a mature follicle in their ovaries. Of the 6 mares treated with oestradiol benzoate, 5 were in this category and 3 showed good oestrous behaviour within 12–18 h after the injection of oestrogen and were mated. The other 2 oestradiol-treated mares showed no change in behaviour despite some improvement in the appearance of the cervix. The 6th mare had no palpable follicle in her ovaries when treated with 10 mg oestradiol benzoate 8 days after injection of a 1·8 mg implant. She nevertheless showed oestrous behaviour and was mated during the next 4 days, although she subsequently failed to ovulate by Day 18 and was not pregnant.

A high conception rate of 70% was achieved in the 100 mares that were mated during the GnRH-induced oestrus (Table 2). Most mares that failed to conceive during this first oestrus did so

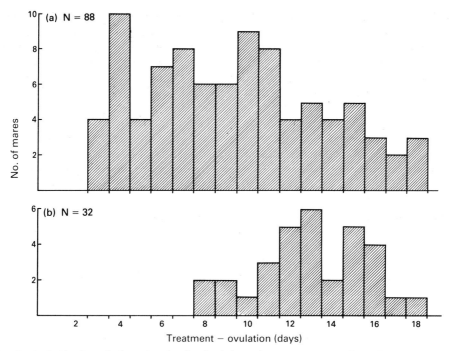

Fig. 2. Individual variations in the interval between treatment and ovulation in (a) 88 anoestrous Thoroughbred mares injected with a single implant of 1·8 mg ICI 118 630 only; (b) 32 mares given a daily oral dose of 27·5 mg allyl trenbolone for 5 days after injection of a single implant of 1·8 mg ICI 118 630.

in subsequent spontaneous or prostaglandin-induced oestrous periods. There was no increase in early pregnancy loss rate, or in later abortions, in any of the GnRH-treated mares, nor any evidence of developmental or other abnormalities in foals born after conception in the GnRH-induced oestrus. None of the treated mares showed any untoward side effects or adverse local reactions around the implant sites.

In the first year of the trial, two barren Thoroughbred mares that appeared not to be responding to an initial implant of 1·8 mg ICI 118 630 were each given a second 1·8 mg implant on Days 9 and 12 respectively. Both these mares persisted in anoestrus for >60 days during which time their ovaries were judged clinically to be very small and completely devoid of follicles.

Discussion

The results of this preliminary clinical trial with the potent GnRH analogue ICI 118 630 support the earlier observations of Evans & Irvine (1976, 1977, 1979) and Johnson (1986) that administration to anoestrous mares of natural GnRH or its synthetic analogues in the correct doses and administration frequency induces a sufficiently large and prolonged release of pituitary LH to stimulate follicular maturation, ovulation and normal luteal function in a high proportion of treated animals. In this regard, GnRH therapy has considerable potential in modern Thoroughbred stud management for shortening seasonal anoestrus in barren and maiden mares and for overcoming anoestrus associated with lactation in early foaling mares. The great value of the low-dose, slow-release formulation of GnRH analogue provided by the ICI 118 630 implants is the simplicity and practicability of their administration coupled with a high degree of efficacy.

A number of features of the present results are of particular interest. First, it remains unclear why 10% of the commercial Thoroughbred mare and 2 of the 15 experimental mares failed to ovulate within 18 days after treatment with a single implant of 1·8 mg ICI 118 630. No obvious clinical features distinguished these non-responders from the majority of animals that did ovulate and there occurred a number of instances in the Thoroughbred group when mares that responded well to treatment had been judged clinically to be in deeper anoestrus than some of the mares which failed to ovulate. Clearly it is not possible to assess accurately the depth of seasonal or lactation-related anoestrus in a heterogeneous population of mares on the basis of clinical estimates alone. Nevertheless, the suspicion remains that the implant failed to function correctly in at least some of the non-responding mares, or that the release rate of 60 µg ICI 118 630/day in these particular animals may have had a suppressive, rather than the desired stimulatory, action on pituitary gonadotrophin release.

The constraints of this study limited the scope and accuracy of the small dose–response trial attempted but the results showed quite clearly that the 0·9 mg implant, releasing only 30 µg ICI 118 630/day, is insufficient. However, it was not considered prudent in the commercial Thoroughbred mares to go above the release rate of 60 µg ICI 118 630/day provided by the 1·8 mg implant in view of the possibility of over-dosing leading to suppression of gonadotrophin release and prolongation of anoestrus in the 3 mares mentioned previously that were given a second implant within the 28-day release life of the first implant. Studies in a range of species, including the rat (Labrie *et al.,* 1978), sheep (Fraser & Lincoln, 1980; Lincoln *et al.,* 1986), rhesus monkey (Sundaram *et al.,* 1982) and human (Schürmeyer *et al.,* 1984) have demonstrated that prolonged administration of high doses of GnRH or its agonists inhibits reproductive function as a result of desensitization of pituitary gonadotrophs against the action of endogenous GnRH.

A second feature of the results of this study is the total of 20 mares that ovulated within 18 days after injection of a GnRH analogue implant yet failed to show sufficiently intense oestrous behaviour or cervical relaxation to be mated. Frequently, a large and softening follicle was palpable in the ovaries of these mares coincidentally with low plasma progesterone concentrations, and yet the normal behavioural and cervical changes associated with the follicular phase of the mare's ovarian cycle remained absent. This syndrome, aptly termed 'silent' ovulation, has been recognized for many years as occurring sporadically in Thoroughbred and other breeds of mares, most commonly in foaling mares at the time of the first post-partum ovulation. During the course of the present trial, however, when serial plasma progesterone measurements were undertaken widely within the general population of non-pregnant mares being teased routinely during the early part of the mating season, it became clear that the syndrome of 'silent' ovulation occurs more frequently in barren and maiden mares than had been considered previously.

The occurrence of 'silent' ovulation appears to be markedly influenced by climatic conditions. In the third year of the present trial (1986), a severe winter followed by an unusually late and cold spring resulted in a very late flush of grass growth in the Newmarket area. 'Silent' ovulations were encountered much more frequently in barren and maiden mares in this year than in the previous 2 years of the trial and, furthermore, the condition occurred in some mares during 2, and occasionally 3, consecutive follicular phases, quite unrelated to treatment with GnRH implants. This observation gives rise to the possibility that the 'silent' ovulations encountered in 17% of the 136 commercial Thoroughbred mares injected with GnRH analogue implants were not an adverse side effect of this type of treatment. Rather, the condition was destined to occur in these particular mares in relation to at least the first ovulation of the new breeding season due to some hitherto unknown metabolic deficiency existing in them at the time. There is clearly a need for further and more detailed investigation of the possible causes of the syndrome of 'silent' ovulation in mares and it may be significant that the striking increase in the incidence of the condition encountered during the present study occurred in the year when grass growth was so markedly delayed.

The great value of a rapid progesterone assay in modern equine stud management is also shown by the present experiments. During the early part of the mating season especially, it is important to categorize non-pregnant mares newly arrived on the stud into those that are already cycling or are in prolonged dioestrus from those still in anoestrus. Regular teasing combined with occasional clinical examinations is not sufficiently accurate for this purpose and only the measurement of raised progesterone concentrations in peripheral plasma will conclusively indicate the existence of luteal tissue in a mare's ovaries. The recent development of rapid and accurate enzyme-linked immunoassay (ELISA) kits, which utilize colorimetric change instead of radioactive uptake as the end point, places progesterone assay availability in the hands of veterinary clinicians and well-managed stud farms.

In conclusion, the results of this study have shown convincingly that slow-release, low-dose formulations of GnRH agonists have considerable potential as efficient therapeutic agents for overcoming anoestrus in mares. Furthermore, the interesting possibility exists that chronic administration of slow-release, high-dose formulations of the same agonists may, by inducing desensitization of pituitary gonadotrophs, cause prolonged sexual quiescence in both male and female horses. The value of such reversible, chemical castration in competition and riding horses is obvious.

We thank Dr P. S. Jackson of ICI Pharmaceuticals Division and Mr D. Morgan of Coopers Animal Health for supplying implants of ICI 118 630 and the managers and staffs of the many Thoroughbred stud farms in Newmarket who co-operated in the trial. The study was financed by The Thoroughbred Breeders' Association of Great Britain.

References

Allen, W.R., Urwin, V., Simpson, D.J., Greenwood, R.E.S., Crowhurst, R.C., Ellis, D.R., Ricketts, S.W., Hunt, M.D.N. & Wingfield-Digby, N.J. (1980) Preliminary studies on the use of an oral progestagen to induce oestrus and ovulation in seasonally anoestrous Thoroughbred mares. *Equine vet. J.* **12**, 141–145.

Dutta, L., Furr, B., Giles, E., Valcaccia, A. & Walpole, A. (1978) Potent agonist and antagonist analogues of luliberin containing an asaglycine residue in position 10. *Biochem. biophys. Res. Commun.* **81**, 382–390.

Evans, M.J. & Irvine, C.H.G. (1976) Measurement of equine Follicle Stimulating Hormone and Luteinising Hormone: response of anestrous mares to gonadotropin releasing hormone. *Biol. Reprod.* **15**, 477–484.

Evans, M.J. & Irvine, C.H.G. (1977) Induction of follicular development, maturation and ovulation by gonadotropin releasing hormone administration to acyclic mares. *Biol. Reprod.* **16**, 452–462.

Evans, M.J. & Irvine, C.H.G. (1979) Induction of follicular development and ovulation in seasonally acyclic mares using gonadotrophin-releasing hormones and progesterone. *J. Reprod. Fert., Suppl.* **27**, 113–121.

Fraser, H.M. & Lincoln, G.A. (1980) Effects of chronic treatment with an LHRH agonist on the secretion of LH, FSH and testosterone in the ram. *Biol. Reprod.* **22**, 269–276.

Garcia, M.C., Freedman, L.J. & Ginther, O.J. (1979) Interaction of seasonal and ovarian factors in the regulation of LH and FSH secretion in the mare. *J. Reprod. Fert., Suppl.* **27**, 103–111.

Jöchle, W., Irvine, C.H.G., Alexander, S.L. & Newby, T.J. (1987) Release of LH, FSH and GnRH into pituitary venous blood in mares treated with a PGF analogue, luprostiol, during the transition period. *J. Reprod. Fert., Suppl.* **35**, 261–267.

Johnson, A.L. (1986) Induction of ovulation in anestrous mares with pulsatile administration of gonadotropin-releasing hormone. *Am. J. vet. Res.* **47**, 983–986.

Labrie, F., Auclair, C., Cusan, L., Kelly, P.A., Pelletier, G. & Ferland, L. (1978) Inhibitory effect of LHRH and its agonists on testicular gonadotrophin receptors and spermatogenesis in the rat. *Int. J. Androl., Suppl.* **2**, 303–318.

Licht, P., Bona Gallo, A., Aggarwal, B.B., Farmer, S.W., Castelino, J.B. & Papkoff, H. (1979) Biological and binding activities of equine pituitary gonadotrophins and pregnant mare serum gonadotrophin. *J. Endocr.* **83**, 311–322.

Lincoln, G.A., Fraser, H.M. & Abbott, M.P. (1986) Blockade of pulsatile LH, FSH and testosterone secretion in rams by constant infusion of an LHRH agonist. *J. Reprod. Fert.* **77**, 587–597.

McLeod, B.J., Haresign, W. & Lamming, G.E. (1982a) The induction of ovulation and luteal function in seasonally anoestrous ewes treated with small-dose multiple injections of LH-RH. *J. Reprod. Fert.* **65**, 215–221.

McLeod, B.J., Haresign, W. & Lamming, G.E. (1982b) Response of seasonally anoestrous ewes to small-dose multiple injections of Gn-RH with and without progesterone pre-treatment. *J. Reprod. Fert.* **65**, 223–230.

McLeod, B.J., Haresign, W. & Lamming, G.E. (1983) Induction of ovulation in seasonally anoestrous ewes by continuous infusion of low doses of GnRH. *J. Reprod. Fert.* **68**, 489–495.

Osborne, V.E. (1966) An analysis of the pattern of ovulation as it occurs in the annual reproductive cycle of the mare in Australia. *Aust. vet. J.* **42**, 149–154.

Palmer, E. (1979) Reproductive management of the mare without detection of oestrus. *J. Reprod. Fert., Suppl.* **27**, 263–270.

Palmer, E., Driancourt, M.A. & Ortavant, R. (1982) Photoperiodic stimulation of the mare during winter anoestrus. *J. Reprod. Fert., Suppl.* **32**, 275–282.

Rutten, D.R., Chaffaux, S., Valon, M., Deletang, F. & de Haas, V. (1986) Progesterone therapy in mares with abnormal oestrous cycles. *Vet. Rec.* **119**, 569–571.

Schürmeyer, T.H., Knuth, U.A., Freischen, C.W., Sandow, J., Bint Akhter, F. & Nieschlag, E. (1984) Suppression of pituitary and testicular function in normal men by constant gonadotropin-releasing hormone agonist infusion. *J. clin. Endocr. Metab.* **59**, 19–24.

Sharp, D.C., Kooistra, L. & Ginther, O.J. (1975) Effects of artificial light on the oestrous cycle of the mare. *J. Reprod. Fert., Suppl.* **23**, 241–246.

Stanley, C.J., Paris, F., Webb, A.E., Heap, R.B., Ellis, S.T., Hamon, M., Worsfold, A. & Booth, J.M. (1986) Use of a new and rapid milk progesterone assay to monitor reproductive activity in the cow. *Vet. Rec.* **118**, 664–667.

Sundaram, K., Connell, K.G., Bardin, C.W., Samojlik, E. & Schally, A.V. (1982) Inhibition of pituitary-testicular function with D-Trp^6- luteinizing hormone-releasing hormone in rhesus monkeys. *Endocrinology* **110**, 1308–1314.

Taylor, T.B., Pemstein, R. & Loy, R.G. (1982) Control of ovulation in mares in the early breeding season with ovarian steroids and prostaglandin. *J. Reprod. Fert., Suppl.* **32**, 219–224.

Urwin, V.E. (1983) The use of heterologous radio-immunoassays for the measurement of follicle-stimulating hormone and luteinizing hormone concentrations in horse and donkey serum. *J. Endocr.* **99**, 199–209.

Urwin, V.E. & Allen, W.R. (1982) Pituitary and chorionic gonadotrophin control of ovarian function during pregnancy in equids. *J. Reprod. Fert., Suppl.* **32**, 371–382.

J. Reprod. Fert., Suppl. 35 (1987), 479–484

Printed in Great Britain
© 1987 Journals of Reproduction & Fertility Ltd

Hormonal changes associated with induced late abortions in the mare

A. Madej, H. Kindahl, C. Nydahl, L.-E. Edqvist and D. R. Stewart*

*Departments of Obstetrics and Gynaecology and Clinical Chemistry, Swedish University of Agricultural Sciences, S-750 07 Uppsala, Sweden, and *Department of Reproduction, University of California at Davis, California 95616, U.S.A.*

Summary. Two mares received PGF-2α twice daily until abortion and 2 mares received a combined treatment with oestradiol benzoate and oxytocin. The mares were about 150 days pregnant. The PG-treated animals aborted after 37 and 61 h, respectively, and the fetuses were expelled in intact fetal membranes. The other 2 mares aborted 13 and 27 h after the first oxytocin injection, respectively, and showed strong uterine contractions and expelled the fetuses in disrupted fetal membranes. Concentrations of 15-ketodihydro-PGF-2α increased both after PG and oxytocin injections and in association with the abortion, but after the PG-induced abortion there was an immediate return to basal levels and after the oxytocin-induced abortion there was a large increase in the concentrations, indicating damage of the uterus. Progesterone and relaxin concentrations followed the placental function and decreased in association with the abortions. Oestrone sulphate values differed in the two groups; the oxytocin-treated animals showed a rapid decrease while the mares treated with PG showed first a marked increase and then a decrease. Concentrations of PMSG appeared to be unaffected by the abortions.

Introduction

There are always demands for induction of abortion at different stages of pregnancy in all domestic species. The main reasons for induction of abortion in the mare during mid- or late gestation are twin pregnancy diagnosed too late or to harvest fetal tissues for virus isolation (Bosu & McKinnon, 1982). Amongst the methods now available for terminating pregnancy in the mare are the use of natural prostaglandin (PG) F-2α (Douglas *et al.*, 1974) or one of its analogues (Squires *et al.*, 1980). Several hormones are involved in pregnancy in the horse and induction of abortions can be used to evaluate the significance of the respective hormones. The findings of Stewart *et al.* (1982a, b) indicated that the placenta is the main source of relaxin activity in the pregnant mare and that oxytocin may be responsible for the increased secretion rate of relaxin at foaling. In late gestation both progesterone and oestrone sulphate are hormones of great importance to maintain pregnancy. Changes in the levels of these hormones could be a good indicator of the condition of the fetus. The aim of the present work was to study sequential hormonal changes during late gestation in mares subjected to abortion and to compare the abortifacient effect of natural PGF-2α and a combination of oestradiol benzoate and oxytocin.

Materials and Methods

Animals. Four Standardbred pregnant mares previously described by Kindahl *et al.* (1982) were used. Two mares, 158 and 173 days of gestation (Nos. 1 and 2, respectively) were given 10 mg PGF-2α (Dinolytic: Upjohn S.A., Puurs, Belgium) intramuscularly at 12-h intervals until abortion occurred. Two mares, 152 and 159 days of gestation (Nos. 3 and 4, respectively) were treated twice with 5 mg oestradiol benzoate (Ovex: AB LEO, Helsinborg, Sweden) intramuscularly at 12-h intervals. Four intravenous injections of 50 i.u. oxytocin (synthetic oxytocin: Ciba-Geigy AG,

Basel, Switzerland) given at 3-h intervals were started 12 h after the last oestradiol treatment. Then, 3 h after the last oxytocin injection, 5 mg oestradiol benzoate were again given. The experiment was carried out in October, at the end of the breeding season.

Blood sampling. Blood samples were collected from the jugular vein into heparinized Vacutainer tubes (Becton-Dickinson). The plasma was removed by centrifugation and stored at $-20°C$ until analyses were performed. The plasma samples were collected daily from 7 days before induction, every 3rd hour from 1 day before the start of treatments until 10 days later, and finally daily for another 14-day period. During the treatments the blood samples were always collected immediately before injection.

Hormone assays. All the radioimmunoassays used had been previously validated for horse plasma (Kindahl *et al.*, 1982; Stewart, 1986).

15-Ketodihydro-PGF-2α, the main initial blood plasma metabolite of PGF-2α was analysed according to Kindahl *et al.* (1976) as slightly modified by Neely *et al.* (1979). The relative cross-reactions of the antibody were 16% with 15-keto-PGF-2α, 4% with 13,14-dihydro-PGF-2α and 0·4% with PGF-2α. The intra-assay coefficient of variation ranged between 6·6 and 11·7% for different ranges of the standard curve, and the inter-assay coefficient of variation was 14% from a pooled sample (mean value 115 pmol/l). The plasma samples were heated in the PG metabolite assay to avoid influence of heat-labile compounds (45°C for 30 min). The detection limit of the assay was 30 pmol/l.

Progesterone concentrations were determined by the method of Kindahl *et al.* (1976). The antibody was raised to an 11α-hydroxyprogesterone-hemisuccinate–bovine serum albumin conjugate (Bosu *et al.*, 1976). The antiserum cross-reacted < 1% for progestagens, oestrogens, androgens and corticoids, except for deoxycorticosterone (3·8%), pregnenolone (3·3%), and 5β-pregnane-3,20-dione (11%). The intra-assay coefficient of variation varied between 5·5 and 18·5% for different ranges of the standard curve. The practical limit of sensitivity was 0·5 nmol/l.

PMSG (horse CG) concentrations were basically determined by the same procedure as described by Menzer & Schams (1979). A highly purified PMSG preparation with a biological activity of 10 000 i.u./mg was used for iodination and as standard. The cross-reactions of horse LH and horse FSH in the assay system were about 100 and 25%, respectively. The inter-assay coefficient of variation was 15·4% (mean value 2320 µg/l). The sensitivity of the assay was 5 µg/l.

Oestrone sulphate concentrations were determined by radioimmunoassay using an antiserum against oestrone glucosiduronate–bovine thyroglobulin raised in rabbits (Wright *et al.*, 1978). The antiserum cross-reacted with oestrone, equilin and equilenin at 62·5, 4·1 and 2·5%, respectively. The intra-assay coefficient of variation was 18·9% at a mean plasma pool concentration of 3·7 nmol oestrone sulphate/l. The sensitivity of the assay was 0·01 pmol/tube.

Relaxin concentrations were determined by homologous radioimmunoassay using horse relaxin, purified from horse placentas, as a radioiodine-labelled tracer and standard (Stewart. 1986). The final antibody dilution was 1:45 000. In the system used there were no cross-reactions with other horse hormones. The sensitivity of the assay system was 23 pg/tube. The intra-assay coefficient of variation was 4·7% in the middle of the standard curve. The respective inter-assay coefficient of variation was 7·7%.

Statistical analysis. Analysis of variance by General Linear Model procedure from Statistical Analysis System (SAS, 1982) was performed to calculate partial correlations between particular hormones.

Results

Mares 1 and 2 treated with PGF-2α aborted after 37 and 61 h, respectively. The number of injections given before abortion occurred was 4 and 6, respectively. The mares showed some abdominal discomfort and sweating during PG injections. The fetuses were aborted in intact fetal membranes and were dead. After abortion, no complications could be noted. The concentrations of 15-ketodihydro-PGF-2α were about 100 pmol/l before any treatment started in Mare 1 (Fig. 1). Within 3 h, after each injection, a peak of PGF-2α metabolite was recorded. The abortion coincided with the peak of PGF-2α metabolite (2092 pmol/l). The duration of that peak was about 9 h. During the PGF-2α injections the plasma concentrations of oestrone sulphate increased from about 1600 to 2500 nmol/l (Fig. 1). An abrupt drop in both oestrone sulphate and progesterone concentrations occurred within 10 h after abortion. During the experimental period the plasma level of PMSG continuously decreased from about 100 µg/l before induction of abortion to about 50 µg/l 20 days later (Fig. 1). Both mares returned to oestrus 6–7 days after abortion, and 5 days later a new episode of progesterone secretion occurred (see Mare 1 in Fig. 1).

Mares 3 and 4, treated with a combination of oestradiol benzoate and oxytocin, aborted after 13 and 27 h, and 4 and 5 injections of oxytocin, respectively. After injections of oxytocin mares were restless, kicked the walls and exhibited strong uterine contractions. The fetuses and fetal membranes were expelled separately. The fetuses were dead and one of them had two broken legs

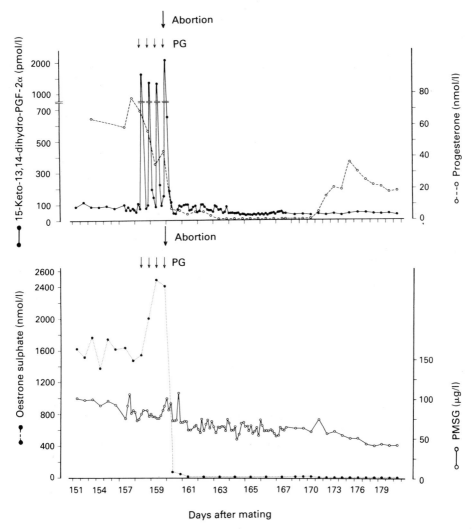

Fig. 1. Peripheral blood plasma concentrations of 15-ketodihydro-PGF-2α and progesterone and oestrone sulphate and PMSG in a Standardbred pregnant mare (No. 1) before, during and after induction of abortion with prostaglandin F-2α (PG).

and an open abdominal wall. Figure 2 depicts the hormonal changes in Mare 3. During the oxytocin injections one distinct peak of the PGF-2α metabolite was seen concomitantly with a drop in progesterone and oestrone sulphate values (Fig. 2). In both mares, PGF-2α metabolite concentrations increased up to 700 pmol/l during the 24 h after abortion. This elevation in the metabolite values lasted about 60 h. Neither of the mares returned to oestrus after the abortion, but new progesterone secretion was seen 2 weeks later. PMSG concentrations were not affected by treatment and instead continuously decreased during the experimental period (e.g. Mare 3, Fig. 2).

Figure 3 depicts relaxin concentrations in Mares 1 and 3 at 3 days before and 3 days after abortion. During the days before abortion relaxin values ranged from 80 to 216 μg/l. Within 1 day after abortion relaxin concentrations decreased to values of about 5 μg/l.

A partial positive correlation between oestrone sulphate and PGF-2α metabolite ($r = 0.37$, $P < 0.05$), between relaxin and oestrone sulphate ($r = 0.77$, $P < 0.05$) and between relaxin and

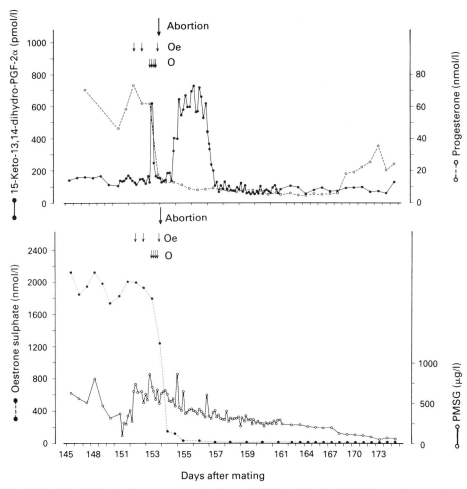

Fig. 2. Peripheral blood plasma concentrations of 15-ketodihydro-PGF-2α and progesterone and oestrone sulphate and PMSG in a Standardbred pregnant mare (No. 3) before, during and after induction of abortion with a combination of oestradiol benzoate (Oe) and oxytocin (O).

PGF-2α metabolite ($r = 0.64$, $P < 0.001$) during treatment and 3 days following abortion was found only in mares given natural PGF-2α.

Discussion

The present results demonstrated that multiple injections of natural prostaglandin F-2α had an abortifacient effect during mid-gestation in the mare. This is in agreement with findings of Douglas *et al.* (1974) who reported that the mean number of PGF-2α injections given before abortion occurred was 3·7. An increase of 15-ketodihydro-PGF-2α in peripheral blood plasma after each PGF-2α injection seemed to be the result of metabolism of this exogenous PGF-2α. However, the duration and magnitude of 15-ketodihydro-PGF-2α peak following the last injection and the abortion occurred might indicate the endogenous release of prostaglandin F-2α. Because of infrequent blood sampling we were not able to see maximal values of the prostaglandin metabolite associated with the abortion. To detect maximal values of 15-ketodihydro-PGF-2α during foaling the blood

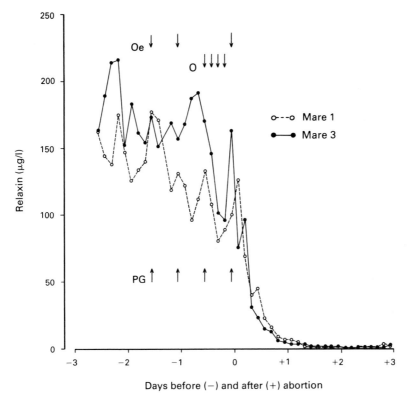

Fig. 3. Peripheral blood plasma concentrations of relaxin in Mares 1 and 3 from 3 days before to 3 days after abortion. Oe = oestradiol benzoate; O = oxytocin; PG = prostaglandin F-2α.

samples have to be collected at 15–30-min intervals (Barnes *et al.*, 1978; Pashen & Allen, 1979; Stewart *et al.*, 1984). There are clearly large differences in 15-ketodihydro-PGF-2α concentrations between the mid-gestation period and before term. In an earlier study (Kindahl *et al.*, 1982) and in the present experiment the basal values of the prostaglandin metabolite during mid-gestation ranged from 100 to 200 pmol/l. In contrast, the basal levels before term reached values of 28 000 pmol/l (Barnes *et al.*, 1978; Stewart *et al.*, 1984).

The sharp release of prostaglandins 2–3 days after oestradiol/oxytocin-induced abortion resembles the release seen following spontaneous foaling (Stewart *et al.*, 1984). However, the magnitude of the former release was less than that of the latter (720 versus 2800 pmol/l). This post-abortion release of prostaglandins was not seen in the prostaglandin-induced abortions. The reason for this difference is not clear but it seems that the oestradiol/oxytocin-induced abortions caused more severe damage to the uterus; the fetal membranes were disrupted and the uterine contractions were more severe. In contrast the prostaglandin-induced abortions were without any complications. Furthermore these animals were also able to express symptoms of oestrus.

Plasma concentrations of relaxin before any treatment began agreed well with the findings of Stewart (1986). Stewart (1986) also reported that during spontaneous and oxytocin-induced foaling the levels of relaxin increased 3-fold and 15-fold, respectively, above the values found during mid-gestation. In contrast, no such dramatic changes were seen in the present study. Whether this indicates a difference in the ability to release relaxin at mid-gestation compared with term merits further investigation.

Pashen (1984) concluded that oestrogens appear to be involved in the mechanisms of prostaglandin production during parturition and reported that the pattern of relaxin production follows that of oestrogens. The correlations between oestrone sulphate and prostaglandin metabolite, relaxin and oestrone sulphate, and finally between relaxin and prostaglandin metabolite found in the present study do not disagree with the suggestions of Pashen (1984). However, no such correlations were found in mares given the combination of oestradiol and oxytocin to induce abortion.

This study was supported by grants from Alex and Eva Wahlströms Stiftelse, Aktiebolaget Trav & Galopp (ATG), Grayson Foundation and the Equine Research Laboratory. We thank Dr D. Schams for the PMSG antibodies and highly purified PMSG and Dr D. C. Collins and Dr K. Wright for antiserum to oestrone sulphate.

References

Barnes, R.J., Comline, R.S., Jeffcott, L.B., Mitchell, M.D., Rossdale, P.D. & Silver, M. (1978) Foetal and maternal concentrations of 13,14-dihydro-15-oxo-prostaglandin F in the mare during late pregnancy and at parturition. *J. Endocr.* **78**, 201–215.

Bosu, W.T.K. & McKinnon, A.O. (1982) Induction of abortion during midgestation in mares. *Can. vet. J.* **23**, 358–360.

Bosu, W.T.K., Edqvist, L.-E., Lindberg, P., Martinsson, K. & Johansson, E.D.B. (1976) The effect of various dosages of lynesterol on the plasma level of oestrogens and progesterone during the menstrual cycle in the rhesus monkey. *Contraception* **13**, 677–684.

Douglas, R.H., Squires, E.L. & Ginther, O.J. (1974) Induction of abortion in mares with prostaglandin F$_{2\alpha}$. *J. Anim. Sci.* **39**, 404–407.

Kindahl, H., Edqvist, L.-E., Granström, E. & Bane, A. (1976) The release of prostaglandin F$_{2\alpha}$ as reflected by 15-keto-13,14-dihydroprostaglandin F$_{2\alpha}$ in the peripheral circulation during normal luteolysis in heifers. *Prostaglandins* **11**, 871–878.

Kindahl, H., Knudsen, O., Madej, A. & Edqvist, L.-E. (1982) Progesterone, prostaglandin F-2α, PMSG and oestrone sulphate during early pregnancy in the mare. *J. Reprod. Fert., Suppl.* **32**, 353–359.

Menzer, C. & Schams, D. (1979) Radioimmunoassay for PMSG and its application to in-vivo studies. *J. Reprod. Fert.* **55**, 339–345.

Neely, D.P., Kindahl, H., Stabenfeldt, G.H., Edqvist, L.-E. & Hughes, J.P. (1979) Prostaglandin release patterns in the mare: physiological, pathophysiological, and therapeutic responses. *J. Reprod. Fert., Suppl.* **27**, 181–189.

Pashen, R.L. (1984) Maternal and foetal endocrinology during late pregnancy and parturition in the mare. *Equine vet. J.* **16**, 233–238.

Pashen, R.L. & Allen, W.R. (1979) The role of the fetal gonads and placenta in steroid production, maintenance of pregnancy and parturition in the mare. *J. Reprod. Fert., Suppl.* **27**, 499–509.

SAS Institute Inc. (1982) SAS User's Guide: Statistics, pp. 139–199. SAS Institute, Cary NC.

Squires, E.L., Hillman, R.B., Pickett, B.W. & Nett, T.M. (1980) Induction of abortion in mares with Equimate: effect on secretion of progesterone, PMSG and reproductive performance. *J. Anim. Sci.* **50**, 490–495.

Stewart, D.R. (1986) Development of a homologous equine relaxin radioimmunoassay. *Endocrinology* **119**, 1100–1104.

Stewart, D.R., Stabenfeldt, G.H. & Hughes, J.P. (1982a) Relaxin activity in foaling mares. *J. Reprod. Fert., Suppl.* **32**, 603–609.

Stewart, D.R., Stabenfeldt, G.H., Hughes, J.P. & Meagher, D.M. (1982b) Determination of the source of equine relaxin. *Biol. Reprod.* **27**, 17–24.

Stewart, D.R., Kindahl, H., Stabenfeldt, G.H. & Hughes, J.P. (1984) Concentrations of 15-keto-13,14-dihydro-prostaglandin F$_{2\alpha}$ in the mare during spontaneous and oxytocin induced foaling. *Equine vet. J.* **16**, 270–274.

Wright, K., Collins, D.C., Musey, P.I. & Preedy, J.R.K. (1978) A specific radioimmunoassay for estrone sulfate in plasma and urine without hydrolysis. *J. clin. Endocr. Metab.* **47**, 1092–1098.

J. Reprod. Fert., Suppl. 35 (1987), 485–492

Effect of *Salmonella typhimurium* endotoxin on PGF-2α release and fetal death in the mare

P. F. Daels, M. Starr, H. Kindahl*, G. Fredriksson*, J. P. Hughes and G. H. Stabenfeldt

Department of Reproduction, School of Veterinary Medicine, University of California, Davis, CA 95616, U.S.A. and *Department of Obstetrics and Gynaecology, College of Veterinary Medicine, Swedish University of Agricultural Sciences, Uppsala, Sweden

Summary. The infusion of *Salmonella typhimurium* endotoxin into pregnant mares resulted in a biphasic release pattern of PGF-2α as determined by 15-keto-13,14-dihydro-PGF-2α concentrations. The initial phase of 1 h duration was followed by accentuated release by 2 h after infusion; concentrations reached basal levels by 6 h. In 7 mares at 23, 26, 29, 33, 36, 53 and 55 days of gestation, fetal death occurred between 36 and 120 h after infusion; 12 mares at 46, 51, 56, 59, 65, 71, 73, 85, 103, 138, 283 and 318 days of gestation did not abort after endotoxin infusion. Luteal activity was compromised in all mares by 9 h after infusion. Progesterone concentrations were consistently lower in mares that aborted (1–2 ng/ml) than in those that did not abort. Mares therefore appear to be vulnerable to fetal loss by a clinical syndrome induced by *Salmonella typhimurium* endotoxin until about 50–60 days of gestation.

Introduction

The occurrence of organisms associated with endotoxin production, the Gram-negative bacteria, is high in animals, being an especially important component of diseases such as enteritis, mastitis, and endometritis. Salmonella-induced abortion in cattle, for example, has been reported from a number of countries (Hinton, 1974). Wrathall *et al.* (1978) were able to induce abortion in pigs by intravenous administration of large doses of endotoxin. Further work has shown that mediators of endotoxin pathophysiology are the prostaglandins (Kessler *et al.*, 1973; Veenendaal *et al.*, 1980). Support for a role of prostaglandins (PGs) in endotoxin action also comes from the finding that the administration of inhibitors of PG synthesis such as aspirin, flurbiprofen, flunixin meglumine and phenylbutazone before endotoxin administration can negate a number of the effects of the endotoxin (Fletcher & Ramwell, 1977; Bottoms *et al.*, 1981; Burrows, 1981; Moore *et al.*, 1981; van Miert *et al.*, 1982).

A rapid and pronounced PG response to endotoxin infusion has been reported in cattle and goats (Fredriksson, 1984; Fredriksson *et al.*, 1985) and endotoxin infusion into pregnant goats caused abortion. Of interest is the finding that endotoxin infusion into hysterectomized goats resulted in the termination of luteal activity. It is therefore obvious that PGF-2α synthesis and release from organ tissues other than the uterus can result in luteolysis. One potential source of PGF-2α production is the pulmonary system (Demling *et al.*, 1981).

The current study was undertaken to determine the effects in the pregnant mare of endotoxin infusion on PGF-2α synthesis and release and on pregnancy maintenance. We wanted to determine whether systemic synthesis and release of PGF-2α would compromise or terminate luteal activity and, if so, would this lead to fetal death. A preliminary study has been reported (Fredriksson *et al.*, 1986).

Materials and Methods

Animals. Twenty-three mares of various breeds (Thoroughbred, Standardbred, Quarter horse, Arabian and crossbred) and ages (3–19 years) were used in the study. Twenty animals were pregnant by natural service with the first day of gestation being the day of ovulation as determined by palpation of the ovaries *per rectum.* Gestation lengths at treatment were 23, 26, 29, 33, 36, 46, 51, 53, 54, 55, 56, 59, 69, 71, 73, 85, 103, 138, 283 and 318 days. Animals were kept initially on grass pasture or in a dry lot and fed oat and alfalfa hay. Animals were maintained in box stalls and fed oat and alfalfa hay from 1 day before to 5 days after endotoxin infusion.

Animals were assessed clinically 1 h before and at hourly intervals until clinical recovery from the endotoxin infusion for the following values: rectal temperature, heart rate, digital pulse, respiratory rate, gut motility, capillary refill time, scleral injection, packed cell volume and general attitude. Additional blood chemistries, including total protein, white cell count and electrolytes, were determined hourly as long as packed cell volume exceeded 45%.

Three animals were given lactated Ringers solution (Travenol Laboratories Inc., Deerfield, IL, U.S.A.) i.v. on the basis of a severe clinical reaction, packed cell volume above 50% and decreased total plasma protein. Fluid administration was stopped as soon as clinical signs of stabilization or improvement were noted. Mare 23 (23 days pregnant) received 13 litres between 3 and 8 h after endotoxin infusion; Mare 53 received 31 litres between 5 and 16 h; and Mare 85 received 14 litres between 6 and 9 h.

Endotoxin. The *Salmonella typhimurium* endotoxin (lipopolysaccharide, LPS) was extracted by the hot phenol–water method from batch-grown cultures of formaldehyde-killed *S. typhimurium* (SH 4809) and the LPS was further purified and characterized as described previously (Svenson & Lindberg, 1978; Lindberg *et al.*, 1983). The endotoxin was stored as a freeze-dried compound. At the start of the project the endotoxin was dissolved in saline and kept at $-20°C$ until 1 h before administration. Three mares (54, 227 and 318 days of gestation) were given a single dose of endotoxin (0·5 µg/kg) i.v. in a preliminary study of endotoxin dosage. Because one of these mares (54 days) showed signs of severe endotoxaemia and shock and subsequently died at 42 h after endotoxin administration, the dose of endotoxin was decreased to 0·33 µg/kg. Sixteen mares (23, 26, 29, 33, 46, 51, 53, 55, 56, 59, 65, 71, 73, 85, 103 and 138 days of gestation) were given a single dose of endotoxin i.v. (0·33 µg/kg). One mare (36 days) was inadvertently given a 0·23 µg/kg dose. Because this mare showed a response to the endotoxin challenge that was indistinguishable from the other mares, based on clinical signs, PGF-2α release and progesterone decline, she was included in the data analysis; the mare also aborted. The 3 non-pregnant mares were given a single dose of endotoxin i.v. (0·33 µg/kg) on Days 7, 8 and 8 after ovulation.

Sampling procedure. Blood samples were taken daily for 10 days, then 10 min and 1 min before the endotoxin administration. Blood samples were obtained at the following intervals after infusion: 10 min for 1 h, hourly for 29 h, every 3 h for 33 h, every 6 h for 72 h and twice daily for 7 days.

Blood was collected in heparinized vacutainers and kept in iced water (0°C). Plasma was obtained by centrifugation within 2 h of collection and stored at $-20°C$ until analysed for hormone concentrations. At 1 h before endotoxin administration an intravenous cannula was placed in the jugular vein and maintained for blood collection for 48 h after endotoxin infusion. The cannulae were flushed with heparinized saline after sample collection. Blood was obtained at other sampling times by venepuncture.

Hormone assays. Luteal response was determined by progesterone analysis of plasma by an enzyme immunoassay (Munro & Stabenfeldt, 1984). The antibody cross-reacts with 11α-hydroxyprogesterone (21·4%), 5α-pregnane-3,20-dione (29·5%), and 20β-dihydroprogesterone (2·4%) while the cross-reaction is <0·5% for 17α-hydroxyprogesterone, oestradiol-17β, cortisol, Δ^4-androstenedione and testosterone. The sensitivity of the assay is 2·5 pg/ml. The inter-assay coefficient of variation ranges from 1·2% to 4·3% at different concentrations of hormone and the intra-assay coefficient of variation from 2·7% to 8·4%.

The release of PGF-2α was monitored by measurement of 15-keto-13,14-dihydro-PGF-2α in plasma by a radioimmunoassay technique previously described (Granström & Kindahl, 1976; Kindahl *et al.*, 1976). The antibody cross-reacts with 15-keto-PGF-2α (16·0%), 13,14-dihydro-PGF-2α (4·0%), and with 15-keto-13,14-dihydro-PGE-2 (1·7%). The detection limit of the assay is 10 pg/ml plasma. The inter-assay coefficient of variation was 14% and the intra-assay coefficient of variation ranged between 6·6 and 11·7% for different areas of the standard curve.

Pregnancy examination. Before the experiment, pregnancy diagnosis was done at regular intervals by rectal palpation and ultrasonography of the uterus. Starting 1 day before the experiment and for 7 days after treatment, or until fetal death, mares were examined daily by rectal palpation and ultrasonography. Fetal death was established to be 12 h after the last positive sign of life of the fetus, i.e. fetal heart beat.

Results

Fetal death

In seven mares (23, 26, 29, 33, 36, 53 and 55 days pregnant) fetal death occurred at 72, 72, 120, 110, 120, 36 and 82 h, respectively, after endotoxin infusion. One of the mares that aborted (36

Table 1. Response of pregnant mares to infusion of *Salmonella typhimurin* endotoxin

Day of pregnancy	Dose (μg/kg body wt)	Status
23	0·33	Aborted 72 h
26	0·33	Aborted 72 h
29	0·33	Aborted 120 h
33	0·33	Aborted 110 h
36	0·23	Aborted 120 h
46	0·33	Pregnant
51	0·33	Pregnant
53	0·33	Aborted 36 h
54	0·5	Aborted 36 h, died at 42 h
55	0·33	Aborted 82 h
56	0·33	Pregnant
59	0·33	Pregnant
65	0·33	Pregnant
71	0·33	Pregnant
73	0·33	Foaled
85	0·33	Foaled
103	0·33	Foaled
138	0·33	Foaled
283	0·5	Foaled
318	0·5	Foaled

days) received 0·23 μg/kg. Ten mares (46, 51, 56, 59, 65, 71, 73, 85, 103 and 138 days pregnant) did not abort after endotoxin infusion: 5 of these mares produced normal foals at term and the other 5 are still in foal.

Of 3 mares given 0·5 μg/kg, the 2 mares at 283 and 318 days of gestation did not abort. Both mares produced normal foals at term. The other mare (54 days) died at 42 h after endotoxin infusion which was preceded by abortion at 36 h. Data from these mares were excluded from all analyses.

Prostaglandin synthesis and release

The composite pattern of PGF-2α synthesis and release after endotoxin infusion is shown for all pregnant mares, both aborting and non-aborting, in Fig. 1. In general the pattern of release of PGF-2α, as determined by measurement of its 15-keto-13,14-dihydro metabolite, followed a biphasic pattern with an initial peak achieved at 30 min after infusion (892 ± 139 pg/ml, mean $\pm$ s.e.m., range 230–2300 pg/ml). Concentrations tended to decrease until 1 h which was followed by a second surge of PGF-2α production which peaked at 2 h (4403 ± 722 pg/ml, range 325–10 528 pg/ml). The 15-keto-13,14-dihydro-PGF2α concentrations were approximately baseline by 6 h after infusion. One mare (53 days) had a third major peak (1596 pg/ml) which was associated with abortion at 36 h after infusion (Fig. 2). This was the only abortion associated with a significant increase in prostaglandin release. One mare (138 days) produced limited amounts of the prostaglandin metabolite (highest concentration, 325 pg/ml) in comparison to other mares.

Total prostaglandin synthesis and release was determined for each mare by measuring the area under the curve (above basal concentrations) for 0–12 h after infusion (Fig. 3). There were no differences between animals that aborted (9118 ± 2657 pg/ml) and those that did not abort (8088 ± 4492 pg/ml).

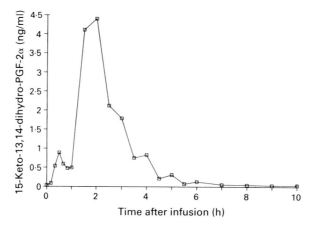

Fig. 1. Composite 15-keto-13,14-dihydro-PGF-2α concentrations in pregnant mares (23–138 days pregnant) 0–10 h after infusion of *Salmonella typhimurium* endotoxin.

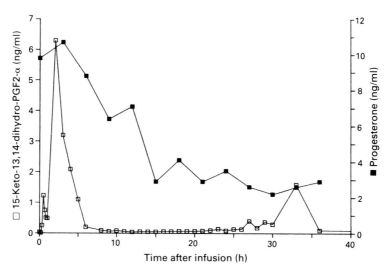

Fig. 2. The response of a pregnant mare (53 days) to infusion of *Salmonella typhimurium* endotoxin. Luteal activity was depressed but not terminated before abortion at 36 h in conjunction with prostaglandin release.

Progesterone patterns

Average concentrations of progesterone following endotoxin infusion for aborting, non-aborting and non-pregnant mares are shown in Fig. 4(a). One non-pregnant mare ovulated at 36 h after treatment during the period of luteolysis and because of the re-establishment of luteal function, the data were not included for non-pregnant mares. All 3 group means were higher at 3 h after infusion than before treatment. Progesterone concentrations were significantly lower ($P < 0.05$) than pre-treatment values at 9.0 ± 2.1 h in aborting mares, 9.4 ± 1.3 h in non-aborting mares and 9.3 ± 0.6 h in non-pregnant mares.

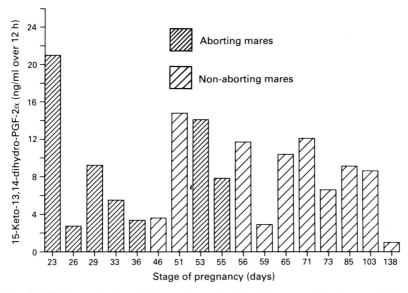

Fig. 3. Total PGF-2α synthesis and release determined by the area under the curve (above basal concentrations) for 12 h after infusion with *Salmonella typhimurium* endotoxin.

Of the 7 aborting mares, 5 had progesterone concentrations between 1 and 2 ng/ml for 27–60 h before fetal death with fetal loss occurring at these concentrations. One mare (53 days) maintained progesterone concentrations > 2 ng/ml before abortion at 36 h (Fig. 2). Abortion in this mare was associated with a major release of prostaglandin. The last mare (29 days) of this group had progesterone concentrations < 2 ng/ml for 59 h followed by values < 1 ng/ml for 48 h before fetal death. This mare began showing oestrus 48 h before fetal death, i.e. in conjunction with a decline in progesterone concentrations to < 1 ng/ml.

While progesterone concentrations declined in non-aborting mares after endotoxin infusion, values remained well above the 1–2 ng/ml range in all mares except one (56 days). Progesterone values in this mare remained between 1 and 2 ng/ml from 32 to 120 h after infusion without the occurrence of abortion (Fig. 4b). The only non-aborting mare that did not have significantly decreased progesterone concentrations post infusion was the mare at 138 days.

Clinical changes

The temperature increase noted in all mares after infusion reached its maximum between 4 and 6 h after infusion with a return to normal at 12 h after infusion. Heart and respiratory rates increased in relation to the body temperature although there was no consistent pattern. All mares displayed acute ileus starting 1 h after infusion and ending between 12 and 18 h after infusion. Ileus was associated with severe diarrhoea that was resolved 12–18 h after infusion. Mares were moderately to severely depressed and anorexic for up to 24 h.

Other signs of endotoxaemia such as changes in the colour of the mucus membrane and capillary refill time were also evident. Leucocytopenia, increased packed cell volume and elevated plasma protein occurred in the first 12 h after infusion and were back to normal values at 24 h after infusion.

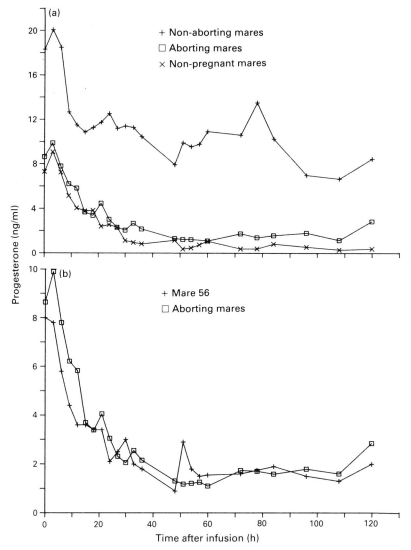

Fig. 4. Plasma progesterone concentrations in (a) non-aborting mares (46–138 days pregnant, N = 10), aborting mares (23–55 days pregnant, N = 7) and non-pregnant mares (N = 2), and (b) in Mare 56 that failed to abort in spite of maintaining progesterone concentrations similar to those of animals that aborted (□, N = 7).

Discussion

One basic question raised by the results of our study involves the reason for fetal death after endotoxin infusion. Amongst various possibilities is that endotoxin could have a direct deleterious effect on the feto–placental unit although there is no evidence that endotoxin *per se* can pass through the endometrium. There was no relationship between the amount of PGF-2α released during the first 12 h after infusion and the incidence of abortion or the duration of pregnancy. The amount of PGF-2α released is therefore not critical to abortion. This does not discount the possibility that PGF-2α has a direct effect on fetal loss.

It is also possible that the depressed luteal activity in which progesterone concentrations remain in the 1–2 ng/ml range for several days can compromise the pregnancy. Shideler *et al.* (1982) indicated that ovariectomized, pregnant mares tended to abort if progesterone concentrations were in the 2 ng/ml range. It is likely that mares with normal pregnancies have occasional values below 2 ng/ml, but this situation is almost certainly transitory occurring over a matter of minutes or a few hours. It is clear that a sustained major release of PGF-2α for a period of 5–6 h, as occurred in this study following endotoxin infusion, results in a long-term depression of luteal activity. It resembles the depression of luteal activity that often occurs in non-pregnant mares with persistent luteal activity because of PGF-2α synthesis and release which is inadequate to effect complete luteal regression (Neely *et al.*, 1979).

It is of interest that the 2 mares in late pregnancy (283 and 318 days) released about the same amount of PGF-2α above baseline, namely, 10 867 and 8026 pg/ml, respectively, as did aborting (9118 ± 2657 pg/ml) and non-aborting mares (8088 ± 4492 pg/ml). The basal concentrations of PGF-2α in the late pregnant mares were several hundred-fold higher than in mares less than 100 days of gestation. These increased baseline values reflect activity of the gravid uterus. Because the amount of PGF-2α released by these late pregnant mares was similar to the amount released earlier in pregnancy, in spite of the greatly increased basal production rate, much of the PGF-2α response to endotoxin must have been by tissues other than those of the reproductive system.

There appears to be a primary interval involved in the response to endotoxin in that PGF-2α synthesis and release tended to peak at 30 min with a decline at 60 min followed by accentuated release during the second hour after infusion. This suggests that organ tissues need a certain exposure period before sustaining a rather large synthesis and release of PGF-2α. Again, the mares in late pregnancy were interesting in that while they released little, if any, PGF-2α above the baseline concentrations during the first hour after infusion, relatively large increases were noted beginning at the second hour.

Prostaglandin synthesis and release was acutely associated with abortion in only one mare (53 days) in which a major surge of prostaglandin was associated with abortion. It could be argued that we missed prostaglandin release in conjunction with abortion at later times because of an inadequate sampling schedule. However, the fetus delivered in conjunction with prostaglandin release was fresh while the other fetuses were found in a decomposed state, suggesting that abortion was passive in these animals in the sense that fetal death occurred before the expulsion of the fetus. In one mare (29 days) fetal death occurred 48 h after progesterone concentrations had declined to values suggestive of complete luteolysis, indicating that the fetus can survive for a considerable period in the absence of progestational support.

It is also not clear as to the effect of the duration of pregnancy on a mare's ability to maintain pregnancy when infused with endotoxin. It appears that the critical period is between Days 45 and 60 in that some mares aborted and some did not during this time. Even though luteal activity was depressed in all mares of < 138 days of gestation, progesterone concentrations were higher in non-aborting mares at the time of endotoxin infusion with values usually continuing to be greater than 4–5 ng/ml throughout the experimental period. One factor could be the formation of accessory corpora lutea which occurs between Days 40 and 60 gestation and extra luteal tissue could be the basis for higher progesterone production in non-aborting mares. It is also clear from the results for one mare (56 days) that progesterone concentrations can be maintained between 1 and 2 ng/ml for some time without fetal loss.

One question we wrestled with concerned the mode of administration of endotoxin, i.e. bolus *vs* continuous infusion. The mode we chose, bolus injection, produced a clinical symptoms similar to those often observed in veterinary clinical practice. The mares showed abdominal discomfort, sweating, depression and diarrhoea with acute symptoms present within 1 h of endotoxin infusion; signs abated by 12–15 h in most animals. The endotoxin infusion took place in the morning and most mares had resumed eating with normal gastrointestinal activity present by evening.

The possibility of pregnancy loss due to endotoxaemia before Day 60 of gestation suggests the importance of avoiding stressful situations during this period. It would appear that the movement of pregnant animals should be minimized before Day 60. If mares develop clinical syndromes suggestive of endotoxaemia before Day 60 of gestation, the question arises as to what therapeutic approaches should be considered. If the animal is thought to be in the early stages of the syndrome, non-steroidal, anti-inflammatory compounds could be useful for modifying prostaglandin synthesis and release. The treatment would have to be early in that prostaglandin synthesis and release was over by about 6 h after infusion. Another therapeutic approach would be to administer progesterone until Day 80 or 90 of gestation, at which time placental progesterone production could be expected to be sufficient to maintain the pregnancy (Holtan *et al.*, 1979). These treatment strategies are currently being investigated.

We thank Sheila MacLachlan and Jeanne Bowers for help in sample collection; Margareta Svensson and Tiina Basu for prostaglandin analysis; Tom Madley for progesterone analysis; Dr S. Svensson, Kardinsha Institutet, Stockholm, for the purified endotoxin; and Margie Merrill for manuscript preparation. The animals were housed at the Animal Research Laboratory, University of California, Davis. Supported by a grant from the Grayson Foundation and the Swedish Council for Agricultural Research and Aktieboraget Trav & Galopp.

References

Bottoms, G.D., Fessler, J.F., Roesel, O.F., Moore, A.B. & Trauenfelder, H.C. (1981) Endotoxin-induced hemodynamic changes in ponies: Effects of flunixin meglumine. *Am. J. vet. Res.* **42**, 1514–1518.

Burrows, G.E. (1981) Therapeutic effect of phenylbutazone on experimental acute *Escherichia coli* endotoxemia in ponies. *Am. J. vet. Res.* **42**, 94–99.

Demling, R., Smith, M., Gunther, R., Flynn, J.T. & Gee, M.H. (1981) Pulmonary injury and prostaglandin production during endotoxemia in conscious sheep. *Am. J. Physiol.* **240**, 4348–4353.

Fletcher, J.R. & Ramwell, P.W. (1977) Modification by aspirin and indomethacin, of the haemodynamic and prostaglandin releasing effects of *E. coli* endotoxin in the dog. *Br. J. Pharmac.* **61**, 175–181.

Fredriksson, G. (1984) Some reproductive and clinical aspects of endotoxins in cows with special emphasis on the role of prostaglandins. *Acta vet. scand.* **25**, 1–13.

Fredriksson, G., Kindahl, H. & Edqvist, L.-E. (1985) Endotoxin-induced prostaglandin release and corpus luteum function in goats. *Anim. Reprod. Sci.* **8**, 109–121.

Fredriksson, G., Kindahl, H. & Stabenfeldt, G. (1986) Endotoxin-induced and prostaglandin-mediated effects on corpus luteum function in the mare. *Theriogenology* **25**, 309–316.

Granström, E. & Kindahl, H. (1976) Radioimmunoassays for prostaglandin metabolites. *Adv. Prostaglandin Thromboxane Res.* **1**, 81–92.

Hinton, M. (1974) *Salmonella dublin* abortion in cattle: studies on the clinical aspects of the condition. *Br. vet. J.* **130**, 556–563.

Holtan, D.W., Squires, E.L., Lapin, D.R. & Ginther, O.J. (1979) Effect of ovariectomy on pregnancy in mares. *J. Reprod. Fert., Suppl.* **27**, 457–463.

Kessler, E., Hughes, R.C., Bennett, E.N. & Nadela, S.M. (1973) Evidence for the presence of prostaglandin-like material in the plasma of dogs with endotoxin shock. *J. lab. clin. Med.* **81**, 85–94.

Kindahl, H., Edqvist, L.-E., Granstrom, E. & Bane, A. (1976) The release of prostaglandin F2α as reflected by 15-keto-13,14-dihydroprostaglandin F2α in the peripheral circulation during normal luteolysis in heifers. *Prostaglandins* **11**, 871–878.

Lindberg, A.A., Sheldon, G.E. & Svenson, S.B. (1983) Induction of endotoxin tolerance with nonpyrogenic O-antigenic oligosaccharide-protein conjugates. *Infect. Immun.* **41**, 883–895.

Moore, J.N., Garner, H.E., Shapland, J.E. & Hatfield, D.G. (1981) Prevention of endotoxin-induced arterial hypoxaemia and lactic acidosis with flunixin meglumine in the conscious pony. *Equine vet. J.* **13**, 95–98.

Munro, C. & Stabenfeldt, G. (1984) Development of a microtitre plate enzyme immunoassay for the determination of progesterone. *J. Endocr.* **101**, 41–49.

Neely, D.P., Kindahl, H., Stabenfeldt, G.H., Edqvist, L.-E. & Hughes, J.P. (1979) Prostaglandin release patterns in the mare: physiological, pathophysiological, and therapeutic responses. *J. Reprod. Fert., Suppl.* **27**, 181–189.

Shideler, R.K., Squires, E.L., Voss, J.L., Eikenberry, D.J. & Pickett, B.W. (1982) Progestagen therapy of ovariectomized pregnant mares. *J. Reprod. Fert., Suppl.* **32**, 459–464.

Svenson, S.B. & Lindberg, A.A. (1978) Immunochemistry of Salmonella O antigen: Preparations of an octasaccharide-bovine serum albumin immunogen representative of Salmonella serogroup B O-antigen and characterization of the antibody response. *J. Immunol.* **180**, 1750–1757.

van Miert, A.S.J.P.A.M., van Duin, C.T.M., Verheijden, J.H.M. & Schotman, A.H.J. (1982) Endotoxin-induced fever and associated haematological and blood biological changes in the goat: the effect of repeated administration and the influence of flurbiprofen. *Res. vet. Sci.* **33**, 248–255.

Veenendaal, G.H., Woutersen-van Nynanten, F.M.A., van Duin, C.Th.M. & van Miert, P.A.M. (1980) Role of circulating prostaglandins in the genesis of pyrogen (endotoxin)-induced ruminal stain in conscious goats. *J. vet. Pharmacol. Therapy* **3**, 59–68.

Wrathall, A.E., Wray, C., Bailey, J. & Wells, D.E. (1978) Experimentally induced bacterial endotoxaemia and abortion in pigs. *Br. vet. J.* **134**, 225–230.

J. Reprod. Fert., Suppl. **35** (1987), 493–495
Printed in Great Britain
© 1987 Journals of Reproduction & Fertility Ltd

Reproductive wastage in the mare and its relationship to progesterone in early pregnancy

D. Forde, L. Keenan, J. Wade*, M. O'Connor† and J. F. Roche

*Faculties of Veterinary Medicine and *Agriculture, University College Dublin, Dublin 4 and †The Irish National Stud, Kildare, Ireland*

Estimates of embryonic and fetal loss in the mare vary tremendously in terms of definition and method of diagnosis. Using ultrasonography, Chevalier & Palmer (1982) found a total pregnancy loss of 9% from first diagnosis of pregnancy on Day 23 but were unable to detect a preferential period of early pregnancy loss, an observation previously made by Bain (1969). Ginther *et al.* (1985a) observed a 10% loss between Days 15 and 50 as compared with the 5% found by Simpson *et al.* (1982) between Days 20 and 45; 14·4% found by Villahoz *et al.* (1985) between Days 15 and 50; and 11·1% by Woods *et al.* (1985a) between Days 14 and 56. Woods *et al.* (1985b) suggested that the embryonic loss rate is high before pregnancy detection by ultrasonography, which would seem to be borne out by the results of Ball *et al.* (1985) who found similar fertilization rates in normal and barren mares but significantly different Day-14 pregnancy rates.

Ovariectomy and progestagen therapy (Holtan *et al.*, 1979; Shideler *et al.*, 1982; Ginther, 1985) have demonstrated the requirement for progesterone for survival of the embryo. It is not clear, however, whether a decline in progesterone concentrations increases the incidence of embryo loss or whether such a decline is due to primary luteal insufficiency (Allen, 1984; Ginther *et al.*, 1985b). The aim of this study was, therefore, to identify the extent and time of reproductive wastage in the mare and its relationship to concentrations of progesterone in serum.

Materials and Methods

During the breeding seasons of 1983/84 and 1984/85 the reproductive performance of 529 Thoroughbred mares (Group 1) on a stud farm was monitored and analysed. Pregnancy diagnosis was carried out using real-time ultrasound echography (Palmer & Driancourt, 1980; Simpson *et al.*, 1982) at 18, 30 and 42 days after the last service (for those mares which remained on the stud farm at those times) and these results were correlated with pregnancy certificates submitted by breeders on 1 October and foaling returns made to the stud book authorities.

In 1985 two replicate groups of 20 Thoroughbred-type mares, aged 3–8 years (Group 2), of which 75% were maiden and 25% barren, were studied through 3 oestrous cycles during which they were bled daily for estimation of progesterone concentrations by radioimmunoassay by the method of Niswender (1973) who also provided the progesterone antiserum (GDN 337). The practical limit of sensitivity of the assay was 0·3 ng/ml while the inter-assay coefficients of variation for the medium and high controls were 14·5% and 10·5% respectively (the value of the low control never exceeded the sensitivity of the assay). The intra-assay CV was 4·4%.

The first observed oestrous cycle was used as a control. During the next oestrus the mares were mated and non-surgical embryo recovery (Squires *et al.*, 1982) was attempted at Days 6–9 (N = 16) or Days 12–14 (N = 22) after ovulation. When the mares returned to oestrus, they were re-mated and slaughtered at Days 2–5 (N = 20) when oviducts were removed *post mortem* and flushed for ova, or Days 15–21 (N = 22, of which 2 were slaughtered in a later cycle). Ultrasound examinations for pregnancy were performed daily from Day 10 in mated mares.

Progesterone values of mares during control, pregnant and non-pregnant cycles were compared using Student's *t* test.

Results and Discussion

Group 1. The cumulative pregnancy rates at Day 18 on 1 October, and the live foal rate for 1983/84 and 1984/85, respectively, were 83% and 88%; 76% and 83%; and 73% and 76%. Late

Table 1. Reproductive wastage in mares after mating in a single cycle

	1983/84	1984/85
No. of cycles studied	458	410
Live foal rate/cycle	39%	40%
Wastage		
Not pregnant on Day 18	51%	48%
Failure to ovulate/short cycle	5%	1%
Return to oestrus before 18 days	33%	26%
Negative ultrasound scan	13%	21%
*Loss between 18 and 24 days	3%	3%
*Loss between 42 days and 1 October	4%	2%
Loss between 1 October and foaling	3%	7%
Total wastage	61%	60%

*Based on mares which remained on the stud farm until at least Day 42 after last service.

Table 2. Extent and time of wastage before Day 18 in experimental mares

Days after ovulation	No. of mares	No. flushed or killed	Pregnant	Not pregnant	Not recovered	Returned to oestrus
15–21	30	22	18	12	4	8
12–14	24	22	15	9	8*	2
6–9	16	16	7	?	9	0
2–5	20	20	11	?	9	0

*1 mare appeared to be pregnant when scanned but no embryo was recovered.

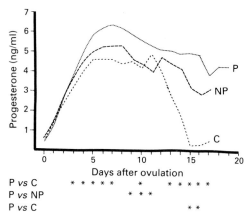

Fig. 1. Serum concentrations of progesterone in cyclic (C), pregnant (P) and non-pregnant (NP) mares up to Day 19 after ovulation. *$P < 0.05$.

embryonic/fetal loss (from Day 18 to foaling) was 10% in Year 1 and 12% in Year 2. The foaling rate for a single cycle in which mares were mated was 40%. The majority of the wastage occurred before 18 days (Table 1).

Group 2. The pregnancy rate in the controlled study for the period 15–21 days (mean, 17 days) after ovulation was 60% and for the period 12–14 days (mean, 13 days) was 62·5%, indicating very little embryonic loss during this period (Table 2). The recovery rate of blastocysts, or 'detected pregnancy rate', at Days 6–9 (mean, Day 7) was 44%. Flushing the oviducts of 20 mares between Days 2 and 5 (mean, Day 4) produced 15 cleaved ova (including 4 sets of twins) out of 20 recovered non-atretic ova from 11 mares for a fertilization rate of 75% (i.e. a pregnancy rate of 55%).

The only difference in progesterone concentrations (Fig. 1) between control (unmated) and non-pregnant mares was at the end of the cycle when luteolysis had occurred in the controls and some of the non-pregnant mares had not yet returned to oestrus. Differences between pregnant and non-pregnant mares were only apparent at Days 9–11 and consequently the data do not support the hypothesis of primary luteal insufficiency as a cause of infertility. The significant difference, on most days of the cycle, between the progesterone concentrations of pregnant mares and the unmated controls can be explained by the finding that 17 out of 21 double ovulations diagnosed during the project occurred at the start of pregnant cycles and double ovulations resulted in a significantly higher ($P < 0.05$) progesterone output than single ovulations.

We conclude that (1) the major source of reproductive wastage in the mare was before Day 12 after ovulation, and (2) there was no association between progesterone concentration and pregnancy status when twin ovulations were taken into account. We suggest that the major source of reproductive wastage in the mare after natural service is a combination of fertilization failure and early death of the fertilized ovum before it enters the uterus.

References

Allen, W.R. (1984) Is your progesterone therapy really necessary? *Equine vet. J.* **16**, 496–498.

Bain, A.M. (1969) Foetal losses during pregnancy in the thoroughbred mare: a record of 2562 pregnancies. *N.Z. Vet. J.* **17**, 155–158.

Ball, B.A., Little, T.V., Hillman, R.B. & Woods, G.L. (1985) Embryonic loss in normal and barren mares. *Proc. Am. Assoc. Equine Pract.* pp. 535–543.

Chevalier, F. & Palmer, E. (1982) Ultrasonic echography in the mare. *J. Reprod. Fert., Suppl.* **32**, 421–430.

Ginther, O.J. (1985) Embryonic loss in mares: nature of loss after experimental induction by ovariectomy or prostaglandin F2. *Theriogenology* **24**, 87–98.

Ginther, O.J., Bergfelt, D.R., Leith, G.S. & Scraba, S.T. (1985a) Embryonic loss in mares: incidence and ultrasonic morphology. *Theriogenology* **24**, 73–86.

Ginther, O.J., Garcia, M.C., Bergfelt, D.R., Leith, G.S. & Scraba, S.T. (1985b) Embryonic loss in mares: pregnancy rate, length of interovulatory intervals, and progesterone concentrations associated with loss during days 11 to 15. *Theriogenology* **24**, 409–417.

Holtan, D.W., Squires, E.L., Lapin, D.R. & Ginther, O.J. (1979) Effect of ovariectomy on pregnancy in mares. *J. Reprod. Fert., Suppl.* **27**, 457–463.

Niswender, G.D. (1973) Influence of the site of conjunction on the specificity of antibodies to progesterone. *Steroids* **22**, 413–424.

Palmer, E. & Driancourt, M.A. (1980) Use of ultrasound echography in equine gynaecology. *Theriogenology* **13**, 203–216.

Shideler, R.K., Squires, E.L., Voss, J.L., Eikenberry, D.J. & Pickett, B.W. (1982) Progesterone therapy of ovariectomized pregnant mares. *J. Reprod. Fert., Suppl.* **32**, 459–464.

Simpson, D.J., Greenwood, R.E.S., Ricketts, S.W., Rossdale, P.D., Sanderson, M. & Allen, W.R. (1982) Use of ultrasonic echography for early diagnosis of single and twin pregnancy in the mare. *J. Reprod. Fert., Suppl.* **32**, 431–439.

Squires, E.L., Imel, K.J., Iuliano, M.F. & Shideler, R.K. (1982) Factors affecting reproductive efficiency in an equine embryo transfer programme. *J. Reprod. Fert., Suppl.* **32**, 409–414.

Villahoz, M.D., Squires, E.L., Voss, J.L. & Shideler, R.K. (1985) Some observations on early embryonic death in mares. *Theriogenology* **23**, 915–924.

Woods, G.L., Baker, C.B., Hillman, R.B. & Schlafer, D.H. (1985a) Recent studies relating to early embryonic death in the mare. *Equine vet. J., Suppl.* **3**, 104–107.

Woods, G.L., Baker, C.B., Bilinski, J., Hillman, R.B., Mank, E.C. & Sluijter, F. (1985b) A field study on early pregnancy loss in Standardbred and Thoroughbred mares. *J. Equine vet. Sci.* **5**, 264–267.

J. Reprod. Fert., Suppl. **35** (1987), 497–498

Clinical and endocrine aspects of early fetal death in the mare

K. Darenius, H. Kindahl and A. Madej

Departments of Obstetrics and Gynaecology and Clinical Chemistry, Swedish University of Agricultural Sciences, S-750 07 Uppsala, Sweden

Early fetal death in the mare is a multifactorial problem in which environmental and managemental factors external to the animal as well as patho-physiological factors in the animal are involved. In the present study efforts have been made to investigate some of these factors.

Sixteen mares which had lost their conceptuses 2–5 times during a period of 6 years were used in a study extending over 4 breeding seasons. The genital organs were examined by means of rectal palpation, ultrasound scanning and vaginoscopy. Cytological, bacteriological and histological examinations were carried out before mating and after early fetal death. Blood samples were collected twice a day beginning one oestrous cycle before mating until 150 days of pregnancy and radioimmunoassay was used to determine peripheral plasma concentrations of progesterone, chorionic gonadotrophin (CG) and oestrone sulphate (Kindahl *et al.*, 1982).

Ten mares had normal pregnancies during the experimental period, 4 mares resorbed or aborted the conceptus once or twice and 2 mares did not conceive. Three of the resorbing/aborting mares were able to carry a foal to term after the pregnancy failure.

The condition of the luminal uterine epithelium before mating, according to clinical, bacteriological and cytological examinations, was good in all mares except 3, which were found to be infected with *Streptococcus zooepidemicus*. After treatment and insemination with minimal contamination technique (Kenney *et al.*, 1975) they all conceived and developed normal pregnancy.

All mares, except two of the resorbing/aborting mares, had a variety of histopathological changes from low grade acute endometritis to advanced chronic degenerative endometritis and pyometra.

In the four mares with early fetal death the embryos/fetuses were confirmed dead on Day 25, Days 37 and 42 (twins), Days 36 and 38 (twice in the same mare) and Day 54. The mares returned to oestrus on Day 59, Day 76, Days 53 and 77, and Day 113, respectively.

When early fetal death occurred during the embryonic stage, the embryonic sac collapsed early after the death of the embryo. The content spread into the uterus making it enlarged and of a doughy consistency. During the fetal stage, however, the sac may remain intact for a long time which makes a correct pregnancy diagnosis by palpation very difficult. The mare in which the fetus was confirmed dead by Day 54 exhibited a 'small for date conceptus' from Day 30. On Day 47 the heart rate was around 60 beats/min, compared with 120–200 beats/min in normal fetuses. From Day 54 no heart activity could be detected. On palpation the vesicular bulge seemed quite normal, grape-fruit sized, well demarcated and fluctuating. This palpatory impression remained for another 59 days when the mare showed oestrus and the fluid of the vesicular bulge had spread out into the uterus. On the same day the cervix was open and greyish, watery uterine content filled the floor of the vagina. With a tube speculum inserted into the vagina and one hand in the rectum a gentle clap on the corpus uterus caused the collapsed vesicular bulge to escape through the speculum.

No primary luteal deficiency, as determined from progesterone concentrations, could be detected in the resorbing/aborting mares as has been reported by Ginther (1985). A different pattern was detected only in the mare which resorbed the conceptus twice and afterwards failed to develop a normal pregnancy: she had irregular interovulatory intervals, showed oestrus and ovulated despite the presence of a growing conceptus.

The production of CG was slightly decreased when the fetus died shortly after the formation of the endometrial cups. Since the presence of this hormone is associated with large individual variation, with mares producing minimal amounts of CG yet carrying the fetus to term (Allen, 1969), it is difficult to determine a minimum level necessary for normal development of pregnancy.

Oestrone sulphate concentrations in normal pregnant mares were comparable to the values reported by other workers during early pregnancy (Terqui & Palmer, 1979; Kindahl *et al.*, 1982). The first detectable rise of conjugated oestrone sulphate around Day 40 is believed to originate from the maternal ovaries and could be seen also in the mares with early fetal death when primary luteal function was maintained. From about Day 80 the fetal gonads begin to enlarge and secrete increasing quantities of dehydroepiandrosterone, which is aromatized in the placenta and appears in the maternal circulation as conjugated oestrone and oestradiol-17β (Pashen & Allen, 1979). Bosu *et al.* (1984) have shown that the fall in oestrone sulphate concentration is very rapid after the death of the conceptus in induced abortions. In the aborting mare in which the fetus was confirmed dead on Day 54, the oestrone sulphate concentrations were significantly lower than in normal pregnant mares from Day 61. The concentration did not return to dioestrous values until Day 104 which coincides with the time when the conceptus lost its spherical outline and darkness of allantoic fluid. By the next scanning it was irregular in outline and had a speckled or granular content.

There may be several explanations of the good breeding results during the trial compared with the previous problems with repeated pregnancy losses of the mares. Good management seems to be essential and many factors may play a part in the breeding results, such as that the mares were in good nutritional condition, were mated only a few times during the height of the ovulatory season and stressful conditions were avoided by keeping them as a group at pasture and by daily uniform routines such as feeding, teasing and blood-sampling.

References

Allen, W.R. (1969) The immunological measurement of pregnant mare serum gonadotrophin. *J. Endocr.* **43**, 593–598.

Bosu, W.T.K., Turner, L. & Franks, T. (1984). Estrone sulphate and progesterone concentrations in the peripheral blood of pregnant mares: clinical implications. *Proc. 10th Int. Congr. Anim. Reprod. & A.I., Urbana*, **2**, 78, Abstr.

Ginther, O.J. (1985) Embryonic loss in mares: incidence, time of occurrence and hormonal involvement. *Theriogenology* **23**, 77–89.

Kenney, R.M., Bergman, R.V., Cooper, W.L. & Morse, G.W. (1975) Minimal contamination techniques for breeding mares: Technique and preliminary findings. *Proc. 21st Ann. Conv. Am. Assoc. equine Pract., Boston*, 327–336.

Kindahl, H., Knudsen, O., Madej, A. & Edqvist, L.-E. (1982) Progesterone, prostaglandin F-2α, PMSG and oestrone sulphate during early pregnancy in the mare. *J. Reprod. Fert., Suppl.* **32**, 353–359.

Pashen, R.L. & Allen, W.R. (1979) The role of the fetal gonads and placenta in steroid production, maintenance of pregnancy and parturition in the mare. *J. Reprod. Fert., Suppl.* **27**, 499–509.

Terqui, M. & Palmer, E. (1979) Oestrogen pattern during early pregnancy in the mare. *J. Reprod. Fert., Suppl.* **27**, 441–446.

J. Reprod. Fert., Suppl. **35** (1987), 499–504

Endometrial histology of early pregnant and non-pregnant mares

L. R. Keenan, D. Forde, T. McGeady, J. Wade* and J. F. Roche

Faculties of Veterinary Medicine and *Agriculture, University College Dublin, Ballsbridge, Dublin 4, Ireland

Introduction

The establishment of early pregnancy in the mare requires a viable blastocyst to enter the uterus, develop *in utero* and initiate the events leading to the successful maternal recognition of pregnancy between Days 14 and 16. Embryonic loss is high in the mare (Ginther, 1985; Woods *et al.*, 1985) and recent results indicate that a high percentage of reproductive wastage occurs before Day 18 (Forde *et al.*, 1987). To increase reproductive efficiency in the mare it is important to determine the various causes of this high wastage. Admission of exogenous agents, such as bacteria, at mating could result in subclinical infection or changes in endometrial function or structure. The aim of this work was to determine histologically whether there were differences in the structure of the endometrium in pregnant and non-pregnant mares that had been mated.

Materials and Methods

Endometrial samples were collected from 40 Thoroughbred mares aged 3–8 years over a period of 3 oestrous cycles using previously described techniques (Ricketts, 1975; Kenney, 1978). During the first cycle the mares were arbitrarily placed in four groups and biopsies were taken between Days 2 and 5, 6 and 9, 12 and 14 and 15 and 21 after ovulation. During the second cycle the mares were mated with a fertile stallion, and their uteri were flushed for embryo recovery and biopsies were taken at Days 6–9 or 12–14 after ovulation. During the third cycle the mares were mated again and slaughtered at Days 2–5 or 15–21 after ovulation. Endometrial samples were taken and the oviducts of the mares at Days 2–5 were flushed for blastocysts and the uteri of the mares at Days 15–21 were flushed for embryos. During each cycle, the mares were teased, bled daily for progesterone determination and had their ovaries palpated or scanned by ultrasound to determine when ovulation occurred. Pregnancy was confirmed by ultrasound from Day 12 onwards in the pregnant mares.

The endometrial tissue was fixed in Bouin's fluid and sections were prepared by routine histological procedures and selectively stained for lymphocytes, neutrophils, eosinophils, mast cells and pigment-bearing macrophages (Drury & Wallington, 1980). The surface and glandular epithelial heights, gland diameters and density and numbers of mobile cells were determined by using a calibrated eyepiece micrometer. Gland activity was observed and the distribution of the mobile cells was estimated. Sections from 60 endometrial samples were divided into three classes according to the status of the mare, i.e. not mated (N = 25), pregnant (N = 26) and mated but no embryo recovered (N = 15). Each class was subdivided into four stages of 1–5, 6–9, 12–14 and 15–21 days after ovulation. Means for each measure were calculated from each section and the data were subjected to an analysis of variance. Duncan's multiple range test was used to determine whether there were significant differences between stages of the oestrous cycle and pregnancy.

Results

The histological observations made are summarized in Table 1. General histological architecture was similar between groups and within categories. The surface epithelium was significantly higher ($P < 0.01$) in pregnant mares in comparison with the other mares at Days 2–5 after ovulation but was not significantly different at other stages (Fig. 1). The superficial and basal endometrial glands in all mares were distributed irregularly, depending on the degree of oedema present, through a moderately vascular lamina propria. Glands of all mares showed increased coiling, tortuosity and

Table 1. Histological observations and measurements, and infiltrations of mobile cells (per $0.5\ cm^2$) of endometria from unmated (C, dioestrous) and pregnant (P) mares and mares that were mated but from which no embryo was recovered (NR)

	Days after ovulation				
	2–5			6–9	
	C (N = 4)	NR (N = 3)	P (N = 5)	C (N = 8)	NR (N = 5)
Surface epithelium					
Type	Columnar/ pseudo-stratified	Columnar/ pseudo-stratified (1 pleo-morphic)	Columnar/ pseudo-stratified (2 pleo-morphic)	Columnar/ pseudo-stratified	Columnar/ (1 pleo-morphic)
Cytoplasm	Slightly vacuolated	Vacuolated/ non-vacuolated	Vacuolated	Vacuolated	Slightly vacuolated
Cytoplasm staining	Pale	Pale/dark	Pale	Pale	Pale
Nuclei	Basal, ovoid	Basal, ovoid	Basal, ovoid (3 elongate)	Basal, ovoid	Basal, ovoid (2 elongate)
Superficial and basal glands					
Configuration	2 Straight/ packed 2 tortuous/ spaced out	2 Tortuous/ spaced out 1 straight/ packed	Tortuous/ spaced out (2 straight/ packed)	Tortuous/ spaced out (1 straight/ packed)	3 Tortuous/ spaced out (2 straight/ packed)
Epithelium	Columnar	Columnar	Columnar	Columnar	Columnar
Cytoplasm	Pale	Pale	Pale	Pale	Pale
Surface epithelium (µm)	15 ± 1^x	18 ± 2^x	$23 \pm 1^{y,c}$	17 ± 1	19 ± 2
Superficial gland diam. (µm)	$47 \pm 3^{x,a}$	51 ± 1	$56 \pm 2^{x,b,c}$	51 ± 2^b	49 ± 4
Superficial gland epithelium ht (µm)	16 ± 1	19 ± 0.5	19 ± 1^b	17 ± 1^b	19 ± 1
Basal gland diam. (µm)	35 ± 1^x	43 ± 1^y	$38 \pm 1^{x,b}$	39 ± 2	41 ± 2
Gland depth (mm)	0.8 ± 0.3	1.1 ± 0.2	1.0 ± 0.2	0.9 ± 0.1	0.9 ± 0.1
Gland density (tubules/0.5 cm²)	39 ± 11^a	68 ± 1	57 ± 13	61 ± 9^a	70 ± 6
Lymphocytes	0.5 ± 0.3	2.0 ± 0.7	0.9 ± 0.3	0.5 ± 0.2^x	1.1 ± 0.6
Neutrophils	—	0.6 ± 0.3	0.1 ± 0.1	0.1 ± 0.1	0.1 ± 0.1
Eosinophils	0.2 ± 0.1	2.3 ± 0.8	1.9 ± 0.6^b	0.2 ± 0.1^x	0.3 ± 0.1^x
Mast cells	1.6 ± 0.4	2.4 ± 0.8	2.2 ± 0.6	2.4 ± 0.5	1.5 ± 0.5
Macrophages	—	—	—	0.02 ± 0.02	0.3 ± 0.3

Values are mean $\pm$ s.e.m.

[a,b,c]Significant differences between groups are indicated by different letter superscripts (at least $P < 0.05$).

[x,y]Significant differences between pregnancy status of mares are indicated by different superscripts (at least $P < 0.05$).

secretory activity up to Day 12 after ovulation but this was less evident in the unmated mares at Days 12–14 after ovulation than at other stages. At Days 15–21 after ovulation the glands of the pregnant and most of the unmated mares were compressed in a rather oedematous lamina propria (Fig. 2). Superficial gland diameter was significantly greater ($P < 0.05$) in all mares at all stages

Table 1 continued

	Days after ovulation						
	12–14				15–21		
P (N = 5)	C (N = 8)	NR (N = 3)	P (N = 8)	C (N = 5)	NR (N = 4)	P (N = 8)
Columnar (1 pleomorphic)	Cuboidal (1 columnar/pseudostratified)	Cuboidal (1 columnar)	3 Columnar/4 cuboidal (1 pleomorphic)	Pleomorphic (1 cuboidal/1 columnar/pseudostratified)	Columnar/cuboidal (1 pleomorphic)	Cuboidal (2 pleomorphic)
Vacuolated (1 non-vacuolated	4 Vacuolated/4 non-vacuolated	Vacuolated/non-vacuolated	Non-vacuolated	Vacuolated/non-vacuolated	Non-vacuolated	Non-vacuolated
Pale (2 dark)	4 Pale/4 dark	Pale/dark	Dark (2 pale)	Pale/dark	Dark	Dark
Elongate (1 basal, ovoid)	Basal, ovoid (2 elongate)	Basal, ovoid	Basal, ovoid (2 elongate)	Basal, ovoid	Basal, ovoid	Basal, ovoid (2 elongate)
3 Tortuous/spaced out (2 straight/packed)	Tortuous/spaced out (2 straight/packed)	Tortuous/spaced out (1 straight/packed)	Tortuous/spaced out (1 straight/packed)	Tortuous/spaced out (2 straight/packed)	Tortuous/spaced out (1 straight/packed)	Straight/packed (2 tortuous/spaced out)
Columnar	Columnar	Columnar	Columnar	Columnar	Columnar	Columnar
Pale/dark	Pale	Pale	Pale/dark	Pale/dark	Pale	Pale/dark
19 ± 2^b	14 ± 2	17 ± 0.4	13 ± 0.4^a	16 ± 2	15 ± 3	13 ± 1^a
53 ± 3^b	41 ± 2^a	47 ± 2	43 ± 2^a	$43 \pm 2^{a,c}$	43 ± 4	42 ± 2^a
17 ± 1^b	14 ± 1^a	16 ± 0.3	14 ± 0.4^a	15 ± 1	16 ± 2	15 ± 1^a
39 ± 1^b	35 ± 2	34 ± 2	$35 \pm 1^{a,b}$	36 ± 1	36 ± 4	32 ± 2^a
1.0 ± 0.1	0.8 ± 0.1	0.8 ± 0.1	0.9 ± 0.1	1.0 ± 0.1	1.0 ± 0.1	1.1 ± 0.1
59 ± 15	85 ± 6^b	74 ± 7	63 ± 10	$65 \pm 5^{a,b}$	71 ± 10	74 ± 7
2.2 ± 0.5^y	1.0 ± 0.4	1.8 ± 0.6	1.0 ± 0.4	0.4 ± 0.1	0.9 ± 0.3	1.6 ± 0.4
0.2 ± 0.05	0.1 ± 0.04	0.1 ± 0.1	0.2 ± 0.1	0.1 ± 0.05	0.1 ± 0.1	0.3 ± 0.1
$1.7 \pm 0.5^{y,b,c}$	0.4 ± 0.2	0.9 ± 0.1	0.3 ± 0.2^a	1.9 ± 1.3	1.9 ± 1.0	$0.6 \pm 0.2^{a,c}$
1.9 ± 0.4	1.9 ± 0.5	3.3 ± 0.5	1.7 ± 0.4	2.5 ± 0.4	2.5 ± 0.7	1.9 ± 0.2
—	2.0 ± 1.3	1.3 ± 1.3	0.02 ± 0.02	0.2 ± 0.1	—	0.4 ± 0.4

than was basal gland diameter. The diameter and epithelial height of the superficial and basal glands in pregnant mares were significantly lower ($P < 0.05$) from Day 12 after ovulation than at earlier stages (Table 1). The basal gland diameter was significantly greater ($P < 0.05$) in mares that were mated but no embryo was recovered than in unmated and pregnant mares at Days 2–5 after

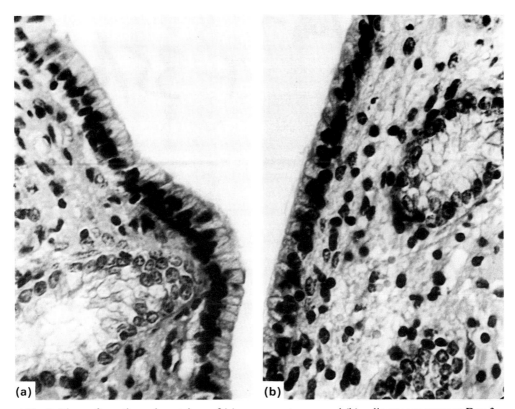

Fig. 1. Biopsy from the endometrium of (a) a pregnant mare and (b) a dioestrous mare at Day 3 after ovulation. × 400.

ovulation. There was no linear correlation of gland size, density and depth of tubules within the stroma with progesterone concentrations. Infiltration of the lamina propria with mobile cells occurred to various degrees in all uteri, and mostly these were concentrated in the stratum compactum. However, mast cells were distributed more frequently through the stratum spongiosum. Macrophages, when present, occurred throughout the lamina propria. Small numbers of randomly distributed lymphocytes were observed in the stratum compactum of all mares, occasionally appearing intra-epithelially and intra-luminally and in small foci in the periglandular stroma and around blood vessels. Significantly increased numbers of lymphocytes and eosinophils were present in pregnant mares at Days 6–9 in comparison with other categories (Table 1). The numbers of eosinophils in mares in which embryos were not recovered were negatively correlated with progesterone concentrations.

Discussion

Endometrial cyclic changes in dioestrous mares have been previously described (Andrews & McKenzie, 1941; Ressang, 1954; Brandt & Manning, 1969; Ricketts, 1975; Kenney, 1978). The results of this study concur broadly with the aforementioned reports for dioestrous mares. The

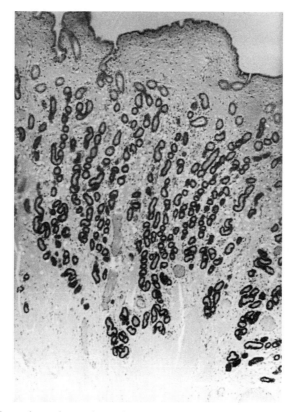

Fig. 2. Biopsy from the endometrium of a pregnant mare at Day 17 after ovulation. × 50.

increased height in surface epithelium, which occurred only in pregnant mares at Days 2–5 after ovulation, appears to be related to pregnancy and not copulatory or seminal effects. This study confirms a reduction in endometrial gland size and epithelial heights in pregnant mares as well as in dioestrous mares as previously reported (Andrews & McKenzie, 1941). Reduction in gland tubule diameter may be explained by increased gland tortuosity and consequent increased tubule density or by increased gland depth within the lamina. However, gland density was significantly increased only in dioestrous mares at Days 12–14 after ovulation and gland depth, while greater in pregnant mares, was not significantly increased. The increase in number of infiltrating lymphocytes and eosinophils in pregnant mares at Days 6–9 after ovulation is interesting although further studies are necessary to establish the practical significance of these findings. The presence of lymphocytes may suggest an immunological response to the embryo during its initial contact with the endometrium. Eosinophils may be concerned in moderating this response.

In conclusion, this study indicates that there are subtle histological distinctions between the endometrial histology of pregnant and non-pregnant mares at very early stages after ovulation. Although most embryo loss in the mare occurs before Day 12 (Forde *et al.*, 1987) we suggest that differences in endometrial structure do not readily explain this early high level of reproductive wastage.

References

Andrews, F.W. & McKenzie, F.F. (1941) Estrus, ovulation and related phenomena in the mare. *Res. Bull. Mo. Agric. exp. Stn*, No. 329.

Brandt, G.W. & Manning, J.P. (1969) Improved uterine biopsy techniques for diagnosing infertility in the mare. *Vet. Med.* **64**, 977–983.

Drury, R.A.B. & Wallington, E.A. (1980) in *Carlton's Histological Technique,* 5th edn. Oxford University Press.

Forde, D., Keenan, L., Wade, J., O'Connor, M. & Roche, J.F. (1987) Reproductive wastage in the mare and its relationship to progesterone in early pregnancy. *J. Reprod. Fert., Suppl.* **35**, 493–495.

Ginther, O.J. (1985) Embryonic loss in mares: Incidence, time of occurrence, and hormonal involvement. *Theriogenology* **23**, 77–89.

Kenney, R.M. (1978) Cyclic and pathological changes of the mare endometrium as detected by biopsy, with a note on early embryonic death. *J. Am. vet. med. Ass.* **172**, 241–262.

Ressang, A. (1954) *Infertility in the mare: a clinical, bacteriological and histopathological investigation.* Thesis, University of Utrecht.

Ricketts, S.W. (1975) The technique and clinical application of endometrial biopsy in the mare. *Equine vet. J.* **7**, 102–107.

Woods, G.L., Baker, C.B., Bilinski, J., Hillman, R.B., Mank, E.C. & Sluijter, F. (1985) A field study on early pregnancy loss in standardbred and thoroughbred mares. *Equine vet. Sci.* **5**, 265–267.

J. Reprod. Fert., Suppl. 35 (1987), 505–506

Printed in Great Britain
© 1987 Journals of Reproduction & Fertility Ltd

Intrauterine inoculation of *Candida parapsilosis* to induce embryonic loss in pony mares

B. A. Ball, S. J. Shin*, V. H. Patten*, M. C. Garcia† and G. L. Woods‡

*Department of Clinical Sciences and the *Diagnostic Laboratory, New York State College of Veterinary Medicine, Cornell University, Ithaca, New York 14853, and †School of Veterinary Medicine, New Bolton Center, University of Pennsylvania, Kennett Square, PA 19348, U.S.A.*

It has been proposed that endometritis causes embryonic loss in mares (Woods *et al.*, 1985). Research on experimentally induced endometritis in the mare has focussed on the non-pregnant mare (Hughes & Loy, 1969; Brown *et al.*, 1979; Evans *et al.*, 1984; Asbury *et al.*, 1984; Williamson *et al.*, 1984; Katila *et al.*, 1984) and not on the pregnant mare. The objectives of this study were to: (1) determine whether intrauterine inoculation with *Candida parapsilosis* would induce embryonic loss in pregnant pony mares and (2) study the induced embryonic losses.

Pony mares were artificially inseminated, and pregnancy was detected by transrectal ultrasonography (Ginther, 1984) at 10–12 days after ovulation. Pregnant mares were randomly assigned to control (single intrauterine inoculation with 1·0 ml sterile phosphate-buffered saline, 12 mares) or treatment groups (single intrauterine inoculation with $6·5 \pm 1·2 \times 10^6$ *C. parapsilosis*, 12 mares). Mares were inoculated 1–3 days after pregnancy was first diagnosed (Days 11–14 after ovulation). Pregnancies were monitored by ultrasound examinations daily from the day of intrauterine inoculation until Day 20 after ovulation and every other day from Day 20 to Day 28 after ovulation. Ultrasound examinations were discontinued when embryonic loss was detected. Mares which did not suffer embryonic loss were given prostaglandin F-2α at Day 28 after ovulation. Serum progesterone concentrations were monitored on the first day of pregnancy detection, the day of intrauterine inoculation and then daily for 4 days or until embryonic loss was detected. Endometrial swabs for culture and cytological examination were taken within 2 days after embryonic loss was detected.

Significantly fewer ($P < 0·01$) embryonic losses occurred in control (4/12) than treated (12/12) mares. The mean interval from intrauterine inoculation until embryonic loss was 5·8 (s.e. 2·8) days for the 4 control mares and 2·1 (s.e. 0·2) days for the 12 treated mares. Before embryonic loss, moderate to marked oedema of the endometrial folds was detected ultrasonically in all 12 treated mares, and free fluid within the uterine lumen was detected in 5 of the treated mares. Slight oedema of the endometrial folds was detected on the first day after inoculation in 9 of the 12 control mares. This oedema resolved after 2–3 days in control mares in which embryonic loss did not occur. After embryonic loss, *C. parapsilosis* was cultured from the uterus of 8/12 treated mares, and cytological evidence of inflammation was detected on the endometrial swab of 10/12 treated mares. Of the 4 control mares that experienced embryonic loss, 2 mares had *Escherichia coli* cultured from the uterus and 3 mares had cytological evidence of inflammation on the endometrial swab. The 8 control mares that did not undergo embryonic loss after inoculation were given prostaglandin F-2α at Day 28 after ovulation, and endometrial swabs were taken after embryonic loss was detected: 1 of these 8 mares had a positive endometrial cytology and *Campylobacter fetus* var. *venerealis* was cultured from her uterus.

There was no significant difference in serum progesterone concentrations between control and treated mares on the first day of pregnancy detection or on the day of intrauterine inoculation (Fig.

---

‡Present address: Department of Veterinary Clinical Medicine and Surgery, Washington, Oregon, Idaho Regional Program of Veterinary Medicine, University of Idaho, Moscow, Idaho 83843, U.S.A.

1). Progesterone concentrations were significantly higher in control mares that did not undergo embryonic loss compared to treated mares at Day $+1$ ($P < 0.05$) and at Days $+2$, $+3$ and $+4$ ($P < 0.01$) after inoculation (Fig. 1). Serum progesterone concentrations were significantly lower ($P < 0.05$) by Day $+3$ after inoculation in the 3 control mares that underwent embryonic loss within 3 days after inoculation than in control mares that did not suffer embryonic loss (Fig. 1).

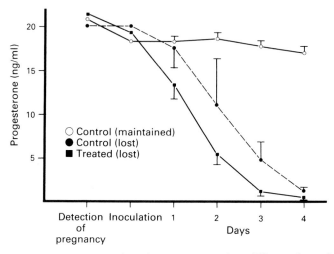

Fig. 1. Serum progesterone concentrations (means $\pm$ s.e.m.) on different days after inoculation for control mares which maintained and lost pregnancies and treated mares which lost pregnancies.

Intrauterine inoculation of *C. parapsilosis* effectively induced embryonic loss in pregnant pony mares. The rapid decline in serum progesterone concentrations after inoculation in the treated mares suggests that uterine-induced luteolysis was a factor in embryonic loss. However, the interval from intrauterine inoculation to embryonic loss in the treated mares appeared shorter than the earlier reported interval for pregnant mares given prostaglandin at Day 12 (2·1 *vs* 6·8 days, respectively) (Ginther, 1985). This apparently reduced interval to embryonic loss may have been related to the direct effect of the inoculated pathogen or the resulting inflammatory response on the embryonic vesicle.

References

Asbury, A.C., Gorman, N.T. & Foster, G.W. (1984) Uterine defense mechanisms in the mare: serum opsonins affecting phagocytosis of *Streptococcus zooepidemicus* by equine neutrophils. *Theriogenology* **21**, 375–385.

Brown, J.E., Corstvet, R.E. & Stratton, L.G. (1979) A study of *Klebsiella pneumoniae* infection in the uterus of the mare. *Am. J. vet. Res.* **40**, 1523–1530.

Evans, M.J., Hamer, J.M., Gason, L.M., Graham, C.S. & Irvine, C.H.G. (1984) Intrauterine inoculation of bacteria in the mare. *Proc. 10th Int. Congr. Anim. Reprod. & A.I., Urbana-Champaign,* **II**, 148, Abstr.

Ginther, O.J. (1984) Ultrasonic evaluation of the reproductive tract of the mare: the single embryo. *J. equine vet. Sci.* **4**, 75–81.

Ginther, O.J. (1985) Embryonic loss in mares: nature of loss after experimental induction by ovariectomy or prostaglandin F₂ alpha. *Theriogenology* **24**, 87–98.

Hughes, J.P. & Loy, R.G. (1969) Investigations on the effect of intrauterine inoculations of *Streptococcus zooepidemicus* in the mare. *Proc. 15th Ann. Conf. Am. Assoc. equine Pract.* **15**, 289–292.

Katila, T., Locke, T.F. & Smith, A.R. (1984) Uterine changes and fertility of mares after an induced Streptococcal endometritis. *Proc. 10th Int. Congr. Anim. Reprod. & A.I., Urbana-Champaign,* **III**, 452, Abstr.

Williamson, P., Penhale, J.W., Munyua, S. & Murray, J. (1984) The acute reaction of the mares' uterus to bacterial infection. *Proc. 10th Int. Congr. Anim. Reprod. & A.I., Urbana-Champaign,* **III**, 477, Abstr.

Woods, G.L., Baker, C.B., Hillman, R.B. & Schlafer, D.H. (1985) Recent studies relating to early embryonic death in the mare. *Equine Vet. J., Suppl.* **3**, 104–107.

J. Reprod. Fert., Suppl. **35** (1987), 507–508

Reproductive function of mares given PGF-2α daily from Day 42 of pregnancy

A. C. Rathwell, A. C. Asbury, P. J. Hansen and L. F. Archbald

Navan Veterinary Centre, P.O. Box 100, Navan Ontario, Canada K0A 2S0

Mares that experience a pregnancy loss during the time that endometrial cups are normally present (about Days 40 to 130 of pregnancy) often fail to resume normal cyclic activity and frequently exhibit anovulatory oestrus, prolonged luteal function, and ovulation without oestrus or anoestrus. The lack of regular oestrous cycles after abortion hinders attempts to re-mate during the same breeding season as an abortion. The present study was designed to determine the effects of daily injections of prostaglandin (PG) F-2α on subsequent oestrous behaviour and ovarian function of pregnant mares that had formed endometrial cups.

Five mares of a light horse breed were teased daily with a stallion to determine oestrus and evaluated daily by rectal palpation for abortion, follicular development and ovulation. The experiment began in August 1985 and was concluded in January 1986. Before the mares became pregnant, reproductive function of each mare was observed after daily PGF-2α administration during the oestrous cycle. During this control period, each mare was injected daily with PGF-2α from Day 6 after ovulation until behavioural oestrus was exhibited. Mares exhibited oestrus at $4\cdot6 \pm 1\cdot0$ days after the first injection. Oestrus lasted an average of $4\cdot2 \pm 1\cdot5$ days and ovulation occurred just before the end of oestrus ($8\cdot6 \pm 2\cdot0$ days after the first injection). Of the 5 mares, 3 conceived at this oestrus and the other two mares conceived at the next oestrus. An immunological test for pregnancy was applied to serum samples from Day 42 of pregnancy to verify formation of endometrial cups.

The mares were then divided into two treatment groups. Group I mares (N = 3) received the same hormonal treatment as during the control period, i.e. PGF-2α from Day 42 until the first day of oestrus after abortion. Mares in Group II (N = 2) were treated similarly except that daily injections of PGF-2α were continued until the second oestrus after abortion. Mares in Group I were mated at the first oestrous period after abortion whereas mares in Group II were mated at the second oestrous cycle after abortion.

Mares in Group I aborted at an average of $3\cdot0 \pm 0$ days after the first injection of PGF-2α and displayed oestrus at $3\cdot7 \pm 1\cdot2$ days from this injection. The oestrous period lasted $4\cdot7 \pm 0\cdot7$ days. Based on rectal palpation and determination of serum progesterone concentrations, all the mares in this group ovulated, at an average of $8\cdot0 \pm 1\cdot2$ days after first injection. All mares were mated at this oestrus, but none became pregnant based on ultrasound examinations at 17–22 days after ovulation.

Both mares in Group II aborted within 3–4 days after the first injection and expressed oestrus at 4 days after first injection. The oestrous period lasted 2–6 days and was accompanied by ovulation 5–9 days after first injection. Both of these mares returned to oestrus 7 days after ovulation. This oestrus lasted 2–3 days and was accompanied by ovulation. The dioestrus after the second ovulation (and after discontinuation of PGF-2α) was not monitored in these 2 mares. Both mares were mated at the second oestrus but neither conceived, based on ultrasound examination at 17 or 18 days after ovulation. There were no significant differences ($P > 0\cdot10$) in interval to first ovulation and oestrus and in duration of first oestrus between the control and post-abortion periods or between Groups I or II.

After the first PGF-2α injection, progesterone concentrations declined in all mares and remained low until after ovulation. For 2 of 3 mares in Group I, values rose after ovulation and

A. C. Rathwell et al.

remained above 1 ng/ml (peak values 8·4–36·4 ng/ml) for the remainder of the sampling period (25–34 days after ovulation). In the third mare, which did not display oestrus after the first post-abortion oestrus, values rose to >4 ng/ml until Day 28 after ovulation, when serum progesterone concentration was 0·33 ng/ml. In contrast, progesterone concentrations increased only slightly after first ovulation in mares of Group II (peak values 1·6–1·8 ng/ml) and began to decline at 5 days after ovulation. After the mares ovulated for a second time, progesterone concentrations again increased (peak values of 6·0–7·0 ng/ml) and remained elevated for the remainder of the sampling period (6–11 days after ovulation).

Overall, mean concentrations of progesterone during the first 7 days of ovulation during the control period (3·1 ± 0·8 ng/ml) were not different ($P > 0·10$) from values after the first post-abortion ovulation (3·6 ± 1·4 ng/ml). Mean progesterone concentrations were lower ($P < 0·05$) for Group II (1·0 ± 0·1 ng/ml) than for Group I (5·4 ± 1·5 ng/ml) mares. Progesterone concentrations during the first 7 days after second ovulation of Group II mares (3·7 ± 0·3 ng/nl) were higher ($P < 0·01$) than after the first ovulation. When endometrial cup function was evaluated by determining serum CG concentrations, all mares still gave positive values at the end of sampling (Days 25–28 after first injection).

These results suggest that (1) daily injections of prostaglandin F-2α can induce abortion in mares at 42 days of pregnancy, (2) abortion is followed by oestrus and ovulation, (3) the endometrial cups do not regress as a result of this treatment, and (4) daily injections of PGF-2α can impair early corpus luteum function.

THE PERINATAL PERIOD
AND THE SUCKING FOAL

Convener

P. D. Rossdale

Chairmen

P. D. Rossdale

P. F. Flood

J. Reprod. Fert., Suppl. **35** (1987), 509–518

Serological and virological investigations of an equid herpesvirus 1 (EHV-1) abortion storm on a stud farm in 1985

J. A. Mumford, P. D. Rossdale*, D. M. Jessett, S. J. Gann, J. Ousey* and R. F. Cook

*Equine Virology Unit, Animal Health Trust, Lanwades Park, Kennett, Newmarket, Suffolk CB8 7PN, and *Beaufort Cottage Stables, High Street, Newmarket, Suffolk CB8 8JS, U.K.*

Summary. An extensive outbreak of EHV-1 abortions occurred on a stud farm in England in 1985. Of the 67 pregnant mares present on the stud farm, 31 were challenged with EHV-1, resulting in the loss of 22 fetuses or foals. Laboratory investigations revealed that the spread of the virus closely followed movement of apparently healthy mares (during the incubation period of the infection). During the outbreak mares were challenged 1–4 months before the expected foaling date. There was no relationship between the gestational age at the time of challenge and the subsequent outcome of infection in terms of abortion. The period between the estimated date of infection and abortion varied between 9 days and 4 months. Virus was isolated from the nasopharynx of 3 apparently healthy foals born during the outbreak.

Introduction

An extensive outbreak of abortions due to an equid herpesvirus (EHV-1) occurred on a stud farm in England in 1985. An intensive epidemiological study was carried out in an attempt to elucidate the origin of the outbreak, the means of spread and the factors affecting the outcome of infections.

EHV-1 abortions are almost always caused by infection of the fetus with subtype 1 strains of EHV-1 which spread via the respiratory route. The virus reaches the fetus via the bloodstream as a cell-associated viraemia. Natural immunity is shortlived (Doll, 1961) and animals can suffer repeated infections (Bryans, 1969) although abortions rarely occur in the same individual in consecutive years. There is ample circumstantial evidence that a state of latency exists for EHV-1 in the horse (Burrows & Goodridge, 1984) and experimental studies have provided evidence to support this assumption (Edington *et al.*, 1985). Respiratory infections in older animals may be completely subclinical. This study highlights the potential danger of the subclinical spread of this virus on a breeding establishment at a time of year when there are large numbers of susceptible pregnant mares present.

Materials and Methods

Movement of mares. The outbreak occurred on a stud farm consisting of two adjoining premises under the same management but staffed largely by two separate groups of personnel (Fig. 1). The two premises were used as (i) a public stud standing 5 stallions, and (ii) a private stud where resident mares and yearlings were accommodated. Both studs had foaling yards. The public stud was used for early foaling mares and the private stud for late foaling mares.

At the beginning of the year (January) it was the stud farm's practice to collect early foaling mares together in the foaling yard of the public stud (Location 2) and group late foaling mares in an American barn in the private stud (Location 3) to release accommodation for visiting barren and maiden mares in an American barn (Location 1) on the public stud.

Management during outbreak. When the first EHV-1 abortion was confirmed and the extent of the infection recognized all movement was stopped and separate personnel allocated to different housing areas. Aborted mares were put into isolation facilities on the stud farm (Location 6) but as these became filled, aborted mares or those expected to abort were moved to isolated premises distant from the stud farm (Location 5).

Sampling. During the course of the outbreak it was possible to obtain sequential serum samples from mares, foals and stallions and occasional nasopharyngeal swabs and heparinized blood samples from mares and foals. Separate personnel carried out sampling in the different housing areas to preclude spread of virus by this means. Tissues from aborted fetuses were also collected for virological examination.

Laboratory tests. Serological tests (complement fixation (CF) and virus neutralization (VN)) and virus isolation attempts from nasopharyngeal swabs, blood and fetal tissues were used to identify infected animals and trace the spread of the infection on the premises.

Complement fixation test. These were microtitre tests as described by Thomson *et al.* (1976). Sera were inactivated at 60°C for 30 min before use. Virus antigens used in the tests were prepared from tissue culture (rabbit kidney cells (RK13), or horse embryonic lung (EEL)) cells infected with representative strains of the two subtypes of EHV-1. Fixation took place overnight at 4°C using 3 CH_{50} units of complement.

Complement fixation antibody produced in response to infection with EHV-1 can be detected about 10 days after exposure, reaching a peak titre at 2–3 weeks. Thereafter antibody may decline relatively rapidly such that base-line levels are reached 10–12 weeks after initial infection (Thomson *et al.*, 1976), although some individuals, particularly mares which have aborted, sustain high levels of antibody for longer periods. Infections can be demonstrated by 4-fold increases in antibody in sera taken in the acute and convalescent phase of infection. Alternatively, when pre-infection samples were not available, high levels of CF antibody of >80 were regarded in this investigation as indicative of recent infection. Mares with a CF titre of <10 have been referred to as seronegative in the text.

When several sequential samples were available from individual horses, it was possible using the CF test to estimate the date of infection with some accuracy.

Restriction endonuclease analysis of EHV-1 DNA. EHV-1 viruses were grown on RK13 cells at a multiplicity of infection of 1 $TCID_{50}$ per 1000 cells. Virus-containing tissue-culture fluids were clarified at 2000 *g* for 15 min and virus was pelleted after centrifugation at 42 000 *g* for 3 h. Virus was resuspended in 0·05 M-Tris–HCl, 0·1 M-NaCl pH 7·4 (TN buffers) and purified using sucrose density gradient. DNA extraction and restriction with BAM HI were as described by Buchman *et al.* (1978).

Virus neutralization test. These microtitre tests were carried out as described by Thomson *et al.* (1976). Dilutions of sera inactivated at 56°C for 30 min were allowed to react with 100 $TCID_{50}$ at 37°C for 60 min and residual virus was titrated in embryonic horse dermis cells. Neutralizing antibody persists for longer than CF antibody after infection and may be used to demonstrate previous exposure in the absence of CF antibody.

Virus isolation. Suspensions (10%) of fetal tissues (lung, liver or thymus), nasopharyngeal swab extracts or leucocyte fractions from heparinized blood were inoculated on to EEL or RK13 cells. Cells showing no cytopathic effect after 7 days at 37°C were frozen at −70°C and then passaged on fresh cell cultures. Virus isolates, as detected by cytopathic effect, were identified by indirect immunofluorescence using specific antisera for EHV-1 and EHV-2 (slow growing herpes virus) (Mumford & Thomson, 1978).

Field observations and laboratory results

Foal losses during 1985 breeding season

The number of animals and their locations in the private and public studs at the beginning of January are given in Table 1. At the beginning of the 1985 season there were 70 pregnant mares on the stud but 3 left the stud before foaling. Additional barren and maiden mares arrived at Location 1 after the first abortion. Figure 1 shows the movement of mares about the stud during the season. The foaling record for the 1985 season were as follows. The 67 mares that foaled at the farm produced 39 healthy foals. Of the 31 mares challenged with EHV-1 before parturition 9 produced healthy foals while 22 mares aborted (i.e. 71% abortion rate). The 36 uninfected pregnant mares produced 30 healthy foals; 6 of these were lost because of twinning, accidents and stillbirths.

Spread of infection

Movement of mares. The primary reason for the extensive spread of infection around the stud was the movement of pregnant mares from the original focus of infection in an American barn

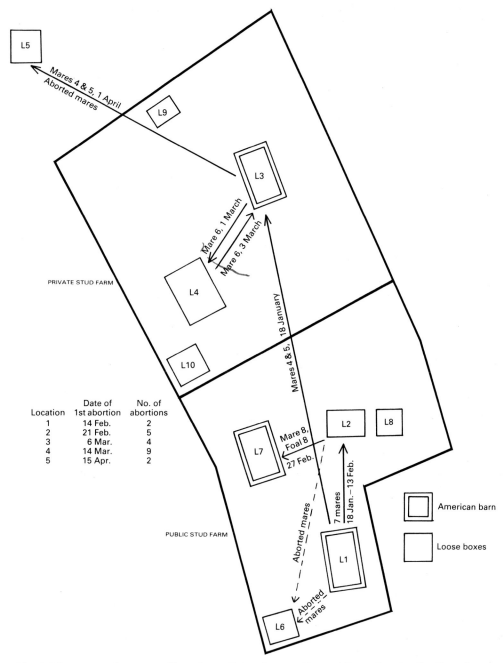

Fig. 1. Movement of mares and locations of abortions on the stud farm. For details of numbers and types of animals in each location at the begining of the season, see Table 1.

(Location 1) during the incubation period of the infection, in the absence of clinical signs and before the first EHV-1 abortion occurred. Abortions occurred in 5 locations as shown in Fig. 1. The first mare (No. 1) aborted in an American barn (Location 1) containing barren, maiden and pregnant mares, and 7 days later 2 abortions occurred in the foaling yard in the public stud (Location 2) some 500 m from Location 1. It was unlikely that the contaminated fetus from the first abortion

Table 1. Location of the animals on the stud farm before 17 January 1985

Location	Type of housing	Stud farm	Animals
1	American barn	Public	19 pregnant mares 21 barren mares 1 maiden mare 1 teaser mare
2	Loose boxes (foaling yard)	Public	12 pregnant mares 3 mares + foals
3	American barn	Private	7 pregnant mares 2 maiden mares
4	Loose boxes (foaling yard)	Private	15 pregnant mares
5	Isolation loose boxes	Public	Empty
6	Isolation loose boxes	—	Empty
7	American barn	Public	6 mares with foals
8	Loose boxes (stallion yard)	Public	5 stallions
9	Loose boxes	Private	8 pregnant mares
10	Loose boxes	Private	5 maiden mares 1 barren mare 1 teaser mare

could have been the source of infection for the 2nd and 3rd abortion mares (Nos 2 and 3) since the distance between the 2 locations precluded easy transfer of virus and the incubation period between infection and abortion is normally > 10–14 days. However, between 18 January and 13 February, i.e. before the first abortion, 7 early foaling mares had been moved from Location 1 to Location 2. Complement fixation tests on sera taken from mares in both locations on 20 February revealed that the 2 mares (including the first to abort) in Location 1 and 4 mares in Location 2 (two of which aborted on 21 February) already had very high levels of CF antibody (> 160), indicating recent infection. This suggested that Mares 1, 2 and 3 had all been exposed to the infection in Location 1 before the first abortion.

The conclusion that the infection started in Location 1 was further strengthened by the finding that 2 late-foaling mares (Nos 4 and 5) moved from Location 1 to Location 3 (an American barn) on 18 January had high levels of CF antibody indicating recent infection when sampled on 27 February, before an abortion occurred in this 3rd location. It was also demonstrated that 4 mares which had been housed in Location 3 since the previous season also had high CF titres in sera on 27 February, indicating recent exposure. This suggested that Mares 4 and 5 were incubating the infection when transferred from Location 1 to Location 3 on 18 January and had subsequently infected in-contact animals in the barn. The first mare aborted in Location 3 on 6 March, 44 days after Mares 4 and 5 arrived from Location 1.

Late-foaling mares housed in the main foaling yard of the private stud (Location 4) were shown to be seronegative by CF tests when sampled on 27 March, i.e. there was no evidence of infection in this group at that time. There had been no direct contact between this group of mares and those known to be infected in Locations 1 and 2 on the public stud. However, on 1 March, just before the results of CF tests became available, a mare (No. 6) in Location 3, was moved to the foaling yard as she was about to deliver twins. Immediately the results of laboratory tests were known this mare was returned to Location 3. Although this mare remained in the foaling yard for no more than 48 h, it was sufficient time to set up a new focus of infection, as demonstrated by CF tests on sera collected on 7 March. The first abortion occurred in the foaling yard on 14 March, 13 days after the introduction of the mare from the infected area, Location 3.

Source of infection. The observations described above suggested that the first identified EHV-1 abortion in Location 1 was not the original source of infection on the farm. Attempts were made to establish other possible sources of infection, e.g. (1) an unidentified EHV-1 abortion before 14 February, (2) introduction of an infected mare onto the stud farm, or (3) resurgence of latent infection in a resident mare.

Unidentified EHV-1 abortion. Before the first EHV-1 abortion two mares had lost twins in Location 1 on 3 January and 31 January respectively. A dead foal was also delivered to a mare in Location 2 on 31 January. Histological examination of tissues from these 3 animals did not reveal any evidence of EHV-1 infection.

Retrospective sera (taken at weekly intervals) dating back to October 1984 were available for both twinning mares. Complement fixation tests demonstrated that there was no evidence of infection in these two mares between October 1984 and the end of January 1985. One of these mares showed a positive test to EHV-1 after the first EHV-1 abortion. Sera taken between 21 February and 27 March from the mare which produced a dead foal in Location 2 on 31 January showed no evidence of infection. Therefore, none of these 3 deliveries could be implicated as the original source of infection.

Introduction of infection with incoming mares. No new non-resident mares had been introduced into Location 1 on the public stud farm between August 1984 and 2–3 days before the first abortion. However, 2 mares resident on the stud farm were sent to the sales in December but returned to the farm unsold and then spent 2 weeks in isolation (Location 5) before rejoining the group in Location 1. Sera taken from these animals at the beginning of February did not indicate recent infection and in fact both animals seroconverted during the outbreak in March and May. Thus neither of these animals could be implicated as the source of infection. Furthermore, retrospective sera available from the twinning mares in Location 1 did not show any serological activity during December or January coincidental with the return of the 2 mares from the sales. New barren and maiden mares which joined the group in February did not arrive in Location 1 in sufficient time to be considered a source of infection for the first mare which aborted.

Resurgence of infection in a resident mare. There had not been an EHV-1 abortion on the stud in the previous 10 years but it is very probable that many mares had been exposed to EHV-1 in their life time and were therefore potential carriers of the virus. While there was no conclusive evidence the possibility exists that an individual in Location 1 which had been on the stud farm for some months shed virus from the respiratory tract, during a resurgence of latent infection. The most likely time for this to have occurred was during the last 2 weeks of January, before Mares 4 and 5 moved to Location 3 and before one or more mares moved to Location 2. There is conclusive evidence that the infection was active in the group in Location 1 before the first abortion on 14 February because a twinning mare in this location which was subsequently moved to Location 2 showed a seroconversion between 31 January and 14 February, demonstrating infection on or before 4 February. Retrospective sera (October–February) from 2 twinning mares in Location 3 on the private stud farm did not show any serological activity until mid-February after Mares 4 and 5 had moved from Location 1. There was therefore no reason to postulate a separate source of infection for this group.

Virus infection on other parts of the stud farm

During the course of the outbreak, animals in other locations on the farm were sampled regularly, i.e. (i) mares and foals in an American barn (Location 7) after foaling in the foaling yard (Location 2), (ii) stallions (Location 8), (iii) pregnant mares (Location 9), and (iv) barren and maiden mares (Location 10). Every precaution was taken to minimize the risk of spread of infection by handlers.

It was shown that one of the foaled mares (Mare 8), moved on 27 February from Location 2 to Location 7, 8 days after the first abortion in Location 2, seroconverted on 11 March having been seronegative at the time of moving. However, none of the other 15 foaled mares or their foals housed in the same barn (Location 7) showed seroconversion.

None of the 5 stallions seroconverted during the outbreak or when mating was resumed after the last abortion on the public stud farm.

A group of pregnant mares on the private stud farm (Location 9) remained seronegative throughout the outbreak and produced healthy foals. There had been no direct contact between these mares and those in Locations 1, 2, 3 and 4 and as soon as the infection was identified in other areas on the private stud farm (Locations 3 and 4) separate handlers were allocated to this group. Similarly, the group of barren and maiden mares (Location 10) remained free of infection throughout the outbreak. Yearling colts and fillies were not sampled.

Variation in incubation period between infection and abortion

By measuring CF antibody in sequential serum samples taken from pregnant mares during the outbreak, it was possible to demonstrate a wide variation in the time between initial infection and abortion in different mares represented by examples given in Fig. 2 (a, b, c). Complement fixation antibody is normally detected 10 days after initial infection. The estimated date of infection was therefore calculated as about 10 days before the date of the first sample showing a significant increase (4-fold) in antibody. Figure 2(a) shows the responses in 4 mares with long periods between initial infection and abortion. Mares 4 and 5 were the 2 mares moved from Location 1 to 3 on 18 January and were the only known direct contact between the area where the initial abortion occurred and Location 3. It was therefore assumed that these mares were shedding virus when moved and set up the new focus of infection in Location 3 by infecting in-contact mares in the barn. The maximum period between infection and abortion was estimated as 130 days for Mare 4 and 85 days for Mare 5 based on this assumption. If the date of infection was calculated only on the basis of available serology (i.e. 10 days before the first date was high CF antibody), the minimum period between infection and abortion would be 88 days and 58 days respectively (see Fig. 2a). For Mare 7, pre- and post-infection serum samples were available and the period between infection and abortion was calculated to be 57 days. Virus was isolated from Mare 8, 38 days before abortion in the presence of detectable CF antibody. In previous field and experimental studies, virus shedding from the nasopharynx has been observed for up to 2 weeks after infection and after the appearance of CF antibody. The period of incubation estimated for this mare (No. 8) was therefore ~48 days.

Figure 2(b) shows serological responses in 3 mares with shorter incubation periods of 20–25 days, whereas Fig. 2(c) demonstrates that some infections showed extremely short incubation periods, with abortion occurring between 9 and 17 days after infection.

Insufficient samples (nasopharyngeal swabs and heparinized blood) were collected from infected pregnant mares to establish the period of virus shedding or the duration of viraemias in infected mares which subsequently aborted. Virus was only recovered from the nasopharynx of one mare 38 days before abortion and a second mare 2 days before abortion.

Infection of foals post partum

EHV-1 infections were demonstrated in 4 foals *post partum*. Virus was recovered from the nasopharynx of 3 foals (7–9 days old) housed in Location 2. One of their dams had high levels of CF and VN antibody at parturition, demonstrating recent exposure, while the other two had no CF antibody but high VN antibody levels, i.e. they had not been recently infected, but retained demonstrable levels of VN antibody from a previous infection. None of the 3 foals showed any clinical signs of respiratory disease while shedding virus, which may indicate that transfer of maternal antibody (VN) provided clinical protection while not preventing infection. Infection was demonstrated in a 4th foal by serology. This foal showed clinical signs of respiratory disease in the 2nd week of life shortly after the dam had seroconverted. Post-suck sera indicated that only poor

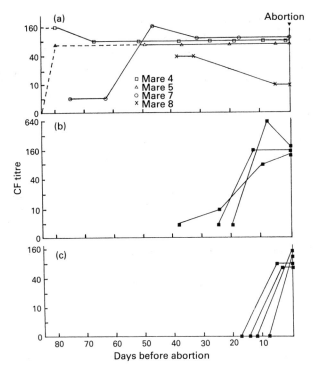

Fig. 2. Variation in incubation period between infection and abortion in mares.

transfer of maternal antibody had occurred in this individual, the serum VN titre in the foal ($10^{1.5}$) being significantly lower than the titre in the serum of the mare ($10^{2.4}$).

Possible factors affecting spread of infection

Other possible factors which may have influenced or contributed to the extensive spread of the infection on these premises were examined.

Virus strain. Isolates recovered from this outbreak were characterized by DNA restriction type to establish whether this virus belonged to a new group which animals on the stud farm may not have encountered previously. The results (not shown) demonstrate that this outbreak was caused by an IP strain. Isolates from 8 individuals involved in the outbreak have been characterized in this manner and all have shown the IP profile. A new variant of EHV-1 subtype 1 was reported by Allen *et al.* (1985) on the basis of having a different electropherotype (IB) from that of the classical IP DNA electropherotype.

Virus neutralizing antibody. It has been suggested that virus neutralizing antibody in the serum contributes to immunity (Bryans, 1969). VN antibody was measured in pregnant and non-pregnant mares housed in areas where the infection was active, as demonstrated by CF antibody responses or virus isolation. Acute-phase sera were examined from mares which became infected and were compared with sera collected at the same time from in-contact individuals which appeared to be resistant to the infection. There was no significant difference in the mean neutralizing antibody titres (VN_{50} $\log_{10}$) of the two groups, 2·27 (s.d. 0·23) for converting mares and 2·24 (s.d. 0·42) for non-converting mares, i.e. susceptibility to the infection could not be associated with the level of neutralizing antibody provided by previous infections.

Possible factors affecting outcome of infection

In those animals that did become infected, various factors were examined in relation to the outcome of infection (i.e. delivery of a healthy foal or an aborted fetus).

Gestational age at the time of infection. Amongst the 31 pregnant mares challenged *pre partum* gestational ages at the time of challenge ranged between 7 and 11 months. The abortions were 2/4 for 7–8 month pregnant mares, 7/8 at 8–9 months, 6/9 at 9–10 months, and 7/10 at 10–11 months.

Virus neutralizing antibody. Although there appeared to be no association between VN antibody and susceptibility to infection, VN antibody was examined in relation to abortion. Sera collected before infection in pregnant mares that aborted were compared with those collected before infection from mares that subsequently became infected but produced healthy foals. There was no significant difference between the mean VN_{50} titres ($\log_{10}$) of the two groups (1·95 (s.d. 1·46) for aborting mares, 2·39 (s.d. 2·3) for non-aborting mares).

Age of pregnant mares. Horses suffer repeated infection throughout life and it has been suggested that this stimulates an increasing level of natural immunity. It would therefore be expected that older animals would be less susceptible to infection and less likely to abort. The outcome of infection in terms of abortion was examined in relation to the age of the mares. There was no obvious association between age and abortion (Fig. 3); in fact 6 of the mares that aborted were $\geqslant 15$ years old whereas all those that produced healthy foals were <15 years old.

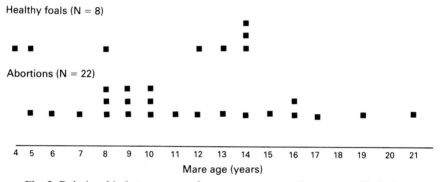

Fig. 3. Relationship between age of pregnant mares and outcome of infections.

Vaccination. Only 6 mares on the stud farm in 1985 had been vaccinated against EHV-1 in the previous year. It was not the farm's policy to vaccinate but in the face of the outbreak some owners requested that their mares should be vaccinated. Only single doses of the inactivated vaccine (Pneumabort-K, Fort Dodge Labs, LA, U.S.A.) were administered to 12 pregnant mares. Serological evidence indicated that 9 of these had experienced the infection before vaccination. Of those vaccinated, 7 aborted and 5 produced healthy foals but 5 of the aborting mares had become infected before vaccination.

Discussion

The co-operation of the stud management and the veterinary practice provided a unique opportunity to investigate an abortion storm on this stud farm in some detail, limited only by the number of different staff available to handle the separate groups of animals when sampling.

Serum samples and virus isolation samples were not always taken at the ideal time to demonstrate infections as the virus spread subclinically and the acute phase of infections could not be identified. However, the information collected provided strong circumstantial evidence to support the theory that the extensive spread of the virus was the result of movement of animals while incubating the infection in the absence of clinical signs.

The original excretion could not be identified but a resurgence of infection in a resident mare was suspected as the source in the absence of the introduction of new animals or circumstances consistent with an aborted fetus being the source of infection.

The CF test proved valuable in differentiating groups of uninfected and infected animals and allowing management practices to contain the spread of infection. The low virus isolation rate from nasal swabs taken from the pregnant mares may indicate that the infectious dose for the respiratory spread of virus is below the limits of detection by conventional virus-isolation techiques, although virus was easily recoverable from foals and aborted fetuses. The observation that healthy young foals 7–10 days old may excrete large quantities of virus in the absence of any clinical signs of respiratory disease demonstrates a hitherto unrecognized source of infection.

High rates of infection were demonstrated in housing areas consisting of individual loose boxes, as well as barns with common air spaces, e.g. in the public and private foaling yards (Locations 2 and 4) infection rates were 60% and 90% respectively whereas the infection rate in the American barn on the private stud (Location 3) was 100%. In the American barn (Location 1), however, where the outbreak began, although the infection rate among the pregnant mares was 70%, only 38% of barren and maiden mares seroconverted; furthermore, the infection did not spread amongst the foaled mares in the American barn at Location 7 (where only 1/16 (6%) of the mares sero-converted). It is tempting to speculate that the state of pregnancy may be immunosuppressive and render the pregnant mares more susceptible than their barren and foaled cohorts.

One of the most interesting observations from the outbreak was the extended period between infection and abortion. This phenomenom raises several questions. Is such a mare potentially infectious to in-contact animals, where does the virus reside during this period? Is a low level viraemia maintained? When does the virus cross the placenta and infect the fetus? Answers to these questions will lead to a better understanding of the pathogenesis of this disease and sounder management practices.

Virus neutralizing antibody did not appear to play a major role in immunity to infection and abortion, and many of the older animals had not developed natural protective immunity. In the light of such observations successful immunization by vaccination must be regarded as difficult to achieve. It is obvious that the present regimen of 3 doses in the 5th, 7th and 9th months of pregnancy recommended for the inactivated vaccine is unlikely to be sufficient in some cases, as infections can occur from the 4th month onwards. Nevertheless, the subclinical nature of this infection before abortion precludes its control by management alone and a vaccine providing stronger immunity than the infection itself is required.

We thank the stud owner and management for full and helpful co-operation; Mrs G. Harrison and Miss T-A. Hammond, for skilled laboratory work at the Equine Virology Unit of the Animal Health Trust; and Coopers Animal Health Limited of Berkhamstead, U.K., for financial assistance to D.M.J.

References

Allen, G.P., Yeargan, M.R., Turtinen, L.W. & Bryans, J.T. (1985) A new field strain of equine abortion virus (equine herpesvirus-1) among Kentucky horses. *Am. J. vet. Res.* **46**, 138–140.

Bryans, J.T. (1969) On immunity to disease caused by equine herpesvirus-1. *J. Am. vet. med. Ass.* **155**, 294–300.

Buchman, T.G., Roizman, B., Adams, G. & Stoner, H. (1978) Restriction endonuclease fingerprinting of herpes simplex virus DNA: a novel epidemiologic tool applied to a nosocomial outbreak. *J. Infect. Dis.* **138**, 488–498.

Burrows, R. & Goodridge, R. (1984) Studies of persistent and latent equid herpesvirus-1 and herpesvirus-3 infections in the Pirbright pony herd. In *Latent herpesvirus Infection in Veterinary Medicine*, pp. 307–319. Martinus Nijhoff, Boston.

Doll, E.R. (1961) Immunization against viral rhino-pneumonitis of horses with live virus propagated in hamsters. *J. Am. vet. med. Ass.* **139**, 1324–1330.

Edington, N., Bridges, C.J. & Huckle, A. (1985) Experimental reactivation of equid herpesvirus 1 following the administration of corticosteroids. *Equine vet. J.* **17**, 369–372.

Mumford, J.A. & Thomson, G.R. (1978) Serological methods for identification of slowly growing herpesviruses isolated from the respiratory tract of horses. In *Proc. 4th Int. Conf. equine infectious Disease, Lyon*, pp. 49–52.

Thomson, G.R., Mumford, J.A., Campbell, J., Griffiths, L. & Clapham, P. (1976) Serological detection of equid herpesvirus 1 infections of the respiratory tract. *Equine vet. J.* **8**, 58–65.

J. Reprod. Fert., Suppl. **35** (1987), 519–528

Plasma concentrations of progestagens, oestrone sulphate and prolactin in pregnant mares subjected to natural challenge with equid herpesvirus-1

J. C. Ousey, P. D. Rossdale, R. S. G. Cash and K. Worthy*

Beaufort Cottage Stables, High Street, Newmarket, Suffolk CB8 8JS, and
Department of Veterinary Clinical Biochemistry, University of Glasgow, Bearsden Road,
Bearsden, Glasgow G61 1QH, U.K.

Summary. Multiparous pregnant mares, on two studfarms, were studied following natural challenge with equid herpesvirus-1 (EHV-1). They were divided into three groups according to serum complement fixation titres: Group A (N = 11) were not challenged and delivered normal foals; Group B (N = 13) were challenged but delivered normal foals; Group C (N = 23) were challenged and delivered infected foals which were stillborn or lived for <31 h. In Groups A and B mean (±s.d.) gestational age at delivery was 343 (±8) and 339 (±8) days respectively, whereas in Group C it was significantly ($P < 0.01$) shorter (294 ± 44 days).

Group C mares had significantly lower pre-partum concentrations of progestagens except for 2 mares in which the values were comparable to those of mares in Groups A and B; one of these 2 mares delivered a live foal which survived for 31 h. Thin-layer chromatography showed that substances eluting in R_F positions similar to 20α-dihydroprogesterone and 5α-pregnane-3,20-dione, were the main progestagen metabolites in plasma at 275–295 and 305–335 days gestation respectively. Incubation of liver, placenta, gonads, kidney and brain revealed no significant difference in the ability of virus-infected and non-infected fetal tissues to metabolize labelled progesterone in the presence of NADPH.

Concentrations of plasma oestrone sulphate *pre partum*, from all groups of mares, were related to gestational age. Plasma prolactin concentrations were elevated at parturition in mares in Groups A and B but lower values were observed in Group C.

We conclude that (1) endocrine patterns in maternal peripheral blood of mares with EHV-1 infected fetuses are similar to those of mares carrying normal fetuses and probably are related to gestational age at the time of delivery; and (2) substantial quantities of progestagen metabolites are found in the plasma of late-term pregnant mares.

Introduction

Epidemiological and virological aspects of abortion in pregnant mares, due to equid herpesvirus-1 (EHV-1) abortion, have been the subject of extensive reports (see Allen & Bryans, 1986) since the disease was first reported in Kentucky (Dimmock & Edwards, 1933). However, there do not appear to have been any reports of endocrinological effects of infection. In contrast, the endocrinology of late gestation in normal pregnancies has been well documented.

Short (1959) could not detect progesterone in the maternal peripheral blood during late pregnancy using an assay with a sensitivity of 4 ng/ml. However, radioimmunoassay (RIA) cross-reacting with progesterone metabolites subsequently revealed higher concentrations increasing towards parturition (Holtan *et al.*, 1975b; Ganjam *et al.*, 1975). Atkins *et al.* (1974) and Holtan *et al.* (1975a) identified these progestagens as 5α-pregnane-3,20-dione (DHP), 3β-hydroxy-5α-pregnan-20-one (3β-ol) and 20α-hydroxy-5α-pregnan-3-one (20α-ol); and Moss *et al.* (1979) showed, in 4 pregnant Pony mares, that DHP was produced by the placenta and a maternal source, 20α-ol was of maternal origin and 3β-ol was formed primarily by the fetus.

Maternal blood oestrogen concentrations are highest in mid-gestation and decrease towards term (Nett *et al.*, 1975). The fetal gonads secrete dehydroepiandrosterone which plays a precursor role in placental oestrogen formation (Raeside *et al.*, 1979). Fetal gonadectomy did not prevent pregnancy continuing to term in 4 Pony mares although 3 foals were 'dysmature', suggesting a contribution by oestrogen to fetal maturation (Pashen & Allen, 1979).

Prolactin concentrations increase in the last weeks of pregnancy in many species (Buttle *et al.*, 1972; Fell *et al.*, 1972; McNeilly & Friesen, 1978) and are partly responsible for the initiation of milk secretion in the mammary glands (Cowie *et al.*, 1980). Reports of prolactin concentration in the mare during pregnancy have failed to detect significant changes (Nett *et al.*, 1975) but attempts to measure prolactin in the horse have used heterologous RIAs based on a sheep prolactin system and it is only recently that homologous RIAs have become available (Worthy *et al.*, 1986).

In this study the effect of EHV-1 infection, on maternal plasma concentrations of progestagens, oestrone sulphate and prolactin, was investigated in mares subjected to natural challenge during 1985/86.

Materials and Methods

Animals and sampling

In total, 47 multiparous Thoroughbred mares on two stud farms were studied after initial abortions diagnosed as being due to EHV-1 infection. The mares were sub-divided into three groups according to serological evidence of viral challenge (Table 1). The time of challenge by EHV-1 was defined according to serum titres of complement fixation as described by Mumford *et al.* (1987).

Once the presence of EHV-1 on the two studfarms had been confirmed, sampling of jugular venous blood for hormone analysis took place until parturition or abortion. Jugular venous blood samples were collected at least twice weekly until Day 3 before expected foaling date when sampling took place daily. Blood samples were collected into vacutainer tubes which were plain or contained lithium heparin. Heparinized samples were centrifuged at 4000 *g* within 2 h of collection and plasma stored at −20°C until time of assay. Serum was harvested and analysed for the presence of antibody titre levels according to methods described by Mumford *et al.* (1987).

Assays of progesterone and its metabolites

Radioimmunoassays. Progestagens were measured by two RIA methods. In the method (i) described by Michel *et al.* (1986), the antiserum cross-reacted with progesterone (100%), 11α-hydroxyprogesterone (76%), DHP (6·2%),

Table 1. Number of pregnant Thoroughbred mares at risk from EHV-1 infection on two stud farms in 1985 and 1986 and details of the outcome of pregnancy

	Stud farm 1	Stud farm 2
Date of first abortion	14 Feb. 1985	9 Sept. 1985
Pregnant mares at risk	34	13
Not challenged*		
Group A	9	2
Challenged*		
Group B (delivered live		
normal foal)	6	7
Group C (delivered		
infected foal)	19	4

*Indicated by acute and convalescent serum titres of complement fixation (Mumford *et al.*, 1987).

corticosterone (2·1%), 17α-hydroxyprogesterone (2·0%), 11-deoxycorticosterone (1·7%), 11-deoxycortisol (0·4%), 5α-pregnen-3β-ol-20-one (0·4%), 5α-pregnan-3β-ol-20-one (0·34%), and 20α-dihydroprogesterone (0·98%). All other compounds tested cross-reacted <0·1%. Inter-assay and intra-assay coefficients of variation were 9·2 and 4·9% respectively.

In the method (ii) described by Barnes *et al.* (1975), the antiserum used cross-reacted with progesterone (100%), 11-deoxycorticosterone (4·0%), DHP (2·0%), pregnenolone (1·5%), 17α-hydroxyprogesterone (0·4%), and testosterone and 20α-dihydroprogesterone (0·35%). All other compounds tested cross-reacted <0·3% (Sheldrick *et al.*, 1980). Inter-assay and intra-assay coefficients of variation were 25·0 and 12·6% respectively.

Thin-layer chromatography. To establish the relative contribution of progestagens to the measurements obtained by the two RIA methods, 4 plasma samples from mares in Groups A and B, taken at 275–295 and again at 305–335 days gestation, and 2 plasma samples from Group C mares, collected 1 and 7 days before abortion (gestational ages: 340 and 300 days respectively), were subjected to thin-layer chromatography (t.l.c.) as follows. Plasma (1 ml) was extracted with 2 × 6 ml diethyl ether. The ether was removed with nitrogen and resuspended in 0·5 ml methylene chloride:benzene (1:1 v/v) spotting mixture. The plasma extracts were run on a thin-layer chromatogram (20 × 20 cm kieselgel 60F-25H; Merck, Darmstadt, West Germany) with a series of reference steroids (progesterone, DHP, 20α-dihydroprogesterone, 3β-hydroxy-5α-pregnan-20-one and 20α-hydroxy-5α-pregnan-3-one). The R_F (relative to front) values of these compounds were compared with those found from the plasma samples using methylene chloride: diethyl ether (70:30 v/v) as the solvent. The positions of the reference steroids were determined by ultraviolet or iodine staining. Those from the plasma samples were present in such small quantities (less than 100 ng/plate) that their positions had to be determined by scraping off 0·5 cm bands along the length of the chromatogram and running them in a progestagen radioimmunoassay (method i) capable of detecting the metabolites by using an antiserum which cross-reacted with those metabolites being investigated.

The specific identity of the progestagens was not clarified further by mass spectrometry or crystallization techniques. Therefore, subsequent reference to metabolites identified by t.l.c. should be taken to mean substances chromatographically identical to these metabolites.

Tissue incubation techniques. To compare the metabolism of progesterone by fetal tissue from viral (*n* = 3) and non-viral (*n* = 3) abortions, incubation studies were performed as follows. Fetal and placental tissues were collected as soon as possible *post mortem* from three Group C mares aborting (TT and VT, gestational ages 292 and 331 days) or delivering a live, infected foal (MEL, gestational age 335 days) and from 3 non-infected Pony fetuses used as controls (gestational age 270–340 days).

The tissues were incubated within 4 h of collection after storage on ice. Tissue (1 cm³) was homogenized with 10 ml 50 mM-potassium phosphate buffer, pH 7·4, using a hand-held PTFE homogenizer. The homogenate (2 ml) was added to 1 ml 2·5 μCi [³H]progesterone in buffer containing 1 mg NADPH. The solution was incubated for 2 h at 37°C with occasional mixing. The incubation was stopped and any steroids extracted by the addition of 2 × 5 ml diethyl ether. The mixture was vortexed for 10 sec and centrifuged at 1000 *g* for 10 min. After decanting the ether and evaporating under nitrogen the steroids were resuspended in methylene chloride:benzene (1:1 v/v) spotting mixture. Thin-layer chromatography was carried out as previously described. Any tritiated steroid compounds were detected using a Panax RT1S-15 (Redhill, Surrey) linked to a servoscribe paper recorder (Smiths Industries, Wembley, Middlesex). Peaks of radioactivity could then be located by aligning the plate with the paper print-out and comparisons made between these peaks and the reference standards.

Assay of oestrone sulphate

Samples from 10 mares, representative of the three groups, were analysed as follows. Aliquants of plasma (100 μl), diluted 1:40 with glass-distilled water, were incubated at 56°C for 1 h with 100 μl sulphatase enzyme obtained from whole digestive juice of the snail *Helix pomatia* (Uniscience, Cambridge) and diluted 1:20 (v/v) with acetate buffer pH 4·6. The oestrone sulphate was completely converted to oestrone. The mixture was extracted using 6 ml diethyl ether and the resulting samples were assayed for oestrone using the protocol detailed by Steranti Research Ltd (St Albans, Herts) employing their anti-oestrone-6-(CMO)-BSA antiserum. To correct the results to oestrone sulphate values, they were multiplied by 1·4. The antiserum cross-reacted with oestrone (100%) and oestradiol-17β (1·4%). All other compounds tested cross-reacted <0·3%. The inter-assay and intra-assay coefficients of variation were 12·2 and 5·7% respectively.

Assay of prolactin

Prolactin samples from 18 mares, representative of the 3 groups, were collected over the last 30 days *pre partum* and analysed by RIA as described by Worthy *et al.* (1986). This antiserum cross-reacted 100% with horse pituitary prolactin and other cross-reactions were <1% for GH, TSH, LH and FSH. The inter-assay and intra-assay coefficients of variation were 17·5 and 5·5% respectively.

Confirmation of outcome of pregnancy

A healthy foal was defined as one delivered alive and showing normal adaptive responses followed by an uneventful history for at least 2 weeks after parturition. An EHV-1 infected foal was defined on clinical grounds of stillbirth or abnormal signs observed immediately after delivery and by the subsequent course of the condition; on gross post-mortem appearance; by the presence of 'Cowdray A' intranuclear inclusion bodies; and by growth of EHV-1 virus in tissue culture (see Mumford *et al.*, 1987).

Statistical analyses

Means and standard deviations or standard errors were applied to all data when numbers were >3. When necessary, data were combined to produce a sufficient number for analysis. Student's *t* test was also applied when appropriate. These methods were according to Armitage (1977).

Results

Gestational age at time of challenge and delivery

Mean ($\pm$ s.d.) gestational age at earliest known time of challenge was 234 $\pm$ 71 days in Group B mares and 268 ($\pm$43) days in Group C mares. Mean ($\pm$s.d.) gestational ages at time of delivery in Groups A and B (343 $\pm$ 8 and 339 $\pm$ 8 days respectively) were not significantly different ($P > 0.05$) and, therefore, the data from these two groups were combined. Mares delivering infected foals (Group C) had a significantly ($P < 0.01$) shorter gestational age (294 $\pm$ 44 days: range 140–341 days; Table 2). Of these, 19 mares delivered fetuses which were stillborn, whereas 4 delivered live foals.

Table 2. Gestational age at delivery in 23 Thoroughbred foals infected with EHV-1 (Group C)

Mare identity	Gestational age at delivery (days)	No. (% of total)	Outcome of delivery
<300 days			
TSN	140		
KS	241		
CRY	243		
DJ	263		
FY	274	10	Dead
CB	286	(43·5%)	fetus
STB	289		
MM	290		
TT	293		
PL	298		
300–320 days			
FN	300		Dead foal
OR	307		Foal lived 0·5 h
CHL	309	6	Dead foal
DA	314	(26·1%)	Dead foal
RB	316		Dead foal
SL	317		Foal lived 1·5 h
320–360 days			
VT	321		Dead foal
LS	324		Dead foal
BD	332	6	Foal lived 18 h
YG	333	(26·1%)	Dead foal
MEL	335		Foal lived 31 h
ALT	341		Dead foal
Unknown			
LAC	Autopsy suggests full term	1 (4·3%)	Dead foal

Endocrinology

The results for Groups A and B were within one standard deviation for all of the parameters measured and therefore these data have been combined on all figures and are shown as Groups A + B.

Progesterone and its metabolites. Progestagen concentrations measured by Method (i) (Fig. 1a) in Groups A + B were <10 ng/ml over the last third of gestation but showed a consistent increase above 10 ng/ml occurring at a mean ($\pm$s.d.) of $22 \cdot 0 \pm 9 \cdot 5$ days (range 40–8 days) before parturition; mean peak values (23 ng/ml) were obtained on Day 3 *pre partum* and concentrations then declined. Group C mares had significantly ($P < 0 \cdot 01$) lower mean progestagen concentrations than did mares in Groups A + B over the last 24 days before delivery, except on Day 4 *pre partum* when mean peak values of $13 \cdot 1$ ng/ml were obtained. This was due to two mares (MEL and VT) in Group C having concentrations of $27 \cdot 5$ and $17 \cdot 5$ ng/ml respectively; these 2 mares were exceptional in Group C because they showed an increase in plasma progestagens for 8 and 12 days respectively *pre partum*.

Mean progestagen concentrations measured by Method (ii) (Fig. 1b) for Groups A + B peaked at $6 \cdot 4$ ng/ml $\pm 1 \cdot 1$, 3 days before delivery and then declined by the day of parturition. However,

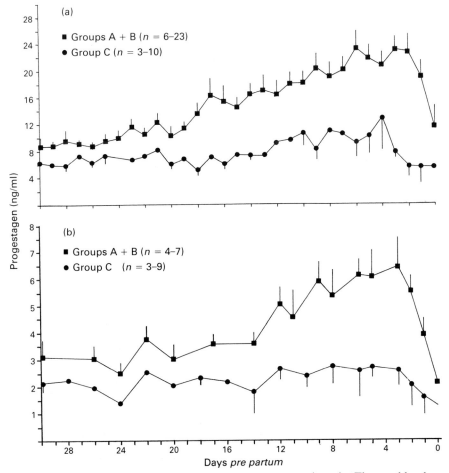

Fig. 1. Mean $\pm$ s.e. maternal plasma progestagen concentrations in Thoroughbred mares during the 30 days *pre partum* by (a) assay method i (Michel *et al.*, 1986) and (b) assay method ii (Barnes *et al.*, 1975). n = number of samples and when numbers were <4, no s.e. is given.

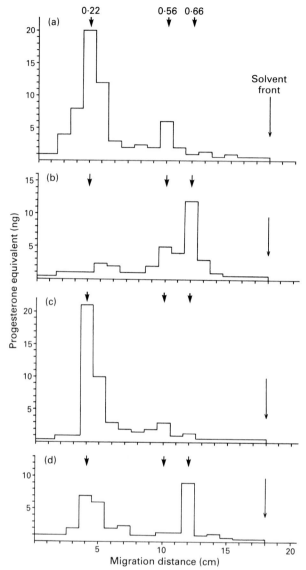

Fig. 2. Maternal plasma progestagen metabolite profile from 4 mares in Groups A and B at (a) 275–295 days and (b) 305–335 days of gestation, (c) a Group C mare at 300 days of gestation (aborted at 307 days) and (d) a Group C mare at 340 days of gestation (aborted at 341 days). Arrows indicate the R_F values for 20α-dihydroprogesterone (0·22), progesterone (0·56) and DHP (0·66).

mean progestagen concentrations in Group C mares remained at < 3 ng/ml, declining to $< 1·5$ ng/ml just before delivery.

Plasma samples obtained from mares in Groups A and B, at 275–295 days gestation, and subjected to t.l.c., separated into a major and lesser peak of 20α-dihydroprogesterone and DHP respectively (Fig. 2a). By 305–335 days gestation, in the same mares, the peak of DHP had increased whereas that of 20α-dihydroprogesterone was undetected (Fig. 2b). In Group C, plasma from one mare (OR) at 300 days gestation resulted in a major and lesser peak of 20α-dihydropro-gesterone (Fig. 2c) and DHP respectively. In a second mare (ALT) examined at 340 days gestation, 1 day before delivery of an infected foal, there was a substantial peak of DHP (Fig. 2d). These

Table 3. Tissue incubation results from virus-infected and non-virus-infected organs*

	Placenta		Liver		Gonad		Kidney		Brain	
	V	NV	V	NV	V	NV	V	NV	V	NV
20α-Dihydroprogesterone	±	−	+	+	−	−	−	−	−	−
3β-Hydroxy-5α-pregnan-20-one	+	+	+	+	±	−	−	−	−	−
20α-Hydroxy-5α-pregnan-3-one	+	+	+	+	−	−	−	−	−	−
Progesterone	+	+	+	+	+	+	+	+	+	+
5α-Pregnane-3,20-dione (DHP)	+	+	+	+	±	±	±	±	−	−

+, >5% conversion of [^{3}H]progesterone; ±, <5% conversion of [^{3}H]progesterone; −, no conversion of [^{3}H]-progesterone.
*V = virus-infected (N = 3; 292, 331 and 335 days of gestation); NV = non-virus infected (N = 2; 270 and 340 days of gestation).

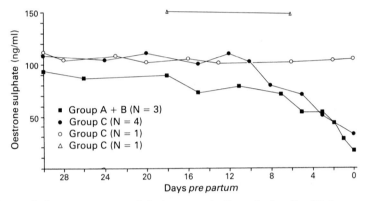

Fig. 3. Maternal plasma oestrone sulphate concentrations during the 30 days *pre partum* in Thoroughbred mares delivering normal and infected fetuses. N = number of mares.

results indicate that patterns of progesterone metabolites in maternal plasma were similar in mares in Groups A + B and C and correlated with advancing gestational age and approach to full term.

There was no significant difference in the ability of virus-infected and non-virus-infected tissues to convert tritiated progesterone, in the presence of NADPH, to the metabolites shown (Table 3). Gonad, kidney and brain only partly metabolized progesterone, <5% conversion for each tissue. Placenta and liver demonstrated much greater metabolic activity and at least 4 progesterone metabolites were detected including DHP (Table 3).

Oestrone sulphate

Maternal plasma concentrations in 4 mares of Groups A + B showed an expected decline towards term (Fig. 3) and a similar pattern was observed in 4 Group C mares which aborted within the full-term period of gestation (320–360 days). In a further 2 Group C mares (DJ and FN) which aborted in mid-gestation (263 and 300 days respectively) concentrations of oestrone sulphate remained elevated compared to those in mares aborting at full term. However, the values in these 2 mares were comparable to those measured at mid-gestation in mares in Groups A + B which carried their pregnancies to term. Therefore, oestrone sulphate concentrations in all groups appeared to be related to gestational age.

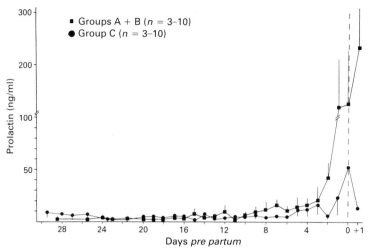

Fig. 4. Mean ± s.e. maternal plasma prolactin concentrations in Thoroughbred mares from 30 days *pre partum* to 1 day *post partum*. *n* = number of samples and when numbers were <4, no s.e. is given.

Prolactin

Prolactin concentrations in mares in Groups A + B tended to increase in all individuals around the time of parturition (Fig. 4). This was associated with maximal mammary development and the presence of 'full term' mammary secretion electrolyte content (Ousey *et al.*, 1984). A similar but smaller increase in mean plasma prolactin concentrations was also observed in Group C mares, although some individuals had no increase before or after abortion.

Discussion

Maternal endocrine patterns in the peripheral blood of mares in late gestation have been the subject of numerous reports since the development of modern RIA techniques. Concentrations of maternal plasma progesterone and its metabolites, leading up to parturition and abortion, have received particular attention because of the relevance of this hormone to the maintenance of pregnancy (Holtan *et al.*, 1975b; Seren *et al.*, 1981; Seamans *et al.*, 1979). Patterns of progestagens (van Niekerk & Morgenthal, 1984) and of electrolyte changes in pre-colostral mammary secretions (Ousey *et al.*, 1984) have also been examined as indicators of fetal well being and readiness for birth respectively. In the present study, with RIA method (i), an increase in plasma progestagen concentrations to > 10 ng/ml in Group A & B mares occurred from 22 ± 9·5 (s.d.) days, and at least 8 days, before delivery. These data provide some indication of impending delivery in Thoroughbreds, based on measurements performed on maternal plasma. The increment appears to be primarily the result of metabolism of progesterone to DHP. In the two RIA methods employed, the antiserum cross-reacted with DHP 6·2 and 2·0% respectively and this difference (a ratio of about 3:1) is reflected in the peak concentrations of progestagen observed *pre partum* (Fig 1). Mean peak concentrations at Day 3 before parturition were 22·5, compared to 6·4 ng/ml respectively. Further, t.l.c. indicated that, in normal mares, 20α-dihydroprogesterone was the major metabolite present in plasma around 300 days gestation but that, by 30 days *pre partum*, this had been replaced by DHP. This is produced from a maternal/placental source (Moss *et al.*, 1979) but the significance of this change in progesterone metabolism close to term is unknown.

Our results indicate that measurement of progesterone by RIA methods alone, cross-reacting even as little as 2·0% with DHP, are not specific enough to identify changes in progesterone itself in maternal plasma in late term, because of the apparently substantial increase in the metabolite at this time.

Out of 23 mares with infected fetuses (Group C), 21 showed no increment in plasma progestagen concentrations. However, these mares had a significantly ($P < 0.01$) shorter gestational length compared with controls (Groups A + B). Incubation experiments in this study suggested that placenta and fetal liver were capable of metabolizing progesterone into DHP *in vitro* as early as 270 days of gestation until term, irrespective of whether the tissues were infected with virus.

Ten Group C mares aborted at less than 300 days gestation (Table 2) and, in all cases, plasma progestagen concentrations were < 10 ng/ml (method i). In these individuals it was not possible, therefore, to determine whether these low concentrations were caused by viral challenge or gestational immaturity. In 6 mares delivering between 300 and 320 days gestation, no increase in pre-partum plasma progestagens occurred. However, 2 of these mares (SL and BD) delivered live foals which survived for 1·5 and 18 h respectively. Of the 6 mares which aborted at full term (320–360 days), 4 did not show a pre-partum increase to above 10 ng/ml. Because delivery occurred within the full-term period, gestational immaturity is unlikely to account for the low concentrations in these mares. However, other studies have demonstrated that gestational age is not the best indicator of readiness for birth and that delivery of a 'premature like' foal can occur within the full term period (Rossdale & Silver, 1982; Ousey *et al.*, 1984; Rossdale *et al.*, 1984). The alternative causal factor responsible for low progestagen concentrations in these mares may have been a direct effect of the virus itself on the placenta, which is the main source of progesterone in the pregnant mare (Holtan *et al.*, 1979). However, the pathology of EHV-1 infection suggests that lesions are found in the fetal lungs and liver but are not recognized consistently in the placenta (Mahaffey, 1968; Prickett, 1969; Jeffcott & Rossdale, 1976). Furthermore, two mares (VT and MEL) had elevated progestagen concentrations *pre partum* which would disagree with this hypothesis. Because of the variability in the timing of progestagen increments, in Groups A and B, they did not provide sufficient evidence to establish whether EHV-1 infection had any disruptive effects on the endocrine system. For this reason, oestrone sulphate was measured because oestrogens may be an index of fetal well being (for review see Casey *et al.*, 1985).

Although it was possible to examine only a limited number of mares in Group C (Fig. 3), patterns corresponded to gestational age rather than to the event of delivery. Mares delivering infected fetuses in the full-term period had decreasing concentrations of oestrone sulphate, whereas those delivering prematurely had high concentrations comparable to those at similar gestational ages in Groups A and B. These results support the hypothesis that the fetus has adequate endocrine function until shortly before delivery.

In Groups A and B, increased concentrations of plasma prolactin at term were associated with onset of lactogenesis and highest peaks (Fig. 4) were found 1–2 days *pre partum* and 1 day *post partum* when mammary secretion electrolyte concentrations were at full-term values (Ousey *et al.*, 1984). These increases in prolactin concentration correspond to those found in Pony mares (Worthy *et al.*, 1986). In mares which delivered infected foals (Group C) the patterns were more variable. Peak values were observed on the day of abortion but even in those mares with highest concentrations (ALT and DA), values were still less than those observed in Groups A + B (Fig. 4).

In conclusion, the results of this study indicate that challenge by EHV-1 does not alter significantly the time course of normal endocrine events taking place in late gestation of Thoroughbred mares. It would appear, therefore, that if mares succumb to the effects of viral challenge, fetal death and abortion follow very rapidly.

We thank the Horserace Betting Levy Board for financial support; Marianne Yelle for assistance in collecting samples; Jenny Mumford and Sheila Gann for serological examinations; Maureen Hamon and Jane Goode for technical assistance in setting up the t.l.c. and oestrone sulphate assays; Marian Silver and Brian Heap for the use of facilities and helpful comments; Sidney Ricketts and Roberto Franco for assisting with post-mortem examinations; and Jan Wade for preparing the manuscript.

References

Allen, G.P. & Bryans, G.T. (1986) Molecular epizootiology, pathogenesis and prophylaxis of equine herpesvirus-1 infection. *Prog. vet. microbiol. Immuno.* **2**, 78–114.

Armitage, P. (1977) *Statistical Methods in Medical Research*, 4th edn. Blackwell Scientific, Oxford.

Atkins, D.T., Sorensen, A.M., Jr & Fleeger, J.L. (1974) 5α-Dihydroprogesterone in the pregnant mare. *J. Anim. Sci.* **39**, 196, Abstr.

Barnes, R.J., Nathanielsz, P.W., Rossdale, P.D., Comline, R.S. & Silver, M. (1975) Plasma progestagens and oestrogens in fetus and mother in late pregnancy. *J. Reprod. Fert., Suppl.* **23**, 617–623.

Buttle, H.L., Forsyth, I.A. & Knaggs, G.S. (1972) Plasma prolactin measured by radioimmunoassay and bioassay in pregnant and lactating goats and the occurrence of placental lactogen. *J. Endocr.* **53**, 483–491.

Casey, M.L., MacDonald, P.C. & Simpson, E.R. (1985) Endocrinological changes of pregnancy. In *Textbook of Endocrinology*, 7th edn, pp. 422–437. Eds J. D. Wilson & D. W. Foster. W.B Saunders, Philadelphia.

Cowie, A.T., Forsyth, I.A. & Hart, I.C. (1980) *Hormonal Control of Lactation*. Monographs of Endocrinology, Springer-Verlag.

Dimmock, W.W. & Edwards, P.R. (1933) Is there a filterable virus of abortion in mares? *Suppl. Ky Agr. Exp. Stn. Bull.* **333**, 297–301.

Fell, L.R., Beck, C., Brown, J.M., Catt, K.J., Cumming, I.A. & Godding, J.R. (1972) Solid-phase radioimmunoassay of ovine prolactin in antibody coated tubes. Prolactin secretion during oestradiol treatment at parturition and during milking. *Endocrinology* **91**, 1329–1336.

Ganjam, V.K., Kenney, R.M. & Flickinger, G. (1975) Plasma progestagens in cyclic, pregnant and postpartum mares. *J. Reprod. Fert., Suppl.* **23**, 441–447.

Holtan, D.W., Ginther, O.J. & Estergreen, V.L. (1975a) 5α-Pregnanes in pregnant mares. *J. Anim. Sci.* **41**, 359, Abstr.

Holtan, D.W., Nett, P.M. & Estergreen, V.L. (1975b) Plasma progestins in pregnant, post partum and cycling mares. *J. Anim. Sci.* **40**, 251–260.

Holtan, D.W., Squires, E.L., Lapin, D.R. & Ginther O.J. (1979) Effect of ovariectomy on pregnancy in mares. *J. Reprod. Fert., Suppl.* **27**, 457–463.

Jeffcott, L.B. & Rossdale, P.D. (1976) Practical aspects of equine virus abortion in the UK. *Vet. Rec.* **98**, 153–155.

Mahaffey, L.W. (1968) Abortion in mares. *Vet. Rec.* **70**, 681–687.

McNeilly, A.S. & Friesen, H.G. (1978) Prolactin during pregnancy and lactation in the rabbit. *Endocrinology* **102**, 1548–1554.

Michel, T.H., Rossdale, P.D. & Cash, R.S.G. (1986) Studies on the efficacy of human chorionic gonadotrophin and gonadotrophin releasing hormone for hastening ovulation in Thoroughbred mares. *Equine vet. J.* **18**, 438–442.

Moss, G.E., Estergreen, V.L., Becker, S.R. & Grant, B.D. (1979) The source of 5α-pregnanes that occur during gestation in mares. *J. Reprod. Fert., Suppl.* **27**, 511–519.

Mumford, J.A., Rossdale, P.D., Jessett, D., Gann, S.J. & Ousey, J.C. (1987) Serological and virological investigations of an equid herpesvirus 1 (EHV-1) abortion storm on a stud farm in 1985. *J. Reprod. Fert., Suppl.* **35**, 509–518.

Nett, T.M., Holtan, D.W. & Estergreen, V.L. (1975) Oestrogens, LH, PMSG and prolactin in serum of pregnant mares. *J. Reprod. Fert., Suppl.* **23**, 457–462.

Ousey, J.C., Dudan, F.E. & Rossdale, P.D. (1984) Preliminary studies of mammary secretions in the mare to assess foetal readiness for birth. *Equine vet. J.* **16**, 259–263.

Pashen, R.L. & Allen, W.R. (1979) The role of the fetal gonads and placenta in steroid production, maintenance of pregnancy and parturition in the mare. *J. Reprod. Fert., Suppl.* **27**, 499–509.

Prickett, M.E. (1969) The pathology of disease caused by equine herpesvirus-1. In *Proc. 2nd Int. Conf. equine infect. Diseases*, pp. 24–33. Eds J.T. Bryans & H. Gerber. Karger, Basel.

Raeside, J.I., Liptrap, R.M., McDonell, W.N. & Milne, F.J. (1979) A precursor role for DHA in a fetoplacental unit for oestrogen formation in the mare. *J. Reprod. Fert., Suppl.* **27**, 493–497.

Rossdale, P.D. & Silver, M. (1982) The concept of readiness for birth. *J. Reprod. Fert., Suppl.* **32**, 507–510.

Rossdale, P.D., Ousey, J.C., Silver, M. & Fowden, A.L. (1984) Studies on equine prematurity. 6: Guidelines for assessment of foal maturity. *Equine vet. J.* **16**, 300–304.

Seamans, K.W., Harms, P.G., Atkins, D.T. & Fleeger, J.L. (1979) Serum levels of progesterone, 5α-dihydroprogesterone and hydroxy-5α-pregnanones in the pre-partum and post partum equine. *Steroids* **33**, 55–63.

Seren, E., Tamanini, C., Gaiani, R. & Bono, G. (1981) Concentrations of progesterone, 17α-hydroxyprogesterone and 20α-dihydroprogesterone in the plasma of mares during pregnancy and at parturition. *J. Reprod. Fert.* **63**, 443–448.

Sheldrick, E.L., Mitchell, M.D. & Flint, A.P.F. (1980) Delayed luteal regression in ewes immunized against oxytocin. *J. Reprod. Fert.* **59**, 37–42.

Short, R.V. (1959) Progesterones in blood. IV: Progesterone in the blood of mares. *J. Endocr.* **19**, 207–210.

van Niekerk, C.H. & Morgenthal, J.C. (1984) Fetal loss and the effect of stress of plasma progestagen levels in pregnant. Thoroughbred mares. *J. Reprod. Fert., Suppl.* **32**, 453–457.

Worthy, K., Escreet, R., Renton, J.P., Eckersall, P.D., Douglas, T.A. & Flint, D.J. (1986) Plasma prolactin concentrations and cyclic activity in pony mares during parturition and early lactation. *J. Reprod. Fert.* **77**, 569–574.

J. Reprod. Fert., Suppl. **35** (1987), 529–533

Printed in Great Britain
© 1987 Journals of Reproduction & Fertility Ltd

Consequence of excess iodine supply in a Thoroughbred stud in southern Brazil

C. A. M. Silva*, H. Merkt†, P. N. L. Bergamo*, S. S. Barros*,
C. S. L. Barros*, M. N. Santos*, H. O. Hoppen†, P. Heidemann‡ and
H. Meyer†

*Faculty of Veterinary Medicine, Federal University of Santa Maria, Brazil; †School of Veterinary Medicine, Hannover, Federal Republic of Germany; and ‡Childrens Clinic, University of Göttingen, Federal Republic of Germany

Summary. Excessive iodine supply of at least 700 mg inorganic iodine in foals and of more than 350 mg iodine in pregnant and lactating mares cause a high incidence of goitres in the newborn and disorders in the long leg bones of foals. Elevated phosphorus and alkaline phosphatase content in the blood may indicate a severe disturbance in the bone metabolism. Of 39 pregnant mares 17 aborted and some of the mares also showed goitres. After withdrawal of the iodine supply normalization took place. Foals born more than 6 weeks later showed normal conditions. The goitres in the mares and the high blood levels of iodine decreased.

Introduction

In a region close to the Breton coast in France, considered to be very rich in iodine, Jacob (1940) observed foals with goitres up to the size of a fist, which he named atypical colloidal goitres.

In 3 Thoroughbred studs, in which the mares received dried seaweed, foals were born weak or dead or died soon after birth (Baker & Lindsay, 1968): their thyroids were extremely enlarged and grossly visible. A few mares also had obvious goitres. The disease disappeared on 2 farms after removal of the iodine from the diet. The iodine level in the diet of each mare was measured and was 48–432 mg/day/mare; this is at least 10 times the normal requirement for pregnant animals of this species.

The present account is of a study of dietary iodine effects in a stud farm in Brazil.

Clinical History and Findings

Animals

In 1983, in a Thoroughbred stud farm in South Brazil two cases of goitres were observed in foals. An iodine deficiency in the diet was assumed. From then onwards all foals were given a daily ration of 50 g salt mixed with 2% potassium iodide (about 700 mg inorganic iodine). Because there was no decrease in the incidence of goitres, the mares were given the same treatment from June 1984 onwards. The daily dosage was 25 g of the above mixture (about 350 mg inorganic iodine).

During February 1985, 17 mares showed nasal discharge and jaundice. An outbreak of piroplasmosis was assumed and in the serum of some mares titres against piroplasms (*B. caballi*) could be found. All mares recovered after a treatment with imidocarb (Imizol: Coopers, Sao Paulo, Brazil). Two mares aborted. Post-mortem examination of the foals revealed no conspicuous findings. In the following months, although 22 foals came to term, one foal was stillborn and showed a striking goitre. Three foals had to be destroyed due to general weakness, and all of them showed goitres, up to a goose egg in size (Fig. 1). A fifth foal had to be destroyed at the age of 3 months because of a fracture of one fore-leg. Moreover, 15 other mares aborted. All of the 15 aborted foals had goitres and the longer the duration of pregnancy the greater the size of the goitres. Only two of the 17 mares which aborted were involved in the outbreak of piroplasmosis in February. No other obvious alterations were found in the post-mortem macroscopic examinations of the aborted foals.

Autopsy

Two of the foals which were destroyed came to the Veterinary Faculty in the University of Sta. Maria for examination. Considerable alterations in the skeleton were found. Particularly the long canon bones of the legs showed an osteopetrosis and a diminished lumen, probably the consequence of a severe disturbance in the bone metabolism.

Biochemistry

Milk samples were taken from 6 mares of the stud and examined for their content of iodine. Samples of blood serum were available from 5 of these 6 mares, 3 other mares and 7 foals. The

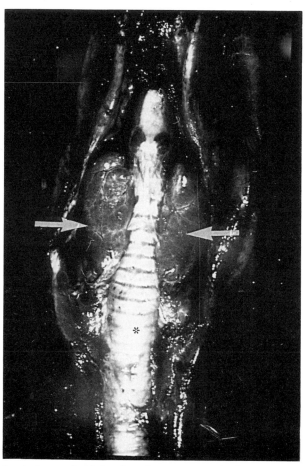

Fig. 1. Thyroid glands (arrows) of the size of a goose egg in foal destroyed due to general weakness in the 2nd week after birth. * = trachea.

serum samples were also examined for iodine, triiodothyronine and thyroxine concentrations (Table 1). Total iodine concentrations were measured in serum and milk samples using an automated wet analysis technique in an Auto Analyser (Technicon) according to Sandell & Kolthoff (1934). Thyroxine was measured by radioimmunoassay (Magic T4, Corning) and triiodothyronine was measured by radioimmunoassay (Ria-Gnost T3, Behring).

For comparison, in some studfarms of the Federal Republic of Germany with no indications of excessive iodine content in the diet, blood or milk samples from 9 mares and blood samples from 4 foals were examined for iodine content (Table 2). About 7 weeks after the withdrawal of the excessive iodine supply in the diet of the Brazilian stud animals, control samples were collected from 7 mares and 5 foals (Table 3).

Results

As shown in Table 1, iodine in the milk of mares suspected of iodine overload averaged 47·3 µg/100 ml, more than 3 times the iodine content in the milk of mares used for comparison (12·4 µg/100 ml; Table 2). In blood samples of the suspected mares the iodine content averaged

Table 1. Content of iodine, thyroxine (T4) and triiodothyronine (T3) in milk and blood samples of 9 mares and blood samples of 7 foals of a Brazilian Thoroughbred stud with excessive iodine supply in the ration

Mare	Date of collection	Iodine content (µg/100 ml)		T4 (µg/100 ml)	T3 (µg/100 ml)
		Milk	Serum		
1	9 Aug. 1985		25·0	<2·5	0·96
1	11 Aug. 1985	20·0			
1a (foal)	9 Aug. 1985		250·0	8·58	0·65
2	9 Aug. 1985		33·0	<2·5	0·45
2	11 Aug. 1985	33·0			
2a (foal)	9 Aug. 1985		225·0	<2·5	3·85
3	9 Aug. 1985		33·0	<2·5	0·67
3	11 Aug. 1985	17·5			
3a (foal)	9 Aug. 1985		195·0	2·58	2·04
4	9 Aug. 1985		30·0	<2·5	0·68
4	11 Aug. 1985	28·0			
4a (foal)	9 Aug. 1985		240·0	4·42	1·74
5	9 Aug. 1985		25·0	<2·5	0·53
5	11 Aug. 1985	138·0			
5a (foal)	9 Aug. 1985		275·0	9·40	0·57
7	9 Aug. 1985		35·0	9·61	0·86
7a (foal)	9 Aug. 1985		73·0	3·23	3·58
8	9 Aug. 1985		20·0	<2·5	0·55
8a (foal)	9 Aug. 1985		75·0	2·63	2·30
9	9 Aug. 1985		70·0	<2·5	0·45
Average Mares (1–9)		47·3	33·8	<2·5	0·64
Foals (1a–8a)			190·4	5·14	2·10

Table 2. Content of iodine, thyroxine (T4) and triiodothyronine (T3) in milk and blood samples of 9 mares and blood samples from 4 foals of some Thoroughbred studs in the Federal Republic of Germany

Mare	Day of collection	Iodine content (µg/100 ml)		T4 (µg/100 ml)	T3 (µg/100 ml)
		Milk	Serum		
A	24 Oct. 1985	32·0	3·5	<2·5	0·25
a (foal)	24 Oct. 1985		3·5	<2·5	2·49
B	24 Oct. 1985	45·0	3·25	<2·5	0·79
b (foal)	24 Oct. 1985		3·75	2·54	0·66
C	24 Oct. 1985	5·75	3·0	<2·5	0·17
c (foal)	24 Oct. 1985		3·0	<2·5	1·11
D	23 Oct. 1985	1·75	3·0	<2·5	0·37
d (foal)	23 Oct. 1985		4·5	5·25	1·90
E	24 Oct. 1985	4·5	4·25	<2·5	0·51
F	24 Oct. 1985	6·5	4·5	<2·5	0·35
G	24 Oct. 1985	10·0	5·75	<2·5	0·13
H	24 Oct. 1985	3·0	4·25	<2·5	0·57
J	24 Oct. 1985	3·0	3·75	2·75	0·91
Average Mares (A–J)		12·4	3·9	<2·5	0·45
Foals (a–d)			3·7	2·8	1·54

33·8 µg/100 ml, exceeding the values used for comparison of 3·9 µg/100 ml, i.e. 8·6-fold. Even more drastic was the difference seen in foals. The average value was 190·4 µg/100 ml serum in the affected foals compared with 3·7 µg/100 ml serum in those used for comparison.

The control samples collected 7 weeks after withdrawal of the excessive iodine supply revealed a considerable decrease of iodine, thyroxine and triiodothyronine content in milk and serum (Table 3). However, the average iodine content in milk and serum was still about twice the average content in the samples collected in the Federal Republic of Germany for comparison.

In 7 of the suspected mares and their foals, serum was also examined for the content of Ca, Mg, phosphorus and alkaline phosphatase, Serum values of mares corresponded well with normal values whilst the blood serum of all foals showed elevated concentrations of phosphorus and alkaline phosphatase, indicative of disturbance of bone metabolism (Table 4).

The foals born in 1983 came into training in 1986. There was a remarkable increase in the frequency of lameness, which appeared to be shifting from front legs to hind legs and back. Two of these animals had goitres. X-ray examination showed different stages of closure of the epiphysial cleft. In addition, cases of osteopetrosis and a diminished lumen in the long canon bones of the legs were seen, similar to those mentioned above, and verified by post-mortem examinations of the foals which were destroyed.

Amongst the 28 foals born in 1984, 3 showed goitres. In most of them there was an accumulation of faults in the posture of the limbs in the following year.

The foals born in 1985 and surviving showed considerable deficiencies in their limb structure: the hind fetlocks particularly were so weak that they touched the ground. In August 1985 iodine addition to the diet was withdrawn. Foals born 6–8 weeks later showed no more goitres and a normal general condition.

The excessive iodine application obviously caused severe losses in the offspring of this stud.

Table 3. Content of iodine in milk samples of 5 mares and content of iodine, thyroxine (T4) and triiodothyronine (T3) in blood samples of 7 mares and 5 foals of a Brazilian Thoroughbred stud collected 7 weeks after withdrawal of the excessive iodine supply in the diet

Mare	Date of collection	Iodine content (μg/100 ml) Milk	Serum	T4 (μg/100 ml)	T3 (μg/100 ml)
1	22 Sep. 1985	15·0			
1	25 Sep. 1985	17·5			
1	27 Sep. 1985		11.25	0.86	0.33
1a (foal)	27 Sep. 1985		8·75	3·69	1·38
2	22 Sep. 1985	25·0			
2	25 Sep. 1985	25·0			
2	27 Sep. 1985		12·5	0·92	0·29
2a (foal)	27 Sep. 1985		13·75	3·47	1·86
3	22 Sep. 1985	25·0			
3	25 Sep. 1985	30·0			
3	27 Sep. 1985		11·25	1·45	0·34
3a (foal)	27 Sep. 1985		10·5	2·67	1·44
4	27 Sep. 1985		3·75	0·81	0·48
7	22 Sep. 1985	17·5			
7	25 Sep. 1985	20·0			
7	27 Sep. 1985		5·5	0·87	0·38
7a (foal)	27 Sep. 1985		9·75	1·87	0·94
8	22 Sep. 1985	11·5			
8	25 Sep. 1985	15·0			
8	27 Sep. 1985		6·25	1·1	0·29
8a (foal)	27 Sep. 1985		10·25	3·36	1·15
9	27 Sep. 1985		3·5	1·38	0·47
Average Mares		22·8	7·7	1·05	0·36
Foals			10·6	3·01	1·35

Table 4. Ca, Mg, phosphorus and alkaline phosphatase content in serum samples collected on 9 August 1985 from Mares 1, 2, 3, 4, 5, 7 and 8 (Table 1) and then foals

	Mares Range	Average	Foals Range	Average	Normal values (range)
Ca (mg/100 ml)	11·2–12·6	12·1	10·55–11·65	11·2	10·0–13·6
Mg (mg/100 ml)	1·74–2·5	1·96	1·66–2·06	1·81	1·58–2·70
Phosphorus (mg/100 ml)	2·11–3·13	2·77	6·81–8·24	7·54	2·5–4·5
Alkaline phosphatase (U/l)	231–307	270	498–1023		350

References

Baker, H.J. & Lindsay, J.R. (1968) Equine goitre due to excess dietary iodide. *J. Am. med. Assoc.* **153**, 1618–1630.

Jacob, Y. (1940) *Contribution à l'étude du goitre: l'hypertrophie des thyroides chez le poulain côtier breton.* Thesis, Faculté de Medicine de Paris, 48 pp.

Sandell, E.B. & Kolthoff, I.M. (1934) Chronometric catalytic method for the determination of microquantities of iodine. *J. Am. chem. Soc.* **56**, 1426.

J. Reprod. Fert., Suppl. 35 (1987), 535–538

The effects of four levels of endophyte-infected fescue seed in the diet of pregnant pony mares

W. E. Loch*, L. D. Swantner* and R. R. Anderson†

*Departments of *Animal Science and †Dairy Science, University of Missouri, Columbia, MO 65211, U.S.A.*

Summary. Mean weight of 20 Quarter Horse placentas on a wet basis was 3.27 ± 0.17 kg. These placentas contained mean dry fat-free tissue, DNA, RNA and collagen weights of 411 ± 24 g, 11.2 ± 0.6 g, 12.4 ± 0.7 g and 210 ± 15 g, respectively. In Ponies and Quarter Horses, there was a trend towards a decrease in these 4 values as gestation length increased.

Analysis of placentas of Ponies fed different amounts of endophyte fungus-infected fescue seed showed that weight of dry fat-free tissue increased with increasing percentages of fescue seed in the diet. Diets containing fescue seed resulted in heavier placentas containing more dry fat-free tissue, DNA, RNA and collagen than placentas from mares fed the control diet containing no fescue seed. In mares fed 45% fescue seed, 3 of the 5 foals born suffered from abnormalities: 2 died and one was saved from suffocation in the placenta. No effects of toxic tall fescue were apparent in foals from mares fed the lower levels of fescue seed (15 or 30%).

Introduction

A summary of studies of the equine placenta by Steven & Samuel (1975) reviewed work relating to the anatomy of the placenta and described the structure of the microcotyledons. The ultrastructural development of the equine placenta was described by Samuel *et al.* (1977). The growth and development of the fetus is dependent upon a placenta of adequate functional capabilities and the health of the placenta, in turn, is dependent on its association with a normal, healthy fetus (Steven *et al.*, 1979).

A limited amount of data is available concerning placental weights for horses. Whitwell & Jeffcott (1975) reported placental weights for Thoroughbred mares as 5.7 ± 0.08 kg and Pony mares as 2.38 ± 0.23 kg.

Tall fescue (*Festua arundinacea*) is a native grass of Europe which was introduced into the United States before 1850. It is an aggressive species that is well adapted to a wide range of soil types and climatic conditions. In the United States, the Kentucky-31 variety of tall fescue is the most prevalent pasture grass in a zone between the northern and southern regions from southwest Missouri eastward to Virginia. It does well in rocky soil and tolerates both wet conditions and drought. Even though it has many useful characteristics, several physiological disorders have been associated with animals grazing tall fescue. In cattle, tall fescue has been associated with necrosis around the coronary band and sloughing of the hoof known as 'fescue foot' (Garner & Harmon, 1973), reduced feed intake, fat necrosis, increased respiratory rate, elevated body temperature and listlessness (Jacobsen & Hatton, 1973). A survey of mare owners by Garrett (1980) revealed agalactia as being the most common abnormality associated with mares grazing fescue. In addition, problems related to thick, tough placentas resulting in suffocation of foals and death of foals at birth, apparently due to respiratory failure, were observed in mares grazing tall fescue (Taylor *et al.*, 1985). The pasture used in this study was found to be infected with an endophyte fungus identified as *Epichloe typhina* or *Acrimonium coenophialum*. Bacon *et al.* (1977b) reported that cattle grazing tall fescue which was heavily infected with *Epichloe typhina* showed signs of fescue toxicity syndrome.

This study was conducted to determine placental weights and the contents of dry fat-free tissue, DNA, RNA and collagen in normal Quarter Horse mares and in placentas from Pony mares fed diets containing different amounts of Kentucky-31 tall fescue seed known to be infected with an endophyte fungus.

Materials and Methods

Animals and treatment. Placentas from 20 mares in the University of Missouri-Columbia Quarter Horse herd were collected and weighed immediately after expulsion. The placentas were then frozen at −20°C until being sampled for analysis.

Pregnant Pony mares were randomly assigned to one of 4 treatment groups at about 3 months before parturition in an experiment designed to test the effects of feeding 4 levels of endophyte fungus-infected fescue seed to pregnant mares. Treatments 1, 2, 3 and 4 consisted of diets containing 0, 15, 30 and 45% Kentucky-31 fescue seed respectively. The seed was analysed microscopically using the method described by Bacon *et al.* (1977a) and it was determined that 61% of the seed was infected with *Epichloe typhina*. The mares were fed at the rate of about 2% of body weight daily divided into 2 equal feedings. The diets containing approximately 50% brome hay and 50% of a concentrate mixture containing the fescue seed. All diets were calculated to contain equal percentages of crude protein, calcium and phosphorus. Vitamins A and D were added to all diets at the rate of 5600 i.u./kg and 1000 i.u./kg, respectively. The mares were maintained in dry lots and group fed.

Placentas were collected immediately after expulsion and weighed. Tissue samples from the pregnant horn, non-pregnant horn, body and umbilical cord were taken and examined histologically. Tissues were examined for lesions, hyperplasia, haemorrhage and other abnormalities.

The placentas were rinsed with distilled water and frozen at −20°C until further samples were taken from the allantochorion in the pregnant horn, non-pregnant horn and body.

Analysis of samples. Placental samples of ∼200 g each were weighed and placed in glass jars containing a solution of chloroform–methanol (2:1 v/v) and left in this solution for 5 days, stirring the solution every 10 h. The samples were then extracted with hot 95% ethanol for 12 h, followed by extraction with diethyl ether for 12 h. The samples were then allowed to dry and the resulting dry fat-free tissue was weighed and ground.

Samples of dry fat-free tissue were analysed for DNA by a modification of a method described by Webb & Levy (1955). DNA was hydrolysed in trichloracetic acid. After hydrolysis, the deoxyribose reacts with *p*-nitrophenylhydrazine to separate it from interfering substances. DNA was then determined colorimetrically in alkaline solution using a Beckman 24 spectrophotometer at 560 nm wavelength.

Samples (25 mg) of dry fat-free tissue were analysed for RNA by a slightly modified procedure described by Albaum & Umbreit (1947).

Collagen was determined in placental samples by analysing samples for hydroxyproline. Collagen contains 14% hydroxyproline (Gustavson, 1956) and this amino acid is found only in collagen. The method used involved hydrolysing 25 mg samples of placental tissue with 6 N-hydrochloric acid and adjusting the pH to 6·9. After colour development, the samples were read in a Beckman 24 spectrophotometer at 557 nm wavelength.

Results and Discussion

Data collected from the Quarter Horse placentas are shown in Table 1.

A regression analysis of dry fat-free tissue weight on gestation length showed a trend towards a

Table 1. Values measured in 20 Quarter Horse placentas

	Mean ± sd weight
Whole wet weight (kg)	3·72 ± 0·17
Dry fat-free tissue (g)	411 ± 24
Deoxyribonucleic acid (g)	11·2 ± 0·6
Ribonucleic acid (g)	12·4 ± 0·7
Collagen (g)	210 ± 15

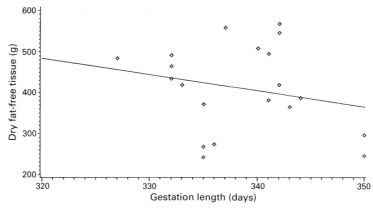

Fig. 1. Regression of dry fat-free tissue weight on gestation length in Quarter Horse mares. Dry fat-free tissue = 1754 + (−3·97 × gestation length).

Table 2. Analysis of placentas from Pony mares fed four levels of endophyte fungus-infected fescue seed

	Treatment			
	1	2	3	4
Fescue seed in diet (%)	0	15	30	45
No. of mares	4	4	4	4
Wet weight (kg)	2·03 ± 0·03[a]	1·41 ± 0·15[b]	1·83 ± 0·17[ab]	2·16 ± 0·13[a]
Dry fat-free tissue (g)	174·2 ± 26[a]	196·6 ± 22[a]	214·3 ± 22[a]	246·9 ± 32[a]
Deoxyribonucleic acid (g)	4·15 ± 0·51[a]	6·18 ± 0·65[b]	6·82 ± 0·13[b]	6·14 ± 0·48[b]
Ribonucleic acid (g)	6·75 ± 1·06[a]	7·74 ± 1·43[a]	7·16 ± 0·72[a]	8·20 ± 1·97[a]
Collagen (g)	88·0 ± 17·8[a]	103·0 ± 15·4[a]	117·3 ± 11·2[a]	120·1 ± 19·0[a]

[a,b]Means on the same line with different superscripts are different ($P < 0.05$).

decrease in this measure with increasing gestation length in the Quarter Horse (Fig. 1) and Pony mares.

Endophyte fungus-infected fescue seed in the diet appeared to cause an increase in the weight of the placenta in the Pony mares based on dry fat-free tissue weight. Each increasing increment of fescue in the diet resulted in a corresponding increase in dry fat-free tissue weight (Table 2).

The mean DNA content of the placentas from Pony mares not consuming fescue seed was 2·3%, while that of mares consuming fescue was 2·94%. This did not prove to be a significant difference ($P > 0.05$), but may be an indication that cell numbers in the placenta are increased by toxic tall fescue. RNA also increased with the addition of fescue seed to the diet, indicating greater cellular activity in those placentas. Collagen content showed a similar pattern to dry fat-free tissue values, which indicates that the placentas from mares consuming high levels of endophyte fungus-infected fescue may contain more connective tissue than others.

No histological abnormalities were observed in any of the placentas with regard to inflammatory lesions, hyperplasia or haemorrhage. All were determined to be normal morphologically.

One foal in Group 4 was born at 301 days gestation encased in the placenta and suffocated. A post-mortem examination showed that the foal had been alive and normal until birth. A second foal in Group 4 was also born encased in the placenta; the attendant cut the placenta open and the foal was alive and normal. A third foal in Group 4 died at birth due to a flattened area in the trachea as it approached the thoracic inlet. The cause was unknown.

Garrett (1980) reported that 15% of mares grazing fescue in Missouri were agalactic at foaling. There appears to be a great deal of variation among mares grazing fescue in the occurrence of agalactia. All the mares in this experiment had sufficient colostrum at foaling to provide adequate immunoglobulin transfer to the foal.

No abnormal retention of placentas was seen in any of the mares in these studies.

Contribution from the Missouri Agriculture Experiment Station Journal Series No. 10159.

References

Albaum, H.G. & Umbreit, W.W. (1947) Differentiation between ribose-3-phosphate and ribose-5-phosphate by means of orcinol-pentose reaction. *J. biol. Chem.* **167**, 369–376.

Bacon, C.W., Porter, J.D., Robbins, J.D. & Luttrell, E.S. (1977a) Epichloe typhina from tall fescue grass. *Appl. Environ. Microbiol.* **34**, 576–581.

Bacon, C.W., Robbins, J.D. & Porter, J.K. (1977b) The systemic infection of toxic fescue grasses by Epichloe typhina. *J. Anim. Sci.* **45** (Suppl. 1), 29, Abstr.

Garner, G.B. & Harmon, B.W. (1973) Experimental pastures and field case data. In *Proc. Fescue Toxicosis Conf., Lexington*, pp. 42–47.

Garrett, L.W. (1980) *Reproductive problems of pregnant mares grazing fescue pastures.* M.S. thesis, Department of Animal Science, University of Missouri.

Gustavson, K.H. (1956) *The Chemistry and Reactivity of Collagen.* Academic Press, New York.

Jacobsen, D.R. & Hatton, R.H. (1973) The fescue toxicity syndrome. In *Proc. Fescue Toxicosis Conf., Lexington*, pp. 4–5.

Samuel, C.A., Allen, W.R. & Steven, D.H. (1977) Studies on the equine placenta. III. Ultrastructure of the uterine glands and the overlying trophoblast. *J. Reprod. Fert.* **51**, 433–437.

Steven, D.H. & Samuel, C.A. (1975) Anatomy of the placental barrier in the mare. *J. Reprod. Fert., Suppl.* **23**, 579–582.

Steven, D.H., Jeffcott, L.B., Mallon, K.A., Ricketts, S.W., Rossdale, P.D. & Samuel, C.A. (1979) Ultrastructural studies on the equine uterus and placenta following parturition. *J. Reprod. Fert., Suppl.* **27**, 579–586.

Taylor, M.C., Loch, W.E. & Ellersieck, M. (1985) Toxicity in pregnant pony mares grazing Kentucky-31 fescue pastures. *Nutr. Rep. Intl.* **31**, 787–795.

Webb, J.M. & Levy, H.B. (1955) A sensitive method for determination of deoxyribonucleic acid in tissues and microorganisms. *J. biol. Chem.* **213**, 107–117.

Whitwell, K.E. & Jeffcott, L.B. (1975) Morphological studies on the fetal membranes of the normal singleton foal at term. *Res. vet. Sci.* **19**, 44–55.

J. Reprod. Fert., Suppl. **35** (1987), 539–545

Effects of inhibiting 3β-hydroxysteroid dehydrogenase on plasma progesterone and other steroids in the pregnant mare near term

A. L. Fowden and M. Silver

The Physiological Laboratory, University of Cambridge, Cambridge CB2 3EG U.K.

Summary. Epostane, a competitive inhibitor of 3β-HSD was administered intravenously to a pregnant mare between 292 and 330 days of gestation at doses of 1–3 mg/kg/min. Plasma progesterone concentrations fell rapidly during epostane infusion in both the artery and uterine vein and remained significantly depressed for 4–5 h after the start of infusion. The venous arterial (V–A) plasma concentration difference in progesterone across the uterus also decreased significantly in response to epostane infusion. There were no significant changes in plasma progesterone or in the V–A concentration difference in control animals infused with vehicle alone. The plasma concentration of total unconjugated oestrogens in the uterine vein was reduced after administration of epostane but remained virtually unchanged in the control experiments. Uterine venous plasma concentrations of PGFM did not change significantly in the control or epostane-treated animals. Arterial plasma cortisol levels fell initially after epostane treatment but then rose to values significantly greater than before infusion. A similar increase in arterial plasma cortisol was observed in the control animals. None of the mares delivered after epostane treatment even at the highest dose. These observations demonstrate that inhibition of 3β-HSD alters steroidogenesis but has little effect on the length of gestation in the pregnant mare.

Introduction

Progesterone is essential for the maintenance of pregnancy and its withdrawal at term is closely associated with the onset of parturition in a number of domestic species (see Thorburn & Challis, 1979). Its production from pregnenolone is controlled by the $\Delta^5$3β-hydroxysteroid dehydrogenase isomerase enzyme complex (3β-HSD: EC 1.1.1.51). Drugs such as trilostane and epostane which inhibit 3β-HSD cause a rapid and prolonged decrease in progesterone which can lead to premature parturition in pregnant animals (Schane *et al.*, 1979; Creange *et al.*, 1981; Taylor *et al.*, 1982).

In common with that of other species, the horse placenta contains 3β-HSD and produces progesterone in sufficient quantities to maintain pregnancy in the absence of the ovary from early in gestation (Ainsworth & Ryan, 1969; Holtan *et al.*, 1979). However, the role of progesterone in the sequence of events preceding delivery in the mare is still not clear (Lovell *et al.*, 1975; Pashen, 1984). Both increases and decreases in maternal progestagens have been reported during the prepartum period in the mare which cannot be related to breed differences or to the subsequent course of labour and delivery (Burns & Fleeger, 1975; Lovell *et al.*, 1975). In the present study, the effects of epostane on the outcome of pregnancy and on progesterone and other plasma steroid concentrations have been investigated in chronically catheterized Pony mares during late gestation.

Materials and Methods

Animals. Nine pregnant mares of mixed breed were catheterized between 240 and 307 days of gestation (term 330–340 days). Food, but not water, was withheld for 12 h before operation and, in mares over 250 days of gestation, the prostaglandin synthetase inhibitor, meclofenamic acid was given before operation as described previously (Silver *et al.*, 1979). Antibiotic (10 ml Streptopen; Glaxovet, Greenford, U.K.) was given for 5 days beginning 1 day before operation.

Operative procedures. Anaesthesia was induced with intravenous thiopentone sodium and maintained with halothane after intubation. The detailed procedure for the induction, maintenance and reversal of anaesthesia was as described by Barnes *et al.* (1978). Catheters were inserted into the aorta and caudal vena cava through the circumflex iliac vessels and into the uterine vein using the surgical procedures described previously (Comline *et al.*, 1975).

Postoperative procedures. Most of the mares were standing and eating hay within 2–3 h of operation and normal feeding patterns were usually restored between 24 and 48 h after operation. A period of 7–10 days was allowed for post-operative recovery before any experiments were begun. Seven live foals were born either spontaneously or by induction at or near term (331·9 ± 3·4 days, N = 7); the remaining 2 foals were delivered prematurely and were probably alive at birth. The possible explanations for this type of premature delivery in chronically catheterized mares have been discussed in detail elsewhere (Silver *et al.*, 1979; Silver & Fowden, 1982).

Experimental procedures. Epostane (4,5-epoxy-17-hydroxy-4, 17-dimethyl-3-oxo-androstane-2-carbonitrile, WIN 32729, Sterling Winthrop, Guildford Surrey, U.K.) was administered intravenously to the pregnant mares between 292 and 330 days of gestation at two dose rates: (1) 300 mg epostane (1–2 mg/kg in ethanol) was given over 15 min (N = 5) or (2) a total of 500 mg epostane (2–3 mg/kg in ethanol) was given over 40 min, with 300 mg being given in the first 15 min (N = 5). One animal was given a total of 900 mg epostane in a 40-min infusion (~ 5 mg/kg i.v.). Control experiments in which ethanol alone was infused were also carried out. Blood samples were taken simultaneously from the maternal artery and uterine vein at 15–30-min intervals for 1 h before and 2 h after the start of the infusion and thereafter at hourly intervals for 6 h. Animals infused with epostane were sampled again at 12 h and 24 h after beginning the infusion when feasible.

Biochemical and statistical analyses. Plasma concentrations of progesterone, cortisol, total unconjugated oestrogens and PGF-2α metabolite (PGFM) were measured in all the samples by the radioimmunoassays described by Barnes *et al.* (1975), Rossdale *et al.* (1982), Barnes *et al.* (1978) and Silver *et al.* (1979) respectively. In the present experiments the following values were obtained for minimum detectable amounts and inter-assay coefficient of variation: progesterone, 30 pg and 8·0%; PGFM, 10–30 pg and 11·6%; cortisol, 30 pg and 7·9%; total unconjugated oestrogen (as oestrone equivalent) 20 pg and 5·8%.

Means and standard errors have been used throughout. Statistical analyses were made according to the methods of Armitage (1971) and statistical significance was assessed by means of the standard or paired-sample *t* test.

Results

The mean initial concentrations of progesterone and the other hormones in the control and epostane-treated animals are shown in Table 1.

Changes in plasma progesterone

Infusion of epostane over 15 min produced a rapid fall in the plasma progesterone concentration in the maternal artery and uterine vein (Fig. 1). Venous progesterone concentrations

Table 1. Initial plasma concentrations of progesterone, total unconjugated oestrogen, PGFM and cortisol in the control and epostane-treated animals

	Progesterone (ng/ml)		Total unconjugated oestrogen (ng/ml)	PGFM (ng/ml)	Cortisol (ng/ml)
	Artery	Uterine vein	Uterine vein	Uterine vein	Artery
Epostane					
1–2 mg/kg	4·1 ± 0·3 (5)	–9·9 ± 0·9 (4)	2·04 ± 0·45 (7)	3·44 ± 0·56 (7)	36·4 ± 4·9 (9)
2–3 mg/kg	4·9 ± 0·9 (5)	–8·9 ± 2·0 (5)			
Control	4·4 ± 0·7 (5)	10·5 ± 1·5 (5)	1·90 ± 0·31 (4)	2·27 (3)	38·3 ± 3·9 (6)

Values are mean ± s.e.m. for the no. of animals in parentheses.

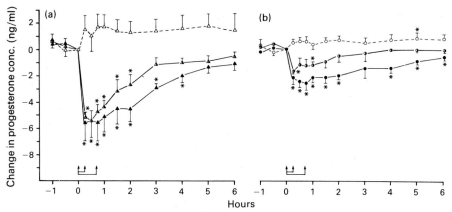

Fig. 1. The mean (±s.e.m.) changes in plasma progesterone in (a) the uterine vein (triangular symbols) and (b) the artery (circular symbols) of pregnant mares infused (arrows) with alcohol alone (N = 9, △○) or epostane at doses of 1–2 mg/kg over 15 min (N = 5, ▲◑) or 2–3 mg/kg over 40 min (N = 5, ▲●). *Significant reduction from 0 h values, $P < 0.01$.

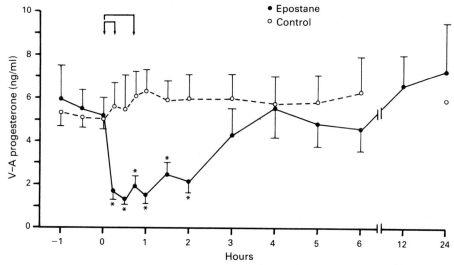

Fig. 2. The mean (±s.e.m.) venous–arterial concentration difference in PGFM across the uterus of pregnant mares infused (arrows) with epostane (N = 8, ●) or alcohol alone (N = 5, ○) over 15 or 40 min. Details of the doses are given in the text. *Significant reduction from 0 h values, $P < 0.05$.

declined to a minimum value 15–30 min after beginning the infusion but had returned to baseline levels by 3 h after infusion (Fig. 1a). The fall in arterial plasma progesterone was smaller and more transient than the venous change (Fig. 1b). Increasing the dose and duration of the epostane infusion prolonged the decrease in plasma progesterone in both the artery and uterine vein, but by only 1–2 h (Fig. 1). Preinfusion concentrations of progesterone were restored by 6 h after infusing the larger dose of epostane (Fig. 1). The maximum decreases in plasma progesterone were similar in the two groups of animals (Fig. 1). There were no significant changes in the plasma arterial or venous concentrations of progesterone in the control animals that were infused with ethanol alone (Fig. 1). Since the two doses of epostane produced essentially similar effects on plasma progesterone, the results from the two groups of experiments have been combined for all subsequent analyses.

The venous arterial (V–A) concentration difference in progesterone across the uterus decreased during epostane infusion and remained significantly depressed for 2 h after beginning the infusion (Fig. 2). No significant change was observed in the V–A progesterone concentration difference during the control infusions (Fig. 2).

Changes in other hormones

The plasma concentration of total unconjugated oestrogens in the uterine vein was reduced after administration of epostane but remained virtually unchanged in the control experiments (Fig. 3). In epostane-treated animals, oestrogen concentrations had fallen significantly by the end of the infusion and did not return to preinfusion values until 6 h afterwards.

Plasma concentrations of PGFM in the uterine vein did not change significantly in the control or epostane-treated animals. Arterial plasma cortisol levels rose in control animals in response to infusion of alcohol alone (Fig. 3). In the epostane-treated animals, plasma cortisol concentrations fell initially and then rose to values similar to those in the controls (Fig. 3).

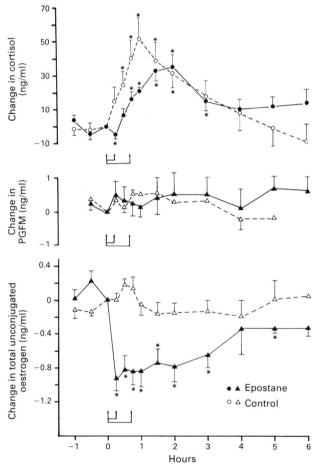

Fig. 3. The mean ($\pm$s.e.m.) changes in the plasma concentrations of uterine venous PGFM and total unconjugated oestrogens and of arterial cortisol in pregnant mares infused (arrows) with epostane (N $\geq$ 4, ●▲) or alcohol alone (N$\geq$4, ○△) over 15 or 40 min. Details of the doses are given in the text. *Significant change from 0 h values, $P < 0.05$.

Outcome of pregnancy

Epostane infusion did not induce delivery in any of the mares even at the highest dose infused (900 mg). The mean interval between the last infusion of epostane and delivery was $13·2 \pm 3·8$ days (N = 9). Even when epostane infusion was combined with fasting, which increases uterine prostaglandin production (Silver & Fowden, 1982), delivery did not occur.

Discussion

The results demonstrate that epostane causes temporary inhibition of 3β-HSD in the pregnant mare near term. An immediate fall in plasma progesterone occurred after epostane treatment, with a decrease in the V–A concentration difference in progesterone across the uterus. Since these changes did not appear to be accompanied by an increase in uterine blood flow, they suggest that the placental production of progesterone was reduced by epostane administration. However, the inhibitory effect of epostane on progesterone synthesis was transient: normal progesterone levels and V–A concentration differences were restored within 6 h of its administration.

In other species, 3β-HSD inhibitors have a more prolonged effect on progesterone synthesis (Schane *et al.*, 1979; Creange *et al.*, 1981; Taylor *et al.*, 1982). In the pregnant ewe, for instance, a single injection of trilostane at a dose comparable to that used in the present study reduced progesterone levels for up to 7 days (Jenkin & Thorburn, 1985). It is not clear why 3β-HSD inhibitors should have such a profound effect on progesterone production in the ewe, and yet have such a transitory action in the mare as the placenta is the source of progesterone in both species during late gestation. Epostane may be less potent in inhibiting 3β-HSD than trilostane but in the ewe both drugs are equally effective in reducing progesterone (Ledger *et al.*, 1985). It therefore seems likely that there are genuine species differences in the sensitivity of the steroidogenic pathways to 3β-HSD inhibitors.

The fall in total unconjugated oestrogens after epostane administration to the pregnant mare suggests that oestrogen synthesis is also dependent on the activity of 3β-HSD. Oestrogens are produced by a feto-placental unit in the mare with the fetal gonads and liver providing C_{19} precursors for biosynthesis to oestrogens in the placenta (Pashen & Allen, 1979). The conversion of dehydroepiandrosterone (DHA) to androstenedione is the biosynthesis step in the oestrogen pathway most likely to be effected by epostane as it is controlled by 3β-HSD (Pashen *et al.*, 1982). Since 3β-HSD is found in a number of sites within the feto-placental unit (Flood & Marrable, 1975), epostane may reduce oestrogen production by inhibiting placental 3β-HSD or by crossing the placenta and inhibiting 3β-HSD in fetal organs such as the adrenals which are known to produce androstenedione from DHA *in vitro* (Pashen *et al.*, 1982). Preliminary observations in other species suggest that epostane can cross the placenta and inhibit adrenal 3β-HSD in the fetus *in utero* (M. Silver & A. L. Fowden, unpublished observations).

Epostane might also be expected to reduce cortisol levels in the mare as 3β-HSD is involved in the initial stages of cortisol synthesis. However, plasma cortisol concentrations increased in the control and epostane-treated animals, indicating that alcohol stimulates the hypothalamic–pituitary–adrenal axis in the pregnant mare. There was an initial fall in cortisol in the epostane-treated animals which suggests that there may have been some inhibition of cortisol production which was subsequently overcome by the effects of alcohol. As epostane is a competitive inhibitor of 3β-HSD, an increase in ACTH in these circumstances might counteract the action of epostane and lead to a delayed rise in plasma cortisol. Epostane does not block glucocorticoid production in response to ACTH in non-pregnant monkeys and it may be that adrenal 3β-HSD is less sensitive to inhibition by epostane than the placental enzyme complex (Creange *et al.*, 1981). In contrast to the mare, infusion of alcohol alone into both sows and ewes near term did not stimulate the production of cortisol (M. Silver & A. L. Fowden, unpublished observations).

None of the mares delivered after epostane treatment even at the highest dose. There were no changes in plasma PGFM nor any other signs of parturient behaviour in the epostane-treated mares. In many species, prostaglandin (PG) synthesis and release depends on the progesterone: oestrogen ratio and hence a rise in oestrogen or a decrease in progesterone can lead to increased PG production and delivery (see Thorburn & Challis, 1979). The progesterone:oestrogen ratio did not change after epostane treatment in the pregnant mare because the progesterone and oestrogen concentrations fell in parallel. However, even at term, there is little apparent change in the ratio although PGFM levels do rise during delivery in the mare (Barnes *et al.*, 1978; Silver *et al.*, 1979; Pashen, 1984). In the pregnant ewe treated with trilostane, the fall in progesterone was severe with only minimal changes in oestrogen and so the decrease in the progesterone:oestrogen ratio was dramatic and resembled the normal pre-partum change (Taylor *et al.*, 1982). However, parturition did not always occur in pregnant ewes treated with epostane before 130 days even though the progesterone:oestrogen ratio was depressed for as long as 7 days (Jenkin & Thorburn, 1985). Since epostane invariably leads to premature delivery in pregnant ewes within 10 days of term (M. Silver, unpublished observations), it appears that other pre-partum changes must occur before 3β-HSD inhibitors can induce labour in the ewe. In monkeys and rats treated with epostane, premature delivery can be induced before term but only when there is a prolonged reduction in progesterone synthesis (Schane *et al.*, 1979; Creange *et al.*, 1981). These observations demonstrate that drugs which inhibit 3β-HSD are unlikely to be effective inducing agents in the pregnant mare, although they do provide a useful means of investigating steroidogenesis during equine pregnancy.

We thank Mrs P. Taylor, Miss T. Grimes and Mr P. Hughes for help during surgery and sampling; Mr D. Clarke and Mr I. Cooper for care of the animals; Mrs J. Knox and Mr K. Richardson for assistance with the biochemical analysis; and Mrs R. Cummings for typing the manuscript. We are indebted to the Horse Racing Betting Levy Board for their financial support.

References

Ainsworth, L. & Ryan, K.J. (1969) Steroid hormone transformations by endocrine organs from pregnant mammals. III Biosynthesis and metabolism of progesterone by the mare placenta *in vitro. Endocrinology* **84**, 91–97.

Armitage, P. (1971) *Statistical Methods in Medical Research.* Blackwell Oxford.

Barnes, R.J., Nathanielsz, P.D., Rossdale, P.D., Comline, R.S. & Silver, M. (1975) Plasma progestagens in fetus and mother in late pregnancy. *J. Reprod. Fert., Suppl.* **23**, 617–623.

Barnes, R.J., Comline, R.S., Jeffcott, L.B., Mitchell, M.D., Rossdale, P.D. & Silver, M. (1978) Foetal and maternal plasma concentrations of 13,14 dihydro-15-oxo-prostaglandin F in the mare during late pregnancy and at parturition. *J. Endocr.* **78**, 201–215.

Burns, S.J. & Fleeger, J.L. (1975) Plasma progestagens in the pregnant mare in the first and last 90 days of gestation. *J. Reprod. Fert., Suppl.* **23**, 435–439.

Comline, R.S., Hall, L.W., Lavalle, R. & Silver, M. (1975) The use of intravascular catheters for long term studies on the mare and fetus. *J. Reprod. Fert., Suppl.* **23**, 583–588.

Creange, J.E., Anzalone, A.J., Potts, G.O. & Schane, H.P. (1981) Win 32,799, A new, potent interceptive agent in rats and rhesus monkeys. *Contraception* **24**, 289–299.

Flood, P.F. & Marrable, A.W. (1975) A histological study of steroid metabolism in the equine fetus and placenta. *J. Reprod. Fert., Suppl.* **23**, 569–573.

Holtan, D.W., Squires, E.L., Lapin, D.R. & Ginther, O.J. (1979) Effect of ovariectomy on pregnancy in mares. *J. Reprod. Fert., Suppl.* **27**, 457–462.

Jenkin, G. & Thorburn, G.D. (1985) Inhibition of progesterone secretion by a 3β hydroxysteroid dehydrogenase inhibitor in late pregnant sheep. *Can. J. Physiol. Pharm.* **63**, 136–142.

Ledger, W.L., Webster, M.A., Anderson, A.B.M. & Turnbull, A.C. (1985) Effect of inhibition of prostaglandin synthesis on cervical softening and uterine activity during ovine parturition resulting from progesterone withdrawal induced by epostane. *J. Endocr.* **105**, 227–233.

Lovell, J.D., Stabenfeldt, G.H., Hughes, J.P. & Evans, J.W. (1975) Endocrine patterns of the mare at term. *J. Reprod. Fert., Suppl.* **23**, 449–456.

Pashen, R.L. (1984) Maternal and foetal endocrinology during late pregnancy and parturition in the mare. *Equine vet. J.* **16**, 233–238.

Pashen, R.L. & Allen, W.R. (1979) The role of the fetal gonads and placenta in steroid production, maintenance of pregnancy and parturition in the mare. *J. Reprod. Fert., Suppl.* **27**, 499–509.

Pashen, R.L., Sheldrick, E.L., Allen, W.R. & Flint, A.P.

(1982) Dehydroepiandrosterone synthesis by the fetal foal and its importance as an oestrogen precursor. *J. Reprod. Fert., Suppl.* **32**, 389–397.

Rossdale, P.D., Silver, M., Ellis, L. & Franenfelder, H. (1982) Response of the adrenal cortex to tetracosatrin ($ACTH_{1-24}$) in the premature and full term foal. *J. Reprod. Fert., Suppl.* **32**, 545–553.

Schane, H.P., Pott, G.O. & Creange, J.E. (1979) Inhibition of ovarian, placenta and adrenal steroidogenesis in the rhesus monkey by trilostane. *Fert. Steril.* **32**, 464–467.

Silver, M. & Fowden, A.L. (1982) Uterine prostaglandin F metabolite production in relation to glucose avail-ability in late pregnancy: the possible influence of diet on the time of delivery in the mare. *J. Reprod. Fert., Suppl.* **32**, 511–519.

Silver, M., Barnes, R.J., Comline, R.S., Fowden, A.L. & Clover, L. (1979) Prostaglandins in maternal and fetal plasma and in allantoic fluid during the second half of gestation in the mare. *J. Reprod. Fert., Suppl.* **27**, 531–539.

Taylor, M.J., Webb, R., Mitchell, M.D. & Robinson, J.S. (1982) Effect of progesterone withdrawal in sheep during late pregnancy. *J. Endocr.* **82**, 85–93.

Thorburn, G.D. & Challis, J.R.G. (1979) Endocrine control of parturition. *Physiol. Rev.* **59**, 863–918.

J. Reprod. Fert., Suppl. **35** (1987), 547–552

The pathogenesis of dystocia and fetal malformation in the horse

M. M. Vandeplassche

Department of Reproduction and Obstetrics, Faculty of Veterinary Medicine, State University, Gent, Belgium

Summary. From a total of 601 severe dystocias in mares, 408 (68%) of the fetuses were in anterior, 95 (16%) in posterior and 98 (16%) in transverse presentation, compared with 99%, 1% and 0·1% respectively for spontaneous parturitions. From the cases with anterior presentation, 151 (37%) showed reflected heads and necks. From the cases with posterior presentation, 47 (50%) presented hip flexions, 25% had hock flexions, and 25% had stretched hind legs, 45 (47%) of the fetuses were in lateral or ventral position, and 28 (30%) of the fetuses were malformed (mainly torticollis and head scoliosis). All 98 cases of transverse presentation were complete or partial bicornual gestations; 34 (35%) fetuses were malformed (mainly wryneck). The best adapted obstetrical method and the results obtained for dam and fetus by means of reposition, embryotomy and Caesarian section are considered. It is concluded that, in mares, posterior and transverse presentations are important causes of fetal malformation and of dystocia.

Introduction

It is surprising how many publications concerning obstetrics in cattle have appeared compared with the sparse literature for equine obstetrics. The aim of this study was to gather a large number of cases of dystocia in mares in which detailed observations have been made, particularly concerning abnormal presentation. It was also intended to compare the obstetrical results for the dam and for the fetus obtained with reposition, embryotomy or Caesarian section.

Materials

From 1941 until 1985 about 1000 mares in dystocia were sent to the clinic by practitioners. Detailed information concerning breed, parity, history, symptoms, diagnosis, treatment and results was obtained from 601 mares in severe dystocia. The majority (63%) were Belgian draught mares; the remainder were warmblood saddle horses (16%), halfblood trotters (11%), Haflinger and ponies (8%) and Thoroughbreds (2%). Of the 601 mares, 186 (31%) were experiencing their first parturition, 102 (17%) were at their second parturition and the rest (52%) were multiparous. There is clearly a higher likelihood of dystocia in primigravid mares.

Observations

Anterior fetal presentation

Of the 601 dystocic mares, 408 (68%) delivered with a foal in anterior presentation: 237 (58%) had a difficult parturition because of the reflected head and, eventually, limbs. Other causes of the dystocia included abnormal posture, malformed front limbs, ventral position, oversized fetus, uterine torsion, inertia uteri, hydrocephalus, prolapse of bladder or rectum, and hydro-allantois. Of the 237 fetuses with reflected head and neck, 85 (36%) suffered from head scoliosis and torticollis (wryneck). The delivery methods are shown in Table 1. That so many of the fetuses with reflected heads were delivered by embryotomy is due to the fact that many fetuses were dead and/or had wryneck malformation: of the foals delivered by fetotomy, 40% needed only 1 section, 30% needed 2 and 30% 3 or more cuts (up to 7 in cases of total embryotomy due to uterine involution, oversized or malformed fetus). However, reposition succeeded in many cases under epidural anaesthesia and when the uterus was relaxed by an injection of isoxsuprin (Duphaspasmin Duphar): the use of eyehooks (even on living foals) often gave valuable help.

Table 1. The delivery methods applied for 237 fetuses with reflected heads, in mares with anterior presentations

	No. of foals	Foals born alive
Reposition	64 (27%)	42
Embryotomy		
(mostly 1 or 2 sections)	154 (65%)	—
Caesarian section	19 (8%)	8
Total	237 (100%)	50 (21%)

Table 2. Osbstetrical methods and survival rate of 95 foals born in posterior presentation from mares in dystocia

	No. of foals	Foals born alive
Reposition and subsequent traction	20 (21%)	4
Embryotomy		
(mean of 2·8 sections)	44 (46%)	—
Caesarian section	12 (13%)	7
Traction	19 (20%)	6
Total	95	17 (18%)

Table 3. Delivery methods and survival rate of foals in transverse presentation born from 98 mares in dystocia

	No. of foals	Foals born alive
Caesarian section	46 (47%)	17
Embryotomy		
(average of 3·3 sections)	47 (47%)	—
Reposition and subsequent traction	5 (6%)	1
Total	98	18 (18%)

Posterior fetal presentation

Of the 95 fetuses (16%) with posterior presentation, 47 (50%) had hip flexion (94% with both limbs), 24 (25%) had hock flexion (90% with both limbs) and 24 (25%) had stretched hindlimbs. It is surprising that 45 (47%) of the fetuses were found in a lateral or in a ventral position, thus complicating fetal expulsion even more. Malformations of the skeleton, mainly of the head and/or limbs, occurred in 28 (30%) of the fetuses. Heavy abdominal straining had caused prolapse of the bladder and/or of the rectum in some mares, while absence of straining was seen in 5 cases of bilateral hip flexions and advanced emphysema of the fetuses. The applied methods for delivery and the obstetrical fetal results are presented in Table 2.

Transverse fetal presentation

Of the 98 mares with transverse fetal presentations, the fetus was completely in both uterine horns in 47 (i.e. no fetal parts had entered the birth tract). Incomplete bicornual gestation was considered to be when one or more limbs, or more exceptionally the head of the fetus, was found well advanced in the birth canal while the trunk was located in both uterine horns. The obstetrical methods applied to deliver these fetuses are presented in Table 3. Reposition is only possible when either the fore-quarters or the hind-quarters of a relatively small fetus have entered the birth canal. In about half of the mares subjected to fetotomy, 3–10 sections were required due to an oversized or malformed fetus, an elongated narrow birth canal or early involution. Several of those fetuses would probably have been better delivered by Caesarian operation. In 34 (35%) of the fetuses malformation of limbs and especially of the head and neck (wryneck) were observed.

Discussion

In a normal population of pregnant mares, 99% foals have an anterior presentation at birth and 1% have a posterior presentation (Harms, 1924; Benesch, 1952; Vandeplassche, 1957; Rozenberger & Tillmann, 1978), while only about 0·1% of mares at term carry their fetus in transverse presentation (Vandeplassche, 1957). When these figures are compared with the findings in the 601 cases of dystocia in this study, it appears that the incidence of posterior presentation is increased by about 16 times and that of transverse presentation by nearly 160 times. In an earlier study comprising 57 cases of dystocia, Tapken (1906) found 8 foals (14%) in posterior presentation.

For a better understanding of the development of fetal presentation one has to investigate what happens in the consecutive stages of pregnancy. Until about 7 months of gestation the amnion and the fetus remain confined to the pregnant uterine horn which is still in a greatly transverse position to the uterine body and the longitudinal axis of the genital tract. The fetal head is directed in 50% of the mares towards the apex of the pregnant horn and for 50% towards the base of that horn. At later stages of gestation, the latter could be considered as an anterior fetal presentation (Vandeplassche, 1957). From 7–8 months of pregnancy on, the amnion and fetus extend into the uterine body and gradually force the transverse position of the pregnant horn into a longitudinal one, resulting in the anatomically and functionally tubular unit of the pregnant horn and the uterine body.

An important observation from 3 months of gestation onwards is the high incidence of twisted umbilical cords (in 78% of the fetuses with an average of 4·4 twists per cord (Vandeplassche, 1957; Vandeplassche & Lauwers, 1986). These twists occur in both parts of the umbilical cord and develop independently one from another, and can only be explained by rotations either of the fetus within the amnion, or/and of the amnion and fetus within the allantois.

Another astonishing fact is that from 8 months pregnancy on, 99% of the fetuses show an anterior presentation with their heads directed towards the cervix, a presentation which persists until parturition. The only possible explanation is that, at about 7–8 months gestation, an extremely preferential and final rotation of the fetuses occurs, leaving only 1% in a longitudinal posterior presentation. In cattle, which normally have 95% anterior presentations, this preferential final rotation seems to fail to a large extent in cases of twin gestation: of a total of 162 twin births, Jöhnk (1951) found both calves in an anterior presentation in 41% of cows, in posterior presentation in 22%, and one in anterior and the other in posterior presentation in 37%. Comparative data for twin parturitions in mares are unfortunately not available in the literature. Reliable information has been obtained from a limited number of 14 twin births close to or at term: both foals were born in anterior presentation in 6 cases, in posterior presentation in 3 cases, and one anterior and the other posterior in 5 cases, indicating a disturbed mechanism of the final rotation. It seems logical that lack of space in the uterus could be an important aetiological factor. The final rotation mechanism also seems to fail in mares suffering from severe hydro-allantois (Vandeplassche *et al.*, 1976); for the 8 published records and another 8 patients since 1976, 5/16 of the fetuses were extracted in posterior presentation. It could be that the always concomitant inertia of the uterine body was not overcome.

Table 4. Pathogenesis of equine dystocia in 95 cases with the fetus in posterior presentation

Defective dilatation of birth tract (cervix, vagina, vulva)
High incidence of abnormal
   position: 47% markedly lateral or ventral position
      (compared with a frequency of 16% for
      fetuses in anterior presentation)
   posture: hip flexion (50%)
      hock flexion (25%)
      stretched hindlimbs (25%)
High percentage of fetal malformations: 30% (head,
  neck, limbs)
Low fetal survival rate (prolonged parturition and
  involution, umbilical cord around a hindlimb)
High incidence of complications: prolapse of bladder
  and/or rectum, inertia uteri, emphysematous fetus

Table 5. Pathogenesis of dystocia observed in 98 mares with the fetus in transverse presentation

Ineffective labour blocking the fetus within both horns
Weak abdominal straining
Defective relaxation of cervix, vagina, vulva
Fetal malformations: 35% of the cases
Oversized fetus (large placenta)
Underdeveloped uterine body

Still more exceptional is the extension of the amnion and fetus into the opposite horn instead of protruding into the uterine body, and resulting in about 0·1% bicornual pregnancies carrying the fetus in a transverse presentation. In such cases, the development of the uterine body varies from normal to almost nothing (Vandeplassche, 1957).

Another intriguing observation is that 147/601 (24%) of the fetuses suffered from torticollis and head scoliosis, occasionally combined with malformation of one or more limbs. By chance, one would expect, for a population of parturient mares, that 99% of the skeletal malformations would occur in fetuses in anterior, 1% in posterior and 0·1% in transverse presentation. However, in the 601 dystocic mares, only 85 (58%) malformations occurred in fetuses in anterior, 28 (19%) in posterior, and 34 (23%) in transverse presentation. This means that posterior and particularly transverse fetal presentations are associated with a tremendously increased incidence of fetal malformations. The question can be asked here whether the abnormal presentations cause the fetal malformations, or whether malformations provoke the abnormal presentations. In the latter, the malformations would have to be considered as caused by a hereditary factor. However, in a recent study, a genetic origin of wryneck in horse fetuses is shown to be unacceptable in most of the cases (Vandeplassche & Lauwers, 1986). What is the mechanism of that dramatic incidence of wry head and neck often combined with malformed front limbs? For fetuses in normal anterior presentation, the head, neck and forelegs become comfortably located in the broad uterine body while for fetuses in posterior or transverse presentation, the massive frontquarters easily become compressed in the narrow apex of a uterine horn with a high risk of malformations (Vandeplassche, 1957). It is logical that fetal malformation could occur if the amnion and fetus have extended from the pregnant uterine horn into the uterine body or into the 2nd horn. The question could be asked whether severe fetal malformations of the head, neck and limbs can develop within the last 3 months of gestation? This relatively short period of time must be sufficient for the development of the lesions, since severely malformed foals born by Caesarian section have recovered, anatomically and functionally within a few weeks after birth (Vandeplassche *et al.*, 1984).

Posterior and transverse presentations of horse fetuses both have to be considered as abnormal (Williams, 1951) because they not only provoke an enormous increase in fetal malformation but also an extremely marked augmentation of dystocia due to a disturbed mechanism of the parturition process. The main aetiological factors for dystocia are summarized in Tables 4 and 5.

The fact that both hindlimbs were involved in 94% of the hip-flexion, and 90% of the hock-flexion, postures indicates that the abnormal posture was not a sign of a simple accident but that it was caused by a failure in the stretching mechanism. However, any explanation for this trouble must remain speculative.

Rather disappointing is the low percentage of foals born alive in posterior (17/95) or in transverse (18/98) presentation. This is mainly caused by the fact that in breech presentations, and particularly in transverse presentations, pressure stimuli are missing and only weak abdominal straining occurs so that the manager often is unaware of the fact that his mare is foaling. An additional loss of time is due to transport to the clinic, with advancing involution and regressing placental blood supply to the fetus. For a fresh living fetus in hock-flexion, reposition is likely to be successful after relaxation of the uterus by tranquillization with or without anaesthesia, elevation of the mare's hindquarters, smearing of the fetus with Vaseline (deep intrauterine infusions of lubricants are contraindicated because there may be a risk of them being swallowed by the fetus). Whenever possible, the delivery should be made by two experienced obstetricians. When reposition trials are not readily successful, Caesarotomy is indicated. The latter method is also justified for living and even for dead fetuses in complete or partial bicornual transverse presentation. It is true that one has to take in consideration the fact that a high percentage of foals are malformed. However, it has been shown that head scoliosis, torticollis and limb malformations can recover completely in vital foals which move intensely especially during the first days *post partum* (Vandeplassche *et al.*, 1984).

References

Benesch, F. (1952) *Lehrbuch der tierärztlichen Geburtshilfe und Gynäkologie.* Urban & Schwarzenberg, Wien.

Harms, F. (1924) *Lehrbuch der tierärztlichen Geburtshilfe,* 6th edn. Eds J. Richter, J. Schmidts & R. Reinhardt. R. Schoetz, Berlin.

Jöhnk, M. (1951) Über fetale Schieflagen und Zwillingsgeburten beim Rind. *Berl. Münch. tierärztl. Wschr.* **64,** 4–9.

Rosenberger, G. & Tillmann, H. (1978) *Tiergeburtshilfe,* 3rd edn. P. Parey, Berlin.

Tapken, A. (1906) Ueber Geburtshilfe beim Pferde. *Mh. prakt. Tierheilk.* **18**, 148.

Vandeplassche, M. (1957) The normal and abnormal presentation, position and posture of the foal-fetus during gestation and at parturition. *Vlaams Diergeneesk. Tijdschr.* **26**, Suppl. 1, 1–68.

Vandeplassche, M., Bouters, R., Spincemaille, J. & Bonte, P. (1976) Dropsy of the fetal sacs in mares. *Vet. Rec.* **99**, 67–69.

Vandeplassche, M., Simoens, P., Bouters, R., De Vos, N. & Verschooten, F. (1984) Aetiology and pathogenesis of congenital torticollis and head scoliosis in the equine fetus. *Eq. vet. J.* **16**, 419–424.

Vandeplassche, M. & Lauwers, H. (1986) The twisted umbilical cord: an expression of kinesis of the equine fetus? *Anim. Reprod. Sci.* **10**, 163–175.

Williams, W.L. (1951) *Veterinary Obstetrics*, 4th edn. E.W. Plimpton Worcester, MA.

Printed in Great Britain

J. Reprod. Fert., Suppl. **35** (1987), 553–564
© 1987 Journals of Reproduction & Fertility Ltd

Electromyographic properties of the myometrium correlated with the endocrinology of the pre-partum and post-partum periods and parturition in pony mares*

G. J. Haluska†, J. E. Lowe‡ and W. B. Currie

Department of Animal Science, and ‡Department of Clinical Sciences, Cornell University, Ithaca, New York 14853-4801, U.S.A.

Summary. A complete set of electromyographic recordings, plasma samples and behavioural observations were collected from 2 mares beginning 7 days *pre partum*, through parturition and into the early post-partum period. During the week *pre partum*, EMG activity was elevated, occurring 26–73% of the time. Activity was least during the day and greatest at night with no significant difference for the hours of the day or between days *pre partum*. During the 24 h before delivery, EMG activity was increased for 7–13 h (55–80%) during the daylight hours. EMG activity decreased 2–4 h immediately preceding delivery of the foal, with an abrupt increase at rupture of the chorioallantois. At delivery, EMG activity consisted of events containing a series of 10–13 discrete bursts of increasing amplitude occurring in rapid succession. After fetal delivery there was a reduction in activity until placental delivery followed by very long (2–22 min) trains of potentials.

Introduction

Numerous accounts of parturition of horses exist in the literature (Wright, 1943; Jeffcott, 1972) as well as in texts (Roberts, 1971; Rossdale & Ricketts, 1980). In many of these sources a gross description of uterine contractility is provided, but no quantitative index exists of uterine activity during the pre-partum, parturient and post-partum periods for the mare.

The evolution of labour culminating in the delivery of the foal is remarkable considering the endocrine milieu of the mare during this period. Progesterone is reported to be increasing in the maternal peripheral circulation (Holtan *et al.*, 1975) and decreasing in the fetal circulation (Barnes *et al.*, 1975) as parturition approaches. However, progesterone does decrease significantly in the peripheral maternal circulation over the last 24 h preceding delivery (Haluska, 1985). Oestrogens continue to decrease from elevated mid-pregnancy values until delivery (Nett *et al.*, 1973), and there is no consistent change in the level of oestradiol-17β during the last week *pre partum* (Haluska, 1985). Relaxin, a known myometrial inhibitor, is elevated during the last half of pregnancy (Stewart & Stabenfeldt, 1981), and concentrations begin to increase above normal pregnancy levels before delivery (Stewart *et al.*, 1982). Concentrations of 13,14-dihydro-15-keto-prostaglandin F-2α (PGFM), reflecting PGF-2α, a myometrial agonist, are elevated in the maternal peripheral circulation during the last 2 weeks *pre partum* (Barnes *et al.*, 1978; Stewart *et al.*, 1984; Haluska, 1985).

*Reprint requests to: W. B. Currie.
†Present address: Oregon Regional Primate Research Center, 505 N.W. 185th Avenue, Beaverton, Oregon 97006, U.S.A.

Both from a managerial point of view and from the study of uterine physiology, a quantitative assessment of uterine contractility in the mare would seem to be essential. Uterine electromyographic (EMG) activity has been shown to be an effective measure of contractile activity in other species at parturition (Krishnamurti *et al.*, 1982). Toutain *et al.* (1983) demonstrated a clear relationship between electrical and mechanical activity in the ewe by using EMG electrodes and implantable strain gauges. This report will present data on myometrial EMG activity in the mare during the pre-partum, parturient and post-partum periods.

Materials and Methods

Animals

Five pony mares of mixed breeding were housed in tie stalls with water available *ad libitum*. They were fed hay at 08:00 and 18:00 h and a concentrate food at 18:00 h daily. The mares were mated to a pony stallion in a controlled environment to ensure insemination. Pregnancies were dated from the last day of standing oestrus (Day 0). All mares received 2 ml of Pneumabort K (Fort Dodge Labs, Inc.) during the 5th, 7th and 9th months of pregnancy as a prophylactic against possible abortion due to equine herpesvirus 1.

Because the mares were mated from the end of the breeding season, a regimen of 16 h light, 8 h darkness was used to ensure continued oestrous cyclicity of the mares until all were pregnant. Lighting consisted of 200-W incandescent bulbs between each of 2 mares (140–346 lux at eye level). Lights, controlled by an automatic timer, were on between 05:00 and 21:00 h daily throughout the experimental period.

Electromyography

Surgery. Each mare had 4 sets of bipolar electrodes surgically implanted in her uterus. Surgery was performed on Days 99, 129, 132, 134 or 163 of pregnancy. The mares received a preliminary i.v. anaesthetic of ketamine (Ketaset: Bristol Laboratories, Syracuse, NY) and xylazine (Rompun: Bayvet, Shawnee, KS) and, after an endotracheal tube was inserted, anaesthesia was maintained during surgery with halothane in oxygen. The uterus was exposed by mid-ventral laparotomy and the pregnant horn was located.

Bipolar electrodes, each consisting of two Teflon-coated multistranded stainless-steel wires (AS 633, Cooner Wire Co., Chatsworth, CA) were sutured into the myometrium and the wires of each set were located 5 mm apart. The electrodes were placed in the pregnant horn near the tubo-uterine junction, above the bifurcation, in the uterine body below the bifurcation, and as close as possible to the cervix. The Teflon coating was removed from 2 mm of the wire that was embedded in the myometrium. The 8 wires, plus an indifferent electrode (free in the peritoneal cavity), were tracked subcutaneously to an area above the flank and exteriorized.

Electromyographic recording. Each electrode set was joined to the appropriate lead off a Grass polygraph. EMG signals were amplified through Grass AC preamplifiers (Model 7P5) and the signals were transcribed to the oscillograph at a chart speed of 10–100 mm/min as circumstance demanded. Signal filters were set at 10 and 35 Hz. Gain varied between electrodes but was constant for each individual electrode throughout the experiment.

The recording area consisted of two stalls in an isolated area of the barn where there was minimal environmental interference. The stalls were separated by a room which housed the polygraph and where active monitoring of the behaviour of the mares was possible without interference.

Electromyographic analysis. About 2 weeks before expected delivery each mare was moved to one of the recording stalls until Day 1 *post partum*. EMG activity was monitored continuously (24 h/day) until parturition. EMG signal analysis was divided into three categories. The primary category consisted of quantification of the duration of voltage burst activity. Individual events were designated as bursts separated by at least 5 sec. Burst activity was subdivided into four types: (a) diffuse activity (single voltage spikes with a time separation ranging from 5 to 30 sec); (b) short bursts (trains of single spikes lasting less than 7 sec); (c) medium bursts (trains of single spikes lasting between 7 and 120 secs); and (d) long bursts (trains of short and medium bursts not separated by 5 sec and lasting more than 120 sec). Periods of inactivity were designated as quiet. The data were expressed as the percentage of time each type of activity occurred during the hours analysed.

The secondary category incorporated an analysis of frequency and amplitude of EMG activity. Frequency was expressed as the number of each type of burst activity per hour. Amplitude of bursts was divided into three subgroups (low, $< 50 \mu V$; intermediate, $50–100 \mu V$; and high, $> 100 \mu V$) and expressed as the number of bursts/h in each subgroup. The tertiary category of analysis examined diurnal fluctuations in percentage activity, amplitude, frequency and duration of bursts. A total of 6 h/day was analysed for Days 6 to 1 *pre partum*. The periods examined were 01: 00–02:00, 05:00–06:00, 09:00–10:00, 13:00–14:00, 17:00–18:00, and 21:00–22:00 h, each of which began 1 h after a blood sample was taken. Beginning 24 h before rupture of the chorio-allantoic membranes, each individual hour was analysed. EMG data were collected during the first day *post partum*.

Statistics

Significant differences between the means of hours of the day and days *pre partum* during the 6 days before the day of parturition were determined by two-way analysis of variance. Linear contrasts using Sheffe's adaptation were made to determine significance between individual or group means.

Results

Of the 5 mares 3 were monitored continuously over the last 7 days *pre partum*, during parturition, and during the early post-partum period. One of the mares, No. 137, delivered her foal unobserved and the precise time of delivery could not be determined. Continuous, reliable EMG recordings were obtained from 2 mares, Nos 203 and 100, along with behavioural observations. Dystocia occurred in Mare 100 and veterinary assistance was required. Mare 203 had a completely normal delivery. The EMG recordings of these 2 mares (Nos 203 and 100) were therefore used to quantify the levels of activity during the pre-partum and parturient stages of pregnancy. Day 0 is the day preceding the night of parturition. Post-partum EMG activity was determined from the recordings of Mares 203 and 100. Diffuse activity showed no consistent pattern and was considered to be of no physiological importance. Short burst activity was always less than 10% total activity and its occurrence mimicked the pattern of % Quiet and so the greatest percentage of short burst activity occurred when the uterus was quiescent.

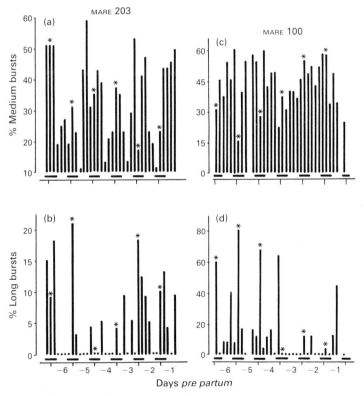

Fig. 1. Myometrial EMG activity for 2 mares during the 6 days before the day of parturition: (a) % medium bursts in Mare 203, (b) % long bursts in Mare 203; (c) % medium bursts in Mare 100; and (d) % long bursts in Mare 100. The stars above the bars indicate the period of 01:00–02:00 h of each day. The horizontal bars above the abscissa indicate night or dark periods.

Days pre partum

The percentage of time that the uterus was quiet, and medium and long burst activity occurred, is illustrated in Fig. 1 for Mares 203 and 100 from Day -6 to Day -1. The uterus of Mare 203 was most quiescent during the hour 13:00–14:00, averaging $64.2 \pm 6.4\%$. The uterus of Mare 100 was least quiescent at 01:00–02:00 h ($24.2 \pm 8.2\%$) and was most quiet at 05:00–06:00 h ($49.2 \pm 4.7\%$). The means of %Q for each of the 6 analysed hours were not significantly different from each other when compared by analysis of variance.

The means of % medium burst and % long burst activity were not significantly different for either mare when the hours of the day were compared. However, there was a trend toward more activity in the early morning hours. For Mare 203, the sum of % medium + long burst activity

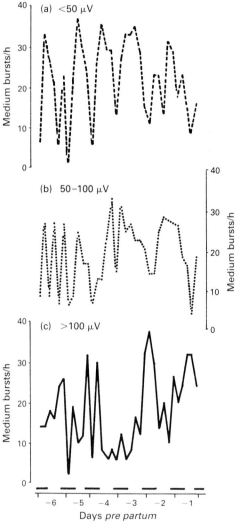

Fig. 2. The amplitude of medium bursts for Mare 100 during the 6 days before the day of parturition. Medium bursts/hour have been divided into low-amplitude bursts ($<50\,\mu V$) intermediate-amplitude bursts ($50–100\,\mu V$), and high-amplitude bursts ($>100\,\mu V$). The horizontal bars above the abscissa indicate night or dark periods.

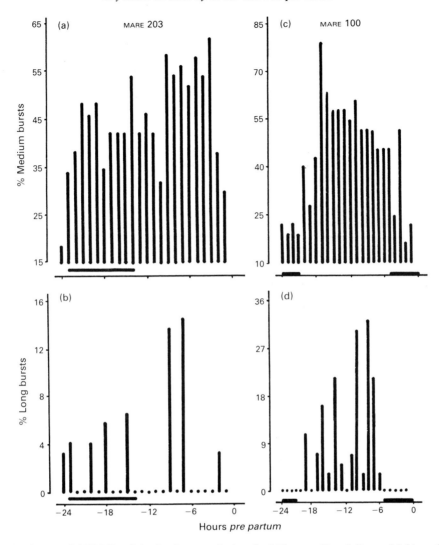

Fig. 3. Myometrial EMG activity for 2 mares during the 24 h preceding delivery: (a) % medium bursts; in Mare 203; (b) % long bursts in Mare 203; (c) % medium bursts in Mare 100; and (d) % long bursts in Mare 100. The bars above the abscissa indicate night or dark periods.

was at its highest level in the early morning hours averaging, $41 \cdot 2 \pm 4 \cdot 5\%$ at 01:00–02:00 h, and $45 \cdot 8 \pm 6 \cdot 0\%$ at 05:00–06:00 h. The myometrium of Mare 100 exhibited greatest activity at 01:00–02:00 h, averaging $73 \cdot 0 \pm 9 \cdot 4\%$ and, in general, was consistently more active than the myometrium of Mare 203, due primarily to higher levels of long-burst activity.

There was a positive correlation between % medium-burst activity and the frequency of medium bursts in Mare 203 ($r = 0 \cdot 88$, $P < 0 \cdot 01$) and in Mare 100 ($r = 0 \cdot 81$, $P < 0 \cdot 01$). The frequency of medium bursts for Mare 203 increased from Day -6 to $61 \cdot 0 \pm 9 \cdot 9$ bursts/h on Day -1. All amplitude sub-categories of medium bursts showed a progressive increase toward the day of parturition. The frequency of medium bursts for Mare 100 reached its highest level on Day -3 ($66 \cdot 3 \pm 1 \cdot 6$ bursts/h) and declined slightly to $57 \cdot 6 \pm 3 \cdot 8$ bursts/h on Day -1. Low, intermediate and high amplitude medium bursts followed the general pattern of total frequency for both mares until Day − 1 when a larger proportion of medium bursts were of high amplitude, as illustrated in Fig. 2 for Mare 100. There was no relationship between percentage or frequency of medium bursts and duration for either mare.

Hours pre partum

Myometrial % medium burst activity for Mare 203 was greatest during the daylight hours on the day of parturition, averaging 54·7 ± 1·4% during the 7 h from 9 to 3 h *pre partum* (Fig. 3a). These hours correspond to 11:00–17:00 h clock time. The % long burst activity was also greatest during the same period, reaching 14·0% at 7 h *pre partum* (Fig. 3b). During the last 2 h before delivery in this mare, there was a decrease in % medium burst activity to 37·5% at 2 h and 28·0% at 1 h *pre partum* (Figs 5a & b).

The % activity for Mare 100 during the last day *pre partum* was similar to that for Mare 203. The greatest % medium burst activity for Mare 100 occurred during the daylight hours of Day 0, averaging 52·9 ± 2·4% during the 13 h from 17 to 5 h *pre partum* corresponding to clock time of 09:00 to 22:00 h (Fig. 3c). At −4 h there was a decrease in % medium burst activity to 21·5% with a transient increase to 48·0% during the 3rd hour *pre partum* and a further drop to 12·0% and 18·4% at 2 and 1 h *pre partum*, respectively. Representative EMG recordings are shown in Fig. 6. The % long burst activity was also greatly increased during the period from 17 to 5 h *pre partum*, reaching 31·8% at 8 h *pre partum* (Fig. 3d). However, in both mares, % long burst activities was not consistently elevated.

There was a strong positive correlation between % activity and frequency of burst for medium and long bursts and an increase in the amplitude of medium burst activity during the period when % medium burst was consistently higher over the last 24 h before delivery (Fig. 4).

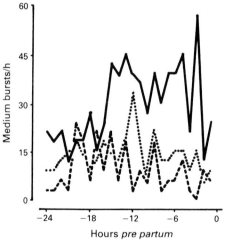

Fig. 4. The amplitude of medium bursts for Mare 100 during the 24 h before delivery of the foal. Medium bursts/h have been divided into low-amplitude bursts, < 50 μV (– – – –), intermediate-amplitude bursts, 50–100 μV (· · · · ·), and high-amplitude bursts, > 100 μV (——).

Parturition

The electrical activity of the myometrium at delivery was not quantified but will be described qualitatively. There was a great difference in the level of activity at delivery in Mares 203 and 100, with greater activity in Mare 100. The length of delivery for Mare 203 was 14 min from rupture of the chorioallantois to fetal delivery and 44 min for Mare 100 which had a malpresentation dystocia.

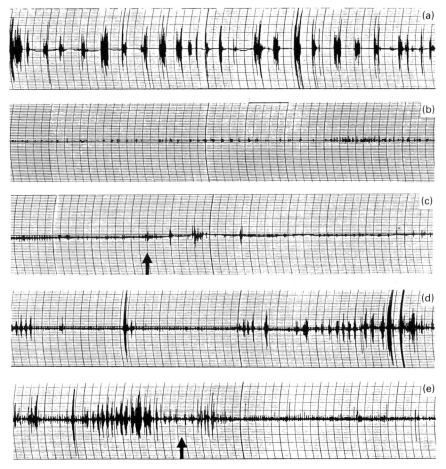

Fig. 5. Myometrial EMG recordings from Mare 203. Electrode position was above the bifurcation in the pregnant horn. Sensitivity, 50 µV/cm; chart speed, 25 mm/min (a, b), 100 mm/min (d, e, f): (a) 3 h before delivery of the foal, % medium bursts = 61·5%; 1 h before delivery, % medium bursts = 28%; (c) at rupture of the chorioallantois (arrow); (d) 2 min before complete delivery; (e) at complete delivery (arrow).

At 5 min before rupture of the chorioallantois of Mare 203, uterine EMG activity was still depressed and resembled the activity that prevailed during the previous 2-h period (Fig. 5b). Figure 5(c) shows that there was little change in activity when the allantoic fluid was released. There was a slight increase in EMG activity preceding the emergence of the front feet of the foal, with a further increase in activity during the passage of the fetal head through the lips of the vulva. All activity to this point had been irregular, following no particular pattern. The activity 2 min before the delivery of the fetus is shown in Fig. 5(d) and consisted of a series of 10–15 bursts separated by 3–6 sec with the first burst in the series having an amplitude of about 25 µV which progressively increased to 300 µV for the last burst of the series. At delivery of the fetus (Fig. 5e) there was a similar series of bursts of activity which were not separated by short periods of inactivity. Immediately after fetal delivery, myometrial EMG activity markedly decreased until after the delivery of the placenta 12 min later.

The level of EMG activity was greater for Mare 100 than for Mare 203 as can be seen in a comparison of Figs 5 and 6. At 8 min before rupture of the chorioallantois (Fig. 6c) the

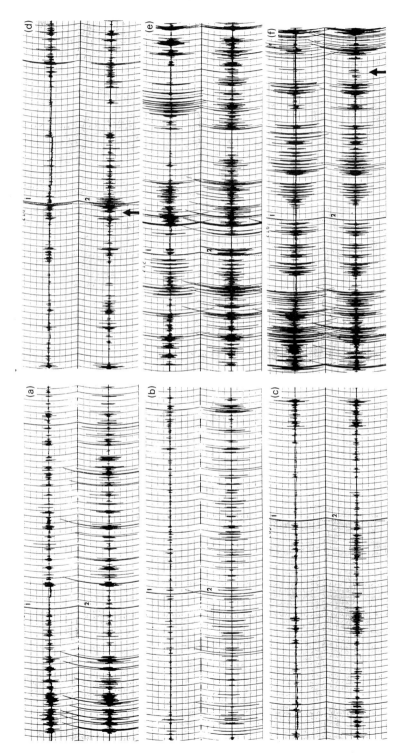

Fig. 6. Myometrial EMG recordings from Mare 100. Electrode positions: (1) near the tubouterine junction in the pregnant horn; (2) above the bifurcation in the pregnant horn. Sensitivity 100 µV/cm, chart speed 25 mm/min (a, b), 50 mm/min (c–f): (a) 5 h before delivery of the foal, % medium bursts 43·2%; (b) 1 h before delivery, % medium bursts = 18·4%; (c) 8 min before rupture of the chorioallantois; (d) at rupture of the chorioallantois (arrow); (e) 12 min after rupture of the chorioallantois as the dystocia was recognized; (f) at fetal delivery (arrow).

myometrium of Mare 100 exhibited relatively erratic activity when compared to the previous hours (Figs 6a & b). At the release of the allantoic fluid (Fig. 6d), there was a transient burst of activity and a subsequent change from erratic-burst activity to the more precise synchronous bursts of increasing amplitude as described for Mare 203 at parturition (Fig. 6e). During the dystocic period, the activity returned to the more erratic pattern seen immediately before the rupture of the chorio-allantois. However, as the foal was pushed back into the uterus to be repositioned, regular bursts of increasing amplitude, synchronous at all electrode sites, returned (Fig. 6f). Immediately following the delivery of the foal, the EMG recordings returned to the pre-delivery state of inactivity until the delivery of the placenta 20 min after fetal delivery.

Hours and days post partum

The percentage of myometrial EMG activity for selected hours *post partum* is listed in Table 1. During the period from fetal to placental delivery there was a great reduction in all activity. During the hours after placental delivery there was a rapid increase in percentage activity due primarily to an increase in % long burst activity. Qualitatively, the activity consisted of long trains of action potentials which ranged from 2 to 22 min in duration and averaged about 5 min. Over the next few days there was a decrease in % long bursts as medium burst activity again became the predominant type of EMG activity.

Table 1. Myometrial EMG activity during the post-partum period in 2 mares

Time relative to delivery	Mare	% quiescent	% short bursts	% medium bursts	% long bursts
Fetal to placental	203	48·7	1·6	49·7	0
	100	77·6	7·8	14·6	0
+1 h	203	34·4	6·0	59·6	0
	100	3·2	0·4	16·6	79·7
+3 h	203	5·9	0·8	32·5	60·8
	100	1·7	0	1·3	97·0
+5 h	203	38·5	5·6	19·2	36·7

Discussion

The increase in myometrial electrical activity during the daylight hours preceding parturition (Stage I of labour) and the drastic decrease in activity during the last few hours before delivery (Stage II of labour) are interesting and novel observations.

In domestic animals, uterine EMG activity consists of long bursts of activity (5–6 min) during pregnancy with a progression to medium bursts (20–30 sec) during labour and delivery. Medium burst activity for Mare 203 was consistently elevated with sporadic episodes of long burst activity for a 7-h period which occurred from 9 to 3 h *pre partum* while elevated activity for Mare 100 lasted for 13 h from −17 to −4 h. There was no change in the duration of bursts during the 24-h period for either mare, but frequency and amplitude of medium bursts increased, i.e. there was an increase in the percentage of time that medium bursts occurred, due to an increase in the frequency of high-amplitude bursts. This is consistent with the period described as labour in rabbits (Csapo & Taketa, 1965), sheep (Toutain *et al.*, 1983), goats (Taverne & Scheerboom, 1984), cows (Zerobin & Sporri, 1972) and pigs (Taverne *et al.*, 1979a). In both mares, this period of consistently elevated activity began during the early daylight hours of the day of parturition (Day 0). The period of labour may be regulated by a photoperiodic cue once the necessary endocrine changes have occurred.

Paradoxically, there is a transient phase of myometrial quiescence separating labour from delivery. This phase of inactivity lasted for 2 h in Mare 203 and for 4 h in Mare 100. It has been shown that there is a surge, above pre-partum levels, of relaxin, a known myometrial inhibitor, which begins before delivery and peaks shortly after delivery (Stewart *et al.*, 1982). Perhaps relaxin is responsible for this transient decrease in spontaneous myometrial activity.

Even though relaxin has been shown to decrease spontaneous myometrial activity in other species, it does not prevent the rat myometrium from responding to oxytocin and PGF-2α, albeit with reduced frequency and amplitude (Cheah & Sherwood, 1981). In fact, relaxin has been shown to enhance the rate of rise in intrauterine pressure cycles of spontaneous and agonist-driven contractions (Downing *et al.*, 1980). This indicates that relaxin may decrease spontaneous activity while co-ordinating agonist-driven smooth muscle contraction of the uterus. It is obvious, from the lack of activity in Fig. 5 that there is little change in myometrial activity even though the concentrations of both PGFM and oxytocin have begun to increase (Haluska, 1985). It may be that relatively high concentrations of these myometrial agonists are necessary to overcome the effect of relaxin on smooth muscle cells. Relaxin increases intracellular cAMP (Judson *et al.*, 1980; Cheah & Sherwood, 1980) and decreases myosin light chain kinase activity of myometrial smooth muscle cells (Nishikori *et al.*, 1983), and Krall & Korenman (1977) have demonstrated that increased intracellular cAMP leads to increased sequestration of Ca^{2+} by the membranous components of the cells. This reduces the transport of Ca^{2+} into the cytoplasm, therefore eliminating an essential component of the contractile process. PGF-2α and oxytocin are thought to reverse this process (Krall & Korenman, 1977), as demonstrated by Torok & Csapo (1976) who reported that a lower concentration of PGF-2α stimulated the post-partum uterus compared to the progesterone dominated pregnant rabbit uterus. Therefore, despite the fact that both PGFM and oxytocin are significantly elevated, the main expulsive effort of the myometrium may not occur until PGF-2α and oxytocin are present in sufficient concentrations to overcome the inhibitory effects of relaxin.

Immediately *post partum*, the myometrium returns to its predelivery state of quiescence. Only after placental delivery does myometrial activity increase again. Relaxin is produced by the horse placenta (Stewart & Stabenfeldt, 1982) and after placental delivery relaxin concentrations rapidly decline (Stewart *et al.*, 1982). With the delivery of the placenta, the source of both the known myometrial inhibitors, relaxin and progesterone, is eliminated. The myometrium responds by exhibiting continuous, low level, erratic electrical activity.

From a qualitative review of the data, a daily rhythm exists in the percentage activity in the mare during the week *pre partum*. The greatest percentage activity occurs during the night and early morning hours. This cycle of activity is similar to the patterns observed in primates (Germain *et al.*, 1982; Taylor *et al.*, 1983). There is no indication of a daily rhythm of EMG activity in other domestic animals. The pattern observed in pregnant ewes (Harding *et al.*, 1982; Krishnamurti *et al.*, 1982; Toutain *et al.*, 1983), goats (Taverne & Scheerboom, 1984; G. J. Haluska & W. B. Currie, unpublished observations), cows (Zerobin & Sporri, 1972), and pigs (Taverne *et al.*, 1979a) consists of regular long bursts of activity separated by periods of almost total inactivity. This pattern changes within a few days of delivery to medium bursts which increase in amplitude and frequency as delivery approaches. During the last week *pre partum* in the mare the predominant type of activity is of medium bursts, which seem to increase as parturition approaches.

In the domestic ruminants, the changes in uterine activity can be related to dynamic changes in fetal corticosteroids, maternal progesterone and oestradiol-17β. However, fetal cortisol increases gradually in the horse with increasing gestational age, with no discrete surge *pre partum* (Nathanielsz *et al.*, 1975; Rossdale *et al.*, 1982; Silver *et al.*, 1984). The concentration of oestradiol does not increase *pre partum*, but mean progesterone concentrations begin to decline and by parturition have decreased significantly (Haluska, 1985). Barnes *et al.* (1975) report a decrease in total plasma progestagens over the last 48 h *pre partum* in the umbilical circulation, indicating that there may be a decrease in progesterone synthesis. Therefore, the only significant endocrine change within the last week *pre partum* is a decrease in progesterone during the 24 h *pre partum*, leading to an increased oestradiol/progesterone ratio. It is perhaps this change which releases the myometrium from inhibitory control and allows the evolution of labour. This interpretation is consistent with the conclusion of Taverne *et al.* (1979b) that myometrial activity in the cyclic mare is well correlated with changes in the concentration of progesterone.

Most mares give birth in relative seclusion at night. Noradrenergic nerve fibres which connect the limbic system (the mediator of stressful emotional input in the central nervous system) and the paraventricular nucleus of the hypothalamus, may allow the mare to respond instinctively to the presence of threats or danger (i.e., human presence and other stimuli) by an inhibition of oxytocin release. Also, central opioids may play a role in the control of oxytocin secretion at parturition (Leng *et al.*, 1985). When 'stress' is eliminated, the central inhibition on oxytocin secretion is released, leading to a surge in PG secretion and a very rapid, controlled delivery as suggested by Stewart *et al.* (1984) and Haluska (1985).

We thank Frank J Michel for technical assistance and Suzanne Bremmer for preparation of the manuscript.

This work was done in partial fulfilment of the requirements of the Ph.D. degree in the field of Animal Science at Cornell University (G.J.H.).

References

Barnes, R.J., Nathanielsz, P.W., Rossdale, P.D., Comline, R.S. & Silver, M. (1975) Plasma progestagens and oestrogens in fetus and mother in late pregnancy. *J. Reprod. Fert., Suppl.* **23**, 617–623.

Barnes, R.J., Comline, R.S., Jeffcott, L.B., Mitchell, M.D., Rossdale, P.D. & Silver, M. (1978) Foetal and maternal plasma concentrations of 13,14-dihydro-15-oxo-prostaglandin F in the mare during late pregnancy and at parturition. *J. Endocr.* **78**, 201–215.

Cheah, S.H. & Sherwood, O.D. (1980) Target tissues for relaxin in the rat: tissue distribution of injected ^{125}I-labelled relaxin and tissue changes in adenosine $3',5'$ monophosphate levels after in vivo relaxin incubation. *Endocrinology* **106**, 1203–1209.

Cheah, S.H. & Sherwood, O.D. (1981) Effects of relaxin on in-vivo uterine contractions in conscious and unrestrained estrogen-treated and steroid untreated ovariectomized rats. *Endocrinology* **109**, 2076–2083.

Csapo, A.I. & Taketa, H. (1965) Effect of progesterone on the electric activity and intrauterine pressure of prepartum and parturient rabbits. *Am. J. Obstet. Gynecol.* **91**, 221–231.

Downing, S.J., Bradshaw, J.M.C. & Porter, D.G. (1980) Relaxin improves the coordination of rat myometrial activity in vivo. *Biol. Reprod.* **23**, 899–903.

Germain, G., Cabrol, D., Visser, A. & Sureau, C. (1982) Electrical activity of the pregnant uterus in the cynomolgus monkey. *Am. J. Obstet. Gynecol.* **142**, 513–519.

Haluska, G.J. (1985) *Electromyographic analysis of the myometrium of the mare correlated with the endocrinology of pregnancy, parturition and the postpartum period.* Ph.D. thesis, Cornell University.

Harding, R., Poore, E.R., Bailey, A., Thorburn, G.D., Jansen, C.A.M. & Nathanielsz, P.W. (1982) Electromyographic activity of the non-pregnant and pregnant sheep uterus. *Am. J. Obstet. Gynecol.* **142**, 448–457.

Holtan, D.W., Nett, T.M. & Estergreen, V.L. (1975) Plasma progestagens in pregnant mares. *J. Reprod. Fert., Suppl.* **23**, 419–424.

Jeffcott, L.B. (1972) Observations on parturition in crossbred pony mares. *Equine vet. J.* **4**, 209–213.

Judson, D.G., Pay, S. & Bhoola, K.D. (1980) Modulation of cyclic AMP in isolated rat uterine tissue slices by porcine relaxin. *J. Endocr.* **87**, 153–159.

Krall, J.F. & Korenman, S.G. (1977) Control of uterine contractility via cyclic AMP dependent protein kinase. In *The Fetus and Birth* (Ciba Fdn Symp.), pp. 319–338. Eds M. O'Connor & J. Knight. Elsevier/Excerpta Medica, Amsterdam.

Krishnamurti, C.R., Kitts, D.D., Kitts, W.D. & Tompkins, J.G. (1982) Myoelectrical changes in the uterus of the sheep around parturition. *J. Reprod. Fert.* **64**, 59–67.

Leng, G., Mansfield, S., Bicknell, R.J., Dean, A.D.P., Ingram, C.D., Marsh, M.I.C., Yates, J.O. & Dyer, R.G. (1985) Central opioids: a possible role in parturition. *J. Endocr.* **106**, 219–224.

Nathanielsz, P.W., Rossdale, P.D., Silver, M. & Comline, R.S. (1975) Studies on fetal, neonatal and maternal cortisol metabolism in the mare. *J. Reprod. Fert., Suppl.* **23**, 625–630.

Nett, T.M., Holtan, D.W. & Estergreen, V.L. (1973) Plasma estrogens in pregnant and post partum mares. *J. Anim. Sci.* **37**, 962–970.

Nishikori, K., Weisbrodt, N.W., Sherwood, O.D. & Sanborn, B.M. (1983) Effects of relaxin on rat uterine myosin light chain kinase activity and myosin light chain phosphorylation. *J. biol. Chem.* **258**, 2568–2474.

Roberts, S.J. (1971) *Veterinary Obstetrics and Genital Disease.* Edwards Brothers, Inc., Ann Arbor.

Rossdale, P.D. & Ricketts, S.W. (1980) *Equine Stud Farm Medicine*, 2nd edn. Balliere Tindal, London.

Rossdale, P.D., Silver, M., Ellis, L. & Frauenfelder, (1982) Response of the adrenal cortex to tetracosactin (ACTH 1-24) in the premature and full-term foal. *J. Reprod. Fert., Suppl.* **32**, 545–553.

Silver, M., Ousey, J.C., Dudan, F.E., Fowden, A.L., Knox, J., Cash, R.S.G. & Rossdale, P.D. (1984) Studies on equine prematurity. 2: Post natal adrenalcortical activity in relation to plasma adrenocorticotrophic hormone and catecholamine levels in term and premature foals. *Equine vet. J.* **16**, 278–286.

Stewart, D.R. & Stabenfeldt, G.H. (1981) Relaxin activity in the pregnant mare. *Biol. Reprod.* **25**, 281–289.

Stewart, D.R. & Stabenfeldt, G.H. (1982) Determination of the source of equine relaxin. *Biol. Reprod.* **27**, 17–24.

Stewart, D.R., Stabenfeldt, G.H. & Hughes, J.P. (1982) Relaxin activity in foaling mares. *J. Reprod. Fert., Suppl.* **32**, 603–609.

Stewart, D.R., Kindahl, H., Stabenfeldt, G.H. & Hughes, J.P. (1984) Concentration of 15-keto-13,14-dihydro-prostaglandin F_2 in the mare during spontaneous and oxytocin induced foaling. *Equine vet. J.* **16**, 270–274.

Taverne, M.A.M. & Scheerboom, J.E.M. (1984) Myometrial electrical activity during pregnancy and parturition in the pigmy goat. *Proc. 10th Int. Congr. Anim. Reprod. & A.I., Urbana-Champaign*, p. 113.

Taverne, M.A.M., Naaktgeboren, C., Elsaesser, F., Forsling, M.L., van der Weyden, G.C., Ellendorff, F. & Smidt, D. (1979a) Myometrial electrical activity and plasma concentrations of progesterone, estrogen, and oxytocin during late pregnancy and parturition in the miniature pig. *Biol. Reprod.* **21**, 1125–1134.

Taverne, M.A.M., van der Weyden, G.C., Fontijne, P., Dieleman, S.J., Pashen, R.L. & Allen, W.R. (1979b) In-vivo myometrial electrical activity in the cyclic mare. *J. Reprod. Fert.* **56**, 521–532.

Taylor, N.F., Martin, M.C., Nathanielsz, P.W. & Seron-Ferre, M. (1983) The fetus determines circadian oscillation of myometrial electromyographic activity in the pregnant rhesus monkey. *Am. J. Obstet. Gynecol.* **146**, 557–567.

Torok, I. & Csapo, A.I. (1976) The effect of progesterone, prostaglandin F2 and oxytocin on the calcium activation of the uterus. *Prostaglandins* **12**, 253–269.

Toutain, P.L., Garcia-Villar, R., Hanzen, C. & Ruckebusch, Y. (1983) Electrical and mechanical activity of the cervix in the ewe during pregnancy and parturition. *J. Reprod. Fert.* **68**, 195–204.

Wright, J.G. (1943) Parturition in the mare. *J. comp. Path.* **53**, 212–219.

Zerobin, K. & Sporri, H. (1972) Motility of the bovine and porcine uterus and fallopian tube. *Adv. vet. Sci. comp. Med.* **16**, 303–354.

J. Reprod. Fert., Suppl. **35** (1987), 565–573

Antepartum evaluations of the equine fetus

C. Adams-Brendemuehl* and F. S. Pipers†

University of Florida Horse Research Center, Ocala, FL 32675, U.S.A.

Summary. Measurements were made by real-time ultrasonography in 14 healthy mares to assess fetal growth and estimate newborn foal weights. Intrauterine fluid volumes were estimated and the placenta was measured and observed for maturational changes. The onset and incidence of echogenic particles in the allantoic fluid were recorded. In the second approach, baseline fetal heart rate, physiological rate variations and number, amplitude and duration of recorded accelerations were measured. Estimates of birth weights were within ± 3.49 kg. Allantoic fluid was evident in all sonographic planes within the uterine cavity. Average vertical axis was 1.9 ± 0.9 cm. The mean thickness of the ventrally located gravid uterine horn and allantochorion was 1.26 ± 0.33 cm. No placental maturational changes were noted. Echogenic particles were observed in the allantoic fluid of all mares recorded within 10 days of foaling. Baseline fetal heart rate was 76 ± 8 beats/min. Spontaneous fetal activity resulted in accelerations in fetal heart rate from 25 to 40 beats/min in amplitude and of 23–36 sec in duration. About 10 accelerations were observed in a 10-min period.

In 17 clinical case mares, abnormal records included decreased fetal activity; lack of fetal growth; haemorrhage and fibrin-like tags in the allantoic fluid; fetal tachycardia, bradycardia, arrhythmia, and cardiac deceleration.

We suggest that ultrasonic fetal measurements provide useful information with regard to the fetal condition, but additional variables are required for more reliable assessments of fetal stress and distress.

Introduction

Antepartum fetal assessments in horses are currently restricted to transrectal ballottement of the gravid uterus. Preliminary research demonstrates that the equine fetus is accessible to interrogation by additional methods (Fraser *et al.*, 1973; Jeffcott & Rossdale, 1979; O'Grady *et al.*, 1981; Pipers & Adams-Brendemuehl, 1984). Organ morphology, fetal number, presentation, and position as well as intermittent fetal heart rate monitoring utilizing two-dimensional, real-time ultra-sonography have been described. Transcutaneous fetal electrocardiography has been reported (Colles *et al.*, 1978) but many problems exist with interpretation and reproducibility (Buss *et al.*, 1980).

Advances in human perinatology in the past 2 decades have contributed to a reduction in both morbidity and mortality. This has been achieved largely through antenatal assessments of fetal well being with identification of high risk patients. Biochemical tests derived from human maternal blood, urine and amniotic fluid assess fetal and placental function. Fetal phospholipids from amniocentesis samples provide reliable information regarding pulmonary maturity. Biophysical parameters utilizing ultrasonic methods assess fetal morphology; growth and motion; placental

*Present address: Ultraview, Inc., P.O. Box 5133, Ocala, FL 32678, U.S.A.
†Present address: Department of Medicine, Tufts University, School of Veterinary Medicine, 200 Westboro Road, N. Grafton, MA 01536, U.S.A.

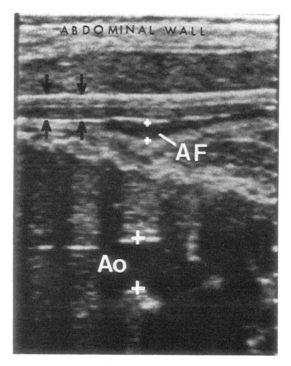

Fig. 1. Transverse scan of fetal thorax illustrates measurement of the aortic diameter (Ao) between the large crosses and the vertical dimension of the allantoic fluid (AF), small crosses. Uterus and allantochorion are also shown (black arrows). 4-MHz linear array transducer.

location and function; and total intrauterine fluid volumes. Non-invasive alternatives to amniocentesis in predicting fetal pulmonary maturity may also be derived indirectly through ultrasound imaging. These include placental classification schemes based on morphological maturation changes (Grannum *et al.*, 1979) and the detection of echogenic particles, i.e. vernix and larger free-floating particles, within the amniotic fluid (Gross *et al.*, 1985). Continuous fetal heart rate monitoring is a widely used technique both in the antepartum period and during labour. Fetal cardiac accelerations in response to fetal movement are considered a sign of fetal well being (Schifrin *et al.*, 1979). The purpose of this paper was to evaluate the use of ultrasonic techniques to assess fetal condition in the horse in late gestation.

Materials and Methods

Animals. Fourteen healthy pregnant mares (Thoroughbreds and Quarter Horses) ranging in age from 10 to 23 years were transabdominally scanned between 300 days gestation to term for an average of 1·5 times weekly. Various types of ultrasound equipment were utilized including an ATL Model 4600 sector scanner with a 3 MHz long-focus and 7·5 MHz transducer capabilities; a Pie Data 560 series, 3 MHz linear array system; and an SDR 1500 Philips equipped with 4 MHz linear array and 7·5 MHz sector transducers. Mare preparation included clipping the ventral abdomen with a No. 40 surgical blade to improve image quality. The clipped area was then cleaned with warm water to remove dirt and debris followed by the application of an ultrasonic coupling gel. Scans were performed without the use of physical or chemical restraint.

Measurements. The fetal presentation, position, and activity level (0 to +3) were observed and described as part of each examination. Fetal aortic measurements were obtained from transverse linear scans where the frame was frozen during systole. Built-in electronic calipers were used to measure the diameter of the great vessel at the level of the caudal border of the fetal heart (Fig. 1). The antepartum aortic measurements were then compared to newborn foal weight, length, girth, hip and shoulder height. Measurements of the allantoic fluid were also achieved from images through the frontal plane of the fetus at approximately the level of the scapulohumeral joint. The vertical distance (cm) was measured from the fetal side of the allantochorion to a fixed location on the fetal thorax just caudal to the shoulder (Fig. 1). The allantoic cavity was scanned for the presence of echogenic floating particles. Measurements were recorded from 7 mares and include the time the particles initially appeared, their density ($+1$ to $+3$), and whether or not swirling of these particles occurred. The uteroplacental unit was routinely scanned using a 7·5 MHz sector transducer for evidence of textural changes. Maturation patterns in human placentas include indentations of the chorionic plate, circular areas of decreased echogenicity, and irregular echogenic calcifications. The combined thickness of the gravid uterine horn and allantochorion was determined from vertical measurements recorded from transverse scans along sagittal or parasagittal planes on the mare's ventral abdomen. Areas of fetal contact were avoided during measurements as compression of the utero-placental unit was observed to occur. Mare weights, *pre* and *post partum*, as well as foal and placental weights were utilized to provide estimates of total intrauterine fluid volumes [(mare's prefoaling wt − post foaling wt) − (placental wt + foaling wt)].

In the second part of the study a fetal Doppler heart rate monitor, the Wakeling FM210, was used to monitor continuously the fetal heart rate. A 3-beat averaging signal was preselected, displayed, and recorded using a dot matrix thermal printer at a speed of 3 cm/min for a period of not less than 10 min per examination. A hand-held event marker provided chart recordings of fetal activity which were determined audibly or manually. The signal processing system consists of a microprocessor-based rate computer with a range of 50–210 beats/min (accuracy ± 1%). The fetal heart was first located with the linear array transducer. The fetal monitor ultrasound transducer (2·3 MHz) was then placed on the mare's abdominal wall directly over the fetal heart. Loss of an audible Doppler signal required relocation of the fetal heart. With experience, interruption of the fetal heart rate recording could often be prevented by slowly moving the Doppler transducer in response to a shifting or fading audible signal until a strong fetal heart beat was relocated. Baseline fetal heart rate, physiological rate variations, and the number, amplitude and duration of accelerations with or without fetal motion were recorded and measured to establish normal fetal cardiac characteristics. Maternal heart rate was recorded at the beginning of each examination and digital pulse pressures were palpated and qualitatively assessed. Digital pulses were graded I–III, with a Grade I pulse considered to be normal, Grade II increased, and Grade III bounding.

Clinical case mares. Antenatal assessments using the biophysical parameters previously described were performed on 17 additional mares (236–346 days gestation) when clinical observations suggested possible obstetrical complications. Clinical findings in these mares included rapid abdominal enlargement, colic, vaginal discharge and uterine torsion, with the most frequent complaint being early mammary development with or without lactation.

Results

Healthy mares

In the 14 healthy mares examined, all fetuses were initially observed in an anterior presentation and found to be lying in a ventrodorsal position in which the fetus most often resided in a left lateral oblique fashion from the maternal sagittal or parasagittal plane. On subsequent scans, some fetuses were found to be temporarily orientated in a transverse position in relation to the long axis of the mare. Actual gestational ages at parturition ranged from 321 to 360 days. All foals in this group appeared physically mature at birth and stood and nursed within 3 h of foaling.

Activity. During the course of each examination, periods of fetal rest and activity were variable ($+1$ to $+3$) with all fetuses exhibiting some form of co-ordinated activity, i.e. flexion and extension of extremities and vertebral column to complex integration of these activities including fetal rotation (≤ 360 degrees) about its long axis. The fetus was observed to be most active from 5 to 72 h before parturition.

Table 1. Correlations of measurements of fetal aortic diameter before birth and foal measurements after birth ($n = 14$)

	Weight (kg)	Length (cm)	Girth (cm)	Hip ht (cm)	Shoulder ht (cm)
Regression coefficient intercept	−19·62	42·08	47·77	40·68	54·19
Slope	29·21	12·43	12·73	23·76	17·81
r	0·78	0·58	0·76	0·87	0·68
Standard error of estimate	3·49	2·53	1·61	2·01	2·79

Weight and size. Fetal aortic diameters were serially recorded from 14 fetuses for a total of 95 measurements. Estimates of newborn foal weights were within ± 3.49 kg with a correlation coefficient of 0.78 ($P < 0.05$). Correlations between recorded fetal aortic dimensions from -7 days to parturition and post-partum parameters demonstrated statistically significant relationships between body weight, girth and hip weight (Table 1).

Allantoic fluid. Allantoic fluid was evident in all sonographic planes within the uterine cavity. The mean vertical distance of the allantoic fluid recorded from a total of 69 scans was 1.9 ± 0.9 cm. Total intrauterine fluid volumes ranged from 8.7 to 41.4 litres for a mean of 24.05 ± 1.53.

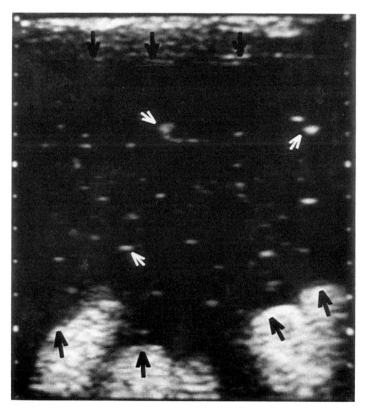

Fig. 2. Ultrasound appearance of echogenic particles in allantoic fluid. Evaluation of allantoic cavity (black arrows) demonstrates echo-dense particles (white arrows) floating within the allantoic fluid. 4-MHz linear array transducer.

Echogenic particles. Echogenic particles were sonographically detected in the allantoic fluid of all mares (Fig. 2). The presence of floating particulate material was first observed 44 days before foaling, whereas the earliest recording of the larger particles was 36 days *pre partum*. Five mares demonstrated increased density with swirling of both particle types within 10 days of parturition (actual gestational ages 321–351 days). In one mare there was no change noted in either the number ($+1$) or appearance of these particles (350 days), while in another mare only an increase in density ($+1$ to $+2$) of both types of particles without swirling (331 days) was observed.

Placenta. No placental textural changes were noted with increasing gestation. The combined thickness of the gravid uterus and allantochorion averaged 1.26 ± 0.33 cm from a total of 75 serial

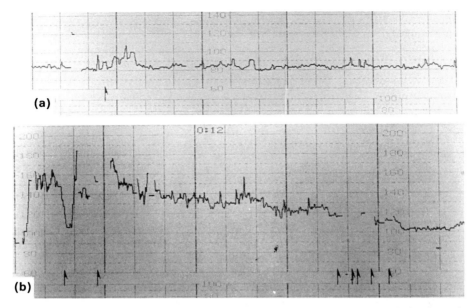

(a)

(b)

Fig. 3. Doppler tracings of fetal heart rate. (a) Normal fetal heart rate pattern in late gestation. Event marker records spontaneous fetal motion preceding heart rate acceleration. (b) Fetal tachycardia; the heart rate pattern returned to normal and parturition occurred 28 h later.

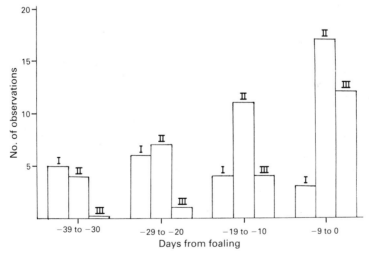

Fig. 4. Qualitative digital pulse pressures in mares from 39 to 0 days before foaling. Pulses were graded I–III.

scans. The post-partum placental weights ranged from 3·6 to 7·7 kg. No significant relationship was found between the utero-placental thickness and either gestational age or post-partum placental weight.

Fetal heart rate. Fetal heart rates and activity were simultaneously recorded for a total of 108 examinations in 14 mares. Continuous monitoring demonstrated that baseline fetal heart rate was 76 ± 8 beats/min. Physiological variations ranged from 7 to 15 beats/min about the baseline.

Spontaneous fetal activity generally resulted in accelerations in the fetal heart rate to 25–40 beats/min in amplitude and of 23–36 sec in duration. An average of 10 accelerations were observed in a 10-min examination period (Fig. 3). Fetal motion without cardiac acceleration was routinely observed while accelerations in the absence of detectable stimuli were recorded only 5% of the time. Three fetuses demonstrated exaggerated heart rate responses. These heart rates ranged from 160 to 192 beats/min in amplitude and 1–6 min in duration. Parturition ensued within 48 h.

Table 2. Antepartum biophysical profiles and outcomes in 17 clinical case mares

Mare no.	Presenting complaint	Gestational age (days) *in utero*	Biophysical profile	Fetal outcome	Mare problems *post partum*
1	Udder development	291	Negative	Negative (foal small, weak; died Day 5)	Placentitis
2	Lactating	310	Positive (FHR 48)	Negative (pneumonia-recovered)	Focal placentitis
3	Twins (1 dead) udder development laminitis	322	Positive (FHR 58)	Negative (dystocia/stillborn necrotizing twin)	Diffuse placentitis
4	Lactating	280	Positive (placenta 0·5 cm)	Negative (weak, pneumonia recovered)	Placentitis
5	Rapid abdominal enlargement	236	Positive (posterior presentation particles)	Negative (aborted)	Placentitis
6	Udder development	291	Positive (FHR 48, no activity)	Negative (seizures/killed)	Unknown
7	Vaginal discharge	270	Positive (arrhythmia, bradycardia)	Negative (stillborn)	Placentitis, calcifications
8	Lactating cervix relaxed	315	Negative	Positive	Focal placentitis
9	Colic	335	Negative	Positive	None
10	MHR 100 beats/min, jugular pulse ventral oedema	335	Negative	Positive	Uterine haemorrhage
11	Uterine torsion	338	Negative	Positive	None
12	Udder development laminitis	327	Positive (FHR 56)	Positive	None
13	Rapid abdominal enlargement	338	Negative	Positive	None
14	Udder development	335	Negative	Positive	None
15	Udder development	346	Negative	Positive	None
16	Udder development	280	Positive (placenta 3·4 cm)	Negative (stillborn)	Unknown
17	Discharge	320	Negative	Negative (unattended foaling, foal retained in amniotic membrane)	Diffuse placentitis

Maternal heart rate and pulse. Maternal heart rates ranged from 44 to 80 beats/min with a mean of 60·6 beats/min from 49 to 0 days before parturition ($n = 80$). All mares exhibited increased pulse pressures, with >90% of the pulses being Grade II or higher within 9 days of foaling (Fig. 4).

Clinical cases

Fetal condition was assessed for 17 additional mares and the biophysical values in these mares were compared to the normal values obtained from healthy mares. The overall biophysical profile was then assigned either a negative or positive rating. A negative profile was defined as one in which all values were within 2 s.d. from the mean. When any value measured was >2 s.d. from the mean, a positive profile was assigned. Abnormal fetal values measured included decreased fetal activity, tachycardia, bradycardia, arrhythmia, and transient cardiac decelerations (Table 2).

When the biophysical profile was negative and within our previously defined normal range, 78% of the mares delivered healthy foals. In contrast, when a positive profile was assigned, 88% of the fetuses had poor outcomes. Two false-negative profiles (22·2%) and one false-positive profile (12·5%) were recorded. One of the false-negative mares (No. 17) was serially examined on 9 occasions and on each a negative profile was recorded. Parturition was unattended and a foal of physically mature appearance was discovered dead within an intact amniotic sac.

Discussion

The horse and human placenta are quite different in both location (placenta diffusa *vs* placenta discoidalis) and feto-maternal exchange (nondeciduate *vs* deciduate). The lack of a large, discrete placental mass in the horse and differences in vascularity may account for the absence of placental textural changes noted in the normal mare group. Heterogeneous textural changes, including calcifications of the intercotyledonary septa and irregular calcifications within the placental substance which may cause acoustical shadowing, are normal morphological changes in the mature human placenta (Fisher *et al.*, 1976). Chronic degenerative placental changes such as infarcts may also involve calcium deposition (Craig, 1977). Placental calcifications with acoustical shadowing were observed in Mare 7 and confirmed on gross inspection of the placenta after abortion.

Differences in types of intrauterine fluid compartments also exist. Allantoic and amniotic fluid-filled sacs are present in the horse while a fluid-filled allantoic compartment is non-existent in humans due to the fusion of the allantochorion and allantoamnionic membranes. Sonographically, the amniotic membrane in the horse is usually found lying close to the fetal trunk with most of the amniotic fluid between the front legs and surrounding the fetal neck and the more dorsally located fetal head. This membrane is easily visualized undulating between the two fluid-filled compartments in this region. Growth-retarded fetuses are frequently found in association with oligohydramnios and show an increased risk of intrapartum asphyxia, stillbirth, and neonatal morbidity and mortality. Qualitative assessments of amniotic fluid volumes alone (Manning *et al.*, 1981) or in combination with other biophysical parameters such as fetal size, weight and total intrauterine volume have been found to be reliable indicators of fetal intrauterine growth retardation.

Vernix, a sebaceous secretion of the skin glands of the human fetus, may be observed in small amounts as echogenic particles suspended within the amniotic fluid in human pregnancies as early as 15 weeks gestation. Increased numbers of vernix and free-floating particles are generally recorded in the third trimester and, with fetal movement, create a dense swirling pattern. Two types of particulate material were sonographically detected in this study floating within the allantoic fluid and appeared to be similar to the vernix and particles described in human amniotic fluid. Hippomanes and smaller debris, long known to be present in the allantoic fluid of equids, are composed of mucoproteins and deposits of minerals, especially calcium phosphates, and are therefore similar to urinary calculi (Ginther, 1979). In this respect, the allantoic fluid in the horse and amniotic fluid in man are alike that in fetal urine exists in both.

Early visualization of echogenic densities within the amniotic fluid in association with human fetal distress has been reported (Khaleghian, 1983). Blood, meconium and infective debris may be additional sources of these echogenic particles (De Vore & Platt, 1986). Echogenic particles were noted in three of the clinical mares (Nos 1, 3 and 5) which had positive biophysical profiles and subsequently aborted. Haemorrhage was grossly observed in the intrauterine fluid of Mare 5.

Human fetal measurements such as biparietal diameter, abdominal circumference and femur length are used to estimate gestational age, fetal size and weight and to monitor individual fetal growth. These measurements become of vital importance when obstetrical complications occur and gestational age is uncertain. Sonographic antepartum fetal measurements in the horse are not without technical limitations. Depth of penetration, focal zones, resolution, field of view, and transducer configuration are all suboptimal for our concerns. Due to the physical size of the horse fetus, most of these measurements are impossible to obtain in late gestation. In spite of these limitations, useful data are available.

In addition to major differences in the actual size of the horse fetus, physiological values (although similar in the nature of their response) are different enough to generate problems in fetal heart rate monitoring. The typical fetal cardiac heart rate monitor operates between 50 and 200 beats/min. Heart rates at both the high and low extremes in this range often produce spurious values due to electronic artefacts resulting in double and half counting. Problems are also encountered when the rate of cardiac acceleration exceeds a preset algorithm installed to differentiate between changes in physiological heart rate and artefact. We attempted to overcome these limitations by modifying the equipment which doubled the size of this 'window'. Since optimal fetal heart rate detection requires the cardiac structures to be within a narrow ultrasonic beam, motion from any number of sources alters this relationship and interferes with data acquisition. As previously stated, experience can increase the clarity of the wave form during monitoring but the lack of convenience and labour-intensive nature of this process can be frustrating.

Finally, the lack of precise clinical, biochemical and histopathological data in the mares with unfavourable pregnancy outcomes makes the interpretation of our biophysical profile more difficult. The fact that a fetus *in utero* appears healthy when at birth it is abnormal may be inconsistent. It appears that additional measurements will be necessary to sort out these dilemmas.

In conclusion, establishing an antepartum biophysical profile in the horse appears to have merit. By augmenting the normal data base, increasing the number of pathological evaluations, acquiring sophisticated clinical and biochemical correlates, and adding equipment modifications, these profiles should become an integral part of antenatal evaluations.

Sponsored in part by the University of Florida Equine Neonatology Study Group and the Florida Thoroughbred Breeders' Association.

References

Buss, D.D., Asbury, A.C. & Chevalier, L. (1980) Limitations in equine fetal electrocardiography. *J. Am. vet. Med. Assoc.* **177**, 174–176.

Colles, C.M., Parkes, R.D. & May, C.J. (1978) Foetal electrocardiography in the mare. *Equine vet. J.* **10**, 32–37.

Craig, M. (1977) Ultrasonic evaluation of the normal and abnormal placenta. *Med. Ultrasound* Aug., 13–20.

De Vore, G.R. & Platt, D.L. (1986) Ultrasound appear-ance of particulate matter in amniotic cavity: vernix or meconium? *J. clin. Ultrasound* **14**, 229–230.

Fisher, C.C., Garrett, W. & Kossoff, G. (1976) Placental aging monitored by gray scale echography. *Am. J. Obstet. Gynecol.* **124**, 483.

Fraser, A.F., Keith, N.W. & Hastie, H. (1973) Summarised observations on the ultrasonic detection of pregnancy and foetal life in the mare. *Vet. Rec.* **92**, 20–21.

Ginther, O.J. (1979) Placentation and embryology. In

Reproductive Biology of the Mare, Ch. 9, pp. 298–300. Naughton & Gunn, Inc., Ann Arbor, Michigan.

Grannum, P.A., Berkowitz, R.L. & Hobbins, J.C. (1979) The ultrasonic changes in the maturing placenta and their relation to fetal pulmonic maturity. *Am. J. Obstet. Gynecol.* **133**, 915–922.

Gross, T.L., Wolfson, R.N., Kuhnert, P.M. & Sokol, R.J. (1985) Sonographically detected free-floating particles in amniotic fluid predict a mature lecithin-sphingomyelin ratio. *J. clin. Ultrasound* **13**, 405–409.

Jeffcott, L.B. & Rossdale, P.D. (1979) A radiographic study of the fetus in late pregnancy and during foaling. *J. Reprod. Fert., Suppl.* **27**, 563–569.

Khaleghian, R. (1983) Echogenic amniotic fluid in the second trimester: a new sign of fetal distress. *J. clin. Ultrasound* **11**, 498–501.

Manning, F.A., Hill, L.M. & Platt, L.D. (1981) Qualitative amnionic fluid volume determination by ultrasound: antepartum detection of intrauterine growth retardation. *Am. J. Obstet. Gynecol.* **139**, 254–258.

O'Grady, J.P., Yeager, C.H., Findleton, L., Brown, J. & Esra, G. (1981) In utero visualization of the fetal horse by ultrasonic scanning. *Equine Practice* **3**, 45–49.

Pipers, F.S. & Adams-Brendemuehl, C.S. (1984) Techniques and applications of transabdominal ultrasonography in the pregnant mare. *Am. J. vet. Med. Assoc.* **185**, 766–771.

Schifrin, B.S., Foye, G., Amato, J., Kates, R. & MacKenna, J. (1979) Routine fetal heart rate monitoring in the antepartum period. *Obstet. Gynecol.* **54**, 21–25.

J. Reprod. Fert., Suppl. **35** (1987), 575–586

Respiratory mechanics and breathing pattern in the neonatal foal

A. M. Koterba* and P. C. Kosch

Department of Physiological Sciences, College of Veterinary Medicine, University of Florida, Gainesville, Florida 32610, U.S.A.

Summary. Breathing pattern, respiratory muscle activation pattern, lung volumes and volume–pressure characteristics of the respiratory system of normal, term, neonatal foals on Days 2 and 7 of age were determined to test the hypothesis that the foal actively maintains end-expiratory lung volume (EEV) greater than the relaxation volume of the respiratory system (Vrx) because of a highly compliant chest wall.

Breathing pattern was measured in the awake, unsedated foal during quiet breathing in lateral and standing positions. The typical neonatal foal breathing pattern was characterized by a monophasic inspiratory and expiratory flow pattern. Both inspiration and expiration were active, with onset of Edi activity preceding onset of inspiratory flow, and phasic abdominal muscle activity detectable throughout most of expiration. No evidence was found to support the hypothesis that the normal, term neonatal foal actively maintains EEV greater than Vrx.

In the neonatal foal, normalized lung volume and lung compliance values were similar to those reported for neonates of other species, while normalized chest wall compliance was considerably lower.

We conclude that the chest wall of the term neonatal foal is sufficiently rigid to prevent a low Vrx. This characteristic probably prevents the foal from having to use a breathing strategy which maintains an EEV greater than Vrx.

Introduction

Two major determinants influencing the pattern of breathing utilized by a particular species are its metabolic requirements and the mechanical characteristics of its respiratory system (Mortola, 1984). As both of these factors differ from the newborn to the adult, it might be expected that the neonatal breathing pattern would differ as well. A general characteristic of the newborn animal is a considerably higher metabolic rate (normalized to body weight) compared to the adult of the same species, which necessitates a high ventilation.

The mechanical characteristics of the chest wall of the neonate may also play an important role in determining the breathing pattern adopted. A high chest wall to lung compliance ratio relative to the adult is a consistent finding in all species of neonates studied to date (Agostoni, 1959; Avery & Cook, 1961; Fisher & Mortola, 1980; Mortola, 1983; Gaultier *et al.*, 1984). Although probably necessary for an uncomplicated delivery through the birth canal, the flexibility of the chest wall may adversely affect both gas exchange and the efficiency of ventilation in a number of ways. First, if tidal volume is increased, chest wall distortion would be expected to increase during inspiration due to an inadequate support of the contracting diaphragm (Mortola, 1983; Mortola *et al.*, 1982). Second, as the relaxation volume of the respiratory system (Vrx) is determined by a balance

---

*Present address: Department of Medical Sciences, Box J-126, JHMHC, College of Veterinary Medicine, Gainesville, Florida 32610, U.S.A.

between the tendency of the lungs to recoil inward and the tendency of the relaxed chest wall to recoil outward, a high chest wall compliance would be expected to predispose to a low Vrx. A low Vrx could be disadvantageous to the neonate in several ways, including a greater tendency for airway collapse at end expiration, thus increasing the work of breathing and decreased O_2 stores because of a small gas reservoir at end expiration (Findley *et al.*, 1983).

There is, however, a considerable body of evidence which suggests that unanaesthetized newborns of several species, including the human infant, do not deflate to Vrx at end expiration but rather actively maintain end-expiratory lung volume (EEV) well above Vrx. This may be accomplished by several different mechanisms. In a system with a high chest wall compliance, the passive time constant, the product of resistance and compliance, should be prolonged. As a result, the expiratory time required for passive emptying of the lungs to Vrx would be increased. A rapid breathing rate with a short expiratory time (T_E) could elevate EEV above Vrx by reducing the time available for complete emptying of the lungs. Two additional strategies potentially useful in maintaining an elevated EEV are inspiratory muscle activity after inspiration and upper airway narrowing during expiration. Both mechanisms exert the same effect on the system; that is, both retard or brake expiratory airflow. The former strategy decreases the rate of lung deflation by counteracting the passive recoil of the respiratory system towards the relaxation volume, while the latter increases expiratory resistance to airflow. Post-inspiratory inspiratory muscle activity has been well documented in both diaphragm and intercostal muscles of the newborn (Bartlett *et al.*, 1973; Gautier *et al.*, 1973; Remmers & Bartlett, 1977; Lopes *et al.*, 1981; Kosch & Stark, 1984). In certain awake newborn animals, including the lamb (Harding *et al.*, 1980), the opossum (Farber, 1978), and puppy (England *et al.*, 1985), laryngeal adductors are active during expiration. In conclusion, in the newborn species studied previously, it appears that a number of strategies are utilized to defend end-expiratory lung volume in a compliant respiratory system.

Little information is available regarding normal respiratory system mechanics in the immature horse. Although Gillespie (1975) speculated that the foal has a respiratory system similar to that of other neonates and thus would be expected to adopt similar strategies to maintain EEV, no evidence in support of this has been published. The only published values of normal lung volumes in foals are those of tidal volume and minute respiratory volume (Rossdale, 1969, 1970; Gillespie, 1975; Stewart *et al.*, 1984). It would certainly seem disadvantageous to the foal to breathe with the adult horse polyphasic pattern involving active deflation to an EEV substantially less than Vrx (Koterba, 1985). From preliminary studies of the breathing pattern of neonatal Pony foals and from one report illustrating airflow tracings against time in two young (5- and 6-months old) horses (Amoroso *et al.*, 1963), it appeared that both inspiration and expiration in the foal were monophasic, and the pattern more closely resembled a typical neonatal breathing pattern than one of an adult horse. No electromyographic (EMG) or pressure measurements were made, however, to confirm that the foals were breathing above or from Vrx.

The purpose of the present study was to test the hypothesis that the awake, quietly breathing neonatal foal uses breathing strategies similar to those utilized by neonates of smaller species to defend lung volume because of a highly compliant chest wall.

Materials and Methods

Ten Thoroughbred and Thoroughbred-cross foals born over a 2-year period at the College of Veterinary Medicine, University of Florida, were utilized for the studies. Mean $\pm$ s.d. body weight at birth was 40 ± 5 kg, and mean gestational age was 334 ± 11 days. All foals were born spontaneously in a pasture; the births were usually unattended, but no obvious abnormalities associated with the birth process were noted.

Respiratory mechanics (under anaesthesia) and breathing pattern (during awake breathing) were measured at 24–36 h and 7–9 days of age. Due to technical problems with the measurement techniques initially, and to an outbreak of equine influenza later, a complete set of data could not be collected on each of these foals during the first week of life. When not being studied, all foals were housed in a large pasture with their mothers.

Before each study, the foal's respiratory system and general health were assessed by physical examination, chest radiography, and a complete blood count.

Anaesthesia protocol. Before induction of anaesthesia, cuffed Silastic (Dow Corning, Midland, MI) endotracheal tubes (Bivona, Cary, IN, U.S.A.) were passed via the external nares into the trachea and secured. Foals were not fasted before induction of anaesthesia.

Pentobarbitone sodium (15 mg/kg body weight) was administered intravenously over a 3–5-min period. Pancuronium sulphate was given intravenously (0·06 mg/kg) to induce respiratory muscle paralysis, and the foals were mechanically ventilated. Additional barbiturate and pancuronium doses were administered to maintain complete respiratory muscle relaxation. After completion of the experiment the foals were allowed to recover and were returned to the mare.

Mechanical measurements. All measurements were made while the anaesthetized foal was maintained in sternal recumbency, with the front legs tucked under the body and the head supported in a horizontal plane with the rest of the body.

Airflow ($\dot{V}$) was measured using a Fleisch no. 2 heated pneumotachograph connected to the endotracheal tube via an adaptor with a side port used for measurement of airway opening pressure (Pao). A transducer measured the pressure drop across the pneumotachograph, proportional to airflow, and this signal was electronically integrated to yield volume. The endotracheal tube and pneumotach were connected to the ventilator during mechanical ventilation and during the passive pressure–volume manoeuvres, to a supersyringe. The dead space of this equipment was 75 ml.

Oesophageal pressure (Pes) was measured with a latex oesophageal balloon filled with 1–2 ml air. It was connected by Teflon PFA tubing to a differential pressure transducer. All pressure systems were calibrated before and after the studies using a water manometer. The balloon was positioned in the caudal part of the oesophagus, a short distance from the cardia.

For the quasistatic pressure–volume (P–V) curves, pressure and volume signals were displayed on the *x*- and *y*-axes, respectively, of a storage oscilloscope and recorded on tape. Transpulmonary pressure (Ptp) was determined by electrical subtraction of Pes from Pao, transthoracic pressure (Pw) was determined by subtraction of the body surface pressure (Pbs) from Pes, and transrespiratory pressure (Prs) was recorded as the difference between Pao and Pbs. The relaxation volume of the respiratory system was defined as the resting volume reached following a large inflation and slow passive deflation. Under anaesthesia and muscle paralysis, functional residual capacity (FRC) or end-expiratory lung volume (EEV) is equivalent to the relaxation volume of the respiratory system (Vrx). Total lung capacity (TLC) was defined as the lung volume at a Ptp of 30 cmH$_2$O, and residual volume (RV) as the lung volume at a Ptp of -30 cmH$_2$O.

Immediately before the generation of each P–V curve, a standard volume history was provided by twice inflating the lungs to TLC. Following these manoeuvres, the lungs were again slowly inflated to TLC, slowly deflated to RV, and then usually allowed to reinflate passively towards FRC. To observe the change in character of the inflation limb of the P–V curve when inflation took place from a low lung volume, in some studies the lungs were reinflated from RV to TLC after completion of the curve. At least 3 complete P–V curves were obtained per study. Inspiratory capacity (IC) and expiratory reserve volume (ERV) were measured from the P–V curves as the lung volume from FRC to TLC, and from FRC to RV, respectively. The lung volume at which the chest wall deflation curve crossed the *y*-axis, the resting position or unstressed volume of the chest wall (OCW), was also measured. Compliance of the lungs (C$_L$) and chest wall (C$_w$) were measured as the slope of the linear part of the deflation limb of the appropriate P–V curve near FRC. The absolute volumes for the RV and TLC were calculated with the addition of the measurements of FRC (see below). All lung volumes were corrected and reported at basal temperature and pressure. Finally, the maximum and minimum value of Pw at TLC and RV, respectively, was recorded, as well as the value at FRC.

The FRC was measured in the paralysed, anaesthetized state using a modification of the closed-system nitrogen-equilibration technique (Robinson *et al.*, 1972) first described by Lundsgaard & Van Slyke (1918). After lung inflation to TLC and passive deflation to FRC, a large syringe and three-way stopcock containing a volume of oxygen similar to the inspiratory capacity of the subject was attached by way of the pneumotach to the endotracheal tube. From FRC, 4 breaths were delivered to the foal using the entire contents of the syringe each time. The fraction of nitrogen in the syringe after the 4 inflations was measured using a nitrogen analyser. FRC was calculated by the following equation:

$$FRC = Vs \times FN_2(end)/[FaN_2 - FN_2(end)]$$

where Vs = volume of oxygen in syringe; FN$_2$(end) = % nitrogen in the syringe at the end of the test; and FaN$_2$ = % nitrogen in the alveoli at the start of the test. Corrections were made for the apparatus dead space, and for the dead space of the valve and tubing of the oxygen syringe. A total of three FRC determinations were made per study and the results were averaged.

Studies of breathing pattern. Before each study, the unrestrained foal's respiratory rate and breathing pattern were observed from a distance to assess the influence of the study conditions on the measured parameters. The foals were removed from the stall and were studied in an enclosed air-conditioned room in a fully padded enclosure with at least 2 people acting as handlers. No tranquilization was utilized in any of the studies.

Airflow ($\dot{V}$), tidal volume (Vt), and airway pressure (Pao), were measured by means of a fibreglass facemask holding one Fleisch No. 4 pneumotachograph. A band of rubber innertube material cemented to the fibreglass was secured to the foal's head with electrical tape to form an airtight seal. The dead space of the mask was 75–100 ml. A transducer measured the pressure drop across the pneumotachograph (proportional to flow) and this signal was electronically integrated to yield volume. Pao was measured from the proximal port of the pneumotach.

Oesophageal (Pes) and gastric (Pga) pressures were measured with 7-cm long Latex balloons sealed over the end of Teflon PFA tubing. Each catheter was connected via a 3-way stopcock to a differential pressure transducer. Both Pga and Pes balloons were routinely filled with 2 ml air during the studies. The frequency responses of the balloon–catheter systems were tested and found sufficient for the conditions of the experiments. All pressure transducers were calibrated using a water manometer system.

Intercostal (Eint) and abdominal (Eabd) EMGs were recorded from paired surface electrodes positioned over the 11th or 12th intercostal space at mid-thoracic level, and over the lower flank area, respectively. The EMG signal from the diaphragm (Edi) was measured using an intraoesophageal electrode system with an inter-electrode distance of 2 cm.

All signals were recorded on tape during periods of quiet breathing, and observations of the behaviour of the foal at the time of the recording were noted on a voice track. Following instrumentation, the foals were maintained in lateral recumbency, and recordings of breathing parameters were made. The majority of the foals relaxed considerably during this time. The foals were subsequently assisted to their feet and all measurements were repeated with the foal standing quietly.

Ventilatory parameters, mechanical timing intervals, and pressure changes were determined by computer analysis. Depending on the study, the number of breaths analysed in each position ranged from 10 to 50. Mechanical inspiratory (T_I) and expiratory (T_E) times were determined from the zero crossover points on the flow tracings. Tidal volume was measured from the volume tracing. The ratio of T_I:T_E, total breath duration (Ttot), T_I:Ttot, instantaneous breathing frequency ($1/$Ttot $\times 60$), and minute ventilation ($\dot{V}_E$; $1/$Ttot $\times 60 \times$ Vt) were calculated breath-by-breath and averaged. Mean inspiratory flow was obtained by division of Vt by T_I. All values were expressed in both absolute terms and as functions of body weight.

The maximum change in Pes between expiration and inspiration (ΔPesmax), and the maximal change observed in Pga (ΔPgamax) during inspiration and expiration were measured and recorded. Transdiaphragmatic pressure (Pdi) was determined by continuous subtraction of Pga from Pes. The time interval between the onset of inspiratory flow and a decrease or change in downward slope of Pdi was determined.

For EMG analysis, taped data were replayed onto a 6-channel pen writer and storage oscilloscope. For each study, the overall pattern of respiratory muscle activation was observed. The presence or absence of persistent intercostal muscle and diaphragm activity into expiration was recorded. In addition, the time intervals between onset of inspiratory flow and onset of both Edi and Eint were recorded for each breath and averaged. The same comparison was made between onset of expiratory flow and Eabd. In selected breaths, the raw Eabd, Edi or Eint signal was electrically added to the flow signal and plotted against volume on the oscilloscope for generation of 'paintbrush' flow–volume loops.

Results

Static mechanics of neonatal foals

Representative quasi-static P–V curves generated in two 2-day-old foals are shown in Fig. 1. The means of the absolute lung volumes, and the mean volumes normalized to body weight and

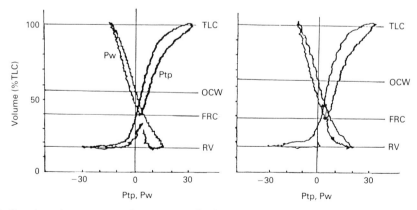

Fig. 1. Quasi-static pressure–volume curves for lung (Ptp) and thorax (Pw) of representative foals at Day 2 of age. The Pw = transthoracic pressure (cmH$_2$O) has signs opposite to that illustrated on scale for Ptp (transpulmonary pressure). Also illustrated are major lung volume subdivisions.

TLC, for foals at Day 2 and Day 7 of age are listed in Table 1. In the newborn foal, TLC averaged 79.5 ± 4.5 ml/kg (mean $\pm$ s.d.), FRC averaged 32 ± 3.2 ml/kg, and RV averaged 12.8 ± 3.8 ml/kg. The volume at the resting position of the chest wall (OCW) in the newborn foal was 46.6 ± 1.7 ml/kg. The ratio of FRC/TLC was $39.2 \pm 2.0\%$. Pressure–volume relationships for the neonatal foal are listed in Table 2. Chest wall compliance was 3.18 ± 0.60 ml/cmH$_2$O/kg on Day 2 of life. C_W/C_L was significantly lower on Day 7 than on Day 2. At Day 2, oesophageal or transthoracic pressure at TLC averaged 10.6 ± 3.9 cmH$_2$O, and was significantly higher on Day 7. A typical P–V curve obtained in the neonatal foal when lung inflation to TLC took place from RV is shown in Fig. 2.

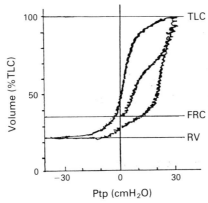

Fig. 2. Pressure–volume characteristics of lungs during inflation from FRC and RV to TLC in a 1-week-old foal. Note high opening pressures required for inflation from RV.

Table 1. Subdivisions of lung volumes in the anaesthetized neonatal foal

Day	n	BW (kg)	TLC	IC	FRC	ERV	RV	OCW
Absolute lung volumes (l)								
2	4–5	40.4	3.23	2.12	1.26	0.78	0.50	1.88
		±5.2	±0.56	±0.42	±0.24	±0.19	±0.14	±0.22
7	6	50.5	4.14	2.56	1.65	0.97	0.68	2.20
		±5.3	±0.43	±0.31	±0.22	±0.24	±0.17	±0.30
Lung volumes normalized to body weight (ml/kg)								
2	4–5	—	79.5	50.9	32.0	19.2	12.8	46.6
			±4.5	±5.5	±3.2	±2.8	±3.8	±1.7
7	6	—	83.2	50.6	33.1	19.0	14.0	44.6
			±4.7	±4.0	±3.2	±3.8	±4.4	±1.2
			FRC/TLC (%)	ERV/TLC	RV/TLC	OCW/TLC		
Lung volumes normalized to total lung capacity (%)								
2	4–5	—	39.2	24.9	14.1	58.8		
			±2.0	±2.8	±2.4	±5.0		
7	6	—	39.8	23.0	16.8	53.5		
			±3.5	±5.4	±5.4	±6.6		

Values are means $\pm$ s.d.
BW = body weight; TLC = total lung capacity; IC = inspiratory capacity; FRC = functional residual capacity; ERV = expiratory reserve volume; RV = residual volume; OCW = resting volume of chest wall.

Table 2. Pressure–volume relationships in the anaesthetized neonatal foal

Day of age	n	C_L (l/cmH$_2$O)	C_L/kg (ml/cmH$_2$O/kg)	C_W (l/cmH$_2$O)	C_W/kg (ml/cmH$_2$O/kg)	C_W/C_L	P_W (TLC) (cmH$_2$O)
2	5	0·15 ±0·04	3·60 ±0·67	0·13 ±0·03	3·18 ±0·60	0·96 ±0·23	10·6 ±3·9
7	7	0·26 ±0·09	5·03 ±1·57	0·13 ±0·03	2·65 ±0·40	0·58* ±0·22	20·0* ±5·9

Values are means ± s.d.
C_L = lung compliance; C_W = chest wall compliance; C_W/C_L = ratio of chest wall to lung compliance; P_W (TLC) = maximum transthoracic pressure at TLC.
*Significantly different from value at Day 2 ($P < 0.05$, Tukey's studentized range test for multiple comparison).

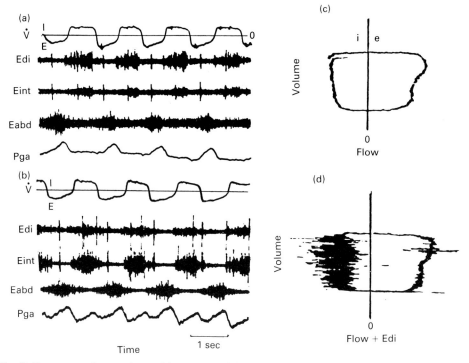

Fig. 3. Representative tracings of immature airflow pattern and EMG activation sequence in the young standing foal. Both inspiration and expiration are essentially monophasic. (a) Tracings of flow (V̇), respiratory muscle EMG activity and gastric pressure (Pga) in a standing 7-day-old foal. Onset of Edi and Eint precede onset of inspiratory flow, and there is no detectable persistence of the signal into expiration. An Eabd signal is present during expiration and Pga peaks towards the end of expiration. (b) Same parameters in another 7-day-old foal, showing similar trends. (c) Flow–volume loop in another standing foal, showing monophasic inspiration and slightly biphasic expiration. (d) Same breath represented by paintbrush flow–volume loop, showing Edi activity throughout most of inspiration.

Breathing studies of awake foals

Table 3 shows the average body weights and ventilatory and pressure parameters in standing foals less than 1 week of age. Table 4 shows the average overall timing parameters in the same foals. Representative tracings of V̇, Pga and EMG activity in two 7-day-old foals are illustrated in Fig. 3.

Table 3. Ventilatory parameters and pressure changes during normal tidal breathing in the standing foal less than 1 week of age

Day of age	n	V_T (l)	V_T/kg (ml/kg)	f (breaths/min)	$\dot{V}_E$ (l/min)	$\dot{V}_E/kg$ (ml/kg)	V_T/T_I (l/min)	$\Delta Pesmax$ (cmH$_2$O)	$\Delta Pgamax$ (cmH$_2$O)	
									Inspiration	Expiration
2	7	0·65 ±0·15	15·8 ±2·5	53·9 ±9·9	34·8 ±12·0	848 ±231	73·9 ±33·0	6·98 ±4·2	1·15 ±0·58	3·32 ±2·31
7	9	0·90 ±0·17	17·6 ±3·0	40·9 ±5·2	33·9 ±7·5	717 ±157	60·6 ±5·7	6·65 ±3·04	1·90 ±1·11	3·26 ±0·65

Values are means ± s.d.
V_T = tidal volume; f = frequency of breathing; $\dot{V}_E$ = minute ventilation; V_T/T_I = mean inspiratory flow; $\Delta Pesmax$ = maximal change in Pes from end expiration to most negative value during inspiration; $\Delta Pgamax$ = maximal change in gastric pressure.

Table 4. Timing parameters in the standing foal during the first week of life

Day of age	T_I (sec)	T_E (sec)	T_{TOT} (sec)	$T_I:T_E$	T_I/T_{TOT}
2	0·51 ±0·15	0·72 ±0·09	1·16 ±0·24	0·79 ±0·15	0·44 ±0·05
7	0·65 ±0·10	0·75 ±0·17	1·45 ±0·24	0·94 ±0·15	0·48 ±0·04

Values are means ± s.d.

Table 5. Timing of onset of EMG parameters relative to onset of inspiratory or expiratory flow in the standing foal

Onset of T_I vs $T_{I(Edi)}$ (sec)	Onset of T_I vs $T_{I(Eint)}$ (sec)	Onset of T_E vs $T_{E(Eabd)}$ (sec)	(% T_E)
+0·056 ± 0·016	+0·012 ± 0·033	−0·036 ± 0·034	5·5
+0·051 ± 0·026	+0·015 ± 0·039	−0·147 ± 0·074	19·4

Means ± s.d. Minus sign denotes that onset of EMG activity lags mechanically event (i.e. beginning of inspiratory or expiratory flow). Plus sign denotes that onset of EMG precedes mechanical event.

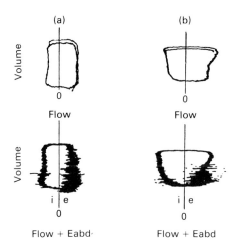

Fig. 4. Paintbrush flow–volume diagrams with Eabd added to flow signal. (a) Box-like, monophasic expiratory flow pattern in a 1-week-old foal and (below) paintbrush flow–volume loop of same breath. Eabd activity is apparent throughout expiration. (b) Flow–volume loop with more pronounced biphasic flow pattern and (below) when Eabd is added to the flow signal, the delay in onset of Eabd relative to onset of expiratory flow is apparent, and probably contributes to the biphasic appearance of flow.

During awake breathing in the standing foal, the most commonly observed pattern was regular in depth and frequency, and inspiration and expiration were essentially monophasic. The configuration of the flow–volume loops varied from essentially box-like to a pattern with slight inflections suggesting the beginning of a biphasic flow pattern (Fig. 3c & d), particularly during expiration. However, no awake neonatal foal showed an adult-like polyphasic pattern of breathing.

Abdominal, intercostal, and diaphragm EMG activities were consistently recorded from all standing foals. Table 5 lists average differences between onset of EMG activity of the 3 respiratory muscles and onset of inspiratory or expiratory airflow. The average onset of Edi and Eint activity preceded the onset of inspiratory flow, but in most foals the signal did not substantially persist into expiration. Phasic abdominal muscle activity during expiration was consistently recorded from every standing foal on Days 2 and 7. The average onset of Eabd lagged the onset of expiratory flow (Table 5) but not nearly to the extent previously observed in the adult horse or in the older foal (Fig. 4b).

Two peaks of Pga, one associated with the last part of inspiration, and one with the last part of expiration, were observed in all standing foals. Pdi reached a minimum value near the end of inspiration in all foals studied. The onset of inspiratory change in Pdi coincided with the onset of inspiratory flow (lagging by 0.02 ± 0.03 sec on Day 2).

Discussion

The values for lung volumes obtained in this study for the anaesthetized 2-day-old neonatal foal are similar to those obtained in a number of other animal species when normalized to body weight (Fisher & Mortola, 1980; Slocombe *et al.*, 1982). Reports of FRC as a percentage of TLC have ranged from 29·8% in neonatal guinea-pigs (Gaultier *et al.*, 1984) to 48% in the 3–8-day-old rat (Fisher & Mortola, 1980), and 51 ± 5% in the neonatal calf (Slocombe *et al.*, 1982). In the puppy, FRC/TLC averaged 30% in one study (Fisher & Mortola, 1980) and 20·1% in an earlier one (Agostoni, 1959). From the configuration of the quasi-static P–V curve of the chest wall (Fig. 1,

Fig. 2), it appears that RV in the neonatal foal, as in other neonates, is determined primarily by airway closure rather than by chest wall stiffness.

The reported values for the resting volume of the chest wall of newborn species have ranged from 39% of TLC in the puppy (Fisher & Mortola, 1980) to $80 \pm 8\%$ of TLC in the calf (Slocombe *et al.*, 1982). However, values for OCW/TLC in the rat, rabbit, cat and pig (Fisher & Mortola, 1980) all have been 61–63%, which is similar to the value of $58\cdot8 \pm 5\%$ reported in this study for the foal.

Lung compliance normalized to body weight (Table 2) was similar to the values reported previously for other neonates (Swyer *et al.*, 1960; Hjalmarson, 1974; Fisher & Mortola, 1980; Gerhardt & Bancalari, 1980; Slocombe *et al.*, 1982).

A soft, flexible chest wall is obviously beneficial for delivery of a mammal through the narrow birth canal (Mortola, 1983), and a high chest wall compliance has been confirmed in all neonatal species examined to date. C_W/kg averaged $4\cdot2 \, ml/cmH_2O/kg$ in term human infants and $6\cdot4 \, ml/cmH_2O/kg$ in premature infants (Gerhardt & Bancalari, 1980). Values in newborn rats, rabbits, cats, dogs and pigs ranged between $6\cdot1 \pm 2\cdot7$ (mean $\pm$ s.d.) $ml/cmH_2O/kg$ (dog) and $12\cdot23 \pm 1\cdot12 \, ml/cmH_2O/kg$ (cat). Slocombe *et al.* (1982) found C_W/kg to be $8\cdot85 \pm 2\cdot44 \, ml/cmH_2O/kg$ (mean $\pm$ s.e.m.) in a group of neonatal calves, and Avery & Cook (1961) found values in two 3-day-old goats to be $6\cdot15$ and $15\cdot4 \, ml/cmH_2O/kg$. The average value obtained for C_W/kg in the 2-day-old foal was only $3\cdot18 \pm 0\cdot60 \, ml/cmH_2O/kg$ (mean $\pm$ s.d.), considerably lower than that recorded for the other neonatal species. C_W/C_L was also low in comparison to other neonatal species. Both parameters appeared to decrease with increasing age, but values at 1 week of age were still considerably greater than reported for the adult horse (Leith & Gillespie, 1971) when values were normalized to body weight.

When normalized for body weight, the mean value for $\dot{V}_E/kg$ was high ($848 \, ml/min/kg$) in the 2-day-old foal relative to that ($162 \, ml/min/kg$) of the adult horse (Koterba, 1985). It is probable that $\dot{V}_E/kg$ was initially elevated at least partly as a result of a high metabolic rate in the newborn foal (Stewart *et al.*, 1984).

The results of the awake breathing studies in the standing position confirmed the observation made in preliminary studies that the pattern of airflow in neonatal foals is essentially monophasic. No awake neonatal foal showed an adult-like polyphasic pattern of breathing.

Analysis of the overall pattern of respiratory muscle activation with respect to airflow revealed important differences between the neonatal and adult horse. In the neonatal foal, onset of Edi consistently preceded the onset of inspiratory flow. Thus, in contrast to the adult horse in which a passive component of inspiration was also observed (Koterba, 1985), but, like the pattern observed in most other species, inspiration in the neonatal foal was an active process. However, in contrast to most other species in which expiration during resting conditions has been described as a primarily passive process, expiration in the standing neonatal foal was a primarily active event. Phasic abdominal EMG activity was always observed in the standing foal. Although there was considerable individual variation, on average, abdominal muscle EMG activity was detected early in expiration in the young foals. Early activation of inspiratory and expiratory respiratory muscles in the younger foals was associated with a flow pattern essentially monophasic in character.

The patterns of change observed in Pga and Pdi during the breathing cycle were consistent with the EMG and airflow data. The configuration of the Pga tracings against time in all neonatal standing foals was very consistent and similar to that described for the adult horse (Koterba, 1985). In the majority of the younger foals, the onset of decline of Pdi to its minimum value during inspiration was closely associated with the onset of inspiratory flow, supporting an early onset of diaphragmatic activation.

In the adult horse, the delay of inspiratory muscle activation relative to the onset of inspiratory flow, the biphasic inspiratory flow pattern, and a mid-expiratory pause in flow were strong indications that EEV was substantially below Vrx (Koterba, 1985). In the younger foal, however, which typically utilized an essentially monophasic airflow pattern, it could not be conclusively established whether the foals were breathing from, above or below Vrx. The major difficulty in determination of the position of EEV relative to Vrx lay in the fact that these individuals persistently used their abdominal muscles phasically during expiration and did not delay onset of inspiratory muscle activity or display a biphasic inspiration. If the expiratory muscles had been relaxed during expiration, analysis of the passive flow–volume curve could have established whether the foals breathed from or above Vrx (Griffiths *et al.*, 1983; Kosch & Stark, 1984; Mortola *et al.*, 1985).

Although conclusive proof cannot be provided, from the data that were acquired in this study, it appears unlikely that the standing neonatal foal typically breathes from above Vrx. First, neonates of other species that have an elevated EEV relative to Vrx routinely utilize mechanisms that effectively prolong the time constant by retarding expiratory airflow. In the standing neonatal foal, obvious upper airway expiratory braking was infrequently observed, and inspiratory muscle activity after inspiration was in fact observed substantially less frequently in the young foals than in the adult horse (Koterba, 1985). Second, the consistent use of the abdominal muscles during expiration would be expected to aid the deflation process; the effective or active time constant in such a system would be expected to be shorter than under passive conditions. Therefore, a complete expiration to or below Vrx should actually be facilitated by such a strategy.

From the results of this study, the neonatal foal was found to utilize a pattern of breathing distinct from the adult horse. Compared to the adult, this pattern was characterized by a higher frequency of breathing, higher airflow rate and minute ventilation on a body weight basis, and a primarily monophasic pattern of both inspiration and expiration. Both phases of the breathing cycle were primarily active in the foal. Most of these findings are similar to those observed in neonates of other smaller species, and most easily explained by the need for the neonate to maintain a high ventilation because of increased metabolic requirements. This need may actually be accentuated in the foal because of the high level of activity it displays from an early age. The active expiration consistently observed in the foals is compatible with a need in the neonatal foal to maintain a high ventilation, as this would be expected to promote movement of air out of the lungs quickly. Whether other neonatal species phasically use their abdomen during expiration has not been well documented.

It appears that the term, standing neonatal foal utilizes few if any of the strategies that are thought to be used by the neonates of other species to actively defend end-expiratory lung volume above Vrx. In the neonatal foal, therefore, the mechanical characteristics of the respiratory system may influence the strategy of breathing to a lesser degree than in other neonates. One obvious explanation for this lies in the results of the studies of static mechanics. The chest wall of the neonatal foal was considerably stiffer than that of neonates of other species, and FRC normalized to body weight or to lung volume was not low. These findings are in contrast to those in other neonates, in which a high chest wall compliance leads to a low Vrx and the need for active mechanisms to defend lung volume. It is probable that the term foal's respiratory system is sufficiently mature from a structural point of view that compensatory mechanisms to defend end-expiratory lung volume are not needed, at least in the standing position.

It should be remembered that the foals included in this study were all term, normal individuals. The breathing strategies utilized by the premature foal or by the foal with respiratory disease may be very different from those reported here: premature and otherwise compromised recumbent neonatal foals may display both markedly irregular breathing patterns and consistently audible grunts, suggestive of upper airway braking.

We thank Dr Kathy Brock for help with the anaesthesia; J. Wozniak, L. Coons and T. Whitlock for assistance during the studies; the Equine Neonatal Study Group for making the foals available; and Linda Rose for help in preparing the manuscript.

References

Agostoni, E. (1959) Volume-pressure relationships of the thorax and lung in the newborn. *J. appl. Physiol.* **14**, 909–913.

Amoroso, E.C., Scott, P. & Williams, K.G. (1963) The pattern of external respiration in the unanesthetized animal. *Proc. R. Soc. B* **159**, 325–347.

Avery, M.E. & Cook, C.D. (1961) Volume-pressure relationships of lungs and thorax in fetal, newborn and adult goats. *J. appl. Physiol.* **16**, 1034–1038.

Bartlett, D., Remmers, J.E. & Gautier, H. (1973) Laryngeal regulation of respiratory airflow. *Respir. Physiol.* **18**, 194–204.

England, S.A., Kent, G. & Stogryn, A.F. (1985) Laryngeal muscle and diaphragmatic activities in conscious dog pups. *Resp. Physiol.* **60**, 95–108.

Farber, J.P. (1978) Laryngeal effects and respiration in the suckling opossum. *Resp. Physiol.* **35**, 189–201.

Findley, L.J., Ries, A.L., Tisi, G.M. & Wagner, P.D. (1983) Hypoxemia during apnea in normal subjects: mechanisms and impact of lung volume. *J. appl. Physiol.* **55**, 1777–1783.

Fisher, J.T. & Mortola, J.P. (1980) Statics of the respiratory system in newborn mammals. *Resp. Physiol.* **41**, 155–172.

Gaultier, C.L., Harf, A., Lorino, A.M. & Atlan, G. (1984) Lung mechanics in growing guinea pigs. *Resp. Physiol.* **56**, 217–228.

Gautier, H., Remmers, J.E. & Bartlett, D., Jr (1973) Control of the duration of expiration. *Respir. Physiol.* **35**, 189–201.

Gerhardt, T. & Bancalari, E. (1980) Chestwall compliance in full-term and premature infants. *Acta paediat. scand.* **69**, 359–364.

Gillespie, J.R. (1975) Postnatal lung growth and function in the foal. *J. Reprod. Fert., Suppl.* **23**, 667–671.

Griffiths, G.B., Noworaj, A. & Mortola, J.P. (1983) End-expiration level and breathing pattern in the newborn. *J. appl. Physiol.* **55**, 243–249.

Harding, R., Johnson, P. & McClelland, M.E. (1980) Respiratory function of the larynx in developing sheep and the influence of sleep state. *Resp. Physiol.* **40**, 165–179.

Hjalmarson, O. (1974) Mechanics of breathing in newborn infants with pulmonary disease. *Acta paediat. scand. Suppl.* **247**, 26–48.

Kosch, P.C. & Stark, A.R. (1984) Dynamic maintenance of end-expiratory lung volume in full-term infants. *J. appl. Physiol.* **57**, 1126–1133.

Koterba, A.M. (1985) *Developmental changes in respiratory mechanics and breathing strategy in the growing horse.* Ph.D. dissertation, University of Florida.

Leith, D.E. & Gillespie, J.R. (1971) Respiratory mechanics of normal horses and one with chronic obstructive lung disease. *Fedn Proc. Fedn Am. Socs exp. Biol.* **30**, 556.

Lopes, J., Muller, N.L., Bryan, M.H. & Bryan, A.C. (1981) Importance of inspiratory muscle tone in maintenance of FRC in the newborn. *J. appl. Physiol.* **51**, 830–834.

Lundsgaard, C. & Van Slyke, D.D. (1918) Studies of lung volume. I. Relation between thorax size and lung volume in normal adults. *J. exp. Med.* **27**, 65–85.

Mortola, J.P. (1983) Some functional mechanical implications of the structural design of the respiratory system in newborn animals. *Am. Rev. Respir. Dis.* **128**, 569–572.

Mortola, J.P. (1984) Breathing pattern in newborns. *J. appl. Physiol.* **56**, 1533–1540.

Mortola, J.P., Fisher, J.T., Smith, B., Fox, G. & Weeks, S. (1982) Dynamics of breathing in infants. *J. appl. Physiol.* **52**, 1209–1215.

Mortola, J.P., Magnante, D. & Saetta, M. (1985) Expiratory pattern of newborn mammals. *J. appl. Physiol.* **58**, 528–533.

Remmers, J.E. & Bartlett, D., Jr (1977) Reflex control of expiratory airflow and duration. *J. appl. Physiol.* **42**, 80–87.

Robinson, N.E., Gillespie, J.R., Berry, J.D. & Simpson, A. (1972) Lung compliance, lung volumes, and single-breath diffusing capacity in dogs. *J. appl. Physiol.* **33**, 808–812.

Rossdale, P.D. (1969) Measurements of pulmonary ventilation in normal newborn thoroughbred foals during the first three days of life. *Br. vet. J.* **125**, 157–161.

Rossdale, P.D. (1970) Some parameters of respiratory function in normal and abnormal foals with special reference to levels of PaO_2 during air and oxygen inhalation. *Res. Vet. Sci.* **11**, 270–276.

Slocombe, R.F., Robinson, N.E. & Derksen, F.J. (1982) Effect of vagotomy on respiratory mechanics and gas exchange in the neonatal calf. *Am. J. vet. Res.* **43**, 1165–1171.

Stewart, J.H., Rose, R.J. & Barko, A.M. (1984) Respiratory studies in foals from birth to seven days old. *Equine vet. J.* **16**, 323–328.

Swyer, P.R., Reiman, R.C. & Wright, J.J. (1960) Ventilation and ventilatory mechanics in the newborn. *J. Pediatr.* **56**, 612–621.

J. Reprod. Fert., Suppl. **35** (1987), 587–592

Bronchoalveolar lavage in the newborn foal

I. K. M. Liu, E. M. Walsh*, M. Bernoco and A. T. W. Cheung*

*Department of Reproduction, School of Veterinary Medicine and *Department of Pediatrics, School of Medicine, University of California, Davis, CA 95616, U.S.A.*

Summary. Two bronchoalveolar lavages, 24 h apart, were performed on 15 foals, ranging in age from 1 to 21 days. In the first lavage, a numerical deficiency in alveolar macrophages was demonstrated in foals up to 2 weeks of age when compared with older (2–3 years of age) horses. Alveolar macrophages obtained from the lungs of 2–3-day-old foals also demonstrated significant impaired chemotactic function. In the second lavage, although an increase of alveolar macrophages was noted, a dramatic increase of polymorphonuclear neutrophil (PMN) mobilization occurred in the foals, thus providing a phagocytic back-up for the alveolar macrophage host defence mechanism. In contrast, PMN response to lower respiratory tract perturbation in older horses was negligible.

The results of this study suggest that the numerical and chemotactic deficiency found in alveolar macrophages of the neonate may play a role in the foal's susceptibility to respiratory disease.

Introduction

Alveolar macrophages are considered to be one of the most important defence mechanisms in the lung (Toews, 1983; Wilson, 1984). Knowledge concerning the role of these macrophages is particularly important in view of the vulnerability of the newborn lung to environmental insult and bacterial invasion. In the neonate, major emphasis has been placed on investigations relating to quantification and functional maturation of alveolar macrophages in several mammalian species. Zeligs *et al.* (1977) clearly demonstrate that the number of lavaged alveolar macrophages in newborn rabbits increases dramatically during the first week of postnatal development and continues to increase up to 4 weeks of age when a plateau is reached. Similar findings were also reported by Sieger *et al.* (1975) utilizing newborn rabbits and human infants. Kazmierowski *et al.* (1976, 1977) demonstrated that repeated lavages (4–24 h after initial lavages) in adult monkeys led to an appreciable rise in the number of polymorphonuclear neutrophils (PMN) in the bronchoalveolar lavage fluid. Cheung *et al.* (1986) confirmed the findings of Kazmierowski *et al.* (1976) in adult monkeys and, in addition, demonstrated a numerical and functional deficiency of alveolar macrophages in newborn monkeys.

Despite the importance and frequency of occurrence of respiratory tract infections in the equine neonate, a paucity of information is available relative to the defence mechanisms of the newborn lung. Sequential bronchoalveolar lavages were performed on 3 foals ranging in age from 6 to 100 days by Zink *et al.* (1982) and low numbers of macrophages were recovered from foals less than 20 days of age; numbers then increased significantly up to 100 days of age. The alveolar macrophages recovered were also capable of ingesting *Corynebacterium equi in vitro* in the presence of immune serum. Subsequently, Zink *et al.* (1985) reported that alveolar macrophages obtained from foals ranging in ages from 6 to 61 days demonstrated significant decrease in phagocytic activity but no differences were found in the killing capacity of the macrophages between the ages of 21 and 61 days.

The purpose of this investigation was to determine the number of alveolar macrophages and polymorphonuclear neutrophils present in the lungs of newborn foals and to investigate the mobilization of cells in response to lower respiratory tract perturbation.

Materials and Methods

Horses. Nineteen (19) clinically healthy horses varying in ages from 1 day to 3 years were utilized. Four (4) foals ranged in ages between 1 and 2 days, seven (7) between 3 and 5 days, two (2) at 2 years and two (2) at 3 years of age. The horses were of Quarter Horse and Thoroughbred breeds. All foals less than 6 months of age were sucking their dams.

Bronchoalveolar lavage. In the newborns, bronchoalveolar lavage was performed using a 1·5-m nasogastric tube with an outer diameter of 7·5 mm and an inner diameter of 5·0 mm. Within the nasogastric tube, a polyethylene tube was inserted with an outer diameter of 4·8 mm and an inner diameter of 3·7 mm. For the adult horses (2–3 years of age), a larger nasogastric tube (2·3 m) was used with an outer diameter of 10·5 mm and an inner diameter of 6·0 mm.

After light anaesthesia with xylazine at 1·0 mg/kg followed by ketamine hydrochloride at 2·2 mg/kg, the horse was placed at left lateral recumbency, the nasogastric tube was inserted via the nostrils, through the pharynx, into the trachea and continued until the tube was lodged into the left caudal lobe of the lung (Fig. 1). A volume of 500 ml phosphate-buffered saline (PBS, 272 mosmol) was used in 9 lavage cycles and the total amount of fluid retrieved was recorded. At the completion of the lavage the animal was allowed to recover normally. For each foal and adult control 2 lavages, 24 h apart, were carried out in identical manner.

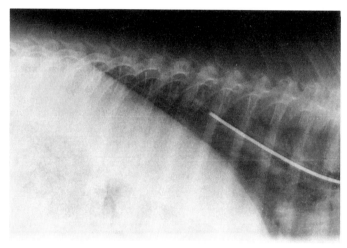

Fig. 1. X-ray radiograph of barium-filled polyethelene catheter to confirm the site of bronchoalveolar lavage.

Harvesting of cells and measurement of phagocytes. The recovered fluid from the bronchoalveolar lavage was centrifuged at 500 *g* for 15 min. The cell pellet was resuspended in 2 ml McCoy's 5a modified culture medium enriched with L-glutamine and 10% heat-inactivated fetal calf serum (Gibco, Santa Clara, CA). Before adjusting the macrophage concentration for migration assay, 3 smears were made for differential cell measurement from the 2 ml McCoy's suspension. The smears were air dried and stained with Wright–Giemsa stain. The number of macrophages, neutrophils and lymphocytes were recorded per 100 cells counted. Cell viability was determined by the Trypan blue exclusion test. In view of the important role played in lung defence mechanisms, total macrophage and neutrophil concentrations of the 2 ml McCoy's suspension was determined by the standard haemocytometer procedure and the original concentration of the cells in the lavage fluid was recorded following adjustment of the total volume retrieved per lavage. The concentration of cells (per group) was expressed as the mean alveolar macrophage or PMN numbers plus the standard error of the mean.

Each newborn foal (as well as adult control) was lavaged twice; the first lavage provided a baseline score for differential macrophage and neutrophil counts. The second lavage, 24 h after the first, gave the count for neutrophil migration arising as a result of lower respiratory tract perturbation.

Chemotaxis assay. Migration responsiveness of alveolar macrophages from the lungs of 11 horses varying in age from 2 days to 3 years was studied by a modification of the Boyden assay (Wilkinson *et al.*, 1982): alveolar macrophages from the first lavage of each test (1×10^6/ml) were placed in the upper portion of a blind well chamber (Neuroprobe, Bethesda, MD) over an 8 μm pore-sized cellulose ester filter. The lower portion of the chamber was filled with 0·2 ml endotoxin-activated plasma (40 μg of *E. coli* lipopolysaccharide (Sigma, St Louis, MO) in 0·4 ml test plasma diluted 1:5 in McCoy's medium as the chemoattractant). Triplicate assays were performed on all samples (with controls). The chambers were incubated at 37°C for 3 h, after which the filters were removed and stained with the modified haematoxylin–Wright procedure. The number of alveolar macrophates per high power field (× 100, oil immersion) which had migrated through the micropore filter was counted. Ten fields under high power were counted per assay of each test sample and the average number of macrophages per field was calculated. Student's *t* test was used in the statistical analysis.

Results

Bronchoalveolar lavage

The lavage tubing was consistently found to be lodged within the segmental bronchi of the left caudal lobe of the lung (Fig. 1). Following 2 lavages, 24 h apart, no symptoms of illness were noted in the recovered horses.

Lavage fluid measurement and cell count

In the first and second lavages, 80% or more of the fluid was consistently recovered. The recovery rates of macrophages, neutrophils and lymphocytes were: 84·22% macrophages, 10·44%

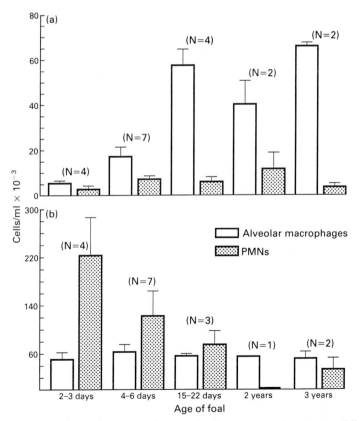

Fig. 2. Concentrations of alveolar macrophage and polymorphonuclear neutrophils (PMN) in foals of different ages at (a) the first lavage and (b) the second lavage.

neutrophils and 5·56% lymphocytes in the first lavage and 47·67% macrophages, 48·67% neutrophils and 3·78% lymphocytes in the second lavage in newborn foals 2 days–2·5 weeks of age. In the
horses which were considered adults (2–3 years of age) and which served as controls, the first lavage
yielded 69·67% macrophages, 1·86% neutrophils and 28·00% lymphocytes. The second lavage
yielded 54·33% macrophages, 14·00% neutrophils and 21·33% lymphocytes. These results suggest
that PMN responsiveness of the lung in the younger foals is significantly greater than in older
horses following lower respiratory tract perturbation.

The concentrations of macrophages and neutrophils recovered from the bronchoalveolar lavage
in foals are shown in Fig. 2. In the first lavage (Fig. 2a), the number of macrophages per ml was
considered minimal ($3·90 \times 10^3 \pm 1·09$) in the first 2 days of the foal's life. A gradual increase in the
number of macrophages ($17·30 \times 10^3 \pm 4·20$) was found in foals ranging from 3 to 5 days of
age. By 14 days of age, the number of macrophages retrieved from the lavage was equivalent
($57·60 \times 10^3 \pm 7·13$) to that recovered from horses considered to be adults ($40·25 \times 10^3 \pm 10·50$
at 2 years and $65·95 \times 10^3 \pm 1·66$ at 3 years). The neutrophil content of the first lavages was
consistently low, ranging from $2·88 \times 10^3 \pm 1·50$ to $11·65 \times 10^3 \pm 7·32$ cells/ml. It appears from
these results that the numbers of macrophages present in the lower respiratory tract of horses reach
their maximum by 2 weeks of age.

The number of macrophages retrieved from the lungs of horses in the second lavage revealed an
increased response from the first lavage at all ages. The most notable increase was found in the
newborn foals (1–2 days of age) (Fig. 2b). PMN response as a result of lower respiratory tract
lavage was most dramatic again in the newborn foals (Fig. 2b). The increase in response continued
in the 4–6-day-old foal group and up to 14–21 days of age. In the adults, the number of PMN found
in the second lavage as a result of lower respiratory tract lavage was considered minimal when
compared to the younger foals.

Chemotactic assay

The ability of alveolar macrophages in newborn foals to migrate through the micropore filter in
the presence of endotoxin-activated plasma was diminished when compared with alveolar macrophages from foals 2 weeks of age and older (Table 1). Although alveolar macrophages obtained
from the lungs of one 3-day-old foal were considered to have functional activity, the overall chemotactic activity of cells obtained from the 2- and 3-day-old foals tested was significantly impaired

Table 1. Migration scores of alveolar macrophages obtained
from the lower respiratory tract of newborn foals and horses of
different ages

Age of foal	Macrophage + endotoxin-activated plasma	Macrophage + medium
2 days	0·1	0·1
2 days	0·0	0·0
3 days	0·5	0·2
3 days	6·1	1·9
2 weeks	16·3	3·3
2·5 weeks	5·1	0·4
4 months	9·6	4·1
4 months	17·5	2·3
6 months	5·4	0·1
1·5 years	3·7	1·3
3 years	3·6	0·3

($P \leqslant 0.05$) when compared with alveolar macrophages obtained from the lungs of older foals and horses.

These preliminary findings suggest a deficiency in the ability of alveolar macrophages in newborn foals to migrate. Further investigations are needed, however, to confirm these findings as the sample size is relatively small.

Discussion

The results obtained from the present investigation closely agree with other studies in several mammalian species with reference to the numerical deficiency of alveolar macrophages present in the newborn lung of rabbits (Zeligs *et al.*, 1977; Sieger, 1978), humans (Sieger *et al.*, 1975), monkeys (Cheung *et al.,* 1986), and horses (Zink & Johnson, 1984).

Although this study demonstrates that the number of alveolar macrophages in newborn foals reaches its peak by 2 weeks of age when compared with horses considered adults, Zink *et al.* (1982) report a biphasic increase in numbers of alveolar macrophages in foals tested within 2 weeks after birth to 100 days of age. Subsequently, Zink & Johnson (1984) reported the increase in number of alveolar macrophages in newborn foals to continue only up to 3 weeks of age. It is clear from the present studies, however, that the newborn foal has a numerical deficiency of alveolar macrophages during the first 2 weeks of life and appears to obtain a full complement of macrophages at a relatively early age. The numbers of lymphocytes and neutrophils also closely agree with those reported by Zink & Johnson (1984), despite the fact that the foals in the present study were younger.

A dramatic increase in the concentration of PMN was found after lower respiratory tract perturbation. This was particularly true in the younger foals in this study, indicating a phagocytic back-up for the alveolar macrophage host defence mechanism. In other mammalian species, however, the opposite appears to be true. Kazmierowski *et al.* (1977) and Cheung *et al.* (1986) demonstrated that, within 4–48 h of bronchoalveolar lavage in adult monkeys, there was a substantial increase of neutrophil migration into the lungs. Conversely, the lungs of newborn monkeys were incapable of mobilizing PMN 24 h after an initial lavage (Cheung *et al.*, 1986). Further investigations are warranted to explain the differences found in horses and non-human primate species.

Other attempts to delineate the role of alveolar macrophages of newborn young against bacterial invasion suggest that microbicidal activity of macrophages is decreased in most 7-day-old or younger animals (Zeligs *et al.*, 1977; Bellanti *et al.*, 1979; Shuit *et al.,* 1982; Weiss *et al.* 1983). This decrease in killing appears to be due to diminished intracellular killing rather than due to defective phagocytosis (Bellanti *et al.*, 1979; Shuit *et al.*, 1982), and the bactericidal activity of the phagocytes becomes comparable to that of adult animals by 3–4 weeks of age (Bellanti *et al.,* 1979; Zeligs *et al.*, 1977; Sieger, 1978; Shuit *et al.*, 1982; Jacobs *et al.*, 1983). Other investigators report chemotactic and phagocytic impairment in 1–2-day-old monkeys (Cheung *et al.*, 1986) while phagocytic function was found to be deficient up to 61 days of age in the horse (Zink *et al.*, 1985). Our preliminary data on the migrational ability of alveolar macrophages suggest a deficiency in chemotactic function in foals 2–3 days of age, with normal activity occurring at 2 weeks of age. Much more data are needed, however, before any conclusions can be drawn.

We have attempted to examine a critical area responsible for the defence mechanism of the lung in newborn foals. We conclude that a numerical deficiency of alveolar macrophages exists in the lung of the newborn foal and that the ability to mobilize PMN into the lung after lower respiratory tract perturbation is significantly greater in newborn foals than that of the adult horse.

This work was supported by the Oak Tree Racing Association, Arcadia, California.

References

Bellanti, J.A., Nerurkar, L.S. & Zeligs, B.J. (1979) Host defenses in the fetus and neonate: studies of the alveolar macrophage during maturation. *Pediatrics* **64**, 726–739

Cheung, A.T.W., Kurland, G., Miller, M.E., Ford, E.W., Ayin, S.A. & Walsh, E.M. (1986) Host defense deficiency in newborn nonhuman primate lungs. *J. med. Primatol.* **15**, 37–47.

Jacobs, R.F., Wilson, C.B., Smith, A.L. & Haas, J.E. (1983) Age-dependent effects of aminobutyryl maramyl dipeptide on alveolar macrophage function in infant and adult Macaca monkeys. *Am. Rev. Respir. Dis.* **128**, 862–867.

Kazmierowski, J.A., Fauci, A.S. & Reynolds, H.Y. (1976) Characterization of lymphocytes in bronchial lavage fluids in monkeys. *J. Immunol.* **116**, 615–618.

Kazmierowski, J.A., Gallin, J.I. & Reynolds, H.Y. (1977) Mechanism for the inflammatory response in primate lungs: demonstration and partial characterization of an alveolar macrophage-derived chemotactic factor with preferential activity for polymorphonuclear leukocytes. *J. clin. Invest.* **59**, 273–281.

Shuit, K.E., Krebs, R.E., Rohn, D. & Steele, V. (1982) Effect of fetal protein malnutrition on the postnatal structure and function of alveolar macrophage. *J. Infect. Dis.* **146**, 498–505.

Sieger, L. (1978) Pulmonary alveolar macrophages in pre- and post-natal rabbits. *J. Reticuloendothel. Soc.* **23**, 389–395.

Sieger, L., Attar, H., Okada, D. & Thiebeault, D. (1975) Alveolar macrophages (A.M.) in rabbit and human fetuses and neonates. *J. Reticuloendothel. Soc.* **18**, 47b, Abstr.

Toews, G.B. (1983) Pulmonary clearance of infectious agents. In *Respiratory Infections: Diagnosis and Management*, pp. 31–39. Ed. J.E. Pennington. Raven Press, New York.

Weiss, R.A., Chanana, A.D. & Joel, D.D. (1983) The development of pulmonary host defense in the neonatal sheep. *J. Reticuloendothel. Soc.* **34**, 190, Abstr.

Wilkinson, P.C., Lackie, J.M. & Allan, R.B. (1982) Methods for measuring leukocyte locomotion. In *Cell Analysis*, vol. 1, pp. 145–193. Ed. N. Catsimpoolas. Plenum Press, New York.

Wilson, C.B. (1984) Lung antimicrobial defenses in the newborn. *Sem. Resp. Med.* **6**, 149–155.

Zeligs, B.J., Nerurkar, L.S., Bellanti, J.A. & Zeligs, J.D. (1977) Maturation of the rabbit alveolar macrophage during animal development. I. Perinatal influx into alveoli and ultrastructural differentiation. *Pediat. Res.* **11**, 197–208.

Zink, M.C. & Johnson, J.A. (1984) Cellular constituents of clinically normal foal bronchoalveolar lavage fluid during postnatal maturation. *Am. J. vet. Res.* **45**, 893–897.

Zink, M.C., Johnson, J.A., Prescot, J.F. & Pascoe, P.J. (1982) The interaction of *Corynebacterium equi* and equine alveolar macrophages *in vitro*. *J. Reprod. Fert., Suppl.* **32**, 491–496.

Zink, M.C., Yager, J.A., Prescott, J.F. & Wilkie, B.N. (1985) In vitro phagocytosis and killing of *Corynebacterium equi* by alveolar macrophages of foals. *Am. J. vet. Res.* **46**, 2171–2174.

J. Reprod. Fert., Suppl. **35** (1987), 593–598

Comparison of systemic and local respiratory tract cellular immunity in the neonatal foal

U. Fogarty and D. P. Leadon

Irish Equine Centre, Johnstown, Naas, Co. Kildare, Ireland

Summary. Blood neutrophils from 10 Thoroughbred and 2 Pony foals were evaluated using in-vitro cellular function tests of chemotaxis, chemiluminescence, phagocytosis and intracellular killing. A comparison of the functional capacities of these cells before and 2–4 days after the ingestion of colostrum indicated an improvement in blood neutrophil chemotaxis and chemiluminescence. Bronchopulmonary lavage was carried out on 9 Thoroughbred and 2 Pony 36-h-old foals. The technique used did not require sedation or anaesthesia. Pulmonary alveolar macrophages were the predominant cell type recovered. When comparisons were made between blood neutrophils and pulmonary alveolar macrophages in the same animal fewer pulmonary alveolar macrophages were phagocytic and there was little if any evidence of intracellular killing by pulmonary alveolar macrophages.

Introduction

The significance of maternally derived immunoglobulin to the newborn foal has been long recognized (Ehrlich, 1892), but the role of systemic and local cellular immunity in neonatal protection is poorly defined. In a study of systemic cellular immunity Coignoul (1983) proposed that the greater susceptibility of young foals to equine herpes virus (EHV)-1 infection may be related to reduced blood neutrophil function. In the young foal immature or defective local respiratory cellular immunity may predispose to infection (Zink & Johnson, 1984; Zink *et al.*, 1985).

The purpose of the present study was to compare local and systemic cellular immunity in the neonatal foal and to monitor the effects of colostrum intake on the systemic cellular response.

Materials and Methods

Foals and methods

The 23 foals used in this study were located on two commercial stud farms, A and B. Foals on Farm A included 10 Thoroughbred and 2 Pony foals. Foals on Farm B included 9 Thoroughbred foals and 1 Pony foal. The mares were foaled indoors between February and June, and all foals stood and sucked within a few hours of birth.

Blood sampling. Blood samples were taken from the 12 foals on Farm A within the first hour *post partum* and again 2–4 days later and from the 10 foals on Farm B and 1 Pony foal on Farm A on the second day *post partum*. Blood samples were taken by jugular venepuncture. Blood for haematology and blood biochemistry was taken into vacutainer EDTA, citrate and plain tubes. Blood (15 ml) for the neutrophil function assays was taken into a sterile vessel containing 100 µl preservative-free sodium heparin (5000 i.u./ml).

Bronchopulmonary lavage. Bronchopulmonary lavage samples were taken from 10 foals on Farm B and 1 of the Pony foals on Farm A. The technique used was a modification of that described by Fogarty *et al.* (1983). Foals were restrained standing, with their necks straight and in an extended position to align the caudal nares with the larynx. The left nostril was swabbed with alcohol and a sterile foal stomach tube (28 cm long) was introduced along the ventral meatus into the upper portion of the trachea. The stomach tube was held in position by an attendant holding the length of suture material attached to the exposed end of the tube. A sterile modified Baker jejunostomy tube (No. 6654, Warne Surgical Products Ltd, Andover, U.K.) was introduced through the foal stomach tube into the lung until it became wedged in a bronchus, where it was gently held in position. When both tubes were in position the foal could

Table 1. White blood cell counts and protein concentrations (mean $\pm$ s.d.) in (a) 12 foals before and 2–4 days after colostrum ingestion and (b) 11 foals sampled during the 2nd day *post partum*

Haematology and biochemistry	(a) Before colostrum ingestion	(a) 2–4 days after colostrum ingestion	(b) 2 days *post partum*
WBC ($\times 10^{-9}$/l)	7·2 $\pm$ 1·1	7·6 $\pm$ 2·0	9·6 $\pm$ 2·5
Neutrophils ($\times 10^{-9}$/l)	4·2 $\pm$ 0·6	5·8 $\pm$ 1·0	6·7 $\pm$ 2·9
Lymphocytes ($\times 10^{-9}$/l)	3·1 $\pm$ 1·1	2·2 $\pm$ 0·9	2·6 $\pm$ 1·3
Monocytes ($\times 10^{-9}$/l)	0·03 $\pm$ 0·06	0·1 $\pm$ 0·1	0·1 $\pm$ 0·1
Eosinophils ($\times 10^{-9}$/l)	0·0	0·05 $\pm$ 0·09	0·01 $\pm$ 0·04
Neutrophil:lymphocyte ratio	1·7 $\pm$ 0·9	3·1 $\pm$ 1·3	3·3 $\pm$ 2·1
Total protein (g/l)	38·4 $\pm$ 2·8	54·5 $\pm$ 9·8	50·7 $\pm$ 10·9
Albumin (g/l)	24·5 $\pm$ 2·6	24·7 $\pm$ 3·2	23·1 $\pm$ 2·9
Globulin (g/l)	13·7 $\pm$ 2·3	28·7 $\pm$ 8·0	27·6 $\pm$ 8·6

breath without difficulty. A total of 180 ml sterile phosphate-buffered saline, pH 7·26 (three 60-ml syringes), was injected through the jejunostomy tube into the lung. The last syringe was left in position for the immediate withdrawal of the lavage fluid. Approximately 40% of the volume introduced was recovered.

Haematology and biochemistry methods

White blood cells were counted using a single Channel Coulter Counter (Coulter Model ZF, Coulter Electronics, Luton, Bedfordshire, U.K.), and the differential count was carried out on smears stained with Diff Quik (Merz and Dade, B. M. Browne, Stillorgan, Dublin, Ireland) using the battlement method of scanning. Total protein was measured by a modification of the Biuret method (Kingsley, 1942) and albumin levels by a modification of the bromocresol green method (Doumas *et al.*, 1971) using a Cobas Mira Discrete Analyser (Roche, B. M. Browne). Globulin concentrations were recorded as the difference between the total protein and serum albumin values.

Blood neutrophil and bronchopulmonary lavage cell preparation

Dextran (0·5 ml, mol. wt 150 000) was added to 15 ml heparinized blood. The blood was allowed to sediment at room temperature for 45 min. The cell-rich plasma was layered onto Ficoll paque (Pharmacia, Uppsala, Sweden) and blood neutrophils were separated by centrifugation at 300 g for 20 min (Boyum, 1968). Blood neutrophils were washed twice with Hanks Balanced Salt Solution (HBSS) and the cell concentration was adjusted to 5 $\times$ 10⁶/ml.

The recovered bronchopulmonary lavage fluid was centrifuged at 300 g for 15 min. The supernatant fluid was decanted and the cell pellet was washed three times with HBSS. The cell concentration was adjusted to 1 $\times$ 10⁶/ml. Cell smears were stained with Wright/Giemsa for differential cell counts and for non-specific esterase activity to identify macrophages (Barka & Anderson, 1963, as modified by Li *et al.*, 1972).

Cell function assays

Chemotaxis. Blood neutrophil chemotaxis was assayed using the method described by Nelson *et al.* (1975). Neutrophils were incubated at 37°C with 5% CO_2 in a central agarose well with HBSS- and *E. coli*- (NCTC 10418) activated autologous serum on either side. Migration was quantified by measuring the linear migration (mm) of the cells towards the activated serum and the HBSS. The chemotactic differential represented increased migration towards the chemoattractant over migration towards the buffer.

Chemiluminescence. Blood neutrophil chemiluminescence was measured in a lumiaggrometer (Coulter Electronics) at 38°C. Neutrophils were stimulated with Zymosan particles, preopsonized in autologous serum, in the presence of a 1/10 dilution of 10^{-3} M-aminophthalhydrazide. Light generation was recorded on an ommiscribe chart recorder (Coulter Electronics). The peak response was measued in mV and the time to reach this response was expressed in min.

Phagocytosis. Bronchopulmonary lavage cell and blood neutrophil phagocytosis and intracellular killing were assayed as described by van Furth & van Zwet (1973). Phagocytosis was assayed by incubating neutrophils and bronchopulmonary lavage cells with *Staphlococcus aureus* (NCTC 6571), in the presence of 20% autologous serum, at

38°C. The phagocytic cell to bacterial cell ratio was 1:1. Samples of the cell suspension were taken at 0, 30 and 60 min. The phagocytic cells were removed by differential centrifugation at 110 *g* for 7 min and the bacteria remaining in the supernatant fluid were diluted and plated on nutrient agar plates. The colony-forming units were counted after overnight incubation at 37°C. Colonies present at 30 and 60 min were expressed as a percentage of those present as 0 min. Smears were made of the phagocytic cells after 30 and 60 min incubation and stained with Wright/Giemsa stain. The percentages of cells that were phagocytic and of those with > 10 bacteria were enumerated.

Intracellular killing. Phagocytic cells were again incubated with *Staph. aureus* as described above for 15 min before the intracellular killing assay. Free bacteria in suspension after 15 min were removed by differential centrifugation and washing. The cells were resuspended with 10% autologous serum and reincubated for 60 min with continuous agitation. Samples of the cell suspension were taken at 0, 30 and 60 min. Cells present were lysed by the addition of sterile distilled water followed by rapid freezing and thawing. The cell lysate was diluted and bacteria present were plated on nutrient agar plates and incubated overnight at 37°C. The colony-forming units present at 30 and 60 min were expressed as a percentage of those present at 0 min.

Statistical analyses

The results were analysed during Student's *t* test and by calculation of Pearson's correlation coefficient (Snedecor & Cochran, 1967).

Results

Foals

All foals included in this study were in normal health and condition throughout the study period. Routine haematology and biochemistry did not indicate disease and colostrum absorption was adequate (Table 1).

Bronchopulmonary lavage

The bronchopulmonary lavage procedure used in this study was easy to perform and no evidence of clinical respiratory disease was detected as a consequence of the procedure. Discomfort experienced by the foals was apparently minimal and many of the foals sucked immediately after completion of the procedure.

A total of $1\cdot5$–$2\cdot8 \times 10^6$ bronchopulmonary lavage cells was recovered with the lavage fluid. Pulmonary alveolar macrophages were the predominant cell type present ($89 \pm 7\cdot5\%$). The percentage of lymphocytes recovered was low ($3\cdot7 \pm 2\%$). Neutrophil numbers varied ($7\cdot6 \pm 7\%$) and in one foal these cells constituted up to 25% of the immune cells present. Epithelial cells were also present and were abundant in some foals.

Cellular function

Blood neutrophils. Blood neutrophils recovered from foals before and again 2–4 days after the ingestion of colostrum exhibited increased chemotaxis and a reduction in the time period in which

Table 2. Blood neutrophil function (mean $\pm$ s.d.) in 12 foals before and after colostrum ingestion

Neutrophil function assay	Before colostrum ingestion	2–4 days after colostrum ingestion	Significance
Chemotaxis (mm)	$0\cdot13 \pm 0\cdot1$	$0\cdot23 \pm 0\cdot1$	$P < 0\cdot05$
Phagocytosis after 60 min (%)	42 ± 29	66 ± 13	N.S.
Killing after 60 min (%)	28 ± 31	54 ± 37	N.S.
Chemiluminescence peak response			
mV	49 ± 17	52 ± 13	N.S.
min	21 ± 8	11 ± 3	$P < 0\cdot001$

Table 3. The percentage (mean ± s.d.) phagocytosis and intracellular killing of *Staphylococcus aureus* by blood neutrophils and bronchopulmonary lavage cells over a 60-min period

	Phagocytosis		Intracellular killing	
	30 min	60 min	30 min	60 min
Blood neutrophils	33 ± 26	64 ± 17	60 ± 28	61 ± 30
Bronchopulmonary lavage cell	38 ± 48	56 ± 40	5 ± 9	8 ± 12

Table 4. The percentage of blood neutrophils and bronchopulmonary lavage (BPL) cells (mean ± s.d.) from 11 foals that contained bacteria and that contained > 10 bacteria/phagocytic cell after incubation for 30 and 60 min with *Staphylococcus aureus*

	Incubation time	
Phagocytic cells	30 min	60 min
Blood neutrophils	77 ± 10	84 ± 10
> 10 bacteria/cell	6 ± 11	22 ± 27
BPL cells	26 ± 13	37 ± 13
> 10 bacteria/cell	0·2 ± 0·6	3·2 ± 7·2

to reach the peak chemiluminescence response (Table 2). Although some foals did exhibit increased phagocytosis and intracellular killing the difference was not significant. There were no significant correlations between increased globulin levels in the sera of young foals and increased chemotaxis, chemiluminescence, phagocytosis or intracellular killing. However, there was a significant correlation ($r = 0·7$) between increased phagocytosis and the reduced timing of the peak chemiluminescence response in colostrum-fed foals.

Blood neutrophil and bronchopulmonary lavage cell comparison. The percentage *Staph. aureus* phagocytosed by blood neutrophils and bronchopulmonary lavage cells increased over the 60-min incubation time (Table 3). The percentages of neutrophils and lavage cells that had phagocytosed bacteria and the number of bacteria per cell also increased over the incubation time (Table 4). However, there was greater individual variation in the phagocytic ability of lavage cells when compared with blood neutrophils. In addition, fewer of the lavage cells were phagocytic and a smaller percentage of them had ingested large numbers of bacteria (Table 4).

Blood neutrophils demonstrated evidence of intracellular killing even though there was considerable individual variation (Table 3), but bronchopulmonary lavage cells demonstrated little, if any, evidence of intracellular killing over the time period studied.

Discussion

The haematological findings in this study were similar to those described by Jeffcott *et al.* (1982). Although the neutrophil to lymphocyte ratio tended to be lower in the first few hours *post partum*, this did increase in older foals.

Several techniques for bronchopulmonary lavage for the foal have been described (Dyer *et al.*, 1983; Zink & Johnson, 1984). These techniques require the use of anaesthesia. The technique used in this study requires neither sedation nor anaesthesia and may be a useful alternative. It is adaptable for use under stud farm conditions and the use of a jejunostomy tube eliminates the possibility of damaging an expensive fibreoptic endoscope.

Pulmonary alveolar macrophages, as in other studies (Dyer *et al.*, 1983; Zink & Johnson, 1984), were the predominant cell type recovered. Variation in the numbers of neutrophils recovered may reflect mild changes in the pulmonary health status of different foals (Hunninghake *et al.*, 1978).

Extracellular humoral factors, for example immunoglobulins and complement, together with functional immune cells are necessary for adequate expression of cellular immunity. Humoral factors have been shown to enhance chemiluminescence, chemotaxis, phagocytosis and intracellular killing (Leijh *et al.*, 1981; Zinkl & Brown, 1982; Johnson *et al.*, 1985; Zink *et al.*, 1985). Colostral proteins present in the serum of the 2–4-day-old foals probably contributed to the enhanced cellular function observed in this study. The lack of correlation between the amount of globulin in the sera of these neonatal foals and improved cellular function may be attributable to differences in the qualitative composition of colostral proteins absorbed. The generation of reactive oxygen radicals and the subsequent production of chemiluminescence, in the presence of luminol, are often but not always associated with phagocytic stimulation of cells (Richards & Renshaw, 1981). In this study, increased phagocytosis appeared to be associated with an enhanced chemiluminescence response.

These findings do not preclude the possibility, however, that the physiological changes occurring within the neonatal period could also have affected cellular function. Plasma cortisol concentrations are raised in the serum of newborn foals and remain elevated over the first 24 h *post partum* (Nathanielsz *et al.*, 1975). Glucocorticoids may enhance or depress cellular function (Guelfi *et al.*, 1985; Frank & Roth, 1986; Webb *et al.*, 1986).

In this study, phagocytosis was assessed by a reduction in viable extracellular bacteria. It is therefore possible that extracellular killing of bacteria in addition to phagocytosis may also have led to a reduction in viable extracellular bacteria. The fact that extracellular killing may be an important pulmonary clearance mechanism (Biggar & Sturgess, 1977; Nugent & Pesanti, 1979) and that only a low percentage of bronchopulmonary lavage cells are phagocytic may indicate that such cells have the ability to kill bacteria outside the cell. The ability of bronchopulmonary lavage cells to phagocytose and/or kill bacteria outside the cell varied considerably between foals. The number of bronchopulmonary lavage cells with ingested bacteria and the number of bacteria per cell were low in comparison to blood neutrophils. Similar findings have been reported for other species (Warshauer *et al.*, 1977; Nugent & Pesanti, 1979; Brody *et al.*, 1982). Zink *et al.* (1985) found that the number of foal pulmonary alveolar macrophages with ingested *R. equi* and the number of *R. equi* per cell could be increased significantly by the addition of immune serum.

The poorly developed bactericidal mechanisms exhibited by bronchopulmonary lavage cells in this study are similar to findings in other species (Sherman *et al.*, 1977; Bellanti *et al.*, 1979; Fogarty, 1984). Zink *et al.* (1985) found little if any evidence of *R. equi* killing by foal pulmonary alveolar macrophages up to 61 days of age.

Physical removal of bacteria from the lung occurs more slowly than inactivation (Rylander, 1968) and so reduced phagocytosis and killing of bacterial challenges may have important implications in neonatal defence mechanisms against infection. Likewise, a delay in the onset of functional systemic cellular immunity may predispose foals to infections immediately *post partum*.

We thank Mr Tom Buckley, Ms Jean Burke and Ms Anna Collins for technical assistance; Ms Bridget McGing M.R.C.V.S. and Mr Brian O'Sullivan M.R.C.V.S. for help in arranging sample collection; and Ms Vivienne Berry and Ms Clare Ludden for typing and preparing the manuscript.

References

Barka, T. & Anderson, P.J. (1963) In *Histochemistry: Theory, Practice and Bibliography*, pp. 260–265. Harper and Row, New York.

Bellanti, J.A., Nerurkar, L.S. & Zeligs, B.J. (1979) Host defenses in the fetus and neonate: studies of the alveolar macrophage during maturation. *Pediatrics* **64**, 726–739.

Biggar, W.D. & Sturgess, J.M. (1977) Role of lysozyme in the microbicidal activity of rat alveolar macrophages. *Infect. Immun.* **16**, 974–982.

Boyum, A. (1968) Isolation of mononuclear cells and granulocytes from human blood. *Scan. J. clin. Lab. Invest.* **21**, 77–89.

Brody, A.R., Roe, M.W., Evans, J.N. & Davis, G.S. (1982) Deposition and translocation of inhaled silica in rats. Quantification of particle distribution, macrophage participation and function. *Lab. Invest.* **47**, 533–542.

Coignoul, F.L. (1983) *Functional and ultrastructural changes in neutrophils in normal and equine herpesvirus 1 subtype 2 infected mares and foals.* Ph.D. thesis, Iowa State University.

Doumas, B.T., Watson, W.A. & Biggs, H.O. (1971) Albumin standards and the measurement of serum albumin with bromocresol green. *Clin. Chem. Acta* **31**, 87–96.

Dyer, R.M., Liggitt, H.D. & Leid, R.W. (1983) Isolation and partial characterisation of equine alveolar macrophages. *Am. J. vet. Res.* **44**, 2379–2384.

Ehrlich, P. (1892) Ueber immunitat durch verebung und saugang. 2. *Hyg. Infektkrankh* **12**, 183–203.

Fogarty, U. (1984) *A study of immune mechanisms in the respiratory tract of the young calf.* Ph.D. thesis, University College, Dublin.

Fogarty, U., Quinn, P.J. & Hannan, J. (1983) Bronchopulmonary lavage in the calf—a new technique. *Irish vet. J.* **37**, 35–38.

Frank, D.E. & Roth, J.A. (1986) Factors secreted by untreated and hydrocortisone treated monocytes which modulate neutrophil function. *Proc. 1st Int. Conf. Vet. Immunol.* (in press).

Guelfi, J.F., Courdouhji, M.K., Alvinerie, M. & Toutain, P.L. (1985) *In vivo* and *in vitro* effects of three glucocorticoids on blood leukocyte chemotaxis in the dog. *Vet. Immunol. Immunopath.* **10**, 245–252.

Hunninglake, G.W., Gallin, J.I. & Fauci, A.S. (1978) Immunologic reactivity of the lung. The *in vivo* and *in vitro* generation of a neutrophil chemotactic factor by alveolar macrophages. *Am. Rev. Respir. Dis.* **177**, 15–23.

Jeffcott, L.B., Rossdale, P.D. & Leadon, D.P. (1982) Haematological changes in the neonatal period of normal and induced premature foals. *J. Reprod. Fert., Suppl.* **32**, 537–544.

Johnson, E.H., Hietala, S. & Smith, B.P. (1985) Chemiluminescence of bovine alveolar macrophages as an indicator of developing immunity in calves vaccinated with aromatic-dependent *Salmonella*. *Vet. Microbiol.* **10**, 451–464.

Kingsley, E.G. (1942) The direct Biuret method for the determination of serum protein as applied to photoelectric and visual colorimetry. *J. Lab. Clin. Med.* **27**, 840.

Leijh, P.C.J., van den Barselaar, M.T., Daha, M.R. & van Furth, R. (1981) Participation of immunoglobulins and complement components in the intracellular killing of *Staphylococcus aureus* and *Escherichia coli* by human granulocytes. *Infect. Immun.* **31**, 714–724.

Li, C.Y., Yan, T. & Crosbie, W.H. (1972) Histochemical characterisation of cellular and structural elements of the human spleen. *J. Histochem. Cytochem.* **20**, 1049–1058.

Nathanielsz, P.W., Rossdale, P.D., Silver, M. & Comline, R.S. (1975) Studies on fetal, neonatal and maternal cortisol. *J. Reprod. Fert., Suppl.* **23**, 625–630.

Nelson, R.D., Quie, P.G. & Simmons, R.L. (1975) Chemotaxis under agarose: a new and simple method for measuring chemotaxis and spontaneous migration of human polymorphonuclear leukocytes and monocytes. *J. Immunol.* **115**, 1650–1656.

Nugent, K.M. & Pesanti, E.L. (1979) Effect of influenza infection on the phagocytic and bactericidal activities of pulmonary macrophages. *Infect. Immun.* **26**, 651–657.

Richards, A.B. & Renshaw, A.W. (1981) Bovine alveolar macrophages. Phagocytic activation of luminol dependent chemiluminescence in bovine pulmonary lavage cells. *Zentbl. VetMed.* **28**, 603–611.

Rylander, R. (1968) Pulmonary defense mechanisms to airborne bacteria. *Acta physiol. scand.* **306**, 1–89.

Sherman, M., Goldstein, E., Lippert, W. & Wennberg, R. (1977) Neonatal lung defense mechanisms: A study of the alveolar macrophage system in neonatal rabbits. *Am. Rev. Respir. Dis.* **116**, 433–440.

Snedecor, G.W. & Cochran, W.G. (1967) *Statistical Methods*, 6th edn. Iowa State University Press, Ames.

van Furth, R. & van Zwet, T.L. (1973) *In vitro* determination of phagocytosis and intracellular killing by polymorphonuclear and mononuclear phagocytes. In *Handbook of Experimental Immunology*, pp. 36.1–36.24. Ed. D. M. Weir. Blackwell Scientific, Oxford.

Warshauer, D., Goldstein, E., Akers, T., Lippert, W. & Kim, M. (1977) Effect of influenza viral infection on the ingestion and killing of bacteria by alveolar macrophages. *Am. Rev. Respir. Dis.* **115**, 269–277.

Webb, D.S.A., Frank, D.E. & Roth, J.A. (1986) Suppression of arachidonic acid metabolism is only partially responsible for the suppression of neutrophil function by glucocorticoids. *Proc. 1st Conf. vet. Immunol.* (in press).

Zink, M.C. & Johnson, J.A. (1984) Cellular constituents of clinically normal foal bronchoalveolar lavage fluid during postnatal maturation. *Am. J. vet. Res.* **45**, 893–897.

Zink, M.C., Yaeger, J.A., Prescott, J.F. & Wilkie, B.M. (1985) *In vitro* phagocytosis and killing of *Corynebacterium equi* by alveolar macrophages of foals. *Am. J. vet. Res.* **46**, 2171–2174.

Zinkl, J.G. & Brown, P.D. (1982) Chemotaxis of horse polymorphonuclear leukocytes to N-formyl-L-methionyl-L-leucyl-L-phenylalanine. *Am. J. vet. Res.* **43**, 613–615.

J. Reprod. Fert., Suppl. **35** (1987), 599–605

Printed in Great Britain
© 1987 Journals of Reproduction & Fertility Ltd

Chemotactic and phagocytic function of peripheral blood polymorphonuclear leucocytes in newborn foals

M. Bernoco, I. K. M. Liu, C. J. Wuest-Ehlert, M. E. Miller and J. Bowers

Department of Reproduction, School of Veterinary Medicine, and Department of Pediatrics, School of Medicine, University of California, Davis, CA 95616, U.S.A.

Summary. Chemotactic and phagocytic responsiveness of peripheral blood polymorphonuclear leucocytes (PMNLs) from 11 foals were analysed immediately after birth (pre-colostral) and at different times after colostrum ingestion. The number of foal PMNLs per microscopic field that had migrated through the filter in chemotaxis and the number of yeast particles ingested per foal PMNL in phagocytosis were significantly lower when tested with foal plasma before colostrum ingestion (chemotaxis, 2.0 ± 0.55 (s.e.m.); phagocytosis, 0.98 ± 0.352) than in tests 4 or more days after colostrum ingestion (chemotaxis, 17.6 ± 3.88; phagocytosis, 3.87 ± 0.410; $P \leqslant 0.005$). A similar functional deficiency was found in foal PMNLs tested with pre-colostral foal plasma when compared to tests of the same cells with mare plasma (chemotaxis, 11.6 ± 2.91; phagocytosis, 3.94 ± 0.269; $P \leqslant 0.005$). PMNLs from neonatal foals responded with normal functions to plasma from normal adult horses. Results of both assays suggest that pre-colostral foal PMNLs are functionally mature but the expression of their chemotactic and phagocytic functions depends on the humoral source of chemotactic and opsonic factors.

Introduction

Investigations into leucocyte function have consistently demonstrated functional impairment of polymorphonuclear leucocytes (PMNL) in the human neonate. PMNLs from cord blood (Pahwa *et al.*, 1977; Tono-oka *et al.*, 1979; Boner *et al.*, 1982), newborn (Miller, 1971; Klein *et al.*, 1977; Miler *et al.*, 1979; Mease *et al.*, 1980; Anderson *et al.*, 1981) and preterm infants (Laurenti *et al.*, 1980) showed significant deficiencies in chemotactic ability when compared with PMNLs from adults. Other investigations further demonstrated diminished intracellular killing with normal phagocytic function (Cohen *et al.*, 1969; Dosset *et al.*, 1969) and diminished deformability (Miller, 1975) of PMNLs in newborn infants. These impairments of PMNL function are considered to be factors that contribute significantly to increased susceptibility of the neonate to microbial invasion (Miller, 1979).

Coignoul *et al.* (1984) reported significantly less random migration of neutrophils obtained from 2–4-month-old pony foals when compared with that of PMNLs from adult ponies. They further indicated that the difference between foals and adults suggests a primary neutrophil deficiency in young foals.

The purpose of this investigation was to determine the functional characteristics of peripheral blood PMNLs obtained from newborn foals before and after the absorption of colostrum from their respective dams.

Materials and Methods

Horses. Eleven randomly selected pregnant mares ranging in age from 4 to 18 years were utilized. Breeds included 3 Quarterhorse, 2 Arabians, 1 Hannoverian, 1 Thoroughbred, 1 Standardbred, 1 Appaloosa, 1 Westphalian and 1 pony. Blood samples were obtained from the foals before (newborn) and 4–12 days after colostrum ingestion. Further blood samples were obtained from 6 of the 11 foals at 1 month and 2·5–5 months of age. Dams were bled at the same time as the foals and served as adult controls.

Blood samples were also obtained from 2 healthy stallions (1 Thoroughbred and 1 Quarterhorse). Their PMNLs were used as control cells, one for chemotaxis and one for the phagocytosis assay.

Six random male horses served as donors for pooled standard positive control plasma which was stored in 0·5 ml aliquants at −70°C.

PMNL isolation. Blood (30–40 ml) was withdrawn into 10-ml Vacutainer tubes with sodium heparin as the anticoagulant. The blood was centrifuged at 400 *g* for 10 min and the plasma was saved. The mixture of 1 part buffy coat and 2 parts saline (9 g NaCl/l) was layered on lymphocyte separation medium (Litton Bionetics, Kensington, MD) and centrifuged at 400 *g* for 25 min (Boyum, 1968). The PMNL-rich buffy coat was collected from the pellet below the lymphocyte separation medium and resuspended in a 2·25% Dextran (average mol. wt 250 000; Sigma, St Louis, MO) solution in saline. After 15 min of sedimentation the supernatant was collected, the cells were washed twice with saline and counted. The final suspension was adjusted to 1×10^6 PMNLs/ml McCoy's 5A medium supplemented with 10% heat-inactivated fetal bovine serum (IFBS) for the chemotaxis assay and to 5×10^6 PMNLs/ml Hanks' Balanced Salt Solution (HBSS) for the phagocytosis assay (both media from GIBCO, Santa Clara, CA).

The PMNL separation technique utilized minimal manipulation of the cells in order to prevent possible stimulation or inhibition of their natural (*in vivo*) functions. The resulting PMNL suspension contained < 5% lymphocytes and a negligible number of erythrocytes.

Chemotaxis assay. A modification of the Boyden assay (Boyden, 1962) in blind-well chambers (Boumsell & Meltzer, 1975) was used to study the chemotactic function of the PMNLs. Endotoxin-activated plasma (EAP) for chemotactic factors was prepared as follows. A sample of 40 μg *Escherichia coli* lipopolysaccharide (Sigma, L3755) dissolved in 20 μl saline (stored frozen) was mixed with 0·4 ml plasma, incubated for 30 min at 37°C, then heat-inactivated for 30 min at 56°C. The chemoattractant, 0·2 ml EAP diluted 1:5 in McCoy's medium with 10% IFBS (GIBCO), or 0·2 ml of the medium alone as control, was placed in the lower well of the blind-well chemotaxis chamber (Neuroprobe Corp., Bethesda, MD). A 3-μm pore-sized polycarbonate filter (Nucleopore Corp., Pleasanton, CA) was secured in place above the well by a tube-shaped cap which formed the upper compartment of the chamber into which $0·2 \times 10^6$ PMNLs in 0·2 ml McCoy's medium with 10% IFBS were delivered. A 45-min incubation period in a humidified incubator at 37°C with 5% CO_2 allowed the PMNLs to migrate through the filter's micropores towards the EAP. The remaining cell suspension was removed by Pasteur pipette from the upper compartment of the chamber and the filter was placed downside up on a microscopic slide and secured by steel clips. The PMNLs attached now to the upper side of the filter were fixed in 80% ethanol, stained with haematoxylin and Wright's stain and mounted for microscopic examination. The PMNLs of 8 consecutive microscopic fields at × 1000 magnification (oil immersion) were counted in two different areas of the filter. Mean numbers of PMNLs per field were calculated from 16 different fields per filter. The chemotactic migration was expressed as the mean number of PMNLs ± the standard error of the means. Triplicate assays were performed for each sample.

Phagocytosis assay. The technique described by Soothill & Harvey (1976) was slightly modified. Heat-killed yeast (*Saccharomyces cerevisiae:* Nabisco Brands, East Hanover, NJ) was suspended in saline at a concentration of 1×10^8 particles/ml saline and stored in 2 ml aliquants at −70°C. Then 1×10^7 yeast particles in 0·1 ml saline were mixed with 0·1 ml 40% plasma in HBSS or 0·1 ml HBSS alone as negative control and incubated with constant rotation for 15 min at 37°C. They were washed 3 times in saline by centrifugation for 2 min at 1000 *g* in a Fisher microfuge. The opsonized yeast pellet was resuspended with 0·2 ml PMNL suspension in HBSS ($= 1 \times 10^6$ PMNLs). The mixture was incubated for 20 min at 37°C with constant rotation and centrifuged for 4 min at 200 *g*. The PMNL–yeast pellet was gently resuspended in 30 μl IFBS, duplicate smears were prepared and stained with Wright's stain. At least 100 PMNLs and their ingested yeast particles were counted at the microscopic magnification of × 1000 (oil immersion). The phagocytic responsiveness was expressed as the mean number of ingested yeast particles per PMNL.

Chemotactic and phagocytic activities of foal and mare PMNLs with foal and mare plasma were simultaneously tested after birth but before colostrum ingestion and at 3 later times after colostrum ingestion. Blood samples obtained from each of the 11 foals tested varied in accordance with their availability: 10 were tested at 4–12 days, 6 at 1 month and 6 at 2·5–5 months of age. All plasma samples from foals and mares were stored at −70°C and retested with control PMNLs. This allowed for simultaneous testing of foal and mare plasma collected at different periods.

Statistical analyses. Analysis of variance and paired *t* tests were used to analyse the data. For chemotaxis, 3 replicate entries were used per test. A probability level of 0·05 was considered significant. Paired *t* tests alone were applied to analyse the data with control cells.

Results

Chemotaxis assay

The means and standard errors of the numbers of PMNLs migrating through the filter in response to foal and mare EAP are shown in Fig. 1. Control PMNLs were additionally tested with standard positive control EAP.

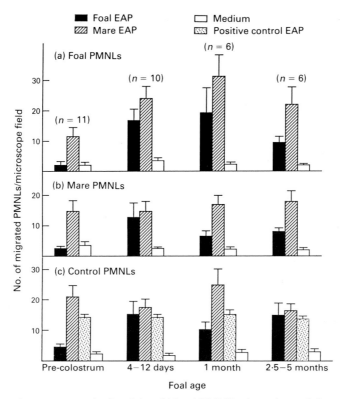

Fig. 1. Numbers (mean ± s.e.m.) of peripheral blood PMNLs that migrated through the micropore filter after exposure to EAP in the chemotaxis assay.

Foal PMNLs (Fig. 1a). The chemotactic response of pre-colostral foal PMNLs to foal EAP was significantly different ($P \leqslant 0.005$) from that of the same cells to mare EAP. Significant differences ($P \leqslant 0.005$) were also found in results with pre-colostral foal cells and EAP when compared with results of foal cells and EAP obtained 4 or more days after the absorption of colostrum. Considering a migration score of 13–20 cells per microscopic field as the normal range for this study, the mean migratory function of foal PMNLs and EAP at 2·5–5 months was below the normal range (9·6). This was caused by lower responses in general and one individual response at 1·94. When that individual EAP was retested with control PMNLs the migration response was 9·53. Foal PMNLs responded to mare EAP with a low migration score before colostrum ingestion (11·6). Their post-colostral migration response to mare EAP was higher overall than to foal EAP.

Mare PMNLs (Fig. 1b). With pre-colostral foal EAP as the chemoattractant, mare PMNLs showed no difference in migration characteristics when compared with pre-colostral foal PMNLs. Both migration responses were considered impaired and the impairment appeared to be plasma related. When tested with post-colostral foal EAP, mare cells elicited some migration inhibition at 1 month (6·45) and 2·5–5 months (8·01). The migration response to autologous EAP was in the normal range (14·72–18·04).

Control PMNLs. Figure 1(c) shows the chemotactic migration pattern of control PMNLs tested with EAP prepared from frozen–thawed foal, mare and standard positive control plasma. Migration results of control cells were significantly different when pre-colostral foal EAP was compared with pre-colostral mare EAP ($P \leqslant 0.005$), with post-colostral foal EAP ($P \leqslant 0.01$) and with positive control EAP ($P \leqslant 0.005$). Results with post-colostral foal EAP were comparable to the results with the standard positive control EAP except for a low mean at 1 month (9·8). Mare EAP overall produced higher migrating control cell numbers than did foal EAP and positive control EAP. When mare EAP was used as the chemoattractant, the migration of foal PMNLs was highest at 1 month (Fig. 1a). A similar peak was noted with control PMNLs and mare EAP.

Because the mare's plasma did not appear able to stimulate normal migratory function of pre-colostral foal cells (Fig. 1a), and yet normal migration activity was apparent with autologous and control cells (Figs 1b, 1c), the possibility of maternal immunological inhibition specifically directed towards the offspring was considered. To determine whether the low migration rate of pre-colostral foal PMNLs towards mare EAP was caused by maternal humoral inhibition or cellular incompetence, 3 foals were tested with the standard positive control EAP for comparison (Table 1). No differences in migration activity were noted when positive control EAP was used as chemoattractant for pre-colostral foal, mare and control PMNLs. Post-colostral foal PMNLs responded with more chemotactic activity than did mare and control PMNLs, especially at 1 month. Pre-colostral mare EAP reflected the slight inhibition mentioned above when tested with pre-colostral foal PMNLs, but resulted in normal to high migration activity with mare and control PMNLs.

These results clearly indicate (a) that PMNLs from newborn foals are fully capable of migrating when exposed to normal EAP, and (b) that the slight inhibition observed in the migration response of pre-colostral foal PMNLs to mare EAP appears to be of maternal humoral origin.

Table 1. Comparison of chemotactic migration responses of foal, mare and control PMNLs when exposed to positive control EAP, foal EAP and mare EAP

			Post-colostrum		
PMNLs	EAP	Pre-colostrum (N = 3)	4-12 days (N = 2)	1 month (N = 3)	2·5–5 months (N = 3)
Foal	Positive control	12·2 ± 3·59	24·3 ± 0·14	41·1 ± 10·07	17·0 ± 5·09
Mare		12·0 ± 3·21	17·3 ± 3·73	11·8 ± 1·66	12·4 ± 1·74
Control		15·8 ± 1·24	16·4 ± 1·75	15·8 ± 1·24	15·8 ± 1·24
Foal	Foal	0·8 ± 0·52	24·1 ± 9·99	30·8 ± 9·05	11·0 ± 1·98
Mare		1·9 ± 0·22	22·8 ± 13·07	5·6 ± 1·42	8·7 ± 0·93
Control		3·1 ± 0·49	25·4 ± 10·53	13·7 ± 2·23	16·7 ± 3·68
Foal	Mare	13·2 ± 6·06	38·1 ± 6·28	41·5 ± 6·96	20·0 ± 5·11
Mare		20·4 ± 5·60	29·6 ± 5·17	15·6 ± 3·89	16·0 ± 2·94
Control		33·1 ± 7·52	31·0 ± 5·49	34·0 ± 7·20	18·9 ± 2·07
Foal	Medium	2·2 ± 0·66	4·1 ± 1·74	4·1 ± 0·50	1·4 ± 0·38
Mare		2·6 ± 0·50	2·2 ± 0·46	3·7 ± 0·70	1·9 ± 0·53
Control		4·2 ± 0·33	3·8 ± 0·15	4·2 ± 0·33	4·2 ± 0·33

Values are mean ± s.e.m.

Phagocytosis assay

Figure 2 shows the numbers of yeast particles ingested per foal (Fig. 2a), mare (Fig. 2b) and control PMNLs (Fig. 2c) after yeast opsonization by foal and mare plasma tested before and after colostrum ingestion. For tests with control PMNLs, yeast was opsonized by frozen–thawed foal, mare and standard positive control plasma.

Foal PMNLs. Pre-colostral foal PMNLs showed considerable phagocytic inhibition after opsonization of yeast particles with pre-colostral foal plasma, but their phagocytic response was above the negative control level (Fig. 2a). Significant differences were found ($P \leqslant 0.0005$) when

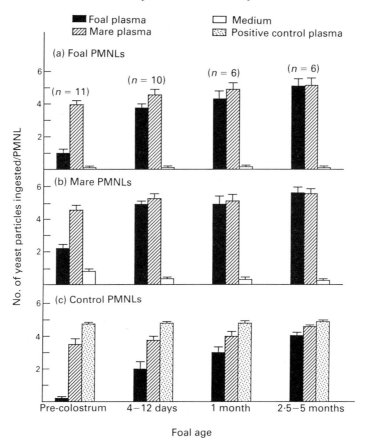

Fig. 2. Numbers (mean ± s.e.m.) of yeast particles ingested per peripheral blood PMNL after yeast opsonization with plasma in the phagocytosis assay. Numbers of phagocytic responses of control PMNL to medium (negative controls) are not reported because they were not higher than 0·01.

phagocytic activity of pre-colostral foal PMNLs with plasma was compared to the activity of the same cells with pre-colostral mare plasma. Comparison of phagocytic function levels of pre-colostral foal PMNLs and plasma with post-colostral foal PMNLs and plasma resulted also in a significant difference ($P \leqslant 0\cdot0005$). A steady functional increase of foal and mare PMNLs after yeast opsonization with their respective plasmas was noted between 4 days and 5 months of age.

Mare PMNLs. Phagocytic responses of mare PMNLs to yeast opsonization with pre-colostral foal, post-colostral foal and autologous plasma were similar to the responses with foal PMNLs, but reflected slightly higher phagocytic activities overall (Fig. 2b). Phagocytic impairment was also noted when pre-colostral foal plasma was exposed to mare PMNLs. Slightly elevated activity of mare PMNLs was also noted in the negative controls when compared with negative control data of foal and control cells in Figs 2(a) and 2(c), indicating non-specific stimulation.

Control PMNLs. The yeast uptake by control PMNLs after opsonization with foal and mare plasma was in general lower than by foal and mare PMNLs. This may be attributed to the freezing–thawing process of the plasma tested. Significant differences in opsonic activity were found when pre-colostral foal plasma was compared with pre-colostral mare plasma ($P \leqslant 0\cdot0005$), with post-colostral foal plasma ($P \leqslant 0\cdot0005$) and with positive control plasma ($P \leqslant 0\cdot0005$). When post-colostral foal and mare plasma were compared with positive control plasma, inhibition of

Table 2. Comparison of phagocytic responses of foal, mare and control PMNLs to yeast opsonization by standard positive control plasma

PMNLs	Pre-colostrum (N = 3)	Post-colostrum		
		4-12 days (N = 2)	1 month (N = 3)	2·5-5 months (N = 3)
Foal	5·05 ± 0·182	5·57 ± 0·035	5·03 ± 0·599	5·46 ± 0·416
Mare	5·39 ± 0·489	5·48 ± 0·629	5·62 ± 0·406	5·83 ± 0·180
Control	4·81 ± 0·183	5·04 ± 0·039	4·81 ± 0·183	4·81 ± 0·183

Values are mean ± s.e.m.
Means of negative controls with medium ranged from 0·00 to 0·13.

opsonic activity was noted particularly with foal plasma. This decrease of opsonic activity seen in foal plasma and to a lesser degree in mare plasma, suggests that foal plasma may be deficient in stimulating phagocytic activity of PMNLs up to at least 1 month of age. Normal chemotactic activity of foal plasma (Fig. 1c), however, was demonstrated during the same period.

Table 2 compares the phagocytic responses of foal, mare and control PMNLs after yeast opsonization with positive control plasma. Pre- and post-colostral foal PMNLs had functional levels similar to mare and control PMNLs at all test times. These results indicate that all cells tested were capable of similar normal phagocytic function when normal control plasma was used as the opsonizing agent.

Discussion

In this study, PMNLs obtained from newborn foals before ingestion of colostrum demonstrated deficient chemotactic and phagocytic activities when tested with autologous plasma. McGuire & Crawford (1973) and Jeffcott (1972) clearly demonstrated the importance of maternal transfer of immunoglobulins via the colostrum to foals immediately after birth. The quantitation of specific immunoglobulin levels was also causally related to diseases associated with the newborn foal (McGuire et al., 1977). The dependence of chemotactic and phagocytic PMNL functions not only upon immunoglobulin but especially on complement-related factors is well known and has been confirmed by many investigators (Klebanoff & Clark, 1978). This study suggests that maternal transfer of immunoglobulins and also other humoral components, e.g. complement, may play a role in activating chemotactic and phagocytic function of PMNLs derived from the newborn foal. Further investigation is needed, however, to determine the functional capacity of PMNLs from colostrum-deprived foals in order to assess the relationship of maternally derived humoral components with the functional characteristics of PMNLs in the newborn foal after ingestion of colostrum.

This investigation further demonstrated that PMNLs tested with plasma from foals ≥ 4 days appeared normal in chemotactic functions. These results are not consistent with the findings of Coignoul et al. (1984) which demonstrate significantly less random migration of neutrophils in 2–4-month-old pony foals when compared with pony mare neutrophils. Phagocytic functions were suppressed, however, when cells from foals up to 1 month of age were tested, suggesting that complete humoral opsonic maturity may occur only after 1 month of age.

Deficiencies in chemotactic and phagocytic functions were also seen when mare and control cells were tested with pre-colostral foal plasma. However, pre-colostral foal cells exhibited normal function responses to positive control plasma. These findings indicate that the cause of the functional deficiency is due to lack of chemotactic and opsonic factors in the pre-colostral plasma and not due to functional immaturity of the newborn foal's PMNLs. The slight inhibition in chemotactic activity of pre-colostral foal PMNLs with EAP prepared from mare plasma has been interpreted as maternal humoral inhibition directed specifically towards the offspring, since mare and control cells responded with normal functions to mare EAP.

It can be concluded that PMNLs derived from foals of all ages before and after ingestion of colostrum are functionally mature. The expression of the cells' functionally maturity depends upon the humoral source of chemotactic and opsonic factors.

This investigation was supported by a grant from the Oak Tree Racing Association, Arcadia, California.

We thank Dr Susan S. Suarez for the preparation of the statistical analysis and assistance with the manuscript, and Mr Kit Smith for technical assistance.

References

Anderson, D.C., Hughes, B.J. & Smith, C.W. (1981) Abnormal mobility of neonatal polymorphonuclear leucocytes. Relationship to impaired redistribution of surface adhesion sites by chemotactic factor or colchicine. *J. clin. Invest.* **68**, 863–874.

Boner, A., Zeligs, B.J. & Bellanti, J.A. (1982) Chemotactic responses of various differential stages of neutrophils from human cord and adult blood. *Infect. Immun.* **35**, 921–928.

Boumsell, L. & Meltzer, M.S. (1975) Mouse mononuclear cell chemotaxis. I. Differential response of monocytes and macrophages. *J. Immunol.* **115**, 1746–1748.

Boyden, S.V (1962) The chemotactic effect of mixtures of antibody and antigen on polymorphonuclear leucocytes. *J. exp. Med.* **115**, 453–466.

Boyum, A. (1968) Isolation of mononuclear cells and granulocytes from human blood. *Scand. J. clin. Lab. Invest.* **21**, Suppl. **97**, 77–89.

Cohen, R., Grush, O. & Kauder, E. (1969) Studies of bactericidal activity and metabolism of the leucocyte in full-term neonates. *J. Pediat.* **75**, 400–406.

Coignoul, F.L., Bertram, T.A., Roth, J.A. & Cheville, N.F. (1984) Functional and ultrastructural evaluation of neutrophils from foals and lactating and non-lactating mares. *Am. J. vet. Res.* **45**, 898–902.

Dosset, J.H., Williams, R.C. & Quie, P.G. (1969) Studies on interaction of bacteria, serum factors and polymorphonuclear leucocytes in mothers and newborns. *Pediatrics* **44**, 49–57.

Jeffcott, L.B. (1972) Passive immunity and its transfer with special reference to the horse. *Biol. Rev.* **47**, 439–464.

Klebanoff, S.J. & Clark, R.A. (1978) *The Neutrophil: Function and Clinical Disorders.* North-Holland Publishing Co., Amsterdam.

Klein, R.B., Fisher, T.J., Gard, S.E., Biberstein, B.S., Rich, K.C. & Stiehm, E. R. (1977) Decreased mononuclear and polymorphonuclear chemotaxis in human newborns, infants, and young children. *Pediatrics* **60**, 467–472.

Laurenti, F., Ferro, R., Marzetti, G., Rossini, M. & Bucci, G. (1980) Neutrophil chemotaxis in preterm infants with infections. *J. Pediat.* **96**, 468–470.

McGuire, T.C. & Crawford, T.B. (1973) Passive immunity in the foal: Measurement of immunoglobulin classes and of specific antibody. *Am. J. vet. Res.* **34**, 1299–1303.

McGuire, T.C., Crawford, T.B., Hallowell, A.L. & Macomber, L.E. (1977) Failure of colostral immunoglobulin transfer as an explanation for most infections and deaths of neonatal foals. *J. Am. vet. med. Ass.* **170**, 1302–1304.

Mease, A.D., Fischer, G.W., Hunter, K.W. & Ruymann, F.B. (1980) Decreased phytohemagglutinin-induced aggregation and C5a-induced chemotaxis of human newborn neutrophils. *Pediat. Res.* **14**, 142–146.

Miler, I., Vondracek, J. & Hromadkova, L. (1979) The decreased in vitro chemotactic activity of polymorphonuclear leucocytes of newborns and infants. *Folia microbiol. Praha* **24**, 247–252.

Miller, M.E. (1971) Chemotactic function in the human neonate. Humoral-cellular aspects. *Pediat. Res.* **5**, 487–492.

Miller, M.E. (1975) Pathology of chemotaxis and random mobility. *Sem. Hematol.* **12**, 59–82.

Miller, M.E. (1979) Phagocytic function in the neonate. Selected aspects. *Paediatrics* **64**, 709–712.

Pawha, S.G., Pahwa, R., Grimes, E. & Smithwick, E. (1977) Cellular and humoral components of monocyte and neutrophil chemotaxis in cord blood. *Pediatr. Res.* **11**, 677–680.

Soothill, J.F. & Harvey, B.A.M. (1976) Defective opsonization. A common immunity deficiency. *Archs dis. Childh.* **51**, 91–99.

Tono-oka, T., Nakayama, M., Uehara, H. & Matsumoto, S. (1979) Characteristics of impaired chemotactic function in cord blood leucocytes. *Pediatr. Res.* **13**, 148–151.

J. Reprod. Fert., Suppl. **35** (1987), 607–614

Sympathoadrenal and other responses to hypoglycaemia in the young foal

M. Silver, A. L. Fowden, J. Knox, J. C. Ousey*, R. Franco* and
P. D. Rossdale*

*Physiological Laboratory, University of Cambridge, Cambridge CB2 3EG and
Beaufort Cottage Stables, Newmarket, Suffolk CB8 8JS, U.K.

Summary. The effects of insulin-induced hypoglycaemia on plasma catecholamines, cortisol and metabolites have been examined in newborn and 7–14-day-old foals. The fall in plasma glucose elicited by the highest dose of insulin (1·0 i.u./kg) given to the neonates was slower in onset and less severe in effect than 0·5 i.u./kg in the older foals. There was a significant inverse correlation between the concentrations of glucose and adrenaline (but not noradrenaline) in plasma once the glucose level had fallen below 2 mmol/l; the adrenergic response to hypoglycaemia was greater in the 7–14-day-old foals than in the neonates. No significant changes in glucose or catecholamines were seen after fasting alone. The adrenocortical response to hypoglycaemia was poor after birth, but significant changes occurred in the older foals with a 3-fold increase in plasma cortisol at 60 min after 0·5 i.u. insulin/kg. There were significant increases in plasma FFA after hypoglycaemia in both groups of animals, but the rise was less pronounced in the neonates. A significant positive correlation was found between plasma adrenaline and FFA values. Hypoglycaemia also resulted in a significant rise in plasma lactate and a slow fall in α-amino nitrogen. These findings show that hypoglycaemia in the foal is followed by stimulation of the adrenal medullary component of the sympathetic system and by activation of the adrenal cortex with a number of consequent metabolic changes. The hypoglycaemic effects of insulin were more intense and the response more rapid in the older foals than in the neonates, which exhibited some degree of insulin resistance.

Introduction

The sympathoadrenal system is particularly active at birth in a wide variety of species (Silver & Edwards, 1980; Lagercrantz, 1983; Sperling *et al.*, 1984). The foal is no exception and high levels of circulating adrenaline and noradrenaline are released in association with neonatal acidaemia, which in turn is related to the degree of hypoxia during delivery (Silver *et al.*, 1984). Among the many other potential hazards of postnatal life which can lead to neonatal distress and sympathoadrenal discharge is hypoglycaemia (Shelley & Nelligan, 1966). The majority of mammals are born with reserves of carbohydrate (liver, muscle) and fat (brown and white adipose tissue) which are mobilized in the first few hours of life (Shelley, 1961; Girard & Ferre, 1982). Indeed, many neonates can withstand many hours without nutrient intake, although their energy stores are normally replenished as soon as lactation and suckling are established. However, if there is lactational failure or maternal rejection hypoglycaemia ensues; it is also a feature of premature delivery when initial energy stores may be low. Whatever the cause, the consequences of hypoglycaemia in the neonate can be severe, especially if prolonged.

Comparatively little is known about the response to and effects of hypoglycaemia in the foal. In the present study the responses of newborn and older foals to low plasma glucose have been examined by testing the effects of different doses of insulin. The sympathoadrenal discharge and the changes in other metabolites and hormones have been studied during the hypoglycaemic episodes and after recovery.

Materials and Methods

Animals. Newborn foals from dated Pony mares were studied within 3–10 h of birth. A few of the mares were induced to deliver close to term by using oxytocin (Rossdale *et al.*, 1982); the majority were allowed to foal spontaneously. The foals were tested again 7–14 days later.

Procedures. Each foal was provided with an indwelling jugular catheter under local anaesthetic; it was then allowed to suck and was immediately muzzled for the remainder of the experiment (2–3 h). Blood samples were taken at half hourly intervals except after the bolus injection of insulin (Novo Actrapid monocomponent porcine insulin, Romford, Essex, U.K.), given 1 h after muzzling, when samples were taken every 15 min for 60 min. The doses of insulin used were as follows: 0·5 or 1·0 i.u./kg in newborn foals; 0·2 or 0·5 i.u./kg in older foals; one 7-day-old animal was given 1·0 i.u./kg, otherwise 4–6 animals were used in each group.

Plasma glucose was monitored using Dextrostix (Miles Laboratories, Slough, U.K.) and the animal's behaviour throughout was noted. The hypoglycaemia was allowed to continue for up to 2 h in the neonates and in the older foals receiving 0·2 i.u. insulin/kg. They were then allowed to feed and a final sample was taken 20–30 min after suckling. The 7–14-day-old foals receiving 0·5 i.u. insulin/kg became unsteady and drowsy by 45 min; they were allowed to suck at 60 min (i.v. glucose was administered to these animals when necessary). In a control series, foals were catheterized, muzzled and sampled at similar intervals.

Treatment of blood samples. Samples of 2–3 ml blood were taken of which 1 ml was placed in a chilled heparinized tube containing EGTA and glutathione for catecholamine assay (Silver *et al.*, 1984) and centrifuged immediately; the supernatant was placed on solid CO_2 and stored in liquid nitrogen until assay. The remainder of the sample was centrifuged in an EDTA-containing tube and the supernatant stored at $-20°C$ for other biochemical analyses. Occasional checks on blood pH and blood gases were made during the course of the experiments.

Biochemical and hormone measurements. Plasma concentrations of glucose, lactate and α-amino nitrogen were determined using established methods adapted for the autoanalyser (Prenton & London, 1967; Comline *et al.*, 1979; Fowden *et al.*, 1982). Free fatty acids (FFA) were measured using a Wako NEFAC kit (Wako Chemicals, GmbH, Neuss, West Germany). Values are expressed as mequiv./l, measured against oleic acid standards. The minimum detectable amount of FFA was 0·02 mequiv./l. The intra- and inter-assay coefficients of variation were 10·2% and 14·8% respectively.

Plasma cortisol concentrations were determined by RIA as described previously (Rossdale *et al.*, 1982). Plasma catecholamine concentrations were measured by high-pressure liquid chromatography (HPLC) using the method developed by Bioanalytical Systems Inc (1985) for the Anachem HPLC system with electrochemical detection. The plasma samples were prepared by adsorption of 300–500 μl onto acid washed alumina (Anton & Sayre, 1962) and 20-μl samples of the 100-μl perchloric acid eluate were injected on to the column. Dihydroxybenzylamine (DHBA) was added as the internal standard to each plasma sample before adsorption. Standard solutions of adrenaline and noradrenaline were also treated in the same way. The limits of sensitivity of the method were 0·1 ng/ml for adrenaline and 0·07 ng/ml for noradrenaline. The intra- and inter-assay coefficients of variation were 5·8% and 7·3% for adrenaline and 5·0% and 6·2% for noradrenaline respectively.

Statistical treatment. Means and standard errors (s.e.) are given throughout and statistical analyses were made according to the methods of Snedecor & Cochrane (1967).

Results

Insulin-induced hypoglycaemia

The changes in plasma glucose concentration elicited by different doses of insulin in newborn and older foals are shown in Fig. 1. The hypoglycaemia seen in the neonates even with the highest dose of insulin (1·0 i.u./kg) was slower in onset and less severe than that in the older group when half the dose was used; even after 2–3 h the newborns showed few overt signs of low plasma glucose. By contrast, 0·5 i.u./kg led to a very rapid decline in glucose concentration both at 7 and 14 days; by 45–60 min the majority of animals showed signs of weakness and instability, but none convulsed. Since there was no difference in the response at 7 and 14 days, the results from the two groups have been combined. The lowest dose of insulin (0·2 i.u./kg) produced much slower changes in plasma glucose, which were similar to those seen in the first few hours of life after 1·0 i.u., although after about 90 min the animals began to recover from the hypoglycaemia.

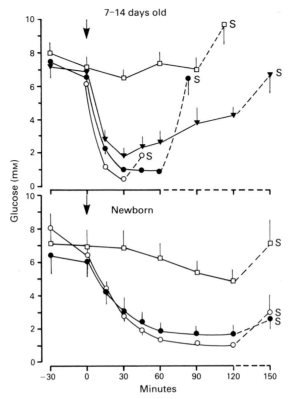

Fig. 1. The effect of different doses of insulin (arrow) on plasma glucose concentrations in newborn and 7–14-day-old foals. Mean values for groups of 4–6 foals given 0·2 (▼), 0·5 (●) or 1·0 (○) i.u. insulin/kg are shown with s.e. values as vertical bars when they exceed symbol size. Control animals (□) were fasted (muzzled) for an equivalent period. S = final value after suckling for the time shown by the broken horizontal axis and broken lines.

All foals were muzzled from 1 h before the insulin was given and Fig. 1 shows that prevention of milk intake alone led to some decrease in plasma glucose concentration over an equivalent period. However, these changes were slight in newborns and 7–14-day-old animals compared with the insulin-induced decreases.

Catecholamine response to hypoglycaemia

Figure 2 shows typical catecholamine responses after insulin given within 10 h of birth and again 7 days later. No change in adrenaline or noradrenaline was detectable until the plasma glucose had fallen below 2 mmol/l. Thereafter a sharp rise in adrenaline concentrations occurred as the glucose levels fell still further. At 7 days the hypoglycaemia was more severe and the adrenaline response correspondingly greater. In neither case was much change in noradrenaline found until the foal was allowed to suck and the adrenaline level had fallen dramatically.

Analysis of the relationship between plasma glucose and circulating adrenaline, using all the data, showed that there was a significant inverse correlation between adrenaline and log plasma glucose when the latter was below 2 mmol/l (Fig. 3). The slope of the line for the 7–14-day animals was significantly steeper than that for the neonates. Above 2 mmol/l (Fig. 3) no correlation was detectable, and there was no significant relationship between noradrenaline and plasma glucose at any level. These results suggest that only the adrenal medulla rather than the whole sympathetic system is stimulated by hypoglycaemia and that the critical plasma glucose required to elicit a response is less than 2 mmol/l.

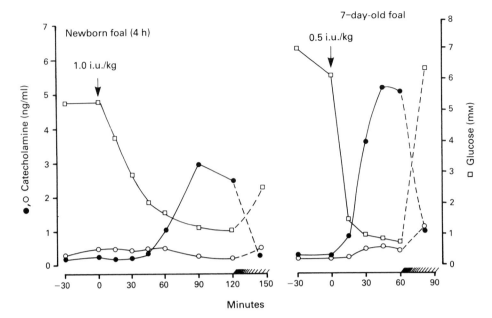

Fig. 2. The changes in plasma glucose (□), adrenaline (●) and noradrenaline (○) in a newborn foal given 1·0 i.u. insulin/kg and 7 days later after 0·5 i.u. insulin/kg at ↓. Foals were allowed to suck from 120 min (newborn) and 60 min (7-day-old).

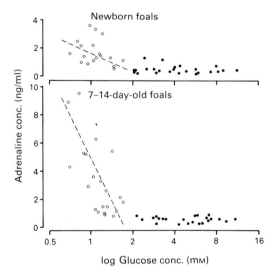

Fig. 3. The relationship between log plasma glucose (x) and plasma adrenaline (y) in newborn and 7–14-day-old foals sampled before and during hypoglycaemia. Below 2·0 mM-glucose (○) the regression line for the newborn foals was $y = -3·2x + 5·7$ with a correlation coefficient (r) of $-0·48$ ($P < 0·05$, $n = 21$). In the 7–14-day group, $y = -19·5x + 27·2$ and $r = -0·65$ ($P < 0·01$, $n = 21$). The difference between the slopes was statistically significant ($P < 0·01$). Above 2 mM-glucose (●) there was no relationship between plasma glucose and adrenaline for either group of foals.

Other changes during hypoglycaemia

A rise in plasma free fatty acids (FFA) was seen in both groups of foals after a slight initial decrease in the first 15–30 min after insulin (Fig. 4). The subsequent increment in FFA was slower in onset and less pronounced in the neonates than in the older foals, but the changes were statistically significant from 30 min after insulin in the former and from 45 min in the latter. These rises reflected the concomitant adrenaline changes in the two groups. There was a significant positive correlation between plasma adrenaline and FFA in the neonates ($r = 0.56$, $P < 0.01$, $n = 27$) and 7–14-day-old foals ($r = 0.89$, $P < 0.001$, $n = 29$). Fasting alone had little effect on FFA levels during the 2–3 h sampling period.

The changes in plasma cortisol were slight in the newborn foals (Fig. 5) except at 2 h after 1·0 i.u. insulin/kg. However, in most of this group the preinjection values were still high after birth, and continued to fall in the fasted controls. At 7–14 days, when basal levels were only 20–25 ng/ml, there was a significant rise by 30 min and a 3-fold increment when plasma glucose was at its lowest at 60 min (Fig. 5; Table 1). Cortisol concentrations remained elevated after suckling in both groups. Fasting alone had little effect on plasma cortisol concentrations.

Other metabolic consequences of the insulin-induced hypoglycaemia included a rise in plasma lactate and a fall in α-amino nitrogen. The latter showed a relatively small but steady decline after

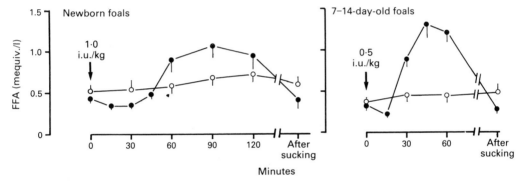

Fig. 4. Mean ± s.e.m. changes in plasma FFA concentrations in newborn and 7–14-day-old foals (N = 5) after insulin (●) at 0 min and during an equivalent period of fasting (○).

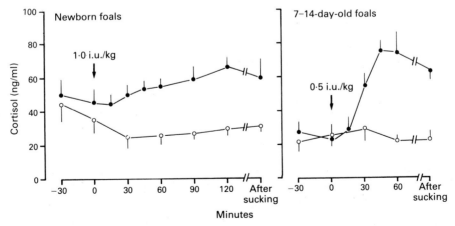

Fig. 5. Mean ± s.e.m. changes in plasma cortisol concentrations in newborn and 7–14-day-old foals (N = 4–6) after insulin (●) at 0 min or fasting alone (○).

insulin administration in both groups of foals and the changes were statistically significant from 60 min in the neonates and from 45 min in the older group (maximum values are given in Table 1). Plasma lactate values were more variable; in the newborns there was an initial lacticacidaemia before insulin was given (Table 1), but little change in lactate levels occurred in either group until minimum glucose values had been attained. Fasting alone had no effect on lactate or α-amino nitrogen over the 2–3 h period (Table 1).

A summary of the maximum changes observed in all the hormones and metabolites measured in this study is given in Table 1 for the newborn and older foals for the periods 0–90 and 0–60 min respectively.

Table 1. Mean concentrations of hormones (ng/ml) and metabolites (mM) in newborn and 7–14-day-old foals after insulin (1·0 and 0·5 i.u./kg, respectively) or fasting

Foals	Metabolite or hormone	After insulin		After fasting	
		0 min	90 min	0 min	90 min
Newborn (N = 5–6)	Adrenaline	0·3 ± 0·1	2·1 ± 0·40**	0·19 ± 0·10	0·17 ± 0·02
	Noradrenaline	0·43 ± 0·10	0·68 ± 0·14	0·48 ± 0·11	0·46 ± 0·12
	Cortisol	46·0 ± 8·3	59·0 ± 7·6	44·1 ± 12·0	29·5 ± 4·5
	Glucose	6·89 ± 0·58	1·0 ± 0·06**	6·89 ± 0·10	4·83 ± 0·63
	Lactate	3·0 ± 0·4	4·6 ± 0·6*	3·0 ± 0·8	2·8 ± 1·0
	FFA	0·41 ± 0·07	1·04 ± 0·14**	0·52 ± 0·06	0·70 ± 0·09
	α Amino N	3·0 ± 0·2	2·3 ± 0·1*	3·7 ± 0·18	3·3 ± 0·15
		0 min	60 min	0 min	60 min
7–14-day-old (N = 5–8)	Adrenaline	0·3 ± 0·1	4·8 ± 1·3**	0·29 ± 0·10	0·22 ± 0·08
	Noradrenaline	0·42 ± 0·17	0·67 ± 0·19	0·39 ± 0·09	0·63 ± 0·18
	Cortisol	23·0 ± 5·8	74·0 ± 12·0**	24·5 ± 5·5	22·1 ± 4·2
	Glucose	6·56 ± 0·62	1·0 ± 0·07**	7·17 ± 0·54	7·11 ± 0·56
	Lactate	1·3 ± 0·1	3·2 ± 0·7*	1·3 ± 0·1	1·1 ± 0·2
	FFA	0·34 ± 0·03	1·21 ± 0·17**	0·36 ± 0·1	0·44 ± 0·09
	α Amino N	3·5 ± 0·1	2·4 ± 0·1**	3·7 ± 0·16	3·5 ± 0·12

Values are mean ± s.e.m.
*$P < 0.05$, **$P < 0.01$ compared with initial value.

Discussion

The adrenal medulla is virtually the only source of circulating adrenaline and is thus clearly implicated in the response to hypoglycaemia in the foal. The present findings show that this response is far more specific than that elicited by delivery (hypoxia or asphyxia) when both adrenaline and noradrenaline are secreted (Silver *et al.*, 1984). The lack of any detectable rise in noradrenaline during hypoglycaemia is perhaps surprising since this hormone is also secreted by the equine adrenal medulla (Comline & Silver, 1971) as well as from sympathetic nerve endings. However, in the present experiments the only changes in noradrenaline were associated with suckling or with excitement and activity.

Early reports of the sympathoadrenal response to hypoglycaemia in both adult and young animals indicated the presence of a small but variable noradrenergic component in the largely adrenergic response (Hökfelt, 1951; Duner, 1954; Crone, 1965). Many of these studies were hampered by poor assay methods, and are difficult to assess in view of the anaesthesia and surgical trauma involved. Bloom *et al.* (1975) investigated the effects of insulin-induced hypoglycaemia in the conscious calf by using an adrenal clamp to sample adrenal venous blood. Doses of insulin equivalent to those given to the 7–14-day-old foals induced a much less severe hypoglycaemia in 3–4-week-old calves despite a previous 12-h fast. The adrenal medullary discharge, which consisted largely of adrenaline, was slow in onset and variable in duration. Whether the poor adrenergic response in the calves compared with that in foals of similar age can be ascribed to species differences or to the effects of surgical implantation of the adrenal clamp 24 h previously is uncertain. However, some preliminary experiments on young lambs, treated in the same way as the foals, have shown that this species is also far less sensitive to insulin. As much as 1–2 i.u./kg could be given to the lamb without any apparent ill effects whereas this dose would lead to collapse in the foal. A possible explanation for these discrepancies may perhaps lie in the amount of glycogen available for mobilization in the different species. However, little is known about the glycogen stores in the foal or about how these reserves are utilized or restored after depletion.

There was a very marked difference between the response of the older foals to hypoglycaemia and that seen in the neonates, not only in the maximum amount of adrenaline released by a given fall in glucose but also in the onset and severity of the hypoglycaemia itself. In addition, none of the newborn foals showed any behavioural signs of hypoglycaemia whereas many of the older foals were drowsy and unstable by 60 min. The comparative insensitivity of the neonates to high doses of insulin was particularly evident in animals tested within 4–6 h of birth. A degree of peripheral antagonism to insulin and resistance to hypoglycaemia has been reported for a number of newborn animals (Edwards, 1964; Shelley & Nelligan, 1966; Silver & Edwards, 1980) and appears to be related to the intense sympathoadrenal stimulation and the high levels of adrenocortical and pancreatic α-cell activity that occur in the immediate post-natal period (Silver & Edwards, 1980; Lagercrantz, 1983; Sperling *et al.*, 1984). In such circumstances pancreatic β-cell activity is minimal and the factors antagonizing the action of insulin may remain dominant for several hours after birth. Certainly when glucose is raised experimentally in newborn lambs there is little or no β-cell response (Shelley, 1979). It is not surprising, therefore, that exogenous insulin is relatively ineffective during this period in many species (Silver & Edwards, 1980), including the foal.

However, when glucose was eventually reduced in the newborn foal, adrenaline and, to a lesser degree, cortisol were both released, albeit in lower concentrations than in the older animals. To what extent the comparatively poor endocrine responses in the newborn can be ascribed to over stimulation in the preceding period after delivery or to incompletely developed glucostatic mechanisms is impossible to determine at present. The adrenal medulla may not be fully mature at this stage because its innervation is incomplete at birth in both the foal and calf (Comline & Silver, 1966; 1971; Silver & Edwards, 1980).

In contrast to the neonate, the rapidity with which the older foals responded to severe hypoglycaemia indicates that homeostatic mechanisms are well developed at this stage although plasma glucose could not be restored after the higher dose of insulin. Nevertheless, the mobilization of FFA and lactate and the increased utilization of amino acids are all metabolic adjustments which help to ameliorate the effects of hypoglycaemia. Adrenaline-induced lacticacidaemia can protect newborn calves against profound hypoglycaemia (Edwards, 1964; Comline & Edwards, 1968) and cerebral uptake of lactate can occur in newborn but not older calves (Gardiner, 1980). There is less information about the role of FFA in hypoglycaemia although the CNS can utilize ketones to a limited extent in some species (Girard & Ferre, 1982). Nevertheless, the increased FFA levels will obviously have a sparing effect on peripheral glucose uptake. Whether the lipolysis is brought about by the high circulating adrenaline (as the close correlation between the two suggests), or whether there is a neural noradrenergic component to the response, cannot be resolved from the present findings.

While the adrenal medullary and cortical responses to hypoglycaemia are clearly important in promoting the metabolic adjustments to low glucose levels, they are not the only endocrine changes induced in these circumstances. Glucagon is also released and, in the calf, is a more sensitive index of hypoglycaemia than sympathoadrenal stimulation (Bloom *et al.*, 1975; Silver & Edwards, 1980). Preliminary findings in the foal have shown that, in addition to the normal postnatal glucagon surge (Sperling *et al.*, 1984), there is also increased α-cell activity once hypoglycaemia is established both in the newborn and older foal.

The wide range of metabolic and endocrine responses elicited by insulin-induced hypoglycaemia in the present study indicates the importance of preventing the development of severe hypoglycaemia in the foal during the weeks immediately after birth. The absence of any deleterious effects from fasting for 2–3 h or more gives no indication of the consequences of deteriorating lactation or inadequate milk intake, which will lead eventually to a depletion of the energy stores. Further studies on the nature of these reserves in the foal and how they are controlled in relation to feeding and fasting are now being undertaken.

We thank the Wellcome Trust and the Horserace Betting Levy Board for generous financial support; the Director and staff of the Equine Research Station for provision of research facilities; and the many members of the Physiological Laboratory, Cambridge, and the E.R.S. for their help in the management of the animals and assistance with the biochemical analyses. J.O. is supported by a training grant from the H.B.L.B. and R.F. was supported by the Wellcome Trust.

References

Anton, A.H. & Sayre, D.F. (1962) A study of the factors affecting the aluminium oxide-trihydroxyindole procedure for the analysis of catecholamines. *J. Pharm. exp. Therap.* **138**, 360–367.

Bioanalytical Systems Inc. (1985) LCEC Applications Note No. 14, West Lafayette, U.S.A.

Bloom, S.R., Edwards, A.V., Hardy, R.N., Malinowska, K.W. & Silver, M. (1975) Endocrine responses to insulin in the young calf. *J. Physiol., Lond.* **244**, 783–803.

Comline, R.S. & Edwards, A.V. (1968) The effects of insulin on the newborn calf. *J. Physiol., Lond.* **198**, 383–404.

Comline, R.S. & Silver, M. (1966) The development of the adrenal medulla of the fetal and newborn calf. *J. Physiol., Lond.* **183**, 305–340.

Comline, R.S. & Silver, M. (1971) Catecholamine secretion by the adrenal medulla of the foetal and newborn foal. *J. Physiol., Lond.* **216**, 659–682.

Comline, R.S., Fowden, A.L. & Silver, M. (1979) Carbohydrate metabolism in the fetal pig during late gestation. *Q. Jl. Exp. Physiol.* **64**, 277–289.

Crone, C. (1965) The secretion of adrenal medullary hormones during hypoglycaemia in intact, decerebrate and spinal sheep. *Acta physiol. scand.* **63**, 213–224.

Duner, H. (1954) The effect of insulin hypoglycaemia on the secretion of adrenaline and noradrenaline from the suprarenal gland of the cat. *Acta physiol. scand.* **32**, 63–68.

Edwards, A.V. (1964) Resistance to hypoglycaemia in the newborn calf. *J. Physiol., Lond.* **171**, 46P–47P. Abstr.

Fowden, A.L., Ellis, L. & Rossdale, P.D. (1982) Pancreatic β cell function in the neonatal foal. *J. Reprod. Fert., Suppl.* **32**, 529–535.

Gardiner, R.M. (1980) The effects of hypoglycaemia on cerebral blood flow in the newborn calf. *J. Physiol., Lond.* **298**, 37–51.

Girard, J. & Ferre, P. (1982) Metabolic and hormonal changes around birth. In *Biochemical Development of the Fetus and Neonate*, pp. 517–551. Ed. C. T. Jones. Elsevier Biomedical Press, Amsterdam.

Hökfelt, B. (1951) Noradrenaline and adrenaline in mammalian tissues. *Acta physiol. scand.* **25**, Suppl. 92.

Lagercrantz, H. (1983) The development and functional role of the sympathoadrenal system in the newborn. In *Neonatal Intensive Care IV*, pp. 67–75. Eds L. Stern, H. Bard & B. Friis-Hansen. Masson, New York.

Prenton, M.A. & London, D.R. (1967) The continuous in vivo monitoring of plasma amino nitrogen. *Technicon Monographs* **2**, 70–75.

Rossdale, P.D., Silver, M., Ellis, L. & Frauenfelder, H. (1982) Response of the adrenal cortex to tetracosactrin (ACTH) in the premature and full term foal. *J. Reprod. Fert., Suppl.* **32**, 545–553.

Shelley, H.J. (1961) Glycogen reserves and their changes at birth. *Br. med. Bull.* **17**, 137–143.

Shelley, H.J. (1979) Transfer of carbohydrate. In *Placental Transfer*, pp. 118–141. Eds G. V. P. Chamberlain & A. W. Wilkinson. Pitman Press, Bath.

Shelley, H.J. & Nelligan, G.A. (1966) Neonatal hypoglycaemia. *Br. med. Bull.* **22**, 34–39.

Silver, M. & Edwards, A.V. (1980) The development of the sympathoadrenal system with an assessment of the role of the adrenal medulla in the fetus and neonate. In *Biogenic Amines in Development*, pp. 147–211. Eds H. Parvez & S. Parvez. Elsevier North-Holland, Amsterdam.

Silver, M., Ousey, J.E., Dudan, F.E., Fowden, A.L., Knox, J., Cash, R.S.G. & Rossdale, P.D. (1984) Studies on equine prematurity 2: post natal adrenocortical activity in relation to plasma adrenocorticotrophic hormone and catecholamine levels in term and premature foals. *Eq. vet. J.* **16**, 278–285.

Snedecor, G.W. & Cochrane, W.G. (1967) *Statistical Methods*, 6th edn. Iowa State Univ. Press, Ames.

Sperling, M.A., Ganguli, S., Leslie, N. & Landt, K. (1984) Fetal-perinatal catecholamine secretion: role in perinatal glucose homeostasis. *Am. J. Physiol.* **247**, E69–E74.

J. Reprod. Fert., Suppl. **35** (1987), 615–622

Fatty acid composition of the plasma lipids of the maternal and newborn horse

J. P. Stammers*, D. P. Leadon and D. Hull*

*The Irish Equine Centre, Johnstown, Naas, Co. Kildare, Ireland, and *Department of Child Health, University Hospital, Queen's Medical Centre, Nottingham NG7 2UH, U.K.*

Summary. The fatty acid composition of the plasma free fatty acid, triacylglycerol and phospholipid fractions was measured in blood and milk samples taken daily from 3 mares and their foals on Days 1–9 *post partum* inclusive, and from a total of 12 mares and foals on Days 22, 30 and 51. A rise in the plasma concentrations of triacylglycerol and phospholipid similar to that well documented in other species occurred in the neonatal period. Alterations in the composition of the foal plasma phospholipid after birth lend support to the view that the placenta rather than the fetus could be responsible for the elongation and desaturation of the essential fatty acids to provide long-chain polyunsaturated fatty acids for the fetus.

Introduction

The newborn of ruminant species have very low concentrations of the essential fatty acids in their tissues and plasma (Leat, 1966; Scott *et al.*, 1967; Noble *et al.*, 1972) and the transfer of fatty acids from the maternal circulation across the placenta of these species is extremely low (Elphick *et al.*, 1979; Leat & Harrison, 1980). In some non-ruminant species essential fatty acids can reach the fetus from maternal free fatty acid or maternal triacylglycerol, e.g. rabbit (Elphick & Hull, 1977a, b; Elphick *et al.*, 1978; Stammers *et al.*, 1983) and guinea-pig (Hershfield & Nemeth, 1968; Thomas & Lowy, 1982). In other non-ruminant mammals, e.g. pig and cat, fatty acid entering the fetal circulation from the placenta is enriched with long-chain polyunsaturated derivatives of essential fatty acids, but the major part of these fatty acids is not derived directly from the maternal free fatty acid compartment (Elphick *et al.*, 1980; Elphick & Hull, 1984). The cat is particularly interesting since it lacks the ability to elongate and desaturate the essential fatty acids (Rivers *et al.*, 1975; Sinclair *et al.*, 1979). In the cat, therefore, long-chain polyunsaturated derivations of essential fatty acids entering the fetal circulation must have been derived directly from the maternal circulation. The only maternal lipid compartment containing the necessary fatty acids is the maternal plasma phospholipids.

The horse placenta is permeable to free fatty acid, an unusual finding in an animal with a complex 6-layered placenta (Silver *et al.*, 1973; J. P. Stammers, D. Hull, D. P. Leadon, L. B. Jeffcott & P. D. Rossdale, unpublished data). Also, long-chain polyunsaturated fatty acids, docosa-hexaenoic acid in particular, are virtually absent from maternal circulating phospholipids and in extremely low concentrations in other maternal lipid compartments. In spite of this, fetal plasma phospholipids are rich in these long-chain polyenoic fatty acids which must therefore be made from their essential fatty acid precursors in fetal tissues, particularly the liver, or placental tissue (unpublished data). If the placenta forms the long-chain polyunsaturated fatty acids for the fetus, then a change in fatty acid composition of circulation lipids would be expected after birth, whereas if the fetal liver were the source, then, providing mares' milk contains adequate linoleic and linolenic acids, the fatty acids in the newborn plasma phospholipids would continue to contain long-chain polyunsaturated fatty acids.

To investigate these possibilities the fatty acid composition of different plasma lipid fractions, including phospholipid, were examined in the foal during the first 50 days of life and were compared with the fatty acid composition of plasma lipids in the maternal circulation over the same period. The fatty acid composition of circulating and tissue lipids in newborn and young ruminants has been extensively examined (Leat, 1966; Noble *et al.,* 1971, 1972; Payne, 1987a, b), but no such information has been reported for the foal.

Materials and Methods

Blood samples were taken daily from 3 mares and their foals between Days 1 and 9 inclusive after birth. Blood samples were also taken from other mare and foal pairs on Days 22 (2 mares and foals), 30 (6 mares and foals) and 51 (4 mares and foals) after birth. The mares and their newly born foals were kept housed for the first 3 days after

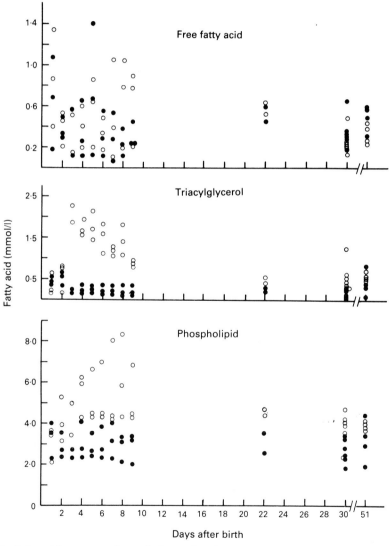

Fig. 1. The fatty acid concentrations of plasma free fatty acid, triacylglycerol and phospholipid fractions in samples from mares (●) and their foals (○) taken during Days 1–51 after birth.

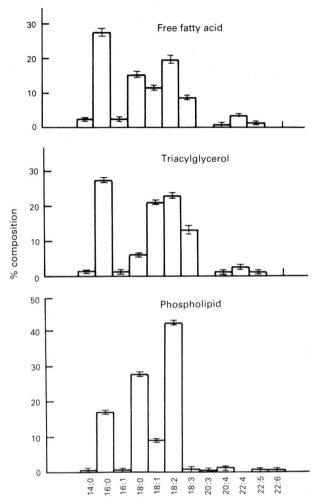

Fig. 2. The mean individual fatty acid composition ($\pm$ s.e.m.) of plasma free fatty acid, triacylglycerol and phospholipid fractions in samples taken from 12 mares on various days between Days 1 and 51 *post partum* ($n = 38$). Each fatty acid is given as follows: 14:0 = myristic acid; 16:0 = palmitic acid; 16:1 = palmitoleic acid; 18:0 = stearic acid; 18:1 = oleic acid; 18:2 = linoleic acid; 18:3: = linolenic acid; 20:3 = eicosatrienoic acid; 20:4 = arachidonic acid; 22:4 = docosatetraenoic acid; 22:5 = docosapentaenoic acid; 22:6 = docosahexaenoic acid.

delivery. Thereafter they were turned out to pasture during the day and housed overnight. Whilst housed, the animals were given hay and water *ad libitum*. Milk samples were taken from all mares on the days when blood samples were collected.

Blood samples were collected from the jugular vein and the plasma was stored on ice until transferred to the laboratory, where it was kept deep frozen before analysis. Milk samples were stored deep frozen. The plasma individual free fatty acid concentrations were determined on samples extracted as described by Dole (1956) and separated by thin-layer chromatography followed by gas–liquid chromatography (Elphick *et al.*, 1980). Plasma phospholipid and triacylglycerol composition and milk triacylglycerol composition were determined by extracting 1 ml plasma or 100 µl milk samples into chloroform/methanol (2:1, v/v) and washing with 0·6 M-phosphate buffer, pH 6·0 (Folch *et al.*, 1957). The extracts were then subjected to thin-layer and gas-liquid chromatography (Elphick & Lawlor, 1977). Precautions were taken to avoid lipid oxidation by using D-L-α-tocopherol as an antioxidant and keeping solutions under nitrogen.

Results

Plasma lipid concentrations

As shown in Fig. 1 free fatty acid concentrations fluctuated in mares and foals for the first 9 days after birth, but towards the end of the study period variations in levels were less and concentrations were similar in mares and foals. Plasma concentrations of triacylglycerol fell slightly during the first 3 days in mares, but thereafter remained remarkably constant for the rest of the study period. In the foals, however, a marked rise in plasma triacylglycerol concentration occurred on the 3rd day after birth, reaching levels about 5 times greater than those in the mare, thereafter declining gradually to similar values to those found in the mother by about Day 22. Maternal plasma phospholipid concentrations were similar throughout the study, whereas foal plasma phospholipids increased during the first 4 days of life. This increase was less rapid than that in the foal plasma triacylglycerol fraction. Foal plasma phospholipid concentrations declined slowly after Day 4 to approach maternal levels by Day 51.

Individual fatty acid composition of plasma lipids

The fatty acid composition of the plasma free fatty acid triacylglycerol and phospholipid fractions of maternal blood during the first 51 days after delivery are given in Fig. 2. Since only

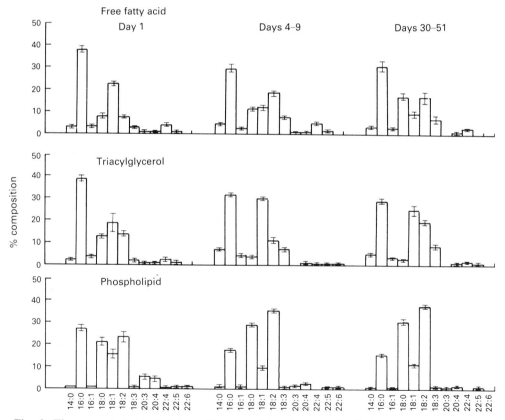

Fig. 3. The mean individual fatty acid composition (± s.e.m.) of plasma free fatty acid, triacylglycerol and phospholipid fractions in samples taken from 3 foals on Day 1($n = 3$) and daily from Days 4 to 9 inclusive ($n = 18$) and from 7 foals between Days 30 and 51 ($n = 10$) after birth.

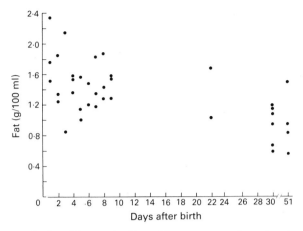

Fig. 4. The percentage fat in milk samples from 12 mares on various days between Days 1 and 51 *post partum*.

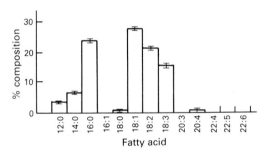

Fig. 5. The individual fatty acid composition of milk fat in all milk samples taken throughout the study (*n* = 38). Since only minor variations in milk fat composition were seen during the study period, the results from different days *post partum* were pooled.

minor changes in composition occurred during this time the results from different days *post partum* are pooled. Figure 3 shows the fatty acid composition of the plasma free fatty acid, triacylglycerol and phospholipid fractions in foal blood on Day 1, Days 4–9 and Days 30–51 after birth. On the first day after birth all three lipid fractions in the foal plasma differed in fatty acid composition from those found in the mother. The foal plasma fatty acid compositions on Days 2 and 3 after birth were intermediate between the values on Days 1 and 4 of life. The above changes occurred gradually during this period.

In the foal plasma free fatty acid compartment, oleic acid decreased and linoleic and linolenic acids increased over the first 4 days of life; the composition then remained similar, except for stearic acid which formed a greater proportion of the plasma free fatty acid compartment from Days 30 to 50 than during the first 9 days after birth. These changes resulted in a profile which resembled that of the maternal plasma free fatty acid compartment.

The foal plasma triacylglycerol fraction also changed in composition; during the first 4 days after birth the proportions of palmitic and stearic acid decreased and those of oleic and linolenic acid increased. Linoleic acid increased between Days 9 and 30; the foal plasma triacylglycerols therefore appeared to change in composition throughout the study period to become similar, but not identical, to those found in the adult.

A progressive fall in palmitic, eicosatrienoic and arachidonic acids together with an increase in stearic and linoleic acids resulted in the foal plasma phospholipids resembling those of the adult by Day 4 of life. Their composition then remained unaltered for the rest of the study period.

Milk lipid composition

Milk samples were taken on each day that blood was sampled from the mares and the percentage fat content analysed (Fig. 4). The fat content of the milk was highest immediately after delivery, at around 1·8 g fat/100 ml, and fell to approximately half this value by Day 30. Despite the fall in total fat content, the fatty acid composition of the milk did not greatly vary from day to day (Fig. 5).

Discussion

Within 2–3 days of delivery the total concentrations of plasma lipids in the maternal circulation are similar to levels reported in the literature for adult horse plasma lipids (Leat & Baker, 1970; Yamamoto *et al.*, 1979). Birth produces a rise in free fatty acid levels in both maternal and newborn circulations due to catecholamine-induced mobilization of lipid stores (Hull, 1979). The rise in plasma concentrations of triacylglycerol and phospholipids seen in the foal over the first few days of life followed by a steady decline towards maternal levels is well documented in other species and occurs over a similar time scale (Friedman & Byers, 1961; Persson, 1974; Potter, 1977; Hamosh *et al.*, 1978). The usual cause for this rise in circulating lipid is cited as a change at birth from the utilization of carbohydrate as the main energy source to that of fat and the relatively high fat content of the milk, coupled with a low triacylglycerol clearing ability of newborn extrahepatic tissues (Hamosh *et al.*, 1978). The fat content of horse's milk is comparatively low (Widdowson, 1976); most of the energy is provided by carbohydrate ($\sim 60\%$). However, fetal levels of plasma triacylglycerol are extremely low (unpublished) and, therefore, despite the low fat content of the milk, the first milk feed is still likely to present the newborn foal with levels of circulating triacylglycerol which it has not previously encountered. However, the changes in composition of the foal plasma triacylglycerol fraction over the first 4 days of life would suggest that the rise in levels of plasma triacylglycerol seen during this period were not just the result of inefficient clearing of dietary fat. Palmitic and oleic acids are the main component fatty acids found in the umbilical cord plasma triacylglycerol fraction during delivery. On the first day after birth, when the foal plasma triacylglycerol levels are already some 5-fold greater than those at delivery, palmitic, oleic, linoleic and stearic acids form the major fatty acid in the foal plasma triacylglycerol fraction. Since stearic acid is virtually absent from mare's milk but does constitute a major fatty acid in the umbilical cord vein plasma free fatty acid fraction at delivery (unpublished) the appearance of this fatty acid would suggest that a considerable proportion of the increased circulating triacylglycerol on the first day of life could be derived from the incorporation of fatty acid mobilized from fetal stores as free fatty acid during parturition. Over the next few days of life, stearic acid levels decline in the foal plasma triacylglycerol fraction as might be expected, whilst the levels of fatty acids which are found in mare's milk increased. The composition on Day 4, when the foal plasma triacylglycerol reached its highest levels, however, still did not fully reflect what might have been expected from the dietary fat, with a relatively greater proportion of palmitic and oleic acids. This would suggest an increased production of endogenous triacylglycerol incorporating fatty acids made by lipogenesis from carbohydrate and other substrates in the liver. Only when the foal plasma triacylglycerol concentrations had fallen to similar levels to those found in the maternal circulation did their fatty acid profile entirely match the fatty acid profile of the dietary lipids.

The fatty acid profiles of both the free fatty acid and phospholipid fractions on the first day of life were similar to those seen in umbilical cord vein plasma during delivery, with the notable exception of some of the long-chain polyunsaturated derivatives of the essential fatty acids.

Arachidonic acid, which formed 6% in the cord plasma free fatty acid fraction, together with docosapentaenoic and docosahexaenoic acids, which formed 2·5% and 4% respectively in the cord plasma phospholipid fraction, had all but disappeared by 1 day after birth. The sudden 'switching-off' of production of these fatty acids at birth lends support to the view that the placenta could be responsible for the desaturation and elongation the essential fatty acids to provide long-chain polyunsaturated fatty acids for the fetus.

Over the first 4 days, increasing amounts of linoleic, linolenic and stearic acids appeared in the foal plasma free fatty acid fraction, resulting in a profile similar to that found in the adult. The proportion of stearic acid continued to increase until by Day 30 the maternal and foal plasma free fatty acids were almost identical in composition. Adipose tissue stores might be expected to expand and fill with lipid in the foal during this period; a large proportion assimilated through lipogenesis from carbohydrate sources tending to favour the production of saturated fatty acids and the rest through uptake of triacylglycerol fatty acid from the circulation.

The foal plasma phospholipid fatty acid profile alters in composition to match the maternal composition by the 4th day after birth. The amounts of the long chain polyunsaturated derivatives, which together comprise over 17% of the plasma phospholipids in the umbilical vein at delivery, are trivial after Day 4 and linoleic acid forms the major phospholipid fatty acid. In other studies in which the fatty acid profiles of plasma lipids have been examined in the horse, despite different compositions in the plasma free fatty acid and triacylglycerol fractions according to the different diets fed to the animals, the phospholipid fatty acid profiles have all been similar to those found in the maternal plasma and foal plasma from Day 4 onwards in this study (Leat & Baker, 1970; Yamamoto *et al.*, 1979; Luther *et al.*, 1981; Bauer & Ransone, 1983). The plasma phosholipid fraction would therefore appear to have a particularly consistent composition in the horse, which is established shortly after birth. The alterations in composition seen in the plasma phospholipid fraction of the newborn foal indicate that the production of the long-chain polyunsaturates ceases at birth, suggesting that the placenta rather than the fetus is the source of these fatty acids *in utero*.

References

Bauer, J.E. & Ransone, W.D. (1983) Fatty acid composition of serum lipids in fasting ponies. *Lipids* **18**, 397–401.

Dole, V.P. (1956) A relation between non-esterified fatty acids in plasma and the metabolism of glucose. *J. clin. Invest* **35**, 150–154.

Elphick, M.C. & Hull, D. (1977a) The transfer of free fatty acids across the rabbit placenta. *J. Physiol., Lond.* **264**, 751–766.

Elphick, M.C. & Hull, D. (1977b) Rabbit placental clearing-factor lipase and transfer to the foetus of fatty acids derived from triglycerides injected into the mother. *J. Physiol., Lond.* **273**, 475–487.

Elphick, M.C. & Hull, D. (1984) Transfer of fatty acid across the cat placenta. *J. devl Physiol.* **6**, 517–525.

Elphick, M.C. & Lawlor, J.P. (1977) Quantitative recovery of free and esterified fatty acids from thin layer plates coated with silicagel. *J. Chromatog.* **130**, 139–143.

Elphick, M.C., Edson, J.L., Lawlor, J.P. & Hull, D. (1978) Source of fetal stored lipids during maternal starvation in rabbits. *Biol. Neonate* **34**, 146–149.

Elphick, M.C., Hull, D. & Broughton-Pipkin, F. (1979) The transfer of fatty acids across the sheep placenta. *J. devl Physiol.* **1**, 31–45.

Elphick, M.C., Flecknell, P., Hull, D. & McFadyen, I.R. (1980) Plasma free fatty acid umbilical venous-arterial concentration differences and placental transfer of ^{14}Cpalmitic acid in pigs. *J. devl Physiol.* **2**, 347–356.

Folch, J., Lees, M. & Sloane Stanley, G.H. (1957) A simple method for the isolation and purification of total lipids from animal tissues. *J. biol. Chem.* **266**, 497–509.

Friedman, M. & Byers, S.O. (1961) Effects of diet on serum lipids of fetal, neonatal and pregnant rabbits. *Am. J. Physiol.* **201**, 611–616.

Hamosh, M., Morton, R.F., Hall, J.R. & Hamosh, P. (1978) Lipoprotein lipase activity and blood triglyceride levels in fetal and newborn rats. *Pediat. Res.* **12**, 1132–1136.

Hershfield, M.S. & Nemeth, A.M. (1968) Placental transport of free palmitic and linoleic acids in the guinea pig. *J. Lipid Res.* **9**, 460–468.

Hull, D. (1979) Fatty acid metabolism before and after birth. In *Nutrition and Metabolism of the Fetus and Infant*, pp. 109–122. Ed. H. K. A. Visser. Martinus Nijhoff, The Hague.

Leat, W.M.F. (1966) Fatty acid composition of the plasma lipids of newborn and maternal ruminants. *Biochem. J.* **98**, 598–603.

Leat, W.M.F. & Baker, J. (1970) Distribution of fatty acids in the plasma lipids of herbivores grazing pasture; a species comparison. *Comp. Biochem. Physiol.* **36**, 153–161.

Leat, W.M.F. & Harrison, F.A. (1980) Transfer of long

chain fatty acids to the fetal and neonatal lamb. *J. devl Physiol.* **2**, 257–274.

Luther, D.G., Cox, H.U. & Dimopoullos, G.T. (1981) Fatty acid composition of equine plasma. *Am. J. vet. Res.* **42**, 91–93.

Noble, R.C., Steele, W. & Moore, J.H. (1971) Postnatal changes in the phospholipid composition of livers from young lambs. *Lipids* **6**, 926–929.

Noble, R.C., Steele, W. & Moore, J.H. (1972) The metabolism of linoleic acid by the young lamb. *Br. J. Nutr.* **27**, 503–508.

Payne, E. (1978a) Fatty acid composition of tissue phospholipids of the foetal calf and neonatal lamb, deer calf and piglet as compared with the cow, sheep, deer and pig. *Br. J. Nutr.* **39**, 45–52.

Payne, E. (1978b) The polyunsaturated fatty acid status of foetal and neonatal ruminants. *Br. J. Nutr.* **39**, 53–59.

Persson, B. (1974) Carbohydrate and lipid metabolism in the newborn infant. *Acta anaesthesol. scand., Suppl.* **55**, 50–57.

Potter, J.M. (1977) Perinatal plasma lipid concentrations. *Aust. N. Z. Jl Med.* **7**, 155–160.

Rivers, J.P.W., Sinclair, A.J. & Crawford, M.A. (1975) Inability of the cat to desaturate essential fatty acids. *Nature, Lond.* **258**, 171–173.

Scott, T.W., Setchell, B.P. & Bassett, J.M. (1967) Characterisation and metabolism of bovine foetal lipids. *Biochem. J.* **104**, 1040–1047.

Silver, M., Steven, D.H. & Comline, R.S. (1973) Placental exchange and morphology in ruminants and the mare. In *Foetal and Neonatal Physiology*, pp. 245–271. Cambridge University Press.

Sinclair, A.J., McLean, J.G. & Monger, E.E. (1979) Metabolism of linoleic acid in the cat. *Lipids* **14**, 932–936.

Stammers, J.P., Elphick, M.C. & Hull, D. (1983) Effect of maternal diet during late pregnancy on fetal lipid stores in rabbits. *J. devl Physiol.* **5**, 395–404.

Thomas, C.R. & Lowy, C. (1982) The clearance and placental transfer of free fatty acids and triglycerides in the pregnant guinea pig. *J. devl Physiol.* **4**, 163–173.

Widdowson, E.M. (1976) Pregnancy and lactation: the comparative point of view. In *Early Nutrition and later Development*, pp. 1–10. Ed. A.W. Wilkinson. Pitman Medical, London.

Yamamoto, M., Tanoka, Y. & Sugano, M. (1979) Serum and liver lipid composition and lecithin; cholesterol acyltransferases in horses, *Equus caballus*. *Comp. Biochem. Physiol.* **62B**, 185–193.

J. Reprod. Fert., Suppl. **35** (1987), 623–628

Printed in Great Britain
© 1987 Journals of Reproduction & Fertility Ltd

Systemic and pulmonary haemodynamics in normal neonatal foals

W. P. Thomas, J. E. Madigan, K. Q. Backus and W. E. Powell

Department of Medicine, School of Veterinary Medicine, University of California, Davis, CA 95616, U.S.A.

Summary. Cardiopulmonary function was studied in 10 full-term healthy foals from birth to 14 days of age. Systemic and pulmonary haemodynamics were recorded in lateral recumbency via indwelling aortic and pulmonary artery catheters. Mean body weight increased from $45\cdot4 \pm 2\cdot4$ kg on Day 1 to $70\cdot6 \pm 6\cdot1$ kg on Day 14. All foals had a continuous murmur of patent ductus arteriosus for 3–6 days. From Day 1 (12 h old) to Day 14, heart rate increased (89 ± 4 to 95 ± 5/min), mean aortic pressure increased ($87\cdot7 \pm 1\cdot9$ to $100\cdot3 \pm 3\cdot2$ mmHg), mean pulmonary artery pressure decreased ($38\cdot6 \pm 4\cdot6$ to $27\cdot4 \pm 3\cdot0$ mmHg), mean right atrial pressure was unchanged ($4\cdot5 \pm 0\cdot5$ to $4\cdot6 \pm 0\cdot9$ mmHg), mean pulmonary artery wedge pressure was unchanged ($7\cdot6 \pm 0\cdot9$ mmHg to $8\cdot1 \pm 0\cdot7$ mmHg), cardiac output increased ($8\cdot03 \pm 0\cdot59$ to $15\cdot88 \pm 1\cdot90$ l/min), cardiac index increased ($180\cdot5 \pm 10\cdot3$ to $222\cdot1 \pm 21\cdot6$ ml/kg/min), stroke volume increased ($90\cdot4 \pm 5\cdot7$ to $164\cdot2 \pm 25\cdot9$ ml), stroke volume index was unchanged ($2\cdot04 \pm 0\cdot10$ to $2\cdot30 \pm 0\cdot35$ ml/kg), pulmonary vascular resistance decreased (314 ± 39 to 104 ± 21 aru), systemic vascular resistance decreased (858 ± 70 to 497 ± 87 aru), and pulmonary/systemic resistance ratio decreased (38 ± 6 to $21 \pm 5\%$). All changes were gradual, although pulmonary artery pressure and pulmonary vascular resistance decreased rapidly in the first 24 h. Catheters were generally well tolerated over several days, indicating their feasibility for studying cardiovascular function in full-term or premature equine neonates.

Introduction

Neonatal respiratory distress syndrome (NRDS) is a major factor adversely affecting survivability of premature foals (Platt, 1973; Rossdale, 1979; Slauson, 1979; Dennis, 1981; Martens, 1982; Sonea, 1985). Compared to its human counterpart, the pathophysiology of NRDS in premature foals is poorly understood. Most research on equine NRDS has focussed on pulmonary function and gas exchange, since hypoxaemia is a consistent feature of the syndrome (Rose *et al.*, 1982, 1983; Rossdale, 1968, 1969, 1970; Stewart *et al.*, 1984a, b). Little attention has been paid to the potential contribution of cardiovascular abnormalities with intracardiac or extracardiac shunting to the systemic hypoxaemia in these foals.

The role of the cardiovascular system in premature or full term foals with neonatal maladjustment or respiratory distress syndrome is unclear (Rossdale, 1979; Martens, 1982; Rose & Stewart, 1983; Sonea, 1985). Patency of the ductus arteriosus is known to persist from 1 to several days after birth in normal term foals (Amoroso *et al.*, 1958; Scott *et al.*, 1975; Fregin, 1982). Premature birth in foals could, as in some human infants, result in delayed ductal closure, delayed involution and maturation of the pulmonary vascular bed, and extracardiac or intrapulmonary shunting, causing or contributing to systemic hypoxaemia (Gootman & Gootman, 1983).

Compared to pulmonary function studies, little information has been published on cardiovascular parameters in normal or premature foals (Rossdale, 1967; Rose & Stewart, 1983;

Lombard *et al.*, 1984; Stewart *et al.*, 1984c). As a prerequisite to the cardiovascular evaluation of neonatal foals with NRDS, we designed a study to establish reference standards for systemic and pulmonary haemodynamics in normal neonates, and to test the feasibility of acute or chronic catheterization for haemodynamic monitoring in newborn foals.

Materials and Methods

Ten full-term, healthy foals of mixed-breed (Quarter horse and Thoroughbred) were studied. Each foal was born without assistance to a healthy mare, and remained in a box stall with the mare between recordings. Each foal had an initial physical examination, complete blood count, venous blood culture, and serum zinc turbidity test for colostral antibody absorption (18 h of age). The first 6 foals were catheterized on Days 1 or 2 and examined daily to Day 6, then every other day to 14 days of age. The other 4 foals were catheterized within 2–4 h of birth and were recorded at 2-h intervals up to 12 h, at 4-h intervals up to 24 h (1 day), then daily to Day 4–6 (3 foals). All foals were examined and weighed daily, and procaine penicillin (20 000 units/kg) and gentamicin (2·2 mg/kg) were given intramuscularly twice daily. All foals had access to their mare at all times.

For catheterization, foals were allowed to suck, then blindfolded and manually restrained in lateral recumbency without sedation. Using local anaesthesia, the jugular vein and carotid artery were isolated. A size 8 French radio-opaque polyethylene catheter was inserted into the carotid artery and the tip was positioned in the ascending aorta. An intravascular sheath was inserted into the jugular vein through which a size 7 French, triple-lumen, balloon-tipped, flow-directed thermodilution catheter (Honeywell) was passed and positioned in the left pulmonary artery. Catheter position was determined by pressure monitoring and confirmed by thoracic radiography. Catheters were sutured to the skin and covered by povidone iodine gauze. Between daily recordings, they were filled with concentrated heparin solution and wrapped with elastic bandage. Before sampling or recording, the catheters were aspirated and flushed with heparinized saline (9 g NaCl/l).

Haemodynamic measurements were recorded in lateral recumbency (for consistency and ease of restraint), while blood gas measurements were made in the lateral recumbent and standing positions. It was difficult to record stable haemodynamics with the foals standing, so this approach was not attempted after the first 2 foals. Arterial and mixed venous blood gases were determined on samples obtained from the aortic and pulmonary artery catheters, respectively. Intravascular pressures were measured using Statham strain-gauge transducers (P23Db) positioned at the midthoracic level. Cardiac output was determined in triplicate by thermodilution, using 5 ml iced saline and an automated cardiac output computer (Honeywell: Model DTCCO-07). Measurements included heart rate, cardiac output, pulmonary artery pressure, aortic pressure, right atrial pressure, pulmonary artery wedge pressure, all recorded on a multichannel physiological recorder (Gould Brush 2600). Also recorded were body weight and heart sounds, and electrocardiograms and echocardiograms were obtained periodically over the 2 weeks. Calculated values included cardiac index, stroke volume index, stroke work index, systemic and pulmonary vascular resistance, and pulmonary/systemic resistance ratio. Results were tabulated and plotted as means ± standard error (s.e.m.). Since the first 6 foals were born during the night, initial data from the 3 foals catheterized on the first morning were arbitrarily combined with the 12-h data on the last 4 foals.

After completion of each study, catheters were removed and the neck was bandaged and cared for until healed.

Results

Results of all measured and calculated values are presented in Table 1. All foals gained weight steadily during this study, averaging 1–2 kg per day, from 45·4 ± 2·4 kg on Day 1 to 70·6 ± 6·1 kg on Day 14. Although a mild fever and localized cellulitis at the catheter exit site occurred in some cases, all foals ate well and tolerated the indwelling catheters well over 1–2 weeks. Gaps in data collection or premature termination of the study in some foals occurred due to catheter malfunction, particularly between 10 and 14 days of age. Nine foals recovered without complications. One foal developed infective carotid arteritis, sepsis, pneumonia and a brain abscess 3 weeks after the study and was killed.

Mean aortic pressure tended to decrease slightly during the first 24 h, but then gradually increased. At the same time, mean pulmonary artery pressure decreased between 2 and 24 h (1 day), with little subsequent change. Throughout the study, mean right atrial pressure varied between 3·1 and 7·1 mmHg, while mean pulmonary artery wedge pressure varied between 7·2 and 11·6 mmHg. There was no clear upward or downward trend over the 2 weeks, but the pulmonary artery wedge pressure was consistently 3–5 mmHg higher than right atrial pressure at each age.

Table 1. Haemodynamics of 10 normal foals

Age	No. of foals	Body wt (kg)	Heart rate (beats/min)	Aortic pressure (mmHg)	Pulmonary artery pressure (mmHg)	Right atrial pressure (mmHg)	Pulmonary artery wedge pressure (mmHg)	Cardiac index (ml/min/kg)	Stroke vol. index (ml/kg)	Stroke work index (g-m/kg)		Vascular resistance (aru)*		PVR/SVR ratio
										Right ventricle	Left ventricle	Pulmonary	Systemic	
2 h	2	45·4 ±2·4	83 ±10	95·7 ±11·9	40·3 ±4·7	3·1 ±0·6	7·5 ±2·5	155·3 ± 8·1	1·89 ±0·13	0·95 ±0·04	2·25 ±0·08	363 ±42	1027 ±174	36 ±2
4 h	4	45·4 ±2·4	84 ± 2	85·0 ± 2·9	41·0 ±2·8	4·4 ±1·2	9·0 ±1·9	197·8 ±25·3	2·35 ±0·26	1·16 ±0·13	2·42 ±0·25	310 ±56	772 ±119	40 ±1
6 h	4	45·4 ±2·4	84 ± 2	91·3 ± 2·4	36·3 ±2·7	5·5 ±1·2	9·0 ±1·8	181·5 ±13·4	2·18 ±0·15	0·90 ±0·08	2·42 ±0·12	274 ±45	865 ±106	32 ±4
8 h	4	45·4 ±2·4	81 ± 3	90·6 ± 1·6	32·4 ±2·4	5·5 ±1·6	7·7 ±1·2	173·9 ±11·6	2·16 ±0·16	0·78 ±0·04	2·44 ±0·19	260 ±39	882 ± 74	29 ±3
10 h	4	45·4 ±2·4	80 ± 1	87·5 ± 1·4	30·3 ±2·6	4·1 ±1·1	7·2 ±1·4	188·1 ±10·4	2·36 ±0·10	0·85 ±0·13	2·58 ±0·12	218 ±32	803 ± 85	28 ±4
12 h	7	44·4 ±1·4	89 ± 4	87·7 ± 1·9	38·6 ±4·6	4·5 ±0·5	7·6 ±0·9	180·5 ±10·3	2·04 ±0·10	0·94 ±0·13	2·22 ±0·13	314 ±39	858 ± 70	38 ±6
16 h	4	45·4 ±2·4	86 ± 4	86·9 ± 2·5	31·4 ±0·8	3·4 ±0·5	7·5 ±1·0	181·5 ±20·7	2·09 ±0·16	0·79 ±0·04	2·24 ±0·12	247 ±41	859 ±123	29 ±1
20 h	4	45·4 ±2·4	84 ± 4	84·1 ± 4·4	26·0 ±2·5	4·5 ±0·5	7·5 ±1·0	195·5 ± 9·8	2·33 ±0·04	0·68 ±0·07	2·43 ±0·17	166 ±22	726 ± 54	23 ±2
1 day	4	45·4 ±2·4	84 ± 3	84·4 ± 3·7	30·5 ±2·3	7·1 ±0·9	9·9 ±1·0	197·3 ±12·0	2·36 ±0·14	0·74 ±0·05	2·38 ±0·18	194 ±37	708 ± 74	27 ±3
2 days	9	48·3 ±2·4	95 ± 6	91·4 ± 4·5	27·8 ±2·3	4·1 ±1·1	8·7 ±0·8	204·1 ±11·8	2·06 ±0·09	0·67 ±0·06	2·34 ±0·17	167 ±24	723 ± 50	23 ±3
3 days	9	49·1 ±2·5	94 ± 7	92·2 ± 3·6	28·1 ±1·3	4·2 ±0·6	8·2 ±1·0	185·6 ±14·3	2·05 ±0·16	0·65 ±0·05	2·30 ±0·16	182 ±19	814 ± 68	22 ±1
4 days	9	50·9 ±2·4	96 ± 5	90·7 ± 3·6	30·0 ±1·8	6·1 ±1·0	9·8 ±1·5	222·8 ±20·3	2·30 ±0·14	0·75 ±0·06	2·58 ±0·23	152 ±20	629 ± 46	26 ±4
5 days	7	55·5 ±2·9	102 ± 9	95·8 ± 3·4	29·9 ±2·6	3·3 ±0·6	7·3 ±1·2	223·4 ±12·6	2·11 ±0·28	0·76 ±0·11	2·66 ±0·20	151 ±20	623 ± 50	25 ±3
6 days	7	57·1 ±2·9	104 ± 8	96·7 ± 5·1	28·7 ±2·3	4·7 ±0·8	10·0 ±2·0	214·5 ±18·1	2·22 ±0·28	0·73 ±0·13	2·77 ±0·25	124 ±14	613 ± 73	20 ±3
8 days	5	60·2 ±3·6	114 ± 9	100·0 ± 5·9	27·3 ±2·3	2·7 ±0·3	8·0 ±1·1	233·9 ±16·3	2·17 ±0·41	0·73 ±0·13	2·55 ±0·40	113 ±14	543 ± 38	21 ±2
10 days	5	63·2 ±3·8	108 ±10	101·3 ± 4·4	27·3 ±2·3	3·5 ±1·5	9·1 ±1·3	205·8 ± 8·5	2·10 ±0·23	0·62 ±0·11	2·56 ±0·15	99 ± 8	596 ± 45	17 ±2
12 days	4	65·8 ±3·9	111 ±10	100·3 ± 4·9	30·7 ±2·3	5·0 ±1·3	11·6 ±2·0	235·2 ±35·1	2·18 ±0·41	0·80 ±0·15	2·61 ±0·50	114 ±20	520 ± 80	23 ±4
14 days	4	70·6 ±6·1	95 ± 5	100·3 ± 3·2	27·4 ±3·0	4·6 ±0·9	8·1 ±0·7	222·1 ±21·6	2·30 ±0·35	0·76 ±0·20	2·80 ±0·27	104 ±21	497 ± 87	21 ±5

Values are mean ± s.e.m.

*Absolute resistance units equivalent to dyne·sec·cm⁻⁵.

As expected, cardiac output increased as the foals grew, but at a faster rate than body weight. As a result, cardiac index, after an initial increase between 2 and 4 h, tended to increase gradually over the 2 weeks up to Day 14. The heart rate also tended to increase from Day 1 to Day 14, with highest values on Days 10 and 12. As a result of parallel changes in cardiac index and heart rate, stroke volume index did not show an upward or downward trend, although there was considerable individual and daily variation in this measurement. Left ventricular stroke work index tended to increase over Days 1–14, while right ventricular stroke work tended to decrease in the first 24 h, but these trends were slight.

Systemic vascular resistance gradually decreased between Days 1 and 14. At the same time, pulmonary vascular resistance decreased sharply in the first 24 h, followed by a more gradual decline to Day 14. As a result, the pulmonary/systemic resistance ratio decreased from a high value of 40% at 4 h to 27% by 24 h, followed by a more gradual decline to Day 14.

Discussion

Like that of other mammalian species, the fetal cardiovascular system of the horse, with its parallel systemic and pulmonary circuits, high pulmonary:systemic vascular resistance ratio, and inter-vascular communications, undergoes dramatic adaptive changes at birth and during the first few weeks of extrauterine life (Dawes, 1961, 1968; Rose & Stewart, 1983). These changes include increased systemic vascular resistance and pressure, decreased pulmonary vascular resistance and pressure, closure of the foramen ovale and ductus arteriosus, and increased systemic arterial oxygen content (Dawes, 1961; Rudolph, 1970). Longitudinal studies in newborn lambs (Dawes, 1961, 1968), dogs (Campbell *et al.*, 1959; Rudolph *et al.*, 1961; Averill *et al.*, 1963; Arango & Rowe, 1971) and other animals (Rudolph, 1970) indicate that, although qualitatively similar, the time course of the transition from fetal to adult haemodynamics varies between species. For example, studies of newborn humans and other animals (Rudolph, 1970) have demonstrated elevated pulmonary artery pressure and pulmonary vascular resistance immediately after birth, with transition to adult values requiring from a few days (dogs) to several weeks (humans, calves). Most of the change occurs within the first few days of life. Systemic arterial pressure is usually low in newborns, with a gradual increase to adult levels over several weeks.

Several investigators have reported on systemic and pulmonary haemodynamics of normal adult horses (Milne *et al.*, 1975; Fregin, 1982). In 1967, Rossdale reported heart rates, heart sounds and electrocardiographic findings in neonatal foals. Lombard *et al.* (1984) and Stewart *et al.* (1984c) have reported echocardiographic measurements, indirect blood pressure measurements, and electrocardiographic findings in foals during the neonatal transition period. However, detailed haemodynamic characterization of newborn foals has not been reported. Our study indicates that the transitional circulation of the normal horse neonate undergoes comparatively rapid changes within the first 24 h, followed by more gradual change over the succeeding 2 weeks. Within a few hours after birth, mean systemic arterial pressure exceeded 80 mmHg, while mean pulmonary artery pressure was about 40 mmHg. Within 24 h, pulmonary artery pressure decreased to about 30 mmHg, which is similar to the 22–30 mmHg values reported for normal adult horses (Fregin, 1982). There was little further decrease in mean pulmonary artery pressure over the 2 weeks, in contrast to the more gradual decline reported in other animals and humans (Rudolph, 1970). Mean systemic arterial pressure increased gradually to about 100 mmHg by 2 weeks, still below reported adult values of 110–130 mmHg (Fregin, 1982). Mean right atrial and pulmonary artery wedge pressures were within the range of reported adult values (4–10 and 5–11 mmHg, respectively) throughout the study.

Heart rates for our foals (80–110/minute) were within the broad normal range reported in previous studies (Rossdale, 1967; Lombard *et al.*, 1984; Stewart *et al.*, 1984c). Cardiac output increased as the foals gained weight, but cardiac index changed little over the 2 weeks. The cardiac index values reported (155–235 ml/kg/min) are higher than those reported for adult Ponies (80–140 ml/kg/min) and substantially higher than those reported for adult horses (56–85 ml/kg/min) (Physick-Sheard, 1985). Similarly, stroke volume increased with age, but stroke volume index did not change. Our stroke volume index values (1·89–2·36 ml/kg) are similar to values reported for adult horses (1·3–2·3 ml/kg) (Physick-Sheard, 1985). The higher cardiac index values in foals were therefore primarily the result of higher heart rates in smaller, younger animals.

Systemic and pulmonary vascular resistance declined during the 2-week study period. The decrease in systemic resistance, however, was gradual over the period, while the pulmonary resistance decreased rapidly in the first 24 h, then declined more gradually over the succeeding 2 weeks. The result was a rapid decrease in the pulmonary/systemic vascular resistance ratio from 35–40% to about 25% within the first 24 h, followed by a more gradual decline over 2 weeks to near 20%. From reported pressure and vascular resistance values (Orr *et al.*, 1975; Physick-Sheard, 1985), the normal pulmonary/systemic resistance ratio in resting adult Ponies and horses can be estimated to be about 20%.

A localized continuous murmur of patent ductus arteriosus was audible in each foal for 3–5 days. In 2 foals, ductal patency and subsequent closure was confirmed by contrast two-dimensional echocardiography. In each case the magnitude of the shunt, estimated visually by the amount of contrast entering the pulmonary artery after aortic injection, appeared to be small.

In addition to the primary objective of establishing normal haemodynamic values for neonatal horses, a secondary objective was to determine the feasibility of using indwelling arterial and pulmonary artery catheters for repeated haemodynamic monitoring of horse neonates. In this study, catheters were well tolerated during the first 5–7 days. Adverse effects included a mild, transient fever and local cellulitis around the neck incision. Percutaneous placement of the right heart thermodilution catheter and systemic pressure monitoring via a peripheral artery would simplify the procedure in hospitalized newborn foals, and should safely allow haemodynamic evaluation for at least 24–48 h. Such monitoring will be essential in determining the role of the pulmonary vasculature, foramen ovale and ductus arteriosus, and cardiac function in the neonatal respiratory disorders of full-term and premature foals.

Nevertheless, several factors potentially affecting the values presented should be considered in interpreting the results of this study. First, the number of foals studied was small and there was considerable individual and daily variability in some values. The effect of the indwelling catheters on these measurements is unknown. In additon, although care was taken to obtain measurements in a steady state of relaxation, such a state could not always be assured. We also chose to use the lateral recumbent position, compared to the standing position usually reported for adult horses and Ponies. Recumbent positioning, which is known to influence blood gas values, may also affect systemic and pulmonary hemodynamics (Hall, 1984). Finally, the effect of the patent ductus arteriosus on the values presented, although probably slight, must be considered. These factors do not alter the relevance of the data to neonatal horse patients, which may be evaluated under similar conditions; rather, they suggest that the data presented should be considered relative normal values under the conditions described.

Supported by grants from the Equine Research Laboratory and the Marcia MacDonald Rivas Endowment Fund, University of California, Davis.

References

Amoroso, E.C., Dawes, G.S. & Mott, J.C. (1958) Patency of the ductus arteriosus in the newborn calf and foal. *Br. Heart J.* **20**, 92–96.

Arango, A. & Rowe, M.I. (1971) The neonatal puppy as an experimental subject. *Biol. Neonate* **18**, 173–182.

Averill, K.H., Wagner, W.W. Jr & Vogel, J.H.K. (1963) Correlation of right ventricular pressure with right ventricular weight. *Am. Heart J.* **66**, 632–635.

Campbell, G.S., Jaques, W.E. & Snyder, D.D. (1959) Maintenance of pulmonary arterial medial hypertrophy in newborn pups. *Surg. Forum* **10**, 688–690.

Dawes, G.S. (1961) Changes in the circulation at birth. *Br. med. Bull.* **17**, 148–153.

Dawes, G.S. (1968) *Foetal and Neonatal Physiology.* Yearbook Medical Publishers, Chicago.

Dennis, S.M. (1981) Perinatal foal mortality. *Comp. Cont. Educ. Pract. Vet.* **3**, S206–S217.

Fregin, G.F. (1982) The cardiovascular system. In *Equine Medicine and Surgery*, 3rd edn, pp. 645–704. Eds R. A. Mansmann, E. S. McAllister & P. W. Pratt. American Veterinary Publications, Inc., Santa Barbara.

Gootman, N.H. & Gootman, P.M. (1983) *Perinatal Cardiovascular Function.* Marcel Dekker, Inc., New York.

Hall, L.W. (1984) Cardiovascular and pulmonary effects of recumbency in two conscious ponies. *Equine vet. J.* **16**, 89–92.

Lombard, C.W., Evans, M., Martin, L. & Tehrani, J. (1984) Blood pressure, electrocardiogram and echocardiogram measurements in the growing pony foal. *Equine vet. J.* **16**, 342–347.

Martens, R.J. (1982) Neonatal respiratory distress: a review with emphasis on the horse. *Comp. Cont. Educ. Pract. Vet.* **4**, S23–S33.

Milne, D.W., Muir, W.W. & Skarda, R.T. (1975) Pulmonary arterial wedge pressure: blood gas tensions and pH in the resting horse. *Am. J. vet. Res.* **36**, 1431–1434.

Orr, J.A., Bisgard, G.E., Forster, H.V., Rawlings, C.A., Buss, D.D. & Will, J.A. (1975) Cardiopulmonary measurements in nonanesthetized, resting normal ponies. *Am. J. vet. Res.* **36**, 1667–1670.

Physick-Sheard, P.W. (1985) Cardiovascular response to exercise and training in the horse. *Vet. Clin. North Am. Equine Pract.* **1**, 383–417.

Platt, H. (1973) Etiological aspects of perinatal mortality in the thoroughbred. *Equine vet. J.* **5**, 116–120.

Rose, R.J. & Stewart, J.H. (1983) Basic concepts of respiratory and cardiovascular function and dysfunction in the full term and premature foal. *Proc. 29th Ann. Conv. Am. Assoc. Equine Pract.* 167–178.

Rose, R.J., Rossdale, P.D. & Leadon, D.P. (1982) Blood gas and acid-base studies in spontaneously delivered, term-induced and induced premature foals. *J. Reprod. Fert., Suppl.* **32**, 521–528.

Rose, R.J., Hodgson, D.R., Leadon, D.P. & Rossdale, P.D. (1983) Effect of intranasal oxygen administration on arterial blood gas and acid base parameters in spontaneously delivered, term induced and induced premature foals. *Res. vet. Sci.* **34**, 159–162.

Rossdale, P.D. (1967) Clinical studies on the newborn Thoroughbred foal. II. Heart rate, auscultation and electrocardiogram. *Br. vet. J.* **123**, 521–532.

Rossdale, P.D. (1968) Blood gas tensions and pH values in the normal thoroughbred foal at birth and in the following 42 h. *Biol. Neonate* **13**, 18–25.

Rossdale, P.D. (1969) Measurements of pulmonary ventilation in normal newborn Thoroughbred foals during the first three days of life. *Br. vet. J.* **125**, 157.

Rossdale, P.D. (1970) Some parameters of respiratory function in normal and abnormal foals with special reference to levels of PaO_2 during air and oxygen inhalation. *Res. vet. Sci.* **11**, 270–276.

Rossdale, P.D. (1979) Neonatal respiratory problems of foals. *Vet. Clin. N. Am. Large Anim. Pract,* **1**, 205–217.

Rudolph, A.M. (1970) The changes in the circulation after birth. Their importance in congenital heart disease. *Circulation* **41**, 343–359.

Rudolph, A.M., Auld, P.A.M., Golinko, R.J. & Paul, M.H. (1961) Pulmonary vascular adjustments in the neonatal period. *Pediatrics* **28**, 28–34.

Scott, E.A., Kneller, S.K. & Witherspoon, D.M. (1975) Closure of ductus arteriosus determined by cardiac catheterization and angiography in newborn foals. *Am. J. vet. Res.* **36**, 1021–1023.

Slauson, D.O. (1979) Naturally occurring hyaline membrane disease syndrome in foals and piglets. *J. Pediatr.* **95**, 889–891.

Sonea, I. (1985) Respiratory distress syndrome in neonatal foals. *Comp. Cont. Educ. pract. Vet.* **7**, S462–S469.

Stewart, J.H., Rose, R.J. & Barko, A.M. (1984a) Respiratory studies in foals from birth to seven days old. *Equine vet. J.* **16**, 323–328.

Stewart, J.H., Rose, R.J. & Barko, A.M. (1984b) Response to oxygen administration in foals: effect of age, duration and method of administration on arterial blood gas values. *Equine vet. J.* **16**, 329–331.

Stewart, J.H., Rose, R.J. & Barko, A.M. (1984c) Echocardiography in foals from birth to three months old. *Equine vet. J.* **16**, 332–341.

J. Reprod. Fert., Suppl. **35** (1987), 629–634
Printed in Great Britain
© 1987 Journals of Reproduction & Fertility Ltd

Transitory changes of hormones in the plasma of parturient pony mares

N. S. Pope, G. F. Sargent, B. S. Wiseman and D. J. Kesler

Physiology Research Laboratory-Reproductive Biotechnology, Department of Animal Sciences, University of Illinois, Urbana, Illinois 61801, U.S.A.

Summary. Frequent blood samples were collected from 8 pony mares before, during and after labour, parturition and placental expulsion and assayed for progesterone, oestradiol, androstenedione and LH concentrations by radioimmunoassay. A significant ($P < 0.05$) decrease in progesterone, oestradiol and in the progesterone:oestradiol ratio was not detected until 0·5 h after foaling. Androstenedione concentrations rose before and peaked at parturition and then declined. A significant ($P < 0.05$) rise in LH was detected 0·5 h after parturition. This LH peak was not detected in one mare and she was the only mare that did not ovulate within the first 20 days *post partum*. These results suggest that: (1) the foal may be an important factor in the production of progesterone and oestradiol by the feto–placental unit; (2) the pituitary is capable of releasing LH immediately after parturition; (3) the parturient rise in LH may be due to removal of negative feedback inhibition by progesterone and/or oestradiol; and (4) the parturient rise in LH at parturition, combined with already elevated concentrations of FSH, may be involved in the rapid growth of follicles *post partum*.

Introduction

The periparturient endocrine patterns of mares have not been well studied. Most foaling occurs at night, between 18:00 and 06:00 h, but the hormonal mechanisms behind this are unknown (Jeffcott, 1972). It has been reported that mares foal in the presence of an increasing progesterone:oestrogen ratio (Lovell *et al.*, 1975), which differs markedly from that which occurs in other farm species, but is similar to reports for women (Llauro *et al.*, 1968). Progestagens and oestrogens decline precipitously in the mare after parturition (Nett *et al.*, 1973; Holtan *et al.*, 1975). High steroid levels have also been found in the umbilical vein (Ganjam *et al.*, 1975) of the fetus, and the newborn foals have higher concentrations of progesterone and oestrogen during the first 24 h *post partum* than do their dams (Lovell *et al.*, 1975). This evidence points to the feto–placental unit as the source of steroids during late pregnancy in the mare.

Corticoid levels in mares have been found to remain unchanged through the peripartum period (Lovell *et al.*, 1975). However, concentrations of androstenedione, an androgen known to be at least partly of adrenal origin, are unknown during this time. Frequent blood sampling during labour, parturition and placental expulsion to determine steroid and gonadotrophin profiles has not been reported. FSH is known to be elevated a few days around parturition and LH has been reported to show no change (Ginther, 1979), but blood samples were taken only on a daily basis in these studies.

Hormone concentrations in blood collected on a more frequent basis are needed to understand clearly physiological changes around parturition and *post partum*. The objective of this study was to determine plasma concentrations of progesterone, oestradiol, androstenedione and LH in parturient mares with the use of frequent (every 30 min) blood sampling throughout labour, parturition and placental expulsion.

Materials and Methods

Experimental animals. Pony mares, aged 5–15 years, were used. The 8 mares were housed at the Physiology Research Laboratory at the University of Illinois, Urbana. Mares were kept in a group, with access to a three-sided shelter, and maintained on free-choice pasture in the summer, supplemented with grass–alfalfa hay, free choice, in the winter. An oats–corn grain mixture was fed at the rate of 0·5–0·9 kg per mare per day. Mares were observed closely for signs of parturition, and confined to a stall when foaling seemed imminent (indicated by the presence of colostrum in the mammary gland, waxing of the teats and relaxation of the pelvic ligaments (Ginther, 1979)).

Blood sampling and hormone measurements. Blood samples were collected by jugular venepuncture into 20-ml heparinized vacuum tubes on a daily basis beginning 45 days before the mare's due date. On the day that parturition seemed impending, blood samples were taken every 30 min from 18:00 h until parturition and placental expulsion had occurred. On Day 1 *post partum*, mares were bled every 30 min from 06:00 to 10:00 h, then daily samples were taken until 30 days after the first day of post-partum oestrus. For purposes of this study, the samples analysed included daily samples beginning 4 days before and until 4 days after parturition, frequent sampling around parturition, and frequent sampling on Day 1 *post partum*. After collection, blood was placed immediately on ice, centrifuged within 4 h and blood plasma was stored at −4°C until assay. Progesterone, oestradiol, androstenedione and LH were measured by radioimmunoassays.

Hormone assays. Progesterone was measured by radioimmunoassay using procedures similar to those reported by Scheffrahn *et al.* (1982). Antiserum was generated in rabbits against 4-pregnen-11α-ol-3,20-dione-11-hemisuccinate–bovine serum albumin. Mean binding exhibited was 49·3%. Recovery aliquants were included by the extraction of [^{3}H]progesterone in pooled mare plasma and recovery was 93%. Cross-reaction of the progesterone antiserum (GS No. 253) was evaluated with androgens, corticosteroids and progestagens. The antiserum showed >2% cross-reactivity only with 5α-pregnan-3,20-dione (4%) and 11α-hydroxyprogesterone (22%). Pools of plasma from a luteal-phase mare were included in each assay and from these values intra-assay and interassay coefficients of variation were calculated to be 5·6% and 14%, respectively. When various volumes of the same pool of plasma (50, 100 and 200 μl) were assayed, the progesterone values were 10·6, 10·7 and 10·6 ng/ml respectively. The sensitivity of the assay, the least amount of progesterone which displaced binding of [^{3}H]progesterone to antibody, was 8 pg ($P < 0.05$).

Oestradiol concentrations in plasma were determined by radioimmunoassay as described by Scheffrahn *et al.* (1982). Antiserum was generated against 1,3,5(10)-oestratrien-3,17β-diol-6-one-6(carboxymethyl-oxime)–bovine serum albumin. Mean binding exhibited was 48·6%. Steroids were extracted from duplicate 500-μl aliquants of plasma. Recovery aliquants were included by the extraction of [^{3}H]oestradiol in 500 μl of mare plasma and recovery was 91·3%. Cross-reactions of the antiserum (No. 842) were evaluated with progesterone, corticosteroids, androgens and oestrogens. The antiserum showed cross-reactivity of >2% only with oestriol (4·9%). Intra-assay and interassay coefficients of variation were calculated from the same plasma pool as 4·1% and 10·9%, respectively. When various volumes of plasma (250, 500, 750 and 1000 μl) were analysed, the oestradiol values were 11·6, 10·2, 10·0 and 10·6 pg/ml, respectively. The sensitivity of the assay was 4 pg ($P < 0.05$).

Plasma testosterone concentrations were quantified using radioimmunoassay procedures reported by Falvo & Nalbandov (1974). Antiserum was generated in rabbits against 4-androsten-11α,17β-diol-3-one-11-hemisuccinate–bovine serum albumin. Recovery aliquants were included by the extraction of [^{3}H]testosterone in mare plasma, with recovery of 89%. Cross-reactivity of the testosterone antiserum (SGS No. 1969-11B) was as follows: 4,6-androstan-dien-17β-ol-3-one (60%); α-androstan-17β-diol-3-one (8·5%), 5α-androstan-3β,17β-diol (5·0%); 4-androsten-3β,17β-diol (3·7%); 4-androsten-3,17-diol (0·9%); 5-androsten-3β,17β-diol (0·3%); 5-androsten-3β-ol-17-one (0·2%). Cross-reaction with progestagens, oestrogens and corticosteroids were <0·01%. When various volumes of plasma (500, 750 and 1000 μl) were assayed, the testosterone values were 15·6, 15·4 and 15·0 pg/ml, respectively. From pools of plasma included in each assay, the intra-assay coefficient of variation was calculated as 5·7%, and the interassay coefficient of variation was 12·9%. The sensitivity of the assay was 8 pg ($P < 0.05$).

Plasma androstenedione concentrations were determined using radioimmunoassay procedures validated for this study. Antiserum was generated in this laboratory against 4-androsten-3,17-dione-3-CMO–bovine serum albumin. Cross-reactivity of the antiserum was as follows: testosterone (0·2%), 5α-dihydrotestosterone (0·2%), dihydroepiandrosterone (0·1%), androsterone (0·7%), and progesterone (0·1%). Cross-reaction with oestrogens and corticosteroids was <0·01%. Final dilution of the antiserum was 1:12 500, and mean binding exhibited was 46·6%. Plasma androstenedione was extracted from duplicate aliquants of 250 μl plasma. After addition of 5 ml diethyl ether, samples were vigorously shaken on a mechanical shaker for 3 min. The tubes were placed in a mixture of solid CO_2 and methanol, and the organic phase decanted into disposable culture tubes. Extracts were dried in a 45°C water bath and then reconstituted with phosphate-buffered saline. Recovery aliquants were included in the assays by addition of [^{3}H]androstenedione in mare plasma, with mean recovery of 92·8%. Standard curves were established for each assay at concentrations of 0, 8, 16, 32, 64, 125, 250 and 500 pg crystalline androstenedione. When various volumes of pooled plasma (250, 500 and 1000 μl) were assayed, concentrations of androstenedione were 475·0, 476·9 and 468·0 pg/ml, respectively. Intra-assay and interassay coefficients of variation from the same pooled plasma were 6·2% and 13·8%, respectively. Samples from one mare were always run within the same assay to reduce within-mare variability. The sensitivity of the assay was 8 pg/ml ($P < 0.05$).

Plasma LH was quantified by double-antibody radioimmunoassay as reported by Scheffrahn *et al.* (1982). The iodination preparation was ovine LH (LER-1056-C2). The anti-ovine LH antiserum (GDN No. 15), diluted 1:40 000, exhibited a mean binding of 25·6%. Duplicate aliquants of 25 μl plasma were assayed. LER-958-1 equine LH was used as the standard. When various volumes of pooled plasma (25, 50 and 100 µl) were assayed, the LH values were 95·3, 106·5 and 93·8 ng/ml, respectively. Intra-assay and interassay coefficients of variation, calculated from the same pooled plasma, were 3·8% and 3·6%, respectively. The sensitivity of the assay was 0·1 ng/ml ($P < 0.05$).

Statistical analysis. Duplicate estimates of hormone concentrations were averaged before analysis. Split-plot analysis of variance for repeated sampling (Gill & Hafs, 1971) was carried out to determine whether hormone concentrations varied over time. Correlations between hormones were done to determine hormone relationships (Steel & Torrie, 1960). Standard errors of the differences between sampling means for each hormone were calculated by methods described by Cochran & Cox (1957). Results were considered significant if a 95% confidence level, the probability of the results occurring by random chance only 5% of the time, was achieved ($P < 0.05$).

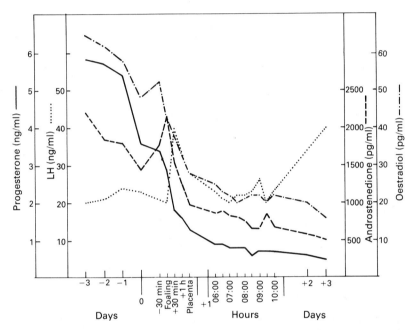

Fig. 1. Hormone concentrations over time in parturient mares. Values are means of 8 mares. The standard error of the difference between the sampling means is 0·6 for progesterone, 4·1 for oestradiol, 149·5 for androstenedione, and 5·9 for LH.

Table 1. Mean (±standard error) progesterone and oestradiol concentrations and their ratio for 8 periparturient pony mares

Time	Progesterone (ng/ml)	Oestradiol (pg/ml)	Ratio (pg progesterone: pg oestradiol)
Day 3 *pre partum*	5·8 ± 3·6	65 ± 26	89
Day 2 *pre partum*	5·7 ± 2·0	62 ± 22	92
Day 1 *pre partum*	5·4 ± 3·3	58 ± 21	93
Day 0 (4–12 h *pre partum*)	3·6 ± 1·9	48 ± 25	75
30 min before foaling	3·4 ± 3·0	52 ± 15	65
Immediately after foaling	2·9 ± 1·9	44 ± 11	66
30 min after foaling	1·8 ± 1·0	38 ± 12	46
1 h after foaling	1·6 ± 0·7	33 ± 11	49
Placental expulsion	1·3 ± 0·6	28 ± 11	47

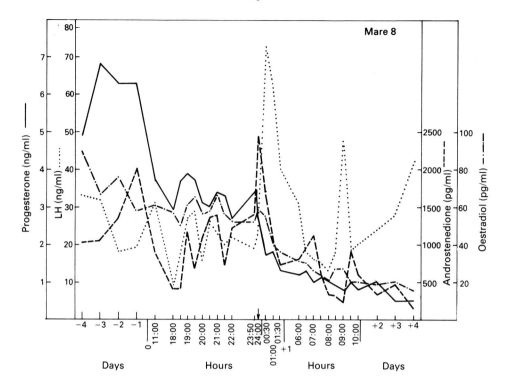

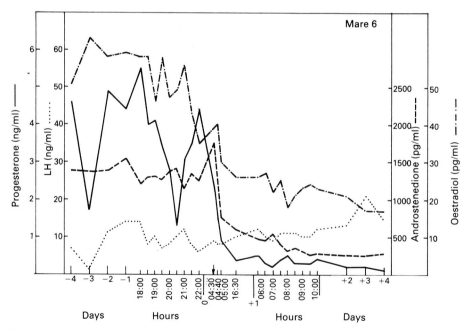

Fig. 2. Hormone concentrations over time in parturient mares: (a) Mare 8 which ovulated on Day 12 *post partum*, and (b) Mare 6 which did not ovulate within 20 days *post partum*. The arrows indicate the time of foaling.

Results and Discussion

Hormone concentrations over time in the 8 mares are shown in Fig. 1. In the analysis of variance, all hormones showed an effect of time.

Progesterone concentrations showed a dramatic decrease between the day before and the day after foaling, beginning before the foal was born (the sample at time 0 represents 4–12 h before foaling). Although there was an apparent decrease in progesterone concentrations from 1 day before foaling to 4–12 h before foaling, it was not significant. A significant decrease was reached at 30 min after foaling, indicating that the foal may be the important factor in the production of progesterone by the feto–placental unit at this time. The same profile was detected for oestradiol, with the most dramatic decline occurring after the foal was born.

It has been stated in the literature that mares foal in the presence of an increasing progesterone: oestrogen ratio (Lovell *et al.*, 1975). Table 1 depicts mean progesterone and oestradiol concentrations in mares in samples taken during the periparturient period. Both progesterone and oestradiol are decreasing throughout this period. When the ratio of progesterone:oestradiol was calculated, there was a decrease over time although due to high between-mare variability *pre partum* this decrease in ratio was not significant until 30 min *post partum*. From this information it can be suggested that the mare may not be as different from other animals as has been assumed in that levels of progesterone are falling faster than levels of oestradiol at the time of parturition, allowing for more myometrial activity, and so uterine contractions can be stimulated.

Androstenedione concentrations rose around parturition peaking at the time of foaling. A decrease in androstenedione then occurred to below pre-partum levels by Day 1 *post partum*. Lovell *et al.* (1975) reported that corticosteroid concentrations did not change during parturition but frequent sampling was not done until after delivery. It is therefore possible that adrenal steroids are involved in parturition in the mare.

A significant rise in LH was seen at 30 min after birth in the present study. This correlated with falling concentrations of progesterone at this time ($r = -0.18$). The rise may be due to removal of negative feedback of progesterone on LH levels, or to some other factor occurring at parturition. This, combined with what is known about the elevated concentrations of FSH for a few days around parturition (Ginther, 1979) may well be responsible for the growth of follicles which occurs rapidly *post partum* in the mare and leads to ovulation within 10–14 days *post partum*. This rise in gonadotrophins and the timing of the first ovulation 10–14 days later corresponds to what is seen at mid-dioestrus in the mare, leading to a wave of follicular growth and ovulation 10–14 days later. The values in Fig. 2 illustrate this suggestion. Mare 8, showed a pronounced rise in LH concentrations associated with foaling. This mare ovulated 12 days after parturition. Mare 6 showed no surge of LH after parturition, and had not ovulated within 20 days after parturition. We would therefore suggest that the LH surge associated with parturition may play an important role in the rapid return to oestrus and subsequent ovulation seen in the post-partum mare.

References

Cochran, W.G. & Cox, G.M. (1957) *Experimental Designs*, 2nd edn. John Wiley, New York.

Falvo, R.E. & Nalbandov, A.V. (1974) Radioimmunoassay of peripheral plasma testosterone in males from eight species using a specific antibody without chromatography. *Endocrinology* **95**, 1466–1468.

Ganjam, V.K., Kenney, R.M. & Flickinger, G. (1975) Plasma progestagens in cyclic, pregnant and post-partum mares. *J. Reprod. Fert., Suppl.* **23**, 441–447.

Gill, J.L. & Hafs, H.D. (1971) Analysis of repeated measurements of animals. *J. Anim. Sci.* **33**, 331–336.

Ginther, O.J. (1979) *Reproductive Biology of the Mare: Basic and Applied Aspects.* McNaughton and Gunn, Inc., Ann Arbor, Michigan.

Holtan, D.W., Nett, T.M. & Estergreen, V.L. (1975) Plasma progestins in pregnant, postpartum and cycling mares. *J. Anim. Sci.* **40**, 251–260.

Jeffcott, L.B. (1972) Observations on parturition in cross-bred pony mares. *Equine vet. J.* **4**, i–v.

Llauro, J.L., Runnebaum, B. & Zander, J. (1968) Progesterone in human peripheral blood before, during and after labor. *Am. J. Obstet. Gynec.* **101**, 867.

Lovell, J.D., Stabenfeldt, G.H., Hughes, J.P. & Evans, J.W. (1975) Endocrine patterns of the mare at term. *J. Reprod. Fert., Suppl.* **23,** 449–456.

Nett, T.M., Holtan, D.W. & Estergreen, V.L. (1973) Plasma estrogens in pregnant and postpartum mares. *J. Anim. Sci.* **37,** 962–970.

Scheffrahn, N.S., Wiseman, B.S., Vincent, D.L., Harrison, P.C. & Kesler, D.J. (1982) Reproductive hormone secretions in pony mares subsequent to ovulation control during late winter. *Theriogenology* **17,** 571–585.

Steel, R.G.D. & Torrie, J.H. (1960) *Principles and Procedures of Statistics.* McGraw-Hill, New York.

J. Reprod. Fert., Suppl. **35** (1987), 635–640

Effect of pulsatile gonadotrophin release on mean serum LH and FSH in peri-parturient mares

K. K. Hines, B. P. Fitzgerald and R. G. Loy

Department of Veterinary Science, University of Kentucky, Lexington, KY 40546-0076, U.S.A.

Summary. Changes in the pattern of LH and FSH in serum were studied in 6 mares foaling during the summer. Samples were collected frequently (every 15 min) for 24 h twice before foaling, -33 ± 2 and -12 ± 2 days, and for 12 h after foaling, on Day 0 and Day 4.

Simultaneous pulses of FSH and LH were observed before foaling ($r^2 = 0.99$). Before foaling, gonadotrophin pulses were infrequent (6 in 264 h of observation). On the day of foaling, LH and FSH pulse frequency increased ($P < 0.005$) with 2–4 pulses per mare. The amplitudes of pulses of LH and FSH were higher before parturition than for those observed on Day 0 ($P < 0.001$). The presence of FSH pulses was associated with an increased mean value of FSH for those days, including Day 0 ($P < 0.02$). The presence of resolvable pulses of LH was not associated with an increase in mean LH values on those days ($P > 0.6$).

The present results demonstrate that LH and FSH release is pulsatile in the peri-parturient mare. In addition, mean values of FSH in serum were elevated when pulses were detected, whereas mean concentrations of LH remained low. These results are consistent with the concept of one hypothalamic releasing hormone for LH and FSH in the mare.

Introduction

Gonadotrophin profiles, based on daily, or less frequent sampling, have been established for the post-partum mare, demonstrating two instances of reciprocal gonadotrophin patterns, in one of which gonadotrophin was elevated while the other was at minimal values (Turner *et al.*, 1979; Loy *et al.*, 1982). The first instance occurred at parturition when elevated concentrations of FSH were observed, relative to pre-partum concentrations, while LH values remained low. The other instance occurred after parturition, when LH concentrations rose, while FSH concentrations declined to minimum values.

During transition into the breeding season, a reciprocal relationship between concentrations of LH and FSH has been found, in which elevated concentrations of FSH and minimal values of LH were present (Freedman *et al.*, 1979). The explanation behind these reciprocal patterns became apparent with the observation of pulsatile release of LH and FSH during the transition into the breeding season (Hines *et al.*, 1986; Fitzgerald *et al.*, 1987). Early in seasonal transition, when mean concentrations of serum FSH were elevated, high-amplitude FSH pulses were detected, accompanied by high-amplitude LH pulses. As ovulation approached and serum FSH concentrations decreased, FSH pulse amplitude also decreased, demonstrating that mean serum concentrations of FSH were influenced by FSH pulse amplitude (Hines *et al.*, 1986). In contrast, mean LH concentrations remained at low values and did not reflect the increase in LH pulse frequency or decrease in pulse amplitude observed during this period (Fitzgerald *et al.*, 1987).

The aim of the present study was to investigate the reciprocal pattern of LH and FSH in the peri-parturient mare by determining the relationship between gonadotrophin pulse amplitude and frequency to mean serum concentrations, and also the identification of factors which may influence pulsatile gonadotrophin release.

Materials and Methods

Six horse mares, which foaled between 4 July and 22 July, were used. Mares were maintained on pasture except during periods of frequent sampling when they were housed in stalls. Foals were kept with their respective dam at all times. Blood samples were collected daily by jugular venepuncture, from 40 days before to 5 days after parturition. In addition, samples were collected frequently via an indwelling jugular cannula (every 15 min for 24 h) twice before foaling (-33 ± 2 and -12 ± 2 days) and (every 15 min for 12 h) twice after parturition on Day 0 (day of foaling) and on Day 4. Samples were collected frequently for 24 h before parturition to ensure that the sampling period was sufficient to allow detection of events if they occurred. A 12-h sampling period was considered to be sufficient on Day 0 and Day 4 after foaling to analyse the changes in gonadotrophin patterns associated with parturition. One mare foaled early, and only one 24-h frequent sampling was performed; this is included in the Day -12 group. Mares were palpated *per rectum* on Days 1 and 5 *post partum* to monitor follicular development. All blood samples were kept for 12–24 h at 4°C, centrifuged and the harvested serum was stored at $-20°C$ until assay.

To determine the response of LH and FSH to a known GnRH stimulus, GnRH (200 μg in 1 ml 0·9% (w/v) NaCl) was injected through the indwelling cannula (3 mares) at the conclusion of the 24-h sampling period on Day -12. Samples were collected every 15 min for an additional 3 h. Three mares served as saline-injected controls. The dose of GnRH was chosen as it had proved to produce a physiological LH response, similar to endogenous pulses (unpublished observations).

Serum concentrations of LH and FSH were measured by radioimmunoassay (RIA). The LH assay, previously validated in our laboratories (Loy *et al.*, 1982), used rabbit anti-ovine LH (GDN No. 15) as the first antibody and sheep anti-rabbit gammaglobulin as the second antibody. Ovine LH (LER 1056-2) was iodionated and the standard preparation was highly purified equine LH (Papkoff No. e98a). The sensitivity of this assay was $0·017 \pm 0·002$ ng/tube, and within- and between-assay variation was 10·1% and 14·8% respectively ($n = 12$ assays).

Serum FSH was also measured by a heterologous system which used anti-human FSH (NIADDK batch No. 6) at an initial dilution of 1:10 000 and was incubated at 4°C for 24 h with 25–200 μl of sample or standard (0·07 to 10 ng/tube, Papkoff equine FSH No. e99b). After addition of ^{125}I-labelled ovine FSH (NIAMDD-OFSH-I-1), incubation was continued for 48 h at 4°C. To precipitate the antigen–antibody complex, sheep anti-rabbit gammaglobulin was added, allowing 24 h incubation before centrifugation. The antiserum specifically bound 10–15% and non-specific binding was $\leqslant 3·5\%$ of the labelled FSH. Serial dilutions of samples and serum pools were parallel to the standard curve. Equine LH (e98a) exhibited $<0·2\%$ cross-reaction, relative to the horse FSH standard. Assay sensitivity was $0·250 \pm 0·035$ ng/tube and intra- and inter-assay variation were both 11% ($n = 12$ assays).

Gonadotrophin pulses were identified by the method described by Fitzgerald *et al.* (1985). Data were subjected to a repeated measures analysis of variance (SAS, 1985). Orthogonal contrasts and protected LSD tests were used to compare means when a significant time effect was detected.

Results

Mean concentrations of LH in serum were low during the pre-partum period and increased ($P < 0·001$) only after parturition (Fig. 1). In contrast, mean concentrations of FSH in serum increased ($P < 0·02$, linear) during late gestation, reaching maximum values on the day of parturition. Elevated concentrations of LH and FSH were observed in the period from Day 0 (day of foaling) to Day 4. Within this period, mean concentrations of LH increased ($P < 0·001$, linear) from $0·31 \pm 0·04$ ng/ml on Day 0, to $2·80 \pm 0·49$ ng/ml on Day 4, while mean concentrations of FSH decreased ($P < 0·001$, linear) from $13·76 \pm 2·55$ ng/ml on Day 0 to $6·70 \pm 0·85$ ng/ml on Day 4.

Pulses of LH and FSH were detected before foaling (Fig. 2). In all mares, pulses of LH and FSH were synchronized before parturition ($r^2 = 0·99$). Gonadotrophin pulses were infrequent before parturition, with only 6 pulses detected in 4 mares during the 264 h of observation.

Both LH and FSH were released in response to exogenous GnRH before foaling (Fig. 2). The amplitude of the induced FSH pulses tended to be greater than the amplitude of the naturally occurring pulses ($33·88 \pm 5·04$ *vs* $23·60 \pm 11·49$ ng/ml, $P < 0·1$). The amplitudes of the induced LH pulses were not different from those of naturally occurring pulses ($0·70 \pm 0·11$ *vs* $0·46 \pm 0·11$ ng/ml, $P > 0·1$).

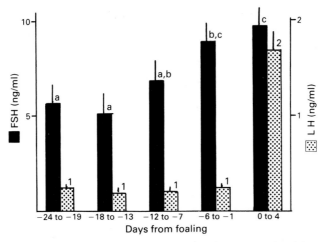

Fig. 1. Changes in mean serum LH and FSH concentrations as analysed by periods. Vertical lines represent the s.e.m. Within hormone, bars with different superscripts are different ($P < 0.005$).

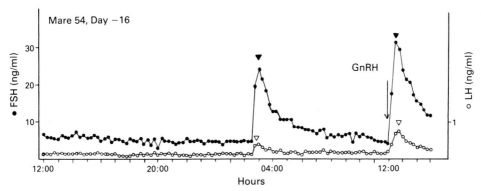

Fig. 2. Changes in serum LH (○—○) and FSH (●—●) concentrations in blood samples collected every 15 min for 27 h in Mare 54, 16 days before foaling. The GnRH (200 µg i.v.) was given at 12:00 h as shown. The occurrence of a pulse is indicated (LH ∇; FSH ▼).

Gonadotrophin pulse frequency increased at parturition ($P < 0.005$). For the 6 mares in this study, on the day of foaling, 10 FSH pulses and 15 LH pulses were detected in 72 h (12 h/mare) of observation. Pulse amplitude appeared to decrease as sampling continued, until pulses often could not be resolved at the conclusion of the sampling period (Fig. 3a). On the day of foaling, in 2/6 mares, all resolvable FSH pulses were accompanied by resolvable LH pulses. In 3/6 mares, LH pulses could be resolved while the change in FSH concentration at the same time could not be resolved as pulses, and in 1/6 mares, FSH pulses were statistically resolved without an accompanying LH pulse.

A change in LH or FSH pulse amplitude or frequency was not detected before parturition ($P > 0.15$). However, at parturition the amplitude of FSH pulses decreased from 19.79 ± 2.6 ng/ml before foaling to 11.77 ± 1.28 ng/ml on Day 0 ($P < 0.002$). Similarly, LH pulse amplitude also decreased at parturition, from 0.38 ± 0.09 ng/ml before, to 0.23 ± 0.02 ng/ml on Day 0 ($P < 0.002$): The amplitudes of the LH pulses on Day 0, while low, were above the limits of assay sensitivity.

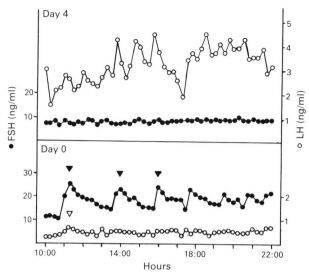

Fig. 3. Changes in serum LH (○—○) and FSH (●—●) concentrations in blood samples collected every 15 min for 12 h from Mare 49 on the day of foaling (Day 0) and on Day 4. The occurrence of a pulse is indicated (LH ▽; FSH ▼).

Distinct pulses of LH could not be resolved by Day 4 *post partum* (Fig. 3b). In all mares, this was the only instance in which mean LH concentrations were elevated above values detected before parturition and on Day 0 ($P < 0.02$). Distinct FSH pulses also could not be resolved at this time; mean FSH concentrations were at the lowest values observed in these mares.

The presence of pulses of FSH was associated with increased mean value of FSH on those days (mean of all samples collected in 24 or 12 h) *vs* days when no pulses were detected ($P < 0.02$). In contrast, the presence of detectable pulses of LH was not associated with an increase in mean LH values on those days ($P > 0.6$).

Discussion

The patterns of gonadotrophins detected in these peri-parturient mares by daily samples were consistent with previous observations (Turner *et al.*, 1979; Fitzgerald *et al.*, 1985). Two instances when the patterns of LH and FSH differed were noted, i.e. elevated FSH values at parturition while LH concentrations in serum remained at minimum values, and, conversely, elevated values of LH in serum 3–4 days *post partum*, accompanied by minimal concentrations of FSH.

An explanation for the reciprocal pattern of LH and FSH at foaling was suggested by the nature of pulsatile gonadotrophin release during this time. Before foaling gonadotrophin pulse frequency was very low, but on the day of parturition, FSH and LH pulse frequency increased. These pulses of FSH were associated with an increase in mean serum FSH values detected on Day 0. However, pulses of LH were not associated with an increase in mean serum LH concentrations. Elevated concentrations of LH were only observed after the day of foaling, when gonadotrophin pulses could no longer be detected in jugular blood samples. These observations are similar to the relationship between mean gonadotrophin concentrations and the presence of pulses during the transition into the breeding season, when changes in mean values of FSH are associated with changes in FSH pulse amplitude (Hines *et al.*, 1986). In contrast, during transition into the breeding season, changes in LH pulse frequency and amplitude were not reflected in changes in mean serum LH concentrations (Fitzgerald *et al.*, 1987).

Before parturition, with the exception of the episodes of gonadotrophin release, within each individual, basal levels of LH and FSH were detected. This may indicate that, during the pre-partum period, the primary mode of gonadotrophin release is through discrete pulses. In addition, it appeared that LH and FSH responded to the same releasing signal because the onset of the pulses were highly synchronized. This contention is further supported by the relative similarity of the response of LH and FSH to exogenous GnRH. If separate releasing factors were responsible for the control of LH and FSH secretion, then the similar response of LH and FSH to exogenous releasing hormone would not be expected.

On the day of foaling, in 4 of 6 mares, pulses of one gonadotrophin could be statistically resolved, while pulses of the other pituitary gonadotrophin could not be. However, this observation may be an artifact of the method used to filter pulsatile gonadotrophin patterns from fluctuations produced by assay variation, and not indicate the action of independent releasing hormones for each gonadotrophin.

In the mare, the pre-partum period is characterized by elevated progesterone and total oestrogen concentrations (Lovell *et al.*, 1975), of fetal and placental origin. It is during this time of elevated concentrations of progestagens and oestrogens in serum that a low gonadotrophin pulse frequency was observed. Similar observations have been reported for the ewe (Goodman & Karsch, 1980; Tamanini *et al.*, 1986). Perhaps in the mare, as in the ewe, steroids may have the ability to modulate gonadotrophin pulse frequency, possibly by altering the release of GnRH. That steroids of fetal and placental origin have a role in controlling gonadotrophin release is further supported by the dramatic increase in gonadotrophin pulse frequency on the day of parturition.

The low levels of gonadotrophins present before foaling resulted in little follicular activity on the day of foaling. However, by 5 days following parturition, all mares possessed detectable follicles. This follicular development was accompanied by an increase in circulating concentrations of LH and a decrease in concentrations of FSH. This suggests that a follicular product may influence serum concentrations of FSH. The involvement of a follicular component in this decrease in concentrations of FSH was also suggested by Turner *et al.* (1979).

In conclusion, the present results demonstrated that LH and FSH release was pulsatile during the peri-partum period in the mare. In addition, mean values of FSH in serum were elevated when pulses were detected, whereas mean values of LH were not. Concentrations of LH were only elevated when discrete pulses could no longer be resolved in jugular blood samples. It is likely that steroids produced by the conceptus influenced gonadotrophin pulse frequency before parturition and ovarian feedback may modulate gonadotrophin release after parturition. Finally, these data are consistent with the concept of one hypothalamic releasing hormone for LH and FSH in the mare.

The investigation reported in this paper (No. 86-4-186) is in connection with a project of the Kentucky Agricultural Experiment station and is published with the approval of the Director. It was supported in part by a grant from the Grayson Foundation Inc.

We thank Kathy Williams for technical assistance; the National Hormone and Pituitary Program, NIADDK, for ovine FSH and anti-human FSH antiserum; Dr H. Papkoff, University of California at San Francisco, for horse LH and FSH standards; Dr G. D. Niswender, Colorado State University, Fort Collins, for anti-ovine LH antiserum; Dr L. Edgerton, University of Kentucky, Lexington, for anti-rabbit gamma-globulin; and Dr F. Bex, Wyeth Pharmaceuticals, for GnRH.

References

Fitzgerald, B.P., l'Anson, H., Legan, S.J. & Loy, R.G. (1985) Changes in patterns of luteinizing hormone secretion before and after the first ovulation in the postpartum mare. *Biol. Reprod.* **33**, 316–323.

Fitzgerald, B.P., Affleck, K.J., Barrows, S.P., Murdoch, W.L., Barker, K.B. & Loy, R.G. (1987) Changes in LH pulse frequency and amplitude in intact mares during the transition into the breeding season. *J. Reprod. Fert.* **79**, 485–493.

Freedman, L.J., Garcia, M.C. & Ginther, O.J. (1979) Influence of photoperiod and ovaries on seasonal reproductive activity in mares. *Biol. Reprod.* **20**, 567–574.

Goodman, R.L. & Karsch, F.J. (1980) Pulsatile secretion of luteinizing hormone: differential suppression by ovarian steroids. *Endocrinology* **107**, 1286–1290.

Hines, K.K., Affleck, K.J., Barrows, S.P., Murdoch, W.L., Barker, K.B., Loy, R.G. & Fitzgerald, B.P. (1986) Evidence that a decrease in follicle stimulating hormone (FSH) pulse amplitude accompanies the onset of the breeding season in the mare. *Biol. Reprod.* **34**, 270, Abstr.

Lovell, J.D., Stabenfeldt, G.H., Hughes, J.P. & Evans, J.W. (1975) Endocrine patterns of the mare at term. *J. Reprod. Fert., Suppl.* **23**, 449–456.

Loy, R.G., Evans, M.J., Pemstein, R. & Taylor, T.B. (1982) Effects of injected steroids on reproductive patterns and performance in post-partum mares. *J. Reprod. Fert., Suppl.* **32**, 199–204.

SAS Institute Inc. (1985) *SAS User's Guide: Version 5 Edition*. SAS Institute Inc., Cary, North Carolina.

Tamanini, C., Crowder, M.E. & Nett, T.M. (1986) Effects of oestradiol and progesterone on pulsatile secretion of luteinizing hormone in ovariectomized ewes. *Acta endocr., Copenh.* **111**, 172–178.

Turner, D.D., Garcia, M.C., Miller, K.F., Holtan, D.W. & Ginther, O.J. (1979) FSH and LH concentrations in periparturient mares. *J. Reprod. Fert., Suppl.* **27**, 547–553.

POSTER ABSTRACTS

J. Reprod. Fert., Suppl. **35** (1987), 641

Printed in Great Britain
© 1987 Journals of Reproduction & Fertility Ltd

Evaluation of the use of fresh, extended, transported stallion semen in the Netherlands

P. J. de Vries*

Clinic for Obstetrics, Gynaecology and AI, Faculty of Veterinary Medicine, State University of Utrecht, Yalelaan 5, 3508 TD Utrecht, The Netherlands

During two seasons experiments were carried out to study the results with the use of transported, fresh, extended stallion semen. After collection (2–3 times/week) gel-free semen was diluted (D-11 extender: distilled water, glucose, fructose, glycine, egg yolk, sulphanilamide, pH 6·90, 280 mosmol) and centrifuged (15 min, 800 g). Sediment was resuspended in D-11 extender and dosages of 500×10^6 total motile spermatozoa in 16-ml polypropylene tubes were made up. Extended semen was slowly cooled to 5°C in constant rotation. On request, extended semen was sent to an insemination station in pre-cooled small, thermo-insulated containers (Sarstedt Inc.; Nümbrecht, F.R.G.) by postal express. After arrival (within 16 h) tubes containing extended semen were stored at 5°C in constant rotation until insemination by a qualified veterinarian. Insemination was performed after preovulatory follicle detection and semen quality control and repeated every 48 h until ovulation. Mares were teased regularly and, when in heat, palpated every other day. Pregnancy was confirmed by rectal palpation. The results of the two experimental seasons are shown in Table 1.

Table 1. Results with stallion semen over two seasons

	No. of stallions	No. of mares	No. of mares pregnant (%)	Inseminations/ pregnancy	Inseminations/ oestrus
1st season	6	79	60 (76)	2·4	1·6
2nd season	13	134	102 (76)	2·5	1·5

I conclude that satisfactory results can be achieved with this system if semen is inseminated within 3 or 4 days of collection. Comparable with the situation using frozen semen, differences in results between stallions were found and seem to be due to differences in susceptibility of semen to the extender. Adjusting the composition of the extender to each stallion may improve results. Hygienic collection and processing of semen are important to prevent unacceptable quantities of bacteria after storage. Problems with transport of the extended semen occur at weekends and holiday periods.

The system is now used commercially in the Netherlands by 3 stallion AI centres and 35 regional insemination stations in the 1986 breeding season.

*Present address: Vierhouten Equine Clinic, Niersenseweg 33, 8076 PW Vierhouten, The Netherlands.

J. Reprod. Fert., Suppl. **35** (1987), 643

Effect of ejaculation frequency on sperm output

J. L. Voss, B. W. Pickett, J. R. Neil and E. L. Squires

Animal Reproduction Laboratory, Colorado State University, Fort Collins, Colorado 80523, U.S.A.

In natural service, it is not uncommon for a stallion to ejaculate 2–4 times daily. Two frequencies of ejaculation commonly used in natural service were studied to determine their effects on sperm output. Ten stallions were subjected to two frequencies of ejaculation in a double switch-back experiment. Stallions were paired based on age, sperm output and testicular size, and assigned to one of two groups. Before beginning the experiment, semen was collected from all stallions daily for 8 days to deplete extra-gonadal sperm reserves. Semen was collected from stallions once or twice per day for 14 days, then collection frequency was switched for another 14 days.

For stallions from which semen was collected twice per day (8 h apart), $1 \cdot 2 \times 10^9$ and $1 \cdot 3 \times 10^9$ spermatozoa were collected in the first and second ejaculates, respectively. After correction for spermatozoa lost in the gel and collection equipment, the number of spermatozoa ejaculated was $3 \cdot 8 \times 10^9$ and $3 \cdot 9 \times 10^9$ for first and second ejaculates, respectively. Total spermatozoa available per day from once and twice frequencies of ejaculation were $3 \cdot 7 \times 10^9$ and $2 \cdot 5 \times 10^9$ ($P < 0 \cdot 001$), respectively, and total spermatozoa ejaculated (after correction for spermatozoa lost in gel and collection equipment) were $7 \cdot 5 \times 10^9$ and $7 \cdot 4 \times 10^9$, respectively. Increasing frequency of ejaculation from once to twice per day did not increase sperm output. Therefore, there are only so many spermatozoa available per day, and for each mare to have an equal opportunity to become pregnant, the intervals between matings must be long enough to ensure that each mare gets a sufficient number of spermatozoa for maximum reproductive efficiency.

J. Reprod. Fert., Suppl. **35** (1987), 645–647

Variation in characteristics of stallion semen caused by breed, age, season of year and service frequency

K. F. Dowsett and W. A. Pattie

Department of Animal Sciences and Production, University of Queensland, St Lucia, Queensland 4067, Australia

The semen characteristics of 168 stallions from 9 breeds, aged from 2 to 26 years, were studied over 4 breeding seasons. Variation due to a range of factors was examined by least-squares analyses of variance with mixed model procedures being used to examine breed effects. Apart from breed, only age, season of year and service frequency had significant effects.

Differences between age groups in semen characteristics are shown in Table 1. Stallions <3 years of age had lowest semen volumes and sperm concentrations which together with the highest proportion of dead spermatozoa resulted in the lowest numbers of live spermatozoa per ejaculate. They also had greater proportions of spermatozoa with abnormal and loose heads, proximal droplets and abnormal tails. Stallions older than 13 years had low sperm concentrations and high percentages of dead spermatozoa but sperm morphology was normal.

Shetland ponies produced largest volumes of gel and gel-free semen with smallest volumes being obtained from Arabs and Ponies respectively (Table 2). The other breeds formed a continuous

Table 1. Means of stallion semen characteristics for various ages

Age (years)	No.	Gel-free vol. (ml)	Gel vol. (ml)	Sperm conc. ($\times 10^{-6}$)	Total sperm no. ($\times 10^{-6}$)	Dead sperm. (%)
1 + 2	28	15·6	0·4	43·38	480·58	30·8
3–13	427	30·2	4·0	147·78	5053·04	18·8
≥14	81	33·7	1·4	83·24	3252·32	22·8
Residual s.d.		0·279	0·638	5·236	33·123	0·25
d.f.	472 for volumes, 496 for others					

Table 2. Means of stallion semen characteristics for various breeds

Breed	No.	Gel-free vol. (ml)	Gel vol. (ml)	Sperm conc. ($\times 10^{-6}$)	Total sperm no. ($\times 10^{-6}$)	Dead sperm. (%)
Thoroughbred	141	28·3	2·7	114·29	5027·47	21·6
Standardbred	111	30·2	3·1	97·24	4737·85	15·4
Arab	73	36·2	1·0	286·82	12661·22	10·1
A.S.H.	73	33·2	5·5	116·06	4854·15	17·1
Quarterhorse	30	23·8	4·0	171·66	5371·53	23·8
Palomino	44	23·8	1·1	138·48	4016·26	21·3
Pony	38	20·8	2·5	104·01	1122·13	24·7
Shetland	8	44·4	13·1	101·25	1720·70	38·5
Appaloosa	18	23·3	2·0	90·42	3331·89	15·8
Residual s.d.		0·279	0·638	5·236	33·123	0·247
d.f.	472 for volumes, 496 for others					

Table 3. Means of stallion semen characteristics for each season

Season of year	No.	Gel-free vol. (ml)	Gel vol. (ml)	Sperm conc. ($\times 10^{-6}$)	Total sperm no. ($\times 10^{-6}$)	Dead sperm (%)
Autumn	80	21·5	1·3	192·27	5505·79	20·2
Winter	102	23·1	1·9	125·16	3776·46	21·4
Spring	232	26·2	4·4	110·19	3726·59	18·5
Summer	122	26·4	6·4	139·45	4369·30	18·7
Residual s.d.		0·279	0·638	5·236	33·123	0·247
d.f.	427 for volumes, 496 for others					

Table 4. Means of stallion semen characteristics for different service frequencies

Service frequency	No.	Gel-free vol. (ml)	Gel vol. (ml)	Sperm conc. ($\times 10^{-6}$)	Total sperm no. ($\times 10^{-6}$)
0–1 h	38	22·2	1·5	69·84	1903·98
1–6 h	47	22·2	2·6	89·92	2318·24
6–24 h	100	30·9	3·2	92·09	3399·80
1–7 days	140	35·5	1·7	163·35	6579·79
>7 days	178	35·1	2·4	193·34	7783·87
1st service	24	42·4	1·0	246·69	9646·98
Residual s.d.		0·279	0·638	5·236	33·123
d.f.	472 for volume, 496 for others				

range between these limits. Arabian stallions had the lowest percentage of non-motile and dead spermatozoa while Shetlands were the only breed with greater than 30% dead, the others having values within accepted normal limits. Arabian stallions had sperm concentrations almost double those of other breeds and with their relatively high volume, this resulted in total sperm numbers that were 3 times greater than those of any other.

There were significant differences between breeds in most morphological characteristics of spermatozoa but they were usually within the range considered normal. Exceptions were found for Palominos and Shetlands whose spermatozoa had high percentages of abnormal tails, abnormal and loose heads. Thoroughbreds had considerably higher percentages of protoplasmic droplets than other breeds. Most breed differences could be considered normal but they should be taken into account when evaluating semen.

Seasonal effects are shown in Table 3. There were no differences in volume of gel-free semen but, because of higher sperm concentrations, total sperm numbers were greatest in summer and autumn. In addition, most dead spermatozoa were present in autumn and winter and these were accompanied by higher proportions of abnormal heads and tails. It seems that the timing of the Australian breeding season in spring and early summer is out of phase with the period of peak semen production.

Gel-free volume and sperm concentration increased with service intervals up to 24 h (Table 4), but there were no increases in morphological abnormalities. It appears that most stallions should only serve twice daily to maintain normal sperm numbers.

In addition to these, the following effects were examined in this study: weather conditions, ambient temperature, time of day, height and weight of stallion, volume and temperature of water in the artificial vagina. None of these factors had any consistent or statistically and biologically meaningful effects.

We conclude that, when evaluating stallion semen in normal stud operations, semen of poor quality may be obtained from immature and aged stallions, from collections made in winter and spring, and from stallions that are required to serve frequently. In these cases, the poor semen quality reflects management problems rather than a pathological condition. Furthermore, some breeds will show consistent differences in some semen characteristics but these will be normal for the breed.

This work was supported by the Queensland Equine Research Foundation.

J. Reprod. Fert., Suppl. **35** (1987), 649–650

Fertility and characteristics of slow-cooled stallion semen

D. H. Douglas-Hamilton, P. J. Burns, D. D. Driscoll and K .M. Viale

Hamilton Thorn Research Associates, 30A Cherry Hill Drive, Danvers, MA 01923, U.S.A.

Transported slow-cooled semen is rapidly becoming accepted as an economical, viable method of horse-breeding. In a field trial during the 1985 season, 4 stallions at Hamilton Farm were mated with 105 mares by transported semen. In the present study we examined relationships between progressive motility and mean velocity of spermatozoa for fresh and stored semen, and the resultant fertility.

Semen was collected and extended for shipment as previously described (Douglas-Hamilton *et al.*, 1984). Briefly, 1 part semen was added to 2 parts glucose + instant non-fat dry milk (Kenney) extender containing 500 mg gentamicin sulphate with 1·2 g NaHCO$_3$ 1. An average of 2·4 × 10^9 progressively motile spermatozoa were packaged into the slow-cooling shipping container (Equitainer) and shipped to the mare; a sample was removed from the rest and stored under standard slow-

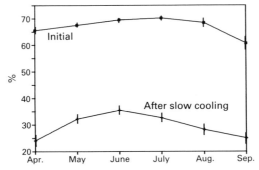

Fig. 1. Mean initial progressive motility (4 stallions) and mean slow-cool (24 h) motility (3 stallions), averaged over each month. One stallion was omitted because it had close to zero slow-cool motility. Standard error is shown.

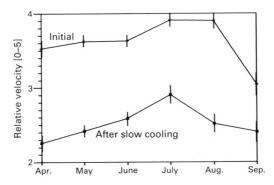

Fig. 2. Mean initial sperm velocity (4 stallions) and slow-cool (24 h) sperm velocity (3 stallions), averaged over each month. One stallion was omitted because it had close to zero slow-cool velocity. Standard error is shown.

cooling conditions (cooling rate at 37°C = −0·25 to −0·3°C/min) in the laboratory refrigerator. Final temperature of 4–8°C was reached after 10 h. After 24–28 h the laboratory sample was examined by placing a drop of semen on a prewarmed slide and allowing it to warm up from 1 min, and the progressive motility and velocity were estimated.

The mares were distributed across the United States and Canada (northeast 60%, southeast 11%, midwest 18%, Pacific 11%): 37% were barren, 23% maiden and 43% foaling. They were examined for pregnancy at 55 days by ultrasound and/or palpation.

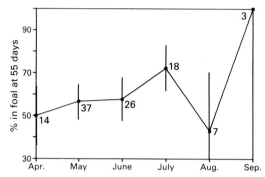

Fig. 3. First-cycle pregnancy rate (at 55 days) for 4 stallions, averaged over month and stallion, using slow-cooled semen transported to 105 mares. Values are mean and s.e. for the no. of mares indicated.

The results are shown in Figs 1–3. The values for first cycle pregnancy for August and September were not significant because of the small sample number. Both slow-cooled sperm velocity and conception rate exhibited peaks in July. Mean first-cycle pregnancy and total seasonal rates were 59% and 83% respectively.

We conclude that the drop in semen characteristics of sperm motility and velocity observed after slow-cool storage does not appear to affect fertility adversely.

References

Douglas-Hamilton, D.H., Osol, R., Osol, G., Driscoll, D. & Noble H. (1984) A field study of the fertility of transported equine semen. *Theriogenology* **22,** 291–304.

J. Reprod. Fert., Suppl. **35** (1987), 651–653

Quantitative assessment of membrane damage in cold-shocked spermatozoa of stallions

P. F. Watson, J. M. Plummer and W. E. Allen*

*Departments of Physiology and of *Surgery & Obstetrics, Royal Veterinary College,
London NW1 0TU, U.K.*

The long-term preservation of semen requires that spermatozoa are subjected to cooling, and several methods use quite rapid cooling rates. However, there is surprisingly little information on the effects of cold shock on stallion spermatozoa (see Watson, 1981). Bogart & Mayer (1950) described stallion spermatozoa as having "little or no inherent resistance" to rapid cooling but, since some 44% of the cells survived cold shock to 0°C, stallion semen may be more resistant than that of other ungulate species. Johnson *et al.* (1980) confirmed that stallion spermatozoa are sensitive to cold shock, but presently no reliable evidence of their relative susceptibility exists.

Our aim was to subject stallion semen to various controlled cold-shock stresses and to quantitate the responses in order to make direct comparison with similar studies in other species.

Materials and methods

Semen was collected by artificial vagina from two pony stallions of proven fertility. Portions (1 ml) of the pooled ejaculates were placed in glass tubes at 30°C and the tubes were subjected to rapid cooling by immersion in water at 16, 8, 4 or 0°C for 5 min. The semen was then prefixed in buffered formaldehyde–glutaraldehyde–picric acid solution at 24°C for 2 h and, after the cells had been washed in buffer, they were post-fixed in buffered 1% osmium tetroxide solution for 1 h (Jones, 1973). A pellet of cells was formed by centrifugation at 1500 *g* for 30 min and was dehydrated and embedded (Plummer & Watson, 1985).

Stained ultrathin sections were examined by transmission electron microscopy in a JEOL 1200EX microscope. At a magnification of × 15 000, 200 longitudinal sperm head profiles were classified directly from the microscope screen according to the degree of damage to the membranes. Profiles of midpieces and tails were also examined for evidence of structural damage.

Results

Sperm heads were categorized into four classes according to the condition of the plasma membrane and the acrosome (Table 1). Although all classes could be identified in any sample irrespective of treatment, the results indicate a progression of structural damage. Both the number of damaged cells and the degree of damage increased with increasing cold stress, as shown by the first three classes (Categories A–C, Table 1). However, the proportion in the class showing the greatest disruption (Category D, Table 1) was apparently unaffected by cold shock. The anterior segment of the acrosome was more labile than the equatorial segment which was unaltered in the majority of spermatozoa subjected even to the greatest cold stress.

The acrosomal contents were very rarely dispersed without prior damage to the plasma membrane. In the process of its disruption, the outer acrosomal membrane appeared initially to undergo the development of undulations, with subsequent distension and 'ballooning', until finally it was considerably displaced from its original position although still apparently continuous. The contents of the acrosome were progressively dispersed during this process.

Table 1. Percentages of spermatozoa with damaged head membranes after various degrees of cold shock from 30°C

Membrane categories*	Cold-shock temperature				
	30°C	16°C	8°C	4°C	0°C
(A) Plasma membrane present, acrosome apparently normal	81	66	63	44	29
(B) Plasma membrane lost, acrosome apparently normal	9	10	12	23	23
(C) Plasma membrane lost, outer acrosomal membrane undulated or 'ballooned', acrosomal contents dispersed	7	16	20	25	41
(D) Plasma and outer acrosomal membrane lost, acrosomal contents dispersed	4	8	5	9	8

*In most cases the equatorial segment of the acrosome was not involved.

Midpiece and tail profiles with intact plasma membranes could readily be found even after severe cold shock, but the plasma membrane of the midpiece was more frequently disrupted than that of the tail. Profiles entirely lacking plasma membranes were very rarely seen. Some mitochondria became electron-lucent with loss of internal structure as a result of cold shock, but apparently normal mitochondrial profiles were frequently evident alongside them. A similar extent of mitochondrial damage was detected whether or not the plasma membrane was ruptured. The axoneme in the midpiece and the tail, together with other tail structures, was apparently unaffected by cold shock.

Discussion

The cellular disruption to stallion spermatozoa induced by cold shock was qualitatively similar to that induced in spermatozoa of other ungulate species, and occurred in a similar temperature range, i.e. below 15°C. As with other species, both the extent of membrane disruption and the proportion of damaged spermatozoa increased with increasing cold stress. The disruption of cell membranes in the head region appeared to be progressive, beginning with the plasma membrane and proceeding to involve the acrosome and its contents. The process did not appear to include the total loss of the outer acrosomal membrane and contents, since the numbers in Category D (Table 1) did not change with treatment and therefore presumably represented a group of degenerate spermatozoa in the ejaculate. However, the extent of the disruption with controlled degrees of cold shock was quantitatively less than that seen with ram or boar spermatozoa subjected to similar stresses (Watson & Plummer, 1985; Plummer *et al.*, 1986).

The acrosomal expansion seen with cold shock suggests an increase in membrane area. Since it appears that biological membranes cannot easily stretch (see McGrath, 1981), such expansion would indicate loss of integrity. While this undoubtedly occurred, in most instances the membrane appeared to maintain its continuity, perhaps suggesting the addition of new membranous material. The possible origin of such material is uncertain, although membrane fragments were frequently observed adjacent to cells.

In contrast to the acrosome, mitochondria appeared to be damaged while the plasma membrane remained intact, perhaps indicating that the damage is a direct result of the cold stress on the mitochondrial cristae, rather than a consequence of a change in intracellular microenvironment resulting from loss of plasma membrane integrity. Since no other structural defects were observed in the midpiece and tail, it is likely that the loss of motility in a proportion of cells with cold shock is related to mitochondrial and/or plasma membrane damage.

It is evident that stallion spermatozoa are susceptible to cold shock and therefore some care is needed in their handling for preservation at reduced temperatures. The inclusion of a protective substance in a diluent as a membrane stabilizer during cold stress is clearly warranted.

References

Bogart, R. & Mayer, D.T. (1950) The effects of egg yolk on the various physical and chemical factors detrimental to spermatozoan viability. *J. Anim. Sci.* **9**, 143–152.

Johnson, L., Amann, R.P. & Pickett, B.W. (1980) Maturation of equine epididymal spermatozoa. *Am. J. vet. Res.* **41**, 1190–1196.

Jones, R.C. (1973) Preparation of spermatozoa for electron and light microscopy. *J. Reprod. Fert.* **33**, 145–149.

McGrath, J.J. (1981) Thermodynamic modelling of membrane damage. In *Effects of Low Temperatures on Biological Membranes,* pp. 335–377. Eds G. J. Morris & A. Clarke. Academic Press, London.

Plummer, J.M. & Watson, P.F. (1985) Ultrastructural localization of calcium ions in ram spermatozoa before and after cold shock as demonstrated by a pyroantimonate technique. *J. Reprod. Fert.* **75**, 255–263.

Plummer, J.M., Robertson, L. & Watson, P.F. (1986) The effects of incubation on membrane damage and calcium movement in boar and ram spermatozoa subjected to cold shock. *Proc. 30th Congr. Int. Union Physiol. Socs* **16**, 497, Abstr.

Watson, P.F. (1981) The effects of cold shock on sperm cell membranes. In *Effects of Low Temperatures on Biological Membranes,* pp. 189–218. Eds G. J. Morris & A. Clarke, Academic Press, London.

Watson, P.F. & Plummer, J.M. (1985) The responses of boar sperm membranes to cold shock and cooling. In *Deep Freezing of Boar Semen,* pp. 113–127. Eds L. A. Johnson & K. Larsson. Swedish University of Agricultural Sciences, Uppsala.

J. Reprod. Fert., Suppl. **35** (1987), 655–656

Printed in Great Britain
© 1987 Journals of Reproduction & Fertility Ltd

An infertility syndrome in stallions associated with low sperm motility

D. B. Galloway

Department of Veterinary Clinical Sciences, University of Melbourne, Prince's Highway, Werribee, Victoria 3030, Australia

From 1975 to 1985, 94 stallions were submitted for reproductive evaluation and 305 semen samples were received from veterinarians for examination at the semen laboratory. Twelve stallions from diverse environments, presented because of reduced fertility, were characterized by absent or low sperm motility and an abnormally high percentage of spermatozoa with attached cytoplasmic droplets (low motility syndrome). Six were Welsh Mountain Ponies 3–5 years of age, 5 were Thoroughbreds from 6 to 10 years of age, and 1 was a 3-year-old Arab. Clinical and reproductive examinations were carried out.

When stallions were first presented, more than 1 semen sample was collected. The ejaculate with the best motility was taken as the most representative and was used in compiling Table 1. One semen sample of good motility ($\geq 60\%$), from each of 60 normal and fertile stallions, was used to define semen characteristics indicative of normal function of the reproductive system. Statistical comparisons were made using Student's *t* test and arc sin transformations for percentages.

Five stallions had normal fertility in 1–3 breeding seasons before diagnosis. For the 3 Thoroughbreds in this group, an average of 44 (range 31–60) mares was served each season, with 75% (range 56–84) becoming pregnant. When examination was requested, an average of 26 (8–40) had been served and had a mean pregnancy rate of 12% (0–37). Two Welsh Mountain Ponies had 'satisfactory' fertility for 2 and 3 previous seasons, respectively, but no mares had become pregnant of 8 and 3 served at the time examination was requested. Low fertility (0–29% pregnant of 2–7 mares served) in both their first and second year of service was the reason for requesting examination of 4 stallions. Low fertility in the 1st year of service (0–17% pregnant of 3–14 mares served) was the reason for examination of 3.

There was no history or clinical evidence of illness or injury in 8 of the horses. Two had suffered severe traumatic accidents to the hind limbs and 1 was surgically treated for colic in the 2 years before diagnosis. The Arab stallion had a chronic upper respiratory tract infection.

The 6 Welsh Mountain ponies were related. A common ancestor appeared in their pedigrees from 2 to 7 times in the 3rd to 6th generation. Two Thoroughbreds were full brothers and another was by the same sire. A third Thoroughbred shared one common ancestor with these 3.

Physical examinations of the reproductive organs revealed no abnormalities other than an intermittently reversed testis in one Thoroughbred. Testicular size was considered consistent with the age and breed of the horses. Epididymides were palpably normal.

Three of the 12 stallions had good libido but a tendency to ejaculatory failure. In 2 Thoroughbreds, management was clearly implicated. In a Welsh Mountain Pony, no cause was found. With perseverance, all 3 yielded semen satisfactory for examination. The other 9 stallions had normal serving behaviour.

Semen motility was uniformly low (Table 1) in stallions with this syndrome. In many samples the immotile spermatozoa were alive as demonstrated by the nigrosin–eosin technique. The mean percentages of abnormal sperm heads and singly bent tails were higher than that for the normal material but not outside the limit of the normal range for the laboratory (20% and 10%, respectively). Spermatozoa with attached cryptoplasmic droplets were significantly increased in 'low motility syndrome' stallions compared with the normal material. In each affected stallion, proximal droplets were over 10% or distal droplets over 20% or both.

Table 1. Characteristics of semen collected from stallions with low motility syndrome (LMS) and from normal stallions

	LMS (N = 12)	Normal (N = 60)
Volume (ml)	38·0 ± 9·0 (3–125)	59·0 ± 6·0 (20–175)
Concentration (× 10⁻⁶/ml)	175·0 ± 24·0 (25–275)	238·0 ± 20·0 (20–450)
Motility (%)	8·0 ± 2·0 (0–20)**	74·0 ± 2·0 (60-100)
% Abnormal heads	12·5 ± 1·5 (6·5–21·5)**	8·0 ± 0·4 (2·5–18·0)
% Proximal droplets	17·1 ± 3·1 (6·0–49·0)**	5·6 ± 0·5 (1·0–21·0)
% Distal droplets	32·0 ± 4·9 (14·0–60·5)**	4·8 ± 0·8 (0·5–33·0)
% Tailless	2·2 ± 0·4 (0·5–4·3)	1·9 ± 0·3 (0–16·0)
% Singly bent tails	5·8 ± 1·5 (0–18·3)*	3·2 ± 0·4 (3·0–19·0)
% Doubly bent tails	1·2 ± 0·7 (0–8·8)	0·7 ± 0·1 (0·2–2·5)
% Coiled tails	0·1 ± 0·05 (0–0·5)	0·4 ± 0·1 (0–2·3)

Values are mean ± s.e.m. and range.
*$P < 0.05$; **$P < 0.001$ compared with normal values.

In 1 stallion examined in the year before diagnosis, when fertility was high, percentages for motility, abnormal heads, proximal droplets and distal droplets were 70, 8, 6 and 2, respectively. In the following year when the complaint of low fertility was investigated, these semen parameters were 10, 14, 31 and 20% respectively. No improvement in semen motility has been observed in follow-up examinations of 1–9 ejaculates from 7 stallions with 'low motility syndrome' from 2 to 25 months after diagnosis. Exhaustive ejaculation was attempted in 3 stallions. In one, motility improved from 0–10% to 50% by the fifth ejaculate on 1 occasion and the third ejaculation on the next day. In two, an improvement of motility was not obtained in 3 ejaculates on 1 day and daily ejaculations for 6 days, respectively. Seminal plasma from a normal stallion did not activate motility in spermatozoa from the 2 affected stallions examined in this way. Spermatozoa from the normal stallion retained motility in the seminal plasma from the affected stallions.

In the low motility syndrome, there is apparently failure of the spermatozoa to acquire the ability to move and fertilize and to mature morphologically in the epididymis. The pattern of sperm abnormalities together with improvement of motility for one stallion subjected to exhaustive ejaculations suggests an epididymal dysfunction.

No cause has been found. Genetic factors could be involved as is likely in spermiostasis of the epididymal head in bulls (Blom & Christensen, 1960), but the evidence is circumstantial. Further work is warranted on the fertility of affected stallions when used intensively close to ovulation for a small number of mares and on the cause, pathogenesis, and prognosis of low motility syndrome.

Reference

Blom, E. & Christensen, N.O. (1960) *Nord VetMed.* **12**, 453–470.

J. Reprod. Fert., Suppl. **35** (1987), 657–658

Printed in Great Britain
© 1987 Journals of Reproduction & Fertility Ltd

Effects of daylength or artificial photoperiod on pituitary responsiveness to GnRH in stallions

C. M. Clay, E. L. Squires, R. P. Amann and T. M. Nett

Animal Reproduction Laboratory, Colorado State University, Fort Collins, CO 80523, U.S.A.

The seasonal influence on reproduction in stallions has been characterized by an increase in testicular size and weight (Pickett *et al.*, 1981; Johnson & Thompson, 1983), sperm production and output (Pickett *et al.*, 1981; Johnson & Thompson, 1983) and libido (Thompson *et al.*, 1977; Pickett *et al.*, 1981) during the spring and summer. In addition to these obvious testicular and behavioural changes there also appears to be a similar annual fluctuation in the serum concentration of gonadotrophin. For luteinizing hormone (LH), this seasonal fluctuation was characterized by higher concentrations in the blood during the spring and summer than during the fall and winter (Thompson *et al.*, 1977). The elevation in serum concentrations of LH in the spring may be a consequence of (a) increased release, (b) increased synthesis, or (c) both (a) and (b). The use of gonadotrophin-releasing hormone (GnRH) should allow a differentiation among these 3 alternatives. Since the primary environmental cue entraining the annual reproductive rhythms of stallions is, apparently, daylength, an experiment was conducted to determine the influence of photoperiod on the pituitary content of LH as assessed by secretion of LH after exogenous GnRH stimulation.

Twenty-one stallions were assigned to 1 of 3 treatments: (a) control, natural daylength; (b) Group S-L, exposed to 8 h light and 16 h dark (8L:16D) for 20 weeks beginning 16 July 1982, then switched to 16L:8D on 2 December and exposed to 16L:8D until March 1984; and (c) Group S-S, exposed to 8L:16D from 16 July 1982 until March 1984. GnRH challenges were conducted at about 8-week intervals and also at 2 weeks before and 2 weeks after the 2 December light shift for stallions in Group S-L. For each challenge, blood samples were collected at 20-min intervals for 2 h before GnRH (dose = 2 µg/kg) and then for 8 h after GnRH. Serum was assayed for concentration of LH (Nett *et al.*, 1975).

In control stallions, baseline serum concentration of LH fluctuated ($P < 0.05$) with season and concomitantly with pituitary content of LH as measured by GnRH-induced releasable LH. Baseline LH was maximal in April and lowest in December. The exposure of Group S-L stallions to 16L:8D after 20 weeks of 8L:16D induced elevated ($P < 0.05$) basal LH values. Baseline LH concentration was higher ($P < 0.05$) for stallions in Group S-L at the December (2 weeks after 16L: 8D) and February (8 weeks after 16L:8D) bleedings than for control or Group S-S stallions. The elevation in basal LH, however, was not accompanied by an increase in releasable LH. After 16 weeks of exposure of Group S-L stallions to 16L:8D, basal LH declined ($P < 0.05$). This decline may have been due to photorefractoriness. The effects of maintaining stallions in Group S-S continuously in 8L:16D were reflected as low basal LH for about 1 year. The depression of basal LH was not reflected as a similar depression in pituitary content of releasable LH. The GnRH-releasable store of LH continued to fluctuate with season in a manner similar to that exhibited by control stallions. After 1 year of exposure to 8L:16D stallions in Group S-S exhibited an elevation ($P < 0.05$) in baseline LH concentration that may have resulted from refractoriness to the inhibitory photoperiod or from the expression of an endogenous cycle.

We suggest that the elevation in serum concentration of LH associated with seasonal reactivation in the stallion probably is due to both an increase in pituitary content of LH and an increase in stimuli for LH release. Furthermore, the effect of increasing photoperiod apparently is induction of LH release, rather than elevation of pituitary content. If this interpretation is correct, the photoinducible mechanism would, presumably, involve an increased stimulation of the anterior pituitary by GnRH.

References

Johnson, L. & Thompson, D.L., Jr (1983) Age-related and seasonal variation in the Sertoli cell population, daily sperm production and serum concentrations of follicle-stimulating hormone, luteinizing hormone and testosterone in stallions. *Biol. Reprod.* **29,** 777–789.

Nett, T.M., Holtan, D.W. & Estergreen, V.L. (1975) Levels of LH, prolactin and oestrogens in the serum of post-partum mares. *J. Reprod. Fert., Suppl.* **23,** 201–206.

Pickett, B.W., Voss, J.L., Squires, E.L. & Amann, R.P. (1981) Management of the stallion for maximum reproductive efficiency. Colorado State University, General Series Bull. 1005.

Thompson, D.L., Jr, Pickett, B.W., Berndtson, W.E., Voss, J.L. & Nett, T.M. (1977) Reproductive physiology of the stallion. VIII. Artificial photoperiod, collection interval and seminal characteristics, sexual behavior and concentrations of LH and testosterone in serum. *J. Anim. Sci.* **44,** 656–664.

J. Reprod. Fert., Suppl. **35** (1987), 659–660

Diagnosis of infertility in mares and stallions by karyotyping

C. R. E. Halnan

Karyotype Research, University of Sydney, New South Wales 2006, Australia

Osborne (1975) reported that live foal rate for Thoroughbred mares was 50% and no improvement had been made utilizing veterinary expertise. Fertility can be improved since conception rates are 70% and live foals related to these about 62% (Laing & Leach, 1975; Sullivan *et al.*, 1975; Dowsett & Pattie, 1982). Early abortions have been estimated at 12% (Sullivan *et al.*, 1975).

The effect of stallion has yet to be determined. Merkt *et al.* (1979) showed that overuse is not likely to be a factor when the harem is below 80 per stallion. The stud books indicate marked differences in foal rate between established stallions by as much as a factor of 2 to 1. Some stallions do not continue to stand after the first season. Differences in sex ratios of foals have been reported (Kent *et al.*, 1985): when the ratio departed from the norm, more offspring were infertile. Stallion performance in these cases has been difficult to evaluate. Stallion farms report conception rates as high as 93%, but the live foal rates are obscure.

Cytogenetic tests by means of karyotypes (Halnan, 1985, 1986) can be used to distinguish chromosomal abnormalities and can give a result within 48 h. Most mares that are infertile after exclusion of other possible causes of infertility are identified by the test, which is possible before they go to stud. The confidence for the prognosis is of the order of 75% before the maiden season and 90% after 1 year at stud. It is estimated that 2% of stallions will have chromosome defects and a presumptive marker has been found in 4 subfertile stallions: The stallions all had a band deletion in autosome 1. One stallion, not successful with only 4 mares, had mosaic autosomal trisomy. The deletion has also been found in mares (unpublished data). If part of the population is heterozygous for a deletion, homozygotic occurrence may be one cause of spontaneous abortion. Stud practice selects on race performance and if this bears an inverse relationship to fertility a possible explanation for the steady foal rate is to hand. This cytogenetic test, performed by graduate scientists, has yet to be automated; it is costly but highly cost effective. Demand for tests is not high and greater demand would lower the cost and increase availability of services. When the operatives are unskilled and the investigation unplanned, the test is unlikely to be successful. The skills are sophisticated and require a year of training and independent capability will need 2–3 years experience. McFeely (1975) postulated that: "In future additional chromosome abnormalities will be found associated with maldevelopment of the equine reproductive system and as a cause of early embryonic mortality". Originally sex chromosome mosaicism, deletion and trisomy were described as predicates for infertility in mares (Chandley *et al.*, 1975; Hughes & Trommerhausen-Smith, 1977). In fact an infertile mare had already been described with 63 chromosomes (Payne *et al.*, 1968). Later, two XY mares were reported, one infertile the other virtually so, as examples of testicular feminization (Kieffer *et al.*, 1976; Sharp *et al.*, 1980). Not all mares with abnormal karyotype fail to reproduce but those which do, do so once and with difficulty. On the other hand a mare with XO:XY:XXY was palpably normal but infertile due to tubal dysgenesis (Halnan, 1985, 1986). A mare with stallion-like characteristics was found to have a deletion in one of the X chromosomes and was infertile.

Chromosomal aberrations have been reported for 3 stallions: a cryptorchid XX (Dunn *et al.*, 1974), an infertile Standardbred with deletion in 13 (Halnan *et al.*, 1982) and a Thoroughbred with a polymorphism of No. 1 and spontaneous abortion in his harem (Haynes & Reisner, 1982).

This means that 20% of all Thoroughbred mares at stud are unlikely to breed for chromosomally identifiable genetic reasons. Removal of that fraction from the mare population would increase the foal rate from 50 to about 62·5%. Identification of infertile stallions should increase the foal rate by about 6%. Subject to the qualification imposed by insufficient data, it is suggested that fertility in horses is perhaps 10% poorer than that of cattle. If the figures are correct, much wasted effort is being expended on attempting to breed infertile mares or stallions without first determining their potential.

References

Chandley, A.C., Fletcher, J.M., Rossdale, P.D., Peace, C.K., Ricketts, S.W., McEnery, R.J., Thorn, J.P., Short, R.V. & Allen, W.R. (1975) Chromosome abnormalities as a cause of infertility in mares. *J. Reprod. Fert., Suppl.* **23**, 377–383.

Dowsett, K.F. & Pattie, W.A. (1982) Characteristics and fertility of stallion semen. *J. Reprod. Fert., Suppl.* **32**, 1–8.

Dunn, H.O., Vaughan, J.T. & McEntee, K. (1974) Bilaterally cryptorchid stallion with female karyotype. *Cornell Vet.* **64**, 265–275.

Halnan, C.R.E. (1985) Equine cytogenetics: role in equine veterinary practice. *Equine vet. J.* **17**, 173–177.

Halnan, C.R.E. (1986) Failure to conceive in a standardbred mare: Karyotype 64XX:64XY. *Equine vet. Sci.* **5**, 161–162.

Halnan, C.R.E., Watson, J.I. & Pryde, L.C. (1982) Prediction of reproductive performance in horses by karyotype. *J. Reprod. Fert., Suppl.* **32**, 627–628.

Haynes, S.E. & Reisner, A.H. (1982) Cytogenetic and DNA analyses of equine abortion *Cytogenet. Cell Genet.* **34**, 204–214.

Hughes, J.P. & Trommershausen-Smith, A. (1977) Infertility in the horse associated with chromosomal abnormalities. *Aust. Vet. J.* **53**, 253–257.

Kent, M.G., Shoffner, L., Bueon, L.C. & Weber, A.F. (1985) Further studies of the cytogenetic aspects and phenotype spectrum present in the XY sex reversal syndromes. *J. Dairy Sci., Suppl.* **68**, 251, Abstr.

Kieffer, N.M., Burns, S.J. & Judge, N.G. (1976) Male pseudohermaphroditism of the testicular ferminizing type in a horse. *Equine vet. J.* **8**, 38–41.

Laing, J.A. & Leech, F.B. (1975) The frequency of infertility in thoroughbred mares. *J. Reprod. Fert., Suppl.* **23**, 307–310.

Osborne, V. (1975) Factors influencing foaling percentages in Australian mares. *J. Reprod. Fert., Suppl.* **23**, 477–483.

McFeely, R.A. (1975) A review of cytogenetics in equine reproduction. *J. Reprod. Fert., Suppl.* **23**, 371–374.

Merkt, H., Jacobs, K-O., Klug, E. & Aukes, E. (1979) An analysis of stallion fertility rates (foals born alive) from the breeding documents of the Landgestut Celle over a 158-year period. *J. Reprod. Fert., Suppl.* **27**, 73–77.

Payne, H.W., Ellsworth, K. & Degroot, A. (1968) Aneuploidy in an infertile mare. *J. Am. vet. Med. Ass.* **153**, 1293–1299.

Sharp, A.J., Wachtel, S.S. & Benirshcke, K. (1980) H-Y antigen in a fertile XY female horse. *J. Reprod. Fert.* **58**, 157–160.

Sullivan, J.J., Turner, P.C., Self, L.C., Gutteridge, H.B. & Bartlett, D.E. (1975) Survey of reproductive efficiency in the quarter-horse and thoroughbred. *J. Reprod. Fert., Suppl.* **23**, 315–318.

J. Reprod. Fert., Suppl. **35** (1987), 661–663

Comparisons of reproductive parameters of two Cape mountain zebra (*Equus zebra zebra*) populations

B. L. Penzhorn* and P. H. Lloyd†

Faculty of Veterinary Science, University of Pretoria, Private Bag X04, Onderstepoort, 0110, and †Cape Department of Nature and Environmental Conservation, Private Bag 5014, Stellenbosch, 7600, Republic of South Africa

The ecology of natural equid populations may offer clues to factors modulating reproduction. Cape mountain zebra (*Equus zebra zebra*) populations were studied in Mountain Zebra National Park (MZNP) (32°15′S, 25°41′E) and De Hoop Nature Reserve (DHNR) (34°26′S, 20°30′E), Cape Province, Republic of South Africa. Monthly distribution of births, age at first foaling, foaling interval and minimum age at last foal were determined. Surviving foals only could be recorded and the figures given here therefore represent minimum values. Eleven surplus stallions were available for research in MZNP; histological sections of their testes were examined. MZNP (6536 ha) is at a high altitude with harsh, dry winters (Penzhorn, 1982); the free-ranging mountain zebra population is maintained at about 220 individuals. DHNR is on the mild coastal plain, with winter rainfall; the expanding mountain zebra population (currently 30 individuals) is semi-free ranging over 3000 ha.

Reproductive lifespan

The mean age at first foaling of DHNR mares was 13 months less than in MZNP (Table 1) and the difference is significant (Satterthwaite modification of Student's *t* test; $P < 0.005$). The DHNR sample was quite small and the range was more restricted than that of the MZNP sample, which casts some doubt on the significance of the differences. First recorded conception of a MZNP filly was at 26 months, while in DHNR it was at 34 months. Two MZNP mares were recorded foaling for the last time at >21 years old. Adult mortality rates are low (Penzhorn, 1984a), which was also found in feral horse populations in Idaho (Eberhardt *et al.*, 1982).

Foaling interval

The means of the two populations are virtually identical (Table 1) (Student's *t* test; $P = 0.8$). Foaling intervals of older and younger mares are similar. On average, mares therefore foal every 2nd year of their reproductive lifespan. In feral horses in North America, at least 50% of the eligible mares were pregnant in consecutive years (Seal & Plotka, 1983).

Foaling season

In both populations foals were born all year round, with a peak in summer. MZNP receives 70% of its rainfall (annual mean 364 mm) in summer; conception occurs later in the season when the spring is dry (Penzhorn, 1985). DHNR is in a predominantly winter rainfall area (annual mean 378 mm). As expected, a peak of conceptions occurred during the southern hemisphere spring/early summer (October–December), when there was ample grazing after the wet winter (Fig. 1).

Onset of spermatogenesis

In MZNP no colt <5 years old succeeded in becoming a herd stallion (Penzhorn, 1984b). Spermatogenesis was absent in bachelor herd colts 24, 29 and 48 months old, but was starting in a

Table 1. Age at first foaling (months) and foaling interval (months) of Cape mountain zebras in the Mountain Zebra National Park (MZNP) and De Hoop Nature Reserve (DHNR)

	Sample	No.	Mean	Range	s.d.
Age at first foaling	MZNP	29	66·5	38–105	17·2
	DHNR	5	53·4	46– 58	5·2
Foaling interval	MZNP	37	22·1	13– 69	6·2
	DHNR	19	21·7	12– 38	7·0

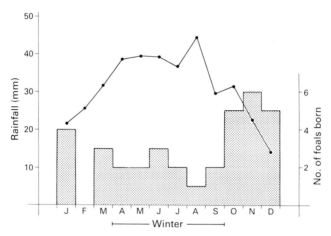

Fig. 1. Mean monthly rainfall (line) and total number of foals born (histogram) in De Hoop Nature Reserve. As the gestation period is ∼ 12 months, this figure also illustrates the monthly distribution of conceptions.

54-month-old bachelor. A DHNR colt running with mares in the absence of adult stallions sired a foal when 42 months old. Domestic horse stallions reach puberty at a younger age (Skinner & Bowen, 1968). The possible psychological inhibitory effect of running with bachelors or being dominated by a mature stallion has to be elucidated.

Conclusions

(1) There are no obvious significant differences in the reproductive patterns of the two populations living under different ecological conditions. The apparent lower mean age at first foaling in the DHNP population may be due to the small sample size, while the greater age of some MZNP mares (e.g. 105 months) may be due to their previous foal(s) not being recorded.

(2) With a low mortality rate and a reproductive lifespan of about 15 years, an average Cape mountain zebra mare could be expected to produce 8–9 foals during her lifetime.

We thank the University of Pretoria for financial support; National Parks Board; Dr P. M. Novellie; Dr N. J. van der Merwe; Mr R. J. Grimbeek; and Mrs I. Cornelius.

References

Eberhardt, L.L., Majorowicz, A.K. & Wilcox, J.A. (1982) Apparent rates of increase for two feral horse herds. *J. Wildl. Mgmt* **46**, 367–374.

Penzhorn, B.L. (1982) Habitat selection by Cape mountain zebras in the Mountain Zebra National Park. *S. Afr. J. Wildl. Res.* **12**, 48–54.

Penzhorn, B.L. (1984a) Observations on mortality of free-ranging Cape mountain zebras *Equus zebra zebra*. *S. Afr. J. Wildl. Res.* **14**, 89–90.

Penzhorn, B.L. (1984b) A long-term study of social organisation and behaviour of Cape mountain zebras *Equus zebra zebra*. *Z. Tierpsychol.* **64**, 97–146.

Penzhorn, B.L. (1985) Reproductive characteristics of a free-ranging population of Cape mountain zebra (*Equus zebra zebra*). *J. Reprod. Fert.* **73**, 51–57.

Seal, U.S. & Plotka, E.D. (1983) Age-specific pregnancy rates in feral horses. *J. Wildl. Mgmt* **47**, 422–429.

Skinner, J.D. & Bowen, J. (1968) Puberty in the Welsh stallion. *J. Reprod. Fert.* **16**, 133–135.

J. Reprod. Fert., Suppl. **35** (1987), 665

Photographic display of various anomalous and aberrant fluid-filled sacs which may confuse diagnostic procedures and the accuracy of ultrasonic echography in the determination of early pregnancy in the mare

Virginia E. Osborne

Department of Veterinary Anatomy, University of Sydney, New South Wales 2006, Australia

Ultrasound echography is now seen to play a broader diagnostic role in equine breeding management than its earlier use in the simple detection of early pregnancy, the determination of the presence of twin conceptuses and the possibilities of failed pregnancies. With advances in equipment type and the technique of its use, echogenic 'black holes' and various shades of the grey scale can offer clinical evaluation of the range of normal and abnormal physiological, cyclic and seasonal vagaries and pathology of the ovary and the entire reproductive tract.

In evidence of this, 36 colour photographs and 6 ultrasound images are offered to picture the array of well-circumscribed or distorted fluid-filled cavities, sacs or cysts which can be found associated with the ovary and the reproductive tract. These may mimic in size and shape the early blastocyst, a deteriorating conceptus or normal ovarian cyclic structures. Of these, the clinician must remain alert.

The ovary may show variations of the seasonal and oestrous cyclic phenomena such as multiple follicles; retained corpora lutea; vascular aberrations from multiple blood-filled follicles to the degree of massive follicular haematomata; or simple submesovarial 'lakes' or haematomata of the superficial venous return; neoplasia.

Associated with the ovary and its ligamentous attachment, residual embryonic anomalies of both the old mesonephric (Wolffian) and the paramesonephric (Müllerian) duct are common. Anterior to the ovary and along the suspensory ligament a scattering of cysts of various dimensions are commonly seen. These are anterior mesonephric remnants. Associated with the cranial pole of the ovary is found almost constantly an epoöphoron entity as a composite, multicystic mass of variable size, representative of the origin of the epididymis in the male. Of paramesonephric origin, ovarian cysts are common, lying within the ovulation fossa.

Anomalies of the oviduct may appear as residual embryological paramesonephric structures such as fimbrial cysts, arising from persistence of the multiple accessory ducts in the formation of the fimbriae of the infundibulum. The rudimentary cephalic tip of the paramesonephric duct persists as a vesicular appendage or hydatid of Morgagni. Aplasia may occur as a failure in the formation of part of the oviduct. One case is shown of a dilated cystic ampulla with no evidence of a fimbriated infundibulum or an external abdominal ostium. Cystic hydrosalpinx also occurs.

Cysts associated with the uterine endometrium are the structures most likely to confuse the diagnosis of early pregnancy or a deteriorating blastocyst. Lymphatic lacunae are common, some as a single, globular unit and others sacculated in a cluster of cysts. Some cysts may actually penetrate through the myometrium to intrude into the peritoneal cavity. One case is presented of a retention cyst dilating the old mesonephric duct within the wall of the oviduct, a continuation of a similar dilated duct as a paroöphoron entity within the mesosalpinx. Solid uterine neoplasia is common.

Cysts within the vagina are found as bilateral dilatations of the old mesonephric pathway as canals of Gartner.

I thank W. R. Allen for photographs of the blastocysts C and D and R. J. Rawlinson for the ultrasound images.

J. Reprod. Fert., Suppl. **35** (1987), 667–668

Fertility and pregnancy loss after delay of foal oestrus with progesterone and oestradiol-17β

R. J. Bell and F. Bristol

Department of Herd Medicine and Theriogenology, Western College of Veterinary Medicine, University of Saskatchewan, Saskatoon, Saskatchewan, Canada S7N 0W0

The gestation length in mares is approximately 340 days (Hintz *et al.*, 1979) and the first post-partum oestrus begins 7–10 days after parturition (Matthews *et al.*, 1967). Despite the risks of lower conception and higher incidence of pregnancy loss associated with breeding at this oestrus (Merkt & Gunzel, 1979; Loy, 1980), many mares are serviced during the first post-partum oestrus in an attempt to maintain a 12-month foaling interval. To determine whether fertility could be improved by a 5-day delay in the onset of the foal oestrus, the first post-partum oestrus was delayed in 58 mares with 5 daily injections of 150 mg progesterone and 10 mg oestradiol-17β beginning on the day of parturition; 58 untreated mares served as controls. All mares were artificially inseminated at the first post-partum oestrus and pregnancy was determined by ultrasound 14 days after ovulation. Pregnancy was monitored by weekly ultrasound examinations until Day 63 of gestation. Blood samples were taken weekly from Days 14 to 63 and at Days 90 and 150 of gestation to determine serum concentrations of CG and progesterone. Mares that did not conceive were re-mated and similarly examined.

The treatment delayed the onset of the first post-partum oestrus by 4·8 days and the number of days to the first ovulation by 6·0 days ($P < 0.001$). The conception rate at foal oestrus was 72% for treated mares and 59% in controls; in control mares that ovulated after 10 days *post partum*, it was 69%. The mean pregnancy rates at the end of the breeding season were the same for both groups (82%). The length and variability of the first post-partum oestrus, gestation length and foaling rate were not affected by the treatment. The mean number of days from parturition to conception (28 days) was not different in treated and control mares, and a 12-month foaling interval was maintained in both groups.

There was no difference in the incidence of pregnancy loss between treated and control mares but it was higher in both these groups (26%) than in mares that conceived at a later oestrus (16%) or in non-lactating mares (8%). The highest incidence (50%) of embryonic loss was observed between Days 14 and 21 of gestation and 78% loss occurred before Day 42. The mean width of the embryonic vesicle on Day 14 in mares that lost their pregnancy between Days 14 and 21 of gestation (18 mm) was significantly smaller ($P < 0.05$) than in mares that remained pregnant (22 mm).

The mean serum progesterone concentration was lower on Day 14 after ovulation in mares that were not pregnant and in mares that resorbed their pregnancy between Days 14 and 21 than in mares that remained pregnant. Changes in serum concentration of progesterone or CG could not be related to embryonic death at any other stage of gestation.

Although a 5-day delay of the onset of the first post-partum oestrus could not be recommended as a routine practice on the basis of this work, it may benefit mares that normally ovulate very soon after parturition or those that foal very late in the breeding season.

Supported by the Canadian Veterinary Research Trust Fund.

References

Hintz, H.F., Hintz, R.L., Lein, D.H. & Van Vleck, L.D. (1979) Length of gestation periods in Thoroughbred mares. *Equine Med Surg.* **3,** 289–292.

Loy, R.G. (1980) Characteristics of postpartum reproduction in mares. *Vet. Clin. North Am., Large Anim. Pract.* **2,** 345–359.

Matthews, R.G., Ropiha, R.T. & Butterfield, R.M. (1967) The phenomenon of foal heat in mares. *Aust. vet. J.* **43,** 579–582.

Merkt, H. & Gunzel A-R. (1979) A survey of early pregnancy losses in West German Thoroughbred mares. *Equine vet. J.* **11,** 256–256.

J. Reprod. Fert., Suppl. **35** (1987), 669–670

Epidemic of *Salmonella abortus equi* abortion in horses and some studies on the development of a vaccine

P. K. Uppal and J. M. Kataria

National Research Centre on Equines, Sirsa Road, Hisar (Haryana), India

Infection with *Salmonella abortus equi* is known to be responsible for abortion in India (Report 1919–20) and is considered to be one of the major causes of equine abortion in this country (Misra *et al.*, 1986). Rossdale & Ricketts (1976) have confirmed that *S. abortus equi* is one of the most important causes of abortion in horses.

To control abortion due to *S. abortus equi* amongst horses Dhanda *et al.* (1955) evolved a formalinized treated potash alum-precipitated vaccine against *S. abortus equi* which was an improvement over the earlier heat-killed vaccine being used in India from 1932–46 which included 12 strains of *S. abortus equi*. Since the duration of immunity with a single dose of the vaccine was short lived for 2 months, repeated vaccinations were advocated in mares.

To overcome this difficulty, it was necessary to search for a better vaccine with a freshly isolated strain from the epidemic of *Salmonella abortus equi* vaccine. In the past, various inactivating agents such as alcohol, acetone, phenol, diethyldithiocarbomate and formalin as well as heat treatment have been used to develop such vaccines.

Amongst these, formalin has been preferred over other inactivating agents for the production of *S. abortus equi* vaccine (Spitznagel & Trainer, 1949; Dhanda *et al.*, 1955; Ueda *et al.*, 1963). Kataria & Uppal (1981) reported that beta-propiolactone and glutaraldehyde could be used as alternative inactivants for the preparation of *S. abortus equi* vaccine. The present study was to compare the protective value and immunological response of *S. abortus equi* vaccine inactivated with formaldehyde or glutaraldehyde or beta-propiolactone which were incorporated with alhydrogel or chrome alum or single oil adjuvant containing lanolin, paraffin or double oil adjuvant containing lanolin, paraffin and Tween 80.

Materials and methods

An epidemic of abortion in one of the Army horse breeding studs resulted in abortion in 60% of the donkey mares, 18% of horse breeding mares, 12% of mountain artillery mares and 6% of general service mules. In this epidemic, histopathological examination eliminated equine *Herpes* virus as well as other pathogens responsible for abortion, but *S. abortus equi* was isolated from 46 samples out of 63 examined. All these isolates were confirmed at the National Salmonella Centre (Veterinary), Indian Veterinary Research Institute, Izatnagar. These were identified as *S. abortus equi* and the strain designated E814, which regularly killed mice after i.p. injection of 0·5 ml inoculum, was used for vaccine preparations.

Highly motile seed culture propagated in nutrient broth was inoculated in roux flasks containing agar. After checking the purity, the growth was washed with physiological saline (9 g NaCl/l) and the suspension was placed in a Brown's opacity tube No. 4. Bacterial suspension was divided into three batches: one was treated with 0·25% formaldehyde, one with 1% beta-propiolactone, and the third with glutaraldehyde. Each batch of the vaccine was incorporated with four different adjuvants. In all, 12 batches of different vaccines were prepared and were tested in mice for protection test and in guinea-pigs for serological response.

Results

The challenge results of various vaccines are shown in Table 1. Not less than 15 mice were used for each batch of the vaccine and 10 controls were included.

Table 1. Results of challenge exposure (no. surviving/no. tested) after vaccination trials in mice

Inactivating agents	Vaccine adjuvants				
	Aluminium hydroxide	0·2% Chrome alum	Single-oil emulsion	Double-oil emulsion	Survival of controls
Formaldehyde	13/15	13/15	11/15	9/15	0/10
Glutaraldehyde	13/15	13/15	11/15	9/15	1/ 9
Beta-propiolactone	14/15	15/15	12/15	12/15	0/10

Table 2. Results* of tube agglutination and indirect haemagglutination test with sera of guinea-pigs immunized with different *Salmonella abortus equi* vaccines after 14 days

Inactivating agents	Test†	Vaccine adjuvants			
		Aluminium hydroxide	0·2% Chrome alum	Single-oil emulsion	Double-oil emulsion
Formaldehyde	O	40	40	20	20
	H	160	320	160	160
	IHA	80	80	20	20
Glutaraldehyde	O	80	160	80	40
	H	320	160	320	320
	IHA	160	160	40	40
Beta-propiolactone	O	160	80	80	40
	H	320	320	320	320
	IHA	640	320	160	160

*Figures represents antibody titre.
†O = Somatic antibody titre; H = flagellar antibody titre; IHA = indirect haemagglutination titre.

Of the various adjuvants used for the preparation of the vaccine, chrome alum or aluminium hydroxide gel showed better results than single-oil or double-oil adjuvant vaccine. Of the inactivants used, beta-propiolactone showed better results than did glutaraldehyde and formalin. These findings were reasonably well supported by the serological response measured by tube agglutination or indirect haemagglutination tests in the sera of guinea-pigs treated with the various vaccines. Groups of 4 guinea-pigs were used for each vaccine and their sera were pooled and were subjected to serological tests before and after 14 days of vaccination (Table 2). None of the sera before vaccination showed any antibody against *S. abortus equi* antigen.

Conclusion

Preliminary studies on the development of an *S. abortus equi* vaccine with a freshly isolated strain from an epidemic in India showed complete protection in mice vaccinated with beta-propiolactone-inactivated chrome alum vaccine, showing its superiority over formalin-treated vaccine. Dhanda *et al.* (1955) reported 86% protection with formalin-treated vaccine in mice.

References

Dhanda, M.R., Lall, J.M. & Singh, M.M. (1955) *Indian J. vet. Sci.* **25**, 245.

Kataria, J.M. & Uppal, P.K. (1981) *Ind. vet. J.* **58**, 839–854.

Misra, V.C., Gupta, R.S. & Verma, R.D. (1968) *Jl Roy. vet. Coll.* **7**, 35–46.

Report (1919–20) *Cattle Livestock Farm*, Hissap, India.

Rossdale, P.D. & Ricketts, S.W. (1976) *Vet. Ann.* **16**, 133–141.

Spitznagel, F. & Trainer, R.Y. (1949) *J. Immunol.* **62**, 229.

Ueda, S., Matsumoto, K. & Soekawa, M. (1963) *Kitamato Archs exp. Med.* **36**, 1–11 and 13–16.

J. Reprod. Fert., Suppl. **35** (1987), 671–673

Printed in Great Britain
© 1987 Journals of Reproduction & Fertility Ltd

The prevalence of *Mycoplasma* spp. and their relationship to reproductive performance in selected horse herds in southern Ontario

V. Bermudez*, R. Miller*, W. Johnson†, S. Rosendal‡ and L. Ruhnke‡

*Departments of *Pathology, †Theriogenology, and ‡Veterinary Microbiology and Immunology, University of Guelph, Guelph, Ontario, Canada N1G 2W1*

Materials and methods

Six breeding farms were asked to participate in this survey and all of them accepted. The survey was carried out between January and July of the 1985 breeding season. All mares on each of the farms were sampled regardless of their reproductive status. A single Culturette (Canlab, McGraw Supply, U.S.A.) swab from the clitoral fossa of each of the 337 mares was taken and, immediately after collection, each swab was placed in a screw-capped plastic microtube and submerged in liquid nitrogen. The swabs were cultured for *Mycoplasma* spp. on agar plates of modified medium B (Erno & Stipkovits, 1973a, b) and for ureaplasmas on complete *Ureaplasma* agar plates (as modified by Ruhnke *et al.*, 1978). For indirect culture semi-solid B medium and ureaplasma broth were used. Inoculated plates of modified medium B were incubated at 37°C in 5% carbon dioxide atmosphere (CO_2 incubator) for up to 9 days. *Ureaplasma* plates were incubated at 37°C using the anaerobic GasPak system (Oxoid Canada Inc., Ottawa). The presence of mycoplasmas was scored as negative if no colonies on plates, low growth (1–10 colonies), moderate (10–100) and heavy (> 100). Pre- and post-ejaculatory urethral swabs and semen cultures were done on some stallions from sampled farms and stallions that came to the College for breeding soundness evaluation in 1985. *Mycoplasma* spp. were identified by using the indirect immunofluorescence test as described by Rosendal & Black (1972). Blocks from positive plates were stored at −70°C for further identification. A questionnaire was used to collect information about the reproductive history of each sampled mare during 1984 and 1985 breeding seasons.

Statistical analysis. A test on the difference between proportions ($Z > 1·96/P < 0·05$) (proportions of *Mycoplasma* spp.) of two binomial distributions was done to determine difference between farms in the proportion of mares colonized by *Mycoplasma* spp. The odds ratio ($< 0·5, > 2$) and χ^2 ($> 3·841$) tests were used to evaluate the strength of association between collected reproductive data and *Mycoplasma* spp. colonization. The conception rate, the number of services (two factorial (5×2) analysis of variance) and oestrus per conception were calculated and the values compared between the mycoplasma-infected and the mycoplasma-free groups. The Mantel–Haenszel technique was used to control statistically the 'farm effect' (confounder) for analysis of the relationship between infertility and colonization of the clitoral fossa by mycoplasmas.

Results and discussion

Mycoplasma spp. were isolated from the clitoral fossa of 114 (33·7%) of 337 mares. Ureaplasmas were not isolated in this study. Table 1 summarizes the prevalence rates by farm and the number of isolates recovered from the clitoral fossa of the 337 mares. The highest prevalence of mycoplasma was found in the pony mares (74·4%) followed by the Thoroughbreds (48·4%) (natural breeding) and Standardbreds (18%) (artificial insemination). *Mycoplasma subdolum* (Lemcke & Kirchhoff, 1979) was more frequently isolated from Standardbred and Thoroughbred mares than *M. equigenitalium* whereas from the pony mares *M. equigenitalium* was more frequent that *M. subdolum*. These results

Table 1. The prevalence rate by farm and the number of isolates recovered from the clitoral fossa of 337 mares

Breed	Farm	M. equigenitalium (Me)	M. subdolum (Ms)	Mixed cultures of Me/Ms	Unidentified Mycoplasma spp.	A. hippikon	Prevalence rate (%)	Total positive mares
Standardbred (199 mares)	1	0	18	0	6	1	13·6	19
	2	1	3	0	0	0	14·3	4
	3	0	13	0	3	0	14·9*	13
Thoroughbred (95 mares)	4	10	17	0	0	0	41·5	27
	5	7	4	8	1	0	63·3	19
Pony (43 mares)	6	18	4	5	5	0	74·4	32

*$P < 0.05$.

Table 2. The list of stallions from Farm 1 and the number of *Mycoplasma*-infected (MI) and clean (C) mares mated and the conception rates (CR) including the overall conception rate by stallion in the breeding season of 1985

Stallions	Total mares mated in sampled population	CR (%)	Mycoplasma-infected			Mycoplasma-clean		
			Pregnant	Non-pregnant	CR (%)	Pregnant	Non-pregnant	CR (%)
F11 (MI)	1	—	1	0	100	0	0	—
F12 (MI)	1	—	1	0	100	0	0	—
F13 (C)	46	80·4	3	0	100	34	9	79·1
F14 (C)	17	64·7	2	2	50	9	4	69·2
F15 (C)	28	71·4	3	2	60	17	6	73·9
F16 (MI)	37	91·9	4	1	80	29	3	90·6

were significantly different ($P < 0.05$) from Farm 3 when it was compared to the other two Standardbred farms (Farms 1 and 2). Mycoplasmas were more frequently recovered from pregnant pony mares ($P < 0.05$) than from barren pony mares. Standardbred and Thoroughbred pregnant and barren mares were more likely to be colonized by mycoplasma than were the maiden mares first mated this season (1985). In the overall population sampled, pregnant mares had a higher prevalence of mycoplasmas than that observed in barren and maiden mares (first mated in the breeding season of 1985). Progesterone during pregnancy could perhaps increase the risk of mycoplasma colonization by affecting local immunity. Eleven virgin mares sampled were free from mycoplasmas. Analysis of the epidemiological data revealed a 'farm effect' as a possible cause of individual farm variation in regard to conception rates, services, and oestrus per conception and probably also the prevalence of mycoplasmas. The overall conception rate (Table 2), services and oestrus per conception values were within normal range. The means of the services per conception differed slightly by farm ($P < 0.07$), but they did not between mycoplasma-infected and clean groups ($P < 0.46$). There was no association between infertility and mycoplasma recovery from the clitoral fossa (adjusted odds ratio = 0·976; $\chi^2 = 0.0005$); nevertheless, it does not mean that the recovery of such organisms from the uterus could not be of significance. Variables such as husbandry, genital sanitation and breeding management are perhaps the most important factors explaining farm variation. It is likely that an artificial insemination system could decrease the risk of a mycoplasma colonization in the genital tract of both mare and stallion. The 'age factor' showed that mares at any age were at risk of being colonized by a mycoplasma between 1 and 24 years of age. An increase in the isolation rate of mycoplasma was observed in mares between 1 and 12 years (peaking at 10–12 years) and was decreased in mares between 13 and 21 years.

Mycoplasmas were isolated from 12 (52·2%) of 23 stallions. The pre-ejaculatory urethral swabs yielded the highest isolation prevalence of mycoplasma (43·5%) as compared to that obtained from post-ejaculatory urethral swabs (30·4%) and semen samples (17·4%). *Mycoplasma subdolum* and *M. equigenitalium* were the most common isolates, but *M. subdolum* was more frequently isolated. In the stallion samples, the parameters of breeding soundness evaluation were within normal range. The epidemiological data and the recovery of mycoplasma in stallions from Farms 1 and 2 did not indicate a straight line association for 'breeding transmission of mycoplasmas'.

In summary, the results of this survey showed that the clitoral fossa in the mare and the urethra in the stallion are the most likely 'ecological niches' of mycoplasmas in the genital tract. *Mycoplasma equigenitalium* and *M. subdolum* isolated from the lower genital tract were not shown to be associated with poor reproductive performance in mares or stallions during the breeding season of 1985. 'Farm effect' was the most likely cause of individual mare variation observed in this study.

We thank the Ontario Racing Commission for funding this project; Mrs Sheila Watson for technical help and kind co-operation in the Mycoplasma Laboratory; and breeding farm owners and veterinarians for allowing mare sampling on their farms.

References

Erno, H. & Stipkovits, X. (1973a) Bovine mycoplasma: cultural and biochemical studies. I. *Acta vet. scand.* **14**, 435–449.

Erno, H. & Stipkovits, X. (1973b) Bovine mycoplasmas: cultural and biochemical studies. II. *Acta vet. scand.* **14**, 450–463.

Lemcke, R.M. & Kirchhoff, H. (1979) *Mycoplasma subdolum*, a new species isolated from horses. *Int. J. Systemic Bacteriol.* **29**, 42–50.

Rosendal, S. & Black, F. (1972) Direct and indirect immunofluorescence of unfixed and fixed mycoplasma colonies. *Acta vet. scand.* **B80**, 615–622.

Ruhnke, H.L., Doig, P.A., MacKay, A.L., Gagnon, A. & Kierstead, M. (1978) Isolation of ureaplasma from bovine granular vulvitis. *Can. J. comp. Med.* **42**, 151–155.

J. Reprod. Fert., Suppl. **35** (1987), 675–677

Dynamic changes which accompany acute and chronic uterine infections in the mare

J. M. Bowen, W. B. Ley, L. Jones*, J. Sutherland*‡, C. de Barros†§,
H. Bergeron†¶, M. Petrites-Murphy†‖, D. Bollinger†, L. Kloppe† and
D. G. Pugh†**

*Virginia–Maryland Regional College of Veterinary Medicine, Virginia Polytechnic Institute &
State University, Blacksburg, Virginia 24061, *Texas Veterinary Medical Diagnostic Laboratory,
and †Large Animal Department, College of Veterinary Medicine, Texas, A&M University,
College Station, Texas 77843, U.S.A.*

The need to be able to differentiate whether or not a bacterium found on culture is significant on a particular occasion is an old one (Asbury, 1984), as many potentially pathogenic bacteria can be inadvertently recovered from the skin of both mares and handlers. Today's improved techniques, such as 'on-farm' plate innoculation, or the extensive use of a transport medium to convey the bacterial swabs back to the laboratory, have increased our need for knowledge of the real patho-logical changes, if any, which are occurring within the uterus of the problem mare (Couto & Hughes, 1984). Endometrial biopsy can be of some use in this regard, but it is traumatic, as evidenced by its effects on the oestrous cycle of the mare, and takes time to prepare and skill to interpret (Ricketts, 1975; Kenney, 1975; Doig *et al.*, 1981).

Endometrial cytology in contrast, is a quick, easy 'on-farm' test, which can be examined immediately, and can be more readily interpreted by the practising veterinarian (Wingfield-Digby, 1978; Couto & Hughes, 1984; Freeman *et al.*, 1986). This poster sets out to demonstrate changes which occur in the course of an intrauterine infection, and differentiates the acute from the chronic reaction (Fig. 1).

Materials and methods

A sterile moistened swab (Culturette: Marion Scientific, Kansas City, MO) is introduced asepti-cally into the lumen of the uterus and allowed to remain in the lumen for about 20 sec. A moistened swab greatly enhances the integrity of the recovered cells in the smear, making their morphology clearer, their identification more positive, and the use of a gelatine-coated glass slide superfluous. The swab is agitated from side to side or up and down during its time within the uterus. Circular motions with the swab should be avoided as this tends to scrape sheets of cells from the uterine wall and give a false impression of the cellular constitution of the uterine lumen. The swab is then withdrawn and rolled slowly but firmly over a glass microscope slide, and after air-drying is stained with a commercial rapid stain, Diff-Quik (American Scientific Products, McGraw Park, IL).

Interpretation

Normal cells from the uterine lumen, the cervix and often the vagina are usually recovered. The morphology of each of these normal cells is shown on the poster. With the introduction of an

---

‡Present address: School of Veterinary Studies, Murdoch University, Perth, Western Australia, Australia.
§Present address: R. Des Saul de Gusmaõ 92, Rio de Janeiro, CEP 22600, Brasil.
¶Present address: 1733 Creekview, DeSoto, Texas, 75115, U.S.A.
‖Present address: Livestock Disease Diagnostic Center, 1429 New Town Pike, Lexington, Kentucky 40511, U.S.A.
**Present address: College of Veterinary Medicine, University of Georgia, Athens, Georgia 30602, U.S.A.

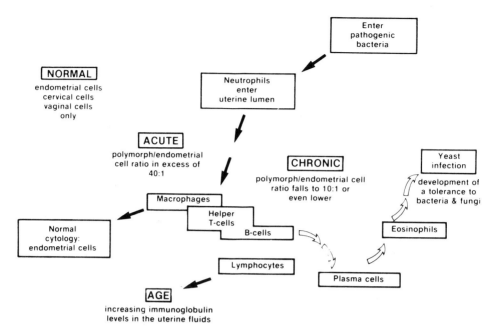

Fig. 1. The dynamic changes which accompany acute and chronic uterine infections in the mare: chart showing the sequence of intrauterine cellular changes which follow the entry of bacteria into the uterus.

infectious agent into the uterus of a normal maiden mare, we find that there is a rapid extravasation of neutrophils into the uterine lumen. These cells are the primary defensive mechanism responding to a chemotactic stimulus. Phagocytosis occurs and many neutrophils die in the process; this serves to stimulate the ingress of further neutrophils into the uterine space. In an acute uterine infection, the ratio of neutrophils to endometrial cells will be in excess of 40:1.

The use of a neutrophil/endometrial cell ratio is a better indicator of the cellular representation on any one slide, as slide preparations and their cellularity vary considerably, and the number of neutrophils per high-powered field is a less consistent quantitative guide.

In a normal mare this massive arrival of neutrophils is sufficient to overcome the invading bacterial organisms, and they are eliminated rapidly. The only other defensive cell that is seen in the uterus during an acute infection is the macrophage, which appears by the 3rd or 4th day after infection; its function is to clear the debris created by the dying neutrophils. By the 6th or 7th day after the introduction of an infectious agent, the normal uterus has cleared this infection, and the cellular contents have returned to normal.

In the problem mare the situation is more complicated. Here the uterus has been repeatedly insulted by an almost continuous flow of bacteria of all sorts entering the uterus. With time and the continuing and changing population of invading bacteria, the secondary defence mechanism, the immune system, is stimulated. Macrophages which have phagocytosed bacteria, will combine with the assistance of 'helper T-cell' lymphocytes to transform 'B-cell' lymphocytes into plasma cells, which are rich in the immunoglobulins (Tizard, 1986a, b). These are essential for the initiation of the 'complement cascade'. Endometrial smears from these mares show fewer neutrophils but a greater number of macrophages with many plasma cells and lymphocytes. The neutrophil/endometrial cell ratio will fall to 10:1 or less, and in the chronically infected mare even to 1:1.

While in a normal mare this would indicate that she was on the way to recovery, in a problem mare the presence of lymphocytes and plasma cells alongside the neutrophils indicates a chronic endometritis.

The sighting of eosinophils in the endometrial smear preparation is an indication of an impairment of the immune system, eventually leading to the development of a tolerance to the persistent presence of bacteria. This is a very poor prognostic indicator.

The use of endometrial cytology in brood mare management allows for: (1) a quicker and more accurate assessment of the uterine environment, (2) the elimination of unnecessary intrauterine therapy, (3) the identification of problem mares which need special treatment if they are to conceive, and (4) an assessment of the efficacy of any treatment regimen.

References

Asbury, A.C. (1984) Endometritis diagnosis in the mare. *Equine vet. Data* **5**, 166.

Couto, M.A. & Hughes, J.P. (1984) Techniques and interpretation of cervical and endometrial cytology in the mare. *J. equine vet. Sci.* **4**, 265–273.

Doig, P.A., McKnight, J.D. & Miller, R.B. (1981) The use of endometrial biopsy in the infertile mare. *Can. vet. J.* **22**, 72–76.

Freeman, K.P., Roszel, J.F. & Slusher, S.H. (1986) Equine endometrial cytologic smear patterns. *Comp. Cont. Ed.* **8**, 349–360.

Kenney, R.M. (1975) Prognostic value of endometrial biopsy of the mare. *J. Reprod. Fert., Suppl.* **23**, 347–348.

Ricketts, S.W. (1975) The technique and clinical application of endometrial biopsy in the mare. *Equine vet. J.* **7**, 102–108.

Tizard, I. (1986a) Basic immunology. 1: The structure and function of antigens. *Vet. Med.* **81**, 55–62.

Tizard, I. (1986b) Basic immunology. 2: Reviewing the properties of antibodies. *Vet. Med.* **81**, 166–178.

Wingfield-Digby, N.J. (1978) The technique and clinical application of endometrial cytology in the mare. *Equine vet. J.* **10**, 167–170.

J. Reprod. Fert., Suppl. **35** (1987), 679

Comparison of histological and cytological recognition of equine endometritis requiring specific therapy

J. F. Roszel, K. P. Freeman and S. H. Slusher

*Oklahoma State University, College of Veterinary Medicine, Stillwater, Oklahoma 74078-0170,
U.S.A.*

We are currently comparing the clinical value of endometrial cytology specimens collected and stained in a manner previously reported (Slusher *et al.*, 1984) with the concurrently collected biopsy tissue specimens. It has become apparent that cytological preparations, in some instances, contain structures difficult to identify or absent in the histological specimens. Amongst our first clinical specimens (~ 350), there were 13 cytological diagnoses of uterine urine pooling (7 cases) and fungal infections (6 cases). All corresponding tissue sections indicated an endometritis, but the aetiological factor was seen in only 1 of the 13 cases. This was a fungal infection in which hyphae were seen with a special fungal stain. This stain would not have been used had it not been for the cytological diagnosis.

In addition to the presence of the aetiological agents in the cytology smears, the components of the specimens representative of urine pooling and fungal infections differed from one another as indicated in Table 1 and from endometritis due to other causes. This was supported by 6 cases with the clinical suspicion of uterine urine pooling which were not substantiated by the cytological findings. On the basis of the cytology report, all 5 cases were treated with medical therapy for bacterial endometritis rather than surgery and all 6 cases responded well to medical treatment.

Table 1. Comparison of components in uterine washings associated with urine pooling and fungal infections (scores from $-$ to $+++$)

Components	Urine pooling	Fungal infections
Urine crystals	$\pm$ to $+++$	$-$
Fungal elements	$-$	$\pm$ to $+++$
Atypical non-secretory endometrial cells	$++$	$\pm$
Atypical secretory endometrial cells	$\pm$	$++$
Atypical vaginal epithelial cells	$\pm$ to $+$	$-$ to $\pm$
Macrophages	$\pm$	$+$
Neutrophils	$+++$	$++$
Erythrocytes	$\pm$	$\pm$
Lymphocytes	$\pm$ to $++$	$\pm$ to $++$
Eosinophils	$+$	$+$
Mucus	$\pm$	$\pm$
Protein background	$\pm$	$\pm$

Fungi were isolated in cultures from 5 of the 6 mares that had fungal elements in the uterine cytology specimens. The cytological findings preceded the results of culture reports by 1–7 days.

Cytological specimens of aspirations from hollow organs contain primarily luminal contents and lining epithelium. The sensitivity of cytology compared to histology in recognizing endometrial urine pooling and fungal infections is presumably due to the superficial involvement of the uterus in such cases.

Reference

Slusher, S.H., Freeman, K.P. & Roszel, J.F. (1984)
Eosinophils in equine uterine cytology and histology
specimens. *J. Am. vet. med. Ass.* **184,** 665–670.

J. Reprod. Fert., Suppl. **35** (1987), 681

Effect of ovarian hormones on opsonins in uterine secretions of ovariectomized mares

E. D. Watson†, C. R. Stokes, J. S. E. David* and F. J. Bourne

Departments of Veterinary Medicine and *Veterinary Surgery, School of Veterinary Science, University of Bristol, Langford House, Langford, Bristol BS18 7DU, U.K.

The bactericidal and phagocytic activity of blood neutrophils suspended in uterine washings and rate of mobilization of neutrophils into the uterine lumen were studied. Ovariectomized mares received daily intramuscular injections of oestradiol benzoate (1 mg; N = 4), progesterone (100 mg; N = 4) or the oily vehicle arachis oil (N = 4) for 14 days. At 7 days after the first injection acute endometritis was induced by intrauterine infusion of a sterile solution of oyster glycogen (1%) and uterine washings (40 ml phosphate-buffered saline) were performed sequentially up to 144 h after infusion. Rate of mobilization of neutrophils into the uterine lumen was similar in all three groups in the first 6 h. However, by 24 h there were significantly more neutrophils present in the washings from progesterone-treated mares ($P < 0.01$) than from the other two groups of mares.

Blood neutrophils suspended in uterine washings from progesterone-treated mares had significantly lower bactericidal activity ($P < 0.001$) against *Streptococcus zooepidemicus* than when suspended in washings from the other two groups. Overall, there was a decrease in bactericidal activity of washings collected between 0 and 24 h after infusion ($P < 0.01$). However, in the oestrogen-treated group bactericidal activity remained high until 144 h after infusion. In progesterone-treated mares bactericidal activity remained very low between 24 and 144 h and 2 of these mares succumbed to uterine infection by 144 h, one with β-haemolytic *Streptococcus* and the other with *Staphylococcus aureus*.

Opsonization of yeast blastopores with uterine washings significantly enhanced phagocytosis compared with opsonization with phosphate-buffered saline ($P < 0.001$). The alternate pathway of complement activation is therefore functional in uterine washings. Yeast opsonization by washings was not affected by hormone treatment or the induction of endometritis.

It is possible that progesterone decreased the specific antibody response to *S. zooepidemicus* or affected concentrations of the bactericidal proteins present in washings. Alternatively, the large numbers of neutrophils present in the uterine lumen of progesterone-treated mares by 24 h after infusion could have increased the breakdown of opsonizing antibodies by release of neutrophil proteases. There was no evidence that hormone treatment influenced activation and function of complement in washings.

This study was supported financially by a grant from the Horserace Betting Levy Board.

†Present address: University of Pennsylvania, Section of Reproductive Studies, New Bolton Center, Kennett Square, PA 19348, U.S.A.

J. Reprod. Fert., Suppl. **35** (1987), 683–684

Investigation of the potential of LHRH or an agonist to induce ovulation in seasonally anoestrous mares with observations on the use of the agonist in problem acyclic mares

B. P. Fitzgerald, K. J. Affleck, R. Pemstein* and R. G. Loy

Department of Veterinary Science, University of Kentucky, Lexington, KY 40546-0076, and
*Hagyard, Davidson & McGee, Lexington, KY 40511, U.S.A.

During the breeding season, a small proportion of horse mares may exhibit reproductive inactivity. This phenomenon has been associated with but not confined to the period *post partum*. Irrespective of reproductive status (i.e. barren, maiden or foaling mare), a paucity of information exists regarding the cause of reproductive inactivity. However, it may be proposed that, in some instances, low circulating gonadotrophin concentrations may be responsible.

Luteinizing hormone-releasing hormone (LHRH) has been used to induce ovulation in seasonally anoestrous mares which, before treatment, exhibit low serum concentrations of luteinizing hormone (LH) (Johnson, 1986a, b). However, a regimen of treatment appropriate for practical application has not been described. Furthermore, it is unknown whether LHRH may be successfully used to induce ovulation in anovulatory mares during the breeding season.

In the present study, dose and frequency of administration of LHRH or agonist were selected to induce an increase in serum LH concentrations comparable to the amplitude and frequency of endogenous pulses of LH reported previously for mares during the transition into the breeding season (Murdoch *et al.*, 1985; Hines *et al.*, 1986; Safir *et al.*, 1987; Fitzgerald *et al.*, 1987). Accordingly, in an initial investigation administration of LHRH (200 µg, i.v.) or LHRH agonist (100 µg, i.v.; Lutrelin, Wyeth Pharmaceuticals, Philadelphia, PA, U.S.A.) every 12 h induced ovulation in 0/4 and 3/3 mares, respectively. In a second study conducted at the same time the following year, 2/4 and 3/4 mares receiving LHRH every 12 h or 6 h, respectively, ovulated within 13–19 days of start of treatment. However, as in the previous year, in 5/5 mares injected with 100 µg LHRH agonist ovulation occurred between 8 and 14 days of treatment. From a total of 10 untreated

Table 1. Results for acyclic mares treated with 100 µg LHRH agonist every 12 h

	Mare class		
	Barren	Maiden	Foaling
No. used	34	37	37
No. ovulating (%)	25 (76)	30 (81)	31 (83)
Mean interval (±s.e.m.) to ovulation (days)	14·5 ± 1·6	9·7 ± 0·8	9·7 ± 1·02
Proportion of mares pregnant	10/23	14/29	13/29
Non-pregnant mares that exhibited oestrous cycles (%)	12 (92)	14 (92)	14 (87)
No. of mares not mated that exhibited oestrous cycles	2*/3	3/3	2*/3
Proportion of non-ovulatory mares that exhibited oestrous cycles	1/8	2/7	0/6
Duration of days of treatment (range) of non-ovulatory mares	7–34	11–31	15–30

*Data unavailable for 1 mare.

control mares, 2 ovulated towards the end of the experimental period. The results suggest that administration of an LHRH agonist, at the dose and frequency used in this study, is a reliable method for inducing ovulation in anoestrous mares.

A second study examined the potential of the LHRH agonist to induce ovulation in problem acyclic mares during the breeding season. Accordingly, as shown in Table 1, an LHRH agonist (100 µg i.v. or i.m.) was administered every 12 h to 37 maiden, 34 barren and 37 foaling mares exhibiting acyclicity. From this total (N = 108), 80% of the mares ovulated. Within each category, the mean interval to occurrence of ovulation was similar and comparable to the range observed in seasonally anoestrous mares (i.e. 9–14 days; Table 1). Conception rates to the agonist-induced ovulation ranged between 43 and 48%. Within each category, 87–93% of mares that did not conceive subsequently exhibited oestrus and were remated in the absence of further treatment.

It is concluded that treatment of seasonally anoestrous mares and problem acyclic mares during the breeding season with an LHRH agonist is a practical, and reliable method of inducing ovulation. Conception rates to the induced ovulation appear satisfactory. More important, however, in the absence of conception a high proportion of mares subsequently exhibit oestrous cycles.

We thank Dr Frederick Bex and Wyeth Pharmaceuticals for donations of LHRH and Lutrelin; collaborating veterinarians from the Central Kentucky region; and the Grayson Foundation, Inc. for financial support. This preliminary report (No. 86-8-000) was in connection with a project of the Kentucky Agriculture Experiment Station and is published with approval of the director.

References

Fitzgerald, B.P., Affleck, K.J., Barrows, S.P., Murdoch, W.L., Barker, K.B. & Loy, R.G. (1987) Changes in LH pulse frequency and amplitude in mares during the transition into the breeding season. *J. Reprod. Fert.* **79**, 485–493.

Hines, K.K., Affleck, K.J., Barrows, S.P., Murdoch, W.L., Barker, K.B., Loy, R.G. & Fitzgerald, B.P. (1986) Evidence that a decrease in follicle stimulating hormone (FSH) pulse amplitude accompanies the onset of the breeding season in the mare. *Biol. Reprod.* **34**, Suppl. 1, 270, Abstr.

Johnson, A.L. (1986a) Induction of ovulation in anestrous mares with pulsatile administration of gonadotrophin-releasing hormone. *Am. J. vet. Res.* **47**, 983–986.

Johnson, A.L. (1986b) Induction of follicular development and ovulation in seasonally anoestrous mares by pulsatile administration of gonadotropin-releasing hormone (GnRH). *Biol. Reprod.* **34**, Suppl. 1, 150, Abstr.

Murdoch, W.L., Affleck, K.J., Barrows, S.P., Loy, R.G. & Fitzgerald, B.P. (1985) Evidence of an increase in luteinizing hormone (LH) pulse frequency in intact mares during the transition from anestrus to the breeding season. *Biol. Reprod.* **32**, Suppl 1, 266, Abstr.

Safir, J.M., Loy, R.G. & Fitzgerald, B.P. (1987) Inhibition of ovulation in the mare by active immunization against LHRH. *J. Reprod. Fert., Suppl.* **35**, 229–237.

J. Reprod. Fert., Suppl. **35** (1987), 685–686

Effects of altered steroid environment on daily concentrations of LH and FSH in intact mares

P. J. Silvia, M. J. Evans*, S. P. Barrows and R. G. Loy

Department of Veterinary Science, University of Kentucky, Lexington, KY 40546-0076, U.S.A. and
*Endocrine Unit, the Princess Margaret Hospital, Christchurch, New Zealand

Methods

Cyclic mares were assigned randomly to 1 of 4 groups; controls (N = 7; Group C), 10 mg PGF-2α (Prostin; Upjohn Co., Kalamazoo, MI) daily for 8 days (N = 6; Group PG), 10 mg oestradiol (in cotton-seed oil) daily for 30 days (N = 7; Group E) or PGF-2α + oestradiol (N = 6; Group PG + E). Injections (i.m.) of PGF-2α and oestradiol began on the day of ovulation (Day 0). Jugular blood samples were taken daily, before each injection, and continued for 6 days after the next ovulation. Serum was harvested and frozen until analysis. Concentrations of LH, FSH and progesterone in serum were quantified by RIA. Effects of treatment on concentrations of LH and FSH were analysed statistically by comparing means of 8-day periods (e.g. Period 1 = Days 0–7, Period 2 = Days 8–15 etc.).

Results and discussion

Relative to controls, the interval from start of the experiment (Day 0) to the next ovulation was increased in Group PG + E. Except for 1 mare, all mares in Group E ovulated by Day 36, yielding a mean interval to ovulation not statistically different from controls (Table 1). Concentrations of progesterone in the two groups receiving PGF-2α were significantly ($P < 0.01$) lower than in the two groups not receiving PG (Table 1). Oestrous behaviour in the Groups E and PG + E during the 30-day treatment period was observed, on average, for 16·7 (55·7%) and 27·8 (92·7%) days respectively. Although individual responses varied, 5/7 mares in Group E ovulated before the end of treatment while 5/6 mares in Group PG + E ovulated >10 days after the end of treatment. Follicular structures in all mares were ≤15 mm in diameter until <10 days before ovulation. The time from Day 0 to the day concentrations of progesterone declined to ≤1 ng/ml was 13·3 and 12·4 days for mares in Groups C and E respectively.

During Period 1 LH was not different ($P > 0.5$) between Groups PG and PG + E or Groups C and E, but means for Groups PG and PG + E tended to be higher ($P = 0.14$). This tendency reached significance during Period 2 ($P < 0.001$) when means (±s.e.m.) were 5·6 ± 1·5, 14·2 ± 2·6, 6·0 ± 1·6 and 20·1 ± 2·9 ng LH/ml for Groups C, PG, E and PG + E, respectively. Concentrations of LH in mares in Group PG + E remained elevated until after the next ovulation. Although not statistically different, the mean for mares in Group PG + E during Period 2 appeared greater than that for mares in Group PG. The area under the curve for LH concentrations associated with the next ovulation, was significantly greater ($P < 0.02$) in Group E mares than in controls. In Group E mares the rise, from baseline, in mean daily concentrations of LH preceded that in controls by about 5 days. After ovulation the pattern and concentration of LH in daily samples for mares in Groups E and C was similar. Mean (±s.e.m.) concentrations of FSH were lower ($P < 0.01$) in Group E (10·1 ± 2·3 ng/ml) during Period 1 than in Groups C and PG (17·7 ± 2·1 and 19·3 ± 1·6 ng/ml, respectively); the value for Group PG + E was intermediate (14·2 ± 1·5 ng/ml). Concentrations of FSH appeared to increase in Group E mares by Period 2 (16·3 ± 2·6 ng/ml) and did not change significantly ($P > 0.1$) in later periods. Daily FSH values in the mares in Group PG ± E rose after Day 0 and remained relatively constant from Day 6 until just before the next ovulation. In 23/26 mares a decreasing trend was observed in concentrations of FSH during the days immediately before ovulation.

Table 1. Mean ($\pm$ s.e.m.) interovulatory interval and concentration of progesterone in serum of intact mares in Groups C (control), PGF-2α), E (oestradiol) and PG + E

Group	Interovulatory interval (days)[†]	Progesterone (ng/ml)[‡]
C	18·7 $\pm$ 1·4[a]	7·79 $\pm$ 0·69[a]
PG	13·3 $\pm$ 1·0[a]	1·10 $\pm$ 0·17[b]
E	30·4 $\pm$ 5·7[b]	7·21 $\pm$ 0·66[a]
	(25·2 $\pm$ 2·5)*	
PG + E	44·5 $\pm$ 4·2[c]	1·24 $\pm$ 0·22[b]

*Mean for group excluding 1 mare (interval of 62 days).
[†]The number of days from Day 0 to the next ovulation.
[‡]Average of values observed over the interval from Day 0 to Day 8.
Within a column means without common letters are different ($P < 0.05$).

A marked inverse relationship between concentrations of progesterone and LH was observed. However, we failed to see the synergistic interaction of progesterone with oestradiol on suppression of LH concentrations, that others have reported for the ovariectomized mare (Garcia & Ginther, 1978). Perhaps low basal concentrations of oestradiol present during dioestrus in control mares is sufficient to act with luteal progesterone, causing maximal suppression of LH, such that addition of exogenous oestradiol was no more effective. Treatment of mares with oestradiol did not seem to alter luteal function compared to controls. Both time to luteolysis and mean concentrations of progesterone released after ovulation were unaffected. Whether the increased area under the curve of LH concentrations associated with the next ovulation observed in Group E mares reflects a positive feedback effect of oestradiol or the fact that the follicular phase (< 1 ng progesterone/ml) in mares in Group E was extended over that seen in controls by about 4–5 days is unclear.

From the results of this study it appears that the initial suppression of FSH in oestradiol-treated mares is not a direct effect of oestradiol but perhaps a combined effect of high luteal progesterone with oestradiol. Under the influence of luteal-phase concentrations of progesterone alone (Group C) FSH rises quickly after ovulation. In the same progesterone environment, addition of exogenous oestradiol (Group E) results in a sluggish or delayed rise in FSH until very near the time of luteolysis. It is also noteworthy that in mares in Group PG + E, in which endogenous progesterone was suppressed to very low (but not always undetectable) levels, concentrations of FSH were intermediate between those of Groups C and E. Burns & Douglas (1981) observed a similar suppression of FSH early after the start of oestradiol treatment in intact mares. Contrary to their findings, we did not observe a significant inhibition of follicular development in the oestradiol-treated mares. However, follicular development was inhibited in mares in Group PG + E until well after the cessation of treatment. This inhibition occurred in the presence of high systemic concentrations of LH and FSH. From these results it appears that treatment of intact mares with PGF-2α and oestradiol resulted in decreased sensitivity of the ovaries to gonadotrophic hormones.

The investigation reported in this paper (No. 86-4175) was supported by the Kentucky Agriculture Experiment Station and is published with approval of the Director.

References

Burns, P.J. & Dougias, R.H. (1981) Effects of daily administration of estradiol-17β on follicular growth, ovulation, and plasma hormones in mares. *Biol. Reprod.* **24**, 1026–1031.

Garcia, M.C. & Ginther, O.J. (1978) Regulation of plasma LH by estradiol and progesterone in ovariectomized mares. *Biol. Reprod.* **19**, 447–453.

J. Reprod. Fert., Suppl. **35** (1987), 687–688

Episodic release of prolactin in the cyclic mare

J. F. Roser, J. O'Sullivan*, J. W. Evans‡, J. Swedlow and H. Papkoff†

*MAb Inc., Mountain View, CA 94043; University of California at *Berkeley and †San Francisco, CA; and ‡Texas A&M University, College Station, TX 77843, U.S.A.*

The physiological role of prolactin in the cyclic mare has not been well defined. Previous attempts to measure prolactin in horses, using daily blood samples or heterologous RIA assay systems, have produced variable and non-significant results. Although several investigators have examined the relationship of LH and oestrogen to the release of prolactin, few studies have been done in the mare. The purpose of this study was to evaluate the episodic nature of prolactin in the cyclic mare in relation to its frequency and amplitude, diurnal rhythm and its correlation with episodic release of LH and oestrogen.

Materials and methods

Eleven mares were bled via a Silastic cannula inserted in the jugular vein on specific days during July. August and September. Collections starting at 08:00 h, occurred every 3 min for 2 h and thereafter every 15 min for 6 h during their oestrous cycle (Day 0 = ovulation; Day 0, N = 8; Day 3, N = 6; Day 7, N = 4; Day 15, N = 8). Serum concentrations of prolactin were quantified by the double-antibody homologous RIA technique described by Roser *et al.* (1984). Serum LH and oestradiol-17β concentrations were determined by RIA methods as described by Geschwind *et al.* (1975) and Edqvist & Johansson (1972) respectively.

Statistical analysis. Amplitude and frequency were determined by the threshold method of Clifton & Steiner (1983). This method, called cycle detection, continuously adjusts the threshold and thereby equalizes the probability of rejected real fluctuations with the probability of accepting artefacts.

Diurnal surges were initially observed in individual graphs and further clarified by processing the data through a running medium smoother. A rise was termed a surge if the rise from baseline was 3 times the assay variation and was sustained for 3 or more observations. Baseline was adjusted to normalize the data

A time-lag analysis was performed to determine patterns of correlation between hormones. In this analysis, correlations at 15-min time intervals were carried out over a 4-h period. Significance was based on the closeness (correlation coefficient = r) of the linear relationship between hormones. In this case any r value above 0·497 was significant at $P = 0·05$.

Results

Pulsatile prolactin patterns tended to decrease in frequency and increase in amplitude from Day 0 to Day 15 (Table 1). A midday diurnal surge ($P < 0·05$) was observed between 12:00 and 14:00 h in all four groups of mares. LH and oestrogen patterns did not demonstrate a similar surge.

Based on a time-lag analysis of each group and individual graphs, there was a significant positive correlation ($P = 0·05$) between oestrogen and prolactin on Day 3. The oestrogen peak appeared to precede the prolactin peak by 1·5–2 h (Figs 1 & 2). A significant correlation was not observed between LH and prolactin.

Table 1. Amplitude and frequency of prolactin peaks per 8-h period

	Groups			
	Day 0	Day 3	Day 7	Day 15
Amplitude	0·86 ± 0·18[c]	0·70 ± 0·17[b]	2·00 ± 0·61[a]	1·61 ± 0·55
Frequency	10·25 ± 3·59	9·00 ± 3·86	6·5 ± 1·19	3·86 ± 1·10

Each value represents the mean ± s.e.m.
a > b or c ($P < 0·05$; Student's *t* test).

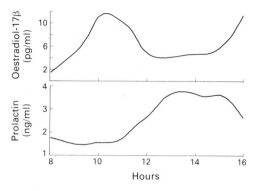

Fig. 1. Concentrations of oestradiol and progesterone in a mare on Day 3.

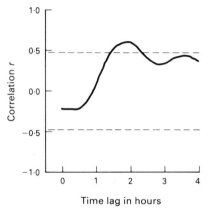

Fig. 2. Correlation of oestrogen with prolactin: statistical analysis of group on Day 3. Broken lines indicate levels of significance at $P = 0.05$.

Discussion

By increasing the frequency of blood sampling and using a homologous RIA for horse prolactin, significant prolactin patterns were observed in the cyclic mare. The frequency and amplitude of prolactin tended to change depending on the time of the oestrous cycle, suggesting that prolactin may be involved in ovarian function in the mare. The mare exhibited a similar mid-day diurnal surge of prolactin regardless of the stage of the ovarian cycle. As in other species, in the mare, oestrogen may have a positive influence in regulating prolactin release during the early luteal phase, whereas LH does not.

References

Clifton, D.K. & Steiner, R.A. (1983) Cycle detection: a technique for estimating the frequency and amplitude of episodic fluctuations in blood hormone and substrate concentrations. *Endocrinology* **112**, 1057–1064.

Edqvist, L.E. & Johansson, E.D.B. (1972) Radioimmunoassay of oestrone and oestradiol in human and bovine peripheral plasma. *Acta endocr., Copenh.* **71**, 716–730.

Geschwind, I.I., Dewey, R., Hughes, J.P., Evans, J.W. & Stabenfeldt, G.H. (1975) Plasma LH levels in the mare during the oestrous cycle. *J. Reprod. Fert., Suppl.* **23**, 207–221.

Roser, J.F., Chang, Y., Papkoff, H. & Li, C.H. (1984) Development and characterization of a homologous radioimmunoassay for equine prolactin. *Proc. Soc. exp. Biol. Med.* **175**, 510–517.

J. Reprod. Fert., Suppl. **35** (1987), 689–690

Non-surgical recovery of follicular fluid and oocytes of mares

E. Palmer, G. Duchamp, J. Bezard, M. Magistrini, W. A. King*,
D. Bousquet* and K. J. Betteridge†

*I.N.R.A. Nouzilly, 37380 Monnaie, France; *C.R.R.A., St-Hyacinthe, Canada J2S 7C6; and
‡University of Guelph, Guelph, Ontario, Canada N1G 2W1*

To collect follicular fluid and oocytes, the following technique was applied to 71 Welsh pony mares. Oestrus was synchronized by vaginal treatment with allyl trenbolone (Regumate) and prostaglandin. Subsequent follicular growth was assessed by echography. Injections of hCG (2000 i.u., i.v.) or crude horse gonadotrophin (CEG, 60 mg, i.v.) were given when the follicle reached 35 mm in diameter.

About 36 h later, the mare was tranquillized (Rompun, 0·08 g, i.v.) and anaesthetized epidurally (Sylvocaine, 20 ml). The ovary was moved by rectal manipulation to determine the site of puncture of the flank which was washed, shaved and cleaned with aseptic fluids. A needle supporting a 3-way tap and two Silastic catheters, previously rinsed with PBS + heparin (100 units/ml), was introduced into the flank. The operator held the ovary *per rectum* with one hand and guided the needle towards the follicle with the other hand. When the follicle was punctured, the golden fluid appeared in the catheter. An assistant connected one empty 60-ml syringe to the catheter, reinjected the first millilitres of fluid to dislodge the oocyte by turbulence and then aspirated all the fluid. The follicle was rinsed with 20 ml PBS + heparin through the second catheter. The liquids were observed with a stereomicroscope to locate the oocyte in the cumulus (a translucid and sticky mass 0·5–2 cm long).

Evaluation of stage of maturation was performed by phase-contrast microscopy after enzymic dispersion of the cumulus or after cytogenetic analysis or morphological study. Different groups of mares (Table 1) received slightly different preparations. The results are presented in Table 1.

Table 1. Collection rate and maturation stage of oocytes after different treatments of mares

Type of mare and treatment	No. of mares	Ovulation before puncture	No. of punctures performed	No. of oocytes recovered	No. of oocytes evaluated	Before M I	M II +PB	M II +PB
Untreated	3	0	3	0	—			
Refractory* + hCG + 36 h	5	0	5	0				
hCG + 36 h	29	5	24	15	7	0	0	7
Pregnant† + hCG + 36 h	2	0	2	2	2	0	0	2
hCG + 38 h	5	3	2	1	1	Undetermined		
CEG + 36 h	21	3	18	12	3	0	0	3
Superovulation‡ + CEG + 36 h	6	1	5	7 (0 + 0 + 2 + 2 + 3)	7	2	2	3
Total	71	12	59	37	20	2	2	15

*Females immunized against hCG (antibodies present).
†Two mares with a large follicle at Day 26–28 of pregnancy, no synchronization and no oestrus.
‡Multiple follicular stimulation by 6 ± 1 daily injections of 20 mg CEG i.m. after synchronization.

From the results we conclude that: (1) non-surgical puncture of the preovulatory follicle through the flank is safe, (2) optimal preparation of mares is hCG or CEG injection 36 h before puncture, (3) recovery rates of fluid and oocytes are high (92% ± 6 of volume, N = 19; and 63% of oocytes, N = 59), (4) metaphase II is reached before ovulation in hCG- and CEG-treated mares, (5) superovulation allows recovery of more oocytes but maturation may be slightly delayed, and (6) some mature oocytes can be collected from pregnant mares.

J. Reprod. Fert., Suppl. **35** (1987), 691–692

Fertilization and pronuclear stages in the horse

A. C. Enders, I. K. M. Liu*, J. Bowers*, C. Cabot* and K. C. Lantz

*Department of Human Anatomy, School of Medicine, and *Department of Reproduction, School of Veterinary Medicine, University of California, Davis, California 95616, U.S.A.*

To determine the events of fertilization and early development in the horse, a timed sequence of ova was obtained after insemination. The first ova to be analysed were in the pronuclear stage and varied in age from 12 to 22 h after insemination.

The time of ovulation was determined by rectal palpation and/or echography of a large (40–50 mm) preovulatory follicle, then every 4 h until ovulation had occurred. After detectable ovulation, the mares were mated naturally to known fertile stallions within 5 h, and the ovaries and oviducts were removed 12–22 h after coitus. The ova were recovered by flushing the oviducts with Dulbecco's phosphate-buffered saline and subsequently were fixed in 2% glutaraldehyde–2% formaldehyde in phosphate buffer, pH 7·2, for 1–2 h, rinsed overnight in cold phosphate buffer, post-fixed in 1% osmium tetroxide, dehydrated in ethanol, and embedded in araldite epoxy resin. Fertilized ova ($n = 8$) were collected at 12, 14, 17, 18, 18·5, 19, 20 and 21 h after insemination. A freshly ovulated, non-fertilized ovum was recovered at 22 h. Freshly ovulated ova were distinguished from the retained ova of previous cycles by the presence of a variable number of cumulus cells and by a distinct interface with the subzonal (perivitelline) space. Although occasional ova from previous cycles were in surprisingly good condition, many showed deterioration and were little more than clustered granules within a zona pellucida.

Examination by transmission electron microscopy revealed that all ova were rich in lipid droplets that were clustered, and the clusters were unevenly distributed throughout the cytoplasm, varying from a hemispheric distribution to a subcortical band. Vacuoles with a dispersed flocculent content were similarly clustered and either occupied a reciprocal position hemispherically or were marginally situated in relation to the lipid. The numerous mitochondria and abundant branched tubular agranular endoplasmic reticulum (aER) were more evenly distributed. Elements of the Golgi complex were also widely distributed as were the rarer annulate lamellae.

In the unfertilized ovum the metaphase plate of the second meiotic division was situated marginally and paralleled the surface. (The spindle was orientated at right angles to the surface.) No centrioles were present, but microtubules formed asters as well as spindle fibres. The spindle was situated in cortical cytoplasm that was devoid of lipid and vacuoles, but a few small vesicles and cortical granules were in this area. This region of the surface of the ovum was devoid of microvilli, and there was a striking accumulation of microfilaments subjacent to the folded cell membrane. Marginal to the region of modified surface, cortical granules were abundant and the surface had typical microvilli.

The more recently ovulated ova retained the antral content in the form of metachromatic material surrounding the zona pellucida, and both younger and older ova showed a pronounced subzonal granular metachromatic layer. In the 12- and 14-h ova, cortical granules though reduced in number were present adjacent to the plasma membrane, and flocculent material was often visualized in plasmalemmal depressions. The microvilli were branched and irregular and coated vesicles were more common between the bases of microvillar clusters. The zona pellucida showed a tripartite structure with the only sperm penetration in the outer zona, suggesting a rapid zonal block to polyspermy.

In the 12-h specimen the male pronucleus had not reached full size and decondensation had not been completed. However, the female pronucleus was typical and the sperm tail was fully within the cytoplasm. In most pronuclear ova the pronuclei were situated some distance apart. In two cases the pronuclei were approximated and in the most advanced stage (21 h) both pronuclei were flattened with slight ruffling at the adjacent borders and the chromosomes aggregated toward these surfaces. Portions of the midpiece of the sperm tail were present and the associated mitochondria were in surprisingly good condition. Even in this ovum the two pronuclei were eccentric rather than central in position. All pronuclei were situated in relatively lipid-free cytoplasm and were associated with aER-rich cytoplasm, vesicle-rich cytoplasm or both. Although the Golgi elements were larger in late pronuclear ova, they did not display any fixed relationship to the pronuclei *per se*.

Except for the uneven distribution of lipid droplets and vacuoles, the pronuclear stage of the horse zygote appears similar to that of the rabbit (Longo & Anderson, 1969), showing somewhat less ruffling of the pronuclear envelope than this species but more than has been reported for the mouse (Van Blerkom & Motta, 1979).

These studies have been supported by grant HD 10342 from the NICHHD, by a grant from the Oak Tree Racing Association, and by the Equine Research Laboratory. We thank S. Schlafke and S. Suarez for assistance.

References

Longo, F.J. & Anderson, E. (1969) Cytological events leading to the formation of the two-cell stage in the rabbit: association of the maternally and paternally derived genomes. *J. Ultrastruct. Res.* **29**, 86–118.

Van Blerkom, J. & Motta, P.M. (1979) *The Cellular Basis of Mammalian Reproduction.* Urban and Schwarzenberg, Baltimore.

J. Reprod. Fert., Suppl. **35** (1987), 693–694

Attempts to produce horse–donkey chimaeras by blastomere aggregation

R. L. Pashen, S. M. Willadsen† and G. B. Anderson*

*Departments of Reproduction, School of Veterinary Medicine and *Animal Science,
University of California, Davis, CA 95616, U.S.A. and †AFRC, Animal Research Station,
Institute of Animal Physiology, 307 Huntingdon Rd, Cambridge, U.K.*

Experimental chimaeras have been produced by embryo manipulation within the same species in sheep, cattle, and mice and between species in mice, sheep and goats (see Willadsen, 1982).

As inter-species embryo transfer between horses (*Equus caballus*) and donkeys (*E. asinus*) has been achieved (W. R. Allen, R. L. Pashen & S. M. Willadsen, unpublished observations; Allen *et al.*, 1985), it was decided to attempt to produce chimaeras between these species.

Methods and results

Embryos at the 2–8-cell stages were collected surgically from the oviducts of mares and jenny donkeys using a technique which has been described previously (Allen & Pashen, 1984).

Individual blastomeres or groups of blastomeres were separated by micromanipulation and then aggregated in an evacuated cow zona pellucida, placed in agar chips and transferred to the ligated oviducts of ewes for in-vivo culture for 3–4 days. The chips were recovered and embryos evaluated for incorporation and continued division of the blastomeres and the production of a discrete embryonic mass. Embryos that developed were transferred surgically to the uteri of mares or jenny donkeys synchronized to have ovulated within $+1$ to -2 days of the original donors.

Altogether, 18 chimaeric embryos were produced and transferred to the oviducts of 6 ewes: 12 (66%) embryos continued to develop and were transferred to jenny donkeys (5) or mares (7) (Table 1). Three jennys and 4 mares failed to return to oestrus and were diagnosed pregnant by rectal palpation. All but 2 mares returned to oestrus by Day 30. Ultrasound scanning showed retarded development of the embryonic vesicle and these mares returned to oestrus by Days 40 and 60 of pregnancy. Chorionic gonadotrophin (CG) was not detected in the blood of either mare.

Discussion

In the present study, several recipients carrying chimaeric embryos failed to return to oestrus at the expected time and were diagnosed pregnant by rectal palpation. Unfortunately, it was not possible to use ultrasound scanning to check the normality of size of the embryonic vesicle and the development of the fetus and fetal heart beat in most of these pregnancies.

In the 2 mares that were scanned, development of the embryonic vesicle was slower than for a normal horse pregnancy and these animals subsequently returned to oestrus by Days 40 and 60. In these mares, the embryos were constructed in such a way that it was likely that donkey cells would have predominated in the trophectoderm and horse cells predominated in the fetus. This prediction can be made as blastomeres of different developmental age were aggregated together. As a rule, blastomeres from embryos of the most advanced developmental stage tend to contribute to the inner cell mass and blastomeres from less advanced embryos contribute mainly to the placenta (Fehilly *et al.*, 1984a, b). In both of the above pregnancies the horse embryos were more advanced developmentally than the donkey embryos. It is possible then that a situation occurred analogous to that occurring with donkey-in-horse pregnancies. In most of these pregnancies, the

Table 1. The make-up of embryos which developed and were transferred to horse or donkey recipients

Embryo	Composition of embryos*		Transferred to†	Pregnant by:	
	Horse	Donkey		Rectal palpation	Ultrasound
1	1/2	1/4	D	*	
2	2/4	1/4	D	*	
3	3/8	1/4	D		
4	1/2	1/4	D		
5	3/8	4/8	D	*	
6	1/2	1/4	M	*	*
7	2/8	1/4	M		
8	4/8	1/4	M	*	
9	2/4	4/8	M	*	*
10	3/8	2/8	M		
11	2/4	2/8	M	*	
12	2/8	2/8	M		

*1/2 denotes 1 blastomere from a 2-cell embryo; 1/4 denotes 1 blastomere from a 4-cell embryo; 2/4 denotes 2 blastomeres from a 4-cell embryo; 2/8 denotes 2 blastomeres from an 8-cell embryo; 3/8 denotes 3 blastomeres from an 8-cell embryo; 4/8 denotes 4 blastomeres from an 8-cell embryo.
†M = mare; D = jenny donkey.

chorionic girdle cells fail to migrate into the maternal endometrium, resulting in failure of endometrial cup development and consequent production of CG (Allen *et al.*, 1985). The fetus is then usually resorbed by Day 80. The possibility that a similar situation occurred in the present chimaeric pregnancies is supported by the fact that CG was not detected in the blood of either mare carrying a chimaeric pregnancy. Other more fundamental reasons, such as differences in rates of embryonic development and incompatibility of cells, could also have prevented further development of the pregnancies.

This work was supported by a Biomedical Research Grant, and a grant from the Equine Research Laboratory, School of Veterinary Medicine, University of California, Davis.

References

Allen, W.R. & Pashen, R.L. (1984) Production of monozygotic (identical) horse twins by embryo micromanipulation. *J. Reprod. Fert.* **71**, 607–613.

Allen, W.R., Kydd, J., Boyle, M.S. & Antczak, D.F. (1985) Between species transfer of horse and donkey embryos: a valuable research tool. *Equine vet. J., Suppl.* **3**, 53–62.

Fehilly, C.B., Willadsen, S.M. & Tucker, E.M. (1984a) Experimental chimaerism in sheep. *J. Reprod. Fert.* **70**, 347–351.

Fehilly, C.B., Willadsen, S.M. & Tucker, E.M. (1984b) Interspecific chimaerism between sheep and goat. *Nature, Lond.* **307**, 634–636.

Willadsen, S.M. (1982) Micromanipulation of embryos of the large domestic species. In *Mammalian Egg Transfer*, pp. 185–210. Ed. C. E. Adams. CRC Press, Boca Raton.

J. Reprod. Fert., Suppl. **35** (1987), 695

Light and ultrastructural studies of morphological alterations in embryos collected from maiden and barren mares

D. H. Schlafer, E. P. Dougherty and G. L. Woods

New York State College of Veterinary Medicine, Cornell University, Ithaca, New York 14853, U.S.A.

Specific mechanisms responsible for early embryonic losses in the mare are poorly understood. It has been shown, however, that mares with histories of subfertility produce not only fewer embryos but those that are produced are of poorer quality than those from 'normal' mares (Woods *et al.*, 1985). The primary objectives of this study were to examine the histomorphology of embryos collected from normal (maiden) mares and embryos from mares with histories of subfertility, to characterize cellular changes in abnormal embryos and to investigate possible causes of early embryonic losses in the mare.

Embryos were collected from each of two groups of mares (33 embryos from maiden mares and 12 embryos from subfertile mares). Embryos were examined by stereomicroscopy and were then fixed and sectioned for light and electron microscopy. Embryo diameters were taken (mean of 2 readings) from photomicrographs of fixed embryos. Thickness of the embryonic capsule of each sectioned embryo was measured using an ocular reticle (expressed as the mean of 3 readings). Statistical analyses were performed using Wilcoxon's Rank Sum Test and the Kruskal–Wallis test.

The histomorphology of all embryos was studied using stereomicroscopy and light level microscopy of stained thick sections. Selected embryos ($n = 21$) were thin sectioned and examined by electron microscopy. Three embryos were lost during processing. Embryo morphology was scored on a scale of 1 to 4 (best to poorest). Criteria used to score embryos included uniformity of cell layers, presence or absence of cells or cellular debris in the perivitelline space or the blastocoele, and presence of degenerative or necrotic cellular changes.

Morphological change evident by stereomicroscopy correlated with cellular pathological changes detected at the light and ultrastructural levels.

Embryos collected from mares with histories of subfertility had more morphological abnormalities than did embryos from maiden mares. Only 4 embryos of 10 (40%) collected from mares with histories of subfertility had good to excellent morphology. In contrast, 26 of 32 (85%) embryos from maiden mares were considered normal morphologically. Marked retardation, cellular degeneration or extensive cellular necrosis was evident in 3 of 10 (30%) embryos from mares with histories of subfertility but only 1 of 32 (3·1%) embryos from maiden mares. Minor alterations in embryonic morphology were detected in 3 of 10, and 5 of 32 embryos from subfertile and maiden mares respectively. Two embryos recovered at different times from one subfertile mare contained numerous hyphae of *Candida tropicalis* invading the embryonic capsule, trophectoderm, endoderm and embryonic disc. Cellular reaction to the invading organisms consisted of necrosis of individual cells, segmented hyperplasia of trophectoderm and thickening of the embryonic capsule.

Reference

Woods, G.L., Baker, C.B., Hillman, R.B. & Schlafer,
 D.H. (1985) Recent studies relating to early embryonic death in the mare. *Eq. vet. J., Suppl.* **3**, 104–107.

J. Reprod. Fert., Suppl. **35** (1987), 697–698

Short-term storage and survival of horse embryos after refrigeration at 4°C

R. L. Pashen

Department of Reproduction, School of Veterinary Medicine, University of California, Davis, CA 95616, U.S.A.

The effectiveness of short-term storage of embryos by refrigeration at 4°C has been investigated in a number of species including rabbits (Hughes & Anderson, 1982) and cattle (Trounson *et al.*, 1976; Kanagawa, 1980; Lindner *et al.*, 1983). In these studies, after cooling, embryos up to the morula stage of development had poor viability but late morula to expanded blastocyst stage embryos survived well both in in-vitro culture and after transfer to synchronized recipients.

The aim of this study was to determine the viability of horse embryos at the late morula (Day 6) to expanded blastocyst (Day 8) stage after short-term storage at 4°C.

Methods and results

Embryos were flushed non-surgically from mares 6–8 days after ovulation and were evaluated for stage of development and morphology. They were then placed in ovum culture medium (OCM; modified Dulbecco's medium (Whittingham, 1971), containing 10% heat-inactivated fetal calf serum), loaded into 0·25-ml plastic straws which were sealed, and placed in a simple water bath in the +4°C compartment of a refrigerator. After storage for 24, 48, 72 or 84 h, they were evaluated for evidence of changes which may have occurred during cooling and then cultured *in vitro* in fresh OCM at 37°C for up to 36 h or transferred non-surgically into recipient mares. Recipient mares ovulated up to 96 h after the original donors.

By morphological assessment, all embryos survived refrigeration for up to 48 h and 13 of 17 (72%) continued to develop *in vitro* after refrigeration for 84 h (Table 1). Seven embryos were transferred non-surgically to recipient mares and 4 pregnancies resulted (Table 2).

Table 1. Development of different embryonic stages after storage for 84 h at 4°C and in-vitro culture in OCM at 37°C for up to 36 h

Stage of development	No. of embryos	
	Before culture	Developing after storage *in vitro*
Late morula	2	2
Early blastocyst	6	5
Early expanding blastocyst	7	5
Expanded blastocyst	2	1

Discussion

Refrigeration at 4°C maintains embryos in a metabolically inactive state. After storage for up to 84 h morula- and blastocyst-stage embryos continued to develop *in vitro*. Although only limited numbers of transfers were done, it appears that pregnancy rates after storage are similar to those obtainable with fresh embryos.

Table 2. Pregnancies achieved after storage of embryos for different periods
and the synchronization between donors and recipients in terms of time of
ovulation and embryo with recipient after storage

Storage period (h)	No. of recipients		Synchronization (days)	
	Transferred to	Pregnant (60 days)	Donor to recipient	Embryo to recipient after storage
48	1	1	-2	0
72	4	2	-3 to -4	0 to -1
84	1	1	-3 to -4	$+0.5$ to -0.5

There are several practical applications of this simple storage technique. First, it allows storage of horse embryos for up to 84 h with little apparent decrease in viability. This is in distinct contrast with results obtained from frozen–thawed horse embryos. Freezing of embryos by currently available methods has limited application for the horse because Day-6 but not Day-7 or -8 embryos give adequate pregnancy rates on transfer after freezing and thawing (Slade *et al.*, 1985; Boyle *et al.*, 1985). This is impractical as recovery rates of embryos on Day 6 are low (< 50%) while Days 7 and 8 are the most efficient times for embryo recovery.

The present method could be used for the maintenance of embryos for shipment over long distances. Embryos could be shipped domestically and internationally in a manner similar to that used for transported chilled semen. Another use would be as a management aid in the sometimes difficult process of synchronization of donors and recipients. Animals could be managed in such a way that recipients could ovulate up to 5 days after the donor. Embryos could be stored while the recipients 'caught up' to the developmental stage of the embryo with transfers done when the difference in synchrony was less than 36 h.

This work was supported by a grant from the Equine Research Laboratory, University of California, Davis.

References

Boyle, M.S., Allen, W.R., Tischner, M. & Czlonkowska, M. (1985) Storage and international transport of horse embryos in liquid nitrogen. *Equine vet. J., Suppl.* **3**, 36–39.

Hughes, M.A. & Anderson, G.B. (1982) Short-term storage of rabbit embryos at 4°C. *Theriogenology* **18**, 275–282.

Kanagawa, H. (1980) One to two day preservation of bovine embryos. *Japanese J. vet. Res.* **28**, 1–6.

Lindner, G.M., Anderson, G.B., BonDurant, R.H. & Cupps, P.T. (1983) Survival of bovine embryos stored at 4°C. *Theriogenology* **20**, 311–319.

Slade, N.P., Takeda, T. & Squires, E.L. (1985) Cryo-preservation of the equine embryo. *Equine vet. J., Suppl.* **3**, 40.

Trounson, A.O., Willadsen, S.M. & Rowson, L.E.A. (1976) The influence of in-vitro culture and cooling on the survival and development of cow embryos. *J. Reprod. Fert.* **46**, 367–370.

Whittingham, D.G. (1971) Survival of mouse embryos after freezing and thawing. *Nature, Lond.* **233**, 125–126.

J. Reprod. Fert., Suppl. **35** (1987), 699–700

Preliminary investigations of the metabolic activity of early horse embryos

D. Rieger, D. Lagneau-Petit* and E. Palmer*

*Centre de recherche en reproduction animale, Université de Montréal, C.P. 5000, St-Hyacinthe, Québec, Canada J2S 7C6 and *I.N.R.A. Station de physiologie de la reproduction, Centre de recherche de Tours, 37380 Nouzilly, France*

The early mammalian embryo undergoes rapid development and growth before implantation or attachment, and therefore must necessarily have significant metabolic requirements. The energy metabolism of mouse and rabbit embryos has received much study, but little is known of the metabolic activity of the embryos of domestic animals. This is particularly true of the horse embryo for which virtually no information is available. The object of this study was to develop a method of measuring the metabolic activity of individual early horse embryos and to use it to evaluate specific aspects of glycolysis.

Materials and methods

Embryos were collected by a non-surgical transcervical flush of the uterus with Dulbecco's phosphate-buffered saline (PBS) containing 1% fetal calf serum (FSC). After recovery from the flushing medium, the embryos were washed four times in culture medium before being cultured at 37°C.

The culture apparatus consisted of microwells (total volume 125 µl) cut to fit into wells of culture plates (Falcon: total volume 2·8 ml). Individual embryos were placed into the inner wells in culture medium which contained 0·5 µCi of a ^{14}C-labelled glycolytic substrate. The outer wells contained 400 µl 1·0 N-NaOH and were capped with Parafilm throughout the culture period. At the end of the culture period, the NaOH was aspirated from the outer wells, mixed with 10 ml scintillation fluid, and counted for $^{14}CO_2$ content. Control wells were included in each trial and contained an equivalent volume of the final wash, and the labelled substrate, but no embryo. A significant metabolism of the labelled substrate was considered to have occurred if the quantity of $^{14}CO_2$ measured was more than 2 standard deviations greater than the mean quantity measured for the control wells.

In the first series of experiments, embryos were collected from mares of mixed breeding at 7·0–10·5 days of gestation (Day 0 = day of ovulation), and cultured in 4 µl droplets of medium (20% FCS in PBS) in 50 Ml paraffin oil, or in 50 or 100 µl medium without oil, with one of D-[1-^{14}C]-glucose (sp. act. 57 mCi/mmol), D-[6-^{14}C]glucose (sp. act. 57 mCi/mmol), or [U-^{14}C]lactate (sp. act. 180 mCi/mmol) (NEN, Boston, MA, U.S.A.).

In the second series of experiments, embryos were collected from Welsh pony mares at Day 7 of gestation, and cultured in 40 µl 10% FCS in PBS with D-[1-^{14}C]glucose (sp. act. 59 mCi/mmol) or D-[6-^{14}C]glucose (sp. act. 58·5 mCi/mmol) (Amersham, Bucks U.K.).

Results and discussion

The first series of experiments included a variety of culture times and volumes, in an attempt to define the optimal conditions for measurement of metabolic activity. Consequently, comparisons cannot be made amongst experiments. However, the results (Table 1) do suggest that the Embden–Meyerhof pathway, the Krebs' cycle, and lactate dehydrogenase are all active from Day 7 to Day 10 in the horse embryo.

The culture conditions used in the second series of experiments were chosen based on the results of the first series and were the same for all embryos, except for the labelled substrate. Although the total concentration of glucose in the medium has not yet been determined, the medium was identical, and the specific activity virtually equal for D-[1-^{14}C]glucose and D-[6-^{14}C]glucose. Consequently, comparisons can be made amongst embryos and between the two labelled substrates.

As for the first series, the results of the second series (Table 2) indicate that the Embden–Meyerhof pathway and the Krebs' cycle are active in the Day-7 horse embryo. For both substrates, the quantity of $^{14}CO_2$ measured was directly proportional to the size of embryo. Regression

Table 1. Metabolism of glycolytic substrates by individual pre-attachment horse embryos (all measurements for individual embryos are significantly different from controls)

Labelled substrate	Embryo		Culture		$^{14}CO_2$ measured	
	Age (days)	Diameter (mm)	Volume (µl)	Time (h)	Embryo (c.p.m.)	Controls (c.p.m.)*
D-[1-^{14}C]Glucose	8·0	0·95	50	17	53 500	102 ± 3
	9·5	3·80	50	1	1178	240 ± 54
	9·5	3·60	100	1	245	112 ± 23
D-[6-^{14}C]Glucose	7·0	0·21	4	3	360	106 ± 14
	8·0	0·31	4	3	698	123 ± 16
	8·0	0·35	4	3	1050	123 ± 16
	8·5	1·03	50	17	17 718	494 ± 162
[U-^{14}C]Lactate	10·5	3·27	50	1	2048	129 ± 33
	7·0	0·17	4	3	677	116 ± 5
	10·0	1·50	50	3	4816	127 ± 17

*Mean ± s.d. (n = 2 or 3).

Table 2. Glucose metabolism by individual Day-7 horse embryos cultured in 40 µl medium for 4 h (the measurements for all embryos are significantly different from the appropriate controls)

Labelled substrate	Embryo		$^{14}CO_2$ measured
	Diameter (mm)	Volume (mm^3)	(c.p.m.)
D-[1-^{14}C]Glucose	0·20	0·004	385
	0·25	0·008	428
	0·55	0·087	2067
	0·55	0·087	2038
	0·60	0·113	3957
	0·75	0·221	10 080
	Controls (n = 10)		66 ± 39*
D-[6-^{14}C]Glucose	0·20	0·004	138
	0·25	0·008	1035
	0·50	0·065	373
	0·60	0·113	921
	0·60	0·113	2054
	0·75	0·221	6112
	Controls (n = 12)		48 ± 18*

*Mean ± s.d.

analyses of $^{14}CO_2$ measured, as a function of embryo volume, yielded slopes of 43 900 c.p.m./mm^3 and 24 000 c.p.m./mm^3, and correlation coefficients (Pearson's product moment) of 0·96 ($P < 0·01$) and 0·88 ($P < 0·05$), for D-[1-^{14}C]glucose and D-[6-^{14}C]glucose, respectively. Given the high and statistically significant correlation coefficients, the slopes of the regression analyses may be used to compare the metabolism of the two terminal carbons. Although a definitive mathematical relationship cannot be determined from these data, the 2-fold higher level of $^{14}CO_2$ produced from D-[1-^{14}C]glucose strongly suggests that the pentose shunt represents a significant portion of the glycolytic activity in the Day-7 horse embryo.

J. Reprod. Fert., Suppl. **35** (1987), 701–702

Comparison of two techniques and three hormone therapies for management of twin conceptuses by manual embryonic reduction

D. R. Pascoe, R. R. Pascoe*, J. P. Hughes, G. H. Stabenfeldt and H. Kindalh†

Department of Reproduction, School of Veterinary Medicine, University of California, Davis, CA 95616, U.S.A.; *Oakey Veterinary Hospital, Hamlyn Road, Oakey, Queensland 4401, Australia; and †Department of Obstetrics and Gynaecology, College of Veterinary Medicine, Swedish University of Agricultural Sciences, 5-750 07 Uppsala, Sweden

Manual rupture of the embryonic vesicle coupled with hormone therapy has been advocated as a method to eliminate one conceptus in mares carrying twin conceptuses. The purposes of this study were to investigate (1) the survival rate of the remaining conceptus when one conceptus was ruptured, (2) the changes in circulating PGFM concentrations after embryonic vesicle rupture, (3) the relationship of the PGFM concentrations to the rupture techniques, (4) the ability of flunixin meglumine to reduce or inhibit uterine prostaglandin release during the embryonic vesicle rupture technique, and (5) the effect of exogenous progesterone on the survival rate of the remaining conceptus.

Mares carrying twin conceptuses between Days 12 and 30 of gestation were divided into four groups: Group 1 (20 mares) were given a placebo (sterile saline i.v. or sesame oil i.m.), Group 2 (32 mares) were given a single dose of progesterone (625 mg, i.m.) in sesame oil, Group 3 (28 mares) received multiple progesterone treatments (625 mg, i.m.) at 6-day intervals until Day 42 of gestation, and Group 4 (20 mares) received single treatments of flunixin meglumine (500 mg, i.v.) and progesterone (625 mg, i.m.). Each group was further divided into two equal subgroups, A and B, according to the age of the embryo. In subgroup A (mobile technique) with embryos of gestation age of 12–16 days, embryonic vesicle rupture was performed after the selected conceptus was moved to the tip of a uterine horn. In subgroup B (fixed technique), gestation age 17–30 days, embryonic vesicle rupture was performed on the selected conceptus *in situ*. An additional 6 mares carrying single conceptuses of gestation ages 12, 14, 16, 20, 25 and 30 days, respectively, were used as subjects for sham embryonic-vesicle rupture.

Both rupture techniques usually caused the release of PGF-2α for a short duration (< 90 min). Concentrations of PGF-2α were directly correlated to the pressure required to effect embryonic vesicle rupture. Embryonic vesicle rupture (mobile technique) of conceptuses $\leqslant 16$ days of age caused 36 of 40 mares to have significant increases in PGFM concentrations, which did not exceed 80 pg/ml. For twin conceptuses between Day 17 and Day 30 of gestation, rupture of the embryonic vesicle (fixed technique) caused 32 of 40 mares to have significant increases in PGFM concentrations (< 172 pg/ml). In the remaining mares in both subgroups, in which embryonic vesicle rupture occurred with relative ease, PGF-2α did not rise above 20 pg/ml. Sham embryonic vesicle rupture attempts caused small PGF-2α releases (PGFM < 96 pg/ml) after each manipulation which rapidly returned to basal levels (10 pg/ml). Flunixin meglumine completely inhibited PGF-2α release after embryonic vesicle rupture, regardless of the technique.

Plasma progesterone concentrations were not significantly decreased ($P > 0{\cdot}01$) after the use of either rupture technique in any group. The use of 17β-hydroxyprogesterone as a single treatment (Group 2) or as a multiple treatment (Group 3) did not significantly alter ($P > 0{\cdot}01$) the endogenous progesterone concentration.

The survival rates for the remaining conceptuses were 10/10 in Group 1A, 10/10 in Group 1B, 16/16 in Group 2B, 10/10 in Group 4A, 10/10 in Group 4B, 15/16 in Group 2A, 13/14 in Group 3B and 12/14 in Group 3A.

In conclusion, the results of this study indicate that either embryonic vesicle rupture technique (mobile or fixed) is an effective method for removing one embryonic vesicle in a mare carrying twin conceptuses between Days 12 and 30 of gestation. Gentle uterine manipulation during the use of either technique was not sufficient to elicit PGF-2α synthesis and release to cause partial or total luteolysis of the corpus luteum. The use of exogenous progesterone was therefore not essential for successful embryo reduction. Similarly, administration of flunixin meglumine, although very effective in preventing the release of PGF-2α from the uterus, was not important for successful embryo reduction. Provided either technique can be carried out with minimal uterine manipulation, the use of exogenous drugs should not be necessary.

Supported by Oak Tree Racing Association, Santa Anita, California, through the Equine Research Laboratory, School of Veterinary Medicine, University of California, Davis.

We thank Coralie Munro and Jeanne Bonura for assay of hormone and drug samples.

J. Reprod. Fert., Suppl. **35** (1987), 703

Shedding of the 'capsule' and proteinase activity in the horse embryo

H.-W. Denker, K. J. Betteridge*† and J. Sirois†

Abteilung Anatomie der RWTH, D-5100 Aachen, West Germany, and †CRRA, Faculté de Médicine Vétérinaire, Université de Montréal, CP 5000, St Hyacinthe, Québec, Canada J2S 7C6

To establish cellular contact with the endometrium, the horse embryo must first shed a glycoprotein 'capsule' (Betteridge *et al.*, 1982; Flood *et al.*, 1982). To investigate whether this process involves a proteinase system analogous to that found in rabbits (Denker, 1977), the uteri from 5 mares have been used in this pilot study. They were collected at surgery or slaughter on Days 20·5, 21·5 and 28·5 after diagnosed ovulation, or on Day 28 and Day 30 as estimated from conceptus diameters measured by ultrasonography. Native cryostat sections were studied with the highly sensitive histochemical gelatin film test for localization of proteinases using a series of inhibitors for identification of the enzyme (Denker, 1974, 1976, 1977). At Day 20·5, the capsule was still intact and surrounded the conceptus completely; at Day 21·5 it was still well preserved but completeness could not be ascertained since this specimen had collapsed; at Day 28 and after, only minor possible remnants were seen. Proteinase activity was considerable in the embryonic membranes and the adjacent endometrium but was much lower in the conceptus-free horn. Two categories of proteinases could be identified: a proteinase with acid pH optimum (cathepsin-like) in the uterine epithelium and embryonic membranes, with markedly increasing activity towards Day 28; and a proteinase with alkaline pH optimum in portions of the trophoblast at Days 20·5 and 21·5. The possible role of the latter enzyme in dissolution of the capsule requires further study of specimens at Days 21–28.

Supported by NSERC Canada.

References

Betteridge, K.J., Eaglesome, M.D., Mitchell, D., Flood, P.F. & Beriault, R. (1982) Development of horse embryos up to twenty-two days after ovulation: observations on fresh specimens. *J. Anat.* **135**, 191–209.

Denker, H.-W. (1974) Protease substrate film test. *Histochemistry* **38**, 331–338; **39**, 193.

Denker, H.-W. (1976) Interaction of proteinase inhibitors with blastocyst proteinases involved in implantation. In *Protides of the Biological Fluids* **23**, pp. 63–68. Ed. H. Peeters. Pergamon Press, Oxford.

Denker, H.-W. (1977) Implantation: the role of proteinases, and blockage of implantation by proteinase inhibitors. *Adv. Anat. Embryol. Cell Biol.* **53**, part 5, 123 pp. Springer-Verlag, Berlin.

Flood, P.F., Betteridge, K.J. & Diocee, M.S. (1982) Transmission electron microscopy of horse embryos 3–16 days after ovulation. *J. Reprod. Fert., Suppl.* **32**, 319–327.

*Present address: Department of Biomedical Sciences, Ontario Veterinary College, University of Guelph, Guelph, Ontario, Canada N1G 2W1.

J. Reprod. Fert., Suppl. **35** (1987), 705–709

Development of Polish-pony foals born after embryo transfer to large mares

M. Tischner

Department of Animal Reproduction, Academy of Agriculture, 30–059 Kraków, Al. Mickiewicza 24/28, Poland

The ultimate quality of a mature horse depends not only on its genetically acquired traits, but also on the method of rearing (environmental influence) during the period of growth. For this reason, many attempts using different methods to manipulate the growing organism of the horse have been made to obtain an animal of desired attributes. Walton & Hammond (1938) initially observed the fundamental influence of the mother in studies on reciprocal crosses between Shire horses × Shetland ponies on the growth and development of the horse. Eventually, with the successful trials of embryo transfer, this issue again became one of interest for many more researchers. Brumby (1960) demonstrated that, in mice, the influence of the mother on the size of the offspring is greater during gestation than after birth, even though the milk capacity of the mother played an important role in the latter. Studies conducted on rabbits similarly showed that factors such as size of mother, sex of offspring and length of gestation (Venge, 1950), have an influential effect on the offspring. Subsequent studies conducted on ewes showed that, through appropriate selection of recipient mothers, it is possible to exert an influence not only on the weight and size of the offspring produced, but also on copper metabolism and type of fleece (Hunter, 1956; Burns, 1972; Wienner *et al.*, 1977). In addition, the study showed maternal effects, embryo effects and maternal–fetal interactions, as key factors.

Comparable studies on transfer in horses have not yet been performed. For this reason, it appeared expedient to conduct an evaluation of foal development of pure-bred Polish ponies born after embryo transfer to heavy-type mares.

Materials and methods

A Tarpan-type Polish pony, mated with a stallion of the same breed, was used as the embryo donor. Average body weight of these horses is in the range of ~400 kg, and height at the withers is 134 cm. Embryos were transferred to heavy-type mares, which have a body weight of ~730 kg (Table 1).

Embryo collection and transfer were carried out by non-surgical means (Allen & Rowson, 1975; Tischner, 1985). Six foals (2 males and 4 females) were born by using this method. Only the fillies were selected for the comparative studies. One pair of fillies, Nos 45 and 46, born in May 1982, were compared with their genetic mothers, and Nos 53 and 54, born in April 1984, were compared with their non-transferred full sisters, born 2·5 months later by the genetic mothers of the transferred foals (Tables 2 & 3).

Environmental conditions, care and feeding were similar for all the horses during the entire observation period. The foals were weaned from their mothers at 7 months of age. Each month, up to 24 months of age, 21 biometric measurements were performed on each foal, and these measurements were repeated every 6 months. The comparative measurements were always carried out by the same individual. Only results of selected measurements that most clearly typify the development of the foals are presented in Tables 1, 2 and 3.

Table 1. Measurements of genetic parents and donor mares used in comparative studies

	Mares							Stallions	
	Donors and genetic mothers				Recipient				
	Polish pony				Half draught-type			Polish pony	
Mare and stallion identification	14	16	20	43	41	42	51	Orzyc	Nurzec
Age measurements (years) taken at	10	8	7	7	8	6	7	10	10
Weight (kg)	420	390	400	380	730	670	780	370	410
Height of withers (cm)	134	127	134	131	161	155	162	132	138
Height at crupper (cm)	136	128	135	131	161	157	165	132	137
Depth of thorax (cm)	63	62	64	58	76	74	78	61	64
Circumference of thorax (cm)	172	172	174	164	204	196	209	172	174
Oblique length of trunk (cm)	143	146	144	139	173	157	171	136	145
Length of front leg (cm)*	78	73	78	76	97	92	92	82	83
Length of forearm (cm)†	40	38	41	39	46	45	44	40	40
Length of shank (cm)‡	28	26	27	27	32	30	32	30	28
Circumference of shank (cm)§	17	17	18	16	23	22	22	18	18
Identification of transferred and control foals (see Tables 2 or 3)	46	52, 54	53, 55	45	45, 54	46	53	43, 45, 46	52, 53, 54, 55

*From the tuber ulner to the ground.
†From the elbow joint to the accessory carpal bone.
‡From the accessory carpal bone to the centre of the fetlock joint.
§Circumference of the upper third of the shank.

Table 2. Comparison of 2 fillies, born after embryo transfer to large mares, with their genetic mothers

	Transferred foal	Genetic mother	Differences	Transferred foal	Genetic mother	Differences
Breed	Polish pony	Polish pony		Polish pony	Polish pony	
Identification No.	46	14		45	43	
Age at measurements (years)	4	10		4	7	
Weight (kg)	400	420	−20	360	380	−20
Height at withers (cm)	139	134	+5	134	131	+3
Height at back (cm)	140	136	+4	135	131	+4
Depth of thorax (cm)	62	63	−1	58	58	—
Circumference of thorax (cm)	170	172	−2	162	164	−2
Oblique length of trunk (cm)	142	143	−1	139	139	—
Length of front leg (cm)	86	78	+8	80	76	+4
Length of forearm (cm)	45	40	+5	43	39	+4
Length of shank (cm)	30	28	+2	29	27	+2
Circumference of shank (cm)	17	18	−1	16	16	—

Table 3. Comparison of 2 of Polish-pony fillies born after embryo transfer to large mares (Nos 53 and 54), with their sisters (Nos 55 and 52), born from the same genetic donor mothers

	Transferred foal	Control foal	Difference	Transferred foal	Control foal	Difference
Foal No.	53*	55*		54†	52*	
Date of birth	14 Apr. 1984	2 July 1984	2·5 months	16 Apr. 1984	29 June 1984	2·5 months
Length of gestation (days)	330	328	+2	330	322	+8
Weight (kg)						
At birth	40	30	+10	46	23·5	+22·5
At 7 months	203	158	+45	240	151	+89
At 16 months	238	237	+1	313	239	+74
At 24 months	293	290	+3	340	284	+56
Height at withers (cm)						
At birth	88	77	+11	90	72	+18
At 7 months	119	110	+9	120	111	+9
At 16 months	126	121	+5	131	124	+7
At 24 months	131	126	+5	134	128	+6
Height at crupper (cm)						
At birth	90	79	+11	92	74	+18
At 7 months	121	113	+8	122	112	+10
At 16 months	129	126	+3	132	125	+7
At 24 months	131	128	+3	135	128	+7
Depth of thorax (cm)						
At birth	29	27	+2	29	25	+4
At 7 months	49	47	+2	52	45	+7
At 16 months	54	52	+2	58	52	+4
At 24 months	57	55	+2	62	58	+4
Circumference of thorax (cm)						
At birth	78	73	+5	78	71	+7
At 7 months	133	126	+7	139	124	+15
At 16 months	138	140	−2	150	139	+11
At 24 months	152	149	+3	160	156	+4
Oblique length of trunk (cm)						
At birth	65	61	+4	66	55	+11
At 7 months	112	107	+5	114	105	+9
At 16 months	120	119	+1	122	115	+7
At 24 months	129	126	+3	140	128	+12
Length of front leg (cm)						
At birth	58	54	+4	60	52	+8
At 7 months	72	69	+3	73	69	+4
At 16 months	80	75	+5	81	77	+4
At 24 months	80	76	+4	82	78	+4
Length of forearm (cm)						
At birth	25	23	+2	25	23	+2
At 7 months	31	30	+1	32	31	+1
At 16 months	40	39	+1	41	39	+2
At 24 months	42	41	+1	42	40	+2
Length of shank (cm)						
At birth	22	20	+2	23	19	+3
At 7 months	25	23	+2	25	23	+2
At 16 months	26	24	+2	28	26	+2
At 24 months	26	24	+2	28	26	+2
Circumference of shank (cm)						
At birth	10·5	9	+1·5	11	9	+2
At 7 months	14·5	14	+0·5	14·5	13·5	+1
At 16 months	15	15	—	17	15	+2
At 24 months	15	15	—	17	15	+2

*First-born offspring of particular mare.
†Second-born offspring of particular mare.

Results and discussion

All of the foal deliveries progressed with ease, and without any complications. Immediately after birth, the transferred foals exhibited great vitality. Several minutes after birth they stood and approached the recipient mare for suckling. These foals were characterized by greater body weight, and long well-developed extremities. At birth, their body weight was greater by 13 kg, that is, 48% greater in comparison to control foals, and were also taller at the withers by an average of 14 cm. The transferred foals maintained this advantage in size to adulthood (Tables 1, 2 & 3), with increases in body weight progressing more intensively to 5 months of age. However, at the time of transition to pasture feeding, differences in body weight increments became less evident, and even a certain standstill was observed in the transferred foals (Table 4). However, specific measurements of foal growth showed advantage in length of limbs, and height at the withers and crupper (average of +5 cm). Bones of the fore limbs and shank were demonstrably longer by 2–5 cm (Tables 2 & 3). On the other hand, dimensions such as head size, neck length, circumference of thorax, and depth of thorax were gradually equalized in both groups of foals. The exception was No. 54, who from birth showed very clear dominance over her genetic sister (No. 52) as well as the other foals. At 2 years of age, this filly attained dimensions usually seen in foals who have completed 3 years of life, while the height of her withers even surpassed that of her mother and other mature horses of this breed. This filly was the second born foal of recipient Mare 41, while the remainder of the fillies were the first-born offspring. At 4 years of age, Fillies 45 and 46 attained an advantage in regard to growth in comparison with their mothers (Table 2), and Fillie 46 even surpassed her father in growth.

All of the fillies reached sexual maturity at the age of 11–12 months and manifested regular oestrous cycles beginning at 16–18 months of age.

The results obtained in this study on the maternal influence on development of transferred foals are in agreement with those reported by Walton & Hammond (1938) for reciprocal crosses. These studies demonstrated that, in horses, nutrition is as important during gestation as it is following birth. The effect can be mediated both by the spacious in-utero environment in the large mare and the higher milk production capacity. In cattle and sheep it was similarly found that there is greater maternal influence on the size of offspring after birth, but these differences level off later through appropriate nourishment (Hammond *et al.*, 1971). In our horses, however, differences in growth were maintained up to maturity. This is most probably connected with the different speed of growth and ossification of bones in the horse. Brzeski (1960) reported that growth and ossification of long bones in horses are most intensive in the second half of pregnancy and a few months after

Table 4. Average increments of body weight of foals during successive months of life

| Month | Daily increase in body weight (g) | | |
	Transferred foals (N = 4)*	Controls (N = 2)†	Difference
I	1290	1198	+92
II	1023	765	+258
III	873	815	+58
IV	908	590	+318
V	578	350	+228
VI	193	300	−107
VII	178	190	−12

*Fillies 46, 45, 53, 54.
†Fillies 52, 55.

birth. The growth plate of the pastern and shank bones ceases at 6–9 months of life, and towards the end of the first year of life there is calcification of the scapulae and cannon bones. Soon after, growth is completed in the lower end of the radius and the tarsal bones. This indicates that, despite genetic makeup, the ultimate height of the horse is decided by nourishment during gestation as well as the later milk capacity of the mare.

Our studies were conducted on a primitive breed of Tarpan-type Polish ponies, which are characterized by late maturity and have dimensions intermediate to those between saddle-horses and Shetland ponies. Embryos were transferred to heavy-type mares, whose dimensions also greatly differ from those of Shire-type horses. Embryo transfers could certainly be used to demonstrate even greater maternal effects with the selection of horses which exhibit much greater differences in height and body weight, as did Walton & Hammond (1938). Moreover, through appropriate selection of recipient mares for embryo transfer amongst pure breeds, it would clearly be possible to influence the ultimate size of the horse.

This work was supported by grant No. CPBP 050 6324.

References

Allen, W.R. & Rowson, L.E.A. (1975) Surgical and non-surgical egg transfer in horses. *J. Reprod. Fert., Suppl.* **23**, 525–530.

Brumby, P.J. (1960) The influence of the maternal environment on growth in mice. *Heredity* **14**, 1–5.

Brzeski, E. (1960) *Studies on horse development and growth.* Ph.D. diss., Acad. Agric. Krakow, 80 pp. In Polish.

Burns, M. (1972) Effects of ova transfer on the birthcoats of lambs. *J. agric. Sci., Camb.* **78**, 1–6.

Hammond, J., Jr., Mason, I.L. & Robinson, T.J. (1971) Quantitative genetics and its application. In *Hammond's Farm Animals*, Vol. 10, pp. 215–232. Edward Arnold, London.

Hunter, G.L. (1956) The maternal influence on size in sheep. *J. agric. Sci., Camb.* **48**, 36–60.

Tischner, M. (1985) Embryo recovery from Polish-pony mares and preliminary observations on foal size after transfer of embryos to large mares. *Equine vet. J., Suppl.* **3**, 96–98.

Venge, O. (1950) Studies of the maternal influence on birth weight in rabbits. *Acta zool., Stockh.* **30**, 1–7.

Walton, A. & Hammond, J. (1938) The maternal effects on growth and conformation in Shire horse–Shetland pony crosses. *Proc. R. Soc. B* **125**, 311–335.

Wienner, G., Wilmut, I. & Field, M.J. (1977) Maternal and lamb breed interactions in the concentration of copper in the tissues and plasma of sheep. In *Proc. 3rd Int. Symp. Trace Elements Metab. Man Anim.*, pp. 469–476. Ed. M. Kirchgessner Inst. fur Ernahurungsphysiologie, Freising-Weihenstephan.

J. Reprod. Fert., Suppl. **35** (1987), 711–714

Plasma steroid concentrations after conception in mares

J. L. Pipkin, D. W. Forrest, G. D. Potter, D. C. Kraemer and J. M. Wilson

Texas Agricultural Experiment Station and Department of Animal Science, Texas A&M University, College Station, Texas 77843, U.S.A.

The transport of the horse embryo into the uterus normally occurs about 6 days (144 h) after ovulation (Oguri & Tsutsumi, 1972). The horse is unique in that unfertilized ova can be retained in the oviduct for variable periods of time (Betteridge & Mitchell, 1974). Ovarian steroid hormones may be involved in controlling the timing of embryo transport through the oviduct (Jansen, 1984). The present study was designed to determine whether progesterone and oestradiol concentrations in peripheral plasma from ovulation to Day 6 after ovulation differ between mares with fertilized or unfertilized ova. Furthermore, the temporal changes in plasma progesterone and oestradiol concentrations were characterized during the first 6 days after ovulation.

Materials and methods

Mares (16 at 2 years old and 2 at 3 years old) of Quarter Horse-type breeding were used in this study. Six mares were sham-inseminated and 10 mares were inseminated to ensure embryo recovery from at least 6 mares. Intrarectal ultrasound examinations (Technicare 210 DX) were performed twice daily beginning at onset of oestrus through ovulation to monitor follicular development and ovulation. Mares were inseminated every other day with 500×10^6 motile spermatozoa extended with bovine serum albumin (BSA) extender until ovulation (inseminated) or infused with 10 ml BSA extender (sham-inseminated). Non-surgical embryo collection (Wilson *et al.*, 1985) was performed on all mares on Day 6 (~ 144 h) after ovulation. Embryos were recovered from 6 inseminated mares, while no ova were recovered from the 6 sham-inseminated mares. Blood samples were obtained by jugular venepuncture twice daily during oestrus. Upon detection of ovulation, samples were collected at 6-h intervals for 4 days and every 2 h for the ensuing 2 days. Prostaglandin F-2α (10 mg; Lutalyse) was administered after embryo recovery and daily blood samples were taken until the mare exhibited oestrus.

Plasma progesterone concentrations were determined by radioimmunoassay (Moseley *et al.*, 1979) using a progesterone antiserum (GDN-337) validated by Gibori *et al.* (1977). The oestradiol antiserum characterized by Korenman *et al.* (1974) was used to quantify concentrations of oestradiol in plasma in the radioimmunoassay described by Moseley *et al.* (1979) without chromatography. Oestradiol concentrations in plasma after ether extraction were similar ($P > 0.05$) to those in plasma that had been extracted and chromatographed.

The 'repeated' option in the General Linear Models (GLM) procedure of the Statistical Analysis System (SAS) was used to evaluate the effects of conception on mean plasma hormone concentrations across time. The 'profile' transformation was used to generate contrasts between adjacent mean concentrations of oestradiol and of progesterone before and after ovulation. Further, the 'contrast' transformation of the 'repeated' option was used to compare mean progesterone concentrations before and after ovulation to mean progesterone concentration at 144 h after ovulation (embryo recovery).

Results

In 6 of 9 collections from inseminated mares 7 embryos and 3 degenerative ova were recovered, while no embryos or ova were recovered in 6 collections from sham-inseminated mares. Mean

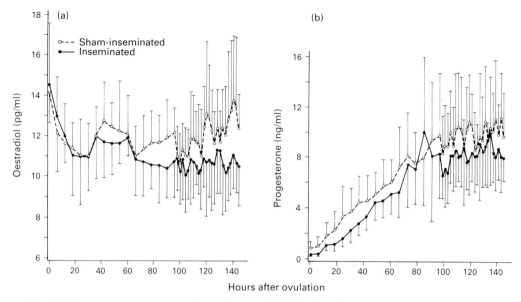

Fig. 1. Mean ± s.e.m. concentrations of oestradiol-17β (a) and progesterone (b) in plasma of inseminated and sham-inseminated mares.

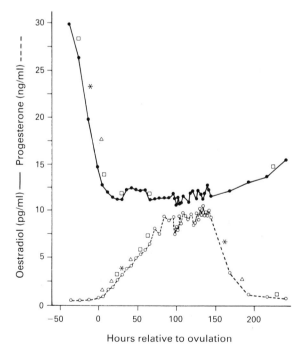

Fig. 2. Analysis with profile contrasts for adjacent mean oestradiol and for adjacent mean progesterone concentrations. □, Denotes difference between adjacent means, $P < 0.05$; △, denotes difference between adjacent means $P < 0.01$; *, denotes difference between adjacent means, $P < 0.001$.

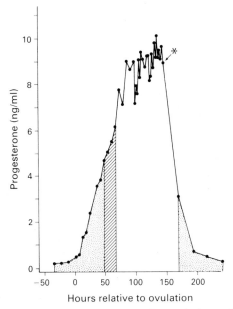

Fig. 3. Contrast analysis for progesterone concentrations relative to Hour 144 after ovulation (∗). ▨ Denotes difference between designated mean and Hour 144 mean, $P < 0.01$; ▢ denotes difference between designated mean and Hour 144 mean, $P < 0.001$.

plasma oestradiol concentrations in both groups of mares were similar ($P > 0.05$, Fig. 1a). Mean plasma progesterone concentrations in inseminated and sham-inseminated mares did not differ ($P > 0.05$) (Fig. 1b). Data from both treatment groups were combined to characterize temporal changes in steroid hormone profiles.

Mean $\pm$ s.e. peak oestradiol concentration (30.1 ± 7.6 pg/ml) occurred 32 ± 6 h before ovulation, and then declined ($P < 0.05$) to the nadir by 6 h after ovulation (Fig. 2). An elevation ($P < 0.05$) in mean concentration occurred at 30 h after ovulation.

Mean progesterone concentration increased ($P < 0.01$) by 12 h after ovulation and continued to rise until 66 h after ovulation (Fig. 3). Mean $\pm$ s.e. peak concentration (13.7 ± 5.6 ng/ml) was attained at 116 ± 5 h after ovulation. Values then declined ($P < 0.001$) by 24 h after embryo collection (PGF-2α administration) and fell below 1 ng/ml by 50 ± 22 h after embryo collection.

Discussion

Conception did not affect the oestradiol or progesterone profile in peripheral plasma during the first 6 days after ovulation in the present study. Changes in secretion of steroid hormones involved in regulating oviducal transport of the ovum therefore may not be dependent upon fertilization of the ovum. It is also possible that changes in oestradiol or progesterone secretion due to the presence of a fertilized ovum may not be detectable in the peripheral circulation in the mare. Both oestrogen (Fuentalba *et al.*, 1982) and progesterone (Forcelledo *et al.*, 1981) are involved in oviducal transport of embryos in the rat.

The timing of the preovulatory peak oestradiol concentration in the present study (24–36 h before ovulation) is comparable with the timing (1–3 days before ovulation) for the peak oestradiol concentration described by Pattison *et al.* (1974). The decrease in oestradiol concentration preceding ovulation in the current study is also consistent with previous reports (Noden *et al.*, 1975; Nelson *et al.*, 1985).

The rapid increase in progesterone concentrations (within 12 h after ovulation) in this study is in agreement with the results reported by Stabenfeldt *et al.* (1972). Progesterone concentrations continued to increase to 3 days after ovulation, and then the elevated progesterone concentrations were maintained until embryo collection on Day 6 after ovulation.

In summary, these results indicate that post-ovulatory progesterone and oestradiol profiles in peripheral plasma in the mare are not affected by the presence of a fertilized ovum within the oviduct. Furthermore, the transport of the embryo from the oviduct into the uterus does not appear to be associated with an acute change in concentration of either hormone in the peripheral circulation in the mare.

We thank Dr G. D. Niswender, Colorado State University, Ft. Collins, Colorado for the gift of anti-progesterone and anti-oestradiol sera. This study was supported in part by Texas Agricultural Experiment Station Project H-6555.

References

Betteridge, K.J. & Mitchell, D. (1974) Direct evidence of retention of unfertilized ova in the oviduct of the mare. *J. Reprod. Fert.* **39**, 145–184.

Forcelledo, M.L., Vera, R. & Croxatto, H.B. (1981) Ovum transport in pregnant, pseudopregnant and cycling rats and its relationship to estradiol and progesterone blood levels. *Biol. Reprod.* **24**, 760–765.

Fuentalba, B., Vera, R., Nieto, M. & Croxatto, H.B. (1982) Changes in nuclear estrogen receptor level in the rat oviduct during ovum transport. *Biol. Reprod.* **27**, 12–16.

Gibori, G., Antczak, E. & Rothchild, I. (1977) The role of estrogen in the regulation of luteal progesterone secretion in the rat after day 12 of pregnancy. *Endocrinology* **100**, 1483–1495.

Jansen, R.P.S. (1984) Endocrine response in the Fallopian tube. *Endocrine Reviews* **5**, 525–551.

Korenman, S.G., Stevens, R.H., Carpenter, L.A., Rabb, M., Niswender, G.D. & Sherman, G.M. (1974) Estradiol radioimmunoassay without chromatography: procedure, validation and normal values. *J. clin. Endocr. Metab.* **38**, 718–720.

Moseley, W.M., Forrest, D.W., Kaltenbach, C.C. & Dunn, T.G. (1979) Effect of norgestomet on peripheral levels of progesterone and estradiol-17β in beef cows. *Theriogenology* **11**, 331–341.

Nelson, E.M., Kiefer, B.L., Roser, J.F. & Evans, J.W. (1985) Serum estradiol-17β concentrations during spontaneous silent estrus and after prostaglandin treatment in the mare. *Theriogenology* **23**, 241–262.

Noden, P.A., Oxender, W.D. & Hafs, H.D. (1975) The cycle of oestrus, ovulation and plasma levels of hormones in the mare. *J. Reprod. Fert., Suppl.* **23**, 189–192.

Oguri, N. & Tsutsumi, Y. (1972) Non-surgical recovery of equine eggs, and an attempt at non-surgical egg transfer in horses. *J. Reprod. Fert.* **31**, 187–195.

Pattison, M.L., Chen, C.L., Kelley, S.T. & Brandt, G.W. (1974) Luteinizing hormone and estradiol in peripheral blood of mares during the estrous cycle. *Biol. Reprod.* **11**, 245–250.

Stabenfeldt, G.H., Hughes, J.P. & Evans, J.W. (1972) Ovarian activity during the estrous cycle of the mare. *Endocrinology* **90**, 1379–1384.

Wilson, J.M., Kreider, J.K., Potter, G.D. & Pipkin, J.L. (1985) Repeated embryo recovery from two-year-fillies. *Proc. 9th Equine Nutr. Physiol. Symp.* 398–401.

J. Reprod. Fert., Suppl. **35** (1987), 715–716

Embryo survival during early gestation in energy-deprived mares

J. T. Potter, J. L. Kreider, G. D. Potter, D. W. Forrest, W. L. Jenkins and J. W. Evans

Texas Agricultural Experiment Station, Department of Animal Science, Texas A&M University, College Station, TX 77843, U.S.A.

Restriction of energy intake may affect concentrations of progesterone and cortisol in the blood of mares. Asa & Ginther (1982) suggested that plasma progesterone concentrations may be affected by stress in mares, and that a significant decrease in progesterone concentration could lead to embryonic mortality. Van Niekerk & Morgenthal (1982) reported that stress in mares increased plasma concentrations of glucocorticoids in the blood.

Nineteen mares of known fertility with condition scores of 5·5 or above were allotted to one of two treatments in a 2×2 factorial experiment. Upon diagnosis of pregnancy at 14–16 days after ovulation by ultrasonography, 9 mares were placed on a control diet that maintained body weight throughout the experiment. Ten mares were placed on a restricted energy diet that was estimated to result in a loss of 0·6–0·9 kg body weight per day.

Pregnancies were verified and embryonic vesicular sizes were measured weekly using ultra-sonography. Blood samples were collected bi-weekly beginning the day of ovulation and continuing through 90 days of gestation. Plasma progesterone concentrations were determined by radioimmunoassay, as described by Kiefer *et al.* (1979). Plasma cortisol concentrations were determined by radioimmunoassay, using a Micromedic Systems Cortisol RIA Kit.

Three mares that were subjected to restricted energy experienced fetal loss during the experiment. Fetal loss in one mare occurred during Week 11 and the last measurable vesicle size was 0·6 cm. The conceptus slowly decreased in size, beginning at Week 8, with total disappearance occurring during Week 11, thus indicating a gradual resorption of the conceptus. Plasma progester-one concentration decreased from 8·4 ng/ml at Week 10 to 2·2 ng/ml at the end of Week 12. A lowered plasma progesterone concentration was therefore detected after the conceptus started to decrease in size and may or may not have been a contributing factor responsible for the initiation of fetal loss. Fetal loss in the second mare occurred between Weeks 7 and 8. The vesicle size at Week 7 was 2·1 cm, with total disappearance of the conceptus occurring by the end of Week 8. Plasma progesterone concentration decreased from 4·9 ng/ml at Week 5 to 1·7 ng/ml at Week 6, and was 2·1 and 1·4 ng/ml at Weeks 7 and 8, respectively. A progesterone deficiency could therefore have caused the fetal loss. The third mare aborted between Weeks 12 and 13, and the expelled fetus was found on Day 87. No change in plasma progesterone concentrations occurred, as they were 17·8 and 18·0 ng/ml at the end of Weeks 12 and 13, respectively. Therefore, the abortion was apparently not associated with a lowered plasma progesterone concentration. A dysfunction of progesterone receptor sites in the pituitary or uterus (Asa & Ginther, 1982) or a non-endocrine related cause (bacterial infection) were thought to be possible reasons for abortion in the third mare.

Mare weights were significantly higher throughout the experiment for the control mares than for the mares fed the restricted energy diet that maintained pregnancy. Mares in the control group maintained a mean body weight of 554 ± 13 kg throughout the experiment. Although the energy-restricted group of mares had significantly lower body weights than the control mares at the beginning of the study, they consistently lost weight at an average of 0·76 kg per day. Mares fed the restricted energy diet and experiencing fetal loss had significantly heavier body weights during Weeks 9–12 than did mares in the restricted energy group which did not experience fetal loss.

Plasma progesterone concentrations were significantly lower for the control group than for the restricted energy group that maintained pregnancy during Weeks 8 through 13. Plasma progesterone concentrations before fetal loss in the group that aborted or experienced fetal loss were not different from those of the energy-restricted group that maintained pregnancy.

Plasma cortisol concentrations were lower for the restricted energy group that maintained pregnancy than for the control group during Weeks 7·5, 8, 10, 11·5 and 12·5. Plasma cortisol concentrations were lower for the restricted energy group that experienced fetal loss than for the energy-restricted mares that maintained pregnancy at Week 1 only. Because of the sampling procedure (bi-weekly) used, one cannot conclude from these results that cortisol had an effect on progesterone concentration.

In summary, mares subjected to energy restriction continually lost body weight throughout the trial. Fetal loss rates were affected ($P < 0.07$) by restriction of energy, and mares fed a restricted energy diet had higher plasma progesterone and lower plasma cortisol concentrations.

References

Asa, C.S. & Ginther, O.J. (1982) Glucocorticoid suppression of oestrus, follicles, LH, and ovulation in the mare. *J. Reprod. Fert., Suppl.* **32,** 247–251.

Kiefer, B.L., Roser, J.F., Evans, J.W., Neely, D.P. & Pacheco, C.A. (1979) Progesterone patterns observed with multiple injections of a PGF-2α analogue in the cyclic mare. *J. Reprod. Fert., Suppl.* **27,** 237–244.

Van Niekerk, C.H. & Morgenthal, J.C. (1982) Fetal loss and the effect of stress on plasma progestagen levels in pregnant Thoroughbred mares. *J. Reprod. Fert., Suppl.* **32,** 453–457.

J. Reprod. Fert., Suppl. **35** (1987), 717–718

Dipstick test for pregnancy (PMSG) and oestrus (LH) in mares

K. C. Kasper, S. Fashandi, T. Le and G. Marr

Monoclonal Antibodies, Inc., 2319 Charleston Rd, Mountain View, CA 94043, U.S.A.

MAb's 30-min dipstick test for pregnant mares' serum gonadotrophin (PMSG) also detects horse luteinizing hormone (LH) semiquantitatively in 2 h. It is a sensitive (as low as 50 ng/ml for purified PMSG, Papkoff), monoclonal antibody-based enzyme-linked immunospecific assay (ELISA). Blue colour forms proportionally to concentration of hormone on a white plastic dipstick. Blood plasma or serum (1419 samples) collected from 210 inseminated mares (3 sites) was tested for PMSG. Occurrence of PMSG test-positive mares after insemination was 0% (0–33 days), 36% (35–39 days), 98% (40–44 days), 100% (49–69 days) and 98% (70–99 days). Detection of PMSG was as early as 34 days until 139 days (latest sample collected) (Table 1). No false positives (369 samples) resulted for 35 mares not conceiving. For LH in plasma, blood was collected daily (April–June) from 19 mares (35 cycles) and quantitated with purified horse LH (Papkoff) by spectrophotometric ELISA in antibody-coated tubes. The first day of oestrus ($3\cdot2 \pm 2\cdot4$ ng/ml) was determined by teasing and ovulation ($8\cdot7 \pm 3\cdot6$ ng/ml) by palpation. The dipstick test correctly identified oestrus compared to teasing, including a silent oestrus (no response to teasing, LH surge, progesterone assay indicated a new corpus luteum). Two mares cycled once, retained their corpus luteum for ≥ 25 days, and maintained basal LH concentrations (< 3 ng/ml). The interval of oestrus to ovulation and LH concentration varied considerably per mare and cycle (Fig. 1). Ovulation could not be predicted. Pregnancy and oestrus were therefore detected with identical test materials by varying the assay protocol.

Table 1. Occurrence of PMSG in pregnant mares in 1050 samples of blood, plasma or serum from 175 pregnant mares

Days after mating	No. of mares				No. of samples			
	−	+	Total*	% +ve	−	+	Total	% +ve
1–9	8	0	8	0	11	0	11	0
10–19	32	0	32	0	48	0	48	0
20–29	106	0	106	0	186	0	186	0
30–34	52	1	53	2	93	1	94	1
35–39	70	37	102	36	116	39	155	25
40–44	7	57	58	98	12	101	113	89
45–49	1	82	82	100	1	101	102	99
50–59	0	44	44	100	0	48	48	100
60–69	0	53	53	100	0	68	68	100
70–79	3	42	43	98	6	85	91	93
80–89	2	32	32	100	7	97	104	93
90–99	1	16	17	94	1	19	20	95
$\geqslant 100$	1	9	10	90	1	9	10	90

*With frequent sampling, the sum of − and + values may exceed actual total of mares, indicating transition from negative to positive in the interval.

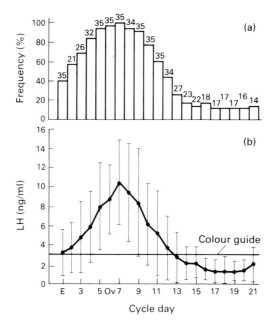

Fig. 1. Oestrus (Ov) distinguished by profile of LH: (a) frequency (%) of dipsticks darker blue than colour guide with total number of samples tested shown above each bar, and (b) LH in plasma (mean ± s.d.); line drawn shows equivalent concentration for colour guide in dipstick assay.

Unusual clinical situations or discrepancies between the original diagnosis (ultrasound or palpation) and MAb's PMSG dipstick test result were noted in 7 of the 210 inseminated mares. Presumed misdiagnosis was concluded for 2 mares when reconfirmation on site was not possible and tests for progesterone and LH were also inconsistent with pregnancy. Persistent endometrial cups after abortion (1 mare), slow development of the cups (1 mare), failure to implant (1 mare) and early fetal loss after a positive ultrasound examination (2 mares) were the reasons for the other 5 mares.

J. Reprod. Fert., Suppl. **35** (1987), 719

Enzyme immunoassay for the qualitative detection of chorionic gonadotrophin, an indicator of pregnancy in mares

T. F. Lock, J. A. DiPietro and K. C. Kasper*

*University of Illinois, College of Veterinary Medicine, Urbana, IL 61801, and
Monoclonal Antibodies Inc., Mountain View, CA 94043, U.S.A.

A monoclonal antibody-based enzyme immunoassay ('Marechek') for the qualitative detection of horse chorionic gonadotropin (CG) was evaluated for accuracy in detection of pregnancy in mares. Chorionic gonadotrophin is produced by the endometrial cups of pregnant mares. These cups form 36–40 days after conception. Horse CG is detectable in the blood between 37 and 42 days of pregnancy. The concentration increases rapidly, usually peaking around Day 60. It then declines and becomes undetectable between 120 and 150 days of pregnancy.

The 'Marechek' test is an enzyme-linked immunospecific dipstick assay. The assay uses two monoclonal antibodies to detect the presence of horse CG: one is coated on the end of a dipstick and the other is linked to an enzyme (alkaline phosphatase). The antibody on the dipstick captures any CG in the sample. Then the enzyme-linked antibody binds to the captured CG, sandwiching it between the antibodies. The dipstick is then placed in a substrate solution. If enzyme is bound, it converts the substrate to a blue product, which is deposited on the end of the dipstick, indicating the presence of CG. If no enzyme is bound, the sample contained no CG and no colour change occurs.

In this test, 349 blood samples were obtained from 144 pregnant mares between 1 and 139 days of pregnancy; 25 blood samples were also collected from 20 non-pregnant mares. Blood was collected in heparinized tubes. Plasma was separated and refrigerated until assays were done. Pregnancy status was confirmed by ultrasound examination.

Of 116 samples collected between Days 1 and 34, only one sample was positive for CG (0·80%). Of 77 samples collected between Days 35 and 39, 35 were positive for CG (45·4%). Of 146 samples collected between Days 40 and 99, 144 were positive for CG (98·6%). Of 10 samples collected between Days 100 and 139, 9 were positive for CG (90%). CG was not detected in any of the samples from the non-pregnant mares. This assay was therefore highly accurate between Days 40 and 99 of pregnancy.

J. Reprod. Fert., Suppl. 35 (1987), 721–723

Printed in Great Britain
© 1987 Journals of Reproduction & Fertility Ltd

Circulating oestrone sulphate concentrations in pregnant mares undergoing ovariectomy

L. H. Kasman, G. H. Stabenfeldt*, J. P. Hughes*, M. A. Couto*
and B. L. Lasley

*San Diego Zoological Society, P.O. Box 551, San Diego, CA 92112, and *Department of Reproduction, School of Veterinary Medicine, U. C. Davis, CA 95616, U.S.A.*

Pregnancy in the domestic horse is associated with a rise in circulating and urinary oestrone sulphate which is detected about 35 days after conception. The source of the oestrone sulphate is somewhat controversial and the proportion of this oestrogen rise derived from the maternal ovaries has yet to be established. Ovariectomy at 35 days after mating resulted in a blunting of the early oestrogen rise, supporting the concept that the maternal ovaries are the major source of oestrogen production (Terqui & Palmer, 1979). Other reports (Flood *et al.*, 1979; Heap *et al.*, 1982) indicate that the products of conception are capable of producing the quantity of oestrogen that is observed in this period of gestation.

The present study was conducted to provide direct evidence regarding the proportion of oestrogens that is contributed by the maternal ovaries in early pregnant mares. To accomplish this,

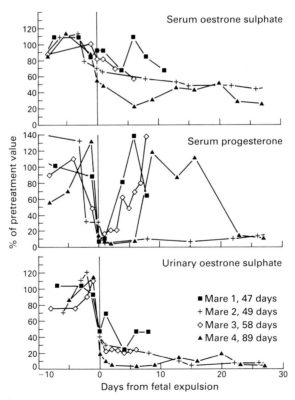

Fig. 1. Concentrations of oestrone sulphate in serum and urine and of progesterone in serum in mares treated with prostaglandin.

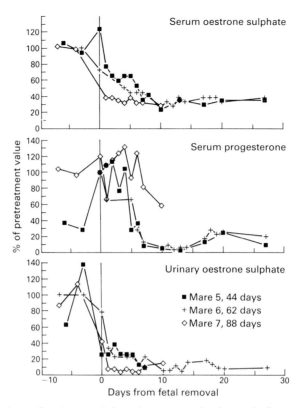

Fig. 2. Concentrations of oestrone sulphate in serum and urine and of progesterone in mares after surgical removal of the fetus.

oestrone sulphate concentrations in mares that were induced to abort or had fetuses removed surgically were compared to those in mares that were ovariectomized but in which pregnancy was maintained by progesterone treatment.

Pregnancy was terminated in 7 mares between Days 44 and 89 by prostaglandin treatment (N = 4) or surgery (N = 3) as reported by Kasman *et al.* (1984). Serum and urinary concentrations of oestrone sulphate were assessed by the method described by Evans *et al.* (1984). Serum progesterone concentrations were also determined by enzyme assay as reported by Munro & Stabenfeldt (1984).

Three additional mares were ovariectomized between Days 51 and 58 of gestation and then treated with 2000 mg progesterone ('Repositol Progesterone': progesterone in 15% benzyl alcohol, 15% ethanol, 45% propylene glycol from Pitman-Moore; Stewart *et al.*, 1982). Daily serum samples were collected 7 days before and 11–15 days after ovariectomy and evaluated for oestrone sulphate values.

The percentage changes from the pre-abortion values of serum oestrone sulphate and progesterone, as well as urinary oestrone sulphate indexed by creatinine are shown in Fig. 1 for 4 mares which were induced to abort by prostaglandin treatment on Days 47, 49, 58 and 89. Serum and urine oestrone sulphate concentrations declined immediately after expulsion of the fetus and in general continued to decline for 10–20 days. Serum progesterone values fell precipitously in all mares, beginning on the day before fetal expulsion but rebounded to pretreatment concentrations within 10 days after expulsion of the fetus in 3 of the 4 mares.

The percentage changes of hormone concentrations from surgical removal of the fetus at Days 44, 62 and 88 of gestation are shown in Fig. 2. As with prostaglandin treatment, serum and urinary

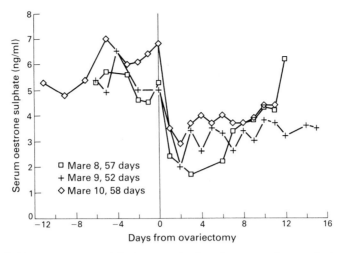

Fig. 3. Serum oestrone sulphate concentrations in mares after ovariectomy.

oestrone sulphate levels declined immediately and continuously after removal of the fetus. By 10 days after surgery serum and urinary oestrone sulphate values were at non-pregnant levels. Unlike the prostaglandin treatment group, serum progesterone concentrations did not fall during the first 4 days after surgery.

Serum oestrone sulphate is depicted for the 3 mares that were ovariectomized between Days 51 and 58 of pregnancy in Fig. 3. Concentrations of serum oestrone sulphate fell precipitously for 2 days after surgery and then plateaued at about 50% of pre-surgery levels. Between Days 3 and 5 after surgery, values in all 3 mares increased to reach about 80% of pretreatment levels by 10 days after surgery.

We conclude that the major source of oestrogen production in early pregnancy is the conceptus.

References

Evans, K.L., Kasman, L.H., Hughes, J.P., Couto, M.A. & Lasley, B.L. (1984) *Theriogenology* **22**, 615–620.

Flood, P.F., Betteridge, K.J. & Irvine, D.S. (1979) Oestrogens and androgens in blastocoelic fluid and cultures of cells from equine conceptuses of 10–22 days gestation. *J. Reprod. Fert., Suppl.* **27**, 413–420.

Heap, R.B., Hamon, M. & Allen, W.R. (1982) Studies on oestrogen synthesis by the preimplantation equine conceptus. *J. Reprod. Fert., Suppl.* **32**, 343–352.

Kasman, L.H., Hughes, J.P., Couto, M.A. & Lasley, B.L. (1984) Changes of circulating and urinary estrogen levels in mares from early pregnancy through induced abortion. *Biol. Reprod.* **30**, Suppl. 1, 68, Abstr. 75.

Munro, C. & Stabenfeldt, G.H. (1984) Development of a microtitre plate enzyme immunoassay for the determination of progesterone. *J. Endocr.* **101**, 41–49.

Stewart, D.R., Stabenfeldt, G.H., Hughes, J.P. & Meagher, D.M. (1982) Determination of the source of equine relaxin. *Biol. Reprod.* **27**, 17–24.

Terqui, M. & Palmer, E. (1979) Oestrogen pattern during early pregnancy in the mare. *J. Reprod. Fert., Suppl.* **27**, 441–446.

J. Reprod. Fert., Suppl. **35** (1987), 725–727

Frequency distribution and daily rhythm of uterine electromyographic epochs of different duration in pony mares in late gestation

F. E. Dudan, J. P. Figueroa, D. A. Frank, B. Elias, E. R. Poore, J. E. Lowe
and P. W. Nathanielsz

*New York State College of Veterinary Medicine, Cornell University, Ithaca, NY 14853-6401,
U.S.A.*

Myometrial activity throughout pregnancy has been described in several species including the horse (Haluska, 1985; Figueroa *et al.*, 1985). In the sheep, epochs of uterine electrical myometrial activity that last longer than 3 min and are associated with low-amplitude increases in intrauterine pressure have been described as 'contractures' (Nathanielsz *et al.*, 1980; Harding *et al.*, 1981). Such contractures have been shown to constitute a distinct and homogeneous population of events, affecting the physiological state of the fetus (Nathanielsz *et al.*, 1980, 1982; Harding *et al.*, 1981).

In late pregnant ponies, direct visual analysis of recording of myometrial activity demonstrated bursts of myometrial activity lasting different lengths of time. However, simple visual analysis may not provide adequate quantification of basal myometrial activity. Therefore, computerized methods have been developed for rigorous pattern analysis of EMG activity (Figueroa *et al.*, 1985).

Aims of study

(1) To determine the frequency distribution of myometrial EMG events of different duration in individual ponies.

(2) To assess whether myometrial activity changes during the last 3 weeks before parturition.

(3) To determine whether myometrial activity displays a daily rhythm during late gestation.

Materials and methods

Five mixed-bred pony mares had 4 sets of multistranded bipolar stainless-steel wires (Cooner AS 632) surgically implanted under general anaesthesia on the uterus, between Days 258 and 282 of gestation. The animals received hay and water *ad libitum*, grain twice daily and were maintained under natural light–dark cycle conditions. The uterine EMG activity of all the ponies was continuously recorded except during 1 h of exercise every 3–4 days. As previously described (Figueroa *et al.*, 1985), a computerized data acquisition storage and analytical system was used for the recording and analysis of uterine EMG. The recorded uterine EMG activity was broken down in sequential periods (from 38 to 118 h) for the analysis. Such periods were then retrospectively classified under intervals expressed in days before parturition. To assess whether myometrial activity was affected by day of gestation, the data obtained from all the ponies were normalized and pooled for each specific interval. The analysis was done on a VAX 11/750 computer, the statistics by one-way analysis of variance (ANOVA) and cosinor analysis.

A major problem in the analysis of the frequency distribution of myometrial EMG events of varied duration is to define the termination of an epoch of myometrial activity. The period of essential quiescence required to terminate an event is called a delimiter and it is an arbitrary definition. For our purpose, it was defined as a preset length of time in which the signal has to fall below the signal:noise threshold designated by a 2-h period running mean (Figueroa *et al.*, 1985).

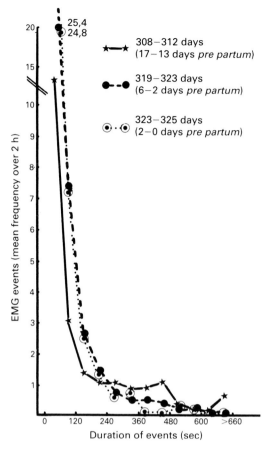

Fig. 1. Frequency distributions of electromyographic events of varied duration at 3 different times before parturition in one pony.

The use of 8-, 16-, 24- and 32-sec delimiters for the analysis did not essentially alter the pattern of the frequency distribution of myometrial EMG events of various durations. Changing the delimiter from 16 to 32 sec produced only a small effect on events of longer duration (≥ 180 sec). In addition, increasing the delimiter from 16 to 32 sec created an initial artificial peak in the distribution of the small events (≤ 180 sec). The subsequent analyses were therefore done using a 16-sec delimiter.

Results

Aim 1. It was shown, using ANOVA, that uterine activity of different durations is not a uniform distribution ($P \leq 0.001$). Comparing our results to those reported in the sheep (Figueroa *et al.*, 1985), the patterns shown by the majority of the frequency distribution suggested that the events might be divided into at least two populations of events, the short (≤ 180 sec) and the long (≥ 180 sec) events. This was based on the mean value given by the end of the exponential function followed by the shorter events, as assessed by visual analysis of 12 frequency distributions taken at random from 4 ponies at different times in late gestation. However, we could not demonstrate a clear-cut separation into 2 mathematically defined populations of events as we have previously demonstrated in the sheep (an exponential model for the short (≤ 180 sec) and a Gaussian model for the long (≥ 180 sec) events: Figueroa *et al.*, 1985). There was a large variability in the patterns of the frequency distribution obtained from the different ponies, for a similar gestation age (expressed as days after ovulation or days before parturition). Furthermore, this variability was also seen in the same individual when frequency distribution patterns were compared from one period to another (Fig. 1).

Aim 2. There was a trend for a decrease in the frequency of the longest events (≥ 180 sec) towards delivery. However, probably because of large individual variability, no statistical significance was observed.

Aim 3. A cosinor analysis was used to assess the presence of a daily rhythmicity of uterine activity. Three periods of EMG recording were tested in 4 ponies during the last month before parturition. At 288–291 and 294–296 days of gestation, the frequency of the short events (≤ 180 sec) recorded in 2 ponies and the frequency of the long events (≥ 180 sec) recorded in one of these ponies (311–313 days) showed daily rhythmicity ($P < 0.025$, 0.020 and 0.005, respectively) with an acrophase at 09:26 h $\pm$ 1.16 h, 00:11 h $\pm$ 1.19 h and 19:14 h $\pm$ 0.57 h respectively. In no instance did uterine activity, expressed as total time active per 2-h period, display any daily rhythm.

Conclusions

Aim 1. Using computerized methods for the analysis of uterine EMG activity, we could not reproduce the findings reported in the sheep (Figueroa *et al.*, 1985). However, our data suggest that there are at least two populations of EMG events in some animals especially at certain stages of gestation (Fig. 1). To characterize further the uterine EMG activity of the pregnant pony, more data and a more detailed analysis of the intrauterine pressure changes are needed.

Aim 2. During the 3 weeks before delivery, no gestational age-related changes could be demonstrated on the frequency of the short (≤ 180 sec) and the long (≥ 180 sec) events or on the total time active per 2-h periods. The lack of statistical significance might be explained by a large variability that could mask individual trends.

Aim 3. A daily rhythm could not be demonstrated as a constant feature in all the animals studied.

We thank Dr Lance Bell, Susan Shell and especially Dr George Haluska for help and comments. The work was supported by the Lalor and the Swiss National Science Foundation.

References

Figueroa, J.P., Mahan, S., Poore, E.R. & Nathanielsz, P.W. (1985) Characteristics and analysis of uterine electromyographic activity in the pregnant sheep. *Am. J. Obstet. Gynecol.* **151**, 524–531.

Haluska, G. (1985) *Electromyographic analysis of the myometrium of the mare correlated with the endocrinology of pregnancy, parturition and the post partum period.* Ph.D. thesis, Cornell University, Ithaca.

Harding, R., Poore, E.R. & Cohn, G.L. (1981) The effect of brief episodes of diminished uterine blood flow on breathing movements, sleep states and heart rate in fetal sheep. *J. devl. Physiol.* **3**, 231–243.

Nathanielsz, P.W., Bailey, A., Poore, E.R., Thorburn, G.D. & Harding, R. (1980) The relationship between myometrial activity and sleep state and breathing in fetal sheep throughout the last third of gestation. *Am. J. Obstet. Gynecol.* **138**, 653–659.

Nathanielsz, P.W., Yu, H.K. & Cabalum, T.C. (1982) Effect of abolition of fetal movement on fetal intravascular PO_2 and incidence of tonic myometrial contractures in the pregnant ewe at 114 to 134 days gestation. *Am. J. Obstet. Gynecol.* **144**, 614–618.

J. Reprod. Fert., Suppl. **35** (1987), 729–730

Printed in Great Britain
© 1987 Journals of Reproduction & Fertility Ltd

GnRH-stimulated release of LH during pregnancy and after parturition

T. M. Nett, C. F. Shoemaker and E. L. Squires

Animal Reproduction Laboratory, Colorado State University, Fort Collins, CO 82523, U.S.A.

The high concentration of oestrogens in the peripheral circulation that occurs during late gestation inhibits synthesis of LH in the anterior pituitary gland (Nilson *et al.*, 1983). This results in a dramatic depletion of pituitary reserves of LH during gestation (Moss *et al.*, 1980, 1985) of most domestic species. Further, the amount of LH secreted is highly correlated with the concentration of LH in the anterior pituitary gland (Crowder *et al.*, 1982; Silvia *et al.*, 1985). This results in a very low rate of secretion of LH during late gestation and during the early post-partum period and is partly responsible for the anoestrus that occurs after parturition. In fact, the blood concentrations of LH during this period are insufficient to induce follicular maturation and ovulation. In many species the duration of this post-partum anoestrous period is greatly extended when the young are allowed to suck their mothers. However, unlike most other domestic species, mares that frequently suckle their foals ovulate soon after parturition. Also, in contrast with other species, the concentrations of oestrone and oestradiol decrease during the last third of pregnancy in the mare (Nett *et al.*, 1973). These observations lead to the hypothesis that the content of LH in the anterior pituitary gland of the mare may increase during late gestation and result in higher levels of LH secretion, leading to normal follicular development and ovulation shortly after parturition.

To test this hypothesis, 10 horse mares were challenged with a maximally stimulatory dose of GnRH (2·0 µg/kg) on Days 240 and 320 of pregnancy and on Day 3 after parturition. We have previously demonstrated that the content of LH in the anterior pituitary gland of mares is highly correlated with the amount of LH released in response to this dosage of GnRH (Silvia *et al.*, 1985). A group of 4 mares was treated with GnRH on Day 2 or 3 of oestrus to serve as controls. The oestrous mares and those on Day 3 after parturition were treated with GnRH at approximately the same time of year (late May or early June). Blood samples were collected at -2, -1, 0, 0·25, 0·5, 0·75, 1, 1·5, 2, 2·5, 3, 3·5, 4, 5, 6, 7 and 8 h relative to the injection of GnRH. Serum was analysed for the concentration of LH by radioimmunoassay (Nett *et al.*, 1975).

Only a minimal release of LH occurred after the injection of GnRH during pregnancy and at Day 3 after parturition in these mares. This was in contrast to a marked release of LH after treatment of oestrous mares with GnRH (Table 1). The smallest release of LH was observed on Day 320 of gestation, a time when the peripheral concentration of oestrogens are reported to be decreasing. The area under the response curve was highly correlated with the basal secretion of LH ($r > 0.99$).

Table 1. Comparative rates of LH secretion

	Baseline (ng/ml)	GnRH response (area under curve)
Pregnant		
Day 240	0·9 ± 0·3	12·5 ± 3·6
Day 320	0·3 ± 0·1	5·7 ± 1·5
Post partum		
Day 3	2·2 ± 0·9	29·1 ± 12·1
Oestrus		
Day 3	59·0 ± 18·5	311·0 ± 54·0

Since both basal secretion of LH and GnRH-induced release of LH are highly correlated with the pituitary content of LH (Silvia *et al.*, 1985), we can conclude that there is very little LH stored in the anterior pituitary of the mare during late pregnancy and early in the post-partum period. Therefore, as in other domestic species, the secretion of LH in the mare during late gestation and after parturition appears to be limited by the amount of LH stored in the anterior pituitary gland. However, the function of the anterior pituitary must recover very rapidly in the mare nursing a foal since enough LH to induce an ovulation is secreted as early as 6–10 days *post partum*.

References

Crowder, M.E., Gilles, P.A., Tamanini, C., Moss, G.E. & Nett, T.M. (1982) Pituitary content of gonadotropins and GnRH-receptors in pregnant, postpartum and steroid-treated OVX ewes. *J. Anim. Sci.* **54,** 1235–1242.

Moss, G.E., Adams, T.E., Niswender, G.D. & Nett, T.M. (1980) Effects of parturition and suckling on concentrations of pituitary gonadotropins, hypothalamic GnRH and pituitary responsiveness to GnRH in ewes. *J. Anim. Sci.* **50,** 496–502.

Moss, G.E., Parfet, J.R., Marvin, C.R., Allrich, R.D. & Diekman, M.A. (1985) Pituitary concentrations of gonadotropins and receptors for GnRH in suckled beef cows at various intervals after calving. *J. Anim. Sci.* **60,** 285–293.

Nett, T.M., Holtan, D.W. & Estergreen, V.L. (1973) Plasma estrogens in pregnant and postpartum mares. *J. Anim. Sci.* **37,** 962–970.

Nett, T.M., Holtan, D.W. & Estergreen, V.L. (1975) Serum levels of oestrogens, luteinizing hormone and prolactin in post-partum mares. *J. Reprod. Fert., Suppl.* **23,** 201–206.

Nilson, J.H., Nejedlik, M.T., Virgin, J.B., Crowder, M. & Nett, T.M. (1983) Estradiol suppresses synthesis of LH by decreasing concentrations of subunit mRNA's in ovine anterior pituitary. *J. biol. Chem.* **25,** 12087–12090.

Silvia, P.J., Squires, E.L. & Nett, T.M. (1985) Changes in the hypothalamic-hypophyseal axis of mares associated with seasonal reproductive recrudescence. *J. Anim. Sci.* **61,** (Suppl. 1), 384, Abstr. 431.

J. Reprod. Fert., Suppl. **35** (1987), 731–732

Plasma prolactin, LH and FSH concentrations after suckling in post-partum mares

J. J. Wiest* and D. L. Thompson, Jr

Department of Animal Science, Louisiana Agricultural Experiment Station, Louisiana State University Agricultural Center, Baton Rouge, Louisiana 70803, U.S.A.

Twelve post-partum mares were used to determine the short-term effects of suckling on plasma concentrations of prolactin, LH and FSH. Mares were of lighthorse type (mainly Quarterhorse) and were used during spring between 7 and 14 days *post partum* (average = 10·2 days). Each foal was separated from its dam for 4 h such that only visual and nose-to-nose contact could occur. An indwelling catheter was placed in one jugular vein of the mare during separation and 4 samples of blood were withdrawn at 15-min intervals. The foal was then reunited with its dam and blood samples were withdrawn at 5, 10, 15, 20, 30, 45 and 60 min relative to first suckling. Prolactin concentrations were determined by a newly-validated radioimmunoassay (Thompson *et al.*, 1986) based on antiserum generated against horse prolactin (supplied by Dr H. Papkoff, Hormone Research Institute, University of California, San Francisco, CA 94143, U.S.A.) and radiolabelled dog prolactin (supplied by Dr A. F. Parlow, Pituitary Hormones and Antisera Center, Harbor-UCLA Medical Center, Torrance, CA 90509, U.S.A.). Concentrations of LH and FSH were determined as described previously (Thompson *et al.*, 1983a, b).

All foals sucked immediately for $5·18 \pm 0·64$ (s.e.m.) min and intermittently thereafter. Averaged over all mares, concentrations of prolactin in plasma increased ($P < 0·05$) within 15 min after suckling and remained elevated ($P < 0·05$) through 60 min (see Fig. 1). However, only 4 of the 12 mares exhibited obvious increases in prolactin concentrations during this time; prolactin concentrations in the remaining mares were constant. Concentrations of LH and FSH in plasma were unchanged over the sampling period for all mares (Fig. 1).

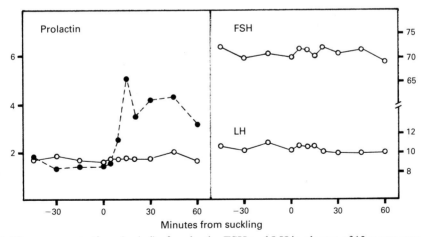

Fig. 1. Mean concentrations (ng/ml) of prolactin, FSH and LH in plasma of 12 mares separated from their foals for 4 h and then allowed to suckle (time 0). Means for the 4 mares that responded with an increase in prolactin concentrations are indicated by the broken line and for the remaining 8 mares by the solid line.

*Present address: Department of Animal Science, Texas A&M University, College Station, Texas 77843, U.S.A.

An effort was made to determine what factor(s) may have contributed to the variation in prolactin response amongst mares. Stage of the oestrous cycle, average concentrations of progesterone, LH and FSH and pre-suckling concentrations of prolactin did not differ between mares that responded with a prolactin rise and those that did not. Moreover, there was no noticeable behavioural difference (such as relative calmness or nervousness) to account for the difference in response.

Prolactin therefore appears to be released during suckling in some but not all mares. In contrast, concentrations of LH and FSH were not affected in any mares, indicating that prolactin and the gonadotrophins are secreted via different mechanisms in the post-partum mare.

References

Thompson, D.L., Jr, Godke, R.A. & Squires, E.L. (1983a) Testosterone effects on mares during synchronization with altrenogest: FSH, LH, estrous duration and pregnancy rate. *J. Anim. Sci.* **56**, 678–686.

Thompson, D.L., Jr, Reville, S.I., Walker, M.P., Derrick, D.J. & Papkoff, H. (1983b) Testosterone administration to mares during estrus: duration of estrus and diestrus and concentrations of LH and FSH in plasma. *J. Anim. Sci.* **56**, 911–918.

Thompson, D.L., Jr, Wiest, J.J. & Nett, T.M. (1986) Measurement of equine prolactin with an equine-canine radioimmunoassay: seasonal effects on the prolactin response to thyrotropin releasing hormone. *Dom. Anim. Endocr.* **3**, 247–252.

J. Reprod. Fert., Suppl. **35** (1987), 733–734

Uterine involution, ovarian activity, and fertility in the post-partum mare

E. Koskinen and T. Katila*

*The State Horse Breeding Institute, 32100 Ypäjä and *College of Veterinary Medicine, 04840 Hautjärvi, Finland*

It is commonly agreed that the foaling rate of mares mated at the first post-partum oestrus is lower than for subsequent oestrous periods. In spite of that it is common practice to mate mares at the first oestrus, since the time from foaling to conception is on average shorter if the mares are mated early in the post-partum period. Many authors have emphasized the importance of evaluation of individual mares when deciding whether to mate at the foal heat or not. The purpose of this study was to determine criteria that could be used when evaluating whether a mare should be mated at the first post-partum oestrus or later.

The course of uterine involution and ovarian function was followed in 34 post-partum periods. The mares were mostly of Finnhorse breed, the age being 12·3 ± 4·97 years (4–23) and the number of foalings 3·9 ± 2·80 (1–13). The mares were examined and sampled on the 2nd and 5th day after parturition and thereafter every other day until the first ovulation had taken place. Biopsies were taken only on Days 2 and 5. The examination included rectal palpation, vaginoscopy and ultrasound scanning. Ovarian activity was monitored also by progesterone determinations. Uterine swabs were examined for the presence of bacteria and neutrophils.

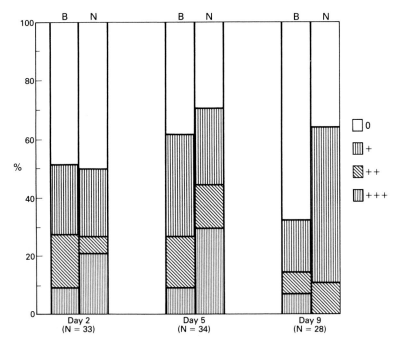

Fig. 1. The occurrence of bacteria (B) and neutrophils (N) in uterine swabs of mares on various days after parturition.

Table 1. The occurrence of neutrophils in uterine biopsies of mares on Days 2 and 5 *post partum*

	Pregnant		Non-pregnant		Total	
	Day 2	Day 5	Day 2	Day 5	Day 2	Day 5
No or only a few neutrophils	9	9	11	5	20	14
Moderate to strong neutrophil infiltration	4	5	9	14	13	19

The mares ovulated for the first time at $11 \cdot 8 \pm 3 \cdot 21$ (6–23) days *post partum*. The conception rates were 47·0 and 66·6% at the 1st and 2nd oestrus, respectively, and the cumulative conception rates were 47·0, 70·6 and 94·1%, respectively for first oestrus, second oestrus and the final rate.

The size of the previously pregnant uterine horn at the last examination before ovulation was the same in mares that conceived at the foal heat and for those that did not (8·8 *vs* 8·4 cm). Some exudate (bloody, grey or purulent mucus) was seen in the vagina in 29% of mares on Day 2, in 56% on Day 5 and in 24% shortly before ovulation. The presence of the discharge had no effect on the conception rate.

More neutrophils were encountered on Day 5 than on Day 2 in uterine swabs (Fig. 1). The conception rate of these mares was not different from that of mares that had no neutrophils in uterine swabs. *Escherichia coli* was the most common organism at 2 days *post partum* (67% of positive samples), but after that their numbers decreased rapidly to 29% on Day 5, and no coliforms were found on Day 9. The proportion of β-haemolytic streptococci increased from 13% on Day 2 to 33% on Day 5 and 67% on Day 9. The next common organisms were coagulase-negative staphylococci and non-haemolytic streptococci. The growth of bacteria was strongest at 5 days *post partum* (Fig. 1).

Neither the type of bacteria nor the amount of bacterial growth on any day was different from the mares that conceived compared with those that did not. The only characteristic that was significantly different ($P < 0 \cdot 05$) for the two groups of mares was the amount of neutrophils in biopsies taken on Day 5 (Table 1). Neutrophils on Day 2 and lymphocytes on either days had no effect on fertility.

J. Reprod. Fert., Suppl. **35** (1987), 733–734

Uterine involution, ovarian activity, and fertility in the post-partum mare

E. Koskinen and T. Katila*

*The State Horse Breeding Institute, 32100 Ypäjä and *College of Veterinary Medicine,
04840 Hautjärvi, Finland*

It is commonly agreed that the foaling rate of mares mated at the first post-partum oestrus is lower than for subsequent oestrous periods. In spite of that it is common practice to mate mares at the first oestrus, since the time from foaling to conception is on average shorter if the mares are mated early in the post-partum period. Many authors have emphasized the importance of evaluation of individual mares when deciding whether to mate at the foal heat or not. The purpose of this study was to determine criteria that could be used when evaluating whether a mare should be mated at the first post-partum oestrus or later.

The course of uterine involution and ovarian function was followed in 34 post-partum periods. The mares were mostly of Finnhorse breed, the age being 12·3 ± 4·97 years (4–23) and the number of foalings 3·9 ± 2·80 (1–13). The mares were examined and sampled on the 2nd and 5th day after parturition and thereafter every other day until the first ovulation had taken place. Biopsies were taken only on Days 2 and 5. The examination included rectal palpation, vaginoscopy and ultra-sound scanning. Ovarian activity was monitored also by progesterone determinations. Uterine swabs were examined for the presence of bacteria and neutrophils.

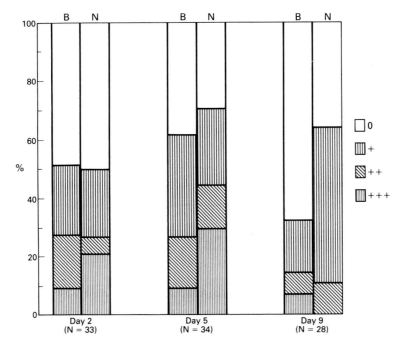

Fig. 1. The occurrence of bacteria (B) and neutrophils (N) in uterine swabs of mares on various days after parturition.

Table 1. The occurrence of neutrophils in uterine biopsies of mares on Days 2 and 5 *post partum*

	Pregnant		Non-pregnant		Total	
	Day 2	Day 5	Day 2	Day 5	Day 2	Day 5
No or only a few neutrophils	9	9	11	5	20	14
Moderate to strong neutrophil infiltration	4	5	9	14	13	19

The mares ovulated for the first time at $11·8 \pm 3·21$ (6–23) days *post partum*. The conception rates were 47·0 and 66·6% at the 1st and 2nd oestrus, respectively, and the cumulative conception rates were 47·0, 70·6 and 94·1%, respectively for first oestrus, second oestrus and the final rate.

The size of the previously pregnant uterine horn at the last examination before ovulation was the same in mares that conceived at the foal heat and for those that did not (8·8 *vs* 8·4 cm). Some exudate (bloody, grey or purulent mucus) was seen in the vagina in 29% of mares on Day 2, in 56% on Day 5 and in 24% shortly before ovulation. The presence of the discharge had no effect on the conception rate.

More neutrophils were encountered on Day 5 than on Day 2 in uterine swabs (Fig. 1). The conception rate of these mares was not different from that of mares that had no neutrophils in uterine swabs. *Escherichia coli* was the most common organism at 2 days *post partum* (67% of positive samples), but after that their numbers decreased rapidly to 29% on Day 5, and no coliforms were found on Day 9. The proportion of β-haemolytic streptococci increased from 13% on Day 2 to 33% on Day 5 and 67% on Day 9. The next common organisms were coagulase-negative staphylococci and non-haemolytic streptococci. The growth of bacteria was strongest at 5 days *post partum* (Fig. 1).

Neither the type of bacteria nor the amount of bacterial growth on any day was different from the mares that conceived compared with those that did not. The only characteristic that was significantly different ($P < 0·05$) for the two groups of mares was the amount of neutrophils in biopsies taken on Day 5 (Table 1). Neutrophils on Day 2 and lymphocytes on either days had no effect on fertility.

J. Reprod. Fert., Suppl. **35** (1987), 733–734

Uterine involution, ovarian activity, and fertility in the post-partum mare

E. Koskinen and T. Katila*

*The State Horse Breeding Institute, 32100 Ypäjä and *College of Veterinary Medicine, 04840 Hautjärvi, Finland*

It is commonly agreed that the foaling rate of mares mated at the first post-partum oestrus is lower than for subsequent oestrous periods. In spite of that it is common practice to mate mares at the first oestrus, since the time from foaling to conception is on average shorter if the mares are mated early in the post-partum period. Many authors have emphasized the importance of evaluation of individual mares when deciding whether to mate at the foal heat or not. The purpose of this study was to determine criteria that could be used when evaluating whether a mare should be mated at the first post-partum oestrus or later.

The course of uterine involution and ovarian function was followed in 34 post-partum periods. The mares were mostly of Finnhorse breed, the age being $12·3 \pm 4·97$ years (4–23) and the number of foalings $3·9 \pm 2·80$ (1–13). The mares were examined and sampled on the 2nd and 5th day after parturition and thereafter every other day until the first ovulation had taken place. Biopsies were taken only on Days 2 and 5. The examination included rectal palpation, vaginoscopy and ultrasound scanning. Ovarian activity was monitored also by progesterone determinations. Uterine swabs were examined for the presence of bacteria and neutrophils.

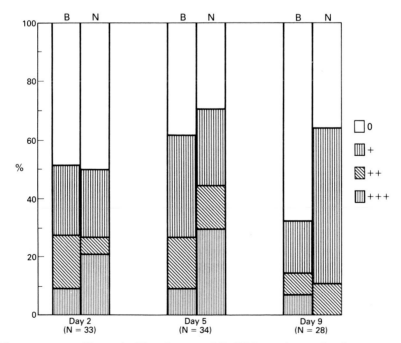

Fig. 1. The occurrence of bacteria (B) and neutrophils (N) in uterine swabs of mares on various days after parturition.

Table 1. The occurrence of neutrophils in uterine biopsies of mares on Days
2 and 5 *post partum*

	Pregnant		Non-pregnant		Total	
	Day 2	Day 5	Day 2	Day 5	Day 2	Day 5
No or only a few neutrophils	9	9	11	5	20	14
Moderate to strong neutrophil infiltration	4	5	9	14	13	19

The mares ovulated for the first time at $11{\cdot}8 \pm 3{\cdot}21$ (6–23) days *post partum*. The conception rates were 47·0 and 66·6% at the 1st and 2nd oestrus, respectively, and the cumulative conception rates were 47·0, 70·6 and 94·1%, respectively for first oestrus, second oestrus and the final rate.

The size of the previously pregnant uterine horn at the last examination before ovulation was the same in mares that conceived at the foal heat and for those that did not (8·8 *vs* 8·4 cm). Some exudate (bloody, grey or purulent mucus) was seen in the vagina in 29% of mares on Day 2, in 56% on Day 5 and in 24% shortly before ovulation. The presence of the discharge had no effect on the conception rate.

More neutrophils were encountered on Day 5 than on Day 2 in uterine swabs (Fig. 1). The conception rate of these mares was not different from that of mares that had no neutrophils in uterine swabs. *Escherichia coli* was the most common organism at 2 days *post partum* (67% of positive samples), but after that their numbers decreased rapidly to 29% on Day 5, and no coliforms were found on Day 9. The proportion of β-haemolytic streptococci increased from 13% on Day 2 to 33% on Day 5 and 67% on Day 9. The next common organisms were coagulase-negative staphylococci and non-haemolytic streptococci. The growth of bacteria was strongest at 5 days *post partum* (Fig. 1).

Neither the type of bacteria nor the amount of bacterial growth on any day was different from the mares that conceived compared with those that did not. The only characteristic that was significantly different ($P < 0{\cdot}05$) for the two groups of mares was the amount of neutrophils in biopsies taken on Day 5 (Table 1). Neutrophils on Day 2 and lymphocytes on either days had no effect on fertility.

J. Reprod. Fert., Suppl. **35** (1987), 735–736
Printed in Great Britain
© 1987 Journals of Reproduction & Fertility Ltd

Relationships among colostral electrolytes, colostral IgG concentrations and absorption of colostral IgG by foals

M. M. LeBlanc and T. Q. Tran

College of Veterinary Medicine, University of Florida, Gainesville, FL 32610, U.S.A.

Measurement of sodium, potassium and calcium concentrations in mammary secretions has been used to assess fetal readiness for birth in Thoroughbred and pony mares (Peaker *et al.*, 1979; Leadon *et al.*, 1984; Ousey *et al.*, 1984). Total calcium concentration has been considered the most important with levels of more than 10 mmol/l indicating impending parturition. It has been suggested that selective concentration of calcium ions by the mammary gland may result in colostrum that enhances active transport mechanisms for intestinal absorption of immunoglobulins (Leadon *et al.*, 1984).

In this study 149 post-partum, presuckle colostrum samples and 96 foal serum samples collected 24 h after parturition were obtained from 4 breeds of horses to determine whether (1) calcium concentration in colostrum was related to foal serum immunoglobulin levels and (2) breed affected colostral electrolyte concentration, colostral IgG concentration, and/or foal serum IgG concentration 24 h after parturition.

Foal serum IgG concentrations as measured by single radial immunodiffusion were highly correlated ($r = 0.70$; $P < 0.0001$) with IgG concentrations of the colostrum each foal ingested (Table 1). No relationships were observed between colostral electrolyte concentrations and foal serum IgG concentrations. Arabian mares had significantly higher ($P < 0.04$) colostral calcium concentrations than did mares of the other breeds. Mean colostral calcium concentrations of post-partum, presuckle colostrum samples in the 4 breeds studied were significantly higher than calcium concentrations previously reported (Leadon *et al.*, 1984). Although methodology for determining calcium concentrations does vary, the O-cresolphthalein complexone method described by Sarker & Chauhan (1967) was used in both studies. Potassium ion concentrations were higher in Arabians than Thoroughbreds ($P < 0.01$). Arabian and Quarter Horse mares had significantly higher colostral IgG concentrations than did Standardbred mares. Arabian foals had significantly higher serum IgG concentrations than Standardbred foals ($P = 0.002$).

Twelve of 36 Thoroughbred mares (33%) prematurely lactated >24 h before foaling. Colostral calcium concentrations were higher in mares that prematurely lactated (Table 2; $P < 0.02$) whereas foal serum IgG concentrations and colostral IgG concentrations were significantly lower (Table 2; $P < 0.001$). Of 31 Standardbred mares 8 (25%) lactated prematurely but significant differences

Table 1. Mean colostral immunoglobulin (Ig) G, colostral calcium concentrations and foal serum IgG concentrations

Breed	No. of colostrum samples	Calcium (mmol/l)	Colostral IgG (g/l)	No. of foal serum samples	Foal IgG (g/l)
Arabian	33	27 ± 1[a]	61 ± 5[a]	22	12 ± 1[a]
Quarter Horse	17	23 ± 2[b]	61 ± 7[a]	10	10 ± 1
Thoroughbred	50	22 ± 1[b]	52 ± 4	34	10 ± 1
Standardbred	49	24 ± 1[b]	40 ± 4[b]	30	8 ± 1[b]

Values are expressed as the mean ± s.e.m.
Means in same column with unlike superscripts differ ($P < 0.04$).

Table 2. Effects of premature lactation on mean colostral calcium, colostral immunoglobulin (Ig) G concentrations and foal serum IgG concentrations

	Thoroughbred		Standardbred	
	Normal	Premature lactation	Normal	Premature lactation
Calcium (mmol/l)	20 ± 1^a	25 ± 2^b	23 ± 1^c	28 ± 3^d
Colostral IgG (g/l)	62 ± 6^a	25 ± 8^b	42 ± 4	30 ± 5
Foal IgG (g/l)	14 ± 1^a	5 ± 1^b	8 ± 1	6 ± 1
No. in group	28	12	23	8

Values are expressed as the mean $\pm$ s.e.m.
Means in same row within breed with unlike superscripts differ ($P < 0.04$).

were seen only in colostral calcium concentrations (Table 2; $P < 0.01$). The higher calcium concentrations seen in mares that prematurely lactated may be due to dilution of their colostrum with milk, since calcium levels are higher in milk than colostrum (Ullrey *et al.*, 1966).

In conclusion, serum IgG concentration in foals is highly correlated with the IgG concentrations of the colostrum ingested. Arabian mares concentrate higher amounts of calcium, potassium and IgG in colostrum than do mares of other breeds studied, whereas Standardbred mares concentrate less IgG in colostrum than do other breeds. Premature lactation affects colostral calcium, colostral IgG concentrations and absorption of colostral immunoglobulins in foals. Colostral calcium concentrations were significantly higher than previously reported. It does not appear that colostral electrolytes are related to absorption of colostral IgG by foals.

References

Leadon, D.P., Jeffcott, L.B. & Rossdale, P.D. (1984) Mammary secretions in normal spontaneous and induced premature parturition in the mare. *Equine vet. J.* **16**, 256–259.

Ousey, J.C., Dudan, F. & Rossdale, P.D. (1984) Preliminary studies of mammary secretions in the mare to assess fetal readiness for birth. *Equine vet. J.* **16**, 259–263.

Peaker, M., Rossdale, P.D., Forsyth, I.A. & Falk, M. (1979) Changes in mammary development and the composition of secretion during late pregnancy in the mare. *J. Reprod. Fert., Suppl.* **27**, 555–561.

Sarker, R.B.C. & Chauham, V.P.S. (1967) A new method for determining microquantities of calcium in biological materials. *Analyt. Biochem.* **20**, 155–160.

Ullrey, D.E., Struthers, R.D. & Henricks, D.G. (1966) Composition of mares' milk. *J. Anim. Sci.* **25**, 217–222.

J. Reprod. Fert., *Suppl.* **35** (1987), 737–738

Printed in Great Britain

A preliminary report on the use of a scoring system for the early identification of sepsis in the equine neonate

B. Brewer and A. Koterba

University of Florida, College of Veterinary Medicine, Gainesville, FL 32610 U.S.A.

Neonatal septicaemia is one of the most common causes of mortality of newborn foals, although frequently the disease remains undiagnosed until post-mortem examination or is missed entirely if no necropsy is performed. A foal which survives the initial septicaemia, if not treated appropriately, will often be plagued later with focal, severe infection (pneumonia, joint ill, enteritis) which may result in death. Septicaemia can be definitively diagnosed *ante mortem* only by the presence of positive blood cultures, but the time lapse involved in obtaining the results (24–48 h) is too lengthy to allow for the early, vigorous treatment necessary to save the patient or prevent serious complications.

No single clinical or historical parameter specifically and consistently identifies the infected neonatal foal and many conditions (neonatal maladjustment syndrome, milk intolerance, prematurity, birth hypoxia and others) present with signs which mimic septicaemia. Newer, expensive antibiotics (with potentially damaging side effects) are available which can, when used for an appropriate duration (2 weeks for septicaemia and up to 1 month for confirmed focal infection), successfully treat the problem and prevent the typical complications including joint infection or osteomyelitis. These antibiotics should not be used irresponsibly for obvious reasons (cost, side effects, development of resistant organisms); furthermore, the duration of the therapy is critical (i.e. 3–5 days is not sufficient even if the foal appears to be recovering clinically). A quick, accurate means of determining the likelihood of septicaemia or focal sepsis in the abnormal neonatal foal is therefore critical and currently does not exist. This abstract describes a weighted scoring system, developed and tested at the University of Florida over a 5-year period which has proved to be extremely useful in defining the need for antibiotic therapy in the abnormal newborn foal ($\leqslant 12$ days of age).

The records of 19 neonatal ($\leqslant 12$ days) foals were retrospectively reviewed. Nine of these foals had confirmed septicaemia (blood culture or post-mortem positive), 2 had confirmed focal infection, 5 had no evidence of infection, and 3 were questionable (thought to be infected but treated and discharged). A weighted scoring system was developed which deliberately included only readily available clinical, laboratory, or historical data.

Foals were examined clinically for the presence or absence of petechiation or scleral injection, fever, hypotonia, coma, depression or convulsions, anterior uveitis, or swollen joints. Laboratory data collected included the segmented and band neutrophil counts, the presence of toxic changes within the neutrophils, fibrinogen values, blood glucose, arterial oxygen tensions, the presence of metabolic acidosis, and the result of the zinc sulphate turbidity test. Historical data regarding the mare, the birth, the placenta and the expected foaling date were also utilized.

This scoring system was then applied prospectively to 187 horse neonates. These neonates included every foal admitted to the University of Florida's Veterinary Medical Teaching Hospital which was $\leqslant 12$ days of age irrespective of the reason for admission (i.e. 'normal' foals admitted with sick mares or foals with limb fractures, but no obvious medical problems were included). Of these 187 patients, 66 had confirmed septicaemia, 18 had confirmed focal infection but were blood culture negative, 64 had no evidence of infection and 39 of the diagnoses were considered questionable since the patients survived and no bacterial organism was isolated. For the purpose of this abstract, the septicaemic and focally infected patients were combined into 1 group ($N = 84$). The second group (N = 64) consisted of foals with no evident infection. Questionable cases were omitted from this particular study.

The sensitivity, specificity, false-positive and false-negative percentages were calculated for the sepsis score and contrasted with the same statistics applied to total white blood cell, segmented neutrophil and band neutrophil counts, the band to neutrophil ratio, the presence or absence of toxic changes, and fibrinogen levels. A Z test was used to determine whether, for each of the above values, the means of the septic *vs* non-septic groups were significantly different.

The sepsis score was much more sensitive and specific for infection, even in very early cases, and had fewer false positives and false negatives than did the total white blood cell, the segmented or band neutrophil counts, the presence or absence of toxic changes, the band to neutrophil ratio, or the fibrinogen levels. Further, it discriminated between the infected and non-infected foals to a far greater degree than did any one of the above measures taken individually.

J. Reprod. Fert., Suppl. **35** (1987), 739–740

Tissue concentrations of carnitine in fetuses, neonates and adult horses

C. V. L. Foster, R. C. Harris and P. D. Rossdale*

*Physiology Unit, Animal Health Trust, Balaton Lodge, Newmarket, Suffolk CB8 7AW, and
Beaufort Cottage Stables, High Street, Newmarket, Suffolk CB8 8JS, U.K.

Carnitine is essential to the transfer of fatty acids across the inner mitochondrial membrane. It is also associated with the regulation of the acetyl CoA/COA ratio within the mitochondrial matrix and with fatty acid synthesis, catabolism of amino acids (Bieber *et al.*, 1982a), thermogenesis, lipolysis, ketogenesis and the shuttling of products of fatty acid chain-shortening out of peroxisomes (Bieber *et al.*, 1982a, b).

Its importance to perinatal metabolism has aroused particular interest. Adaptation from intrauterine to extrauterine life involves metabolic changes. After parturition, when energy intake is low, there is an increased dependence on internal fat stores for energy. Maintenance of body temperature is vital for survival, particularly in premature animals of low bodyweight. Carnitine appears to stimulate lipolysis, ketogenesis and thermogenesis and, therefore, an adequate supply in the first few days of life may be critical.

In many species neonates, especially those born prematurely, have lower tissue carnitine concentrations than do adults. This may be caused by a decreased capacity for carnitine biosynthesis because the last enzyme in the biosynthetic pathway has a low activity. Studies indicate that tissue stores after birth are derived largely from milk. Carnitine is widely considered to be essential to premature neonates, those on feed supplements low in carnitine and those in which carnitine deficiency may accompany some form of pathophysiology.

To establish whether carnitine levels in horse fetuses and foals are lower than adult concentrations and, therefore, whether supplementation might be indicated for premature foals, various tissues obtained at *post mortem* were divided into three groups and analysed for total carnitine concentration (Table 1).

After collection, tissues were freeze-dried, dissected free of connective tissue and powdered. Total carnitine was extracted by alkaline digestion of the muscle powder in 0·5 mmolKOH/l for 1 h at 60°C. Extracts were neutralized with HCl. Carnitine was assayed enzymatically using the DTNB method of Marquis & Fritz (1964).

Table 1. Mean content and s.d. of total carnitine in tissues obtained from fetuses, foals and adults at *post mortem*

| | Total carnitine (mmol/kg) | | | | | | | | |
| | Fetuses | | | Foals | | | Adults | | |
	No.	Mean	s.d.	No.	Mean	s.d.	No.	Mean	s.d.
Middle gluteal (deep)	5	25·7	7·0	7	22·2	5·9	6	25·8	6·6
Middle gluteal (superficial)	5	19·9	5·3	7	20·4	7·7	6	19·5	2·5
Triceps	6	22·3	6·1	5	20·8	2·5	4	28·0	2·1
Semitendinosus	6	24·2	8·5	6	21·2	3·5	6	18·8	5·5
Diaphragm	5	16·5	7·4	7	20·2	7·8	6	30·1	9·7
Heart	5	9·7	3·4	5	8·7	1·7	6	10·3	4·1

The animals used were of various ages, breeds and sex. Gestational ages of fetuses ranged from 6 to 10 months. Cause of death in fetuses and foals varied, including abortion, dystocia, joint-ill and still-birth. Foal ages ranged from 0 to 8 days except for one which was destroyed at 9 months. Adults were between 1 and 23 years of age.

In this preliminary study there was a trend towards lower carnitine concentrations in triceps, and diaphragm of fetuses and foals compared to adults, but this was statistically significant ($P < 0.05$) only in diaphragm. Mean concentrations in the three groups are presented in Table 1.

Despite the similarity in the mean muscle content of carnitine between the groups, the lowest individual values were generally found in neonates, this being particularly marked in a 290-day pony fetus which died after the mare underwent surgery. Values were: middle gluteal (deep) 17·2; middle gluteal (superficial) 17·1; triceps 16·4; semitendinosus 16·2; diaphragm 4·2 and heart 5·2 mmol/kg d.m.

As the size of the survey increases it should be possible to determine the proportion of the population with low carnitine levels which might benefit from supplementation. Future studies will investigate the importance of biosynthesis or placental transfer in determining tissue carnitine concentrations.

The study was supported by a grant from the Sigma Tau Pharmaceutical Company, Rome, Italy. We thank Jennifer Ousey and Jennifer Stewart for providing some of the post-mortem specimens.

References

Bieber, L.L., Emaus, R., Valkner, K. & Farrell, S. (1982a) Possible functions of short-chain and medium-chain carnitine acyltransferases. *Fedn Proc. Fedn Am. Socs exp. Biol.* **41**, 2858–2862.

Bieber, L.L., Valkner, K. & Farrell, S. (1982b) Carnitine acyltransferases of liver peroxisomes. *Ann. N.Y. Acad. Sci.* **386**, 395–396.

Marquis, N.R. & Fritz, I.B. (1964) Enzymological determination of free carnitine concentrations in rat tissues. *J. Lipid Res.* **5**, 184–187.

SUMMATION

J. Reprod. Fert., Suppl. **35** (1987), 741–749

Calgary encapsulated: a personal summation of the Fourth International Symposium on Equine Reproduction

K. J. Betteridge

Department of Biomedical Sciences, Ontario Veterinary College, University of Guelph, Guelph, Ontario, Canada N1G 2W1

Introduction

Carl Hartman, in one of the earliest supplements to the *Journal of Reproduction and Fertility,* said that Research should spell FUN (Hartman, 1967). He would surely have approved of ISER spelling FUN for the fourth time here at Calgary and, for that, we delegates owe a big 'thank you' to the International Committee and to the Local Arrangements Committee. Once more they have brought together a broader spectrum of reproductive biologists and clinicians than is usual and have shown that the resulting cross-fertilization of ideas and approaches to problems really does produce concepts with hybrid vigour. Partly for this reason, this Symposium, like its predecessors, has produced papers and discussions that have relevance well beyond the titled session in which they have been presented and so I have been given the task of condensing all that we have heard and discussed over the past week into a 'take-home message'. That is a daunting assignment! However, I believe that Eric Lamming's summary of the Sydney meeting (Lamming, 1982) served a very useful purpose in leaving us with an inventory of challenges requiring further investigation, and so I hope that this summation, with reference to the first, will help us gauge the progress that has been made over the intervening 4 years, and to think about where we go from here.

I have divided the summary into three overlapping sections; the normal reproductive processes, their manipulation, and clinical and kindred aspects of the subject. Within the first section, I have followed the same physiological progression from conception to birth that we have used all week.

The fact that we have enjoyed 126 presentations (84 oral and 42 by poster) from 18 countries will explain that not all can be mentioned with equal force in a summary and emphasis has been based on my effort to make a cohesive report, not on any judgement of the relative merits of the communications. Presentations are referred to simply by the name of the first author so that they can be traced with reference to the author index of this volume.

The normal processes

The domesticated horse as an equid

The major thrust of Lamming's (1982) summary concerned the perils, for those of us who deal with domesticated horses, of forgetting or ignoring two factors that are essential to satisfactory reproductive performance and survival in wild or feral equids. These are the seasonal responses of males and females to the environment, and the intricate patterns of social behaviour that underlie sexual behaviour. This message has again come through loud and clear here at Calgary and several papers heightened our awareness of how horses and their cousins should reproduce, and how efficiently they will reproduce, given the chance.

The symposium's most exotic contribution (Penzhorn) vividly illustrated the effects of environment on the whole pattern of reproduction in two different populations of zebras living under different rainfall conditions in Southern Africa. Those of us who manage horses need to understand the mechanisms through which the environment and diet affect reproductive efficiency and there is no doubt that wild and feral equine populations have much to teach us, besides being of great ethological interest in their own right.

Another instalment of the observations being made of pasture breeding on the Canadian prairies (Bristol) described foaling rates of 90% or more for groups of about 300 mares during two successive seasons. Although this was in a population that had already been heavily culled for infertility, such figures must surely eventually inspire an enlightened (or cavalier) Thoroughbred owner to try to pasture breeding and thereby respond to one of Lamming's (1982) challenges. The indisputable success of a return to a more natural mating system makes one wonder whether beneficial effects would also be seen if foaling was allowed to occur in family groups, as happens in the wild but never on the farm. As a prelude, we could do with some observational work on parturition and perinatal behaviour in wild and feral populations; a tall order in view of the mare's penchant for foaling in solitude.

As to seasonality, three studies (Magistrini; Kenney; Dowsett) from both hemispheres agreed in general that the stallion is not at his best in the winter. This might be because baseline concentrations of LH are then at their lowest in stallions (Clay) and geldings (Hoffman). This suggests that the LH concentrations probably reflect seasonal variations in LHRH pulse frequency and are obviously independent of gonadal steroids. However, another paper underlined the difficulties of interpreting the 'concentrations' of hormones measured in immunological, receptor and biological assays (Alexander), so we must be careful in drawing 'cause and effect' conclusions from these endocrinological studies. That the concentrations and affinities of testicular receptors for LH remain constant throughout the year (Evans) also fits in with the concept that seasonal variation in testicular function may be controlled centrally rather than peripherally. Marked seasonal effects on plasma prolactin concentrations have been demonstrated in the stallion (Thompson) as well as in the mare (Worthy). The patterns in both sexes resemble those seen in the sheep (a short-day breeder) and therefore reflect a response to changes in photoperiod rather than the onset of sexual activity. It seems that we can exclude the possibility that the thyroid gland plays an essential role in the control of seasonal breeding because seasonality is seen even after thyroidectomy (Lowe).

The stallion and his gametes

Iconoclastic papers are always refreshing. The ritual of washing the stallion's penis before use (seldom practised in the wild) was shown to be contraindicated at the Sydney meeting (Bowen *et al.*, 1982) and this view was reinforced here at Calgary by the demonstration (Rath) that chlorhexidine can be absorbed from such washings, with an associated decline in semen quality. This may be caused by adverse effects on plasma membranes, even though the horse spermatozoon, in comparison with those of other domestic ungulates, is relatively resistant to such damage (P. F. Watson). Were there shades of iconoclasm, too, in the study (Rousset) which showed that an accurate estimation of a stallion's fertility in terms of live foals born would take 10 seasons and 25 mares per season? This does not invalidate attempts to find better means of assessing the fertilizing potential of individual ejaculates, of course, and we learned that seminal proteins are under investigation for that, so far without conclusive results (Amann), and also that chemotactic chambers (Strzemienski) and certain biophysical methods (Rousset) may hold some promise.

Two posters clearly demonstrated that there is nothing to be gained (in terms of sperm output) from using a stallion more than once (Voss) or twice (Dowsett) a day. This is well below the service frequency observed in free-ranging conditions when mares oestrous cycles were artificially synchronized (Bristol, 1982). Since those services were associated with a high pregnancy rate, perhaps fertility and sperm count are not as closely related as one might think, at least for natural service.

The graphic report of the effects of psychoneurotropic agents on sexual behaviour in stallions (McDonnell) reminded us that the 'stresses' imposed by domestication and management are very different from those encountered in the wild.

The mare and her gametes

We learned a good deal about the gonadotrophins (Alexander; Manning) and the endocrinology of ovulation in the mare. Four studies of gonadotrophin concentrations in relation to folliculogenesis and ovulation were presented. From one (Silvia) it would appear that oestrogens can suppress follicular development despite high circulating concentrations of both LH and FSH, suggesting that the effect is exerted at the ovarian level. Three studies of LH in the peri- and post-partum period reached different conclusions. In one (Pope) a rise in LH concentrations 30 min after parturition seemed important to the occurrence of an early post-partum ovulation but in two others (Hines; Nett) the low basal concentrations of LH found during late pregnancy persisted for the first 2 or 3 days of the post-partum period. The reports of an increase in pulse frequency for LH as the mare passes from anoestrus into the breeding season that were presented in Sydney have been confirmed. Furthermore, the control of these pulses by LHRH has been inferred (Safir) although a push–pull cannulation study did not demonstrate pulses of LHRH release that could be correlated with those of LH unless the pulse was initiated by an injected stimulatory agent (Sharp).

Prolactin seems to have come into its own as an equine hormone. Besides the seasonal studies already mentioned, differences in its pulsatile secretion that depend on the stage of the cycle were reported. Early in the cycle, the pulses of prolactin release may be associated with oestrogen pulses (Roser); near the time of luteolysis the rise in prolactin concentrations appear to be associated with those of PGFM, although there is no evidence that prolactin is required for luteolysis (Worthy). We also learned that TRH can release prolactin in mares, just as it can in stallions (Thompson), and may even be useful in the treatment of agalactia (Lothrop).

Another new hormonal candidate for a role in luteolysis is oxytocin which, in pharmacological doses, has been shown to bring about more PGF-2α release when injected at the time of normal luteolysis than when given earlier or later in the cycle (Goff), correlating well with a separate study that showed circulating concentrations of the peptide to be at their highest at Day 15 (Tetzke).

A description of the basic behavioural features of the oestrous cycle of the donkey reminded us that oestrous cycles involve more than oscillating hormones (Henry). It became very evident at this Symposium that interest in this unusual beast is on the increase.

Turning now to the follicle and its all-important contents, we heard that the follicle destined to ovulate is 'selected' as early as Day 14 of the cycle, about 7 days before it actually ovulates (Fay), much earlier than in the domestic ruminants, for example. Since access to large volumes of follicular fluid by relatively simple puncture through the flank is quite practicable in the mare (Vogselsang; Palmer), the horse may be useful for investigation of the factors controlling selection and the properties of the selected follicle which are believed to repress others. However, the puncture technique has not been developed to gather follicular fluid but to harvest oocytes mature enough to fertilize *in vitro*. Although successful in-vitro fertilization was not reported at this meeting, reports of oocyte structure and maturation and of sperm capacitation are obviously leading in that direction (Vogelsang; Palmer; Okólski).

Fertilization and embryonic development

I relished the description of the places in which spermatozoa appear to lie in wait for the egg (Boyle) and to survive for several days, much longer than is possible in the ruminants. Fred Day would have appreciated these renewed Cambridge efforts to explain his old clinical and experimental observations of the early 40s (Day, 1942).

Nobody has yet explained the phenomenon of differential transport of fertilized and unfertilized ova in the mare but we saw that fertilization was without detectable effect on systemic progesterone and oestradiol concentrations (Pipkin). We also saw beautiful micrographs illustrating the structure of oocytes during the final processes of maturation (Palmer) and of pronuclear zygotes which are impossible to distinguish from unfertilized oocytes during the first 24 h in the fresh condition (Enders), a point of considerable importance for those interested in in-vitro fertilization or in the determination of fertilization rates.

Presentations concerning uterine embryos included original descriptions of X-chromosome inactivation (Romagnano), metabolic activity (Rieger), the differing steroidogenic roles of two groups of cells in the Day-14 conceptus (Marsan), and the ultrastructure of embryos before and after glycerol treatment and freezing (Wilson). The rise in oestrone sulphate concentrations in the maternal circulation soon after attachment was shown to be due in large part to oestrogen production by the conceptus rather than the mare's ovaries (Kasman). Furthermore, there is additional indirect evidence that the source of the oestrogens later in gestation is placental (Jeffcott; Kindahl).

A graphic example of the evident relationship between structure and function in the early conceptus was provided by the morphological differences found in those obtained from normal and subfertile mares (Schlafer). Since these included striking effects on the capsule, and indications of its reaction to insult, it would seem important to continue investigation of the structure, function and fate (Denker) of this envelope.

The uterine environment, besides being strongly influenced by ovarian steroids (McDowell), is very responsive to the presence of a conceptus by Day 11–13 because, by that stage, the uterus has ceased to release much PGF-2α in response to an injection of oxytocin (Goff). Although a time for the 'recognition of pregnancy' has been brought forward to Day 11–13 as a result of this study (Goff), others amongst us feel that this event may be more of a continuum in the horse than in the ruminants, for example.

Maintenance of pregnancy

A standard-bearer for the cause of adopting a comparative and interdisciplinary approach to the study of reproduction is the investigation of the immunological significance of equine endometrial cups. Using techniques ranging from interspecies embryo transfer (Allen) to tissue typing with monoclonal antibodies (Antczak), the Cambridge–Cornell axis is making great strides in this field. Here at Calgary they showed that interactions between maternal lymphocytes and cup cells cause alterations in lymphocyte function that can be assumed to be of importance to the maintenance of pregnancy even if we do not yet understand how (Crump; Allen). Is there any analogous significance to the observation of considerable lymphocytic infiltration of pregnant, but not of non-pregnant, uteri much sooner after ovulation, Days 6–9 (Keenan)?

It is remarkable that horse CG should be so thoroughly understood structurally (Stewart) and yet, from the gonadotrophic point of view, remain so entirely enigmatic as to function (if any) in equids. The endometrial cups which persist after PGF-2α-induced abortion do prevent spontaneous luteolysis (Rathwell) but the success of pregnancies in the face of extremely variable levels of horse CG has been clear for years, so it would be rash to assume that this luteostatic effect is the prime function of horse CG in normal pregnancy. The genetic influence on horse CG production, first demonstrated in hybrids and then confirmed by interspecies embryo transfer, has now been shown to occur during normal horse–horse pregnancies and to affect not only quantity but also biological and chemical quality of the gonadotrophin (Manning). Methods for detecting horse CG for the purposes of pregnancy diagnosis have progressed through ELISA systems to simple dipstick methods (Kasper; Lock).

The phagocytic role of neutrophils in making the uterus a suitable place for an embryo to live was the subject of no fewer than four communications, perhaps in response to Lamming's (1982) exhortations of 4 years ago. Neutrophils pour into the uterus soon after serum proteins do in response to a variety of provocations, reaching a peak at about 6 h after the stimulus (Williamson). Progesterone does not reduce the number of cells that invade (E. Watson) but in normal mares, it does reduce or delay their phagocytic activity (Evans; E. Watson). There is no reason to think that susceptibility to uterine infection can be blamed on defective phagocytosis (Asbury).

The mare does not need a thyroid gland to become and to remain pregnant (Lowe) and attempts to correct hypothyroidism during pregnancy may do more harm than good (Silva). Unlike sows and ewes, at least, the mare can also temporarily dispense with progesterone during late pregnancy without entering labour prematurely (Fowden).

Fertilization failure and pregnancy loss

From the precirculated abstracts, I thought that we were destined for disagreement concerning the importance of fertilization failure as a cause of reproductive wastage. However, the round-table discussants seemed to agree that fertilization rates in horses are usually high (Woods; Forde; Enders).

In general agreement with previous work, two large scale surveys of brood mares showed remarkable consistency in the pregnancy rate (50–55%) at Days 14–18 and rate of loss of those pregnancies before term (11–13%) (Woods; Forde). There is also a consensus that most losses occur in the first 4–6 weeks of pregnancy (Woods; Forde; Bell). The abortion rate after 3 months was found to vary considerably (3–18%) from year to year in pasture-mated mares (Bristol) and this was related to stress induced by management procedures. In another study, and in agreement with the experience of many practitioners in the field, a history of embryonic loss or abortion did not seem to prejudice a mare's chances of subsequently producing a foal (Darenius), although this would not be true of those animals in which the infertility was due to a cytogenetic abnormality (Bowling; Hurtgen; Halnan) or a recent uterine infection (Woods).

As to the causes of embryonic loss, one common thread running through several papers was the adverse effect of uterine inflammation on pregnancy rates, whether occurring spontaneously (Adams; Woods) or subsequent to experimental intrauterine infusions of yeast or even, to a lesser extent, of sterile saline (Ball) which could explain the very low rate of success of embryo reinsertion (Sirois).

There is certainly room for development of a more subtle indicator of impending pregnancy failure than complete luteolysis. Progesterone determinations rarely indicate complete CL regression before embryonic loss (Sirois; Daels) although some decline in plasma progesterone concentrations was to be seen after loss induced by experimental endotoxaemia and in most cases of loss associated with uterine inflammation (Adams; Daels). Furthermore, the Irish Thoroughbred work revealed significant differences in progesterone concentrations in the blood of mares according to pregnancy status as early as 9–13 days after ovulation (Forde).

Another poster (Potter) underlined the importance of nutrition in the maintenance of pregnancy.

Ultrasound continues to advance the study of early pregnancy and embryonic mortality and, as gestation progresses, to be useful in assessing the well-being of the fetus (Adams-Brendemuehl).

Parturition and lactation

We were reminded that the foal is the most mature of the domestic animals at birth and we heard the first report of a newly discovered cortisol-binding protein which is apparently exclusive to the horse fetus and neonate and which may serve to compensate for the relatively low levels of cortisol-binding globulin in horses compared with other perinates (Irvine). The probable function of this protein is to increase the plasma cortisol pool in the fetus and the rate of transfer of cortisol from dam to fetus, and later to dampen the effects of the post-natal surge of cortisol on the foal at birth.

There were two presentations describing comprehensive studies of the electromyographic activity of the mare's uterus during late gestation and at parturition (Dudan; Haluska). One of them (Haluska) correlated the muscular activity with changes in hormonal concentrations in the maternal circulation and arrived at a useful working hypothesis as to how the mare exerts so much control over the time of birth. This she probably accomplishes by central restraint of the release of oxytocin, the final trigger for delivery. The forces unleashed by that trigger have always posed an enormous problem for the obstetrician who has to understand them and work with them when things go wrong, so it was informative to hear a masterly classification of such dystocias (Vandeplassche).

There were only three reports on the subject of lactation (Lothrop; Wiest; LeBlanc) and I feel that I should reiterate Lamming's (1982) comment that this is disappointing in view of the importance of the subject. That importance was well illustrated by the description of the marked decrease in IgG content of milk available to the newborn foal that occurs if lactation begins 24 h prematurely (LeBlanc).

Adaptation of the foal to extrauterine life

It is in this section of the Symposium that many of we reproductive biologists find our horizons considerably extended into the world of non-reproductive physiology embracing such vital subjects as repiratory mechanics and cardiopulmonary function (Koterba; Madigan). This is interesting for us but I wonder whether we can provide the informed critique of these topics that our colleagues deserve and might wish. Nevertheless, I shall not be able to look at a newborn foal again without remembering that it was at Calgary that I first learned of the physiological significance of the rapid assumption of sternal recumbency.

Three studies of the newborn foal (Liu; Fogarty; Bernoco) brought us back to the polymorpho-nuclear leucocytes (PMNLs or neutrophils). They, as well as accompanying macrophages, tend to be functionally deficient when recovered from the respiratory tract or peripheral circulation of newborn foals, suggesting that the deficiencies could account for the newborn's high susceptibility to respiratory disease.

Endocrinological studies outside the usual range of reproduction showed us that hypoglycae-mia in the newborn foal probably stimulates only the adrenal component of the sympatho–adrenal system (Silver) and that the placenta, rather than the fetus, is probably responsible for the elonga-tion and desaturation of the fatty acids that are essential to the fetus (Leadon). The interesting hypothesis that a deficiency of carnitine in newborn foals could cause malaise by impeding transport of the fatty acids to mitochondria has not so far been borne out (Foster).

Manipulation of the reproductive process

Molecular biology and reproduction

The quality, behaviour and physiology of every horse we ever deal with is, in the last analysis, a reflection of the DNA within its chromosomes. Appropriately for us, the new manipulators of this DNA are usually called 'gene jockeys' but their growing influence on all aspects of reproduction is only just beginning to be felt in the horse world although we were ushered into the era of molecular biology with the very first paper of this symposium (Stewart). It is thanks to molecular biology's additions to decades of biochemical research that we now find ourselves at an unusual juncture in endocrinology where more seems to be known about the structure of the gonadotrophins than about their biological activities. The notion of gonadotrophins existing in different isoforms, for example (Alexander), will have been new to most of us and the use of bioengineered gonadotrophin molecules in the battery of assays now available should soon unravel some more of the relation-ships between their structure and function. This could make 'designer drugs' available for superovulating other species, too.

It will be interesting to see whether the traditionally conservative horse breeding industry takes advantage of this 'new biology' or prefers to insulate itself from the changes that it could make possible. Paradoxically, forensic necessities could preclude any attempt at insularity; it may be my perverse mind, but after hearing that horse growth hormone (GH) is being bioengineered (Stewart) and that human athletes are now misusing human GH instead of steroids to improve their performance, I do not think that one has to be a Dick Francis to conjure up some rather alarming scenarios for the Sport of Kings. We heard of precedents for a forensic stimulus to research at this meeting; the detailed work on the synthesis of C-18 steroids in the horse testis (Smith).

On the scientific front, I think that it is worth remembering that although the use of recombinant DNA technology in genetic research amplifies the power of cytogenetic analysis, it does not outmode the cytogenetic approach. Four presentations at this meeting have illustrated the value of cytogenetics in both clinical and research applications (Hurtgen; Halnan; Bowling; Romagnano) and it is a fair bet that the complementariness of the new and the old will be much in evidence by the time we meet again. In the same vein, any forseeable method of creating transgenic horses or sexing embryos is going to require as much expertise in the handling of gametes, zygotes, embryos and recipient mares as in recombining DNA; again, the fields must progress in parallel. These points may seem obvious to us but they bear some stressing when research priorities have to be set and scarce resources allocated.

Control of seasonal breeding and of ovulation

The efficiency of pharmacological induction of ovulation in the anoestrous mare has advanced remarkably since the Sydney Symposium due to the development of new analogues of LHRH (Sanderson; Fitzgerald) and of new delivery systems (Hyland). The most practicable and satisfactory would certainly seem to be the slow-release implant (Sanderson). The ovulation-inducing action of certain PG analogues during oestrus (Savage) or during the period of transition from anoestrus to the breeding season (Jöchle) also evidently involves stimulation of LHRH release (Jöchle).

The old problem of reduced or abolished ovulatory response to hCG when used repeatedly can be circumvented by treatment with crude horse gonadotrophins (Duchamp).

It is a pity that we have to miss out on the remarkable progress being made in controlling the sheep's breeding season with melatonin implants. This is because it is relatively easy to add melatonin to the circulation and thus fool the animal into acting as though days were getting shorter, but quite another thing to remove melatonin to suggest that spring is on the way. Some ingenuity will be necessary to accomplish this, so it will be interesting to see what our next Symposium will have to offer on this topic.

In the longer term, one wonders whether genetic selection for reproductive photoresponsiveness will be applicable to horses. This has recently been shown to be feasible in deer mice (Desjardins *et al.*, 1986) which share with horses the feature of considerable individual variation within the species as far as reproductive response to photoperiod is concerned. A generation interval measured in years rather than weeks will make repetition of the *Peromyscus* work in equids much more difficult. Nonetheless, it is worth considering, particularly as we move into an era in which there could be considerable incentive to identify the gene(s) responsible for photoresponsiveness in other species of farm animal as well.

Artificial insemination

There were several reports on the conservation and use of fresh semen over prolonged periods. Two (de Vries; Douglas-Hamilton) described how extended semen could be cooled, shipped at 5–8°C in insulated containers, and give good results when inseminated within 3 or 4 days of collection. A third confirmed the benefits of slow cooling to that temperature in order to maintain motility for similar periods of time but also found it advantageous to remove the original seminal fluid and to store either in a closed bag with no atmosphere or in an atmosphere of 5% CO_2 (Heiskanen). There is still some question as to which non-glycolysable sugars are best to use in semen diluents (Arns). The field use of frozen semen was reported to be giving satisfactory results in both Czechoslovakia (Müller) and South Africa (Volkmann) and we heard that stallions which are 'poor freezers' cannot be changed into 'good freezers' just by changing the season of frequency of semen collection (Magistrini).

Embryo manipulation and transfer

The viabilities of embryos resulting from spontaneous double ovulations and from single ovulations are the same (Squires) and the Colorado group has also been able to more than double the normal embryo recovery rate (to 2 per donor) by using horse gonadotrophins to boost the ovulation rate to almost 4 per donor, increasing the efficiency of embryo transfer accordingly.

The birth of foals from progesterone-treated ovariectomized recipient mares has been a significant development and the extension of the pilot study in Pennsylvania has produced further encouraging results (Hinrichs). As was to be expected, substitution of orally-active progestagens for daily injections of progesterone, and of cyclic mares for mares without ovaries has now been investigated. Simply feeding altrenogest permits the use of recipients that ovulated from 3 days before to 3 days after the donor (Parry-Weeks) and this 7-day span can be extended to 12 days with suitable PG–oestrogen–progesterone treatment (Pool). The indication that horse embryos are inhibited in their development whilst retaining their viability for at least 84 h in a simple medium at 4°C also promises well as an alternative way of making use of recipients that ovulate after the donor (Pashen).

Chimaera production by micromanipulation has been extended to equids but it remains to be seen whether the cause of the failure of pregnancies initiated with horse–donkey chimaeras is indeed a lack of girdle cell migration (Willadsen).

At the Polish poster on transfers between small and large breeds of horse (Tischner), I sensed the ghost of John Hammond clattering by in a cart drawn by a pair of 48-year old Shetland–Shire crosses (Walton & Hammond, 1938; Hammond, 1962).

Clinical and kindred aspects of reproduction

Twinning

The management of the 5–10% twin pregnancies in Thoroughbreds remains an important part of studfarm veterinary practice and, although there is no doubt that it has been considerably simplified since early detection of the two conceptuses by ultrasound became possible, the problem has not been obviated by the 'natural reduction mechanism' by which one conceptus often apparently kills the other (R. R. Pascoe). Excellent survival of the conceptus remaining after manual crushing of the other one 12–30 days after ovulation was reported here and there is no advantage in employing any supplementary hormone therapy (D. R. Pascoe).

The foal heat

The debate over whether or not to breed at the first post-partum oestrus still rages, though large numbers of wild and feral horses seem indifferent to the question. Although conception rates are lower and embryonic loss rates higher when the foal heat is used for breeding, about half of the matings are fertile (Saltiel; Koskinen). Some indication of the chances of conception may be obtained from determining the number of neutrophils in uterine biopsies taken on Day 5 *post partum* but this, of course, is something of a statistical exercise. Delaying the first post-partum oestrus for 5 days therapeutically has been shown to offer no advantage for increasing overall fertility but it may be useful in mares that usually ovulate sooner than 10 days after parturition (Bell).

Diagnostic methods

The use of ultrasonic echography is no longer confined to females; we heard of its usefulness for examining the accessory sex glands in the stallion (Little). Some potential pitfalls for the novitiate using the system in the mare were well illustrated (Osborne).

Two posters contended that intrauterine cytology can be at least as useful as histological examination of biopsies for identifying acute and chronic endometritis (Bowen; Roszel) and a third (Brewer) described a scoring system to identify septicaemia in newborn foals and thereby counteract the indiscriminate (or delayed) use of antibiotic therapy.

Anaerobic bacteria have been shown to be capable of establishing themselves as opportunistic pathogens in the uterus (Ricketts) and two species of *Mycoplasma* have been isolated from about one third of over 300 mares surveyed in Southern Ontario, although there was no obvious effect of the organism on their fertility (Bermudez).

Stallions are thought to play a significant role in the dissemination of equine arteritis virus infection which may be spread entirely venereally (Timoney). A new infertility syndrome in stallions themselves has been described; it is associated with low sperm motility and may involve an epididymal dysfunction (Galloway).

An extensive serological, virological and endocrinological study of an EHV-1 abortion storm on an English studfarm provided important epidemiological information and an interesting comparison of how hormonal changes associated with abortion from this cause resemble, and differ from, those seen at normal term delivery (Mumford; Ousey). The account of the abortion storm also underlined just how devastating an outbreak of the disease can be. Infectious abortion of different aetiology, and vaccination against it in India, were the subject of a poster (Uppal).

We also heard of a careful investigation of abortions caused by grazing an endophyte-infected fescue grass, apparently as a result of an alkaloid produced by the fungus (Loch), although an alternative explanation in terms of a high oxalate content of similar pastures in Brazil was advanced from the floor.

Conclusions

I have no conclusions. We need time after this intensive week to digest, to ruminate a little (if you will forgive the term), and to arrive at our own personal conclusions in the fullness of time. That is the joy of these meetings—each one spawns ideas for the next and so, as we all go home and prepare to bring those ideas to fruition for our next Symposium in 4 years' time, may we all, in Professor Hans Merkt's memorable words:

'Carry spoons in our pockets ready for the day that it rains soup'.

References

Bowen, J.M., Tobin, N., Simpson, R.B., Ley, W.B. & Ansari, M.M. (1982) Effects of washing on the bacterial flora of the stallion's penis. *J. Reprod. Fert., Suppl.* **32,** 41–45.

Bristol, F. (1982) Breeding behaviour of a stallion at pasture with 20 mares in synchronized oestrus. *J. Reprod. Fert., Suppl.* **32,** 71–77.

Day, F.T. (1942) Survival of spermatozoa in the genital tract of the mare. *J. agric. Sci., Camb.* **32,** 108–111.

Desjardins, C., Bronson, F.H. & Blank, J.L. (1986) Genetic selection for reproductive photoresponsiveness in deer mice. *Nature, Lond.* **322,** 172–173.

Hammond, J. (1962) An interview, curriculum vitae, selected publications. *J. Reprod. Fert,* **3,** 2–13.

Hartman, C.G. (1967) Research should spell FUN. *J. Reprod. Fert., Suppl.* **2,** 1–10.

Lamming, G.E. (1982) A summation. *J. Reprod. Fert., Suppl.* **32,** 611–621.

Walton, A. & Hammond, J. (1938) The maternal effects on growth and conformation in Shire horse–Shetland pony crosses. *Proc. R. Soc. B* **125,** 311–335.

INDEX

751

SUBJECTS